The Finite Element Method

The Finite Element Method

Fundamentals and Applications in Civil, Hydraulic, Mechanical and Aeronautical Engineering

Bofang Zhu
Professor, China Institute of Water Resources and Hydropower Research
and
Academician, Chinese Academy of Engineering
Beijing, China

WILEY

清華大學出版社
TSINGHUA UNIVERSITY PRESS

This edition first published 2018 by John Wiley & Sons Singapore Pte. Ltd under exclusive licence granted by Tsinghua University Press for all media and languages (excluding simplified and traditional Chinese) throughout the world (excluding Mainland China), and with non-exclusive license for electronic versions in Mainland China.

Registered Offices
John Wiley & Sons, Inc., 111 River Street, Hoboken, NJ 07030, USA
John Wiley & Sons Singapore Pte. Ltd, 1 Fusionopolis Walk, #07-01 Solaris South Tower, Singapore 138628

Editorial Office
1 Fusionopolis Walk, #07-01 Solaris South Tower, Singapore 138628

For details of our global editorial offices, customer services, and more information about Wiley products visit us at www.wiley.com.

Wiley also publishes its books in a variety of electronic formats and by print-on-demand. Some content that appears in standard print versions of this book may not be available in other formats.

Library of Congress Cataloging-in-Publication Data applied for

Hardback ISBN: 9781119107316

Cover design: Wiley
Cover Image: © Magnilion/Gettyimages

Set in 10/12pt WarnockPro by SPi Global, Chennai, India
Printed in Singapore by C.O.S. Printers Pte Ltd

10 9 8 7 6 5 4 3 2 1

Contents

Preface

The finite element method (FEM) is so powerful that many very complicated engineering problems can be solved by it. This book is primarily written for engineers. It introduces the basic principles and applications of FEM. It may also be used as textbook in universities and colleges.

The first purpose of this book is to make an easy read for engineers, so the physical ideas are enhanced and the basic principles and computing methods are introduced in an easy but accurate way.

The second purpose of this book is to be of practical value to engineers, so the formulas that can be used to analyze problems in practical engineering are given in detail.

Thus, there are three distinguishing characteristics of this book: (1) it is easy to read; (2) the theory and computing formulas of finite element method are complete; (3) it is of practical use to readers, especially to engineers and professors and engineering students.

Before the publication of the first edition of this book in Chinese in 1979, the predicted readers were engineers, but after publication it was noticed that it was well received not only by engineers but also by professors and students in universities and colleges. It is now not only a widely accepted reference book for engineers but also widely used as textbook for professors and students in universities and colleges in China.

According to the Information Center of Chinese Academy of Science *The Finite Element Method, Theory, and Applications* (in Chinese) is one of the most well-received 10 books in China in water resources and hydropower domain.

Now the new book in English will be published, I hope it will be well received not only by engineers working in practical engineering project but also by professors and students.

About the Author

Bofang Zhu, an academician of the Chinese Academy of Engineering and a famous scientist of hydraulic structures and solid mechanics in China, was born on October 17, 1928, in Yujiang County, Jiangxi Province. In 1951, he graduated in civil engineering from Shanghai Jiao Tong University and then participated in the design of the first three concrete dams in China (Foziling dam, Meishan dam, and Xianghongdian dam). In 1957, he was transferred to the China Institute of Water Resources and Hydropower Research where he was engaged in the research work of high concrete dams. He was awarded China National Outstanding Young Scientist in 1984 and was elected the academician of the Chinese Academy of Engineering in 1995. He is now the consultant of the technical committee of the Ministry of Water Resources of China, a member of the consultant group of the three very high dams in the world: the Xiaowan dam, the Longtan dam, and the Baihetan dam. He was a member of the eighth and the ninth Chinese People's Political Consultative Conference, the board chairman of the Institute of Computer Application of China Civil Engineering Society, and a member of the standing committee of the China Civil Engineering Society and the standing committee of the China Hydropower Engineering Society.

He is the founder of the theory of thermal stresses of mass concrete, the shape optimization of arch dams, the simulating computation of concrete dam, and the theory of creep of mass concrete in the word.

He has established a perfect system of the theory of thermal stress and temperature control of mass concrete, including two basic theorems of creep of nonhomogeneous concrete structures; the law of variation and the methods of computation of the thermal stresses of arch dams, gravity dams, docks, sluices, tunnels, and various massive concrete structures; the method of computation of temperature in reservoirs and pipe cooling, thermal stress in beams on foundation, cold wave, heightening of gravity dam; and the methods and criteria for control of temperatures. He proposed the idea of "long-time thermal insulation as well as comprehensive temperature control" that ended the history of "no concrete dam without crack" and some concrete dams without crack that had been first constructed in China in recent years, including the Sanjianghe concrete arch dam and the third stage of the famous Three Gorges concrete gravity dam.

He proposed the mathematical model and methods of solution for shape optimization of arch dams, which was realized for the first time in the world and up to now had been applied to more than 100 practical dams, resulting in 10–30% saving of dam concrete, and the efficiency of design was raised a great deal.

He had a series of contributions to the theory and applications of the finite element method (FEM).

He proposed a lot of new methods for finite element analysis, including the compound element, different time increments in different regions, the equivalent equation of heat conduction for pipe cooling, and the implicit method for computing elastocreeping stresses by FEM.

He developed the method of simulating computation of high concrete dams by FEM. All factors, including the course of construction, the variation of ambient temperatures, the heat hydration of cement, the change of mechanical and thermal properties with age of concrete, and the pipe cooling, precooling, and surface insulation can be considered in the analysis of the stress state. If the tensile stress is larger than the allowable value, the methods of temperature control must be changed until the maximum tensile stress is not bigger than the allowable value. Thus cracks will not appear in the dam. Experience shows that this is an important contribution in dam technology.

He proposed the equivalent stress for FEM and its allowable values that had been adopted in the design specifications of arch dams in China; thus the condition for substituting the trial load method by FEM is provided.

The instrumental monitoring can give only the displacement of some particular points but cannot give the stress field and the coefficient of safety of concrete dams. In order to overcome this defect, he proposed a new method of numerical monitoring by FEM that can give the stress field and the coefficient of safety and raise the level of safety control of concrete dam and that had begun to be applied in practical projects in China.

A new idea for semimature age of concrete has been proposed by him. The crack resistance of concrete may be promoted by changing its semimature age.

A vast amount of scientific research works had been conducted under his direction for a series of important concrete dams in China, such as Three Gorges, Xiaowan, Longtan, Xiluodu, Sanmenxia, Liujiaxia, Xing'anjiang, and so on. Fourteen results of his scientific research were adopted in the design specifications of gravity dams, arch dams, docks, and hydraulic concrete structures.

He has published 10 books: *Theory and Applications of the Finite Element Method* (1st ed. in 1979, 2nd ed. in 1998, 3rd ed. in 2009), *Thermal Stresses and Temperature Control of Mass Concrete* (1st ed. in 1999, 2nd ed. in 2012), *Thermal Stresses and Temperature Control of Hydraulic Concrete Structures* (1976), *Theory and Applications of Structural Optimization* (1984), *Design and Research of Arch Dams* (2002), *Collected Works on Hydraulic Structures and Solid Mechanics* (1988), *Selected Papers of Academician Bofang Zhu* (1997), *New Developments in Theory and Technology of Concrete Dams* (2009), and *Thermal Stresses and Temperature Control of Mass Concrete (in English)* (2014). He has published more than 200 scientific papers.

He was awarded the title of China National Outstanding Young Scientist in 1984, the China National Prize of Natural Science in 1982 for his research work in thermal stresses in mass concrete, the China National Prize of Scientific Progress in 1988 for his research work in the optimum design of arch dams, and the China National Prize of Scientific Progress in 2001 for his research works in simulating computation and thermal stresses. He became ICOLD (International Commission on Large Dams) Honorary Member.

1

Introduction to Finite Element Method and Matrix Analysis of Truss

This chapter first introduces the basic conception of finite element method. The basic principles of truss analysis are similar to finite element method but easier to be understood, so the matrix analysis of truss is introduced later as an introduction to the finite element method.

1.1 Introduction to Finite Element Method

A truss is shown in Figure 1.1(a) with all nodes pin jointed and each element is a member only bearing axial force. A frame is shown in Figure 1.1(b) with all nodes rigid jointed and each element is a member bearing bending moment, shearing force, and axial force.

A beam is shown in Figure 1.1(c). All of the above three types of structures may be analyzed by structural mechanics and the theory of strength of materials. The basic assumption of them is the plane section assumption; in other words, the plane perpendicular to the central axis of the member before deformation remains to be a plane after deformation. For a rectangular high beam with relatively high ratio of the height H to length L (H/L), as shown in Figure 1.1(d), the plane section assumption cannot be applied. The calculation must be made according to the theory of elasticity that is actually a complicated problem even though the shape is simple.

Figure 1.2 shows some engineering structures. Figure 1.2(a) shows a gravity dam on the rock foundation. The dam body is nearly a triangle. However, there are slopes on both upstream and downstream boundaries. The mechanical and thermal properties of the dam body and the base rock are different. Figure 1.2(b) shows a double-curvature arch dam, which is a parabolic shell with varied curvature and thickness, supported on the base rock. Figure 1.2(c) shows an underground cavern in rock foundation. Figure 1.2(d) shows a massive concrete block in the construction of the concrete dam. The block is great in volume with concrete placed layer by layer. Generally, a layer of concrete with thickness of 1.5–3.0 m is placed every 6–10 days. Due to different ages, the modulus of elasticity, creep, and heat of hydration are all different in each layer.

For the various types of actual engineering structures shown in Figure 1.2, it is obviously impossible to work out the theoretical solutions by means of theory of elasticity. Numerical method is probably the only solution for stress calculation. Previously, attempts have been made to analyze such complicated structures by finite difference method. For example, for plane problems, the structural sections are divided into

The Finite Element Method: Fundamentals and Applications in Civil, Hydraulic, Mechanical and Aeronautical Engineering, First Edition. Bofang Zhu.

Companion website: www.wiley.com/go/zhufem17

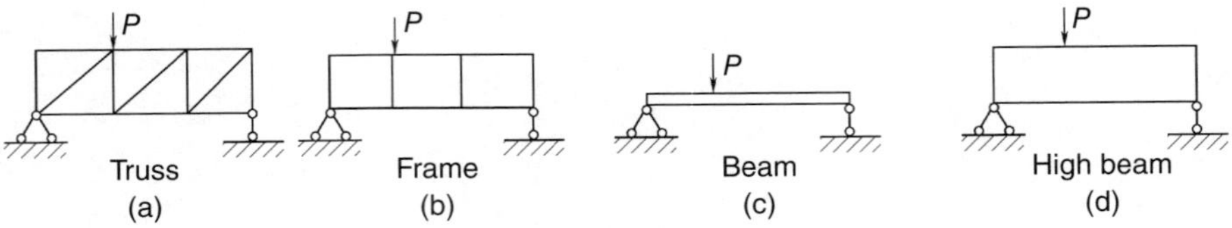

Figure 1.1 Truss, frame, beam, and high beam.

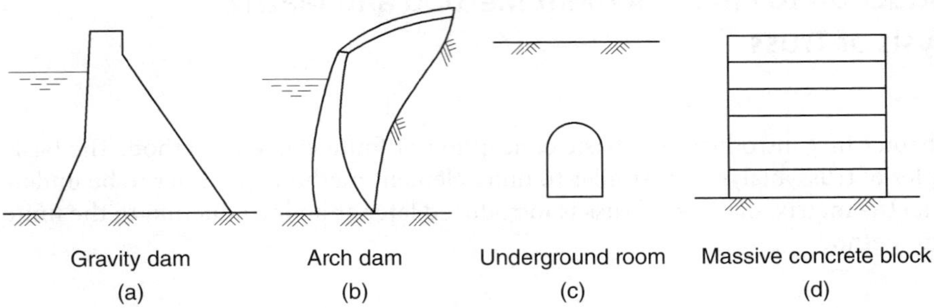

Figure 1.2 Practical engineering structures.

rectangular meshes, and the differential equations of equilibrium are transformed into finite difference equations. But the rectangular computing mesh is difficult to adapt to the boundary of the true structure, so it is rarely applied in the analysis of practical complicated structures.

The finite element method divides the original structure into finite elements, as shown in Figure 1.3.

The elements are a series of triangles of different size and shape; thus the computing mesh can adapt to the boundary of the true structure. Furthermore, different elements

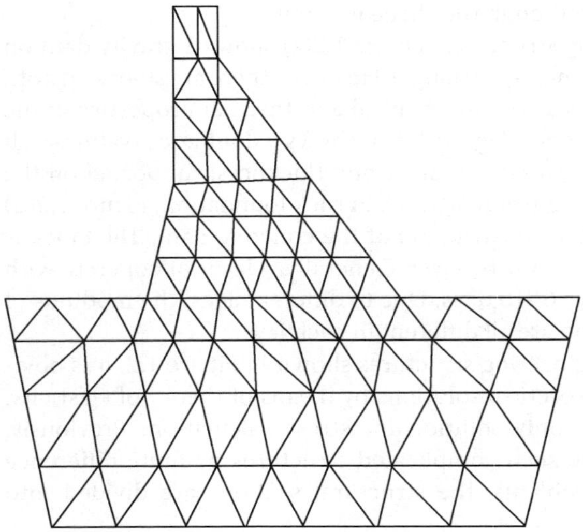

Figure 1.3 Cross section of gravity dam discreted into triangular elements.

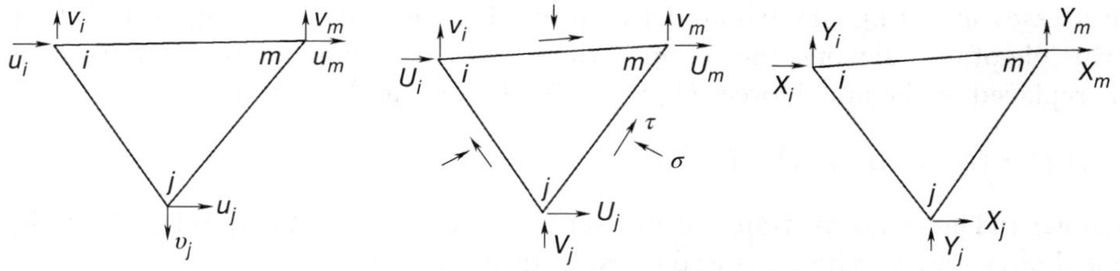

Figure 1.4 Nodal displacements, equivalent nodal forces, and nodal loads of an element.

may possess different material properties, for example, some elements represent concrete and others represent rock or soil. Thus, the finite element method can be used to analyze practical complicated structures in engineering on computers. It is now widely applied in civil, hydraulic, mechanical, and aeronautical engineering.

For example, a gravity dam on the rock foundation shown in Figure 1.3 is analyzed as plane problem. It may be divided into definite number of triangular elements, as shown in Figure 1.4. Each element has three nodes, and each node has two displacement components (horizontal displacement u_i and vertical displacement v_i). So each element has six nodal displacements:

$$\{\delta\}^{\mathrm{e}} = \begin{bmatrix} u_i & v_i & u_j & v_j & u_m & v_m \end{bmatrix}^{\mathrm{T}}$$

If the element is sufficiently small, the variation of the displacements in the element may be expressed approximately as follows:

$$\left.\begin{aligned} u &= N_i u_i + N_j u_j + N_m u_m \\ v &= N_i v_i + N_j v_j + N_m v_m \end{aligned}\right\} \tag{1.1}$$

where N_i is the linear function of x and y; at node i, $N_i = 1$ and $N_j = N_m = 0$; at node j, $N_j = 1$ and $N_i = N_m = 0$; and u_i and v_i are, respectively, the horizontal and vertical displacement of node i. The strain components are

$$\left.\begin{aligned} \varepsilon_x &= \frac{\partial N_i}{\partial x}u_i + \frac{\partial N_j}{\partial x}u_j + \frac{\partial N_m}{\partial x}u_m \\ \varepsilon_y &= \frac{\partial N_i}{\partial y}v_i + \frac{\partial N_j}{\partial y}v_j + \frac{\partial N_m}{\partial y}v_m \\ \gamma_{xy} &= \frac{\partial N_i}{\partial y}u_i + \frac{\partial N_j}{\partial y}u_j + \frac{\partial N_m}{\partial y}u_m + \frac{\partial N_i}{\partial x}v_i + \frac{\partial N_j}{\partial x}v_j + \frac{\partial N_j}{\partial x}v_m \end{aligned}\right\} \tag{1.2}$$

With strain components available and by the generalized Hooke's law, the stress components in the element may be computed and expressed by

$$\{\sigma\} = \begin{Bmatrix} \sigma_x \\ \sigma_y \\ \tau_{xy} \end{Bmatrix} = \frac{E}{1-\mu}\begin{bmatrix} 1 & \mu & 0 \\ \mu & 1 & 0 \\ 0 & 0 & (1-\mu)/2 \end{bmatrix}\begin{Bmatrix} \varepsilon_x \\ \varepsilon_y \\ \gamma_{xy} \end{Bmatrix} = [D]\{\varepsilon\} \tag{1.3}$$

The stresses are acting on the boundaries of the element as shown in Figure 1.4(b). By the principle of virtual work, the action of the stresses on the boundary of an element can be replaced by the nodal forces $U_i, V_i, U_j, V_j, U_m, V_m$ and expressed by

$$\{F\}^{\mathrm{e}} = [U_i, V_i, U_j, V_j, U_m, V_m]^{\mathrm{T}}$$

As shown in Figure 1.4, as strain is expressed by the displacement of the element node, the nodal force may also be expressed by nodal displacements as

$$\{F\}^{\mathrm{e}} = [k]^{\mathrm{e}}\{\delta\}^{\mathrm{e}} \tag{1.4}$$

The loads undertaken by the element such as boundary force, volume force, and temperature variation may also be transferred into nodal loads X_i, Y_i on node i based on virtual work principle, as shown in Figure 1.4(c).

Assuming that the total number of elements around node i is "m", the equilibrium equation of node i may be given as follows:

$$\sum_e U_i = \sum_e X_i, \quad \sum_e V_i = \sum_e Y_i \tag{1.5}$$

where U_i is the horizontal nodal force of each element, X_i is the horizontal nodal load, and $\sum_e$ refers to the sum of the elements around node i.

Expressing the nodal force in Eq. (1.5) by nodal displacements as Eq. (1.4), we get the nodal equilibrium equation expressed by nodal displacements as follows:

$$[K]\{\delta\} = \{P\} \tag{1.6}$$

where $[K]$ refers to the stiffness matrix and $\{P\}$ refers to the load vector. The inverse of the previous formula for nodal displacement is $\{\delta\} = [K]^{-1}\{P\}$. With nodal displacement available, the strain and stress of each element may be computed.

The finite element method possesses the following superiorities:

1) The original continuous medium is provided with infinite degrees of freedom and is impossible to be solved on computers. After being divided into finite elements, the nodal displacements are taken as unknown quantities. The number of unknown quantities is finite and it may be solved on computers.
2) It may be used to analyze various types of structures with complicated shapes in actual engineering.
3) Different materials may be used for different elements.
4) The nonlinear problems may be calculated, including material nonlinearity and geometrical nonlinearity (great deformation) problems.

The finite element method is now the most powerful numerical method that can be used to analyze any complicated structure in engineering. Figure 1.5 shows the finite element mesh of a fighter aircraft [1]; of course, it is not computed by the triangular plane elements, and it is computed by the complex elements that will be described in Chapters 7, 10, and 11.

Though the finite element method is originally developed to solve structural problems, due to a lot of researches, it is now available to solve various field problems such as temperature fields, flow fields, and electromagnetic fields.

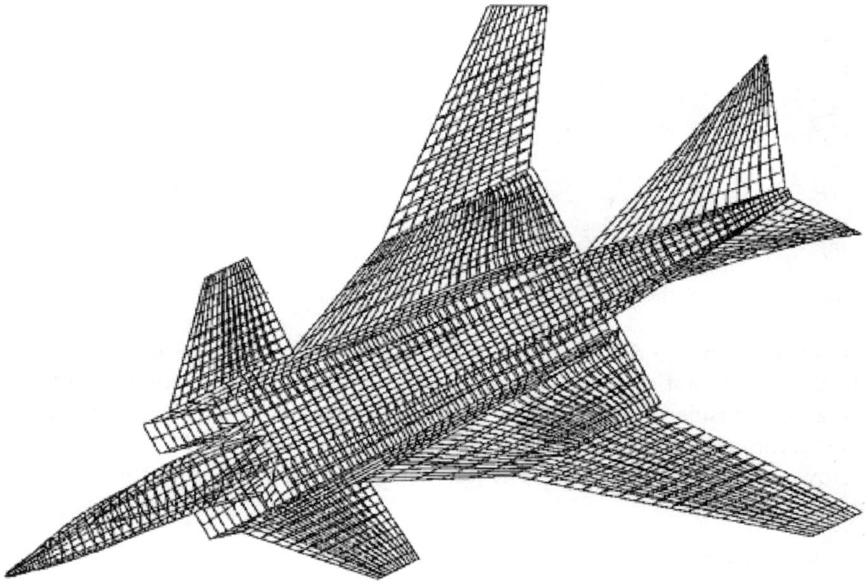

Figure 1.5 Finite element mesh of an aircraft.

1.2 Truss Analysis Overview

With the invention of computers, in order to make use of this innovative and powerful computation tool, matrix method is first applied on truss and frame [2, 3] and then generalized into continuous medium. The continuous medium is transformed into a group of finite elements, with the nodal displacements as unknown variables. An equilibrium equation set with finite number of degrees of freedom is derived, which can be solved by computer. This is finite element method [4–6] in a narrow sense. The finite element method of continuous medium and the matrix analysis of truss and frame share the same basic conceptions, that is, regarding a structure as the integration of finite elements and connecting such elements on finite nodes. The only difference is that, for truss and frame, the original members such as bars, beams, and pillars may be directly taken as elements, but there are no such natural elements for continuous mediums. Therefore, the original structure must be artificially divided into finite blocks, which are taken as the elements for calculation. With the assistance of truss, the conception of finite element method may be explained in a relatively vivid manner. Therefore, in the following part of this chapter, the analysis of truss by matrix method is introduced first as an introduction to the analysis of continuous medium by the finite element method. Of course, the framed structure matrix analysis itself is of great application values in actual engineering.

When solving any mechanical problems related to statically indeterminate structures, it is necessary to consider equilibrium conditions, deformation conditions, and physical conditions. The key for solution is usually the accurate analysis of deformation conditions.

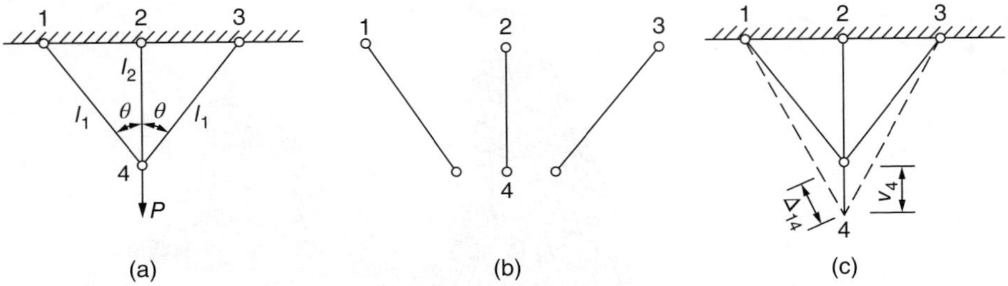

Figure 1.6 Truss. (a) Truss, (b) element, and (c) displacement.

For a truss shown in Figure 1.6, under the action of load P, what is the axial force for each bar? This is a statically indeterminate problem. The three members – 14, 24, and 34 – may be considered as the three "elements" forming the truss and the entire truss as the "combination" of the three elements. The axial force N for each element is the unknown variable to be solved. The geometric and physical properties for each member are:

$$\text{Length}: \quad l_{14} = l_{34} = l_1, \quad l_{24} = l_2$$
$$\text{Cross-sectional area}: \quad A_{14} = A_{34} = A_1, \quad A_{24} = A_2$$
$$\text{Modulus of elasticity}: \quad E_{14} = E_{34} = E_1, \quad E_{24} = E_2$$

As there are three elements intersecting in node 4, it is impossible to compute the axial force for each element just by the equilibrium conditions, and the deformation conditions must be analyzed.

According to the theory of strength of materials, under the action of the axial force N of the uniform section member, the deflection on the member end is

$$\Delta = \frac{Nl}{AE} \tag{a}$$

Due to symmetry, the horizontal displacement of node 4 is 0. Set the vertical displacement of node 4 as v_4; as nodes 1, 2, and 3 are fixed, the deflection for each element may be calculated as follows:

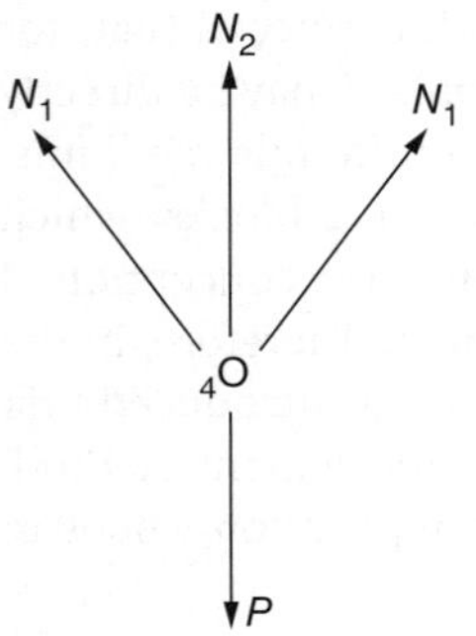

Figure 1.7 Equilibrium of node 4.

$$\Delta_{14} = v_4 \cos\theta, \quad \Delta_{24} = v_4, \quad \Delta_{34} = v_4 \cos\theta \tag{b}$$

Now, take the equilibrium conditions of node 4 into consideration (Figure 1.7); the equilibrium requirement for y direction is

$$2N_{1y} + N_{2y} = P \tag{c}$$

where N_{1y}, N_{2y} are, respectively, the projections of the axial force of each member on y-axis.

According to formulas (a) and (b),

$$\left.\begin{aligned} N_{1y} &= N_1 \cos\theta = \frac{E_1A_1}{l_1}\Delta_{14}\cos\theta = \frac{E_1A_1\cos^2\theta}{l_1}v_4 = k_1v_4 \\ N_{2y} &= N_2 = \frac{E_2A_2}{l_2}\Delta_{24} = \frac{E_2A_2}{l_2}v_4 = k_2v_4 \end{aligned}\right\} \tag{d}$$

where k_1 and k_2 are the stiffness coefficients of the element, which are the member end forces in the y-direction when the displacement of node 4 is 1. The numerical value depends on the geometric and physical properties of the member.

For example, the stiffness coefficient of element 14 is

$$k_{14} = k_1 = \frac{E_1A_1\cos^2\theta}{l_1}$$

The stiffness coefficient of element 24 is

$$k_{24} = k_2 = \frac{E_2A_2}{l_1}$$

Substitute the formula (d) into formula (c):

$$(2k_1 + k_2)v_4 = P$$

Then the displacement of node 4 may be solved as

$$v_4 = \frac{P}{2k_1 + k_2} \tag{e}$$

where $2k_1 + k_2$ is the global stiffness coefficient of the structure.

Based on the solved displacement, the axial force for each element may be computed as

$$N_{14} = N_{34} = N_1 = \frac{N_{1y}}{\cos\theta} = \frac{k_1v_4}{\cos\theta} = \frac{k_1P}{(2k_1 + k_2)\cos\theta}$$

$$N_{24} = N_2 = k_2v_4 = \frac{k_2P}{2k_1 + k_2}$$

The approach solves the nodal displacement by the equilibrium equation with the displacement as the basic unknown variable and then deducts reversely the internal force of each element. This is called the "displacement method."

According to the previous example, the finite element method may be briefed as follows: (1) The original structure is replaced by a group of elements that are connected at the nodes. (2) Take the displacement components of all the nodes as unknown variables and establish an equation set of equilibrium. (3) The inverse of the equilibrium equation will give the displacements of the nodes. (4) Compute the internal force for each element by the nodal displacements.

The internal force of each element may also be regarded as the basic unknown variable for solutions, which is called the "force method." However, the force method is more complicated compared with the displacement method for matrix calculation. Therefore, it is rarely applied in actual work, while the displacement method is widely applied in practical computation of structures.

1.3 Stiffness Matrix of Horizontal Bar Element

The stiffness coefficient of each member in the previous section is a special case because the member is fixed at one end. In the general case, as the truss shown in Figure 1.8, there may be displacements on both ends of elements. To explain in an evolutionary manner, in this section the horizontal straight bar *ij* will be researched first, as shown in Figure 1.9.

There is a horizontal displacement on each end of the element, that is, u_i and u_j, so the element has two degrees of freedom, and the stiffness matrix should be in a second-order form.

The stress condition of the member may be analyzed in two statuses:

Status 1: $u_i = u_i, u_j = 0$. At this time, node *j* is fixed.

$$\text{Element strain：} \quad \varepsilon = -\frac{u_i}{l}$$

$$\text{Element stress：} \quad \sigma = E\varepsilon = -\frac{Eu_i}{l}$$

In the theory of strength of material, the tensile stress is regarded as positive, while in finite element method, as shown in Figure 1.8, the rightward nodal force is regarded as positive. Therefore, a negative sign is added in the following formula:

$$\text{Nodal force at left end of element：} \quad U_i = -A\sigma = \frac{AE}{l}u_i$$

$$\text{Nodal force at right end of element：} \quad U_j = A\sigma = -\frac{AE}{l}u_j$$

Status 2: $u_i = 0, u_j = u_j$; status 2 is opposite to status 1.

$$\text{Element strain：} \quad \varepsilon = -\frac{u_j}{l}$$

$$\text{Element stress：} \quad \sigma = \frac{Eu_j}{l}$$

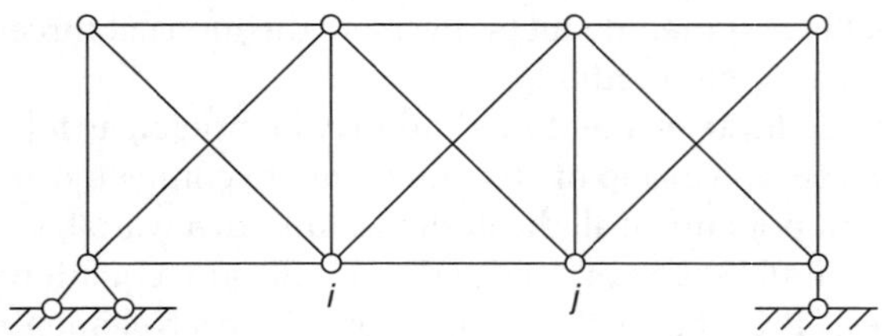

Figure 1.8 Truss.

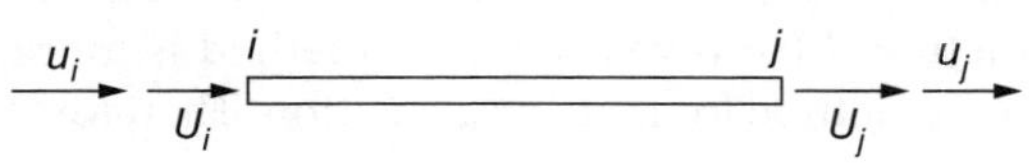

Figure 1.9 Horizontal bar element.

Nodal force at left end of element : $U_i = -A\sigma = -\frac{AE}{l}u_j$

Nodal force at right end of element : $U_j = A\sigma = \frac{AE}{l}u_j$

By combining the results of the previous two statuses, the element nodal forces are given by

$$U_i = \frac{AE}{l}u_i - \frac{AE}{l}u_j$$

$$U_j = -\frac{AE}{l}u_i + \frac{AE}{l}u_j$$

in matrix form:

$$\begin{Bmatrix} U_i \\ U_j \end{Bmatrix} = \frac{AE}{l}\begin{bmatrix} 1 & -1 \\ -1 & 1 \end{bmatrix}\begin{Bmatrix} u_i \\ u_j \end{Bmatrix} = [k]^{\mathrm{e}}\begin{Bmatrix} u_i \\ u_j \end{Bmatrix} \tag{1.7}$$

where

$$[k]^{\mathrm{e}} = \frac{AE}{l}\begin{bmatrix} 1 & -1 \\ -1 & 1 \end{bmatrix} \tag{1.8}$$

$[k]^{\mathrm{e}}$ is named as stiffness matrix of element.

The element axial force may be expressed as

$$N = \frac{AE}{l}\begin{bmatrix} -1 & 1 \end{bmatrix}\begin{Bmatrix} u_i \\ u_j \end{Bmatrix} = [S]\begin{Bmatrix} u_i \\ u_j \end{Bmatrix} \tag{1.9}$$

where

$$[S] = \frac{AE}{l}\begin{bmatrix} -1 & 1 \end{bmatrix} \tag{1.10}$$

Generally, the axial force is regarded as generalized stress. Therefore, matrix $[S]$ is also called the element stress matrix.

Actually, at node i and node j, besides the horizontal displacements, vertical displacements may also occur (however under small deformation conditions, the vertical node displacements bear no influence on the internal force of the bars). We may extend the stiffness matrix of the element into a fourth-order form. Therefore, the element nodal force is

$$\begin{Bmatrix} U_i \\ V_i \\ U_j \\ V_j \end{Bmatrix} = \frac{AE}{l}\begin{bmatrix} 1 & 0 & -1 & 0 \\ 0 & 0 & 0 & 0 \\ -1 & 0 & 1 & 0 \\ 0 & 0 & 0 & 0 \end{bmatrix}\begin{Bmatrix} u_i \\ v_i \\ u_j \\ v_j \end{Bmatrix} \tag{1.11}$$

In this formula, the vertical node displacement v_i and v_j as well as the vertical nodal force V_i and V_j are introduced, and accordingly some zeros are added into the stiffness matrix.

The general formula for the axial force of the horizontal bar element is

$$\text{Element axial force :} \quad N = \frac{AE}{l}\begin{bmatrix} -1 & 0 & 1 & 0 \end{bmatrix}\begin{Bmatrix} u_i \\ v_i \\ u_j \\ v_j \end{Bmatrix} = [S]\{\delta\}^{\mathrm{e}} \tag{1.12}$$

where $\{\delta\}^{\mathrm{e}} = \begin{bmatrix} u_i & v_i & u_j & v_j \end{bmatrix}^{\mathrm{T}}$.

1.4 Stiffness Matrix of Inclined Bar Element

For an inclined bar element as shown in Figure 1.10, the nodal displacement at node i is

$$\{\delta_i\} = \begin{Bmatrix} u_i \\ v_i \end{Bmatrix}$$

where u_i and v_i are, respectively, the horizontal and vertical nodal displacement components.

The nodal force at node i is

$$\{F_i\} = \begin{Bmatrix} U_i \\ V_i \end{Bmatrix}$$

where U_i and V_i are, respectively, the horizontal and vertical components of the nodal force.

It is stipulated that the symbols for nodal displacement u_i and v_i as well as the nodal force U_i and V_i should be positive when the direction is the same as coordinate axes x and y, or negative otherwise.

First, analyze the strain–displacement relationship of the inclined bar. Set the length of the bar element as l, as may be derived by geometrical relationship:

$$l^2 = (x_j - x_i)^2 + (y_j - y_i)^2 \tag{a}$$

Taking differential derivative on both sides of formula (a), we get

$$l\,\mathrm{d}l = (x_j - x_i)(\mathrm{d}x_j - \mathrm{d}x_i) + (y_j - y_i)(\mathrm{d}y_j - \mathrm{d}y_i)$$

Dividing both ends of the previous formula, respectively, by l, then

$$\mathrm{d}l = \alpha(\mathrm{d}x_j - \mathrm{d}x_i) + \beta(\mathrm{d}y_j - \mathrm{d}y_i) \tag{b}$$

where

$$\alpha = \cos\theta = \frac{x_j - x_i}{l}, \quad \beta = \sin\theta = \frac{y_j - y_i}{l} \tag{c}$$

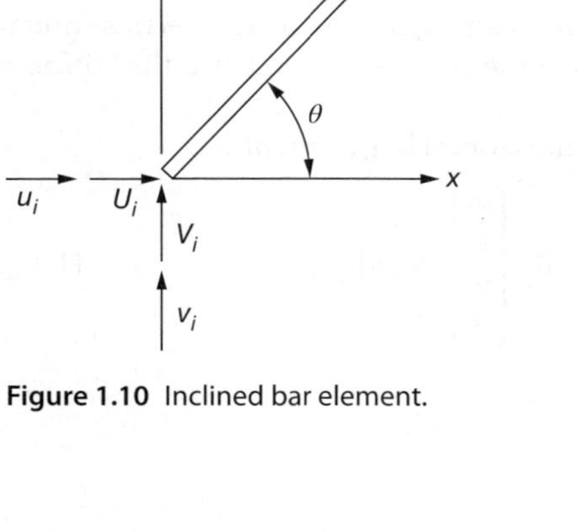

Figure 1.10 Inclined bar element.

After the member is deformed under stress, the coordinates of node i will be changed from (x_i, y_i) to $(x_i + u_i, y_i + v_i)$, namely,

$$\mathrm{d}x_i = u_i, \quad \mathrm{d}y_i = v_i$$

Similarly,

$$\mathrm{d}x_j = u_j, \quad \mathrm{d}y_j = v_j$$

Substituting the previous formula into formula (b) and dividing it by l, then we get the strain of the bar element:

$$\varepsilon = \frac{\mathrm{d}l}{l} = \frac{\alpha}{l}(-u_i + u_j) + \frac{\beta}{l}(-v_i + v_j) \tag{d}$$

The axial force of the inclined bar element is

$$N = AE\varepsilon = \frac{AE}{l}[\alpha(-u_i + u_j) + \beta(-u_j + v_j)] \tag{1.13}$$

It is stipulated that the tensile force is positive for axial force N.

The absolute value of the nodal force, respectively, is equal to the horizontal and vertical components of axial force N, namely,

$$\left.\begin{array}{ll} U_i = -N\cos\theta, & V_i = -N\sin\theta \\ U_j = N\cos\theta, & V_j = N\sin\theta \end{array}\right\} \tag{e}$$

Substitution of formula (1.13) into formula (e) yields the nodal force of the inclined bar element:

$$\{F\}^e = \begin{Bmatrix} U_i \\ V_i \\ U_j \\ V_j \end{Bmatrix} = \frac{AE}{l}\begin{bmatrix} \alpha^2 & \alpha\beta & -\alpha^2 & -\alpha\beta \\ \alpha\beta & \beta^2 & -\alpha\beta & -\beta^2 \\ -\alpha^2 & -\alpha\beta & \alpha^2 & \alpha\beta \\ -\alpha\beta & -\beta^2 & \alpha\beta & \beta^2 \end{bmatrix}\begin{Bmatrix} u_i \\ v_i \\ u_j \\ v_j \end{Bmatrix} = [k]^e\{\delta\}^e \tag{1.14}$$

where $[k]^e$ is the stiffness matrix of the inclined bar element.

Formula (1.14) may also be expressed by partitioned matrix as follows:

$$\begin{Bmatrix} F_i \\ F_j \end{Bmatrix} = \begin{bmatrix} k_{ii} & k_{ij} \\ k_{ji} & k_{jj} \end{bmatrix}\begin{Bmatrix} \delta_i \\ \delta_j \end{Bmatrix} \tag{1.15}$$

where

$$\left.\begin{array}{l} \{F_i\} = \begin{Bmatrix} U_i \\ V_i \end{Bmatrix}, \quad \{F_j\} = \begin{Bmatrix} U_j \\ V_j \end{Bmatrix}, \quad \{\delta_i\} = \begin{Bmatrix} u_i \\ v_i \end{Bmatrix}, \quad \{\delta_j\} = \begin{Bmatrix} u_j \\ v_j \end{Bmatrix} \\ [k_{ii}] = \dfrac{AE}{l}\begin{bmatrix} \alpha^2 & \alpha\beta \\ \alpha\beta & \beta^2 \end{bmatrix}, \quad [k_{ij}] = [k_{ji}] = \dfrac{AE}{l}\begin{bmatrix} -\alpha^2 & -\alpha\beta \\ -\alpha\beta & -\beta^2 \end{bmatrix} \\ [k_{jj}] = \dfrac{AE}{l}\begin{bmatrix} \alpha^2 & \alpha\beta \\ \alpha\beta & \beta^2 \end{bmatrix}, \quad \alpha = \cos\theta, \ \ \beta = \sin\theta \end{array}\right\} \tag{1.16}$$

According to formula (1.14), the element stiffness matrix is symmetrical. It is not an accidental phenomenon. Instead, it is a necessary result of the reciprocal theory of work in the structural mechanics.

Based on formula (1.13), the stress matrix of the inclined bar element is expressed as

$$[S] = \frac{AE}{l}\begin{bmatrix} -\alpha & -\beta & \alpha & \beta \end{bmatrix} \tag{1.17}$$

1.5 Coordinate Transformation

In Section 1.4, the stiffness matrix $[k]$ of the inclined bar element is deducted from the strain–displacement relationship of the bar. It may also be directly derived from the stiffness matrix of the horizontal bar element of the formula (1.11) according to the coordinate transformation relationship.

As shown in Figure 1.7, besides the global coordinates (x, y) of the structure, take a local coordinate $(\overline{x}, \overline{y})$, of which $\overline{x}$ axis is parallel to the axial line of bar element ij, while $\overline{y}$-axis is perpendicular to the axial line of the bar element. From formula (1.11), for the nodal force of the horizontal bar element, the nodal force of the bar element ij in the local coordinate system $(\overline{x}, \overline{y})$ may be given as follows:

$$\{\overline{F}\} = \begin{Bmatrix} \overline{U}_i \\ \overline{V}_i \\ \overline{U}_j \\ \overline{V}_j \end{Bmatrix} = \frac{AE}{l} \begin{bmatrix} 1 & 0 & -1 & 0 \\ 0 & 0 & 0 & 0 \\ -1 & 0 & 1 & 0 \\ 0 & 0 & 0 & 0 \end{bmatrix} \begin{Bmatrix} \overline{u}_i \\ \overline{v}_i \\ \overline{u}_j \\ \overline{v}_j \end{Bmatrix} \tag{1.18}$$

or

$$\{\overline{F}\} = [\overline{k}]^{e}\{\overline{\delta}\} \tag{1.19}$$

According to Figure 1.11, the transformation relationship between local coordinate and global coordinate is

$$\left.\begin{aligned} \overline{U}_i &= U_i \cos\theta + V_i \sin\theta \\ \overline{V}_i &= -U_i \sin\theta + V_i \cos\theta \\ \overline{U}_j &= U_j \cos\theta + V_j \sin\theta \\ \overline{V}_j &= -U_j \sin\theta + V_j \cos\theta \end{aligned}\right\} \tag{a}$$

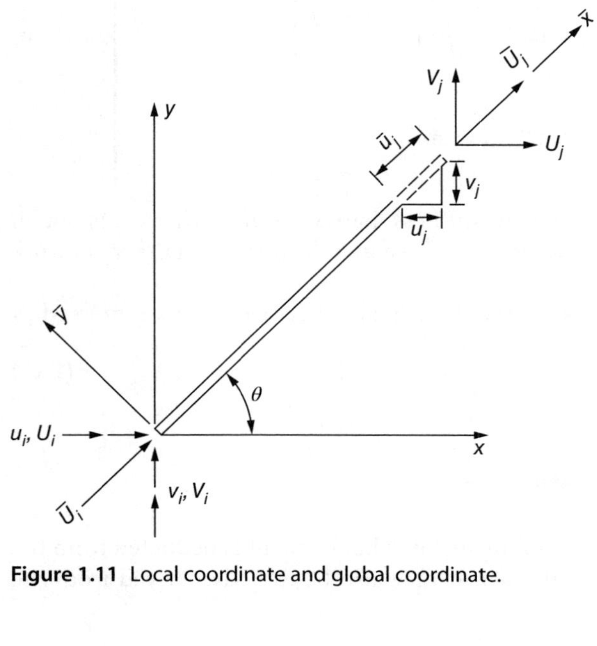

Figure 1.11 Local coordinate and global coordinate.

If in a matrix form

$$\begin{Bmatrix} \overline{U}_i \\ \overline{V}_i \\ \overline{U}_j \\ \overline{V}_j \end{Bmatrix} = \begin{bmatrix} \alpha & \beta & 0 & 0 \\ -\beta & \alpha & 0 & 0 \\ 0 & 0 & \alpha & \beta \\ 0 & 0 & -\beta & \alpha \end{bmatrix} \begin{Bmatrix} U_i \\ V_i \\ U_j \\ V_j \end{Bmatrix} \tag{b}$$

or

$$\{\overline{F}\} = [\lambda]\{F\} \tag{1.20}$$

where

$$[\lambda] = \begin{bmatrix} \alpha & \beta & 0 & 0 \\ -\beta & \alpha & 0 & 0 \\ 0 & 0 & \alpha & \beta \\ 0 & 0 & -\beta & \alpha \end{bmatrix} \tag{1.21}$$

$$\alpha = \cos\theta, \quad \beta = \sin\theta$$

in which $[\lambda]$ is the transformation matrix.

For nodal displacement, there is also a similar relationship between the local coordinate and the global coordinate:

$$\{\overline{\delta}\} = [\lambda]\{\delta\} \tag{1.22}$$

Now, try to transfer the stiffness matrix of the bar element in the local coordinate system into the stiffness matrix in the global coordinate system. Substitution of formulas (1.20) and (1.18) into formula (1.19) yields

$$[\lambda]\{F\} = [\overline{k}]^{\mathrm{e}}[\lambda]\{\delta\}$$

Multiplying $[\lambda]^{-1}$ on both sides of the previous formula, we have

$$\{F\} = [\lambda]^{-1}[\overline{k}]^{\mathrm{e}}[\lambda]\{\delta\}$$

As $[\lambda]$ is an orthogonal matrix, $[\lambda]^{-1} = [\lambda]^{\mathrm{T}}$, so

$$\{F\} = [\lambda]^{\mathrm{T}}[\overline{k}]^{\mathrm{e}}[\lambda]\{\delta\}$$

or

$$\{F\} = [k]^{\mathrm{e}}\{\delta\} \tag{1.23}$$

where

$$[k]^{\mathrm{e}} = [\lambda]^{\mathrm{T}}[\overline{k}]^{\mathrm{e}}[\lambda] \tag{1.24}$$

in which matrix $[k]$ is the stiffness matrix of the element in the global coordinate system.

Formula (1.24) is a basic formula to transfer the stiffness matrix $[\overline{k}]^{\mathrm{e}}$ of the element in local coordinate system into the stiffness matrix of the element in global coordinate system. Therefore, it is easy to prove that the result of $[k]^{\mathrm{e}}$ obtained by substituting $[\overline{k}]^{\mathrm{e}}$ and $[\lambda]$ into formula (1.24) is the same as formula (1.14).

1.6 Nodal Equilibrium Equation and Global Stiffness Matrix

Let us take node i out of a truss, as shown in Figure 1.12, and assume that there are three bar elements around node i, which, respectively, are ij, im, and ip. The horizontal and vertical loads on node i are, respectively, X_i and Y_i.

According to equilibrium of forces, the nodal force of the bar elements and the nodal force on the node are equal in value but opposite in direction. Taking bar ij, for example, the nodal force on the element bar is $[U_{ij}, V_{ij}]^{\mathrm{T}}$, while the nodal force on node i is $[-U_{ij}, -V_{ij}]^{\mathrm{T}}$.

To analyze the equilibrium condition of the node, node i is taken as the free body, as shown in Figure 1.12(c). The loads (X_i, Y_i) on node i and the nodal force on each bar element must be in equilibrium. The equilibrium equation of node i in the horizontal and vertical direction should be

$$
\begin{aligned}
&X_i - U_{ij} - U_{im} - U_{ip} = 0 \\
&Y_i - V_{ij} - V_{im} - V_{ip} = 0
\end{aligned}
$$

or

$$
\left.
\begin{aligned}
\sum_{e=j,m,p} U_{ie} = X_i \\
\sum_{e=j,m,p} V_{ie} = Y_i
\end{aligned}
\right\}
\tag{1.25}
$$

where Σ refers to the sum of the elements around node i.

Obviously, elements unrelated to node i are not included in the previous sum formula.

According to formula (1.15), the nodal force on the node by bar element ij is

$$
\{F_{ij}\} = \begin{Bmatrix} U_{ij} \\ V_{ij} \end{Bmatrix} = [k_{ii}]\{\delta_i\} + [k_{ij}]\{\delta_j\}
$$

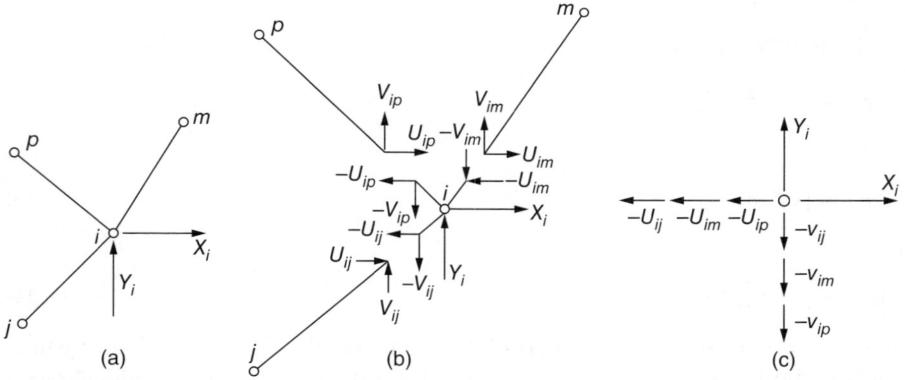

Figure 1.12 Equilibrium of node i.

The nodal load is

$$\{P_i\} = \begin{Bmatrix} X_i \\ Y_i \end{Bmatrix}$$

Substitution of the previous formula into formula (1.25) yields the equilibrium equation of node i:

$$\left(\sum [K_{ii}]\right)\{\delta_i\} + [k_{ij}]\{\delta_j\} + [k_{im}]\{\delta_m\} + [k_{ip}]\{\delta_p\} + \cdots = \{P_i\}$$

or

$$\left(\sum [k_{ii}]\right)\{\delta_i\} + \sum [k_{ij}]\{\delta_j\} = \{P_i\} \tag{1.26}$$

Each node is provided with a pair of equilibrium equations as mentioned earlier. After writing down the equilibrium equations one by one for all nodes $i = 1, 2, \dots, N$, we get the $2N$ order system of linear equations of equilibrium as follows (N means the number of nodes of the structure):

$$[K]\{\delta\} = \{P\} \tag{1.27}$$

where

$$\{\delta\} = [\delta_1, \delta_2, \dots, \delta_N]^{\mathrm{T}}$$
$$\{P\} = [P_1, P_2, \dots, P_N]^{\mathrm{T}}$$

in which $\{\delta\}$ is the vector formed by all nodal displacements, $\{P\}$ is the vector formed by all nodal loads, and $[K]$ is the global stiffness matrix of the structure. Equation (1.27) is the system of equilibrium equations for the structural nodes.

According to formula (1.26), the elements of $[K]$ may be calculated as follows:

$$K_{rs} = \sum_e k_{ij} \tag{1.28}$$

where $\sum_e$ is the sum of all the elements intersecting at node i, suffix rs means that K_{rs} is located in the row r and line s of the global stiffness matrix $[K]$, and suffix ij means that k_{ij} is located in line i and row j of the element stiffness matrix. According to the node number and freedom degree order, rs and ij may be determined easily, as detailed in the following sections. The physical meaning of the global stiffness coefficient K_{rs} is the rth nodal force due to element deformation of the sth degree of freedom.

After the node displacement is worked out by equilibrium equation (1.27), internal force of each element may be computed by the stress matrix.

1.7 Treatment of Boundary Conditions

Generally, the nodes on the boundary represent two cases. One is that the nodes on the boundaries are free in deformation, for example, nodes 5, 6, 7, 8, and so on in Figure 1.13. At this time, the load on such nodes may just be set as zero. Or, if there are external loads on node 3, the nodal load of such node may be set equal to the prescribed load Q. The

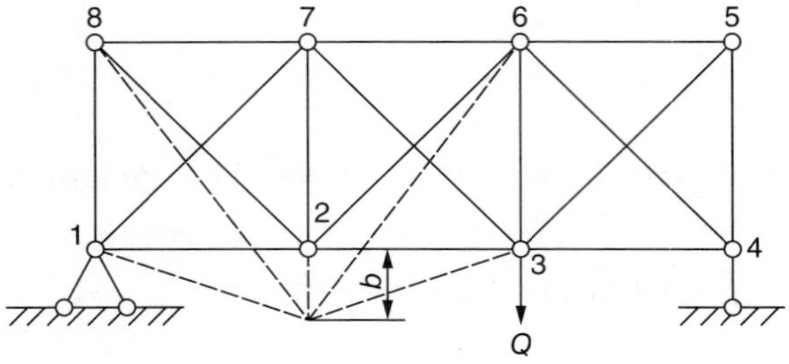

Figure 1.13 Truss.

treatment of such boundary conditions is relatively simple. The other case is that the values of node displacement are stipulated for the nodes on the boundaries, for example, the truss shown in Figure 1.8; it is required that

$$u_1 = v_1 = v_4 = 0, \quad v_2 = b \tag{a}$$

Then, whether we can directly put the formula (a) with prescribed displacement value into equilibrium equation $[P] = \{K\}(\delta)$. It is impossible if direct method is applied for solution, because all the nodal displacements by direct method are treated as unknown variables.

Now, rearrange the set of equilibrium equations (1.27) as follows:

$$\begin{bmatrix} K_{aa} & K_{ab} \\ K_{ab}^{\mathrm{T}} & K_{ab} \end{bmatrix} \begin{Bmatrix} \delta_a \\ \delta_b \end{Bmatrix} = \begin{Bmatrix} P_a \\ P_b \end{Bmatrix} \tag{b}$$

where $\{\delta_b\}$ is the known node displacement, for example, the node displacements expressed in formula (a), and (δ_a) is the unknown nodal displacement. Accordingly, $\{P_a\}$ is the known nodal load and $\{P_b\}$ is the unknown reaction of the support. According to the matrix multiplication, it may be drawn from formula (b) that

$$[K_{aa}]\{\delta_a\} + [K_{ab}]\{\delta_b\} = \{P_a\} \tag{c}$$

$$[K_{ab}]^{\mathrm{T}}\{\delta_a\} + [K_{bb}]\{\delta_b\} = \{P_b\} \tag{d}$$

As long as the given displacement $\{\delta_b\}$ is sufficient to block the rigid body movement of the structure, the submatrix $[K_{aa}]$ will be nonsingular. Solve the unknown nodal displacement $\{\delta_a\}$ in accordance with formula (c) as follows:

$$\{\delta_a\} = [K_{aa}]^{-1}(\{P_a\} - [K_{ab}]\{\delta_b\}) \tag{e}$$

We substitute formula (e) into formula (d) for the reaction of the support:

$$\{P_b\} = ([K_{aa}] - [K_{ab}]^{\mathrm{T}}[K_{aa}]^{-1}[K_{ab}])\{\delta_b\} + [K_{ab}]^{\mathrm{T}}[K_{aa}]^{-1}\{P_a\} \tag{f}$$

On condition that all the supports are rigid and the given node displacement $\{\delta_b\} = \{0\}$, formulas (c) and (d) may be simplified as

$$[K_{aa}]\{\delta_a\} = \{P_a\} \tag{g}$$

$$[K_{ab}]^{\mathrm{T}}\{\delta_a\} = \{P_b\} \tag{h}$$

Under such circumstance, it is easy to deduct formula (g) from formula (b). In the global stiffness matrix $[K]$, we first delete each line corresponding to $\{P_b\}$ and then delete each row corresponding to $\{\delta_b\}$ from the remaining part. Then $[K_{aa}]$ and formula (g) will be deducted. After solving $\{\delta_a\}$ from formula (g), substitute them into formula (h) for the reaction of the support $\{P_b\}$.

In the aforementioned sections, the steps to solve the set of equilibrium equations are explained. However, in finite element method, there are generally hundreds or even thousands of unknown variables. Generally, computer is used for solution. The nodal points of the given displacement and the given load generally appear in a staggered manner. Therefore, it is relatively complicated in procedure to partition a large-sized matrix $[K]$ in the manner of formula (b). Generally, to facilitate program design, the order of the matrix is recommended to remain unchanged; in other words, the numbers of lines and rows of stiffness matrix $[K]$ should not be reduced. The treatment methods of such boundary conditions will be detailed in the following part.

We set the equilibrium equation of the structure as

$$\begin{bmatrix} k_{1,1} & k_{1,2} & k_{1,3} & k_{1,4} & \cdots & k_{1,16} \\ k_{2,1} & k_{2,2} & k_{2,3} & k_{2,4} & \cdots & k_{2,16} \\ k_{3,1} & k_{3,2} & k_{3,3} & k_{3,4} & \cdots & k_{3,16} \\ k_{4,1} & k_{4,2} & k_{4,3} & k_{4,4} & \cdots & k_{4,4} \\ \vdots & \vdots & \vdots & \vdots & \cdots & \vdots \\ k_{16,1} & k_{16,2} & k_{16,3} & k_{16,14} & \cdots & k_{16,16} \end{bmatrix} \begin{Bmatrix} u_1 \\ v_1 \\ u_2 \\ v_2 \\ \vdots \\ v_8 \end{Bmatrix} = \begin{Bmatrix} X_1 \\ Y_1 \\ X_2 \\ Y_2 \\ \vdots \\ Y_8 \end{Bmatrix} \tag{i}$$

First, for the realization of the condition $u_1 = 0$, we change formula (i) as follows:

1) In stiffness matrix $[K]$, all the coefficients in the first line and first row are changed to zero except coefficient $k_{1,1}$ corresponding to u_1 on the primary diagonal.
2) Setting X_1 as 0 in the load matrix, then formula (i) is changed into

$$\begin{bmatrix} k_{1,1} & 0 & 0 & 0 & \cdots & 0 \\ 0 & k_{2,2} & k_{2,3} & k_{2,4} & \cdots & k_{2,16} \\ 0 & k_{3,2} & k_{3,3} & k_{3,4} & \cdots & k_{3,16} \\ 0 & k_{4,2} & k_{4,3} & k_{4,4} & \cdots & k_{4,16} \\ \vdots & \vdots & \vdots & \vdots & & \vdots \\ 0 & k_{16,2} & k_{16,3} & k_{16,4} & \cdots & k_{16,16} \end{bmatrix} \begin{Bmatrix} u_1 \\ v_1 \\ u_2 \\ v_2 \\ \vdots \\ v_8 \end{Bmatrix} = \begin{Bmatrix} 0 \\ Y_1 \\ X_2 \\ Y_2 \\ \vdots \\ Y_8 \end{Bmatrix} \tag{j}$$

Solution of formula (j) will satisfy automatically the condition of $u_1 = 0$. Analogy may be made similarly for conditions such as $v_1 = v_4 = 0$.

Then for the realization of condition $v_2 = b$, we assume a status that all the nodal displacements are zero except $\bar{v}_2 = -b$. Then, the node displacement vector is

$$\{\bar{\delta}\} = \begin{bmatrix} 0 & 0 & 0 & -b & 0 & \cdots & 0 \end{bmatrix}^{\mathrm{T}}$$

The node load vector is

$$\{\bar{P}\} = [K]\{\bar{\delta}\} = \begin{bmatrix} -k_{1,4}b & -k_{2,4}b & \cdots & -k_{16,4}b \end{bmatrix}^{\mathrm{T}}$$

Superposing such condition onto the original equation, we get the new equilibrium equation as follows:

$$[K](\{\delta\} + \{\bar{\delta}\}) = \{P\} + \{\overline{P}\}$$

Then we set the boundary to a condition meeting the requirements of the following formula:

$$v_2 + \bar{v}_2 = v_2 - b = 0$$

With reference to the treatment of $u_1 = 0$ mentioned earlier, the following treatment should be made for $v_2 = b$:

1) All coefficients in line 4 and row 4 corresponding to v_2 are changed into zero in stiffness matrix $[K]$ with only coefficient $k_{4,4}$ remained on the diagonal.
2) Change the load corresponding to v_2 as $Y_2 = k_{4,4}b$ in the load array, and change all other loads to the corresponding items in $\{P\} + \{\overline{P}\}$ as follows:

$$X_1 - k_{1,4}b, \quad Y_1 - k_{2,4}b, \ldots$$

In other words, we change the equilibrium equation into the following form for $v_2 = b$:

$$\begin{bmatrix} k_{1,1} & k_{1,2} & k_{1,3} & 0 & k_{1,5} & \cdots \\ k_{2,1} & k_{2,2} & k_{2,3} & 0 & k_{2,5} & \cdots \\ k_{3,1} & k_{3,2} & k_{3,3} & 0 & k_{3,5} & \cdots \\ 0 & 0 & 0 & k_{4,4} & 0 & \cdots \\ k_{5,1} & k_{5,2} & k_{5,3} & 0 & k_{5,5} & \cdots \\ \vdots & \vdots & \vdots & \vdots & \vdots & \end{bmatrix} \begin{Bmatrix} u_1 \\ v_1 \\ u_2 \\ v_2 \\ u_3 \\ \vdots \end{Bmatrix} = \begin{Bmatrix} X_1 - k_{1,4}b \\ Y_1 - k_{2,4}b \\ X_2 - k_{3,4}b \\ k_{4,4}b \\ X_3 - k_{5,4}b \\ \vdots \end{Bmatrix} \tag{k}$$

The solution of formula (k) will automatically realize the condition of $v_2 = b$.

To further simplify the calculation, the following approximate method may be applied:

1) The stiffness coefficient $k_{4,4}$ corresponding to v_2 on the diagonal is changed to an extreme great value, for example, $k_{4,4} \times 10^8$, namely,

$$k_{4,4} \to k_{4,4} \times 10^8$$

2) The nodal loads corresponding to v_2 is changed to $k_{4,4} \times 10^8 \times b$, namely,

$$Y_2 \to k_{4,4} \times 10^8 \times b$$

All the other coefficients remain unchanged; in other words, the equilibrium equation (i) is changed to the following form:

$$\begin{bmatrix} k_{1,1} & k_{1,2} & k_{1,3} & k_{1,4} & k_{1,5} & \cdots \\ k_{2,1} & k_{2,2} & k_{2,3} & k_{2,4} & k_{2,5} & \cdots \\ k_{3,1} & k_{3,2} & k_{3,3} & k_{3,4} & k_{3,5} & \cdots \\ k_{4,1} & k_{4,2} & k_{4,3} & k_{4,4} \times 10^8 & k_{4,5} & \cdots \\ k_{5,1} & k_{5,2} & k_{5,3} & k_{5,4} & k_{5,5} & \cdots \\ \vdots & \vdots & \vdots & \vdots & \vdots & \end{bmatrix} \begin{Bmatrix} u_1 \\ v_1 \\ u_2 \\ v_2 \\ u_3 \\ \vdots \end{Bmatrix} = \begin{Bmatrix} X_1 \\ Y_1 \\ X_2 \\ k_{4,4} \times 10^8 \times b \\ X_3 \\ \vdots \end{Bmatrix} \tag{l}$$

In the previous set of equations, if either side of the fourth formula is divided by 10^8, except $k_{4,4}v_2$, all other items are approximate to zero on the left and $k_{4,4}b$ on the right. Therefore, to solve in accordance with the aforementioned formula, v_2 will be quite approximate to b, namely,

$$v_2 \approx b$$

The aforementioned two treatment methods are applicable for the solution of various types of matrices, especially the second method, which may greatly facilitate the design of program.

Under many circumstances, the reaction of the support is required to be computed. To this end, the reaction required may be analyzed just by substituting the solved node displacements into the unmodified equilibrium equation of the support. For example, to compute the reactions of node 1, R_{x1} and R_{y1}, according to the first line and second line in formula (i), the equilibrium condition of support 1 may be expressed as

$$\left.\begin{aligned} R_{X1} &= X_1 - k_{1,1}u_1 - k_{1,2}v_1 - k_{1,3}u_2 - \cdots - k_{1,16}v_8 \\ R_{y1} &= Y_1 - k_{2,1}u_1 - k_{2,2}v_1 - k_{2,3}u_2 - \cdots - k_{2,16}v_8 \end{aligned}\right\} \tag{m}$$

By substituting the solved nodal displacements into formula (m), we will get the reaction R_{x1} and R_{y1} of the support.

Now, a truss shown in Figure 1.14 is taken as an example to explain the calculation methods. The supporting condition is

$$\begin{aligned} &u_1 = v_1 = u_4 = v_4 = 0 \\ &u_3 = b \end{aligned} \tag{n}$$

The truss consists of six bar elements in total, with the dimension and angle of inclination of each element listed in Table 1.1.

First, we calculate the stiffness matrix of the element according to formula (1.14):

$$[k]^e = \frac{AE}{l}\begin{bmatrix} \alpha^2 & \alpha\beta & -\alpha^2 & -\alpha\beta \\ \alpha\beta & \beta^2 & -\alpha\beta & -\beta^2 \\ -\alpha^2 & -\alpha\beta & \alpha^2 & \alpha\beta \\ -\alpha\beta & -\beta^2 & \alpha\beta & \beta^2 \end{bmatrix}$$

Figure 1.14 Truss.

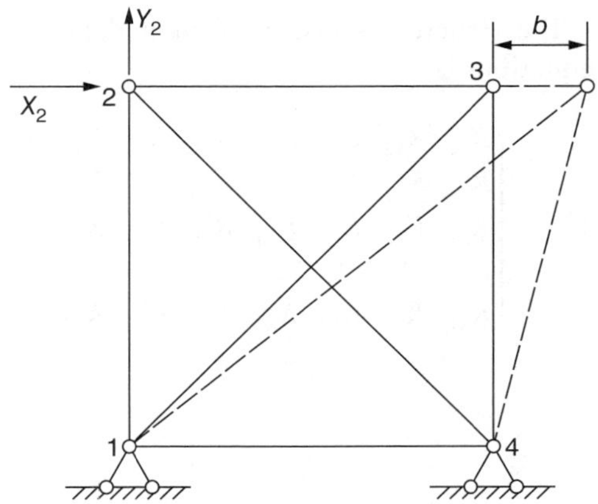

Table 1.1 Element dimension.

Bar element	*i* Point	*j* Point	Area	Length	Modulus of elasticity	Angle of inclination $\theta(*)$	$\alpha = \cos\theta$	$\beta = \sin\theta$	α^2	β^2	$\alpha\beta$
1-2	1	2	A	l	E	90	0	1	0	1	0
1-3	1	3	A	$\sqrt{2}l$	E	45	$1/\sqrt{2}$	$1/\sqrt{2}$	1/2	1/2	1/2
1-4	1	4	A	l	E	0	1	0	1	0	0
2-3	2	3	A	l	E	0	1	0	1	0	0
2-4	2	4	A	$\sqrt{2}l$	E	315	$1/\sqrt{2}$	$-1/\sqrt{2}$	1/2	1/2	−1/2
3-4	3	4	A	l	E	270	0	−1	0	1	0

By substituting the data in Table 1.1 into the previous formula, the stiffness matrix of each bar element is obtained as follows:

$$[k]^{\mathrm{e}}_{12} = \frac{AE}{l}\begin{bmatrix} 0 & 0 & 0 & 0 \\ 0 & 1 & 0 & -1 \\ 0 & 0 & 0 & 0 \\ 0 & -1 & 0 & 1 \end{bmatrix}$$

$$[k]^{\mathrm{e}}_{13} = \frac{AE}{\sqrt{2}l}\begin{bmatrix} 0.5 & 0.5 & -0.5 & -0.5 \\ 0.5 & 0.5 & -0.5 & -0.5 \\ -0.5 & -0.5 & 0.5 & 0.5 \\ -0.5 & -0.5 & 0.5 & 0.5 \end{bmatrix}$$

$$[k]^{\mathrm{e}}_{14} = \frac{AE}{l}\begin{bmatrix} 1 & 0 & -1 & 0 \\ 0 & 0 & 0 & 0 \\ -1 & 0 & 1 & 0 \\ 0 & 0 & 0 & 0 \end{bmatrix}$$

The structure consists of eight degrees of freedom. The set of nodal equilibrium equations is

$$\begin{pmatrix} K_{11} & K_{12} & K_{13} & K_{14} & K_{15} & K_{16} & K_{17} & K_{18} \\ K_{21} & K_{22} & K_{23} & K_{24} & K_{25} & K_{26} & K_{27} & K_{28} \\ K_{31} & K_{32} & K_{33} & K_{34} & K_{35} & K_{36} & K_{37} & K_{38} \\ \vdots & \vdots & \vdots & \vdots & \vdots & \vdots & \vdots & \vdots \\ K_{81} & K_{82} & K_{83} & K_{84} & K_{85} & K_{86} & K_{87} & K_{88} \end{pmatrix} \begin{Bmatrix} u_1 \\ v_1 \\ u_2 \\ \vdots \\ v_4 \end{Bmatrix} = \begin{Bmatrix} X_1 \\ Y_1 \\ X_2 \\ \vdots \\ Y_4 \end{Bmatrix}$$

According to formula (1.6), the elements of the global stiffness matrix may be computed according to the following formula:

$$k_{rs} = \sum k_{ij}$$

where $\sum$ is the sum of all elements intersecting on node i.

For example, the three bar elements (1-2, 1-3, 1-4) are intersecting on node 1. In the calculation of the global stiffness coefficient of node 1, the corresponding stiffness coefficients related to the three elements should be substituted into the previous formula for sum, for example:

$$K_{11} = (k_{11})_{12} + (k_{11})_{13} + (k_{11})_{14} = \frac{AE}{\sqrt{2}l}(0 + 0.5 + \sqrt{2}) = \frac{AE}{\sqrt{2}l} \times 1.914$$

$$K_{12} = (k_{12})_{12} + (k_{12})_{13} + (k_{12})_{14} = \frac{AE}{\sqrt{2}l}(0 + 0.5 + 0) = \frac{AE}{\sqrt{2}l} \times 0.5$$

$$K_{13} = (k_{13})_{12} = 0$$

$$K_{14} = (k_{14})_{12} = 0$$

$$K_{15} = (k_{13})_{13} = \frac{AE}{\sqrt{2}l} \times (-0.5)$$

$$K_{16} = (k_{14})_{13} = \frac{AE}{\sqrt{2}l} \times (-0.5)$$

$$K_{17} = (k_{13})_{14} = \frac{AE}{l} \times (-1)$$

$$K_{18} = (k_{14})_{14} = 0$$

In aforementioned formulas, $(k_{13})_{14}$ is the third element in line 1 of the element stiffness matrix for bar 14. The items mentioned earlier are elements in line 1 of the global stiffness matrix $[K]$. Explanations will be made in the following part on the calculation of such coefficients.

K_{11} is the first nodal force U_1 caused by the element displacement $u_1 = 1$ of the first degree of freedom u_1. According to Figure 1.13, when $u_1 = 1$, nodal U_1 will be generated in the three elements of 12, 13, and 14. Therefore, K_{11} may be computed by summing up the stiffness coefficient k_{11} of the three elements.

Another example is K_{16}, which is the nodal force U_1 caused by the sixth degree of freedom $v_3 = 1$ of the structure. Elements 12 and 14 have nothing to do with v_3 and are excluded out of the calculation. v_3 is the sixth degree of freedom for the structure but the fourth degree of freedom for element 13; therefore, $K_{16} = (k_{14})_{13}$.

The rest may be deduced by analogy for other elements in the global stiffness matrix $[K]$, and then we get the set of equilibrium equations for all the nodes as follows:

$$\frac{AE}{\sqrt{2}l}\begin{bmatrix} 1.914 & 0.5 & 0 & 0 & -0.5 & -0.5 & -1.414 & 0 \\ 0.5 & 1.914 & 0 & -1.414 & -0.5 & -0.5 & 0 & 0 \\ 0 & 0 & 1.914 & -0.5 & -1.414 & 0 & -0.5 & 0.5 \\ 0 & -1.414 & -0.5 & 1.914 & 0 & 0 & 0.5 & -0.5 \\ -0.5 & -0.5 & -1.414 & 0 & 1.914 & 0.5 & 0 & 0 \\ -0.5 & -0.5 & 0 & 0 & 0.5 & 1.914 & 0 & -1.414 \\ -1.414 & 0 & -0.5 & 0.5 & 0 & 0 & 1.914 & -0.5 \\ 0 & 0 & 0.5 & -0.5 & 0 & -1.414 & -0.5 & 1.914 \end{bmatrix}$$

$$\times \begin{Bmatrix} u_1 \\ v_1 \\ u_2 \\ v_2 \\ u_3 \\ v_3 \\ u_4 \\ v_4 \end{Bmatrix} = \begin{Bmatrix} X_1 \\ Y_1 \\ X_2 \\ Y_2 \\ X_3 \\ Y_3 \\ X_4 \\ Y_4 \end{Bmatrix} \tag{o}$$

Now, the boundary conditions expressed in formula (n) are taken into consideration. In accordance with the approximate methods mentioned earlier, we multiply stiffness coefficients $k_{11}, k_{22}, k_{55}, k_{77}, k_{88}$ on the primary diagonal corresponding to u_1, v_1, u_3, u_4, v_4, and so on by 10^8 in the global stiffness matrix $[K]$ and then change corresponding loads X_1, Y_1, X_4, Y_4 in load vector $\{P\}$ to zero and X_3 to $k_{35} \times 10^8 \times b$, and then we get the equilibrium equation as formula (p):

$$\frac{AE}{\sqrt{2}l}\times\begin{bmatrix} 1.914\times10^8 & 0.5 & 0 & 0 & -0.5 & -0.5 & -1.414 & 0 \\ 0.5 & 1.914\times10^8 & 0 & -1.414 & -0.5 & -0.5 & 0 & 0 \\ 0 & 0 & 1.914 & -0.5 & -1.414 & 0 & -0.5 & 0.5 \\ 0 & -1.414 & -0.5 & 1.914 & 0 & 0 & 0.5 & -0.5 \\ -0.5 & -0.5 & -1.414 & 0 & 1.914\times10^8 & 0.5 & 0 & 0 \\ -0.5 & -0.5 & 0 & 0 & 0.5 & 1.914 & 0 & -1.414 \\ -1.414 & 0 & -0.5 & 0.5 & 0 & 0 & 1.914\times10^8 & -0.5 \\ 0 & 0 & 0.5 & -0.5 & 0 & -1.414 & -0.5 & 1.914\times10^8 \end{bmatrix}$$

$$\times \begin{Bmatrix} u_1 \\ v_1 \\ u_2 \\ v_2 \\ u_3 \\ v_3 \\ u_4 \\ v_4 \end{Bmatrix} = \begin{Bmatrix} 0 \\ 0 \\ X_2 \\ Y_2 \\ 1.914\times10^8\times b \\ Y_3 \\ 0 \\ 0 \end{Bmatrix} \tag{p}$$

Solving formula (p) for the nodal displacements in accordance with formula (n) and substituting the solved nodal displacements into formula (1.13), the axial forces of each element will be obtained.

It is clear from this example that after the global stiffness matrix is established, only some algebraic operations are required. Therefore, the key to solve the problem is to establish the global stiffness matrix.

Bibliography

1 Rao, S.S. (2001) *The Finite Element Method in Engineering*, Elsevier, New York.
2 Langefors, B. (1952) Analysis of elastic structures by matrix transformation, with special regard to semimonocoque structures. *J. Aerosol Sci.*, **19** (7), 451–458.
3 Denke, P.H. (1954) *A Matrix Method of Structural Analysis*. Proceedings of the Second U. S. National Congress Applied Mechanics. ASME, pp. 445–457.
4 Turner, M.J., Clough, R.W., Martin, H.C. and Topp, L.J. (1956) Stiffness and deflection analysis of complex structures. *J. Aerosol Sci.*, **23**, 805–823.
5 Clough, R.W. (1960) *The Finite Element in Plane Stress Analysis*. Proceedings of the Second ASCE Conference on Electronic Computation.
6 Argyris, J.H. (1954) Energy theorems and structural analysis. *Aircr. Eng.*, **26**, 341–356, 383–387, 394; 1955, **27**: 42–58, 80–94, 125–134, 145–158.
7 Ghali, A. and Neville, A.M. (1992) *Structural Analysis*, Intex Educational Publisher, Scranton, PA.

2

Plane Problems in Theory of Elasticity

2.1 Discretization of Continuous Medium

In Chapter 1 we apply the matrix displacement method to analyze the truss; every member is treated as an element that is connected to one another on nodes. Nodal displacements are taken as the basic unknown quantities for analysis. As the number of nodes is limited, the number of the nodal equilibrium equations is also limited; thus linear algebra (or matrix) equations may be used for analysis on the computer.

In a continuous medium, interconnected points are infinite with an infinite number of degrees of freedom (DOFs), making it difficult for numerical solutions. The finite element method extends the application of the matrix analysis method for the truss to continuous medium: the original continuous medium is replaced by the combination of a finite number of elements. Thus a group of elements are interconnected on a finite number of nodes containing a finite number of DOFs, making it possible to be analyzed on a computer.

As shown in Figure 2.1(a), between A and B, using a truss to bear load P, every member can be taken as an element and every element connects to each other only by public nodes. It is easy to obtain internal forces of members by the equilibrium conditions of nodes. Using a plate to bear load P, this plate has an infinite number of DOFs.

Now we imitate the analysis of truss and divide the plate into some triangular elements by a group of dash lines. Assume that every element is interconnected only on public nodes as shown in Figure 2.1(d). Take nodal displacements as unknown quantities and establish a set of algebraic equations by the equilibrium conditions of nodes. The solution of the equilibrium equations will give the nodal displacements of the structure and hence the stresses within elements. So judging superficially, the computation of plates is almost the same as that of trusses; however, there remains a significant difference. For trusses, elements are connected only on nodes with no other connections; thus the solution described in Chapter 1 leads to exact outcomes. If we further divide each member of truss into several elements and make calculation more carefully, we will get the same results. As for the plate, displacements and stresses are originally continuous on the common boundaries of adjacent elements. Now assume that the elements are only interconnected on public nodes, the computed displacements and stresses on the common boundaries of adjacent elements may be discontinuous and lead to errors. As shown in Figure 2.2, two adjacent elements share the same displacement only on common nodes; displacement differences possible on common boundaries are presented as

The Finite Element Method: Fundamentals and Applications in Civil, Hydraulic, Mechanical and Aeronautical Engineering, First Edition. Bofang Zhu.

Companion website: www.wiley.com/go/zhufem17

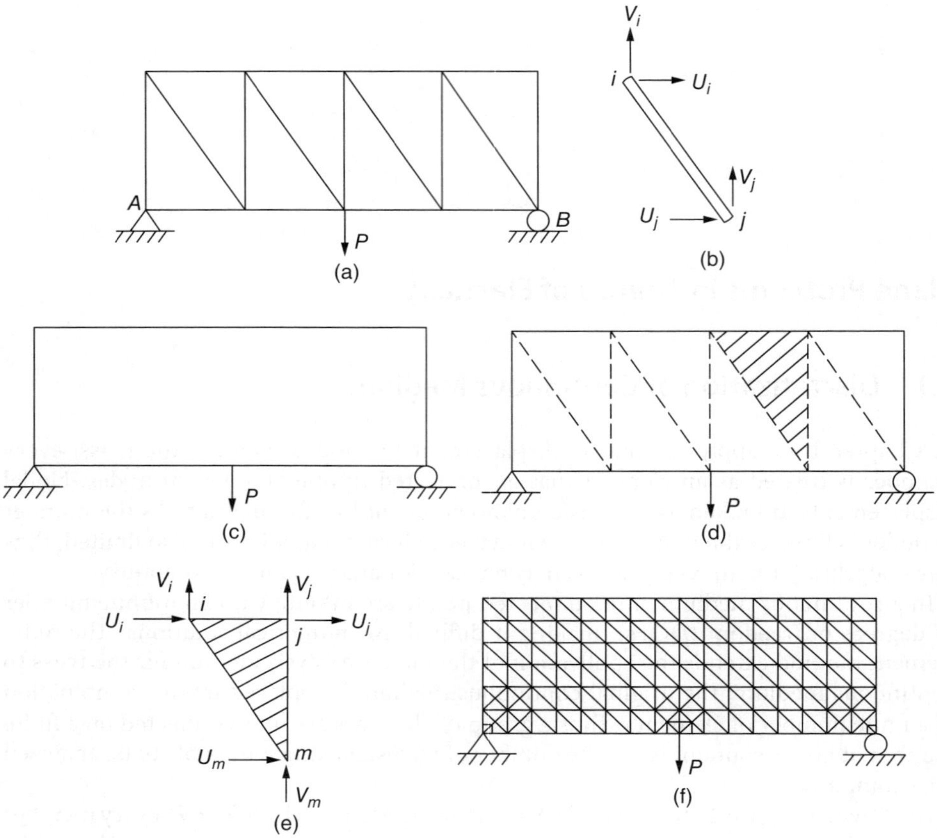

Figure 2.1 Discretization of a structure. (a) Truss, (b) element of truss, (c) plate, (d) plate replaced by the combination of several elements, (e) plane element, and (f) typical division of elements.

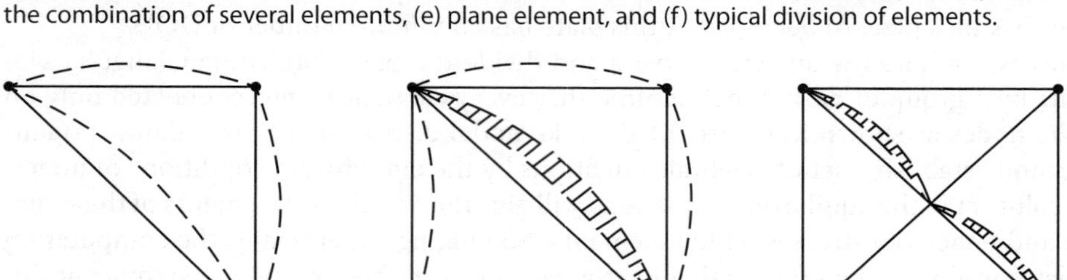

Figure 2.2 Displacements of structure. (a) Element displacement, (b) displacement difference on common boundaries of adjacent elements, and (c) intensified grid (reduced displacement difference).

the shaded area of the figure. If we thicken the computational grid and have the elements divided more smaller, the displacement differences will also be reduced.

Therefore, in order to ensure the necessary precision, intensive computational grids must be adopted, and in stress-concentrated areas, for example, near fulcrums or centralized forces, local grids should also be intensified. In addition, in some cases,

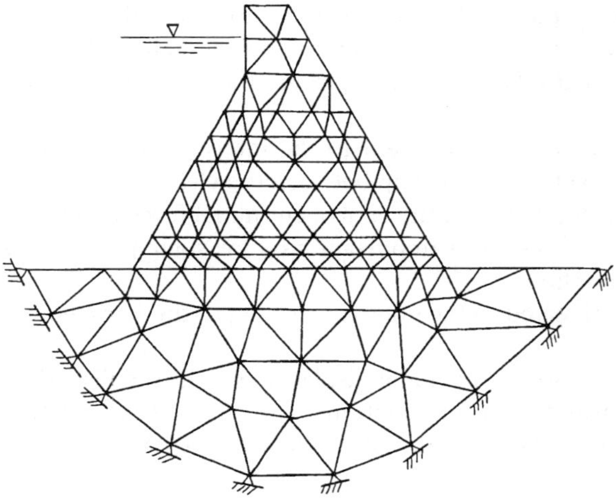

Figure 2.3 Discretization of the buttress dam.

some other methods concerning element design may be taken to improve the accuracy of calculation. For example, when we solve 3-node triangular elements by the displacement method, the element displacement function is generally a linear function of the coordinates, so on the common boundaries of adjacent elements, the displacement is continuous but the stress is discontinuous. Therefore using the finite element method to calculate continuous medium, the results obtained are not exact but approximate solutions. This is different from the truss. However, the application of high-speed large-capacity computers can make the computational grids so dense as to ensure adequate calculation precision for engineering needs.

In the application of the finite element method, a 2D continuous medium is replaced with combination of a finite number of 2D elements; a three-dimensional (3D) continuous medium is replaced with combination of a finite number of 3D elements. Of course, within these elements, all material properties of the original medium are maintained.

The finite element analysis for continuous medium contains three basic aspects: discretization of the medium, calculation of element properties, and structural analysis of element combination.

As for 2D continuous medium, taking the buttress dam on the rock base as shown in Figure 2.3, for example, the analysis steps by the finite element method are as follows:

1) The original medium is divided into a finite number of triangular elements by virtual straight lines. These lines are boundaries of elements and the intersection of several straight lines is called the node.
2) Assuming that the elements are interconnected on nodes and the nodal displacements are taken as the basic unknown quantities.
3) A function is given that can uniquely represent the displacement of any point within the element by the displacements of 3 nodes; this function is called a displacement function.

4) By a displacement function, the nodal displacement can be used to uniquely represent the strain of any point within the element; then by the generalized Hooke's law, the nodal displacement can be used to uniquely represent the stress of any point within the element.
5) By the energy principles the nodal force equivalent to the internal stress state of the element is derived; and then by the relationship between the element stress and nodal displacement, the relationship between the equivalent nodal force and nodal displacement is established. This is the most important step for solving the stress problems by the finite element method.
6) The loads of every element are relocated to the nodes according to the static equivalence principle.
7) A static equilibrium equation represented by nodal displacements is established on every node and we obtain a linear system of equations; solution of this equation set will give the nodal displacements; then the stress of every element can be obtained.

To provide a step-by-step exploration, in this chapter we will only introduce the easiest but most frequently used triangular element to solve plane problems. The element displacement changes linearly and the strain and stress within the element are constants. The subsequent chapters will expound plane elements in other various forms.

2.2 Displacement Function

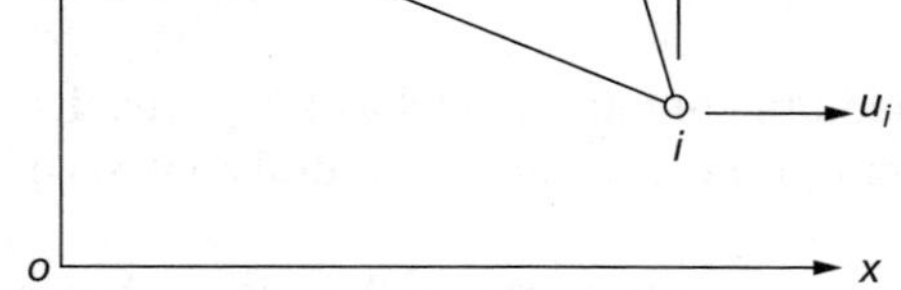

Figure 2.4 Triangular elements in plane problems.

Figure 2.4 shows a typical triangular element. The 3 nodes are i, j, m, which are arranged in a counterclockwise order. Every node has two displacement components, namely,

$$\{\delta_i\} = \begin{Bmatrix} u_i \\ v_i \end{Bmatrix} \tag{2.1}$$

The six nodal displacement components of every element can be expressed as a vector, which is

$$\{\delta\}^e = \begin{Bmatrix} \delta_i \\ \delta_j \\ \delta_m \end{Bmatrix} \tag{2.2}$$

If only the nodal displacements are available, we cannot directly get the strain and stress within the element. Therefore in order to represent the strain and stress in the element with nodal displacements, it is necessary to assume that the displacement component of any point within the element is a certain function of the coordinates.

Now it is assumed that displacement components within the element are linear functions of the coordinates, that is to say,

$$u = \beta_1 + \beta_2 x + \beta_3 y, \quad v = \beta_4 + \beta_5 x + \beta_6 y \tag{2.3}$$

The six coefficients β_i in the above formula can be represented by nodal displacements. Substitution of the coordinates of points i, j, m into formula (2.3) gives

$$u_i = \beta_1 + \beta_2 x_i + \beta_3 y_i, \quad u_j = \beta_1 + \beta_2 x_j + \beta_3 y_j, \quad u_m = \beta_1 + \beta_2 x_m + \beta_3 y_m \tag{a}$$

$$v_i = \beta_4 + \beta_5 x_i + \beta_6 y_i, \quad v_j = \beta_4 + \beta_5 x_j + \beta_6 y_j, \quad v_m = \beta_4 + \beta_5 x_m + \beta_6 y_m \tag{b}$$

By Cramer's rule in the linear algebra, we can solve the coefficients β_1, β_2, β_3 from formula (a) and β_4, β_5, β_6 from formula (b) and then substitute them back into formula (2.3), and then we will get the following element displacement function:

$$\begin{aligned} u &= \frac{1}{2A}\{(a_i + b_i x + c_i y)u_i + (a_j + b_j x + c_j y)u_j + (a_m + b_m x + c_m y)u_m\} \\ v &= \frac{1}{2A}\{(a_i + b_i x + c_i y)v_i + (a_j + b_j x + c_j y)v_j + (a_m + b_m x + c_m y)v_m\} \end{aligned} \tag{2.4}$$

$$\left.\begin{aligned} a_i &= x_j y_m - x_m y_j, \quad & b_i &= y_j - y_m, \quad & c_i &= x_m - x_j \\ a_j &= x_m y_i - x_i y_m, \quad & b_j &= y_m - y_i, \quad & c_j &= x_i - x_m \\ a_m &= x_i y_j - x_j y_i, \quad & b_m &= y_i - y_j, \quad & c_m &= x_j - x_i \end{aligned}\right\} \tag{2.5}$$

$$A = \frac{1}{2}\begin{bmatrix} 1 & x_i & y_i \\ 1 & x_j & y_j \\ 1 & x_m & y_m \end{bmatrix} \tag{2.6}$$

According to the analytic geometry, A equals the area of triangle ijm. To make the area A obtained not negative, as previously mentioned, i, j, m must be in the counterclockwise order.

To simplify the expression of displacement functions, they are described as

$$N_i = \frac{a_i + b_i x + c_i y}{2A}, \quad N_j = \frac{a_j + b_j x + c_j y}{2A}, \quad N_m = \frac{a_m + b_m x + c_m y}{2A} \tag{2.7}$$

Substitution of Eq. (2.7) into formula (2.4) yields a simple expression of the displacement function as follows:

$$u = N_i u_i + N_j u_j + N_m u_m, \quad v = N_i v_i + N_j v_j + N_m v_m \tag{2.8}$$

N_i, N_j, N_m are functions of coordinates, reflecting the displacement patterns of the element and known as shape functions of the *element displacement*.

Element displacements expressed by formula (2.8) can also be transformed into the following matrix form:

$$\{r\} = \begin{Bmatrix} u \\ v \end{Bmatrix} = [N]\{\delta\}^{\mathrm{e}} = \begin{bmatrix} IN_i & IN_j & IN_m \end{bmatrix}\{\delta\}^{\mathrm{e}} \tag{2.9}$$

where $I = \begin{bmatrix} 1 & 0 \\ 0 & 1 \end{bmatrix}$ is a second-order unit matrix.

According to the displacement function (2.8), the displacement changes linearly on the element boundary. Since two adjacent elements have the same nodal displacement on their common nodes, the two elements will have the same displacement on their common boundary; in other words, the displacement function selected ensures the continuity of the displacement between two adjacent elements.

2.3 Element Strain

As a plane problem, the three strain components ε_x, ε_y, γ_{xy} in the element can be expressed by matrix as follows:

$$\{\varepsilon\} = \begin{Bmatrix} \varepsilon_x \\ \varepsilon_y \\ \gamma_{xy} \end{Bmatrix} = \begin{Bmatrix} \dfrac{\partial u}{\partial x} \\ \dfrac{\partial v}{\partial y} \\ \dfrac{\partial u}{\partial y} + \dfrac{\partial v}{\partial x} \end{Bmatrix} \tag{2.10}$$

By substituting the displacement function (2.8) into formula (2.10), we have

$$\{\varepsilon\} = \begin{bmatrix} \dfrac{\partial N_i}{\partial x} & 0 & \dfrac{\partial N_j}{\partial x} & 0 & \dfrac{\partial N_m}{\partial x} & 0 \\ 0 & \dfrac{\partial N_i}{\partial y} & 0 & \dfrac{\partial N_j}{\partial y} & 0 & \dfrac{\partial N_m}{\partial y} \\ \dfrac{\partial N_i}{\partial y} & \dfrac{\partial N_i}{\partial x} & \dfrac{\partial N_j}{\partial y} & \dfrac{\partial N_j}{\partial x} & \dfrac{\partial N_m}{\partial y} & \dfrac{\partial N_m}{\partial x} \end{bmatrix} \begin{Bmatrix} u_i \\ v_i \\ u_j \\ v_j \\ u_m \\ v_m \end{Bmatrix}$$

$$= \frac{1}{2A} \begin{bmatrix} b_i & 0 & b_j & 0 & b_m & 0 \\ 0 & c_i & 0 & c_j & 0 & c_m \\ c_i & b_i & c_j & b_j & c_m & b_m \end{bmatrix} \{\delta\}^{\mathrm{e}} \tag{a}$$

or

$$\{\varepsilon\} = [B]\{\delta\}^{\mathrm{e}} \tag{2.11}$$

Matrix $[B]$ can be written in partitions as

$$[B] = \begin{bmatrix} B_i & B_j & B_m \end{bmatrix} \tag{2.12}$$

and its submatrix is

$$[B_i] = \frac{1}{2A} \begin{bmatrix} b_i & 0 \\ 0 & c_i \\ c_i & b_i \end{bmatrix} \quad (i, j, m) \tag{2.13}$$

The sign (i, j, m) behind the formula above means this formula actually represents three formulas and the other two formulas will be obtained by rotation of corner connection i, j, m thereafter. This book will continue the usage of this mark afterward to save space.

The elemental area A and coefficient b_i, c_i, and so on are all constants; as a result, the elements of matrix $[B]$ are constants. Thus elements of strain $\{\varepsilon\}$ are also constants, that is to say, in every element, the strain components ε_x, ε_y, γ_{xy} are all constants.

2.4 Initial Strain

Initial strain refers to the strain not related to stress and caused by temperature changes, shrinkage, crystal growth, and other factors:

$$\{\varepsilon_0\} = \begin{Bmatrix} \varepsilon_{x0} \\ \varepsilon_{y0} \\ \gamma_{xy0} \end{Bmatrix} \tag{2.14}$$

In general, the initial strain within the element is a function of the coordinates. When the element is small enough, the initial strain within the element may take an average value that is also a constant. This is consistent with the strain in the element specified in Eq. (2.11) – it is also a constant. Take the temperature deformation, for example, and let the temperature within the element be $T(x,y)$. The average temperature will be adopted to calculate the initial strain:

$$\overline{T} = \frac{1}{A}\iint T(x,y)\,\mathrm{d}x\,\mathrm{d}y$$

When $T(x,y)$ is a linear function of x and y, from the above formula, we can get

$$\overline{T} = \frac{T_i + T_j + T_m}{3}$$

in which T_i, T_j, T_m are, respectively, the temperature of nodes i, j, m.

As for nonlinear temperatures, the above formula may still be used approximately and the error caused herefrom is in the same order as that caused by a linear displacement function.

As for plane stress problems, the initial strain caused by temperatures $\overline{T}$ is

$$\{\varepsilon_0\} = \begin{Bmatrix} \alpha\overline{T} \\ \alpha\overline{T} \\ 0 \end{Bmatrix} \tag{2.15}$$

in which α is the coefficient of linear expansion.

As temperature changes will not cause shear deformation in isotropic medium, $\gamma_{xyo} = 0$.

As for plane strain problems, the initial strain caused by temperatures $\overline{T}$ is

$$\{\varepsilon_0\} = (1+\mu)\begin{Bmatrix} \alpha\overline{T} \\ \alpha\overline{T} \\ 0 \end{Bmatrix} \tag{2.16}$$

in which μ is Poisson's ratio of the material.

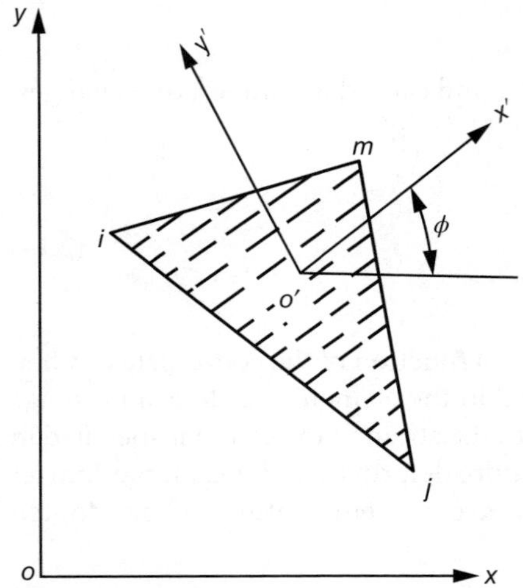

Figure 2.5 Elements in layered materials.

For layered anisotropic materials, the linear expansion coefficient may change with directions. As shown in Figure 2.5, let x' and y' be the main direction of layered materials, and the linear expansion coefficient in the x' and y' direction is, respectively, α_1 and α_2, and the initial strain of plane stress problems in the local coordinate system (x', y') is

$$\{\varepsilon_0'\} = \begin{Bmatrix} \varepsilon_{x'0} \\ \varepsilon_{y'0} \\ \gamma_{x'y'0} \end{Bmatrix} = \begin{bmatrix} \alpha_1 \overline{T} \\ \alpha_2 \overline{T} \\ 0 \end{bmatrix} \tag{2.17}$$

To obtain the initial strain in the global coordinate system (x, y), the following transformation is necessary:

$$\{\varepsilon_0'\} = [\theta]^{\mathrm{T}} \{\varepsilon_0\} \tag{2.18}$$

For the angle ϕ shown in Figure 2.5, it is easy to obtain

$$[\theta] = \begin{bmatrix} \cos^2\phi & \sin^2\phi & -2\sin\phi\cos\phi \\ \sin^2\phi & \cos^2\phi & 2\sin\phi\cos\phi \\ \sin\phi\cos\phi & -\sin\phi\cos\phi & \cos^2\phi - \sin^2\phi \end{bmatrix} \tag{2.19}$$

After transformation, the shear component of the initial strain in the global coordinate system (x, y) may not be zero.

2.5 Element Stress

After solving the elemental strain, we can easily get the elemental stress by the generalized Hooke's law, which will be introduced as follows.

2.5.1 Isotropic Body: Plane Stress

By the generalized Hooke's law, for plane stress problems of the isotropic body, the strain components are given by the following formulas:

$$\varepsilon_x = \frac{\sigma_x}{E} - \frac{\mu\sigma_y}{E} + \varepsilon_{x0}$$

$$\varepsilon_y = \frac{\sigma_y}{E} - \frac{\mu\sigma_x}{E} + \varepsilon_{y0}$$

$$\gamma_{xy} = \frac{2(1+\mu)}{E}\tau_{xy} + \gamma_{xy0}$$

From the above formulas we can solve the stress components and obtain

$$\sigma_x = \frac{E}{1-\mu^2}(\varepsilon_x - \varepsilon_{x0} + \mu\varepsilon_y - \mu\varepsilon_{y0})$$

$$\sigma_y = \frac{E}{1-\mu^2}(\mu\varepsilon_x - \mu\varepsilon_{x0} + \varepsilon_y - \varepsilon_{y0})$$

$$\tau_{xy} = \frac{E}{2(1+\mu)}(\gamma_{xy} - \gamma_{xy0}) = \frac{E}{1-\mu^2}\frac{1-\mu}{2}(\gamma_{xy} - \gamma_{xy0})$$

Combine all of the above three formulas and express them with a matrix equation as

$$\begin{Bmatrix} \sigma_x \\ \sigma_y \\ \tau_{xy} \end{Bmatrix} = \frac{E}{1-\mu^2}\begin{bmatrix} 1 & \mu & 0 \\ \mu & 1 & 0 \\ 0 & 0 & \dfrac{1-\mu}{2} \end{bmatrix}\begin{Bmatrix} \varepsilon_x - \varepsilon_{x0} \\ \varepsilon_y - \varepsilon_{y0} \\ \gamma_{xy} - \gamma_{xy0} \end{Bmatrix}$$

or

$$\{\sigma\} = [D](\{\varepsilon\} - \{\varepsilon_0\}) \tag{2.20}$$

$$[D] = \frac{E}{1-\mu^2}\begin{bmatrix} 1 & \mu & 0 \\ \mu & 1 & 0 \\ 0 & 0 & \dfrac{1-\mu}{2} \end{bmatrix} \tag{2.21}$$

in which $[D]$ is the elasticity matrix, expressed by the elastic constants E and μ.

Substitution of formula (2.11) into formula (2.20) gives

$$\{\sigma\} = [S]\{\sigma\}^e - [D]\{\varepsilon_0\} \tag{2.22}$$

$$[S] = [D][B] = \begin{bmatrix} S_i & S_j & S_m \end{bmatrix} \tag{2.23}$$

$$[S_i] = [D][B_i] = \frac{E}{2(1-\mu^2)A}\begin{bmatrix} b_i & \mu c_i \\ \mu b_i & c_i \\ \dfrac{1-\mu}{2}c_i & \dfrac{1-\mu}{2}b_i \end{bmatrix} \quad (i, j, m)$$

in which $[S]$ is the stress matrix.

According to formula (2.22), the element stress may be determined by the nodal displacements.

As the displacement function is linear, the strain component and stress component in every element are all constants. In a variable stress field, the adjacent elements always have different stresses, so on the common boundary of two elements, the stress will go through sudden change. But as the elements are smaller and smaller, this sudden change will be sharply reduced, not affecting the solution by the finite element method and leading to correct outcomes.

2.5.2 Isotropic Body: Plane Strain

Due to $\varepsilon_z = 0$, for plane strain problems, there exists the normal stress σ_z besides the stress components σ_x, σ_y, τ_{xy}. Assuming the initial stress is caused by temperature changes, the stress components in the element are given by the following formula:

$$\left.\begin{aligned} \varepsilon_x &= \frac{1}{E}(\sigma_x - \mu\sigma_y - \mu\sigma_z) + \alpha\overline{T} \\ \varepsilon_y &= \frac{1}{E}(\sigma_y - \mu\sigma_z - \mu\sigma_x) + \alpha\overline{T} \\ \gamma_{xy} &= \frac{2(1+\mu)}{E}\tau_{xy} \end{aligned}\right\} \tag{a}$$

Besides, there is also

$$\varepsilon_z = \frac{1}{E}(\sigma_z - \mu\sigma_x - \mu\sigma_y) + \alpha\overline{T} = 0 \tag{b}$$

After obtaining σ_z from formula (b) and substituting it into formula (a) to compute the three stress components σ_x, σ_y, τ_{xy}, we will the obtain matrix $[D]$ for plane strain problem as given by Eq. (2.24), and the initial strain $\{\varepsilon_0\}$ is shown in formula (2.16):

$$[D] = \frac{E(1-\mu)}{(1+\mu)(1-2\mu)}\begin{bmatrix} 1 & \dfrac{\mu}{1-\mu} & 0 \\ \dfrac{\mu}{1-\mu} & 1 & 0 \\ 0 & 0 & \dfrac{1-2\mu}{2(1-\mu)} \end{bmatrix} \tag{2.24}$$

By replacing E, μ, α in the plane stress formulas with $E/(1-\mu^2)$, $\mu/(1-\mu)$, $(1+\mu)\alpha$, we will get the corresponding plane strain formulas. On the contrary, by replacing E, μ, α in the plane strain formulas with $E(1+2\mu)/(1+\mu)^2$, $\mu/(1+\mu)$, $(1+\mu)\alpha/(1+2\mu)$, we will get the corresponding plane stress formulas. The general plane computing program can be set by plane stress (or plane strain) formula. In case there is any need to solve plane strain (or plane stress) problems, just make the above substitution with E, μ, α.

2.5.3 Anisotropic Body

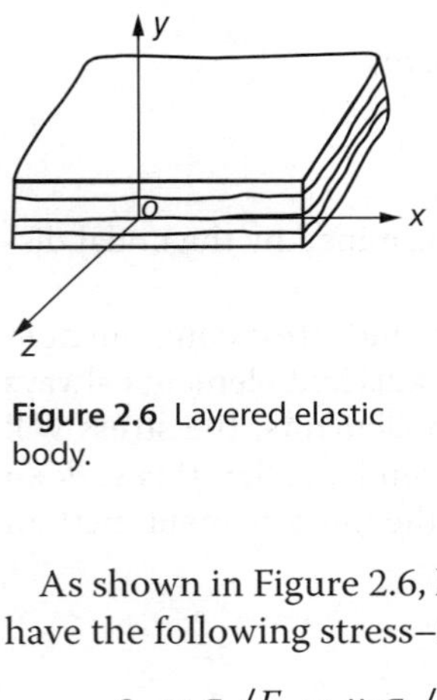

Figure 2.6 Layered elastic body.

For the most common anisotropic body with no elastic symmetry relations, 21 independent elastic constants are necessary to describe the 3D stress–strain relationship.

In order to be analyzed as plane problems, any point within the elastic body must have an elastic symmetry plane, and all the elastic symmetry planes must be parallel to each other. In general cases, there are six independent elastic constants.

What is of the most interest in real projects is the layered elastic body, namely, transversely isotropic body. It is elastic isotropic within the layered planes. This type of material has only five independent elastic constants.

As shown in Figure 2.6, let the y-axis be normal to the layered plane, and then we will have the following stress–strain relationship:

$$\varepsilon_x = \sigma_x/E_1 - \mu_2\sigma_y/E_2 - \mu_1\sigma_z/E_1, \quad \gamma_{xy} = \tau_{xy}/G_2$$

$$\varepsilon_y = -\mu_2\sigma_x/E_2 + \sigma_y/E_2 - \mu_2\sigma_z/E_2, \quad \gamma_{yz} = \tau_{yz}/G_2$$

$$\varepsilon_z = -\mu_1\sigma_x/E_1 - \mu_2\sigma_y/E_2 + \sigma_z/E_1, \quad \gamma_{zx} = 2(1+\mu_1)\tau_{zx}/E_1$$

in which E_1, μ_1 are elastic constants within the layered plane and E_2, G_2, μ_2 are elastic constants normal to the layered plane.

Let

$$E_1/E_2 = n, \quad G_2/E_2 = m$$

The elasticity matrix for plane stress problems of the layered elastic body is

$$[D] = \frac{E_2}{(1 - n\mu_2^2)} \begin{bmatrix} n & n\mu_2 & 0 \\ n\mu_2 & 1 & 0 \\ 0 & 0 & m(1 - n\mu_2^2) \end{bmatrix} \tag{2.25}$$

The elasticity matrix for plane strain problems of the layered elastic body is

$$[D] = \frac{E_2}{(1 + \mu_1)(1 - \mu_1 - 2n\mu_2^2)} \times \begin{bmatrix} n(1 - n\mu_2^2) & n\mu_2(1 + \mu_1) & 0 \\ n\mu_2(1 + \mu_1) & (1 - \mu_1^2) & 0 \\ 0 & 0 & m(1 + \mu_1)(1 - \mu_1 - 2n\mu_2^2) \end{bmatrix} \tag{2.26}$$

As shown in Figure 2.5, when there is an angle between the plane and x-axis, transformation is necessary to obtain the elasticity matrix $[D]$ in the global coordinate system (x, y).

Let the elasticity matrix in the local coordinate system (x', y') be $[D']$, stress be $\{\sigma'\}$, and strain be $\{\varepsilon'\}$. Then the stress–strain relationship is

$$\{\sigma'\} = [D']\{\varepsilon'\} \tag{c}$$

The stress in the global coordinate system (x, y) is

$$\{\sigma\} = [\theta]\{\sigma'\} \tag{d}$$

in which $[\theta]$ is a transformation matrix, as expressed in formula (2.19).

Substitute formula (c) into formula (d) to obtain

$$\{\sigma\} = [\theta][D']\{\varepsilon'\}$$

Then substitute $\{\varepsilon'\} = [\theta]^{\mathrm{T}}\{\varepsilon\}$ into the above formula to obtain

$$\{\sigma\} = [\theta][D'][\theta]^{\mathrm{T}}\{\varepsilon\}$$

Then we will see

$$[D] = [\theta][D'][\theta]^{\mathrm{T}} \tag{2.27}$$

in which $[D]$ is an elasticity matrix in the global coordinate system.

2.6 Equivalent Nodal Force and Element Stiffness Matrix

Under the action of external loads, there are displacements and stresses in the elements of the planar continuous structure as shown in Figure 2.3. Now we take an element *ijm* from it, as shown in Figure 2.7; according to the linear displacement assumptions, there will be uniform stress component $\sigma_x, \sigma_y, \tau_{xy}$ in the element. On the three boundaries of the element, there are distributed forces p that maintain balance with elemental stresses.

In order to be able to use concepts of structural mechanics for solving the problems in the theory of elasticity, the distributed forces acting on the boundaries of the element are replaced by the equivalent concentrated forces acting on the nodes and call them

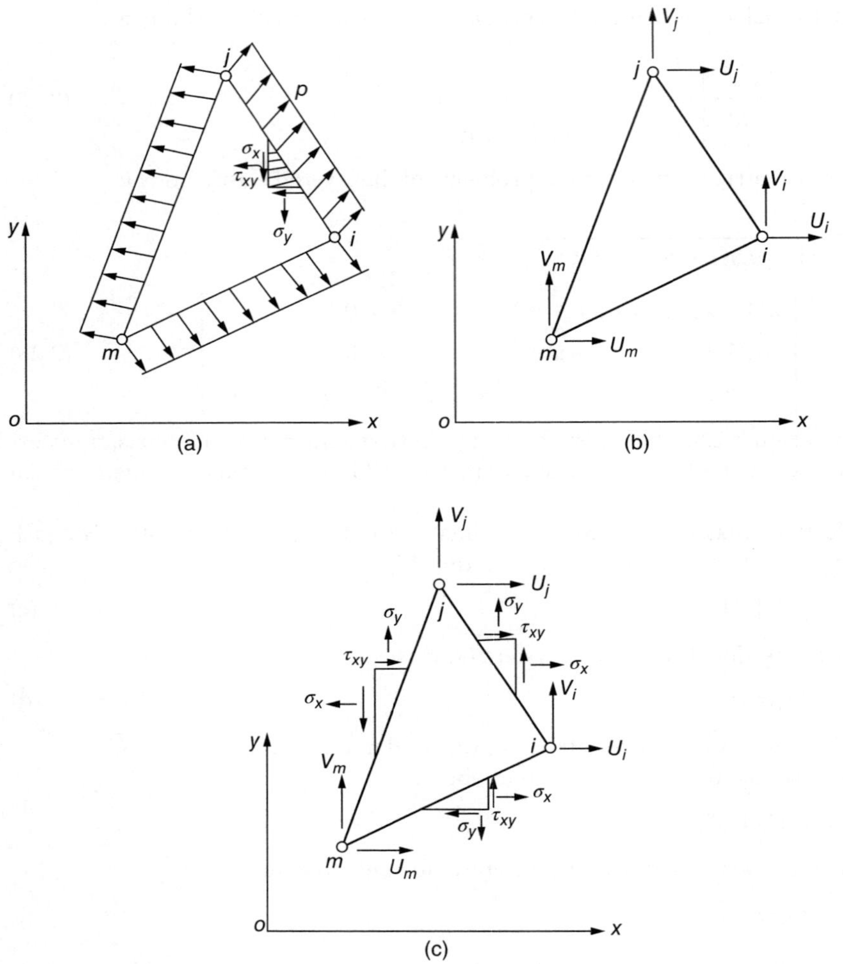

Figure 2.7 Equivalent nodal force. (a) Element boundary force, (b) equivalent nodal force, and (c) equivalent nodal force and element boundary stress.

the nodal forces. The direction of nodal force is consistent with the direction of nodal displacement. For example, on node i, the nodal forces are

$$\{F_i\} = \begin{Bmatrix} U_i \\ V_i \end{Bmatrix}$$

Now compute these equivalent nodal forces by the static equivalence principle. Refer to Figure 2.7(c); the horizontal resultant acting on edge ij is

$$X_{ij} = \sigma_x t(y_j - y_i) + \tau_{xy} t(x_i - x_j)$$

where t is the thickness of element.

The horizontal resultant acting on edge mi is

$$X_{mi} = \sigma_x t(y_i - y_m) + \tau_{xy} t(x_i - x_m)$$

Now let $x_{ij}/2$ and $x_{mi}/2$ be removed to the adjacent node i to obtain the horizontal nodal force as follows:

$$U_i = \frac{1}{2}(X_{ij} + X_{mi}) = \frac{t}{2}\sigma_x(y_j - y_m) + \frac{t}{2}\tau_{xy}(x_m - x_j)$$
$$= \frac{t}{2}(b_i\sigma_x + c_i\tau_{xy})$$

Similarly, the vertical nodal force acting on point i can be obtained as follows:

$$V_i = \frac{t}{2}(c_i\sigma_y + b_i\tau_{xy})$$

With similar ways we can get the nodal forces acting on node j and node m. Thus the nodal forces of the element can be expressed with matrices as follows:

$$\{F\}^e = \begin{pmatrix} U_i \\ V_i \\ U_j \\ V_j \\ U_m \\ V_m \end{pmatrix} = \frac{t}{2}\begin{bmatrix} b_i & 0 & c_i \\ 0 & c_i & b_i \\ b_j & 0 & c_j \\ 0 & c_j & b_j \\ b_m & 0 & c_m \\ 0 & c_m & b_m \end{bmatrix}\begin{Bmatrix} \sigma_x \\ \sigma_y \\ \tau_{xy} \end{Bmatrix} \tag{2.28}$$

By expression (2.13) for matrix $[B]$, formula (2.28) can be simplified as

$$\{F\}^e = tA[B]^T\{\sigma\} \tag{2.29}$$

In case with no initial strain, the stress can be expressed by nodal displacements as follows:

$$\{\sigma\} = [D]\{\varepsilon\} = [D][B]\{\delta\}^e$$

Substitution of the above formula into formula (2.29) will give

$$\{F\}^e = [B]^T[D][B]tA\{\delta\}^e$$

Let

$$\{k\}^e = [B]^T[D][B]tA \tag{2.30}$$

Then

$$\{F\}^e = [k]^e\{\delta\}^e \tag{2.31}$$

Formula (2.31) demonstrates the relationship between nodal forces and nodal displacements of the element. Matrix $[k]^e$ is called the element stiffness matrix.

According to formula (2.12), matrix $[B]$ can be written in the partitioned form as follows:

$$[B] = \begin{bmatrix} B_i & B_j & B_m \end{bmatrix}$$

By substituting the above equation into Eq. (2.30), the stiffness matrix can also be expressed in the partitioned form in the following:

$$[k]^e = \begin{bmatrix} k_{ii} & k_{ij} & k_{im} \\ k_{ji} & k_{jj} & k_{jm} \\ k_{mi} & k_{mj} & k_{mm} \end{bmatrix} \tag{2.32}$$

in which $[k_{rs}]$ is a 2×2 submatrix, namely,

$$[k_{rs}] = [B_r]^{\mathrm{T}}[D][B_s]tA = \begin{bmatrix} k_{rs}^1 & k_{rs}^2 \\ k_{rs}^3 & k_{rs}^4 \end{bmatrix} \tag{2.33}$$

Formula (2.33) is easy for calculation.

As for isotropic plane stress problems, by substituting the expression (2.21) of elastic matrix $[D]$ into formula (2.33), we get

$$[k_{rs}] = \frac{Et}{4(1-\mu^2)A}\begin{bmatrix} b_r b_s + (1-\mu)c_r c_s/2 & \mu b_r c_s + (1-\mu)c_r b_s/2 \\ \mu c_r b_s + (1-\mu)b_r c_s/2 & c_r c_s + (1-\mu)b_r b_s/2 \end{bmatrix} \tag{2.34}$$

As for isotropic plane strain problems, E in the above formula should be replaced by $E/(1-\mu^2)$ and μ by $\mu/(1-\mu)$, and then we will get

$$[k_{rs}] = \frac{E(1-\mu)t}{4(1+\mu)(1-2\mu)A} \times \begin{bmatrix} b_r b_s + (1-2\mu)c_r c_s/2(1-\mu) & \mu b_r c_s/(1-\mu) + (1-2\mu)c_r b_s/2(1-\mu) \\ \mu c_r b_s/(1-\mu) + (1-2\mu)b_r c_s/2(1-\mu) & c_r c_s + (1-2\mu)b_r b_s/2(1-\mu) \end{bmatrix} \tag{2.35}$$

The element stiffness matrix $[k]$ of the plane stress and plane strain of isotropic bodies is as follows:

$$[k]^{\mathrm{e}} = H\begin{bmatrix} b_i^2 + qc_i^2 & rb_ic_i & b_ib_j + qc_ic_j & pb_ic_j + qb_jc_i & b_ib_m + qc_ic_m & pb_ic_m + qb_mc_i \\ & c_i^2 + qb_i^2 & pb_jc_i + qb_ic_j & c_ic_j + qb_ib_j & pb_mc_i + qb_ic_m & c_ic_m + qb_ib_m \\ & & b_j^2 + qc_j^2 & rb_jc_j & b_jb_m + qc_jc_m & pb_jc_m + qb_mc_j \\ & \text{symmetrical} & & c_j^2 + qb_j^2 & pb_mc_j + qb_jc_m & c_jc_m + qb_jb_m \\ & & & & b_m^2 + qc_m^2 & rb_mc_m \\ & & & & & c_m^2 + qb_m^2 \end{bmatrix} \tag{2.36}$$

Plane strain : $H = \dfrac{E(1-\mu)t}{4(1+\mu)(1-2\mu)A}$, $p = \dfrac{\mu}{1-\mu}$, $q = \dfrac{1-2\mu}{2(1-\mu)}$, $r = \dfrac{1}{2(1-\mu)}$

Plane stress : $H = \dfrac{Et}{4(1-\mu^2)A}$, $p = \mu$, $q = \dfrac{1-\mu}{2}$, $r = \dfrac{1+\mu}{2}$

In the finite element method, the equivalent nodal force is an important concept. To deepen the understanding, we will deduce its calculation formula in another way.

For the element ijm as shown in Figure 2.8, the distributed forces acting on element boundaries have been replaced by six nodal forces; the internal stresses of the element are σ_x, σ_y, τ_{xy}. Three straight lines are drawn from the elemental centroid b to the midpoint a, c, d of the three edges to trisect the element. One part is taken out as shown in Figure 2.8(b); now the distributed force acting on edge abc is p, and its resultant force

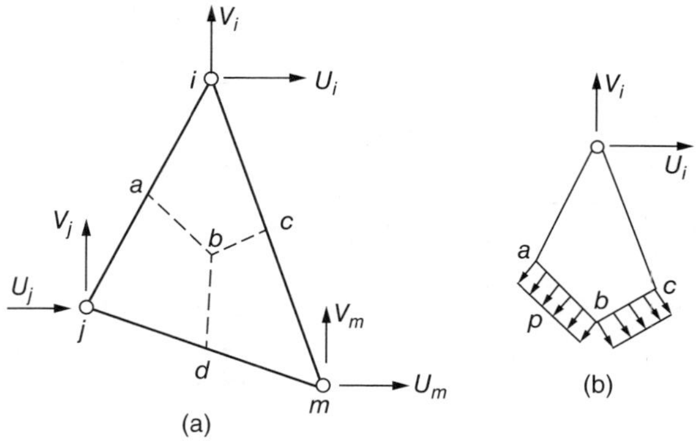

Figure 2.8 Equivalent nodal force.

will keep balance with the nodal force U_i and V_i. Let the horizontal component of the distributed force be p_x; do the line integration along abc and we will get

$$U_i = t\int_{abc} p_x \, ds = t\{(y_a - y_c)\sigma_x + (x_c - x_a)\tau_{xy}\} \tag{a}$$

in which x_a, y_a are the coordinates of point a.

As a and c are midpoints of two edges,

$$x_a = \frac{x_i + x_j}{2}, \quad y_a = \frac{y_i + y_j}{2}, \quad x_c = \frac{x_i + x_m}{2}, \quad y_c = \frac{y_i + y_m}{2}$$

By substituting the above formula into formula (a), we get

$$U_i = \frac{t}{2}(b_i\sigma_x + c_i\tau_{xy}) \tag{b}$$

Formula (b) is identical to formula (2.28) obtained previously. Similarly, the other five nodal forces can be obtained and lead to formula (2.29), and then by substituting $\{\sigma\} = [D]\,[B]\,\{\delta\}^e$, we obtain the element stiffness matrix $[k]^e$ as formula (2.30).

If taking the variational approach to solve the nodal force, we will get the same formula, but the deducing process is more abstract. This chapter will deduce the formula of nodal forces by primary methods at first, which is helpful for readers to establish clear mechanical concepts. The variational approach will be expounded in Chapter 3.

Now we compute a simple case. Let a plane stress element ijm be as shown in Figure 2.9; its thickness is t, Poisson's ratio is $\mu = 0.30$, and the nodal coordinates are $x_i = 100\,\text{cm}$, $y_i = 0$, $x_j = 100\,\text{cm}$, $y_j = 60\,\text{cm}$, $x_m = y_m = 0$. By formulas (2.5) and (2.6), $b_i = 60\,\text{cm}$, $c_i = -100\,\text{cm}$, $b_j = 0$, $c_j = 100\,\text{cm}$, $b_m = -60\,\text{cm}$, $c_m = 0$, $A = 3000\,\text{cm}^2$.

By formula (2.22), we will obtain the following relationship between the element stresses and nodal displacements:

$$\begin{Bmatrix} \sigma_x \\ \sigma_y \\ \tau_{xy} \end{Bmatrix} = \frac{E}{2(1-\mu^2)A}\begin{bmatrix} 60 & -30 & 0 & 30 & -60 & 0 \\ 18 & -100 & 0 & 100 & -18 & 0 \\ -35 & 21 & 35 & 0 & 0 & -21 \end{bmatrix}\begin{Bmatrix} u_i \\ v_i \\ u_j \\ v_j \\ u_m \\ v_m \end{Bmatrix}$$

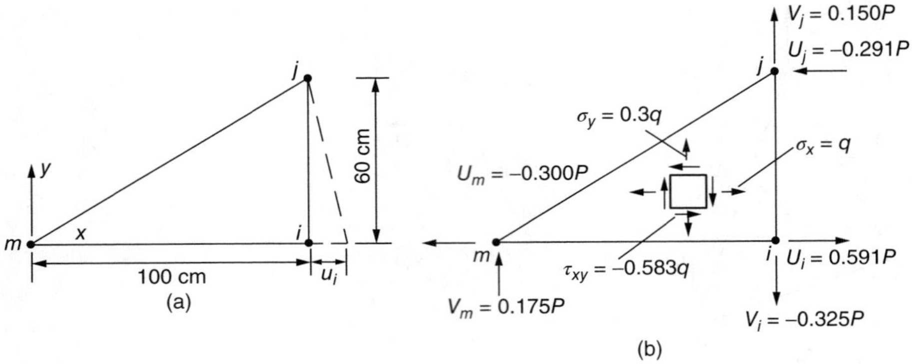

Figure 2.9 Nodal forces of the element. (a) Element dimension and (b) stress and nodal forces caused by u_i.

By formula (2.36), we will obtain the following relationship between the elemental nodal forces and nodal displacements:

$$\begin{Bmatrix} U_i \\ V_i \\ U_j \\ V_j \\ U_m \\ V_m \end{Bmatrix} = \frac{Et}{1-\mu^2}\begin{bmatrix} 0.591 & -0.325 & -0.291 & 0.150 & -0.300 & 0.175 \\ -0.325 & 0.938 & 0.175 & -0.833 & 0.150 & -0.105 \\ -0.291 & 0.175 & 0.291 & 0 & 0 & -0.175 \\ 0.150 & -0.833 & 0 & 0.833 & -0.150 & 0 \\ -0.300 & 0.150 & 0 & -0.150 & 0.300 & 0 \\ 0.175 & -0.105 & -0.175 & 0 & 0 & 0.105 \end{bmatrix}\begin{Bmatrix} u_i \\ v_i \\ u_j \\ v_j \\ u_m \\ v_m \end{Bmatrix} \tag{d}$$

Now assuming except $u_i \neq 0$, the remaining nodal displacements are all zero; by formula (c) we can see that the stresses within the element are

$$\sigma_x = q, \quad \sigma_y = 0.30q, \quad \tau_{xy} = -0.583q$$

where

$$q = \frac{60Eu_i}{2(1-\mu^2)A} = \frac{0.01Eu_i}{1-\mu^2}$$

By formula (d), it is known that the elemental nodal forces are

$$[U_i \ V_i \ U_j \ V_j \ U_m \ V_m]^{\mathrm{T}} = P[0.591 \ -0.325 \ -0.291 \ 0.150 \ -0.300 \ 0.175]^{\mathrm{T}}$$

$$P = \frac{Eu_it}{1-\mu^2}$$

2.7 Nodal Loads

There are two types of loads acting on the structure: one is the concentrated loads and the other is distributed loads. As for the concentrated loads, their points of application should be made to be the nodes when dividing grids. As for distributed loads, they should be replaced by equivalent nodal loads for easy calculation. Replacement of this type of loads should be done according to the static equivalence principle; only in this

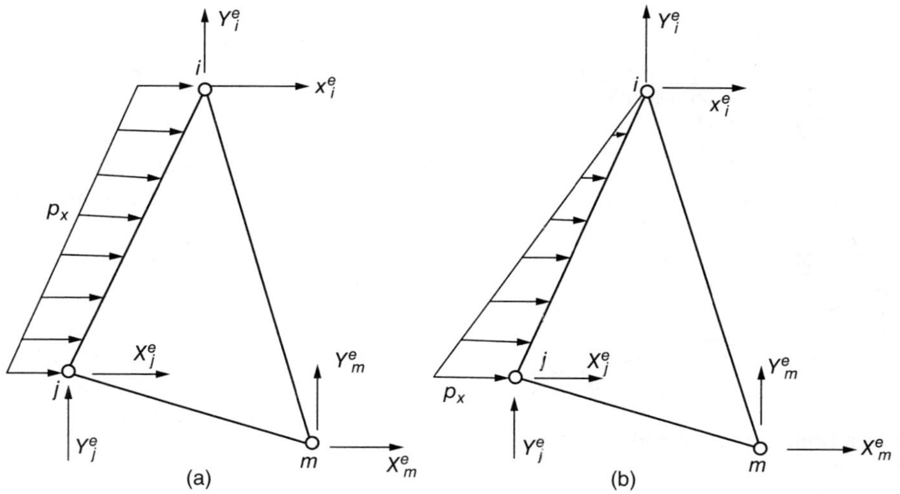

Figure 2.10 Equivalent nodal loads of boundary forces. (a) Uniform boundary force and (b) triangularly distributed boundary force.

way the stress errors caused by replacement will be limited locally and not affect the global stress state.

2.7.1 Equivalent Nodal Loads of Distributed Boundary Forces

As shown in Figure 2.10(a), let e be the element close to the boundary and the evenly distributed horizontal boundary force acting on edge ij be p_x. Their resultant is $p_x tl$, in which l is the length of edge ij and t the elemental thickness. The acting point of the resultant is midpoint of edge ij. By virtual displacement principle, it is easy to prove the equivalent nodal load is

$$\{P\}^{\mathrm{e}} = \begin{bmatrix} X_i^{\mathrm{e}} & Y_i^{\mathrm{e}} & X_j^{\mathrm{e}} & Y_j^{\mathrm{e}} & X_m^{\mathrm{e}} & Y_m^{\mathrm{e}} \end{bmatrix}^{\mathrm{T}} = \frac{p_x tl}{2} \begin{bmatrix} 1 & 0 & 1 & 0 & 0 & 0 \end{bmatrix}^{\mathrm{T}} \tag{2.37}$$

Another example is shown in Figure 2.10(b); on edge ij there is a triangularly distributed horizontal boundary force, the intensity of which on point j is p_x. Their resultant will be $p_x tl/2$ and the point of application of the resultant is $l/3$ away from point j and $2l/3$ from point i. By virtual displacement principle, it is easy to prove the equivalent nodal load is

$$\{P\}^{\mathrm{e}} = \frac{p_x tl}{2} \begin{bmatrix} \frac{1}{3} & 0 & \frac{2}{3} & 0 & 0 & 0 \end{bmatrix}^{\mathrm{T}} \tag{2.38}$$

2.7.2 Nodal Loads of Uniform Volume Force

Given a homogeneous and uniformly thick element ijm, a horizontal volume force q_x and vertical volume force q_y are acting on its volume, of which the resultant is

$$Q_x = q_x tA, \quad Q_y = q_y tA \tag{a}$$

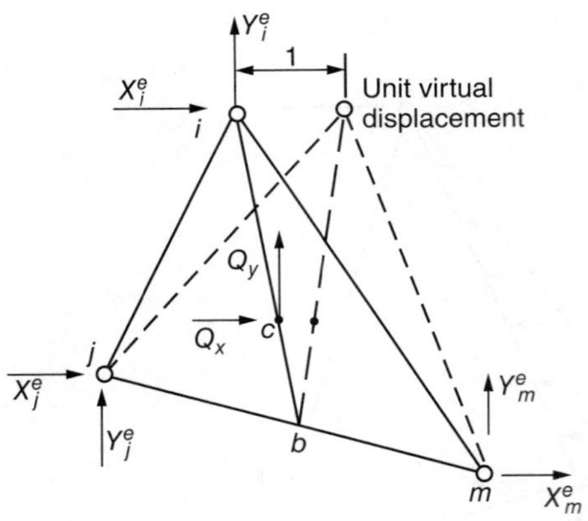

Figure 2.11 Equivalent nodal loads of uniform volume force.

in which A is area of the element and t is thickness. The point of application of the resultant is the elemental center c as shown in Figure 2.11.

At first compute the horizontal load X_i^e of point i. Assume that unit virtual displacement occurs at node i along the horizontal direction and the other 2 nodes j and m do not move. According to the linear displacement assumption, point b will not move and center c will move by 1/3 unit horizontally and by 0 vertically. By the static equivalent principle, the virtual work of loads of the element should equal to that of X_i^e, namely,

$$X_i^e \times 1 = Q_x \times \frac{1}{3} + Q_y \times 0$$

Hereby it is known that $X_i^e = Q_x/3$; in a similar way other nodal loads can be obtained. Finally the load vector is as follows:

$$\{P\}^e = \frac{1}{3}\begin{bmatrix} Q_x & Q_y & Q_x & Q_y & Q_x & Q_y \end{bmatrix}^T \tag{2.39}$$

2.7.3 Nodal Loads Due to Potential of Volume Force

In some cases, the volume force can be expressed by volume force potential ϕ as follows:

$$q_x = -\frac{\partial \phi}{\partial x}, \quad q_y = -\frac{\partial \phi}{\partial y} \tag{2.40}$$

Let the volume force potential of 3 nodes of element ijm be ϕ_i, ϕ_j, ϕ_m; the volume force potential within the element can be represented by shape functions as follows:

$$\phi = N_i\phi_i + N_j\phi_j + N_m\phi_m$$

where the shape function N_i is given by formula (2.7). By formula (2.40) the volume force can be given as follows:

$$\begin{Bmatrix} q_x \\ q_y \end{Bmatrix} = \frac{-1}{2A}\begin{bmatrix} b_i & b_j & b_m \\ c_i & c_j & c_m \end{bmatrix}\begin{Bmatrix} \phi_i \\ \phi_j \\ \phi_m \end{Bmatrix}$$

So by formula (2.39), we can get the nodal load as

$$\{P\}^e = -\frac{t}{6}\begin{bmatrix} b_i & b_j & b_m \\ c_i & c_j & c_m \\ b_i & b_j & b_m \\ c_i & c_j & c_m \\ b_i & b_j & b_m \\ c_i & c_j & c_m \end{bmatrix}\begin{Bmatrix} \phi_i \\ \phi_j \\ \phi_m \end{Bmatrix} \tag{2.41}$$

2.7.4 Nodal Loads Caused by Initial Strain

If there exists an initial strain $\{\varepsilon_0\}$ in element e, then the stress calculation formula is

$$\{\sigma\} = [D](\{\varepsilon\} - \{\varepsilon_0\}) = [D][B]\{\delta\}^{\mathrm{e}} - [D]\{\varepsilon_0\}$$

Substitution of the above equation into formula (2.29) gives the following elemental nodal force:

$$\{F\}^{\mathrm{e}} = [B]^{\mathrm{T}}[D][B]tA\{\delta\}^{\mathrm{e}} - [B]^{\mathrm{T}}[D]tA\{\varepsilon_0\}$$

That is,

$$\{F\}^{\mathrm{e}} = [k]\{\delta\}^{\mathrm{e}} - [B]^{\mathrm{T}}[D]tA\{\varepsilon_0\}$$

Compared with formula (2.31), it can be seen that the second term in the right part of the above formula did not exist previously; it is the nodal force produced by the initial strain. Change its sign and we will get the following nodal load caused by the initial strain:

$$\{P\}^{\mathrm{e}}_{\varepsilon_0} = [B]^{\mathrm{T}}[D]\{\varepsilon_0\}tA \tag{2.42}$$

Substitution of the expression of matrix $[B]$ and $[D]$ into formula (2.42) will give

$$\{P\}^{\mathrm{e}}_{\varepsilon_0} = A_0 \begin{bmatrix} b_i & \beta b_i & \gamma c_i \\ \beta c_i & c_i & \gamma b_i \\ b_j & \beta b_j & \gamma c_j \\ \beta c_j & c_j & \gamma b_j \\ b_m & \beta b_m & \gamma c_m \\ \beta c_m & c_m & \gamma b_m \end{bmatrix} \begin{Bmatrix} \varepsilon_{x0} \\ \varepsilon_{y0} \\ \gamma_{xy0} \end{Bmatrix} \tag{2.43}$$

$$\text{Plane strain problem}: \quad A_0 = \frac{E(1-\mu)t}{2(1+\mu)(1-2\mu)}, \quad \beta = \frac{\mu}{1-\mu}, \quad \gamma = \frac{1-2\mu}{2(1-\mu)}$$

$$\text{Plane stress problem}: \quad A_0 = \frac{Et}{2(1-\mu^2)}, \quad \beta = \mu, \quad \gamma = \frac{1-\mu}{2}$$

Substitution of formula (2.15) or formula (2.16) into formula (2.43) gives the following equivalent nodal load of temperature deformation:

$$\{P\}^{\mathrm{e}}_{\varepsilon_0} = A_1 \begin{bmatrix} b_i & c_i & b_j & c_j & b_m & c_m \end{bmatrix}^{\mathrm{T}} \tag{2.44}$$

$$\text{Plane strain problem}: \quad A_1 = \frac{Et\alpha\overline{T}}{2(1-2\mu)}$$

$$\text{Plane stress problem}: \quad A_1 = \frac{Et\alpha\overline{T}}{2(1-\mu)}$$

All the above formulas apply to the isotropic body. Substitution of expression $[D]$ and $\{\varepsilon_0\}$ of the anisotropic body into formula (2.42) will give the equivalent nodal loads of the anisotropic body.

2.8 Nodal Equilibrium Equation and Global Stiffness Matrix

Taking a node as a free body, under the action of nodal forces and nodal loads, every node must be in equilibrium, for example, taking node i from our analyzed structure as

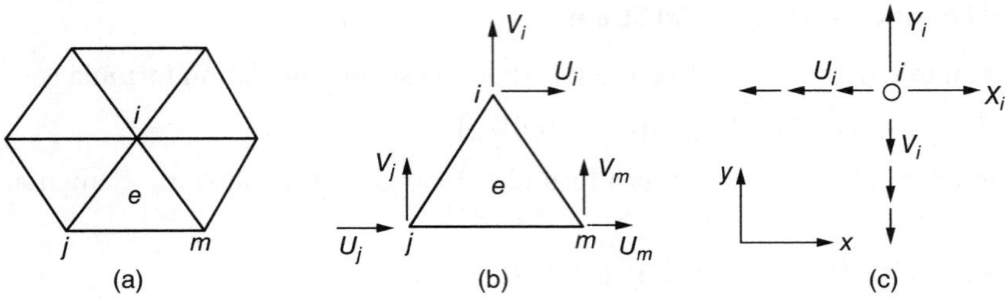

Figure 2.12 Nodal equilibrium. (a) Node *i*, (b) element e, and (c) equilibrium of node *i*.

shown in Figure 2.12(a). There are six elements around node i. Take the element e from it, and then the element is subjected to the action of the nodal forces from node i, j, m as follows:

$$\{F\}^{\mathrm{e}} = \left[F_i \; F_j \; F_m\right]^{\mathrm{T}} = \left[U_i \; V_i \; U_j \; V_j \; U_m \; V_m\right]^{\mathrm{T}}$$

Conversely, nodes i, j, m are also under the action of nodal force from this element, which are equal in value and opposite in direction with the above nodal force. As shown in Figure 2.12, node i is under the action of U_i and V_i imposed by element e in the negative direction. Similarly, other elements surrounding node i impose the same type of forces on node i.

Node i also bears nodal loads transferred from all elements around the node:

$$\{P_i\} = \sum_e \{P_i\}^{\mathrm{e}} = \begin{Bmatrix} X_i \\ Y_i \end{Bmatrix}$$

According to the equilibrium condition of node i in the horizontal and vertical direction, the following equilibrium equation may be set up:

$$\sum U_i = X_i, \quad \sum V_i = Y_i$$

in which Σ is the sum of all elements around node i.

The above equilibrium equation may also be represented as

$$\sum \{F_i\} = \{P_i\} \tag{2.45}$$

Substituting $\{F\}^{\mathrm{e}} = [k]\{\delta\}^{\mathrm{e}}$ into formula (2.45), we get the nodal equilibrium equation represented by nodal displacements as

$$[K]\{\delta\} = \{P\} \tag{2.46}$$

By deduction similar to Section 1.5, elements of the global stiffness matrix $[K]$ can be obtained as follows:

$$K_{rs} = \sum k_{ij} \tag{2.47}$$

If there are n nodes in the structure, the equilibrium equation set (2.46) will be linear equation set in $2n$ order. Solving this equation set to obtain nodal displacements, by formula (2.20), the elemental stress can be computed. Therefore, the key for the problem lies in the establishment of the global stiffness matrix $[K]$.

According to Figure 2.8, the nodal equilibrium equation can be set up in another approach as follows.

As shown in Figure 2.13, connect the centroid of every element to the midpoints of every edge with lines to form a contour Γ around node i. According to Section 2.6 formula (a), let the horizontal and vertical component of boundary force acting on contour Γ be p_x and p_y, and then the equilibrium equation about node i can be written as

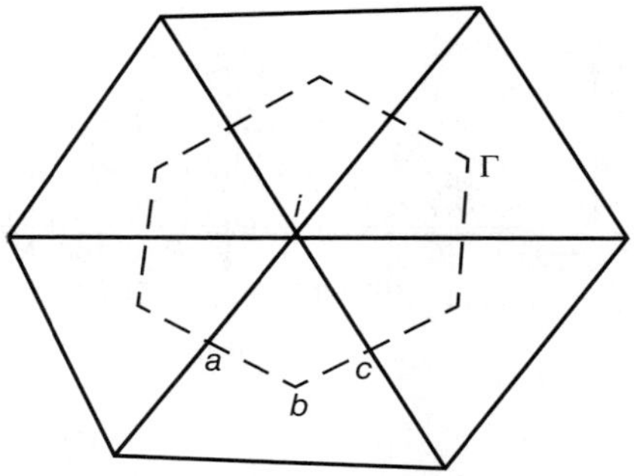

Figure 2.13 Nodal equilibrium.

$$\left.\begin{aligned}\sum U_i &= \int tp_x \,\mathrm{d}s = X_i \\ \sum V_i &= \int tp_y \,\mathrm{d}s = Y_i\end{aligned}\right\} \tag{a}$$

in which p_x and p_y can be given by the stress of every element and X_i and Y_i are nodal loads.

Let the angle between the outer normal n of the contour Γ and horizontal axis be ϕ; according to equilibrium conditions, there is

$$p_x = \sigma_x \cos\phi + \tau_{xy}\sin\phi, \qquad p_y = \sigma_y \sin\phi + \tau_{xy}\cos\phi \tag{b}$$

Substitution of the above formula into formula (a) will give a nodal equilibrium equation directly represented by the contour integral of elemental stress as follows:

$$\left.\begin{aligned}\int_\Gamma t(\sigma_x \cos\phi + \tau_{xy}\sin\phi)\,\mathrm{d}s &= X_i \\ \int_\Gamma t(\sigma_y \sin\phi + \tau_{xy}\cos\phi)\,\mathrm{d}s &= Y_i\end{aligned}\right\} \tag{2.48}$$

Then by substituting $\{\sigma\} = [D]\{\varepsilon\} = [D][B]\{\delta\}^e$ into formula (2.48), we will get the nodal equilibrium equation (2.46).

Now take a simple example to show the establishment of equilibrium equations. As shown in Figure 2.14(a), let a triangular plate with stiffeners bear load P. Adopt the triangular plane stress element for the plate and bar elements for the stiffeners, so this is a composite structure. In order to calculate the stress more accurately, it is necessary to use a more intensive computing grid. Now to offer a simple example, we divide it into two triangular plane stress elements. Due to the symmetry, only one-half needs to be calculated as shown in Figure 2.14(b). There are a total of four elements – 1 triangular plane stress element and three bar elements as shown in Figure 2.14(c). Besides dimensions shown in the figure, the other data are $E = 200$ GPa, $\mu = 0.30$.

First of all, compute the stiffness matrix of each element as follows:

Element Ⓐ triangular plate $\dfrac{Et}{1-\mu^2} = 2.2 \times 10^6\ \mathrm{N/cm}$

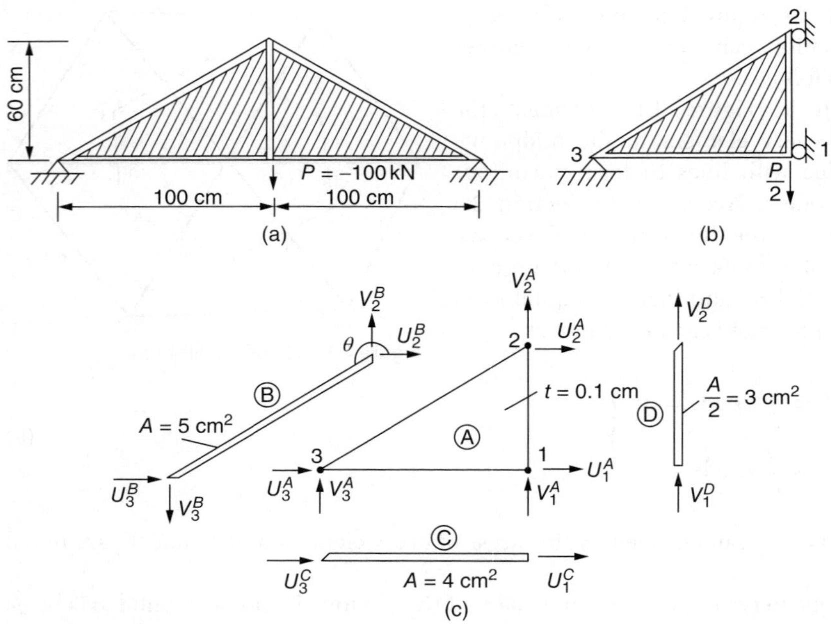

Figure 2.14 Figure for a simple example.

Substituting this value into Section 2.6 formula (d), we get

$$
\begin{array}{c}
\\
\begin{bmatrix} U_1^{\mathrm{A}} \\ V_1^{\mathrm{A}} \\ U_2^{\mathrm{A}} \\ V_2^{\mathrm{A}} \\ U_3^{\mathrm{A}} \\ V_3^{\mathrm{A}} \end{bmatrix}
\end{array}
= 10^6
\begin{array}{c}
\begin{array}{cccccc} u_1 & v_1 & u_2 & v_2 & u_3 & v_3 \end{array} \\
\begin{bmatrix}
1.300 & -0.714 & -0.640 & 0.330 & -0.660 & 0.385 \\
-0.714 & 2.06 & 0.385 & -1.832 & 0.330 & -0.231 \\
-0.640 & 0.385 & 0.640 & 0 & 0 & -0.385 \\
0.330 & -1.832 & 0 & 1.832 & -0.330 & 0 \\
-0.660 & 0.330 & 0 & -0.330 & 0.660 & 0 \\
0.385 & -0231 & -0.385 & 0 & 0 & 0.231
\end{bmatrix}
\end{array}
\begin{Bmatrix} u_1 \\ v_1 \\ u_2 \\ v_2 \\ u_3 \\ v_3 \end{Bmatrix}
$$

Above the element stiffness matrix are displacements corresponding to all columns, so that it is easy to locate corresponding positions when superimposing stiffness coefficients afterward.

Element Ⓑ: member element 2–3

$$l = \sqrt{100^2 + 60^2} = 116.7\,\text{cm}, \qquad \frac{AE}{l} = \frac{5 \times 20 \times 10^6}{116.7} = 0.858 \times 10^6\,\text{N/cm}$$

$$\alpha = \cos\theta = -0.858, \quad \beta = \sin\theta = -0.515$$

By formula (1–8), we can get

$$\left\{\begin{array}{c} U_2^{\mathrm{B}} \\ V_2^{\mathrm{B}} \\ U_3^{\mathrm{B}} \\ V_3^{\mathrm{B}} \end{array}\right\} = 10^6 \begin{array}{c} \begin{array}{cccc} u_2 & v_2 & u_3 & v_3 \end{array} \\ \begin{bmatrix} 0.630 & 0.378 & -0.630 & -0.378 \\ 0.378 & 0.227 & -0.378 & -0.227 \\ -0.630 & -0.378 & 0.630 & 0.378 \\ -0.378 & -0.227 & 0.378 & 0.227 \end{bmatrix} \end{array} \left\{\begin{array}{c} u_2 \\ v_2 \\ u_3 \\ v_3 \end{array}\right\}$$

Element Ⓒ, bar element 1–3

$$\frac{AE}{l} = 0.800 \times 10^6\,\mathrm{N/cm}, \quad \alpha = -1, \quad \beta = 0$$

$$\left\{\begin{array}{c} U_1^{\mathrm{C}} \\ U_3^{\mathrm{C}} \end{array}\right\} = 10^6 \begin{array}{c} \begin{array}{cc} u_1 & u_3 \end{array} \\ \begin{bmatrix} 0.800 & -0.800 \\ -0.800 & 0.800 \end{bmatrix} \end{array} \left\{\begin{array}{c} u_1 \\ u_3 \end{array}\right\}$$

Element Ⓓ, bar element 1–2

$$\frac{AE}{l} = 10^6\,\mathrm{N/cm}, \quad \alpha = 0, \beta = 1$$

$$\left\{\begin{array}{c} V_1^{\mathrm{D}} \\ V_2^{\mathrm{D}} \end{array}\right\} = 10^6 \begin{array}{c} \begin{array}{cc} v_1 & v_2 \end{array} \\ \begin{bmatrix} 1 & -1 \\ -1 & 1 \end{bmatrix} \end{array} \left\{\begin{array}{c} v_1 \\ v_2 \end{array}\right\}$$

In order to establish the global stiffness matrix and equilibrium equation set, add up the above four matrix equations to get

$$10^6 \begin{bmatrix} 2.10 & -0.714 & -0.640 & 0.330 & -1.460 & 0.385 \\ -0.714 & 3.06 & 0.385 & -2.832 & 0.330 & -0.231 \\ -0.640 & 0.385 & 1.270 & 0.378 & -0.630 & -0.763 \\ 0.330 & -2.832 & 0.378 & 3.059 & -0.708 & -0.227 \\ -1.460 & 0.330 & -0.630 & -0.708 & 2.09 & 0.378 \\ 0.385 & -0.231 & -0.763 & -0.227 & 0.378 & 0.458 \end{bmatrix} \left\{\begin{array}{c} u_1 \\ v_1 \\ u_2 \\ v_2 \\ u_3 \\ v_3 \end{array}\right\} = \left\{\begin{array}{c} X_1 \\ Y_1 \\ X_2 \\ Y_2 \\ X_3 \\ Y_3 \end{array}\right\} \tag{c}$$

In the following, we will consider the boundary conditions by methods provided in Chapter 1. By Figure 2.14(b), it is known that

$$u_1 = u_2 = u_3 = v_3 = 0 \tag{d}$$

From the equilibrium equation set (c), delete equations 1, 3, 5, and 6 corresponding to the above zero nodal displacements. Then delete the terms corresponding to the above quantities from the remaining two equations, leading to an equilibrium equation with v_1 and v_2 as the unknown quantities as follows:

$$10^6 \begin{bmatrix} 3.06 & -2.832 \\ -2.832 & 3.059 \end{bmatrix} \left\{\begin{array}{c} v_1 \\ v_2 \end{array}\right\} = \left\{\begin{array}{c} -50{,}000 \\ 0 \end{array}\right\}$$

By solving the above equation, we get

$$v_1 = -0.1131\,\text{cm}, \quad v_2 = -0.1047\,\text{cm} \tag{e}$$

Now the nodal displacement is available, by $\{\sigma\} = [S]\{\delta\}^e$, it is easy to get the elemental stress. By substituting the known nodal displacements into equations 1, 3, 5, and 6 of the equilibrium equation (c), we can obtain the reaction forces on the support.

2.9 Establish the Global Stiffness Matrix by the Coding Method

Previously we set up the global stiffness matrix by a direct method, that is, selecting and combining relevant elemental stiffness coefficient k_{ij} to obtain the global stiffness matrix $[K]$. Now we will describe how to use the coding method to automatically establish the global stiffness matrix $[K]$ of the plane problem.

At present, the element stiffness matrix is in 6×6 order, which is

$$[k]^e = \begin{bmatrix} k_{11} & k_{12} & k_{13} & k_{14} & k_{15} & k_{16} \\ k_{21} & k_{22} & k_{23} & k_{24} & k_{25} & k_{26} \\ \vdots & \vdots & \vdots & \vdots & \vdots & \vdots \\ k_{61} & k_{62} & k_{63} & k_{64} & k_{65} & k_{66} \end{bmatrix}$$

The subscript "ij" of the stiffness coefficient k_{ij} denotes it lies in row i and column j, of which the physical meaning is the ith nodal force caused by elemental deformation of the jth DOF of the element. We use 1, 2, 3, …, 6 to represent the six DOFs and corresponding six numbers of nodal forces.

If there are n nodes in the structure, the total number of DOF is $2n$; the serial numbers of them are as follows:

$$1,2,3,4,\ldots,2n-1,2n \tag{a}$$

Take one element e from the structure as shown in Figure 2.15, and its 3 nodes are numbered as n_i, n_j, n_m. The two DOFs of node n_i are coded among the structural DOF coding (a) as $2n_i - 1$ and $2n_i$, and among the elemental six DOFs coding as 1 and 2, as shown in Table 2.1.

Using Table 2.1 it is easy to determine the relationship between the elemental stiffness coefficient and global stiffness coefficient. Taking the elemental stiffness coefficient k_{25}, for example, if we take $2n_i$ from column 2 and $2n_m - 1$ from column 5 of the coding table, then the relation between the element stiffness coefficient and the corresponding global stiffness coefficient is $k_{25} \to K_{2n_i,2n_m-1}$, namely,

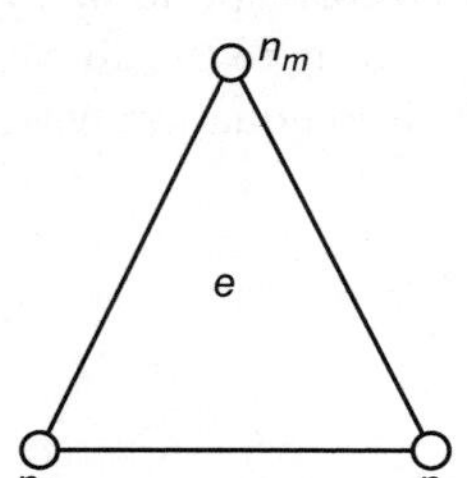

Figure 2.15 Element e.

$$k_{25} \to K_{2n_i,2n_m-1}$$

$$k_{11} \to K_{2n_i-1,2n_i-1}$$

In the same way

$$k_{23} \to K_{2n_i,2n_j-1}$$

Table 2.1 Nodal DOF coding table of element e.

Nodal DOF	u_i	v_i	u_j	v_j	u_m	v_m
Code in the element (i, j)	1	2	3	4	5	6
Code in the structure (r, s)	$2n_i - 1$	$2n_i$	$2n_j - 1$	$2n_j$	$2n_m - 1$	$2n_m$

After superimposing stiffness coefficients of all elements into corresponding global stiffness coefficients according to the coding table, we will obtain the global stiffness matrix. By the method of Section 1.7 to treat the boundary conditions, solutions are available.

For the problems analyzed in the following chapters, the global stiffness matrix can be established in the similar way, which will not be repeated later.

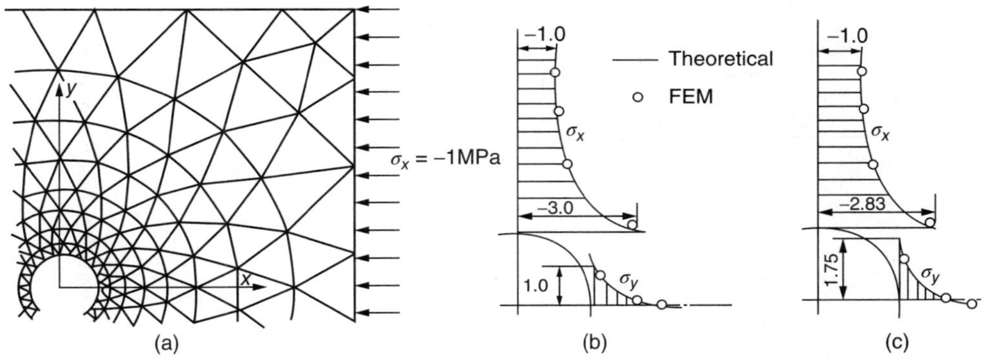

Figure 2.16 Stress concentration of the circular hole. (a) Computing grid near the circular hole, (b) stresses in an isotropic body, and (c) stresses in an anisotropic body, $E_x = 1$, $E_y = 3$, $G_{xy} = 0.42$, $\mu_1 = 0.1$, $\mu_2 = 0$.

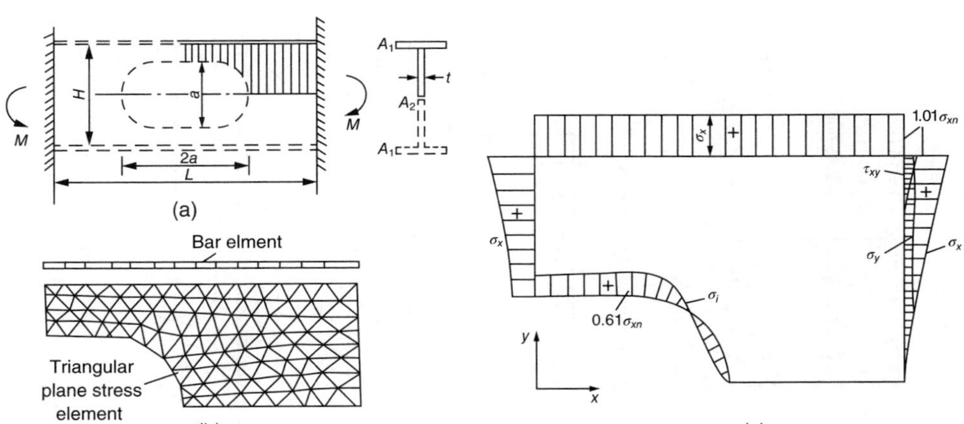

Figure 2.17 Stresses in I beam with a hole in web. (a) I beam, (b) computing grid, and (c) computing results. $A_1 = A_2$, $L/H = 1.6$, $t/H = 1/30$, $a/H = 0.40$.

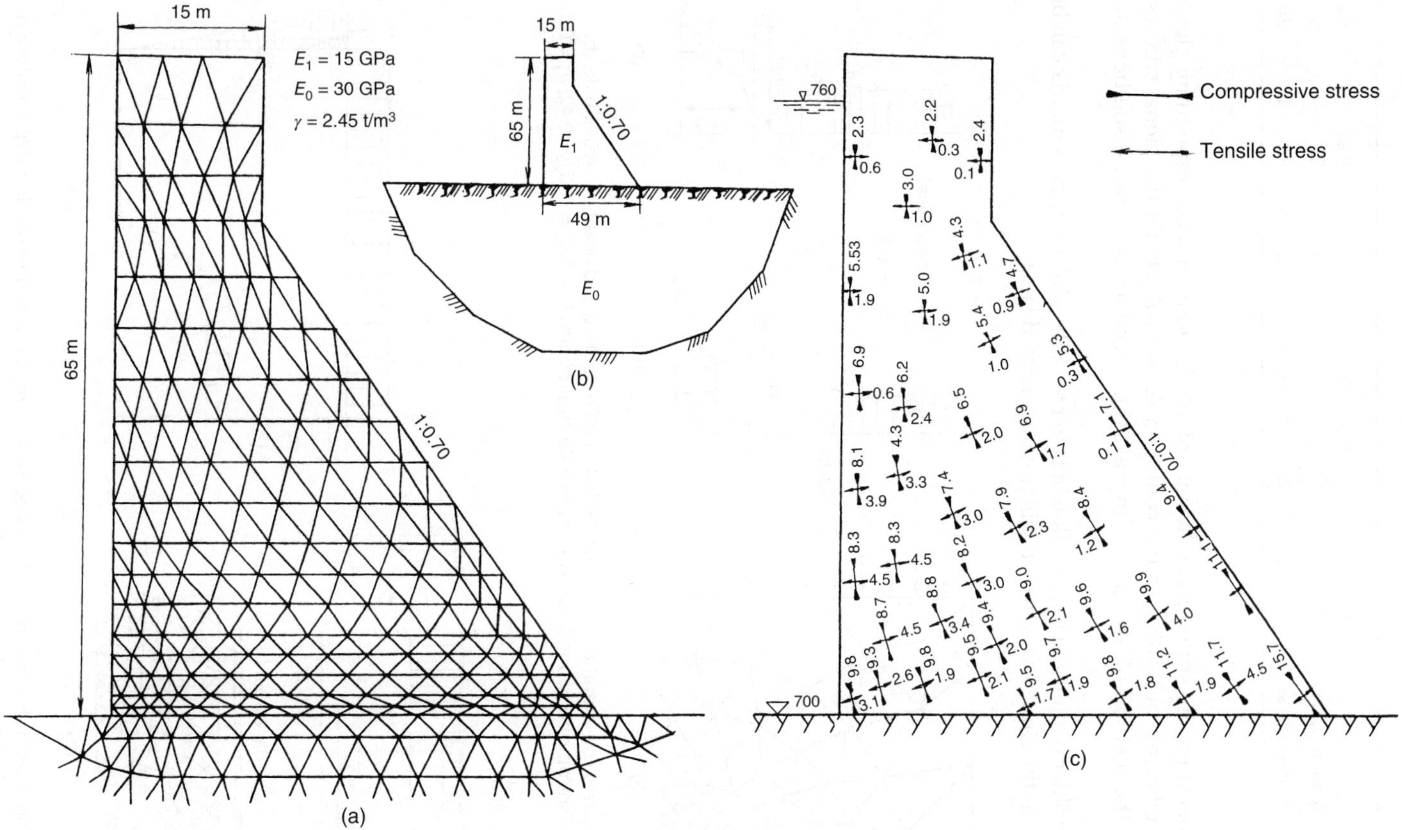

Figure 2.18 Stresses of the gravity dam (0.1 MPa). (a) Computing grid (only a part of the base is shown), (b) dimension of the gravity dam, and (c) calculated stresses (water pressure + self-weight).

2.10 Calculation Example

It is proved by theory and practice that if the dimension of the element is sufficient small, the finite element method of the above plane problem will lead to correct solutions. The convergence criteria will be discussed in Chapter 4. This section will introduce some calculation examples.

2.10.1 Stress Concentration near the Circular Hole

Figure 2.16(a) represents a computing grid near the circular hole under action of the uniform stress $\sigma_x = -1\,\text{MPa}$. As stress concentration occurs near the hole, the grid near the hole will gradually intensified. Figure 2.16(b) and (c) demonstrates calculation results of the isotropic and anisotropic body. The curve in the figure is a theoretic answer to the circular hole in infinite domain, and the small circles are the calculation outcome by the finite element method, both of which are perfectly consistent.

2.10.2 Stress Analysis of I Beam with a Hole in Web

In ship structures, there are holes in the web of I beam. Figure 2.17(a) represents an I beam with a holed web, and Figure 2.17(b) represents the computing grid. Take 1/4 for computation due to symmetry. The web adopts triangular plane stress elements and the flange adopts bar element (if the flange is very wide, it can also adopt triangular plane stress elements). The computing results are shown in Figure 2.17(c), in which σ_{xn} is the maximum fiber stress calculated by the beam theory.

2.10.3 Stress Analysis of the Concrete Gravity Dam

Figure 2.18 gives a stress analysis of a concrete gravity dam on rock foundation under water pressure and concrete weight. This dam is 65 m high and 49 m wide in base. The modulus of elasticity of rock foundation is $E_0 = 30\,\text{GPa}$, modulus of elasticity of concrete (considering creep effects) is $E_1 = 15\,\text{GPa}$, water density is $1\,\text{t/m}^3$, and concrete density is $2.45\,\text{t/m}^3$. The computing results were given by the writer and Professor Song Jingting.

Bibliography

1 Turner, M.J., Clough, R.W., Martin, H.C. and Topp, L.J. (1956) Stiffness and deflection analysis of complex structures. *J. Aero. Sci.*, **23**, 805–823.

2 Clough, R.W. (1960) *The Finite Element in Plane Stress Analysis.* Proceeding of the Second ASCE Conference on Electronic Computation.

3 Argyris, J.H. (1954) Energy theorems and structural analysis. *Aircraft Eng*, **26**, 347–356, 383–387, **394**; 1955, **27**:42–58, 80–94, 125–134, 145–158.

4 Zienkiewicz, O.C. and Taylor, R.L. (2009) *Finite Element Method for Solid and Structural Mechanics*, Elsevier, New York, London.

5 Holland, I. and Bell, K. (1972) *Finite Element Methods in Stress Analysis*, Tapir, Trondheim (Norway).

6 Timoshenko, S. and Goodier, J.N. (1951) *Theory of Elasticity*, 2nd edn, McGraw-Hill, New York.

3

Element Analysis

It can be seen from the previous two chapters that the core of the finite element method is to establish the element stiffness matrix, which after appropriate combination will leave us a set of equilibrium equations and a few algebraic operations to complete. During calculation of elastic plane problems, we apply the intuitive approach to set up the element stiffness matrix, which is easy to understand and helpful for beginners to establish clear mechanical concepts. However, there are some defects in this intuitive approach. On one hand, it is difficult to depend on it for establishing the stiffness matrices of complex elements; on the other hand, it cannot provide the proof of convergence. We'll explain in this chapter how to use the energy principle to set up basic formulas about elements and in the next chapter how to apply the energy principle in overall analysis of the structure.

At first we'll introduce the principle of virtual displacement and its application in element analysis. Among all the energy principles, the virtual displacement principle is the easiest for application and most widely used. Then we'll move on to application of the minimum potential energy principle, minimum complementary energy principle, and Reissner energy principle in element analysis.

3.1 Principle of Virtual Displacement

The so-called virtual displacement can be any infinitely small displacement, which must be continuous within the structure and meet the kinematic boundary conditions on structural boundaries. Take a cantilever beam, for example, whose virtual displacement and slope at fixed ends must be zero.

Consider the body in Figure 3.1; it is under the action of external forces F_1, F_2, …, which may be expressed as

$$\{\overline{F}\} = \left[F_1\ F_2\ F_3\ \cdots\right]^{\mathrm{T}}$$

Under these external forces, the stresses in the body are

$$\{\sigma\} = \left[\sigma_x\ \sigma_y\ \sigma_z\ \tau_{xy}\ \tau_{yx}\ \tau_{xx}\right]^{\mathrm{T}}$$

Now suppose that there are virtual displacements in the body; at the points of application of every external force, the virtual displacements in the direction of each external

The Finite Element Method: Fundamentals and Applications in Civil, Hydraulic, Mechanical and Aeronautical Engineering, First Edition. Bofang Zhu.

Companion website: www.wiley.com/go/zhufem17

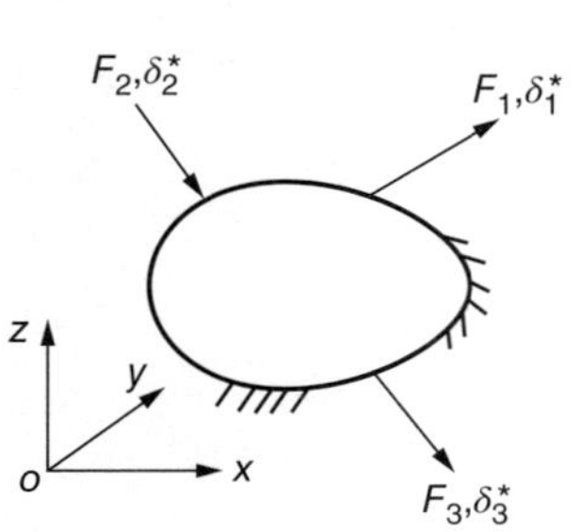

Figure 3.1 Boundary conditions of the solid.

force are δ_1^*, δ_2^*, δ_3^*, …, which can be expressed as

$$\{\delta^*\} = \begin{bmatrix}\delta_1^* & \delta_2^* & \delta_3^* & \cdots\end{bmatrix}^{\mathrm{T}}$$

The virtual strains caused by the virtual displacements are

$$\{\varepsilon^*\} = \begin{bmatrix}\varepsilon_x^* & \varepsilon_y^* & \varepsilon_x^* & \gamma_{xy}^* & \gamma_{yx}^* & \gamma_{xx}^*\end{bmatrix}^{\mathrm{T}}$$

When the virtual displacements occur, the external forces already act on the body, and during the process of virtual displacement, the external forces remains unchanged. Thus the virtual work done by external forces on virtual displacements is

$$\delta V = F_1\delta_1^* + F_2\delta_2^* + F_3\delta_3^* + \cdots = \{\delta^*\}^{\mathrm{T}}\{\overline{F}\}$$

Within the unit volume of the body, the virtual strain energy of the stress on virtual strain is

$$\sigma_x\varepsilon_x^* + \sigma_y\varepsilon_y^* + \sigma_z\varepsilon_z^* + \tau_{xy}\gamma_{xy}^* + \tau_{yz}\gamma_{yx}^* + \tau_{xx}\gamma_{zx}^* = \{\varepsilon^*\}^{\mathrm{T}}\{\sigma\}$$

The virtual strain energy of the whole body is

$$\delta U = \iiint \{\varepsilon^*\}^{\mathrm{T}}\{\sigma\}\,\mathrm{d}x\,\mathrm{d}y\,\mathrm{d}z$$

The virtual displacement principle demonstrates that if the body is in equilibrium before virtual displacements occur, the virtual work done by external forces at the time of virtual displacements occurring is equal to the virtual strain energy of the body, $\delta V = \delta U$, namely,

$$\{\delta^*\}^{\mathrm{T}}\{F\} = \iiint \{\varepsilon^*\}^{\mathrm{T}}\{\sigma\}\,\mathrm{d}x\,\mathrm{d}y\,\mathrm{d}z \tag{3.1}$$

Now we take a plane stress problem, for example, to prove the above assertion. We consider a plate of unit thickness with no volume force or initial stress for simplicity (it is not difficult to extend it to general cases). As shown in Figure 3.2, the boundary may be divided into two parts. The displacement equals to zero on S_u and on S_σ is acting boundary force p, of which the components are p_x and p_y. The boundary conditions on this part may be written as

$$\sigma_x l_x + \tau_{xy} l_y = p_x, \quad \sigma_y l_y + \tau_{xy} l_x = p_y \tag{a}$$

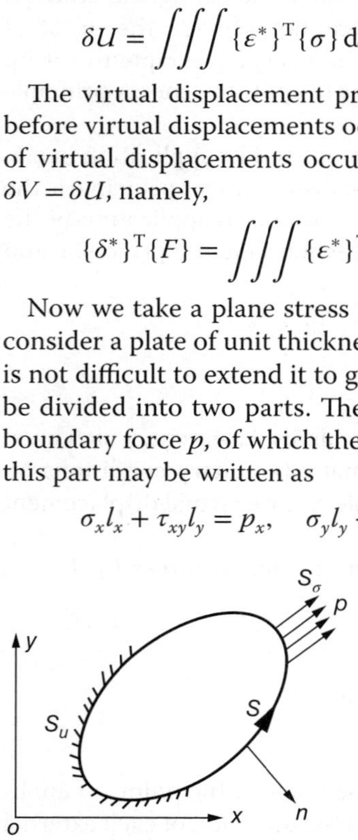

Figure 3.2 Boundary conditions on S_u and S_σ.

Under the action of external force p, there are stresses σ_x, σ_y, τ_{xy} in the body. It is known by the theory of elasticity that these stresses satisfy the following equilibrium equations:

$$\frac{\partial\sigma_x}{\partial x} + \frac{\partial\tau_{xy}}{\partial y} = 0, \quad \frac{\partial\sigma_y}{\partial y} + \frac{\partial\tau_{xy}}{\partial x} = 0 \tag{b}$$

Now suppose the body produces virtual displacements u^*, v^*, the corresponding virtual strains are

$$\varepsilon_x^* = \frac{\partial u^*}{\partial x}, \quad \varepsilon_y^* = \frac{\partial v^*}{\partial y}, \quad \gamma_{xy}^* = \frac{\partial u^*}{\partial y} + \frac{\partial v^*}{\partial x} \tag{c}$$

Because of the above virtual strain, the virtual strain energy caused by stresses within the body is

$$\begin{aligned}\delta U &= \iint \{\varepsilon^*\}\{\sigma\}\,\mathrm{d}x\,\mathrm{d}y \\ &= \iint \left[\sigma_x \frac{\partial u^*}{\partial x} + \sigma_y \frac{\partial v^*}{\partial y} + \tau_{xy}\left(\frac{\partial u^*}{\partial y} + \frac{\partial v^*}{\partial x}\right)\right]\mathrm{d}x\,\mathrm{d}y \end{aligned} \tag{d}$$

As

$$\iint \frac{\partial(\sigma_x u^*)}{\partial x}\,\mathrm{d}x\,\mathrm{d}y = \iint u^* \frac{\partial \sigma_x}{\partial x}\,\mathrm{d}x\,\mathrm{d}y + \iint \sigma_x \frac{\partial u^*}{\partial x}\,\mathrm{d}x\,\mathrm{d}y$$

so the first term on the right part of formula (d) is

$$\iint \sigma_x \frac{\partial u^*}{\partial x}\,\mathrm{d}x\,\mathrm{d}y = \iint \frac{\partial(\sigma_x u^*)}{\partial x}\,\mathrm{d}x\,\mathrm{d}y - \iint u^* \frac{\partial \sigma_x}{\partial x}\,\mathrm{d}x\,\mathrm{d}y$$

In the same way, if we do calculations for the other two terms on the right part of the formula and substitute them into formula (d), we will get

$$\begin{aligned}\delta U = &-\iint \left[\left(\frac{\partial \sigma_x}{\partial x} + \frac{\partial \tau_{xy}}{\partial y}\right) u^* + \left(\frac{\partial \sigma_y}{\partial y} + \frac{\partial \tau_{xy}}{\partial x}\right) v^*\right]\mathrm{d}x\,\mathrm{d}y \\ &+ \iint \left[\frac{\partial}{\partial x}(\sigma_x u^*) + \frac{\partial}{\partial x}(\tau_{xy} v^*) + \frac{\partial}{\partial y}(\sigma_y v^*) + \frac{\partial}{\partial y}(\tau_{xy} u^*)\right]\mathrm{d}x\,\mathrm{d}y\end{aligned}$$

By equilibrium equation (b), we know the first integral on the right of the formula above should be equal to zero; thus the expression of virtual strain energy δU may be simplified as

$$\delta U = \iint \left[\frac{\partial}{\partial x}(\sigma_x u^* + \tau_{xy} v^*) + \frac{\partial}{\partial y}(\sigma_y v^* + \tau_{xy} u^*)\right]\mathrm{d}x\,\mathrm{d}y \tag{e}$$

By Green's theorem, it is known that

$$\delta U = \iint \left(\frac{\partial Q}{\partial x} + \frac{\partial P}{\partial y}\right)\mathrm{d}x\,\mathrm{d}y = \int (Ql_x + Pl_y)\,\mathrm{d}s \tag{f}$$

in which l_x, l_y are direction cosines of the boundary normal.

Let

$$Q = \sigma_x u^* + \tau_{xy} v^*, \quad P = \sigma_y v^* + \tau_{xy} u^*$$

By formulas (e) and (f), we know the virtual strain energy δU can be expressed as

$$\delta U = \int \left[(\sigma_x u^* + \tau_{xy} v^*)\, l_x + (\sigma_y v^* + \tau_{xy} u^*) l_y\right]\mathrm{d}s$$

or

$$\delta U = \int \left[(\sigma_x l_x + \tau_{xy} l_y)\, u^* + (\sigma_y l_y + \tau_{xy} l_x) v^*\right]\mathrm{d}s \tag{g}$$

According to the boundary condition, on boundary S_u the displacement equals to zero, so $u^* = 0$, $v^* = 0$, and the S_u part of linear integral on the right of formula (g) should be equal to zero. On boundary S_σ, $\sigma_x l_x + \tau_{xy} l_y = p_x$, $\sigma_y l_y + \tau_{xy} l_x = p_y$, by substituting them into formula (g), we obtain

$$\delta U = \int_{S\sigma} (p_x u^* + p_y v^*)\, \mathrm{d}s \tag{h}$$

The right part of formula (h) represents the virtual work δV on the virtual displacement done by external forces p_x and p_y acting on the boundary of the body, so

$$\delta U = \delta V \tag{3.2}$$

Thus the virtual displacement principle is proved.

It is worth noting that during the proving process we did not cite the stress–strain relationship of the material, so the virtual displacement principle applies not only to linear materials but also to nonlinear materials.

3.2 Element Displacement

Figure 3.3 shows the triangular element for solving plane stress problems, tetrahedral element for solving space stress problems, and triangular element for solving plate bending problems. Use $\{\delta_i\}$ to represent the displacement of node i, as

$$\text{Plane stress problems :} \quad \{\delta_i\} = \begin{Bmatrix} u_i \\ v_i \end{Bmatrix}$$

$$\text{Space stress problem :} \quad \{\delta_i\} = \begin{Bmatrix} u_i \\ v_i \\ w_i \end{Bmatrix}$$

$$\text{Plate bending problems :} \quad \{\delta_i\} = \begin{Bmatrix} w_i \\ \theta_{xi} \\ \theta_{yi} \end{Bmatrix}$$

in which u_i, v_i, w_i are the linear displacement components of node i in the x-, y-, and z-directions, respectively, and θ_{xi}, θ_{yi} are angular displacement of node i around the x- and y-axes, respectively.

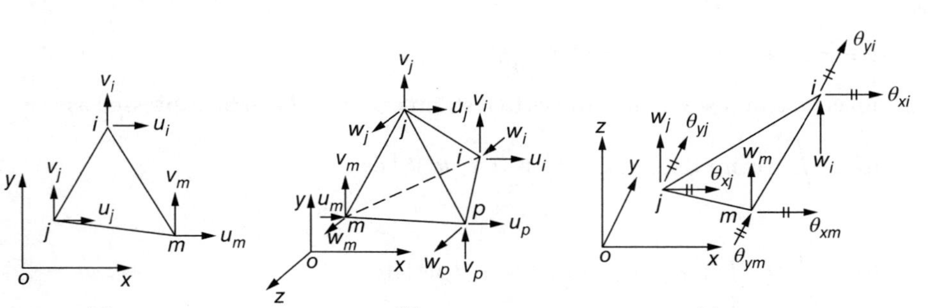

Figure 3.3 Several elements to solve the stresses. (a) Plane stress element, (b) space stress element, and (c) plate bending element.

Use $\{\delta\}^e$ to represent the vector consisting of displacements of all nodes of the element e, that is,

$$\{\delta\}^e = \left[\delta_i \ \delta_j \ \delta_m \ \cdots\right]^T \tag{3.3}$$

Displacement of any point (x, y, z) in the element can be represented as below:

$$\{r\} = [N]\{\delta\}^* = \left[N_i \ N_j \ N_m \ \cdots\right] \begin{Bmatrix} \delta_i \\ \delta_j \\ \delta_m \\ \vdots \end{Bmatrix} \tag{3.4}$$

in which N_i is a shape function.

The shape function has the following properties:

On node i : $N_i(x_i, y_i, z_i) = I$ (unit matrix)

On other nodes : $N_i(x_i, y_i, z_i) = N_i(x_m, y_m, z_m) = 0 \cdots$

where N_i is function of coordinates, of which the specific form will be given in the following chapters.

3.3 Element Strain and Stress

When the displacement of any point within the element is known, appropriate differential operation can help us to obtain the strain of any point within the element. Usually it can be expressed as follows:

$$\{\varepsilon\} = [B]\{\delta\}^\varepsilon \tag{3.5}$$

For example, for plane stress problems, matrix $[B]$ is shown in formulas (2.12) and (2.13).

Let $\{\varepsilon_0\}$ be the initial strain caused by factors such as temperature changes and contraction and $\{\sigma_0\}$ be the initial stress already existing in the element before applying loads. According to the generalized Hooke's theorem, the elemental stress may be represented as below:

$$\{\sigma\} = [D](\{\varepsilon\} - \{\varepsilon_0\}) + \{\sigma_0\} \tag{3.6}$$

in which $[D]$ is the elasticity matrix.

Substitute formula (3.5) into formula (3.6) and rewrite it as

$$\{\sigma\} = [S]\{\delta\}^e - [D]\{\varepsilon_0\} + \{\sigma_0\} \tag{3.7}$$

in which

$$[S] = [D][B] \tag{3.8}$$

3.4 Nodal Force and Element Stiffness Matrix

In order to establish the equilibrium equations of the structure, we use the equivalent concentrated force acting on the elemental nodes to replace the stress on element boundaries, which are called nodal force. The number and direction of nodal forces must be consistent with the nodal displacement. Use $\{F_i\}$ to represent the nodal force of node i to have

$$\text{Plane stress problems}: \quad \{F_i\} = \begin{Bmatrix} U_i \\ V_i \end{Bmatrix}$$

$$\text{Pace stress problems}: \quad \{F_i\} = \begin{Bmatrix} U_i \\ V_i \\ W_i \end{Bmatrix}$$

$$\text{Plate bending problems}: \quad \{F_i\} = \begin{Bmatrix} W_i \\ M_{\theta xi} \\ M_{\theta yi} \end{Bmatrix}$$

in which U_i, V_i, W_i are nodal forces acting on node i in the x-, y-, z-directions, respectively, and $M_{\theta xi}$, $M_{\theta yi}$ are concentrated moments around the x- and y-axes.

Use $\{F\}^e$ to represent the vectors composed of all nodal forces of element e:

$$\{F\}^e = \begin{bmatrix} F_i & F_j & F_m & \cdots \end{bmatrix}^T \tag{3.9}$$

Now use the virtual work principle to deduce the expression of nodal forces. Supposing there is a virtual displacement $\{r^*\}$ in element e, the corresponding virtual nodal displacement is $\{\delta^*\}^e$; according to formula (3.4), we have

$$\{r^*\} = [N]\{\delta^*\}^e$$

By formula (3.5), the virtual strain occurring within the element is

$$\{\varepsilon^*\} = [B]\{\delta^*\}^e$$

The virtual work done by nodal forces is equal to the sum of products of every nodal force component multiplied by corresponding virtual nodal displacement component:

$$u_i^* U_i + v_i^* V_i + w_i^* W_i + \cdots$$

which may be represented by matrix as

$$\delta V = \left(\{\delta^*\}^e\right)^T \{F\}^e$$

Within the whole element, the virtual strain energy of the stress on virtual strain $\{\varepsilon^*\}$ is

$$\delta U = \iiint \{\varepsilon^*\}^T \{\sigma\}\,dx\,dy\,dz$$

Substituting $\{\varepsilon^*\} = [B]\{\delta^*\}^e$ into the formula above, we get

$$\delta U = \left(\{\delta^*\}^e\right)^T \iiint [B]^T \{\sigma\}\,dx\,dy\,dz$$

By formula (3.2), $\delta U = \delta V$, so

$$\left(\{\delta^*\}^e\right)^T \{F\}^e = \left(\{\delta^*\}^e\right)^T \iiint [B]^T \{\sigma\}\,dx\,dy\,dz$$

The formula above must be tenable for any virtual displacement. As the virtual displacement is arbitrary, matrix $(\{\delta^*\}^e)^T$ is also arbitrary. The matrices multiplied with $(\{\delta^*\}^e)^T$ on both sides of the formula above should be equivalent. Then we obtain

$$\{F\}^e = \iiint [B]^T\{\sigma\}\,dx\,dy\,dz \tag{3.10}$$

The multiple integration on the right of the formula above is done within the volume of the whole element e. For cases without initial strains or initial stresses, the stress may be represented as follows:

$$\{\sigma\} = [D]\{\varepsilon\} = [D][B]\{\delta\}^e$$

Substituting the above formula into formula (3.10), as elements of $\{\delta^*\}^e$ are constants and can be put outside of the integration sign, then we'll get

$$\{F\}^e = \iiint [B]^T[D][B]\,dx\,dy\,dz\{\delta\}^e$$

Let

$$[k]^e = \iiint [B]^T[D][B]\,dx\,dy\,dz \tag{3.11}$$

then

$$\{F\}^e = [k]^e\{\delta\}^e \tag{3.12}$$

Formula (3.12) establishes the relationship between nodal forces and nodal displacements. Matrix $[k]^e$ is called the element stiffness matrix, which depends on the shape, size, direction, and elastic constant of the element, and is irrelevant to the location; that is to say, it will not change with the translation of the element or coordinate axis.

3.5 Nodal Load

In order to analyze the stress of continuous medium by the structural mechanics method, all distributed loads must be substituted with equivalent nodal loads. Now use $\{P_i\}$ to represent the equivalent nodal loads of node i, and then

Plane stress problems： $\{P_i\} = \begin{Bmatrix} X_i \\ Y_i \end{Bmatrix}$

Space stress problems： $\{P_i\} = \begin{Bmatrix} X_i \\ Y_i \\ Z_i \end{Bmatrix}$

Plate bending problems： $\{P_i\} = \begin{Bmatrix} Z_i \\ R_{\theta xi} \\ R_{\theta yi} \end{Bmatrix}$

in which X_i, Y_i, Z_i, are the concentrated loads acting on node i in the x-, y-, z-directions, respectively, and $R_{\theta xi}$, $R_{\theta yi}$ are the concentrated moment around the x- and y-axes, respectively.

Use $\{P\}^{\mathrm{e}}$ to represent the vector composed of all nodal loads of element e, namely,

$$\{P\}^{\mathrm{e}} = \left[P_i \ P_j \ P_m \ \cdots\right]^{\mathrm{T}} \tag{3.13}$$

Below we'll use the virtual displacement method to derive a number of nodal load formulas.

3.5.1 Distributed Volume Force

Let the volume force acting within unit volume be

$$\{q\} = \begin{Bmatrix} q_x \\ q_y \\ q_z \end{Bmatrix}$$

When virtual displacement $\{r^*\}$ occurs within the element, the work done by volume force $\{q\}$ is

$$\iiint \{r^*\}^{\mathrm{T}}\{q\}\,\mathrm{d}x\,\mathrm{d}y\,\mathrm{d}z = \left(\{\delta^*\}^{\mathrm{e}}\right)^{\mathrm{T}} \iiint [N]^{\mathrm{T}}\{q\}\,\mathrm{d}x\,\mathrm{d}y\,\mathrm{d}z$$

It should be equal to the work done by the equivalent nodal load, that is,

$$\left(\{\delta^*\}^{\mathrm{e}}\right)^{\mathrm{T}}\{P\}_q^{\mathrm{e}} = \left(\{\delta^*\}^{\mathrm{e}}\right)^{\mathrm{T}} \iiint [N]^{\mathrm{T}}\{q\}\,\mathrm{d}x\,\mathrm{d}y\,\mathrm{d}z$$

As virtual displacement $\{\delta^*\}$ is arbitrary, by the formula above we obtain the equivalent nodal load of volume force $\{q\}$ as follows:

$$\{P\}_q^{\mathrm{e}} = \iiint [N]^{\mathrm{T}}\{q\}\,\mathrm{d}x\,\mathrm{d}y\,\mathrm{d}z \tag{3.14}$$

Multiple integration on the right part of formula (3.14) should be done within the whole volume of element e.

3.5.2 Distributed Surface Force

Let element e be near the boundary and on its boundary S is acting as the distributed surface force $\{p\}$

$$\{p\} = \begin{Bmatrix} p_x \\ p_y \\ p_z \end{Bmatrix}$$

When virtual displacement $\{r^*\} = [N]\{\delta^*\}^{\mathrm{e}}$ happens in the element, the work done by surface force $\{p\}$ is

$$\int_S \{r^*\}^{\mathrm{T}}\{p\}\,\mathrm{d}S = \left(\{\delta^*\}^{\mathrm{e}}\right)^{\mathrm{T}} \int_S [N]^{\mathrm{T}}\{p\}\,\mathrm{d}S$$

It must be equal to the work done by the equivalent nodal load; hereby we obtain

$$\{P\}_p^{\mathrm{e}} = \int_S [N]^{\mathrm{T}}\{p\}\,\mathrm{d}S \tag{3.15}$$

The surface integration on the right part of formula (3.15) is done on surface S under the action of distributed loads.

3.5.3 Initial Strain and Initial Stress

If within element e there are initial strain $\{\varepsilon_0\}$ and initial stress $\{\sigma_0\}$, substitute formula (3.6) into formula (3.10) and obtain

$$\{F\}^e = \iiint [B]^T[D][B]\,dx\,dy\,dz\{\delta\}^e - \iiint [B]^T[D]\{\varepsilon_0\}\,dx\,dy\,dz + \iiint [B]^T\{\sigma_0\}\,dx\,dy\,dz$$

The second and third terms on the right part of the formula above are nodal forces caused by the initial strain and initial stress. Change their signs and obtain the nodal load $\{P\}^e_{\varepsilon_0}$ caused by the initial strain and nodal load $\{P\}^e_{\sigma_0}$ caused by the initial stress as follows:

$$\{P\}^e_{\varepsilon_0} = \iiint [B]^T[D]\{\varepsilon_0\}\,dx\,dy\,dz \tag{3.16}$$

$$\{P\}^e_{\sigma_0} = -\iiint [B]^T\{\sigma_0\}\,dx\,dy \in z \tag{3.17}$$

3.6 Application Examples of the Principle of Virtual Displacements: Beam Element

Now we shall use the principle of virtual displacement to derive the beam element stiffness matrix, as an example.

For the beam element 12 shown in Figure 3.4, the nodal force $\{F\}^e$ and nodal displacement $\{\delta\}^e$ are, respectively,

$$\{F\}^e = \begin{Bmatrix} V_1 \\ M_1 \\ V_2 \\ M_2 \end{Bmatrix}, \quad \{\delta\}^e = \begin{Bmatrix} v_1 \\ \phi_1 \\ v_2 \\ \phi_2 \end{Bmatrix}$$

Let

$$L_1 = 1 - \frac{x}{l}, \quad L_2 = \frac{x}{l} \tag{3.18}$$

Obviously, L_1, L_2 have the following features:

$$\left.\begin{array}{lll} \text{On node 1}: & L_1 = 1, & L_2 = 0 \\ \text{On node 2}: & L_1 = 0, & L_2 = 1 \end{array}\right\} \tag{3.19}$$

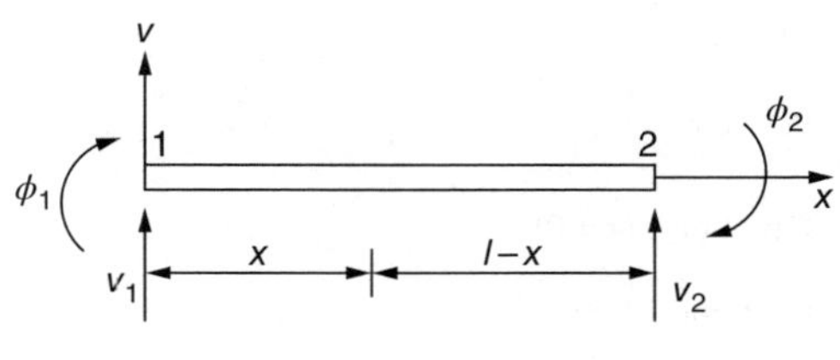

Figure 3.4 Beam element.

Let $\psi = \psi(L_1, L_2)$. Then

$$\frac{d\psi}{dx} = \frac{\partial \psi}{\partial L_1}\frac{\partial L_1}{\partial x} + \frac{\partial \psi}{\partial L_2}\frac{\partial L_2}{\partial x} = \frac{1}{l}\left(\frac{\partial \psi}{\partial L_2} - \frac{\partial \psi}{\partial L_1}\right) \tag{3.20}$$

Because $L_1 + L_2 = 1$, it is not difficult to prove the integrals of L_1 and L_2 may be calculated as follows:

$$\int_0^1 L_1^a L_2^b \, dx = l\int_0^1 (1 - L_2)L_2^b \, dL_2 = \frac{a!b!l}{(a+b+1)!} \tag{3.21}$$

where

$$a! = 1 \cdot 2 \cdot 3 \cdots (a-1) \cdot a$$
$$0! = 1$$

in which L_1, L_2 are coordinates represented by the ratio of the length, thus called length coordinates.

In order to conduct element analysis, we at first must use a function to represent the deflection of any point of beam element 12. Now we use shape functions and nodal displacements to represent the deflection v of any point within the element as follows:

$$v = N_1 v_1 + N_2\phi_1 + N_3 u_2 + N_4\phi_2 = \begin{bmatrix} N_1 & N_2 & N_3 & N_4 \end{bmatrix} \begin{Bmatrix} v_1 \\ \phi_1 \\ v_2 \\ \phi_2 \end{Bmatrix} = [N]\{\delta\}^e \tag{3.22}$$

$$[N] = \begin{bmatrix} N_1 & N_2 & N_3 & N_4 \end{bmatrix} \tag{3.23}$$

$$\left.\begin{aligned} N_1 &= L_1^2(3 - 2L_1), \quad N_2 = -L_1^2 L_2 l \\ N_3 &= L_2^2(3 - 2L_2), \quad N_4 = -L_1 L_2^2 l \end{aligned}\right\} \tag{3.24}$$

in which N_1, N_2, N_3, N_4 are shape functions of the beam element.

The reason for adopting the above shape function is to make v and dv/dx on both ends of the element to meet the supposed boundary conditions. Calculate the derivative of N_i and by formula (3.20) we can obtain

$$\left.\begin{aligned} \frac{dN_1}{dx} &= -\frac{6L_1L_2}{l}, \quad \frac{dN_2}{dx} = -L_1(L_1 - 2L_2) \\ \frac{dN_3}{dx} &= \frac{6L_1L_2}{l}, \quad \frac{dN_4}{dx} = -L_2(L_2 - 2L_1) \end{aligned}\right\} \tag{3.25}$$

It is easy to find that, on node 1 and node 2, N_i and its derivative will take the following values:

Position	N_1	N_2	N_3	N_4	$\frac{dN_1}{dx}$	$\frac{dN_2}{ds}$	$\frac{dN_3}{dx}$	$\frac{dN_4}{dx}$
On node 1 ($x = 0$)	1	0	0	0	0	−1	0	0
On node 2 ($x = l$)	0	0	1	0	0	0	0	−1

Calculate the derivative by formula (3.22) and get

$$\frac{dv}{dx} = \frac{dN_1}{dx}v_1 + \frac{dN_2}{dx}\phi_1 + \frac{dN_3}{dx}v_2 + \frac{dN_4}{dx}\phi_2 \tag{3.26}$$

According to values listed above, we can see

$$\text{On node 1}: \quad v = v_1, \quad \frac{dv}{dx} = -\phi_2$$

$$\text{On node 2}: \quad v = v_2, \quad \frac{dv}{dx} = -\phi_2$$

They meet the boundary conditions we supposed for the deflection curve of the beam. By formula (3.24) we can see N_i is a cubic polynomial of x, so the displacement formula (3.22) of the beam represented by shape functions is essentially the same as the displacement curve of the beam directly represented by cubic polynomials. But the usage of shape functions can make displacement v represented directly by nodal displacement $\{\delta\}^e$, which brings convenience to deduction of the element stiffness matrix and nodal load in the next step.

Then calculate the derivative of Eq. (3.26) and get

$$\frac{d^2v}{dx^2} = \{\varepsilon\} = [B]\{\delta\}^e \tag{3.27}$$

where

$$[B] = \begin{bmatrix} \frac{d^2N_1}{dx^2} & \frac{d^2N_2}{dx^2} & \frac{d^2N_3}{dx^2} & \frac{d^2N_4}{dx^2} \end{bmatrix} \tag{3.28}$$

Use the bending moment to represent the general stress of the beam, so

$$\{\sigma\} = EI\frac{d^2v}{dx^2} = [D]\{\varepsilon\} = [D][B]\{\delta\}^e$$

in which $[D] = EI$.

By the virtual work principle, from formula (3.11) we can obtain the element stiffness matrix by the following formula:

$$[k]^e = \int_0^l [B]^T[D][B]\,dx$$

Substitute formula (3.28) into the above formula and we can see the element k_{rs} of row r, column s in $[k]^e$ is

$$k_{rs} = EI\int_0^l \frac{d^2N_r}{dx^2}\frac{d^2N_s}{dx^2}\,dx \quad (r, s = 1,2,3,4) \tag{3.29}$$

Hereby we get the beam element stiffness matrix as below:

$$[k]^e = \frac{EI}{l^2}\begin{bmatrix} 12 & -6l & -12 & -6l \\ -6l & 4l^2 & 6l & 2l^2 \\ -12 & 6l & 12 & 6l \\ -6l & 2l^2 & 6l & 4l^2 \end{bmatrix} \tag{3.30}$$

As formula (3.22) used to represent the deflection curve of the beam is a cubic polynomial consistent with the theoretical solution, the element stiffness matrix obtained is identical to that obtained by the structural mechanics. If the displacement function adopted is approximate, the element stiffness matrix obtained will also be approximate.

3.7 Strain Energy and Complementary Strain Energy

Under the action of external forces, the stress $\{\sigma\}$ and strain $\{\varepsilon\}$ will appear within the body. Suppose the external forces are increased gradually from zero, the stress and strain will also increase from zero gradually. During this increasing process, the work done by stress within unit volume is called strain energy density denoted by U. For simplicity, we don't take initial stress and initial strain into consideration here (it is not difficult to extend it to general cases including initial stress and initial strain). By Figure 3.5 we know that the strain energy density $\overline{U}$ is the area right of the $\sigma - \varepsilon$ curve. For general space problems, it can be calculated as follows:

$$\overline{U} = \int_0^{\varepsilon_x} \sigma_x \, d\varepsilon_x + \int_0^{\varepsilon_y} \sigma_y \, d\varepsilon_y + \int_0^{\varepsilon_z} \sigma_z \, d\varepsilon_z + \int_0^{\gamma_{xy}} \tau_{xy} \, d\gamma_{xy} + \int_0^{\gamma_{yz}} \tau_{yz} \, d\gamma_{yz} + \int_0^{\gamma_{zx}} \tau_{zx} \, d\gamma_{zx} = \int \{\sigma\}^T d(\varepsilon) \tag{3.31}$$

For a linear elastic body, substituting the stress–strain relationship $\{\sigma\} = [D]\{\varepsilon\}$ into formula (3.31), after integration we get

$$\overline{U} = \int \{\varepsilon\}^T [D] d\{\varepsilon\} = \frac{1}{2}\{\varepsilon\}^T [D]\{\varepsilon\} \tag{3.32}$$

Integration within the whole volume of the body will give the strain energy of the body as below:

$$U = \frac{1}{2}\iiint \{\varepsilon\}^T [D]\{\varepsilon\} \, dx \, dy \, dz \tag{3.33}$$

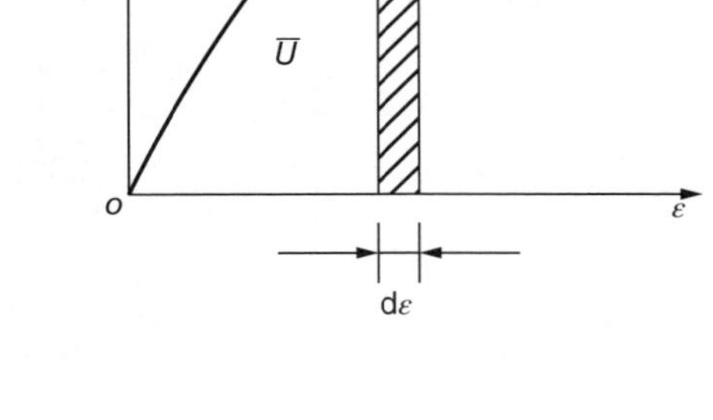

Figure 3.5 Stress, strain, and strain energy.

In Figure 3.5, the area left of the stress–strain curve is defined as complementary strain energy density denoted as $\overline{U}^*$, which may be deemed as the complementary area of rectangle $\sigma\varepsilon$. The complementary strain energy density should be calculated by the following formula:

$$\overline{U}^* = \int_0^{\sigma_x} \varepsilon_x \, \mathrm{d}\sigma_x + \int_0^{\sigma_y} \varepsilon_y \, \mathrm{d}\sigma_y + \int_0^{\sigma_z} \varepsilon_z \, \mathrm{d}\sigma_z + \cdots + \int_0^{\gamma_{zx}} \gamma_{zx} \, \mathrm{d}\tau_{zx} = \int \{\varepsilon\}^{\mathrm{T}} \mathrm{d}\{\sigma\} \tag{3.34}$$

For a linear elastic body, substituting the stress–strain relationship $\{\varepsilon\} = [D]^{-1}\{\sigma\}$ into formula (3.34), after integration we get

$$\overline{U}^* = \int \{\sigma\}^{\mathrm{T}}[D]^{-1} \mathrm{d}\{\sigma\} = \frac{1}{2}\{\sigma\}^{\mathrm{T}}[D]^{-1}\{\sigma\} \tag{3.35}$$

If we do integration within the whole volume, we will obtain the complementary stress energy of the body as follows:

$$U^* = \frac{1}{2}\iiint \{\sigma\}^{\mathrm{T}}[D]^{-1}\{\sigma\} \, \mathrm{d}x \, \mathrm{d}y \, \mathrm{d}z \tag{3.36}$$

3.8 Principle of Minimum Potential Energy

The potential energy Π_{p} of the body is defined as the difference between the strain energy U and the external force potential V, that is,

$$\Pi_{\mathrm{p}} = U - V \tag{3.37}$$

in which the strain energy U may be calculated by formula (3.33) and the external force potential can be calculated by the following formula:

$$V = \sum F\delta + \iiint \{r\}^{\mathrm{T}}\{q\} \, \mathrm{d}x \, \mathrm{d}y \, \mathrm{d}z + \int_{S_\sigma} \{r_b\}^{\mathrm{T}}\{\overline{p}\} \, \mathrm{d}s \tag{3.38}$$

On the right part of the above equation, the first term is the potential of concentrated force F, the second term is the potential of volume force $\{q\}$, the third term is the potential of surface force $\{p\}$, S_σ is the surface under action of the surface force, $\{r_b\}$ is the displacement on surface S_σ, and $\{\overline{p}\}$ is the preset surface force.

The minimum potential energy principle can be described as follows: among all the displacements meeting the boundary conditions, those meeting the equilibrium condition can make the potential energy Π_{p} of the body to take the stationary value, that is,

$$\delta\Pi_{\mathrm{p}} = \delta U - \delta V = 0 \tag{3.39}$$

For a linear elastic body, the potential energy takes the minimum value, that is,

$$\delta^2\Pi_{\mathrm{p}} = \delta^2 U - \delta^2 V \geq 0 \tag{3.40}$$

The minimum potential energy principle can be proved by virtual displacements. If there is virtual displacement $\delta(r) = \delta \begin{pmatrix} u & v & w \end{pmatrix}$, in the body and corresponding virtual strain is $\delta\{\varepsilon\}$, the virtual strain energy of the whole body is

$$\iiint \{\sigma\}^{\mathrm{T}}\delta\{\varepsilon\} \, \mathrm{d}x \, \mathrm{d}y \, \mathrm{d}z$$

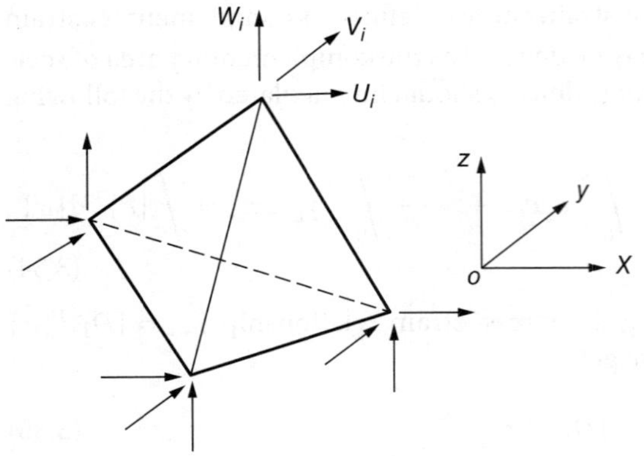

Figure 3.6 Element.

By substituting $\{\sigma\} = [D]\{\varepsilon\}$ into the above equation, we get

$$\iiint \{\varepsilon\}^{\mathrm{T}}[D]\delta\{\varepsilon\}\,\mathrm{d}x\,\mathrm{d}y\,\mathrm{d}z = \delta U$$

It is known by formula (3.33) that the left of the above formula equals to the first variation of the strain energy. Similarly, by formula (3.38) we know the virtual work done by external forces on virtual displacement $\delta(r)$ equals to the first variation δV of the external force potential. According to the virtual displacement principle $\delta U - \delta V = 0$, it is known that

$$\delta\Pi_{\mathrm{p}} = 0$$

By the principle of minimum potential energy, we can derive the element stiffness matrix and nodal load. Considering one element shown in Figure 3.6, on the node is acting nodal force $\{F\}^{\mathrm{e}}$, the element node displacement is $\{\delta\}^{\mathrm{e}}$, and element strain is $\{\varepsilon\} = [B]\{\delta\}^{\mathrm{e}}$. Substitution of this formula into formula (3.33) yields

$$U = \frac{1}{2}\{\delta\}^{\mathrm{eT}}\left(\iiint [B]^{\mathrm{T}}[D][B]\,\mathrm{d}x\,\mathrm{d}y\,\mathrm{d}z\right)\{\delta\}^{\mathrm{e}}$$

or

$$U = \frac{1}{2}\{\delta\}^{\mathrm{eT}}[k]^{\mathrm{e}}\{\delta\}^{\mathrm{e}} \tag{3.41}$$

in which

$$[k]^{\mathrm{e}} = \iiint [B]^{\mathrm{T}}[D][B]\,\mathrm{d}x\,\mathrm{d}y\,\mathrm{d}z \tag{a}$$

By formula (3.38), the external force potential of the element nodal force is

$$V = \{\delta\}^{\mathrm{eT}}\{F\}^{\mathrm{e}}$$

Then the potential energy of the element is

$$\Pi_{\mathrm{p}} = \frac{1}{2}\{\delta\}^{\mathrm{eT}}[k]^{\mathrm{e}}\{\delta\}^{\mathrm{e}} - \{\delta\}^{\mathrm{eT}}\{F\}^{\mathrm{e}} \tag{3.42}$$

By the minimum potential energy principle, $\delta \Pi_{\mathrm{p}} = 0$, so

$$\frac{\partial \Pi_{\mathrm{p}}}{\partial \{\delta\}^{\mathrm{e}}} = 0$$

thus

$$\{F\}^{\mathrm{e}} = [k]^{\mathrm{e}} \{\delta\}^{\mathrm{e}} \tag{3.43}$$

The above formula establishes the relationship between the nodal force and nodal displacement.

From the physical consideration, the strain energy must be a positive quantity, that is,

$$U = \frac{1}{2} \{\delta\}^{\mathrm{eT}} [k]^{\mathrm{e}} \{\delta\}^{\mathrm{e}} \geq 0$$

and the nodal displacement $\{\delta\}^{\mathrm{e}}$ is arbitrary, so the element stiffness matrix $[k]^{\mathrm{e}}$ is positive definite. Hereby it can be inferred that the second variation of potential is not negative. Since the first variation of potential energy equals to zero and second variation is not negative, we can assert the potential takes the minimum value.

By substituting $\{r\} = [N]\{\delta\}^{\mathrm{e}}$ into the formula (3.38) of external force potential, we obtain the potential of volume force $\{q\}$ and surface force $\{\overline{p}\}$ as

$$V = \{\delta\}^{\mathrm{eT}} \iiint [N]\{q\}\, \mathrm{d}x\, \mathrm{d}y\, \mathrm{d}z + \{\delta\}^{\mathrm{eT}} \int_{s} [N]^{\mathrm{T}} \{\overline{p}\}\, \mathrm{d}S$$

So the element potential is

$$\begin{aligned} \Pi_{\mathrm{p}} &= U - V \\ &= \frac{1}{2} \{\delta\}^{\mathrm{eT}} [k]^{\mathrm{e}} \{\delta\}^{\mathrm{e}} - \{\delta\}^{\mathrm{eT}} \iiint [N]\{q\}\, \mathrm{d}x\, \mathrm{d}y\, \mathrm{d}z - \{\delta\}^{\mathrm{eT}} \int_{s} [N]^{\mathrm{T}} \{\overline{p}\}\, \mathrm{d}S \end{aligned}$$

According to the minimum potential principle, $\partial \Pi_{\mathrm{p}} / \partial \{\delta\}^{\mathrm{e}} = 0$. Thus we obtain

$$[k]^{\mathrm{e}} \{\delta\}^{\mathrm{e}} = \{P\}_{q}^{\mathrm{e}} + \{P\}_{p}^{\mathrm{e}}$$

$$\{P\}_{q}^{\mathrm{e}} = \iiint [N]^{\mathrm{T}} \{q\}\, \mathrm{d}x\, \mathrm{d}y\, \mathrm{d}z \tag{3.14}$$

$$\{P\}_{p}^{\mathrm{e}} = \int_{S} [N]^{\mathrm{T}} \{\overline{p}\}\, \mathrm{d}S \tag{3.15}$$

in which $\{P\}_{q}^{\mathrm{e}}$ is the nodal load caused by volume force and $\{P\}_{p}^{\mathrm{e}}$ is the nodal load caused by surface force.

From the deduction above, we can see the element stiffness matrix derived by the minimum potential energy principle is identical to the result given by the virtual displacement principle. The two principles are the same in essence. The virtual displacement principle is easy for use, while the minimum potential principle can help us to determine the convergence and lower-limit nature of the solution.

When depending on the virtual displacement principle or the minimum potential energy principle for solution, it is necessary to suppose the element displacement function with equal displacements along the common boundary of adjacent elements. This type of element is called displacement-consistent element or consistent element in brief. The unknown quantity during solving is the nodal displacement.

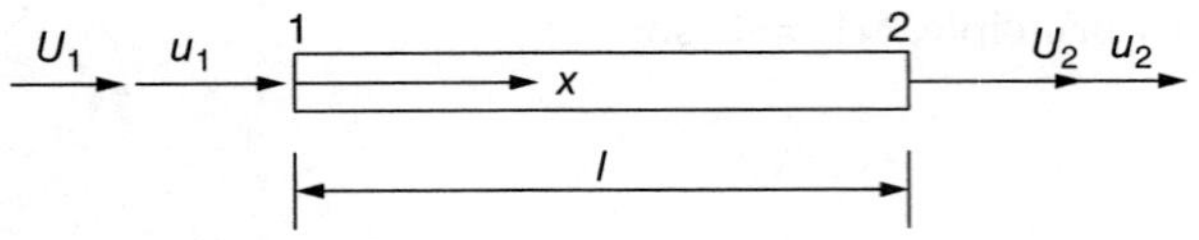

Figure 3.7 Bar element.

Now we take the bar element shown in Figure 3.7 to illustrate how to compute the element stiffness matrix by the potential energy principle. The element displacement is expressed by the following formula:

$$u = L_1u_1 + L_2u_2 = \left(1 - \frac{x}{l}\right)u_1 + \frac{x}{l}u_2$$

The element strain is

$$\varepsilon = \frac{du}{dx} = -\frac{1}{l}u_1 + \frac{1}{l}u_2 = [B]\{\delta\}^e$$

in which $[B] = \begin{bmatrix} -\frac{1}{l_1} \frac{1}{l} \end{bmatrix}$.

The stress in the element is $\sigma = E\varepsilon$, so under present conditions the elastic matrix is

$$[D] = E$$

By formula (a), the element stiffness matrix is

$$[k]^e = \int_0^l [B]^T[D][B]A\,dx = \frac{AE}{l}\begin{Bmatrix} -1 \\ 1 \end{Bmatrix}[-1\ 1] = \frac{AE}{l}\begin{bmatrix} 1 & -1 \\ -1 & 1 \end{bmatrix}$$

The strain energy is

$$U = \frac{1}{2}\{\delta\}^{eT}[k]^e\{\delta\}^e = \frac{AE}{2l}(u_1^2 - 2u_1u_2 + u_2^2)$$

The external force potential is

$$V = U_1u_1 + U_2u_2$$

So the potential energy of the element is

$$\Pi_p = U - V = \frac{AE}{2l}(u_1^2 - 2u_1u_2 + u_2^2) - U_1u_1 - U_2u_2$$

According to the minimum potential energy principle,

$$\frac{\partial \Pi_p}{\partial u_1} = 0, \quad \frac{\partial \Pi_p}{\partial u_2} = 0$$

thus we obtain

$$\begin{Bmatrix} U_1 \\ U_2 \end{Bmatrix} = \frac{AE}{l}\begin{bmatrix} 1 & -1 \\ -1 & 1 \end{bmatrix}\begin{Bmatrix} u_1 \\ u_2 \end{Bmatrix} = [k]^e\begin{Bmatrix} u_1 \\ u_2 \end{Bmatrix}$$

The above formula illustrates the relationship between the nodal force and nodal displacement of bar element.

3.9 Minimum Complementary Energy Principle

If on a certain part of boundary S_u of the body is preset, the displacement $\{\overline{r_b}\} = \left[\overline{u_b}\ \overline{v_b}\ \overline{w_b}\right]$ and let the boundary reaction force (including support reaction force) acting on the boundary be $\{p\} = \left[p_x\ p_y\ p_z\right]^{\mathrm{T}}$. Then the potential energy of boundary reaction force may be calculated by the following formula:

$$V^* = \int_{S_u} \left(p_x\overline{u_b} + p_y\overline{v_b} + p_z\overline{w_b}\right) \mathrm{d}s = \int_{S_u} \{P\}^{\mathrm{T}}\{\overline{r_b}\}\,\mathrm{d}s \tag{3.44}$$

The complementary energy Π_c of the body is defined as difference between the complementary strain energy U^* and potential V^* of preset boundary reaction force on that part of boundary S_u, which is

$$\Pi_c = U^* - V^* \tag{3.45}$$

By formula (3.36), the complementary strain energy is

$$U^* = \frac{1}{2}\iiint \{\sigma\}^{\mathrm{T}}[D]^{-1}\{\sigma\}\,\mathrm{d}x\,\mathrm{d}y\,\mathrm{d}z$$

So the complementary energy of the body may be determined by the following formula:

$$\Pi_c = \frac{1}{2}\iiint \{\sigma\}^{\mathrm{T}}[D]^{-1}\{\sigma\}\,\mathrm{d}x\,\mathrm{d}y\,\mathrm{d}z - \int_{S_u} \{p\}^{\mathrm{T}}\{\overline{r_b}\}\,\mathrm{d}S \tag{3.46}$$

The minimum complementary energy principle may be described as follows: among all stress states meeting the equilibrium requirements within the body and preset stress conditions on the boundary, only the stress state that further meets the stress–strain relationship within the body and preset boundary displacement condition on the boundary can make the complementary energy of the body to take the stationary value; that is to say,

$$\delta\Pi_c = \delta U^* - \delta V^* = 0 \tag{3.47}$$

For a linear elastic body, Π_c takes the minimum value, which is

$$\delta^2\Pi_c = \delta^2 U^* - \delta^2 V^* \geq 0 \tag{3.48}$$

The minimum complementary energy principle may be proved by the virtual stress principle. On the premise of keeping the displacement of the body unchanged, supposing any tiny virtual stress $\delta\{\sigma\}$ occurs, it should meet the preset boundary stress condition. By steps similar to Section 3.8, we can prove that the complementary energy takes the minimum value.

The minimum complementary energy principle has a basic difference from the minimum potential energy principle as the latter corresponds to the equilibrium condition of the structure, while the former corresponds to the deformation compatibility condition of the structure. The latter takes displacement as the variable quantity, while the former takes force as the variable quantity.

During element analysis, if the displacement function of the element is preset, the element stiffness matrix can be derived by the minimum potential energy principle. If the stress state of the element is preset, the element flexibility matrix can be derived by the minimum complementary energy principle.

Suppose the element stress $\{\sigma\}$ can be expressed by nodal force $\{F\}^{\mathrm{e}}$ as follows:

$$\{\sigma\} = [\rho]\{F\}^{\mathrm{e}} \tag{3.49}$$

The value taken by the above formula on the element boundary is just the boundary force $\{p\}$ of the element denoted as

$$\{p\} = [\lambda]\{F\}^{\mathrm{e}} \tag{3.50}$$

It should be pointed out that except bar elements and beam elements, for ordinary continuous medium, it is difficult to represent the internal stress of the elements like formula (3.49). Substitution of formulas (3.49) and (3.50) into formula (3.46) will give

$$\Pi_{\mathrm{c}} = \frac{1}{2}\{F\}^{x\mathrm{T}}[f]\{F\}^{x} - \{F\}^{\mathrm{eT}}\{\overline{\delta}\} \tag{3.51}$$

$$[f] = \iiint \{\rho\}^{\mathrm{T}}[D]^{-1}\{\rho\}\,\mathrm{d}x\,\mathrm{d}y\,\mathrm{d}z \tag{3.52}$$

$$\{\overline{\delta}\} = \iint_{S_u} [\lambda]^{\mathrm{T}}\{\overline{r_b}\}\,\mathrm{d}S \tag{3.53}$$

where $[f]$ is the element flexibility matrix and $\{\overline{\delta}\}$ is the displacement of element nodes.

By the minimum complementary energy principle, $\delta\Pi_{\mathrm{c}} = 0$, so

$$\frac{\partial\Pi_{\mathrm{c}}}{\partial\{F\}^{\mathrm{e}}} = 0$$

By formula (3.51), we get

$$[f]\{F\}^{\mathrm{e}} = \{\overline{\delta}\} \tag{3.54}$$

If formula (3.54) is used for overall analysis of the structure, it is called the matrix force method. At first choose the redundant force, and then establish a set of equations with nodal forces as the unknown quantities in light of the deformation compatibility condition, which leads to the solution of nodal forces. The matrix force method is far more difficult to realize than the matrix displacement method on computers, so it is rarely used. When following the minimum complementary energy for solution, the stress field assumed should satisfy the equilibrium equation within the element. Stresses on common boundaries of adjacent elements don't have to be continuous, but should meet the equilibrium condition. This type of element is called equilibrium element, and the unknown quantity during solving is the nodal force.

3.10 Hybrid Element

From the previous section we can see that if the element stress is preset, the element flexibility matrix can be deduced by the minimum complementary energy principle. If it is applied in overall analysis, it is the matrix force method, but it is difficult for actual application. Besides, if the element displacement is preset, the element stiffness matrix can be deduced by the minimum potential energy principle; if it is applied in overall analysis, it is the matrix displacement method and is now most widely used. But the requirements of the displacement functions are strict, it requires that not only the displacement is continuous within the element but also the displacement between adjacent elements on element boundaries and certain derivatives are consistent. In some cases, for example, calculation of stresses on thin plates, thin shells, and crack tips, building a compatible displacement function is quite difficult. To solve this problem, several hybrid elements are developed.

One is the displacement hybrid element, within which we adopt a displacement function and on its boundary adopt another independent stress or displacement function and use the minimum potential energy principle for solution. The other is the stress hybrid element, within which we assume a stress field and adopt another compatible displacement field on its boundary and use the minimum complementary energy principle for solution.

Below we'll expound the stress hybrid element (Figure 3.8), as this type of hybrid element is more valuable in practice.

When solving by the displacement method, we take the nodal displacement as the parameter; when solving by the force method, we take the nodal force as the parameter. When using the above two methods for solution, we only adopt one set of parameters; however, in case of the stress hybrid elements, we adopt two sets of parameters – one set is to represent the element stress and the other to represent the element boundary displacement.

Within the element, we suppose there is a stress field $\{\sigma\}$ meeting the equilibrium condition, which may be represented by generalized parameter $\{\beta\}$ as follows:

$$\{\sigma\} = [\Omega]\{\beta\} \tag{3.55}$$

Let the above formula take values on the element boundaries, and we can obtain the boundary force on element boundaries as follows:

$$\{p\} = [\rho]\{\beta\} \tag{3.56}$$

in which $\{\rho\} = [\Omega_b]$ is obtained by $[\Omega]$ taking values on element boundaries.

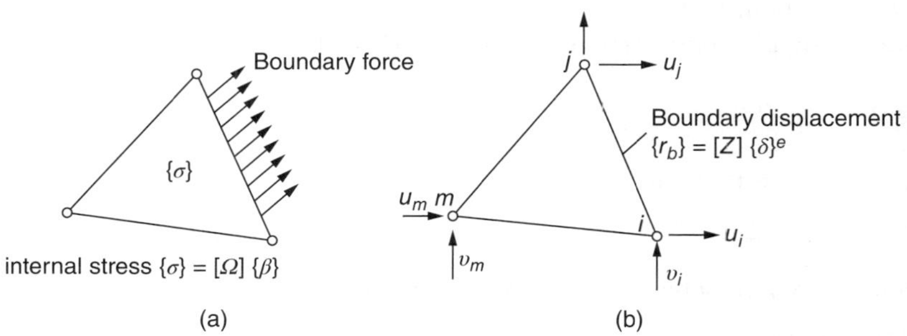

Figure 3.8 Stress hybrid element. (a) Use generalized parameter (β) to represent the internal stress $\{\sigma\}$ and boundary stress $\{p\}$ of the element (b) use nodal displacement $\{\delta\}^e$ to represent the element boundary displacement $\{r_b\}$.

The displacement on element boundaries can be represented by nodal displacement $\{\delta\}^e$ as follows:

$$\{r_b\} = [Z]\{\delta\}^e \tag{3.57}$$

The interpolating function $[Z]$ here is applied only to the boundaries other than the whole element, so it is easy to build $[Z]$ that maintains consistency between adjacent elements. In the displacement compatible element,

$$\{r\} = [N]\{\delta\}^e$$

in which $[N]$ is defined for the whole element. It is difficult to build $[N]$ keeping consistency between adjacent elements. This is the major difference between the two kinds of elements.

In order to deal with the two independent fields, the expression of complementary energy should be modified as

$$\Pi_c^s = \frac{1}{2}\iiint \{\sigma\}^T[D]^{-1}\{\sigma\}\,dx\,dy\,dz - \int_S \{p\}^T\{r_b\}\,dS \tag{3.46a}$$

in which S is all the boundaries of the element and $\{p\}$ is the boundary force consistent with the internal stress field.

The difference between the above formula and the ordinary complementary energy Π_c is that $\{r_b\}$ is represented by parameters of nodal displacements, so the unknown quantities in Π_c^s include not only nodal forces but also nodal displacements, but the usual complementary energy Π_e includes only the nodal force parameter.

Substitution of expression of $\{\sigma\}$, $\{p\}$, $\{r_b\}$ into formula (3.46a) yields

$$\Pi_c^s = \frac{1}{2}\{\beta\}^T[H]\{\beta\} - \{\beta\}^T[Q]\{\delta\}^e \tag{3.58}$$

where

$$[H] = \iiint \{\Omega\}^T[D]^{-1}\{\Omega\}\,dx\,dy\,dz \tag{3.59}$$

$$[Q] = \int_s [\rho]^T[Z]\,dS \tag{3.60}$$

is calculated by volume integration of the element and $[Q]$ is calculated by surface integration of the element boundaries.

By the minimum complementary energy principle, $\delta\Pi_c^s = 0$, so

$$\frac{\partial \Pi_c^s}{\partial\{\beta\}} = 0$$

So by formula (3.58) we arrive at

$$[H]\{\beta\} - [Q]\{\delta\}^e = 0$$

Thus the generalized parameter $\{\beta\}$ can be obtained as follows:

$$\{\beta\} = [H]^{-1}[Q]\{\delta\}^e \tag{3.61}$$

We substitute the above formula into formula (3.58) and get

$$\Pi_c = -\frac{1}{2}\{\delta\}^{eT}[k]^e\{\delta\}^e \tag{3.62}$$

in which

$$[k]^e = [Q]^T[H]^{-1}[Q] \tag{3.63}$$

where $[k]^e$ is the element stiffness matrix (refer to Section 4.3).

By overall analysis of the structure, the displacement $\{\delta\}$ can be obtained; by formula (3.61) the generalized parameter $\{\beta\}$ can be obtained; then by formula (3.55), the element stress can be determined.

Because the final matrix equation takes the nodal displacement as the unknown quantities, the stress hybrid element is easy to use. Furthermore, compared with the displacement hybrid element, the stress hybrid element can provide a more precise estimate of the stress.

3.11 Hybrid Element Example: Plane Rectangular Element

Now we take the plane rectangular element, for example, to explain how to analyze the stress hybrid element. As shown in Figure 3.9, of the rectangular element, the length is a and the height is b. Place the coordinate origin on node 1 and let the coordinate axes coincide with the edges of element.

The axes are superposed on the two edges of the element. First, suppose that the internal stress field of the element is as follows:

$$\begin{aligned} \sigma_x &= \beta_1 + \beta_2 y \\ \sigma_y &= \beta_3 + \beta_4 x \\ \tau_{xy} &= \beta_5 \end{aligned} \tag{3.64}$$

That is,

$$\{\sigma\} = \begin{bmatrix} 1 & y & 0 & 0 & 0 \\ 0 & 0 & 1 & x & 0 \\ 0 & 0 & 0 & 0 & 1 \end{bmatrix} \begin{Bmatrix} \beta_1 \\ \beta_2 \\ \beta_3 \\ \beta_4 \\ \beta_5 \end{Bmatrix} = [\Omega]\{\beta\} \tag{3.65}$$

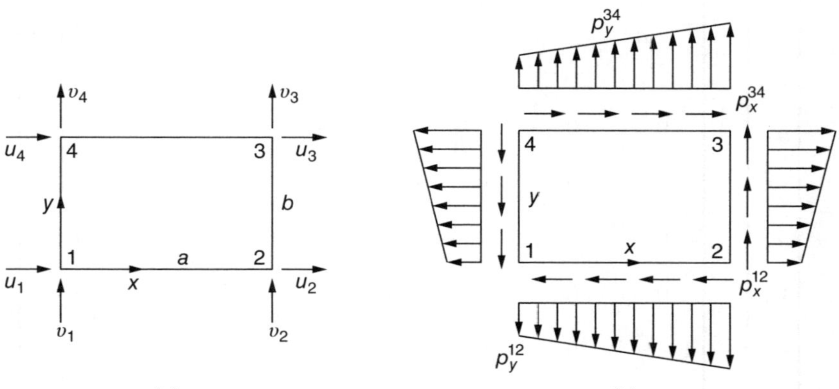

Figure 3.9 Rectangular hybrid element. (a) Nodal displacement and (b) element stress.

Obviously, the above stress field satisfies the plane problem equilibrium equation below:

$$\frac{\partial \sigma_x}{\partial x} + \frac{\partial \tau_{xy}}{\partial y} = 0, \qquad \frac{\partial \sigma_y}{\partial y} + \frac{\partial \tau_{xy}}{\partial x} = 0$$

By formula (3.64) we can determine the boundary forces acting on all boundaries of the element, such as:

$$\text{On boundary 12}: \quad p_x^{12} = -\beta_5, \quad p_y^{12} = -\beta_5 - \beta_4 x$$

$$\text{On boundary 23}: \quad p_x^{23} = \beta_1 + \beta_2 y, \quad p_y^{23} = \beta_5$$

Similarly, we can determine the boundary force on the other two boundaries, thus,

$$\{p\} = \begin{Bmatrix} p_x^{12} \\ p_y^{12} \\ p_x^{23} \\ p_y^{23} \\ p_x^{34} \\ p_y^{34} \\ p_x^{41} \\ p_y^{41} \end{Bmatrix} = \begin{bmatrix} 0 & 0 & 0 & 0 & -1 \\ 0 & 0 & -1 & -x & 0 \\ 0 & 0 & 0 & 0 & 1 \\ 0 & 0 & 0 & 0 & 1 \\ 0 & 0 & 1 & x & 0 \\ -1 & -y & 0 & 0 & 0 \\ 0 & 0 & 0 & 0 & -1 \end{bmatrix} \begin{Bmatrix} \beta_1 \\ \beta_2 \\ \beta_3 \\ \beta_4 \\ \beta_5 \end{Bmatrix} = [\rho][\beta] \tag{3.66}$$

Supposing the displacement changes linearly along the element boundary, thus, the boundary displacement can be represented by the nodal displacement. For example, on boundary 23, the displacement is

$$u_{23} = \left(1 - \frac{y}{b}\right) u_2 + \frac{y}{b} u_3$$

$$v_{23} = \left(1 - \frac{y}{b}\right) v_2 + \frac{y}{b} v_3$$

For the other three boundaries, we can write similar formulas, thus, the element boundary displacement $\{r_b\}$ is obtained as follows:

$$\{r_b\} = \begin{Bmatrix} u_{12} \\ v_{12} \\ u_{23} \\ v_{23} \\ u_{34} \\ v_{34} \\ u_{41} \\ v_{41} \end{Bmatrix} = \begin{bmatrix} 1-\frac{x}{a} & 0 & \frac{x}{a} & 0 & 0 & 0 & 0 & 0 \\ 0 & 1-\frac{x}{a} & 0 & \frac{x}{a} & 0 & 0 & 0 & 0 \\ 0 & 0 & 1-\frac{y}{b} & 0 & \frac{y}{b} & 0 & 0 & 0 \\ 0 & 0 & 0 & 1-\frac{y}{b} & 0 & \frac{y}{b} & 0 & 0 \\ 0 & 0 & 0 & 0 & \frac{x}{a} & 0 & 1-\frac{x}{a} & 0 \\ 0 & 0 & 0 & 0 & 0 & \frac{x}{a} & 0 & 1-\frac{x}{a} \\ 1-\frac{y}{b} & 0 & 0 & 0 & 0 & 0 & \frac{y}{b} & 0 \\ 0 & 1-\frac{y}{b} & 0 & 0 & 0 & 0 & 0 & \frac{y}{b} \end{bmatrix} \begin{Bmatrix} u_1 \\ v_1 \\ u_2 \\ v_2 \\ u_3 \\ v_3 \\ u_4 \\ v_4 \end{Bmatrix}$$

$$= [Z]\{\delta\}^{\mathrm{e}} \tag{3.67}$$

By formulas (3.65)–(3.67), we substitute matrix $[\Omega]$, $\{\rho\}$, $[Z]$ into formulas (3.59) and (3.60) and obtain $[H]$ and $[Q]$. Then we substitute them into formula (3.63) and obtain the element stiffness matrix like formula (3.68). This is the stiffness matrix of plane stress problems; for plane strain problems, the modulus of elasticity and Poisson's ratio should be replaced properly.

$$[k] = \frac{Et}{16 s_1 s_2 ab}\begin{bmatrix} u_1 & v_1 & u_2 & v_2 & u_3 & v_3 & u_4 & v_4 \\ s_1a^2+s_4b^2 & & & & & & & \\ s_2ab & s_1b^2+s_4a^2 & & & \text{symmetrical} & & & \\ s_1a^2-s_4b^2 & s_3ab & s_1a^2+s_4b^2 & & & & & \\ -s_3ab & s_5a^2-s_1b^2 & -s_2ab & s_1b^2+s_4a^2 & & & & \\ -s_1a^2-s_5b^2 & -s_2ab & s_5b^2-s_1a^2 & s_3ab & s_1b^2+s_4a^2 & & & \\ -s_2ab & -s_1b^2-s_5a^2 & -s_3ab & s_1b^2-s_4a^2 & s_2ab & s_1b^2+s_4a^2 & & \\ s_5b^2-s_1a^2 & -s_3ab & -s_1a^2-s_5b^2 & s_2ab & s_1a^2-s_4b^2 & & & \\ s_3ab & s_1b^2-s_4a^2 & s_2ab & -s_1b^2-s_5a^2 & -s_3ab & s_5a^2-s_1b^2 & -s_2ab & s_1b^2+s_4a^2 \end{bmatrix} \tag{3.68}$$

in which

$$s_1 = \frac{1-\mu}{2}, \quad s_2 = \frac{1+\mu}{2}, \quad s_3 = \frac{1-3\mu}{2}, \quad s_4 = \frac{4-\mu^2}{3}, \quad s_5 = \frac{2+\mu^2}{3}$$

and t is the plate thickness.

3.12 Mixed Energy Principle

The functional Π_p in the minimum potential energy principle is represented by displacement field, and the functional Π_c in the minimum complementary energy principle is represented by stress field. However, in the mixed energy principle, the functional Π_R will be represented by both displacement and stress. When applying the mixed energy principle to the finite element method, we can assume a stress field and a displacement field, respectively, for every element.

By Figure 3.5 we know the sum of the strain energy density and complementary strain energy density is equal to the area of rectangular, so

$$\overline{U} + \overline{U}^* = \{\sigma\}^{\mathrm{T}}\{\varepsilon\}$$

By integration in the whole volume, we obtain

$$U + U^* = \iiint \{\sigma\}^{\mathrm{T}}\{\varepsilon\}\,\mathrm{d}x\,\mathrm{d}y\,\mathrm{d}z$$

$$U = \iiint \{\sigma\}^{\mathrm{T}}\{\varepsilon\}\,\mathrm{d}x\,\mathrm{d}y\,\mathrm{d}z - U^*$$

in which U is the strain energy and U^* is the complementary strain energy.

By substituting the above formula into equation $\Pi_p = U - V$, we get

$$\Pi_R = \iiint \{\sigma\}^{\mathrm{T}}\{\varepsilon\}\,\mathrm{d}x\,\mathrm{d}y\,\mathrm{d}z - U^* - V$$

By the expression (3.36) of the complementary strain energy U^*, we obtain

$$\Pi_R = \iiint \left(\{\sigma\}^{\mathrm{T}}\{\varepsilon\} - \frac{1}{2}\{\sigma\}^{\mathrm{T}}[D]^{-1}\{\sigma\}\right)\mathrm{d}x\,\mathrm{d}y\,\mathrm{d}z - V \tag{3.69}$$

in which Π_R is called the Reissner functional.

The stain $\{\varepsilon\}$ in the formula is the derivative of the displacement field, so the above functional includes not only the stress field but also the displacement field. In the potential $\Pi_p = U - V$, the strain energy U includes only strain $\{\varepsilon\}$, and V includes only the external force potential on boundary S_σ. In the complementary energy $\Pi_e = U^* - V^*$, the complementary strain energy U^* includes only $\{\delta\}$, and V^* includes only the potential of basic reaction force on boundary S_u. Now the Reissner functional includes both the stress $\{\sigma\}$ and the strain $\{\varepsilon\}$, so V in formula (3.69) includes both the external force potential on boundary S_σ and the potential of boundary reaction on boundary S_u. The following formula may be adopted for calculation:

$$V = \int_{S_\sigma} \{r\}^{\mathrm{T}}\{\overline{p}\}\,\mathrm{d}S + \int_{S_u} \{p\}^{\mathrm{T}}\{r - \overline{r}\}\,\mathrm{d}S \tag{3.70}$$

in which $\{\overline{p}\}$ is the preset force on boundary S_σ and $\{\overline{r}\}$ is the preset displacement on boundary S_u.

The mixed energy principle demonstrates that functional Π_R takes the stationary value, namely,

$$\delta\Pi_R = 0 \tag{3.71}$$

It takes only the stationary value other than extreme value. When Π_R takes the variations corresponding to stress $\{\sigma\}$ or displacement $\{r\}$, it doesn't offer the maximum value or minimum value.

For a hybrid element, we may suppose a stress (or displacement) field within the element and another displacement (or stress) field on the element boundary. Application of the above mixed energy principle necessitate a simultaneous assumption of a stress field and displacement field within the element.

After structure discretization, we may use formula $\{\sigma\} = [\rho]\{F\}$ to represent the element stress by nodal forces and use formula $\{\varepsilon\} = [B]\{\delta\}$ to represent the element strain by nodal displacements. By $\{P\} = [\lambda]\{F\}$ we can represent the surface force on boundary S_u with the nodal force $\{F\}$, and by formula $\{r\} = [Z]\ \{\delta\}$ we can represent the displacement on boundary S_σ with nodal displacements. By substituting them into formulas (3.69) and (3.70), we obtain

$$\Pi_R = -\frac{1}{2}\{F\}^{\mathrm{T}}[\alpha_1]\{F\} + \{F\}^{\mathrm{T}}[\alpha_2]\{\delta\} - \{\delta\}^{\mathrm{T}}\{\overline{F}\} + \{F\}^{\mathrm{T}}\{\overline{\delta}\} \tag{3.72}$$

in which

$$[\alpha_1] = \iiint [\rho]^{\mathrm{T}}[D][\rho]\,\mathrm{d}x\,\mathrm{d}y\,\mathrm{d}z$$

$$[\alpha_2] = \iiint [\rho]^{\mathrm{T}}[B]\,\mathrm{d}x\,d$$

$$\left[\overline{F}\right] = \int_{S_u} [Z]^{\mathrm{T}}\{\overline{p}\}\,\mathrm{d}S$$

$$\left[\overline{\delta}\right] = \int_{S_u} [\lambda]^{\mathrm{T}}\{\overline{r}\}\,\mathrm{d}S$$

Let the functional Π_R takes the stationary value, that is,

$$\frac{\partial \Pi_R}{\partial \{F\}} = 0, \quad \frac{\partial \Pi_R}{\partial \{\delta\}} = 0$$

hereby we obtain

$$\left.\begin{aligned} -[\alpha_1]\{F\} + \{\delta\} + \{\bar{\delta}\} = 0 \\ \{\alpha_2\}^{\mathrm{T}}\{F\} - \{\bar{F}\} = 0 \end{aligned}\right\} \tag{3.73}$$

In formula (3.73), the unknown quantities include nodal displacement $\{\delta\}$ and nodal force $\{F\}$.

The mixed energy principle is more frequently applied in the finite element analysis of plate and shells; detailed examples are provided in Chapters 10 and 21.

3.13 Composite Element

Sometimes several simple elements may be combined into a composite element. As shown in Figure 3.10, we combine four triangles into a rectangle or a polygon. In establishing the stiffness matrix of composite element, we can eliminate all internal nodes and take only nodes on boundary of composite element into consideration during global equilibrium of the structure; thus the order and bandwidth of the overall equilibrium equation are all decreased and lead to reduced computer storage capacity and calculation time.

The node equilibrium equation of composite element can be written by partitioned matrix as follows:

$$\begin{bmatrix} K_{aa} & K_{ab} \\ K_{ba} & K_{bb} \end{bmatrix} \begin{Bmatrix} \delta_a \\ \delta_b \end{Bmatrix} = \begin{Bmatrix} P_a \\ P_b \end{Bmatrix} \tag{3.74}$$

in which $\{\delta_a\}$ is the internal node displacement to be removed, $\{\delta_b\}$ is the element boundary node displacement left, $\{P_a\}$ is the internal node load, and $\{P_b\}$ is the load of boundary nodes.

By expanding formula (3.74), we obtain

$$[K_{aa}]\{\delta_a\} + [K_{ab}]\{\delta_b\} = \{P_a\} \tag{a}$$

$$[K_{ba}]\{\delta_a\} + [K_{bb}]\{\delta_b\} = \{P_b\} \tag{b}$$

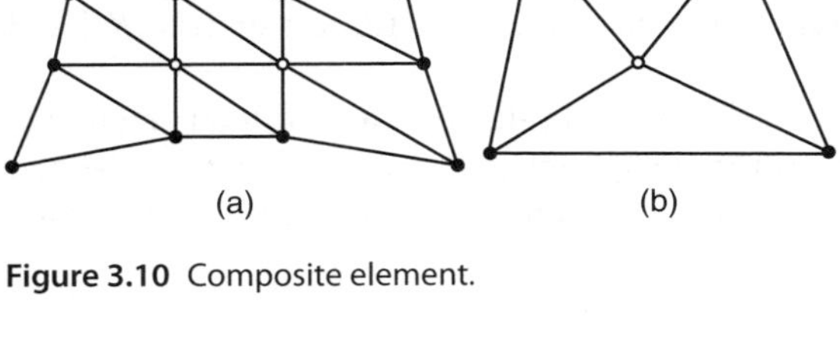

Figure 3.10 Composite element.

By formula (a) we can solve $\{\delta_a\}$ as

$$\{\delta_a\} = [C] - [T]\{\delta_b\} \tag{3.75}$$

in which

$$[C] = [K_{aa}]^{-1}\{P_a\} \tag{3.76}$$

$$[T] = [K_{aa}]^{-1}[K_{ab}] \tag{3.77}$$

By substituting formula (3.75) into formula (b), we get a set of equilibrium equations concerning element boundary nodes as follows:

$$[K^*]\{\delta_b\} = \{P^*\} \tag{3.78}$$

in which

$$[K^*] = [K_{bb}] - [K_{ba}][T] \tag{3.79}$$

$$[P^*] = [P_b] - [K_{ba}]\{C\} \tag{3.80}$$

where $[K_{ba}][T]$ is revision of the element stiffness matrix due to release of internal nodes and $[K_{ba}]\{C\}$ is the load on element boundary nodes transferred from internal nodes.

The stress within the element is

$$\{\sigma\} = \begin{bmatrix} S_a & S_b \end{bmatrix} \begin{Bmatrix} \delta_a \\ \delta_a \end{Bmatrix} + \{\sigma_0\} \tag{3.81}$$

in which $\{\sigma_0\}$ is initial stress.

By substituting formula (3.75) into formula (3.81), we will remove the internal DOF $\{\delta_a\}$ and obtain

$$\{\sigma\} = [S^*]\{\delta_b\} + \{\sigma_0^*\} \tag{3.82}$$

in which

$$[S^*] = [S_b] - [S_a][T] \tag{3.83}$$

$$\{\sigma_0^*\} = \{\sigma_0\} + [S_a]\{C\} \tag{3.84}$$

It can be seen from formula (3.82) that if $[S^*]$ and $\{\sigma_0^*\}$ are formed during forming of $[K^*]$, the element stress can be directly calculated by element boundary node displacement $\{\delta_b\}$, so there is no need to calculate the internal node displacement $\{\delta_a\}$ in the stress calculation phase.

Adoption of matrix signs above is to clarify basic concepts, but in computer programs multiplication and inversion of matrices are complicated. It is more convenient in actual computation to adopt Gauss elimination method; that is to say, in the element equilibrium equation set (3.74), we put the internal DOF on the front (or rear), use Gauss elimination method to remove $\{\delta_a\}$ to get the equilibrium equation (3.78) represented by boundary DOF $\{\delta_b\}$, and at last obtain the reduced element stiffness matrix $[K^*]$ and element node load vector $\{P^*\}$.

It can be found that the difference between composite elements and ordinary elements is that the displacement function of ordinary element is a smooth function defined for the whole element, while that of composite element is a function smooth in each segment.

Bibliography

1 Washizhu, K. (1975) *Variational Methods in Elasticity and Plasticity*, 2nd edn, Pergamon Press, Oxford.
2 Zienkiewicz, O.C. and Taylor, R.L. (2009) *Finite Element Method for Solid and Structural Mechanics*, 6th edn, Elsevier, New York, London.
3 Bathe, K.J. (1982) *Finite Element Procedures in Engineering Analysis*, Prentice Hall, Englewood Cliffs.
4 Strang, G. and Fix, G. (1973) *An Analysis of the Finite Element Method*, Prentice Hall, Englewood Cliffs.
5 Clough, R.W. (1960) *The Finite Element Method in Plane Stress Analysis*. Proceedings of the Second ASCE Conference on Electronic Computation, September 1960, Pittsburgh, PA.

4

Global Analysis

The previous chapter illustrated how to establish the stiffness matrix, flexibility matrix, and nodal load of an element. This chapter will show how to use those relations to analyze the whole structure. To go from easy to difficult, first, structural mechanics will be used to establish node equilibrium equation; second, the global stiffness matrix will be derived according to the principle of minimum potential energy; third, the convergence of solution and method of substructure will be illustrated; and finally, the principle of minimum complementary energy, which is rarely used in global analysis, will just be briefly introduced.

4.1 Nodal Equilibrium Equation

In Chapters 1 and 2, the structural mechanics method was used to establish the node equilibrium equations of truss and plane problems of the theory of elasticity. The node equilibrium equation of general cases can be built by the same method.

After the continuum was discretized by finite element method, for any node, as node i, the nodal loads from the elements around it are

$$\{P_i\} = \sum_e \{P_i\}^e$$

in which $\sum_e$ means the sum of all the elements around node i.

Let $\sum_e \{F_i\}$ be the nodal forces acting on node i by the elements around it; the equilibrium equation of node i can be presented as:

$$\sum_e \{F_i\} = \{P_i\} \sum_e \{F_i\} = \{P_i\} \tag{a}$$

For example, for the spatial problem in the theory of elasticity, the force in x, y, and z directions of any nodes should all be kept in equilibrium; then, the Eq.(a) means

$$\sum_e U_i = X_i, \quad \sum_e V_i = Y_i, \quad \sum_e W_i = Z_i$$

For shell structure, there are three moments that should be kept in equilibrium in node i except for the three forces in x, y, and z directions. Therefore, for each node i, there are

The Finite Element Method: Fundamentals and Applications in Civil, Hydraulic, Mechanical and Aeronautical Engineering, First Edition. Bofang Zhu.

Companion website: www.wiley.com/go/zhufem17

another three equations:

$$\sum_e M_{\theta xi} = R_{\theta xi}, \quad \sum_e M_{\theta yi} = R_{\theta yi}, \quad \sum_e M_{\theta xi} = R_{\theta xi}$$

Substituting $\{F\}^{\mathrm{e}} = [k]\{\delta\}^{\mathrm{e}}$ into Eq. (a), the equilibrium equation of node i, which is presented by nodal displacement, will be gained. Establishing the equilibrium equation for each node, then the equilibrium equations of the whole structure is

$$[K]\{\delta\} = \{P\} \tag{4.1}$$

in which $[K]$ is the global stiffness matrix, $\{\delta\}$ is the vector composed of all the nodal displacements, and $\{P\}$ is the vector composed of all the nodal loads.

The elements of global stiffness matrix $[K]$ can be calculated as

$$K_{rs} = \sum_e K_{ij} \tag{4.2}$$

By the coding method (refer to Section 2.9), the global stiffness matrix can be established by the element stiffness matrix.

After resolving the displacement $\{\delta\}$ by equilibrium equation (4.1), the strain of each element can be calculated by $\{\varepsilon\} = [B]\{\delta\}$; then, the stress of each element can be calculated by

$$\{\sigma\} = [D]\{\varepsilon\}$$

According to Eqs (3.6) and (3.11):

$$[k]^{\mathrm{e}} = \int_{V_{\mathrm{e}}} [B]^{\mathrm{T}}[D][B]\,\mathrm{d}V$$

$$\{\sigma\} = [D][B]\{\delta\}$$

Using the two equations above, the equilibrium equation (4.1) can be changed to the following form:

$$\int_V [B]^{\mathrm{T}}\{\sigma\}\,\mathrm{d}V = \{P\} \tag{4.1a}$$

The integral on the left side of the Eq. (4.1a) is the sum of the integral of volume V_{e} of each related element, which is

$$\int_V ()\mathrm{d}V = \sum_e \int_{V_{\mathrm{e}}} ()\mathrm{d}V$$

4.2 Application of the Principle of Minimum Potential Energy

The previous chapter has shown how to analyze an element by the principle of minimum potential energy; this chapter will further explain how to make the global analysis of structure by the principle of minimum potential energy.

As Figure 4.1 shows, supposing that the structure has been divided into m elements, according to Eq. (3.41), the strain energy of the element j will be

$$U_j = \frac{1}{2}\{\delta_j\}^{\mathrm{T}} [k_j] \{\delta_j\}$$

in which $[k_j]$ is the stiffness matrix of element j and $\{\delta_j\}$ is the nodal displacements of element j.

Because the strain energy is scalar, the strain energy of the whole structure can be gained by adding up the strain energy of m elements.

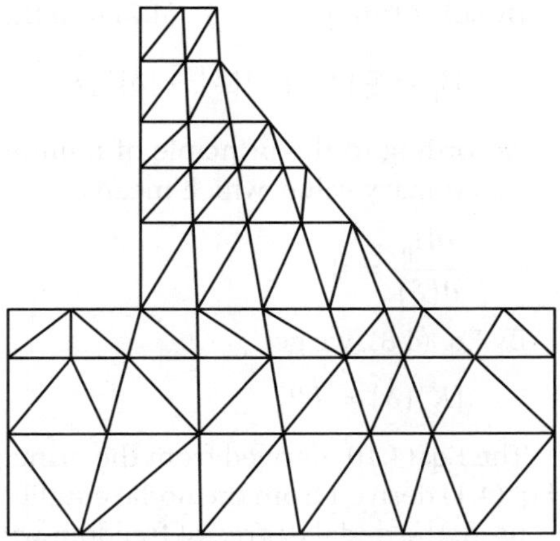

Figure 4.1 Structure.

$$U = \sum_{j=1}^{m} U_j$$
$$= \frac{1}{2}\sum \{\delta_j\}^{\mathrm{T}} [k_j] \{\delta_j\} \qquad (4.3)$$

Let

$$\{s\} = \begin{bmatrix}\delta_1 & \delta_2 & \cdots & \delta_m\end{bmatrix}^{\mathrm{T}}$$
$$[k_s] = \begin{bmatrix} k_1 & 0 & 0 \\ 0 & \ddots & 0 \\ 0 & 0 & k_m \end{bmatrix}$$

where $\{s\}$ is to arrange the nodal displacement ($\{\delta_1\}$, $\{\delta_2\}$, …) of each element in order, in which the same terms have not combined, and $[k_s]$ is the unassembled global stiffness matrix, which means the submatrix on the main diagonal of $[k_s]$ is the stiffness matrix ($[k_1]$, $[k_2]$, $[k_m]$) of each element and the other submatrix is zero. Then, the Eq. (4.3) can be rewritten as

$$U = \frac{1}{2}\{S\}^{\mathrm{T}}[K_s]\{S\} \qquad (4.4)$$

The relationship of nodal displacement vectors $\{\delta\}$ and column $\{s\}$ of the whole structure is

$$\{s\} = [A]\{\delta\}$$

in which $[A]$ is transformation matrix.

Substitution of the equation above into Eq. (4.4) yields

$$U = \frac{1}{2}\{\delta\}^{\mathrm{T}}[K]\{\delta\} \qquad (4.5)$$

$$[K] = [A]^{\mathrm{T}}[K_s][A] \qquad (4.6)$$

where $[K]$ is the global stiffness matrix of the structure.

Because all the loads $\{P\}$ are acting on the nodes, the potential of the load is

$$V = \{\delta\}^{\mathrm{T}}\{P\} \qquad (4.7)$$

The potential energy of the structure is

$$\Pi_{\mathrm{p}} = U - V$$

By substituting Eq. (4.5), (4.7) into the equation above, we get

$$\Pi_p = \frac{1}{2}\{\delta\}^T[K]\{\delta\} - [\delta]^T\{P\} \tag{4.8}$$

According to the principle of minimum potential energy, the potential energy takes the stationary value, which means

$$\frac{\partial \Pi_p}{\partial\{\delta\}} = 0 \tag{4.9}$$

By Eq. (4.8), we get

$$[K]\{\delta\} = \{P\} \tag{4.10}$$

The Eq. (4.10) derived from the principle of minimum potential energy is equal to the Eq. (4.1) derived from the nodal equilibrium conditions. But if the higher order element is used, the nodal degree of freedom will contain the higher derivative of displacement, and the corresponding nodal force will have no explicit physical meanings. In this situation, it is hard to build the global equation by the nodal equilibrium conditions, but there is no difficulty to build the global equilibrium equation by the principle of minimum potential energy.

4.3 The Low Limit Property of the Solution of Minimum Potential Energy

By discretizing the continuum with finite element method, we can find the approximate solution of displacement with the principle of minimum potential energy, which will be less than the exact solution, and this displacement solution can be called the lowest limit solution.

It is not hard to prove the above conclusion. As Figure 4.2 shows, supposing there is only a concentrated force P_i that acts on the body, this force is gradually increasing from zero to P_i. Supposing the displacement is δ_i in point i, in the direction of P_i, the work of force P_i during the loading process will be $P_i\delta_i/2$, which equals the strain energy U of the objects.

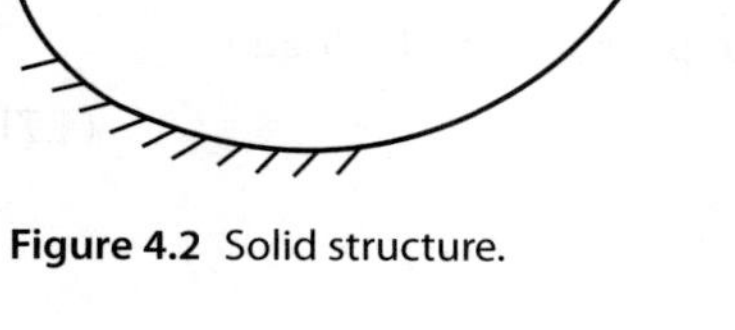

Figure 4.2 Solid structure.

$$U = \frac{1}{2}P_i\delta_i$$

The potential energy of external force is

$$V = P_i\delta_i$$

So when the body is in equilibrium, its potential energy is

$$\tilde{\Pi}_p = U - V = \frac{1}{2}P_i\delta_i - P_i\delta_i = -\frac{1}{2}P_i\delta_i = \min$$

In the equation above, δ_i is the exact solution and $\tilde{\Pi}_p$ is the minimum value according to the minimum potential principle.

If the approximate solution is $\tilde{\delta}_i$ calculated with finite element method, the approximate solution of potential energy is

$$\tilde{\Pi}_p = -\frac{1}{2}P_i\tilde{\delta}_i$$

Since the exact potential energy $\tilde{\Pi}_p$ is the minimum value, any other approximate solution $\prod_p$ of the potential energy should not be less than $\prod_p$, which means

$$-\frac{1}{2}P_i\tilde{\delta}_i \geq -\frac{1}{2}P_i\delta_i$$

Since both sides of the equation above have minus sign, after simplification we can get

$$\tilde{\delta}_i \leq \delta_i \tag{4.11}$$

which shows that the approximate solution of displacement calculated according to the principle of minimum potential energy after discretizing the body with finite element method will not be more than the exact displacement solution. Note that the direction of displacement δ_i and $\tilde{\delta}_i$ should be along with the direction of the external force P_i.

Why is the approximate solution of displacement calculated according to the principle of minimum potential energy less than the exact solution of displacement? When calculating with the principle of minimum potential energy, the element displacement function should be assumed first. Those displacement functions are continuous, but not exact. The element taken from the body as a part of the continuum should have infinite degrees of freedom, which becomes finite degrees of freedom after applying displacement functions. The deformation capacity of element has been limited by displacement functions, which means the element stiffness has been increased. So the calculated approximate solution of displacement is less than the exact solution.

When calculating by the principle of minimum complementary energy, the stress is balanced on the border of the adjacent elements, but the displacement is not continuous, which means the deformation capacity of the object will increase and the object will be softer. The approximate solution calculated by the principle of minimum complementary energy will be larger than or equal to the exact solution, so it has the property of the upper limit.

When calculating with hybrid elements, it is not sure whether the approximate solution is larger or less than the exact solution. In general, the value is between the solution calculated by the principle of minimum potential energy and the solution calculated by the minimum complementary energy.

4.4 The Convergence of Solutions

When calculated by the finite elements, the continuum that has infinite degrees of freedom is replaced by a computing model that has only finite degrees of freedom. Then, will the solution converge to the exact solution? The answer is that if the calculating method meets certain conditions, the calculating result will converge to the exact solution.

When solving the problem with displacement function, the following conditions should be satisfied to make the calculation result converge to the exact solution:

1) The rigid body displacement does not produce strains. Obviously, this condition is necessary; otherwise, the translation and rotation of the body will also produce strains.
2) The displacement functions should reflect the constant strain of the element. When the size of element infinitely shrinks, the stains of element will approach to a constant; therefore, the displacement function of element should contain constant strain terms.
3) The displacement function should assure that the strain on the contact surface of adjacent elements is finite.

Condition 1 can be regarded as a particular condition of condition 2, because the rigid body displacement is a particular condition of constant strain – the strain is zero.

When calculating on the basis of displacement (by the principle of minimum potential energy) with finite element method, only the internal energy (strain energy) will be calculated, but the energy on the contact surface of the adjacent elements will not be computed. Because the thickness of the contact surface is zero, the work will be zero if the strain on the contact surface is finite; otherwise, the work will not be zero, and therefore there will be errors if ignoring it.

If the function and its less-than nth derivative (including nth) are continuous, it is called C_n continuous. The function u shown in Figure 4.3 is continuous, but its first derivative is not continuous at x_1, where its second derivative is infinite. So u is a C_0 continuous function.

In the plane and spatial stress problem, strain is the first derivative of displacement. The condition of convergence demands that the strain on the interface of adjacent

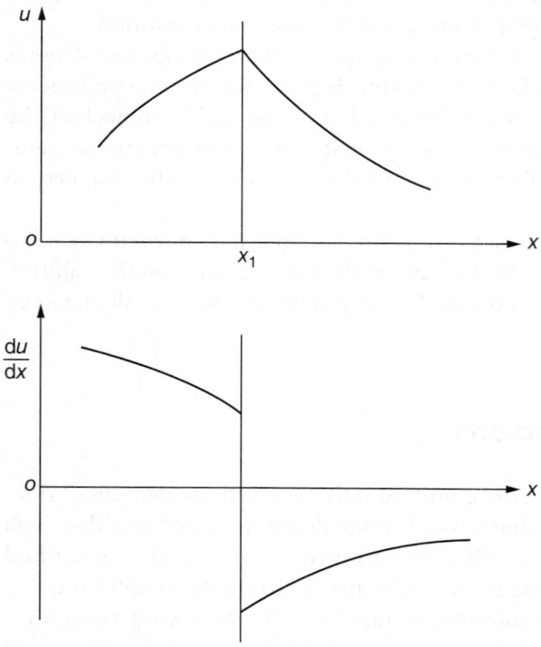

Figure 4.3 C_0 continuous function.

elements is finite, which means the displacement is continuous and the displacement function is C_0 continuous. In the problems of plate and shell, stress is the second derivative of displacement, so the displacement and its first derivative on the interface of elements are continuous, which means the displacement function is C_1 continuous.

The following is an analysis using the two dimensional stress problem as example. According to Eq. (2.3), the displacement functions of a triangular element are

$$u = \beta_1 + \beta_2 x + \beta_3 y, \quad v = \beta_4 + \beta_5 x + \beta_6 y$$

The equation above can be transformed as follows:

$$u = \left(\beta_1 - \frac{\beta_5 - \beta_3}{2} y\right) + \frac{\beta_5 + \beta_3}{2} y + \beta_2 x$$

$$v = \left(\beta_4 - \frac{\beta_5 - \beta_3}{2} x\right) + \frac{\beta_5 + \beta_3}{2} y + \beta_6 y$$

The terms within the brackets on the right side of the two equations above represent the rigid body displacements, in which $u = \beta_1$ and $v = \beta_4$ are the rigid movement of elements along x and y directions, respectively, while $u = -\frac{(\beta_5 - \beta_3)y}{2}$ and $v = \frac{(\beta_5 - \beta_3)x}{2}$ are the rigid rotation around z-axis. According to Eq. (2.10), these rigid body displacements do not produce strain within the element.

By substituting Eq. (2.3) into Eq. (2.10), we can get

$$\varepsilon_x = \beta_2, \quad \varepsilon_y = \beta_6, \quad \varepsilon_{xy} = \beta_3 + \beta_5$$

which reflect the normal strain and shearing strain of the constant value.

In addition, the displacement component shown by Eq. (2.3) is a linear function of coordinate in each element; of course, it changes linearly on the public boundary of an adjacent element. Because there is the same nodal displacement in the public nodes of adjacent elements, the displacement is continuous on the whole public border.

So, the displacement function Eq. (2.3) satisfies the three conditions mentioned above.

The actual structure has infinite degrees of freedom. But, in finite element method, we use n elements to represent the original structure, which means the calculation model just has finite degrees of freedom. It is intuitive that if n is larger, which means more elements, the result will be closer to the true solution. If the biggest scale of element in the calculation model is h, then the smaller the h, the closer the result is to the true solution.

As Figure 4.4 shows, the uniform cross-sectional bar bears an axial uniformly distributed load, and the element stiffness matrix will be as the following equation when calculated by finite element method:

$$[k] = \frac{EA}{h} \begin{bmatrix} 1 & -1 \\ -1 & 1 \end{bmatrix} \tag{a}$$

in which A is the area of cross section and E is modulus of elasticity.

The equilibrium condition at node i will be as the following:

$$\frac{EA}{h}[-u_{i-1} + u_i] + \frac{EA}{ah}[u_i - u_{i+1}] = \frac{ph(1+a)}{2}$$

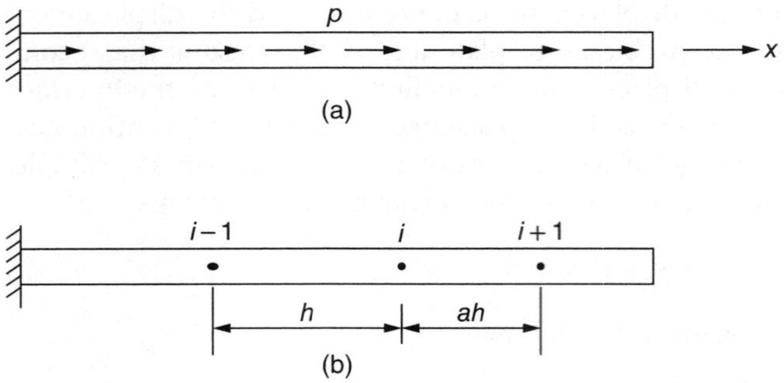

Figure 4.4 Sketch of a bar of equal cross-sectional. (a) Uniform load and (b) nodes.

After making the Taylor expansion of displacement u in the neighborhood of node i, we can get

$$u_{i+1} = u_i + u_i'(ah) + u_i''\frac{(ah)^2}{2} + u_i'''\frac{(ah)^3}{6} + u_i^{\mathrm{IV}}\frac{(ah)^4}{24} + \cdots$$

$$u_{i-1} = u_i - u_i'h + u_i''\frac{h^2}{2} + u_i'''\frac{h^3}{6} + u_i^{\mathrm{IV}}\frac{h^4}{24} - \cdots$$

By substituting the two equations into Eq. (a), we can get the differential equation corresponding to the balance equation of finite element method.

$$u_i'' - \frac{h}{3}(1-a)u_i''' + \frac{h^2}{12} \times \frac{1+a^3}{1+a}u_i^{\mathrm{IV}} + \cdots + \frac{p}{EA} = 0 \tag{b}$$

The accurate equilibrium differential equation at node i is

$$u_i'' + \frac{p}{EA} = 0 \tag{c}$$

Comparing Eq. (b) with Eq. (c), we can see that when h approaches to zero, the solution of finite element method is approaching to the accurate result. When $a \neq 1$, the convergence rate is linear. When $a = 1$, the convergence rate is quadratic.

4.5 Analysis of the Substructure

When analyzing a large-scale complicated structure, because of the large amount of elements, the equation set may be so huge that it will be beyond the storage capacity of the computer. We can divide the original structure into several regions and call each region a substructure. The substructures connect with each other on their public boundary. First, we analyze the substructures, eliminating their internal degrees of freedom by static condensation; then, we analyze the whole structure, in which we can only consider the constrained boundary of structure and the degrees of freedom on the public boundary of adjacent substructures. The scale of problem is much less than the original structure. We can also set substructures into substructures, which can be called multiple substructures.

4.5.1 Multiple Substructures

We use Figure 4.5 as a simple example to explain the multiple substructure method. Figure 4.5(a) shows the original structure with seven first-layer substructures divided by dash lines; Figure 4.5(b) shows the substructures in the second layer; and Figure 4.5(c) shows the substructures in the third layer. First, the third-layer substructure will be calculated by condensing its internal degrees of freedom. Second, the second-layer substructure will be calculated by condensing its internal degrees of freedom. Last, the original structure will be calculated; in this part, there are only degrees of freedom on the constrained boundary and the public boundaries of adjacent first-layer substructures.

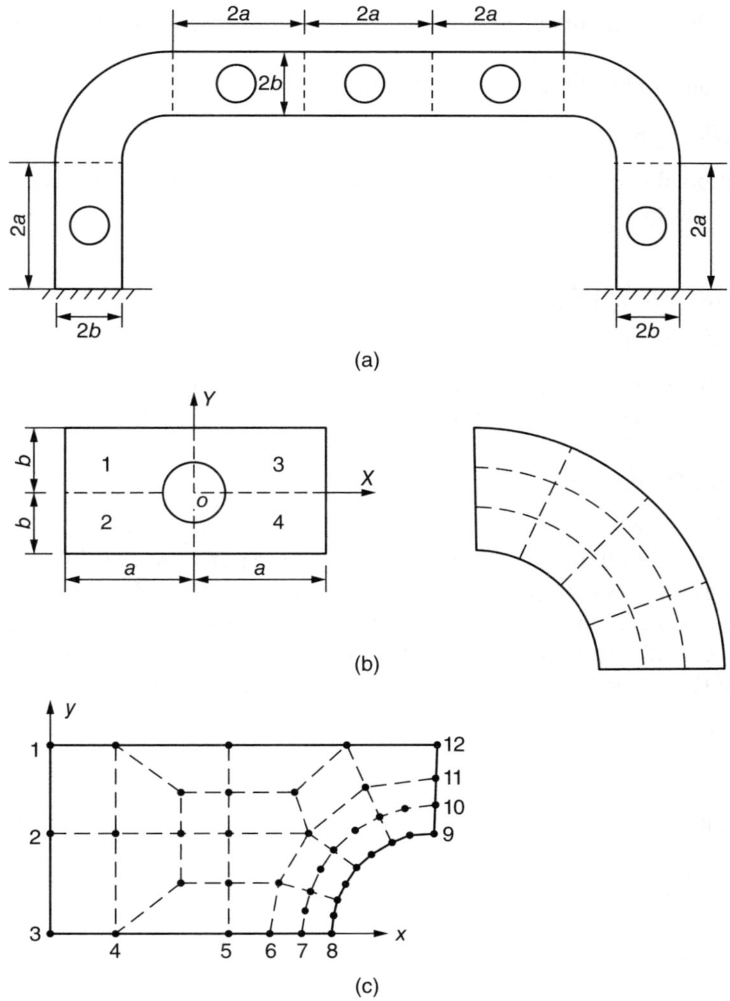

Figure 4.5 Multiple substructures. (a) Original structure and the first-layer substructure; (b) the second-layer substructure; (c) the third-layer substructure.

4.5.2 Condensation of the Internal Degrees of Freedom of Substructures

By using proper nodal number, the equilibrium equation of the substructure may be expressed as follows:

$$\begin{bmatrix} K_{bb} & K_{bi} \\ K_{ib} & K_{ii} \end{bmatrix} \begin{pmatrix} \delta_b \\ \delta_i \end{pmatrix} = \begin{Bmatrix} P_b \\ P_i \end{Bmatrix} \tag{4.12}$$

in which $\{\delta_b\}$ is the nodal displacement vector on the interfaces and constrained surfaces of substructures (the constrained surface is also interface – the interface of structure and foundation), and $\{\delta_i\}$ is the nodal displacement vector of the internal structure and non-constrained surfaces.

In Figure 4.5(c), the displacements of nodes 1–12 belong to $\{\delta_b\}$, while those of the others belong to $\{\delta_i\}$.

The following equation can be obtained from the second equation of Eq. (4.12):

$$\{\delta_i\} = [K_{ii}]^{-1}(\{P_i\} - [K_{ib}]\{\delta_b\})$$

Substitute the equation above into the first equation of Eq. (4.12); after condensation we get the following equations:

$$[K_b^*]\{\delta_b\} = \{P_b^*\}$$
$$[K_b^*] = [K_{bb}] - [K_{bi}][K_{ii}]^{-1}[K_{ib}]$$
$$\{P_b^*\} = \{P_b\} - [K_{bi}][K_{ii}]^{-1}\{P_i\}$$

After condensation, the internal degrees of freedom of the substructures are all eliminated. Only the degrees of freedom of the interface $\{\delta_b\}$ are left.

4.5.3 Coordinate Transformation

In the substructure of the second layer shown in Figure 4.5(b), the shape and sizes of substructures 1, 2, 3, and 4 are all the same, but the directions are not. If every substructure takes the local coordinate system (x, y) shown in Figure 4.5(c), their stiffness matrix will be the same, which means it only needs to calculate once. But from local coordinate system (x, y) to global coordinate system (X, Y), the nodal displacements, nodal loads, and stiffness matrix of the element all need a coordinate transformation, which means

$$\{\overline{\delta}\} = [\lambda]^{\mathrm{T}}\{\delta_b\} \tag{4.13}$$

$$\{\overline{P}_b\} = [\lambda]^{\mathrm{T}}\{P_b{}^*\} \tag{4.14}$$

$$\{\overline{K}_b\} = [\lambda]^{\mathrm{T}}[K_b^*][\lambda] \tag{4.15}$$

$$[\lambda] = \begin{bmatrix} L & 0 & \cdots & 0 \\ 0 & L & \cdots & 0 \\ \vdots & \vdots & \cdots & \vdots \\ 0 & 0 & \cdots & L \end{bmatrix} \tag{4.16}$$

$$[L] = \begin{bmatrix} \cos(x, X) & \cos(x, Y) \\ \cos(y, X) & \cos(y, Y) \end{bmatrix} \tag{4.17}$$

in which $\{\overline{\delta}_b\}$, $\{\overline{P}_b\}$, and $\{\overline{K}_b\}$ are terms in the global coordinate system (X, Y); $\{\delta_b\}$, $\{P_b^*\}$, and $\{K_b^*\}$ are terms in local coordinate system (x, y); and (x, X), (x, Y), (y, X), and (y, Y) are the included angles from X-, Y-axis to x-, y-axes, respectively; counterclockwise direction is positive.

From Figure 4.5, we can see that the included angles of local coordinate system (x, y) of four second-layer substructures and global coordinate system (X, Y) of the first-layer substructures are as follows:

Substructure 1: $(x,X)_1 = (y,Y)_1 = 0$
Substructure 2: $(x,X)_2 = 0, (y,Y)_2 = \pi$
Substructure 3: $(x,X)_3 = \pi, (y,Y)_3 = 0$
Substructure 4: $(x,X)_4 = (y,Y)_4 = \pi$

Meanwhile $(y,X) = (y,Y) + {}^{\pi}/_{2}$, $(x,Y) = {}^{3\pi}/_{2} + (x,X)$
So, the coordinate transformation matrices of four substructures are

$$[L_1] = \begin{bmatrix} 1 & 0 \\ 0 & 1 \end{bmatrix}, \quad [L_2] = \begin{bmatrix} 1 & 0 \\ 0 & -1 \end{bmatrix}, \quad [L_3] = \begin{bmatrix} -1 & 0 \\ 0 & 1 \end{bmatrix}, \quad [L_4] = \begin{bmatrix} -1 & 0 \\ 0 & -1 \end{bmatrix}$$

Bibliography

1 Bathe, K.J. (1982) *Finite Element Procedures in Engineering Analysis*, Prentice-Hall, Englewood Cliffs.

2 Zienkiewicz, O.C., Taylor, R.L. and Zhu, J.Z. (2008) *The Finite Element Method, Its Basis and Fundamentals*, Elsevier, New York.

3 Xucheng, W. and Min, S. (1988) *The Basic Principle and Numerical Method of Finite Element Method*, Tsinghua University Press, Beijing.

4 Desai, C.S. and Abel, J.F. (1978) *An Introduction to Finite Element Method*. Jiangbonan, Yinzeyong translate, Science Press, Beijing.

5 Strang, G. and Fix, G. (1973) *An Analysis of the Finite Element Method*, Prentice-Hall, Englewood Cliffs.

5

High-Order Element of Plane Problem

In Chapter 2, we introduced the constant strain triangular element, one of the first elements proposed in the finite element method. As it is simple, it is still in application. However, the strain and stress in the element are both constants, while the stresses in engineering structures vary sharply with coordinates, so intensive elements must be laid when applying the constant strain triangular element to realize good calculation accuracy, which causes a big number of nodes and large equations. In this chapter we will introduce the rectangular and triangular high-order element. The element displacement function adopts quadratic and higher polynomials. The strain and stress within the element are variable, which can well reflect the variation of actual stresses. Adoption of a small number of elements may obtain good calculation results.

5.1 Rectangular Elements

In Chapter 3 we introduced the rectangular stress hybrid element. Now we will move to rectangular elements solved by displacement. Rectangular elements with the edge of $2a$ and $2b$ are shown in Figure 5.1. For simplicity, put the origin of coordinates on the element centroid, and take two symmetry axes parallel to the two edges as the x-and y-axis. The element displacement functions are set as

$$\left.\begin{aligned} u &= \beta_1 + \beta_2 x + \beta_3 y + \beta_4 xy \\ v &= \beta_5 + \beta_6 x + \beta_7 y + \beta_8 xy \end{aligned}\right\} \tag{a}$$

The strains in the element are

$$\begin{aligned} \varepsilon_x &= \frac{\partial u}{\partial x} = \beta_2 + \beta_4 y \\ \varepsilon_y &= \frac{\partial v}{\partial y} = \beta_7 + \beta_8 x \\ \gamma_{xy} &= \frac{\partial u}{\partial y} + \frac{\partial v}{\partial x} = (\beta_3 + \beta_6) + \beta_4 x + \beta_8 y \end{aligned} \tag{b}$$

From this, the normal strain ε_x is constant in the x-direction and is a linear function in the y-direction; the normal strain ε_y is constant in the y-direction and is a linear function in the x-direction; the shear strain γ_{xy} is linear functions in both x- and

The Finite Element Method: Fundamentals and Applications in Civil, Hydraulic, Mechanical and Aeronautical Engineering, First Edition. Bofang Zhu.

Companion website: www.wiley.com/go/zhufem17

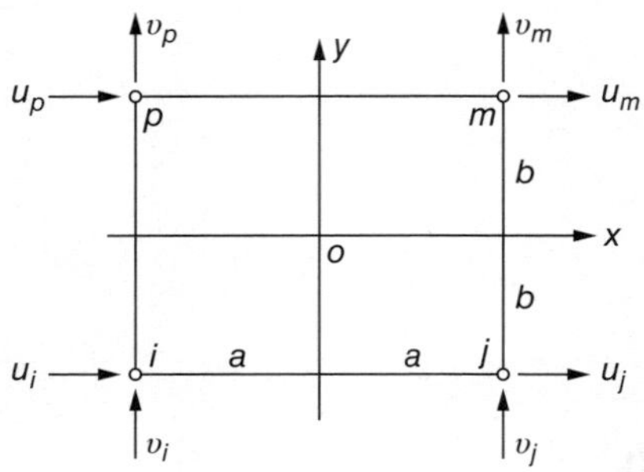

Figure 5.1 4-Node rectangular element.

y-directions. Coefficients $\beta_2, \beta_7, \beta_3, \beta_6$ represent constant strains, and β_1, β_5 reflect rigid-body displacement. From Eq. (a), we can see on the element boundary $x = \pm a$ and $y = \pm b$, displacement components are in linear variation, thus ensuring the continuity of the displacement of any two adjacent elements at their common boundary. It can be seen that the displacement function formulas (a) meet the sufficient conditions for convergence of the solution. By means of the coefficients β_1–β_8 in Eq. (a), the displacements of nodes i, j, m, and p are given as follows:

$$
\begin{aligned}
u_i &= \beta_1 - a\beta_2 - b\beta_3 + ab\beta_4 \\
u_j &= \beta_1 + a\beta_2 - b\beta_3 - ab\beta_4 \\
u_m &= \beta_1 + a\beta_2 + b\beta_3 + ab\beta_4 \\
u_p &= \beta_1 - a\beta_2 + b\beta_3 - ab\beta_4
\end{aligned}
$$

Using Cramer's rule, from the above four equations, $\beta_1, \beta_2, \beta_3, \beta_4$ can be solved. In a similar way, we can solve $\beta_5, \beta_6, \beta_7, \beta_8$. Substituting these eight coefficients back into formula (a), we obtain

$$
\left.\begin{aligned}
u &= N_i u_i + N_j u_j + N_m u_m + N_p u_p \\
v &= N_i v_i + N_j v_j + N_m v_m + N_p v_p
\end{aligned}\right\} \tag{5.1}
$$

$$
\left.\begin{aligned}
N_i &= \frac{1}{4}(1-\xi)(1-\eta), \quad & N_j &= \frac{1}{4}(1+\xi)(1-\eta) \\
N_m &= \frac{1}{4}(1+\xi)(1+\eta), \quad & N_p &= \frac{1}{4}(1-\xi)(1+\eta) \\
\xi &= x/a, & \eta &= y/b
\end{aligned}\right\} \tag{5.2}
$$

which can also be represented in matrix as

$$
\{r\} = \begin{Bmatrix} u \\ v \end{Bmatrix} = [N]\{\delta\}^{\mathrm{e}} \tag{5.3}
$$

$$
[N] = \begin{bmatrix} IN_i & IN_j & IN_m & IN_p \end{bmatrix} \tag{5.4}
$$

$$
\{\delta\}^{\mathrm{e}} = \begin{bmatrix} u_i & v_i & u_j & v_j & u_m & v_m & u_p & v_p \end{bmatrix}^{\mathrm{T}} \tag{5.5}
$$

where I is the unit matrix and N_i, N_j, N_m, N_p are all shape functions of the rectangular element. Element strains can also be represented by node displacements as

$$
\{\varepsilon\} = \begin{Bmatrix} \dfrac{\partial u}{\partial x} \\ \dfrac{\partial v}{\partial y} \\ \dfrac{\partial u}{\partial y} + \dfrac{\partial v}{\partial x} \end{Bmatrix} = \begin{Bmatrix} \sum \dfrac{\partial N_i}{\partial x} u_i \\ \sum \dfrac{\partial N_i}{\partial x} v_i \\ \sum \left(\dfrac{\partial N_i}{\partial y} u_i + \dfrac{\partial N_i}{\partial x} v_i \right) \end{Bmatrix} = [B]\{\delta\}^{\mathrm{e}} \tag{5.6}
$$

$$[B]=\begin{bmatrix}\frac{\partial N_i}{\partial x} & 0 & \frac{\partial N_j}{\partial x} & 0 & \frac{\partial N_m}{\partial x} & 0 & \frac{\partial N_p}{\partial x} & 0\\ 0 & \frac{\partial N_i}{\partial y} & 0 & \frac{\partial N_j}{\partial y} & 0 & \frac{\partial N_m}{\partial y} & 0 & \frac{\partial N_p}{\partial y}\\ \frac{\partial N_i}{\partial y} & \frac{\partial N_i}{\partial x} & \frac{\partial N_j}{\partial y} & \frac{\partial N_j}{\partial x} & \frac{\partial N_m}{\partial y} & \frac{\partial N_m}{\partial x} & \frac{\partial N_p}{\partial y} & \frac{\partial N_p}{\partial x}\end{bmatrix} \tag{5.7}$$

Substitution of the shape function (5.2) into formula (5.7) yields

$$[B]=\frac{1}{4ab}\begin{bmatrix}-(b-y) & 0 & b-y & 0 & b+y & 0 & -(b+y) & 0\\ 0 & -(a-x) & 0 & -(a+x) & 0 & a+x & 0 & a-x\\ -(a-x) & -(b-y) & -(a+x) & b-y & a+x & b+y & a-x & -(b+y)\end{bmatrix} \tag{5.8}$$

Substituting the above equation into formula (3.11) and integrating every term of the matrix, we obtain the stiffness matrix of the plane element as follows:

$$[k]=HEt\begin{bmatrix}\beta+r\alpha & & & & & & & \\ m & \alpha+r\beta & & & & \text{symmetrical} & & \\ -\beta+\frac{1}{2}r\alpha & s & \beta+r\alpha & & & & & \\ -s & \frac{\alpha}{2}-r\beta & -m & \alpha+r\beta & & & & \\ -\frac{\beta}{2}-\frac{1}{2}r\alpha & -m & \frac{\beta}{2}-r\alpha & s & \beta+r\alpha & & & \\ -m & \frac{\alpha}{2}-\frac{r\beta}{2} & -s & -\alpha+\frac{r\beta}{2} & m & \alpha+r\beta & & \\ \frac{\beta}{2}-r\alpha & -s & -\frac{\beta}{2}-\frac{r\alpha}{2} & m & -\beta+\frac{r\alpha}{2} & s & \beta+r\alpha & \\ s & -\alpha+\frac{r\beta}{2} & m & -\frac{\alpha}{2}-\frac{r\beta}{2} & -s & \frac{\alpha}{2}-r\beta & -m & \alpha+r\beta\end{bmatrix} \tag{5.9}$$

Plane stress : $H=\dfrac{1}{1-\mu^2},\quad r=\dfrac{1-\mu}{2},\quad s=\dfrac{1-3\mu}{8},\quad m=\dfrac{1+\mu}{8},$

$\alpha=\dfrac{a}{3b},\quad \beta=\dfrac{b}{3a}$

Plane strain : $H=\dfrac{1-\mu}{(1+\mu)(1-2\mu)},\quad r=\dfrac{1-2\mu}{2(1-\mu)},\quad s=\dfrac{1-4\mu}{8(1-\mu)},$

$m=\dfrac{1}{8(1-\mu)},\quad \alpha=\dfrac{a}{3b},\quad \beta=\dfrac{b}{3a}$

By substituting $[N]$ into formula (3.14) and (3.15), it is easy to determine the nodal load formula. For example, for self-weight, the load of every node is 1/4 of the element weight. Another example is that when a boundary of the element is exposed to the triangular distribution of surface force such as water pressure, 1/3 of the resultant may be placed on the node with 0 water pressure and 2/3 on the other node.

For the 8-node rectangular element shown in Figure 5.2(a), the displacement function can be expressed as

$$u=\beta_1+\beta_2x+\beta_3y+\beta_4xy+\beta_5x^2+\beta_6y^2+\beta_7x^2y+\beta_8xy^2$$
$$v=\beta_9+\beta_{10}x+\cdots+\beta_{16}xy^2$$

Along the element boundary, the curve of displacement is a parabola. On every boundary there are 3 nodes, which ensure that the displacement of every common boundary is continuous. The internal strain in the element is a quadratic function of coordinates. Similar to the derivation for 4-node rectangular elements above, it is easy to determine the stiffness matrix and nodal load of this type of element.

For the 12-node rectangular element shown in Figure 5.2(b), the displacement variation along the element boundary is a cubic function of coordinates, and the element strain is also a cubic function of coordinates. For the 6-node rectangular element shown in Figure 5.2(c), the displacement function should be chosen as follows: along the upper and lower horizontal boundary, the displacement varies according to a quadratic function of x; along the left and right vertical boundary, the displacement varies according to a linear function of y.

As the order of displacement functions is high, the stresses in the element are variable. However, when adopting high-order elements, the number of elements is always small and the sizes of elements are always large. Boundaries of these elements are not easy to match the boundaries of actual structures; thus their application is rare.

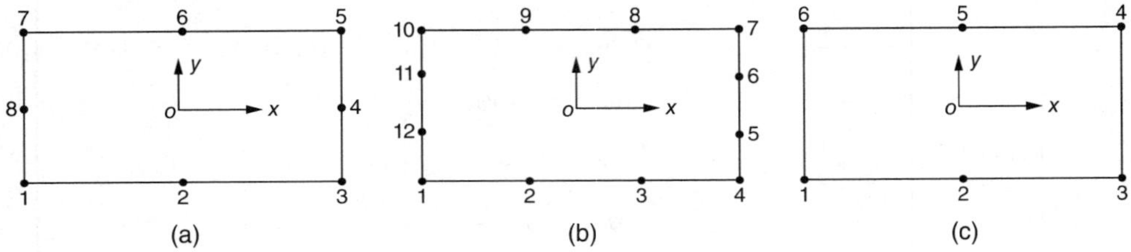

Figure 5.2 Other forms of rectangular elements. (a) 8-Node rectangular element; (b) 12-node rectangular element; and (c) 6-node rectangular element.

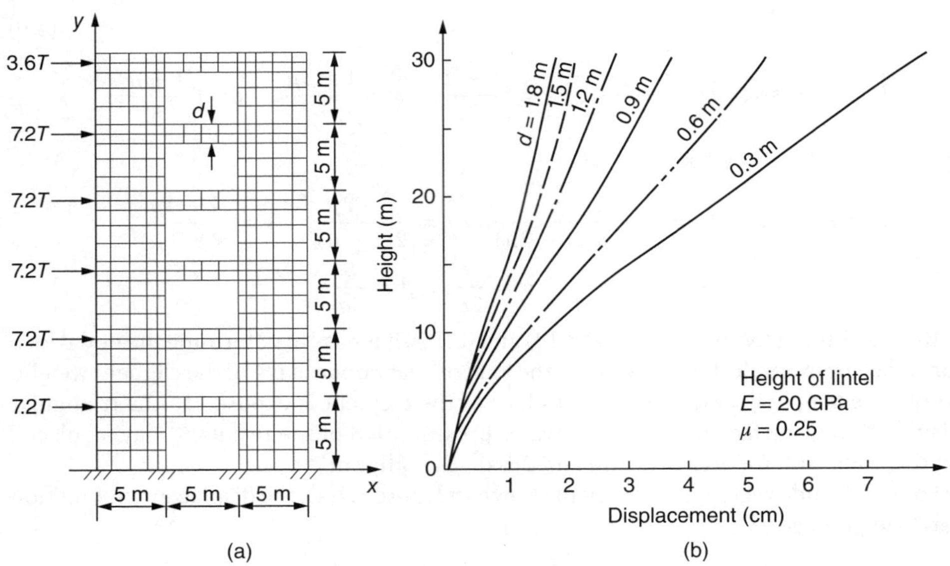

Figure 5.3 Analysis of lateral displacement of shear wall of six-story buildings. (a) Load, size, and computational grid and (b) left lateral displacement of the wall.

4-Node rectangular elements have easy calculations and are still commonly used. In order to match irregular boundaries, sometimes they are used together with constant strain triangular elements. The rectangular elements are adopted in regular parts and the triangular elements in irregular parts.

Figure 5.3 presents the grid and calculated results of the analysis of shear wall lateral displacement of six-story buildings using 4-node rectangular elements. The figure shows the lintel stiffness has an important influence on the shear wall lateral displacement.

5.2 Area Coordinates

As for high-order triangular element, if still using orthogonal coordinates to define the shape function, the formula about the stiffness matrix will be very complicated. It can be greatly simplified when turning to area coordinates.

As shown in Figure 5.4, the position of an arbitrary point P in the triangular element 123 may be fixed by the following three ratios:

$$L_1 = \frac{A_1}{A}, \quad L_2 = \frac{A_2}{A}, \quad L_3 = \frac{A_3}{A} \tag{5.10}$$

where A is the area of the triangular element 123 and A_1, A_2, A_3 are, respectively, the area of triangular $P23$, $P31$, and $P12$.

The three ratios are called the area coordinates of point P. Clearly, the three area coordinates are not independent of each other, $A_1 + A_2 + A_3 = A$; thus

$$L_1 + L_2 + L_3 = 1 \tag{5.11}$$

It is easy to see the area coordinates of the 3 nodes are as follows:

Node 1: $L_1 = 1, L_2 = L_3 = 0$
Node 2: $L_2 = 1, L_3 = L_1 = 0$
Node 3: $L_3 = 1, L_1 = L_2 = 0$.

The following are equations expressing the three edges of a triangular by area coordinates:

Edge 23: $L_1 = 0$
Edge 31: $L_2 = 0$
Edge 12: $L_3 = 0$.

The area coordinates of the centroid of the triangle are

$$L_1 = L_2 = L_3 = 1/3$$

As seen from Figure 5.4, all points along a line parallel to edge 23 have the same area coordinates L_1, and L_1 is just equal to the ratio of the distance from the line to edge 23 and that from node 1 to edge 23. The isolines of L_1 shown in Figure 5.4 are all straight lines parallel to one another.

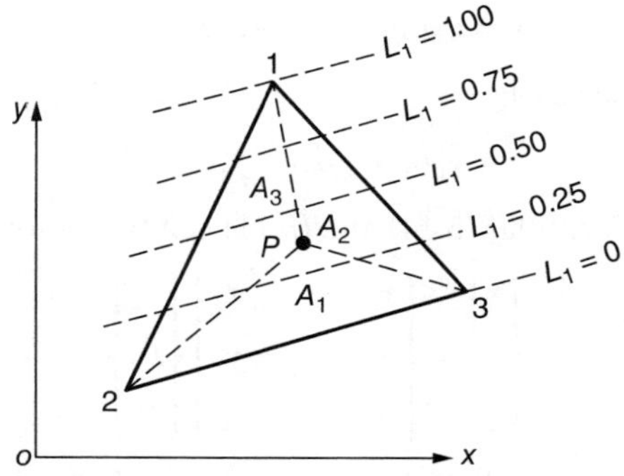

Figure 5.4 Area coordinates.

Set the coordinate of point P as (x, y), and the area of triangular $P23$ is

$$A_1 = \frac{1}{2}\begin{vmatrix} 1 & x & y \\ 1 & x_2 & y_2 \\ 1 & x_3 & y_3 \end{vmatrix}$$

$$= \frac{1}{2}\{(x_2y_3 - x_3y_2) + (y_2 - y_3)x + (x_3 - x_2)y\} \tag{a}$$

Just the same as Chapter 2, let

$$a_1 = x_2y_3 - x_3y_2, \quad b_1 = y_2 - y_3, \quad c_1 = x_3 - x_2 \tag{b}$$

Thus formula (a) becomes

$$A_1 = \frac{1}{2}(a_1 + b_1x + c_1y) \quad (1, 2, 3)$$

Substitution of the above formula into formula (5.10) yields the formula expressing area coordinates with the orthogonal coordinates:

$$\left.\begin{aligned} L_1 &= \frac{(a_1 + b_1x + c_1y)}{2A} \\ L_2 &= \frac{(a_2 + b_2x + c_2y)}{2A} \\ L_3 &= \frac{(a_3 + b_3x + c_3y)}{2A} \end{aligned}\right\} \tag{5.12}$$

Compare formula (5.12) with formula (2.7), it can be seen the shape functions N_i, N_j, and N_m of the triangular elements with constant strain are area coordinates L_1, L_2, L_3.

Formula (5.12) can also be expressed by matrix as follows:

$$\begin{Bmatrix} L_1 \\ L_2 \\ L_3 \end{Bmatrix} = \frac{1}{2A}\begin{bmatrix} a_1 & b_1 & c_1 \\ a_2 & b_2 & c_2 \\ a_{31} & b_3 & c_3 \end{bmatrix}\begin{Bmatrix} 1 \\ x \\ y \end{Bmatrix} \tag{5.13}$$

Multiply each of formula (5.12) with x_1, x_2, x_3, respectively, and then sum up by formula (b) and the following equation can be obtained:

$$x = x_1L_1 + x_2L_2 + x_3L_3 \tag{5.14a}$$

Similarly,

$$y = y_1L_1 + y_2L_2 + y_3L_3 \tag{5.14b}$$

Formulas (5.14) and (5.11) can be combined and represented in the matrix form as

$$\begin{Bmatrix} x \\ y \\ 1 \end{Bmatrix} = \begin{bmatrix} x_1 & x_2 & x_3 \\ y_1 & y_2 & y_3 \\ 1 & 1 & 1 \end{bmatrix}\begin{Bmatrix} L_1 \\ L_2 \\ L_3 \end{Bmatrix} \tag{5.15}$$

Formula (5.15) demonstrates that polynomials represented by the Cartesian coordinates may also be same-order polynomials represented by the area coordinates, but the area coordinates have the following advantages: they are independent of the shape and orientation of the triangle and their integral calculations are very simple.

When calculating the derivative of functions of area coordinates L_1, L_2, L_3 with respect to Cartesian coordinates (x, y), the following formula may be applied:

$$\left.\begin{aligned}\frac{\partial}{\partial x} &= \frac{\partial L_1}{\partial x}\frac{\partial}{\partial L_1} + \frac{\partial L_2}{\partial x}\frac{\partial}{\partial L_2} + \frac{\partial L_3}{\partial x}\frac{\partial}{\partial L_3}\\ &= \frac{1}{2A}\left(b_1\frac{\partial}{\partial L_1} + b_2\frac{\partial}{\partial L_2} + b_3\frac{\partial}{\partial L_3}\right)\\ \frac{\partial}{\partial y} &= \frac{\partial L_1}{\partial y}\frac{\partial}{\partial L_1} + \frac{\partial L_2}{\partial y}\frac{\partial}{\partial L_2} + \frac{\partial L_3}{\partial y}\frac{\partial}{\partial L_3}\\ &= \frac{1}{2A}\left(c_1\frac{\partial}{\partial L_1} + c_2\frac{\partial}{\partial L_2} + c_3\frac{\partial}{\partial L_3}\right)\end{aligned}\right\} \tag{5.16}$$

The integral of power functions of the area coordinates in the triangular element is as follows:

$$I = \iint_A L_1^a L_2^b L_3^c \, \mathrm{d}x \, \mathrm{d}y \tag{c}$$

Let L_1 and L_2 be independent variables, and then

$$\mathrm{d}x \, \mathrm{d}y = |J| \, \mathrm{d}L_1 \, \mathrm{d}L_2 = 2A \, \mathrm{d}L_1 \, \mathrm{d}L_2 \tag{d}$$

Thus the integral formula (c) may be written as

$$I = 2A\int_0^1 \left(\int_0^{1-L_1} L_2^b L_3^c \, \mathrm{d}L_2\right) L_1^a \, \mathrm{d}L_1 \tag{e}$$

The above integral can be calculated using the following formula:

$$\int_0^p x^b (p-x)^c \, \mathrm{d}x = \frac{b!c!}{(b+c+1)!} p^{b+c+1} \tag{f}$$

This formula can be verified by integration by parts, and from this, it can be seen that

$$\int_0^{1-L_1} L_2^b L_3^c \, \mathrm{d}L_2 = \int_0^{1-L_1} L_2^b (1-L_1-L_2)^c \, \mathrm{d}L_2 = \frac{b!c!}{(b+c+1)!}(1-L_1)^{b+c+1}$$

Substitute it into formula (e) and obtain

$$I = 2A\frac{b!c!}{(b+c+1)!}\int_0^{1_1} L_1^a (1-L_1)^{b+c+1} \mathrm{d}L_1 = 2A\frac{a!b!c!}{(a+b+c+2)!}$$

That is,

$$\iint_A L_1^a L_2^b L_3^c \, \mathrm{d}x \, \mathrm{d}y = 2A\frac{a!b!c!}{(a+b+c+2)!} \tag{5.17}$$

For example,

$$\left.\begin{aligned}
&\iint_A L_1\,\mathrm{d}x\,\mathrm{d}y = \frac{A}{3} \quad (1,2,3)\\
&\iint_A L_1^2\,\mathrm{d}x\,\mathrm{d}y = \frac{A}{6} \quad (1,2,3)\\
&\iint_A L_1 L_2\,\mathrm{d}x\,\mathrm{d}y = \frac{A}{12} \quad (1,2,3)
\end{aligned}\right\} \tag{5.18}$$

5.3 High-Order Triangular Element

The calculations of the 3-node triangular element introduced in Chapter 2 are simple, but the shape function adopted is linear, namely,

$$N_1 = L_1, \quad N_2 = L_2, \quad N_3 = L_3$$

Thus the stress and strain within the element are constants. Now we introduce several improved triangular elements.

5.3.1 6-Node Quadratic Triangular Element

As shown in Figure 5.5, each element has 6 nodes including 3 corner points and 3 edge midpoints with 12 degrees of freedom (DOFs). The displacement function can be set as the complete quadratic polynomial. The stress of the element is linear, no longer a constant.

Take the following displacement function:

$$u = \beta_1 + \beta_2 x + \beta_3 y + \beta_4 x^2 + \beta_5 xy + \beta_6 y^2 \tag{a}$$

$$v = \beta_7 + \beta_8 x + \beta_9 y + \beta_{10} x^2 + \beta_{11} xy + \beta_{12} y^2 \tag{b}$$

The 12-node displacement components u_i, v_i can just determine the 12 coefficients in the above formula. On the element boundary, each displacement component varies as a parabola. On every common boundary, there are three common nodes, which can just guarantee the displacement continuity of two adjacent elements. Besides, β_1, β_2, β_3 in formula (a) and β_7, β_8, β_9 in formula (b) reflect the rigid-body displacement and constant strain. So the displacement function satisfies the convergence conditions for solution.

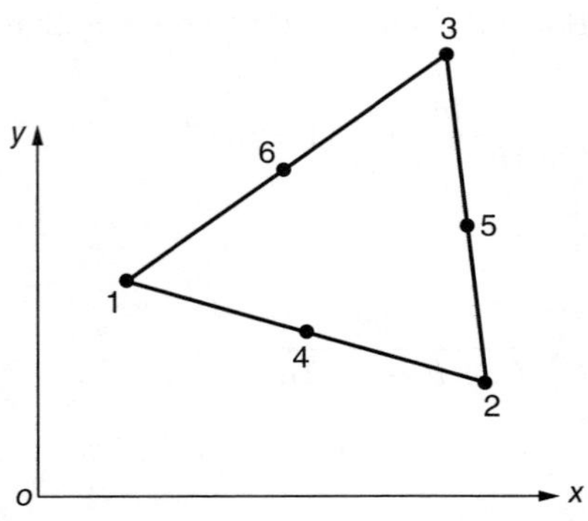

Figure 5.5 6-Node triangular element.

When the displacement functions are represented by Cartesian coordinates as formulas (a) and (b), the calculations of coefficients β_1–β_{12} and the stiffness matrix and nodal loads are very tedious, and the use of area coordinates can make it greatly simplified.

The quadratic shape functions represented by area coordinates are as follows:

$$\left.\begin{array}{ll}\text{Corner point}: & N_1 = (2L_1 - 1)L_1 \quad (1, 2, 3) \\ \text{Edge midpoint}: & N_4 = 4L_1L_2 \quad (4, 5, 6)\end{array}\right\} \tag{5.19}$$

The above shape functions may represent the displacements as follows:

$$u = \sum_{i=1}^{6} N_i u_i, \quad v = \sum_{i=1}^{6} N_i v_i \tag{5.20}$$

Or

$$\{r\} = \begin{Bmatrix} u \\ v \end{Bmatrix} = [N]\{\delta\}^e \tag{5.21}$$

Where

$$[N] = \begin{bmatrix} N_1 & 0 & N_2 & 0 & N_3 & 0 & N_4 & 0 & N_5 & 0 & N_6 & 0 \\ 0 & N_1 & 0 & N_2 & 0 & N_3 & 0 & N_4 & 0 & N_5 & 0 & N_6 \end{bmatrix}$$

$$\{\delta\}^e = \begin{bmatrix} u_1 & v_1 & u_2 & v_2 & u_3 & v_3 & u_4 & v_4 & u_5 & v_5 & u_6 & v_6 \end{bmatrix}^T$$

Then by the general formula in Chapter 3, we can calculate the stiffness matrix, nodal loads, and so on.

5.3.2 10-Node 3-Order Triangular Element

The element displacements are expressed by complete cubic polynomials:

$$\begin{aligned} u &= \beta_1 + \beta_2 x + \beta_3 y + \beta_4 x^2 + \beta_5 xy + \beta_6 y^2 \\ &\quad + \beta_7 x^3 + \beta_8 x^2 y + \beta_9 xy^2 + \beta_{10} y^3 \\ v &= \beta_{11} + \beta_{12} x + \beta_{13} y + \beta_{14} x^2 + \beta_{15} xy + \beta_{16} y^2 \\ &\quad + \beta_{17} x^3 + \beta_{18} x^2 y + \beta_{19} xy^2 + \beta_{20} y^3 \end{aligned}$$

There are 20 coefficients that need 20 nodal displacements to be determined, so 10 nodes are required. Besides three corner points, take two third points on each edge and the element centroid as shown in Figure 5.6. On each element edge, displacement is in a cubic-curve distribution. Now on every edge there are 4 nodes that can ensure the displacement continuity between adjacent elements. Coefficients β_1, β_2, β_3, and β_{11}, β_{12}, β_{13} reflect the rigid-body displacement and constant strain. So the above displacement functions can meet convergence conditions for solution.

In order to simplify calculation, adopt three-order shape functions represented by area coordinates as follows:

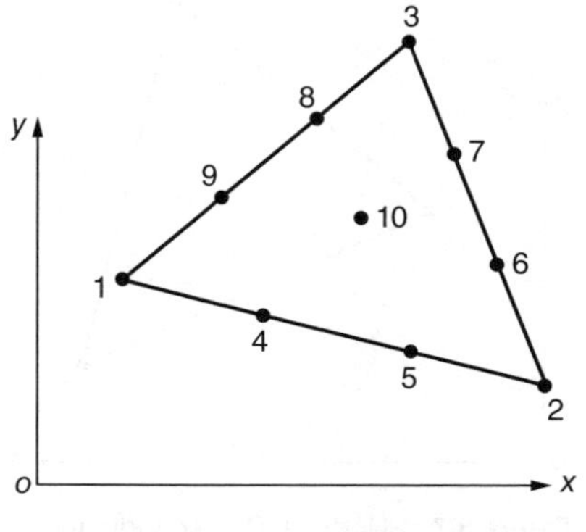

Figure 5.6 10-Node triangular element.

Corner point :

$$N_1 = \frac{1}{2}(3L_1 - 1)(3L_1 - 2)L_1 \quad (1, 2, 3)$$

$$\text{Edge third point}: \quad N_4 = \frac{9}{2}L_1L_2(3L_1 - 1) \quad (4,5,6,7,8,9) \tag{5.22}$$

$$\text{Centroid}: \quad N_{10} = 27L_1L_2L_3$$

The displacement functions are taken as

$$u = \sum_{i=1}^{10} N_i u_i, \quad v = \sum_{i=1}^{10} N_i v_i \tag{5.23}$$

By the general formula of stress analysis in Chapter 3, we can compute the stiffness matrix, nodal loads, and so on. By matrix computation, the internal node can be removed and we will get a 9-node 18 DOF element. The method to remove internal DOF in composite element in Section 3.13 may be referred to.

5.3.3 3-Node 18 DOF Triangular Element

As shown in Figure 5.7, take three corner points as the specific points of the element. For every node, take the displacements and their derivatives as parameters, and then there are 6 DOFs on every node and 18 DOF for the whole element, which are

$$\{\delta\}^e = \begin{bmatrix}\delta_1 & \delta_2 & \delta_3\end{bmatrix}^T$$

$$\{\delta_1\} = \begin{bmatrix} u_1 & \frac{\partial u_1}{\partial x} & \frac{\partial u_1}{\partial y} & v_1 & \frac{\partial v_1}{\partial x} & \frac{\partial v_1}{\partial y}\end{bmatrix}^T \quad (1,2,3)$$

A complete three-order polynomial represented by area coordinates includes the following 10 items:

$$1, L_1, L_2, L_1L_2, L_2L_3, L_3L_1, L_1L_2^2 - L_2L_1^2, L_2L_3^2 - L_3L_2^2, L_3L_1^2 - L_1L_3^2, L_1L_2L_3$$

The element displacement function should include the 10 items listed above, but the last item, $L_1L_2L_3$, takes 0 on the three edges of the element, so it represents the internal DOF. When generating virtual displacement, the work done by the boundary force on the virtual displacement determined by internal DOF is zero; hence internal DOF may be removed by matrix computation when deducing the element stiffness matrix. In the end, there are 18 DOFs left for every element. See Ref. [1] for more details. Along the element boundary, the displacement varies in a cubic curve. There are two displacements and two displacement derivatives; in total there are four parameters just enough to ensure the displacement continuity of adjacent elements. Within the element, the strain changes according to a quadratic function of coordinates. On nodes, adjacent elements share not only the same displacement component but also the same strain component and even the same stress component in the case of homogeneous materials. On the edge, the derivative of the displacement is usually not continuous, so the strain and stress are not continuous.

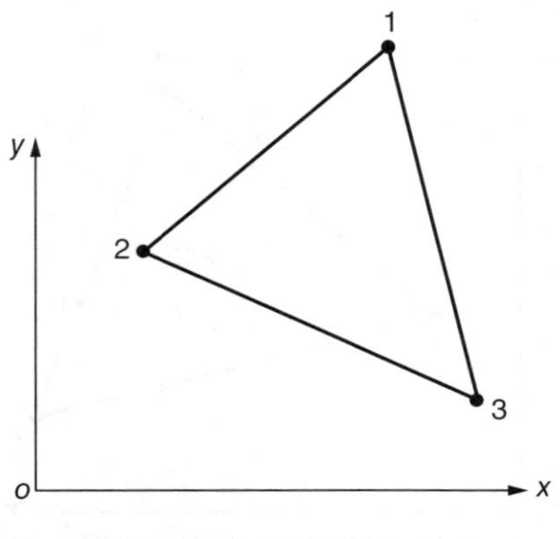

Figure 5.7 3-Node 18 DOF triangular element.

Figure 5.8 compares calculation results by this type of element and constant strain triangular element. The stress results of the constant strain element shown in the figure are averages of stresses of the two elements (like element 1 and 2 in the figure) of a square. Although five rows of nodes are laid in the beam height direction, the deflection error still reaches 14%. From Figure 5.8(c), high computational accuracy has already realized using two 18 DOF triangular elements: the deflection error is 0.6% and the normal stress error is 2.3%.

As 18 parameters are concentrated in 3 nodes, this type of element has another advantage: less nodes and small band width of the overall stiffness matrix (compared to 9-node 18 DOF triangular element).

Figure 5.9 shows calculation results of a cantilever beam with unit thickness. At the free end of the beam, the loads are in a parabolic distribution and assigned to every nodes according to the work equivalent principle. The figure shows vertical displacements of the point A calculated by four different elements. The calculation accuracy of the rectangular hybrid elements is high. When DOF is less than 6, only one layer of elements in the height direction is taken and good results are still obtained.

In short, as the internal stress is not constant, high-order elements can well adapt to the variable stress field of the structure. Using a few elements can obtain good results. However, the stiffness matrix of high-order elements is complicated, and more calculation time is needed to form the structural stiffness matrix.

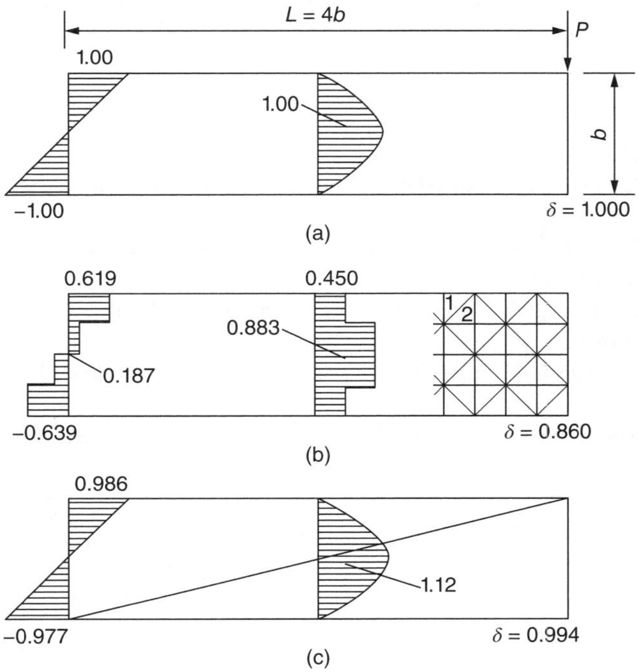

Figure 5.8 Comparison of constant strain element with high-order element of the cantilever. (a) Beam theory; (b) constant strain element (128 elements and 170 equations); and (c) 3-node 18 DOF elements (2 elements and 24 equations).

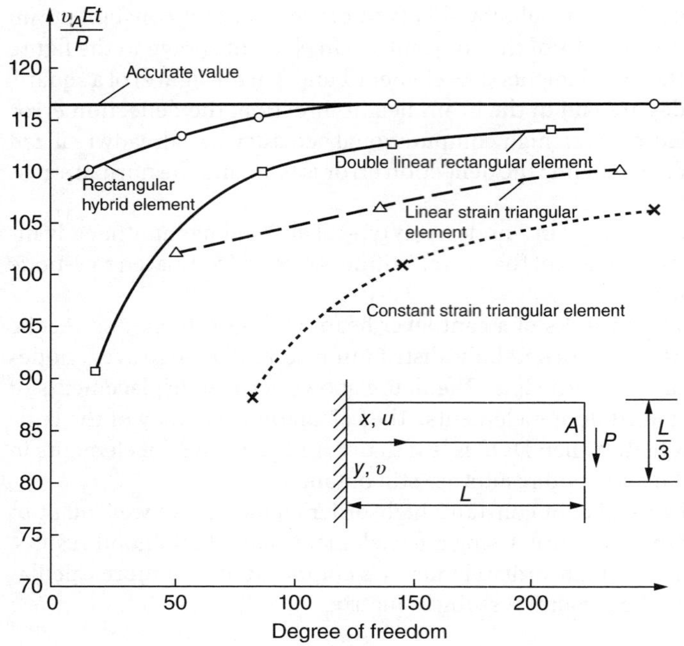

Figure 5.9 Comparison of calculation results for triangular element and rectangular element of the cantilever beam.

Bibliography

1 Holland, I. and Bell, K. (1972) *Finite Element Methods in Stress Analysis,* Tapir, Trondheim (Norway).
2 Norrie, D.H. and de Vries, G. (1973) *The Finite Element Methods, Fundamentals and Applications,* Academic Press, London.
3 Zienkiewicz, O.C. and Taylor, R.L. (2000) *The Finite Element Method,* 5th edn, Butterworth-Heinemann, Oxford.
4 ASCE (1969) *Proceedings of Symposium on Application of Finite Element Methods in Civil Engineering,* ASCE.
5 Whiteman, J.R. (1973) *The Mathematics of Finite Elements and Applications,* Academic Press, New York.

6

Axisymmetrical Problems in Theory of Elasticity

If the geometry, constraint conditions, and loads are symmetrical about an axis, as z-axis, then all the displacements, strains, and stresses are symmetrical about the axis. This problem is called an axisymmetrical stress problem, which often appears in the engineering of shaft, pressure vessel, and machine manufacturing.

The cylindrical coordinates (r, θ, z) are more convenient for the axisymmetrical problem. If the axis of symmetry of the elastic body is taken as z-axis, all the stresses, strains, and displacements are independent of θ and are functions of r and z. At any point, there are only two displacement components, namely, the radial displacement u in the r direction and the axial displacement w in the axial direction z. Due to symmetry, the displacement in the θ direction is zero.

The elements adopted in the axisymmetrical problem are annular rings with triangular cross section ijm (or other shape) as shown in Figure 6.1. The elements are connected by annular hinges, the intersection of which with the rz plane as i, j, m are called nodes. Due to symmetry, it is necessary to analyze only one cross section.

If the geometry of the body is axisymmetrical but the loads are not axisymmetrical, the loads may be expressed by Fourier series. Then the problem may be resolved into two problems, one symmetrical and one antisymmetrical problem.

6.1 Stresses Due to Axisymmetrical Loads

If all the geometry, constraint conditions, and loads of an elastic body are axisymmetrical, it is necessary to analyze only one cross section.

6.1.1 Displacement Function

Take the cross section ijm of an annular element as shown in Figure 6.2, and the displacements at node i are

$$\{\delta_i\} = \begin{Bmatrix} u_i \\ w_i \end{Bmatrix}$$

The element nodal displacements are

$$\{\delta\}^e = \begin{bmatrix} \delta_i & \delta_j & \delta_m \end{bmatrix}^{\mathrm{T}}$$

The Finite Element Method: Fundamentals and Applications in Civil, Hydraulic, Mechanical and Aeronautical Engineering, First Edition. Bofang Zhu.

Companion website: www.wiley.com/go/zhufem17

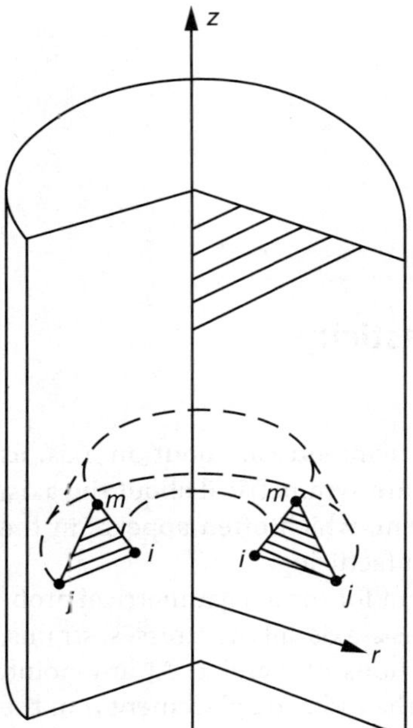

Figure 6.1 Triangular elements of axisymmetrical elastic body.

Adopting the linear displacement functions as follows

$$u = \beta_1 + \beta_2 r + \beta_3 z, \quad w = \beta_4 + \beta_5 r + \beta_6 z$$

then we get the displacement expressions

$$\begin{aligned} u &= N_i u_i + N_j u_j + N_m u_m, \\ w &= N_i w_i + N_j w_j + N_m w_m \end{aligned} \tag{6.1}$$

where $N_i = (a_i + b_i r + c_i z)/2A \ (i, j, m)$

$$a_i = r_j z_m - r_m z_j, \quad b_i = z_j - z_m,$$

$$c_i = -r_j + r_m \quad (i, j, m)$$

$$2A = \begin{vmatrix} 1 & r_i & z_i \\ 1 & r_j & z_j \\ 1 & r_m & z_{ni} \end{vmatrix}$$

Equation (6.1) may also be expressed by matrices as follows:

$$\{r\} = \begin{Bmatrix} u \\ w \end{Bmatrix} = [N]\{\delta\}^e$$

$$= \begin{bmatrix} N_i & 0 & N_j & 0 & N_m & 0 \\ 0 & N_i & 0 & N_j & 0 & N_m \end{bmatrix} \{\delta\}^e \tag{6.2}$$

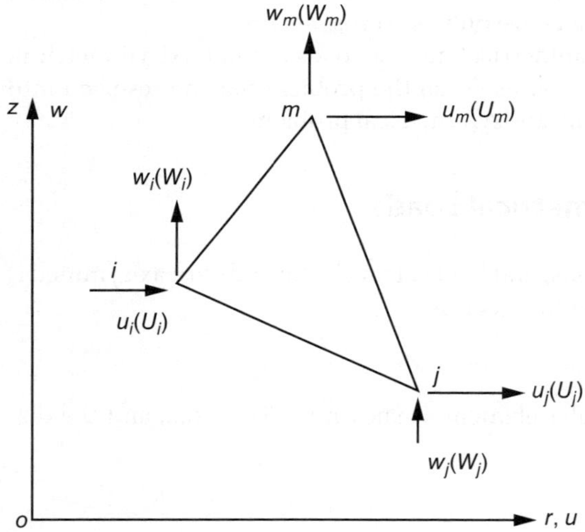

Figure 6.2 Nodal displacements and nodal forces for triangular axisymmetrical element.

6.1.2 Element Strains

As shown in Figure 6.3, for axisymmetrical stress problem, at any point there are 4 strain components: the radial normal strain ε_r, the tangential normal strain ε_θ, the axial normal strain ε_z, and the shearing strain γ_{rz} in rz plane.

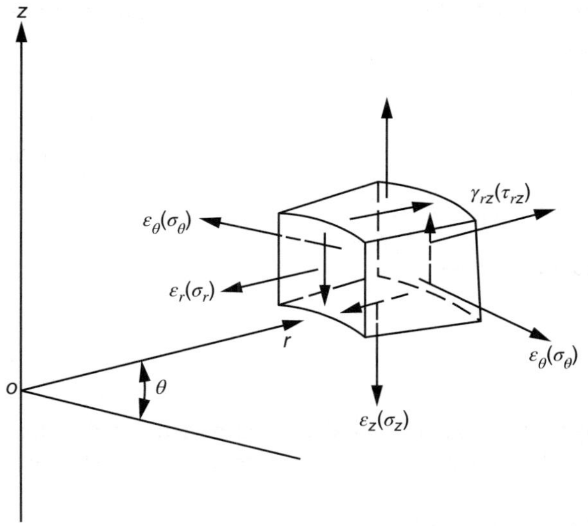

Figure 6.3 Stresses and strains in axisymmetrical elastic body.

Due to axisymmetry, $\gamma_{r\theta} = \gamma_{\theta z} = 0$. According to the geometrical relation, the strains may be expressed by displacements in the following:

$$\{\varepsilon\} = \begin{Bmatrix} \varepsilon_r \\ \varepsilon_\theta \\ \varepsilon_z \\ \gamma_{rz} \end{Bmatrix} = \begin{Bmatrix} \dfrac{\partial u}{\partial r} \\ \dfrac{u}{r} \\ \dfrac{\partial w}{\partial z} \\ \dfrac{\partial w}{\partial r} + \dfrac{\partial u}{\partial z} \end{Bmatrix} \tag{6.3}$$

Substitution of Eq. (6.2) into Eq. (6.3) yields

$$\{\varepsilon\} = [B]\{\delta\}^{\mathrm{e}} = \begin{bmatrix} B_i & B_j & B_m \end{bmatrix} \{\delta\}^{\mathrm{e}} \tag{6.4}$$

$$[B_i] = \begin{bmatrix} \dfrac{\partial N_i}{\partial r} & 0 \\ \dfrac{N_i}{r} & 0 \\ 0 & \dfrac{\partial N_i}{\partial z} \\ \dfrac{\partial N_i}{\partial z} & \dfrac{\partial N_i}{\partial r} \end{bmatrix} = \frac{1}{2A} \begin{bmatrix} b_i & 0 \\ h_i & 0 \\ 0 & c_i \\ c_i & b_i \end{bmatrix} \quad (i, j, m) \tag{6.5}$$

$$h_i = \frac{a_i}{r} + b_i + c_i \frac{z}{r}$$

For axisymmetrical problem, there may be four independent initial strain components, namely,

$$\{\varepsilon_0\} = \begin{Bmatrix} \varepsilon_{r0} \\ \varepsilon_{\theta 0} \\ \varepsilon_{z0} \\ \gamma_{rz0} \end{Bmatrix} \tag{6.6}$$

The initial strains caused by temperature change are

$$\{\varepsilon_0\} = \begin{bmatrix} \alpha T & \alpha T & \alpha T & 0 \end{bmatrix}^{\mathrm{T}}$$

6.1.3 Element Stress

For axisymmetrical problem, at any point there are four stress components: radial normal stress σ_r, circumferential normal stress σ_θ, axial normal stress σ_z, and shearing stress τ_{rz}.

The relation between stresses and strains is given by

$$\{\sigma\} = \begin{Bmatrix} \sigma_r \\ \sigma_\theta \\ \sigma_z \\ \tau_{rz} \end{Bmatrix} = [D](\{\varepsilon\} - \{\varepsilon_0\}) \tag{6.7}$$

where $[D]$ is the elasticity matrix.

For laminar material, if z-axis is normal to the layer, the stress–strain relation is as follows:

$$\left.\begin{aligned} \varepsilon_r &= \frac{\sigma_r}{E_1} - \frac{\mu_1\sigma_\theta}{E_1} - \frac{\mu_2\sigma_z}{E_2} \\ \varepsilon_\theta &= \frac{\sigma_\theta}{E_1} - \frac{\mu_1\sigma_r}{E_1} - \frac{\mu_2\sigma_z}{E_2} \\ \varepsilon_z &= \frac{\sigma_z}{E_2} - \frac{\mu_2\sigma_r}{E_2} - \frac{\mu_2\sigma_\theta}{E_2} \\ \gamma_{rz} &= \frac{\tau_{rz}}{G_2} \end{aligned}\right\} \tag{6.8}$$

From the inverse of Eq. (6.8), we get the elasticity matrix as follows:

$$[D] = \frac{E_2}{(1+\mu_1)(1-\mu_1-2n\mu_2^2)} \times \begin{bmatrix} 1-\mu_1^2 & n\mu_2(1+\mu_1) & n\mu_2(1+\mu_1) & 0 \\ & n(1-n\mu_2^2) & (\mu_1+n\mu_2^2)n & 0 \\ \text{symmetrical} & & n(1-n\mu_2^2) & 0 \\ & & & m(1+\mu_1)(1-\mu_1-2n\mu_2^2) \end{bmatrix} \tag{6.9}$$

$$n = \frac{E_1}{E_2}, \quad m = \frac{G_2}{E_2} \tag{6.10}$$

For isotropic body, let

$$E_1 = E_2 = E, \quad n = 1$$

$$\mu_1 = \mu_2 = \mu$$

$$m = \frac{G_2}{E_2} = \frac{G}{E} = \frac{1}{2}(1+\mu)$$

$$[D] = \frac{E(1-\mu)}{(1+\mu)(1-2\mu)} \begin{bmatrix} 1 & \frac{\mu}{1-\mu} & \frac{\mu}{1-\mu} & 0 \\ & 1 & \frac{\mu}{1-\mu} & 0 \\ & & 1 & 0 \\ \text{Symmetrical} & & & \frac{1-2\mu}{2(1-\mu)} \end{bmatrix} \tag{6.11}$$

6.1.4 Element Stiffness Matrix

The element stiffness matrix may be derived from Eq. (3.10). By volume integration of the whole circular ring, we get

$$[k] = 2\pi \iint [B]^{\mathrm{T}}[D][B] r \,\mathrm{d}r\,\mathrm{d}z \tag{6.12}$$

in which the matrix $[B]$ is given by Eq. (6.5). As there are variables r and z in the function of integration, the integral of the right part of the above equation cannot be gotten by simple method. In order to avoid the complicated operation of integration and the trouble of $r = 0$ at the axis of symmetry, the following approximate method may be used. Let r and z in each element be approximately equal to constants as follows:

$$r = \frac{r_i + r_j + r_m}{3}, \quad z = \frac{z_i + z_j + z_m}{3} \tag{6.13}$$

The error will be small when the computing mesh is not too large. As the function of integration is constant, from Eq. (6.12), we get

$$[k] = 2\pi r A [B]^{\mathrm{T}}[D][B] = \begin{bmatrix} k_{ii} & k_{ij} & k_{im} \\ k_{ji} & k_{jj} & k_{jm} \\ k_{mi} & k_{mj} & k_{mm} \end{bmatrix} \tag{6.14}$$

in which

$$[k_{rs}] = 2\pi r A [B_r]^{\mathrm{T}}[D][B_s] \tag{6.15}$$

where A is the area of triangle ijm and $[k_{rs}]$ is submatrix.

Substitution of the formulas of $[B]$ and $[D]$ into Eq. (6.15) will yield $[k_{rs}]$, for example, for isotropic body:

$$[k_{rs}] = g_3 \begin{bmatrix} b_r b_s + h_r h_s + g_1(b_r h_s + h_r b_s) + g_2 c_r c_s & g_1(b_r c_s + h_r c_s) + g_2 c_r b_s \\ g_1(c_r b_s + c_r h_s) + g_2 b_r c_s & c_r c_s + g_2 b_r b_s \end{bmatrix}$$

$$(r, s = i, j, m) \tag{6.16}$$

in which

$$g_1 = \frac{\mu}{(1-\mu)}, \quad g_2 = \frac{(1-2\mu)}{2(1-\mu)}, \quad g_3 = \frac{\pi E(1-\mu)r}{2(1+\mu)(1-2\mu)A}$$

6.1.5 Nodal Loads

For axisymmetrical problem, the nodal loads are acting on the circular rings. For example, let r be the radius of node, and then the loads acting on unit length of the ring are $\overline{R}$ (radial) and $\overline{Z}$ (axial), and the nodal loads must be $2\pi r\overline{R}$ in radial direction and $2\pi r\overline{Z}$ in axial direction.

Let the volumetric forces (gravity force, centrifugal force, etc.) in unit volume be

$$\{q\} = \begin{Bmatrix} q_r \\ q_z \end{Bmatrix}$$

From Eq. (3.13), the nodal loads are

$$\{P\}_q^{\mathrm{e}} = 2\pi \iint [N]^{\mathrm{T}} \begin{Bmatrix} q_r \\ q_z \end{Bmatrix} r\,\mathrm{d}r\,\mathrm{d}z \tag{6.17}$$

When the volumetric forces are constant, r and z in the function of integration may approximately take its mean value; hence

$$\{P_i\}_q^{\mathrm{e}} = \{P_j\}_q^{\mathrm{e}} = \{P_m\}_q^{\mathrm{e}} = 2\pi \begin{Bmatrix} q_r \\ q_z \end{Bmatrix} \frac{rA}{3} \tag{6.18}$$

If the volumetric force is not constant, such as the centrifugal force of a rotating machine, it is best to conduct precise integration by Eq. (6.17).

From Eq. (3.15), the nodal loads due to initial strain $\{\varepsilon_0\}$ may be given by

$$\{P\}_{\varepsilon_0}^{\mathrm{e}} = 2\pi \iint [B]^{\mathrm{T}}[D]\{\varepsilon_0\} r\,\mathrm{d}r\,\mathrm{d}z$$

When the initial strain $\{\varepsilon_0\}$ are constant in an element and r and z approximately take their mean values, we get

$$\{P_i\}_{\varepsilon_0}^{\mathrm{e}} = 2\pi rA[B_i]^{\mathrm{T}}[D]\{\varepsilon_0\} \tag{6.19}$$

6.2 Antisymmetrical Load

If the geometrical shape of an elastic body is axisymmetrical, but the load is not axisymmetrical, such as the wind load and earthquake load, we may express the load in θ direction by Fourier series. Taking advantage of the axisymmetrical shape of the body, the three-dimensional problem can be transformed into a set of two-dimensional problems that are easier to solve.

Resolve the load into radial load R, circumferential load F, and axial load Q, and express them by Fourier series as follows:

$$\left.\begin{aligned} R &= \sum R_n^{\mathrm{s}} \cos n\theta + \sum R_n^{\mathrm{t}} \sin n\theta \\ F &= -\sum F_n^{\mathrm{s}} \sin n\theta + \sum F_n^{\mathrm{t}} \cos n\theta \\ Q &= \sum Q_n^{\mathrm{s}} \cos n\theta + \sum Q_n^{\mathrm{t}} \sin \theta \end{aligned}\right\} \tag{6.20}$$

where the superscript “s” means axisymmetry and “t” means antisymmetry.

The displacement components may also be expressed by Fourier series in the following:

$$\left.\begin{aligned} u &= \sum u_n^{\mathrm{s}} \cos n\theta + \sum u_n^{\mathrm{t}} \sin n\theta \\ v &= -\sum v_n^{\mathrm{s}} \sin n\theta + \sum v_n^{\mathrm{t}} \cos n\theta \\ w &= \sum w_n^{\mathrm{s}} \cos n\theta + \sum w_n^{\mathrm{t}} \sin n\theta \end{aligned}\right\} \tag{6.21}$$

in which $u_n^{\mathrm{s}}, v_n^{\mathrm{s}}, w_n^{\mathrm{s}}$ are the Fourier coefficients of axisymmetric displacements, $u_n^{\mathrm{t}}, v_n^{\mathrm{t}}, w_n^{\mathrm{t}}$ are the Fourier coefficients of antisymmetric displacements, and all of them are functions of coordinates r and z.

In order to discrete the problem, $u_n^{\mathrm{s}}, \ldots, w_n^{\mathrm{t}}$ are expressed by nodal parameters. The computing mesh in rz plane is composed by triangular elements as shown in Figure 6.2. Let

$$\{u_n^{\mathrm{s}}\} = \begin{Bmatrix} u_{ni}^{\mathrm{s}} \\ u_{nj}^{\mathrm{s}} \\ u_{nm}^{\mathrm{s}} \end{Bmatrix}, \quad \{u_n^{\mathrm{t}}\} = \begin{Bmatrix} u_{ni}^{\mathrm{t}} \\ u_{nj}^{\mathrm{t}} \\ u_{nm}^{\mathrm{t}} \end{Bmatrix}, \ldots \tag{6.22}$$

where $u_{ni}^{\mathrm{s}}, u_{nj}^{\mathrm{s}}, u_{nm}^{\mathrm{s}}, \ldots$ are nodal parameters.

By means of these nodal parameters and the shape functions of element, $u_n^{\mathrm{s}}, v_n^{\mathrm{s}}, \ldots$ can be expressed as follows:

$$\left.\begin{aligned} u_n^{\mathrm{s}} &= \begin{bmatrix} N_i & N_j & N_m \end{bmatrix} \begin{Bmatrix} u_{ni}^{\mathrm{s}} \\ u_{nj}^{\mathrm{s}} \\ u_{nm}^{\mathrm{s}} \end{Bmatrix} = [N]\{u_n^{\mathrm{s}}\} \\ v_n^{\mathrm{s}} &= [N]\{v_n^{\mathrm{s}}\} \\ w_n^{\mathrm{s}} &= [N]\{w_n^{\mathrm{s}}\} \end{aligned}\right\} \tag{6.23}$$

Substitution of Eq. (6.23) into Eq. (6.21) yields the formulas for displacement components in the following:

$$\left.\begin{aligned} u &= \sum [N]\{u_n^{\mathrm{s}}\} \cos n\theta + \sum [N]\{u_n^{\mathrm{t}}\} \sin n\theta \\ v &= -\sum [N]\{v_n^{\mathrm{s}}\} \sin n\theta + \sum [N]\{v_n^{\mathrm{t}}\} \cos n\theta \\ w &= \sum [N]\{w_n^{\mathrm{s}}\} \cos n\theta + \sum [N]\{w_n^{\mathrm{t}}\} \sin n\theta \end{aligned}\right\} \tag{6.24}$$

In the case of nonaxisymmetry, the strains of the elastic body are

$$\{\varepsilon\} = \begin{Bmatrix} \varepsilon_r \\ \varepsilon_z \\ \varepsilon_\theta \\ \gamma_{rz} \\ \gamma_{z\theta} \\ \gamma_{r\theta} \end{Bmatrix} = \begin{Bmatrix} \dfrac{\partial u}{\partial r} \\ \dfrac{\partial w}{\partial z} \\ \dfrac{u}{r} + \dfrac{1}{r}\dfrac{\partial v}{\partial \theta} \\ \dfrac{\partial u}{\partial z} + \dfrac{\partial w}{\partial r} \\ \dfrac{\partial v}{\partial z} + \dfrac{1}{r} + \dfrac{\partial w}{\partial \theta} \\ \dfrac{1}{r}\dfrac{\partial u}{\partial \theta} + \dfrac{\partial v}{\partial r} - \dfrac{v}{r} \end{Bmatrix} \tag{6.25}$$

Substituting Eq. (6.24) into Eq. (6.25), we get

$$\{\varepsilon\} = \sum [B_n^{\mathrm{s}}]\{\delta_n^{\mathrm{s}}\} + \sum [B_n^{\mathrm{t}}]\{\delta_n^{\mathrm{t}}\} \tag{6.26}$$

in which

$$\left.\begin{aligned} &[B_n^{\mathrm{s}}] = \begin{bmatrix} B_{ni}^{\mathrm{s}} & B_{nj}^{\mathrm{s}} & B_{nm}^{\mathrm{s}} \end{bmatrix} \\ &[B_n^{\mathrm{t}}] = \begin{bmatrix} B_{ni}^{\mathrm{t}} & B_{nj}^{\mathrm{t}} & B_{nm}^{\mathrm{t}} \end{bmatrix} \\ &\{\delta_n^{\mathrm{s}}\} = \begin{Bmatrix} \delta_{ni}^{\mathrm{s}} \\ \delta_{nj}^{\mathrm{s}} \\ \delta_{nm}^{\mathrm{s}} \end{Bmatrix}, \quad \{\delta_{ni}^{\mathrm{s}}\} = \begin{Bmatrix} u_{ni}^{\mathrm{s}} \\ v_{ni}^{\mathrm{s}} \\ w_{ni}^{\mathrm{s}} \end{Bmatrix}, \ldots \\ &\{\delta_n^{\mathrm{t}}\} = \begin{Bmatrix} \delta_{ni}^{\mathrm{t}} \\ \delta_{nj}^{\mathrm{t}} \\ \delta_{nm}^{\mathrm{t}} \end{Bmatrix}, \quad \{\delta_{ni}^{\mathrm{t}}\} = \begin{Bmatrix} u_{ni}^{\mathrm{t}} \\ v_{ni}^{\mathrm{t}} \\ w_{ni}^{\mathrm{t}} \end{Bmatrix}, \ldots \end{aligned}\right\} \tag{6.27}$$

$$[B_{ni}^{\mathrm{s}}] = \begin{bmatrix} \dfrac{\partial N_i}{\partial r}\cos n\theta & 0 & 0 \\ 0 & 0 & \dfrac{\partial N_i}{\partial z}\cos n\theta \\ \dfrac{N_i}{r}\cos n\theta & -\dfrac{nN_i}{r}\cos n\theta & 0 \\ \dfrac{\partial N_i}{\partial z}\cos n\theta & 0 & \dfrac{\partial N_i}{\partial r}\cos n\theta \\ 0 & -\dfrac{\partial N_i}{\partial z}\sin n\theta & -\dfrac{nN_i}{r}\sin n\theta \\ -\dfrac{nN_i}{r}\sin n\theta & \left(\dfrac{\partial N_i}{\partial r} - \dfrac{N_i}{r}\right)\sin n\theta & 0 \end{bmatrix} \quad (i,j,m) \tag{6.28}$$

$$[B_{ni}^{t}] = \begin{bmatrix} \frac{\partial N_i}{\partial r}\sin n\theta & 0 & 0 \\ 0 & 0 & \frac{\partial N_i}{\partial z}\sin n\theta \\ \frac{N_i}{r}\sin n\theta & -\frac{nN_i}{r}\sin n\theta & 0 \\ \frac{\partial N_i}{\partial z}\sin n\theta & 0 & \frac{\partial N_i}{\partial r}\sin n\theta \\ 0 & \frac{\partial N_i}{\partial z}\cos n\theta & \frac{nN}{r}\cos n\theta \\ \frac{nN_i}{r}\cos n\theta & \left(\frac{N_i}{r} - \frac{\partial N_i}{\partial r}\right)\cos n\theta & 0 \end{bmatrix} \quad (i, j, m) \tag{6.29}$$

In the right part of Eq. (6.26), the first term represents the axisymmetric strain and the second term, the antisymmetric strain.

The strain energy of the elastic body is

$$U = \frac{1}{2}\int \{\varepsilon\}^{\mathrm{T}}[D]\{\varepsilon\}\,\mathrm{d(vol)}$$

By substituting the formula for strain (6.26) into the above equation and integrating in the volume of the element, due to the orthogonality of triangular functions, when $l \neq n$, there is no coupling between the lth term and the nth term; thus

$$U = \sum U_n \tag{6.30}$$

$$U_n = \frac{1}{2}\{\delta_n^{\mathrm{s}}\}^{\mathrm{T}}[k_n^{\mathrm{s}}]\{\delta_n^{\mathrm{s}}\} + \frac{1}{2}\{\delta_n^{t}\}^{\mathrm{T}}[k_n^{\mathrm{t}}]\{\delta_n^{\mathrm{t}}\} \tag{6.31}$$

$$\left.\begin{aligned} [k_n^{\mathrm{s}}] &= \iiint [B_n^{\mathrm{s}}]^{\mathrm{T}}[D][B_n^{\mathrm{s}}]r\,\mathrm{d}\theta\,\mathrm{d}r\,\mathrm{d}z \\ [k_n^{\mathrm{t}}] &= \iiint [B_n^{\mathrm{t}}]^{\mathrm{T}}[D][B_n^{\mathrm{t}}]r\,\mathrm{d}\theta\,\mathrm{d}r\,\mathrm{d}z \end{aligned}\right\} \tag{6.32}$$

where $r\,\mathrm{d}\theta\,\mathrm{d}r\,\mathrm{d}z = \mathrm{d(vol)}$

$[k_n^{\mathrm{s}}]$ and $[k_n^{\mathrm{t}}]$ are the axisymmetrical and antiaxisymmetrical nth stiffness coefficients.

From Eq. (6.24), the nth displacement component is expressed in the following:

$$\{r_n\} = \begin{Bmatrix} u_n \\ v_n \\ w_n \end{Bmatrix} = [N_n^{\mathrm{s}}]\{\delta_n^{\mathrm{s}}\} + [N_n^{\mathrm{t}}]\{\delta_n^{\mathrm{t}}\} \tag{6.33}$$

in which $[N_n^{\mathrm{s}}]$ and $[N_n^{\mathrm{t}}]$ are 3×9 matrix of shape functions:

$$\left.\begin{aligned} [N_n^{\mathrm{s}}] &= \begin{bmatrix} N_i\cos n\theta & 0 & 0 & N_j\cos n\theta & \cdots & 0 \\ 0 & -N_i\sin n\theta & 0 & 0 & \cdots & 0 \\ 0 & 0 & N_i\cos n\theta & 0 & \cdots & N_m\cos n\theta \end{bmatrix} \\ [N_n^{\mathrm{t}}] &= \begin{bmatrix} N_i\sin n\theta & 0 & 0 & N_j\sin n\theta & \cdots & 0 \\ 0 & N_i\cos n\theta & 0 & 0 & \cdots & 0 \\ 0 & 0 & N_i\sin n\theta & 0 & \cdots & N_m\sin n\theta \end{bmatrix} \end{aligned}\right\} \tag{6.34}$$

Substitution of Eq. (6.20) into Eq. (3.14) yields the nth axisymmetrical nodal load $\{P_n^{\rm s}\}$ and antiaxisymmetrical nodal load $\{P_n^{\rm t}\}$ as follows:

$$\left.\begin{aligned}\{P_n^{\rm s}\} &= \iiint [N_n^{\rm s}]^{\rm T}\begin{Bmatrix} R_n^{\rm s}\cos n\theta \\ -F_n^{\rm s}\sin n\theta \\ Q_n^{\rm s}\cos n\theta \end{Bmatrix} r\,{\rm d}\theta\,{\rm d}r\,{\rm d}z \\ \{P_n^{\rm t}\} &= \iiint [N_n^{\rm t}]^{\rm T}\begin{Bmatrix} R_n^{\rm t}\cos n\theta \\ F_n^{\rm t}\cos n\theta \\ Q_n^{\rm t}\cos n\theta \end{Bmatrix} r\,{\rm d}\theta\,{\rm d}r\,{\rm d}z \end{aligned}\right\} \tag{6.35}$$

There are some trigonometric functions in the formulas for nodal loads and stiffness matrix expressed as

$$\int_0^{2\pi} \sin^2 n\theta\,{\rm d}\theta = \int_0^{2\pi} \cos^2 n\theta\,{\rm d}\theta = \pi$$

After integration, there will be no circumferential coordinate θ in the stiffness matrix and nodal loads, and the original three-dimensional problem is converted into two independent two-dimensional problems:

$$[K_n^{\rm s}]\{\delta_n^{\rm s}\} = \{P_n^{\rm s}\}$$
$$[K_n^{\rm t}]\{\delta_n^{\rm t}\} = \{P_n^{\rm t}\}$$

Although the computing mesh is two-dimensional, there are three displacement parameters and three loading parameters at each node, which is different from the plane problems in the theory of elasticity.

In order to ensure that the global stiffness matrix is nonsingular, three restraint conditions are necessary when $n=1$; only the axial displacement of rigid body must be restrained when $n>1$; the rigid-body rotation around the axis of symmetry and the rigid-body translation along the axis of symmetry must be restrained when $n=0$.

Bibliography

1 Clough, R.W. and Rashid, Y.R. (1965) Finite element analysis of axi-symmetric solids. *ASCE Eng. Mech. Div. J.*, **91**, 71.

2 Wilson, E.L. (1965) Structural analysis of axi-symmetric solids. *AIAA J.*, **3**, 2269–2274.

3 Timoshenko, S. and Goodier, J.N. (1951) *Theory of Elasticity*, 2nd edn, McGraw-Hill, New York.

7

Spatial Problems in Theory of Elasticity

In the previous chapters, we explain how to solve the elastic plane problems and axisymmetric problems. However, some structures with complex shapes are hard to be simplified as plane or axisymmetric problems and must be treated as spatial problems. For spatial problems, isoparametric elements are currently used (see Chapter 8). To provide a step-by-step exploration, this chapter will first describe the tetrahedral elements.

7.1 Constant Strain Tetrahedral Elements

Figure 7.1 represents a tetrahedral element with four corner points i, j, m, and p as the nodes. This is the first and simplest spatial element.

7.1.1 Displacement Function

There are three nodal displacement components in every node

$$\{\delta_i\} = \begin{Bmatrix} u_i \\ v_i \\ w_i \end{Bmatrix} \tag{7.1}$$

There are a total of 12 nodal displacement components for every element, which can be represented as a vector.

$$\{\delta\}^{\mathrm{e}} = \left[\delta_i \; \delta_j \; \delta_m \; \delta_p\right]^{\mathrm{T}} \tag{7.2}$$

It is assumed that the displacement components of any point within the element are linear functions of the coordinates, namely,

$$\left.\begin{aligned} u &= \beta_1 + \beta_2 x + \beta_3 y + \beta_4 z \\ v &= \beta_5 + \beta_6 x + \beta_7 y + \beta_8 z \\ w &= \beta_9 + \beta_{10} x + \beta_{11} y + \beta_{12} z \end{aligned}\right\} \tag{a}$$

The Finite Element Method: Fundamentals and Applications in Civil, Hydraulic, Mechanical and Aeronautical Engineering, First Edition. Bofang Zhu.

Companion website: www.wiley.com/go/zhufem17

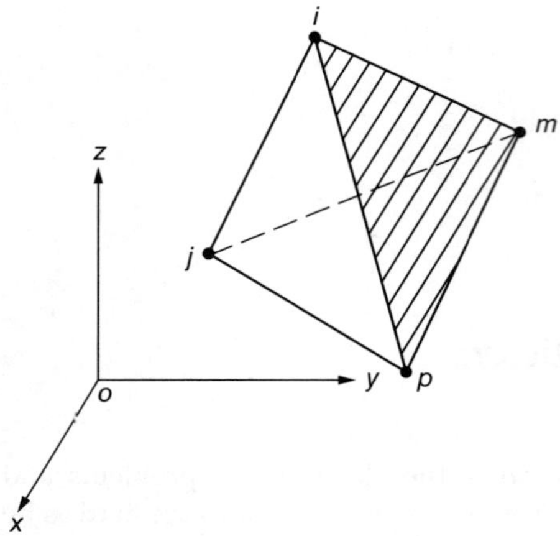

Figure 7.1 Constant strain tetrahedron elements.

Substituting the coordinates of node i, j, m, and p into the first equation of formula (a), we get

$$\left.\begin{aligned} u_i &= \beta_1 + \beta_2 x_i + \beta_3 y_i + \beta_4 z_i \\ u_j &= \beta_1 + \beta_2 x_j + \beta_3 y_j + \beta_4 z_j \\ u_m &= \beta_1 + \beta_2 x_m + \beta_3 y_m + \beta_4 z_m \\ u_p &= \beta_1 + \beta_2 x_p + \beta_3 y_p + \beta_4 z_p \end{aligned}\right\} \quad \text{(b)}$$

Inverse of formula (b) gives the coefficient β_1, β_2, β_3, and β_4 and substituting them into formula (a) yields

$$u = N_i u_i + N_j u_j + N_m u_m + N_p u_p \quad (7.3a)$$

In the same way

$$v = N_i v_i + N_j v_j + N_m v_m + N_p v_p \quad (7.3b)$$

$$w = N_i w_i + N_j w_j + N_m w_m + N_p w_p \quad (7.3c)$$

$$\left.\begin{aligned} N_i &= \frac{(a_i + b_i x + c_i y + d_i z)}{6V} \\ N_j &= \frac{(a_j + b_j x + c_j y + d_j z)}{6V} \\ N_m &= \frac{(a_m + b_m x + c_m y + d_m z)}{6V} \\ N_p &= \frac{(a_p + b_p x + c_p y + d_p z)}{6V} \end{aligned}\right\} \quad (7.4)$$

$$V = \frac{1}{6}\begin{vmatrix} 1 & x_i & y_i & z_i \\ 1 & x_j & y_j & z_j \\ 1 & x_m & y_m & z_m \\ 1 & x_p & y_p & z_p \end{vmatrix} \quad (7.5)$$

$$\left.\begin{aligned} a_i &= \begin{vmatrix} x_j & y_j & z_j \\ x_m & y_m & z_m \\ y_p & y_p & z_p \end{vmatrix}, \quad b_i = -\begin{vmatrix} 1 & y_j & z_j \\ 1 & y_m & z_m \\ 1 & y_p & z_p \end{vmatrix} \\ c_i &= -\begin{vmatrix} x_j & 1 & z_j \\ x_m & 1 & z_m \\ x_p & 1 & z_p \end{vmatrix}, \quad d_i = -\begin{vmatrix} x_j & y_j & 1 \\ x_m & y_m & 1 \\ y_p & y_p & 1 \end{vmatrix} \end{aligned} \quad (i,j,m,p)\right\} \quad (7.6)$$

in which V is the volume of tetrahedron $ijmp$.

To make the tetrahedron volume V not negative, the element node i, j, m, and p must be in a certain order. In the right-handed coordinate system, when rotating in the $i \rightarrow j \rightarrow m$ direction, the right-hand spiral must advance toward p, as shown in Figure 7.1.

With formula (7.3), the element displacements can be represented by the nodal displacement components as follows:

$$\{r\} = \begin{Bmatrix} u \\ v \\ w \end{Bmatrix} = [N]\{\delta\}^{\mathrm{e}} = \left[IN_i \ IN_j \ IN_m \ IN_p\right] \{\delta\}^{\mathrm{e}} \tag{7.7}$$

in which I is a three-order unit matrix.

As the displacement function is linear, on the contact surface of adjacent elements, the displacement is obviously continuous.

7.1.2 Element Strain

In spatial stress problems, there are six strain components in every point. By the theory of elasticity it is known that the strain matrix can be defined as follows:

$$\begin{aligned}\{\varepsilon\} &= \left[\varepsilon_x \ \varepsilon_y \ \varepsilon_z \ \gamma_{xy} \ \gamma_{yz} \ \gamma_{zx}\right]^{\mathrm{T}} \\ &= \left[\frac{\partial u}{\partial x} \ \frac{\partial v}{\partial y} \ \frac{\partial w}{\partial z} \ \frac{\partial u}{\partial y}+\frac{\partial v}{\partial x} \ \frac{\partial v}{\partial z}+\frac{\partial w}{\partial y} \ \frac{\partial w}{\partial x}+\frac{\partial u}{\partial z}\right]^{\mathrm{T}}\end{aligned} \tag{7.8}$$

By substituting formula (7.3) into formula (7.8), we obtain

$$\{\varepsilon\} = [B]\{\delta\}^{\mathrm{e}} = [B_iB_jB_mB_p]\{\delta\}^{\mathrm{e}} \tag{7.9}$$

The submatrix $[B_i]$ equals to the following 6×3 matrix:

$$[B_i] = \frac{1}{6V}\begin{bmatrix} b_i & 0 & 0 \\ 0 & c_i & 0 \\ 0 & 0 & d_i \\ c_i & b_i & 0 \\ 0 & d_i & c_i \\ d_i & 0 & b_i \end{bmatrix} \quad (i, j, m, p) \tag{7.10}$$

As the elements in matrix $[B]$ are constants, the element strain components are also all constants.

The initial strain of the element is

$$\{\varepsilon_0\} = \left[\varepsilon_{x0} \ \varepsilon_{y0} \ \varepsilon_{z0} \ \gamma_{xy0} \ \gamma_{yz0} \ \gamma_{zx0}\right]^{\mathrm{T}} \tag{7.11}$$

For example, the initial strain caused by temperature change T in the isotropic body is

$$\{\varepsilon_0\} = \left[\alpha T \ \alpha T \ \alpha T \ 0 \ 0 \ 0\right]^{\mathrm{T}} \tag{7.12}$$

in which α is the coefficient of linear expansion.

Using formula (a) and formula (7.8), it is known that the coefficients β_1, β_5, and β_9 in formula (a) represent the rigid body movements; β_2, β_7, and β_{12} represent constant normal strains; and the other six coefficients reflect the constant shear strain and rotation of a rigid body. So the 12 coefficients in formula (a) fully reflect the rigid body displacements and constant strains of the element. In addition, as the displacement function is linear and can ensure the displacement continuity between adjacent elements, the displacement functions (a) meet the convergence criteria.

7.1.3 Element Stress

The element stress can be represented by nodal displacements as follows:

$$\begin{aligned}\{\sigma\} &= \left[\sigma_x\ \sigma_y\ \sigma_z\ \tau_{xy}\ \tau_{yz}\ \tau_{zx}\right]^{\mathrm{T}} \\ &= [D][B]\{\delta\}^{\mathrm{e}} - [D]\{\varepsilon_0\} + \{\sigma_0\}\end{aligned} \tag{7.13}$$

For the isotropic body, the elasticity matrix $[D]$ is determined by formula (7.14):

$$[D] = \frac{E(1-\mu)}{(1+\mu)(1-2\mu)}\begin{bmatrix} 1 & \frac{\mu}{1-\mu} & \frac{\mu}{1-\mu} & 0 & 0 & 0 \\ & 1 & \frac{\mu}{1-\mu} & 0 & 0 & 0 \\ & & 1 & 0 & 0 & 0 \\ & \text{symmetrical} & & \frac{1-2\mu}{2(1-\mu)} & 0 & 0 \\ & & & & \frac{1-2\mu}{2(1-\mu)} & 0 \\ & & & & & \frac{1-2\mu}{2(1-\mu)} \end{bmatrix} \tag{7.14}$$

For the transversely isotropic material shown in Figure 7.2, there are the following stress–strain relationships:

$$\begin{aligned}
\varepsilon_x &= \frac{\sigma_x}{E_1} - \mu_1\frac{\sigma_y}{E_1} - \mu_2\frac{\sigma_z}{E_2} \\
\varepsilon_y &= -\mu_1\frac{\sigma_x}{E_1} + \frac{\sigma_y}{E_1} - \mu_2\frac{\sigma_z}{E_2} \\
\varepsilon_z &= -\mu_2\frac{\sigma_x}{E_2} - \mu_2\frac{\sigma_y}{E_2} + \frac{\sigma_z}{E_2} \\
\gamma_{xy} &= \frac{2(1+\mu_1)}{E_1}\tau_{xy} \\
\gamma_{yz} &= \frac{\tau_{yz}}{G_2}, \quad \gamma_{zx} = \frac{\tau_{zx}}{G_2}
\end{aligned}$$

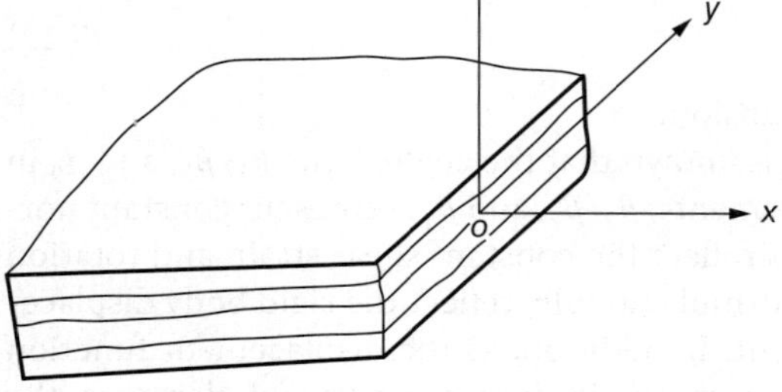

Figure 7.2 Transversely isotropic body.

Solving the six stress components by the above formula, we get the elasticity matrix of this type of layered material as follows:

$$[D] = \frac{E_2}{(1+\mu_1)p}\begin{bmatrix} n(1-\mu_2 n) & n(\mu_1+\mu_2 n) & n\mu_2(1+\mu_1) & 0 & 0 & 0 \\ n(\mu_1+\mu_2 n) & n(1-n\mu_2^2) & n\mu_2(1+\mu_1) & 0 & 0 & 0 \\ n\mu_2(1+\mu_1) & n\mu_2(1+\mu_1) & 1-\mu_1^2 & 0 & 0 & 0 \\ 0 & 0 & 0 & \frac{np}{2} & 0 & 0 \\ 0 & 0 & 0 & 0 & m(1+\mu_1)p & 0 \\ 0 & 0 & 0 & 0 & 0 & m(1+\mu_1)p \end{bmatrix} \tag{7.15}$$

where

$$n = \frac{E_1}{E_2}, \quad m = \frac{G_2}{E_2}, \quad p = 1 - \mu_1 - 2n\mu_2^2$$

As the strains are constant, in this type of element the stresses are also constant.

7.1.4 Stiffness Matrix of the Element

By substituting the matrix $[B]$ of expression (7.9) into formula (3.10), we can compute the element stiffness matrix $[K]$. As the elements of matrix $[B]$ are constants, the computing is easy. The computing formula is as follows:

$$[k] = [B]^{\mathrm{T}}[D][B]V$$

or

$$[k] = \begin{bmatrix} k_{ii} & -k_{ij} & k_{im} & -k_{ip} \\ -k_{ji} & k_{jj} & -k_{jm} & k_{jp} \\ k_{mi} & -k_{mj} & k_{mm} & -k_{mp} \\ -k_{pi} & k_{pj} & -k_{pm} & k_{pp} \end{bmatrix} \tag{7.16}$$

The submatrix $[k_{rs}]$ is calculated by the following formula:

$$[k_{rs}] = [B_r]^{\mathrm{T}}[D][B_s]V$$

For the isotropic body

$$[k_{rs}] = \frac{E(1-\mu)}{36(1+\mu)(1-2\mu)V} \times \begin{bmatrix} b_r b_s + g_2(c_r c_s + d_r d_s) & g_1 b_r c_s + g_2 c_r b_s & g_1 b_r d_s + g_2 d_r b_s \\ g_1 c_r b_s + g_2 b_r b_s & c_r c_s + g_2(b_r b_s + d_r d_s) & g_1 c_r d_s + g_2 d_r c_s \\ g_1 d_r b_s + g_2 b_r d_s & g_1 d_r c_s + g_2 c_r d_s & d_r d_s + g_2(b_r b_s + c_r d_s) \end{bmatrix} \quad (r, s = i, j, m, p) \tag{7.17}$$

where $g_1 = \mu/(1-\mu)$, $g_2 = (1-2\mu)/2(1-\mu)$.

For the transversely isotropic body

$$[k_{rs}] = \frac{E_2}{36(1+\mu_1)pV} \times \begin{bmatrix} g_1 b_r b_s + g_6 c_r c_s + g_7 d_r d_s & g_2 b_r c_s + g_6 c_r b_s & g_3 b_r d_s + g_7 d_r b_s \\ g_2 c_r b_s + g_6 b_r c_s & g_4 c_r c_s + g_6 b_r b_s + g_7 d_r d_s & g_3 c_r d_s + g_7 d_r c_s \\ g_3 d_r b_s + g_7 b_r d_s & g_3 d_r c_s + g_7 c_r d_s & g_5 d_r d_s + g_7 (c_r c_s + b_r b_s) \end{bmatrix} \quad (7.18)$$

where

$$g_1 = n(1-\mu_2 n), \quad g_2 = n(\mu_1 + \mu_2 n)$$
$$g_3 = n\mu_2(1+\mu_1), \quad g_4 = n(1-n\mu_2^2)$$
$$g_5 = 1-\mu_1^2, \quad g_6 = \frac{np}{2}, \quad g_7 = m(1+\mu_1)p$$

7.1.5 Nodal Load

Let the volume force be $\{q\} = \begin{bmatrix} q_x & q_y & q_z \end{bmatrix}^{\mathrm{T}} \{q\} = [q_x q_y q_z]^{\mathrm{T}}$, in which q_x, q_y, and q_z are constants. Using formula (3.14) the nodal load of point i can be obtained as follows:

$$\{P_i^{\mathrm{e}}\}_q = \begin{Bmatrix} X_i^{\mathrm{e}} \\ Y_i^{\mathrm{e}} \\ Z_i^{\mathrm{e}} \end{Bmatrix}_q = \frac{V}{4} \begin{Bmatrix} q_x \\ q_y \\ q_z \end{Bmatrix} \quad (i,j,m,p)$$

in which V is the volume of the element.

In other words, the volume forces in three directions are all assigned averagely onto the four nodes of the element.

If the element e is close to the boundary and one of its boundary surfaces like ijm bears the linear surface load,

$$\{p\} = \begin{bmatrix} p_x & p_y & p_z \end{bmatrix}^{\mathrm{T}}$$

The intensities on node i, j, and m are $\begin{bmatrix} p_x^i & p_y^i & p_z^i \end{bmatrix}^{\mathrm{T}}$, respectively. Using formula (3.15) the nodal load can be obtained as follows:

$$X_{ip}^{\mathrm{e}} = \frac{1}{6} A_{ijm} \left(p_x^i + \frac{1}{2} p_x^j + \frac{1}{2} p_x^m \right) \quad \begin{pmatrix} i,j,m \\ x,y,z \end{pmatrix}$$

in which A_{ijm} is the area of the boundary surface ijm.

The nodal load caused by the initial strain can be calculated by the following formula:

$$\{P\}_{\varepsilon_0}^{\mathrm{e}} = [B]^{\mathrm{T}}[D]\{\varepsilon_0\}V$$

For instance, let the average temperature of the element be $\overline{T} = (T_i + T_j + T_m + T_p)/4$, by the above formula the nodal load caused by temperature deformation can be given as follows:

$$\{P\}_{\varepsilon_0}^{\mathrm{e}} = \frac{E\alpha\overline{T}}{6(1-2\mu)} \begin{bmatrix} b_i & c_i & d_i & -b_j & -c_j & -d_j & b_m & c_m & d_m & -b_p & -c_p & -d_p \end{bmatrix}^{\mathrm{T}}$$

7.2 Volume Coordinates

For high-order tetrahedral elements for spatial problems, adoption of volume coordinates can simplify the calculation formula. As shown in Figure 7.3, in the tetrahedral element 1234, the position of any point P can be determined by the following ratio:

$$L_1 = \frac{V_1}{V}, \quad L_2 = \frac{V_2}{V}, \quad L_3 = \frac{V_3}{V}, \quad L_4 = \frac{V_4}{V} \tag{7.19}$$

$$V = \frac{1}{6}\begin{vmatrix} 1 & 1 & 1 & 1 \\ x_1 & x_2 & x_3 & x_4 \\ y_1 & y_2 & y_3 & y_4 \\ z_1 & z_2 & z_3 & z_4 \end{vmatrix} \tag{a}$$

in which V is volume of tetrahedron 1234; V_1, V_2, V_3, and V_4 are volumes of tetrahedron P234, P412, and P123, respectively; and L_1, L_2, L_3, and L_4 are volume coordinates of point P.

Since $V_1 + V_2 + V_3 + V_4 = V$,

$$L_1 + L_2 + L_3 + L_4 = 1 \tag{b}$$

The following relationship exists between orthogonal coordinates and volume coordinates:

$$\begin{Bmatrix} 1 \\ x \\ y \\ z \end{Bmatrix} = \begin{bmatrix} 1 & 1 & 1 & 1 \\ x_1 & x_2 & x_3 & x_4 \\ y_1 & y_2 & y_3 & y_4 \\ z_1 & z_2 & z_3 & z_4 \end{bmatrix} \begin{Bmatrix} L_1 \\ L_2 \\ L_3 \\ L_4 \end{Bmatrix} \tag{7.20}$$

By the inverse of the above formula, the volume coordinates can be represented by orthogonal coordinates as follows:

$$\begin{Bmatrix} L_1 \\ L_2 \\ L_3 \\ L_4 \end{Bmatrix} = \frac{1}{6V}\begin{bmatrix} V_1 & a_1 & b_1 & c_1 \\ V_2 & a_2 & b_2 & c_2 \\ V_3 & a_3 & b_3 & c_3 \\ V_4 & a_4 & b_4 & c_4 \end{bmatrix} \begin{Bmatrix} 1 \\ x \\ y \\ z \end{Bmatrix} \tag{c}$$

in which a_i, b_i, and c_i are projection areas of surface i (surface opposite to corner point i) on the x, y, and z planes, respectively.

In fact, a_i, b_i, and c_i are complement minors corresponding to coordinates x_i, y_i, and z_i of the matrix in formula (7.20), for example,

$$a_3 = -\begin{vmatrix} 1 & 1 & 1 \\ y_1 & y_2 & y_4 \\ z_1 & z_2 & z_4 \end{vmatrix}$$

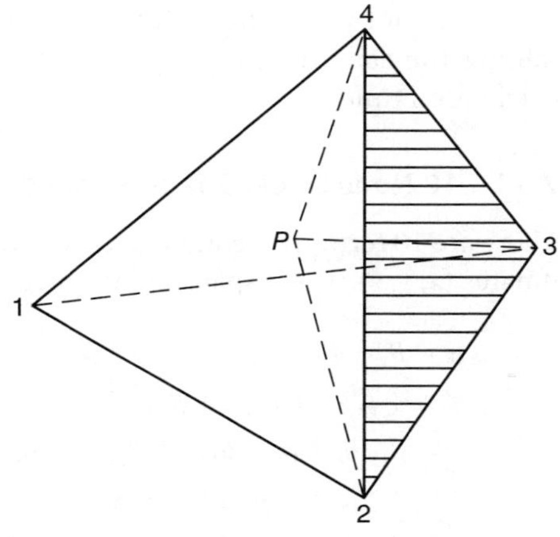

Figure 7.3 Volume coordinates.

When calculating the integral of the power function of volume coordinates on the tetrahedral element, the following formula can be used:

$$\iiint_V L_1^a L_2^b L_3^c L_4^d \,\mathrm{d}x\,\mathrm{d}y\,\mathrm{d}z = 6V \frac{a!b!c!d!}{(a+b+c+d+3)!} \tag{7.21}$$

To compute the derivative of the power function of volume coordinates with respect to orthogonal coordinates, the following formula may be used:

$$\left.\begin{aligned} \frac{\partial}{\partial x} &= \sum \frac{\partial L_i}{\partial x}\frac{\partial}{\partial L_i} = \frac{1}{6V}\left(a_1\frac{\partial}{\partial L_1} + a_2\frac{\partial}{\partial L_2} + a_3\frac{\partial}{\partial L_3} + a_4\frac{\partial}{\partial L_4}\right) \\ \frac{\partial}{\partial y} &= \sum \frac{\partial L_i}{\partial y}\frac{\partial}{\partial L_i} = \frac{1}{6V}\left(b_1\frac{\partial}{\partial L_1} + b_2\frac{\partial}{\partial L_2} + b_3\frac{\partial}{\partial L_3} + b_4\frac{\partial}{\partial L_4}\right) \\ \frac{\partial}{\partial z} &= \sum \frac{\partial L_i}{\partial z}\frac{\partial}{\partial L_i} = \frac{1}{6V}\left(c_1\frac{\partial}{\partial L_1} + c_2\frac{\partial}{\partial L_2} + c_3\frac{\partial}{\partial L_3} + c_4\frac{\partial}{\partial L_4}\right) \end{aligned}\right\} \tag{7.22}$$

For constant strain tetrahedral elements described in the preceding section, the shape function adopted can also be represented by volume coordinates as follows:

$$N_1 = L_1, \quad N_2 = L_2, \quad N_3 = L_3, \quad N_4 = L_4$$

7.3 High-Order Tetrahedral Elements

The stress field in practical engineering structure often changes rapidly with the coordinates. The stress components in the constant strain tetrahedral elements are all constants, which are difficult to adapt to stress fields undergoing sharp changes. To ensure a necessary precision, intensive computing grids must be adopted and leads to numerous nodes and colossal sets of equations. If we turn to the high-order displacement function, the stress in the element is changing; we can get necessary calculation precision with less elements and DOFs and smaller equation sets. Of course, the stiffness matrix of high-order element is more complex to form, requiring more time for calculation. But past experiences tell us that on the premise of maintaining the same calculation precision, higher-order elements will reduce the total calculation time.

7.3.1 10-Node Linear Strain Tetrahedral Elements

There are 10 terms in a complete quadratic polynomial represented by orthogonal coordinates (x, y, z). The displacement functions are taken as

$$\begin{aligned} u &= \beta_1 + \beta_2 x + \beta_3 y + \beta_4 z + \beta_5 x^2 + \beta_6 y^2 + \beta_7 z^2 \\ &\quad + \beta_8 xy + \beta_9 yz + \beta_{10} zx \\ v &= \beta_{11} + \beta_{12} x + \beta_{13} y + \beta_{14} z + \beta_{15} x^2 + \beta_{16} y^2 + \beta_{17} z^2 \\ &\quad + \beta_{18} xy + \beta_{19} yz + \beta_{20} zx \\ w &= \beta_{21} + \beta_{22} x + \beta_{23} y + \beta_{24} z + \beta_{25} x^2 + \beta_{26} y^2 + \beta_{27} z^2 \\ &\quad + \beta_{28} xy + \beta_{29} yz + \beta_{30} zx \end{aligned}$$

Calculating the derivatives by the above formula, we can get the strain components in the element as follows:

$$\varepsilon_x = \frac{\partial u}{\partial x} = \beta_2 + 2\beta_5 x + \beta_8 y + \beta_{10} z$$

$$\vdots$$

$$\gamma_{xy} = \frac{\partial u}{\partial y} + \frac{\partial v}{\partial x} = (\beta_3 + \beta_{12}) + (\beta_8 + 2\beta_{15})x + (2\beta_6 + \beta_{18})y + (\beta_9 + \beta_{20})z$$

$$\vdots$$

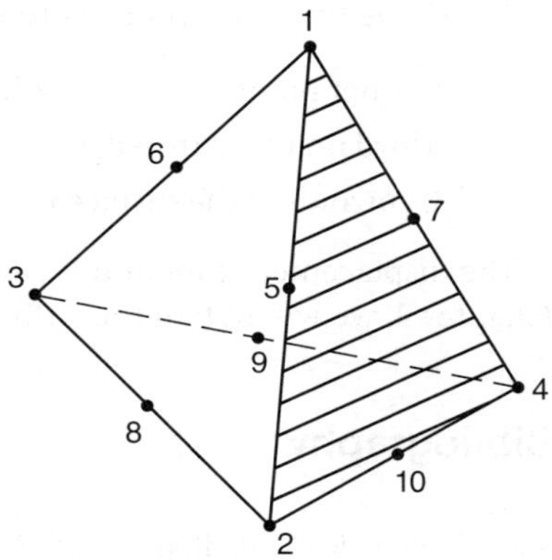

Figure 7.4 10-node linear strain tetrahedral elements.

It is clear that the strain components of the element are linear functions of coordinates. The displacement function includes 30 coefficients, which required 30 parameters to be determined. Now take 10 nodes, that is, four corner points and the midpoints of six edges as shown in Figure 7.4. For every node there are three displacement components as parameters that are sufficient to determine the coefficients in the displacement function.

For high-order elements, direct adoption of Cartesian orthogonal coordinates will cause a lot of calculations. So we turn to volume coordinates below, and the displacement functions of the element are taken as

$$u = \sum_{i=1}^{10} N_i u_i, \quad v = \sum_{i=1}^{10} N_i v_i,$$

$$w = \sum_{i=1}^{10} N_i w_i \tag{7.23}$$

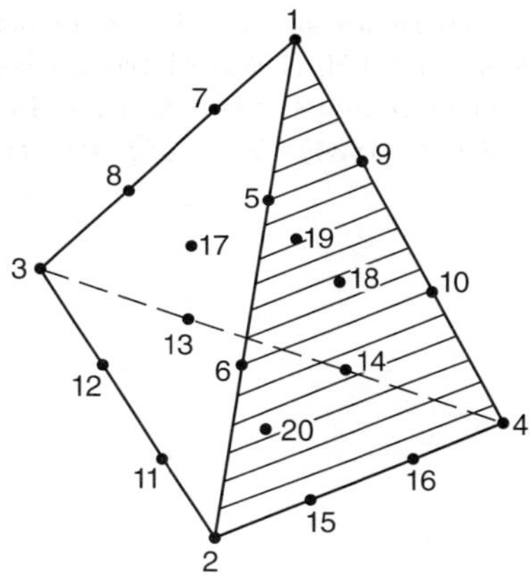

Figure 7.5 20-Node tetrahedron element.

in which u_i, v_i, and w_i are the displacement components of node i; N_i is a two-order shape function represented by volume coordinates as follows:

$$\text{Corner points}: N_1 = (2L_1 - 1)L_1 \quad (1, 2, 3, 4) \tag{7.24}$$

$$\text{Midpoints of the edge}: N_5 = 4L_1 L_2 \quad (5, 6, 7, 8, 9, 10)$$

Using the general formulas in Chapter 3, we can obtain the calculation formula for the element stiffness matrix and nodal loads.

7.3.2 20-Node Tetrahedral Element

In the orthogonal coordinate system, there are 20 terms in total in a complete cubic polynomial. There are 20 terms also in the displacement function of the tetrahedral element if adopting a complete cubic polynomial, so 20 nodes are needed. Now we take four corner points, third points of the six edges and centroids of four surfaces as shown in Figure 7.5. The displacement function is a cubic polynomial, and the elemental strain and stress is a quadratic function of the coordinates.

The shape functions represented by volume coordinates are

$$\left.\begin{array}{ll}\text{Corner points}: & N_1 = \frac{1}{2}(3L_1 - 1)(3L_1 - 2)L_1 \quad (1, 2, 3, 4) \\ \text{Third points of the edge}: & N_5 = \frac{9}{2}L_1 L_2 (3L_1 - 1) \quad (5, \ldots, 16) \\ \text{Centroids of the surface}: & N_{18} = 27L_1 L_2 L_3 \quad (17, 18, 19, 20)\end{array}\right\} \tag{7.25}$$

The displacement function is similar to formula (7.23). Using the general formulas in Chapter 3, we will obtain the stiffness matrix, nodal loads, and other formulas.

Bibliography

1 Gallagher, R.H., Padlog, J. and Bijlaard, P.P. (1962) Stress analysis of heated complex shapes. *ARS J.*, **32**, 700–707.
2 Melosh, R.J. (1963) Structural analysis of solids. *Proc. ASCE*, **89** (ST4), 205–223.
3 Argyris, J.H. (1965) Three dimensional anisotropic and inhomogeneous media-matrix analysis for small and large displacements. *Ingenieur Archiv.*, **34**, 33–55.
4 Argyris, J.H. (1965) Matrix analysis of three dimensional elastic media-small and large displacements. *AIAA J.*, **3** (1), 45–51.
5 Xu, Z. (1984) *Theory of Elasticity*, (vols 1 and 2), Higher Education Press, Beijing.

8

Shape Function, Coordinate Transformation, Isoparametric Element, and Infinite Element

The boundaries of elements provided in the previous chapters are all lines or planes. To fit more complex structural shapes, we have to increase the number of elements, which leads to more degrees of freedoms (DOFs) and more data and information required to be input, causing longer calculation time and more preparations in advance. To solve this problem, complex shape elements are helpful. This chapter will demonstrate how to use coordinate transformation to establish elements with complex shapes and the corresponding calculation methods.

8.1 Definition of Shape Functions

It can be seen from the previous chapters that when using the finite element method, after the element shape and corresponding shape function is determined, the remaining calculations may be done according to standard steps and general formulas, which are relatively simple. In the following sections, we will also see the geometrical shapes of some complex elements established by shape functions. Therefore shape functions are very important in the finite element method.

A shape function is a continuous function of coordinates defined within the element, which should meet the following conditions:

1)
$$\left.\begin{array}{ll}\text{On node } i: & N_i = 1 \\ \text{On other nodes}: & N_i = 0\end{array}\right\} \tag{8.1}$$

2) It can ensure the continuity of unknown quantities (u, v or x, y) defined by it between adjacent elements.
3) It should include linear terms, so that the element displacement defined by it can meet the condition of constant strain.
4) It should meet the following equations:

$$\sum N_i = 1 \tag{8.2}$$

so the element defined by it can reflect the movement of rigid body.

The Finite Element Method: Fundamentals and Applications in Civil, Hydraulic, Mechanical and Aeronautical Engineering, First Edition. Bofang Zhu.

Companion website: www.wiley.com/go/zhufem17

Now we describe the necessity of the previous equations. Define the element displacement with shape functions as follows:

$$u = \sum N_i u_i, \quad v = \sum N_i v_i, \quad w = \sum N_i w_i \tag{a}$$

in which u_i, v_i, w_i are displacements of node i.

Let the element take rigid-body movement along the horizontal (or other) direction $u = 1$, and then all nodes and any point within the element will take unit movements along the horizontal direction, namely,

$$u = \sum N_i \times 1 = 1$$

so we have formula (8.2).

The higher the order of the shape function is, the more complicated is the shape of the element and the less elements are needed for a problem, leading to lower-order equation set and less time for solving. But once the shape function is higher ordered, calculations to establish the stiffness matrix become more complex. So for every specific problem, there is a shape function with the most suitable order that makes the total calculation time most economic. This is always determined by calculation experiences.

8.2 One-Dimensional Shape Functions

As one-dimensional elements have accurate solutions, Chapter 1 did not adopt the general finite element method for analysis. In practical engineering, one-dimensional elements always appear simultaneously with two- or three-dimensional (3D) elements in a continuous medium. For example, reinforcing steel bars in the reinforced concrete structure, thin lining of shafts, edge stiffener rings of openings, flanges of flanged beams, and so on may all be deemed as one-dimensional elements, which are working together with other two-dimensional or 3D elements. In this case, we should apply a uniform method for analysis.

As shown in Figure 8.1, now we use coordinate ξ and the element adopted is a straight line: $-1 \le \xi \le +1$. Later by coordinate transformation, we can get curved elements of different lengths and shapes in the global coordinate system (x, y, z). The current coordinate ξ is local coordinate, also called natural coordinates. The elements shown in Figure 8.1 are called parent elements, and the curved elements obtained by coordinate transformation in the global coordinate system are called child elements. Shape functions are defined by natural coordinates in parent elements. One-dimensional shape functions are as follows (Figure 8.1):

1) Linear element (2-node)

$$N_1 = \frac{1-\xi}{2}, \quad N_2 = \frac{1+\xi}{2} \tag{8.3}$$

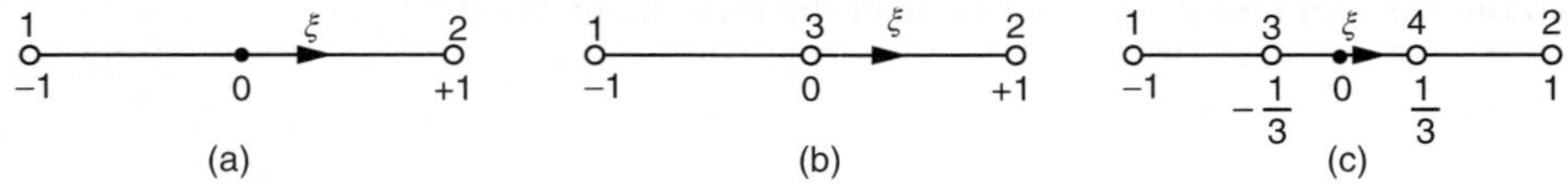

Figure 8.1 One-dimensional parent element. (a) Linear element, (b) quadratic element, and (c) cubic element.

2) Quadratic element (3-node)

$$N_1 = -\frac{(1-\xi)\xi}{2}, \quad N_2 = \frac{(1+\xi)\xi}{2}, \quad N_3 = 1-\xi^2 \tag{8.4}$$

3) Cubic element (4-node)

$$\left.\begin{aligned} N_1 &= \frac{(1-\xi)(9\xi^2-1)}{16}, \quad N_2 = \frac{(1+\xi)(9\xi^2-1)}{16} \\ N_3 &= \frac{9(1-\xi^2)(1-3\xi)}{16}, \quad N_4 = \frac{9(1-\xi^2)(1+3\xi)}{16} \end{aligned}\right\} \tag{8.5}$$

By the Lagrange polynomial, a one-dimensional shape function can be expressed in the following uniform form:

$$N_i^n = \frac{(\xi-\xi_1)(\xi-\xi_2)\cdots(\xi-\xi_{i-1})(\xi-\xi_{i+1})\cdots(\xi-\xi_n)}{(\xi_i-\xi_1)(\xi_i-\xi_2)\cdots(\xi_i-\xi_{i-1})(\xi_i-\xi_{i+1})\cdots(\xi_i-\xi_n)} \tag{a}$$

It is not difficult to verify that the previous shape functions satisfy formula (8.1) and formula (8.2).

8.3 Two-Dimensional Shape Function

A two-dimensional parent element is the 2×2 square in (ξ,η) plane:

$$-1 \le \xi \le +1, \quad -1 \le \eta \le +1$$

As shown in Figure 8.2, the origin of coordinates is placed on the centroid of the element. The element boundaries are four straight lines: $\xi = \pm1, \eta = \pm1$. The number of nodes should correspond to the order of the shape function to ensure the continuity of unknown quantities defined by shape functions between adjacent elements. Therefore, for a linear, quadratic, and cubic shape function, there are, respectively, 2, 3, and 4 nodes on every edge of the element. In addition to four corner points, other nodes are placed on the bisection or trisection points of every edge as shown in Figure 8.2. Two-dimensional shape functions are as follows:

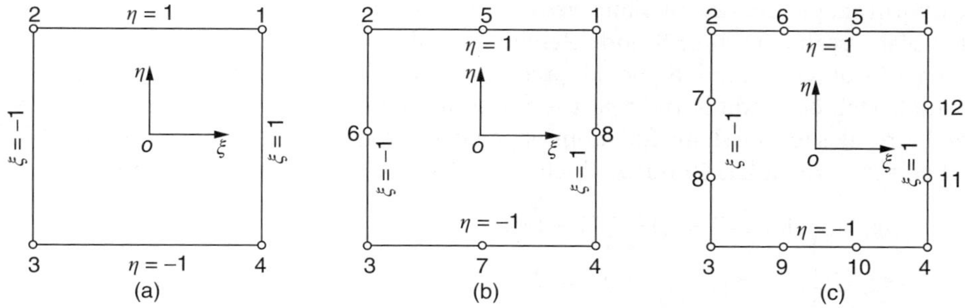

Figure 8.2 Two-dimensional parent element. (a) Linear element, (b) quadratic element, and (c) cubic element.

1) Linear element (4-node)

$$\left.\begin{aligned} N_1 &= \frac{(1+\xi)(1+\eta)}{4}, \quad N_2 = \frac{(1-\xi)(1+\eta)}{4} \\ N_3 &= \frac{(1-\xi)(1-\eta)}{4}, \quad N_4 = \frac{(1+\xi)(1-\eta)}{4} \end{aligned}\right\} \tag{8.6}$$

Quoting new variables,

$$\xi_0 = \xi_i\xi, \quad \eta_0 = \eta_i\eta \tag{8.7}$$

in which ξ_i, η_i is the coordinate of point.

Thus the shape function shown in formula (8.6) can be represented in a combined way as

$$N_i = \frac{(1+\xi_0)(1+\eta_0)}{4} \quad (i = 1,2,3,4) \tag{8.8}$$

Obviously, on the four boundaries of the element, the shape function is linear.

2) Quadratic element (8-node)

$$\left.\begin{aligned} &\text{Corner points:} && N_i = \frac{1}{4}(1+\xi_0)(1+\eta_0)(\xi_0+\eta_0-1) \quad (i = 1,2,3,4) \\ &\text{Midpoints of the edge:} && N_i = \frac{1}{2}(1-\xi^2)(1+\eta_0) \quad (i = 5,7) \\ & && N_i = \frac{1}{2}(1-\eta^2)(1+\xi_0) \quad (i = 6,8) \end{aligned}\right\} \tag{8.9}$$

On each edge of the element, the shape function is a quadratic function.

3) Cubic element (12-node)

$$\left.\begin{aligned} &\text{Corner points:} && N_i = \frac{1}{32}(1+\xi_0)(1+\eta_0)[9(\xi^2+\eta^2)-10] \ (i = 1,2,3,4) \\ &\text{Third points of the edge:} && N_i = \frac{9}{32}(1+\xi_0)(1-\eta^2)(1+9\eta_0) \ (i = 7,8,11,12) \\ & && N_i = \frac{9}{32}(1+\eta_0)(1-\xi^2)(1+9\xi_0) \ (i = 5,6,9,10) \end{aligned}\right\} \tag{8.10}$$

On each edge of the element, the shape function is a cubic shape function.

4) Uniform shape function of plane elements with 4–9 nodes

For plane problems, 4- or 8-node elements are always preferred. But sometimes elements in other forms may be adopted. For example, in Figure 8.3(a), the left is a 5-node element and the right part is a 8-node element. Table 8.1 presents the general formulas for the shape functions of plane elements with 4–9 nodes. For example, when the element has 5 nodes, which are 1, 2, 3, 4, and 8, the shape functions are

$$\begin{aligned} N_1 &= \frac{1}{4}(1+\xi)(1+\eta) - \frac{1}{4}(1+\xi)(1-\eta^2), \quad N_2 = \frac{1}{4}(1-\xi)(1+\eta) \\ N_3 &= \frac{1}{4}(1-\xi)(1-\eta), \quad N_4 = \frac{1}{4}(1+\xi)(1-\eta) - \frac{1}{4}(1+\xi)(1-\eta^2) \\ N_5 &= \frac{1}{2}(1+\xi)(1-\eta^2) \end{aligned}$$

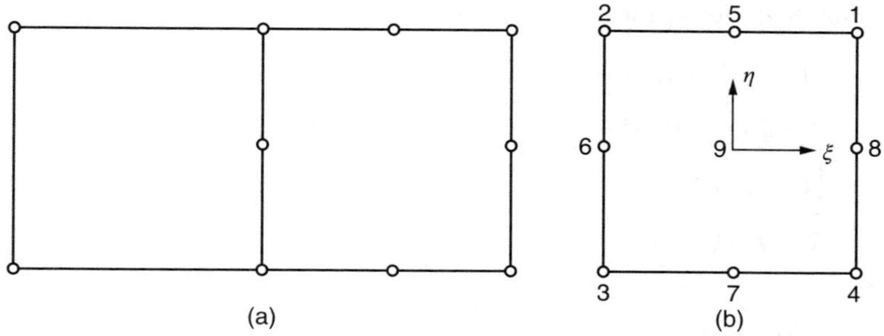

Figure 8.3 Plane element.

Table 8.1 General shape functions of 4–9-node plane elements (refer to Figure 8.3b).

N_i	Shape function	The following terms are included only when there is node i				
		$i = 5$	$i = 6$	$i = 7$	$i = 8$	$i = 9$
N_1	$\frac{1}{4}(1+\xi)(1+\eta)$	$-\frac{1}{2}N_5$			$-\frac{1}{2}N_8$	$-\frac{1}{4}N_9$
N_2	$\frac{1}{4}(1-\xi)(1+\eta)$	$-\frac{1}{2}N_5$	$-\frac{1}{2}N_6$			$-\frac{1}{4}N_9$
N_3	$\frac{1}{4}(1-\xi)(1-\eta)$		$-\frac{1}{2}N_6$	$-\frac{1}{2}N_7$		$-\frac{1}{4}N_9$
N_4	$\frac{1}{4}(1+\xi)(1-\eta)$			$-\frac{1}{2}N_7$	$-\frac{1}{2}N_8$	$-\frac{1}{4}N_9$
N_5	$\frac{1}{2}(1-\xi^2)(1+\eta)$					$-\frac{1}{2}N_9$
N_6	$\frac{1}{2}(1-\xi)(1-\eta^2)$					$-\frac{1}{2}N_9$
N_7	$\frac{1}{2}(1-\xi^2)(1-\eta)$					$-\frac{1}{2}N_9$
N_8	$\frac{1}{2}(1+\xi)(1-\eta^2)$					$-\frac{1}{2}N_9$
N_9	$(1-\xi^2)(1-\eta^2)$					

It is easy to verify that all the previous types of shape functions meet the four requirements on shape functions as described in Section 8.1. Now take the quadratic shape function represented by formula (8.9), for example, to offer explanation.

On node i, $\xi_0 = \eta_0 = 1$, so $N_i = 1$, and on other nodes, $N_i = 0$, meeting the condition (1).

On the four edges of the element, the shape function is quadratic, and every edge has 3 nodes sufficient to ensure the continuity of unknown quantities defined by shape functions between adjacent elements, thus meeting the condition (2).

By expanding formula (8.9), we can see it contains the linear terms ξ and η. The linear combination of these functions can fully reflect any linear changes of unknown quantities defined by shape functions, thus meeting the condition (3).

Substitute formula (8.9) into formula (8.2) to obtain

$$\begin{aligned}\sum N_i &= \frac{1}{4}\{(1-\xi)(1-\eta)(-\xi-\eta-1)+(1+\xi)(1-\eta)(\xi-\eta-1)\\&\quad+(1-\xi)(1+\eta)(-\xi+\eta-1)+(1+\xi)(1+\eta)(\xi+\eta-1)\\&\quad+\frac{1}{2}(1-\xi^2)(1-\eta)+(1-\xi^2)(1+\eta)(1-\eta^2)(1-\xi)\\&\quad+(1-\eta^2)(1+\xi)\}\\&=1\end{aligned}$$

Obviously it meets the condition (4).

5) Shape function of triangular elements
   The shape function of triangular elements is represented by area coordinates. For example, the shape function of a 3-node triangular element can be represented as follows:

$$N_i = L_i \quad (i = 1,2,3)$$

   in which L_i is the area coordinate.
   Shape functions of various plane elements are provided in Table 8.2.

8.4 Three-Dimensional Shape Function

A 3D parent element is the $2\times2\times2$ regular hexahedron in the (ξ,η,ζ) coordinate system, in which

$$-1 \le \xi \le +1, \quad -1 \le \eta \le +1, \quad -1 \le \zeta \le +1$$

As shown in Figure 8.4, the origin of coordinates is placed on the centroid of the element, and the element boundaries are six planes: $\xi = \pm1,\ \ \eta = \pm1,\ \ \zeta = \pm1$. The elemental nodes are placed on corner points and equidistant points of the edges. 3D shape functions are as follows:

1) Linear element (8-node)

$$N_i = \frac{1}{8}(1+\xi_0)(1+\eta_0)(1+\zeta_0) \tag{8.11}$$

2) Quadratic element (20-node)

$$\left.\begin{aligned}&\text{Corner points}: && N_i = \frac{1}{8}(1+\xi_0)(1+\eta_0)(1+\xi_0)(\xi_0+\eta_0+\xi_0-2)\\&\text{Typical mid points of the edge}: && \xi_i = 0,\quad \eta_i = \pm1,\quad \xi_i = \pm1\\& && N_i = \frac{1}{4}(1-\xi^2)(1+\eta_0)(1+\xi_0)\end{aligned}\right\} \tag{8.12}$$

Table 8.2 Shape functions of various plane elements ($\xi_0 = \xi_i \xi, \eta_0 = \eta_i \eta$).

Name of the element	Shape of the element (parent element)	Degree of freedom	Shape function	Features of the element
3-Node triangular planar element	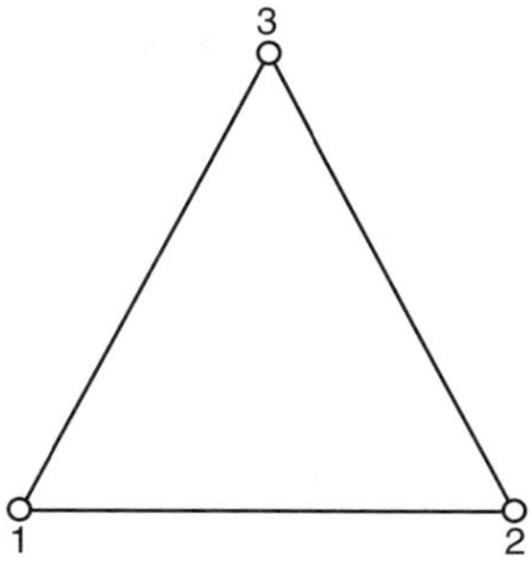	u, v	$N_1 = L_1$ $N_2 = L_2$ $N_3 = L_3$ (L_i is area coordinate)	The element is easy to calculate and adaptable to complex geometric shapes with sharp corners
6-Node triangular planar element	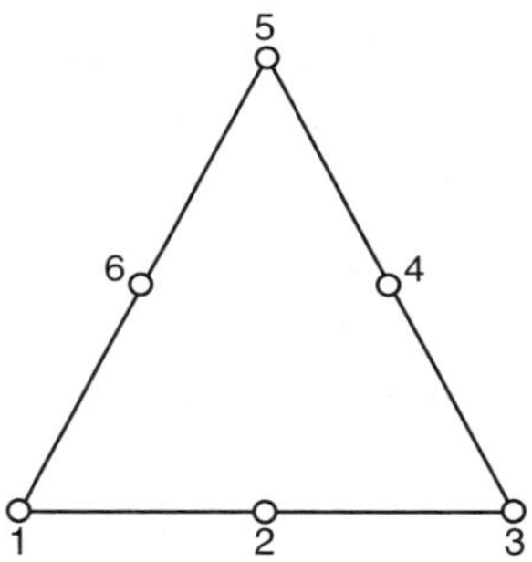	u, v	$N_1 = (2L_1 - 1)L_1,$ $N_2 = 4L_1L_2$ $N_3 = (2L_2 - 1)L_2,$ $N_4 = 4L_2L_3$ $N_5 = (2L_3 - 1)L_3,$ $N_6 = 4L_3L_1$	The boundary is flexible and adaptable to complex geometric shapes with sharp corners
10-Node triangular planar element	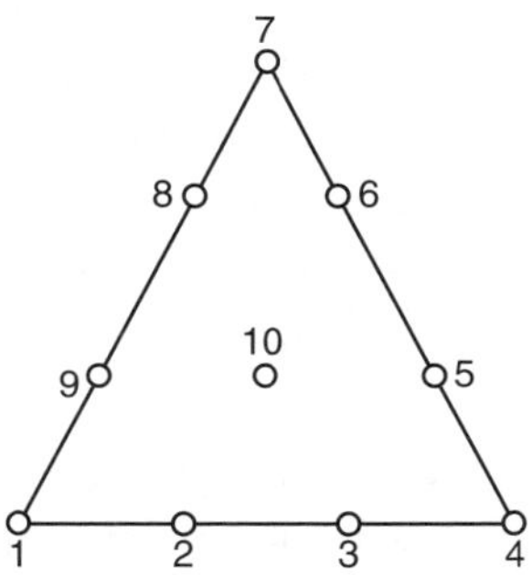	u, v	$N_1 = \frac{1}{2}(3L_1 - 1)(3L_1 - 2)L_1$ $N_2 = \frac{9}{2}L_1L_2(3L_1 - 1)$ $N_3 = \frac{9}{2}L_1L_2(3L_2 - 1)$ $N_4 = \frac{1}{2}(3L_2 - 1)(3L_2 - 2)L_4$ $N_5 = \frac{9}{2}L_2L_3(3L_1 - 1)$ $N_6 = \frac{9}{2}L_3L_2(3L_3 - 1)$ $N_7 = \frac{1}{2}(3L_3 - 1)(3L_3 - 2)L_2$ $N_8 = \frac{9}{2}L_3L_1(3L_1 - 1)$ $N_9 = \frac{9}{2}L_3L_1(3L_1 - 1)$ $N_{10} = 27L_1L_2L_3$	The boundary is flexible and adaptable to complex geometric shapes with sharp corners

(Continued)

Table 8.2 (Continued)

Name of the element	Shape of the element (parent element)	Degree of freedom	Shape function	Features of the element
4-Node planar isoparametric element	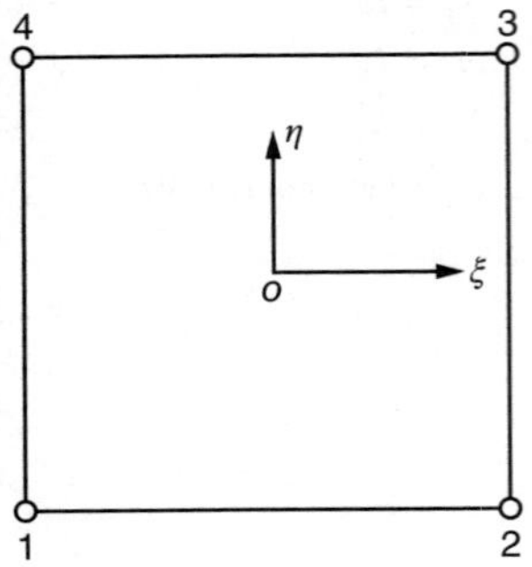	u, v	$N_i = \frac{1}{4}(1+\xi_0)(1+\eta_0)$ $(i = 1,2,3,4)$	Easy to calculate
8-Node planar isoparametric element	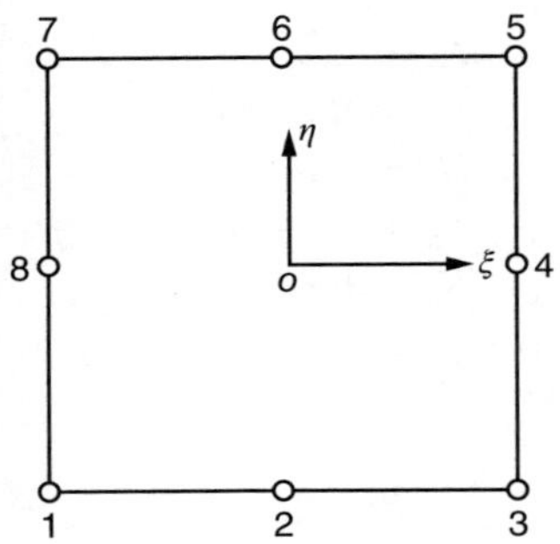	u, v	$N_i = \frac{1}{4}(1+\xi_0)(1+\eta_0) \times (\xi_0 - \eta_0 - 1)$ $(i = 1,3,5,7)$ $N_i = \frac{1}{2}(1-\xi^2)(1+\eta_0)$ $(i = 2,6)$ $N_i = \frac{1}{2}(1-\eta^2)(1+\xi_0)$ $(i = 4,8)$	The boundary is flexible, and simulation of cracks is possible by moving the edge midpoint to the 1/4 position
12-Node planar isoparametric element	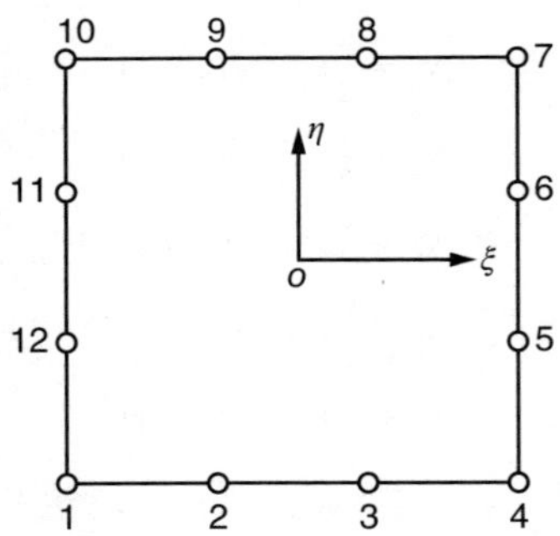	u, v	$N_i = \frac{1}{32}(1+\xi_0)(1+\eta_0) \times [-10 + 9(\xi^2+\eta^2)]$ $(i = 1,4,7,10)$ $N_i = \frac{9}{32}(1+\xi_0)(1+\eta^2) \times (1+9\eta_0)$ $(i = 5,6,11,12)$ $N_i = \frac{9}{32}(1+\eta_0)(1-\xi^2) \times (1+9\xi_0)$ $(i = 2,3,8,9)$	The boundary is flexible and can be used for fracture analysis

Table 8.2 (Continued)

Name of the element	**Shape of the element (parent element)**	**Degree of freedom**	**Shape function**	**Features of the element**
6-Node planar isoparametric element	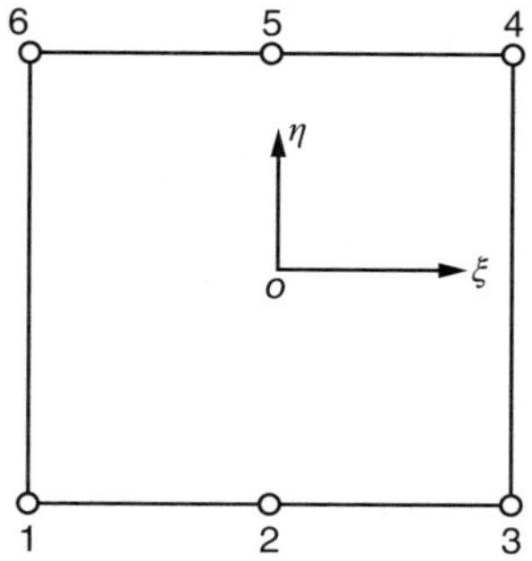	u, v	$N_i = \frac{1}{4}(1+\xi_0)(1+\eta_0)$ $(i = 1,3,4,6)$ $N_i = \frac{1}{2}(1+\xi^2)(1+\eta_0)$ $(i = 2,5)$	The upper and lower boundaries are flexible and applicable to transition between linear elements and quadratic elements
8-Node planar isoparametric element	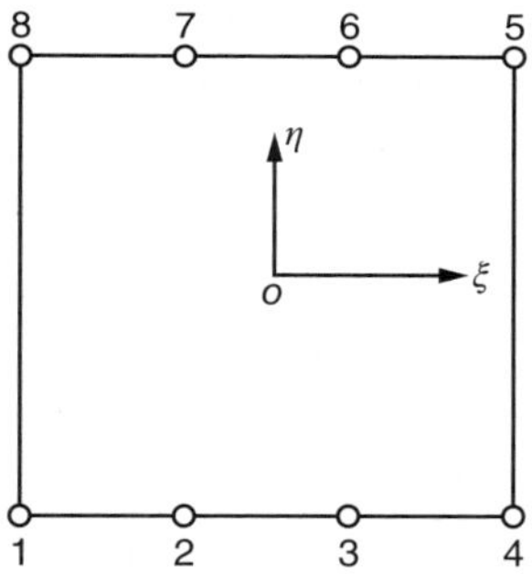	u, v	$N_i = \frac{1}{32}(1+\xi_0)(1+\eta_0) \times (-1+9\xi^2)$ $(i = 1,4,5,8)$ $N_i = \frac{9}{32}(1-\xi^2)(1+9\xi_0) \times (1+\eta_0)$ $(i = 2,3,6,7)$	The upper and lower boundaries are flexible and applicable to transition between linear elements and cubic elements
7-Node plane transition element	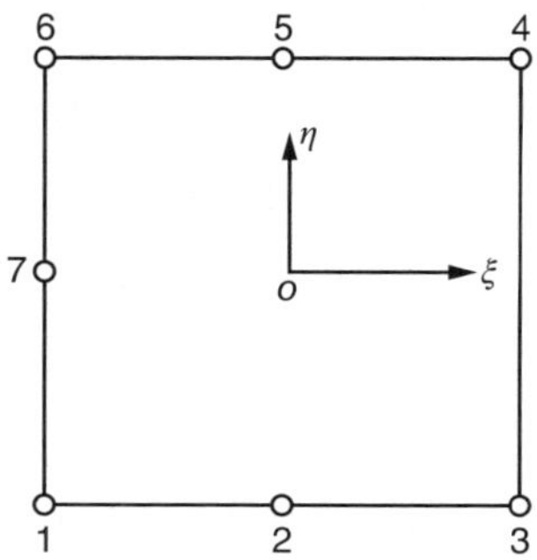	u, v	$N_1 = \frac{1}{4}(1-\xi)(1-\eta) \times (1+\xi+\eta)$ $N_2 = \frac{1}{2}(1-\eta)(1-\xi^2)$ $N_3 = \frac{1}{4}\xi(1+\xi)(1-\eta)$ $N_4 = \frac{1}{4}\xi(1+\xi)(1+\eta)$ $N_5 = \frac{1}{2}(1+\eta)(1-\xi^2)$ $N_6 = -\frac{1}{4}(1-\xi)(1+\eta) \times (1+\xi-\eta)$ $N_7 = \frac{1}{2}(1-\xi)(1-\eta^2)$	The three edges are flexible to be applied in transition between linear elements and quadratic elements

(Continued)

Table 8.2 (Continued)

Name of the element	Shape of the element (parent element)	Degree of freedom	Shape function	Features of the element
9-Node plane transition element	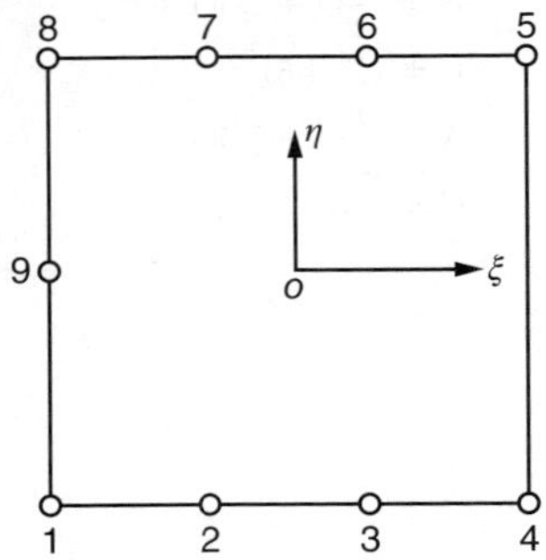	u, v	$N_i = \frac{1}{32}(1+\xi_0)$ $\times[-9(1-\xi^2)+\eta_0$ $(1-9\xi^2)+8\eta^2]$ $(i = 1,8)$ $N_i = \frac{1}{32}(1+\xi_0)(-1+9\xi^2)$ $\times(1+\eta_0)\ \ (i = 4,5)$ $N_i = \frac{9}{32}(1-\xi^2)(1+9\xi_0)$ $\times(1+\eta_0)\ \ (i = 2,3,6,7)$ $N_9 = \frac{1}{2}(1-\xi)(1-\eta^2)$	The three edges are flexible to be applied in transition between several types of elements
10-Node planar transition element	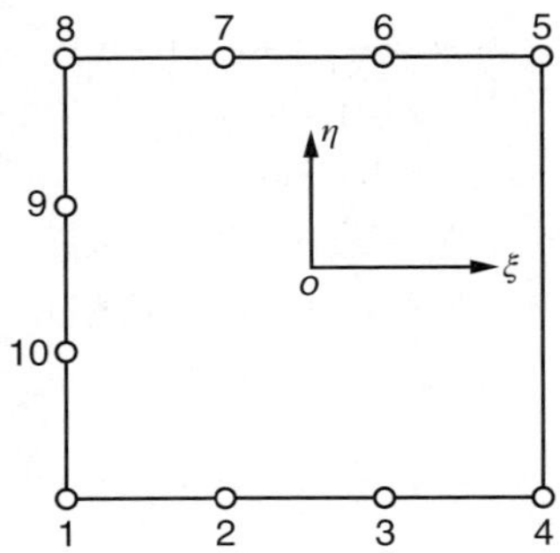	u, v	$N_i = \frac{1}{32}(1+\xi_0)(1+\eta_0)$ $\times[-10+9(\xi^2+\eta^2)]$ $(i = 1,8)$ $N_i = \frac{1}{32}(1+\xi_0)(1+\eta_0)$ $\times(-1+9\xi^2)\ \ (i = 4,5)$ $N_i = \frac{9}{32}(1-\xi^2)(1+9\xi_0)$ $\times(1+\eta_0)\ \ (i = 2,3,6,7)$ $N_i = \frac{9}{32}(1+\xi_0)(1-\eta^2)$ $\times(1+\eta_0)\ \ (i = 9,10)$	The three edges are flexible to be applied in transition between several types of elements
6-Node plane transition element	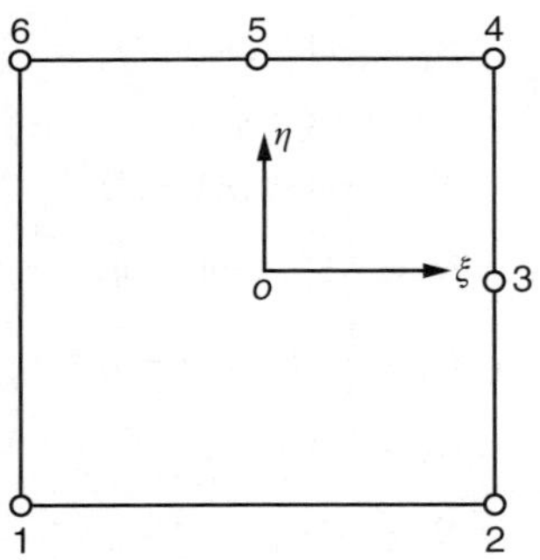	u, v	$N_1 = \frac{1}{4}(1-\xi)(1-\eta)$ $N_2 = -\frac{1}{4}\eta(1+\xi)(1-\eta)$ $N_3 = \frac{1}{2}(1+\xi)(1-\eta^2)$ $N_4 = \frac{1}{4}(1+\xi)(1+\eta)$ $\times(\xi+\eta-1)$ $N_5 = \frac{1}{2}(1+\eta)(1-\xi^2)$ $N_6 = -\frac{1}{4}\xi(1-\xi)(1+\eta)$	The three edges are flexible to be applied in transition between two types of elements

Table 8.2 (Continued)

Name of the element	Shape of the element (parent element)	Degree of freedom	Shape function	Features of the element
9-Node plane isoparametric element	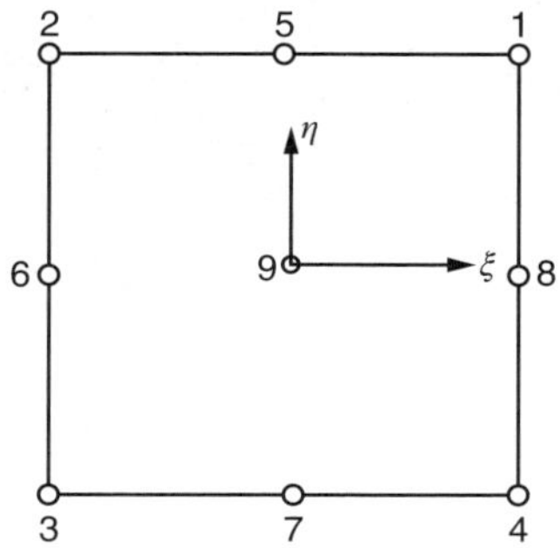	u, v	$N_i = \frac{1}{4}(1+\xi_0)(1+\eta_0)$ $(\xi_0+\eta_0-1)$ $+\frac{1}{4}(1-\xi^2)(1-\eta^2)$ $(i = 1,2,3,4)$ $N_i = \frac{1}{2}(1-\xi^2)(1-\eta^2)$ $-\frac{1}{2}(1-\xi^2)(1-\eta^2)$ $(i = 5,7)$ $N_i = \frac{1}{2}(1-\eta^2)(1+\xi_0)$ $-\frac{1}{2}(1-\xi^2)(1-\eta^2)$ $(i = 6,8)$ $N_9 = \frac{1}{2}(1-\xi^2)(1-\eta^2)$	Generally used for analysis of plates and shells

3) Cubic element (32-node)

$$\left.\begin{array}{ll} \text{Corner points :} & N_i=\frac{1}{64}(1+\xi_0)(1+\eta_0)(1+\zeta_0)[9(\xi^2+\eta^2+\zeta^2)-19] \\ \text{Typical third points of the edge :} & \xi_i = \pm\frac{1}{3}, \quad \eta_i = \pm 1, \quad \zeta_i = \pm 1 \\ & N_i = \frac{9}{64}(1-\xi^2)(1+9\xi_0)(1+\eta_0)(1+\zeta_0) \end{array}\right\} \tag{8.13}$$

4) Universal shape function of spatial elements (8–20 nodes)
For space problems, 8- or 20-node elements are always adopted. But sometimes elements in other forms may be adopted. In the following we will give the universal shape functions of 8–20-node spatial elements (refer to Figure 8.4(b)):

$$N_1 = g_1 - \frac{1}{2}(g_9 + g_{12} + g_{17}), \quad N_2 = g_2 - \frac{1}{2}(g_9 + g_{10} + g_{18})$$

$$N_3 = g_3 - \frac{1}{2}(g_{10} + g_{11} + g_{19}), \quad N_4 = g_4 - \frac{1}{2}(g_{11} + g_{12} + g_{20})$$

$$N_5 = g_5 - \frac{1}{2}(g_{13} + g_{16} + g_{17}), \quad N_6 = g_6 - \frac{1}{2}(g_{13} + g_{14} + g_{18})$$

$$N_7 = g_7 - \frac{1}{2}(g_{14} + g_{16} + g_{19}), \quad N_8 = g_8 - \frac{1}{2}(g_{15} + g_{16} + g_{20})$$

$$N_j = g_j \quad (j = 9,10,\ldots,20)$$

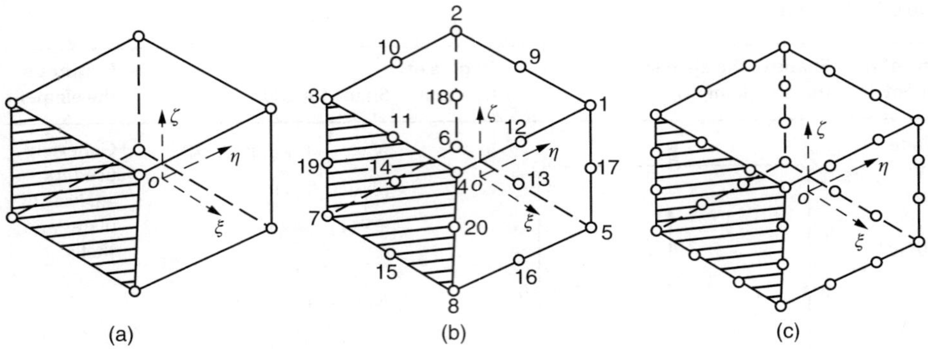

Figure 8.4 Three-dimensional parent element. (a) Linear element (8 nodes), (b) quadratic element (20 nodes), and (c) cubic element (32 nodes).

If there is no node i, $g_i = 0$; otherwise

$$\left.\begin{aligned} g_i &= f(\xi,\xi_i)f(\eta,\eta_1)f(\zeta,\zeta_i) \\ f(t,t_i) &= \frac{1}{2}(1+t_it) \quad (t_i = \pm 1; t = \xi,\eta,\zeta) \\ f(t,t_i) &= (1-t^2) \quad (t_i = 0; t = \xi,\eta,\zeta) \end{aligned}\right\} \tag{8.14}$$

5) Tetrahedral elements
   The shape function of tetrahedral element is represented by volume coordinates. For example, the shape function of 4-node polyhedral element can be represented as follows:

   $$N_i = L_i(i = 1,2,3,4)$$

   in which L_i is the volume coordinate.
6) Triangular prism element
   In order to represent the shape function of triangular prism element, coordinates ζ is used in the direction parallel to the seamed edge, and area coordinate L_i is used within the triangular plane normal to the seamed edge, as detailed in Table 8.3.

Table 8.3 presents shape functions of various space elements.

8.5 Coordinate Transformation

The parent elements described previously are simple and regular in the geometric shapes and easy to calculate but difficult to be used for various complex structural shapes encountered in practical engineering. To solve this issue, coordinate transformation can be used to transform the geometrically simple parent elements in the (ξ, η, ζ) coordinate system into geometrically complex elements with curved boundaries. Elements after transformation are called child elements and can adapt to complex shapes of various actual structures. After this treatment, the element acquires dual natures: on one hand, the geometric features and loads of child elements all originate

Table 8.3 Shape functions of various space elements ($\xi_0 = \xi_i\xi, \eta_0 = \eta_i\eta, S_0 = \zeta_i\zeta$).

Name of the element	Shape of the element (parent element)	Degree of freedom	Shape function	Features of the element
8-Node space isoparametric element	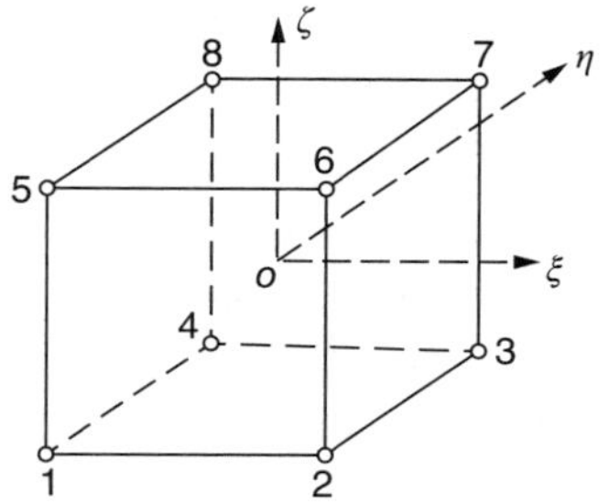	u, v, w	$N_i = \frac{1}{8}(1+\xi_0)(1+\eta_0)(1+\zeta_0)$ $(i = 1, 2, \ldots, 8)$	Easy to calculate, poor adaptability to complex shapes
20-Node space isoparametric element	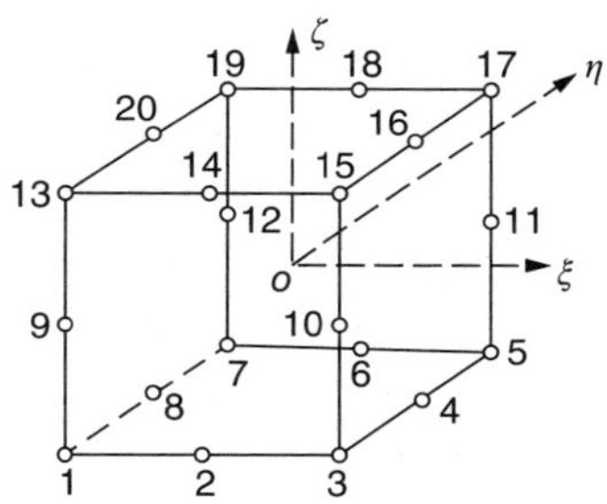	u, v, w	$N_i = \frac{1}{8}(1+\xi_0)(1+\eta_0) \times (1+\zeta_0)(\xi_0+\eta_0+\xi_0-2)$ $(i = 1,3,5,7,13,15,17,19)$ $N_i = \frac{1}{4}(1-\xi^2)(1+\eta_0) \times (1+\xi_0)$ $(i = 2,6,14,18)$ $N_i = \frac{1}{4}(1-\eta^2)(1+\xi_0) \times (1+\zeta_0)$ $(i = 4,8,16,20)$ $N_i = \frac{1}{4}(1-\zeta^2)(1+\xi_0) \times (1+\eta_0)$ $(i = 9,10,11,12)$	The boundary can be a curved surface; the element is applicable to fracture analysis
32-Node space isoparametric element	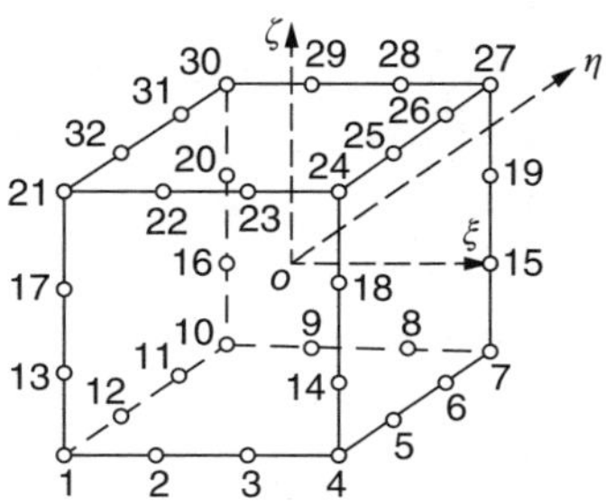	u, v, w	$N_i = \frac{1}{64}(1+\xi_0)(1+\eta_0) \times (1+\xi_0) \times [9(\xi^2+\eta^2+\zeta^2)-19]$ $(i = 1,4,7,10,21,24,27,30)$ $N_i = \frac{9}{64}(1-\xi^2)(1+9\xi_0) \times (1+\eta_0)(1+\xi_0)$ $(i = 2,3,8,9,22,23,28,29)$ $N_i = \frac{9}{64}(1-\eta^2)(1+9\eta_0) \times (1+\xi_0)(1+\zeta_0)$ $(i = 5,6,11,12,25,26,31,32)$ $N_i = \frac{9}{64}(1-\zeta^2)(1+9\zeta_0) \times (1+\xi_0)(1+\eta_0)$ $(i = 13,14,15,16,17,18,19,20)$	The boundary can be a curved surface; the element is applicable to fracture analysis

(Continued)

Table 8.3 (Continued)

Name of the element	Shape of the element (parent element)	Degree of freedom	Shape function	Features of the element
16-Node space isoparametric element	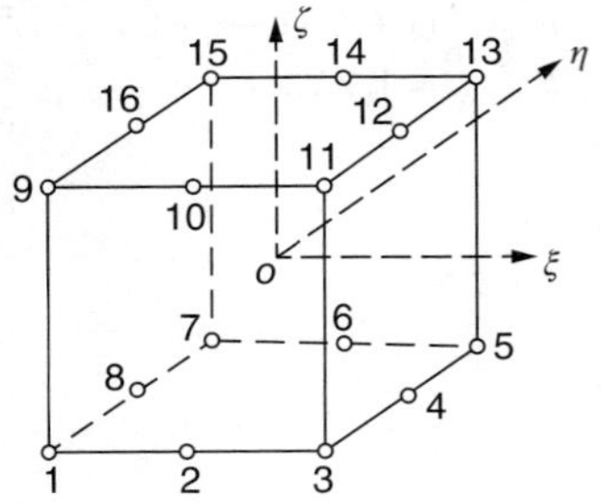	u, v, w	$N_i = \frac{1}{8}(1+\xi_0)(1+\eta_0)$ $(1+\zeta_0)\times(\xi_0+\eta_0-1)$ $N_i = \frac{1}{8}(1+\xi_0)(1+\eta_0)$ $(1+\zeta_0)\times(\xi_0+\eta_0-1)$ $(i = 1,3,5,7,9,11,13,15)$ $N_i = \frac{1}{4}(1-\xi^2)(1+\eta_0)$ $(1+\zeta_0)$ $(i = 2,6,10,14)$ $N_i = \frac{1}{4}(1-\eta^2)(1+\xi_0)$ $(1+\zeta_0)$ $(i = 4,8,12,16)$	Upper and lower surfaces can be curved
24-Node space isoparametric element	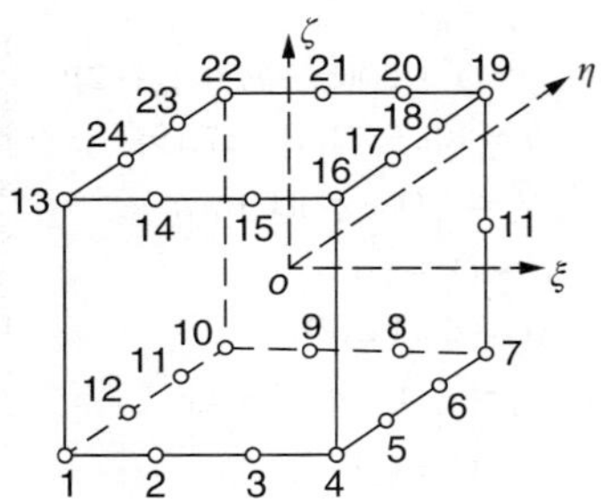	u, v, w	$N_i = \frac{1}{64}(1+\xi_0)(1+\eta_0)$ $(1+\zeta_0)\times[-10+9(\xi^2+\eta^2)]$ $N_i = \frac{1}{64}(1+\xi_0)(1+\eta_0)$ $(1+\xi_0)\times[-10+9(\xi^2+\eta^2)]$ $(i = 1,4,7,10,13,16,19,22)$ $N_i = \frac{9}{64}(1-\eta^2)(1+\xi_0)$ $(1+9\eta_0)(1+\xi_0)$ $(i = 5,6,11,12,17,18,23,24)$ $N_i = \frac{9}{64}(1-\xi^2)(1+9\xi_0)$ $(1+\eta_0)(1+\xi_0)$ $(i = 2,3,8,9,14,15,20,21)$	Upper and lower surfaces can be curved
18-Node space transition element	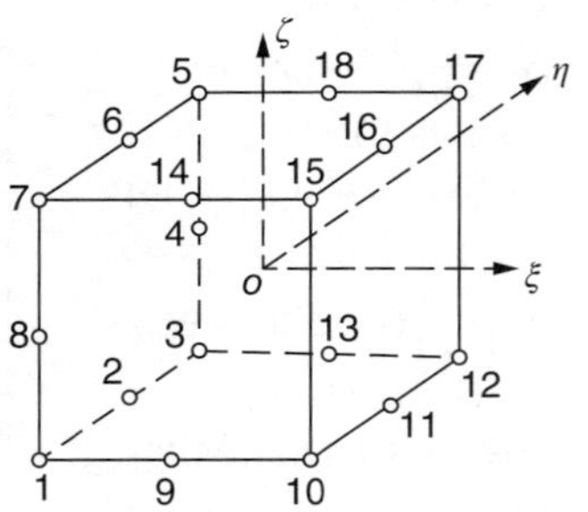	u, v, w	$N_i = \frac{1}{8}(1+\xi_0)(1+\eta_0)$ $(1+\zeta_0)\times(\xi_0+\eta_0+\zeta_0-2)$ $N_i = \frac{1}{8}(1+\xi_0)(1+\eta_0)$ $(1+\zeta_0)\times(\xi_0+\eta_0+\zeta_0-2)$ $(i = 1,3,5,7)$ $N_i = \frac{1}{4}(1-\eta^2)(1+\xi_0)$ $(1+\zeta_0)$ $(i = 2,6)$ $N_i = \frac{1}{4}(1-\zeta^2)(1+\xi_0)$ $(1+\eta_0)$ $(i = 4,8)$ $N_i = \frac{1}{8}(1+\xi_0)(1+\eta_0)$ $(1+\zeta_0)\times(\xi-1-\eta_0)$ $(i = 10,12,15,17)$ $N_i = \frac{1}{4}(1-\xi^2)(1+\eta_0)$ $(1+\zeta_0)$ $(i = 9,13,14,18)$ $N_i = \frac{1}{4}(1-\eta^2)(1+\xi_0)$ $(1+\zeta_0)$ $(i = 11,16)$	The boundary can be a curved surface; the element is applicable between linear elements and quadratic elements

Table 8.3 (Continued)

Name of the element	Shape of the element (parent element)	Degree of freedom	Shape function	Features of the element
28-Node space transition element	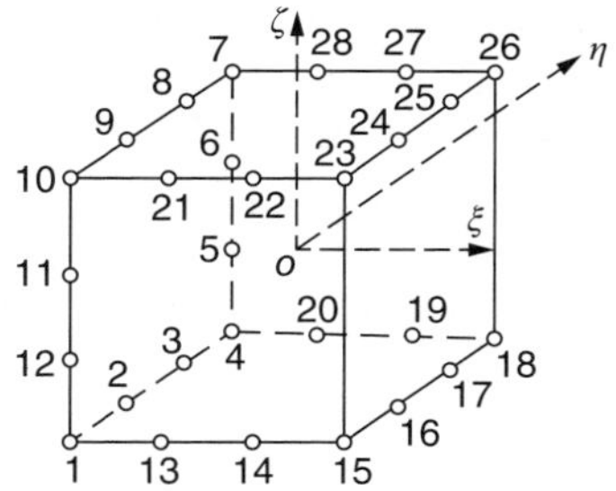	u, v, w	$N_i = \frac{1}{64}(1+\xi_0)(1+\eta_0)$ $(1+\zeta_0)$ $\times[9(\xi^2+\eta^2+\zeta^2)-19]$ $(i = 1,4,7,10)$ $N_i = \frac{1}{64}(1-\eta^2)(1+\xi_0)$ $\times(1+9\eta_0)(1+\zeta_0)$ $(i = 2,3,8,9)$ $N_i = \frac{9}{64}(1-\zeta^2)(1+\xi_0)$ $\times(1+\eta_0)(1+9\zeta_0)$ $(i = 5,6,11,12)$ $N_i = \frac{9}{64}(1-\xi^2)(1+9\xi_0)$ $\times(1+\eta_0)(1+\zeta_0)$ $(i = 13,14,19,20,21,22,27,28)$ $N_i = \frac{1}{64}(1+\xi_0)(1+\eta_0)$ $(1+\zeta_0)\times[9(\xi^2+\eta^2)-10]$ $(i = 15,18,23,26)$ $N_i = \frac{9}{64}(1+\xi_0)(1+\zeta_0)$ $\times(1+9\eta_0)(1-\eta^2)$ $(i = 16,17,24,25)$	The boundary can be a curved surface; the element is applicable to transition between linear elements and cubic elements
4-Node linear tetra-hedron element	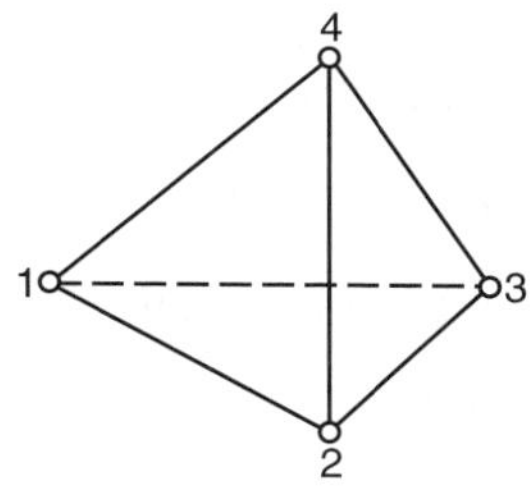	u, v, w	$N_1 = L_1$ $N_2 = L_2$ $N_3 = L_3$ $N_4 = L_4$ (L_i is the volume coordinate)	Easy to calculate and adaptable to complex geometric shapes. The stress within the element is a constant with lower precision
10-Node quadratic tetrahedral element	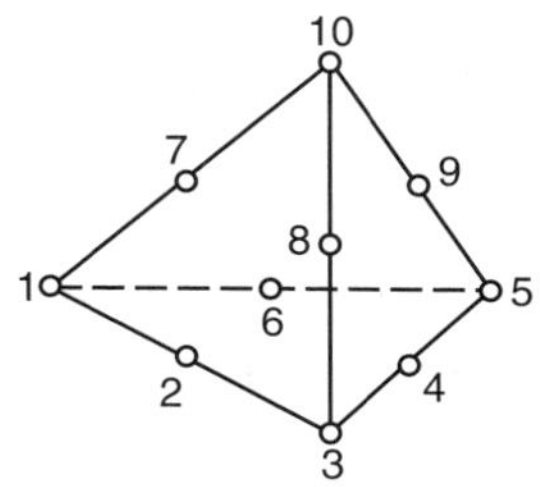	u, v, w	$N_1 = L_1(2L_1-1)$, $N_2 = 4L_1L_2$ $N_3 = L_2(2L_2-1)$, $N_4 = 4L_2L_3$ $N_5 = L_3(2L_3-1)$, $N_6 = 4L_3L_1$ $N_7 = 4L_1L_4$, $N_8 = 4L_2L_4$ $L_9 = 4L_3L_4$, $N_{10} = L_4(2L_4-1)$	The boundary can be a curved surface; the element is adaptable to complex geometric shapes with sharp corners

(Continued)

Table 8.3 (Continued)

Name of the element	Shape of the element (parent element)	Degree of freedom	Shape function	Features of the element
20-Node cubic tetrahedral element	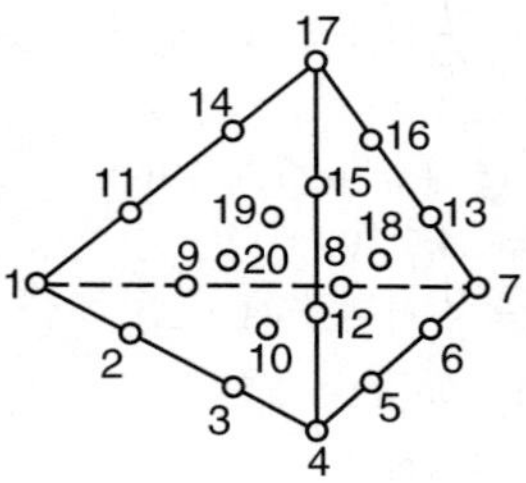	u, v, w	$N_1 = \frac{1}{2}(3L_1 - 1)(3L_1 - 2)L_1$ $N_2 = \frac{9}{2}L_1L_2(3L_2 - 1)$ $N_3 = \frac{9}{2}L_1L_2(3L_2 - 1)$ $N_4 = \frac{1}{2}(3L_2 - 1)(3L_2 - 2)L_2$ $N_5 = \frac{9}{2}L_2L_3(3L_2 - 1)$ $N_6 = \frac{9}{2}L_2L_3(3L_3 - 1)$ $N_7 = \frac{1}{2}(3L_3 - 1)(3L_3 - 2)L_3$ $N_8 = \frac{9}{2}L_2L_3(3L_3 - 1)$ $N_9 = \frac{9}{2}L_3L_1(3L_1 - 1)$ $N_{10} = 27L_1L_2L_3$ $N_{11} = \frac{9}{2}L_1L_4(3L_1 - 1)$ $N_{12} = \frac{9}{2}L_2L_4(3L_2 - 1)$ $N_{13} = \frac{9}{2}L_3L_4(3L_3 - 1)$ $N_{14} = \frac{9}{2}L_1L_4(3L_4 - 1)$ $N_{15} = \frac{9}{2}L_2L_4(3L_5 - 1)$ $N_{16} = \frac{9}{2}L_2L_4(3L_4 - 1)$ $N_{17} = \frac{1}{2}(3L_4 - 1)(3L_4 - 2)L_4$ $N_{18} = 27L_2L_1L_4$ $N_{19} = 27L_1L_3L_4,$ $N_{20} = 27L_1L_2L_4$	The boundary can be a curved surface; the element is adaptable to complex geometric shapes with sharp corners
6-Node linear triangular prism element	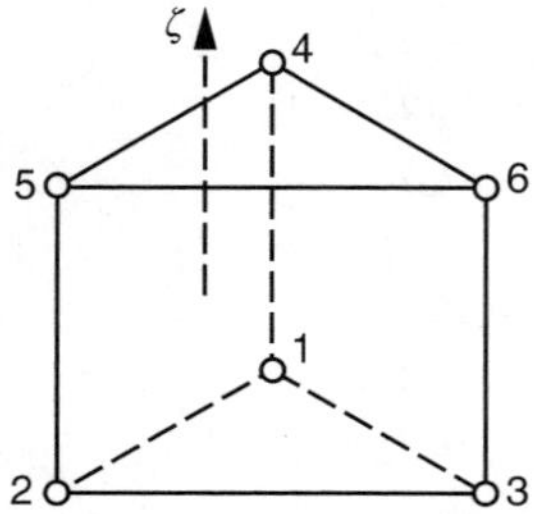	u, v, w	$N_i = \frac{1}{2}L_1(1 + \zeta_0) \quad (i = 1,4)$ $N_i = \frac{1}{2}L_2(1 + \zeta_0) \quad (i = 2,5)$ $N_i = \frac{1}{2}L_3(1 + \zeta_0) \quad (i = 3,6)$ (L_i is area coordinate)	Adapt to complex geometric shapes with sharp corners

Table 8.3 (Continued)

Name of the element	Shape of the element (parent element)	Degree of freedom	Shape function	Features of the element
15-Node quadratic triangular prism element	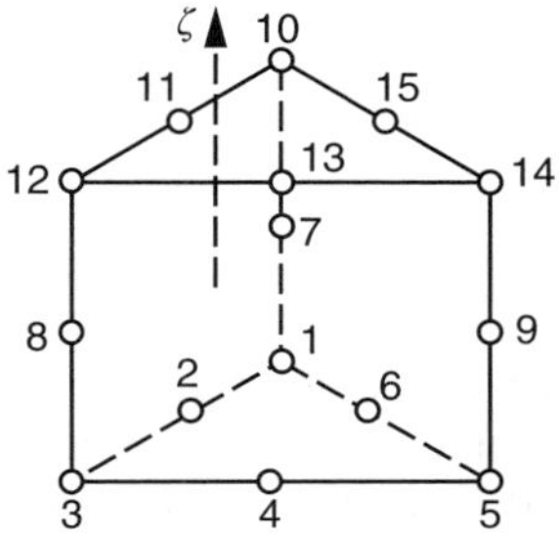	u, v, w	$N_i = \frac{1}{2}L_1(2L_1 - 1)(1 + \zeta_0)$ $-\frac{1}{2}L_1(1 - \zeta^2) \quad (i = 1,10)$ $N_i = \frac{1}{2}L_2(2L_2 - 1)(1 + \zeta_0)$ $-\frac{1}{2}L_2(1 - \zeta^2) \quad (i = 3,12)$ $N_i = \frac{1}{2}L_3(2L_3 - 1)(1 + \zeta_0)$ $-\frac{1}{2}L_3(1 - \zeta^2)$ $N_i = \frac{1}{2}L_1(2L_1 - 1)(1 + \zeta_0)$ $-\frac{1}{2}L_1(1 - \zeta^2) \quad (i = 1,10)$ $N_i = \frac{1}{2}L_2(2L_2 - 1)(1 + \zeta_0)$ $-\frac{1}{2}L_2(1 - \zeta^2) \quad (i = 3,12)$ $N_i = \frac{1}{2}L_3(2L_3 - 1)(1 + \zeta_0)$ $-\frac{1}{2}L_3(1 - \zeta^2) \quad (i = 5,14)$ $N_i = 2L_1L_2(1 + \zeta_0) \quad (i = 2,11)$ $N_i = 2L_2L_3(1 + \zeta_0) \quad (i = 4,13)$ $N_i = 2L_3L_1(1 + \zeta_0) \quad (i = 6,15)$ $N_7 = L_1(1 - \zeta^2)$ $N_8 = L_2(1 - \zeta^2)$ $N_9 = L_1(1 - \zeta^2)$	The boundary can be a curved surface; the element is adaptable to complex geometric shapes with sharp corners
12-Node triangular prism element	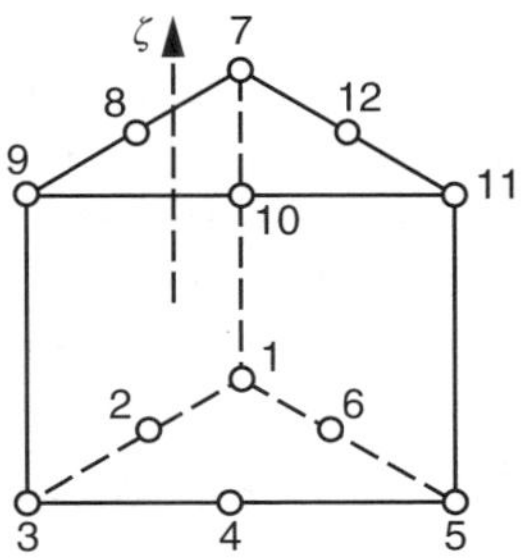	u, v, w	$N_i = \frac{1}{2}L_1(2L_1 - 1)(1 + \zeta_0)$ $(i = 1,7)$ $N_i = \frac{1}{2}L_2(2L_2 - 1)(1 + \zeta_0)$ $(i = 3,9)$ $N_i = \frac{1}{2}L_3(2L_3 - 1)(1 + \zeta_0)$ $(i = 5,11)$ $N_i = 2L_1L_2(1 + \zeta_0) \; (i = 2,8)$ $N_i = 2L_2L_3(1 + \zeta_0) \; (i = 4,10)$ $N_i = 2L_3L_1(1 + \zeta_0) \; (i = 6,12)$	The upper and lower surfaces can be curved surfaces; the element is adaptable to complex geometric shapes with sharp corners

from actual structures and fully reflect the actual situation; on the other hand, a lot of calculations are done within the parent elements. Its shape is simple and regular, easy to calculate and circulate, and particularly applicable for calculations on the computer. Therefore it has two advantages.

In order to conduct coordinate transformation, a one-to-one correspondence must be set up between the coordinate system (ξ, η, ζ) and coordinate system (x, y, z). In the finite element method, this correspondence is set up by shape functions.

8.5.1 Plane Coordinate Transformation

In the global coordinate system, coordinates of any point within the element can be represented by shape functions as follows:

$$\left.\begin{aligned} x &= \sum N_i x_i = N_1 x_1 + N_2 x_2 + \cdots \\ y &= \sum N_i y_i = N_1 y_1 + N_2 y_2 + \cdots \end{aligned}\right\} \tag{8.15}$$

in which $N_i(x_i, y_i)$ are shape functions represented by local coordinates; $N_i(x_i, y_i)$ is the global coordinate of node i.

Formula (8.15) is the transformation formula of plane coordinates.

Figure 8.5 demonstrates coordinate transformation of one-dimensional element. The original straight lines are, respectively, transformed into a straight line, quadratic curve, and cubic curve, as the shape functions N_i in the transformation formula (8.14) are, respectively, linear, quadratic, and cubic functions of ξ.

Figure 8.6 demonstrates the plane coordinate transformation of two-dimensional linear elements. The parent element is a square and child element is an arbitrary quadrilateral.

The coordinate transformation formula is

$$\begin{aligned} x &= \frac{x_1}{4}(1-\xi)(1-\eta) + \frac{x_2}{4}(1+\xi)(1-\eta) + \frac{x_3}{4}(1-\xi)(1+\eta) + \frac{x_4}{4}(1+\xi)(1+\eta) \\ y &= \frac{y_1}{4}(1-\xi)(1-\eta) + \frac{y_2}{4}(1+\xi)(1-\eta) + \frac{y_3}{4}(1-\xi)(1+\eta) + \frac{y_4}{4}(1+\xi)(1+\eta) \end{aligned} \tag{a}$$

Let $\xi = 1$ and we will obtain

$$x = \frac{x_2}{2}(1-\eta) + \frac{x_4}{2}(1+\eta), \quad y = \frac{y_2}{2}(1-\eta) + \frac{y_4}{2}(1+\eta)$$

which is equation of straight line 24. Similarly let $\xi = -1,\ \eta = 1,\ \eta = -1$ and we will obtain other three edges, respectively: in formula (a) let $\xi = \eta = 0$ to obtain $x = (x_1 + x_2 + x_3 + x_4)/4,\ y = (y_1 + y_2 + y_3 + y_4)/4$, which is the centroid of the quadrilateral. Use the two straight lines bisecting four edges to divide the child element, take the centroid of the child element as the original point ($\xi = \eta = 0$), and draw ξ axis

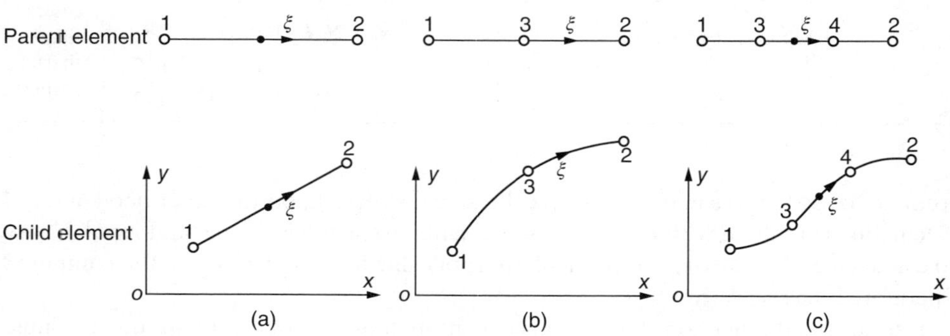

Figure 8.5 Plane coordinate transformation of one-dimensional element. (a) Linear element, (b) quadratic element, and (c) cubic element.

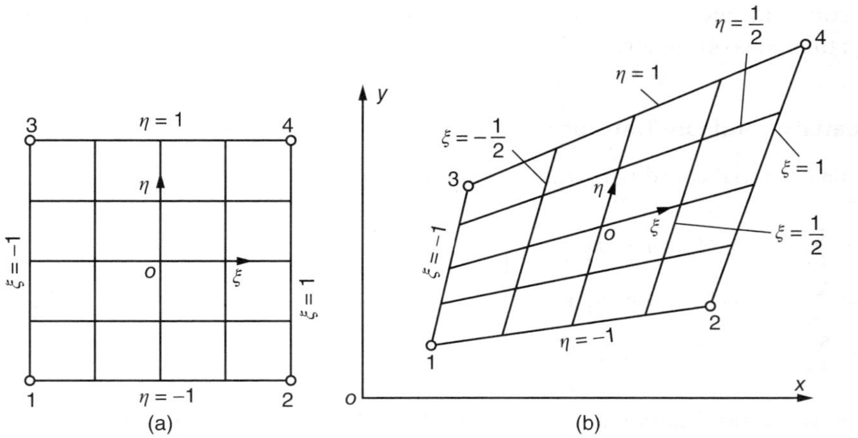

Figure 8.6 Plane coordinates transformation of two-dimensional linear elements. (a) Parent element and (b) child element.

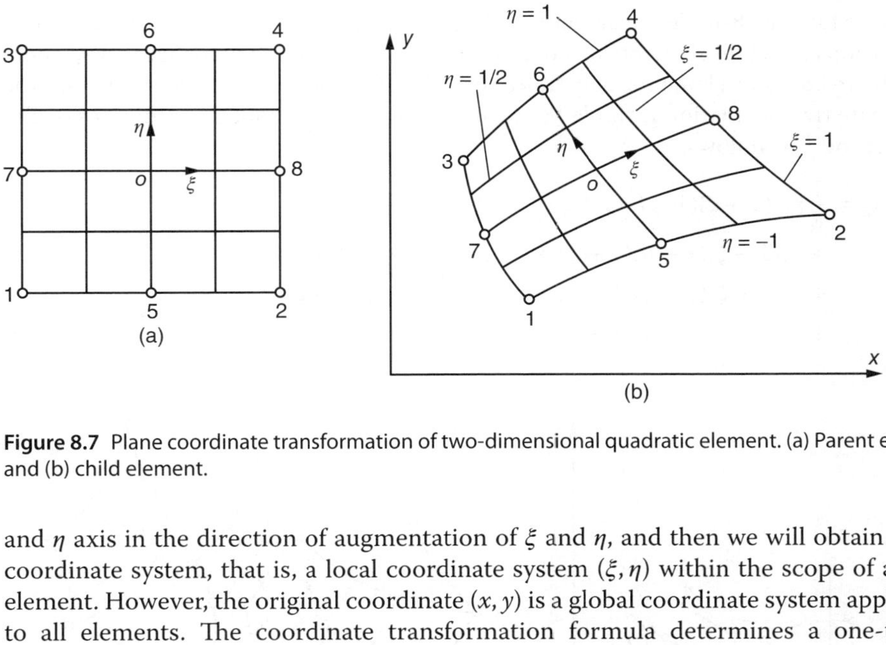

Figure 8.7 Plane coordinate transformation of two-dimensional quadratic element. (a) Parent element and (b) child element.

and η axis in the direction of augmentation of ξ and η, and then we will obtain a new coordinate system, that is, a local coordinate system (ξ, η) within the scope of a child element. However, the original coordinate (x, y) is a global coordinate system applicable to all elements. The coordinate transformation formula determines a one-to-one relationship between the local coordinate (ξ, η) and global coordinates (x, y).

Figure 8.7 shows the plane coordinate transformation of two-dimensional quadratic element.

The four edges of child element are all quadratic curves, and the local coordinates (ξ, η) are curvilinear coordinates.

Now we explore the coordinate continuity of adjacent child elements on common boundaries. Taking quadratic elements, for example, two adjacent elements are all quadratic curves on the common boundary. Since they have identical coordinates on three common nodes, they have identical coordinates along the whole common boundary; thus the coordinates are continuous.

8.5.2 Spatial Coordinate Transformation

The formula of spatial coordinate transformation is as follows:

$$\left.\begin{aligned} x &= \sum N_i x_i = N_1 x_1 + N_2 x_2 + \cdots \\ y &= \sum N_i y_i = N_1 y_1 + N_2 y_2 + \cdots \\ z &= \sum N_i z_i = N_1 z_1 + N_2 z_2 + \cdots \end{aligned}\right\} \tag{8.16}$$

in which N_i is a shape function and x_i, y_i, z_i are the global coordinates of node i.

After spatial coordinate transformation, the original straight line will become a spatial curve, and the original plane will become a spatial curved surface. The regular hexahedron parent element will become a hexahedron child element with curved edges and curved surfaces, as shown in Figure 8.8.

Now we take the 8-node linear element, for example, to explore the coordinate continuity of adjacent child elements on common boundaries. As shown in Figure 8.9, where a and b are adjacent elements, substitute the shape function in formula (8.11) into the coordinate transformation formula (8.16) and obtain the horizontal coordinates of elements a and b as follows:

$$\begin{aligned} x_a = \frac{1}{8}\{ & x_1(1+\xi)(1-\eta)(1+\zeta) + x_2(1+\xi)(1+\eta)(1+\zeta) \\ & + x_3(1-\xi)(1-\eta)(1+\zeta) + x_4(1-\xi)(1+\eta)(1+\zeta) \\ & + x_5(1+\xi)(1-\eta)(1-\zeta) + x_6(1+\xi)(1+\eta)(1-\zeta) \\ & + x_7(1-\xi)(1-\eta)(1-\zeta) + x_8(1-\xi)(1+\eta)(1-\zeta)\} \end{aligned} \tag{b}$$

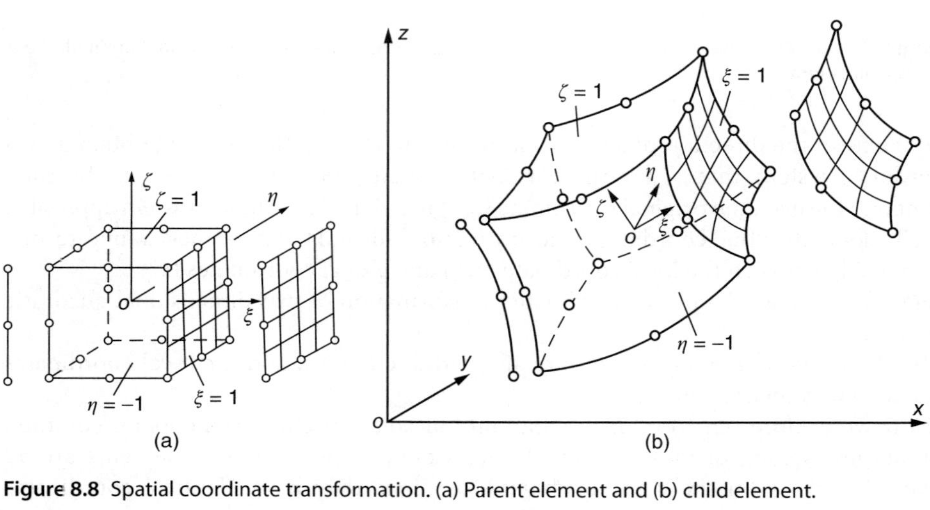

Figure 8.8 Spatial coordinate transformation. (a) Parent element and (b) child element.

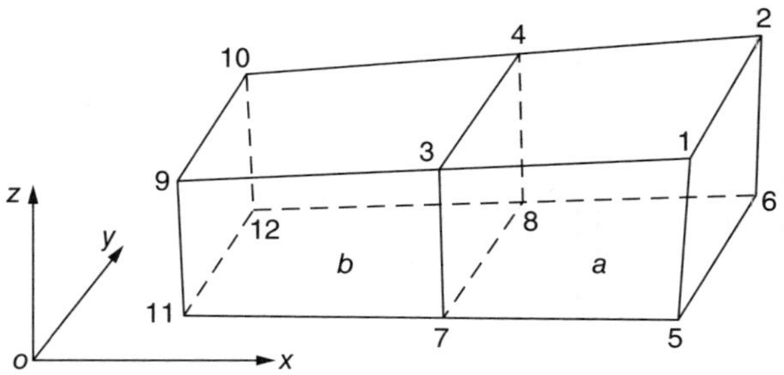

Figure 8.9 Child element.

$$
\begin{aligned}
x_b = \frac{1}{8}\{ & x_3(1+\xi)(1-\eta)(1+\zeta) + x_4(1+\xi)(1+\eta)(1+\zeta) \\
& + x_9(1-\xi)(1-\eta)(1+\zeta) + x_{10}(1-\xi)(1+\eta)(1+\zeta) \\
& + x_7(1+\xi)(1-\eta)(1-\zeta) + x_9(1+\xi)(1+\eta)(1-\zeta) \\
& + x_{11}(1-\xi)(1-\eta)(1-\zeta) + x_{12}(1-\xi)(1+\eta)(1-\zeta)\}
\end{aligned} \tag{c}
$$

The interface between two elements is 3478; in the formula let $\xi = -1$ to obtain

$$
\begin{aligned}
x_a(\xi = -1) = \frac{1}{4}\{ & x_3(1-\eta)(1+\zeta) + x_4(1+\eta)(1+\zeta) + x_7(1-\eta)(1-\zeta) \\
& + x_8(1+\eta)(1-\zeta)\}
\end{aligned}
$$

In formula (c) let $\xi = +1$ to obtain

$$
\begin{aligned}
x_b(\xi = +1) = \frac{1}{4}\{ & x_3(1-\eta)(1+\zeta) + x_4(1+\eta)(1+\zeta) + x_7(1-\eta)(1-\zeta) \\
& + x_8(1+\eta)(1-\zeta)\}
\end{aligned}
$$

Comparing the previous two formulas, it is clear that on the interface between elements a and b, there are

$$x_a(\xi = -1) \equiv x_b(\xi = +1)$$

It can also be proved that y and z are also continuous on the interface.

8.6 Displacement Function

The displacement of the element can be represented by shape functions as follows:

$$
\begin{aligned}
u &= \sum N_i u_i = N_1 u_1 + N_2 u_2 + \cdots \\
v &= \sum N_i v_i = N_1 v_1 + N_2 v_2 + \cdots \\
w &= \sum N_i w_i = N_1 w_1 + N_2 w_2 + \cdots
\end{aligned}
$$

or

$$\{r\} = \begin{Bmatrix} u \\ v \\ w \end{Bmatrix} = \begin{bmatrix} N_1 & 0 & 0 & N_2 & 0 & 0 & \cdots \\ 0 & N_1 & 0 & 0 & N_2 & 0 & \cdots \\ 0 & 0 & N_1 & 0 & 0 & N_2 & \end{bmatrix} \begin{Bmatrix} u_1 \\ v_1 \\ w_1 \\ u_2 \\ v_2 \\ w_2 \\ \vdots \end{Bmatrix} = [N]\{\delta\}^{\mathrm{e}} \tag{8.17}$$

where $\{\delta_i\}^{\mathrm{e}} = \begin{bmatrix} u_1 & v_1 & w_1 & u_2 & \cdots \end{bmatrix}^{\mathrm{T}}$ in which N_i is a shape function and $\{\delta_i\}^*$ are nodal displacements.

By comparing formulas (8.16) and (8.17), we can see the coordinate transformation and element displacement function all used the shape function N_i, which could be a linear, quadratic, cubic, or high-order function of the local coordinate (ξ, η, ζ). If the shape functions used by elemental coordinate transformation and displacement function have the same order, the number of nodes used to determine the elemental shape should equal to that of nodes used to determine the element displacement. This type of element is called isoparametric element. If the shape function used for coordinate transformation has a higher order than that for the displacement function, the nodes in coordinate transformation should exceed those used to determine the element displacement. This type of element is called hyper-parameter element. Conversely, if the shape function used for coordinate transformation has a lower order than that for the displacement function, it is called sub-parameter element. In the isoparametric element, coordinate transformation and displacement function always adopt the same nodes.

According to Section 4.4, in order that the calculation results by the finite element method converge to correct solutions, the displacement function must include the rigid-body displacement and constant strain; for space problems, the displacement function must contain the following linear terms:

$$u = \beta_1 + \beta_2 x + \beta_3 y + \beta_4 z \tag{a}$$
$$v = \beta_5 + \beta_6 x + \beta_7 y + \beta_8 z \tag{b}$$
$$w = \beta_9 + \beta_{10} x + \beta_{11} y + \beta_{12} z \tag{c}$$

Now take formula (a) for analysis. On node I, it is tenable that

$$u_i = \beta_1 + \beta_2 x_i + \beta_3 y_i + \beta_4 z_i \quad (i = 1, 2{,}3, \ldots) \tag{d}$$

Substituting formula (d) into formula (8.16), we get

$$\begin{aligned} u &= \sum N_i u_i \\ &= N_1(\beta_1 + \beta_2 x_1 + \beta_3 y_1 + \beta_4 z_1) + N_2(\beta_1 + \beta_2 x_2 + \beta_3 y_2 + \beta_4 z_2) + \cdots \\ &= \beta_1 \sum N_i + \beta_2 \sum N_i x_i + \beta_3 \sum N_i y_i + \beta_4 \sum N_i z_i \\ &= \beta_1 + \beta_2 x + \beta_3 y + \beta_4 z \end{aligned}$$

So in order that formula (a) is tenable for any coefficients $\beta_1, \beta_2, \beta_3, \beta_4$, it is required that

$$\sum N_i x_i = x, \quad \sum N_i y_i = y, \quad \sum N_i z_i = z \tag{e}$$

where

$$\sum N_i = 1 \tag{f}$$

For isoparametric elements, formula (e) is the coordinate transformation formula (8.16), and formula (f) is the conditional expression of rigid-body displacement (8.2). As the shape functions provided in this chapter are all coincided with formula (8.2), the displacement function (8.17) is coincided with the conditions for rigid-body displacement and constant strain.

Similar to the demonstration about coordinate continuity between adjacent elements in the previous section, it is easy to prove when the various shape functions provided in this chapter are used in displacement function (8.17), the displacement between adjacent elements is also continuous.

Therefore, according to the convergence conditions listed in Section 4.4, for child elements after coordinate transformation, the displacement function (8.17) will offer a solution converging to a correct one.

8.7 Element Strain

Strains of space problems can be represented as follows:

$$\{\varepsilon\} = \begin{Bmatrix} \varepsilon_x \\ \varepsilon_y \\ \varepsilon_z \\ \gamma_{xy} \\ \gamma_{yz} \\ \gamma_{xz} \end{Bmatrix} = \begin{Bmatrix} \dfrac{\partial u}{\partial x} \\ \dfrac{\partial v}{\partial y} \\ \dfrac{\partial w}{\partial z} \\ \dfrac{\partial u}{\partial y} + \dfrac{\partial v}{\partial x} \\ \dfrac{\partial v}{\partial z} + \dfrac{\partial w}{\partial y} \\ \dfrac{\partial w}{\partial x} + \dfrac{\partial u}{\partial z} \end{Bmatrix} \tag{8.18}$$

By substituting displacement function (8.17) into formula (8.18), we obtain

$$\{\varepsilon\} = [B]\{\delta\}^* = \left[B_1\ B_2\ B_3\ \cdots\right]\{\delta\}^* \tag{8.19}$$

where

$$[B_i] = \begin{bmatrix} \dfrac{\partial N_i}{\partial x} & 0 & 0 \\ 0 & \dfrac{\partial N_i}{\partial y} & 0 \\ 0 & 0 & \dfrac{\partial N_i}{\partial z} \\ \dfrac{\partial N_i}{\partial y} & \dfrac{\partial N_i}{\partial x} & 0 \\ 0 & \dfrac{\partial N_i}{\partial z} & \dfrac{\partial N_i}{\partial y} \\ \dfrac{\partial N_i}{\partial z} & 0 & \dfrac{\partial N_i}{\partial x} \end{bmatrix} \tag{8.20}$$

The shape function $N_i(\xi, \eta, \zeta)$ is given by the local coordinates. In light of the partial differential rule, it is known that

$$\frac{\partial N_i}{\partial \xi} = \frac{\partial N_i}{\partial x}\frac{\partial x}{\partial \xi} + \frac{\partial N_i}{\partial y}\frac{\partial y}{\partial \xi} + \frac{\partial N_i}{\partial z}\frac{\partial z}{\partial \xi} \tag{a}$$

In the same way we get $\frac{\partial N_i}{\partial \eta}$ and $\frac{\partial N_i}{\partial \zeta}$, bring them together to obtain

$$\begin{Bmatrix} \frac{\partial N_i}{\partial \xi} \\ \frac{\partial N_i}{\partial \eta} \\ \frac{\partial N_i}{\partial \zeta} \end{Bmatrix} = \begin{bmatrix} \frac{\partial x}{\partial \xi} & \frac{\partial y}{\partial \xi} & \frac{\partial z}{\partial \xi} \\ \frac{\partial x}{\partial \eta} & \frac{\partial y}{\partial \eta} & \frac{\partial z}{\partial \eta} \\ \frac{\partial x}{\partial \zeta} & \frac{\partial y}{\partial \zeta} & \frac{\partial z}{\partial \zeta} \end{bmatrix} \begin{Bmatrix} \frac{\partial N_i}{\partial x} \\ \frac{\partial N_i}{\partial y} \\ \frac{\partial N_i}{\partial z} \end{Bmatrix} = [J] \begin{Bmatrix} \frac{\partial N_i}{\partial x} \\ \frac{\partial N_i}{\partial y} \\ \frac{\partial N_i}{\partial z} \end{Bmatrix} \tag{8.21}$$

As the shape function is given by local coordinates, the left part of the previous formula can be computed according to $N_i(\xi, \eta, \zeta)$. In addition, by the coordinate transformation formula (8.16), we can get matrix $[J]$, which is called Jacobian matrix:

$$[J] = \begin{bmatrix} \frac{\partial x}{\partial \xi} & \frac{\partial y}{\partial \xi} & \frac{\partial z}{\partial \xi} \\ \frac{\partial x}{\partial \eta} & \frac{\partial y}{\partial \eta} & \frac{\partial z}{\partial \eta} \\ \frac{\partial x}{\partial \zeta} & \frac{\partial y}{\partial \zeta} & \frac{\partial z}{\partial \zeta} \end{bmatrix} = \begin{bmatrix} \sum \frac{\partial N_i}{\partial \xi} x_i & \sum \frac{\partial N_i}{\partial \xi} y_i & \sum \frac{\partial N_i}{\partial \xi} z_i \\ \sum \frac{\partial N_i}{\partial \eta} x_i & \sum \frac{\partial N_i}{\partial \eta} y_i & \sum \frac{\partial N_i}{\partial \eta} z_i \\ \sum \frac{\partial N_i}{\partial \zeta} x_i & \sum \frac{\partial N_i}{\partial \zeta} y_i & \sum \frac{\partial N_i}{\partial \zeta} z_i \end{bmatrix}$$

$$= \begin{bmatrix} \frac{\partial N_1}{\partial \xi} & \frac{\partial N_2}{\partial \xi} & \cdots & \frac{\partial N_m}{\partial \xi} \\ \frac{\partial N_1}{\partial \eta} & \frac{\partial N_2}{\partial \eta} & \cdots & \frac{\partial N_m}{\partial \eta} \\ \frac{\partial N_1}{\partial \zeta} & \frac{\partial N_2}{\partial \zeta} & \cdots & \frac{\partial N_m}{\partial \zeta} \end{bmatrix} \begin{bmatrix} x_1 & y_1 & z_1 \\ x_2 & y_2 & z_2 \\ \vdots & \vdots & \vdots \\ x_m & y_m & z_m \end{bmatrix} = [j][X] \tag{8.22}$$

$$[j] = \begin{bmatrix} \frac{\partial N_1}{\partial \xi} & \frac{\partial N_2}{\partial \xi} & \cdots & \frac{\partial N_m}{\partial \xi} \\ \frac{\partial N_1}{\partial \eta} & \frac{\partial N_2}{\partial \eta} & \cdots & \frac{\partial N_m}{\partial \eta} \\ \frac{\partial N_1}{\partial \zeta} & \frac{\partial N_2}{\partial \zeta} & \cdots & \frac{\partial N_m}{\partial \zeta} \end{bmatrix}, \quad [X] = \begin{bmatrix} x_1 & y_1 & z_1 \\ x_2 & y_2 & z_2 \\ \vdots & \vdots & \vdots \\ x_m & y_m & z_m \end{bmatrix} \tag{8.23}$$

After inverting matrix $[J]$, by formula (8.21), we can obtain the derivative of the shape function in the global coordinates as follows:

$$\begin{Bmatrix} \frac{\partial N_i}{\partial x} \\ \frac{\partial N_i}{\partial y} \\ \frac{\partial N_i}{\partial z} \end{Bmatrix} = [J]^{-1} \begin{Bmatrix} \frac{\partial N_i}{\partial \xi} \\ \frac{\partial N_i}{\partial \eta} \\ \frac{\partial N_i}{\partial \zeta} \end{Bmatrix} \tag{8.24}$$

Substitution of formula (8.24) into formulas (8.20) and (8.19) will give the elemental strain.

From formulas (8.19), (8.20), and (8.24), it is evident that Jacobian matrix exerts significant effects on the elemental strain. Next we will calculate the Jacobian matrices of three elements shown in Figure 8.10.

Element *a*: We substitute the coordinates of 4 nodes into formula (8.15) and simplify it to obtain

$$x = 3\xi, \quad y = 2\eta$$

By formula (8.22), we can get

$$[J] = \begin{bmatrix} 3 & 0 \\ 0 & 2 \end{bmatrix}$$

Element *b*: We substitute the coordinates of nodes into formula (8.15) and get

$$\begin{aligned} x = \frac{1}{4}\Bigg\{ & \left(3 + \frac{1}{2\sqrt{3}}\right)(1+\xi)(1+\eta) + \left(-3 + \frac{1}{2\sqrt{3}}\right)(1-\xi)(1+\eta) \\ & + \left(-3 - \frac{1}{2\sqrt{3}}\right)(1-\xi)(1-\eta) + \left(3 - \frac{1}{2\sqrt{3}}\right)(1+\xi)(1-\eta)\Bigg\} \end{aligned}$$

$$\begin{aligned} y &= \frac{1}{4}\left\{\frac{1}{2}(1+\xi)(1+\eta) + \frac{1}{2}(1-\xi)(1+\eta) - \frac{1}{2}(1-\xi)(1-\eta) - \frac{1}{2}(1+\xi)(1-\eta)\right\} \\ &= \frac{1}{2}\eta \end{aligned}$$

By formula (8.22), we can get

$$[J] = \begin{bmatrix} 3 & 0 \\ \dfrac{1}{2\sqrt{3}} & \dfrac{1}{2} \end{bmatrix}$$

Element *c*:

$$\begin{aligned} x &= \frac{1}{4}\{(1+\xi)(1+\eta) - (1-\xi)(1+\eta) - (1-\xi)(1-\eta) + (1+\xi)(1-\eta)\} \\ &= \xi \end{aligned}$$

$$\begin{aligned} y &= \frac{1}{4}\left\{\frac{4}{5}(1+\xi)(1+\eta) + \frac{1}{4}(1-\xi)(1+\eta) - \frac{3}{4}(1-\xi)(1-\eta) - \frac{3}{4}(1+\xi)(1-\eta)\right\} \\ &= \frac{1}{4}(\xi + 3\eta + \xi\eta) \end{aligned}$$

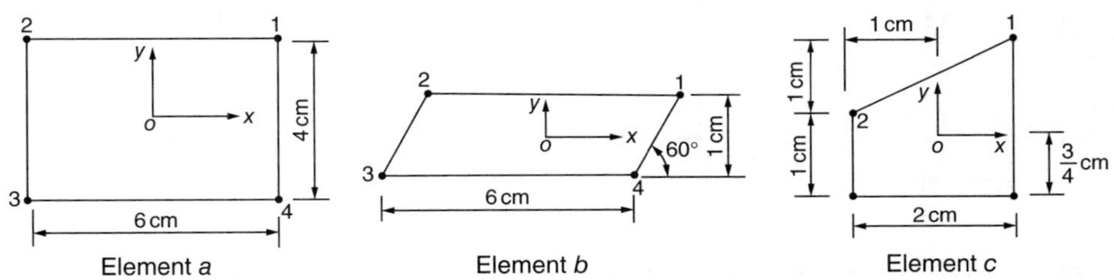

Figure 8.10 Calculation example.

By formula (8.22), we can get

$$[J] = \frac{1}{4}\begin{bmatrix} 4 & (1+\eta) \\ 0 & (3+\xi) \end{bmatrix}$$

It is clear that the Jacobian matrix of 2×2 square element is an identity matrix: $[J] = [I]$. Therefore, the Jacobian matrix $[J]$ of ordinary element represents changes in the geometric shape from the 2×2 square element. For elements a and b, this sort of changes is constants on any points (ξ, η), so the Jacobian matrices $[J]$ of the two elements are also constants.

Obviously, the variation rule of strain within the element depends on the order of shape function. Now take plane problems, for example, to explain the variation of strain within curve side element. Although the shape function is a simple polynomial of local coordinates, it is difficult to represent the displacement function of curve side element with an explicit function of global coordinates due to coordinate transformation. So it is really an issue to determine the variation of strain within curve side element in general situations. For a brief determination of variation, we may simplify the problem and have a look at Figure 8.11 that demonstrates how the strain is changing within the rectangular element and let the coordinate axis be the two symmetry axes of the rectangle. Obviously, magnify the parent element by a and b times in the two directions of ξ, η, and we will get this child element. The global coordinates is related with local coordinates in the following pattern:

$$\xi = \frac{x}{a}, \quad \eta = \frac{y}{b} \tag{b}$$

By substituting formula (b) into the shape function (8.6) of 4-node linear element and making comparison with the shape function (5.2) of 4-node rectangular element, we can see they are identical. It is clear from formula (b) of Section 5.1 that in this type of element, the normal strain ε_x is a constant in the x direction and a linear function of coordinates in the y direction, ε_y is a constant in the y direction and a linear function of coordinates in the x direction, and the shear strain γ_{xy} is linear function of coordinates x and y.

Then by substituting formula (b) into the shape function (8.9) and displacement function (8.17) of 8-node quadratic element, it is evident that after merging and sorting, the displacement function of this type of element will take the following form:

$$u = \beta_1 + \beta_2 x + \beta_3 y + \beta_4 xy + \beta_5 x^2 + \beta_6 y^2 + \beta_7 x^2 y + \beta_8 xy^2$$
$$v = \beta_9 + \beta_{10} x + \beta_{11} y + \beta_{12} xy + \beta_{13} x^2 + \beta_{14} y^2 + \beta_{15} x^2 y + \beta_{16} xy^2$$

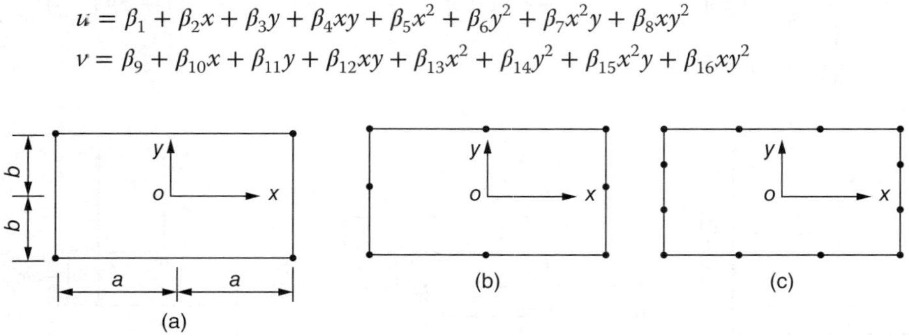

Figure 8.11 Plane element. (a) 4-node element, (b) 8-node element, and (c) 12-node element.

With the previous formula we can obtain the strain component within the element as follows:

$$\varepsilon_x = \frac{\partial u}{\partial x} = \beta_2 + \beta_4 y + 2\beta_5 x + 2\beta_7 xy + \beta_8 y^2$$

$$\varepsilon_y = \frac{\partial v}{\partial y} = \beta_{11} + \beta_{12} x + 2\beta_{14} y + \beta_{15} x^2 + 2\beta_{16} xy$$

$$\begin{aligned}\gamma_{xy} = \frac{\partial u}{\partial y} + \frac{\partial v}{\partial x} &= (\beta_3 + \beta_{10}) + (\beta_4 + 2\beta_{13})x + (2\beta_6 + \beta_{12})y \\ &+ \beta_1 x^2 + 2(\beta_8 + \beta_{15})xy + \beta_{16} y^2\end{aligned}$$

Thus for the 8-node rectangular element shown in Figure 8.11(b), the normal strain ε_x is linear in the x direction and quadratic in the y direction, the normal strain ε_y is linear in the y direction and quadratic in the x direction, and the shear strain γ_{xy} is quadratic in both x and y directions.

Similarly, for the 12-node rectangular element shown in Figure 8.11(c), the displacement function is

$$\begin{aligned}u &= \beta_1 + \beta_2 x + \beta_3 y + \beta_4 xy + \beta_5 x^2 + \beta_6 y^2 + \beta_7 x^2 y + \beta_8 xy^2 + \beta_9 x^3 + \beta_{10} y^3 \\ &+ \beta_{11} x^3 y + \beta_{12} xy^3 \\ v &= \beta_{13} + \beta_{14} x + \beta_{15} y + \beta_{16} xy + \beta_{17} x^2 + \beta_{18} y^2 + \beta_{19} x^2 y + \beta_{20} xy^2 + \beta_{21} x^3 + \beta_{22} y^3 \\ &+ \beta_{23} x^3 y + \beta_{24} xy^3\end{aligned}$$

Solving the derivative by formula, we can get the strain component. In this type of element, the normal strain ε_x is quadratic in the x direction and cubic in the y direction, ε_y is quadratic in the y direction and cubic in the x direction, and the shear strain γ_{xy} is cubic in both x and y directions.

Deducing by analog, it is easy to obtain the distribution rule of strain in spatial elements. For example, in a $2a \times 2b \times 2c$ hexahedral linear element with 8 nodes, the normal strain ε_x is a constant in the x direction and linear in the y and z directions, ε_y is a constant in the y direction and linear in the x, y direction, the shear strain is linear in all the three directions, and so on.

For another example, in a 20-node hexahedron quadratic element $2a \times 2b \times 2c$, the normal strain ε_x is linear in the x direction and quadratic in the y direction, the normal strain ε_y is linear in the y direction and quadratic in the x and z directions, the shear strain is quadratic in all the three directions, and so on.

8.8 Stiffness Matrix

According to the general formula (3.1), the stiffness matrix of the element is

$$[k] = \iiint [B]^{\mathrm{T}}[D][B]\,\mathrm{d}x\,\mathrm{d}y\,\mathrm{d}z \tag{8.25}$$

Let $[B] = \begin{bmatrix} B_1 & B_2 & \cdots & B_i & \cdots & B_m \end{bmatrix}$.

After partition of the matrix, the submatrix is

$$[k_{rs}] = \iiint [B_r]^{\mathrm{T}}[D][B_s]\,\mathrm{d}x\,\mathrm{d}y\,\mathrm{d}z \tag{8.26}$$

The derivative calculation in the right part of the previous formula is done in the global coordinate system (x, y, z). By formula (8.62), we can see

$$\mathrm{d}x\,\mathrm{d}y\,\mathrm{d}z = |J|\mathrm{d}\xi\,\mathrm{d}\eta\,\mathrm{d}\zeta \tag{8.27}$$

in which $[J]$ is determinant of Jacobian matrix $[J]$. Substituting formulas (8.20), (8.24), and (8.27) into formula (8.26), we obtain

$$\begin{aligned}[k_{rs}] &= \int_{-1}^{1}\int_{-1}^{1}\int_{-1}^{1}[B_r]^{\mathrm{T}}[D][B_s]|J|\mathrm{d}\xi\,\mathrm{d}\eta\,\mathrm{d}\zeta = \int_{-1}^{1}\int_{-1}^{1}\int_{-1}^{1}G_{rs}(\xi,\eta,\zeta)\mathrm{d}\xi\,\mathrm{d}\eta\,\mathrm{d}\zeta\\ G_{rs}(\xi,\eta,\zeta) &= [B_r]^{\mathrm{T}}[D][B_s]|J|\end{aligned} \tag{8.28}$$

The integral calculation in the right part of formula (8.28) is done in the parent element other than a complexly shaped child element; thus the upper and lower limits of the derivative are very simple. Linear and quadratic elements will have responding single and double integrals.

If the child element is in a simple geometric shape, for example, a two-dimensional rectangle or 3D regular hexahedron, we can compute the integral in right of formula (8.28) and obtain the stiffness matrix in an explicit expression. Generally child elements have complex shapes, and it is difficult to represent $G_{rs}(\xi,\eta,\zeta)$ with an explicit expression. Numerical integration is required to obtain the value of $[k_{rs}]$.

The calculation steps are as follows:

1) Use formula (8.23) to calculate $[j]$ that is shared by all elements.
2) Substitute the nodal coordinates into formula (8.22) to obtain $[J]$. Solve its inversion and obtain $[J]^{-1}$.
3) Calculate $[\partial N_i/\partial x, \partial N_i/\partial y, \partial N_i/\partial z]^{\mathrm{T}}$ by formula (8.24).
4) Calculate $[B_i]$ by formula (8.20).
5) Calculate $[k_{rs}]$ by formula (8.28).

For example, for a 4-node planar linear element, the shape function is formula (8.6). By formula (8.23) we can get

$$[j] = \begin{bmatrix}\dfrac{\partial N_1}{\partial\xi} & \cdots & \dfrac{\partial N_4}{\partial\xi}\\ \dfrac{\partial N_2}{\partial\eta} & \cdots & \dfrac{\partial N_4}{\partial\eta}\end{bmatrix} = \frac{1}{4}\begin{bmatrix}1+\eta & -1-\eta & -1+\eta & 1-\eta\\ 1+\xi & 1-\xi & -1+\xi & -1-\xi\end{bmatrix} \tag{a}$$

By formula (8.22) we get

$$\begin{aligned}[J] &= \begin{bmatrix}\dfrac{\partial x}{\partial\xi} & \dfrac{\partial y}{\partial\xi}\\ \dfrac{\partial x}{\partial\eta} & \dfrac{\partial y}{\partial\eta}\end{bmatrix} = \frac{1}{4}\begin{bmatrix}1+\eta & -1-\eta & -1+\eta & 1-\eta\\ 1+\xi & 1-\xi & -1+\xi & -1-\xi\end{bmatrix}\begin{bmatrix}x_1 & y_1\\ \vdots & \vdots\\ x_4 & y_4\end{bmatrix}\\ &= \frac{1}{4}\begin{bmatrix}(1+\eta)x_1-(1+\eta)x_2-(1-\eta)x_3+(1-\eta)x_4\\ (1+\xi)x_1+(1-\xi)x_2-(1-\xi)x_3-(1+\xi)x_4\end{bmatrix.}\\ &\qquad\begin{matrix}(1+\eta)y_1-(1+\eta)y_2-(1-\eta)y_3+(1-\eta)y_4\\ (1+\xi)y_1+(1-\xi)y_2-(1-\xi)y_3-(1+\xi)y_4\end{matrix}\Bigg]\end{aligned} \tag{b}$$

By formula (8.24) it is proved that

$$\begin{Bmatrix} \dfrac{\partial N_i}{\partial x} \\ \dfrac{\partial N_i}{\partial y} \end{Bmatrix} = [J]^{-1} \begin{Bmatrix} \dfrac{\partial N_i}{\partial \xi} \\ \dfrac{\partial N_i}{\partial \eta} \end{Bmatrix} \tag{c}$$

By formula (8.20) it is proved that

$$[B_i] = \begin{bmatrix} \dfrac{\partial N_i}{\partial x} & 0 \\ 0 & \dfrac{\partial N_i}{\partial y} \\ \dfrac{\partial N_i}{\partial y} & \dfrac{\partial N_i}{\partial x} \end{bmatrix} = [J]^{-1} \begin{bmatrix} \dfrac{\partial N_i}{\partial \xi} & 0 \\ 0 & \dfrac{\partial N_i}{\partial \eta} \\ \dfrac{\partial N_i}{\partial \eta} & \dfrac{\partial N_i}{\partial \xi} \end{bmatrix} \tag{d}$$

In formulas (b) and (d), let $\xi = \xi_j, \eta = \eta_j$ to obtain

$$[J_i] = [J]_{\xi=\xi_j,\eta=\eta_j}$$

$$[B_{ij}] = [B_i]_{\xi=\xi_j,\eta=\eta_j}$$

Substitution into formula (8.28) will give

$$G_{rs}(\xi_j, \eta_j) = [B_{rj}]^{\mathrm{T}}[D][B_{sj}]|J_j|$$

Then we can use the Gaussian integral formula to compute $[k_{rs}]$.

8.9 Nodal Loads

By the general formula (3.13), the nodal load produced by volume force $\{q\} = \left[q_x\ q_y\ q_z\right]^{\mathrm{T}}$ is

$$\{P\}_q^{\mathrm{e}} = \iiint [N]^{\mathrm{T}}\{q\}\mathrm{d}x\,\mathrm{d}y\,\mathrm{d}z = \int_{-1}^{1}\int_{-1}^{1}\int_{-1}^{1} [N]^T\{q\}|J|\mathrm{d}\xi\,\mathrm{d}\eta\,\mathrm{d}\zeta \tag{8.29}$$

$$[N] = \begin{bmatrix} N_1 & 0 & 0 & N_2 & 0 & 0 & \cdots \\ 0 & N_1 & 0 & 0 & N_2 & 0 & \cdots \\ 0 & 0 & N_1 & 0 & 0 & N_2 & \cdots \end{bmatrix}_{3\times 3m} \tag{a}$$

in which m is the number of elemental nodes.

As the shape function and Jacobian matrix $[J]$ are all represented by natural coordinates ξ, η, ζ, the integrated term in formula (8.29) is a function of ξ, η, ζ. Nodal loads produced by initial strain $\{\varepsilon_0\}$ and initial stress $\{\sigma_0\}$ are also attributed to similar integrals. These integrals are always computed by the Gauss integral method.

The nodal load produced by distributed force $\{p\}$ acting on the element boundary Ω should be calculated as follows:

$$\{P\}_p^{\mathrm{e}} = \iint_{\Omega} [N]^{\mathrm{T}}\{p\}\,\mathrm{d}\Omega \tag{8.30}$$

in which $[N]$ is the shape function matrix of the surface, which is a 3×3 s-order matrix. s is the number of nodes on the elemental surface.

Let the surface Ω under action of $\{p\}$ be the $\zeta = 1$ surface. In the coordinate transformation formula, let $\zeta = 1$ to obtain the equation of Ω as follows:

$$x = \sum N_i(\zeta = 1)x_i, \quad y = \sum N_i(\zeta = 1)y_i, \quad z = \sum N_i(\zeta = 1)z_i \tag{b}$$

By formulas (8.60) and (8.61), we can get

$$\mathrm{d}\Omega = A\,\mathrm{d}\xi\,\mathrm{d}\eta \tag{8.31}$$

$$A = \sqrt{\left(\frac{\partial x}{\partial \xi}\frac{\partial y}{\partial \eta} - \frac{\partial x}{\partial \eta}\frac{\partial y}{\partial \xi}\right)^2 + \left(\frac{\partial y}{\partial \xi}\frac{\partial z}{\partial \eta} - \frac{\partial y}{\partial \eta}\frac{\partial z}{\partial \xi}\right)^2 + \left(\frac{\partial z}{\partial \xi}\frac{\partial x}{\partial \eta} - \frac{\partial z}{\partial \xi}\frac{\partial x}{\partial \eta}\right)^2} \tag{c}$$

The pressure intensity of any point on surface Ω can be represented as

$$p = [\overline{N}]\{p\}^{\mathrm{e}} \tag{8.32}$$

$$[\overline{N}] = \left[N_1 \ N_2 \ \cdots \ N_i\right]$$

$$[p]^{\mathrm{e}} = \left[p_1 \ p_2 \ \cdots \ p_n\right]^{\mathrm{T}}$$

which are the pressure intensities of s surface nodes.

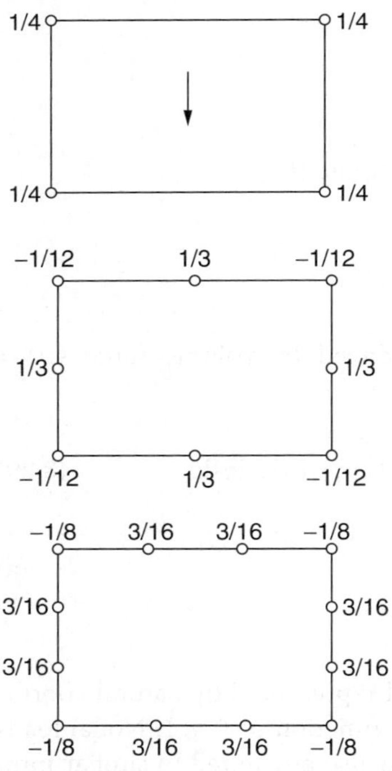

Figure 8.12 Nodal load of the rectangular element caused by the uniform volume force.

Let the direction cosine of the normal line of any point on surface Ω be l, m, n, and then the components of surface pressure in x, y, z directions are

$$\{p\} = \begin{Bmatrix} p_x \\ p_y \\ p_z \end{Bmatrix} = \begin{Bmatrix} l \\ m \\ n \end{Bmatrix} p = \{L\}[\overline{N}]\{p\}^{\mathrm{e}} \tag{d}$$

$$\{L\} = \left[l \ m \ n\right]^{\mathrm{T}}$$

By substituting formula (8.31) and formula (d) into formula (8.30), we will obtain the nodal load caused by the pressure acting on the $\zeta = 1$ surface as follows:

$$\{P\}_p^{\mathrm{e}} = [f][p]^{\mathrm{e}} \tag{8.33}$$

$$[f] = \int_{-1}^{1}\int_{-1}^{1} [N]^{\mathrm{T}}\{L\}[\overline{N}]A\,\mathrm{d}\xi\,\mathrm{d}\eta \tag{8.34}$$

Generally the surface Ω has a complex shape, and the previous formula should also be calculated by numerical integration.

In short, for curved elements, nodal loads caused by volume forces and surface forces are all calculated by the computer according to formulas (8.29) and (8.30) and cannot be assigned simply. Only in case that the element is very simply shaped that nodal loads can be obtained by simple calculation.

Figure 8.12 represents the nodal load caused by uniform volume forces in regular rectangular elements, which demonstrates that the nodal load is consistent with that by

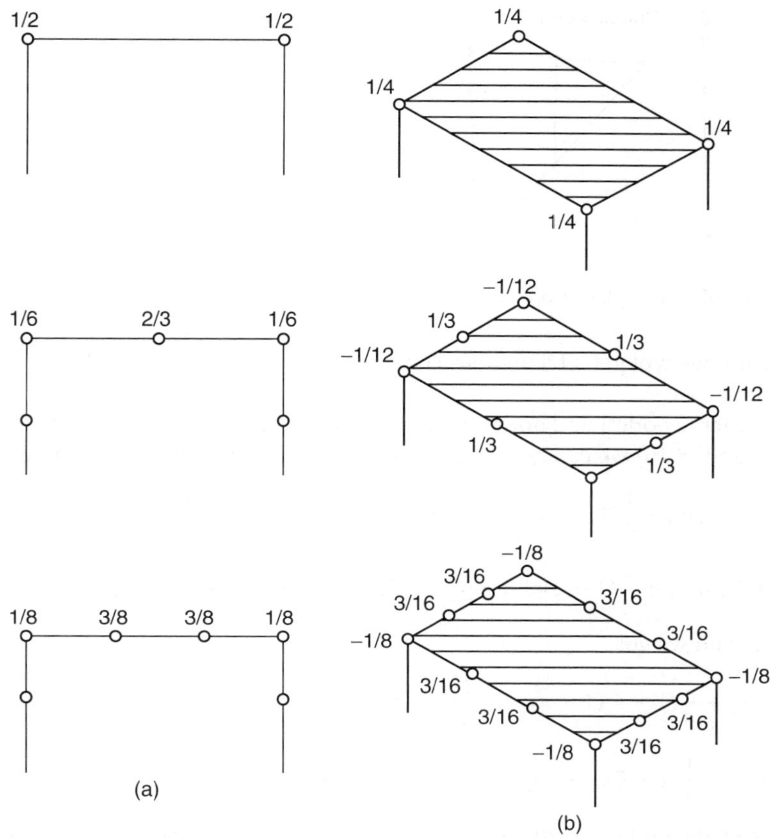

Figure 8.13 Nodal load caused by the uniform surface force acting on top of the plane element and space element (a) plane element and (b) space element.

intuitive distribution only in the case of linear elements. Both quadratic and cubic elements have negative nodal loads on corner points, which is inconsistent with people's intuitive outcomes. However, it offers a more precise stress and strain than nodal loads by intuitive distribution. Figure 8.13 demonstrates the nodal load caused by the uniform surface force.

8.10 Degradation of Isoparametric Elements

Isoparametric elements are highly adaptable to irregular geometric shapes; however in certain particularly complex situations, we need to adopt triangular (plane problems) or sphenoid (space problems) elements. This section will explain how to degrade an isoparametric element into a triangular or sphenoid element [1].

8.10.1 Degradation of 4-Node Plane Isoparametric Elements

For a 4-node plane isoparametric element, by merging 2 adjacent nodes, we will obtain a triangular element. For example, in Figure 8.14, just let node 3 and node 4 share the

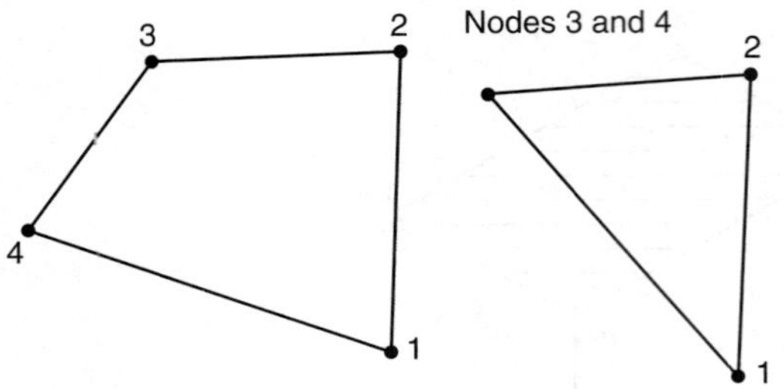

Figure 8.14 Degradation of 4-node plane isoparametric elements.

same coordinates, and we will get a triangular element. Next we will illustrate it with a calculation example.

In Figure 8.15, we merge node 1 and node 2 to get a constant strain triangular element. By formulas (8.6) and (8.15) we can get the coordinates within the 4-node element as

$$x = \frac{1}{4}(1+\xi)(1+\eta)x_1 + \frac{1}{4}(1-\xi)(1+\eta)x_2 + \frac{1}{4}(1-\xi)(1-\eta)x_3 + \frac{1}{4}(1+\xi)(1-\eta)x_4$$

$$y = \frac{1}{4}(1+\xi)(1+\eta)y_1 + \frac{1}{4}(1-\xi)(1+\eta)y_2 + \frac{1}{4}(1-\xi)(1-\eta)y_3 + \frac{1}{4}(1+\xi)(1-\eta)y_4$$

Let $x_1 = x_2$, $y_1 = y_2$ and we get

$$x = \frac{1}{2}(1+\eta)x_2 + \frac{1}{4}(1-\xi)(1-\eta)x_3 + \frac{1}{4}(1+\xi)(1-\eta)x_4$$

$$y = \frac{1}{2}(1+\eta)y_2 + \frac{1}{4}(1-\xi)(1-\eta)y_3 + \frac{1}{4}(1+\xi)(1-\eta)y_4$$

Substituting the specific nodal coordinates shown in Figure 8.15 into the previous two formulas, we have

$$x = \frac{1}{2}(1+\xi)(1-\eta)$$

$$y = 1 + \eta$$

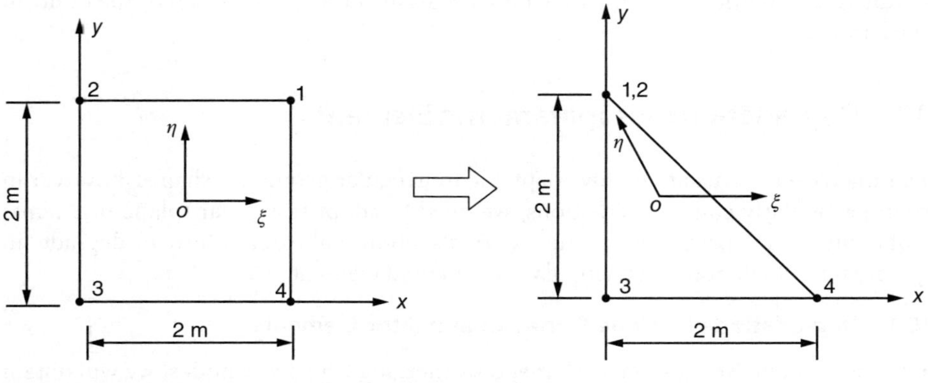

Figure 8.15 Degradation of a 4-node plane element into a 3-node element.

Then it is known

$$\frac{\partial x}{\partial \xi} = \frac{1}{2}(1-\eta), \quad \frac{\partial y}{\partial \xi} = 0$$

$$\frac{\partial x}{\partial \eta} = -\frac{1}{2}(1+\xi), \quad \frac{\partial y}{\partial \eta} = 1$$

By formula (8.22), we can get

$$[J] = \begin{bmatrix} \dfrac{\partial x}{\partial \xi} & \dfrac{\partial y}{\partial \xi} \\ \dfrac{\partial x}{\partial \eta} & \dfrac{\partial y}{\partial \eta} \end{bmatrix} = \frac{1}{2}\begin{bmatrix} 1-\eta & 0 \\ -(1+\xi) & 2 \end{bmatrix}$$

$$[J]^{-1} = \begin{bmatrix} \dfrac{2}{1-\eta} & 0 \\ \dfrac{1+\xi}{1-\eta} & 1 \end{bmatrix} \tag{a}$$

The displacement of any point within the element is

$$u = \frac{1}{2}(1+\eta)u_2 + \frac{1}{4}(1-\xi)(1-\eta)u_3 + \frac{1}{4}(1+\xi)(1-\eta)u_4$$

$$v = \frac{1}{2}(1+\eta)v_2 + \frac{1}{4}(1-\xi)(1-\eta)v_3 + \frac{1}{4}(1+\xi)(1-\eta)v_4$$

Then it is known that

$$\frac{\partial u}{\partial \xi} = -\frac{1}{4}(1-\eta)u_3 + \frac{1}{4}(1-\eta)u_4, \quad \frac{\partial v}{\partial \xi} = -\frac{1}{4}(1-\eta)v_3 + \frac{1}{4}(1-\eta)v_4$$

$$\frac{\partial u}{\partial \eta} = \frac{1}{2}u_2 - \frac{1}{4}(1-\xi)u_3 - \frac{1}{4}(1+\xi)u_4, \quad \frac{\partial v}{\partial \eta} = \frac{1}{2}v_2 - \frac{1}{4}(1-\xi)v_3 - \frac{1}{4}(1+\xi)v_4$$

By formula (8.24) it is proved that

$$\begin{Bmatrix} \dfrac{\partial u}{\partial x} \\ \dfrac{\partial u}{\partial y} \end{Bmatrix} = \begin{bmatrix} \dfrac{2}{1-\eta} & 0 \\ \dfrac{1+\xi}{1-\eta} & 1 \end{bmatrix} \begin{bmatrix} 0 & 0 & -\dfrac{1-\eta}{4} & 0 & \dfrac{1-\eta}{4} & 0 \\ \dfrac{1}{2} & 0 & -\dfrac{1-\xi}{4} & 0 & -\dfrac{1+\xi}{4} & 0 \end{bmatrix} \begin{Bmatrix} u_2 \\ v_2 \\ u_3 \\ v_3 \\ u_4 \\ v_4 \end{Bmatrix} = \begin{bmatrix} 0 & 0 & -\dfrac{1}{2} & 0 & \dfrac{1}{2} & 0 \\ \dfrac{1}{2} & 0 & -\dfrac{1}{2} & 0 & 0 & 0 \end{bmatrix} \begin{Bmatrix} u_2 \\ v_2 \\ u_3 \\ v_3 \\ u_4 \\ v_4 \end{Bmatrix}$$

Similarly it is known that

$$\begin{Bmatrix} \dfrac{\partial v}{\partial x} \\ \dfrac{\partial v}{\partial y} \end{Bmatrix} = \begin{bmatrix} 0 & 0 & 0 & -\dfrac{1}{2} & 0 & \dfrac{1}{2} \\ 0 & \dfrac{1}{2} & 0 & -\dfrac{1}{2} & 0 & 0 \end{bmatrix} \begin{Bmatrix} u_2 \\ v_2 \\ u_3 \\ v_3 \\ u_4 \\ v_4 \end{Bmatrix}$$

Thus the strain of any point within the element is

$$\{\varepsilon\} = \begin{Bmatrix} \varepsilon_x \\ \varepsilon_y \\ \gamma_{zy} \end{Bmatrix} = \begin{bmatrix} 0 & 0 & -\frac{1}{2} & 0 & \frac{1}{2} & 0 \\ 0 & \frac{1}{2} & 0 & -\frac{1}{2} & 0 & 0 \\ \frac{1}{2} & 0 & -\frac{1}{2} & -\frac{1}{2} & 0 & \frac{1}{2} \end{bmatrix} \begin{Bmatrix} u_2 \\ v_2 \\ u_3 \\ v_3 \\ u_4 \\ v_4 \end{Bmatrix} \tag{b}$$

We can see that the strain of any point within the element is irrelevant with and it is a constant strain triangular element.

The previous computing example is degradation of a square element into a right-angled triangular element, so as to simplify the calculation. In fact adopting the same steps, we can prove that for any 4-node plane element, only merging 2 adjacent nodes will produce a constant strain triangular element.

From formula (a) it is clear that for the previous computing example, when $\eta = +1$ the Jacobian matrix $[J]$ is singular, but this singularity disappeared during calculation of strains. This means that in a degraded element, it is not proper to calculate strains at the merging point of 2 nodes. In fact, degraded elements have constant strains, so we can calculate the stress at the centroid of the element ($\xi = 0, \eta = 0$).

8.10.2 Degradation of an 8-Node Space Isoparametric Element

As shown in Figure 8.16, for an 8-node space isoparametric element, just merging 2 or more adjacent nodes will degrade it to a 4–7-node element. In the next we will illustrate it with a calculation example.

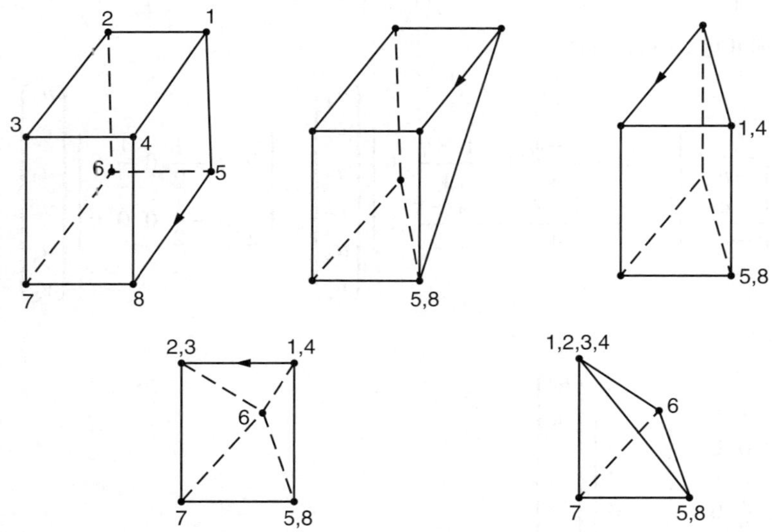

Figure 8.16 Degradation of an 8-node space isoparametric element.

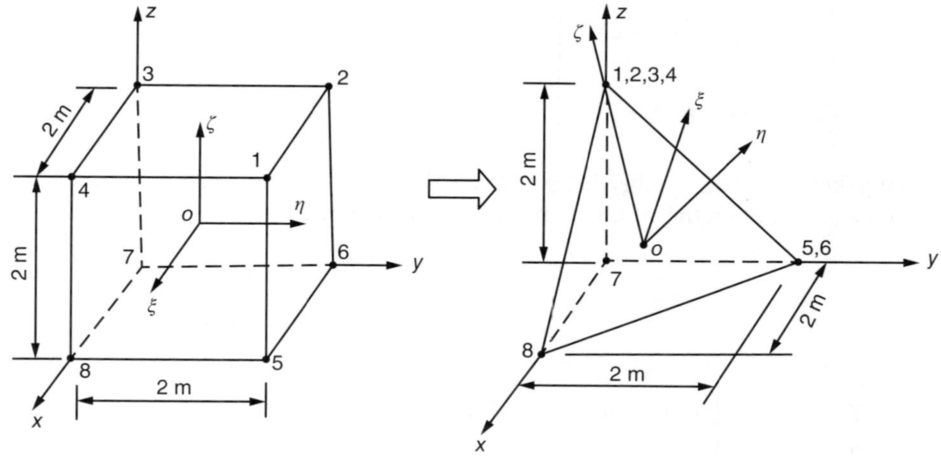

Figure 8.17 Degradation of an 8-node space element into a tetrahedral element.

As shown in Figure 8.17, we merge nodes 1, 2, 3, and 4 into one node and nodes 5 and 6 into one node, and then we get a tetrahedral element. Do calculation just as the previous plane example. First use the shape function (8.11) to express the coordinates of any point within the element like formula (8.16), and then substitute the coordinates of every node of the tetrahedral element into it to get

$$x = \frac{1}{4}(1+\xi)(1-\eta)(1-\zeta), \quad y = \frac{1}{2}(1+\eta)(1-\zeta), \quad z = 1+\zeta$$

Then we will see

$$[J] = \frac{1}{4}\begin{bmatrix} (1-\eta)(1-\zeta) & 0 & 0 \\ -(1+\xi)(1-\zeta) & 2(1-\zeta) & 0 \\ -(1+\xi)(1-\eta) & -2(1+\eta) & 4 \end{bmatrix}$$

$$[J]^{-1} = \begin{bmatrix} \dfrac{4}{(1-\eta)(1-\zeta)} & 0 & 0 \\ \dfrac{2(1+\xi)}{(1-\eta)(1-\zeta)} & \dfrac{2}{1-\zeta} & 0 \\ \dfrac{2(1+\xi)}{(1-\eta)(1-\zeta)} & \dfrac{1+\xi}{1-\zeta} & 1 \end{bmatrix} \tag{c}$$

Use the shape function (8.11) to represent the displacement, then let $u_1 = u_2 = u_3 = u_4, u_5 = u_6, v_1 = v_2, u_3 = u_4, \ldots$, and get

$$\left.\begin{aligned} u &= f_4u_4 + f_5u_5 + f_7u_7 + f_8u_8 \\ v &= f_4v_4 + f_5v_5 + f_7v_7 + f_8v_8 \\ w &= f_4w_4 + f_5w_5 + f_7w_7 + f_8w_8 \end{aligned}\right\} \tag{d}$$

where

$$f_4 = \frac{1}{2}(1+\zeta), \quad f_5 = \frac{1}{4}(1+\eta)(1-\zeta)$$

$$f_7 = \frac{1}{8}(1-\xi)(1-\eta)(1-\zeta)$$

$$f_8 = \frac{1}{8}(1+\xi)(1-\eta)(1-\zeta)$$

Solving $\partial u/\partial\xi, \partial u/\partial\eta, \ldots, \partial w/\partial\zeta$ by formula (d), then we use formula (8.24) and $[J]^{-1}$ in formula (c) to obtain the strain of any point in the element as follows:

$$\begin{Bmatrix} \dfrac{\partial u}{\partial x} \\ \dfrac{\partial v}{\partial y} \\ \dfrac{\partial w}{\partial z} \\ \dfrac{\partial u}{\partial y}+\dfrac{\partial v}{\partial x} \\ \dfrac{\partial v}{\partial z}+\dfrac{\partial w}{\partial y} \\ \dfrac{\partial u}{\partial z}+\dfrac{\partial w}{\partial x} \end{Bmatrix} = \begin{bmatrix} 0 & 0 & 0 & 0 & 0 & 0 & -\frac{1}{2} & 0 & 0 & \frac{1}{2} & 0 & 0 \\ 0 & 0 & 0 & 0 & \frac{1}{2} & 0 & 0 & -\frac{1}{2} & 0 & 0 & 0 & 0 \\ 0 & 0 & \frac{1}{2} & 0 & 0 & 0 & 0 & 0 & -\frac{1}{2} & 0 & 0 & 0 \\ 0 & 0 & 0 & \frac{1}{2} & 0 & 0 & -\frac{1}{2} & -\frac{1}{2} & 0 & 0 & \frac{1}{2} & 0 \\ 0 & \frac{1}{2} & 0 & 0 & 0 & \frac{1}{2} & 0 & -\frac{1}{2} & -\frac{1}{2} & 0 & 0 & 0 \\ \frac{1}{2} & 0 & 0 & 0 & 0 & 0 & -\frac{1}{2} & 0 & -\frac{1}{2} & 0 & 0 & -\frac{1}{2} \end{bmatrix} \begin{Bmatrix} u_4 \\ v_4 \\ w_4 \\ \cdots \\ u_5 \\ v_5 \\ w_5 \\ \cdots \\ u_7 \\ v_7 \\ w_7 \\ \cdots \\ u_8 \\ v_8 \\ w_8 \end{Bmatrix} \tag{e}$$

By formula (e) it is evident that the strain within a degraded element is a constant.

8.10.3 Degradation of High-Order Elements

As explained before, for a linear element, whether it is two-dimensional or 3D problem, just merging adjacent nodes, we can get a degraded element with no need to make any modification of the shape function of the element, quite easy for application. It is different for high-order elements. After degradation, the shape function of the element must be modified [2].

In the following text we will take the plane problem, for example, to illustrate this point. Figure 8.18(a) is an 8-node square element. We merge nodes 1, 2, and 5 and get a triangular element. As shown in Figure 8.18(b), it is an equilateral triangle. In this case, the element must be isotropic in space, which means that variation of the displacements u and v within the element must be the same near every corner point and midpoint of edges. But if merging nodes 1, 2, and 5 of the 8-node element but not modifying the shape function of the element, the triangular element obtained does not meet the previous requirements on isotropism. To meet this requirement, we must modify the shape functions of nodes 3, 4, and 7 and will obtain the following shape function of

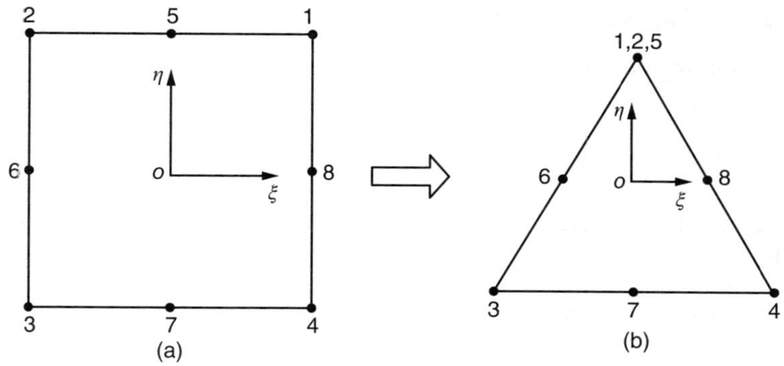

Figure 8.18 Degradation of an 8-node square element into a 6-node equilateral triangle element. (a) Square and (b) equilateral triangle.

the triangular element:

$$\left.\begin{aligned}
N_1^* &= \frac{1}{2}(1+\eta) - \frac{1}{2}(1-\eta^2) \\
N_3^* &= \frac{1}{4}(1-\xi)(1-\eta) - \frac{1}{4}(1-\eta^2)(1-\xi) - \frac{1}{4}(1-\xi^2)(1-\eta) + \Delta N \\
N_4^* &= \frac{1}{4}(1+\xi)(1-\eta) - \frac{1}{4}(1-\xi^2)(1-\eta) - \frac{1}{4}(1-\eta^2)(1+\xi) + \Delta N \\
N_6^* &= \frac{1}{2}(1-\eta^2)(1-\xi) \\
N_7^* &= \frac{1}{2}(1-\xi^2)(1-\eta) - 2\Delta N \\
N_8^* &= \frac{1}{2}(1-\eta^2)(1+\xi)
\end{aligned}\right\} \tag{f}$$

where

$$\Delta N = \frac{1}{8}(1-\xi^2)(1-\eta^2) \tag{g}$$

8.11 Numerical Integration

In order to compute the stiffness matrix and nodal load of curved elements, it is necessary to calculate the integral as that of formula (8.28). Usually the integrand $G(\xi,\eta,\zeta)$ is very complicated, and it is difficult to let its integral be represented by an explicit expression. So we always adopt the numerical integration method to calculate the integral value, that is to say, select certain points in the element as the integral points, calculate the values of the integrand $G(\xi,\eta,\zeta)$ on these integral points, and obtain the integral value by these values.

There are two approaches for numerical integration: one adopts equispaced integral points, for example, Simpson method, and the other adopts unequally spaced integral points, for example, the Gauss method. By the finite element method, we always choose the Gauss method because the integrand is complex; by Gauss method, a few integral points will ensure a relatively high precision of integration.

8.11.1 One-Dimensional Gauss Quadrature Formula

We are required to compute the following integral value within the interval [−1, 1] (Figure 8.19):

$$I = \int_{-1}^{1} f(\xi)\,\mathrm{d}\xi \tag{a}$$

We choose the integral points as $\xi_1, \xi_2, \ldots, \xi_n$ and then calculate the integral value by the following formula:

$$\begin{aligned} I &= \int_{-1}^{1} f(\xi)\,\mathrm{d}\xi = H_1 f(\xi_1) + H_2 f(\xi_2) + \cdots + H_n f(\xi_n) \\ &= \sum_{i}^{n} H_i f(\xi_i) \end{aligned} \tag{8.35}$$

where H_i is weight coefficient. H_i and ξ_i are so chosen to obtain the maximum precision. In order to achieve the highest precision, the integral points ξ_i should be roots of Legendre polynomial $L_n(\xi)$, and the weighted coefficients H_i should be determined by the following formula:

$$H_i = \frac{2}{(1-\xi_i^2)[L_n'(\xi_i)]^2} \tag{b}$$

For easy calculation, Table 8.4 presents the values of integral points ξ_i and weighting factors H_i. In the case of n integral points, the Gauss quadrature formula (8.35) can produce accurate results for any polynomial $f(\xi)$ no higher than $2n-1$ in the order.

Now take another example. Let $n=4$. In Table 8.1 we know $\xi_1 = -0.86113$, $\xi_2 = -0.33998$, $\xi_3 = 0.33998$, $\xi_4 = 0.86113$, $H_1 = 0.34785$, $H_2 = 0.65214$, $H_3 = 0.65214$, and $H_4 = 0.34785$, substitute them into formula (8.35), and obtain

$$\begin{aligned} I = {} & 0.34785 f(-0.86113) + 0.65214 f(-0.33998) + 0.65214 f(0.33998) \\ & + 0.34785 f(0.86113) \end{aligned}$$

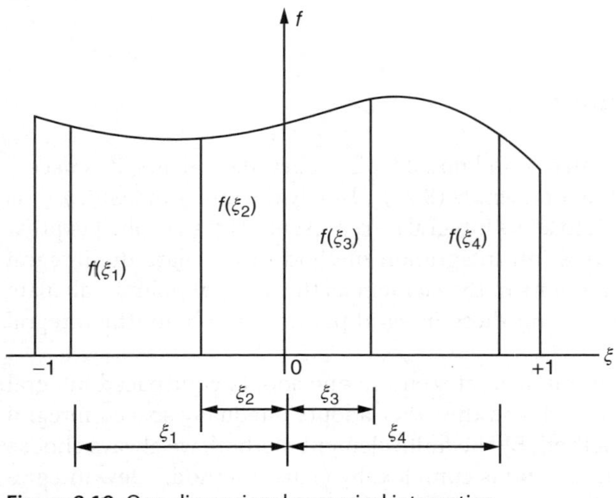

Figure 8.19 One-dimensional numerical integration.

Table 8.4 Integral point coordinates and weighting coefficient of the Gauss quadrature formulas $\left[\int_{-1}^{1} f(\xi)\,d\xi = \sum_{i=1}^{n} H_i f(\xi_i)\right]$.

$\pm\xi_i$		H_i
0.5773502692	$n=2$	1.0000000000
0.7745966692	$n=3$	0.5555555556
0.0000000000		0.8888888889
0.8611363116	$n=4$	0.3478548451
0.3399810436		0.6521451549
0.9061798459	$n=5$	0.2369268851
0.5384693101		0.4786286705
0.0000000000		0.5688888889

8.11.2 Two-Dimensional and Three-Dimensional Gauss Quadrature Formulas

Calculate the double integrals as follows:

$$I = \int_{-1}^{1}\int_{-1}^{1} f(\xi,\eta)\,d\xi\,d\eta \tag{8.36}$$

First let η be a constant, and calculate the integral in the ξ direction:

$$\int_{-1}^{1} f(\xi,\eta)\,d\xi = \sum_{j=1}^{n} H_j f(\xi_j,\eta) = \phi(\eta)$$

Then calculate the integral in the η direction and obtain

$$I = \int_{-1}^{1} \phi(\eta)\,d\eta = \sum_{i=1}^{n} H_i\phi(\eta_i) = \sum_{i=1}^{n} H_i \sum_{j=1}^{n} H_j f(\xi_j,\eta_i) = \sum_{i=1}^{n}\sum_{j=1}^{n} H_i H_j f(\xi_j,\eta_i) \tag{8.37}$$

Similarly, for the triple integral, it is proved that

$$I = \int_{-1}^{1}\int_{-1}^{1}\int_{-1}^{1} f(\xi,\eta,\zeta)\,d\xi\,d\eta\,d\zeta = \sum_{m=1}^{m}\sum_{j=1}^{m}\sum_{i=1}^{m} H_i H_j H_m f(\xi_i,\eta_j,\zeta_m) \tag{8.38}$$

The foregoing calculations all adopt the same number of integral points in every direction, which is not necessary. Sometimes different integral points may be adopted in different directions.

Figure 8.20 illustrates layout of 3×3 integral points in a plane square area. As $n=3$, the five-order polynomial can be achieved in the accuracy in every direction. For 3D elements, there will be n^3 integral points.

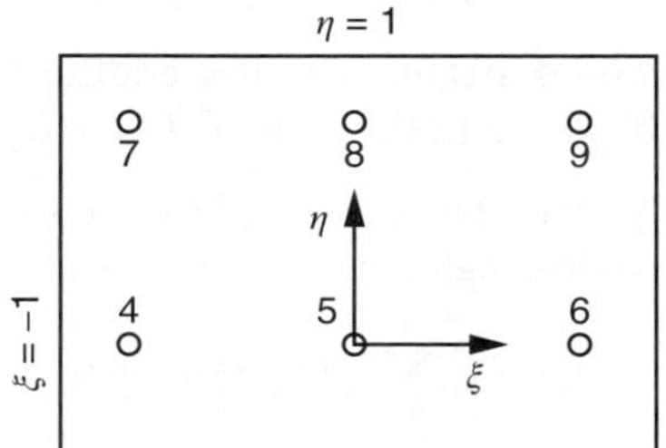

Figure 8.20 Square area. Integral points for $N=3$.

For plane problems, the calculation amount approximately equals to n^2; for space problems, the calculation amount reaches nearly n^3.

Based on experiences, on the premise of ensuring convergence, the number of integral points n should be as small as possible.

8.12 Selection of the Numerical Integration Order

In an isoparametric element, the stiffness matrix is obtained by numerical integration with large calculation amount and much computer work time. Therefore, the principle to choose the order of numerical integration is reducing the order as much as possible on the premise of ensuring necessary calculation precision in order to save costs.

8.12.1 Conditions for Nonsingularity of the Global Stiffness Matrix [*K*]

The condition for the global equilibrium equation $[K]\{\delta_1\} = \{P\}$ having a unique nonzero solution is the existence of inverse matrix of the stiffness matrix, that is to say, after introducing boundary conditions, $[K]$ must be nonsingular, and the conditions for this nonsingularity is $|K| \neq 0$, or $[K]$ is full rank. Let the structural DOF be N in consideration of boundary constraint conditions, and then $[K]$ is an N-order square matrix. The conditions for an equilibrium equation set having a unique nonzero solution are

$$r(K) = N \tag{a}$$

in which $r(K)$ is rank of matrix $[K]$.

As for the rank of matrices, there are two basic rules as follows:

1) The rank rule for matrix multiplication is

$$r(ABC) \leq \min\left[r(A)\ r(B)\ r(C)\right] \tag{b}$$

which means that after multiplication of matrix $[A]$, $[B]$, $[C]$, the rank $r\ (ABC)$ is not bigger than the smallest one of $r(A)$, $r(B)$, $r(C)$.

2) The rank rule for matrix adding is

$$r(A + B + C) \leq r(A) + r(B) + r(C) \tag{c}$$

which means that after addition of matrix $[A]$, $[B]$, $[C]$, the rank $r\ (A + B + C)$ is no bigger than the sum of $r(A)$, $r(B)$, $r(C)$.

By formulas (8.28) and (8.38), we know the element stiffness matrix is calculated by the following formula:

$$[K]^e = \sum_{i=1}^{g} h_i [B]^T [D][B]|J| \tag{d}$$

in which h_i is the Gauss integral coefficient; g is the number of Gauss integral points within the element; $[D]$ is an elasticity matrix, which is a $d \times d$ square matrix with rank $r(D) = d$; d is the strain components calculated in every point as well as the number of independent relations of every Gauss point, the plane problem $d = 3$, axisymmetric problem $d = 4$, and 3D problem $d = 6$; $[B]$ is a strain matrix and $d \times e$ matrix; and e is the DOF of elemental node.

Usually $d < e$. Thus the rank of $[B]$ is $r(B) = d$. Considering Equation (b), the rank of $[B]^T[D][B]$ is less than or equal to d. According to formula (c), the rank of element stiffness matrix $[K]$ is less than or equal to dg. Supposing there are m elements, by formula (c), the rank of global stiffness matrix $[K]$ is

$$r(K) \leq mgd = S \tag{e}$$

On every integral point there are d independent relations and there are total mg integral points, so S is total number of independent relations. By formula (a), the necessary condition of an equilibrium equation set having a unique nonzero solution is

$$N = r(K) \leq mgd = S \tag{f}$$

It means the number of DOFs is not greater than the number of independent relations. If $N > S$, the stiffness matrix $[K]$ is singular and there is no unique solution to the equilibrium equation set.

In the following text we will take plane elasticity problems, for example, adopt linear and quadratic elements and select one-point and four-point Gauss integrals, respectively. Now the number of strain components $d = 3$. On every integral point there are three independent relations, so the total number of independent relations S is three times that of the Gauss integral points. Every node has two DOFs, and the total number of structural DOFs N equals to two times the number of nodes minus the number of constraints.

As shown in Figure 8.21, for a one-point integral linear element, $[K]$ is singular in any of (a), (b), (c). The exception is case (d), in which only after constraining all the 3 nodes on the left, the necessary condition for nonsingularity of the matrix is met. Just because of this, one-point integral linear element is rarely adopted. For a 2×2 integral quadratic element, except case (a) that has only one element, other cases all meet the necessary conditions of nonsingularity of the matrix, so it is more frequently applied in practice.

8.12.2 Integral Order Ensuring the Calculation Precision

As explained previously, for a one-dimensional Gauss integral represented by formula (8.35), if taking n integral points, we can get accurate outcomes for any polynomial $f(\xi)$ with order not higher than $2n-1$.

Taking the integral of an element stiffness matrix, for example, by formula (8.28), we can see that the integrand is

$$G(\xi, \eta, \zeta) = [B]^T[D][B]|J|$$

If the highest order of ξ in $G(\xi, \eta, \zeta)$ is m, then the integral points in the ξ direction can be set as

$$n \geq (m+1)/2 \tag{g}$$

For example, let the coordinate transformation formula of a 4-node rectangular element $2a \times 2b$ be

$$x = a\xi, \quad y = b\eta$$

Then

$$\xi = x/a, \quad \eta = y/b$$

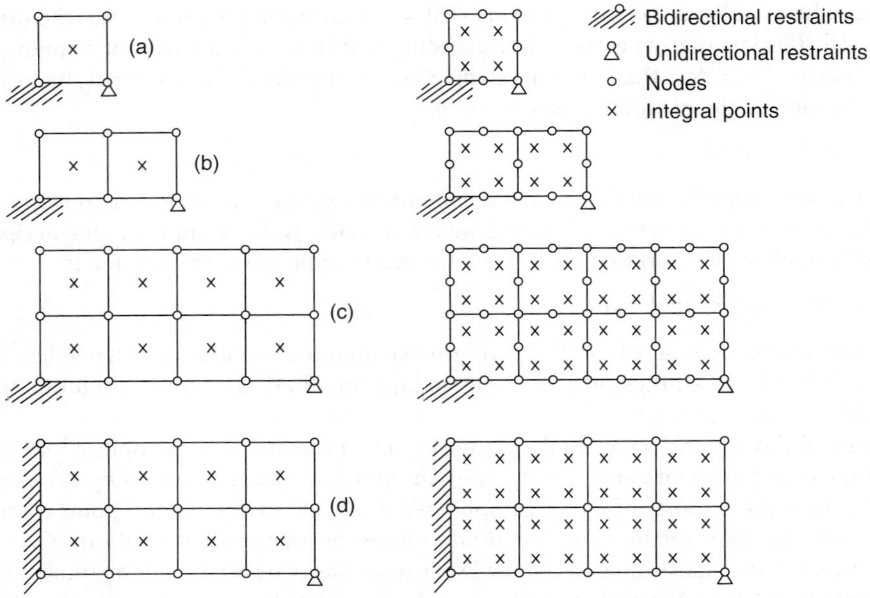

	Linear element	Quadratic element
	N S	N S
(a)	4×2−3 = 5>1×3 = 3 Singular	2×8−3 = 15>4×3 = 12 Singular
(b)	6×2−3 = 9>2×3 = 6 Singular	13×2−3 = 23<8×3 = 24
(c)	15×2−3 = 27>8×3 = 24 Singular	37×2−3 = 71<32×3 = 96
(d)	15×2−7 = 23<8×3 = 24	37×2−11 = 63<32×3 = 96

Figure 8.21 Singularity examination of stiffness matrices of elastic plane problems (if $N > S$, singular).

Substitute the previous formula into the two-dimensional linear shape function (8.6). After merging we obtain the displacement function as

$$u = \beta_1 + \beta_2 x + \beta_3 y + \beta_4 xy$$

$$v = \beta_5 + \beta_6 x + \beta_7 y + \beta_8 xy$$

Thus we get three strain components as

$$\varepsilon_x = \beta_2 + \beta_6 b\eta, \quad \varepsilon_y = \beta_7 + \beta_8 a\xi, \quad \gamma_{xy} = (\beta_3 + \beta_6) + \beta_4 a\xi + \beta_8 b\eta$$

Besides

$$J = \begin{bmatrix} a & 0 \\ 0 & b \end{bmatrix}, \quad |J| = ab = \text{constant}$$

In summary, we have

$$G(\xi, \eta) = G(\xi^2, \xi\eta, \eta^2)$$

Table 8.5 Integration orders of plane element stiffness matrices.

	Element	Reliable integration orders	Commonly used reduced integration
4-Node rectangular element	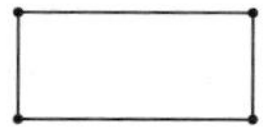	2×2	—
4-Node arbitrary quadrilateral element		2×2	—
8-Node rectangular element		3×3	2×2
8-Node curved-edge element	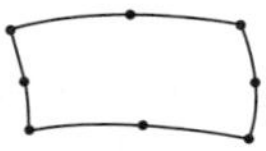	3×3	2×2
9-Node rectangular element	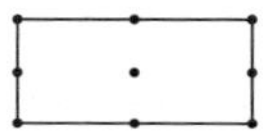	3×3	2×2
9-Node curved-edge element	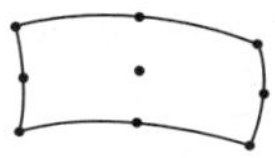	3×3	2×2
16-Node rectangular element	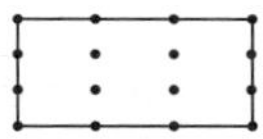	4×4	3×3
16-Node curved-edge element	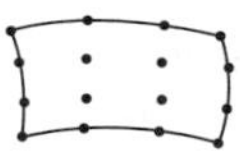	4×4	3×3

In the two directions of ξ, η, $rn = 2$, so using the 2×2 integration scheme, we can obtain accurate integral values.

Generally, is not a constant and the coordinate transformation formula is also complex. It is hard to determine the order of accurate integration. In fact, the integral order of irregularly shaped elements is also chosen in light of regularly shaped elements, as shown in Table 8.5. For rectangular elements, adoption of reliable integration order listed in the table leads to accurate values of the element stiffness matrix; for curved elements, it leads to approximate values. But experiences tell us that the integration orders offered are sufficient.

8.12.3 Reduced Integration and Selected Integration

In the early stage of adoption of curved-edge elements, 3×3 Gauss integration is always used for 8-node plane curved-edge elements, and $3\times3\times3$ Gauss integration is always used for 20-node 3D curved elements. Researches in recent years reveal that for the two types of the previous elements, respective usage of 2×2 and $2\times2\times2$ Gauss integration will remarkably improve the elements' feature and save calculation amount [16].

As for the significance of reduced integration, the following explanation may help your understanding: taking two-dimensional curved-edge elements, for example, it can be proved that adoption of 2×2 Gauss integration to calculate the stiffness matrix integral $\iint [B]^{\mathrm{T}}[D][B]\,\mathrm{d}x\,\mathrm{d}y$ will lead to an outcome tantamount to the $\iint [\tilde{B}]^{\mathrm{T}}[D][\tilde{B}]\,\mathrm{d}x\,\mathrm{d}y$ integral values, in which elements in $[\tilde{B}]$ are obtained through smoothing by least square(s) of elements in strain matrix $[B]$. After smoothing, some false strain values are filtered and a more soft and efficient element comes into being.

The integration order should not be reduced too much. If so, it cannot reflect the various states and features of the element, which may bring down the rank of element stiffness matrix and make the global stiffness matrix singular or ill conditioned. Taking the plane linear element, for example, the precise integral order is 2×2. The one-point integral plan shown in Figure 8.21 is a reduced integration, making the global stiffness matrix singular.

When adopting reduced integration, the following conditions should be met as much as possible:

1) Do not reduce the rank of element stiffness matrix, that is to say, the rank of element stiffness matrix obtained by reduced integration should not be lower than that obtained by precise Gauss integration.
2) The element should contain the constant strain state.

Condition (1) here is to ensure a unique solution for the global equilibrium equation set; condition (2) is to ensure the convergence of solution. Reduced integration inconsistent with the previous conditions may result in false outcomes, which should be noted by users.

Table 8.5 lists the normal integration order and reduced integration order of plane element.

Another approach is the selected integration method, which means adoption of different integration orders for different objects when computing the element stiffness matrix. For instance, in a thin-plate bending element, we adopt the normal integration order for bending strain energy and one reduced order for shearing strain energy.

The problem mentioned previously is integration of element stiffness matrix. As for Gauss integration of node load, the usual adoption is the order identical to that of the element stiffness matrix. As for integration order of two types of mass matrices, for the concentrated mass matrices, only the element volume must be obtained precisely, while the consistent mass matrix requires higher integration order than the element stiffness matrix. This is because in calculation of the consistent mass matrix, we directly adopt the displacement interpolation function and in calculation of the element stiffness matrix, we adopt the derivative of the displacement function. Comparing formulas (22.4) and (22.5) with the stiffness matrix formula (3.11), it is easy to understand the previous assertion.

8.13 Stress Refinement and Stress Smoothing

8.13.1 Stress Refinement

The stress computation formula is

$$\{\sigma\} = [D](\{\varepsilon\} - \{\varepsilon_0\}) + \{\sigma_0\}$$

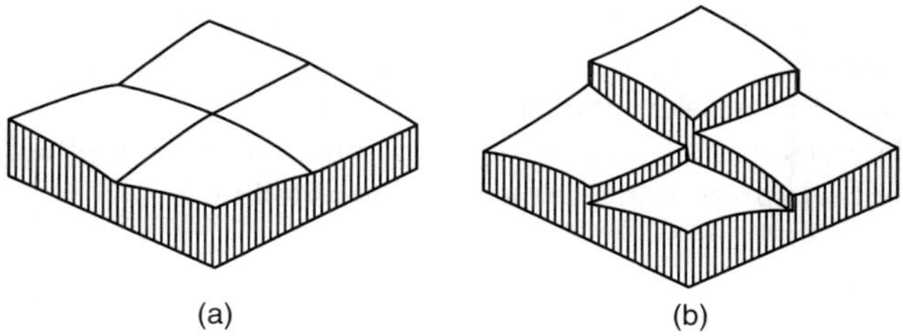

Figure 8.22 Smoothed and unsmoothed stresses. (a) smoothed stress and (b) unsmoothed stress.

The displacement is continuous along the boundary of elements, but the derivatives of displacement are often discontinuous, so the stress on the boundary of elements is also often discontinuous, as shown in Figure 8.22(b). For the curved-side elements by using numerical integration, the experience shows that the computed stress on an integral point has the best accuracy, while that on a node has the worst accuracy. It is not difficult to explain this phenomenon, because the accuracy of an interpolation function on the edge near the interpolation area is usually poor. Therefore, in a shape function, the derivatives and the stress inside elements have higher accuracy than those on the boundary of the elements.

Usually, much attention is paid to the stresses on edges and nodes in engineering. In order to overcome the discontinuity, poor accuracy, and other disadvantages of the stresses on edges and nodes, the stresses of several elements around a node are averaged to obtain the nodal stress $\{\sigma_i\}$ (the stresses cannot be averaged if the materials are different or the thickness mutates). The stress at any point inside an element is interpolated with a shape function as follows:

$$\{\sigma\} = \sum N_i\{\sigma_i\} \tag{a}$$

Here, N_i is a shape function.

The stress computed in equation (a) is continuous throughout the structure, and the corresponding nodal load may be different from the originally given load $\{P\}$, which means that there will be imbalance force on the node. In order to eliminate the imbalance force, the displacement of structure can be adjusted for once, and increments of displacement are calculated in the following equation, that is,

$$[K]\{\Delta\delta\} = \{P\} - \sum \iiint [B]^{\mathrm{T}}\{\sigma\}\,\mathrm{d}x\,\mathrm{d}y\,\mathrm{d}z \tag{b}$$

Here, the right side represents the nodal imbalance force. If the stress is accurate, the imbalance force is zero, $\{\Delta\delta\} = 0$. Otherwise, $\{\Delta\delta\}$ can be computed by equation (b), and the real displacement of the structure is $\{\delta\} + \{\Delta\delta\}$. Thus for computing the more accurate stress $\{\sigma\} + \{\Delta\sigma\}$, iterative computation can be conducted according to equation (b) when necessary, until a satisfactory result is obtained. As $[K]$ is decomposed in advance, so the amount of calculation is not heavy.

8.13.2 Stress Smoothing

Another way of overcoming the stress discontinuity, poor accuracy, and other disadvantages is stress smoothing, and the best way is using the least squares method for the smoothing in the whole area, but this method causes a heavy workload. Thus, a more practical method is using the least squares method for smoothing in each element and then taking the average of the stresses in relevant elements on a node.

We take the smoothed nodal stresses $\sigma_1, \sigma_2, \ldots, \sigma_p$ as unknowns, and the smoothed stress at any point in an element is expressed in a shape function as follows:

$$\tilde{\sigma}(\xi,\eta) = \sum_{i=1}^{p} \tilde{N}_i \tilde{\sigma}_i \tag{8.39}$$

Here, $\tilde{N}_i$ is a shape function of the stress for smoothing and can be different in order from the shape function N_i in the stiffness matrix of the computation element (e.g., N_i can be the second order, while $\tilde{N}_i$ can be the first order). At any point in the element, the error between the smoothed and unsmoothed stresses is

$$e(\xi,\eta) = \sigma(\xi,\eta) - \tilde{\sigma}(\xi,\eta) \tag{8.40}$$

Here, the unsmoothed stress $\sigma(\xi,\eta)$ can be computed by the stress–displacement relationship as follows:

$$\sigma(\xi,\eta) = [D][B]\{\delta\}^*$$

Our task is to find the values $\tilde{\sigma}_1, \tilde{\sigma}_2, \ldots, \tilde{\sigma}_n$ when the minimum value is taken in the functional

$$\phi = \iint e^2(\xi,\eta)\,\mathrm{d}x\,\mathrm{d}y \tag{8.41}$$

In order to take the minimum value in the functional ϕ, we must have

$$\frac{\partial\phi}{\partial\tilde{\sigma}_i} = 0 \quad (i = 1,2,\ldots,p) \tag{8.42}$$

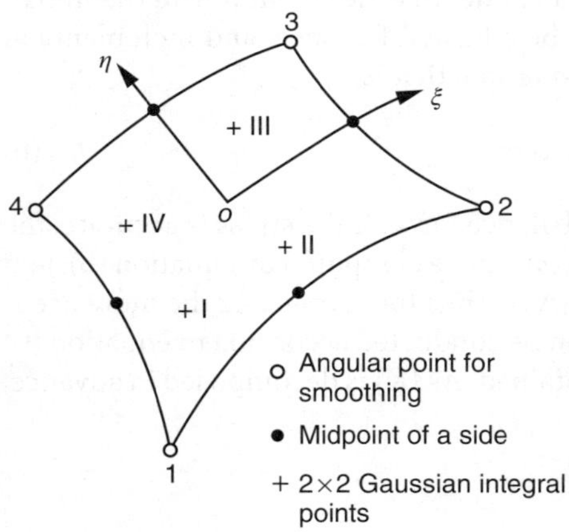

Figure 8.23 Two-dimensional curved-side element.

Here, p is the number of angular points. Equations (8.40) and (8.41) are substituted in the previous equation, and we have the equation set

$$\frac{\partial\phi}{\partial\tilde{\sigma}_i} = \iint (\sigma - \tilde{\sigma})\tilde{N}_i\,\mathrm{d}x\,\mathrm{d}y = 0 \quad (i = 1,2,\ldots,p) \tag{8.43}$$

This equation set contains a total of p equations, and the smoothed nodal stress calculated in equation (8.43) is σ_i, and then, this nodal stress is substituted in equation (8.39) to obtain the smoothed stress at any point inside the element.

Now, take a two-dimensional curved-side element as an example, and see Figure 8.23. We take the smoothed stresses $\tilde{\sigma}_1, \tilde{\sigma}_2, \tilde{\sigma}_3, \tilde{\sigma}_4$

of four angular points as unknowns and use the following linear shape function for smoothing:

$$\left.\begin{aligned} \tilde{N}_1 &= \frac{(1-\xi)(1-\eta)}{4}, \quad \tilde{N}_2 = \frac{(1+\xi)(1-\eta)}{4} \\ \tilde{N}_3 &= \frac{(1+\xi)(1+\eta)}{4}, \quad \tilde{N}_4 = \frac{(1-\xi)(1+\eta)}{4} \end{aligned}\right\} \tag{c}$$

The smoothed stress is expressed as equation (d):

$$\tilde{\sigma}(\xi,\eta) = \tilde{N}_1\tilde{\sigma}_1 + \tilde{N}_2\tilde{\sigma}_2 + \tilde{N}_3\tilde{\sigma}_3 + \tilde{N}_4\tilde{\sigma}_4 \tag{d}$$

Equation (d) is substituted in equation (8.43), and we have the following equation set:

$$\left.\begin{aligned} \iint (\sigma-\tilde{\sigma})\tilde{N}_1|J|\,\mathrm{d}\xi\,\mathrm{d}\eta = 0 \\ \iint (\sigma-\tilde{\sigma})\tilde{N}_2|J|\,\mathrm{d}\xi\,\mathrm{d}\eta = 0 \\ \iint (\sigma-\tilde{\sigma})\tilde{N}_3|J|\,\mathrm{d}\xi\,\mathrm{d}\eta = 0 \\ \iint (\sigma-\tilde{\sigma})\tilde{N}_4|J|\,\mathrm{d}\xi\,\mathrm{d}\eta = 0 \end{aligned}\right\} \tag{e}$$

Equation set (e) is an algebraic equation set with $\tilde{\sigma}_1, \tilde{\sigma}_2, \tilde{\sigma}_3, \tilde{\sigma}_4$ as unknowns, and these equations can be solved after computing the coefficient in each equation by using the Gaussian integration. However, we can use a more convenient solution. Since $\tilde{N}_i$ is a linear shape function, as long as the 2×2 Gaussian integral is used, there will be a total of four integral points, namely,

$$\text{I.}\ \left(-\frac{1}{\sqrt{3}}, -\frac{1}{\sqrt{3}}\right), \quad \text{II.}\ \left(\frac{1}{\sqrt{3}}, -\frac{1}{\sqrt{3}}\right)$$

$$\text{III.}\ \left(\frac{1}{\sqrt{3}}, \frac{1}{\sqrt{3}}\right), \quad \text{IV.}\ \left(-\frac{1}{\sqrt{3}}, \frac{1}{\sqrt{3}}\right)$$

For the four integral points, let the smoothed stress $|\tilde{\sigma}|$ be equal to the unsmoothed stress σ, namely,

$$\left.\begin{aligned} \tilde{\sigma}\left(-\frac{1}{\sqrt{3}}, -\frac{1}{\sqrt{3}}\right) = \sigma_{\mathrm{I}} \\ \tilde{\sigma}\left(\frac{1}{\sqrt{3}}, -\frac{1}{\sqrt{3}}\right) = \sigma_{\mathrm{II}} \\ \tilde{\sigma}\left(\frac{1}{\sqrt{3}}, \frac{1}{\sqrt{3}}\right) = \sigma_{\mathrm{III}} \\ \tilde{\sigma}\left(-\frac{1}{\sqrt{3}}, \frac{1}{\sqrt{3}}\right) = \sigma_{\mathrm{IV}} \end{aligned}\right\} \tag{f}$$

Therefore, we have $\sigma - \tilde{\sigma} = 0$ for each of the four integral points. When we use the 2×2 integrals to obtain the coefficient in each equation in equation set (e), the solution to each equation will be zero, which means that equation set (f) and equation set (e) are equivalent. The smoothed stresses $\tilde{\sigma}_1, \tilde{\sigma}_2, \tilde{\sigma}_3, \tilde{\sigma}_4$ of angular points solved by equation (f) are as follows:

$$\begin{Bmatrix} \tilde{\sigma}_1 \\ \tilde{\sigma}_2 \\ \tilde{\sigma}_3 \\ \tilde{\sigma}_4 \end{Bmatrix} = \begin{bmatrix} a & b & c & b \\ b & a & b & c \\ c & b & a & b \\ b & c & b & a \end{bmatrix} \begin{Bmatrix} \sigma_{\mathrm{I}} \\ \sigma_{\mathrm{II}} \\ \sigma_{\mathrm{III}} \\ \sigma_{\mathrm{IV}} \end{Bmatrix} \tag{8.44}$$

where

$$a = 1 + \frac{\sqrt{3}}{2}, \quad b = -\frac{1}{2}, \quad c = 1 - \frac{\sqrt{3}}{2} \tag{8.45}$$

The $\sigma_{\mathrm{I}}, \sigma_{\mathrm{II}}, \sigma_{\mathrm{III}}, \sigma_{\mathrm{IV}}$ are unsmoothed stresses on four Gaussian integral points I, II, III, and IV. As $\tilde{N}_i$ is a linear shape function, $\tilde{N}_i(\xi, \eta)$ is a linear function of ξ or η on the boundary of the elements, and therefore, the smoothed stress of the midpoint of each side can be obtained by averaging the smoothed stresses of adjacent angular points. The stresses of all other points can be calculated in equation (8.39).

Thus, for the two-dimensional element, we first compute the unsmoothed stresses $\sigma_{\mathrm{I}}, \sigma_{\mathrm{II}}, \sigma_{\mathrm{III}}, \sigma_{\mathrm{IV}}$ on four Gaussian integral points. The smoothed stresses $\tilde{\sigma}_1, \tilde{\sigma}_2, \tilde{\sigma}_3, \tilde{\sigma}_4$ on four angular points can be computed in equation (8.44), and the average of the stresses of adjacent angular points is taken to obtain the smoothed stress of the midpoint of the side.

The smoothed stresses calculated by such a method are computed on the common boundary of all elements, and different elements may have different values, so we can take the average of the stresses of all the elements.

For the 3D element, as shown in Figure 8.24, a trilinear shape function (8.11) is used for smoothing according to the least squares method, and similar results can be obtained. The smoothed stress $\tilde{\sigma}_i$ of eight angular points can be computed in the following equation:

$$\begin{Bmatrix} \tilde{\sigma}_1 \\ \tilde{\sigma}_2 \\ \tilde{\sigma}_3 \\ \tilde{\sigma}_4 \\ \tilde{\sigma}_5 \\ \tilde{\sigma}_6 \\ \tilde{\sigma}_7 \\ \tilde{\sigma}_8 \end{Bmatrix} = \begin{bmatrix} a & b & c & b & b & c & d & c \\ b & a & b & c & c & b & c & d \\ c & b & a & b & d & c & b & c \\ b & c & b & a & c & d & c & b \\ b & c & d & c & a & b & c & b \\ b & b & c & d & b & a & b & c \\ d & c & b & c & c & b & a & b \\ c & d & c & b & b & c & b & a \end{bmatrix} \begin{Bmatrix} \sigma_{\mathrm{I}} \\ \sigma_{\mathrm{II}} \\ \sigma_{\mathrm{III}} \\ \sigma_{\mathrm{IV}} \\ \sigma_{\mathrm{V}} \\ \sigma_{\mathrm{VI}} \\ \sigma_{\mathrm{VII}} \\ \sigma_{\mathrm{VIII}} \end{Bmatrix} \tag{8.46}$$

where

$$a = \frac{5+3\sqrt{3}}{4}, \quad b = -\frac{\sqrt{3}+1}{4}, \quad c = \frac{\sqrt{3}-1}{4}, \quad d = \frac{5-3\sqrt{3}}{4}$$

Here, $\sigma_{\mathrm{I}}, \sigma_{\mathrm{II}}, \ldots, \sigma_{\mathrm{VIII}}$ are unsmoothed stresses of eight Gaussian integral points $2\times2\times2$.

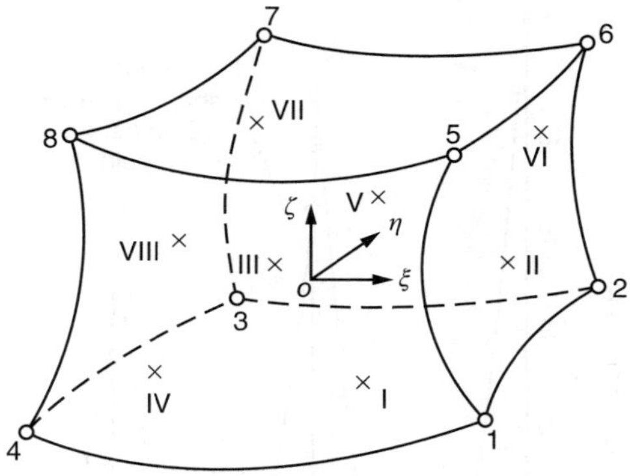

Figure 8.24 Three-dimensional curved-side element.

The smoothed stress of the midpoint of each side can be obtained by averaging the smoothed stresses of adjacent angular points, and the smoothed stress at any point inside an element can be calculated in the following equation:

$$\tilde{\sigma}(\xi, \eta, s) = \sum_{i=1}^{s} \tilde{N}_i \tilde{\sigma}_i \tag{g}$$

where $\tilde{N}_i$ is a linear shape function expressed in equation (8.11).

Now, we summarize the stress smoothing method of curved-side elements as follows:

1) Computing the unsmoothed stresses $\sigma_{\mathrm{I}}, \sigma_{\mathrm{II}}, \ldots$ on all integral points
2) Calculating the smoothed stresses $\tilde{\sigma}_1, \tilde{\sigma}_2, \tilde{\sigma}_3, \tilde{\sigma}_4, \ldots$ on all nodes by equation (8.44) or (8.46)
3) Taking the average of the stresses of related elemental nodes on the common node of different elements as the nodal stress
4) Taking the average of the stresses of adjacent nodes as the stress of the midpoint of each side

Here is an example to illustrate the state of stress before and after smoothing. As shown in Figure 8.25, a cantilever was calculated, and the height-to-length ratio was $h/l = 1/20$. It bore a uniform load and was divided into four 8-node rectangular elements. The horizontal normal stress σ_x was calculated by the 2×2 Gaussian integration, and the numerical values were very consistent with the theoretical solutions before and after smoothing. However, the unsmoothed shear stress was very far from the theoretical solution, while the smoothed shear stress was identical with the theoretical solution.

8.14 Elemental Form and Layout

For an isoparametric element, as a result of the coordinate transformation, the variation of strain in the element depends on the order of shape function and relates to the elemental shape and the position of the node of each side, as explained in the following.

8.14.1 Effect of the Elemental Shape on Strain

For the child element as shown in Figure 8.26, as long as the four apex angles of the element remain at 90°, a great change in the value of the side length a/b does not affect the

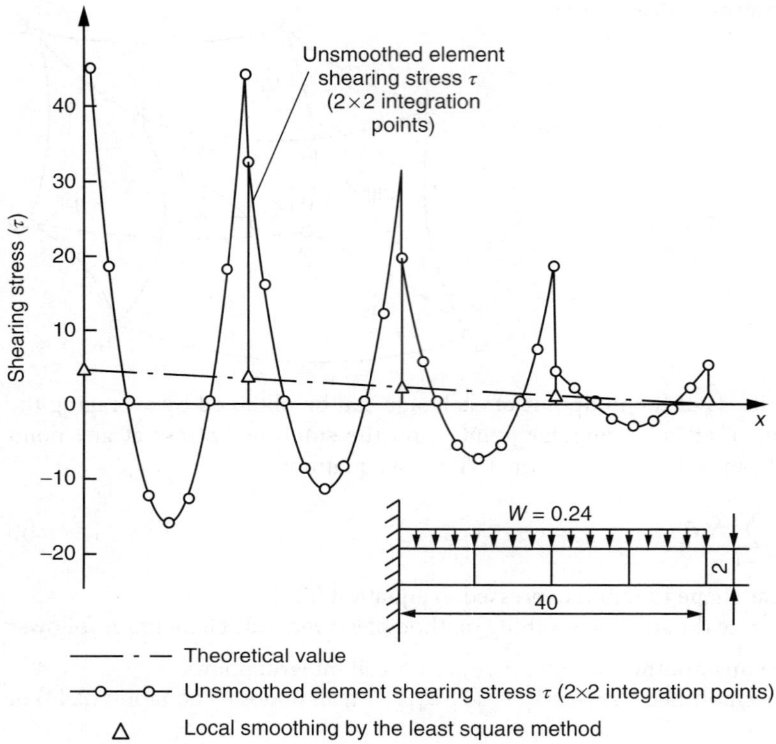

Figure 8.25 Shearing stresses on the neutral axis of a beam.

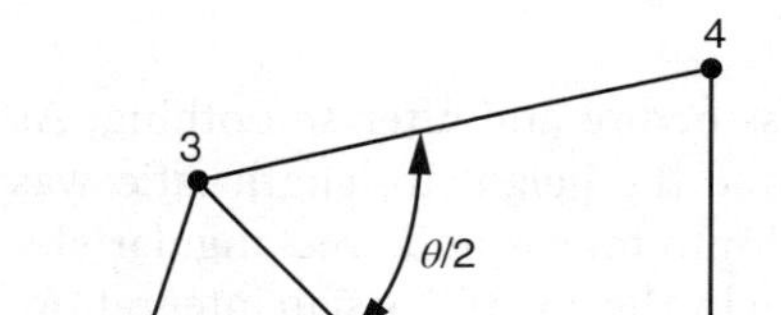

Figure 8.26 Quadrilateral element.

computation results. However, if the apex angles deviate from 90° farther, the computation results may be affected remarkably. For example, for the quadrilateral element shown in Figure 8.26, the determinant of the Jacobian matrix of node 3 is

$$|J| = a^2 ctg\frac{\theta}{2}$$

When θ approaches 180° (in Figure 8.27), $ctg\frac{\theta}{2} \rightarrow 0$, the Jacobian matrix cannot be inverted, and the strain at node 3 cannot be obtained. Even if the strain near node 3 can be calculated, the calculation error will be big.

Thus, in the case of arranging the computational grid of curved-side elements, the apex angles of each element cannot be close to 180° and should generally remain at about 90° as far as possible. The computational grids shown in Figure 8.27 (a–c) are bad.

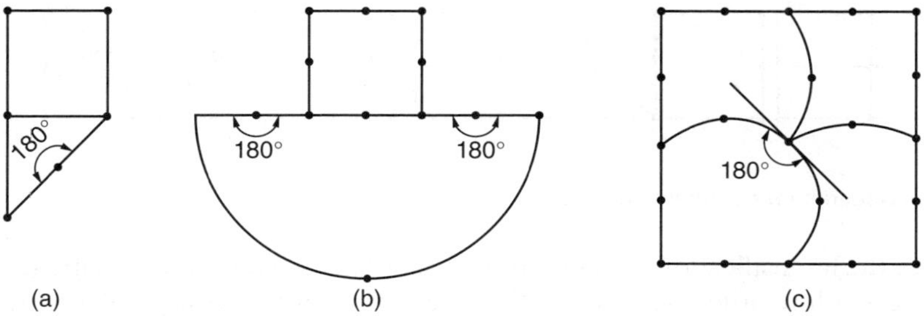

Figure 8.27 Inappropriate computational grids ($\theta = 180°$).

8.14.2 Effect of Edge Node Spacing on Strain

In a parent element, the nodes on each edge are in equidistant distribution, and in a child element, the nodes on each edge should also be in equidistant distribution. If the difference of nodal spacing on the same edge is very large, the computation results will be influenced. For example, Section 22.7 illustrated the situation that the strain at the angular point 1 tends to infinity when the midpoint of an edge in the quadratic isoparametric element moves from the normal position to 1/4 of the side length.

Therefore, in the case of arranging the computational grid of isoparametric elements, attention must be paid to two points: (1) four apex angles should be less than 180° and should preferably be close to 90°, and (2) the midpoint of a side should be arranged inside the middle 1/3 of the side, as close to the midpoint of the side as possible, as shown in Figure 8.28.

8.14.3 Intensification of Computing Mesh of Isoparametric Elements

The stress within the structure is uneven. Elements can be sparse in the area with a small stress gradient and should be dense in the area with a large stress gradient. Figure 8.29 (a)

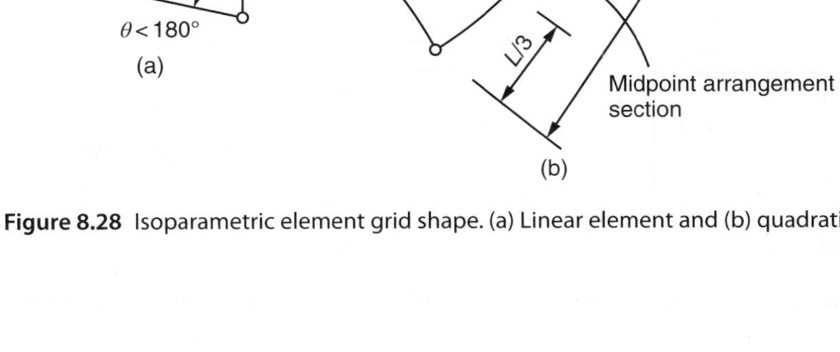

Figure 8.28 Isoparametric element grid shape. (a) Linear element and (b) quadratic element.

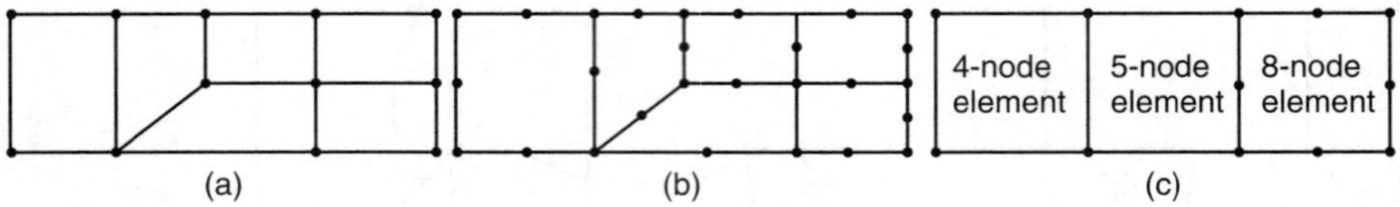

Figure 8.29 Planar isoparametric element encryption.

and (b) indicates the intensification of computing mesh of the 4- and 8-node elements. Another way is using low-order elements in the area with a small stress gradient and high-order elements in the area with large stress gradient, and the middle is connected by a transition element, as shown in Figure 8.29(c).

8.15 Inconsistent Elements

Consider the rectangular element $2a \times 2b$. The origin of the coordinates lies on its center, and the accurate solution of the displacement under the effect of pure bending is

$$u = cxy, \quad v = \frac{c}{2}(a^2 - x^2) + \frac{\mu c}{2}(b^2 - y^2) \tag{a}$$

The solution of stress is $\sigma_x = cEy, \tau_{xy} = \sigma_y = 0$; the deformation is shown in Figure 8.30(a).

By using a 4-node planar linear isoparametric element for simulation, the displacement function is

$$u = \sum_{i=1}^{4} N_i u_i, \quad v = \sum_{i=1}^{4} N_i v_i$$

$$N_i = (1 + \xi_i \xi)(1 + \eta_i \eta)/4 \tag{b}$$

For the state of pure bending stress, the obtained displacement is

$$u = c'xy, \quad v = 0 \tag{c}$$

The stress is $\sigma_x = Ec'y, \sigma_y = -\mu Ec'y, \tau_{xy} = Gc'x$; the deformation is shown in Figure 8.30(b), and false shearing strains appear in the element.

The displacement function represented by equation (b) contains only the items 1, ξ, η, and $\xi\eta$, and it is not a complete quadratic polynomial equation, namely, there is no item ξ^2 or η^2. This is the reason for the error.

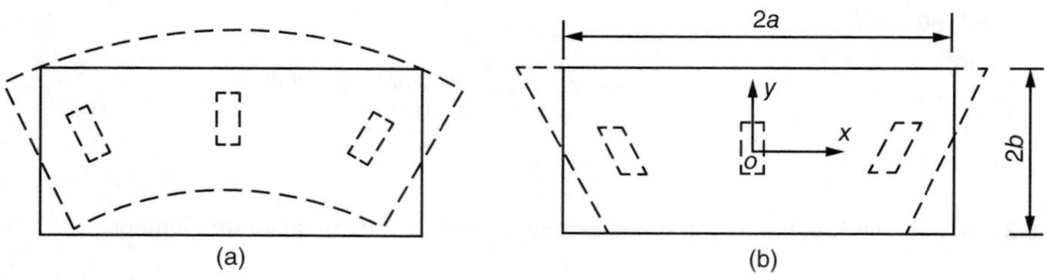

Figure 8.30 Rectangular element displacement. (a) Pure bending mode and (b) linear displacement mode.

In order to improve the calculation accuracy of elements, Wilson proposed to take the element displacement function as follows [18]:

$$u = \sum_{i=1}^{4} N_i u_i + \alpha_1(1 - \xi^2) + \alpha_2(1 - \eta^2)$$

$$v = \sum_{i=1}^{4} N_i v_i + \alpha_3(1 - \xi^2) + \alpha_4(1 - \eta^2) \tag{8.47}$$

Here, the shape function N_i is the same as equation (b), and a_1–a_4 are four new DOFs. On the four nodes, $a_i = 0$, so a_i is the internal degree of freedom. After the four items are added, the shape function is a complete quadratic polynomial and can improve the calculation accuracy, but the displacement is not continuous between the adjacent elements – thus they are called inconsistent elements. Now, the elements have a total of 12 DOFs, and the internal DOFs a_1–a_4 are eliminated by static condensation in the element calculation stage in order not to increase the number of variables in the overall balance equation set.

Equation (8.47) is expressed in a matrix form, that is,

$$\{r\} = \begin{Bmatrix} u \\ v \end{Bmatrix} = [N]\{\delta\}^{\mathrm{e}} + [\overline{N}]\{\alpha\}^{\mathrm{e}}$$

$$[\overline{N}] = \begin{bmatrix} 1-\xi^2 & 1-\eta^2 & 0 & 0 \\ 0 & 0 & 1-\xi^2 & 1-\eta^2 \end{bmatrix}$$

$$\{\alpha\}^{\mathrm{e}} = \begin{bmatrix} \alpha_1 & \alpha_2 & \alpha_3 & \alpha_4 \end{bmatrix}^{\mathrm{T}} \tag{d}$$

Here, $[N], \{\delta\}^{\mathrm{e}}$ are the shape function matrix and the nodal displacement array in a conventional 4-node isoparametric element.

The element strain is

$$\{\varepsilon\} = [B]\{\delta\}^{\mathrm{e}} + [\overline{B}]\{\alpha\}^{\mathrm{e}} \tag{e}$$

By the principle of minimum potential energy, we can have

$$\begin{bmatrix} K_{\delta\delta}^{\mathrm{e}} & K_{\delta a}^{\mathrm{e}} \\ K_{a\delta}^{\mathrm{e}} & K_{aa}^{\mathrm{e}} \end{bmatrix} \begin{Bmatrix} \delta^{\mathrm{e}} \\ \alpha^{\mathrm{e}} \end{Bmatrix} = \begin{Bmatrix} P^{\mathrm{e}} \\ 0 \end{Bmatrix}$$

$$[K_{\delta\delta}^{\mathrm{e}}] = \int [B]^{\mathrm{T}}[D][B]\,\mathrm{d}V$$

$$[K_{\delta a}^{\mathrm{e}}] = [K_{a\delta}]^{\mathrm{T}} = \int [B]^{\mathrm{T}}[D][\overline{B}]\,\mathrm{d}V \tag{f}$$

$$[K_{aa}^{\mathrm{e}}] = \int [B]^{\mathrm{T}}[D][\overline{B}]\,\mathrm{d}V \tag{g}$$

From the second equation in (f), we can obtain

$$\{\alpha\}^{\mathrm{e}} = -[K_{aa}^{\mathrm{e}}]^{-1}[K_{a\delta}^{\mathrm{e}}]\{\delta^{\mathrm{e}}\} \tag{h}$$

Equation (h) is substituted into the first equation of (f) to eliminate $\{\alpha\}^{\mathrm{e}}$, and we have

$$[K]^{\mathrm{e}}\{\delta\}^{\mathrm{e}} = \{P\}^{\mathrm{e}} \tag{i}$$

$$[K]^{\mathrm{e}} = [K_{\delta\delta}^{\mathrm{e}}] - [K_{\delta a}^{\mathrm{e}}][K_{aa}^{\mathrm{e}}]^{-1}[K_{a\delta}^{\mathrm{e}}] \tag{j}$$

The inconsistent elements cannot strictly meet the convergence conditions as described in Section 4.4 due to the discontinuous displacement between the elements, but an explanation will be given in the next section that the computation results are still convergent as long as a patch test is passed. According to the patch test results of inconsistent elements, when an element is a rectangle or parallelogram, $|J|$ is a constant, and the inconsistent element can pass a patch test; an arbitrary quadrilateral inconsistent element cannot pass a patch test. Wilson proposed to take the numerical value at the center of an element ($\xi = \eta = 0$) for $\partial x/\partial\xi$ $\partial x/\partial\eta$ $\partial y/\partial\xi$ $\partial y/\partial\eta$ while calculating $[K^e_{\delta a}]$ in order to replace different partial derivatives at each point so that inconsistent elements can pass a patch test in an arbitrary quadrilateral element. Thus, the element $|J|$ is still a constant and can pass a patch test. The experience showed good effect.

Figure 8.31 indicates a numerical example, where two calculation grids and two element modes are used: conventional 4-node linear consistent and inconsistent elements, with the computation results shown in Table 8.6. According to this table, inconsistent elements have higher calculation accuracy.

In the displacement function of 8-node isoparametric 3D elements, a similar inconsistent displacement mode is added, and the displacement function with 8-node

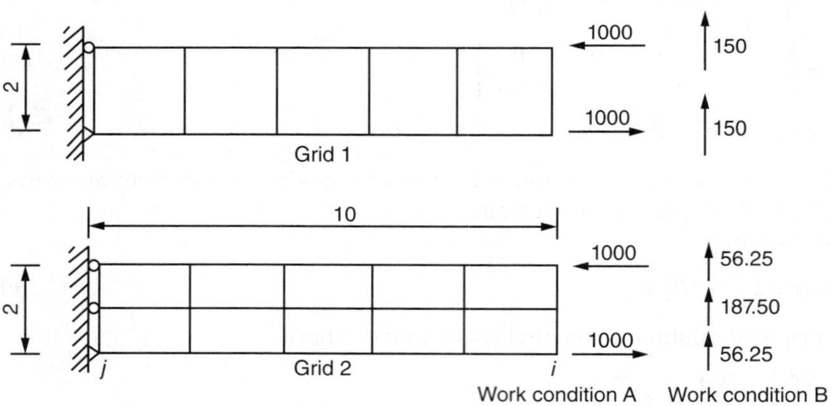

Figure 8.31 Numerical example of comparison between consistent and inconsistent elements.

Table 8.6 Comparison between computing results of consistent and inconsistent element.

		Point *i* displacement		Point *i* bending stress	
		Work condition *A*	Work condition *B*	Work condition *A*	Work condition *B*
Theory of beams		10.00	103.0	300	4050
Consistent quadrilateral element	Grid 1	6.81	70.1	218.2	2945
	Grid 2	7.06	72.3	218.8	2954
Inconsistent quadrilateral element	Grid 1	10.00	101.5	300.0	4050
	Grid 2	10.00	101.3	300.0	4050

inconsistent 3D elements can be obtained as follows:

$$u = \sum_{i=1}^{8} N_i u_i + a_1(1-\xi^2) + a_2(1-\eta^2) + a_3(1-\zeta^2)$$
$$v = \sum_{i=1}^{8} N_i v_i + a_4(1-\xi^2) + a_5(1-\eta^2) + a_6(1-\zeta^2)$$
$$w = \sum_{i=1}^{8} N_i w_i + a_7(1-\xi^2) + a_8(1-\eta^2) + a_9(1-\zeta^2) \tag{8.48}$$

In equation (8.48), the internal DOFs a_1–a_9 are eliminated by static condensation in the element calculation stage, and the numerical value at the center of the element is taken for $\partial x/\partial \xi, \ldots, \partial z/\partial \zeta$ etc. while calculating the $[K^{e}_{\delta a}]$ in order to pass the patch test.

8.16 Patch Test

As explained in Section 4.4, when the displacement method is used to find the solution, three conditions must be met to make the computation results converge to the genuine solution:

1) The rigid-body displacement of an element does not generate strain.
2) The displacement function should contain the constant strain of an element.
3) The strain should be limited on the contact surface of adjacent elements.

The first and second conditions are collectively called completeness conditions, and the third condition is called the continuity condition.

In fact, some elements do not meet the continuity condition, and the computation results still converge to the genuine solution, and an example is the inconsistent element mentioned in the preceding section. It can also be found in the following text that for thin-plate elements established according to the Kirchhoff thin-plate theory, the deflection should be continuous and the first-order partial derivatives of the deflection should be continuous normally between elements. In fact, due to the difficulty in mathematical treatment, the first-order partial derivatives of the deflection in many shell elements were discontinuous between adjacent elements, but the computation results converged. Therefore, people want to ask: Can the continuity condition be relaxed, and to what extent can this condition be relaxed?

When the displacement between elements is discontinuous, the strain on the contact surface is infinite. If integration is conducted across the contact surface, limited or unlimited strain energy may be produced. However, in fact, in the finite element method, calculation is conducted on Gaussian integral points, and the strain energy on the contact surface is not included. This is a characteristic of the finite element method.

According to the research results, the two conditions of rigid-body displacement and constant strain must be met at any time; otherwise, the computation results will not converge. If the continuity condition is satisfied at the same time, the computation results will be in monotonous convergence. If the continuity condition is not met, the computation results will not be in monotonous convergence. However, if the completeness conditions are met at the elemental level and also at the overall level after elemental combination, in other words, if the inconsistent displacement between elements does not undermine the overall completeness, the computation results are not in monotonous convergence, but in convergence.

If the completeness conditions have been met at the elemental level, how will the completeness conditions be further tested at the overall level? Clearly, so long as the rigid-body displacement conditions are satisfied at the elemental level, they will surely be met at the overall level. The problem is how to test the constant strain conditions at the overall level. Therefore, Irons [3, 4] proposed a patch test method, with which several elements form a patch to check whether the constant strain conditions are met.

For consistent elements, a patch test is not required theoretically, but in fact, a patch test is often used to check whether the programming is correct.

As shown in Figure 8.32, the patch test method is as follows:

1) Test A
   All nodes are given known exact nodal displacement, and then check whether the node i satisfies the following equilibrium equation:

$$\sum K_{ij}\delta_j - P_i = 0$$

2) Test B
   All the nodes on the boundary are given exact nodal displacement, and the displacement of the node i is calculated in the following equation:

$$\delta_i = K_{ii}^{-1}\left(P_i - \sum K_{ji}\delta_j\right)$$

   Then, the calculated value and the exact displacement of the node i are compared.
3) Test C
   Only the nodal DOFs necessary for eliminating the rigid-body displacement are fixed on the boundary, and nodal loads calculated according to the exact solution are applied to other nodes, and then the equilibrium equation is solved:

$$[K]\{\delta\} - \{P\} = 0$$

Then, the obtained nodal displacement and the exact solution are compared.

If the previous three tests are passed, the computation results surely converge to the genuine solution.

For a plane problem, the theoretical solution can be

$$u = ax + b, \quad v = cx + d \tag{a}$$

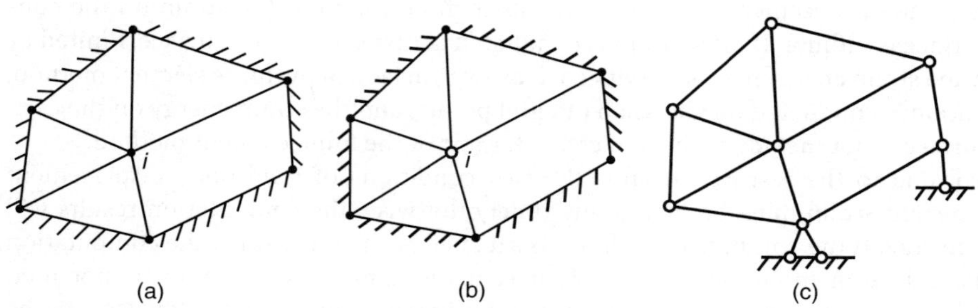

Figure 8.32 Patch test method. (a) Test A, (b) test B, and (c) test C.

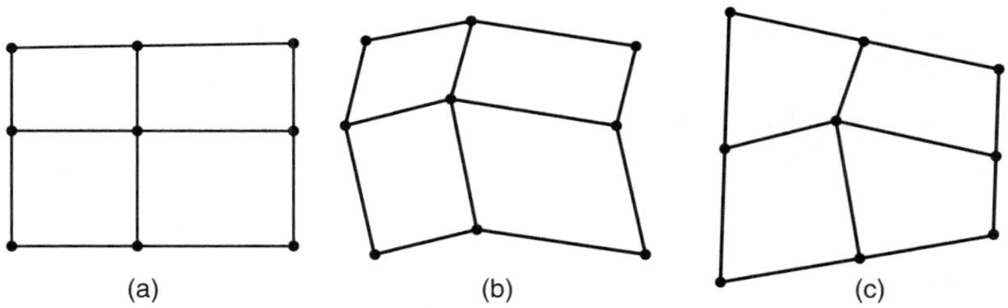

Figure 8.33 Patch test of 4-node inconsistent element. (a) Rectangular element, (b) parallelogram element, and (c) irregular element.

Here, a, b, c, and d are constants.

At this time, the corresponding strain is

$$\varepsilon_x = a, \quad \varepsilon_y = c, \quad \gamma_{xy} = 0$$

If the theoretical solution is

$$u = a + by, \quad v = c + dx \tag{b}$$

the corresponding strain is

$$\varepsilon_x = 0, \quad \varepsilon_y = 0, \quad \gamma_{xy} = b + d$$

Suppose $E = 1000$, $\mu = 0.3$, and take

$$u = 0.002x, \quad v = -0.0006y \tag{c}$$

The corresponding stresses are

$$\sigma_x = 2, \quad \sigma_y = -0.6, \quad \tau_{xy} = 0$$

For the 4-node inconsistent elements expressed in equation (8.47), three grids shown in Figure 8.33 were used for a patch test, and the test results were as follows: for rectangular and parallelogram elements, the tests A, B, and C passed; for irregular elements, none of the three tests (A, B, and C) passed. It is thus obvious that passing a patch test relates to the geometry of an element.

It is normally unnecessary to conduct a patch test for consistent elements, but when the order-reducing integration method is adopted, the elemental calculation accuracy may decline, and sometimes a patch test is needed.

The grid shown in Figure 8.34 was used for a patch test of 4-node consistent elements, and the theoretical solution used formula (c); the test results were as follows: When the 2×2 Gaussian integral was used, the patch tests A, B, and C passed. When the 1×1 Gaussian integral (order-reducing integral) was adopted, the test

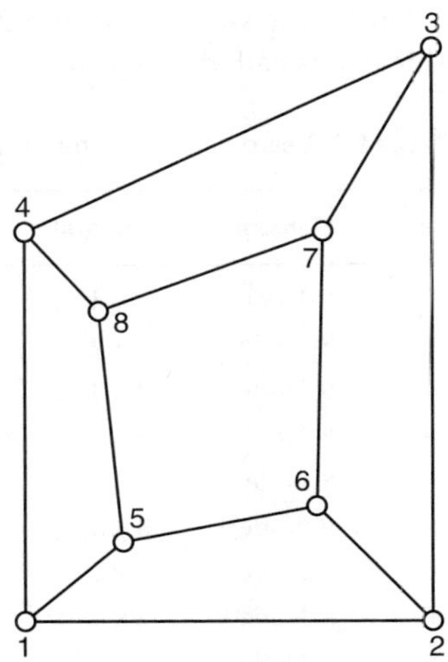

Figure 8.34 Patch test of 4-node consistent element.

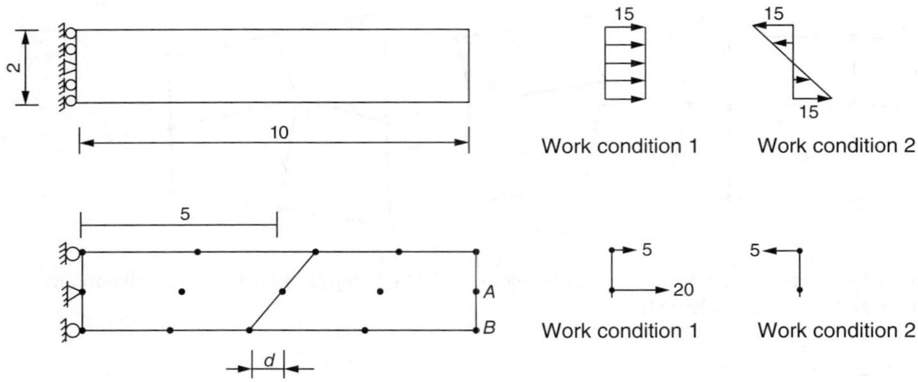

Figure 8.35 Patch test of 8- and 9-node isoparametric elements.

C failed. Therefore, it should be especially careful to use the order-reducing integration method.

As shown in Figure 8.35, consideration was given to the patch test that included an 8-node isoparametric element and a 9-node Lagrange element. Under the work condition 1, test C was conducted. For the 8-node element, the 2×2 (reduction of order) and the 3×3 (conventional) Gaussian integrals passed the test; for the 9-node element, only the 3×3-order Gaussian integral passed the test, and the 2×2 order-reducing integral failed.

The computation results under the work condition 2 were shown in Table 8.7. It is clear from the table that for the rectangular elements with $d = 0$, the computation results of the three cases were very good. For the nonrectangular elements with $d \neq 0$, the computation results of the 9-node 3×3 element were very good. For the 8-node element, the computation results were poor by using the 3×3 Gaussian integral, and the computation accuracy improved significantly by using the 2×2 order-reducing integral.

Under the work condition 2, the displacement was not linear, but in a high order, so this test is called a high-order patch test.

Table 8.7 Beam bending computation results ($E = 100$, $\mu = 0.3$).

d	Element	Integral	u_A	u_B	v_B
0	8-Node	3×3	0.750	0.150	0.75225
	8-Node	2×2	0.750	0.150	0.75225
	9-Node	3×3	0.750	0.150	0.75225
1	8-Node	3×3	0.7448	1490	0.74572
	8-node	2×2	0.750	0.150	0.75100
	9-Node	3×3	0.750	0.150	0.75225
2	8-Node	3×3	0.6684	0.1333	0.66364
	8-Node	2×2	0.750	0.150	0.75225
	9-Node	3×3	0.750	0.150	0.75225
	Exact value		0.750	0.150	0.75225

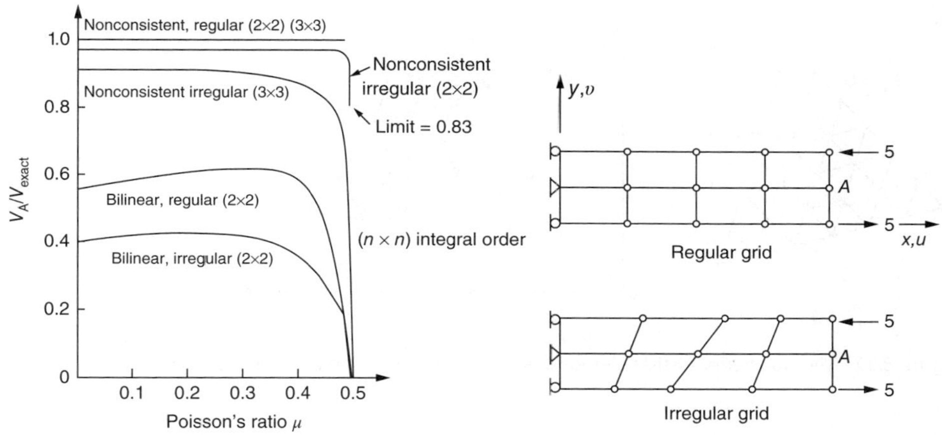

Figure 8.36 Consistent and inconsistent elements, high-order patch test in 4-node plane strain quadrilateral.

The high-order patch test can be used to test the "robustness" of elements. If the computation results are not sensitive to physical parameters, these elements are called robust elements. In problems of plane strain, when Poisson's ratio is close to 0.5, the strain energy per unit volume is higher by several magnitudes than that per unit shear strain. Therefore, the computation accuracy is very poor for some elements when Poisson's ratio is close to 0.5.

Figure 8.36 shows the computation results of the plane strain problem when Poisson's ratio is close to 0.5 in two 4-node (consistent and inconsistent) elements. It is clear from the figure that when Poisson's ratio is close to 0.5, the calculation accuracy reduces significantly. The calculation accuracy of inconsistent elements is higher.

8.17 Triangular, Tetrahedral, and Prismatic Curved-Side Elements

In high-order triangular and tetrahedral elements as mentioned in Sections 5.3 and 7.3, due to the use of a high-order displacement function, the elemental stress varies and is no longer constant, but the boundary of the elements is a straight line or a plane. In order to better adapt to the irregular geometry of the structure, it is also possible to obtain the child elements whose boundary is a curve or a curved surface through coordinate transformation.

The triangular curved-side element is shown in Figure 8.37, and its coordinate conversion formula is as follows:

$$x = \sum N_i x_i, \quad y = \sum N_i y_i \tag{8.49}$$

where N_i is a shape function expressed in the area coordinates and x_i, y_i are the global coordinates of node i.

The local coordinates of the child elements are the area coordinates.

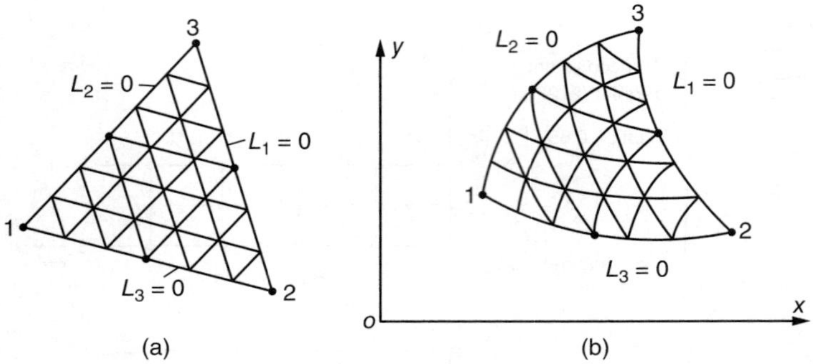

Figure 8.37 Triangular curved-side element. (a) Parent element and (b) child element.

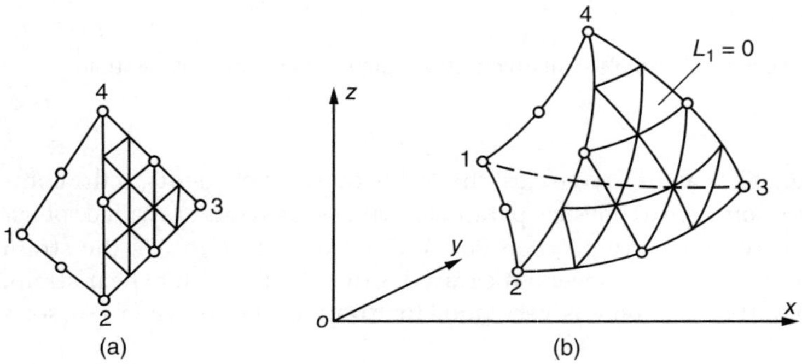

Figure 8.38 Tetrahedral curved-surface element. (a) Parent element and (b) child element.

The displacement u, v of any point in the elements is expressed by shape function as the formula (5.20).

The tetrahedral curved-side element is shown in Figure 8.38, and the coordinate conversion formula is

$$x = \sum N_i x_i, \quad y = \sum N_i y_i, \quad z = \sum N_i z_i \tag{8.50}$$

Here, N_i is a shape function expressed in the area coordinates.

The displacement u, v, w of any point in the element is expressed by shape function as the formula (7.23).

While computing the Jacobian matrix $[J]$, it must be noted that the area coordinates and the volume coordinates are not mutually independent and should satisfy the formulas (5.11) and $L_1 + L_2 + L_3 + L_4 = 1$. In order to solve this problem, the volume coordinates are taken as an example. Suppose

$$\begin{aligned} &\xi = L_1, \quad \eta = L_2, \quad \zeta = L_3, \\ &1 - \xi - \eta - \zeta = L_4 \end{aligned} \tag{8.51}$$

The shape function N_i is expressed by L_1, L_2, L_3, and L_4, and according to the partial differentiation rule,

$$\frac{\partial N_i}{\partial \xi} = \frac{\partial N_i}{\partial L_1}\frac{\partial L_1}{\partial \xi} + \frac{\partial N_i}{\partial L_2}\frac{\partial L_2}{\partial \xi} + \frac{\partial N_i}{\partial L_3}\frac{\partial L_3}{\partial \xi} + \frac{\partial N_i}{\partial L_4}\frac{\partial L_4}{\partial \xi} \tag{a}$$

After formula (8.51) is substituted into equation (a), we have

$$\frac{\partial N_i}{\partial \xi} = \frac{\partial N_i}{\partial L_1} - \frac{\partial N_i}{\partial L_4}$$

Similarly, other derivatives can be solved. Thus, according to the coordinate conversion formula (8.50), the Jacobian matrix can be obtained as follows:

$$[J] = \begin{bmatrix} \frac{\partial x}{\partial \xi} & \frac{\partial y}{\partial \xi} & \frac{\partial z}{\partial \xi} \\ \frac{\partial x}{\partial \eta} & \frac{\partial y}{\partial \eta} & \frac{\partial z}{\partial \eta} \\ \frac{\partial x}{\partial \zeta} & \frac{\partial y}{\partial \zeta} & \frac{\partial z}{\partial \zeta} \end{bmatrix} \quad (8.52)$$

Then, $\frac{\partial N_i}{\partial x}$ and the like can be computed by formula (8.53):

$$\begin{Bmatrix} \frac{\partial N_i}{\partial x} \\ \frac{\partial N_i}{\partial y} \\ \frac{\partial N_i}{\partial z} \end{Bmatrix} = [J]^{-1} \begin{Bmatrix} \frac{\partial N_i}{\partial \xi} \\ \frac{\partial N_i}{\partial \eta} \\ \frac{\partial N_i}{\partial \zeta} \end{Bmatrix} \quad (8.53)$$

For curved-side elements, the stiffness matrix is difficult to be expressed in explicit formulation, and it is necessary to use the numerical integration. For example, for the integration, in a triangle we have

$$I = \int_0^1 \int_0^{1-L_1} f(L_1 L_2 L_3)\,\mathrm{d}L_1\,\mathrm{d}L_2 = \sum W_i f(L_1 L_2 L_3) \quad (8.54)$$

The weight coefficient W_i and integral points are shown in Table 8.8.

A prismatic curved-side element can sometimes be used, as shown in Figure 8.39. The area coordinates L_1, L_2, and L_3 are adopted within the triangle, and the coordinate ξ is used in the height direction. The parent element is a triangular prism; the longitudinal side is 2 in length, and the top and bottom surfaces are two same triangles perpendicular to the longitudinal side. Five surfaces of the child element are curved faces, and the coordinate conversion formula is

$$x = \sum_{i=1}^{15} N_i x_i, \quad y = \sum_{i=1}^{15} N_i y_i, \quad z = \sum_{i=1}^{15} N_i z_i \quad (8.55)$$

The displacement function is

$$u = \sum_{i=1}^{15} N_i u_i, \quad v = \sum_{i=1}^{15} N_i v_i, \quad w = \sum_{i=1}^{15} N_i w_i \quad (8.56)$$

where N_i is a shape function as follows:

Lower corner point: $N_1 = \frac{1}{2}(1-\zeta)(2L_1-1)L_1 - \frac{1}{2}L_1(1-\zeta^2)$ (1,2,3)
Lower edge midpoint: $N_4 = 2L_1L_2(1-\zeta)$ (4,5,6)

Table 8.8 Triangular and tetrahedral numerical integration parameters.

Order	Schematic plan	Error	Integral point	Area and volume coordinates $L_1L_2L_3$	Weight coefficient W_i
Linear	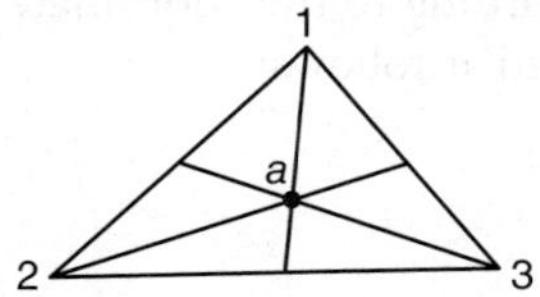	$0\ (h^2)$	a	1/3, 1/3, 1/3	1/2
Second	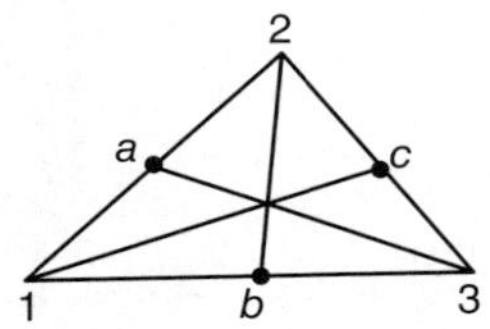	$0\ (h^3)$	a b c	1/2, 1/2, 0 0, 1/2, 1/2 1/2, 0, 1/2	1/6 1/6 1/6
Third	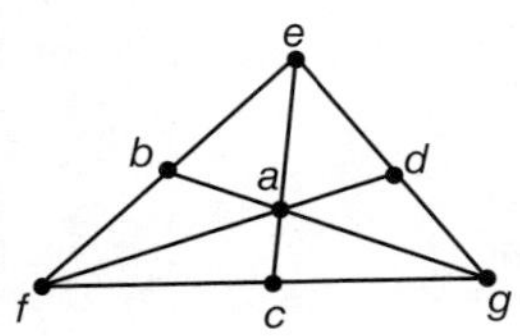	$0\ (h^4)$	a b c d e f g	1/3, 1/3, 1/3 1/2,1/2,0 0,1/2,1/2 1/2, 0,1/2 1, 0,0 0, 1,0 0, 0,1	27/120 8/120 (b, c, d) 3/120 (e, f, g)
Linear	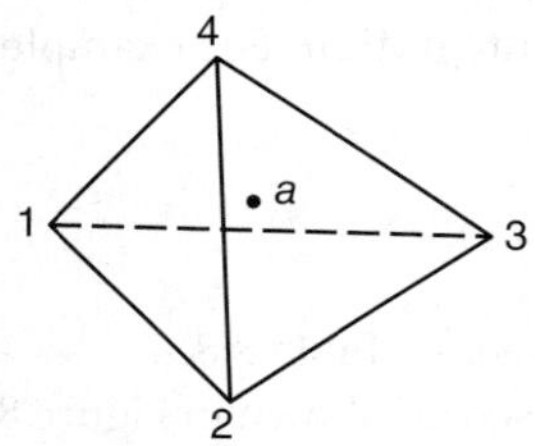	$0(h^2)$	a	1/4, 1/4, 1/4, 1/4	1
Second	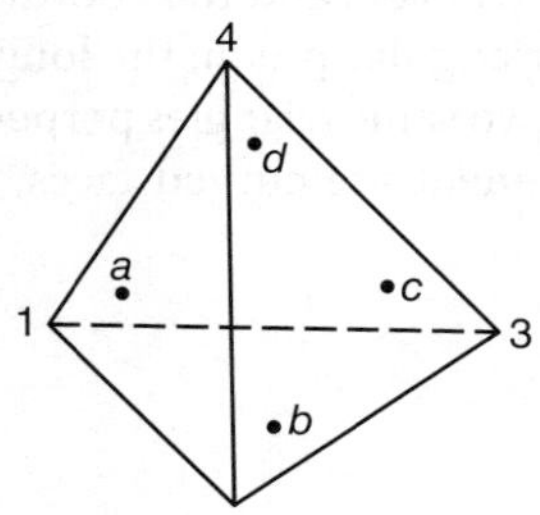	$0\ (h^3)$	a b c d	$\alpha, \beta, \beta, \beta$ $\beta, \alpha, \beta, \beta$ $\beta, \beta, \alpha, \beta$ $\beta, \beta, \beta, \alpha$ $\alpha = 0.58541020$ $\beta = 0.13819660$	1/4 1/4 1/4 1/4
Third	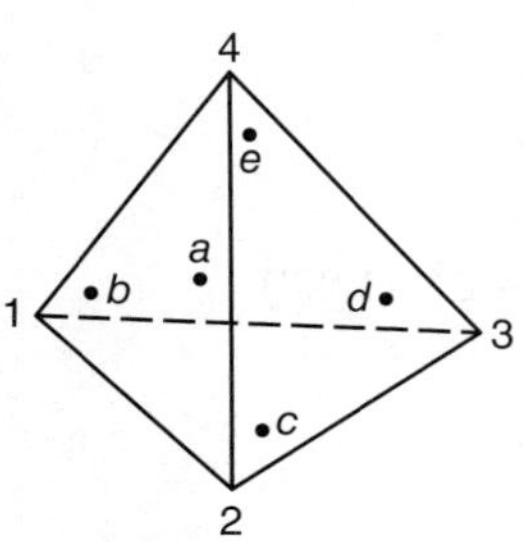	$0\ (h^4)$	a b c d e	1/4,1/4,1/4,1/4 1/3,1/6,1/6,1/6 1/6,1/3,1/6,1/6 1/6,1/6,1/3,1/6 1/6,1/6,1/6,1/3	4/5 9/20 9/20 9/20 9/20

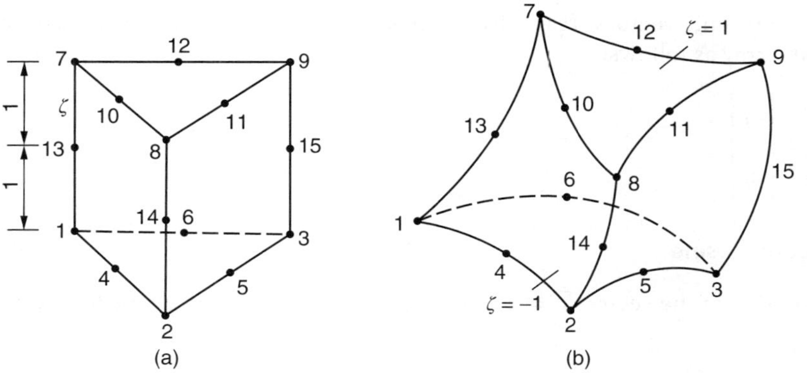

Figure 8.39 Prismatic curved-surface element. (a) Parent element and (b) child element.

Upper corner point: $N_7 = \frac{1}{2}(1+\zeta)(2L_1 - 1)L_1 - \frac{1}{2}L_1(1-\zeta^2)$ (7,8,9)
Upper edge midpoint: $N_{10} = 2(1+\zeta)L_1L_3$ (10,11,12)
Side edge midpoint: $N_{13} = L_1(1-\zeta^2)$ (13,14,15)

Here, L_i is the area coordinates.

This prismatic curved-surface element can match with a 20-node isoparametric element for mixed use in order to better fit an irregular boundary.

8.18 Vector Computation in Isoparametric Elements

The following is a brief description of some vector computation used in curved-side elements, which explains how to calculate the area and the volume of a curved-face element.

Vectors can be expressed by components in x, y, z directions. For example, the vector $\vec{V}_{01}$ shown in Figure 8.40 can be expressed as follows:

$$\vec{V}_{01} = \vec{i}x_1 + \vec{j}y_1 + \vec{k}z_1$$

Figure 8.40 Vector.

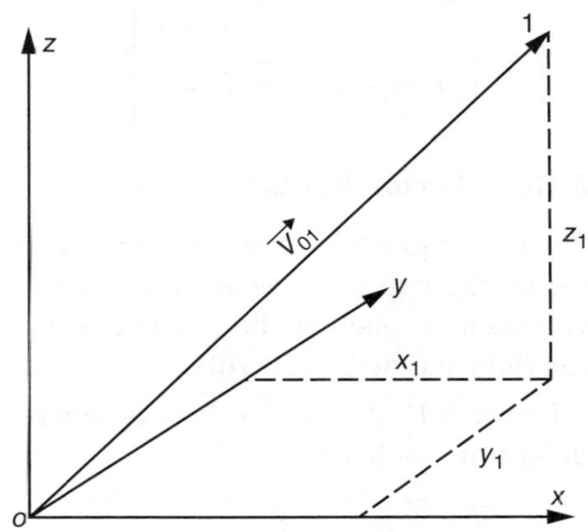

Here, $\vec{i}$, $\vec{j}$, $\vec{k}$ are unit vectors, respectively, in x, y, z directions. They can also be expressed in an array as follows:

$$\{V_{01}\} = \begin{Bmatrix} x_1 \\ y_1 \\ z_1 \end{Bmatrix}$$

8.18.1 Direction Cosine

The direction cosine of the vector $\vec{a} = \vec{i}\,a_x + \vec{j}\,a_y + \vec{k}\,a_z$ is computed by the following equation:

$$\left.\begin{aligned} l &= \cos(\vec{a}, x) = \frac{a_x}{a} \\ m &= \cos(\vec{a}, y) = \frac{a_y}{a} \\ n &= \cos(\vec{a}, z) = \frac{a_z}{a} \end{aligned}\right\}$$

where

$$a = \sqrt{a_x^2 + a_y^2 + a_z^2}$$

Here, $(\vec{a}, x)$ is an included angle between the vector $\vec{a}$ and the axis x.

8.18.2 Scalar Product

The scalar product of two vectors is defined as the product of a vector's length and another vector's projected length on the former vector and can be calculated in the equation as follows:

$$\vec{a}\cdot\vec{b} = ab\cos\theta = \vec{b}\cdot\vec{a} \tag{8.57}$$

Here, θ is the angle between $\vec{a}$ and $\vec{b}$; a and b are the length of $\vec{a}$ and $\vec{b}$, respectively.

According to the previous definition, we have

$$\left.\begin{aligned} \vec{i}\cdot\vec{i} &= \vec{j}\cdot\vec{j} = \vec{k}\cdot\vec{k} = 1 \\ \vec{i}\cdot\vec{j} &= \vec{j}\cdot\vec{k} = \vec{k}\cdot\vec{i} = 0 \end{aligned}\right\} \tag{d}$$

8.18.3 Vector Product

The vector product of two vectors is defined as another vector that is orthogonal to the plane where the two vectors are and its size equals the product of the length of the two vectors multiplied by the sine of the angle between the two vectors; its direction meets the right-handed screw rule.

Figure 8.41 shows $\vec{a}\times\vec{b} = \vec{c}$, where $|\vec{c}| = ab\sin\theta$, and according to the previous definition, we have

$$\vec{a}\times\vec{b} = \vec{b}\times\vec{a}$$

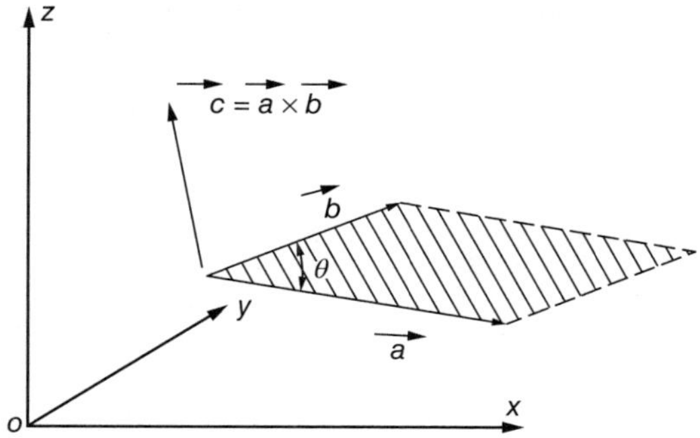

Figure 8.41 Vector product of two vectors.

From Figure 8.41, the length of $\vec{c}$ equals the area of the parallelogram with $\vec{a}$ and $\vec{b}$ as the sides.

According to the previous definition, we get

$$\left.\begin{array}{l}\vec{i}\times\vec{i}=\vec{j}\times\vec{j}=\vec{k}\times\vec{k}=0\\ \vec{i}\times\vec{j}=\vec{k},\vec{j}\times\vec{k}=\vec{i},\vec{k}\times\vec{i}=\vec{j}\end{array}\right\}\tag{e}$$

The vector product of the vector $\vec{a}=\vec{i}\,a_x+\vec{j}\,a_y+\vec{k}\,a_z$ and the vector $\vec{b}=\vec{i}\,b_x+\vec{j}\,b_y+\vec{k}\,b_z$ (or called cross product) can be expressed in the following equation:

$$\vec{a}\times\vec{b}=\begin{bmatrix}\vec{i}&\vec{j}&\vec{k}\\ a_x&a_y&a_z\\ b_x&b_y&b_z\end{bmatrix}=\vec{i}\,(a_yb_z-a_zb_y)+\vec{j}\,(a_zb_x-a_xb_z)+\vec{k}\,(a_xb_y-a_yb_x)\tag{8.58}$$

8.18.4 Infinitesimal Area in Curvilinear Coordinate System

Let ξ and η be the curvilinear coordinates in a plane, $\overrightarrow{\mathrm{d}\xi}$ be a vector tangent to the curve $\eta=\bar{k}_1$, and $\overrightarrow{\mathrm{d}\eta}$ be a vector tangent to the curve $\xi=\bar{k}_1$. Referring to Figure 8.42, we get

$$\overrightarrow{\mathrm{d}\xi}=\vec{i}\,\frac{\partial x}{\partial\xi}\mathrm{d}\xi+\vec{j}\,\frac{\partial y}{\partial\xi}\mathrm{d}\xi$$

$$\overrightarrow{\mathrm{d}\eta}=\vec{i}\,\frac{\partial x}{\partial\eta}\mathrm{d}\eta+\vec{j}\,\frac{\partial y}{\partial\eta}\mathrm{d}\eta$$

Let $\vec{c}=\overrightarrow{d\xi}\times\overrightarrow{d\eta}$, and in equation (8.58), we have

$$\vec{c}=\overrightarrow{\mathrm{d}\xi}\times\overrightarrow{\mathrm{d}\eta}=\begin{vmatrix}\vec{i}&\vec{j}&\vec{k}\\ \frac{\partial x}{\partial\xi}\mathrm{d}\xi&\frac{\partial y}{\partial\xi}\mathrm{d}\xi&0\\ \frac{\partial x}{\partial\eta}\mathrm{d}\eta&\frac{\partial y}{\partial\eta}\mathrm{d}\eta&0\end{vmatrix}=\vec{i}\cdot 0+\vec{j}\cdot 0+\vec{k}\begin{vmatrix}\frac{\partial x}{\partial\xi}&\frac{\partial x}{\partial\xi}\\ \frac{\partial x}{\partial\eta}&\frac{\partial x}{\partial\eta}\end{vmatrix}\mathrm{d}\xi\,\mathrm{d}\eta$$

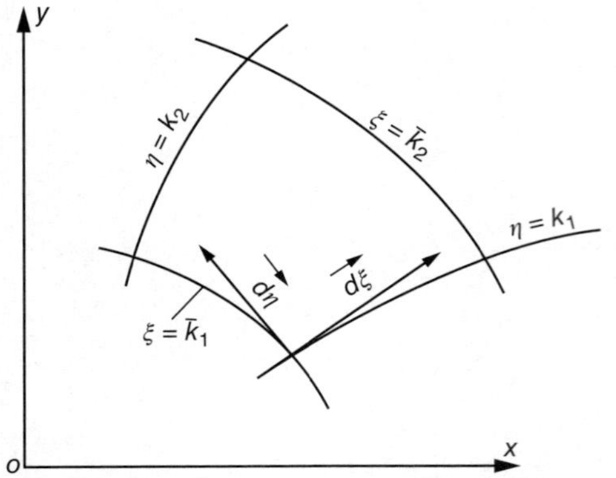

Figure 8.42 Planar curvilinear coordinates.

The infinitesimal area of the parallelogram with $\overrightarrow{d\xi}$ and $\overrightarrow{d\eta}$ as the sides equals the length of the vector $\vec{c}$, so in the previous equation, we get

$$\mathrm{d}A = |\vec{c}\,| = \begin{vmatrix} \dfrac{\partial x}{\partial \xi} & \dfrac{\partial x}{\partial \eta} \\ \dfrac{\partial y}{\partial \xi} & \dfrac{\partial y}{\partial \eta} \end{vmatrix} \mathrm{d}\xi \mathrm{d}\eta = |J| \mathrm{d}\xi \mathrm{d}\eta \tag{8.59}$$

8.18.5 Infinitesimal Area of Spatial Curved Surface

In Figure 8.43, there are curvilinear coordinates $\xi = \text{constant}$ and $\eta = \text{constant}$ in the spatial curved surface, and draw two vectors $\overrightarrow{d\xi}$ and $\overrightarrow{d\eta}$ in their tangential direction as follows:

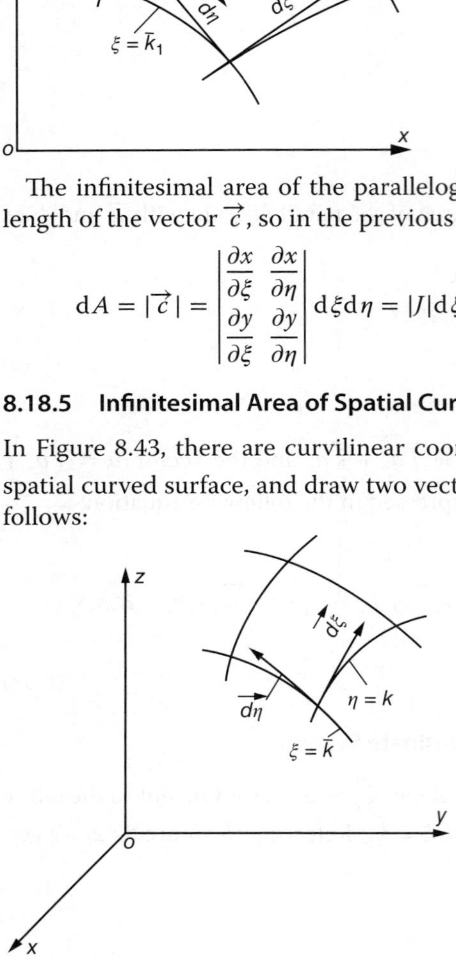

Figure 8.43 Spatial curved surface.

$$\overrightarrow{\mathrm{d}\xi} = \left(\vec{i}\,\frac{\partial x}{\partial \xi} + \vec{j}\,\frac{\partial y}{\partial \xi} + \vec{k}\,\frac{\partial z}{\partial \xi} \right) \mathrm{d}\xi$$

$$\overrightarrow{\mathrm{d}\eta} = \left(\vec{i}\,\frac{\partial x}{\partial \eta} + \vec{j}\,\frac{\partial x}{\partial \eta} + \vec{k}\,\frac{\partial z}{\partial \eta} \right) \mathrm{d}\eta$$

Draw the vector $\vec{c}$ orthogonal to $\overrightarrow{d\xi}$ and $\overrightarrow{d\eta}$, and we get

$$\vec{c} = \overrightarrow{\mathrm{d}\xi} \times \overrightarrow{\mathrm{d}\eta} = \begin{vmatrix} \vec{i} & \vec{j} & \vec{k} \\ \dfrac{\partial x}{\partial \xi} & \dfrac{\partial y}{\partial \xi} & \dfrac{\partial z}{\partial \xi} \\ \dfrac{\partial x}{\partial \eta} & \dfrac{\partial y}{\partial \eta} & \dfrac{\partial z}{\partial \eta} \end{vmatrix} \mathrm{d}\xi \, \mathrm{d}\eta$$

The infinitesimal area of the parallelogram with $\overrightarrow{d\xi}$ and $\overrightarrow{d\eta}$ as the sides equals the length of $\vec{c}$, so the infinitesimal area of the spatial curved surface is obtained, namely,

$$\mathrm{d}\Omega = |\overrightarrow{\mathrm{d}\xi} \times \overrightarrow{\mathrm{d}\eta}| = A \, \mathrm{d}\xi \, \mathrm{d}\eta \tag{8.60}$$

$$A = \left\{ \left(\frac{\partial x}{\partial \xi}\frac{\partial y}{\partial \eta} - \frac{\partial x}{\partial \eta}\frac{\partial y}{\partial \xi} \right)^2 + \left(\frac{\partial y}{\partial \xi}\frac{\partial z}{\partial \eta} - \frac{\partial y}{\partial \eta}\frac{\partial z}{\partial \xi} \right)^2 + \left(\frac{\partial z}{\partial \eta}\frac{\partial x}{\partial \xi} - \frac{\partial z}{\partial \xi}\frac{\partial x}{\partial \eta} \right)^2 \right\}^{\frac{1}{2}} \tag{8.61}$$

8.18.6 Spatial Infinitesimal Volumes

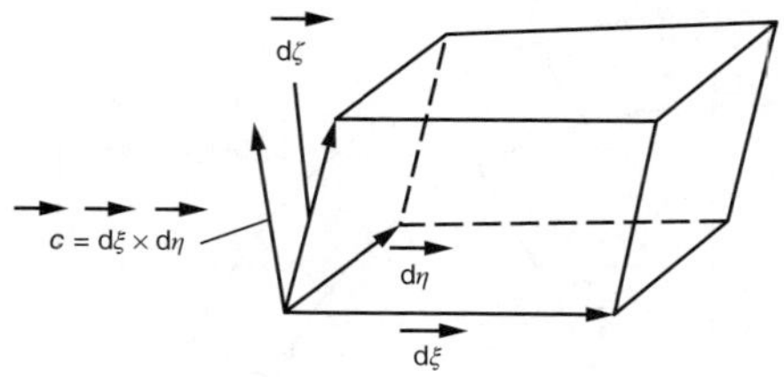

Figure 8.44 Infinitesimal volume.

Considering the spatial curvilinear coordinates $\xi, \eta, \zeta, \xi =$ constant, $\eta =$ constant, and $\zeta =$ constant are three curved surfaces, draw the vector $\overrightarrow{\mathrm{d}\xi}$ along the tangential direction to the intersection of the two curved surfaces $\eta =$ constant and $\zeta =$ constant, and similarly, draw $\overrightarrow{\mathrm{d}\eta}$ and $\overrightarrow{\mathrm{d}\xi}$. As shown in Figure 8.44, draw the vector $\vec{c} = \overrightarrow{\mathrm{d}\xi} \times \overrightarrow{\mathrm{d}\eta}$; the length of $\vec{c}$ equals the area of the quadrilateral with $\overrightarrow{\mathrm{d}\xi}$ and $\overrightarrow{\mathrm{d}\eta}$ as the sides, and the direction of $\vec{c}$ is orthogonal to $\overrightarrow{\mathrm{d}\xi}$ and $\overrightarrow{\mathrm{d}\eta}$. Then, form the scalar product of $\vec{c}$ and $\overrightarrow{\mathrm{d}\zeta}$, which is equal to the length of $\vec{c}$ multiplied by the length of $\overrightarrow{\mathrm{d}\zeta}$ and by the cosine of their included angle. Obviously, it is the volume of the elementary hexahedron with $\overrightarrow{\mathrm{d}\xi}, \overrightarrow{\mathrm{d}\eta}, \overrightarrow{\mathrm{d}\zeta}$ as the edges in the space, that is,

$$\mathrm{d}V = \overrightarrow{\mathrm{d}\zeta} \cdot (\overrightarrow{\mathrm{d}\xi} \times \overrightarrow{\mathrm{d}\eta}) = \left(\vec{i}\frac{\partial x}{\partial \zeta} + \vec{j}\frac{\partial y}{\partial \zeta} + \vec{k}\frac{\partial z}{\partial \zeta}\right)\mathrm{d}\zeta$$

$$\times \left\{\vec{i}\begin{vmatrix} \frac{\partial y}{\partial \xi} & \frac{\partial z}{\partial \xi} \\ \frac{\partial y}{\partial \eta} & \frac{\partial z}{\partial \eta} \end{vmatrix} + \vec{j}\begin{vmatrix} \frac{\partial z}{\partial \xi} & \frac{\partial z}{\partial \xi} \\ \frac{\partial z}{\partial \eta} & \frac{\partial x}{\partial \eta} \end{vmatrix} + \vec{k}\begin{vmatrix} \frac{\partial x}{\partial \xi} & \frac{\partial y}{\partial \xi} \\ \frac{\partial x}{\partial \eta} & \frac{\partial y}{\partial \eta} \end{vmatrix}\right\}\mathrm{d}\xi\,\mathrm{d}\eta$$

From the previous equation and equation (d), we get the infinitesimal volume as follows:

$$\mathrm{d}V = |J|\,\mathrm{d}\xi\,\mathrm{d}\eta\,\mathrm{d}\zeta \tag{8.62}$$

$$|J| = \begin{vmatrix} \frac{\partial x}{\partial \xi} & \frac{\partial x}{\partial \eta} & \frac{\partial x}{\partial \zeta} \\ \frac{\partial y}{\partial \xi} & \frac{\partial y}{\partial \eta} & \frac{\partial y}{\partial \zeta} \\ \frac{\partial z}{\partial \xi} & \frac{\partial z}{\partial \eta} & \frac{\partial z}{\partial \zeta} \end{vmatrix}$$

$|J|$ is the determinant of the Jacobian matrix.

8.19 Numerical Examples of Isoparametric Elements

The following are numerical examples of isoparametric curved-side elements.

Figure 8.45 shows the stress results of a round tube under the effect of the internal water pressure and the internal and external temperature differences that the author and Mr. Song Jingting computed by 8-node rectangular curved-side elements. Only 1/4 of the elements were used for the computation because of the symmetry. Five elements were divided, and a quadratic shape function was adopted. As shown in the figure, the computed value was in good agreement with the theoretical solution.

Figure 8.46 shows the stress results of a rotating sphere under the effect of the centrifugal force that were calculated by isoparametric spatial curved-face elements [19].

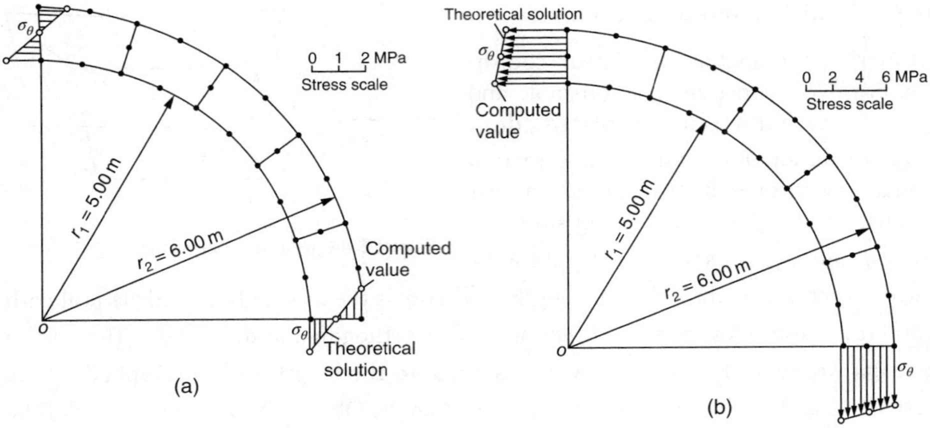

Figure 8.45 Comparison between computed stress and theoretical solution of round tube. (a) Circumferential stress under the effect of internal and external temperature differences (internal temperature of 10 °C, the external temperature of 0 °C, $E = 20$ GPa, $\mu = 0.16$, $\alpha = 1 \times 10^{-5}(1/°C)$ and (b) circumferential stress under the effect of uniform internal water pressure (internal water pressure of 1 MPa).

Only 1/8 was taken out, and seven elements were used because of the symmetry, and the computed value was also in good agreement with the theoretical solution.

Figure 8.47 shows the computation results of the cantilevers with different length-to-height ratios, and four 8-node planar quadratic isoparametric elements were used.

The figure gives the comparison of the effect of the order-reducing integration. It can thus be seen that the use of the order-reducing integration (2×2) decreased the amount of computation and increased the computation accuracy. According to the figure, when the length-to-height ratio of the beam was very large, there would be an ill-conditioned equation; when 3D finite elements were used to calculate thin-walled structure, a similar ill-conditioned equation might also appear. Therefore, bar elements should be used for the slender member bar structure during computation, and thin-plate and thin-shell elements should be used for the thin-walled structure during computation to reduce the amount of computation and avoid the occurrence of an ill-conditioned equation.

8.20 Infinite Elements

Sometimes, it is necessary to analyze the stresses in the infinite field or the semi-infinite field. In this case, we may adopt the infinite elements [5, 6]. An infinite element is a radial strip; one end is connected with a finite element, and the other end tends to infinity. The displacement is zero at infinity.

Figure 8.48 shows a gravity dam foundation; finite elements are used in the near field, and infinite elements in the far field.

8.20.1 Two-Dimensional Infinite Elements

In plane problems, 4- or 5-node infinite elements are commonly used, as shown in Figure 8.49, and they can be connected to 4- or 8-node finite elements, respectively.

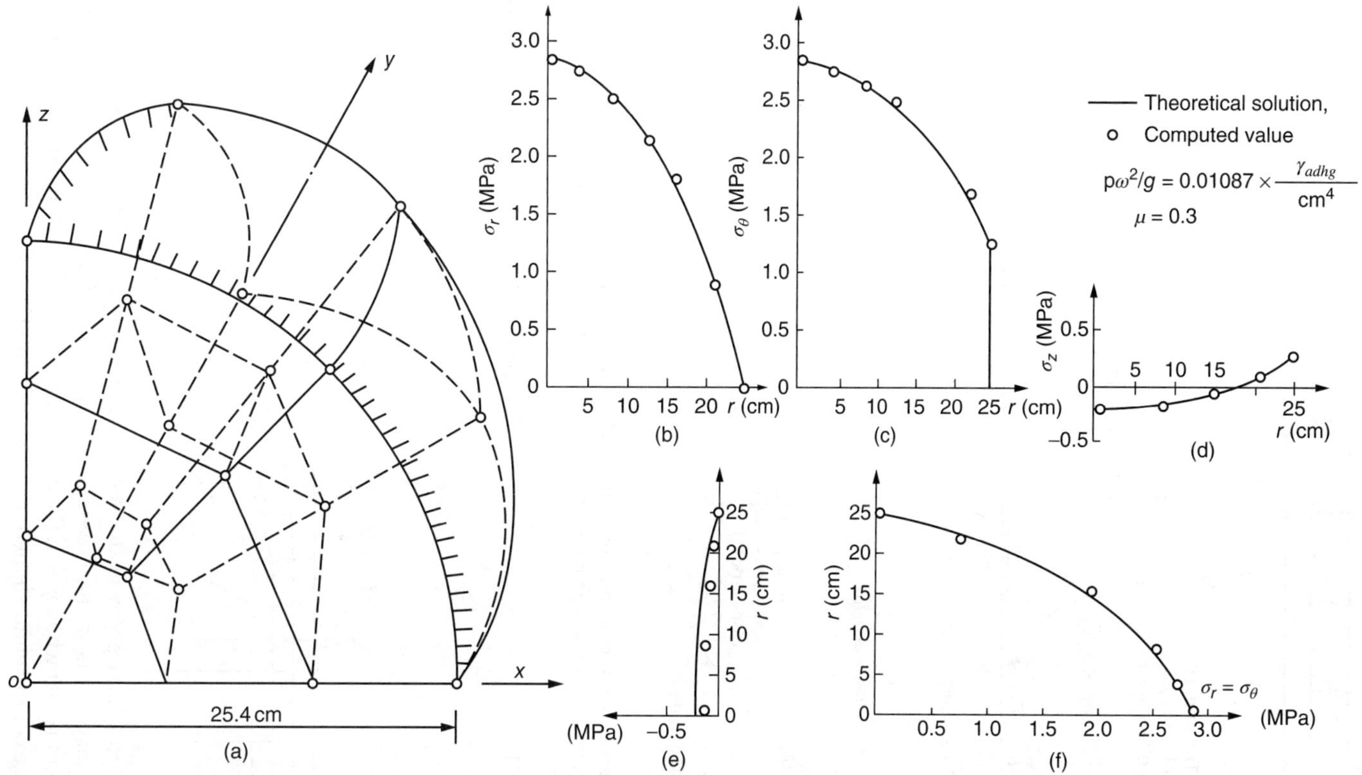

Figure 8.46 The use of seven quadratic isoparametric spatial elements to compute a rotating sphere. (a) Computational grid, (b) $\sigma_r(z = 0)$, (c) $\sigma_\theta(z = 0)$, (d) $\sigma_z(z = 0)$, (e) $\sigma_z(r = 0)$, (f) $\sigma_r = \sigma_\theta(r = 0)$.

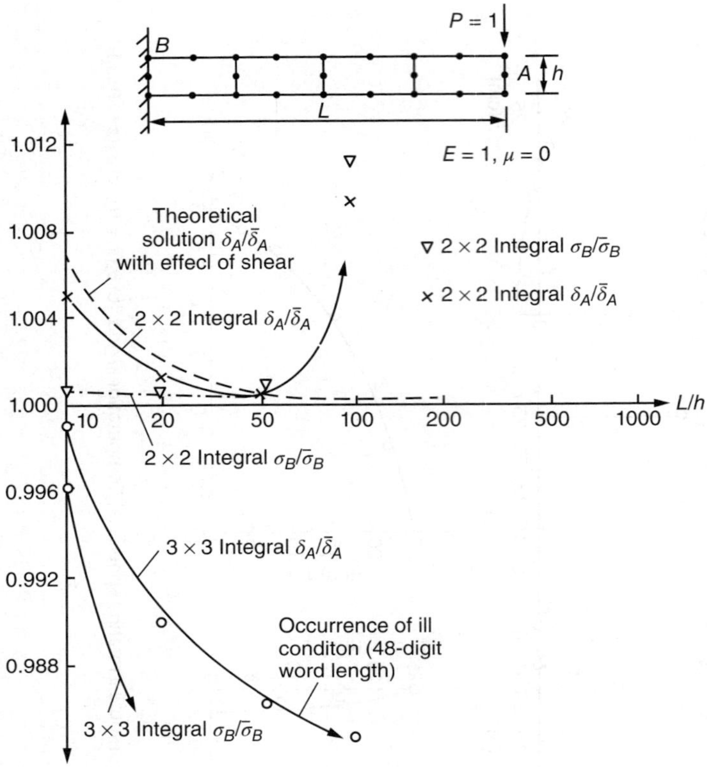

Figure 8.47 Effect of order-reducing integration while computing cantilevers with different length-to-height ratios by using quadratic isoparametric elements.

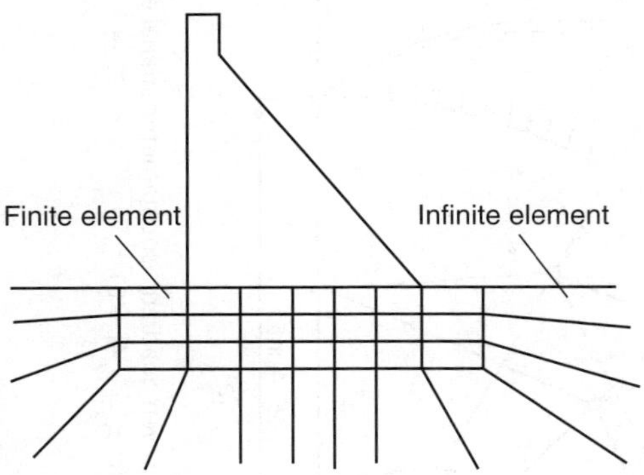

Figure 8.48 Application of infinite elements.

Consideration is first given to the geometric mapping; the parent element is a 2×2 square and is mapped as an infinite element by the shape function $N_i\ (\xi, \eta)$. In addition, the requirement is that when $\eta \to 1$, the coordinates of the child element tends to infinity. The coordinates of the child element are computed in the following equation:

$$x = \sum N_i x_i, \quad y = \sum N_i y_i \tag{a}$$

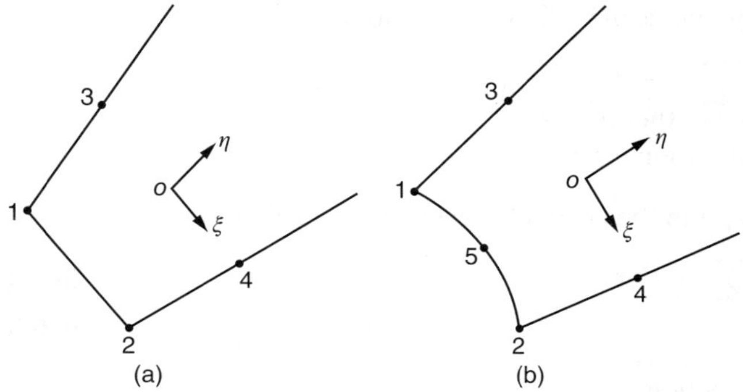

Figure 8.49 Two-dimensional infinite elements: (a) 4-node infinite element and (b) 5-node infinite element.

For 4-node infinite elements, a shape function can be taken as follows:
When $\eta \le 0$,

$$\left.\begin{aligned} N_1 &= -\frac{1}{2}(1-\xi)\eta, \quad N_2 = -\frac{1}{2}(1+\xi)\eta \\ N_3 &= \frac{1}{2}(1-\xi)(1+\eta), \quad N_4 = \frac{1}{2}(1+\xi)(1+\eta) \end{aligned}\right\} \tag{8.63}$$

When $\eta > 0$,

$$\left.\begin{aligned} N_1 &= -\frac{1}{2}(1-\xi)\eta \Big/ (1-\eta), \quad N_2 = -\frac{1}{2}(1+\xi)\eta \Big/ (1-\eta) \\ N_3 &= \frac{1}{2}(1-\xi)[1+\eta/(1-\eta)], \quad N_4 = \frac{1}{2}(1+\xi)[1+\eta/(1-\eta)] \end{aligned}\right\} \tag{8.64}$$

For 5-node infinite elements, a shape function can be taken as follows:
When $\eta \le 0$,

$$\left.\begin{aligned} N_1 &= \frac{1}{2}\xi\eta(1-\xi), \quad N_2 = -\frac{1}{2}\xi\eta(1+\xi) \\ N_3 &= \frac{1}{2}(1-\xi)(1+\eta), \quad N_4 = \frac{1}{2}(1+\xi)(1+\eta) \\ N_5 &= -(1-\xi^2)\eta \end{aligned}\right\} \tag{8.65}$$

When $\eta > 0$,

$$\left.\begin{aligned} N_1 &= -\frac{1}{2}(1-\xi)\eta/(1-\eta) \\ N_2 &= -\frac{1}{2}(1+\xi)\eta/(1-\eta) \\ N_3 &= \frac{1}{2}(1-\xi)[1+\eta/(1-\eta)] \\ N_4 &= \frac{1}{2}(1+\xi)[1+\eta/(1-\eta)] \\ N_5 &= 0 \end{aligned}\right\} \tag{8.66}$$

The previous shape function meets the following conditions:

1) Any point in the element, $\sum N_i = 1$.
2) $N_i = 1$ at node i, and $N_i = 0$ at other nodes.
3) Both x and y tend to infinity when $\eta = 1$.

The displacement of any point in the element is expressed as follows:

$$u = \sum M_i u_i, \quad v = \sum M_i v_i \tag{8.67}$$

$$M_i = M_i^0 f_i(r) \tag{8.68}$$

where M_i is a displacement function.

M_i^0 can take $N_i\ (\xi, \eta)$ when $\eta < 0$, and $f_i\ (r)$ is an attenuation function and should meet the following condition:

$$\lim_{r \to \infty} f_i(r) = 0 \tag{b}$$

For example, $f_i(r)$ can take

$$f_i(r) = (r_i/r)^a \tag{8.69}$$

or

$$f_i(r) = \exp(1 - r/r_i) \tag{8.70}$$

Here, r is the distance from any point to the selected attenuation center and can be called attenuation radius (the attenuation center is often chosen near the geometric center of the computational domain); r_i is the attenuation radius of node i; the index a is equal to or larger than 1, usually taking 1 or 2.

It is thus clear from the aforementioned that the shape function and the displacement function of infinite elements are different, so infinite elements are not isoparametric elements.

The elemental stiffness matrix and the load array are calculated in the same way as curved-side finite elements, omitted here.

8.20.2 Three-Dimensional Infinite Elements

3D infinite elements are shown in Figure 8.50, the natural coordinates are ξ, η, ζ, and when $\eta = 1$, the coordinates of the child element are infinity.

The coordinates of any point in the child element is expressed as follows:

$$x = \sum N_i x_i, \quad y = \sum N_i y_i, \quad z = \sum N_i z_i \tag{8.71}$$

For 8-node infinite elements, the shape function is as follows:

When $\eta \le 0$,

$$\left.\begin{aligned} N_i &= -\frac{1}{4}(1 + \xi_i \xi)(1 + \zeta_i \zeta)\eta \quad (i = 1,2,3,4) \\ N_i &= -\frac{1}{4}(1 + \xi_i \xi)(1 + \zeta_i \zeta)(1 + \eta) \quad (i = 5,6,7,8) \end{aligned}\right\} \tag{8.72}$$

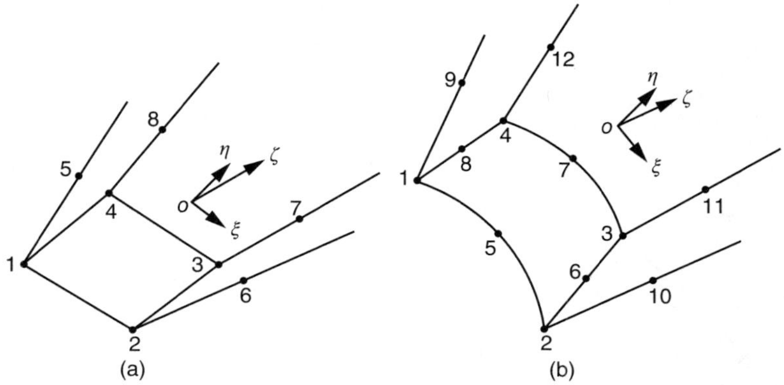

Figure 8.50 Three-dimensional infinite elements: (a) 8-node infinite element and (b) 12-node infinite element.

When $\eta > 0$,

$$\left.\begin{aligned} N_i &= -\frac{1}{4}(1+\xi_i\xi)(1+\zeta_i\zeta)\eta/(1-\eta) \quad (i=1,2,3,4) \\ N_i &= \frac{1}{4}(1+\xi_i\xi)(1+\zeta_i\zeta)\left(1+\frac{\eta}{1+\eta}\right) \quad (i=5,6,7,8) \end{aligned}\right\} \tag{8.73}$$

For 12-node infinite elements, the shape function is as follows:
When $\eta \le 0$,

$$\left.\begin{aligned} N_i &= -\frac{1}{4}(1+\xi_i\xi)(1+\zeta_i\zeta)\eta(\xi_i\xi+\zeta_i\zeta-\eta-2) \quad (i=1,2,3,4) \\ N_i &= \frac{1}{4}(1+\xi_i\xi)(1+\zeta_i\zeta)(1+\eta^2) \quad (i=9,10,11,12) \\ N_i &= -\frac{1}{2}(1-\xi^2)(1+\zeta_i\zeta)\eta \quad (i=5,7) \\ N_i &= -\frac{1}{2}(1+\xi_i\xi)(1-\zeta^2)\eta \quad (i=6,8) \end{aligned}\right\} \tag{8.74}$$

When $\eta > 0$,

$$\left.\begin{aligned} N_i &= -\frac{1}{4}(1+\xi_i\xi)(1+\zeta_i\zeta)\eta/(1-\eta) \quad (i=1,2,3,4) \\ N_i &= \frac{1}{4}(1+\xi_i\xi)(1+\zeta_i\zeta)/(1-\eta) \quad (i=9,10,11,12) \\ N_i &= 0 \quad (i=5,6,7,8) \end{aligned}\right\} \tag{8.75}$$

The displacement of any point in the element can be expressed as

$$\begin{aligned} u &= \sum M_i u_i, \quad v = \sum M_i v_i, \\ w &= \sum M_i w_i \end{aligned} \tag{8.76}$$

where M_i is a displacement function, expressed in Equation (8.68). M_i^0 takes $N_i\ (\xi, \eta, \zeta)$ when $\xi < 0$; the displacement is zero when $r \to \infty$.

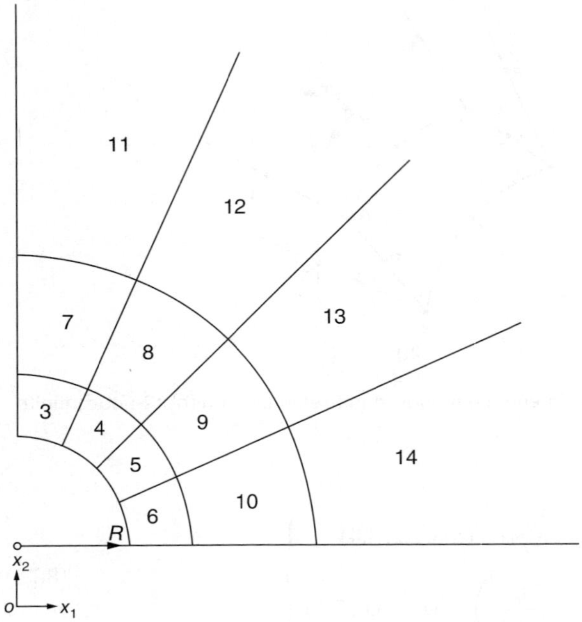

Figure 8.51 Computational grid of infinite element example.

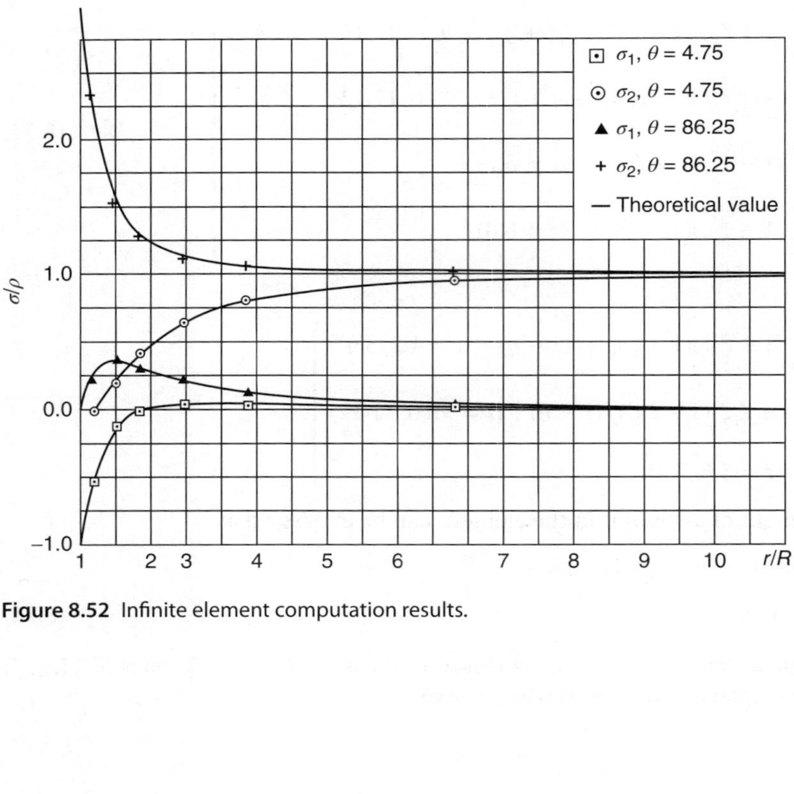

Figure 8.52 Infinite element computation results.

Experience showed that the computation accuracy of infinite elements was quite good.

The following is a numerical example of a plane problem [6]. The infinite field has a round hole, and there is the initial stress $\sigma_x = p = -1000$ in the field before the opening. After the excavation, the hole surface becomes free, thus causing stress and displacement near the hole. Because of the symmetry, only 1/4 is needed for analysis. The computational grid is shown in Figure 8.51, where there are two layers of 8-node isoparametric elements inside and one layer of infinite elements outside. The computation results are shown in Figure 8.52, which shows that the computed stresses are very close to the theoretical solution.

Bibliography

1 Bathe, K.J. (1982) *Finite Element Procedures in Engineering Analysis*, Prentice-Hall, Englewood Cliffs.

2 Newton, R.E. (1973) Degeneration of brick-type isoparametric elements. *Int. J. Numer. Methods Eng.*, 7, 579–581.

3 Irons, B.M. (1966) *Numerical Integration Applied to Finite Element Methods.* Conference on Use of Digital Computers in Structural Engineering, University of Newcastle.

4 Irons, B.M. and Zienkiewicz, O.C. (1968) *The Isoparametric Finite System – A New Concept in Finite Element Analysis.* Proceedings, Conference on Recent Advances in Stress Analysis, Royal Aeronautical Society, London.

5 Beer, G. and Meek, J.L. (1981) Infinite domain elements. *Int. J. Numer. Methods Eng.*, **17** (1), 43–52.

6 Irons, B.M. and Razzaque, A. (1972) in *Mathematical Foundations of the Finite Element Method* (ed. A.K. Ariz), Academic Press, pp. 557–587.

7 Irons, B.M. (1966) Engineering application of numerical integration in stiffness method. *AIAA J.*, **14**, 2035–2037.

8 Irons, B.M. (1971) Quadrature rules for brick based finite elements. *Int. J. Numer. Methods Eng.*, **3**, 293–294.

9 Hellen, T.K. (1972) Effective quadrature rules for quadratic solid isoparametric finite elements. *Int. J. Numer. Methods Eng.*, **4**, 597–600.

10 Cowper, G.R. (1973) Gaussian quadrature formulas for triangles. *Int. J. Numer. Methods Eng.*, 7, 405–408.

11 Fried, I. (1974) Numerical integration in the finite element method. *Comput. Struct.*, **4**, 921–932.

12 Zienkiewicz, O.C. and Taylor, R.L. (2009) *Finite Element Method for Solid and Structural Mechanics*, Elsevier, New York, London.

13 Kardestuncer, H. and Norrie, D.H. (1987) *Finite Element Handbook*, McGraw-Hill, New York.

14 Bettess, P. (1977) Infinite elements. *Int. J. Numer. Methods Eng.*, **11** (1), 53–64.
15 Wilson, E.L., Taylor, R.L., Doherty, W. and Ghaboussi, J. (1973) in *Numerical and Computer Methods in Structural Mechanics* (eds S.T. Fenves *et al.*), Academic Press, New York.
16 Hinton, E. and Cambell, J.S. (1974) Local and global smoothing of discontinuous finite element functions using a least square method. *Int. J. Numer. Methods Eng.*, **8**, 461–480.

9

Comparison and Application Instances of Various Planar and Spatial Elements

In previous chapters, we described various types of planar and spatial elements. Which elements should be used in practical work? This problem relates to computation accuracy, computation time and preparation work, and other factors. In this chapter, we will compare various planar and spatial elements and explain how to select elements in practical work.

Since the finite element method was invented, many complex problems in practical engineering were solved. In this chapter, we will introduce some application instances of planar and spatial elements.

9.1 Comparison and Selection of Various Planar Elements

The shape functions of various planar elements are listed in Table 8.2.

Figure 9.1 shows the computing grid of cantilevers under the action of concentrated load, which is calculated by three different types of triangular elements [1]. Three computational grids have the same number of nodes. According to Section 4.3, the deflection computed by the displacement method is less than the true value, so the greater the displacement is, the higher the accuracy is. It is clear from Figure 9.1 that 9-node triangular elements have the best results, followed by 6-node triangular elements; 3-node isostrain triangular elements have very poor results.

Figure 9.2 shows the computed results of the cantilevers by constant strain triangular elements, rectangular elements, and isoparametric elements [2]. It is evident that the computation error is quite large by using a row of rectangular elements or two rows of triangular elements, but a row of 8-node isoparametric elements can give satisfactory computation results.

As shown in Figure 5.8, the error is still relatively great in computing the cantilevers under the action of concentrated loads by using four rows of constant strain triangular elements.

According to the above computation results, for the thin structures in which bending is the principal action such as concrete cutoff walls in the soil base, tunnel lining, and so on, it is not suitable to adopt isostrain triangular elements. This is because these structures are relatively thin and the arrangement for more than five rows of elements in the thickness direction is relatively difficult, while the computation error is relatively great for fewer than four rows of elements. It would be best to use 8-node isoparametric

The Finite Element Method: Fundamentals and Applications in Civil, Hydraulic, Mechanical and Aeronautical Engineering, First Edition. Bofang Zhu.

Companion website: www.wiley.com/go/zhufem17

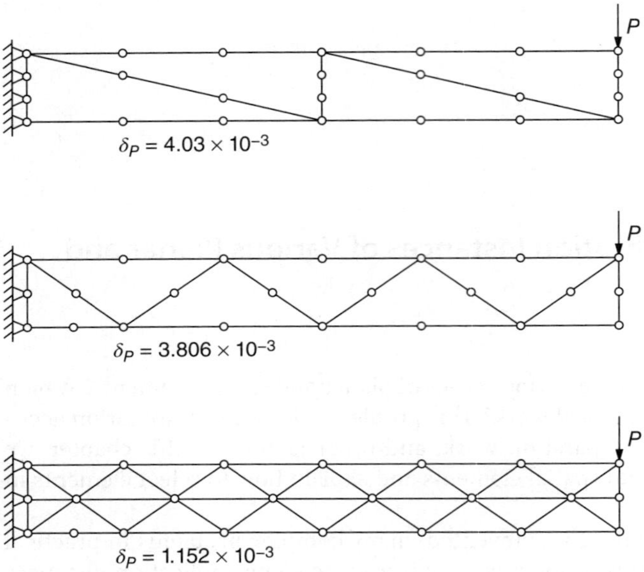

Figure 9.1 Cantilever deflection computation.

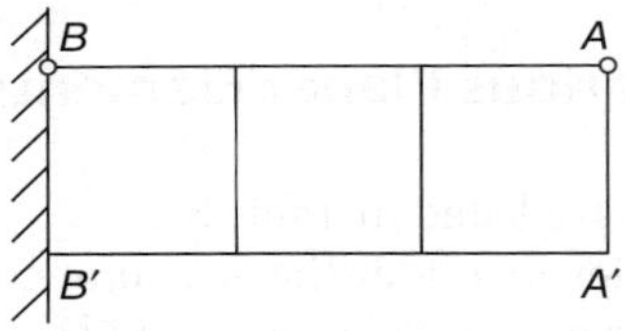

Element \ Deflection and stress \ Load	Point-*A* vertical force		*AA′* section couple	
	δ_A	σ_B	δ_A	σ_B
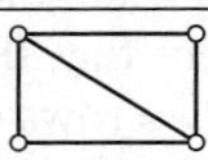	0.26	0.19	0.22	0.22
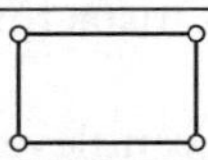	0.65	0.56	0.67	0.67
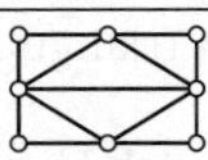	0.53	0.51	0.52	0.55
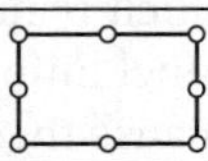	0.99	0.99	1.00	1.00
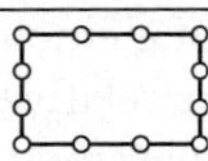	1.00	1.00	1.00	1.00
Exact value	1.00	1.00	1.00	1.00

Figure 9.2 Comparison of computation results of a cantilever by various planar elements.

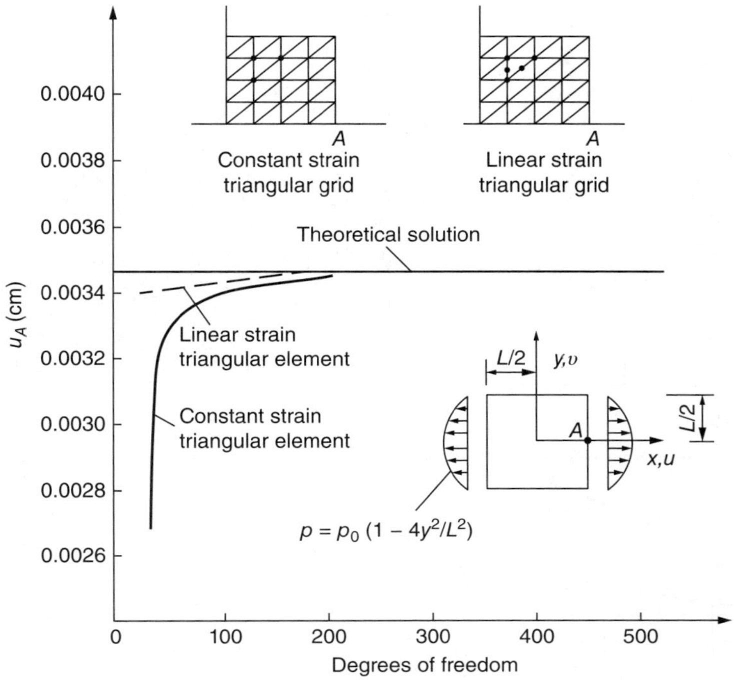

Figure 9.3 Square plate computed by two kinds of triangular elements.

elements or high-order triangular elements. Similarly, for arch dams and other spatial structures, it is better to adopt 20-node isoparametric elements or high-order tetrahedral elements rather than constant strain tetrahedral elements.

Figure 9.3 shows the square plates of equal thickness under the action of boundary forces in parabolic distribution. In his book *Theory of Elasticity*, Timoshenko gave a minimum work solution to this problem. In order to make a comparison, the literature [3] used constant strain triangular elements and linear strain triangular elements for computation. Only one-fourth of the plate was taken because of the symmetry; the figure indicates the computational grid of the two types of elements, and on this basis, the other grids were obtained by intensification. The figure shows the horizontal displacement u_A of point A, which indicates that quite high computation accuracy can be obtained by using not many linear strain elements. The stress accuracy also has a similar trend, and linear strain triangular elements have higher accuracy.

Figure 9.4 illustrates the results of a hollow cylinder computed by the author and Song Jingting by two different kinds of elements: only one-fourth was calculated because of the symmetry. Four layers of 8-node isoparametric elements were arranged in the thickness direction, with a total of 79 nodes; eight layers of isostrain triangular elements were arranged in the thickness direction, with a total of 103 nodes. It is clear from the figure that the stress values calculated with isoparametric elements were completely on the theoretical solution curve, having very high computation accuracy. The stress values computed with isostrain triangular elements fluctuated near the theoretical solution curve, having slightly poor computation accuracy, but were still available.

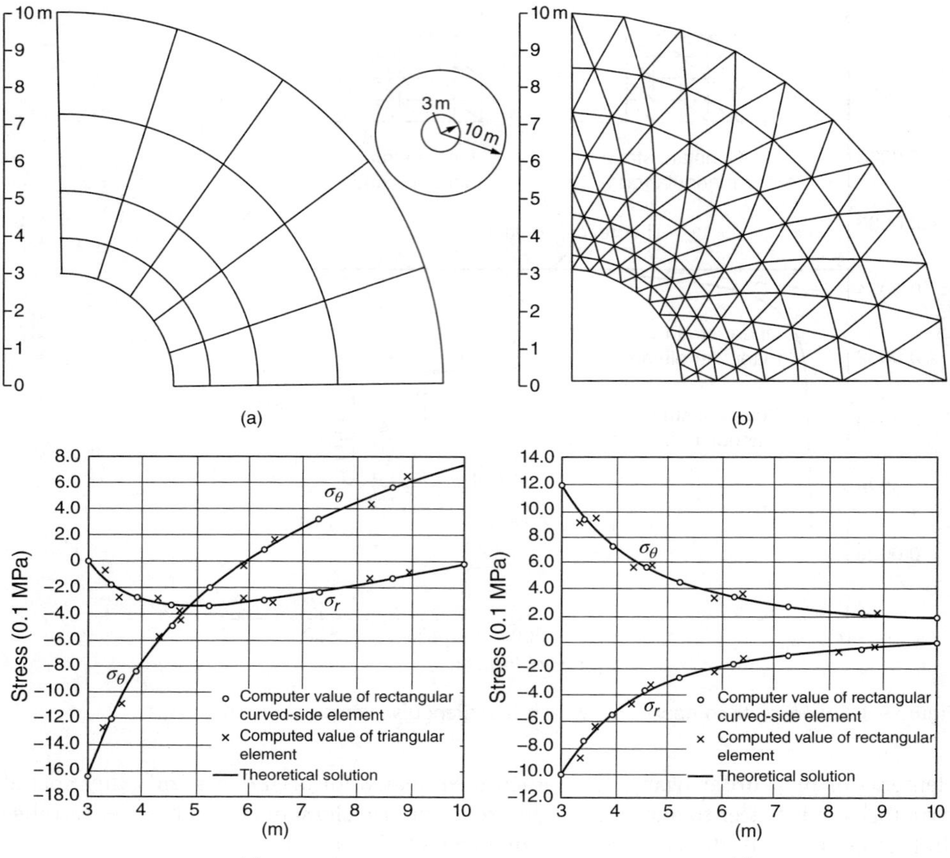

Figure 9.4 Comparison of computed values and theoretical solutions of a thick-walled round tube by two kinds of elements. (a) Computational grid of isoparametric elements (79 nodes), (b) computational grid of isostrain triangular elements (103 nodes), (c) stress caused by the internal and external temperature difference (10 °C on inner surface, 0 °C on outer surface, and (d) stress caused by the internal water pressure of 1 MPa.

In order to achieve the same accuracy, the number of high-order elements can be reduced, and the total degrees of freedom decrease accordingly. Thus, there are a smaller amount of information preparation and less time of solving the equation set by the computer. The high-order elements have relatively great bandwidth, and larger storage capacity is needed for the same degree of freedom. A more significant drawback of high-order elements is the complexity of the elemental stiffness matrix, and the computer spends much time during the formation of the stiffness matrix. Especially for isoparametric elements, their stiffness matrices must be computed by numerical integration, and the computer consumes more time for the formation of the stiffness matrices. However, in most cases, these disadvantages can be remedied by reduction in the degree of freedom.

9.2 Comparison and Selection of Various Spatial Elements

Various spatial elements are listed in Table 8.3 and can be roughly divided into three types, namely, tetrahedral elements, hexahedral elements, and isoparametric elements, represented by symbols T_n, H_n, and I_n, respectively. Among them, n is the number of degrees of freedom of elemental nodes, for example, T_{12} meaning the tetrahedral element with 12 degrees of freedom, H_{24} meaning the hexahedral element with 24 degrees of freedom, and I_{60} being the isoparametric element with 60 degrees of freedom.

Hexahedral elements have a regular shape and are difficult to adapt to the complex shape of the engineering structure; they are rarely applied at present.

In tetrahedral elements, T_{12} is the element that was first proposed and is now being applied because of its relatively simple stiffness matrix and its capability of adapting to the complex geometry. However, the internal strain of the element is constant, and intensive elements must be used to obtain better stress results. Otherwise, the computation accuracy is very poor, as shown in Figure 7.11. In all types of tetrahedral elements, T_{48} is the element with better result, and the elemental strain is a quadratic function, with higher computation accuracy; each node uses 12 parameters. Compared with other tetrahedral high-order elements, T_{48} has fewer nodes, and the bandwidth of the overall stiffness matrix decreases accordingly. It can better adapt to the complex geometry than hexahedral elements but is inferior to isoparametric curved-face elements.

Isoparametric elements have high computation accuracy and can adapt to complex geometry, so they are widely applied. Currently, I_{60} and I_{24} are the most widely applied elements. The prismatic curved-face element I_{45} is mainly used to match the I_{60} in order to better adapt to irregular boundaries.

To analyze and compare the computation accuracy and effect of various spatial elements, the literature [4] used several different types of elements to compute the short cantilever shown in Figure 9.5(a) and the simply supported slab shown in Figure 9.5(b). The elements used are as follows:

$5T_{12}$, five constant strain T_{12} elements, formed a brick-shaped 8-node hexahedral element.

$6T_{12}$, six constant strain T_{12} elements, first constituted a triangular prism, and then two triangular prisms formed a brick-shaped 8-node hexahedron.

I_{24} was an 8-node isoparametric element

$5T_{30}$, five linear strain tetrahedral T_{30} elements, formed a 26-node hexahedral element, where six nodes were located in the center of six surfaces and were eliminated through matrix operation and 20 nodes remained.

I_{60} was a 20-node isoparametric element.

The above five different types of elements were used for computation according to the grids shown in Figure 9.5(a) and (b), and the results were listed in Table 9.1. It is clear from the table that the isoparametric element I_{60} had very high computation accuracy for the two structures. The isoparametric element I_{24} had satisfactory computation accuracy of the cantilever but showed relatively poor computation accuracy of thin plates. The elements $5T_{12}$ and $6T_{12}$ that were made up of five or six constant strain elements showed very poor computation accuracy of the two structures. The element $5T_{30}$ that was formed by five linear strain elements had comparatively good computation accuracy, but its surface was a plane, not as good as the isoparametric element I_{60}

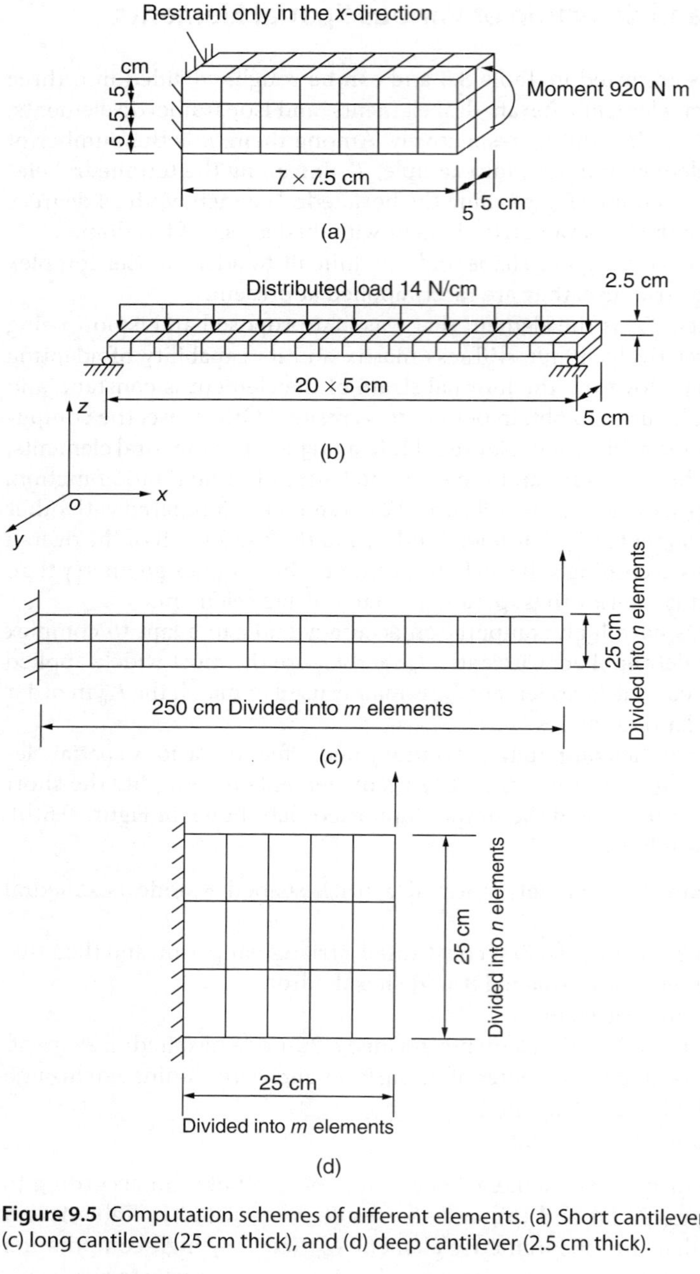

Figure 9.5 Computation schemes of different elements. (a) Short cantilever, (b) simply supported slab, (c) long cantilever (25 cm thick), and (d) deep cantilever (2.5 cm thick).

that could fit the complex shape of the structure. In addition, the stiffness matrix of the isoparametric element was isotropic, while according to the computation, the stiffness matrix of the regular hexahedral element that was formed by $5T_{30}$ elements was slightly different in three directions.

According to the above computation results, in two kinds of isoparametric elements, I_{60} seemed better, but the results in Table 9.3 were calculated by different elements under

Table 9.1 Comparison of computation results of spatial elements.

Number of nodes	Element number	Description	Short cantilever			Simply supported slab		
			Maximum stress (0.1 MPa)		Maximum deflection (10^{-4} cm)	Maximum stress (0.1 MPa)		Maximum deflection (cm)
			σ_x	σ_z		σ_x	σ_z	
	$5T_{12}$	Combination of five constant strain elements	Not computed	Not computed	1.54	Not computed	Not computed	0.0786
8-Node	$6T_{12}$	Combination of six constant strain elements	1.056	0.084	1.39	Not computed	Not computed	0.0662
	I_{24}	Isoparametric elements	2.22	0.099	1.86	788	0.38	0.236
20-Node	$5T_{30}$	Combination of five linear strain elements	2.35	0	2.08	—	—	—
	I_{60}	Isoparametric elements	2.35	0	2.08	1690	0.74	0.616
Theory of beams			2.35	0	2.08	1690	0.70	0.616

the same grid conditions. The I_{60} element had higher accuracy but consumed more computation time due to its complexity. From the point of view of engineering, comparison should be conducted according to the computation accuracy obtained from unit time. Obviously, in order to carry out this comparison, different computational grids should be adopted for different elements. Thus, the long cantilever shown in Figure 9.5(c) and the deep cantilever shown in Figure 9.5(d) were calculated by different grids. In the length and thickness directions, m and n elements were divided, respectively. The computation results are shown in Figure 9.6, in which the computation time (IBM 7094 computer) is taken as the abscissa and the beam end deflection as the vertical coordinate. The element obtaining greater deflection in the same time is superior. It is clear from the figure that for the long cantilever, the I_{60} isoparametric element had the best performance, but for the deep cantilever with greater shear effect, the I_{24} element was better. However, only one layer of elements was taken in the y-direction (perpendicular to the paper surface) in the above computation, not giving full play to the advantages of the I_{60} element. When the y-direction has multilayer elements, the I_{60} and I_{24} elements may show smaller differences in the massive structures like deep cantilevers.

In short, for spatial problems, the 20-node isoparametric element I_{60} can well reflect the plate bending action; when taking only one layer of elements in the thickness direction, it can calculate the structures with relatively significant bending action, such as arch dams, thick-walled pressure vessels, and so on. However, for the massive structures with relatively significant internal shear stress, the 8-node isoparametric element I_{24} may be more effective. If there is a phenomenon of stress concentration, tetrahedral elements are still a form worthy of consideration because we can use dense grids to adapt to a dramatically changing stress field.

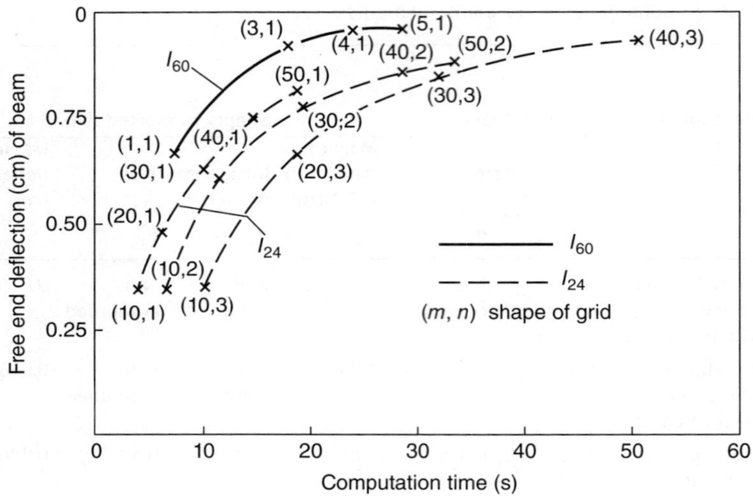

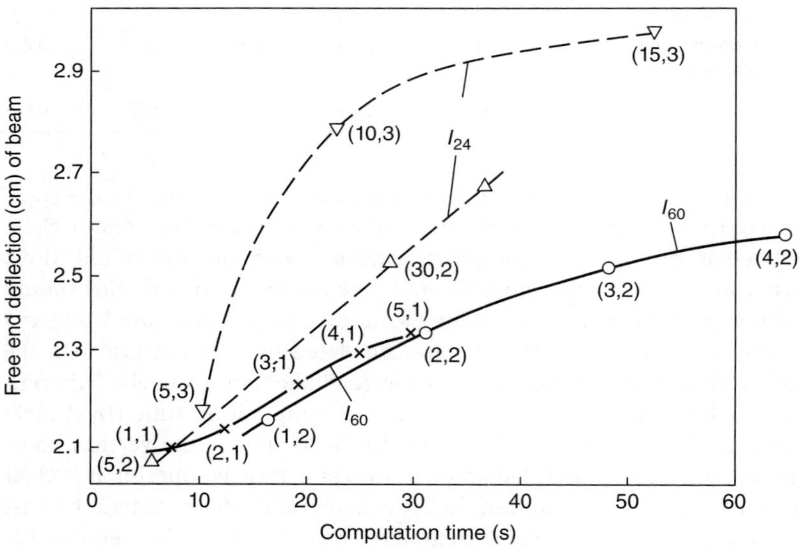

Figure 9.6 Comparison of computational efficiency of different elements. (a) Long cantilever and (b) deep cantilever.

According to Figure 8.26, when the structure is very thin, such as penstocks in a hydropower station, the use of spatial elements for computation may cause an ill-conditioned equation. It would be best to adopt thin-plate and thin-shell elements for computation for such thin-walled structures, as shown in the following sections in detail.

In the following sections, we will give some application instances of the finite element method.

9.3 Analysis of Stresses in Arch Dam

Arch dams are variable-thickness and variable-radius doubly curved shells built in rocky canyons, and their stress analysis is a relatively difficult issue due to their complex shapes. In the engineering design, the trial-load method was mainly adopted in the past, and the use of the finite element method increased gradually in recent years. Here, we introduce some typical results by using spatial finite elements for the analysis of arch dams [5] and compare these results with the results calculated by the traditional trial-load method. In the following sections, we will also describe some computation results of arch dams by thin- and thick-shell finite elements.

The canyon is symmetric. The dam is 120 m high, the crest 5.35 m thick, the dam base 23.35 m thick, and the radius of the crest arch ring 170.80 m, with the central angle of 106°; the radius of the arch ring of the dam base is 73.4 m, with the central angle of 80°. The chord–height ratio and the arc–height ratio of the crest are 2.27 and 2.67, respectively, and the thickness–height ratio of the dam base is 0.1943. The dam shape is shown in Figure 9.7(a). Only the water pressure up to the crest was considered; the modulus of elasticity of the dam body E is 20,000 MPa, and Poisson's ratio μ is 0.15.

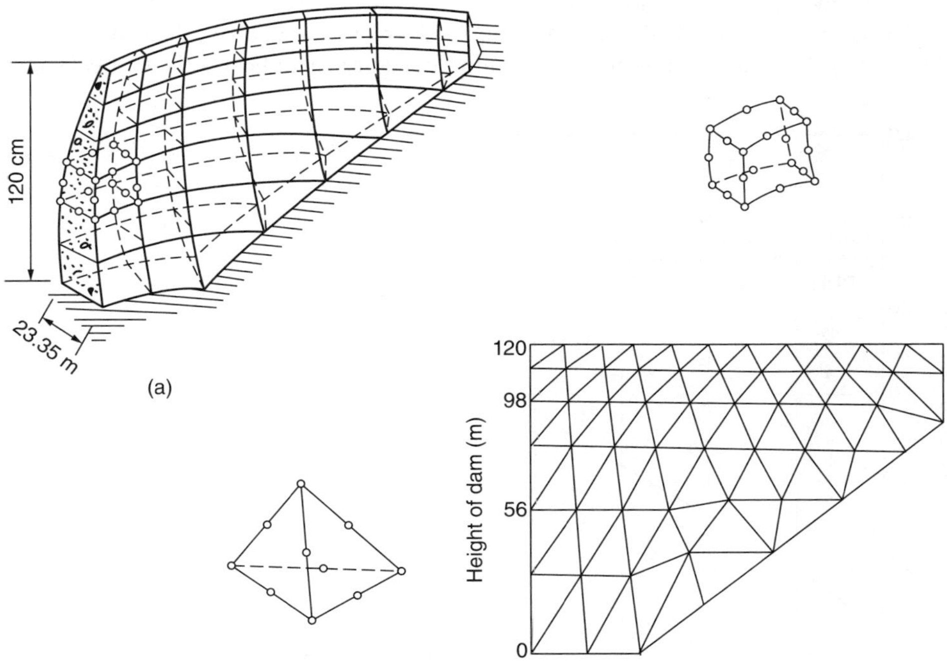

Figure 9.7 Computational grid of doubly curved arch dam. (a) 20-Node curved-surface element grid and (b) 10-node tetrahedral element grid (middle surface).

9.3.1 Comparison of Different Computation Methods

The dam was calculated by the following three different methods:

1) *20-Node spatial curved-surface elements*
   One layer of elements was used in the thickness direction, and only half of them were taken for computation because of the symmetric valley; there were a total of 32 elements with 272 nodes used.
2) *10-Node linear strain tetrahedral elements*
   Three 10-node tetrahedral elements first formed a triangular prism, and then two triangular prisms constituted a brick-shaped hexahedral element; two layers of combined hexahedral elements were used in the thickness direction of the dam body. The computational grid on the middle surface was shown in Figure 9.7(b). The whole dam had a total of 1015 nodes, and there were a total of 2800 unknowns after considering the conditions on the base boundary and the symmetric faces.
3) *Trial-load method*
   The loads on the arches and cantilevers are fully adjusted to make the consistence of the radial displacement, tangential displacement, and torsion.

The influence of base deformation on the arch dam stress is a comparatively complicated issue, and for the purpose of simplification, the base was assumed to be rigid, in order to compare the computation results of the dam stress by these three methods.

Figure 9.8 shows the radial displacement of the arch crown. The radial displacement calculated by the three methods was very close below the elevation of 70 m, and three displacement curves gradually separated above the elevation of 70 m; the displacement obtained by the trial-load method was minimum. Figure 9.9 indicates the vertical stress

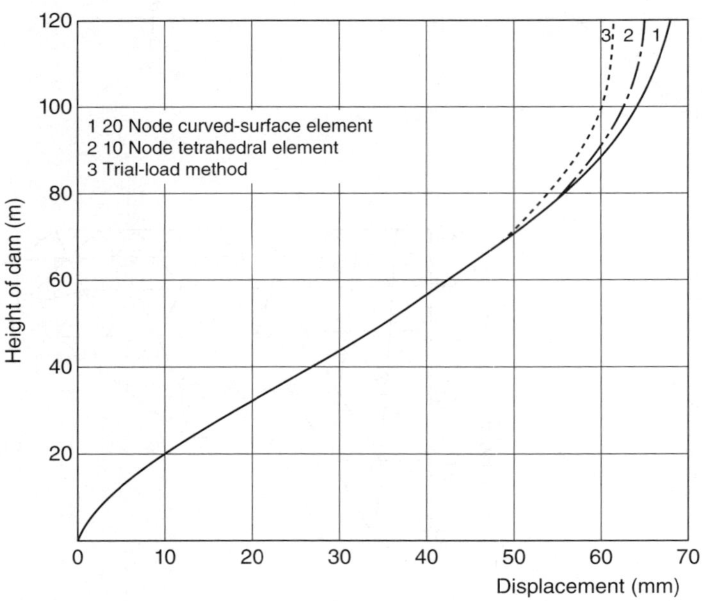

Figure 9.8 Radial displacement of crown section.

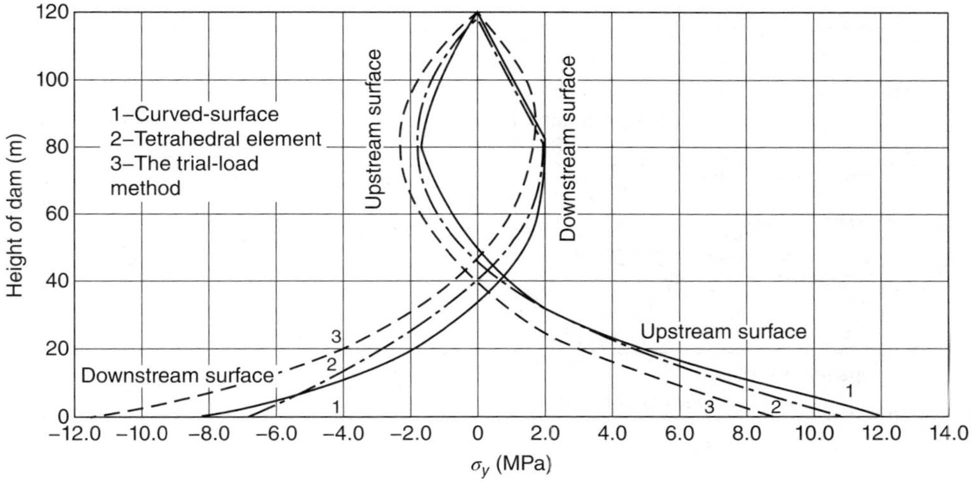

Figure 9.9 Vertical stress on the crown section.

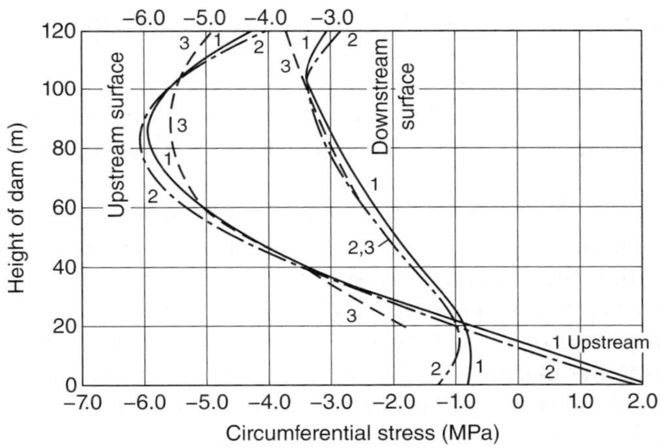

Figure 9.10 Circumferential stress of the arch ring of crown section. (1) Curved-face elements; (2) tetrahedral elements; and (3) trial-load method.

on the crown section. The vertical stresses computed by two finite element methods were quite close, while those calculated by the trial-load method had a big difference. Figure 9.10 compares the circumferential stresses of the arch ring on the crown section. The stress was relatively great on the crest by the trial-load method; the circumferential stress was maximum at the elevation of 80 m, where the stress was slightly smaller on the contrary by the trial-load method. The circumferential stress of the dam base was not given under the trial-load method. Figure 9.11 shows the changes in the circumferential stress of the arch ring. The stress values computed by two finite element methods were very close, and the stresses calculated by the trial-load method had a big difference. Figure 9.11(b) compares the changes in the circumferential stress of the arch ring at the elevation of 80 m. Overall, the stress values by the three methods were quite close,

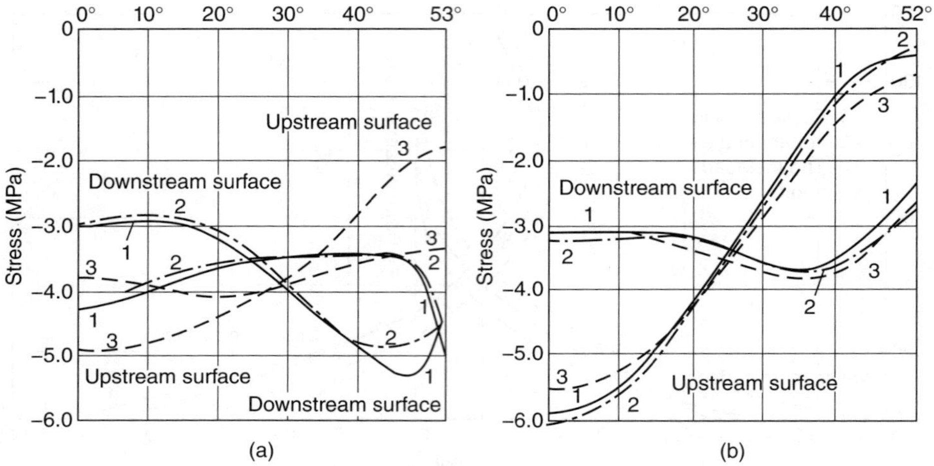

Figure 9.11 Circumferential stress of arch ring. (a) Crest; (b) elevation of 80 m. (1) curved-face elements, (2) tetrahedral elements; and (3) trial-load method.

but the maximum compressive stress value calculated by the trial-load method was slightly smaller.

According to the above displacement and stress results, the computation results by two kinds of finite elements were very close to each other, but only 272 nodes of the curved-face elements were used, and 1015 nodes of the linear strain tetrahedral elements were used.

There were much fewer nodes and elements of the curved-face elements used, so the amount of computation, preparation, and result organization was much smaller. According to the later computation, the computation results were very satisfactory even if only nine 20-node curved-face elements were used. Compared with 32 elements, the maximum error in displacement was only 2%. It is thus clear that the application of 20-node curved-face elements to the analysis of arch dam stress is very appropriate.

The displacement and stress of the crest and the vertical stress of the dam base differed greatly between the trial-load method and the finite element method, and the rest was basically the same. Since the computing time of the trial-load method is less than the finite element method, the trial-load method can yet be regarded as an available computation tool during the preliminary design stage.

9.3.2 The Effect of Foundation Deformation on the Displacement and Stress of Arch Dam

The extent of effect of foundation deformation on the arch dam stress is an issue of concern in the engineering; the literature [5] analyzed this problem and gave the computational grid by using 20-node spatial curved-face elements, as shown in Figure 9.12; four cases were computed as follows:

1) The base was rigid, $E_f = \infty$.
2) The modulus of elasticity of the *foundation* was the same as that of the dam concrete, $E_f = E_c$.

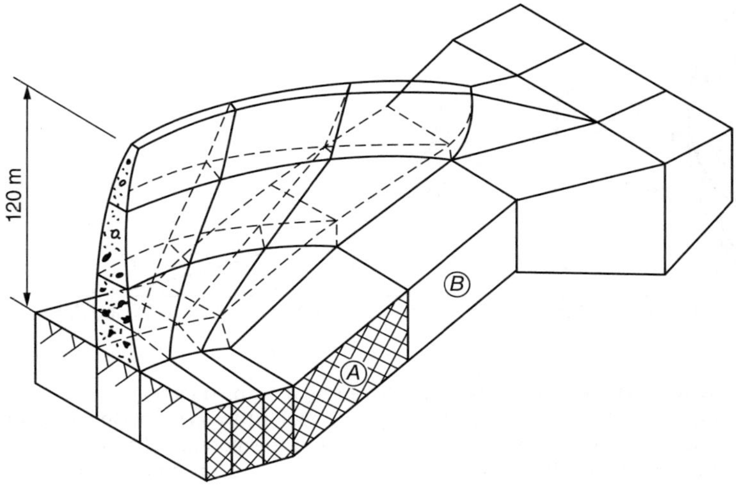

Figure 9.12 Arch dam and foundation.

3) The modulus of elasticity of the *foundation* was 1/10 of that of the dam concrete, $E_f = E_c/10$.
4) The *foundation* was not homogeneous (1): for the river bed part (area A), $E_f = E_c/8$; for the bank slope part (area B), $E_f = E_c/2$.

Figure 9.13 shows the radial displacement of the downstream face of the arch crown; apparently, the modulus of elasticity of the *foundation* had great influence on the dam displacement. Figures 9.14–9.16 indicate the effect of the *foundation* deformation on the dam stress, and it is thus clear that the changes in the modulus of elasticity of the *foundation* affected significantly the vertical stress, namely, the stress on the cantilever.

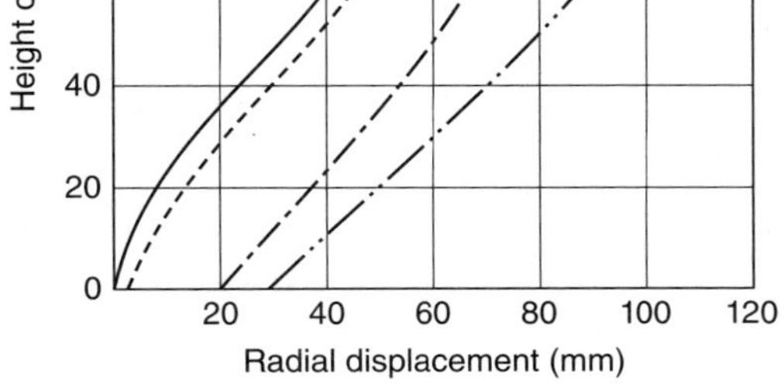

Figure 9.13 Radial displacement of downstream face on the crown section with different elastic ratios of the foundation.

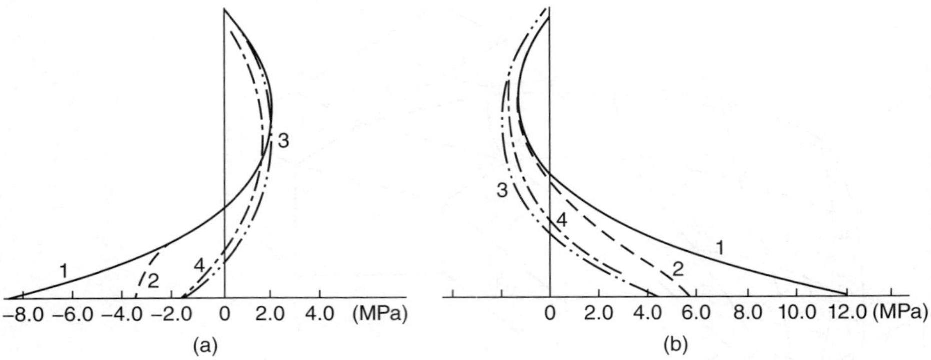

Figure 9.14 Effect of foundation deformation on the vertical stress on the crown section $1 - E_f = \infty$, $2 - E_f = E_t$, $3 - E_f = E_t/10$, $4 - E_f(\text{region A}) = E_f/8, E_f$ (region B) $= E_f/2$. (a) Downstream face of the arch crown; (b) upstream face of the arch crown. , where E_f is the modulus of elasticity of the foundation, E_c is the modulus of elasticity of concrete and $E_c = 20{,}000$ MPa for computation, and Poisson's ratio of the dam body and the foundation is 0.15.

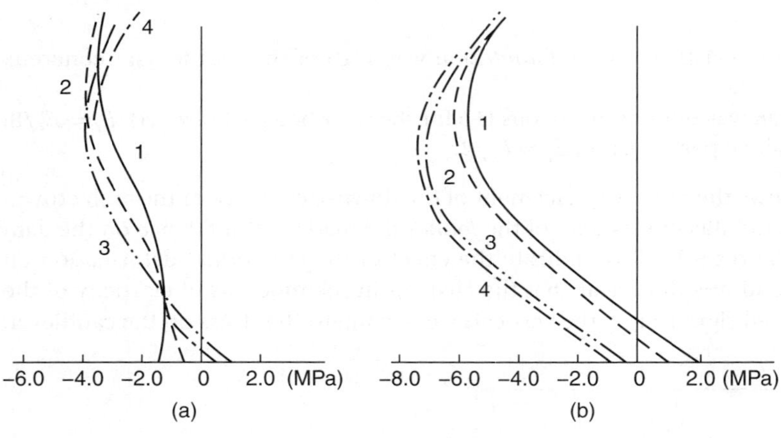

Figure 9.15 Effect of foundation deformation on the circumferential stress of the arch ring on the crown section $1 - E_f = \infty$, $2 - E_f = E_c$, $3 - E_f = E_c/10$, $4 - E_f(\text{region A}) = E_c/8, E_f$ (region B) $= E_c/2$. (a) Downstream surface of the arch crown and (b) upstream surface of the arch crown.

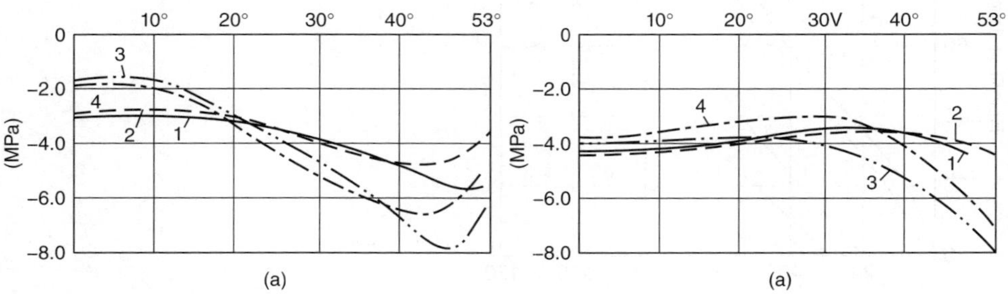

Figure 9.16 Effect of foundation deformation on the circumferential stress of the arch ring of the crest $1 - E_f = \infty$, $2 - E_f = E_c$, $3 - E_f = E_c/10$, $4 - E_f(\text{region A}) = E_c/8$, $E_f(\text{region B}) = E_c/2$.
(a) Circumferential stress of the downstream surface and (b) circumferential stress of the upstream surface.

For a relatively soft *foundation*, the vertical stress of the dam *foundation* is only 1/3–1/4 of the rigid *foundation*. According to Figure 9.16, the *foundation* deformation also made relatively large effect on the circumferential stress of the abutment but had quite small influence on the circumferential stress of the arch crown.

9.4 Analysis of Stress in Buttress Dam

Contraction joints are usually used to separate all buttresses of the dam, so each buttress has independent action. In analyzing the stress in the buttress dam, buttresses are often regarded as variable-thickness flat plates, and plane stress finite elements are adopted to compute the overall stress in buttresses. Then, head flakes of unit thickness are cut in the direction perpendicular to the upstream dam surface, and planar finite elements are used to calculate the head stress under the effect of the water pressure.

The state of stress inside the dam is obtained by analyzing two-dimensional finite elements twice. However, the buttress dam is actually a spatial structure, and there is a great difference between the sizes of the head and buttress. The extent of error produced according to the analysis of plane problems also interest designers. The literature [6] analyzed this issue and used 20-node isoparametric elements to compute a buttress dam with the height of 62.5 m and the span of 17 m. Only half of the dam was taken for the analysis because of the symmetry, and the computational grid of the dam body and the principal stresses are shown in Figure 9.17. The computation included a portion of the foundation (not shown in the figure), and a layer of relatively thin elements were arranged on the foundation surface in order to better reflect the effect of foundation deformation on the dam stress. For comparison, the same grid was adopted for two-dimensional analysis. Figure 9.18 shows the principal stress on the

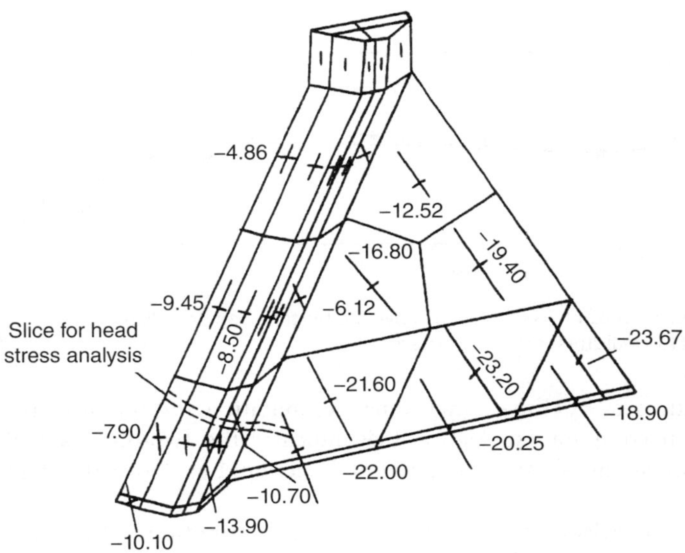

Figure 9.17 Analysis of three-dimensional stress in buttress dam (principal stress: 0.1 MPa).

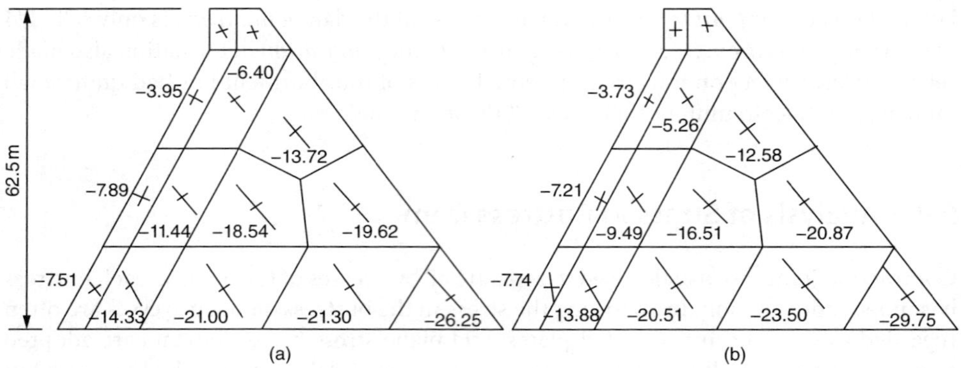

Figure 9.18 Principal stresses in buttress dam due to water pressure and deadweight (0.1 MPa). (a) Three-dimensional analysis, principal stress on middle surface and (b) principal stress in two-dimensional analysis.

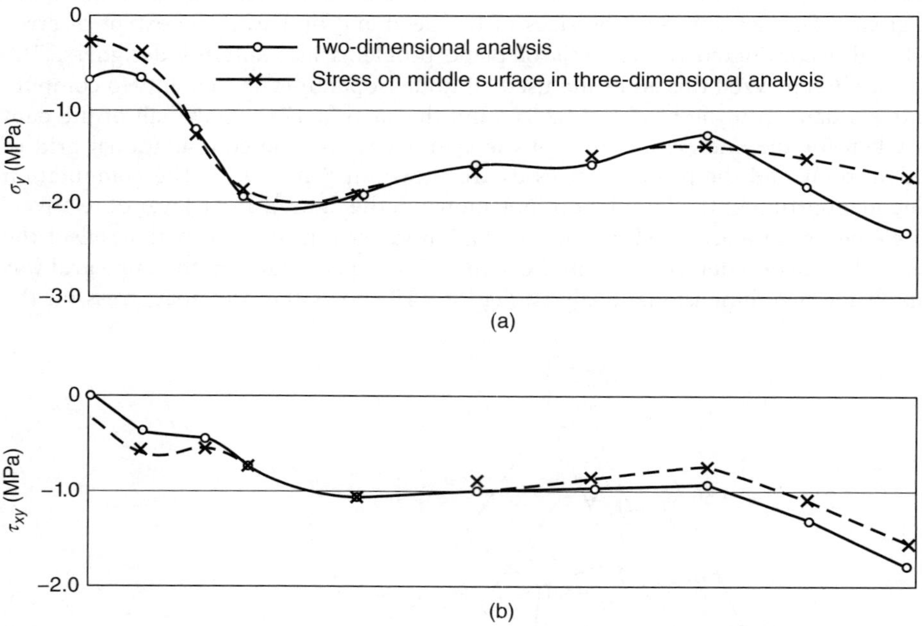

Figure 9.19 Stress distribution in horizontal section 1 m above the buttress dam foundation (water pressure and deadweight). (a) Normal stress s (b) shear stress.

middle surface of the buttresses by three-dimensional computation and the principal stress by two-dimensional computation. Figure 9.19 indicates the two-dimensional and three-dimensional stress distribution at horizontal section 1 m above the dam foundation.

Figure 9.20 shows the comparison of the horizontal stress distribution in the head. The following can be obtained from the figure: (1) By regarding the buttresses as variable-thickness flat plates and using two-dimensional finite elements for computation, the obtained state of stress was generally similar in the law of distribution to

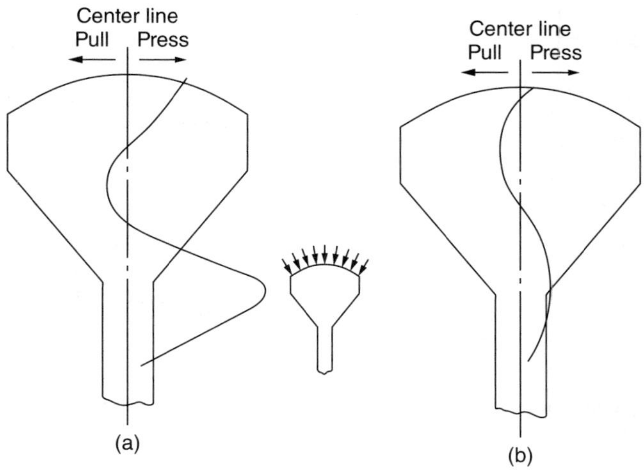

Figure 9.20 Horizontal stress in the head of buttress dam. (a) Two-dimensional analysis (water pressure) and (b) three-dimensional analysis, 17 m away from the dam foundation (water pressure and deadweight).

the three-dimensional analysis results, but there were some discrepancies in the stress values at the heel and the toe of the dam (e.g., Figure 9.19 shows that the absolute value was slightly larger by comparing the obtained vertical compressive stress on the heel of the dam in the three-dimensional analysis and the two-dimensional stress). (2) For the horizontal stress in the head, the stress graphs had a great difference in the two- and three-dimensional results (of course, the loads were different), but the maximum values of the tensile stresses relating to design were relatively close.

9.5 Analysis of Spatial Effect of Gravity Dam

In a gravity dam, generally there are vertical and horizontal joints during construction; it is reasonable to assume that the deadweight of the dam body is undertaken separately by each dam block during stress analysis, but if the vertical and horizontal joints have been grouted before storing water, the dam body will have spatial effect under water pressure. As some water loads are transferred to both banks through the spatial effect, part of the dam body in the river bed bears slightly fewer water loads. Figure 9.21 shows some computation results by using 20-node curved-face elements [6]. According to the figure, the spatial effect of the gravity dam was closely related to the value of the length–height ratio B/H of the dam body. When the B/H was greater than 4, the spatial effect was small; when the B/H was less than 2, the spatial effect was very significant.

9.6 Analysis of Spatial Effect of Earth Dam

Earth dams are built in canyons, and there is no joint in the dam body, so they belong to spatial structures. However, the computation will be very heavy if spatial finite elements are used for computation. The computation should be conducted according to plane

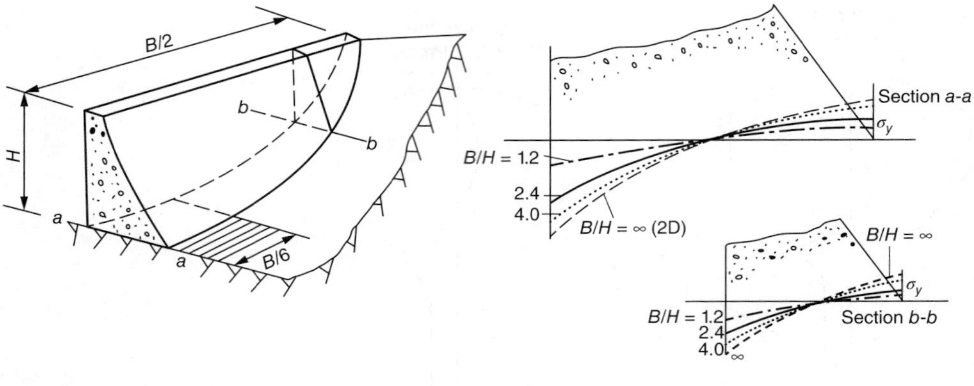

Figure 9.21 Spatial effect of gravity dam (the vertical stress σ_y on two dam sections) $B/H = \infty$ represents two-dimensional analysis.

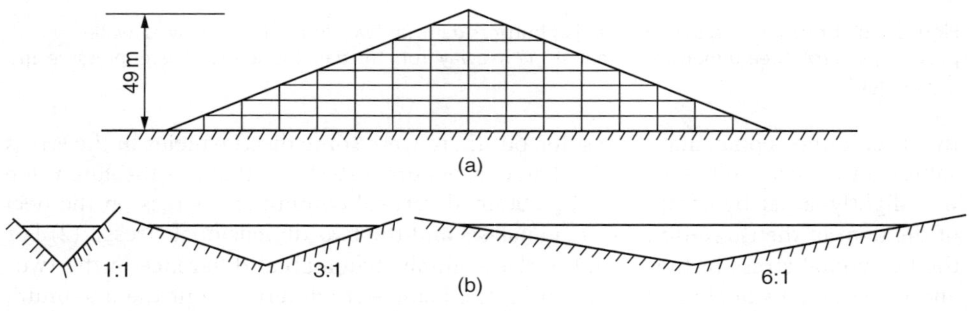

Figure 9.22 Earth dam section. (a) Cross section and (b) longitudinal section.

problems as far as possible if the computation error is not very great. Now, we compare and analyze the computation results of an earth dam, respectively, according to a plane problem and a spatial problem under the conditions of different valley slopes [7].

The cross section of the earth dam is shown in Figure 9.22; the upstream and downstream dam slopes were 2.5:1, and the height of the dam was 49 m, and the valley banks were symmetrical. We analyzed three different bank slopes, namely, 1:1, 3:1, and 6:1. Eight-node isoparametric elements were used to analyze the spatial problem, and the internal degree of freedom was increased to improve the computation accuracy. Quadrilateral elements were adopted to analyze the plane problem and could degenerate into triangular elements at the boundary. In order to simulate the construction process of the earth dam, eight layers were divided, and the incremental method was used for computation. The modulus of elasticity of the soil body E was 9.7 MPa, the soil density γ was 2000 kg/m^3, and Poisson's ratio μ was 0.40. The comparison of the computation results is shown in Tables 9.2–9.4.

The following can be seen from the above computation results:

1) When the valley slope was relatively gentle, such as 3:1 or gentler, if the transversal section of the earth dam was calculated according to a plane strain problem, relatively reasonable results could be obtained. When the valley slope was relatively steep, such as 1:1, the results computed according to plane strain were inaccurate due to the significant spatial effect of the dam body.

Table 9.2 Computation results (plane strain/three-dimensional computation) × 100% of transverse section of earth dam.

	1:1		3:1		6:1	
	Average	Range	Average	Range	Average	Range
Maximum principal stress, σ_1	113	100–129	102	100–113	101	98–109
Minimum principal stress, σ_3	98	79–125	96	81–111	97	88–100
Maximum shear stress, τ_{max}	138	108–225	112	100–150	108	100–150
Vertical displacement, v	136	91–156	106	85–114	100	85–105
Horizontal displacement, u	268	75–435	120	80–149	105	85–120

Table 9.3 Computation results I (plane stress/three-dimensional computation) × 100% of longitudinal section of earth dam.

	1:1		3:1		6:1	
	Average	Range	Average	Range	Average	Range
Maximum principal stress, σ_1	109	80–127	110	102–115	111	100–115
Minimum principal stress, σ_3	77	20–109	84	60–109	85	63–100
Maximum shear stress, τ_{max}	149	107–185	149	111–179	149	126–181
Vertical displacement, v	160	122–196	173	138–217	173	140–224
Horizontal displacement, u	220	130–1300	228	139.400	224	139.400

Table 9.4 Computation results II (plane strain/three-dimensional computation) × 100% of longitudinal section of earth dam.

	1:1		3:1		6:1	
	Average	Range	Average	Range	Average	Range
Maximum principal stress, σ_1	111	91–115	110	98–115	110	93–115
Minimum principal stress, σ_3	122	85–130	124	100–136	123	100–135
Maximum shear stress, τ_{max}	94	70–117	91	74–105	90	83–107
Vertical displacement, v	97	78–113	98	75–120	97	80–124
Horizontal displacement, u	117	72–166	118	75–233	115	71–200

2) The longitudinal section of the earth dam had unreliable computation results according to plane stress, and the obtained computation results of the three bank slopes were not good.
3) The longitudinal section of the earth dam could be calculated according to a plane strain problem, and the obtained computation results of the three different valleys were good; the maximum computation error was in σ_3, and the calculated σ_3 according to a plane strain problem was greater by 24% than the three-dimensional computation value.

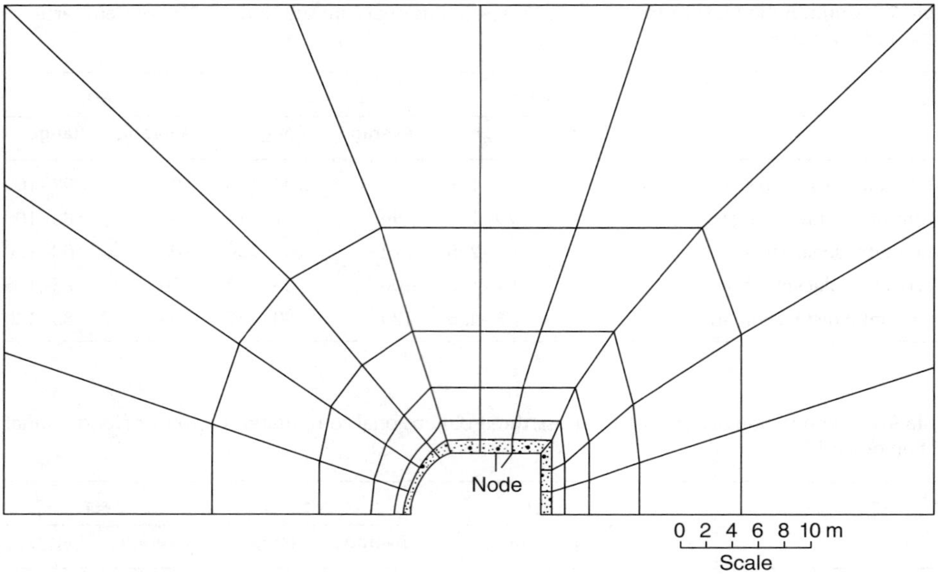

Figure 9.23 Computational grid of tunnel lining. Rock: $E_0 = 6900$ MPa, $\mu_0 = 0.16$. Concrete: $E_1 = 18{,}000$ MPa, $\mu_1 = 0.16$.

9.7 Analysis of Stress on Tunnel Lining

The computation of the stress of tunnel lining buried in the rock is a very complex contact problem in the theory of elasticity. In the past engineering design, the rock was first assumed to have a certain resistance coefficient, and then the lining was computed according to the method of structural mechanics. This computation method could not reflect the real situation of the interaction between the rock and the lining. The use of the finite element method can correctly calculate the state of stress in the lining. Since the tunnel lining is relatively thin, it is hard to arrange more than five layers of constant strain elements in the thickness direction. If only two or three rows of constant strain elements are arranged, there will be a significant computation error. Therefore, it is preferable to use quadratic isoparametric elements; they have higher computation accuracy, and their boundary is a quadratic curve, which is able to fit the boundary of the tunnel lining.

Figure 9.23 shows the grid by which the author and Song Jingting jointly used 8-node planar isoparametric elements to compute a horseshoe-shaped hydraulic tunnel. Only half of the structure was taken for the computation because of the symmetry. The modulus of elasticity of the rock was 6900 MPa, and that of the concrete lining was 18,000 MPa. Two loads were calculated: one was the internal water pressure, and the other was the rock pressure and the deadweight of the lining. The stress computation results are shown in Figure 9.24.

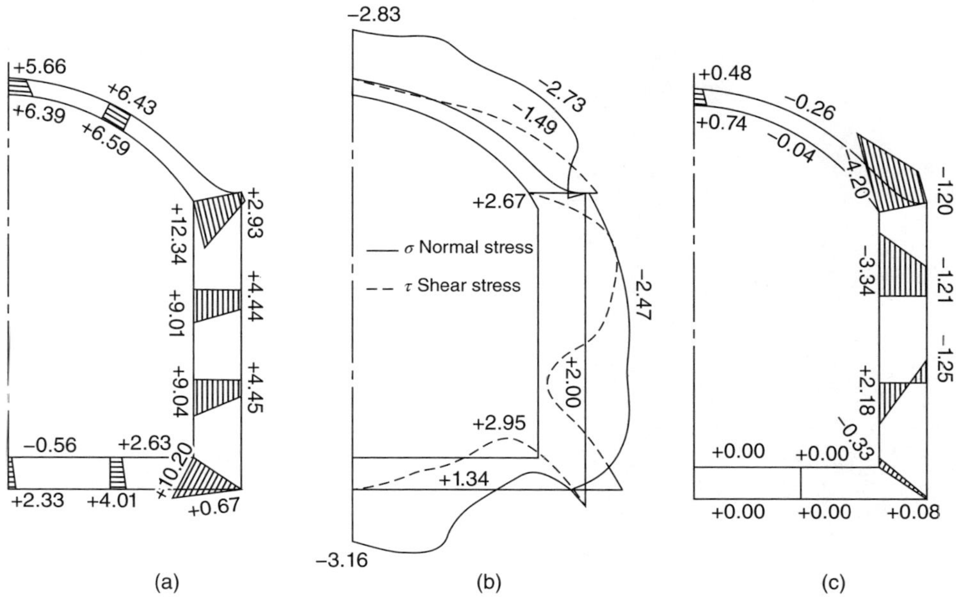

Figure 9.24 Tunnel lining stress computation results (the tensile stress is positive, 10^5 Pa). (a) Stresses in cross section under the effect of internal water; (b) stresses on contact surface under the effect of internal water; and (c) stresses in cross section under the effect of rock pressure and deadweight of lining.

Bibliography

1 Zienkiewicz, O.C. and Taylor, R.L. (2009) *Finite Element Method for Solid and Structural Mechanics*, 6th edn, Elsevier, New York, London.
2 Holland, I. and Bell, K. (1972) *Finite Element Methods in Stress Analysis*, Tapir, Trondheim (Norway).
3 Gallagher, R.H. (1975) *Finite Element Analysis Fundamentals*, Prentice-Hall, Englewood Cliffs.
4 ASCE (1969) *Symposium on Applications of Finite Element Methods in Civil Engineering*. November 13–14, Nashville, Tennessee, United States. ASCE.
5 (1968) Proc. Symp. Arch Dams, Inst. Civ. Eng.
6 Naylor, D.J., Stagg, K.G. and Zienkiewicz, O.C. (1975) *Criteria and Assumptions for Numerical Analysis of Dams.*
7 Lefebvre, G., Duncan, J.M. and Wilson, E.L. (1973) Three dimensional finite element analysis of dams. *Proc. ASCE*, **SM7**, 495–507.

10

Elastic Thin Plate

Elastic thin plates are widely used in engineering. Some theoretical solutions and tables are available for relatively simple situations, such as thin plates with equal thickness, single span, no large orifice, and regular shape (rectangular, circular). But for some complex situations occurring in engineering, such as plates with variable thickness, multi span, large orifice, irregular shape, and supported by elastic beam or column, theoretical method is incapable to solve these problems. Now using the finite element method, they can be solved quickly.

When solving the bending problem of thin plates by the displacement method, it is required that not only the deflection w is continuous, but the first derivative of w is also continuous in the common boundary of adjacent elements. But it is not easy to do so. Therefore, in the finite element method for solving bending problems of thin plates, besides the elements solved by the displacement method, hybrid elements and mixed elements also gain prominence. In later years, an element with curved boundary is developed whose deflection and rotation are interpolated, respectively. It is very effective.

10.1 Bending of Elastic Thin Plate

Under a load perpendicular to the surface of the plate, the thin plate bends. If the deflection w compared with thickness is small, the following assumptions can be used in the analysis of the bending problem of plate:

1) Normal stress in the thickness direction is negligible, and the thickness does not change after deformation.
2) After bending, the normal line of the plate remained normal to the middle surface of the thin plate.
3) For the points of the middle surface, there is no displacement parallel to the middle surface of the plate.

Using the assumptions above, all the stress and strain components can be expressed by deflection w. As shown in Figure 10.1, let middle surface be xy surface and the z-axis perpendicular to it. According to the first assumption, we can get

$$\varepsilon_z = \frac{\partial w}{\partial z} = 0$$

The Finite Element Method: Fundamentals and Applications in Civil, Hydraulic, Mechanical and Aeronautical Engineering, First Edition. Bofang Zhu.

Companion website: www.wiley.com/go/zhufem17

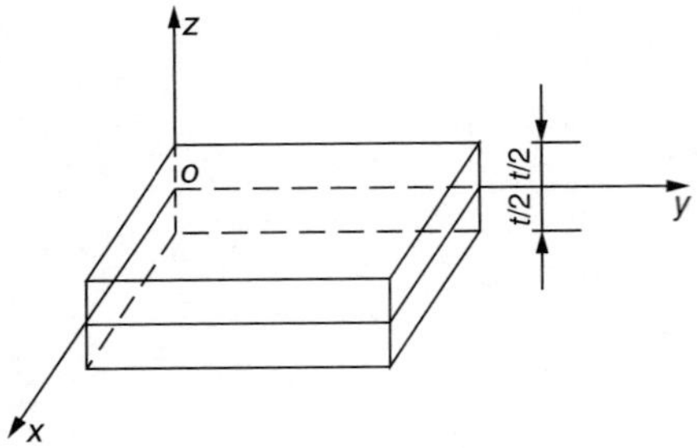

Figure 10.1 Thin plate.

So w is independent on z and $w = w(x, y)$ is obtained, that is to say, all the points on each normal line of the middle surface have the same displacement w.

According to the second assumption, after bending, normal lines of the plate remain perpendicular to the tangents of the elastic surface in the direction of x or y.

There is no shearing strains, namely, $\gamma_{yz} = 0$, $\gamma_{zx} = 0$, that is,

$$\frac{\partial v}{\partial z} + \frac{\partial w}{\partial y} = 0, \quad \frac{\partial w}{\partial x} + \frac{\partial u}{\partial z} = 0$$

From the equations above, we can get

$$\frac{\partial v}{\partial z} = -\frac{\partial w}{\partial y}, \quad \frac{\partial u}{\partial z} = -\frac{\partial w}{\partial x} \tag{a}$$

But from formula $w = w(x, y)$, it is clear that $\partial w/\partial x$ and $\partial w/\partial y$ don't change with variable z. We integrate formula (a) with z and get

$$v = -z\frac{\partial w}{\partial y} + f_1(x, y), \quad u = -z\frac{\partial w}{\partial x} + f_2(x, y) \tag{b}$$

in which $f_1(x, y)$ and $f_2(x, y)$ are arbitrary functions.

By the third assumption, we can get $(u)_{z=0} = (v)_{z=0} = 0$. From the formula (b), $f_1(x, y) = f_2(x, y) = 0$ is obtained. So formula (b) is simplified as

$$v = -z\frac{\partial w}{\partial y}, \quad u = -z\frac{\partial w}{\partial x}$$

From the formulas above, we can get three strain components of all the points in the thin plate:

$$\begin{aligned}
\varepsilon_x &= \frac{\partial u}{\partial x} = -z\frac{\partial^2 w}{\partial x^2} \\
\varepsilon_y &= \frac{\partial v}{\partial y} = -z\frac{\partial^2 w}{\partial y^2} \\
\gamma_{xy} &= \frac{\partial u}{\partial y} + \frac{\partial v}{\partial x} = -2z\frac{\partial^2 w}{\partial x \partial y}
\end{aligned} \tag{c}$$

In the case of small deformations, $\partial^2 w/\partial x^2$ and $\partial^2 w/\partial y^2$ represent the curvature in x- and y-direction of the elastic surface. And $\partial^2 w/\partial x\partial y$ represents the torsion in x- and y-direction. These three terms entirely determine the strain components of each point of the plate. So they can be described as deformations of the thin plate and expressed by matrix as

$$\{\psi\} = \left[-\frac{\partial^2 w}{\partial x^2} \quad -\frac{\partial^2 w}{\partial y^2} \quad -2\frac{\partial^2 w}{\partial x \partial y}\right]^{\mathrm{T}} \tag{10.1}$$

By the formula (c), the strain $\{\varepsilon\}$ of each point in the plate can be expressed with deformation of plate $\{\psi\}$ as

$$\{\varepsilon\} = z\{\psi\} \tag{10.2}$$

Since the normal stress can be neglected, the strain components of each point in the isotropic plate can be expressed with stress as

$$\varepsilon_x = \frac{\sigma_x}{E} - \mu\frac{\sigma_y}{E}, \quad \varepsilon_y = -\mu\frac{\sigma_x}{E} + \frac{\sigma_y}{E}, \quad \gamma_{xy} = \frac{2(1+\mu)}{E}\tau_{xy}$$

From the three formulas above, stresses are solved as follows:

$$\sigma_x = \frac{E}{1-\mu^2}(\varepsilon_x + \mu\varepsilon_y), \quad \sigma_y = \frac{E}{1-\mu^2}(\mu\varepsilon_x + \varepsilon_y), \quad \tau_{xy} = \frac{E}{2(1+\mu)}\gamma_{xy}$$

By substitution of formula (c) into the above equation, stress components of each point in the plate can be expressed with the deflection w as follows:

$$\begin{aligned}
\sigma_x &= -\frac{E}{1-\mu^2}z\left(\frac{\partial^2 w}{\partial x^2} + \mu\frac{\partial^2 w}{\partial y^2}\right)\\
\sigma_y &= -\frac{E}{1-\mu^2}z\left(\mu\frac{\partial^2 w}{\partial x^2} + \frac{\partial^2 w}{\partial y^2}\right)\\
\tau_{xy} &= -\frac{E}{1+\mu}z\frac{\partial^2 w}{\partial x\partial y}
\end{aligned} \tag{d}$$

Now we calculate the internal forces of plate synthesized by the stress components. As shown in Figure 10.2, taking out a small hexahedron from the plate, it has a unit width in x- and y-direction. In a cross section perpendicular to the x-axis, normal stress is σ_x. From formula (d), we know it is proportional to z, so it can only synthesize a moment. Expressing the moment of unit width as M_x, we get

$$M_x = \int_{-t/2}^{t/2} z\sigma_x \, \mathrm{d}_z$$

Substituting the expression of σ_x in the formula (d) and integrating with z, then we get

$$M_x = -\frac{Et^3}{12(1-\mu^2)}\left(\frac{\partial^2 w}{\partial x^2} + \mu\frac{\partial^2 w}{\partial y^2}\right) \tag{e}$$

Similarly, the shear stress in the cross section will synthesize a torque:

$$M_{xy} = \int_{-t/2}^{t/2} z\tau_{xy} \, \mathrm{d}_z = -\frac{Et^3}{12(1+\mu)}\frac{\partial^2 w}{\partial x\partial y} \tag{f}$$

In a cross section perpendicular to the y-axis, σ_y and τ_{yx} can synthesize a moment and a torque, namely,

$$\begin{aligned}
M_y &= \int_{-t/2}^{t/2} z\sigma_y \, \mathrm{d}_z\\
&= -\frac{Et^3}{12(1-\mu^2)}\left(\mu\frac{\partial^2 w}{\partial x^2} + \frac{\partial^2 w}{\partial y^2}\right)
\end{aligned} \tag{g}$$

$$M_{yx} = \int_{-t/2}^{t/2} z\tau_{yx} \, \mathrm{d}_z = \int_{-t/2}^{t/2} z\tau_{xy} \, \mathrm{d}_z = M_{xy}$$

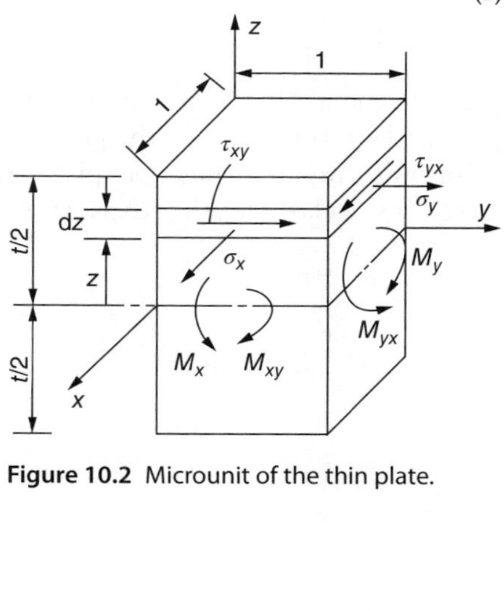

Figure 10.2 Microunit of the thin plate.

Combining the internal forces of the thin plate, denoting as

$$\{M\} = \begin{Bmatrix} M_x \\ M_y \\ M_{xy} \end{Bmatrix} \tag{10.3}$$

Substitution of the three formulas (e), (f), (g) into the above equation yields

$$\{M\} = \frac{Et^3}{12(1-\mu^2)} \begin{Bmatrix} -\frac{\partial^2 w}{\partial x^2} - \mu\frac{\partial^2 w}{\partial y^2} \\ -\mu\frac{\partial^2 w}{\partial x^2} - \frac{\partial^2 w}{\partial y^2} \\ -(1-\mu)\frac{\partial^2 w}{\partial x \partial y} \end{Bmatrix} = \frac{Et^3}{12(1-\mu^2)} \begin{bmatrix} 1 & \mu & 0 \\ \mu & 1 & 0 \\ 0 & 0 & \frac{1-\mu}{2} \end{bmatrix} \begin{Bmatrix} -\frac{\partial^2 w}{\partial x^2} \\ -\frac{\partial^2 w}{\partial y^2} \\ -2\frac{\partial^2 w}{\partial x \partial y} \end{Bmatrix}$$

Using formula (10.1), the formula above can be abbreviated as

$$\{M\} = [D]\{\psi\} \tag{10.4}$$

$$[D] = D\begin{bmatrix} 1 & \mu & 0 \\ \mu & 1 & 0 \\ 0 & 0 & \frac{1-\mu}{2} \end{bmatrix} \tag{10.5}$$

where

$$D = \frac{Et^3}{12(1-\mu^2)} \tag{10.6}$$

in which $[D]$ is the elasticity matrix of the bending problem of isotropic elastic thin plate and t is the thickness of the plate.

Using the three formulas (e), (f), (g) and eliminating w, we get

$$\sigma_x = \frac{12M_x}{t^3}z, \quad \sigma_y = \frac{12M_y}{t^3}z, \quad \tau_{xy} = \frac{12M_{xy}}{t^3}z$$

Or for short

$$\{\sigma\} = \frac{12}{t^3}z\{M\} \tag{10.7}$$

The formula (10.7) expresses the relationship between stresses of each point in the plate and internal forces of the plate.

For anisotropic–orthotropic plate, when x- and y-axes are parallel to the principal directions of elasticity, the stress–strain relationship is as follows:

$$\varepsilon_x = \frac{1}{E_x}(\sigma_x - \mu_1\sigma_y)$$

$$\varepsilon_y = \frac{1}{E_y}(\sigma_y - \mu_2\sigma_x)$$

$$\gamma_{xy} = \frac{1}{G}\tau_{xy}$$

The relationship between moment and curvature is

$$\left.\begin{aligned} M_x &= -D_x\left(\frac{\partial^2 w}{\partial x^2} + \mu_2\frac{\partial^2 w}{\partial y^2}\right)\\ M_y &= -D_y\left(\frac{\partial^2 w}{\partial y^2} + \mu_1\frac{\partial^2 w}{\partial x^2}\right)\\ M_{xy} &= -2D_{xy}\frac{\partial^2 w}{\partial x\partial y}\end{aligned}\right\} \tag{10.8}$$

Therefore, the elasticity matrix is

$$[D] = \begin{bmatrix} D_x & D_1 & 0\\ D_1 & D_y & 0\\ 0 & 0 & D_{xy}\end{bmatrix} \tag{10.9}$$

$$\begin{aligned} D_x &= \frac{E_x t^3}{12(1-\mu_1\mu_2)}\\ D_y &= \frac{E_y t^3}{12(1-\mu_1\mu_2)}\\ D_{xy} &= \frac{Gt^3}{12}\\ D_1 &= \mu_2 D_x\end{aligned} \tag{10.10}$$

If x- and y-axes are not parallel to the principal directions of elasticity, conversion should be performed. Let the local coordinate x', y' be parallel to the principal direction of elasticity; the angle between x- and x'-axis is ϕ. In the global coordinate system (x, y), the elasticity matrix of the thin plate is as follows:

$$[D] = [\theta][D'][\theta]^{\mathrm{T}}$$

in which $[\theta]$ is the transformation matrix, and it is described as formula (2.19). $[D']$ is an elasticity matrix in the local coordinate system, and it is described as formula (10.9).

As shown in Figure 10.3, in the application of the finite element method to solve the bending problem of thin plate, some discrete elements replaced the original continuous thin plate. Each element connects to each other only at the node. Because moments should be transferred between adjacent elements, the nodes are considered as rigid joints.

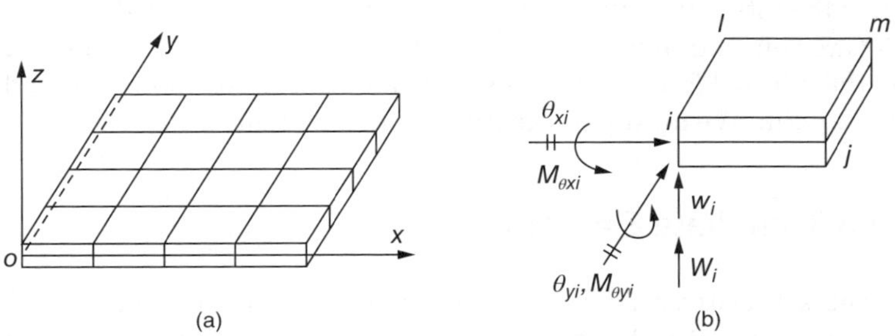

Figure 10.3 The thin plate is divided into rectangular element. (a) Thin plate (b) rectangular element.

Each node has three displacement components, namely, the deflection w, the angular displacement θ_x rolling around the x-axis, and the angular displacement θ_y rolling around the y-axis. The deflection w along the positive direction of the z-axis is positive. If the vector according to the right-hand rule is along the positive direction of the axis, the angular displacement is positive, as shown in Figure 10.3(b). In small deformation conditions, by the geometric relation, we can get $\theta_x = \partial w/\partial y$, $\theta_y = -\partial w/\partial x$ so the displacement of node i can be expressed as

$$\{\delta_i\} = \begin{Bmatrix} w_i \\ \theta_{xi} \\ \theta_{yi} \end{Bmatrix} = \begin{Bmatrix} w_i \\ (\partial w/\partial y)_i \\ -(\partial w/\partial x)_i \end{Bmatrix}$$

The corresponding nodal forces are a concentrated force and two moments, namely,

$$\{F_i\} = \begin{Bmatrix} W_i \\ M_{\theta xi} \\ M_{\theta yi} \end{Bmatrix}$$

The equivalent nodal loads are also a concentrated force and two moments, namely,

$$\{P_i\} = \begin{Bmatrix} Z_i \\ R_{\theta xi} \\ R_{\theta yi} \end{Bmatrix}$$

Using the displacement method, three equilibrium equations can be established at each node as follows:

$$\sum_e W_i = Z_i$$

$$\sum_e M_{\theta xi} = R_{\theta xi}$$

$$\sum_e M_{\theta yi} = R_{\theta yi}$$

In which the first formula expresses the equilibrium of force along the direction of w, the remaining two formulas express the equilibrium of moment along the direction of θ_x and θ_y, respectively. Using the principle of virtual work, the nodal forces can be expressed by the nodal displacements. Therefore the nodal equilibrium equations above can also be described by the nodal displacement, so they can be used to solve the nodal displacements and then the internal stresses of each element can be calculated.

The nodal moments are concentrated moments, and their dimensions are N · m. But in the formula (10.3), the internal forces of the thin plate are distributed moments, and their dimensions are N · m/m. We must pay attention to the difference.

10.2 Rectangular Thin Plate Element

The rectangular element is a relatively simple kind of thin plate elements, so it's introduced first. As shown in Figure 10.4, the four edges of the element are parallel to the x- and y-axes, respectively, and its plane size is $2a \times 2b$.

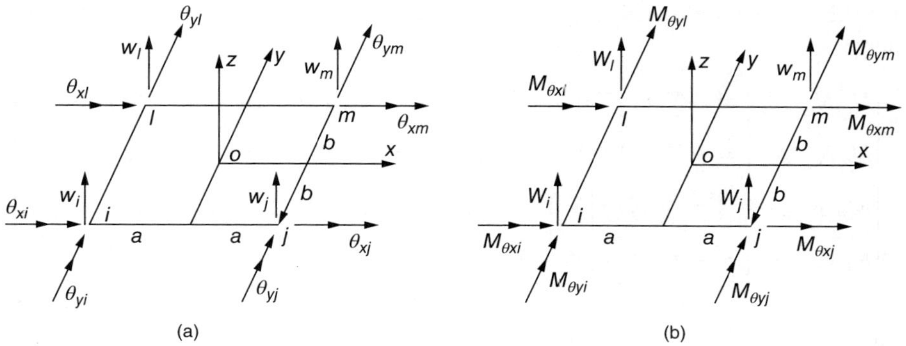

Figure 10.4 The nodal displacement and force of the rectangular element.

10.2.1 Displacement Function

As mentioned above, the deformation and the internal stress depend on the deflection w of the middle surface entirely. A rectangular thin plate element has 12 degrees of freedom, so the expression of the deflection w should contain 12 parameters; now we try to take the following quartic polynomial

$$w = \beta_1 + \beta_2 x + \beta_3 y + \beta_4 x^2 + \beta_5 xy + \beta_6 y^2 + \beta_7 x^3 + \beta_8 x^2 y + \beta_9 xy^2 + \beta_{10} y^3 + \beta_{11} x^3 y + \beta_{12} xy^3 \tag{a}$$

As shown in Figure 10.4, the displacement of nodes of the element is

$$\{\delta\}^e = \begin{bmatrix}\delta_i & \delta_j & \delta_m & \delta_l\end{bmatrix}^T = \begin{bmatrix}w_i & \theta_{xi} & \theta_{yi} & w_j & \theta_{xj} & \theta_{yj} & w_m & \theta_{xm} & \theta_{ym} & w_l & \theta_{xl} & \theta_{yl}\end{bmatrix}^T \tag{10.11}$$

By substitution of the coordinates of node i, $x = a, y = b$, into formula (a) and its derivative, we get

$$\left.\begin{aligned} w_i &= \beta_1 - a\beta_2 - b\beta_3 + a^2\beta_4 + ab\beta_5 + b^2\beta_6 - a^3\beta_7 - a^2 b\beta_8 - ab^2\beta_9 \\ &\quad -b^3\beta_{10} + a^3 b\beta_{11} + ab^3\beta_{12} \\ \theta_{xi} &= \left(\frac{\partial w}{\partial y}\right)_i = \beta_3 - a\beta_5 - 2b\beta_6 + a^2\beta_8 + 2ba\beta_9 + 3b^2\beta_{10} - a^3\beta_{11} - 3ab^2\beta_{12} \\ -\theta_{yi} &= \left(\frac{\partial w}{\partial x}\right)_i = \beta_2 - 2a\beta_4 - b\beta_5 + 3a^2\beta_7 + 2ab\beta_8 + b^2\beta_9 - 3a^2 b\beta_{11} - b^3\beta_{12} \end{aligned}\right\} \tag{b}$$

There are also three equations similar to formula (b) at the nodes of j, m, l. By solving the 12 equations, β_1–β_{12} are obtained. Then by substitution of them into formula (a), we get

$$w = N_i w_i + N_{xi}\theta_{xi} + N_{yi}\theta_{yi} + N_j w_j + N_{xj}\theta_{xj} + N_{yj}\theta_{yj} + N_m w_m + N_{xm}\theta_{xm} + N_{ym}\theta_{ym} + N_l w_l + N_{xl}\theta_{xl} + N_{yl}\theta_{yl} \tag{10.12}$$

in which the shape functions $N_i, N_{xi}, N_{yi}, \ldots, N_{yl}$ are all the quartic polynomials of x and y, namely,

$$\left.\begin{aligned}
\left[N_i\ N_{xi}\ N_{yi}\right] &= \tfrac{1}{16}X_1Y_1\left[X_1Y_1 - X_2Y_2 + 2X_1X_2 + 2Y_1Y_2,\ 2bY_1Y_2,\ -2aX_1X_2\right]\\
\left[N_j\ N_{xj}\ N_{yj}\right] &= \tfrac{1}{16}X_2Y_1\left[X_2Y_1 - X_1Y_2 + 2X_1X_2 + 2Y_1Y_2,\ 2bY_1Y_2,\ 2aX_1X_2\right]\\
\left[N_m\ N_{xm}\ N_{ym}\right] &= \tfrac{1}{16}X_2Y_2\left[X_2Y_2 - X_1Y_1 + 2X_1X_2 + 2Y_1Y_2,\ -2bY_1Y_2,\ 2aX_1X_2\right]\\
\left[N_l\ N_{xl}\ N_{yl}\right] &= \tfrac{1}{16}X_1Y_2\left[X_1Y_2 - X_2Y_1 + 2X_1X_2 + 2Y_1Y_2,\ -2bY_1Y_2,\ -2aX_1X_2\right]
\end{aligned}\right\} \tag{10.13}$$

in which

$$X_1 = 1 - \frac{x}{a},\quad X_2 = 1 + \frac{x}{a},\quad Y_1 = 1 - \frac{y}{b},\quad Y_2 = 1 + \frac{y}{b}$$

It is not difficult to verify that the shape functions above have the following features. At the node of i,

$$N_i = \frac{\partial N_{xi}}{\partial y} = -\frac{\partial N_{yi}}{\partial x} = 1 \quad \text{and} \quad N_{xi} = N_{yi} = \frac{\partial N_i}{\partial x} = \frac{\partial N_i}{\partial y} = 0$$

the other nodes, N_i, N_{xi}, N_{yi} and their first derivative are zero. Other shape functions $N_j, N_{xj}, \ldots$ also have the similar features.

The displacement function (10.12) can also be written in matrix form, namely,

$$w = [N]\{\delta\}^e \tag{10.14}$$

$$[N] = \left[N_i\ N_{xi}\ N_{yi}\ N_j\ N_{xj}\ N_{yj}\ N_m\ N_{xm}\ N_{ym}\ N_l\ N_{xl}\ N_{yl}\right]$$

As mentioned above, the displacements of the thin plate depend entirely on the displacements of the middle surface. There is no displacement in the x- and y-direction, only the deflection w in the z-direction is generated. In the formula (a), the first three terms $\beta_1 + \beta_2 x + \beta_3 y$ represent the displacement of the middle surface, namely, the rigid displacement of the thin plate, in which β_1 represents the movement of the thin plate in the z-direction, $-\beta_2$ and β_3 represent the rotation around y- and x-axes, respectively.

By substitution of $\beta_4 x^2 + \beta_5 xy + \beta_6 y^2$ in the formula (a) into formula (10.1), we get

$$-\frac{\partial^2 w}{\partial x^2} = -2\beta_4,\quad -\frac{\partial^2 w}{\partial y^2} = -2\beta_6,\quad -2\frac{\partial^2 w}{\partial x \partial y} = -2\beta_5$$

Namely, three constant strains of the thin plate are obtained. It is obvious that the displacement function (a) reflects the rigid body displacement and constant strain of the thin plate.

Then we analyze the continuity of the displacement between adjacent elements. In the bending problem of the thin plate, the strain is the second derivative of the deflection w, so on the contact surface of adjacent elements, the deflection w and its first derivative should be continuous. Using the displacement function (a), between the adjacent elements, the deflection w is completely continuous, but its derivative is not completely continuous. Taking the edge of ij as an example, y is a constant and the deflection w is a cubic polynomial, so the four nodal displacements $w_i, w_j, \theta_{yi} = -(\partial w/\partial x)_i, \theta_{yj} = -(\partial w/\partial x)_j$ can be determined completely. Because the two adjacent elements taking the ij edge as the common boundary have the same four nodal displacements mentioned above at the points i and j and the deflection functions of the two elements will be exactly the same cubic curve at the edge of ij, thus the continuity of the deflection w between the two elements is ensured. At the edge of ij, $\theta_x = \partial w/\partial y$ is also a cubic polynomial of x, so four constants are also needed to determine it completely. But now only two nodal displacements $\theta_{xi} = (\partial w/\partial y)_i$ and $\theta_{xj} = (\partial w/\partial y)_j$ can limit it partly, so the two adjacent elements have the same θ_x only at the nodes of i and j, while not the whole edge of ij. But the actual calculation results show that when the size of the element becomes smaller gradually, the calculation result is able to converge to the correct answer.

10.2.2 Stiffness Matrix

By substitution of the displacement function (10.12) into formula (10.1), we get the strain of the thin plate element as follows:

$$\{\psi\} = [B]\{\delta\}^e \tag{10.15}$$

where $[B]$ is a 3×12 matrix

$$[B] = -\begin{bmatrix} \dfrac{\partial^2 N_i}{\partial x^2} & \dfrac{\partial^2 N_{xi}}{\partial x^2} & \dfrac{\partial^2 N_{yi}}{\partial x^2} & \cdots & \dfrac{\partial^2 N_{yl}}{\partial x^2} \\ \dfrac{\partial^2 N_i}{\partial y^2} & \dfrac{\partial^2 N_{xi}}{\partial y^2} & \dfrac{\partial^2 N_{yi}}{\partial y^2} & \cdots & \dfrac{\partial^2 N_{yl}}{\partial y^2} \\ 2\dfrac{\partial^2 N_i}{\partial x \partial y} & 2\dfrac{\partial^2 N_{xi}}{\partial x \partial y} & 2\dfrac{\partial^2 N_{yi}}{\partial x \partial y} & \cdots & 2\dfrac{\partial^2 N_{yl}}{\partial x \partial y} \end{bmatrix} \tag{10.16}$$

Then by substitution of formula (10.15) into formula (10.14), we get the stress of the thin plate element as follows:

$$\{M\} = [D][B]\{\delta\}^e \tag{10.17}$$

Now we will prove that the universal formula (3.10) still can be used to calculate the stiffness matrix of the thin plate element. Assuming that in the thin plate element there is the nodal virtual displacement $\{\delta^*\}^e$, the corresponding virtual strain of the points in the plate is

$$\{\varepsilon^*\} = z\{\Psi^*\} = z[B]\{\delta^*\}^e \tag{c}$$

The virtual work generated by the stress $\{\sigma\}$ per unit volume is $\{\varepsilon^*\}^T\{\sigma\}$. After integration in the entire element, it should be equal to the virtual work that the nodal force $\{F\}^e$ has done, namely,

$$(\{\delta^*\}^e)^T\{F\}^e = \iiint \{\varepsilon^*\}^T\{\sigma\}\, dz\, dx\, dy$$

By substitution of formula (c) and (10.17) into the formula above, we get

$$(\{\delta^*\}^e)^T\{F\}^e = \iint \left(\int_{-t/2}^{t/2} \frac{12z^2}{t^3}\, dz\right)\{\Psi^*\}^T\{M\}\, dx\, dy = \iint \{\Psi^*\}^T\{M\}\, dx\, dy$$

Then by substitution of $\{\Psi^*\} = [B]\{\delta^*\}^e$ and formula (10.17) into the formula above and simplifying it, we get

$$\{F\}^e = [k]^e\{\delta\}^e \tag{10.18}$$

where

$$\{k\}^{\mathrm{e}} = \iint [B]^{\mathrm{T}} [D][B]\,\mathrm{d}x\,\mathrm{d}y \tag{10.19}$$

The matrix $[k]^{\mathrm{e}}$ is the stiffness matrix of the thin plate element, and it is in line with the general formula (3.10). By the formula (10.13), we get the second derivative of the shape function; by substitution of it into formula (10.16), we get the matrix $[B]$. Then by substitution of it into formula (10.19) and through integration, we get the explicit expression of the stiffness matrix $[k]^{\mathrm{e}}$.

10.2.3 Nodal Load

The equivalent load of each node of the element is

$$\{P\}^{\mathrm{e}} = \left[P_i\ P_j\ P_m\ P_l\right]^{\mathrm{T}} \tag{10.20}$$

Assuming that the distributed load q acts on the surface of the thin plate, by the virtual work principle, we know the nodal load can be calculated by the formula below:

$$\{P\}^{\mathrm{e}} = \iint [N]^{\mathrm{T}} q\,\mathrm{d}x\,\mathrm{d}y \tag{10.21}$$

in which $[N]$ is seen in the formula (10.14). When the load distributes evenly, q is a constant, by the formula (10.21), we get

$$\{P\}^{\mathrm{e}} = q\int_{-a}^{a}\int_{-b}^{b} [N]^{\mathrm{T}}\,\mathrm{d}x\,\mathrm{d}y = q\int_{-a}^{a}\int_{-b}^{b} \left[N_i\ N_{xi}\ N_{yi}\ N_j\ \cdots\ N_{yl}\right]^{\mathrm{T}}\mathrm{d}x\,\mathrm{d}y$$

By substitution of formula (10.13) into the formula above and by integration, we get (Figure 10.5)

$$\begin{aligned}\{P\}^{\mathrm{e}} &= \left[Z_i\ R_{\theta xi}\ R_{\theta yi}\ Z_j\ R_{\theta xj}\ R_{\theta yj}\ Z_m\ R_{\theta xm}\ R_{\theta ym}\ Z_l\ R_{\theta xl}\ R_{\theta yl}\right]^{\mathrm{T}} \\ &= 4qab\left[\frac{1}{4}\ \frac{b}{12}\ -\frac{a}{12}\ \frac{1}{4}\ \frac{b}{12}\ \frac{a}{12}\ \frac{1}{4}\ -\frac{b}{12}\ \frac{a}{12}\ \frac{1}{4}\ -\frac{b}{12}\ -\frac{a}{12}\right]^{\mathrm{T}}\end{aligned} \tag{10.22}$$

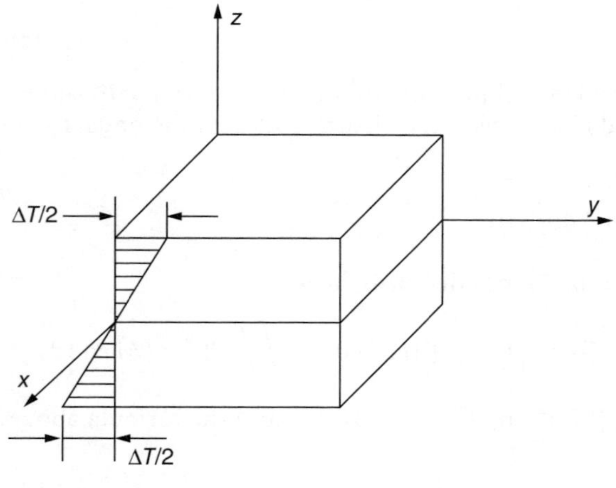

Figure 10.5 The temperature distribution in the plate.

Thus, for the thin plate element, the nodal load also contains the moment load besides the normal load.

Assuming that the difference of the temperature on the top and bottom surfaces of the thin plate is ΔT and the temperature of the middle surface is zero, the initial strain of each point in the plate due to the temperature change is

$$\{\varepsilon_0\} = \begin{Bmatrix} \varepsilon_{x0} \\ \varepsilon_{y0} \\ \gamma_{xy0} \end{Bmatrix} = z \begin{Bmatrix} \dfrac{\alpha\Delta T}{2} \\ \dfrac{\alpha\Delta T}{2} \\ 0 \end{Bmatrix} = z\{\Psi^0\}$$

Compared with formula (10.2), it's known that the initial deformation of the thin plate caused by the temperature change is

$$\{\Psi^0\} = \begin{Bmatrix} \dfrac{\alpha\Delta T}{2} \\ \dfrac{\alpha\Delta T}{2} \\ 0 \end{Bmatrix} \tag{10.23}$$

Then the internal forces of the plate should be calculated by the formula below:

$$\{M\} = [D](\{\Psi\} - \{\Psi^0\}) \tag{10.24}$$

Using the virtual work principle, it's deduced that the nodal load caused by the temperature change should be calculated by the formula below:

$$\{P\}^{\mathrm{e}}_{\varepsilon_0} = \iint [B]^{\mathrm{T}}[D]\{\Psi^0\}\,\mathrm{d}x\,\mathrm{d}y \tag{10.25}$$

10.2.4 Example

By the two simple examples below, we will explain the convergence of the bending of plate with the finite element method.

10.2.4.1 Square Thin Plate Supported by Four Edges

For the square plate with four edges fixed, the computing grid is shown in Figure 10.6(a). Under the uniform distributed load, the deflection w along the center line and its bending moment M_x are shown in Figure 10.6(b) and (c).

For the square plate under the uniform distributed load or central concentrated load, with four edges clamped or simply supported, the comparison between the deflection w of the center point obtained by the finite element method and the exact solution are shown in Table 10.1, where L is the length of the plate, $D = \frac{Et^3}{12(1-\mu^2)}$ and $\mu = 0.30$ in calculation.

10.2.4.2 Square Thin Plate Supported by Corner Points

For the square plate under the uniform distributed load and supported by the pillars at the corner, the comparison of the deflection and the bending moment obtained by the finite element method and the finite difference method is shown in Table 10.2.

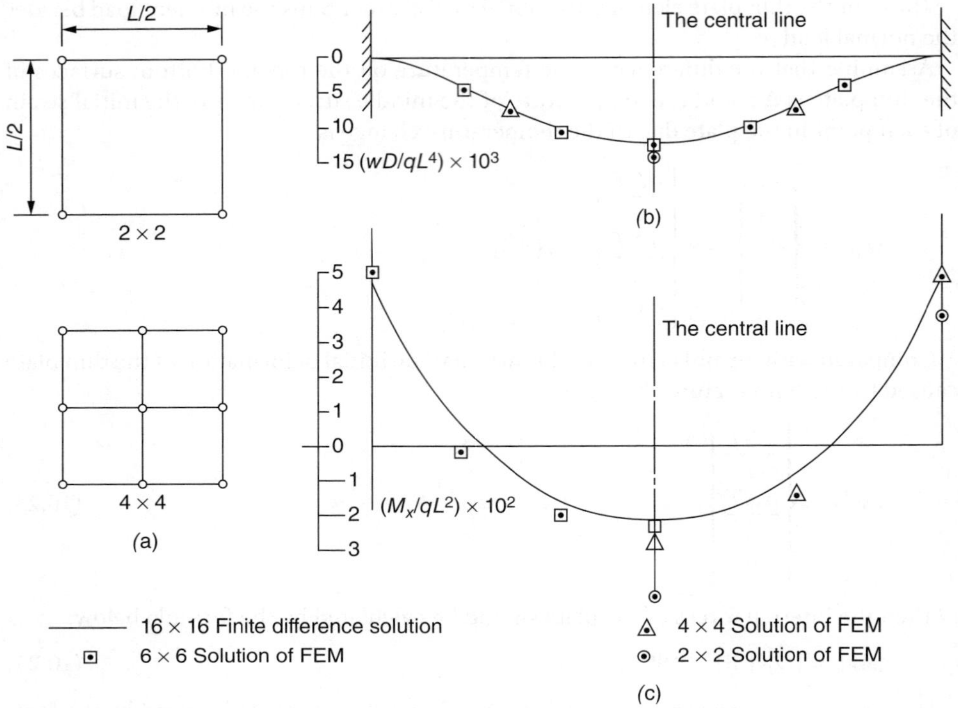

Figure 10.6 The square plate with four edges clamped under the uniform distributed load q (the rectangular element). (a) The computing grid, (b) the deflection, and (c) the bending moment.

Table 10.1 The deflection of the center point of the square thin plate (rectangular element).

Grid	The number of nodes	Four edges simply supported		Four edges clamped	
		Uniformly distributed load	Concentrated load	Uniformly distributed load	Concentrated load
		$w_{max}D/qL^4$	$w_{max}D/PL^2$	$w_{max}D/qL^4$	$w_{max}D/PL^2$
2×2	9	0.003446	0.013784	0.001480	0.005919
4×4	25	0.003939	0.012327	0.001403	0.006134
8×8	81	0.004033	0.011829	0.001304	0.005803
16×16	28	0.004056	0.011671	0.001275	0.005672
Exact solution	—	0.004062	0.01160	0.00126	0.00560

Table 10.2 The deflection and the bending moment of the square thin plate with the corner points supported (rectangular element).

Grid	The center of the plate		The middle points of edges	
	$w_{max}D/qL^4$	M/qL^2	$w_{max}D/qL^4$	M/qL^2
2×2	0.0176	0.095	0.0126	0.139
4×4	0.0232	0.108	0.0165	0.149
6×6	0.0244	0.109	0.0173	0.150
Finite difference solution	0.0265	0.109	0.0170	0.140

From the examples above, it is evident that when calculating the elastic thin plate by the finite element method, the convergence of the solution is very good. Even for the thin plate supported on the corner points, there is stress concentration near the corner; the calculated deflection and bending moment are good.

10.3 Triangular Thin Plate Element

Triangular thin plate element can better adapt to the complex boundary shape, so its practical value is bigger. The triangular element with three nodes as shown in Figure 10.7 is often used.

10.3.1 Displacement Function

Each node has three displacement components, namely, the deflection w, the angular displacement θ_x rolling around the x-axis, and the angular displacement θ_y rolling

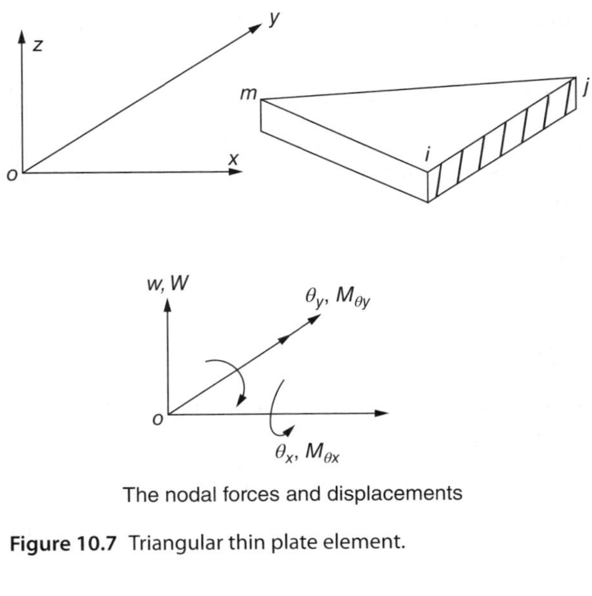

Figure 10.7 Triangular thin plate element.

around the y-axis. There are 9 nodal displacements of 3 nodes, so the displacement function should include 9 terms, but a complete cubic polynomial contains 10 items as follows:

$$\beta_1 + \beta_2 x + \beta_3 y + \beta_4 x^2 + \beta_5 xy + \beta_6 y^2 + \beta_7 x^3 + \beta_8 x^2 y + \beta_9 xy^2 + \beta_{10} y^3$$

One item should be deleted from the formula above. In the formula, the first three items represent the rigid body displacement, and the three quadratic items represent constant deformation (curvature). They can't be deleted in order to ensure the convergence of solution; only one item can be removed from the four cubic items behind. If we delete any one form the four cubic items behind, the symmetry of the expression will be lost. $\beta_8 = \beta_9$ can keep the symmetry of the formula, but in some cases, for example, when two edges of the triangle parallel to the x- and y-axis, it's impossible to solve the coefficients in the formula. If we use area coordinate, the contradiction can be solved.

The linear, quadratic and cubic expression of the area coordinates have the following items, respectively.

$$\text{The linear expression}: \; L_i, L_j, L_m \tag{a}$$

$$\text{The quadratic expression}: \; L_iL_j, L_jL_m, L_mL_i, {L_i}^2, {L_j}^2, {L_m}^2 \tag{b}$$

$$\text{The cubic expression}: \; L_iL_jL_m, {L_i}^2L_j, {L_j}^2L_m, {L_m}^2L_i, L_i{L_j}^2, L_j{L_m}^2, L_m{L_i}^2, {L_i}^3, {L_j}^3, {L_m}^3 \tag{c}$$

Based on formulas (8.4) and (8.6), the complete linear polynomial of x and y expressed with the area coordinates is

$$\beta_1 L_i + \beta_2 L_j + \beta_3 L_m$$

The complete quadratic polynomial of x and y should include three items of formula (a) at least and other three items of formula (b). It includes a total of six items, for example,

$$\beta_1 L_i + \beta_2 L_j + \beta_3 L_m + \beta_4 L_iL_j + \beta_5 L_jL_m + \beta_6 L_mL_i$$

The complete cubic polynomial of x and y should include four items of formula (c) at least and other six items of formula (a) and (b). It includes a total of ten items, for example,

$$L_i, L_j, L_m, {L_i}^2L_j, {L_j}^2L_m, {L_m}^2L_i, L_i{L_j}^2, L_j{L_m}^2, L_m{L_i}^2, L_iL_jL_m \tag{d}$$

The combination of the 10 items above can be used as the displacement function, but the nodal displacement only has nine items, so an independent item should be reduced; the last item has the following features at the node of i, j, and m.

$$L_iL_jL_m = \frac{\partial}{\partial x}(L_iL_jL_m) = \frac{\partial}{\partial y}(L_iL_jL_m) = 0$$

Now we put it into other cubic items and take the displacement function:

$$w = \beta_1 L_1 + \beta_2 L_2 + \beta_3 L_3 + \beta_4 (L_2^2 L_1 + cL_1L_2L_3) + \cdots + \beta_9 (L_1^2 L_2 + cL_1L_2L_3)$$

Through some algebraic operation, in order to satisfy the constant deformation of the thin plate, we can only take $c = 1/2$, so we get

$$w = \beta_1 L_1 + \beta_2 L_2 + \beta_3 L_3 + \beta_4 \left(L_2^2 L_1 + \frac{1}{2}L_1L_2L_3\right) + \cdots + \beta_9 \left(L_1^2 L_2 + \frac{1}{2}L_1L_2L_3\right) \tag{10.26}$$

Substitute into formula (10.26) the corresponding nodal displacements:

$$w_i, \quad \theta_{xi} = \left(\frac{\partial w}{\partial y}\right)_i, \quad \theta_{yi} = -\left(\frac{\partial w}{\partial x}\right)_i$$

Using the nodal displacements, the coefficients β_1–β_9 of formula (10.26) can be solved. Then by their substitution into formula (10.26) and simplification, the displacement function is as follows:

$$\begin{aligned} w &= N_i w_i + N_{xi}\theta_{xi} + N_{yi}\theta_{yi} + N_j w_j + N_{xj}\theta_{xj} + N_{yj}\theta_{yj} + N_m w_m + N_{xm}\theta_{xm} + N_{ym}\theta_{ym} \\ &= \left[N_i \ N_{xi} \ N_{yi} \ N_j \ N_{xj} \ N_{yj} \ N_m \ N_{xm} \ N_{ym}\right] \{\delta\}^e \\ &= [N]\{\delta\}^e \end{aligned} \tag{10.27}$$

in which

$$\left.\begin{aligned} N_i &= L_i + L_i^2 L_j + L_i^2 L_m - L_i L_j^2 - L_i L_m^2 \\ N_{xi} &= b_j L_i^2 L_m - b_m L_i^2 L_j + \tfrac{1}{2}(b_j - b_m) L_i L_j L_m \\ N_{yi} &= c_j L_i^2 L_m - c_m L_i^2 L_j + \tfrac{1}{2}(c_j - c_m) L_i L_j L_m \end{aligned}\right\} \quad (i, j, m) \tag{10.28}$$

Using the formula (8.8), it's obvious that the shape functions and their derivatives above have the properties listed in the Table 10.3.

By the properties listed in the table above, it's known that the displacement function w meets the following conditions:

$$\text{At the node } i: \quad w = w_i, \quad \frac{\partial w}{\partial y} = \theta_{xi}, \quad -\frac{\partial w}{\partial x} = \theta_{yi} \quad (i, j, m)$$

By substitution of (10.28) into the displacement function (10.27) and by proper treatment, we get

$$\begin{aligned} w = {} & w_i L_i + w_j L_j + w_m L_m + (c_m \theta_{yj} + b_m \theta_{xj}) L_i L_j \\ & + (b_j \theta_{xi} + c_j \theta_{yi}) L_i L_m + (b_i \theta_{xm} + c_i \theta_{ym}) L_j L_m + \cdots \end{aligned}$$

The first six items of the formula above is a complete quadratic polynomial, so the displacement function can meet the condition of the rigid body displacements and constant strain of the element. In addition, along three boundaries of the element, the deflection w is the cubic function of the length s, two endpoints (nodes) of each boundary have two parameters (w and $\partial w/\partial s$), respectively. There are four same parameters of each boundary, thus it's ensured that two adjacent elements have the same displacement along the common boundary. The normal slope $\partial w/\partial n$ of the boundary of the element is also the cubic function of the length s, but only two nodes have the same slope, so the normal

Table 10.3 The properties of the shape functions.

	N_i	$\frac{\partial N_i}{\partial y}$	$-\frac{\partial N_i}{\partial x}$	N_{xi}	$\frac{\partial N_{xi}}{\partial y}$	$-\frac{\partial N_{xi}}{\partial x}$	N_{yi}	$\frac{\partial N_{yi}}{\partial y}$	$-\frac{\partial N_{yi}}{\partial x}$
i	1	0	0	0	1	0	0	0	1
j, m	0	0	0	0	0	0	0	0	0
								(i, j, m)	

slope of adjacent elements is controlled partly and the continuity of the normal slope can't be fully guaranteed.

10.3.2 Stiffness Matrix and Nodal Load

By substitution of the displacement function (10.27) into formula (10.1), the deformation of the element $\{\Psi\} = [B]\{\delta\}^{e}$ can be also obtained, where $[B]$ is a 3×9 matrix

$$[B] = -\begin{bmatrix} \dfrac{\partial^2 N_i}{\partial x^2} & \dfrac{\partial^2 N_{xi}}{\partial x^2} & \dfrac{\partial^2 N_{yi}}{\partial x^2} & \cdots & \dfrac{\partial^2 N_{ym}}{\partial x^2} \\ \dfrac{\partial^2 N_i}{\partial y^2} & \dfrac{\partial^2 N_{xi}}{\partial y^2} & \dfrac{\partial^2 N_{yi}}{\partial y^2} & \cdots & \dfrac{\partial^2 N_{ym}}{\partial y^2} \\ 2\dfrac{\partial^2 N_i}{\partial x \partial y} & 2\dfrac{\partial^2 N_{xi}}{\partial x \partial y} & 2\dfrac{\partial^2 N_{yi}}{\partial x \partial y} & \cdots & 2\dfrac{\partial^2 N_{ym}}{\partial x \partial y} \end{bmatrix} \tag{e}$$

By the formula (10.4), the stress of the element $\{M\} = [D][B]\{\delta\}^{e}$ can be obtained, and the stiffness matrix is still calculated by the formula (10.19).

Using the integration formula (5.17) of the area coordinates, it is easy to express the stiffness matrix. But the expression is longer; actually it is also very convenient to use the numerical integration. Because the integrand is a quadratic expression, accurate results can be obtained using the integration formula of three points in Table 8.1.

Using the integration formula of the area coordinate, it's easy to calculate the nodal loads by the formula (10.21). When the thin plate element is subjected to a uniform normal load, the nodal load is as follows:

$$\begin{aligned} \{P\}^{e} &= \left[P_i \ P_j \ P_m\right]^{T} \\ &= qA\left[\frac{1}{3} \cdot \frac{b_j - b_m}{24} \cdot \frac{c_j - c_m}{24} \ \ \frac{1}{3} \cdot \frac{b_m - b_i}{24} \cdot \frac{c_m - c_i}{24} \ \ \frac{1}{3} \cdot \frac{b_i - b_j}{24} \cdot \frac{c_i - c_j}{24}\right] \end{aligned}$$

where A is the area of the triangle.

10.3.3 Smoothing Curvature

The triangular thin plate element above, due to its simple calculation, is still widely used so far. But the normal slope of the adjacent elements is not entirely consistent; only when the computation grid is regular, the results can converge to the correct answer. For some irregular grid, the accuracy of the calculation is very poor; in order to overcome this shortcoming, the literature [16] suggests smoothing curvature for the element.

For smoothing curvature of the element, it is sufficient to smooth the derivative of the shape function, namely, replacing the primary derivative by the smoothed one. The second derivative N_i'' of the shape function N_i on the angular point is often singular; only using high-order numerical integration, a good convergence can be obtained. In order to get the element of simple calculation and good convergence, we should replace N_i'' by smoothed derivative $\overline{N_i''} = \alpha_1 + \alpha_2 x + \alpha_3 y$, which is a linear function of coordinates; using the least square method, the coefficient α_j of the derivative $\overline{N_i''}$ can be determined as follows:

$$\frac{\partial}{\partial \alpha_j}\left[\int_A (N_i'' - \overline{N}_i'')^2 \, \mathrm{d}A\right] = 0 \tag{10.29}$$

Because $\overline{N}_i''$ is a linear polynomial, when calculating the stiffness matrix of the element, only three integral points are needed, namely, three midpoints of edges.

10.3.4 Example

10.3.4.1 The Square Plate Bearing Concentrated and Distributed Loads

For the square plate with four edges simply supported or fixed under the action of concentrated load P or distributed load q, because of symmetry, only a quarter of the plate needs to be taken out. The concentrated load is acting on the center of the plate. The computing grid is shown in Figure 10.8. The deflection w_0 of the center of the plate is shown in Table 10.4. The moments M_x of the elements on the centerline of the plate under the uniformly distributed load are shown in Figure 10.9. The figure shows that M_x given by the smoothed elements are more close to the theoretical solution.

10.3.4.2 The Distortion of the Square Plate

As shown in Figure 10.10, the angular points A, B, and D of the square plate are simply supported, and the angular point C bears the concentrated load $P = 50N$. According to the exact solution of the theory of the thin plate, the bending moment of the plate is

$$M_x = 0, \quad M_y = 0, \quad M_{xy} = 25\,\mathrm{N} \cdot \mathrm{m/m}$$

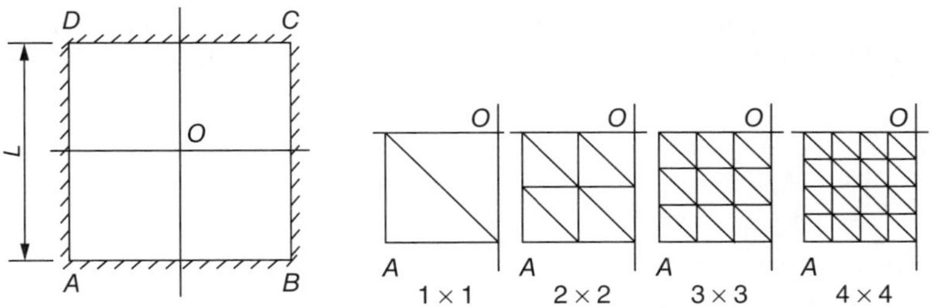

Figure 10.8 The square plate bearing concentrated and distributed loads.

Table 10.4 The value k of the plate with edges supported simply or fixed (triangular thin plate element).

	The concentrated load P acting on the center of the plate $k = Dw_0/PL^2$				**The distributed load q $k = Dw_0/qL^4$**			
	Simply supported edges		**Fixed edges**		**Simply supported edges**		**Fixed edges**	
	Smoothed	**Unsmoothed**	**Smoothed**	**Unsmoothed**	**Smoothed**	**Unsmoothed**	**Smoothed**	**Unsmoothed**
1×1	0.00984	0.01302	0.00261	0.00521	0.00400	—	0.000870	0.00172
2×2	0.01088	0.01176	0.00503	0.00589	0.00402	0.00413	0.001234	0.00157
3×3	0.01126	0.01211	0.00533	0.00583	0.00404	0.00413	0.001248	0.00135
4×4	0.01141	0.01165	0.00545	0.00572	0.00405	0.00405	0.001256	0.00134
Accurate k	0.01160		0.00560		0.00406		0.00126	

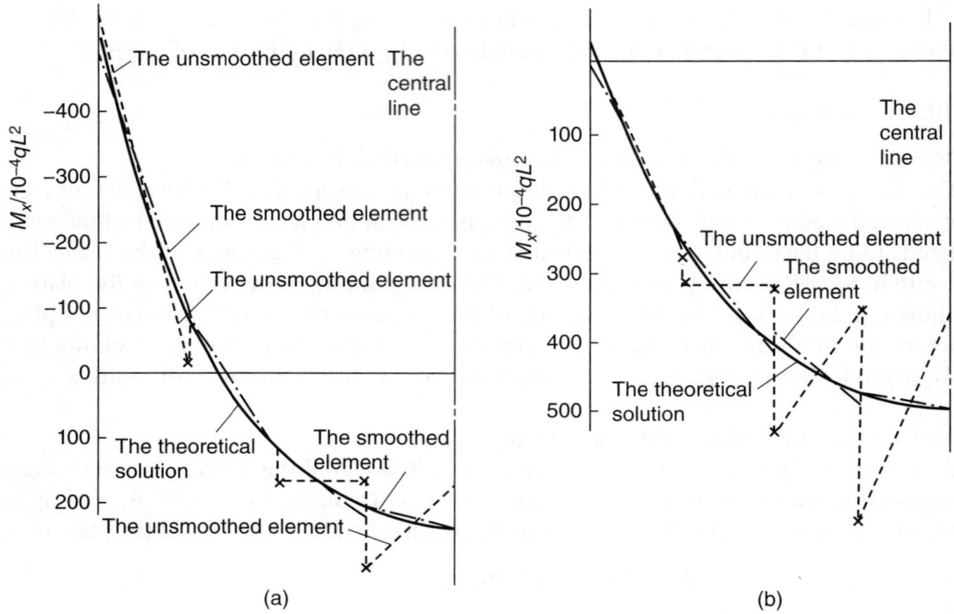

Figure 10.9 The distribution of M_x of the elements on the centerline of the square plate under the uniformly distributed load. (a) The square plate with four edges fixed and (b) the square plate with four edges simply supported.

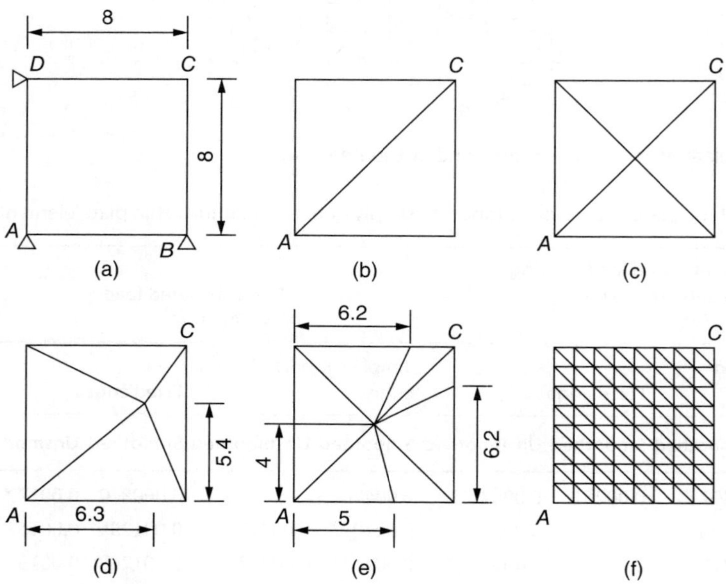

Figure 10.10 The computing grid of the distorted square plate.

This problem can be used to test whether the thin plate element can accurately reflect the constant state of stresses. The computing grid is shown in Figure 10.10. The results show that for the regular grid, even with the coarse grid 1×1, both smoothed and not smoothed elements can give the accurate value of stress. For the irregular grid, the unsmoothed element even with the dense grid, also can't give the correct value of stress.

For the grid shown in Figure 10.10(d) and (e), the errors of the centroid stress given by the unsmoothed element are 12.2% and 30.0%, respectively. Even if using the average values of the nodes, the error also can't be reduced.

It is noticed that the grid of Figure 10.10(e) is denser than the grid (c); the coarse grid (c) gives the correct value of stress, while the error of the denser grid (e) is 30%. This phenomenon is not normal of course. For any practical problems, even with very dense grid, as long as the grid is irregular, if unsmoothing, even in the area of uniform stress, considerable error will also be obtained and after smoothing, for all the grids shown in Figure 10.10, accurate stress values can be obtained.

10.4 Plate Element with Curved Boundary and Deflection and Rotation Defined Respectively

The thin plate element mentioned above is based on the thin plate theory of Kirchhoff. When calculating the deflection, the influence of shearing deformation is ignored. According to the theory of Kirchhoff, between the adjacent elements, the deflection and its derivative should be continuous, but in fact only the deflection is continuous, the derivative of the deflection is not completely continuous. In order to solve this contradiction, two methods are proposed; the first method is to use the hybrid element and the second one is to use the thin plate theory of Mindlin in which the shearing deformation is considered and the deflection and rotation are expressed individually. It now appears that the calculation of the thin plate based on the second method is relatively simple and has good precision. It can also adapt to the irregular shape by the coordinate transformation. So the practical value is bigger.

According to Mindlin thin plate theory, when calculating the deflection of the plate, the influence of shearing deformation is considered. The deflection of the plate and the angular displacement of the normal line of the middle surface are independent of each other and interpolated respectively, so the continuity of the deflection and the angular displacement on the boundary of the adjacent elements is guaranteed. In order to be easier for the reader to understand, here we introduce a beam element considering the influences of shearing deformation first.

10.4.1 Beam Element Considering the Shearing Deformation

First we analyze the influence of the shearing deformation. Assuming that the deflection generated by the shearing force is w_1, for any cross section of the beam, there is

$$\gamma = \frac{\mathrm{d}w_1}{\mathrm{d}x} = \frac{kV}{AG} \tag{a}$$

in which V is the shearing force, A is the sectional area, G is the shear modulus, and k is the coefficient considering the uneven distribution of the shearing stress.

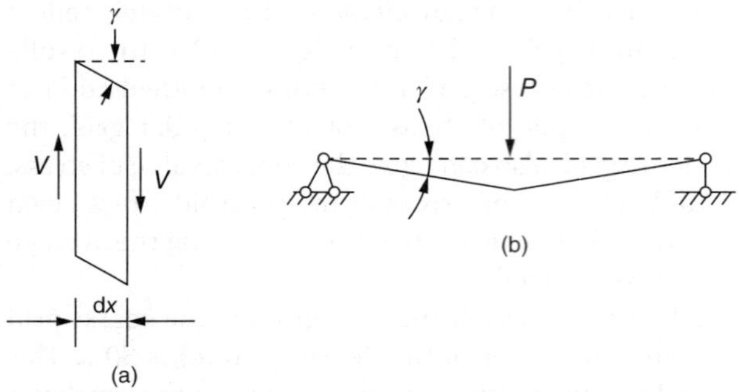

Figure 10.11 The deformation caused by the shear force. (a) The parallel movement between the adjacent sections. (b) The deflection of the simple supported beam.

By differentiating the formula above, we get

$$\frac{\mathrm{d}^2 w_1}{\mathrm{d}x^2} = \frac{k}{AG}\frac{\mathrm{d}V}{\mathrm{d}x} \tag{b}$$

Assuming that the deflection caused by the bending moment is w_2, by the theory of strength of materials, we get

$$\frac{\mathrm{d}^2 w_2}{\mathrm{d}x^2} = \frac{M}{EJ} \tag{c}$$

where M is the bending moment and J is the moment of inertia of the section.

Combining formulas (b) and (c), considering the bending moment and shearing force, the deflection w of the beam should satisfy the following equation:

$$\frac{\mathrm{d}^2 w}{\mathrm{d}x^2} = \frac{M}{EJ} + \frac{k}{AG}\frac{\mathrm{d}V}{\mathrm{d}x} \tag{d}$$

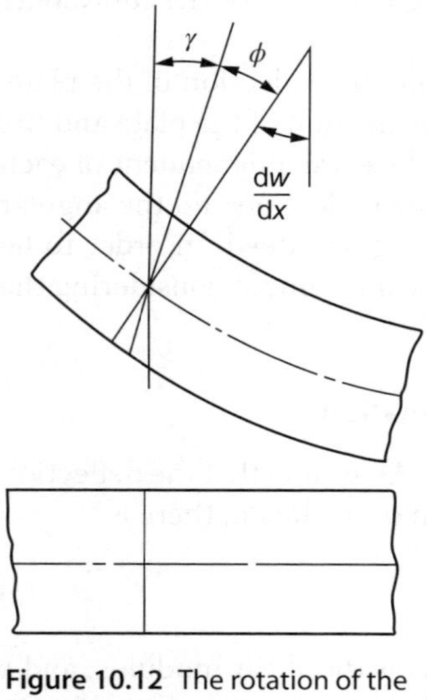

Figure 10.12 The rotation of the beam's section.

Thus, the deflection of the beam is composed of two parts, one is caused by the bending moment and the other is caused by the shearing force.

As shown in Figure 10.11(a), the shearing force only causes the parallel movement between two adjacent sections and does not cause the rotation of the cross sections. The deflection caused by the shearing force is the result of the parallel movement of the cross section, not due to the rotation of the cross section. As shown in Figure 10.11(b), for the simply supported beam bearing concentrated load, the deflection produced by the shearing deformation is two straight lines. At the point of application of force P, $\mathrm{d}w/\mathrm{d}x$ is discontinuous.

The rotation of the beam section caused by the bending moment is shown in Figure 10.12. Considering the effect of the shearing deformation, after bearing the load, the original cross section vertical to the neutral axis of the beam is not vertical to the neutral axis, and its rotation angle is

$$\phi = \frac{\mathrm{d}w}{\mathrm{d}x} - \gamma \tag{10.30}$$

Therefore dividing the beam into elements, on the contact surface between the adjacent elements, w and ϕ should be continuous and $\mathrm{d}w/\mathrm{d}x$ does not need to be continuous. In fact, if a concentrated load is acting on the node, the shearing strain on both sides of the contact surface is different, so $\mathrm{d}w/\mathrm{d}x$ on the contact surface is discontinuous.

In sum, considering the effect of the shearing deformation, the deflection w of the beam and the rotation ϕ of the section of the beam are independent of each other.

After considering the effect of the shearing deformation, the potential energy of the beam is

$$\Pi = \frac{1}{2}\int_0^L EJ\left(\frac{\mathrm{d}\phi}{\mathrm{d}x}\right)^2 \mathrm{d}x + \frac{k}{2}\int_0^L GA\left(\frac{\mathrm{d}w}{\mathrm{d}x} - \phi\right)^2 \mathrm{d}x - \int_0^L qw\,\mathrm{d}x \tag{10.31}$$

in which, q is the distributed load on the beam.

In the formula (10.31), the former two items on the right side represent the strain energy generated by the bending and shearing deformation; the third item represents the potential energy of the load.

For the uniform beam, by the condition of stationary value $\delta\Pi = 0$, we get

$$EJ\int_0^L \frac{\mathrm{d}\phi}{\mathrm{d}x}\delta\left(\frac{\mathrm{d}\phi}{\mathrm{d}x}\right)\mathrm{d}x + kGA\int_0^L \left(\frac{\mathrm{d}w}{\mathrm{d}x} - \phi\right)\delta\left(\frac{\mathrm{d}w}{\mathrm{d}x} - \phi\right)\mathrm{d}x - \int_0^L q\delta w\,\mathrm{d}x = 0 \tag{10.32}$$

Let's deduce some relevant formulas of the beam element considering the shearing deformation. The deflection w and the rotation angle ϕ of any point in the beam element expressed with the shape function are as follows:

$$w = \sum_{i=1}^{s} N_i w_i, \quad \phi = \sum_{i=1}^{s} N_i \phi_i \tag{10.33a}$$

where s is the number of nodes of the element.

The formula (10.33a) can also be expressed with matrices as follows:

$$w = [N_w]\{\delta\}, \quad \phi = [N_\phi]\{\delta\} \tag{10.33b}$$

$$\left.\begin{aligned} [N_w] &= \begin{bmatrix} N_1 & \cdots & N_s & 0 & \cdots & 0 \end{bmatrix} \\ [N_\phi] &= \begin{bmatrix} 0 & \cdots & 0 & N_1 & \cdots & N_s \end{bmatrix} \\ [\delta] &= \begin{bmatrix} w_1 & \cdots & w_s & \phi_1 & \cdots & \phi_s \end{bmatrix} \end{aligned}\right\} \tag{10.34}$$

By differentiating formula (10.33) with respect to x, we get

$$\frac{\partial w}{\partial x} = [B_w]\{\delta\}, \quad \frac{\partial \phi}{\partial x} = [B_\phi]\{\delta\} \tag{10.35}$$

in which

$$\left.\begin{aligned}
[B_w] &= [J]^{-1}\left[\frac{\partial N_1}{\partial \xi}\ \cdots\ \frac{\partial N_s}{\partial \xi}\ 0\ \cdots\ 0\right]\\
[B_\phi] &= [J]^{-1}\left[0\ \cdots\ 0\ \frac{\partial N_1}{\partial \xi}\ \cdots\ \frac{\partial N_s}{\partial \xi}\right]\\
[J] &= \frac{\partial x}{\partial \xi}
\end{aligned}\right\} \tag{10.36}$$

By the formula (10.32), the stiffness matrix $[K]$ and the nodal load $[P]$ of the element are obtained as follows:

$$[K] = \int_{-1}^{1} [B_\phi]^{\mathrm{T}} EJ [B_\phi] |J|\, \mathrm{d}\xi + k \int_{-1}^{1} [B_w - N_\phi]^{\mathrm{T}} GA [B_w - N_\phi] |J|\, \mathrm{d}\xi \tag{10.37}$$

$$\{P\} = \int_{-1}^{1} [N_w]^{\mathrm{T}} q |J|\, \mathrm{d}\xi \tag{10.38}$$

For example, the beam element with three nodes shown in Figure 10.13, the coordinate transformation formula is

$$x = \sum_{i=1}^{3} N_i x_i$$

Three nodal coordinates are $x_1 = 0,\ x_2 = L,\ x_3 = L/2$. Substituting them into the formula above, we get

$$x = \frac{L}{2}(1 + \xi)$$

Therefore the shape functions expressed with x are

$$N_1 = \frac{2x^2}{L^2} - \frac{3x}{L} + 1, \quad N_2 = \frac{2x^2}{L^2} - \frac{x}{L}, \quad N_3 = \frac{4x}{L} - \frac{4x^2}{L^2}$$

The derivatives of the shape functions are

$$\frac{\partial N_1}{\partial x} = \frac{4x}{L} - \frac{3}{L}, \quad \frac{\partial N_2}{\partial x} = \frac{4x}{L^2} - \frac{1}{L}, \quad \frac{\partial N_3}{\partial x} = \frac{4}{L} - \frac{8x}{L}$$

Then we get

$$[B_\phi] = \left[0\ \ 0\ \ 0\ \ \frac{\partial N_1}{\partial x}\ \ \frac{\partial N_2}{\partial x}\ \ \frac{\partial N_3}{\partial x}\right]$$

$$[B_w - N_\phi] = \left[\frac{\partial N_1}{\partial x}\ \ \frac{\partial N_2}{\partial x}\ \ \frac{\partial N_3}{\partial x}\ \ -N_1\ \ -N_2\ \ -N_3\right]$$

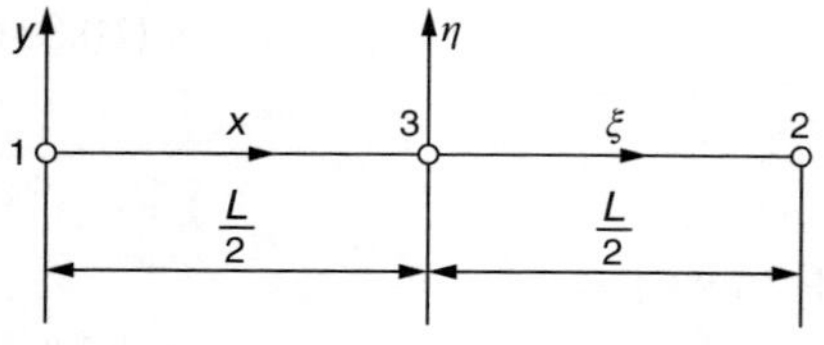

Figure 10.13 Beam element of three nodes.

By substitution into formula (10.37), we get the stiffness matrix of element.

The computing experience shows that if we use the element with 3 or 4 nodes, whether it is a beam with medium height or small height, the calculation results are satisfactory. But if we use the beam element with 2 nodes, when the height of the beam is very small, the stiffness matrix will be too large and a larger calculation error will result. This phenomenon is called element locking.

In order to solve the problem of element locking, two ways can be used. One way is to integrate with lower order. The other more efficient way is to use the discrete Kirchhoff's theory, namely, on several discrete points of the element, making

$$\frac{\mathrm{d}w}{\mathrm{d}x} = \phi \tag{10.39}$$

Through such treatment, the characteristics of the element will be improved.

10.4.2 Curved Plate Element with the Deflection and Rotation Interpolated Respectively

According to the Mindlin elastic plate theory, during calculation, we use the following assumptions:

1) The deflection w of the plate is small.
2) After deformation the normal line of the middle surface of the plate is still a line, but it doesn't have to be perpendicular to the deformed middle face.
3) The normal stress perpendicular to the middle surface can be ignored.

According to the assumptions above, the displacement of any point (x, y, z) of the plate is calculated as follows (Figure 10.14):

$$u = z\phi_y(x, y), \quad v = -z\phi_x(x, y), \quad w = w(x, y) \tag{10.40}$$

The strain in the plane of the plate is

$$\{\varepsilon\} = \begin{Bmatrix} \varepsilon_x \\ \varepsilon_y \\ \gamma_{xy} \end{Bmatrix} = z \begin{Bmatrix} \dfrac{\partial \phi_y}{\partial x} \\ -\dfrac{\partial \phi_x}{\partial y} \\ \dfrac{\partial \phi_y}{\partial y} - \dfrac{\partial \phi_x}{\partial x} \end{Bmatrix} = z\{\Psi\} \tag{10.41}$$

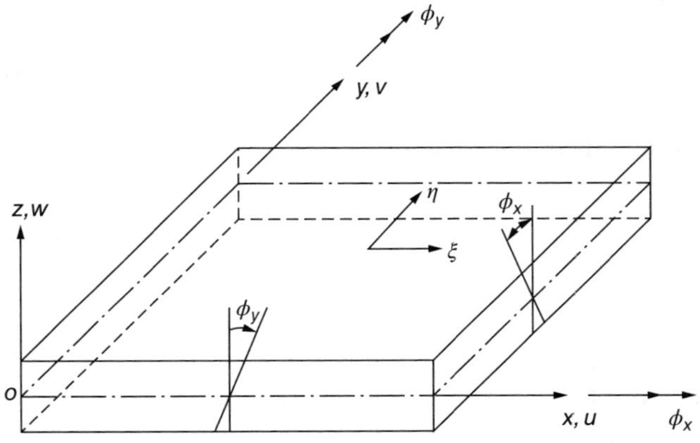

Figure 10.14 The plate element considering the shear deformation.

where

$$\{\Psi\} = \left[\frac{\partial \phi_y}{\partial x}, -\frac{\partial \phi_x}{\partial y}, \frac{\partial \phi_y}{\partial y} - \frac{\partial \phi_x}{\partial x}\right]^{\mathrm{T}} \tag{10.42}$$

Assuming that the shearing strain along the thickness in the plate is constant and it is calculated by the formula below:

$$\{\gamma\} = \begin{Bmatrix} \gamma_{yz} \\ \gamma_{zx} \end{Bmatrix} = \begin{Bmatrix} \dfrac{\partial w}{\partial y} - \phi_x \\ \dfrac{\partial w}{\partial x} + \phi_y \end{Bmatrix} \tag{10.43}$$

For the elastic isotropic plate, the stress should be calculated according to the formula below:

$$\{\sigma\} = \begin{Bmatrix} \sigma_x \\ \sigma_y \\ \tau_{xy} \end{Bmatrix} = z\frac{E}{1-\mu^2}\begin{bmatrix} 1 & \mu & 0 \\ \mu & 1 & 0 \\ 0 & 0 & \dfrac{1-\mu}{2} \end{bmatrix}\begin{Bmatrix} \dfrac{\partial \phi_y}{\partial x} \\ -\dfrac{\partial \phi_x}{\partial y} \\ \dfrac{\partial \phi_y}{\partial y} + \dfrac{\partial \phi_x}{\partial x} \end{Bmatrix} \tag{e}$$

$$\{\tau\} = \begin{Bmatrix} \tau_{yz} \\ \tau_{zx} \end{Bmatrix} = \frac{kE}{2(1+\mu)}\begin{Bmatrix} \dfrac{\partial w}{\partial y} - \phi_x \\ \dfrac{\partial w}{\partial x} + \phi_y \end{Bmatrix} \tag{f}$$

The moment of the plate is

$$\{M\} = \begin{Bmatrix} M_x \\ M_y \\ M_{xy} \end{Bmatrix} = \int_{-t/2}^{t/2} z\{\sigma\}\,\mathrm{d}z \tag{g}$$

The transverse shear of the plate is

$$\{Q\} = \begin{Bmatrix} Q_{yz} \\ Q_{zx} \end{Bmatrix} = \int_{-t/2}^{t/2} \{\tau\}\,\mathrm{d}z \tag{h}$$

By substitution of formulas (e) and (f) into formulas (g) and (h), we get

$$\{M\} = [D_b]\{\Psi\} \tag{10.44}$$

$$\{Q\} = [D_s]\{\gamma\} \tag{10.45}$$

where

$$[D_b] = \frac{Et^3}{12(1-\mu^2)}\begin{bmatrix} 1 & \mu & 0 \\ \mu & 1 & 0 \\ 0 & 0 & \dfrac{1-\mu}{2} \end{bmatrix}, \quad [D_s] = \frac{kEt}{2(1+\mu)}\begin{bmatrix} 1 & 0 \\ 0 & 1 \end{bmatrix} \tag{10.46}$$

The potential energy of the plate element is

$$\begin{aligned}\Pi &= \frac{1}{2}\int_A (\{M\}^{\mathrm{T}}\{\Psi\} + \{Q\}^{\mathrm{T}}\{\gamma\})\,\mathrm{d}A - \int_A qw\,\mathrm{d}A \\ &= \frac{1}{2}\int_A (\{\Psi\}^{\mathrm{T}}[D_b]\{\Psi\} + \{\gamma\}^{\mathrm{T}}\{D_s\}\{\gamma\})\,\mathrm{d}A - \int_A qw\,\mathrm{d}A\end{aligned} \tag{10.47}$$

The coordinate of any point of the middle surface expressed with the shape function is as follows:

$$x = \sum_{i=1}^{s} N_i x_i, \quad y = \sum_{i=1}^{s} N_i y_i \tag{10.48}$$

in which s is the number of nodes of the element.

The deflection and rotation angle of any point of the plate expressed with the nodal deflection w_i and nodal rotation angles ϕ_{xi} and ϕ_{yi} is as follows:

$$w = \sum_{i=1}^{s} N_i w_i, \quad \phi_x = \sum_{i=1}^{s} N_i \phi_{xi}, \quad \phi_y = \sum_{i=1}^{s} N_i \phi_{yi} \tag{10.49}$$

By substitution of formula (10.49) into formulas (10.42) and (10.43), we get

$$\{\Psi\} = [B_b]\{\delta\}, \quad \{\gamma\} = [B_s]\{\delta\} \tag{10.50}$$

where

$$[B_b] = \begin{bmatrix} 0 & 0 & \dfrac{\partial N_1}{\partial x} & \cdots & 0 & 0 & \dfrac{\partial N_s}{\partial x} \\ 0 & -\dfrac{\partial N_1}{\partial y} & 0 & \cdots & 0 & -\dfrac{\partial N_s}{\partial y} & 0 \\ 0 & -\dfrac{\partial N_1}{\partial x} & \dfrac{\partial N_1}{\partial y} & \cdots & 0 & -\dfrac{\partial N_s}{\partial x} & \dfrac{\partial N_s}{\partial y} \end{bmatrix} \tag{10.51}$$

$$[B_s] = \begin{bmatrix} \dfrac{\partial N_1}{\partial y} & -N_1 & 0 & \cdots & \dfrac{\partial N_s}{\partial y} & -N_s & 0 \\ \dfrac{\partial N_1}{\partial x} & 0 & N_1 & \cdots & \dfrac{\partial N_s}{\partial x} & 0 & N_s \end{bmatrix} \tag{10.52}$$

$$[\delta] = \begin{bmatrix} w_1 & \phi_{x1} & \phi_{y1} & \cdots & w_s & \phi_{xs} & \phi_{ys} \end{bmatrix}^{\mathrm{T}} \tag{10.53}$$

Then the stiffness matrix of the element is

$$[K] = \int_{-1}^{1}\int_{-1}^{1} ([B_b]^{\mathrm{T}}[D_b][B_b] + [B_s]^{\mathrm{T}}[D_s][B_s])|J|\,\mathrm{d}\xi\,\mathrm{d}\eta \tag{10.54}$$

The nodal load of the element is

$$\{P\} = \int_{-1}^{1}\int_{-1}^{1} [N]^{\mathrm{T}} q|J|\,\mathrm{d}\xi\,\mathrm{d}\eta \tag{10.55}$$

where

$$[N] = \begin{bmatrix} N_1 & 0 & 0 & N_2 & 0 & 0 & \cdots \end{bmatrix} \tag{10.56}$$

For this kind of plate element, because the deflection and rotation angle are independent of each other, the condition of continuous deformation is easily satisfied. And due to the coordinate transformation with the shape function, it can adapt to the irregular geometry. Experience shows that if we use the high-order element with 9 nodes, the accuracy and efficiency of calculation are good, but if we use the low-order element with 4 nodes, when calculating the thin plate, the stiffness of the plate is too big and the calculation result is not accurate. In this case, we can use the discrete Kirchhoff's theory, namely, on some selected discrete points, making the shearing deformation equal to zero; the character of the element is improved.

10.5 The Plate on Elastic Foundation

The characteristic of the plate on elastic foundation is that the lower surface of the plate is attached to the foundation. This is a hybrid structure and the key to calculation is to determine the counterforce of the foundation correctly.

10.5.1 Plate on Winkler Foundation

Assuming that the foundation counterforce is directly proportional to the deflection of the plate, namely,

$$p = \beta w \tag{a}$$

where p is the foundation counterforce, β is the resistance coefficient of the foundation, and w is the subsidence of the foundation surface, which is equal to the deflection of the plate.

By substitution of $w = [N]\{\delta\}^e$ into formula (a), we get

$$p = \beta[N]\{\delta\}^e \tag{b}$$

where $\{\delta\}^e$ is the nodal displacement of the plate element.

Now we consider the equilibrium of the plate element. The normal load q is on the upper surface, and the foundation counterforce p is on the lower surface. According to the formula (10.21), the nodal equilibrium condition can be written as

$$\begin{aligned}[k]\{\delta\}^e &= \iint [N]^T q \,\mathrm{d}x\,\mathrm{d}y - \iint [N]^T p \,\mathrm{d}x\,\mathrm{d}y \\ &= \iint [N]^T q \,\mathrm{d}x\,\mathrm{d}y - \beta \iint [N]^T [N] \,\mathrm{d}x\,\mathrm{d}y \{\delta\}^e\end{aligned}$$

The second item on the right side of the formula above represents the influence of the foundation counterforce. Because the direction of p is opposite to that of w, we take a minus sign. Moving the item of the foundation counterforce to the left of the formula above, we get

$$([k] + [k_0])\{\delta\}^e = \{P\}^e \tag{c}$$

$$[k_0] = \beta \iint [N]^T [N] \,\mathrm{d}x\,\mathrm{d}y \tag{10.57}$$

in which $\{P\}^e$ is the nodal load.

By the formula (c), it's known that when calculating the plate on Winkler foundation, it is only necessary to replace the original stiffness matrix of the element with $[k] + [k_0]$ where $[k_0]$ is calculated by the formula (10.57) and $[k]$ is still calculated by the formula (10.59). Winkler assumption simplifies the calculation, but it assumes that the foundation is consisted of independent springs. It is inconsistent with reality.

10.5.2 Plate on Elastic Half Space

As shown in Figure 10.15, we regard the foundation as an elastic half space, divide the thin plate into rectangular elements, and connect each node with the corresponding points of the foundation by a bar. Assuming that the foundation counterforce on the area of each bar allocated (the shading shown in Figure 10.15) is evenly distributed, then the displacement of the foundation can be calculated.

Assuming that the force of the bar is P_i at the node of i and it is uniformly distributed in the area ab, so around the node of i, the pressure of the foundation is P_i/ab (Figure 10.16).

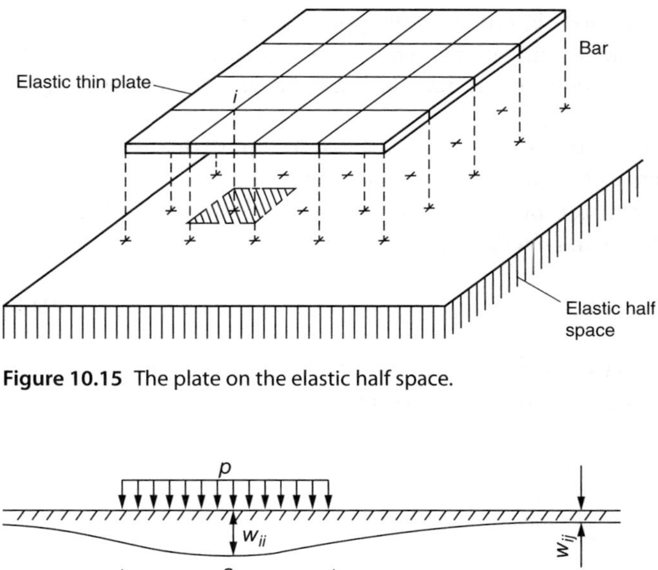

Figure 10.15 The plate on the elastic half space.

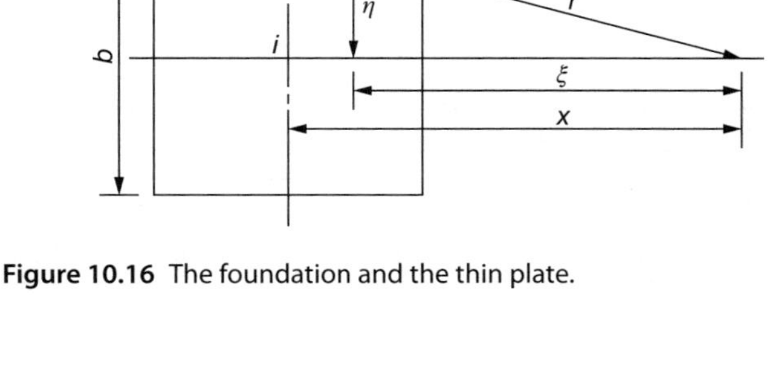

Figure 10.16 The foundation and the thin plate.

According to Bousinesq's formula, the vertical displacement at another point j on the surface caused by the concentrated force F at the point of i on the surface of the elastic half space is as follows:

$$w_{ji} = \frac{F(1-\mu_0^2)}{\pi E_0 r} \tag{10.58}$$

where E_0 and μ_0 are the modulus of elasticity and Poisson's ratio of the foundation, respectively, and r is the distance between point i and point j.

Integrating in the rectangle area ab by the formula (10.58), the vertical deflection of the center of the rectangular under uniformly distributed load P_i/ab is as follows:

$$\begin{aligned} w_{ii} &= 2\int_{\xi=0}^{\xi=a/2} 2\int_{\eta=0}^{\eta=b/2} \frac{P_i(1-\mu_0^2)\,\mathrm{d}\xi\,\mathrm{d}\eta}{ab\pi E_0\sqrt{\xi^2+\eta^2}} \\ &= \frac{P_i(1-\mu_0^2)}{ab\pi E_0} f_{ii} \end{aligned} \tag{10.59}$$

Coefficient values of f_{ii} are shown in Table 10.5.

If the point j is outside of the loading area, it also can be integrated of course. But if we calculate directly by the formula (10.58), the approximate displacement can be obtained. In this case, F is the load acting on the whole rectangular area, namely, P_i, r is the distance between point j and point i. The accurate and approximate values of f_{ji} are listed in the Table 10.6.

With the table, we know that even for $x = a$, the error of the approximate calculation is only 4% and the error decreases rapidly with the increase of x.

Under the action of all the contact force $\{P\} = \begin{bmatrix} P_1 & P_2 & \cdots & P_n \end{bmatrix}^{\mathrm{T}}$, the vertical displacement of the foundation surface can be written as

$$\{w\} = \frac{(1-\mu_0^2)}{\pi E_0 a}[f]\{P\} \tag{10.60}$$

in which $[f]$ is the flexibility matrix of the foundation; its diagonal elements are calculated by the formula (10.59) and the other elements by the formula (10.58).

Table 10.5 Coefficient f_{ii}.

b/a	2/3	1	2	3	4	5
f_{ii}	4.265	3.525	2.406	1.867	1.543	1.322

Table 10.6 Coefficient f_{ji}.

x/a	1	2	3	4	5	6	8	10
Exact value	1.038	0.505	0.333	0.251	0.200	0.167	0.125	0.100
Approximate value	1.000	0.500	0.333	0.250	0.200	0.167	0.125	0.100

Reversing the formula (10.60), we get

$$\{P\} = \frac{\pi E_0 a}{(1-\mu_0^2)}[k_0]\{w\} \tag{10.61}$$

where $[k_0] = [f]^{-1}$.

In the nodal displacement $\{\delta\}$ of the plate, in addition to the vertical displacement, there are angular displacements θ_x and θ_y. Generally it is assumed that the foundation counterforce has nothing to do with the angular displacement of the surface. So outside the elements of $\{k_0\}$, filling in some zeros, we can get the foundation counterforce as follows:

$$\{P\}_f = [k_f]\{\delta\}$$

Assuming that the load, which the plate is subjected to, is $\{P\}_q$, minus the foundation counterforce, we get the effective load of the plate, namely,

$$\{P\}_q - \{P\}_f = [K]\{\delta\}$$

Hence, we get

$$([K]+[K_f])\{\delta\} = \{P\}_q \tag{10.62}$$

By the formula above $\{\delta\}$ can be solved, then the internal stress of the plate can be calculated.

The calculation results of the square plate under uniformly distributed load are shown in Figure 10.17. $\gamma = 180\pi(E_0/E_p)(a/t)^3$ represents the relative stiffness, E_0 is the

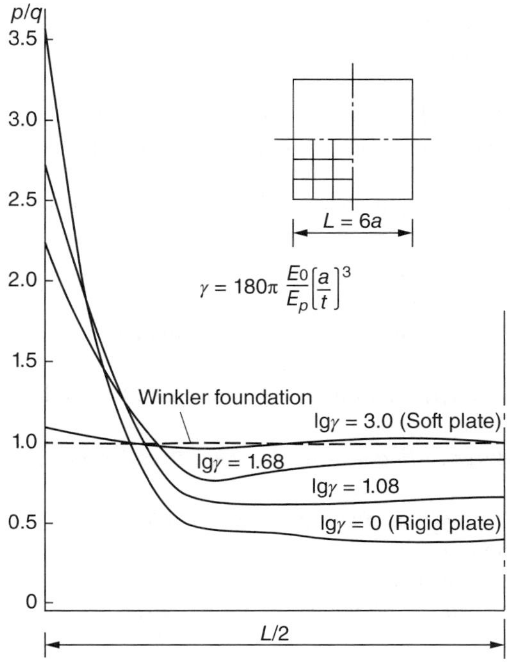

Figure 10.17 The distribution of the contact stress of the central line under uniformly distributed load.

modulus of elasticity of the foundation, and E_p is the modulus of elasticity of the plate. During the calculation, we assume that the foundation and the plate have the same value of $1 - \mu^2$.

By the figure it is clear that when the stiffness of the plate is very big, the foundation counterforce on the edge can achieve a large numerical value (the theoretical value is infinity), and when the plate is very soft, the foundation counterforce is equal to the uniformly distributed load. If we use the Winkler assumption, whatever the stiffness of the plate is, the foundation counterforce is equal to the uniformly distributed load; this fact indicates the defects of the Winkler assumption.

Bibliography

1 Hughes, T.J.R. and Hinton, E. (1986) *Finite Element Methods for Plate and Shell Structures*, Pineridge Press, Swansea.

2 Clough, R.W. and Tocher, J.L. (1965) Finite element stiffness matrices for analysis of plates in bending, in *Proceedings of the Conference on Matrix Methods in Structural Mechanics, Air Force Institute of Technology*, Wright Patterson A. F. Base, Ohio.

3 Bazeley, G.P., Cheung, Y.K., Irons, B.M. and Zienkiewicz, O.C. (1965) Triangular elements in bending-conforming and non-conforming solutions, in *Proceedings of the Conference on Matrix Methods in Structural Mechanics, Air Force Institute of Technology*, Wright Patterson A. F. Base, Ohio.

4 Zienkiewicz, O.C. and Cheung, Y.K. (1964) The finite element method for analysis of elastic isotropic and orthotropic slabs. *Proc. Inst. Civ. Eng*, **28**, 471–488.

5 Pian, T.H.H. and Tong, P. (1969) Basis of finite element methods for solid continua. *Int. J. Numer. Meth. Eng.*, **1**, 3–28.

6 Alwood, R.J. and Cornes, G.M.M. (1969) A polygonal finite element for plate bending problems using assumed stress approach. *Int. J. Numer. Meth. Eng.*, **1**, 135–150.

7 Herrmann, L.R. (1967) Finite element bending analysis of plates. *J. Eng. Mech. Div. Proc. ASCE*, **93** (EM5), 13–26.

8 Mindlin, R.D. (1951) Influence of rotary inertia and shear on flexural motion of isotropic elastic plates. *J. Appl. Mech.*, **18**, 31–38.

9 Batoz, J.L., Bathe, K.J. and Ho, L.W. (1980) A study of three-node triangular plate bending elements. *Int. J. Numer. Meth. Eng.*, **15**, 1771–1812.

10 Bathe, K.J. and Bolouzchi, S. (1980) A geometric and material nonlinear plate and shell element. *J. Comput. Struct*, **11**, 23–48.

11 Bathe, K.J. (1982) *Finite Element Procedures in Engineering Analysis*, Prentice-Hall, Englewood Cliffs.

12 Zienkiewicz, O.C. and Taylor, R.L. (2009) *Finite Element Method for Solid and Structural Mechanics*, Elsevier, New York, London.

13 Cheung, Y.K. and Zienkiewicz, O.C. (1965) Plates and tanks on elastic foundations and application of finite element method. *Int. J. Solids Struct.*, **1**, 451–461.

14 Zienkiewicz, O.C., Xu, Z. and Zeng, L.F. (1993) Linked interpolation for Reissner–Mindlin plate elements – Part I: a simple quadrilateral. *Int. J. Numer. Meth. Eng.*, **36** (18), 3043–3056.

15 Taylor, R.L. and Auricchio, F. (1993) Linked interpolation of Reissner–Mindlin plate elements – Part II: a simple triangle. *Int. J. Numer. Meth. Eng.*, **36** (18), 3057–3066.

16 Razzaque, A.Q. (1973) Program for triangular elements with derivative smoothing. *Int. J. Numer. Meth. Eng.*, **6** (3), 333–343.

11

Elastic Thin Shell

There are two different approaches to analyze the elastic shell by finite element method. One is replacing the original shell by a folded plate system consisting of plate elements in which the stress of the thin shell is equal to the sum of the plane stress state and thin-plate bending stress state. The other approach is the direct application of curved-surface element and deduction of the element stiffness matrix by shell theory.

First, we will introduce the computation approach by replacing the shell with the folded plate. As shown in Figure 11.1, generally the shell is replaced by a combination of triangular or rectangular thin-plate elements. The triangular element is provided with relatively wide application as it is adaptable to complicated shapes of the shell. Then, we will introduce the curved-surface shell element.

11.1 Element Stiffness Matrix in Local Coordinate System

The stress state of the elastic shell may be regarded as the combination of the plane stress state and the bending stress state. Therefore, the stiffness matrix of the shell element may also be computed by combining the stiffness matrices of both stress states. As shown in Figure 11.2, for the local coordinate system, the x- and y-axes lie in the plane of the element.

For the plane stress state, as may be seen in Chapter 2, the stress state of the element fully depends on the displacements u and v of the nodes. Taking triangular element, for example, the relation between the nodal force and nodal displacement of the element is shown as follows:

$$\begin{Bmatrix} F_i^p \\ F_f^p \\ F_m^p \end{Bmatrix} = \left[k^p\right] \begin{Bmatrix} \delta_i^p \\ \delta_f^p \\ \delta_m^p \end{Bmatrix} \tag{11.1}$$

$$\left\{\delta_i^p\right\} = \begin{Bmatrix} u_i \\ v_i \end{Bmatrix} \quad (i, j, m)$$

$$\left\{F_i^p\right\} = \begin{Bmatrix} U_i \\ V_i \end{Bmatrix} \quad (i, j, m)$$

The Finite Element Method: Fundamentals and Applications in Civil, Hydraulic, Mechanical and Aeronautical Engineering,
First Edition. Bofang Zhu.

Companion website: www.wiley.com/go/zhufem17

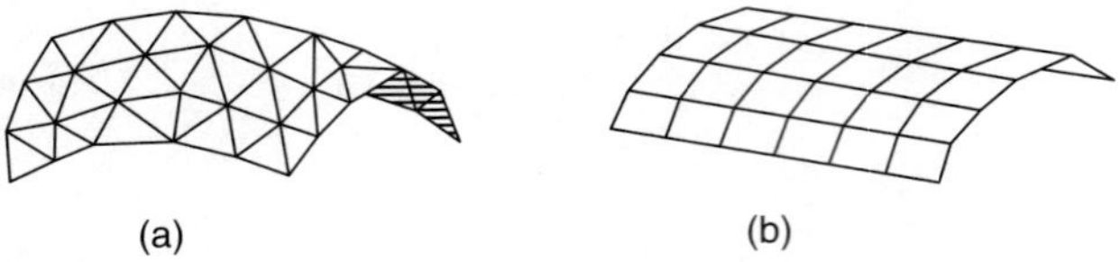

Figure 11.1 Replacement of shell by folded plate. (a) Arbitrary shell consisting of triangular elements; (b) cylindrical shell consisting of rectangular elements.

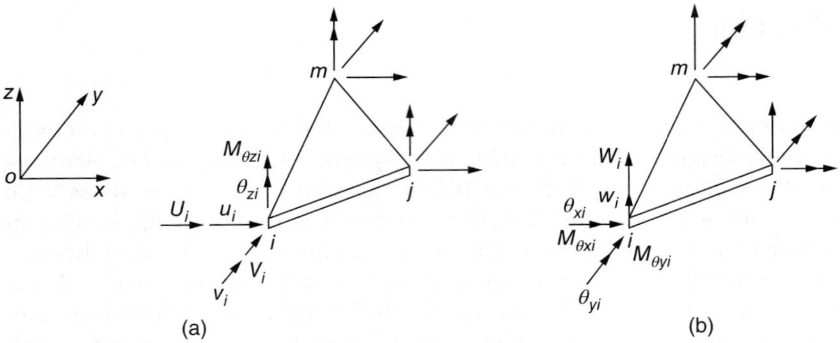

Figure 11.2 Nodal force and nodal displacement of shell element. (a) Plane stress state; (b) bending stress state.

The nodal angular displacements $\theta_{zi}, \theta_{zj}, \theta_{zm}$ bear no influence on the nodal forces and therefore may be neglected. The corresponding nodal forces $M_{\theta zi}, M_{\theta zj}, M_{\theta zm}$ do not exist either.

For the bending stress state, the element stress state depends on the linear displacement w of the node along z direction and the angular displacement θ_x around x-axis and θ_y around the y-axis; the relation between the nodal force and nodal displacement is expressed as follows:

$$\begin{Bmatrix} F_i^{\mathrm{b}} \\ F_f^{\mathrm{b}} \\ F_m^{\mathrm{b}} \end{Bmatrix} = \left[k^{\mathrm{b}}\right] \begin{Bmatrix} \delta_i^{\mathrm{b}} \\ \delta_f^{\mathrm{b}} \\ \delta_m^{\mathrm{b}} \end{Bmatrix}$$

$$\left\{\delta_i^{\mathrm{b}}\right\} = \begin{Bmatrix} w_i \\ \theta_{xi} \\ \theta_{yi} \end{Bmatrix} \quad (i, j, m)$$

$$\left\{F_i^{\mathrm{b}}\right\} = \begin{Bmatrix} W_i \\ M_{\theta xi} \\ M_{\theta yi} \end{Bmatrix} \quad (i, j, m) \tag{11.2}$$

By combination of the plane stress and bending stress, the nodal displacement and nodal force of the element are shown as follows:

$$\{\delta_i\} = \begin{Bmatrix} u_i \\ v_i \\ w_i \\ \theta_{xi} \\ \theta_{yi} \\ \theta_{zi} \end{Bmatrix}, \quad \{F_i\} = \begin{Bmatrix} U_i \\ V_i \\ W_i \\ M_{\theta xi} \\ M_{\theta yi} \\ M_{\theta zi} \end{Bmatrix} \tag{11.3}$$

Though the angular displacement θ_{zi} bears no influence on the stress state, to facilitate the further transformation of the stiffness matrix of the local coordinate system into the global coordinate system and the assemblage of the global stiffness matrix, we specifically include θ_{zi} into the node displacement and include a virtual bending moment $M_{\theta zi}$ into the nodal force accordingly.

The relation between the nodal force and the nodal displacement of the element may be expressed as

$$\begin{Bmatrix} F_i \\ F_j \\ F_m \end{Bmatrix} = [k] \begin{Bmatrix} \delta_i \\ \delta_j \\ \delta_m \end{Bmatrix} \tag{11.4}$$

or

$$\{F\}^{\mathrm{e}} = [k]\{\delta\}^{\mathrm{e}} \tag{11.5}$$

where $[k]$ refers to the element stiffness matrix under the combined stress state.

The nodal force $\{F_i^{\mathrm{p}}\}$ of the plane stress state and the nodal displacement $\{\delta_i^{\mathrm{b}}\}$ of the bending stress state bear no mutual influence. The nodal force $\{F_i^{\mathrm{b}}\}$ of the bending stress state and the node displacement $\{\delta_i^{\mathrm{p}}\}$ of the plane stress state bear no mutual influence either. Therefore, the submatrix of the element stiffness matrix under combined stress state may be expressed as

$$[k_n] = \left[\begin{array}{cc|ccc|c} \multicolumn{2}{c|}{\multirow{2}{*}{$\left[k_n^p\right]$}} & 0 & 0 & 0 & 0 \\ & & 0 & 0 & 0 & 0 \\ \hline 0 & 0 & & & & 0 \\ 0 & 0 & & \left[k_n^b\right] & & 0 \\ 0 & 0 & & & & 0 \\ \hline 0 & 0 & 0 & 0 & 0 & 0 \end{array}\right] \tag{11.6}$$

where $[k_{rs}^{\mathrm{p}}]$ and $[k_{rs}^{\mathrm{b}}]$ refer to the corresponding submatrix of the plane stress problem and thin-plate bending problem, respectively.

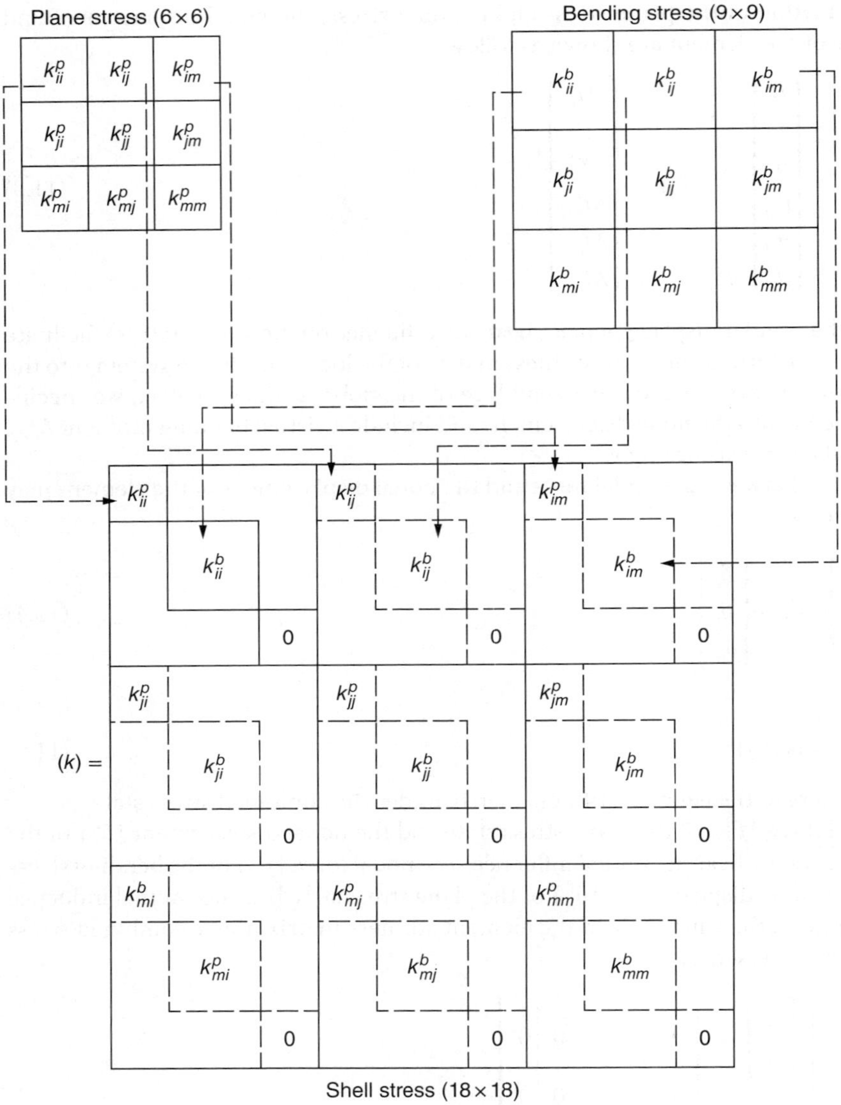

Figure 11.3 Shell element stiffness matrix by combination of stiffness matrices of plane stress and thin-plate bending.

Therefore, by combining the triangular element stiffness matrix of the plane stress problem and thin-plate bending problem, the triangular element stiffness matrix under combined stress state may be obtained, as shown in Figure 11.3. This combination approach is also applicable to the rectangle shell element, with the only difference that the stiffness matrix is in the order of 24×24.

11.2 Coordinate Transformation: Global Stiffness Matrix

As mentioned above, the element stiffness matrix is deduced in the local coordinate system of the element, taking the middle surface of the element as xy plane and z-axis being perpendicular to the middle surface of the element. To establish the global stiffness matrix for the shell, the element stiffness coefficients of different planes must be assembled at the nodes. Therefore, it is necessary to determine a global coordinate system, and the stiffness matrices of each element in the local coordinate system should be transformed into the global coordinate system.

Express global coordinates by x', y', z' and local coordinates still by x, y, z. In the local coordinate system, the displacement and nodal force of node i are still expressed as $\{\delta_i\}$ and $\{F_i\}$, as shown in formula (11.3). And in the global coordinate system, the displacement and node force of node i are expressed as $\{\delta'_i\}$ and $\{F'_i\}$:

$$\begin{aligned}\{\delta'_i\} &= [u'_i v'_i w'_i \theta'_{xi} \theta'_{yi} \theta'_{zi}]^{\mathrm{T}}\\ \{F'_i\} &= [U'_i V'_i W'_i M'_{\theta xi} M'_{\theta yi} M'_{\theta zi}]^{\mathrm{T}}\end{aligned} \tag{11.7}$$

The transformation of the nodal displacement and nodal force between the two coordinate systems (Figure 11.4) may be computed by the following formula:

$$\{\delta_i\} = [\lambda]\{\delta'_i\}, \{F_i\} = [\lambda]\{F'_i\} \tag{11.8}$$

$$[\lambda] = \begin{bmatrix} \cos(x,x') & \cos(x,y') & \cos(x,z') & 0 & 0 & 0\\ \cos(y,x') & \cos(y,y') & \cos(y,z') & 0 & 0 & 0\\ \cos(z,x') & \cos(z,x') & \cos(z,z') & 0 & 0 & 0\\ 0 & 0 & 0 & \cos(x,x') & \cos(x,y') & \cos(x,z')\\ 0 & 0 & 0 & \cos(y,x') & \cos(y,y') & \cos(y,z')\\ 0 & 0 & 0 & \cos(z,x') & \cos(z,x') & \cos(z,z') \end{bmatrix} \tag{11.9}$$

where (x, x') is the included angle between x-axis and x'-axis.

Expressing the displacement and node force of the element in the global coordinates as $\{\delta'\}^{\mathrm{e}}$ and $\{F'\}^{\mathrm{e}}$, respectively, from transform formula (11.8), we have

$$\begin{aligned}\{\delta\}^{\mathrm{e}} &= [L]\{\delta'\}^{\mathrm{e}},\\ \{F\}^{\mathrm{e}} &= [L]\{F'\}^{\mathrm{e}}\end{aligned} \tag{11.10}$$

$$[L] = \begin{bmatrix} \lambda & 0 & 0\\ 0 & \lambda & 0\\ 0 & 0 & \lambda \end{bmatrix} \tag{11.11}$$

From the latter formula of formula (11.10),

$$\{F'_i\}^{\mathrm{e}} = [L]^{-1}\{F\}^{\mathrm{e}}$$

Substituting the formula of the displacement and nodal force of the element in the local coordinate system $\{F'\}^{\mathrm{e}} = [k]\{\delta\}^{\mathrm{e}}$ into the abovementioned formula, we get

$$\{F'_i\}^{\mathrm{e}} = [L]^{-1}[k]\{\delta\}^{\mathrm{e}} = [L]^{-1}[k][L]\{\delta'\}^{\mathrm{e}}$$

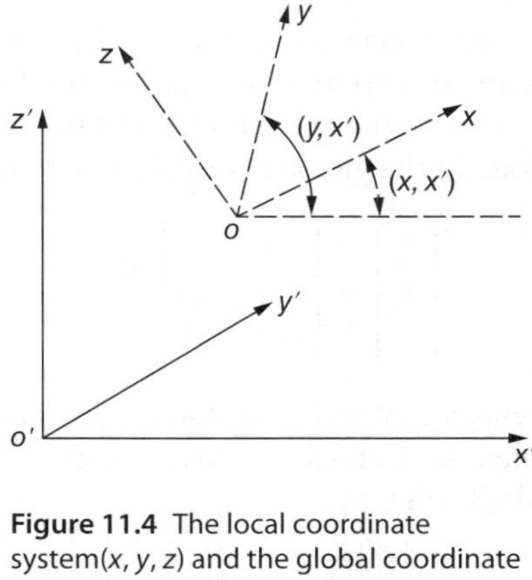

Figure 11.4 The local coordinate system(x, y, z) and the global coordinate system(x', y', z').

It may be seen that $[L]^{-1}[k][L]$ is the stiffness matrix in the global coordinate system, namely,

$$[k'] = [L]^{-1}[k][L]$$

As $[L]$ is an orthogonal matrix, so $[L]^{-1} = [L]^{\mathrm{T}}$; thus

$$[k'] = [L]^{\mathrm{T}}[k][L] \tag{11.12}$$

This is the transformation formula of the stiffness matrix. With this formula, the element stiffness matrix $[k']$ in the global coordinate system may be computed by the element stiffness matrix $[k]$ in the local coordinate system.

Expressing the stiffness matrix into partitioned form, the submatrix of $[k']$ may be computed as follows:

$$[k'_{rs}] = [\lambda]^{\mathrm{T}}[k_{rs}][\lambda] \tag{11.13}$$

where $[k_{rs}]$ is calculated in the local coordinate system by formula (11.6).

After the element stiffness matrix in the global coordinate system is worked out, the results will be assembled according to the processes stated above. The global stiffness matrix $[k]$ will be easily obtained.

To establish a nodal equilibrium equation, the nodal load must also be computed. It is easy to calculate the element nodal load by equalizing the work of the element load to the work of nodal load. Generally, under the distributed load, the element node load involves the force and the moment component. Actually, when the element is relatively small, a relatively simple approach may be applied to work out the element nodal load, that is, neglecting the moment component and allocating the resultant force of the external loads on the element evenly to each of the node of the element. For example, on the condition that the shell surface is bearing normal distributed loads, the load intensity on the three nodes of the triangular element is p_i, p_j, p_m, respectively, and the resultant force of the loads on the element will be

$$R = \frac{A}{3}(p_i + p_j + p_m)$$

where A refers to the area of the element. Assume it is positive when the load has the same direction with z-axis of the local coordinate system. Allocate the resultant force evenly to the three nodes and transform it into the global coordinate system. The node load in the global coordinate system will be obtained as

$$\begin{bmatrix} X_i \\ Y_i \\ Z_i \end{bmatrix} = \begin{Bmatrix} \cos(x', z) \\ \cos(y', z) \\ \cos(z', z) \end{Bmatrix} \frac{R}{3} \tag{11.14}$$

where x', y', z' is the global coordinate system. After assembling the nodal loads of each element to form the load vector of the shell $\{P'\}$, the node equilibrium equation of the shell will be

$$[K]\{\delta'\} = \{P'\} \tag{11.15}$$

The displacements of all nodes in the global coordinate system may be solved by the abovementioned formula. With the transformation formula $\{\delta\}^{\mathrm{e}} = [L]\{\delta'\}^{\mathrm{e}}$, we can compute the displacement of each node in the local coordinate system and then compute the stress of each element with the stress matrix in the local coordinate system.

A specific circumstance must be noted, if all elements around one node are within the same plane, and as the stiffness coefficient along θ_{zi} direction has already been set equal to zero in formula (11.6), the sixth equilibrium equation of such node (in θ_z direction) in the local coordinate system will be

$$0 = 0 \tag{11.16}$$

On the condition that the global coordinate system is consistent with the local coordinate system, obviously the determinant of the stiffness matrix will be $|K| = 0$. Therefore, formula (11.15) does not satisfy the unique solution conditions. If the global coordinate system is inconsistent with the local coordinate system, six equilibrium equations that are apparently correct may be obtained for the node after transformation, and due to formula (11.16), the row vectors of determinant $|K|$ are in linear correlation, so $|K| = 0$. Formula (11.15) still does not satisfy the unique solution conditions.

To solve the problem, for such nodes with corresponding elements lying in the same plane, the node equilibrium equation may be established in the local coordinate system with the equilibrium equation $0 = 0$ in the direction of θ_z deleted. Therefore, the unique solution requirements will be met. However, such approach is relatively complicated in program design.

Another approach is giving an arbitrary stiffness coefficient $k_{\theta z}$ on such special nodes, and the equilibrium equation of such node in the direction of θ_z in the local coordinate system will be

$$k_{\theta z}\theta_{zi} = 0 \tag{11.17}$$

By coordinate transformation, the node equilibrium equation in the global coordinate system can meet the unique solution requirements, and θ_{zi} is included in the solved node displacements. θ_{zi} bears no influence on the element stress and no relationship to equilibrium equation for other nodes, so the arbitrary value may be given to $k_{\theta z}$ and will not influence the computation results.

11.3 Direction Cosine of Local Coordinate

As shown in the previous section, to establish a node equilibrium equation, the direction cosine matrix $[\lambda]$ of the local coordinate of each element will be required, as shown in formula (11.9).

In Figure 11.5, the global coordinates of the three nodes in triangular element ijm will be $x'_i, y'_i, z'_i, x'_j, \ldots, z'_m$. Set node i as the origin of the local coordinate system and side ij as x-axis. y-axis is located within the element plane and is perpendicular to side ij, and z-axis is perpendicular to the element plane. If side ij is regarded as a vector, it may

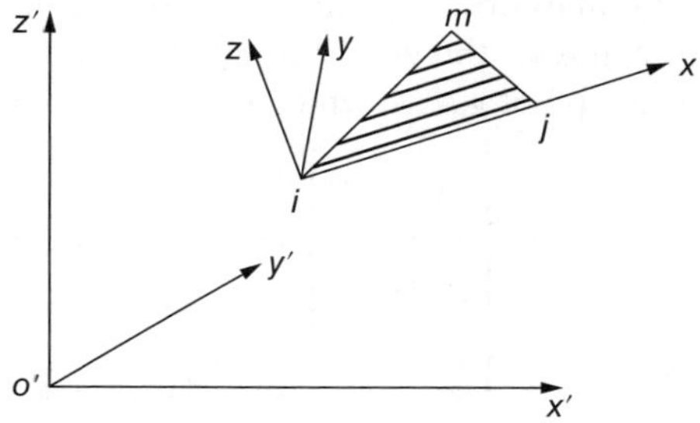

Figure 11.5 Local coordinate of triangular element.

be expressed by its three components in the global coordinate system as

$$V_{ij} = \begin{Bmatrix} x'_{ji} \\ y'_{ji} \\ z'_{ji} \end{Bmatrix} \tag{11.18}$$

where $x'_{ij} = x'_j - x'_i(x', y', z')$. Dividing the three components by the length of the vector, we get the direction cosine of the vector, which is the direction cosine of the x-axis of the local coordinate in the global coordinate system. So the unit vector is deduced:

$$v_x = \begin{Bmatrix} \cos(x, x') \\ \cos(x, y') \\ \cos(x, z') \end{Bmatrix} = \frac{1}{l_{ij}} \begin{Bmatrix} x'_{ji} \\ y'_{ji} \\ z'_{ji} \end{Bmatrix} \tag{11.19}$$

$$l_{ij} = \left[\left(x'_{ji}\right)^2 + \left(y'_{ji}\right)^2 + \left(z'_{ji}\right)^2 \right]^{1/2}$$

z-axis is perpendicular to the triangular plane. With the nature that the vector product of two vectors is perpendicular to the plane where the two vectors are located, the vector V_z perpendicular to the triangular plane will be

$$V_z = V_{ij} \times V_{im} = \begin{Bmatrix} y'_{ji} z'_{mi} - z'_{ji} y'_{mi} \\ z'_{ji} x'_{mi} - x'_{ji} z'_{mi} \\ x'_{ji} y'_{mi} - y'_{ji} x'_{mi} \end{Bmatrix}$$

The length of vector V_z is twice of the triangular area, namely,

$$\begin{aligned} l &= \left[\left(y'_{ji} z'_{mi} - z'_{ji} y'_{mi}\right)^2 + \left(z'_{ji} x'_{mi} - x'_{ji} z'_{mi}\right)^2 + \left(x'_{ji} y'_{mi} - y'_{ji} x'_{mi}\right)^2 \right]^{1/2} \\ &= 2A \end{aligned}$$

Therefore, the direction cosine of z-axis is

$$v_z = \begin{Bmatrix} \cos(z, x') \\ \cos(z, y') \\ \cos(z, z') \end{Bmatrix} = \frac{1}{2A} \begin{Bmatrix} y'_{ji} z'_{mi} - z'_{ji} y'_{mi} \\ z'_{ji} x'_{mi} - x'_{ji} z'_{mi} \\ x'_{ji} y'_{mi} - y'_{ji} x'_{mi} \end{Bmatrix} \tag{11.20}$$

Similarly, the direction cosine of y-axis may be obtained. As the y-axis is perpendicular to plane xz, the vector product of unit vector v_z and vector v_x is the unit vector in the direction of y-axis, which is

$$v_y = \begin{Bmatrix} \cos(y, x') \\ \cos(y, y') \\ \cos(y, z') \end{Bmatrix} = v_z \times v_x$$

$$= \begin{Bmatrix} \cos(z, y')\cos(x, z') - \cos(x, y')\cos(z, z') \\ \cos(z, z')\cos(x, x') - \cos(x, z')\cos(z, x') \\ \cos(z, x')\cos(x, y') - \cos(x, x')\cos(z, y') \end{Bmatrix} \tag{11.21}$$

As the length of v_y equals 1, There is no need to further divide it by its length. To this end, all elements in the direction cosine matrix $[\lambda]$ in local coordinate are worked out.

In the abovementioned computation, we assume that the x-axis of the local coordinate is along side ij of the element. Of course this is not the only method. There are other approaches available to define the local coordinate system. For example, in the computation of shells for arch dam, cooling tower, and oil tank, to facilitate the sorting of the results, it is preferred to set plane $x'y'$ of the global coordinate system into the horizontal level and the x-axis of the local coordinate system parallel to plane $x'y'$. The z-axis of the local coordinate is still perpendicular to the triangular plane. Therefore, formula (11.20) may be used first to compute the direction cosine along z-axis. As the x-axis of local coordinate is perpendicular to z'-axis, the direction cosine along x-axis is

$$v_x = \begin{Bmatrix} \cos(x, x') \\ \cos(x, y') \\ 0 \end{Bmatrix} \tag{11.22}$$

The sum of the squares of the cosines of all directions equals 1, so

$$\cos^2(x, x') + \cos^2(x, y') = 1 \tag{11.23}$$

In addition, the scalar product of v_x and v_z should be equal to zero, that is,

$$\cos(x, x')\cos(z, x') + \cos(x, y')\cos(z, y') = 0 \tag{11.24}$$

Solving the abovementioned two equations for $\cos(x, x')$ and $\cos(x, y')$, then we obtain v_x accordingly. With v_x and v_z worked out, v_y may be computed.

For the rectangle element, the scope of application is only restricted to cylinder or box-type shells, and the boundary of the shell must be parallel or perpendicular to the generating line of the cylindrical surface. Therefore, the x-axis of the global coordinate system may be set along the generating line of the cylindrical surface, while the x-axis of the local coordinate system of each element parallel to x'-axis, as shown in Figure 11.6.

Obviously, the direction cosine along x-axis is

$$\cos(x, x') = 1, \quad \cos(x, y') = 0, \quad \cos(x, z') = 0 \tag{11.25}$$

The y-axis is parallel to side pi. Therefore, by the coordinates of node p and node i in the global coordinate system, the direction cosine of y-axis may be computed as follows:

$$\left.\begin{aligned} &\cos(z, x') = 0 \\ &\cos(y, y') = \frac{y'_p - y'_i}{l_{pi}} \\ &\cos(y, z') = \frac{z'_p - z'_i}{l_{pi}} \end{aligned}\right\} \tag{11.26}$$

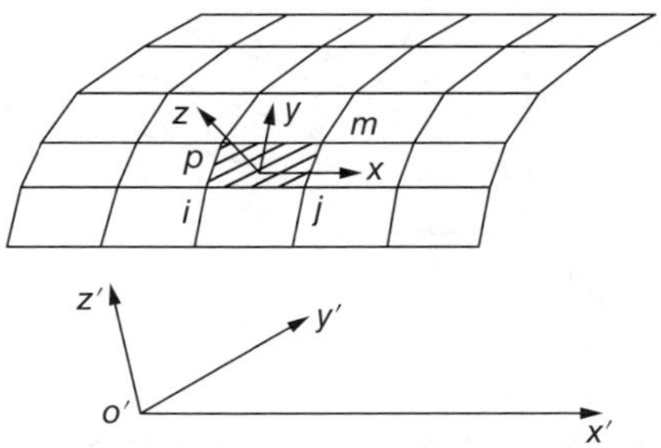

Figure 11.6 Local coordinate of the rectangular element.

where l_{pi} is the length of side pi, namely,

$$l_{pi} = \left[\left(z'_p - z'_i\right)^2 + \left(y'_p - y'_i\right)^2\right]^{1/2}$$

Similarly, the direction cosine of z-axis may be computed as

$$\left.\begin{aligned} \cos(z, x') &= 0 \\ \cos(z, z') &= \frac{z'_p - z'_i}{l_{pi}} \\ \cos(y, z') &= \frac{y'_p - y'_i}{l_{pi}} \end{aligned}\right\} \tag{11.27}$$

11.4 Curved-Surface Shell Element

In the previous sections, the actual shell is substituted by folded plate structure. As long as the mesh is sufficiently dense, the computation precision can meet the engineering requirements. However, if the curved-surface element is directly applied, it may reflect the actual geometry of the shell, and better computation results may be obtained.

As shown in Figure 11.7, a number of "s" nodes are allocated on the middle surface of the shell. The coordinates of any node on the middle surface may be expressed as

$$x = \sum_{i=1}^{s} N_i x_i, \quad y = \sum_{i=1}^{s} N_i y_i, \quad z = \sum_{i=1}^{s} N_i z_i \tag{a}$$

where s is the number of nodes of the element and N_i is a two-dimensional shape function, as shown in Table 8.2.

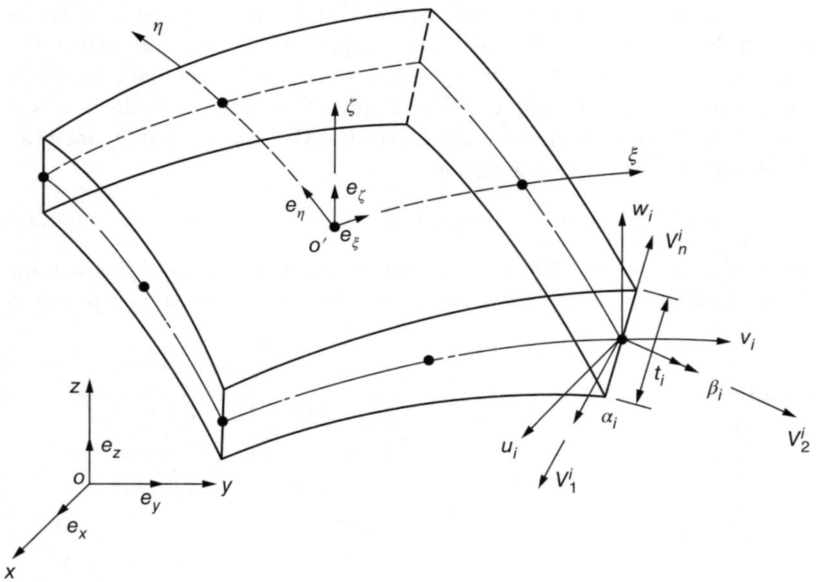

Figure 11.7 Nine-node curved-surface shell element.

Coordinates of any node in the element before transformation are as follows:

$$x^0(\xi,\eta,\zeta) = \sum_{i=1}^{s} N_i(\xi,\eta)x_i^o + \frac{\xi}{2}\sum_{i=1}^{s} N_i(\xi,\eta)t_i\vec{V}_{nx}^{io}$$

$$y^0(\xi,\eta,\zeta) = \sum_{i=1}^{s} N_i(\xi,\eta)y_i^o + \frac{\xi}{2}\sum_{i=1}^{s} N_i(\xi,\eta)t_i\vec{V}_{ny}^{io}$$

$$z^0(\xi,\eta,\zeta) = \sum_{i=1}^{s} N_i(\xi,\eta)z_i^o + \frac{\xi}{2}\sum_{i=1}^{s} N_i(\xi,\eta)t_i\vec{V}_{nz}^{io} \tag{b}$$

where $x^\circ(\xi, \eta, \zeta)$, $y^\circ(\xi, \eta, \zeta)$, $z^\circ(\xi, \eta, \zeta)$ are the coordinates of any point before transformation; $x_i^\circ, y_i^\circ, z_i^\circ$ are the coordinates of the node i before transformation; $\vec{V}_{nx}^{i_0}$, $\vec{V}_{ny}^{i_0}$, $\vec{V}_{nz}^{i_0}$ refer to the component of unit vector $\vec{V}_n^{i_0}$ along x, y, and z, respectively; $\vec{V}_n^{i_0}$ refers to the unit vector of the node i along the normal line of the middle surface of the shell before transformation; and t_i refers to the thickness at node i.

Assume that the normal line of the middle surface before deformation is still a straight line after deformation (not necessarily still perpendicular to the middle surface) and the unit vector along the straight line is V_n^{i1}, the coordinates of any node after deformation will be

$$\left.\begin{aligned} x(\xi,\eta,\zeta) &= \sum_{i=1}^{s} N_i x_i + \frac{\zeta}{2}\sum_{i=1}^{s} N_i t_i V_{nx}^{i1} \\ y(\xi,\eta,\zeta) &= \sum_{i=1}^{s} N_i y_i + \frac{\zeta}{2}\sum_{i=1}^{s} N_i t_i V_{ny}^{i1} \\ z(\xi,\eta,\zeta) &= \sum_{i=1}^{s} N_i z_i + \frac{\zeta}{2}\sum_{i=1}^{s} N_i t_i V_{nz}^{i1} \end{aligned}\right\} \tag{c}$$

Subtracting equation (b) from equation (c), we get the displacement components of any point within the element as

$$\left.\begin{aligned} u(\xi,\eta,\zeta) &= \sum_{i=1}^{s} N_i u_i + \frac{\zeta}{2}\sum_{i=1}^{s} N_i t_i V_{nx}^{i} \\ v(\xi,\eta,\zeta) &= \sum_{i=1}^{s} N_i v_i + \frac{\zeta}{2}\sum_{i=1}^{s} N_i t_i V_{ny}^{i} \\ w(\xi,\eta,\zeta) &= \sum_{i=1}^{s} N_i w_i + \frac{\zeta}{2}\sum_{i=1}^{s} N_i t_i V_{nz}^{i} \end{aligned}\right\} \tag{11.28}$$

$$V_n^i = V_n^{i1} - V_n^{i0} \tag{11.29}$$

V_{nx}^i, V_{ny}^i, V_{nz}^i are the three components of V_n^i; actually they are the increments of the direction cosine of the normal line of the middle surface. They may be expressed by the rotation of node i. However, there is no unique expression method. A relatively effective approach is defining two vectors V_1^{i0} and V_2^{i0} perpendicular to V_n^{i0}. At first set V_1^{i0} is the unit vector simultaneously perpendicular to y-axis and V_n^{i0}, that is,

$$V_1^{i0} = \frac{e_y \times V_n^{i0}}{|e_y \times V_n^{i0}|} \tag{11.30}$$

where e_y is the unit vector along y-axis (under the special circumstance where V_n^{i0} is parallel to y-axis, set $V_1^{i0} = e_x$), and then make V_2^{i0} perpendicular to V_n^{i0} and V_1^{i0}, namely,

$$V_2^{i0} = V_n^{i0} \times V_1^{i0} \tag{11.31}$$

Let α_i and β_i be the rotations of the middle surface normal line V_n^{i0} around vector V_1^{i0} and V_2^{i0}, respectively, as the values of α_i and β_i are relatively small, so

$$V_n^i = -V_2^{i0}\alpha_i + V_1^{i0}\beta_i \tag{11.32}$$

when $V_1^{i0} = e_x$, $V_2^{i0} = e_y$, and $V_n^{i0} = e_z$, the abovementioned relationship is easy to be proved. As such vectors are tensors, the abovementioned vectors should be valid under common conditions. Substituting the abovementioned formula into formula (11.28), then

$$\left.\begin{aligned} x^\circ(\xi,\eta,\zeta) &= \sum_{i=1}^{s} N_i u_i + \frac{\zeta}{2}\sum_{i=1}^{s} N_i t_i(-V_{2x}^{i0}\alpha_i + V_{1x}^{i0}\beta_i) \\ y^\circ(\xi,\eta,\zeta) &= \sum_{i=1}^{s} N_i v_i + \frac{\zeta}{2}\sum_{i=1}^{s} N_i t_i(-V_{2y}^{i0}\alpha_i + V_{1y}^{i0}\beta_i) \\ z^\circ(\xi,\eta,\zeta) &= \sum_{i=1}^{s} N_i w_i + \frac{\zeta}{2}\sum_{i=1}^{s} N_i t_i(-V_{2z}^{i0}\alpha_i + V_{1z}^{i0}\beta_i) \end{aligned}\right\} \tag{11.33}$$

By computing the partial derivatives of u with respect to ξ, η, ζ, we have

$$\begin{Bmatrix} \dfrac{\partial u}{\partial \xi} \\ \dfrac{\partial u}{\partial \eta} \\ \dfrac{\partial u}{\partial \zeta} \end{Bmatrix} = \sum_{i=1}^{s} \begin{bmatrix} \dfrac{\partial N_i}{\partial \xi}[1 \;\; \zeta g_{1x}^i \;\; \zeta g_{2x}^i] \\ \dfrac{\partial N_i}{\partial \eta}[1 \;\; \zeta g_{1x}^i \;\; \zeta g_{2x}^i] \\ N_i\,[0 \;\; g_{1x}^i \;\; g_{2x}^i] \end{bmatrix} \begin{Bmatrix} u_i \\ \alpha_i \\ \beta_i \end{Bmatrix} \quad (u, v, w; x, y, z) \tag{11.34}$$

$$g_{1x}^i = -\frac{1}{2}t_i V_2^{io}, \quad g_{2x}^i = \frac{1}{2}t_i V_1^{io} \tag{11.35}$$

where $(u, v, w; x, y, z)$ indicates that replacing u in order by v, w, and x *in order by* y, z, and then the partial derivative $\partial v/\partial\xi, \partial v/\partial\eta, \partial v/\partial\zeta$ and $\partial w/\partial\xi, \partial w/\partial\eta, \partial w/\partial\zeta$ for v and w may be computed. The following relationship may be applied to work out the partial derivatives of the displacement with respect to x, y and z:

$$\begin{Bmatrix} \dfrac{\partial}{\partial x} \\ \dfrac{\partial}{\partial y} \\ \dfrac{\partial}{\partial z} \end{Bmatrix} = [J]^{-1} \begin{Bmatrix} \dfrac{\partial}{\partial \xi} \\ \dfrac{\partial}{\partial \eta} \\ \dfrac{\partial}{\partial \zeta} \end{Bmatrix} \tag{11.36}$$

in which $[J]$ refers to the Jacobian Matrix.

Substituting formula (11.34) into formula (11.36), we have

$$\begin{Bmatrix} \dfrac{\partial u}{\partial x} \\ \dfrac{\partial u}{\partial y} \\ \dfrac{\partial u}{\partial z} \end{Bmatrix} = \sum_{i=1}^{s} \begin{bmatrix} \dfrac{\partial N_i}{\partial x} & g_{1x}^i G_x^i & g_{2x}^i G_x^i \\ \dfrac{\partial N_i}{\partial y} & g_{1x}^i G_y^i & g_{2x}^i G_y^i \\ \dfrac{\partial N_i}{\partial z} & g_{1x}^i G_z^i & g_{2x}^i G_z^i \end{bmatrix} \begin{Bmatrix} u_i \\ \alpha_i \\ \beta_i \end{Bmatrix} \tag{11.37}$$

in which

$$\frac{\partial N_i}{\partial x} = J_{11}^{-1}\frac{\partial N_i}{\partial \xi} + J_{12}^{-1}\frac{\partial N_i}{\partial \eta}$$

$$G_x^i = \zeta\left(J_{11}^{-1}\frac{\partial N_i}{\partial \xi} + J_{12}^{-1}\frac{\partial N_i}{\partial \eta}\right) + J_{13}^{-1}N_i$$

where J_{ij}^{-1} is the (i, j) element of $[J]^{-1}$. Similarly, the partial derivatives of v and w may be computed.

Then we have

$$[\varepsilon] = [B]\{\delta\} \tag{11.38}$$

$$[\varepsilon] = \begin{bmatrix} \varepsilon_x & \varepsilon_y & \varepsilon_z & \gamma_{xy} & \gamma_{yz} & \gamma_{zx} \end{bmatrix}^{\mathrm{T}}$$

$$\{\delta\} = \begin{bmatrix} u_1 & v_1 & w_1 & \alpha_1 & \beta_1 & \cdots & u_s & v_s & w_s & \alpha_s & \beta_s \end{bmatrix}^{\mathrm{T}}$$

Assuming the normal stress along the normal line of the middle surface of the shell $\sigma_\zeta = 0$, the stress–strain relationship of any point within the shell will be

$$[\sigma] = [D_{\mathrm{sh}}]\{\varepsilon\} \tag{11.39}$$

$$[D_{\mathrm{sh}}] = [Q_{\mathrm{sh}}]^{\mathrm{T}} \left(\frac{E}{1-\mu^2} \begin{bmatrix} 1 & \mu & 0 & 0 & 0 & 0 \\ & 1 & 0 & 0 & 0 & 0 \\ & & 0 & 0 & 0 & 0 \\ & & & \dfrac{1-\mu}{2} & 0 & 0 \\ & \text{symmetrical} & & & \dfrac{1-\mu}{2} & 0 \\ & & & & & \dfrac{1-\mu}{2} \end{bmatrix} \right) [Q_{\mathrm{sh}}] \tag{11.40}$$

Matrix $[Q_{\mathrm{sh}}]$ transforms the stress–strain relationship from the local coordinate system ξ, η, ζ to the global coordinate system (x, y, z). *The elements of* $[Q_{\mathrm{sh}}]$ *may be expressed by the direction cosine of* ξ, η, ζ *in the* (x, y, z) *coordinate system*:

$$[Q_{\mathrm{sh}}] = \begin{bmatrix} l_1^2 & m_1^2 & n_1^2 & l_1 m_1 & m_1 n_1 & n_1 l_1 \\ l_2^2 & m_2^2 & n_2^2 & l_2 m_2 & m_2 n_2 & n_2 l_2 \\ l_3^2 & m_3^2 & n_3^2 & l_3 m_3 & m_3 n_3 & n_3 l_3 \\ 2l_1 l_2 & 2m_1 m_2 & 2n_1 n_2 & l_1 m_2 + l_2 m_1 & m_1 n_2 + m_2 n_1 & n_1 l_2 + n_2 l_1 \\ 2l_2 l_3 & 2m_2 m_3 & 2n_2 n_3 & l_2 m_3 + l_3 m_2 & m_2 n_3 + m_3 n_2 & n_2 l_3 + n_3 l_2 \\ 2l_3 l_1 & 2m_3 m_1 & 2n_3 n_1 & l_3 m_1 + l_1 m_3 & m_3 n_1 + m_1 n_3 & n_3 l_1 + n_1 l_3 \end{bmatrix} \tag{11.41}$$

where

$$\begin{aligned} l_1 &= \cos(e_{x,} e_\xi), & m_1 &= \cos(e_{y,} e_\xi), & n_1 &= \cos(e_{z,} e_\xi) \\ l_2 &= \cos(e_{x,} e_\eta), & m_2 &= \cos(e_{y,} e_\eta), & n_2 &= \cos(e_{x,} e_\eta) \\ l_3 &= \cos(e_{x,} e_\xi), & m_3 &= \cos(e_{y,} e_\xi), & n_3 &= \cos(e_{z,} e_\xi) \end{aligned}$$

The element stiffness matrix is as follows:

$$[k] = \int_{-1}^{1}\int_{-1}^{1}\int_{-1}^{1} [B]^{\mathrm{T}}[D_{\mathrm{sh}}][B]|J|\mathrm{d}\xi\eta\mathrm{d}\xi \tag{11.42}$$

Generally, it is necessary to conduct the numerical integration along the three directions in the computation of the element stiffness matrix of the shell, and $[Q_{\mathrm{sh}}]$ and $[D_{\mathrm{sh}}]$ for each integral point should be recomputed. However, under some specific circumstances where element of the same type is provided with the same $[D_{\mathrm{sh}}]$, computation is required only once, and then reutilization will be available.

The shell element is provided with the following superiorities:

1) Accurately representing the shapes of various types of complicated shell
2) Fully meeting the requirements of deformation continuity

Experience has shown that the computation precision is satisfactory by application of a 9-node shell element.

11.5 Shell Supported or Reinforced by Curved Beam

In actual engineering, sometimes thin shells supported or reinforced by curved beams will be applied. Generally, the beam and shell are rigidly connected, and, under the action of the loads, the beam and shell will deform simultaneously. For such mixed structure, it is quite difficult to solve by means of theory. The finite element method is an effective tool to solve such structural problem.

Figure 11.8 is a case of thin shell supported by beams. The shell and beam are in rigid connection. The shell is divided into triangular elements. The supporting beam *ij* next to element *ijm* is also regarded as one element, a beam element. When establishing the equilibrium equation for node *i* and node *j*, besides the nodal force of the shell element, the nodal force for such beam element must also be considered. Therefore, it is necessary to establish the relationship between the node force $\{F\}^{\mathrm{e}}$ and node displacement $\{\delta\}^{\mathrm{e}}$ of element *ij*, which is

$$\{F\}^{\mathrm{e}} = [k]^{\mathrm{e}}\{\delta\}^{\mathrm{e}} \tag{11.43}$$

$$\{\delta\}^{\mathrm{e}} = \left[u_i \ v_i \ w_i \ \theta_{xi} \ \theta_{yi} \ \theta_{zi} \ u_j \ v_j \ w_j \ \theta_{xi} \ \theta_{yi} \ \theta_{zi}\right]^{\mathrm{T}}$$

$$\{F\}^{\mathrm{e}} = \left[U_i \ V_i \ W_i \ M_{\theta xi} \ M_{\theta yi} \ M_{\theta zi} \ U_j \ V_j \ W_j \ M_{\theta xj} \ M_{\theta yj} \ M_{\theta zj}\right]^{\mathrm{T}}$$

The beam element is provided with 12 degrees of freedom. The element stiffness matrix $[k]^{\mathrm{e}}$ is in 12×12 order. The influence of the shearing deformation is omitted, and setting $b_y = b_z = 0$, we get the element stiffness matrix in the local coordinate

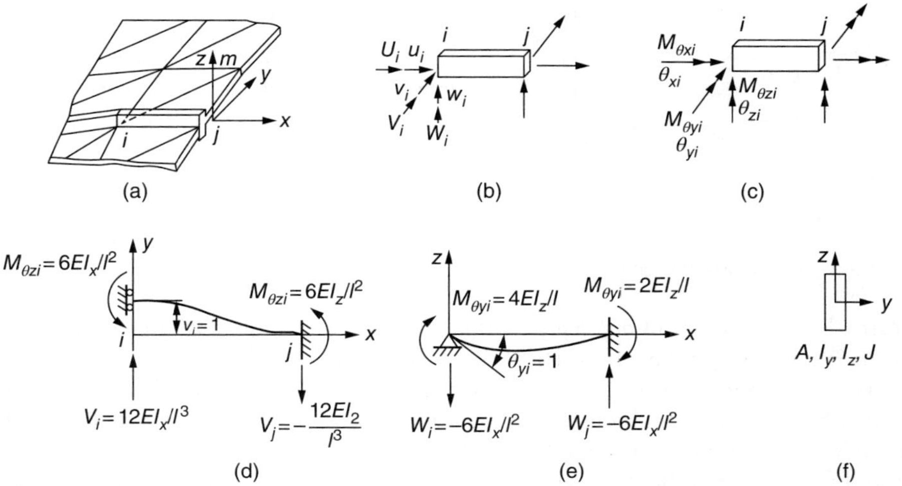

Figure 11.8 Thin shell supported by curved beam.

system as formula (11.44):

$$
[k]^{\mathrm{e}} = \frac{EI_y}{l^3}
\begin{bmatrix}
al^2 & 0 & 0 & 0 & 0 & 0 & -al^2 & 0 & 0 & 0 & 0 & 0 \\
0 & 12b & 0 & 0 & 0 & 6bl & 0 & -12l & 0 & 0 & 0 & 6bl^2 \\
0 & 0 & 12 & 0 & -6l & 0 & 0 & 0 & -12l & 0 & -6l & 0 \\
0 & 0 & 0 & cl^2 & 0 & 0 & 0 & 0 & 0 & -d^2 & 0 & 0 \\
0 & 0 & -6bl & 0 & 4l^2 & 0 & 0 & 0 & 6l & 0 & 2l^2 & 0 \\
0 & 6bl & 0 & 0 & 0 & 4bl^2 & 0 & -6bl & 0 & 0 & 0 & 2bl^2 \\
-al^2 & 0 & 0 & 0 & 0 & 0 & al^2 & 0 & 0 & 0 & 0 & 0 \\
0 & -12b & 0 & 0 & 0 & -6bl & 0 & 12b & 0 & 0 & 0 & -6bl \\
0 & 0 & -12l & 0 & 6l & 0 & 0 & 0 & 12 & 0 & 0 & 0 \\
0 & 0 & 0 & -d^2 & 0 & 0 & 0 & 0 & 0 & cl^2 & 0 & 0 \\
0 & 0 & -6l & 0 & 2l^2 & 0 & 0 & 0 & 6l & 0 & 4l^2 & 0 \\
0 & 6bl & 0 & 0 & 0 & 2bl^2 & 0 & -bl & 0 & 0 & 0 & 4bl^2
\end{bmatrix}
\tag{11.44}
$$

where

$$a = \frac{A}{I_y}, \quad b = \frac{I_x}{I_y}, \quad c = \frac{GJ}{EI_y}$$

GJ is the torsional rigidity of the beam.

The local coordinate system is inconsistent with the global coordinate system. We may compute the element stiffness matrix $[k']^{\mathrm{e}}$ in the global coordinate system by formula $[k']^{\mathrm{e}} = [L]^{\mathrm{T}}[k]^{\mathrm{e}}[L]$.

In the abovementioned computation, the neutral axis of the beam element is located on the middle surface of the shell as shown in figure 11.8. In the actual engineering

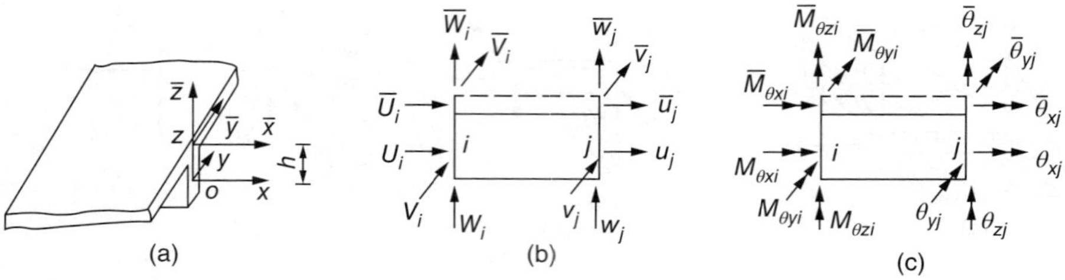

Figure 11.9 Shell supported by beam.

structure, the neutral axis of beam element is often not on the shell middle surface, as shown in Figure 11.9(a). Let the vertical distance between the beam element neutral axis and the shell middle surface be h, the local coordinate system of the beam element is xyz, of which x-axis is the neutral axis of the beam. To facilitate the establishment of the node equilibrium equation, take an additional coordinate system $\bar{x}\bar{y}\bar{z}$, in which $\bar{x}$-axis and $\bar{y}$-axis are parallel to x- and y-axes, respectively, and $\bar{x}$-axis is located in the middle surface of the shell; the $\bar{z}$-axis coincides with z-axis. If the assumption of the plane section is still applicable for the beam element, the element stiffness matrix $[k]$ in the local coordinate system may still be computed by formula (11.44).

The nodal forces and nodal displacements of both local coordinate systems are shown in Figure 11.9(b) and (c). According to the plane assumption, the nodal displacements $\{\delta\}^{\mathrm{e}}$ and $\{\bar{\delta}\}^{\mathrm{e}}$ of the two local coordinate systems conform to the following relationship:

$$\{\delta\}^{\mathrm{e}} = [C]\{\bar{\delta}\}^{\mathrm{e}} \tag{11.45}$$

$$[C] = \begin{bmatrix} \beta & 0 \\ 0 & \beta \end{bmatrix}$$

$$[\beta] = \begin{bmatrix} 1 & 0 & 0 & 0 & -h & 0 \\ 0 & 1 & 0 & h & 0 & 0 \\ 0 & 0 & 1 & 0 & 0 & 0 \\ 0 & 0 & 0 & 1 & 0 & 0 \\ 0 & 0 & 0 & 0 & 1 & 0 \\ 0 & 0 & 0 & 0 & 0 & 1 \end{bmatrix}$$

According to the conditions of equilibrium of forces, the nodal forces $\{F\}^{\mathrm{e}}$ and $\{\bar{F}\}^{\mathrm{e}}$ of the two local coordinate systems are given by the following formula:

$$\{\bar{F}\}^{\mathrm{e}} = [C]^{\mathrm{T}}\{F\}^{\mathrm{e}}$$

Substituting formula (11.43) and formula (11.45) into the above formula, then

$$\{\bar{F}\}^{\mathrm{e}} = [C]^{\mathrm{T}}\{F\}^{\mathrm{e}} = [C]^{\mathrm{T}}[k]\{\delta\}^{\mathrm{e}} = [C]^{\mathrm{T}}[k][C]\{\bar{\delta}\}^{\mathrm{e}}$$

$$\{\bar{F}\}^{\mathrm{e}} = [\bar{k}]\{\bar{\delta}\}^{\mathrm{e}} \tag{11.46}$$

$$[\bar{k}] = [C]^{\mathrm{T}}[k][C] \tag{11.47}$$

Computing the element stiffness matrix $[\overline{k}]$ of the local coordinate system $\overline{x}\overline{y}\overline{z}$ with formula (11.47) and the element stiffness matrix $[k']$ in the global coordinate system by $[k'] = [L]^T[\overline{k}][L]$, then the nodal equilibrium equation of the shell may be established.

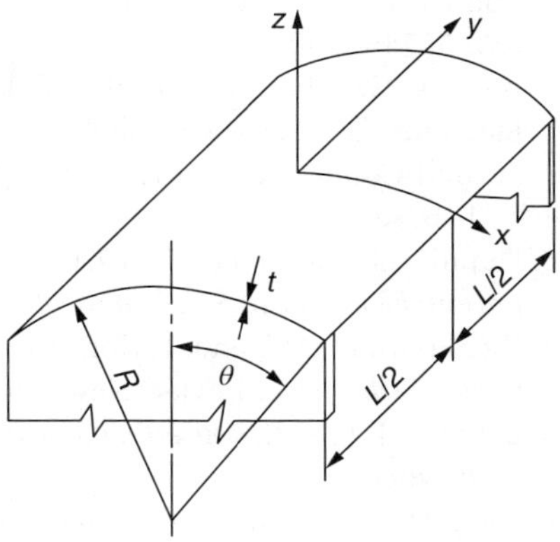

Figure 11.10 Cylindrical shell roof ($L = 15.2$ m, $R = 7.6$ m, $t = 7.6$ cm, deadload $= 4400$ N/m^2).

11.6 Example

For a cylindrical shell roof with dimension shown in Figure 11.10, the stress under dead load is computed with the rectangle elements. Due to the symmetry, only a proportion of 1/4 is required to be taken out for computation. Take $\mu = 0$ in calculation to facilitate comparison with the theoretical solution. The computation results are shown in Figure 11.11.

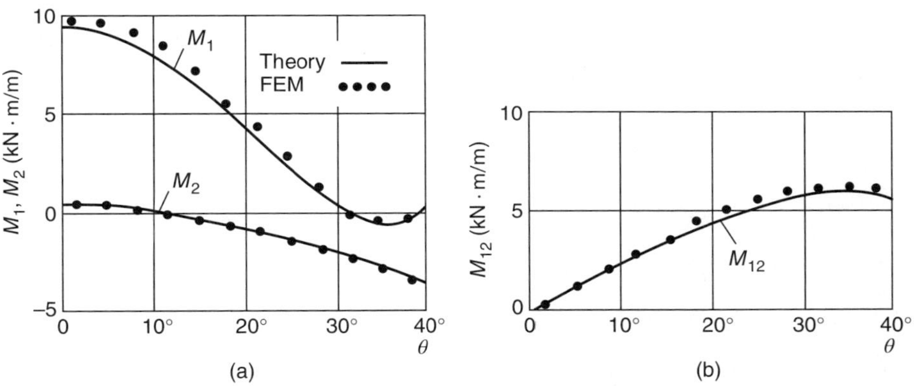

Figure 11.11 Computation results of cylindrical shell roof. (a) Transverse bending moment M_1 and longitudinal bending moment M_2 of the central section. (b) Twisting moment M_{12} of the supporting edge.

Bibliography

1 Flugge, W. (1960) *Stresses in Shells*, Springer-Verlag, New York.
2 Argyris, J.H. (1965) Matrix displacement analysis of anisotropic shells by triangular elements. *Journal of the Royal Aeronautical Society*, **69**, 801–805.
3 Clough, R.W. and Johnson, C.P. (1968) A finite element approximation for the analysis of thin shells. *Journal of Solids and Structures*, **4**, 43–60.

4 Bogner, F.K., Fox, R.L. and Schmi, L.A. (1967) A cylindrical shell element. *AIAA Journal*, **5**, 745–750.

5 Bonnes, G., Dhatt, G., Giroux, Y.M. and Robichand, L.P.A.. (1968) Curved Triangular Elements for Analysis of Shells. Air Force Institute of Technology, Proceedings of the Second Conference Matrix Methods in Structural Mechanics. Ohio: Wright-Patterson A. F. Base.

6 Dupuis, G. and Goel, J.J. (1970) A curved finite element for thin elastic shells. *International Journal of Solids and Structures*, **6**, 987–996.

7 Zienkiewicz, O.C. and Taylor, R.L. (2009) *Finite Element Method for Solid and Structural Mechanics*, Elsevier, New York, London.

8 Bathe, K.J. (1982) *Finite Element Procedures in Engineering Analysis*, Prentice-Hall, Englewood Cliffs.

9 Yang, H.T.Y., Saigal, S. and Liaw, D.G. (1990) Advances of thin shell finite elements and some applications version I. *Computers and Structures*, **35** (4), 481–504.

12

Axisymmetric Shell

Axisymmetrical shells are extensively used in actual projects. The analysis is greatly simplified due to axial symmetry. The element will be unidimensional if both the geometrical shape and load of the shell are axisymmetrical.

Grafton and Strome made the first study of finite element method [1] for the axisymmetrical shell by using a linear element. Zienkiewicz and other researchers put forward the method for analyzing axisymmetrical shell [2–4].

12.1 Linear Element

As shown in Figure 12.1, the axisymmetrical shell is divided along the direction of meridian line into several linear elements, each of which is a truncated cone shell. The angle Φ is constant in each element. On the basis of Kirchhoff assumption, under the action of axisymmetrical load, four strain components are obtained by the following formula [5]:

$$\{\varepsilon\} = \begin{Bmatrix} \varepsilon_s \\ \varepsilon_\theta \\ x_s \\ x_\theta \end{Bmatrix} = \begin{Bmatrix} du/ds \\ (w\cos\varphi + u\sin\phi)/r \\ -d^2w/ds^2 \\ -(\sin\phi/r)dw/ds \end{Bmatrix} \tag{12.1}$$

Four internal forces are obtained as follows:

$$\{\sigma\} = \begin{Bmatrix} N_S \\ N_\theta \\ M_s \\ M_\theta \end{Bmatrix} = [D]\{\varepsilon\} \tag{12.2}$$

The elasticity matrix $[D]$ for the isotropic shell is as follows:

$$[D] = \frac{Et}{1-\mu^2}\begin{bmatrix} 1 & \mu & 0 & 0 \\ \mu & 1 & 0 & 0 \\ 0 & 0 & t^2/12 & \mu t^2/12 \\ 0 & 0 & \mu t^2/12 & t^2/12 \end{bmatrix} \tag{12.3}$$

in which t stands for the shell thickness.

As shown in Figure 12.2, the linear element has 2 nodes, either of which has the nodal displacement parameter under the action of axisymmetrical load as follows:

$$\{\delta_i\} = \begin{bmatrix} \overline{u}_i & \overline{w}_i & \beta_i \end{bmatrix}^T \tag{12.4}$$

The Finite Element Method: Fundamentals and Applications in Civil, Hydraulic, Mechanical and Aeronautical Engineering, First Edition. Bofang Zhu.

Companion website: www.wiley.com/go/zhufem17

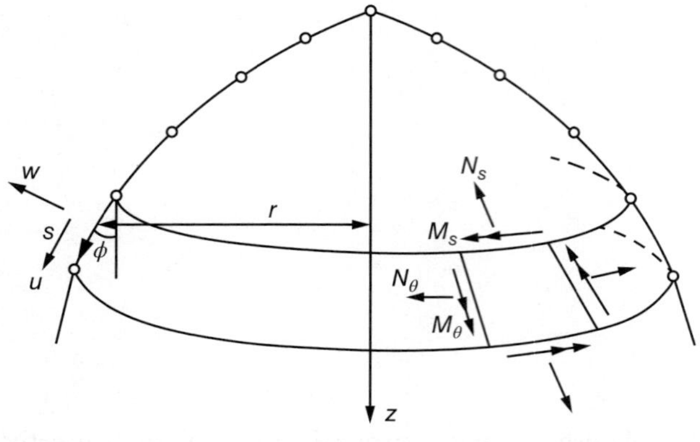

Figure 12.1 Internal force and displacement of the axisymmetrical shell under the action of axisymmetrical load.

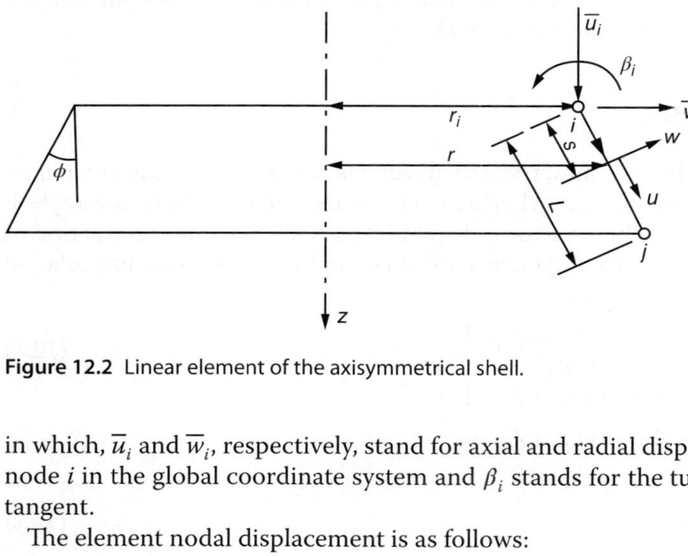

Figure 12.2 Linear element of the axisymmetrical shell.

in which, $\overline{u}_i$ and $\overline{w}_i$, respectively, stand for axial and radial displacement components of node i in the global coordinate system and β_i stands for the turn angle of a meridional tangent.

The element nodal displacement is as follows:

$$\{\delta\}^e = \left[\overline{u}_i \ \overline{w}_i \ \beta_i \ \overline{u}_j \ \overline{w}_j \ \beta_j\right]^T \tag{12.5}$$

In the local coordinate system, radial displacement u and normal displacement w of any point on the middle plane of the shell can be expressed by s as follows:

$$u = a_1 + a_2 s, \quad w = a_3 + a_4 s + a_5 s^2 + a_6 s^3 \tag{a}$$

u and w shall meet the following boundary conditions:

$$\text{When} \quad s = 0: \quad u = u_i, \quad w = w_i, \quad \frac{dw}{ds} = \left(\frac{dw}{ds}\right)_i$$

$$\text{When} \quad s = L: \quad u = u_j, \quad w = w_j, \quad \frac{dw}{ds} = \left(\frac{dw}{ds}\right)_j$$

After the coefficients $\alpha_1, \alpha_2, \ldots, \alpha_s$ are calculated from the aforementioned boundary conditions, they are substituted in formula (a), with such a formula obtained as follows:

$$\begin{Bmatrix} u \\ w \end{Bmatrix} = [N_i' \ \ N_j'] \left[u_i \ w_i \ \left(\frac{\mathrm{d}w}{\mathrm{d}s}\right)_i \ u_j \ w_j \left(\frac{\mathrm{d}w}{\mathrm{d}s}\right)_j \right]^{\mathrm{T}} \tag{12.6}$$

in which

$$\begin{aligned}
[N_i'] &= \begin{bmatrix} 1-\xi & 0 & 0 \\ 0 & 1-3\xi^2+2\xi^3 & L(\xi-2\xi^2+\xi^3) \end{bmatrix} \\
[N_j'] &= \begin{bmatrix} \xi & 0 & 0 \\ 0 & 3\xi^2-2\xi^3 & L(-\xi^2+\xi^3) \end{bmatrix} \\
\xi &= s/l
\end{aligned} \tag{12.7}$$

where L is the length of the element meridian line.

From Figure 12.2 we can know the following relation for node i:

$$\begin{Bmatrix} u_i \\ w_i \\ (\mathrm{d}w/\mathrm{d}s)_i \end{Bmatrix} = \begin{bmatrix} \cos\theta & \sin\theta & 0 \\ -\sin\theta & \cos\theta & 0 \\ 0 & 0 & 1 \end{bmatrix} \begin{Bmatrix} \overline{u}_i \\ \overline{w}_i \\ \beta_i \end{Bmatrix} = [\lambda]\{\delta_i\} \tag{12.8}$$

By substituting formula (12.8) into formula (12.6), we can obtain formula (12.9) in the following:

$$\begin{Bmatrix} u \\ w \end{Bmatrix} = \left[N_i'\lambda \ \ N_j'\lambda\right] \begin{Bmatrix} \delta_i \\ \delta_j \end{Bmatrix} = [N]\{\delta\}^{\mathrm{e}} \tag{12.9}$$

By substituting formula (12.9) into formula (12.1), we can obtain formula (12.10) as follows:

$$\{\varepsilon\} = [B]\{\delta\}^{\mathrm{e}} = \left[B_i'\lambda \ \ B_j'\lambda\right]\{\delta\}^{\mathrm{e}} \tag{12.10}$$

$$\begin{aligned}
[B_i'] &= \begin{bmatrix} -1/L & 0 & 0 \\ (1-\xi)\sin\phi/r & (1-3\xi^2+2\xi^3)\cos\phi/r & L(\xi-2\xi^2+\xi^3)\cos\phi/r \\ 0 & 6(1-2\xi)/L^2 & (4-6\xi)/L \\ 0 & 6\xi(1-\xi)\sin\phi/rL & (-1+4\xi-3\xi^2)\sin\phi/r \end{bmatrix} \\
[B_j'] &= \begin{bmatrix} 1/L & 0 & 0 \\ \xi\sin\phi/r & (3\xi^2-2\xi^3)\cos\phi/r & L(-\xi^2+\xi^3)\cos\phi/r \\ 0 & -6(1-2\xi)/L^2 & 2(1-3\xi)/L \\ 0 & -6\xi(1-\xi)/rL & (2\xi-3\xi^2)\sin\phi/r \end{bmatrix}
\end{aligned} \tag{12.11}$$

Now it is not difficult to obtain an element stiffness matrix and a load array by using the fundamental formulas in Chapter 3. From Figure 12.2 we can know that

$$\mathrm{d}A = 2\pi r\mathrm{d}s = 2\pi rL\mathrm{d}\xi$$

in which ξ varies from 0 to 1.

The element stiffness matrix is calculated by the formula (12.12) as follows:

$$[K^{\mathrm{e}}] = \int_0^1 2\pi rL[B]^{\mathrm{T}}[D][B]\mathrm{d}\xi \tag{12.12}$$

Formula (12.12) can be expressed in a partitioned form:

$$[K^e] = \begin{bmatrix} K_{ii} & K_{ij} \\ K_{ji} & K_{jj} \end{bmatrix} \tag{12.13}$$

in which

$$[K_{ij}] = 2\pi L[\lambda]^T \left(\int_0^1 [B_i^t][D][B_j^t] r d\xi \right) [\lambda] \tag{12.14}$$

Suppose lateral distributed load $\{p\} = \left[p_u \ p_w\right]^T$, in which p_u and p_w, respectively, stand for a radial load component and a normal load component, thereupon the nodal load vector is as follows:

$$\{P_i\} = 2\pi L[\lambda]^T \int_0^1 [N_i']^T \{p\} r d\xi \tag{12.15}$$

$$\{P_i\} = \left[P_{\bar{u}_i},\ P_{\bar{w}_i},\ P_{\beta_i}\right]$$

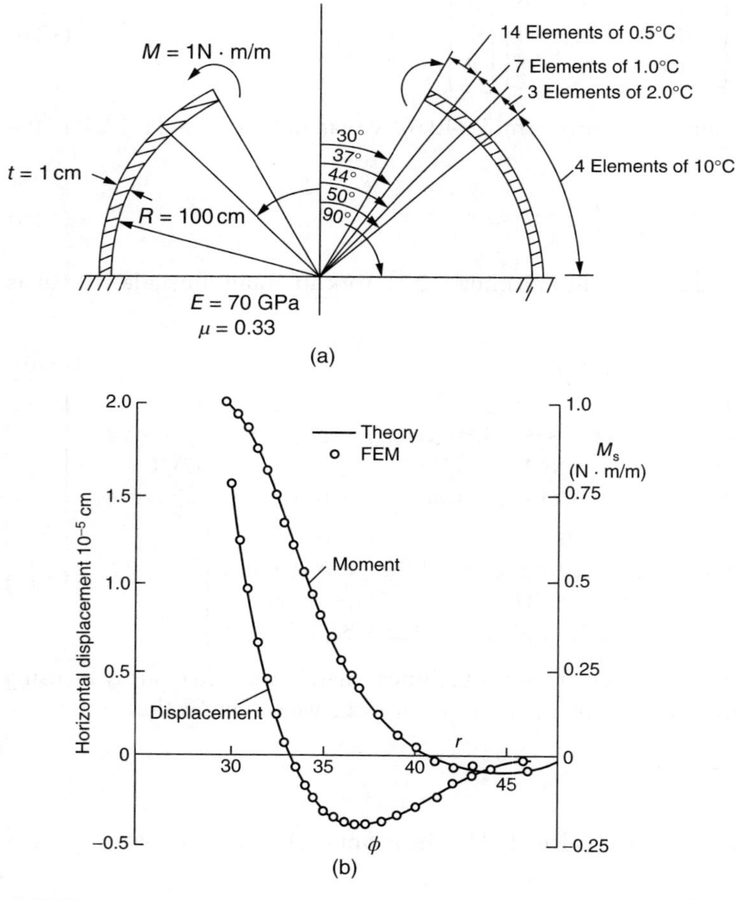

Figure 12.3 Linear element example (hemispheric shell).

In formula (12.15), $P_{\overline{u}_i}$, $P_{\overline{w}_i}$, and P_{β_i}, respectively, stand for the forces along the direction of z and r and the moment along the direction of β_i.

In formulas (12.12)–(12.15), r can be expressed as

$$r = r_i + \xi L \sin\phi$$

Formulas (12.14) and (12.15) can be calculated by Gaussian quadrature method.

Figure 12.3 shows an example [1]. A hemispheric shell with a center bore whose edge is subjected to the action of a bending moment is divided into 28 elements at uneven intervals. The computed result accords with the analytical solution.

12.2 Curved Element

A curved element can be used in order to improve computational accuracy of the element. Figure 12.4 shows a three-node curved element. Coordinates r and z on middle surface of the shell are expressed by shape functions as follows:

$$r = \sum_{i=1}^{3} N_i r_i, \quad z = \sum_{i=1}^{3} N_i z_i \tag{12.16}$$

in which

$$N_1 = -\xi(1-\xi)/2, \quad N_2 = \xi(1+\xi)/2, \quad N_3 = 1-\xi^2$$

Suppose the turn angle and the linear displacement of any point are independent, displacements $\overline{u}, \overline{w}$ in the global coordinate system (r, z) and turn angle β are expressed by shape functions as follows:

$$\overline{u} = \sum_{i=1}^{3} N_i \overline{u}_i, \quad \overline{w} = \sum_{i=1}^{3} N_i \overline{w}_i, \quad \beta = \sum_{i=1}^{3} N_i \beta_i \tag{12.17}$$

The nodal displacement of node i is as follows:

$$\{\delta_i\} = \begin{bmatrix} \overline{u}_i & \overline{w}_i & \beta_i \end{bmatrix}^{\mathrm{T}} \tag{12.18}$$

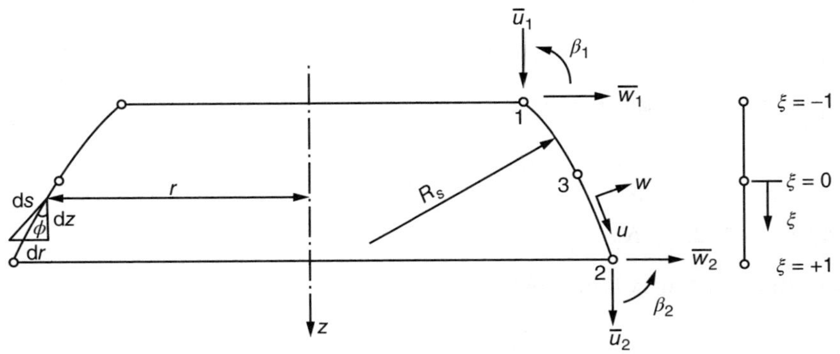

Figure 12.4 Curved element.

Let

$$J = \frac{ds}{d\xi} = \sqrt{\left(\frac{dr}{d\xi}\right)^2 + \left(\frac{dz}{d\xi}\right)^2} = \sqrt{\left(\sum \frac{dN_i}{d\xi} r_i\right)^2 + \left(\sum \frac{dN_i}{d\xi} z_i\right)^2} \tag{12.19}$$

From Figure 12.4 we can know that

$$\cos\phi = \frac{dz}{ds} = \frac{dz/d\xi}{ds/d\xi} = \frac{1}{J}\sum_{i=1}^{3} \frac{dN_i}{d\xi} z_i$$

$$\sin\phi = \frac{dr}{ds} = \frac{dr/d\xi}{ds/d\xi} = \frac{1}{J}\sum_{i=1}^{3} \frac{dN_i}{d\xi} r_i$$

The radial curvature $1/R_s$ of a curved element is not zero. Under the action of axisymmetrical load, the relation of stress–displacement is as follows:

$$\{\varepsilon\} = \begin{Bmatrix} \varepsilon_s \\ \varepsilon_\theta \\ x_s \\ x_\theta \\ \gamma \end{Bmatrix} = \begin{Bmatrix} du/ds + w/R_s \\ (u\sin\phi + w\cos\phi)/r \\ -d\beta/ds \\ -(\sin\phi/r)\beta \\ dw/ds - u/R_s - \beta \end{Bmatrix} \tag{12.20}$$

In formula (12.20), γ stands for transverse shear strain; u, w, and β stand for displacement in a local coordinate system.

The relation between u, w, and β and $\overline{u}_i$, $\overline{w}$, β in the global coordinate system is as follows:

$$\begin{Bmatrix} u \\ w \\ \beta \end{Bmatrix} = \begin{bmatrix} \cos\phi & \sin\phi & 0 \\ -\sin\phi & \cos\phi & 0 \\ 0 & 0 & 1 \end{bmatrix} \begin{Bmatrix} \overline{u} \\ \overline{w} \\ \beta \end{Bmatrix} \tag{12.21}$$

By substituting formulas (12.17) and (12.21) into formula (12.20), we obtain the next formula (12.22):

$$\{\varepsilon\} = [B_1 \;\; B_2 \;\; B_3]\,[\delta_1 \;\; \delta_2 \;\; \delta_3]^T \tag{12.22}$$

$$[B_i] = \begin{bmatrix} -\sin\phi N_i/R_s + (\cos\phi/J)dN_i/d\xi & \cos\phi N_i/R_s + (\sin\phi/J)dN_i/d\xi & 0 \\ 0 & N_i/r & 0 \\ 0 & 0 & -(dN_i/d\xi)/J \\ 0 & 0 & -\sin\phi N_i/r \\ -\cos\phi N_i/R_s - (\sin\phi/J)dN_i/d\xi & -\sin\phi N_i/R_s + (\cos\phi/J)dN_i/d\xi & -N_i \end{bmatrix} \tag{12.23}$$

in which

$$dN_i/d\xi = \xi - 1/2, \quad dN_2/d\xi = \xi + 1/2, \quad dN_3/d\xi = -2\xi$$

The element stiffness matrix is given by

$$[K^e] = 2\pi \int_{-1}^{1} [B]^T[D][B] rJ d\varepsilon = \begin{bmatrix} k_{11} & k_{12} & k_{13} \\ k_{21} & k_{22} & k_{23} \\ k_{31} & k_{32} & k_{33} \end{bmatrix} \tag{12.24}$$

where

$$[K_{ij}] = 2\pi \int_{-1}^{1} [B_i]^{\mathrm{T}}[D][B_j] rJ \mathrm{d}\xi \tag{12.25}$$

In formula (12.24), $[D]$ stands for an elasticity matrix.

Experience shows that a satisfactory result can be obtained by using two-point Gaussian quadrature.

For an isotropic shell, $[D]$ can be expressed by the following equation:

$$[D] = \frac{Et}{1-\mu^2}\begin{bmatrix} 1 & \mu & 0 & 0 & 0 \\ \mu & 1 & 0 & 0 & 0 \\ 0 & 0 & t^2/12 & \mu t^2/12 & 0 \\ 0 & 0 & \mu t^2/12 & t^2/12 & 0 \\ 0 & 0 & 0 & 0 & k(1-\mu)/2 \end{bmatrix} \tag{12.26}$$

in which $k = 5/6$.

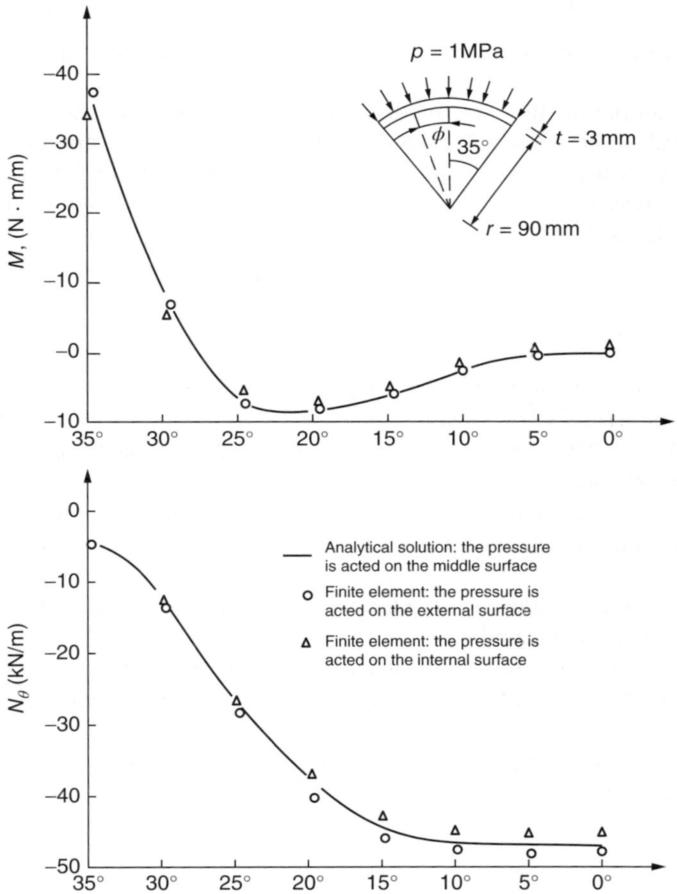

Figure 12.5 Curved element example (a spherical head cover).

The nodal load vector of the element can be calculated by the following formula:

$$\{P_i^e\} = 2\pi \int_{-1}^{1} N_i\{p\}rJ\mathrm{d}\xi \tag{12.27}$$

in which

$$\{p\} = \begin{bmatrix} p_{\overline{u}} & p_{\overline{w}} & \beta \end{bmatrix}^{\mathrm{T}}$$

In case the load is normal uniformly distributed p, the following formulas apply:

$$p_{\overline{u}} = -p \sin\phi, \quad p_{\overline{w}} = p \cos\phi, \quad p_\beta = 0$$

Formula (12.27) is calculated by supposing that the load is acted on middle surface of the shell. In case of consideration of influence of actual load on internal or external surfaces, the following formula (12.28) applies:

$$\{P_i^e\} = 2\pi \int_{-0}^{1} N_i\{p\}r^*J^*\mathrm{d}\xi$$
$$r^* = r \pm (t\cos\phi)/2, J^* = J(1 \pm t/2R_s) \tag{12.28}$$

In (12.28), "+" and "−," respectively, are used for loads acted on the external surface and the internal surface.

Figure 12.5 shows an example [5] that ten quadratic curved elements are used for calculating a spherical head cover, $t/R = 3.3\%$, the theoretical solution when pressure acting on middle surface is shown as the solid line; the results calculated by finite element method have a maximum stress difference of 7% when an identical pressure is acted on the internal or external surfaces.

Bibliography

1 Grafton, P.E. and Strome, D.R. (1963) Analysis of axi-symmetric shells by the direct stiffness method. *Journal of AIAA*, **1**, 2342–2347.
2 Jones, R.E. and Strome, D.R. (1966) Direct stiffness method of analysis of shells of revolution utilising curved elements. *Journal of AIAA*, **4**, 1519–1525.
3 Stricklin, J.A., Navaratna, D.R. and Pian, T.H.H. (1966) Improvements in the analysis of shells of revolution by matrix displacement method (curved elements). *Journal of AIAA*, **4**, 2069–2072.
4 Zienkiewicz, O.C. and Taylor, R.L. (2009) *Finite Element Method for Solid and Structural Mechanics*, Elsevier, New York, London.
5 Wang Y. (1983) Analysis of Temperature and Stress for a Rotational Symmetrical Structure, a Master's Thesis from Department of Engineering Mechanics, Tsinghua University.
6 Timoshenko, S. and Woinowsky-Krieger, S. (1959) *Theory of Plates and Shells*, 2nd edn, New York, McGraw Hill.

13

Problems in Fluid Mechanics

The problems in fluid mechanics are often encountered in engineering practice. Generally speaking, there are two kinds of problems of fluid mechanics: the first kind is the seepage flow through porous media, and the second kind is the flow of air or water. Due to the irregular geometries of engineering structures and the complicated boundary conditions, these problems can hardly be solved through classical analytical methods, but the finite element method can lead to good solutions. This chapter will explain how to solve the problems in fluid mechanics with finite element method.

13.1 Relation between Stress and Strain for Newtonian Fluids

13.1.1 Stress–Strain Relations for Solids

According to Hooke's law, the strains of elastic solid are given by

$$\left.\begin{aligned} \varepsilon_x &= \tfrac{1}{E}[\sigma_x - \mu(\sigma_y + \sigma_z)] \\ \gamma_{xy} &= \tfrac{\tau_{xy}}{G} \end{aligned}\right\} \tag{13.1}$$

Let

$$e = \varepsilon_x + \varepsilon_y + \varepsilon_z = \frac{\partial u}{\partial x} + \frac{\partial v}{\partial y} + \frac{\partial w}{\partial z} \tag{13.2}$$

$$\overline{\sigma} = \frac{\sigma_x + \sigma_y + \sigma_z}{3} \tag{13.3}$$

Substituting Eq. (13.1) into Eq. (13.2), we get

$$e = \frac{3\overline{\sigma}(1-2\mu)}{E} \quad \text{or} \quad \overline{\sigma} = \frac{Ee}{3(1-2\mu)} \tag{13.4}$$

Thus

$$\begin{aligned} \varepsilon_x &= \frac{1}{E}[\sigma_x - \mu(\sigma_x + \sigma_y + \sigma_z) + \mu\sigma_x] \\ &= \frac{1+\mu}{E}\sigma_x - \frac{3\mu}{E}\overline{\sigma} \end{aligned} \tag{13.5}$$

Substitution of Eq. (13.4) into Eq. (13.5) yields

$$\sigma_x = \frac{E}{1+\mu}\varepsilon_x + \frac{3\mu}{1+\mu}\overline{\sigma} = \frac{E}{1+\mu}\varepsilon_x + \frac{\mu Ee}{(1+\mu)(1-2\mu)} \tag{13.6}$$

The Finite Element Method: Fundamentals and Applications in Civil, Hydraulic, Mechanical and Aeronautical Engineering, First Edition. Bofang Zhu.

Companion website: www.wiley.com/go/zhufem17

As $E/(1+\mu) = 2G$, the above equation becomes

$$\sigma_x = 2G\varepsilon_x + \frac{2G\mu}{1-2\mu}e \tag{13.7}$$

Subtracting $\overline{\sigma}$ from both sides of the above equation, we have

$$\sigma_x - \overline{\sigma} = 2G\varepsilon_x + \frac{2G\mu}{1-2\mu}e - \overline{\sigma} \tag{13.8}$$

Substituting Eq. (13.4) into Eq. (13.8), we get

$$\sigma_x - \overline{\sigma} = 2G\left(\varepsilon_x - \frac{e}{3}\right) \tag{13.9a}$$

Similarly, we have

$$\sigma_y - \overline{\sigma} = 2G\left(\varepsilon_y - \frac{e}{3}\right) \tag{13.9b}$$

$$\sigma_z - \overline{\sigma} = 2G\left(\varepsilon_z - \frac{e}{3}\right) \tag{13.9c}$$

$$\tau_{xy} = G\gamma_{xy} \tag{13.9d}$$

$$\tau_{yz} = G\gamma_{yz} \tag{13.9e}$$

$$\tau_{zx} = G\gamma_{zx} \tag{13.9f}$$

13.1.2 Stress–Rate and Strain Relations for Fluid

Experimental results indicate that the stresses in a fluid are proportional to the rate of strain instead of strain itself. The relation between stress and rate of strain of fluid is given by

$$\sigma_x - \overline{\sigma} = 2\eta\left(\frac{\partial \varepsilon_x}{\partial t} - \frac{1}{3}\frac{\partial e}{\partial t}\right)$$

$$\frac{\partial \varepsilon_x}{\partial t} = \frac{\partial}{\partial t}\left(\frac{\partial u}{\partial x}\right) = \frac{\partial}{\partial x}\left(\frac{\partial u}{\partial t}\right) = \frac{\partial v_x}{\partial x} \tag{a}$$

$$\frac{\partial e}{\partial t} = \frac{\partial}{\partial t}(\varepsilon_x + \varepsilon_y + \varepsilon_z) = \frac{\partial v_x}{\partial x} + \frac{\partial v_y}{\partial y} + \frac{\partial v_z}{\partial z} \tag{b}$$

where v_x, v_y, v_z are the velocities in x, y, z directions,

$$v_x = \frac{\partial u}{\partial t}, \quad v_y = \frac{\partial v}{\partial t}, \quad v_z = \frac{\partial w}{\partial t}$$

Taking $\overline{\sigma} = -p$, where p is the mean fluid pressure, thus the stress rate and strain relation for fluid are given by

$$\sigma_x = -p + 2\eta\frac{\partial v_x}{\partial x} - \frac{2}{3}\eta\left(\frac{\partial v_x}{\partial x} + \frac{\partial v_y}{\partial y} + \frac{\partial v_z}{\partial z}\right) \tag{13.10a}$$

$$\sigma_y = -p + 2\eta\frac{\partial v_y}{\partial y} - \frac{2}{3}\eta\left(\frac{\partial v_x}{\partial x} + \frac{\partial v_y}{\partial y} + \frac{\partial v_z}{\partial z}\right) \tag{13.10b}$$

$$\sigma_z = -p + 2\eta\frac{\partial v_z}{\partial z} - \frac{2}{3}\eta\left(\frac{\partial v_x}{\partial x} + \frac{\partial v_y}{\partial y} + \frac{\partial v_z}{\partial z}\right) \tag{13.10c}$$

$$\tau_{xy} = \eta\left(\frac{\partial v_y}{\partial x} + \frac{\partial v_x}{\partial y}\right) \tag{13.10d}$$

$$\tau_{yz} = \eta \left(\frac{\partial v_z}{\partial y} + \frac{\partial v_y}{\partial z} \right) \tag{13.10e}$$

$$\tau_{zx} = \eta \left(\frac{\partial v_x}{\partial z} + \frac{\partial v_z}{\partial x} \right) \tag{13.10f}$$

where η is the viscosity of fluid.

13.2 Equation of Motion

By applying Newton's second law to a differential volume $dxdydz$ of fluid, we get the following equation:

$$\left. \begin{aligned} \rho X + \frac{\partial \sigma_x}{\partial x} + \frac{\partial \tau_{yx}}{\partial y} + \frac{\partial \tau_{zx}}{\partial z} &= \rho a_x \\ \rho Y + \frac{\partial \tau_{xy}}{\partial x} + \frac{\partial \sigma_y}{\partial y} + \frac{\partial \tau_{zy}}{\partial z} &= \rho a_y \\ \rho Z + \frac{\partial \tau_{xz}}{\partial x} + \frac{\partial \tau_{yz}}{\partial y} + \frac{\partial \sigma_z}{\partial z} &= \rho a_z \end{aligned} \right\} \tag{13.11}$$

where X, Y, Z are the body forces in x, y, z directions and a_x, a_y, a_z are the accelerations in x, y, z directions. Equations (13.11) are general and applicable to any fluid with gravitational-type body forces. For Newtonian fluid, substitution of Eq. (13.10) into Eq. (13.11) yields the equations of motion in x, y, z directions as

$$\begin{aligned} \rho a_x = \rho X - \frac{\partial p}{\partial x} + \frac{\partial}{\partial x} \left[2\eta \frac{\partial v_x}{\partial x} - \frac{2}{3}\eta \left(\frac{\partial v_x}{\partial x} + \frac{\partial v_y}{\partial y} + \frac{\partial v_z}{\partial z} \right) \right] \\ + \frac{\partial}{\partial y} \left[\eta \left(\frac{\partial v_y}{\partial x} + \frac{\partial v_x}{\partial y} \right) \right] + \frac{\partial}{\partial z} \left[\eta \left(\frac{\partial v_x}{\partial z} + \frac{\partial v_z}{\partial x} \right) \right] \end{aligned} \tag{13.12a}$$

$$\begin{aligned} \rho a_y = \rho Y - \frac{\partial p}{\partial y} + \frac{\partial}{\partial x} \left[\eta \left(\frac{\partial v_x}{\partial y} + \frac{\partial v_y}{\partial x} \right) \right] \\ + \frac{\partial}{\partial y} \left[2\eta \frac{\partial v_y}{\partial y} - \frac{2}{3}\eta \left(\frac{\partial v_x}{\partial x} + \frac{\partial v_y}{\partial y} + \frac{\partial v_z}{\partial z} \right) \right] + \frac{\partial}{\partial z} \left[\eta \left(\frac{\partial v_y}{\partial z} + \frac{\partial v_z}{\partial y} \right) \right] \end{aligned} \tag{13.12b}$$

$$\begin{aligned} \rho a_z = \rho Z - \frac{\partial p}{\partial x} + \frac{\partial}{\partial x} \left[\eta \left(\frac{\partial v_z}{\partial x} + \frac{\partial v_x}{\partial z} \right) \right] + \frac{\partial}{\partial y} \left[\eta \left(\frac{\partial v_y}{\partial z} + \frac{\partial v_z}{\partial y} \right) \right] \\ + \frac{\partial}{\partial z} \left[2\eta \frac{\partial v_z}{\partial z} - \frac{2}{3}\eta \left(\frac{\partial v_x}{\partial x} + \frac{\partial v_y}{\partial y} + \frac{\partial v_z}{\partial z} \right) \right] \end{aligned} \tag{13.12c}$$

Equations (13.12) are called the Navier–Stokes equations for compressible fluids. For incompressible fluid, $\partial v_x/\partial x + \partial v_y/\partial y + \partial v_z/\partial z = 0$, the Navier–Stokes equations of motion become

$$\left. \begin{aligned} a_x &= X - \frac{1}{\rho}\frac{\partial p}{\partial x} + \frac{\eta}{\rho} \left(\frac{\partial^2 v_x}{\partial x^2} + \frac{\partial^2 v_x}{\partial y^2} + \frac{\partial^2 v_x}{\partial z^2} \right) \\ a_y &= Y - \frac{1}{\rho}\frac{\partial p}{\partial x} + \frac{\eta}{\rho} \left(\frac{\partial^2 v_y}{\partial x^2} + \frac{\partial^2 v_y}{\partial y^2} + \frac{\partial^2 v_y}{\partial z^2} \right) \\ a_z &= Z - \frac{1}{\rho}\frac{\partial p}{\partial z} + \frac{\eta}{\rho} \left(\frac{\partial^2 v_z}{\partial x^2} + \frac{\partial^2 v_z}{\partial y^2} + \frac{\partial^2 v_z}{\partial z^2} \right) \end{aligned} \right\} \tag{13.13}$$

13.3 Continuity Equation

Considering a differential volume of size dxdydz, the flux of mass per second for the three directions x, y, z, respectively, is

$$-\frac{\partial}{\partial x}(\rho v_x)\cdot \mathrm{d}y\mathrm{d}z, \quad -\frac{\partial}{\partial y}(\rho v_y)\cdot \mathrm{d}x\mathrm{d}z, \quad \text{and} \quad -\frac{\partial}{\partial z}(\rho v_z)\cdot \mathrm{d}x\mathrm{d}y$$

From the principle of conservation of matter, the sum of these must be equal to the time rate of change of mass,

$$\frac{\partial}{\partial t}(\rho \mathrm{d}x\mathrm{d}y\mathrm{d}x)$$

Cancel dxdydz from all the terms, we get the continuity equation of fluid as follows:

$$\frac{\partial \rho}{\partial t}+\frac{\partial}{\partial x}(\rho v_x)+\frac{\partial}{\partial y}(\rho v_y)+\frac{\partial}{\partial z}(\rho v_z)=0 \tag{13.14}$$

where ρ is the mass density and v_x, v_y, and v_z are the x, y, z components of velocity.

If the fluid is incompressible, $\partial\rho/\partial t=0$, and the continuity equation for both steady and unsteady flows becomes

$$\frac{\partial v_x}{\partial x}+\frac{\partial v_y}{\partial y}+\frac{\partial v_z}{\partial z}=0 \tag{13.15}$$

13.4 Energy Equation

The energy equation represents the conservation of energy and is dependent on the temperature distribution in space. It is expressed as follows:

$$\lambda\left(\frac{\partial^2 T}{\partial x^2}+\frac{\partial^2 T}{\partial y^2}+\frac{\partial^2 T}{\partial z^2}\right)+\frac{\partial Q}{\partial t}+\Phi=\frac{\partial}{\partial x}(\rho v_x)+\frac{\partial}{\partial y}(\rho v_y)+\frac{\partial}{\partial z}(\rho v_z)$$
$$+\frac{\rho}{2}\cdot\frac{\partial}{\partial t}(v_x^2+v_y^2+v_z^2)+\rho C_v\frac{\partial T}{\partial t} \tag{13.16}$$

where λ is the conductivity, C_v is the specific heat, Q is the heat generated in the fluid per unit volume, and Φ is the dissipation function, that is, time rate of energy dissipated due to action of viscosity:

$$\Phi=-\frac{2}{3}\eta\left(\frac{\partial v_x}{\partial x}+\frac{\partial v_y}{\partial y}+\frac{\partial v_z}{\partial z}\right)+2\eta\left[\left(\frac{\partial v_x}{\partial x}\right)^2+\left(\frac{\partial v_y}{\partial y}\right)^2+\left(\frac{\partial v_z}{\partial z}\right)^2\right]$$
$$+\eta\left[\left(\frac{\partial v_z}{\partial y}+\frac{\partial v_y}{\partial z}\right)^2+\left(\frac{\partial v_x}{\partial z}+\frac{\partial v_z}{\partial x}\right)^2+\left(\frac{\partial v_y}{\partial x}+\frac{\partial v_x}{\partial y}\right)^2\right] \tag{13.17}$$

13.5 State and Viscosity Equations

The variations of density ρ and viscosity η with pressure p and temperature T can be expressed by the state and viscosity equations:

$$\rho=\rho\,(p,T) \tag{13.18}$$

$$\eta=\eta\,(p,T) \tag{13.19}$$

For a general three-dimensional flow problem, there are seven unknowns: v_x, v_y, v_z, ρ, p, η, and T, which are determined by seven governing Eqs (13.12a), (13.12b), (13.12c), (13.14), (13.16), (13.18), and (13.19). The solution of the complete set of equations is difficult. However, in many practical situations, the governing equations get simplified considerably and the solution of the problems is possible.

13.6 Fundamental Equations for Steady Seepage Flow and Their Discretization

The steady seepage flow is defined as the flow with stable seepage head and velocity, that is, these two variables are not time dependent. Solution of steady seepage will lead to various engineering applications. This section is to solve this problem with finite element methods.

13.6.1 Generalized Darcy Law

The water head is formulated as

$$\phi = z + \frac{p}{\gamma} \tag{13.20}$$

where γ is the unit weight of fluid, p is the water pressure, and z is the height from certain base and upward is positive for z axis.

For the anisotropic medium, each component of the seepage velocity is proportional to not only corresponding component of hydraulic gradient but also other components of hydraulic gradient. According to generalized Darcy law, each component of the seepage velocity can be formulated as

$$\left.\begin{aligned} v_x &= -k_{xx}\frac{\partial \phi}{\partial x} - k_{xy}\frac{\partial \phi}{\partial y} - k_{xz}\frac{\partial \phi}{\partial z} \\ v_y &= -k_{yx}\frac{\partial \phi}{\partial x} - k_{yy}\frac{\partial \phi}{\partial y} - k_{yz}\frac{\partial \phi}{\partial z} \\ v_z &= -k_{zx}\frac{\partial \phi}{\partial x} - k_{zy}\frac{\partial \phi}{\partial y} - k_{zz}\frac{\partial \phi}{\partial z} \end{aligned}\right\} \tag{13.21a}$$

It can also be written as the matrix form:

$$\{v\} = -[k]\{\phi'\} \tag{13.21b}$$

Here

$$\{v\} = \begin{bmatrix} v_x & v_y & v_z \end{bmatrix}^{\mathrm{T}} \tag{13.22}$$

$$\{\phi'\} = \begin{bmatrix} \dfrac{\partial \phi}{\partial x} & \dfrac{\partial \phi}{\partial y} & \dfrac{\partial \phi}{\partial z} \end{bmatrix}^{\mathrm{T}} \tag{13.23}$$

$$[k] = \begin{bmatrix} k_{xx} & k_{xy} & k_{xz} \\ k_{yx} & k_{yy} & k_{yz} \\ k_{zx} & k_{zy} & k_{zx} \end{bmatrix} \tag{13.24}$$

where $[k]$ is defined as seepage matrix.

For three-dimensional problems, $[k]$ has nine components in total and six independent ones as $k_{ij} = k_{ji}$ for symmetric requirements. For plane problems, $[k]$ has three independent components.

Components k_{ij} in seepage matrix should be transformed when the coordinate system is changed. This coordinate transformation from ij to rs is formulated as

$$k_{rs} = \sum_{i=1}^{3} \sum_{j=1}^{3} l_{ri} l_{sj} k_{ij} \tag{13.25}$$

where l_{ri} is the cosine of the angle between r axis and i axis.

It can be found that Eq. (13.25) follows the coordinate transformation rules for tensors; hence, $[k]$ is also called seepage tensor.

For plane problems, when coordinate xoy is rotated to $x'oy'$ by angle α, Eq. (13.25) can be rewritten in coordinate $x'oy'$ (Figure 13.1) as

$$\begin{Bmatrix} k_{x'x'} \\ k_{y'y'} \\ k_{x'y'} \end{Bmatrix} = \begin{bmatrix} \cos^2\alpha & \sin^2\alpha & \sin 2\alpha \\ \sin^2\alpha & \cos^2\alpha & -\sin 2\alpha \\ -\frac{1}{2}\sin 2\alpha & \frac{1}{2}\sin 2\alpha & \cos 2\alpha \end{bmatrix} \begin{Bmatrix} k_{xx} \\ k_{yy} \\ k_{xy} \end{Bmatrix} \tag{13.26}$$

Let $k_{x'y'} = 0$ after rotating coordinate $x'oy'$; then we could get

$$\alpha = \frac{1}{2}\tan^{-1}\left(\frac{2k_{xy}}{k_{xx} - k_{yy}}\right) \tag{13.27}$$

Here, α indicates the principal direction of the seepage tensor; $k_{x'x'}$ and $k_{y'y'}$ are the principal permeability coefficients.

For two-dimensional (2D) problem, the seepage tensor has three independent components, that is, two principal permeability coefficients and one principal direction; for three-dimensional problems, it has six independent components, that is, three principal permeability coefficients and three principal directions.

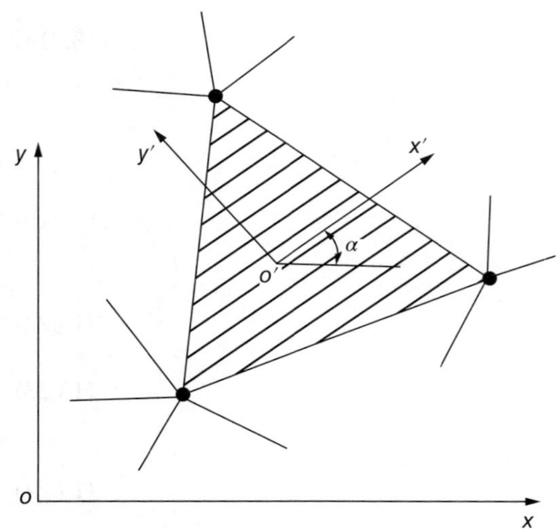

Figure 13.1 Anisotropic plane problem.

13.6.2 Fundamental Equations

For incompressible fluids, the continuity equation is

$$\frac{\partial v_x}{\partial x} + \frac{\partial v_y}{\partial y} + \frac{\partial v_z}{\partial z} - Q = 0 \tag{13.28}$$

where Q is the internal source.

Substituting Eq. (13.21a) into Eq. (13.28), we obtain the fundamental equation that water head ϕ has to satisfy in the area R:

$$\frac{\partial}{\partial x}\left(k_{xx}\frac{\partial \phi}{\partial x} + k_{xy}\frac{\partial \phi}{\partial y} + k_{xz}\frac{\partial \phi}{\partial z}\right) + \frac{\partial}{\partial y}\left(k_{yx}\frac{\partial \phi}{\partial x} + k_{yy}\frac{\partial \phi}{\partial y} + k_{yz}\frac{\partial \phi}{\partial z}\right)$$
$$+\frac{\partial}{\partial z}\left(k_{zx}\frac{\partial \phi}{\partial x} + k_{zy}\frac{\partial \phi}{\partial y} + k_{zz}\frac{\partial \phi}{\partial z}\right) - Q = 0 \tag{13.29}$$

Here, the water head also has to satisfy the boundary conditions. The following two types of boundary conditions are usually encountered:

1) Water head is known on the boundary B.
   On boundary B

$$\phi = \phi_b \tag{13.30}$$

2) Unit seepage flow by area is known on boundary C, that is, the normal velocity v_n is known.
   On boundary C

$$l_x v_x + l_y v_y + l_z v_z = v_n \tag{13.31}$$

Here, l_x, l_y, and l_z denote the directional cosines of outer normal of the boundary. The detailed formulations of v_x, v_y, and v_z refer to Eq. (13.21a).

The water head ϕ in seepage field can be obtained through fundamental Eq. (13.28) and boundary conditions (13.29) and (13.30). Based on the variational principle, it is equivalent to the extreme-value problem of the following functional.

If function $\phi(x, y, z)$ satisfies $\phi = \phi_b$ on the boundary b and makes the following functional reach the extreme value:

$$I(\phi) = \iiint_R \left\{ \frac{1}{2}\left[k_{xx}\left(\frac{\partial \phi}{\partial x}\right)^2 + k_{yy}\left(\frac{\partial \phi}{\partial y}\right)^2 + k_{zz}\left(\frac{\partial \phi}{\partial z}\right)^2\right.\right.$$
$$\left.\left. + 2k_{xy}\frac{\partial \phi}{\partial x}\frac{\partial \phi}{\partial y} + 2k_{yz}\frac{\partial \phi}{\partial y}\frac{\partial \phi}{\partial z} + 2k_{zx}\frac{\partial \phi}{\partial z}\frac{\partial \phi}{\partial x}\right] - Q\phi \right\} \mathrm{d}x\mathrm{d}y\mathrm{d}z$$
$$+ \iint_C v_n \phi \mathrm{d}s \tag{13.32}$$

According to the Euler equation, $\phi(x, y, z)$ should satisfy the continuity Eqs (13.29) and (13.31) on the boundary C; thus, $\phi(x, y, z)$ is the water head function to be solved. In the process of solution, we have to give $\phi = \phi_b$ on the boundary B, that is, Eq. (13.30).

13.6.3 Discretization of the Problems

The area R to be solved is divided into finite elements. Assuming that the nodes of each element are $i, j, m, \ldots$, the water head at each node is $\phi_i, \phi_j, \phi_m, \ldots$, and the shape function of each element is $N_i, N_j, N_m, \ldots$, we get the water head at any point in the element with the shape function as follows:

$$\phi^e(x,y,z) = \begin{bmatrix} N_i & N_j & N_m & \ldots \end{bmatrix} \begin{Bmatrix} \phi_i \\ \phi_j \\ \phi_m \\ \vdots \end{Bmatrix} = [N]\{\phi\}^e \tag{13.33}$$

Substituting Eq. (13.33) into Eq. (13.23) and Eq. (13.21b), we obtain

$$\{\phi'\} = \begin{bmatrix} \frac{\partial \phi}{\partial x} & \frac{\partial \phi}{\partial y} & \frac{\partial \phi}{\partial z} \end{bmatrix}^{\mathrm{T}} = [B]\{\phi\} \tag{13.34}$$

$$\{v\} = \begin{bmatrix} v_x & v_y & v_z \end{bmatrix}^{\mathrm{T}} = -[k][B][\phi] \tag{13.35}$$

Here

$$[B] = \begin{bmatrix} \frac{\partial N_i}{\partial x} & \frac{\partial N_j}{\partial x} & \frac{\partial N_m}{\partial x} & \cdots \\ \frac{\partial N_i}{\partial y} & \frac{\partial N_j}{\partial y} & \frac{\partial N_m}{\partial y} & \cdots \\ \frac{\partial N_i}{\partial z} & \frac{\partial N_j}{\partial z} & \frac{\partial N_m}{\partial z} & \cdots \end{bmatrix}^{\mathrm{T}} \tag{13.36}$$

Taking element e as one subarea ΔR in area R, we can get the functional in this subarea as follows:

$$I(\phi) = \iiint_{\Delta R} \left\{ \frac{1}{2} \left[k_{xx} \left(\frac{\partial \phi}{\partial x} \right)^2 + k_{yy} \left(\frac{\partial \phi}{\partial y} \right)^2 + k_{zz} \left(\frac{\partial \phi}{\partial z} \right)^2 \right. \right. \\ \left. \left. +2k_{xy} \frac{\partial \phi}{\partial x} \frac{\partial \phi}{\partial y} + 2k_{yz} \frac{\partial \phi}{\partial y} \frac{\partial \phi}{\partial z} + 2k_{zx} \frac{\partial \phi}{\partial z} \frac{\partial \phi}{\partial x} \right] - Q\phi \right\} \mathrm{d}x\mathrm{d}y\mathrm{d}z + \iint_{\Delta c} v_n \phi \mathrm{d}s \tag{13.37}$$

where the second term in the right part of Eq. (13.37) means the surface integral along the boundary C, which only needs to be considered in the elements close to the boundary C.

Differentiating the terms behind the sign of integration in the above equation, we get

$$\frac{\partial I^e}{\partial \phi_i} = \iiint_{\Delta R} \left[k_{xx} \frac{\partial \phi}{\partial x} \frac{\partial}{\partial \phi_i} \left(\frac{\partial \phi}{\partial x} \right) + k_{yy} \frac{\partial \phi}{\partial y} \frac{\partial}{\partial \phi_i} \left(\frac{\partial \phi}{\partial y} \right) + k_{zz} \frac{\partial \phi}{\partial z} \frac{\partial}{\partial \phi_i} \left(\frac{\partial \phi}{\partial z} \right) \right. \\ + k_{xy} \frac{\partial \phi}{\partial x} \frac{\partial}{\partial \phi_i} \left(\frac{\partial \phi}{\partial y} \right) + k_{xy} \frac{\partial \phi}{\partial y} \frac{\partial}{\partial \phi_i} \left(\frac{\partial \phi}{\partial x} \right) + k_{yz} \frac{\partial \phi}{\partial y} \frac{\partial}{\partial \phi_i} \left(\frac{\partial \phi}{\partial z} \right) \\ + k_{yz} \frac{\partial \phi}{\partial z} \frac{\partial}{\partial \phi_i} \left(\frac{\partial \phi}{\partial y} \right) + k_{zx} \frac{\partial \phi}{\partial z} \frac{\partial}{\partial \phi_i} \left(\frac{\partial \phi}{\partial x} \right) + k_{zx} \frac{\partial \phi}{\partial x} \frac{\partial}{\partial \phi_i} \left(\frac{\partial \phi}{\partial z} \right) \\ \left. -Q \frac{\partial p}{\partial \phi_i} \right] \mathrm{d}x\mathrm{d}y\mathrm{d}z + \iint_{\Delta c} v_n \frac{\partial \phi}{\partial \phi_i} \mathrm{d}s \tag{13.38}$$

According to Eq. (13.33), in the element e, there are

$$\frac{\partial \phi}{\partial x} = \frac{\partial N_i}{\partial x} \phi_i + \frac{\partial N_j}{\partial x} \phi_j + \frac{\partial N_m}{\partial x} \phi_m + \ldots$$

$$\frac{\partial}{\partial \phi_i}\left(\frac{\partial \phi}{\partial x}\right) = \frac{\partial N_i}{\partial x}$$
$$\frac{\partial \phi}{\partial \phi_i} = N_i$$
$$\vdots$$

Substituting these into Eq. (13.38), we get

$$\begin{Bmatrix} \dfrac{\partial I^{\mathrm{e}}}{\partial \phi_i} \\ \dfrac{\partial I^{\mathrm{e}}}{\partial \phi_j} \\ \dfrac{\partial I^{\mathrm{e}}}{\partial \phi_m} \\ \vdots \end{Bmatrix} = \frac{\partial I^{\mathrm{e}}}{\partial \{\phi\}^{\mathrm{e}}} = [H]^{\mathrm{e}}\{\phi\}^{\mathrm{e}} - \{F\}^{\mathrm{e}} \tag{13.39}$$

$$[H]^{\mathrm{e}} = \iiint_{\Delta R} [B]^{\mathrm{T}}[k][B]\mathrm{d}x\mathrm{d}y\mathrm{d}z \tag{13.40}$$

$$\{F\}^{\mathrm{e}} = \iiint_{\Delta R} [N]^{\mathrm{T}} Q\mathrm{d}x\mathrm{d}y\mathrm{d}z - \iint_{\Delta c} [N]^{\mathrm{T}} v_n \mathrm{d}s \tag{13.41}$$

where $[H]^{\mathrm{e}}$ is the elemental conductivity matrix, which is corresponding to the stiffness matrix in structural analyses.

The components of $[H]^{\mathrm{e}}$ and $\{F\}^{\mathrm{e}}$ are computed as follows:

$$\begin{aligned} H_{ij}^{\mathrm{e}} = \iiint_{\Delta R} \Bigg[& k_{xx}\frac{\partial N_i}{\partial x}\frac{\partial N_j}{\partial x} + k_{yy}\frac{\partial N_i}{\partial y}\frac{\partial N_j}{\partial y} + k_{zz}\frac{\partial N_i}{\partial z}\frac{\partial N_j}{\partial z} \\ & + k_{xy}\left(\frac{\partial N_i}{\partial x}\frac{\partial N_j}{\partial y} + \frac{\partial N_i}{\partial y}\frac{\partial N_j}{\partial x}\right) + k_{yz}\left(\frac{\partial N_i}{\partial y}\frac{\partial N_j}{\partial z} + \frac{\partial N_i}{\partial z}\frac{\partial N_j}{\partial y}\right) \\ & + k_{zx}\left(\frac{\partial N_i}{\partial z}\frac{\partial N_j}{\partial x} + \frac{\partial N_i}{\partial x}\frac{\partial N_j}{\partial z}\right)\Bigg] \mathrm{d}x\mathrm{d}y\mathrm{d}z \end{aligned} \tag{13.42}$$

$$\{F\}^{\mathrm{e}} = \iiint_{\Delta R} N_i Q\mathrm{d}x\mathrm{d}y\mathrm{d}z - \iint_{\Delta C} N_i v_n \mathrm{d}s \tag{13.43}$$

Compared with structural analyses, the first term in the right part of Eq. (13.43) is equivalent to the nodal force induced by body force. The second term is equivalent to the nodal force induced by surface force on the boundary.

Assembling the terms of all the elements, we get the following equation for all the nodes in the area:

$$\frac{\partial I^{\mathrm{e}}}{\partial \{\phi\}^{\mathrm{e}}} = [H]\{\phi\} - \{F\} = 0$$

That is,

$$[H]\{\phi\} = \{F\} \tag{13.44}$$
$$H_{ij} = \sum_e H_{ij}^{\mathrm{e}}, \quad F_i = \sum_e F_i^{\mathrm{e}}$$

Then, the value ϕ of each node can be solved and the flow velocity of each node can be computed by Eq. (13.35).

As there is no derivative term of permeability coefficient k in the functional $I(\phi)$ in Eq. (13.37), the above solution suits both constant coefficient and variable coefficient problem. Different coefficient can be specified in different element, and the coefficient can also be variable in even one element. Of course, the variation of the coefficient should be considered in the integration procedure.

Difference could be found in the solution of anisotropic seepage problem and that of the anisotropic structural problem. The elemental conductivity matrix $[H^e]$ suits any local coordinates, as it is only linked with ϕ according to Eq. (13.40). Therefore, a different local coordinate can be used in each element and the transformation is not necessary for the integration of global matrix $[H]$.

There are two difficult points in seepage analyses: free surface calculation and effect of the drainage holes. The following sections will explain how to overcome these difficulties.

13.7 Free Surface Calculation for Seepage Analysis

As shown in Figure 13.2, it is an earth-filled dam. Boundary condition on the upstream surface is set as $\phi = H$ (constant); on downstream surface, $\phi = h$; online cd, $\phi = z$ for $p = 0$; and on the impermeable rock surface, it must satisfy Eqs (13.30) and (13.31) and these boundary conditions can be treated with the above methods. For free surface, as no flow cross it and $p = 0$, the following two equations can be written:

$$\phi = z$$

$$v_{\mathrm{n}} = 0$$

Here, the location of the free surface and the seepage point at the downstream surface are unknown. The following iterative analytical methods can be used to solve this problem.

13.7.1 Method of Mesh Revision

First, a free surface is assumed according to engineering experience, and it is treated as a boundary. Restricted by $v_n = 0$, analysis is carried out to check whether $\phi = z$ can be satisfied. If not, the location of free surface and seepage point should be revised. Generally, coordinate z of the new free surface can be set as ϕ of the previous analysis. Then, a second iterative analysis is performed. Usually, five or six iterations will be enough.

This method is used early in the past, which has a big disadvantage. The computing mesh is revised and global conductivity matrix should be computed with the new free surface in every iteration step. It leads to very low computation efficiency.

13.7.2 Method of Revision of the Conductivity Matrix

The solving area is divided into two subareas. In R_1 above free surface, no seepage velocity exists, while there is positive velocity in the subarea R_2 below free surface. Therefore, k can be set 0 in R_1 and the real permeability coefficient k only needs to be set in R_2.

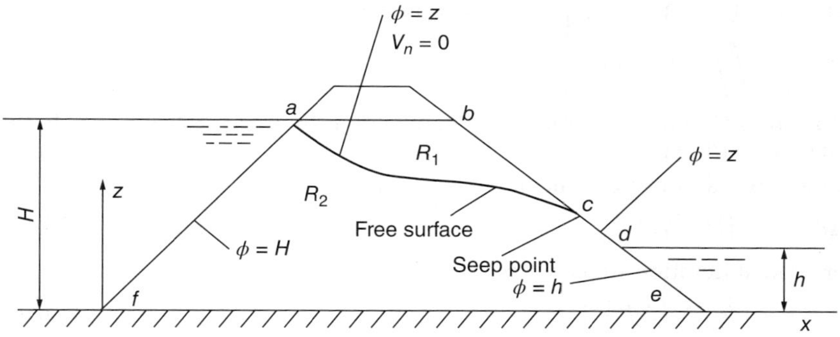

Figure 13.2 Earth dam with free surface.

For analytical stability, a small positive value is usually set in R_1. It is recommended by Bathe:

$$\left.\begin{array}{ll} k = k & \text{when } \phi \geq z \\ k = k/1000 & \text{when } \phi < z \end{array}\right\} \tag{13.45}$$

Therefore, only permeability coefficient k needs to be revised and a fixed mesh can be used. The flowchart is as follows. First, mesh is discretized in the whole solving area (Figure 13.2a–f). Then, the global conductivity matrix $[H]$ is formed with the assumption that real k is set in the whole area. After solving water head ϕ, the permeability coefficient k is set for the each element according to Eq. (13.45). Then, refresh $[H]$ and solve new water head ϕ and so forth. The iteration can be stopped when ϕ^n and ϕ^{n-1} solved by two adjacent iterations are close to each other.

Using isoparametric elements, we could have the free surface cross one element due to the large size of elements. As the component is calculated with the Gauss integral method, we could check and revise each Gauss point with Eq. (13.45). Revise k as $k/1000$ when $\phi < z$.

The advantage of this method is using fixed mesh; however, the conductivity matrix $[H]$ has to be revised and inversed in each iteration step, which still leads to very long computation time.

13.7.3 Residual Velocity Method

Residual velocity method is proposed by Desai [1]; it is often called the residual discharge method. The author chooses residual velocity method as the better name.

Taking the earth-filled dam (Figure 13.2) as example, we could have the computation flowchart as follows:

1) The computing mesh is discretized in the whole solving area *abcdef* with the following boundary conditions:
   line *af*: $\phi = H$. line *abc*: $v_n = 0$. line *cd*: $\phi = z$. line *de*: $\phi = h$. line *ef*: $v_n = 0$.
   Solve the first approximate value $\{\phi^1\}$ with Eq. (13.44).
2) Obtain the rth approximate location of free surface based on Eqs (a) and (b).
3) Calculate the rth residual velocity normal to the free surface. As we should have $v_n = 0$ on free surface, let

$$v_n = -v_n^r \tag{a}$$

Substituting it to Eq. (13.41), we get the right term as

$$\{F^{r+1}\} = \iint [N]^{\mathrm{T}} v_n^r \mathrm{d}s \tag{13.46}$$

The right term of (13.46) is the surface integral along the free surface; v_n has to be computed by Eq. (13.31).

4) The increment of water head is calculated by (13.44):

$$\{\Delta\phi^{r+1}\} = [H]^{-1}\{F^{r+1}\} \tag{13.47}$$

5) The water head at iteration step $r+1$ is

$$\{\phi^{r+1}\} = \{\phi^r\} + \{\Delta\phi^{r+1}\} \tag{13.48}$$

6) Iterative computation of step (2)–(5) will not stop until $\Delta\phi^{r+1}$ is small enough.

With residual velocity method, conductivity matrix $[H]$ only needs to be decomposed once, and only substitution is needed for computing water head increment with Eq. (13.47). Thus, high computing efficiency can be expected.

Next, we will compute the free surface location with $z=\phi$.

For linear elements, as Figure 13.3(a), $\phi(1)<z(1)$ and $\phi(2)>z(2)$, the free surface location is

$$z = z(2) + \frac{[z(1) - z(2)]\Delta_2}{\Delta_2 - \Delta_1} \tag{b}$$

where

$$\Delta_1 = \phi(1) - z(1), \quad \Delta_2 = \phi(2) - z(2)$$

For quadratic elements, as Figure 13.3(b), $\Delta = \phi - z$ can be written as follows with η.

$$\Delta(\eta) = \phi(\eta) - z(\eta) = \frac{1}{2}\eta^2(\Delta_1 + \Delta_2 - 2\Delta_3) + \frac{1}{2}\eta(\Delta_1 - \Delta_2) + \Delta_3 \tag{c}$$

At free surface, $\Delta(\eta)=0$, and it can be derived from Eq. (c) as follows:

$$(\Delta_1 + \Delta_2 - 2\Delta_3)\eta^2 + (\Delta_1 - \Delta_2)\eta + 2\Delta_3 = 0 \tag{d}$$

η can be solved from above equation and the free surface location is

$$z = N_1(\eta)z(1) + N_2(\eta)z(2) + N_3(\eta)z(3) \tag{e}$$

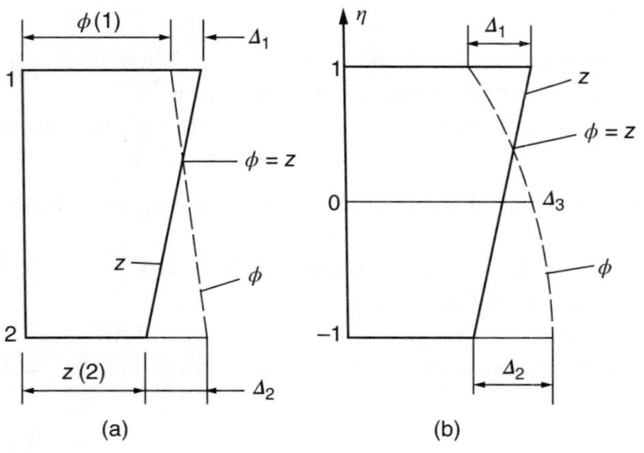

Figure 13.3 Free surface location. (a) Linear element; (b) second-order element.

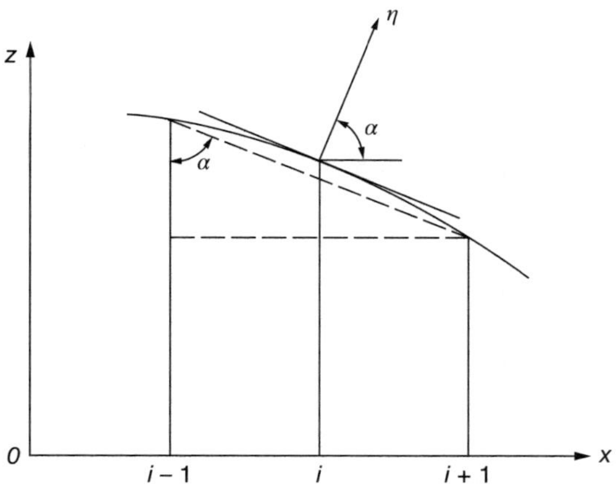

Figure 13.4 Free surface in two-dimensional problem.

To compute the right term $\{F^{r+1}\}$ from Eq. (13.46), it is necessary to know the normal direction and infinitesimal area dS. For two-dimensional free surface as shown in Figure 13.4, the infinitesimal length Δs and direction cosine $\cos\alpha$ is calculated as follows:

$$\Delta s \cong \frac{1}{2}\sqrt{\Delta x^2 + \Delta z^2} \tag{f}$$

$$\cos\alpha \cong \frac{\Delta z}{\sqrt{\Delta x^2 + \Delta z^2}} \tag{g}$$

$$\Delta x = x(i+1) - x(i-1), \quad \Delta z = z(i+1) - z(i-1)$$

For three-dimensional free surface, its location can also be computed according to Eqs (b) and (d); as shown in Figure 13.5, the tangential plane to the free surface for infinitesimal area ΔS is quadrilateral $ABCD$, and its area can be computed as follows:

$$\Delta S = [(p-a)(p-b)(p-c)(p-d) - abcd\cos^2\beta]^{1/2} \tag{h}$$

$$p = \frac{a+b+c+d}{2}, \quad \beta = \frac{\angle A + \angle C}{2}$$

Here, a, b, c, and d are the length of the four sides, and $\angle A$ and $\angle C$ are the angles.

The normal directional cosine of the free surface can be computed with the formulation similar to Eq. (g).

Also, the shape function can be used to compute the elemental area and normal direction. Once the free surface location is obtained, the coordinate of the free surface can be written as follows with $N_i(\xi, \eta)$:

$$x = \sum N_i(\xi,\eta)x_i, \quad y = \sum N_i(\xi,\eta)y_i, \quad z = \sum N_i(\xi,\eta)z_i \tag{i}$$

Here, x_i, y_i, z_i are the coordinates on the known free surface.

Assuming the mapping of free surface infinitesimal area $ABCD$ on its mother element is $\Delta\xi\Delta\eta$, as shown in Figure 13.5(a), we could get the area of $ABCD$ as

$$\Delta S = a\Delta\xi\Delta\eta \tag{j}$$

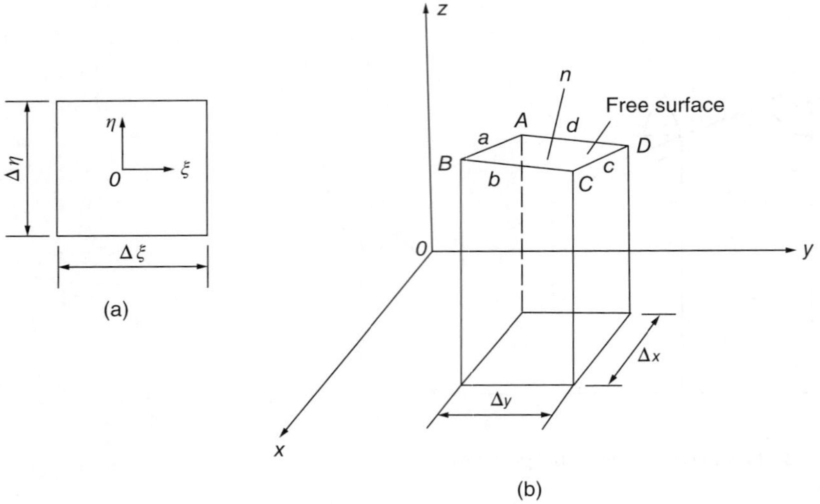

Figure 13.5 Free surface for three-dimensional problem.

And the normal direction n is

$$l_x = \cos(n,x) = a_x/a, \quad l_y = \cos(n,y) = a_y/a, \quad l_z = \cos(n,z) = a_z/a \tag{k}$$

$$\text{Here} \quad \left.\begin{aligned} a_x &= \frac{\partial y}{\partial \xi}\frac{\partial z}{\partial \eta} - \frac{\partial y}{\partial \eta}\frac{\partial z}{\partial \xi}, \quad a_y = \frac{\partial z}{\partial \xi}\frac{\partial x}{\partial \eta} - \frac{\partial z}{\partial \eta}\frac{\partial x}{\partial \xi} \\ a_z &= \frac{\partial x}{\partial \xi}\frac{\partial y}{\partial \eta} - \frac{\partial x}{\partial \eta}\frac{\partial y}{\partial \xi}, \quad a = \sqrt{a_x^2 + a_y^2 + a_z^2} \end{aligned}\right\} \tag{l}$$

13.7.4 Initial Velocity Method

This method is proposed by Zhang with the basis of Gell's work [3]. It is originally called initial discharge method, and the author renamed as initial velocity method. To clarify this concept more clearly, we elaborate the rationale and algorithm in following new approach.

As a matter of fact, the fundamental equation $[H]\{\phi\} = \{F\}$ represents the continuity equation. When internal source $Q = 0$ and boundary normal velocity $v_n = 0$, the right term $\{F\} = 0$. According to Eq. (13.40) and $\{v\} = -[k][B]\{\phi\}$, the fundamental Eq. (13.44) can be rewritten as follows:

$$\int_R [B]^{\mathrm{T}}\{v\}\mathrm{d}V = 0 \tag{13.49}$$

Here, R is the solving area, $\{v\}$ is the velocity vector, and $\mathrm{d}V$ is the infinitesimal volume. The left term of Eq. (13.49) is a volume integral.

Taking the earth-filled dam shown in Figure 13.2 as example, the nodal water head ϕ^0 and velocity $\{v^0\}$ is solved from Eq. (13.49) with the mesh for the whole area *abcdef*. In the subarea R_1 above the free surface, velocity should equal zero. As discussed, the revision of conductivity matrix lets permeability coefficient $k = 0$ in subarea R_1 $(\phi < z)$

and performs the next iteration. Instead of modifying permeability coefficient, an initial velocity term $-\{v^0\}$ is added into Eq. (13.49) as

$$\left.\begin{aligned} &\text{when } \phi < z \text{ (area } R_1), -\{v^0\} = [k][B]\{\phi^0\} \\ &\text{when } \phi \geq z \text{ (area } R_2), -\{v^0\} = 0 \end{aligned}\right\} \tag{13.50}$$

Then, the velocity in the subarea R_1 above the free surface can be revised to zero. Hence, the fundamental equation is modified as

$$\int_R [B]^{\mathrm{T}}(\{v\} - \{v^0\})\mathrm{d}V = 0 \tag{13.51}$$

It can be written as the iterative format:

$$\int_{R_1+R_2} [B]^{\mathrm{T}}\{v^{n+1}\}\,\mathrm{d}V = \int_{R_1} [B]^{\mathrm{T}}\{v^n\}\mathrm{d}V \tag{13.52}$$

Here, n means the iteration steps; R_1 means the subarea above free surface; and R_2 means the subarea below free surface.

It can also be rewritten as

$$[H]\{\phi^{n+1}\} = \{F^{n+1}\} \tag{13.53}$$

$$\{F^{n+1}\} = \int_R [B]^{\mathrm{T}}\{v^n\}\mathrm{d}V = [H]\{\phi^n\} \tag{13.54}$$

Because $\{\varphi^n\}$ can satisfy Eq. (13.49), the right term of above equation will be zero if all the elements around node i locate inside subarea R_1. Thus, not all the nodes inside R_1 have the nonzero right term, only the elements and nodes corresponding to free surface have. Essentially speaking, this method is the same as the residual velocity method. The velocity across free surface is forced to be zero in residual velocity method, while here velocity inside the element close to free surface is forced to be zero.

As a matter of fact, Gauss integral method is used for the integration of Eq. (13.52). Therefore, the number of Gauss points in the elements close to free surface should be large enough according to the above discussion. In each element, the number of Gauss point should be larger than 3 and the controlled area of each Gauss point should be small enough.

Taking the case of Figure 13.2 as an example, we could use the following steps:

1) First perform mesh discretization in the whole area *abcdef*. Boundary *abc* is treated with $v_n = 0$. $\{\phi^1\}$ is obtained from $[H]\{\phi\} = 0$.
2) Compute the right term according to Eq. (13.54) for the Gauss points in subarea R_1 ($\phi < z$).
3) Inverse the matrix $[H]$, and, according to (13.53), the new water head can be computed: $\{\phi^{n+1}\} = [H]^{-1}\{F^{n+1}\}$.
4) Stop computation when the $\{\phi^{n+1}\}$ is close enough to $\{\varphi^n\}$; otherwise, go on with iteration.

Generally speaking, residual velocity method and initial velocity method are better in the above four methods. The conductivity matrix is only decomposed once and only simple substitution is needed for the following iteration. The high computation efficiency can be expected. The comparison of these two methods shows that the computation time for each iteration step is almost the same and the total computation time relies on the iteration steps. According to the principle of computation, the convergence could be faster for residual velocity method.

13.8 Substitution of the Curtain of Drainage Holes by the Seeping Layer for Seepage Analysis

The use of drainage holes is an important measure to reduce seepage pressure in engineering practice. However, it is difficult to simulate this effect because of the singularity of the drainage holes in the seepage field. Early studies let the nodal water head equal that at the drainage hole, which hints the linear distribution of the water head between the nodes. It might overestimate the drainage effect as the real water head is nonlinearly distributed. To simulate the real water head, small size of finite elements has to be used for those around the drainage hole. The general diameter of the drainage hole is 0.05–0.10 m, and the same size of finite elements has to be used. Compared with the usual size of foundation elements 5–20 m with no drainage hole, a large number of finite elements should be used. To overcome this difficulty, the author proposed a hybrid element method [4], which may suit the analysis for a single dam section, but still encounter difficulties for a whole dam analysis. In Ref. [8] the author proposed the method of equivalent seeping layer in which the effect of drainage holes are modeled by the equivalent coefficient of permeability; thus the effect of drainage holes may be computed with normal three-dimensional seepage program. It is introduced as follows:

As shown in Figure 13.6(a), a curtain of drainage holes is set in the foundation. The spacing is $2b$; the diameter is r_w. The equivalent seeping layer is shown in Figure 13.6(b); its thickness is e and it has the same depth as the holes. The same horizontal permeability coefficient is used as the base rock, while the vertical permeability coefficient k_z is so determined that the layer will have the same total drainage effect as the rock with drainage holes. It can be computed with a general code based on the normal three-dimensional seepage analytical program.

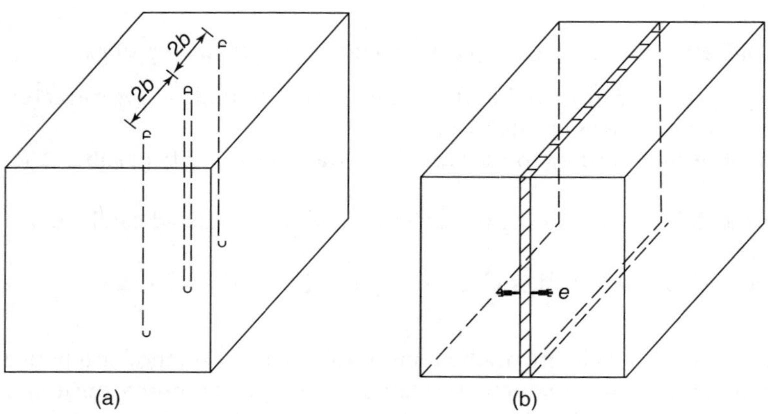

Figure 13.6 Curtain of drainage holes and the equivalent seeping layer.

The water head function is

$$H = z + \frac{p}{\gamma}$$

Here, z is the vertical coordinate, p is the water pressure, and γ is unit weight of water.

The equivalent coefficient k_z is computed as follows. A three-dimensional mesh is used with dam section, foundation, a grouting curtain, and a drainage hole; its thickness is set as the half of the drainage hole spacing. (A general program can be used here.)

The top boundary conditions of seeping layer: the exit of the equivalent seeping layer is set on the foundation surface. Let the pressure of layer i at the exit equal the average pressure along the dam axis on the foundation surface from three-dimensional analysis, that is,

$$H_i = z_0 + \frac{p_{\mathrm{mi}}}{\gamma} \tag{13.55}$$

where H_i is the equivalent water head of the layer i on the foundation surface; z_0 is the vertical coordinate of the foundation surface; and p_{mi} is the line average pressure of the drainage holes along the dam axis on the foundation surface.

The vertical permeability coefficient k_{zi} is selected to force the average water head in the region L_i equal the corresponding average water head from three-dimensional analysis.

$$\overline{H}_i = H_i' \tag{13.56}$$

Here, $\overline{H}_i$ is the average water head in the region L_i with the equivalent method, while H_i is the average water head in the region L_i with the simulation of the drainage holes.

Using (13.56) to compute k_z, we could consider not only the drainage hole size and spacing but the effect of the grouting curtain and complicated geometrical permeability characteristics. Both single row of holes and multi-rows of holes can be calculated.

When the vertical equivalent permeability coefficient is obtained, the equivalent seeping layer method can be used for a whole dam analysis; the mesh sensitivity is no longer affected by the drainage hole diameter and spacing. Both the preprocess and the computation can be effectively simplified.

Example 1 The cross section of an arch dam is shown in Figure 13.7(a): its bottom width is 50 m, the width of the grouting curtain is 12 m and the depth is 100 m, the diameter of the drainage hole is 0.10 m, and the spacing is 3.0 m and the depth is 50 m. The upstream water level is 500 m, while the downstream water level is 350 m. The foundation surface level is 300 m, and the bottom level of the drainage gallery is 310 m. The permeability coefficient of the rock k_0 is 1.0 m/day, and the permeability coefficient of the grouting curtain k_1 is 0.1 m/day.

First, three-dimensional computing mesh as shown in Figure 13.7(b) is used, including drainage holes in the foundation. Due to symmetry, the thickness is set as 1.5 m, the length is 450 m, and the height is 300 m. The boundary conditions are as follows: top of the foundation, upstream $H = 500$ m, downstream $H = 350$ m, bottom of the dam $\partial H/\partial z = 0$, and another five boundaries are impermeable. The internal height of the drainage hole is $H = 310$ m; fine mesh is used around the hole and transit to the normal foundation element. The computing result shows that the average water head along the

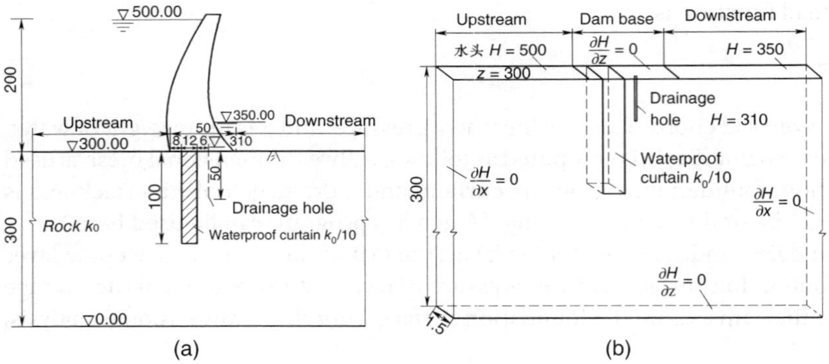

Figure 13.7 Seepage in the arch dam foundation. (a) Cross section of arch dam; (b) 3D computation model and boundary conditions of foundations.

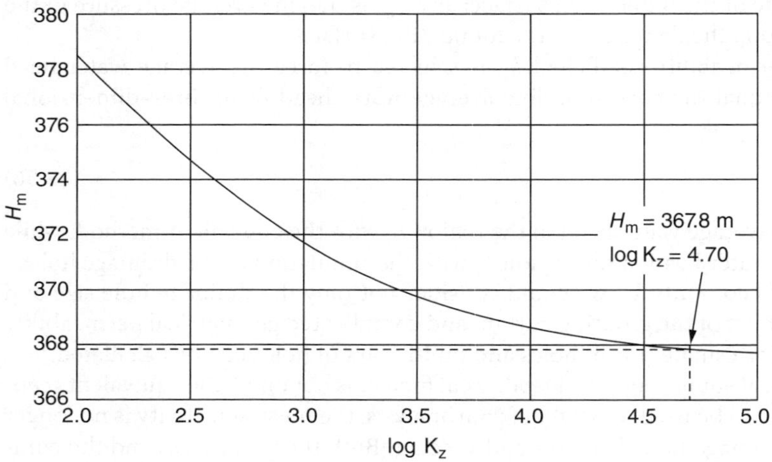

Figure 13.8 Relation between average head H_m at dam bottom and vertical permeability coefficient of the equivalent seeping layer k_x.

thick direction (dam axis) at the top of the drainage hole is $H = 312.58$ m and average water head at the dam bottom surface is $H_m = 367.8$ m.

Second, as shown in Figure 13.7(b), a seeping layer with 50 m depth and 0.10 m thickness is used to be equivalent to the original hole. The permeability coefficient of the rock k_0 is 1.0 m/day, and the permeability coefficient of the grouting curtain k_1 is 0.1 m/day. The horizontal permeability coefficient of the seeping layer is 1.0 m/day and the vertical equivalent coefficient k_z is set as a series of numbers. The relation curve between average water head at dam bottom surface H_m and log k_z can be obtained as Figure 13.8. From the cross-point between the curve and $H_m = 367.8$ m, we could get log $k_z = 4.70$, that is, the vertical equivalent coefficient $k_z = 50{,}000$ m/day. Then, the three-dimensional analysis for a whole dam is performed with the value $k_z = 50{,}000$ m/day for the vertical equivalent permeability coefficient of the seeping layer. When several rows of drainage holes are

Table 13.1 Water head distribution along water flow direction on foundation surface (m).

Distance to dam heel *x*	Three-dimensional analysis results			Results of equivalent layer
	Water head on the symmetric plane in the water flow direction	Sectional water head across the hole along the water flow direction	Average water head along the thick direction (across river)	
0.00	500.0	500.0	500.0	500.0
8.00	473.8	473.8	473.8	474.0
20.00	319.9	319.9	319.9	319.9
25.00	313.8	313.6	313.7	313.7
26.00	313.3	310.0	312.6	312.6
27.00	313.8	313.5	313.6	313.7
30.00	317.1	317.1	317.1	317.2
40.00	329.5	329.5	329.5	329.5
50.00	350.0	350.0	350.0	350.0

used and the geometrical conditions are complicated, several subareas can be divided along the direction of the water flow. Different k_z can be used for different subareas.

Computing results: The H distribution along the water flow direction on the foundation surface is shown in Table 13.1 and Figure 13.9. When the distance is larger than 1.5 m to the drainage hole, the water heads by two methods are close to each other. It can be found in Table 13.1 that, on the whole dam foundation surface, the average water heads along the thick direction by two methods are also close to each other. In a word, on

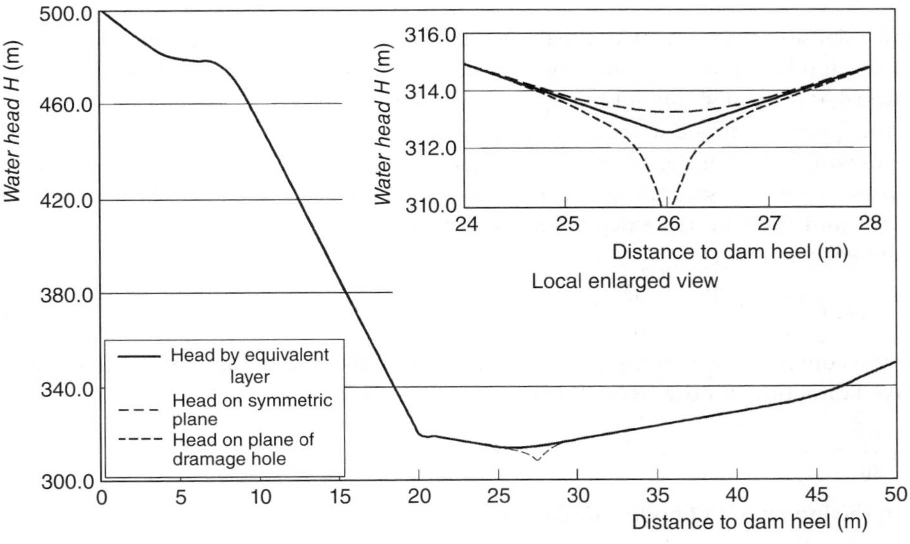

Figure 13.9 Longitudinal distribution of water head at foundation base.

the whole dam foundation surface, the average water heads by two methods are close to each other,and the maximum error is 0.20 m. Compared with the water head difference between upstream and downstream levels of 150 m, the relative error is only 0.13%. It implies that the two methods will give not only the same vertical uplift pressure but also the same uplift moment. As a matter of fact, the distribution of water heads in the large area of the foundation is also close to each other; the detail can be referred to Ref. [8].

Summary: Using the equivalent seeping layer instead of the drainage holes for analysis, the method will give the same drainage effect on foundation surface and whole area of foundation; it proves the applicability of this method.

13.9 Unsteady Seepage Flow

Taking the earth dam with saturation line inside as example (Figure 13.2), the seepage is steady when the upstream level is stable. If not, the unsteady seepage will appear, and the water head, velocity, and saturation line will vary with time.

Assuming the fluid is incompressible and there is no internal source, let $Q=0$ in Eq. (13.28), and the fundamental equation is

$$\frac{\partial v_x}{\partial x} + \frac{\partial v_y}{\partial y} + \frac{\partial v_z}{\partial z} = 0 \tag{13.57a}$$

In this case, the free surface will not coincide with the stream line, that is, it will not satisfy Eq. (13.46) and only follow the zero pressure condition, that is, $\phi = \mathrm{z}$. The location of free surface will be time dependent. Assuming the initial location of the free surface is known, we could solve ϕ according to the free surface boundary $\phi = \mathrm{z}$ and Eq. (13.44). Here, the moving speed of free surface along the normal direction is assumed as v_n; according to the flow boundary conditions on free surface, we could have

$$sv_n = v_x l_x + v_y l_y + v_z l_z \tag{13.57b}$$

Here, s denotes the porosity of the foundation soil. It could be saturation deficit when the saturation line goes upward, while it is specific yield when the saturation line goes downward. If the water content by capillary action is neglected, it presents soil porosity. l_x, l_y, and l_z are the directional cosines of the normal.

The moving speed of the saturation line can be calculated. The nodes at free surface should move with the saturation line. Assuming the nodal coordinate on the free surface are $\bar{x}, \bar{y}, \bar{z}$ and they are time dependent, the moving speed of the free surface can be formulated as follows:

$$v_n = l_x \frac{\mathrm{d}\bar{x}}{\mathrm{d}t} + l_y \frac{\mathrm{d}\bar{y}}{\mathrm{d}t} + l_z \frac{\mathrm{d}\bar{z}}{\mathrm{d}t} \tag{13.58}$$

For the convenience of computation, it is assumed the nodes could move along only one fixed direction, for example, if it is y direction, we could have

$$\frac{\mathrm{d}\bar{y}}{\mathrm{d}t} = \frac{v_n}{l_y} \tag{13.59}$$

Then the increment of node coordinate is

$$\Delta\bar{y} = \frac{\mathrm{d}\bar{y}}{\mathrm{d}t}\Delta t \tag{13.60}$$

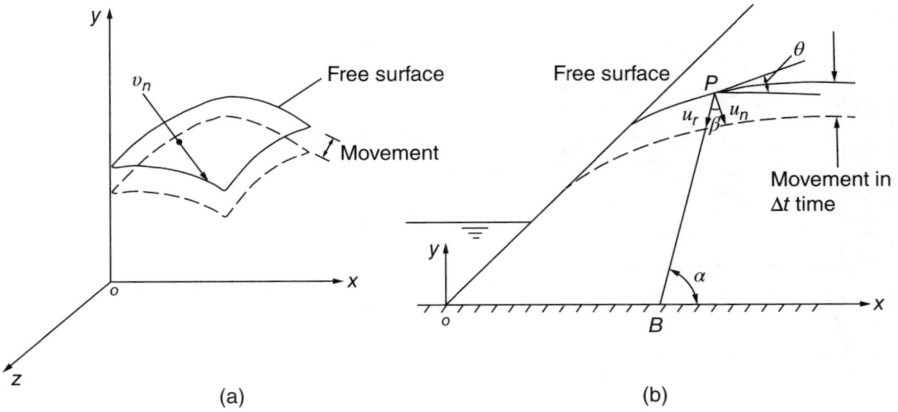

Figure 13.10 Free surface movement in earth dam. (a) Space problem; (b) plane problem.

For the plane problem in Figure 13.10, if it is assumed the nodes could only move along line PB and the angle between PB and x axis is α, the moving distance along the normal direction is

$$u_n = v_n \Delta t \tag{13.61}$$

The moving distance along PB is

$$u_r = \frac{u_n}{\cos \beta} \tag{13.62}$$

Here,

$$\beta = \frac{\pi}{2} - \alpha + \theta$$

The increment of nodal coordinate in time Δt is

$$\Delta \overline{x} = u_r \cos \alpha, \quad \Delta \overline{y} = u_r \sin \alpha$$

Then the new location of the free surface can be obtained. Next, iterative analysis can be performed till the real free surface location and water head function ϕ are solved. Obviously, the computing mesh should be modified every time the new surface location is refreshed.

As Figure 13.11, taking this earth dam as an example, the upstream water level suddenly decreases 30 m from the maximum, and isoparametric elements are used to compute the time-dependent surface location [6]. The analysis shows good agreement with experimental results [6].

13.10 Dynamic Water Pressure during Earthquake

This section introduces the dynamic water pressure applied on the water-retaining structures during earthquake. If the viscosity and convection is neglected when the fluid is moving with a small amplitude and let $\eta = 0$ in Eq. (13.12), the additional pressure

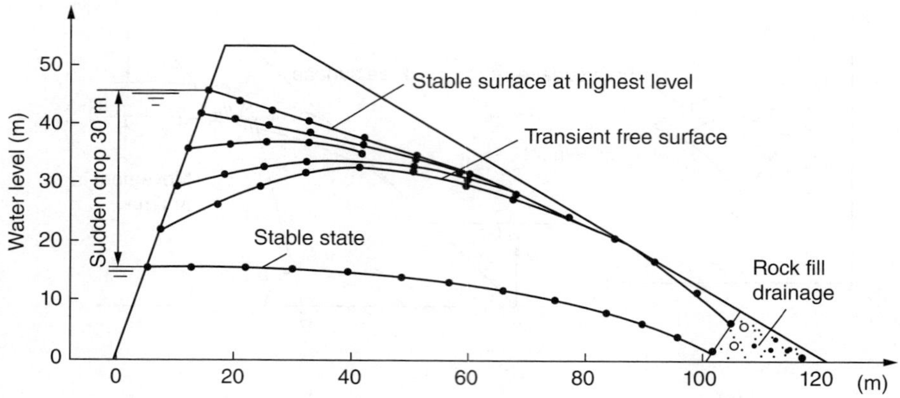

Figure 13.11 Unstable seepage in the earth dam after rapid drop-down.

p meets the following equations according to the equilibrium of an infinitesimal volume:

$$\frac{\partial p}{\partial x} = -\rho \frac{\partial v_x}{\partial t}, \quad \frac{\partial p}{\partial y} = -\rho \frac{\partial v_y}{\partial t}, \quad \frac{\partial p}{\partial z} = -\rho \frac{\partial v_z}{\partial t} \tag{a}$$

where ρ is the density; v_x, v_y, and v_z are the velocities of the mass point along x, y, and z direction, respectively.

The continuous equation is

$$\frac{\partial v_x}{\partial x} + \frac{\partial v_y}{\partial y} + \frac{\partial v_z}{\partial z} = -\frac{1}{K}\frac{\partial p}{\partial t} \tag{b}$$

where K is the bulk modulus.

Computing the differential quotient of Eq. (b) with respect to time t, and substituting Eq. (a) to it, we obtain the basic equation as follows:

$$\frac{\partial^2 p}{\partial x^2} + \frac{\partial^2 p}{\partial y^2} + \frac{\partial^2 p}{\partial z^2} = \frac{\rho}{K}\frac{\partial^2 p}{\partial t^2} \tag{13.63}$$

Neglecting the compressibility of the water, we get

$$\frac{\partial^2 p}{\partial x^2} + \frac{\partial^2 p}{\partial y^2} + \frac{\partial^2 p}{\partial z^2} = 0 \tag{13.64}$$

The boundary conditions at free surface are

$$p = 0 \tag{13.65}$$

As it is assumed no friction in the fluid, there is no restriction of the relative speed between fluid and solid on the fluid–solid interface in the tangential direction. However, the acceleration of fluid and solid has to be the same at the normal direction of boundary because there is no separation between them. According to Eq. (a), the condition of the boundary should be

$$\frac{\partial p}{\partial n} = -\rho \frac{\partial^2 \delta_n}{\partial t^2} \tag{13.66}$$

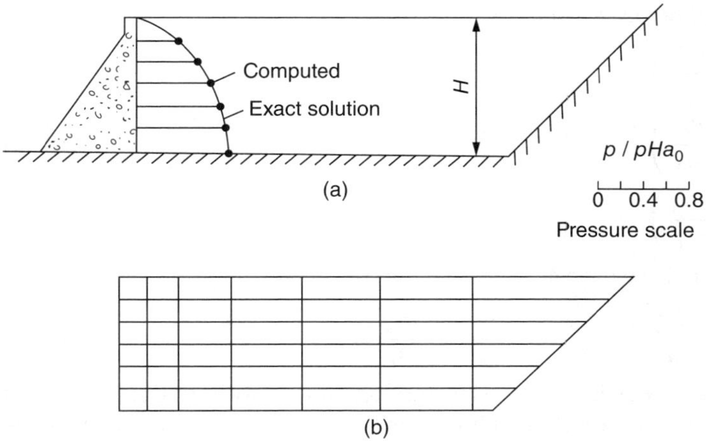

Figure 13.12 Dynamic water pressure on dam surface. (a) Computation results; (b) computation mesh.

In this equation, n is the normal direction at the interface with its positive direction pointing outside the fluid; δ_n is the normal components of boundary displacement.

Equation (13.66) is similar with Eq. (13.31).

Because Eqs (13.64)–(13.66) are similar with Eqs (13.29)–(13.31), dynamic water pressure can be obtained by this equation:

$$[H]\{p\} = \{F\} \tag{13.67}$$

In the above equation, $\{p\}$ is the node pressure; $[H]$ and $\{F\}$ can be gained according to Eqs (13.42) and (13.43).

Figure 13.12 is an example for computation of dynamic water pressure. Supposing the dam body is rigid, the body will move horizontally with the ground during earthquake, and the acceleration is a_0; the computing grid is shown as Figure 13.12(b) by applying quadrilateral element. The result is shown as Figure 13.12(a).

13.11 Inviscid Fluid Flow Formulated by Potential Function Φ

For incompressible and nonviscous fluid, the velocity of flow on any part can be expressed by potential function Φ if the flow is not rotational; this type of flow is defined as potential flow. Potential flow has a wide range of application in engineering, including the overflow of dam, outflow of orifice, and subsonic flow around aircraft.

13.11.1 Basic Equations

If the fluid is incompressible, the velocity of flow on any part should match the continuity equation below:

$$\frac{\partial v_x}{\partial x} + \frac{\partial v_y}{\partial y} + \frac{\partial v_z}{\partial z} = 0 \tag{13.68}$$

In this equation, v_x, v_y, v_z are the components of the flow velocity of the fluid in x, y, z directions, respectively.

The rotation that fluid revolves around x, y, z axis is

$$\omega_x = \frac{1}{2}\left(\frac{\partial v_z}{\partial y} - \frac{\partial v_y}{\partial z}\right)$$
$$\omega_y = \frac{1}{2}\left(\frac{\partial v_x}{\partial z} - \frac{\partial v_z}{\partial x}\right) \tag{13.69}$$
$$\omega_z = \frac{1}{2}\left(\frac{\partial v_y}{\partial x} - \frac{\partial v_x}{\partial y}\right)$$

If the flow is irrotational, we have $\omega_x = \omega_y = \omega_z = 0$; it means

$$\frac{\partial v_z}{\partial y} - \frac{\partial v_y}{\partial z} = 0, \quad \frac{\partial v_x}{\partial z} - \frac{\partial v_z}{\partial x} = 0, \quad \frac{\partial v_y}{\partial x} - \frac{\partial v_x}{\partial y} = 0 \tag{13.70}$$

Therefore, irrotational flow of incompressible fluid should satisfy both Eqs (13.67) and (13.70). For such flow, we can use potential function Φ to express the components of the flow velocity as below:

$$v_x = \frac{\partial \phi}{\partial x}, \quad v_y = \frac{\partial \phi}{\partial y}, \quad v_z = \frac{\partial \phi}{\partial z} \tag{13.71}$$

Apparently, the potential function Φ defined like this must also satisfy the conditional expression of irrotational flow (13.70). If the fluid is also incompressible, the continuity equations below should be met:

$$\frac{\partial^2 \phi}{\partial x^2} + \frac{\partial^2 \phi}{\partial y^2} + \frac{\partial^2 \phi}{\partial z^2} = 0 \tag{13.72}$$

Along the boundary, the normal flow velocity and tangential flow velocity are, respectively,

$$v_n = \frac{\partial \phi}{\partial n}, \quad v_s = \frac{\partial \phi}{\partial s} \tag{13.73}$$

where n stands for normal of the boundary and s stands for the tangent of the boundary.

When the flow is parallel to x axis and has the uniform flow velocity U, the potential function can be expressed as

$$\phi = Ux + \text{constant} \tag{13.74}$$

Equation (13.74) can also sometimes be used to express the boundary condition of the infinity. Of course, in practical computation, a section far enough is taken to represent the infinity.

Thus, the potential flow formulated by potential function Φ can be shown as below.

In the area R where the boundary is $C = C_1 + C_2$; the potential function Φ satisfies the Laplace equation:

$$\nabla^2 \phi = \frac{\partial^2 \phi}{\partial x^2} + \frac{\partial^2 \phi}{\partial y^2} + \frac{\partial^2 \phi}{\partial z^2} = 0$$

On the boundary C_1, Φ takes the known value:

$$\phi = g \tag{13.75}$$

On the boundary of C_2, the normal flow of velocity has the following value:

$$\frac{\partial \phi}{\partial n} = \beta \tag{13.76}$$

Similar with the problem of steady seepage, the potential flow can also be expressed as the extreme value of functional below:

The function Φ on the boundary C_1 takes the known value, $\Phi = g$, and it makes the functional below reach the extreme value:

$$I(\phi) = \iiint_R \frac{1}{2}\left[\left(\frac{\partial\phi}{\partial x}\right)^2 + \left(\frac{\partial\phi}{\partial y}\right)^2 + \left(\frac{\partial\phi}{\partial z}\right)^2\right] dxdydz + \iint_{C_2} \beta\phi ds = \min \tag{13.77}$$

Equations (13.77) and (13.33) are alike, and the calculation equations are also similar for discretized finite elements. Thus the detail is omitted.

Based on the knowledge of hydrodynamics, if the flow is nonviscous, the energy at any point of flow space is a constant. For example, in the gravitational field, there is

$$\frac{V^2}{2g} + \frac{p}{\gamma} + z = \text{constant} = H$$

$$V^2 = v_x^2 + v_y^2 + v_z^2$$

where p is the pressure, H is the total head, γ is the unit weight of the fluid, and V is the velocity of flow.

This is the Bernoulli equation, with which the pressure p at any point could be calculated after figuring the velocity of the flow.

According to the difference of solving region and boundary condition, the problem of potential flow can be classified into three types:

1) Potential flow in single connected region
2) Potential flow in multiple connected region
3) Potential flow with free surface (some parts of the boundary are free surface).

The solution of the single connected region is not difficult and does not need special technique; while some skill is necessary for solving the other two problems. The problem of flow with the free surface will be discussed in next chapter, whereas what follows in this chapter is the method of calculation for the multiple connected field.

Considering the planar space of potential flow, the circulation Γ is defined as the integral of tangential velocity of flow v_s along any closed curve C as below:

$$\Gamma = \int_C v_s ds \tag{13.78}$$

According to character of potential flow, if there are no objects in C, then $\Gamma = 0$. If there are objects inside C, the circulation usually will not be zero. Only under some special circumstances, for example, the flow circling the objects is symmetrical, the circulation $\Gamma = 0$. In this case, there is no lift in the flow and ϕ is continuous in the full space.

When the circulation $\Gamma \neq 0$, there is lift of flow. Suppose the velocity of the coming flow is U, based on the famous Kutta–Joukowski lift theorem, the lift that the objects suffering is

$$L = -\rho U\Gamma$$

where L is the lift, ρ is the density, and Γ is the circulation around the object.

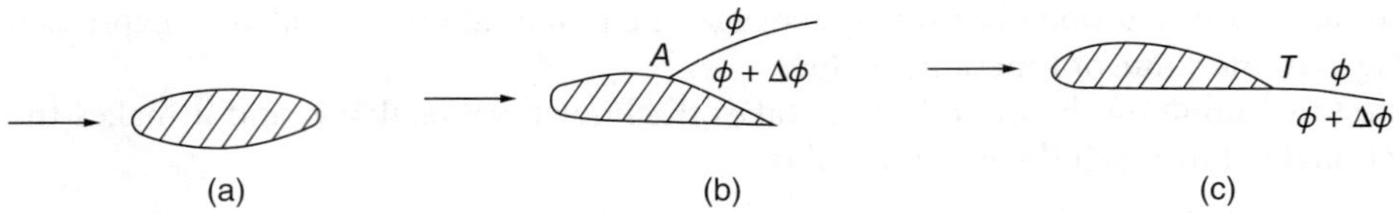

Figure 13.13 Two-dimensional potential flow around objects. (a) Flow without lift; (b,c) flow with lift.

Consider a random point A on the surface of the object and substitute $v_s = \partial\varphi/\partial s$ into Eq. (13.78), and then take the integration around the whole surface. We will get

$$\Gamma = \int_{A_-}^{A_+} \frac{\partial \phi}{\partial s} \mathrm{d}s = \phi_{A_+} - \phi_{A_-} = \Delta\phi$$

It is obvious that there is bound $\Delta\phi$ in the potential function at the surface, for example, point A, on the lift object. In fact, on the whole curve crossing A, there is the bound $\Delta\phi = \Gamma$, shown as Figure 13.13(b). This curve could be arranged in a random way. It will be easy to understand if the bound appeared at the end of the wing, just like the Figure 13.13(c). In this situation, the curve actually is the vortex line at the end.

13.11.2 The Flow around Objects without Lift

As shown like Figure 13.14, it is required to analyze the flow around the non-lifting object, and it is known that there is uniform velocity U for the fluid at the infinity; but a finite solving region R is always used in the actually calculation. A rectangular zone is often taken at the parallel direction of the flow, where the boundary is far enough from the object. Suppose the flow at the boundary is not disturbed, with the result that U is known, shown as Figure 13.14. Along the object's boundary, the normal velocity must be zero; thus under present conditions, as all the conditions of the boundary are given by $\partial\varphi/\partial s$, the following answer is not unique. When solving it by the finite element method, the global stiffness matrix $[K]$ is singular. In order to overcome this difficulty,

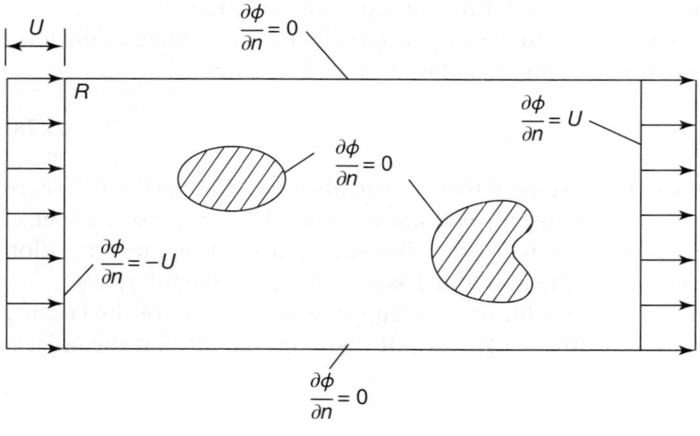

Figure 13.14 Flow around non-lifting objects.

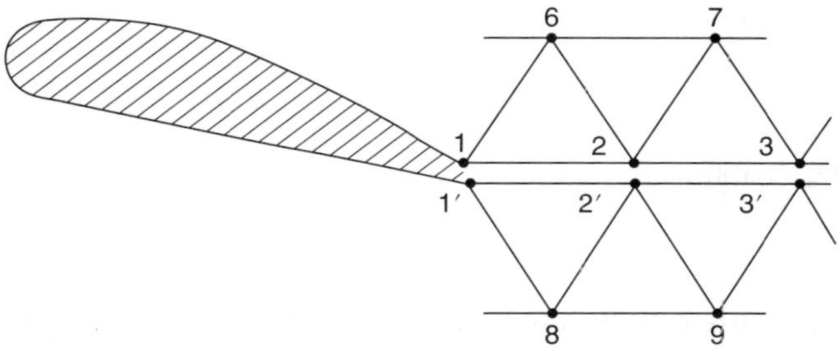

Figure 13.15 Flow around lifting objects.

we can pick up any node in the solution area and let ϕ take any known value at this node. For example, let $\phi = 0$ at a node on the boundary.

13.11.3 The Flow around Objects with Lift

Now we will discuss the two-dimensional potential flow around objects with lift. As shown in Figure 13.15, the flow field was divided along the trailing vortex line into simply connected regions. Taking 123 and 1′2′3′ as boundaries, the finite elements are established, respectively. For points 2 and 2′, there are

$$\left.\begin{aligned} H_{21}\phi_1 + H_{22}\phi_2 + H_{23}\phi_3 + H_{26}\phi_6 + H_{27}\phi_7 = F_2 \\ H_{21}\varphi_1 + H_{22}\varphi_2 + H_{23}\varphi_3 + H_{28}\varphi_8 + H_{29}\varphi_9 = F_2' \end{aligned}\right\} \tag{13.79}$$

Due to the outflow volume along the boundary, 123 is equal to the volume along the boundary 1′2′3′; from Eq. (13.22), it can be seen that $F_2' = -F_2$. Besides, $\varphi_{1'} = \varphi_1 + \Gamma, \varphi_{2'} = \varphi_2 + \Gamma, \varphi_{3'} = \varphi_3 + \Gamma$. Considering these conditions, equation of point 2 can be obtained as below when the two equations in (13.79) are merged:

$$\sum H_i\phi_i = -(H_{21}' + H_{22}' + H_{23}')\Gamma \tag{13.80}$$

In this way, point 2 also becomes an inner point; with appearance of the right-hand side. Point 3's situation is similar.

Based on the Kutta condition, the end points 1 and 1′ are stationary points. Hence two equations can be established as below:

$$\left.\begin{aligned} H_{11}\phi_1 + H_{12}\phi_2 + \textstyle\sum H_i\phi_i = 0 \\ H_{11}'(\phi_1 + \Gamma) + H_{12}'(\phi_2 + \Gamma) + \textstyle\sum H_i\phi_i = 0 \end{aligned}\right\} \tag{13.81}$$

In fact, two equations were established at the tail point under the Kutta conditions, and an extra equation may be used to solve Γ.

13.12 Potential Flow Formulated by Stream Function ψ

The plane potential flow can also be formulated by stream function ψ.

13.12.1 Basic Equations

The continuous equation of the plane potential flow is

$$\frac{\partial v_x}{\partial x} + \frac{\partial v_y}{\partial y} = 0 \tag{13.82}$$

The rotation that is circling z axis is

$$\omega_z = \frac{1}{2}\left(\frac{\partial v_y}{\partial x} - \frac{\partial v_x}{\partial y}\right)$$

Thus the condition of irrotational flow is

$$\frac{\partial v_y}{\partial x} - \frac{\partial v_x}{\partial y} = 0 \tag{13.83}$$

Now the stream function ψ is used to represent velocity component v_x and v_y as follows:

$$v_x = \frac{\partial \psi}{\partial y}, \quad v_y = -\frac{\partial \psi}{\partial x} \tag{13.84}$$

It is obvious that stream function ψ satisfies the continuous Eq. (13.82). Then substitute Eq. (13.84) into Eq. (13.83) for irrotational flow:

$$\frac{\partial^2 \psi}{\partial x^2} + \frac{\partial^2 \psi}{\partial y^2} = 0 \tag{13.85}$$

So the stream function has to satisfy Laplace's equation, too.

The normal velocity v_n and the tangential velocity v_s along the boundary are

$$v_n = \frac{\partial \psi}{\partial s}, \quad v_s = -\frac{\partial \psi}{\partial n}$$

The integral along the boundary is

$$\psi = \int \frac{\partial \psi}{\partial s} ds = \int v_n ds \tag{13.86}$$

If the normal direction of the velocity is known, the value of ψ at the boundary could be gained from Eq. (13.86). Let α be the known tangential direction of the velocity and the condition of the boundary can be obtained from the equation below:

$$\frac{\partial \psi}{\partial n} = -\alpha \tag{13.87}$$

For the flow parallel to the x axis with uniform velocity U, the stream function can be expressed as

$$\psi = Uy + \text{constant} \tag{13.88}$$

In solving the potential flow problem by stream function, the boundary condition is generally given by Eq. (13.86) or (13.88).

13.12.2 The Flow around Objects without Lift

Supposing far away from the object the fluid has uniform velocity U, the value of ψ at the outer boundary C_0 can be given according to Eq. (13.88) as shown in the Figure 13.16. Along the boundary C_1, $\partial\psi/\partial s = 0$ so ψ = constant. However, this constant is unknown, which causes the difficulty of the solution. In order to overcome it, Ref. [7] suggests to take

$$\psi(x, y) = \psi_0(x, y) + b\psi_1(x, y) \tag{13.89}$$

In this equation, b is an undetermined coefficient.

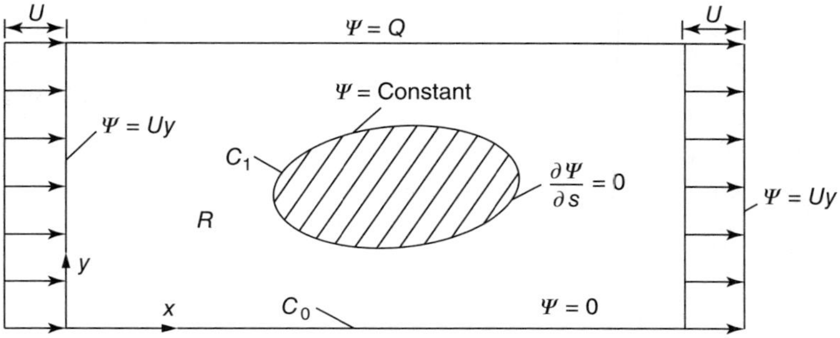

Figure 13.16 Flow around a non-lifting object.

ψ_0 and ψ_1 satisfy the following conditions, respectively:

1) In the area R

$$\nabla^2\psi_0 = 0$$

At C_0

$$\psi_0 = g(x, y) \tag{13.90}$$

At C_1

$$\psi_0 = 0$$

2) In the area R

$$\nabla^2\psi_1 = 0$$

At C_0

$$\psi_1 = 0 \tag{13.91}$$

At C_1

$$\psi_1 = 1$$

ψ_0 and ψ_1 can be obtained by the finite element method. Then we return to Eq. (13.89); at the point near the boundary C_0 inside the area R the ψ value of which is known, ψ_0 and ψ_1 are calculated according to Eq. (13.89). Then the constant b can be gained.

If there are n objects in the flow field, the method of solution is similar.

Let

$$\psi = \psi_0 + b_1\psi_1 + b_2\psi_2 + \cdots + b_n\psi_n \tag{13.92}$$

where $b_1, b_2, \ldots, b_n$ are n undetermined constants and $\psi_1, \psi_2, \ldots, \psi_n$ should satisfy the following conditions:

1) In the area R

$$\nabla^2\psi_0 = 0$$

At C_0

$$\psi_0 = g(x, y) \tag{13.93}$$

At $C_1, C_2, \ldots, C_n$

$$\psi_0 = 0$$

2) In the area R
   At C_1

$$\psi_1 = 1 \tag{13.94}$$

   At $C_0, C_2, C_3, \ldots, C_n$

$$\psi_1 = 0$$
$$\nabla^2\psi_1 = 0$$

3) In the area R

$$\nabla^2\psi_2 = 0$$

   At C_2

$$\psi_2 = 1 \tag{13.95}$$

   At $C_0, C_1, C_3, \ldots, C_n$

$$\psi_2 = 0$$

   (2) In the area R

$$\nabla^2\psi_n = 0$$

   At C_n

$$\psi_n = 1 \tag{13.96}$$

   At $C_0, C_1, \ldots, C_{n-1}$

$$\psi_n = 0$$

It is not difficult to solve $\psi_1, \psi_2, \ldots, \psi_n$ with the finite element method. Substituting them to Eq. (13.92), compute the ψ value at n points near boundary C_0 to form n linear equations the solution of which will give the n constants $b_1, b_2, \ldots, b_n$.

13.12.3 The Flow around Objects with Lift

When there are some lifting objects such as wings in the flow field, the Kutta condition must be considered in the calculation. Refer to Figure 13.17, and let U be inflow velocity, C_1 be the wing boundary, and C_0 be the boundary that is far away from the wing where the flow is not disturbed; the problem to be solved is

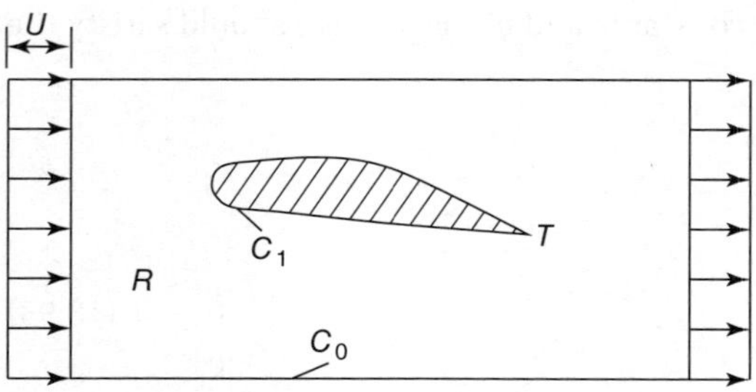

Figure 13.17 Flow around object with lift.

In the area R

$$\nabla^2\psi = 0$$

At the boundary C_0

$$\psi = Uy$$

At the boundary C_1

$$\psi = \text{constant}$$

The Kutta condition, at the leaving edge T

$$v_x = v_y = 0$$

To solve this problem, we can also use superposition method:
Let

$$\psi = \psi_0 + b\psi_1 \tag{13.97}$$

where ψ_0 and ψ_1 satisfy Eqs (13.90) and (13.91), respectively. Undetermined coefficient b can be solved by the Kutta condition. At the boundary of the wing C_1, $\psi_0 = 0$, $\psi_1 = 1$, and $\psi_n = b =$ constant, thus along the C_1, there is

$$\psi_n = \frac{\partial\psi}{\partial s} = 0$$

The normal velocity of the flow is also equal to zero at the leave edge T. If there is another velocity component equal to zero in the different direction of the normal n, the velocity at the point T will be zero. Taking this direction for x direction, there will be

$$v_{x\mathrm{T}} = \left.\frac{\partial\psi}{\partial y}\right|_{\mathrm{T}} = \left.\frac{\partial\psi_0}{\partial y}\right|_{\mathrm{T}} + b\left.\frac{\partial\psi_1}{\partial y}\right|_{\mathrm{T}} = 0 \tag{13.98}$$

When Eq. (13.98) is established, the Kutta condition is satisfied, and then coefficient b can be obtained from this equation.

Figure 13.18 shows the streamline distribution around a wing that gained by the superposition method, in which P_1 and P_2 are two stationary points.

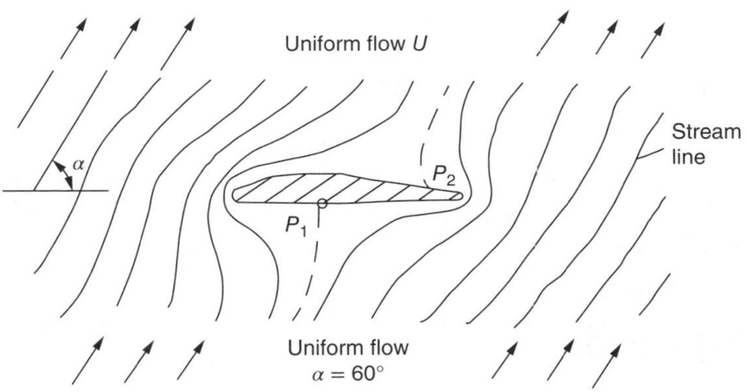

Figure 13.18 Flow around a wing.

13.13 Flow on the Free Surface

In practical engineering, flow on the free surface is a common problem, for example, the overflow on the dam surface and orifice outflow. The difficulty of this problem is that the exact location of the free surface is hard to know in advance. The free surface is a flow line; the normal velocity along the free surface is zero and so does the pressure. Similar with the problem of the dam seepage surface that was mentioned before, this problem can also be solved by iteration method, that is, presuppose a free surface to solve Laplace's equation and then check whether the conditions on the free surface are satisfied. If not, we may modify the position of the free surface and resolve the problem. Usually, we might repeat the calculation five or six times to get a satisfactory result.

The flow with the free surface is a variational problem with variable boundary. According to this theory, Professor Xueqing Xu in China has solved the problem of horizontal flow with the free surface; calculating two or three times could reach a satisfactory result as in Figure 13.19(a). Based on Prof. Xu's research, we solved the problem of flow with inclined free surface as shown in Figure 13.19(b) – by stream function ψ in a simpler and straighter way.

Referring to Figure 13.19(b), on section AB there is the uniform velocity U, $\psi = Uy$ along AB. Along the boundary $BCDE$, because $\partial\psi/\partial s = 0$, we have $\psi = 0$. The value of ψ can also be given, supposing the velocity is uniform along EF. Hence ψ is known along the boundary of $ABCDEF$. On the free surface AF, it is a streamline, so $\psi = Q$ (flow) and the pressure $p = 0$. That is to say the free surface should satisfy two conditions at the same time.

The problem to be solved is

In the region R

$$\frac{\partial^2\psi}{\partial x^2} + \frac{\partial^2\psi}{\partial y^2} = 0$$

On the boundary $ABCDEF$

$$\psi = g(x, y)$$

On free surface AF

$$\psi = Q$$

$$p = 0$$

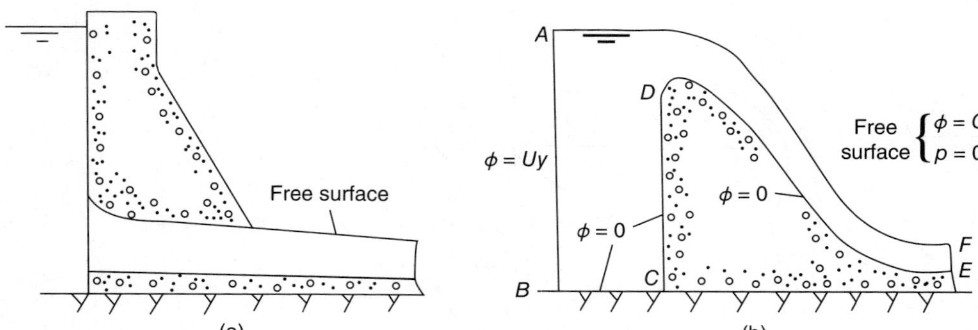

Figure 13.19 Flow with free surface. (a) Free surface boundary of flow at gentle slope; (b) free surface boundary of flow at steep slope.

The question above is equal to the variational problem below: to find the value for function $\psi(x, y)$ at all the boundaries and make the functional

$$I(\psi) = \iint_R \frac{1}{2}\left[\left(\frac{\partial\psi}{\partial x}\right)^2 + \left(\frac{\partial\psi}{\partial y}\right)^2\right] \mathrm{d}x\mathrm{d}y \tag{13.99}$$

to take the minimum value. The value of ψ is given in advance and at the same time $p = 0$ on the free surface.

One layer of trapezoidal-surface element is applied near the free surface, and triangular elements are applied at the other parts, of which the node coordinates are given in advance. While at node i, we calculate node parameter ψ and the thickness d of the trapezoidal-surface element at the same time, where d was considered as a floating and unknown value.

Refer to Figure 13.20, we considered node b on the free surface and the corresponding node i at Γ. At node i, the following equation can be got by setting functional $I(\psi)$ to take minimum value:

$$\sum_{e} \frac{\partial I^{e}}{\partial \psi_i} = 0 \tag{13.100}$$

Equation (13.100) sums all the elements around i. At node b, stream function was given a known value $\psi = Q$ and must satisfy the condition below due to the free surface:

$$p = 0 \tag{13.101}$$

By this equation we can compute the thickness d_i of the surface element.

In the plane xy, the Bernoulli equation for 2D flow is

$$\frac{V^2}{2g} + y + \frac{p}{\gamma} = \text{constant} = H \tag{13.102}$$

$$V^2 = v_x^2 + v_y^2 = \left(\frac{\partial\psi}{\partial x}\right)^2 + \left(\frac{\partial\psi}{\partial y}\right)^2 \tag{13.103}$$

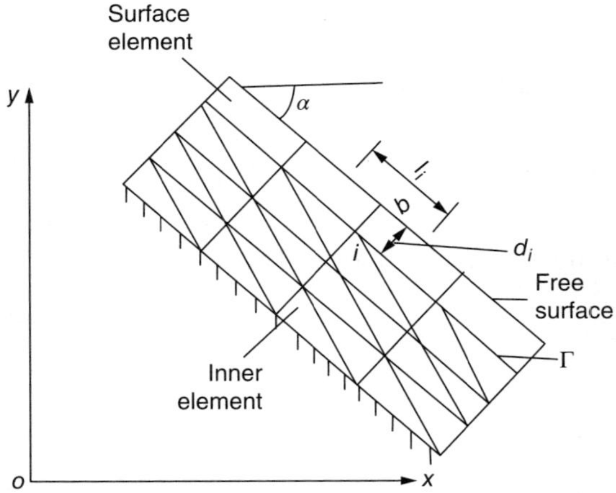

Figure 13.20 Free surface of overflow boundary.

In Eq. (13.102), H is the total head, γ is the unit weight, and V is the velocity of flow. On the free surface, for $p=0$, from the Bernoulli Eq. (13.102), we can get

$$V = -\sqrt{2g(H-y)} \tag{13.104}$$

According to Figure 13.20, the ordinate of point b is

$$y_b = y_i + d_i \cos \alpha_i$$

Substituting in Eq. (13.104), there will be

$$V = -\sqrt{2g(H-y_i)}\left(1 - \frac{d_i \cos \alpha_i}{H - y_i}\right)^{1/2}$$

Expand the right end above by the binomial theorem and omit the higher terms for d_i is smaller when compared with $H-y_i$. Then there is

$$V = -\sqrt{2g(H-y_i)}\left[1 - \frac{d_i \cos \alpha_i}{2(H - y_i)}\right] \tag{13.105}$$

Equation (13.105) represents the Bernoulli equations on the free surface.

Go back to the conditions where functional $I(\psi)$ takes minimization. Equation (13.100) sums for all the elements around node i. There are two types of elements, which are triangular elements and trapezoidal-surface elements. The formula for calculating the value of $\partial I^e/\partial \psi_i$ in triangular elements can be developed from shape function Eq. (2.7). The calculation formula for $\partial I^e/\partial \psi_i$ in the trapezoidal-surface element is given below.

On the free surface, the normal velocity $v_n = 0$, so

$$V = v_s = -\frac{\partial \psi}{\partial n} = -\frac{\psi_b - \psi_i}{d_i} \tag{13.106}$$

hence

$$\frac{\partial V}{\partial \psi_i} = \frac{1}{d_i} \tag{13.107}$$

Suppose there is a uniform velocity V for the surface element within the scope of l_i near point b, we substituted Eq. (13.103) into Eq. (13.99), and then will get

$$I^e = \iint\limits_{\Delta R} \frac{1}{2} V^2 \mathrm{d}x\mathrm{d}y$$

Evaluate $\partial I^e/\partial \psi_i$ by the equation above, and there will be

$$\frac{\partial I^e}{\partial \psi_i} = \frac{\partial I^e}{\partial V}\frac{\partial V}{\partial \psi_i} = \frac{\partial}{\partial V}\left[\iint\limits_{\Delta R} \frac{1}{2} V^2 \mathrm{d}x\mathrm{d}y\right]\frac{\partial V}{\partial \psi_i}$$

Then substitute Eq. (13.107) into the equation above, there will be

$$\frac{\partial I^e}{\partial \psi_i} = V l_i \tag{13.108}$$

where

$$l_i = \frac{l_1 + l_2}{2}$$

l_1 and l_2 are the length of the left and right surface element.

Substitute Eq. (13.105) into Eq. (13.108), we will get

$$\frac{\partial I^e}{\partial \psi_i} = -\sqrt{2g(H-y_i)}\left[1 - \frac{d_i \cos \alpha_i}{2(H-y_i)}\right] l_i \tag{13.109}$$

Then substitute Eq. (13.106) into Eq. (13.105) and omit higher terms of $d/(H-y)$, and we will get

$$\psi_i - \psi_b = -\sqrt{2g(H-y_i)}d_i \tag{13.110}$$

The equation above represents the Bernoulli equations on the free surface. At the node b, stream function is given the known value $\psi = Q$; d_i is unknown and the governing equation is $p=0$ on the free surface, which is Eq. (13.110). At the node i, ψ_i is unknown and the governing equation is Eq. (13.106).

In the actual calculation, we suppose curve Γ at first; then calculate the value of ψ at all nodes and the value of d_i at the free boundary. If d_i is more than the predetermined value when compared with the depth of the water, for example, 0.03%, we will let $y_i = y_i + d_i \cos \alpha_i$ and recalculate it until d_i satisfies the requirement.

If there is $\alpha_i = 0$ in Eq. (13.109), we will get Xu's formula. Prof. Xu had calculated a case in which d_i was 0.0996 times the depth of water in the first result and 0.00164 times in the second results, which reached the prescribed accuracy 0.03.

Free surface calculation for seepage analysis in Section 13.72 is applicable to the current situation. But the method in this section shows a new way of thinking and some computing skills.

In order to provide the primary position of the curve Γ for the first calculation, we could suppose the velocity distributed evenly along the depth. Let y be the channel bottom height (Figure 13.21); α is the angle between the channel bottom and horizontal line; S is the water depth that is perpendicular to the channel bottom; the flow is Q and the total heal is H. Based on the Bernoulli equation, there will be

$$\frac{1}{2g}\left(\frac{Q}{S}\right)^2 + y + S\cos\alpha = H$$

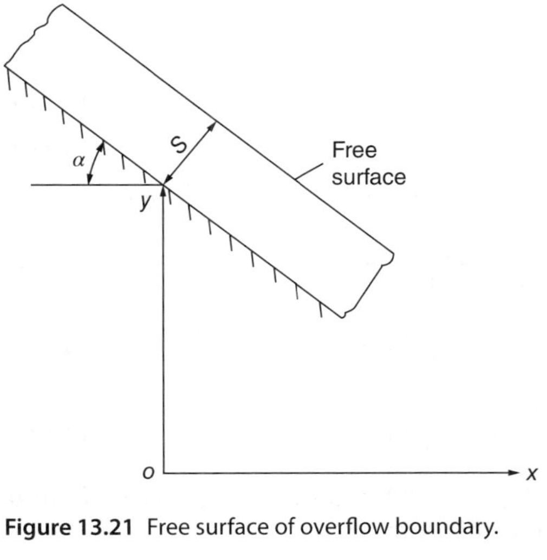

Figure 13.21 Free surface of overflow boundary.

Multiplying both sides of the equation above by S^2, we get

$$S^3 \cos \alpha - (H - y)S^2 + \frac{Q^2}{2g} = 0 \tag{13.111}$$

Once S was got from Eq. (13.111), approximate free surface line will be obtained.

13.14 Viscous and Non-Newtonian Flow

13.14.1 Solution of the Stokes Equation

The basic equations governing the two-dimensional steady incompressible flow of fluid are as follows:

$$\frac{\partial p}{\partial x} + \rho \left(v_x \frac{\partial v_x}{\partial x} + v_y \frac{\partial v_x}{\partial y} \right) = X + \eta \left(\frac{\partial^2 v_x}{\partial x^2} + \frac{\partial^2 v_x}{\partial y^2} \right) \tag{13.112}$$

$$\frac{\partial p}{\partial y} + \rho \left(v_x \frac{\partial v_y}{\partial x} + v_y \frac{\partial v_y}{\partial y} \right) = Y + \eta \left(\frac{\partial^2 v_y}{\partial x^2} + \frac{\partial^2 v_y}{\partial y^2} \right) \tag{13.113}$$

The continuity equation is

$$\frac{\partial v_x}{\partial x} + \frac{\partial v_y}{\partial y} = 0 \tag{13.114}$$

where X and Y are the components of body force. When the convective terms (terms involving ρ) in Eqs (13.112) and (13.113) are neglected, we obtain the Stokes equations:

$$\eta \left(\frac{\partial^2 v_x}{\partial x^2} + \frac{\partial^2 v_x}{\partial y^2} \right) - \frac{\partial p}{\partial x} + X = 0 \tag{13.115}$$

$$\eta \left(\frac{\partial^2 v_y}{\partial x^2} + \frac{\partial^2 v_y}{\partial y^2} \right) - \frac{\partial p}{\partial y} + Y = 0 \tag{13.116}$$

To solve the above equations by finite element method, in each element, the unknowns p, v_x, and v_y are expressed as

$$p^{\mathrm{e}} = \sum N_i^{p}(x, y) P_i \tag{13.117}$$

$$v_x^{\mathrm{e}} = \sum N_i^{vx}(x, y) V_{xi} \tag{13.118}$$

$$v_y^{\mathrm{e}} = \sum N_i^{vy}(x, y) V_{yi} \tag{13.119}$$

Using Galerkin's weighted residual method, we get the following equations:

$$\iint N_i(x, y) \left[\eta \left(\frac{\partial^2 v_x}{\partial x^2} + \frac{\partial^2 v_x}{\partial y^2} \right) - \frac{\partial p}{\partial x} + \rho X \right] \mathrm{d}A = 0 \tag{13.120}$$

$$\iint N_i(x, y) \left[\eta \left(\frac{\partial^2 v_y}{\partial x^2} + \frac{\partial^2 v_y}{\partial y^2} \right) - \frac{\partial p}{\partial y} + \rho Y \right] \mathrm{d}A = 0 \tag{13.121}$$

$$\iint N_i \left[\frac{\partial v_x}{\partial x} + \frac{\partial v_y}{\partial y} \right] \mathrm{d}A = 0 \tag{13.122}$$

Substituting Eqs (13.117)–(13.119) into Eqs (13.120)–(13.122), we get a set of linear equations, the solution of which will give the unknowns p_i, v_{xi}, and v_{yi}.

13.14.2 Solution of the Navier–Stokes Equations

To solve the Navier–Stokes equations (13.12) by the weighted residual method, the following set of equations are to be solved:

$$\iint N_i(x,y)\left[\rho X - \frac{\partial p}{\partial x} + 2\eta\frac{\partial}{\partial x}\left(\frac{2}{3}\frac{\partial v_x}{\partial x} - \frac{\partial v_y}{\partial y} - \frac{\partial v_z}{\partial z}\right)\right.$$
$$\left. +\eta\frac{\partial}{\partial y}\left(\frac{\partial v_y}{\partial x} + \frac{\partial v_x}{\partial y}\right) + \eta\frac{\partial}{\partial z}\left(\frac{\partial v_x}{\partial z} + \frac{\partial v_z}{\partial x}\right)\right] \mathrm{d}A = 0 \qquad (13.123)$$

$$\iint N_i(x,y)\left[\rho Y - \frac{\partial p}{\partial y} + \eta\frac{\partial}{\partial x}\left(\frac{\partial v_x}{\partial y} + \frac{\partial v_y}{\partial x}\right)\right]$$
$$+2\eta\left(\frac{2}{3}\frac{\partial v_y}{\partial y} - \frac{\partial v_x}{\partial x} - \frac{\partial v_z}{\partial z}\right) + \eta\frac{\partial}{\partial z}\left(\frac{\partial v_y}{\partial z} - \frac{\partial v_y}{\partial y}\right)\Bigg] \mathrm{d}A = 0$$
$$(13.124)$$

$$\iint N_i(x,y)\left(\frac{\partial v_x}{\partial x} + \frac{\partial v_y}{\partial y}\right)\mathrm{d}A = 0 \qquad (13.125)$$

Substituting Eqs (13.117)–(13.119) to Eqs (13.123)–(13.125), we will have a set of linear equations, the solution of which will give the unknowns (Figure 13.22).

Yamada and Ito considered the problem of flow past a circular cylinder to illustrate the previous analysis [11]. As indicated in Figure 13.23, the infinite field of flow is confined by a circle of radius $r/a = 8.0$ (or 20.0) and the region is divided into finite elements. The boundary conditions at these radii are specified as $V = 1.0$ or 5.0 cm/s and $p = 0$ for a Reynolds number of 30.0. The boundary conditions on the surface of the cylinder ($r = a$) are taken as $v_x = v_y = 0$. The velocity distribution obtained by solving the Stokes equations is given by Figure 13.24. For the Navier–Stokes equations, the velocity distribution becomes unsymmetrical due to the inclusion of convective terms as shown in Figure 13.24.

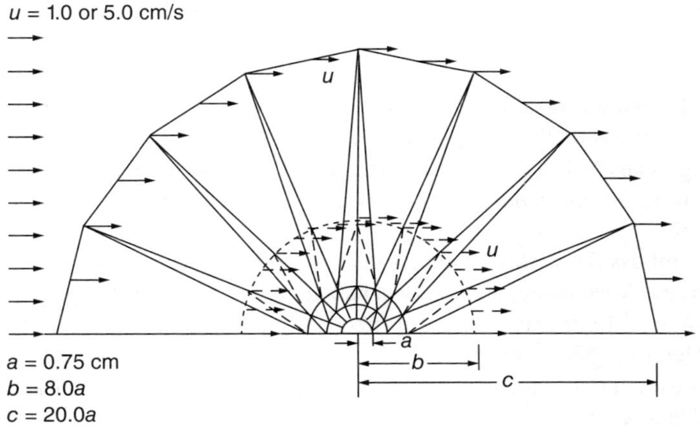

Figure 13.22 Boundary conditions and finite element mesh for the analysis of the flow past a circular cylinder.

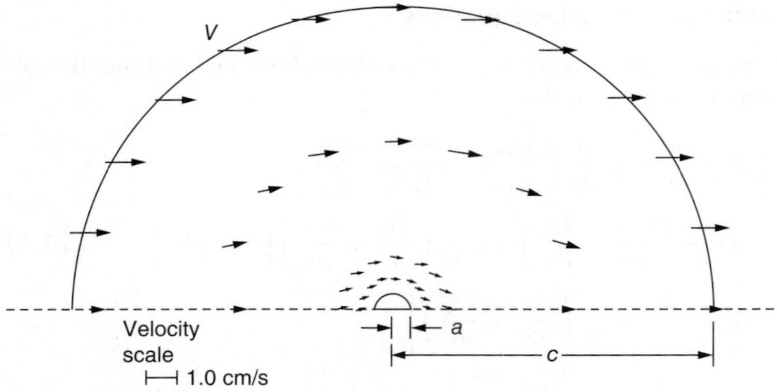

Figure 13.23 Velocity solution of the Stokes equations.

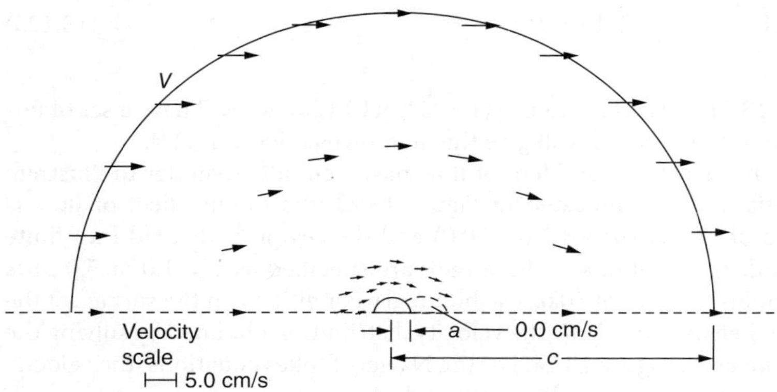

Figure 13.24 Velocity solution of Navier–Stokes equations ($R_e = 30$).

Bibliography

1 Desai, C.S. (1976) Finite element residual schemes for unconfined flow. *International Journal for Numerical Methods in Engineering*, **10** (6), 1415–1418.

2 Bathe, K.J. and Khoshgoftaar, M.R. (1979) Finite element free surface seepage analysis without mesh iteration. *International Journal for Numerical Analysis Methods in Geomechanics*, **3**, 13–22.

3 Gell, V.K. (1984) Der Einfluss der Sickrestromung im Untergrund auf die Berechnung der Spannungen und Verformungen von Bogenstaumauern. Veroffentlichungen des Institutes fur Grundbau, Bodenmechanik, Felsmechanik und Verkchrswasserbau der RWTH Aachen. Herausgegeben von Prof. Dr. Ing. W. Wittke, Heft II.

4 Zhu, B. (1982) Hybrid elements considering effect of drainage holes in seepage field. *Journal of Hydraulic Engineering*, **9**, 32–41.

5 Whiteman, J.R. (1973) *The Mathematics of Finite Elements and Applications*, Academic Press, New York.

6 Desai, C.S. (1972) Seepage Analysis of Earth Banks under Drawdown. *Journal of the Geotechnical Engineering Division, ASCE*, **SM11**, 1143–1162.
7 Norrie, D.H. and de Vries, G. (1973) *The Finite Element Method, Fundamentals and Applications*, Academic Press, London.
8 Zhu, B., Li, Y., Xu, P. and Zhang, G. (2007) Substitution of draining holes by seeping layer. *Water Resources and Hydropower Engineering*, **10**, 24–28.
9 Zienkiewicz, O.C., Taylor, R.L. and Nithiarasu, P. (2005) *The Finite Element Method for Fluid Dynamics*, Elsevier, New York.
10 Xu, X. (1978) Finite element method for hydraulics problems with free surface. *Science and Technology of Hydraulic and Water Transportation*, **2**, 1–17.
11 Y. Yamada, K. Ito, Finite element analysis of steady fluid and metal flow. *Finite Elements in Fluids*, **1**, RH Gallagher, Wiley, London, 1975.
12 Rao, S.S. (2011) *The Finite Element Method in Engineering*, Elsevier, New York.

14

Problems in Conduction of Heat in Solids

Experience shows that the finite difference method is suitable for the solution of one dimensional temperature field and the finite element method is more suitable for the solution of two-dimensional (2D) and three-dimensional (3D) temperature field in practical engineering.

14.1 Differential Equation: Initial and Boundary Conditions for Conduction of Heat in Solids

Let $T(x, y, z, \tau)$ be the temperature of point (x, y, z) at time τ; according to the theory of heat conduction in solids, $T(x, y, z, \tau)$ must satisfy the following differential equation:

$$\frac{\partial T}{\partial \tau} = a\left(\frac{\partial^2 T}{\partial x^2} + \frac{\partial^2 T}{\partial y^2} + \frac{\partial^2 T}{\partial z^2}\right) + \frac{\partial \theta}{\partial \tau} \tag{14.1}$$

where τ is time, a is the diffusivity, and θ is the adiabatic temperature rise of the solid, for example, the temperature rise of concrete due to the hydration heat of cement. The initial condition is when $\tau = 0$

$$T = T_0(x, y, z) \tag{14.2}$$

As shown in Figure 14.1, the surface of the solid is divided into two parts: C and C'. The boundary condition of the problem is as follows:

$$\left.\begin{aligned} &\text{on boundary } C': \quad T = T_{\mathrm{B}} \\ &\text{on boundary } C: \quad -\lambda\left(l_x\frac{\partial T}{\partial x} + l_y\frac{\partial T}{\partial y} + l_z\frac{\partial T}{\partial z}\right) = \beta(T - T_{\mathrm{a}}) \end{aligned}\right\} \tag{14.3}$$

where l_x, l_y, l_z are the direction cosines of the outward normal to the surface, T_{B} is the given surface temperature, T_{a} is the air temperature, β is the surface conductance of the solid or the coefficient of surface heat transfer, and λ is the conductivity of the solid.

The relation between a and λ, c, ρ of the solid is

$$a = \frac{\lambda}{c\rho} \tag{14.4}$$

where a is the diffusivity, λ is the conductivity, c is the specific heat, and ρ is the density.

The Finite Element Method: Fundamentals and Applications in Civil, Hydraulic, Mechanical and Aeronautical Engineering, First Edition. Bofang Zhu.

Companion website: www.wiley.com/go/zhufem17

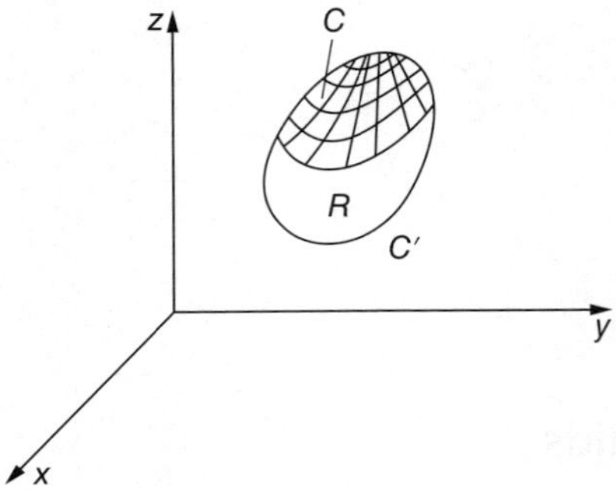

Figure 14.1 A space problem.

14.2 Variational Principle for Conduction of Heat in Solids

14.2.1 Euler's Equation

As shown in Figure 14.1, consider the functional

$$I(T) = \iiint_R F(T, T_x, T_y, T_z)\,\mathrm{d}x\mathrm{d}y\mathrm{d}z + \iint_C G(T)\mathrm{d}s \tag{14.5}$$

where $T_x = \partial T/\partial x$, $T_y = \partial T/\partial y$, $T_z = \partial T/\partial z$ and the boundary conditions of temperature T are

$$\left.\begin{aligned} &\text{on boundary } C' \quad T = T_\mathrm{b} \\ &\text{on boundary } C \quad l_x\frac{\partial T}{\partial x} + l_y\frac{\partial T}{\partial y} + l_z\frac{\partial T}{\partial z} = -\frac{\beta}{\lambda}(T - T_\mathrm{a}) \end{aligned}\right\} \tag{14.6}$$

in which l_x, l_y, l_z are the direction cosines of the outward normal to the surface and T_a is the air temperature. It has been proved that when the functional $I(T)$ takes the minimum value, namely,

$$\delta(I) = 0 \tag{14.7}$$

the following equations are valid:

$$\frac{\partial F}{\partial T} - \frac{\partial}{\partial x}\left(\frac{\partial F}{\partial T_x}\right) - \frac{\partial}{\partial y}\left(\frac{\partial F}{\partial T_y}\right) - \frac{\partial}{\partial z}\left(\frac{\partial F}{\partial T_z}\right) = 0 \tag{14.8}$$

$$\frac{\partial G}{\partial T} + l_x\frac{\partial F}{\partial T_x} + l_y\frac{\partial F}{\partial T_y} + l_z\frac{\partial F}{\partial T_z} = 0 \tag{14.9}$$

These are the Euler's equations of the variational problem Eqs. (14.5)–(14.7).

14.2.2 Variational Principle of Problem of Heat Conduction

Consider the unsteady temperature $T(x, y, z, t)$ that satisfies the following equations: in region R

$$\frac{\partial^2 T}{\partial x^2} + \frac{\partial^2 T}{\partial y^2} + \frac{\partial^2 T}{\partial z^2} + \frac{1}{a}\left(\frac{\partial \theta}{\partial \tau} - \frac{\partial T}{\partial \tau}\right) = 0 \tag{14.10}$$

when $\tau = 0$

$$T = T_0(x, y, z) \tag{14.11}$$

on surface C' when $\tau > 0$,

$$T = T_\mathrm{b} \tag{14.12}$$

on surface C when $\tau > 0$,

$$l_x\frac{\partial T}{\partial x} + l_y\frac{\partial T}{\partial y} + l_z\frac{\partial T}{\partial z} + \frac{\beta}{\lambda}(T - T_\mathrm{a}) = 0 \tag{14.13}$$

Now take the function F and G as follows:

$$\left.\begin{aligned} F &= \frac{1}{2}\left[\left(\frac{\partial T}{\partial x}\right)^2 + \left(\frac{\partial T}{\partial y}\right)^2 + \left(\frac{\partial T}{\partial z}\right)^2\right] - \frac{1}{a}\left(\frac{\partial \theta}{\partial \tau} - \frac{\partial T}{\partial \tau}\right)T \\ G &= \frac{\beta}{\lambda}\left(\frac{1}{2}T^2 - T_{\mathrm{a}}T\right) \end{aligned}\right\} \tag{14.14}$$

Substituting F and G into Eq. (14.5), we get

$$\begin{aligned} I(T) = &\iiint_R \left\{\frac{1}{2}\left[\left(\frac{\partial T}{\partial x}\right)^2 + \left(\frac{\partial T}{\partial y}\right)^2 + \left(\frac{\partial T}{\partial z}\right)^2\right] - \frac{1}{a}\left(\frac{\partial \theta}{\partial \tau} - \frac{\partial T}{\partial \tau}\right)T\right\} \mathrm{d}x\mathrm{d}y\mathrm{d}z \\ &+ \iint_C \frac{\beta}{\lambda}\left(\frac{1}{2}T^2 - T_{\mathrm{a}}T\right)\mathrm{d}s \end{aligned} \tag{14.15}$$

From Eq. (14.14), we get the partial derivatives of F and G:

$$\frac{\partial F}{\partial T} = -\frac{1}{a}\left(\frac{\partial \theta}{\partial \tau} - \frac{\partial T}{\partial \tau}\right),\ \frac{\partial F}{\partial T_x} = \frac{\partial T}{\partial x},\ \frac{\partial F}{\partial T_y} = \frac{\partial T}{\partial y}$$

$$\frac{\partial F}{\partial T_z} = \frac{\partial T}{\partial z},\ \frac{\partial}{\partial x}\left(\frac{\partial F}{\partial T_x}\right) = \frac{\partial^2 T}{\partial x^2},\ \frac{\partial}{\partial y}\left(\frac{\partial F}{\partial T_y}\right) = \frac{\partial^2 T}{\partial y^2}$$

$$\frac{\partial}{\partial z}\left(\frac{\partial F}{\partial T_z}\right) = \frac{\partial^2 T}{\partial z^2},\ \frac{\partial G}{\partial T} = \frac{\beta}{\lambda}(T - T_{\mathrm{a}})$$

Substituting them into Eqs (14.8) and (14.9), we have in region R

$$\begin{aligned} &\frac{\partial F}{\partial T} - \frac{\partial}{\partial x}\left(\frac{\partial F}{\partial T_x}\right) - \frac{\partial}{\partial y}\left(\frac{\partial F}{\partial T_y}\right) - \frac{\partial}{\partial z}\left(\frac{\partial F}{\partial T_z}\right) \\ &= -\frac{1}{a}\left(\frac{\partial \theta}{\partial \tau} - \frac{\partial T}{\partial \tau}\right) - \frac{\partial^2 T}{\partial x^2} - \frac{\partial^2 T}{\partial y^2} - \frac{\partial^2 T}{\partial z^2} = 0 \end{aligned} \tag{14.16}$$

on boundary C

$$\begin{aligned} &\frac{\partial G}{\partial T} + l_x\frac{\partial F}{\partial T_x} + l_y\frac{\partial F}{\partial T_y} + l_z\frac{\partial F}{\partial T_z} \\ &= \frac{\beta}{\lambda}(T - T_{\mathrm{a}}) + l_x\frac{\partial T}{\partial x} + l_y\frac{\partial T}{\partial y} + l_z\frac{\partial T}{\partial z} = 0 \end{aligned} \tag{14.17}$$

Eqs (14.16) and (14.17) are identical with Eqs (14.10) and (14.13). Hence the following conclusion is derived: if the temperature $T(x, y, z, \tau)$ takes the initial temperature T_0 when $\tau = 0$ takes the boundary temperature T_{b} on the surface C' and enforce the functional $I(T)$ expressed by Eq. (14.15) to take the minimum value, then $T(x, y, z, \tau)$ will satisfy Eq. (14.10) in region R and Eq. (14.13) on surface C, namely, $T(x, y, z, \tau)$ is the unsteady temperature solution that we seek.

14.3 Discretization of Continuous Body

Temperatures are the unknown variables in the solution of temperature field. There are infinite points in a continuous body, and every point has a temperature; hence there

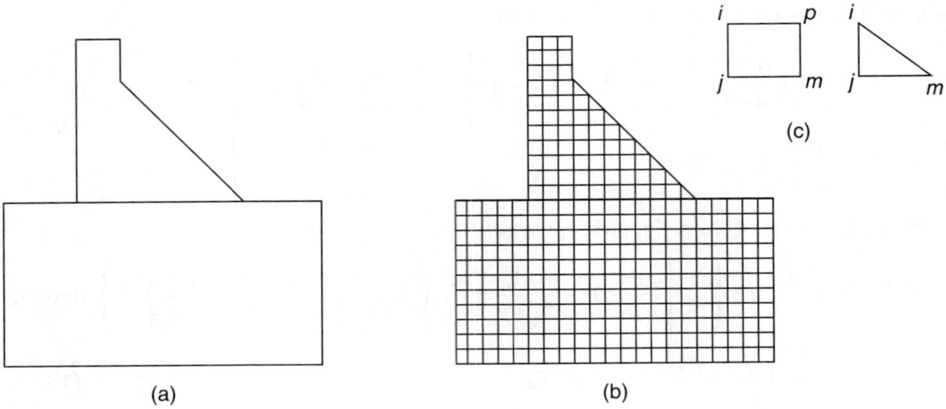

Figure 14.2 Discretization of continuous structure by finite elements: (a) original structure and (b) structure after discretization.

are infinite unknown variables in the temperature field of a continuous body, which is difficult to be solved by numerical method. In the finite element method, the continuous body is substituted by a group of elements, which are connected at finite number of nodes as shown in Figure 14.2. The temperatures in the element are expressed by the temperatures of the nodes with shape functions. Only temperatures of the finite number of nodes are unknown variables, and the original problem with infinite unknown variables is replaced by a new problem with finite unknown variables, which can be solved by numerical methods on the computer.

14.4 Fundamental Equations for Solving Unsteady Temperature Field by FEM

Consider a 3D unsteady temperature field, and the equation of heat conduction, the initial and boundary conditions are given by Eqs (14.1)–(14.3). According to the variational principle and Eq. (14.14), this heat conduction problem is identical with the minimum value problem of the functional $I(T)$: $T(x,y,z,\tau) = T_0(x,y,z)$ when $\tau = 0$ and $T(x,y,z,\tau) = T_b$ on the first kind of boundary C' and enforce the following functional $I(T)$ to take the minimum value.

$$I(T) = \iiint_R \left\{ \frac{1}{2}\left[\left(\frac{\partial T}{\partial x}\right)^2 + \left(\frac{\partial T}{\partial y}\right)^2 + \left(\frac{\partial T}{\partial z}\right)^2\right] - \frac{1}{a}\left(\frac{\partial \theta}{\partial \tau} - \frac{\partial T}{\partial \tau}\right) T \right\} \mathrm{d}x\mathrm{d}y\mathrm{d}z + \iint_C \frac{\beta}{\lambda}\left(\frac{1}{2}T^2 - T_a T\right)\mathrm{d}s \tag{14.18}$$

The first term of the right part of Eq. (14.18) is the space integral in the region of solution R, while the second term is the area integral of the boundary C. As shown in Figure 14.2, the solution region R is divided into finite number of elements. Suppose that the nodes of element e are $i, j, m, \ldots, p$, the temperatures of nodes are $T_i(\tau)$, $T_j(\tau)$, $T_m(\tau), \ldots, T_p(\tau)$, and the temperature of any point of the element is expressed by the temperatures of nodes as follows:

$$
\begin{aligned}
T^{\mathrm{e}}(x, y, z, \tau) &= N_i T_i + N_j T_j + N_m T_m + \cdots + N_p T_p \\
&= [N_i, N_j, N_m \cdots] \begin{Bmatrix} T_i \\ T_j \\ T_m \\ \vdots \end{Bmatrix} = [N]\{T\}^{\mathrm{e}}
\end{aligned}
\tag{14.19}
$$

where the shape function $N_i(\xi, \eta, \zeta)$ is a function of the local coordinates ξ, η, ς and the nodal temperature $T_i(\tau)$ is a function of time.

The rate of change of temperature of any point in the element is

$$
\begin{aligned}
\frac{\partial T}{\partial \tau} &= N_i \frac{\partial T_i}{\partial \tau} + N_j \frac{\partial T_j}{\partial \tau} + N_m \frac{\partial T_m}{\partial \tau} + \cdots \\
&= [N_i, N_j, N_m + \cdots] \begin{Bmatrix} \dot{T}_i \\ \dot{T}_j \\ \dot{T}_m \\ \vdots \end{Bmatrix} = [N]\{\dot{T}\}^{\mathrm{e}}
\end{aligned}
\tag{14.20}
$$

Considering element e as a subregion ΔR of the solution region R, the value of the functional in the subregion is

$$
\begin{aligned}
I^{\mathrm{e}}(T) = \iiint_{\Delta R} \Bigg\{ & \frac{1}{2}\left[\left(\frac{\partial T}{\partial x}\right)^2 + \left(\frac{\partial T}{\partial y}\right)^2 + \left(\frac{\partial T}{\partial z}\right)^2\right] \\
& - \frac{1}{a}\left(\frac{\partial \theta}{\partial \tau} - \frac{\partial T}{\partial \tau}\right) T \Bigg\} \mathrm{d}x\mathrm{d}y\mathrm{d}z + \iint_{\Delta C} \frac{\beta}{\lambda}\left(\frac{1}{2}T^2 - T_{\mathrm{a}} T\right) \mathrm{d}s
\end{aligned}
\tag{14.21}
$$

Differentiating Eq. (14.21) in the sign of integral, we obtain

$$
\begin{aligned}
\frac{\partial I^{\mathrm{e}}}{\partial T_i} = \iiint_{\Delta R} \Bigg(& \frac{\partial T}{\partial x}\frac{\partial}{\partial T_i}\left(\frac{\partial T}{\partial x}\right) + \frac{\partial T}{\partial y}\frac{\partial}{\partial T_i}\left(\frac{\partial T}{\partial y}\right) + \frac{\partial T}{\partial z}\frac{\partial}{\partial T_i}\left(\frac{\partial T}{\partial z}\right) \\
& + \frac{1}{a}\left(\frac{\partial T}{\partial \tau} - \frac{\partial \theta}{\partial \tau}\right)\frac{\partial T}{\partial T_i}\Bigg) \mathrm{d}x\mathrm{d}y\mathrm{d}z + \iint_{\Delta C} \frac{\beta}{\lambda}\left(T\frac{\partial T}{\partial T_i} - T_{\mathrm{a}}\frac{\partial T}{\partial T_i}\right) \mathrm{d}s
\end{aligned}
$$

Substitution of Eq. (14.19) into the above equation yields

$$
\begin{aligned}
\frac{\partial I^{\mathrm{e}}}{\partial T_i} = {} & h_{ii}^{\mathrm{e}} T_i + h_{ij}^{\mathrm{e}} T_j + h_{im}^{\mathrm{e}} T_m + \cdots r_{ii}^{\mathrm{e}} \frac{\partial T_i}{\partial \tau} + r_{ij}^{\mathrm{e}} \frac{\partial T_j}{\partial \tau} + r_{im}^{\mathrm{e}} \frac{\partial T_m}{\partial \tau} + \cdots - f_i^{\mathrm{e}} \frac{\partial \theta}{\partial \tau} \\
& + g_{ii}^{\mathrm{e}} T_i + g_{ij}^{\mathrm{e}} T_j + g_{im}^{\mathrm{e}} T_m + \cdots - p_i^{\mathrm{e}} T_a
\end{aligned}
\tag{14.22}
$$

where

$$\left.\begin{aligned}
h_{ij}^{e} &= \iiint\limits_{\Delta R} \left(\frac{\partial N_i}{\partial x}\frac{\partial N_j}{\partial x} + \frac{\partial N_i}{\partial y}\frac{\partial N_j}{\partial y} + \frac{\partial N_i}{\partial z}\frac{\partial N_j}{\partial z} \right) \mathrm{d}x\mathrm{d}y\mathrm{d}z \\
f_i^{e} &= \frac{1}{a}\iiint\limits_{\Delta R} N_i \mathrm{d}x\mathrm{d}y\mathrm{d}z \\
g_{ij}^{e} &= \frac{\lambda}{\beta}\iint\limits_{\Delta C} N_i N_j \mathrm{d}s \\
p_i^{e} &= \frac{\lambda}{\beta}\iint\limits_{\Delta C} N_i \mathrm{d}s \\
r_{ij}^{e} &= \frac{1}{a}\iiint\limits_{\Delta R} N_i N_j \mathrm{d}x\mathrm{d}y\mathrm{d}z
\end{aligned}\right\} \tag{14.23}$$

in which g_{ij}^{e} and p_i^{e} are area integrals on the second kind of boundary C.

When the elements are small enough, the original functional may be replaced by the sum of the functional of the elements, namely,

$$I(T) \cong \sum_e I^{e}(T) \tag{14.24}$$

In order to enforce the functional $I(T)$ to take the minimum value, it is necessary that

$$\frac{\partial I}{\partial T_i} \cong \sum_e \frac{\partial I^{e}}{\partial T_i} = 0 \tag{14.25}$$

Substituting Eq. (14.21) into the above equation, we get

$$\begin{aligned}
&\sum_e \left(h_{ii}^{e} T_i + h_{ij}^{e} T_j + h_{im}^{e} T_m + \cdots + r_{ii}^{e}\frac{\partial T_i}{\partial \tau} + r_{ij}^{e}\frac{\partial T_j}{\partial \tau} + r_{im}^{e}\frac{\partial T_m}{\partial \tau} + \cdots \right. \\
&\qquad \left. - f_i \frac{\partial \theta}{\partial \tau} + g_{ii}^{e} T_i + g_{ij}^{e} T_j + g_{im}^{e} T_m + \cdots - p_i^{e} T_a \right) = 0
\end{aligned} \tag{14.26}$$

which may be transformed into the following equation by the symbol of matrix:

$$[H]\{T\} + [R]\left\{\frac{\partial T}{\partial \tau}\right\} + \{F\} = 0 \tag{14.27}$$

where

$$\left.\begin{aligned}
H_{ij} &= \sum_e (h_{ij}^{e} + g_{ij}^{e}) \\
R_{ij} &= \sum_e r_{ij}^{e} \\
F_i &= \sum_e \left(-f_i \frac{\partial \theta}{\partial \tau} - p_i^{e} T_a \right)
\end{aligned}\right\} \tag{14.28}$$

in which $\sum_e$ indicates the sum of all the elements relevant to the node i.

The Eq. (14.27) is valid at any time τ, so it is valid at $\tau = \tau_n$ and $\tau = \tau_{n+1}$, namely,

$$[H]\{T_n\} + [R]\left\{\frac{\partial T}{\partial \tau}\right\}_n + \{F_n\} = 0 \tag{14.29}$$

$$[H]\{T_{n+1}\} + [R]\left\{\frac{\partial T}{\partial \tau}\right\}_{n+1} + \{F_{n+1}\} = 0 \tag{14.30}$$

Now let

$$\Delta T_n = T_{n+1} - T_n = \Delta\tau_n\left[(1-s)\left(\frac{\partial T}{\partial \tau}\right)_n + s\left(\frac{\partial T}{\partial \tau}\right)_{n+1}\right] \tag{14.31}$$

According to the value of s, these are the following cases:

1) For $s = 0$, $\Delta T_n = \Delta\tau_n\left(\frac{\partial T}{\partial \tau}\right)_n$, forward difference, explicit solution
2) For $s = 1$, $\Delta T_n = \Delta\tau_n\left(\frac{\partial T}{\partial \tau}\right)_{n+1}$, backward difference, implicit solution
3) For $s = 1/2$, $\Delta T_n = \frac{1}{2}\Delta\tau_n\left[\left(\frac{\partial T}{\partial \tau}\right)_n + \left(\frac{\partial T}{\partial \tau}\right)_{n+1}\right]$, midpoint difference, implicit solution

Experience shows that the backward difference method ($s = 1$) is the better one in the two implicit methods.

From Eq. (14.31), we have

$$\left\{\frac{\partial T}{\partial \tau}\right\}_{n+1} = \frac{1}{s\Delta\tau_n}[\{T_{n+1}\} - \{T_n\}] - \frac{1-s}{s}\left\{\frac{\partial T}{\partial \tau}\right\}_n \tag{14.32}$$

Substituting into Eq. (14.30), we get

$$[H]\{T_{n+1}\} + [R]\left(\frac{1}{s\Delta\tau_n}[\{T_{n+1}\} - \{T_n\}] - \frac{1-s}{s}\left\{\frac{\partial T}{\partial \tau}\right\}_n\right) + \{F_{n+1}\} = 0 \tag{14.33}$$

From Eq. (14.29)

$$-[R]\left\{\frac{\partial T}{\partial \tau}\right\}_n = [H]\{T_n\} + \{F_n\}$$

Substituting into Eq. (14.33), the fundamental equation for computing the unsteady temperature field by FEM is derived in the following:

$$\left([H] + \frac{1}{s\Delta\tau_n}[R]\right)\{T_{n+1}\} + \left(\frac{1-s}{s}[H] - \frac{1}{s\Delta\tau_n}[R]\right)\{T_n\} + \frac{1-s}{s}\{F_n\} + \{F_{n+1}\} = 0 \tag{14.34}$$

in which, $\{T_n\}, \{F_n\}$, and $\{F_{n+1}\}$ are known and $\{T_{n+1}\}$ is unknown; thus the above equation is a linear equation system about $\{T_{n+1}\}$ the solution of which will give the temperatures $\{T_{n+1}\}$ of all the nodes at time $\tau = \tau_{n+1}$.

14.5 Two-Dimensional Unsteady Temperature Field, Triangular Elements

In order to give the readers a clear idea, it will be explained in the following how triangular elements are used to compute the 2D unsteady temperature field.

As shown in Figure 14.3, the solution region R is divided into a set of elements each of which has three nodes. Taking one element e whose nodes are i, j, and m, the temperature in the element e is expressed in the following: (Figure 14.4)

$$T^e(x, y, y, \tau) = N_i(x, y)T_i(\tau) + N_j(x, y)T_j(\tau) + N_m(x, y)T_m(\tau) \tag{14.35}$$

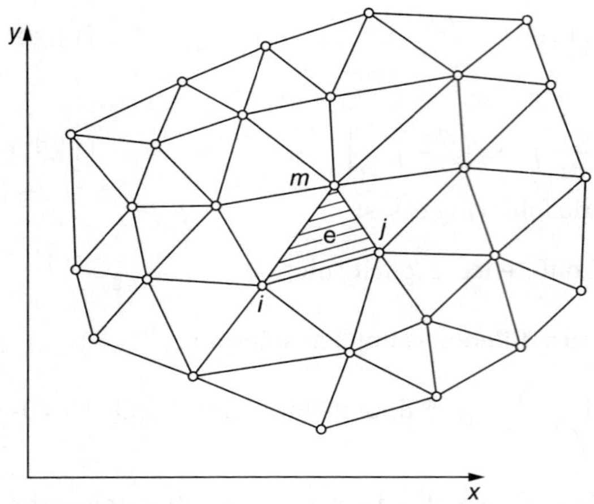

Figure 14.3 Discretization of the solution region.

in which N_i, N_j, N_m are the following shape functions

$$\left.\begin{aligned} N_i &= \frac{1}{2A}(a_i + b_i x + c_i y) \\ N_j &= \frac{1}{2A}(a_j + b_j x + c_j y) \\ N_m &= \frac{1}{2A}(a_m + b_m x + c_m y) \end{aligned}\right\} \tag{14.36}$$

Where

$$a_i = x_j y_m - x_m y_j, \quad b_i = y_j - y_m, \quad c_i = x_m - x_j \tag{14.37}$$

and A is the area of the triangle ijm.

$$A = \frac{1}{2}\begin{vmatrix} 1 & x_i & y_i \\ 1 & x_j & y_j \\ 1 & x_m & y_m \end{vmatrix}$$

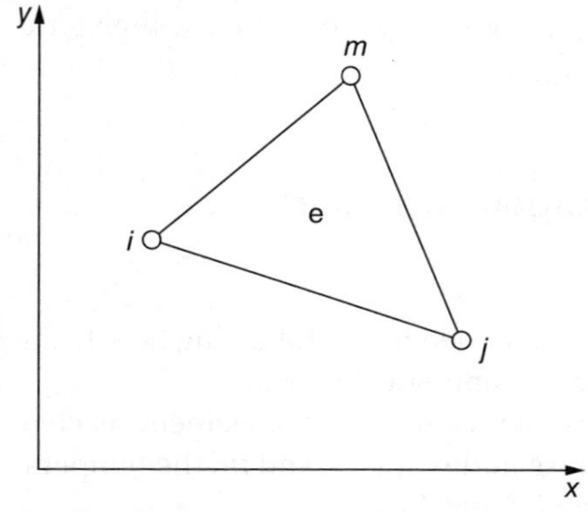

Figure 14.4 Element e.

Substituting Eq. (14.36) into Eq. (14.23), we get

$$\begin{aligned} h_{ii}^{e} &= \frac{b_i^2 + c_i^2}{4A}, h_{ij}^{e} = \frac{b_i b_j + c_i c_j}{4A}, \\ h_{im}^{e} &= \frac{b_i b_m + c_i c_m}{4A} \\ r_{ii}^{e} &= \frac{A}{6a}, r_{ij}^{e} = r_{im}^{e} = \frac{A}{12a} \\ f_i^{e} &= \frac{A}{3a}, p_i^{e} = \overline{\beta} L/2 \\ g_{ii}^{e} &= \overline{\beta} L/3, g_{ij}^{e} = g_{im}^{e} = \overline{\beta} L/6, \overline{\beta} = \beta/\lambda \end{aligned} \tag{14.38}$$

Now we have obtained all the coefficients in Eq. (14.23). Let

$$\frac{\partial \theta}{\partial \tau} = \frac{\Delta \theta}{\Delta \tau} \tag{14.39}$$

Then the Eq. (14.34) has established the solution of which will give the nodal temperatures $\{T_{n+1}\}$ for $\tau = \tau_{n+1}$.

14.6 Isoparametric Elements

Today the isoparametric elements are extensively used in practice to raise the efficiency of computing.

14.6.1 Two-Dimensional Isoparametric Elements

The isoparametric elements are constructed by transformation of coordinates. The 2D linear isoparametric element is shown in Figure 14.5. The element in the local coordinates is a 2×2 square, the origin of the local coordinates (ξ, η) is the center of element, and the four boundaries of element are $\xi = \pm 1$ and $\eta = \pm 1$. In the Cartesian coordinates, the element is a quadrilateral. The coordinates of any point in the element are expressed by

$$\left.\begin{aligned} x &= \sum N_i x_i = N_1 x_1 + N_2 x_2 + \cdots \\ y &= \sum N_i y_i = N_1 y_1 + N_{2y} x_2 + \cdots \end{aligned}\right\} \tag{14.40}$$

where $N_i(\xi, \eta)$ is the shape function expressed by local coordinates ξ, η, and (x_i, y_i) are the global coordinates of node i. Equation (14.40) is the coordinate transform formula for 2D isoparametric element.

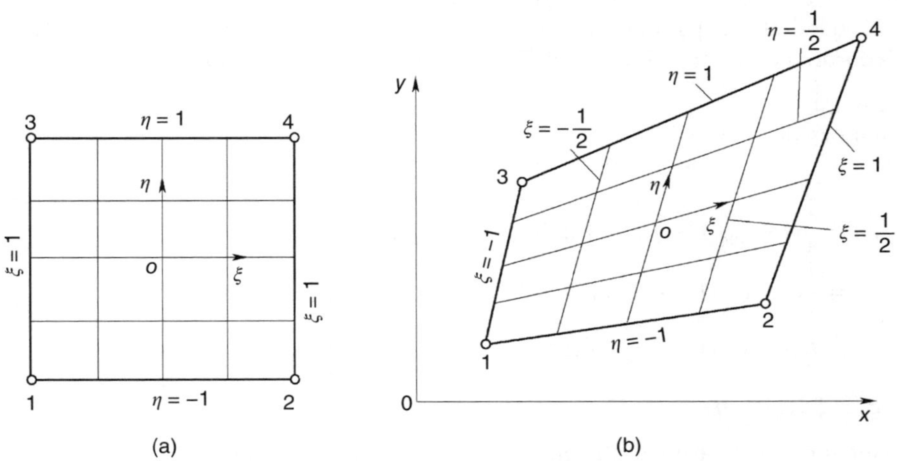

Figure 14.5 Two-dimensional linear isoparametric element: (a) local coordinates and (b) Cartesian coordinates.

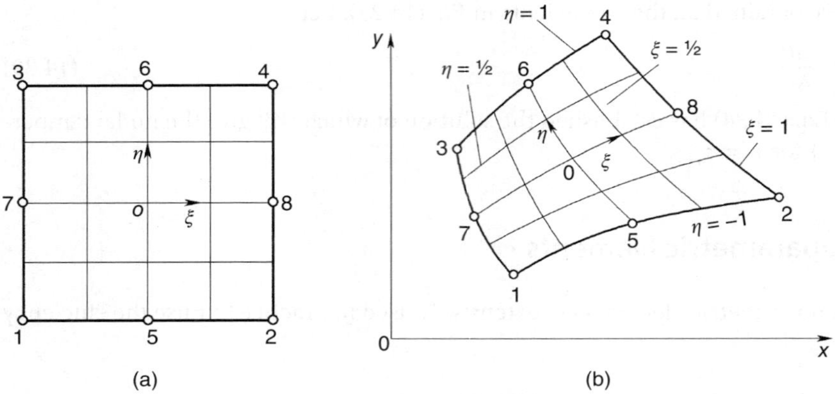

Figure 14.6 Two-dimensional quadratic isoparametric element: (a) local coordinates and (b) Cartesian coordinates.

The shape functions for 2D linear isoparametric element are

$$\begin{aligned} N_1 &= \frac{1}{4}(1-\xi)(1-\eta) \\ N_2 &= \frac{1}{4}(1+\xi)(1-\eta) \\ N_3 &= \frac{1}{4}(1-\xi)(1+\eta) \\ N_4 &= \frac{1}{4}(1+\xi)(1+\eta) \end{aligned} \tag{14.41}$$

The temperature at any point of the element in the Cartesian coordinate system is

$$T = \sum N_i T_i = N_1 T_1 + N_2 T_2 + \cdots \tag{14.42}$$

where $N_i(\xi, \eta)$ is the shape function expressed by local coordinates (ξ, η) and T_i is the temperature at node i in the Cartesian coordinate.

The 2D quadratic isoparametric element with eight nodes is shown in Figure 14.6. The coordinates are transformed by Eq. (14.40) with shape functions as follows:

Corner point
Midpoint of side

$$\left.\begin{aligned} N_i &= \frac{1}{4}(1+\xi_0)(1+\eta_0)(\xi_0+\eta_0-1), \quad i = 1,2,3 \\ N_i &= \frac{1}{2}(1-\xi^2)(1+\eta_0), \quad i = 5,6 \\ N_i &= \frac{1}{2}(1-\eta^2)(1+\xi_0), \quad i = 7,8 \\ \xi_0 &= \xi_i\xi, \eta_0 = \eta_i\eta \end{aligned}\right\} \tag{14.43}$$

The shape functions are quadratic on the four sides of the element.

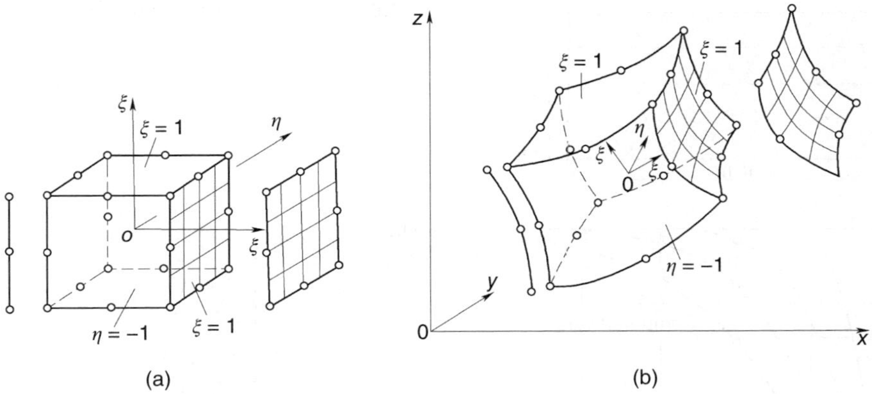

Figure 14.7 Three-dimensional quadratic isoparametric element: (a) local coordinates and (b) Cartesian coordinates.

14.6.2 Three-Dimensional Isoparametric Elements

A 3D quadratic isoparametric element with 20 nodes is shown in Figure 14.7. The coordinates of any point of the element in the Cartesian coordinates are

$$\left.\begin{aligned} x &= \sum N_i x_i = N_1 x_1 + N_2 x_2 + \cdots \\ y &= \sum N_i y_i = N_1 y_1 + N_2 y_2 + \cdots \\ z &= \sum N_i z_i = N_1 z_1 + N_2 z_2 + \cdots \end{aligned}\right\} \tag{14.44}$$

where x_i, y_i, z_i are the Cartesian coordinates of node i and $N_i(\xi, \eta, \zeta)$ is the shape function expressed by local coordinates ξ, η, ζ as follows:

Side

$$\left.\begin{aligned} &\text{corner point} && N_i = \frac{1}{8}(1+\xi_0)(1+\eta_0)(1+\zeta_0)(\xi_0+\eta_0+\zeta_0-2) \\ &\text{typical midpoint of side} && (\xi_i = 0, \eta_i = \pm 1, \zeta_i = \pm 1) \\ & && N_i = \frac{1}{4} - (1-\xi^2)(1+\eta_0)(1+\zeta_0) \end{aligned}\right\} \tag{14.45}$$

The temperature in the element in the Cartesian coordinates is still expressed by Eq. (14.42), but the shape functions are expressed by ξ, η, ζ as Eq. (14.45).

14.7 Computing Examples of Unsteady Temperature Field

Example 1 Cooling of rectangular prism

A rectangular prism of infinite length, the cross section is 10 m × 4 m, the initial temperature is $T_0 = 20\,°\text{C}$, the sides AD and BC are insulated, the sides AB and CD kept at

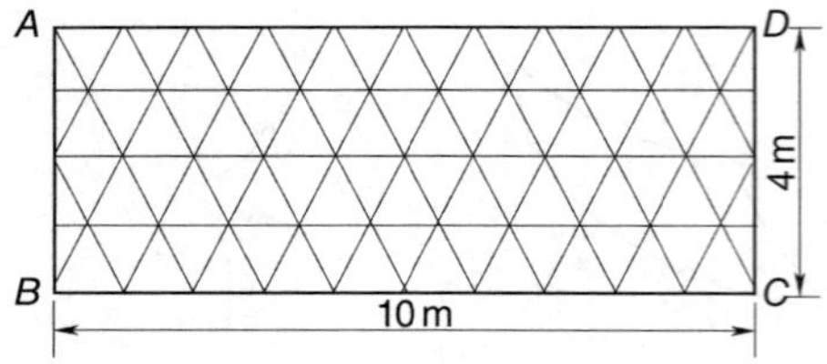

Figure 14.8 Example, net of computation.

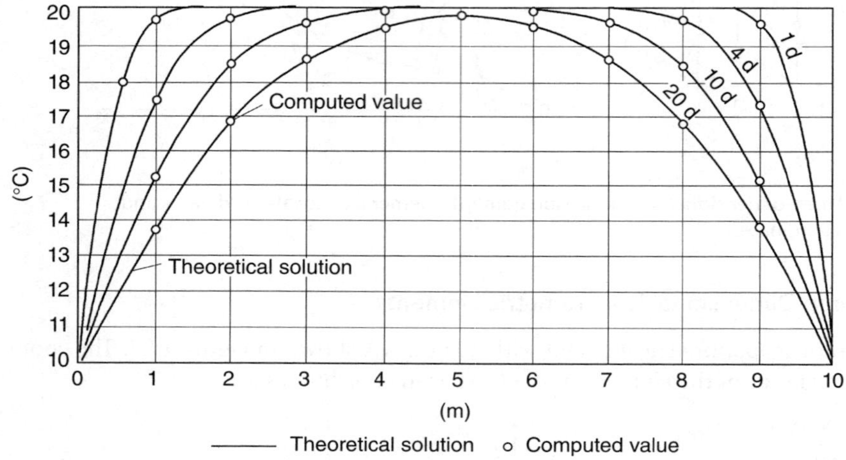

Figure 14.9 Example 1 comparison of the results given by FEM and the theoretical solution.

$T_b = 10\,°\text{C}$, the diffusivity $a = 0.10\,\text{m}^2/\text{d}$, the net of computation is shown in Figure 14.8, and the results computed by FEM are shown in Figure 14.9. The computed results all lie on the curves of theoretical solutions.

Example 2 Cooling of concrete dam

A concrete gravity dam is shown in Figure 14.10(a). The height of dam is $H = 65$ m, the width of the base of dam is $L = 49$ m, the initial temperature of the dam is 20 °C and that of the rock foundation is 0 °C, and the boundary temperature is also 0 °C. The process of cooling is computed by FEM, and the computed results are shown in Figure 14.10. It is clear from Figure 14.10(e) that the temperature in the interior of dam is still 2 °C after cooling for eight years.

14.8 Temperature Field of Mass Concrete with Pipe Cooling

14.8.1 Concrete Cylinder Cooled by Water Pipe

The temperature of mass concrete generally is controlled by water flowing in pipes embedded in concrete: the radius b or diameter D of the concrete cylinder is given by

$$b = \sqrt{1.07 s_1 s_2/\pi} = 0.5836\sqrt{s_1 s_2}, \quad D = 2b = 1.1672\sqrt{s_1 s_2} \tag{14.46}$$

where s_1, s_2 are the horizontal and vertical spacing of pipes, considering the effect of cooling of a rectangular prism with sides s_1 and s_2 is lower than that of a circular cylinder, so the coefficient $k = 1.07 > 1.00$ is used in the formula.

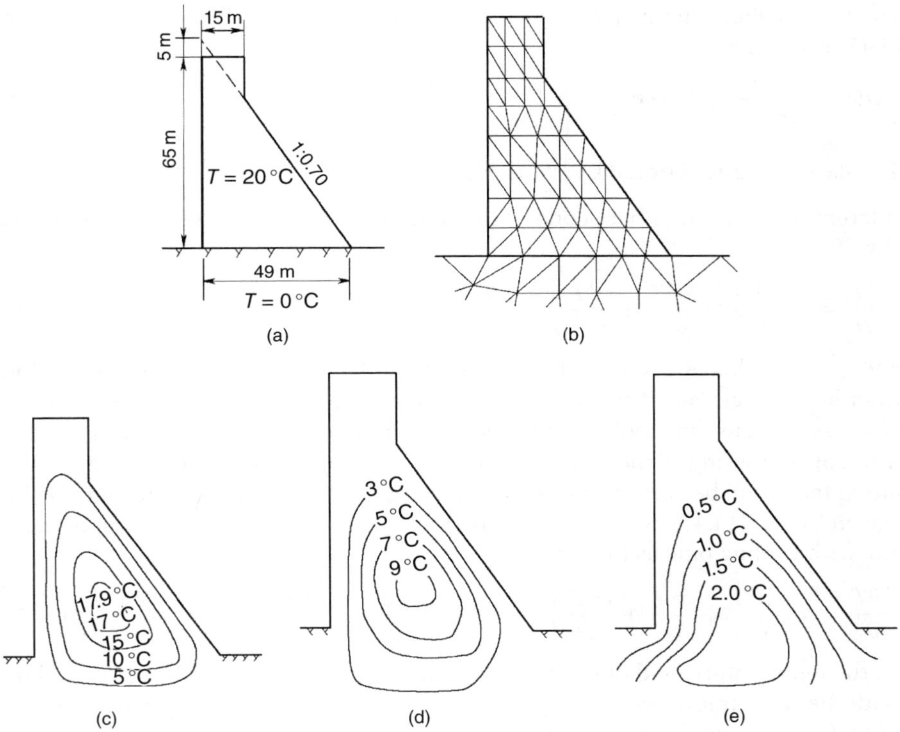

Figure 14.10 Example 2 cooling of concrete dam (a) dimensions of dam and initial temperature T_0, (b) net of computing (only part of net is shown for the foundation), (c) temperature distribution after cooling of one year, (d) temperature distribution after cooling of three years, and (e) temperature distribution after cooling for eight years.

The solution of concrete cylinder cooled by water pipe is an infinite series, but it converges so quickly that only the first term is needed in practical computation. See Ref. [3].

The mean temperature of a concrete cylinder cooled by water pipe is

$$T_1(t) = T_w + (T_0 - T_w)\phi(\tau) + \theta_0\psi(\tau) \tag{14.47}$$

where $T(t)$ is mean temperature of concrete cylinder with diameter D and length L, T_w is temperature of water at the inlet of the cooling pipe, and T_0 is initial temperature of concrete

$$\left.\begin{aligned} &\phi(t) = e^{-k_2 z} = e^{-p\tau} \\ &k_2 = 2.09 - 1.35\xi + 0.320\xi^2 \\ &\xi = \frac{\lambda L}{c_w \rho_w q_w}, \quad z = \frac{ga\tau}{D^2} \\ &g = \frac{\ln 100}{\ln(b/c) + (\lambda/\lambda_1)\ln(c/r_0)} \\ &p = k_2 ga/D^2 \end{aligned}\right\} \tag{14.48}$$

Where λ is conductivity of concrete, λ_1 is conductivity of cooling pipe, c, r_0 are outer and inner radius of the cooling pipe, and c_w, ρ_w, q_w are specific heat, density, and discharge of cooling water.

When the adiabatic temperature rise of concrete is $\theta(t) = \theta_0(1 - e^{-m\tau})$, $\psi(\tau)$ in Eq. (14.47) is given by

$$\psi(\tau) = \frac{m}{m-p}(e^{-p\tau} - e^{-m\tau}) \tag{14.49}$$

14.8.2 Mass Concrete Cooled by Water Pipe

The differential equation of heat conduction of mass concrete without cooling pipe is given by

$$\frac{\partial T}{\partial \tau} = a\left(\frac{\partial^2 T}{\partial x^2} + \frac{\partial^2 T}{\partial y^2} + \frac{\partial^2 T}{\partial z^2}\right) + \frac{\partial \theta}{\partial \tau} \tag{14.50}$$

where $\theta(\tau)$ is the adiabatic temperature rise due to internal source of heat, such as the hydration heat of cement. If the above equation is used to compute the temperature field of mass concrete with cooling pipes, as the radius of cooling pipe is about 1 cm, the dimension of computing elements near the cooling pipe must also be about 1 cm, so the computing mesh will be very intense with a huge number of DOF. Substituting $\theta(\tau)$ by $T_1(\tau)$ given by Eq. (14.47), we get the differential equation of heat conduction of mass concrete with cooling pipe as follows:

$$\frac{\partial T}{\partial \tau} = a\left(\frac{\partial^2 T}{\partial x^2} + \frac{\partial^2 T}{\partial y^2} + \frac{\partial^2 T}{\partial z^2}\right) + (T_0 - T_w)\frac{\partial \phi}{\partial \tau} + \theta_0\frac{\partial \psi}{\partial \tau} \tag{14.51}$$

Thus, the temperature field of mass concrete with cooling pipes can be computed by FEM with the conventional computing mesh; the effect of pipe cooling is considered by $\phi(\tau)$ and $\psi(\tau)$. The calculation is simplified remarkably.

14.8.3 Mass Concrete Cooled by Water Pipe with Precise $\theta(\tau)$

Experience shows that the formula

$$\theta(\tau) = \theta_0(1 - e^{-m\tau})$$

does not tally well with the experimental results. It would be better to use the following more precise formula:

$$\theta(\tau) = \theta_1(1 - e^{-m_1\tau}) + \theta_2(1 - e^{-m_2\tau}) \tag{14.52}$$

In this case, the differential equation of heat conduction of mass concrete with cooling pipe will be

$$\frac{\partial T}{\partial \tau} = a\left(\frac{\partial^2 T}{\partial x^2} + \frac{\partial^2 T}{\partial y^2} + \frac{\partial^2 T}{\partial z^2}\right) + (T_0 - T_w)\frac{\partial \varphi}{\partial \tau} + \theta_1\frac{\partial \psi_1}{\partial \tau} + \theta_2\frac{\partial \psi_2}{\partial \tau} \tag{14.53}$$

where

$$\left.\begin{aligned} \psi_1(\tau) &= \frac{m_1}{m_1 - p}(e^{-p\tau} - e^{-m_1\tau}) \\ \psi_2(\tau) &= \frac{m_2}{m_2 - p}(e^{-p\tau} - e^{-m_2\tau}) \end{aligned}\right\} \tag{14.54}$$

Bibliography

1 Zhu, B. (2014) *Thermal Stresses and Temperature Control of Mass Concrete*, Elsevier, New York, London.
2 Carslaw, H.S. and Jaeger, J.C. (1986) *Conduction of Heat in Solids*, 2nd edn, Oxford University Press, Oxford.
3 U. S. Bureau of Reclamation (1949) *Cooling of Concrete Dams*, U. S. Bureau of Reclamation, Denver.
4 Zhu, B. (1957) The effect of pipe cooling in mass concrete with internal source of heat. *Journal of Hydraulic Engineering*, **4**, 87–106.
5 Zhu, B. and Jiangbo, C. (1989) Finite element analysis of pipe cooling in mass concrete, a three dimensional problem. *Journal of Construction Engineering and Management ASCE.*, **115** (14), 487–498.
6 Zhu, B. (1991) Equivalent equation of heat conduction considering the effect of pipe cooling. *Journal of Hydraulic Engineering*, **3**, 28–34.
7 Zhu, B. (1999) Effect of cooling by water flowing in nonmetal pipes embedded in mass concrete. *Journal of Construction Engineering and Management, ASCE*, **125** (1), 61–68.

15

Methods for Nonlinear Finite Element Analysis

So far, what we discussed are all linear problems, but a large number of nonlinear problems will be encountered in engineering practice. Take solid mechanics, for example, the following two assumptions are adopted in linear elasticity: ① the stress–strain relationship of material is linear, as shown in Eq. (3.6), which assumes that materials obey Hooke's law, and ② the strain–displacement relationship is linear, as shown as Eq. (3.5), which is valid only for small displacement.

A large number of problems are consistent with these assumptions; thus, the linear method has been widely applied. But there are many problems that do not follow the above assumptions. For example, when the steel stress exceeds its proportional limit, the stress–strain relationship is nonlinear, and the stress–strain relations of soils and rocks are also nonlinear. These are called material nonlinear problems. For structures such as beams, plates, and shells in the post-buckling state, the stress–strain relationship becomes nonlinear due to its large displacement; these are called geometrical nonlinear problems.

In addition, nonlinear heat conduction problems will be encountered when thermal diffusivity and internal heat sources of materials are related to temperature. Nonlinearity in fluid mechanics occurs when viscosity coefficient is related to velocity of flow or when Darcy's law is no longer applicable as a result of turbulence.

This chapter describes the solving methods for nonlinear finite element analysis. Despite that there are various kinds of nonlinear finite element problems, their solving methods are similar. Hence, the problem of material nonlinearity is taken here as an example to describe how to solve nonlinear problems with finite element methods.

When the stress–strain relationship of materials is nonlinear, the stiffness matrix will not be a constant, but related to strain and displacement, and labeled as $[K(\delta)]$. Then, global balance equations of structures are a set of nonlinear equations as follows:

$$\{\psi\} = [K(\delta)]\{\delta\} - \{P\} = 0 \tag{15.1}$$

The methods for solving nonlinear problems can be divided into three categories: the incremental method, the iterative method, and the mixed method. The incremental method divides the load into a set of increments applied once each time. The stiffness matrix is assumed to be a constant in each load increment but have different values related to the stress–strain relationships in different load increments. The iterative method applies all the load value in each iteration process but gradually modifies displacement and strain to meet the nonlinear stress–strain relationships.

The Finite Element Method: Fundamentals and Applications in Civil, Hydraulic, Mechanical and Aeronautical Engineering, First Edition. Bofang Zhu.

Companion website: www.wiley.com/go/zhufem17

The mixed method uses both the incremental method and the iterative method, which also divides load into small increments, but the number of increments is less than that in the incremental method, and iterative computations will be performed in each increment.

15.1 Incremental Method

When nonlinear problems are analyzed with the incremental method, the load is divided into a series of small increments that may be equal or unequal. When one load increment is applied in each step, equations are assumed to be linear, and the stiffness matrix $[K]$ is constant, which may have different values in different load increment. The displacement increment $\{\Delta\delta\}$ is obtained in each step when one load increment $\{\Delta P\}$ applied, and the displacement $\{\delta\}$ is obtained by accumulation $\{\delta\} = \sum \Delta\delta$.

The incremental method approximates the nonlinear problem with a series of linear problems, which actually means that a nonlinear curve is replaced by a broken line.

Divide the load into m increments, so the total load is

$$\{P\} = \sum_{j=1}^{m} \{\Delta P_j\}$$

After applying the ith increment, the load is

$$\{P_i\} = \sum_{j=1}^{i} \{\Delta P_j\} \tag{15.2}$$

Each load increment generates one displacement increment $\{\Delta\delta_i\}$ and one stress increment $\{\Delta\sigma_j\}$, so after applying ith load, the displacement and the stress are

$$\{\delta_i\} = \sum_{j=1}^{i} \{\Delta\delta_j\} \tag{15.3}$$

$$\{\sigma_i\} = \sum_{j=1}^{i} \{\Delta\sigma_j\} \tag{15.4}$$

The following methods may be used to calculate the displacement increment $\{\Delta\delta_i\}$ and the stress increment $\{\Delta\sigma_i\}$ for the load increment $\{\Delta P_i\}$.

15.1.1 Method of Starting Point Stiffness

As shown in Figure 15.1 (the figure indicates a single-variable problem, but all equations in this chapter can be used in multivariable problems), assuming that the stress $\{\sigma_{i-1}\}$ at the end of step $i-1$ is already obtained, the elasticity matrix $[D_{i-1}]$ at the end of step $i-1$ can be determined according to $\{\sigma_{i-1}\}$ and stress–strain relationship, so that the stiffness matrix $[K_{i-1}]$ at the end of step $i-1$ can be calculated. Then, assuming that the stiffness matrix within step i remains unchanged and is approximately equal to $[K_{i-1}]$, the displacement increment $\{\Delta\delta_i\}$ at step i can be calculated according to the following equations:

$$[K_{i-1}]\{\Delta\delta_i\} = \{\Delta P_i\}, \quad i = 1,2,3,\ldots,m \tag{15.5}$$

where

$$[K_{i-1}] = [K_{i-1}\{\Delta\delta_{i-1}\}]$$

Since the stiffness matrix $[K_{i-1}]$ is calculated by the value of the stress–strain relationship curve at the starting point of step i, it is called the method of starting point stiffness.

After $\{\delta_i\}$ is calculated by Eqs. (15.3) and (15.5), $[K_i]$ can be computed. If $\{\delta_i\}$ is the exact solution, it should satisfy Eq. (15.1). In fact, $\{\delta_i\}$ is just an approximate solution, and after substituting it into Eq. (15.1), we get

$$\{\psi_i\} = [K_i]\{\delta_i\} - \{P_i\} \neq 0 \tag{15.6}$$

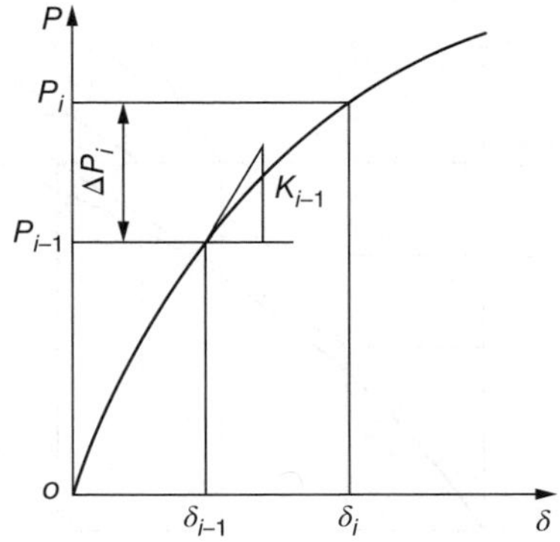

Figure 15.1 Method of starting point stiffness.

$\{\Psi_i\}$ is called unbalanced force, which represents the calculation error. The calculation error will be accumulated during the calculation process. In order to improve accuracy, $\{\Psi_i\}$ could be deducted from the nodal loads in the next calculation as follows:

$$[K_i]\{\Delta\delta_{i+1}\} = \{\Delta P_{i+1}\} - \{\psi_i\} \tag{15.7}$$

$$\{\delta_{i+1}\} = \{\delta_i\} + \{\Delta\delta_{i+1}\} \tag{15.8}$$

With this improved algorithm, we make a correction to the deviation of calculated values $\{\delta_i\}$ from the true solution $\{\delta\}$ of Eq. (15.1) at each step, so it is called self-correction method. The basic equations are Eqs. (15.6)–(15.8).

15.1.2 Method of Midpoint Stiffness

The method of starting point stiffness aforementioned is equivalent to Euler method of numerical integration for differential equations; its calculations are simple but rough, with lower accuracy. To improve the accuracy, a natural idea is to use the average stiffness in the calculation at each step: firstly, calculate preliminary $\{\Delta\delta_i^*\}$ and $\{\delta_i^*\}$ with $[K_{i-1}]$ according to Eq. (15.5), and then the stiffness matrix $[K_i]$ at the end of step i is calculated according to this preliminary displacement $\{\delta_i^*\}$ and the stress–strain relationship. Thus the average stiffness matrix at step i is calculated as follows:

$$[\overline{K}_i] = \frac{1}{2}([K_{i-1}] + [K_i]) \tag{a}$$

Then, the displacement increment at step i can be calculated according to the following equation:

$$[\overline{K}_i]\{\Delta\delta_i\} = \{\Delta P_i\} \tag{b}$$

Using this method, the accuracy is improved, but Eq. (a) shows that the amount of computer storage is increased. Therefore, using the following method of midpoint stiffness is more appropriate. As shown in Figure 15.2, half of the load increment $\frac{1}{2}\{\Delta P_i\}$ is applied; with the stiffness matrix $[K_{i-1}]$ at the end of step $i-1$, the temporary displacement increment $\{\Delta\delta_{i-1/2}^*\}$ is calculated by the following equation:

$$[K_{i-1}]\{\Delta\delta_{i-1/2}^*\} = \frac{1}{2}\{\Delta P_i\} \tag{c}$$

where the variable with * is temporary.

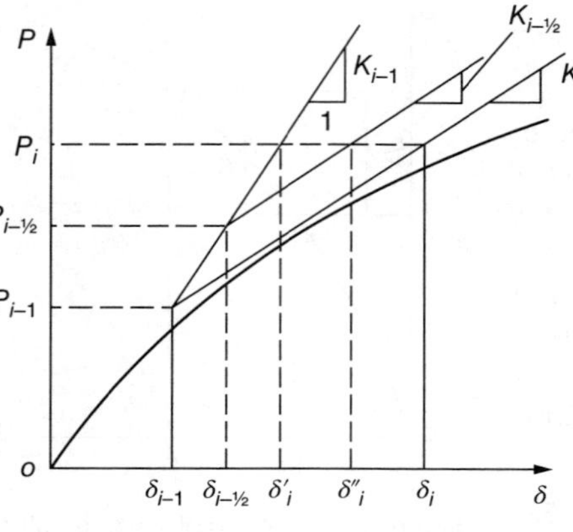

Figure 15.2 The comparison of midpoint stiffness method and starting point stiffness method.

The midpoint displacement is obtained from the following:

$$\{\delta^*_{i-1/2}\} = \{\delta_{i-1}\} + \{\Delta\delta^*_{i-1/2}\} \tag{d}$$

The midpoint stiffness matrix $[K_{i-1/2}]$ is calculated according to $\{\delta^*_{i-1/2}\}$ and the stress–strain relationship, and then the displacement increment $\{\Delta\delta_i\}$ at step i is calculated by the following equation:

$$[K_{i-1/2}]\{\Delta\delta_i\} = \{\Delta P_i\} \tag{15.9}$$

The difference between the method of midpoint stiffness and the method of starting point stiffness can be seen in Figure 15.2 in which δ_i is obtained by the method of midpoint stiffness and δ'_i is the displacement obtained by the method of starting point stiffness. δ''_i is the displacement obtained by the method of starting point stiffness in two steps. The computation amount of δ_i and δ''_i are basically the same, but the accuracy of δ_i is higher. The method of midpoint stiffness is equivalent to Runge–Kutta method for solving the differential equations.

In the calculation process of the midpoint stiffness method, self-correcting method can also be used, in which Eq. (15.9) is changed into the following equation:

$$[K_{i-1/2}]\{\Delta\delta_i\} = \{\Delta P_i\} - \{\psi_{i-1}\} \tag{15.10}$$

In general, $\{\delta_i\}$ is assumed to be the deflection already calculated, and $\{\delta^*_{i+1}\} = \{\delta_i\} + \{\Delta\delta^*_{i+1}\}$ is a temporary variable calculated approximately, taking

$$\{\delta_{i+\theta}\} = (1-\theta)\{\delta_i\} + \theta\{\Delta\delta^*_{i+1}\} \tag{15.11}$$

Calculate $[K_{i+\theta}]$ by $\{\delta_{i+\theta}\}$, and then calculate the deflection at the end of this step by the following equation:

$$\{\delta_{i+1}\} - \{\delta_i\} = [K_{i+\theta}]^{-1}(\{\Delta P_{i+1}\} - \{\psi_i\}) \tag{15.12}$$

Obviously, taking $\theta = 1/2$, it is the method of midpoint stiffness, while taking $\theta = 0$, it is the method of starting point stiffness, where $\{\Psi_i\}$ is self-correction term.

Now take a simple example to illustrate the accuracy of the approximation method aforementioned. Assuming that there is a nonlinear spring, and its stiffness $K(\delta)$ is not constant, but is given by the following equation,

$$K(\delta) = K_0\left(1 - \frac{\delta}{a}\right)$$

While the deflection δ increases, the stiffness $K(\delta)$ decreases; when the deflection $\delta = a$, the stiffness tends to zero, meaning that at this time the spring is no longer able to carry loads. If the load applied to spring is P, then the balance equation is

$$K(\delta)\delta = K_0\left(1 - \frac{\delta}{a}\right)\delta = P \tag{e}$$

$\delta(P)$ is solved by the above equation and noted that $\delta(0) = 0$, so the exact solution of deflection is

$$\delta(P) = \frac{1}{2}\left(a - \sqrt{a^2 - \frac{4aP}{K_0}}\right) \tag{f}$$

Differentiating Eq. (e), we get the following differential equation:

$$\frac{\mathrm{d}\delta}{\mathrm{d}P} = \frac{1}{K_0(1 - 2\delta/a)} \tag{g}$$

This is a first-order nonlinear ordinary differential equation of δ, and exact solution as Eq. (f) can also be obtained from the integral for the last equation. Using the numerical method for solving, rewriting (g) as a difference equation, we obtain

$$\frac{\delta(P + \Delta P) - \delta(P)}{\Delta P} = \frac{1}{K_0(1 - 2\delta/a)}$$

or

$$\delta(P + \Delta P) = \delta(P) + \frac{\Delta P}{K_0(1 - 2\delta/a)} \tag{h}$$

Note

$$\delta_i = \delta(P_i), \quad \Delta P = \Delta P_i$$

Then

$$\delta_{i+1} = \delta_i + \frac{\Delta P_i}{K_0(1 - 2\delta_i/a)} \tag{i}$$

This is equivalent to the equation of the starting point stiffness method.

From Eq. (g), at the place of $\delta(P_i) = \delta_i$, the gradient of the deflection curve is

$$\delta_i' = \frac{1}{K_0(1 - 2\delta_i/a)} \tag{j}$$

$\delta_{i+1/2}$ obtained from the gradient aforementioned is

$$\delta_{i+1/2} = \delta_i + \frac{0.5\Delta P_i}{K_0(1 - 2\delta_i/a)} \tag{k}$$

At the place of $\delta_{i+1/2}$, the gradient of deflection curve is

$$\delta_{i+1/2}' = \frac{1}{K_0(1 - 2\delta_{i+1/2}/a)} \tag{l}$$

δ_{i+1} obtained is as follows:

$$\delta_{i+1} = \delta_i + \frac{\Delta P_i}{K_0(1 - 2\delta_{i+1/2}/a)} \tag{m}$$

The above equation is equivalent to the equation of the midpoint stiffness method.

Table 15.1 The results of nonlinear spring deflection δ ($a = 4$, $K_0 = 20$).

Load P	Exact value	Method of starting point stiffness	Method of midpoint stiffness	Load P	Exact value	Method of starting point stiffness	Method of midpoint stiffness
2	0.103	0.100	0.103	10	0.586	0.563	0.585
4	0.211	0.205	0.211	12	0.735	0.702	0.734
6	0.327	0.317	0.326	14	0.905	0.856	0.902
8	0.451	0.435	0.450	15	1.106	1.031	1.097

The results calculated with three different methods are shown in Table 15.1. The accuracy of approximate calculation can be seen roughly from this table. Of course, the calculation accuracy is not only related to the calculation methods but also related to the nonlinear degree of the problem and the calculation steps and other factors. In this example, if reducing the length of the steps, the calculation accuracy of the starting point stiffness method can also be improved.

15.2 Iterative Method

When we solve nonlinear problems with iterative method, all the load is applied once, and then adjust the displacement gradually to satisfy the basic equation (15.1).

15.2.1 Direct Iterative Method

One of the simplest ways to solve the nonlinear equation (15.1) is the direct iterative method. Firstly, give an initial approximate solution $\{\delta_0\}$, for example, let $\{\delta_0\} = 0$. $[K(\delta_0)] = [K_0]$ is obtained by the stress–strain relationship, and the first improved approximate solution can be computed by Eq. (15.1):

$$\{\delta_1\} = [K_0]^{-1}\{P\}$$

Repeating this process, the equations for the $(n+1)$th approximate solution calculated from the nth approximate solution are

$$\left.\begin{aligned}[K_n] &= [K(\delta_n)] \\ \{\delta_{n+1}\} &= [K_n]^{-1}\{P\}\end{aligned}\right\} \tag{15.13}$$

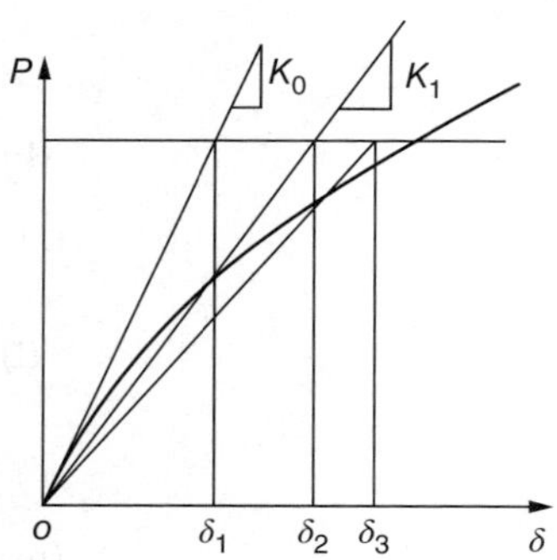

Figure 15.3 Direct iterative method.

The calculation process is shown in Figure 15.3. Repeat the above calculations until the two calculation results are sufficiently close. The scale of judgment may be

$$\|\Delta\delta\| = \max\Delta\delta_i \tag{15.14}$$

or

$$\|\Delta\delta\| = \sqrt{\{\Delta\delta\}^{\mathrm{T}}\{\Delta\delta\}} \tag{15.15}$$

The criteria of convergence may be

$$\|\Delta\delta_n\| \leq \alpha\|\Delta\delta_n\| \tag{15.16}$$

where α is a small number given in advance. When the above equation is satisfied, the calculation is thought to be converged, and iteration is terminated.

The unbalance force $\{\Psi_n\}$ computed by Eq. (15.6) may be also regarded as a measure of deviation from the equilibrium condition; accordingly the criterion of convergence is

$$\|\psi_n\| \leq \beta\|P\| \tag{15.17}$$

where β is a small number given in advance.

As shown in Figure 15.3, the secant stiffness matrix is used at each step in the direct iterative method.

15.2.2 Newton Method

Firstly, consider the nonlinear equation of a single variable x:

$$f(x) = 0 \tag{a}$$

Make the Taylor expansion at x_0, remain only the linear terms, and get the linear approximate equation of $f(x) = 0$ near x_0 to obtain

$$f(x_0) + f'(x_0)(x - x_0) = 0 \tag{b}$$

If $f'(x_0) \neq 0$, the solution of the above equation (Figure 15.4) is

$$x_1 = x_0 - \frac{f(x_0)}{f'(x_0)} \tag{c}$$

Repeating the above process, we get the $(n+1)$th approximate solution of $f(x) = 0$:

$$x_{n+1} = x_n - \frac{f(x_n)}{f'(x_n)} \tag{d}$$

This is the famous Newton–Raphson method, called Newton method for short.

Now back to our nonlinear equations,

$$\{\psi\} = [K(\delta)]\{\delta\} - \{P\} = \{F(\delta)\} - \{P\} = 0 \tag{15.18}$$

Make the Taylor expansion for $\{\Psi\}$ at $\{\delta\} = \{\delta_n\}$, and remain only the linear terms, and then we get

$$\{\psi\} = \{\psi_n\} + [K_t^n](\{\delta\} - \{\delta_n\}) = 0$$

From this we obtain the $(n+1)$th approximate solution as follows:

$$\{\delta_{n+1}\} = \{\delta_n\} - [K_t^n]^{-1}\{\psi_n\} \tag{15.19}$$

where $[K_t^n]$ is tangent stiffness matrix.

Figure 15.5 shows the convergence of Newton method schematically.

The convergence of Newton method is good. But in some cases (such as the ideal plastic and

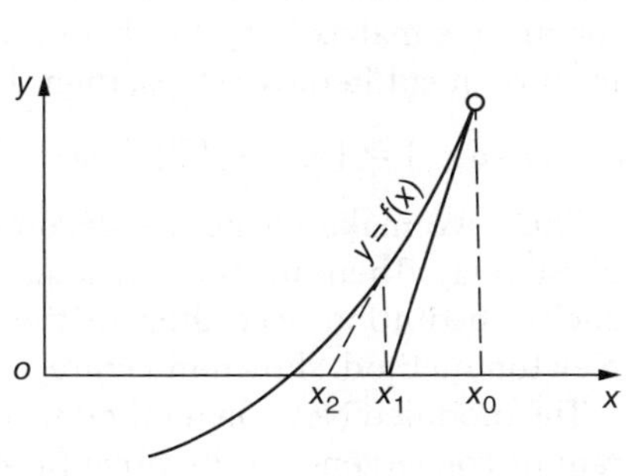

Figure 15.4 Newton method.

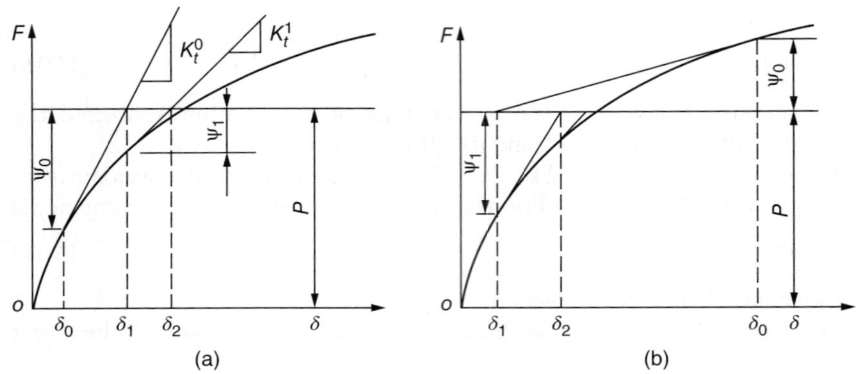

Figure 15.5 Newton method. (a) δ_0 is less than the true solution. (b) δ_0 is larger than the true solution.

the softening plastic problems), in the calculation process, $[K_t^n]$ may be singular or ill-conditioned, so that the inversion of $[K_t^n]$ is difficult to get. To overcome this difficulty, we can introduce a damping factor μ^n to make $[K_t^n] + \mu^n[I]$ become non-singular or make its ill condition weakened, where $[I]$ is identity matrix. Then, replace Eq. (15.19) with the following iterative equation:

$$\{\delta_{n+1}\} = \{\delta_n\} - ([K_t^n] + \mu^n[I])^{-1}\{\psi_n\} \tag{15.20}$$

In Eq. (15.20), μ^n is used to change the main diagonal elements of the matrix $[K_t]$. But when μ^n gets larger, the convergence rate will be slower. When $\mu^n \to 0$, Eq. (15.20) tends to Eq. (15.19), and the convergence rate is the fastest.

The criterion of convergence in Newton method is the same as in the direct iterative method, which is not repeated here. When using the Newton method or the direct iterative method for solving, $[K_t^n]$ and $[K^n]$ must be recalculated, and their inverses must be calculated in each iterative step; therefore, these two methods are called variable stiffness methods.

15.2.3 Modified Newton Method

For large problems, the formation of the stiffness matrix and the computation of its inverse matrix are time consuming. In Newton method, the stiffness matrix must be reestablished, and the inverse matrix must be calculated in each iterative step, so the calculation time is not economical. If we establish the stiffness matrix $[K_t^0]$ and compute the inverse matrix $[K_t^0]^{-1}$ only in the first iterative step, and the inverse matrix is used in subsequent iterative steps, then the nth iterative equation becomes

$$\{\delta_{n+1}\} = \{\delta_n\} - [K_t^0]^{-1}\{\psi_n\} \tag{15.21}$$

Thus, we make triangular decomposition on $[K_t^0]$ and store up in the calculation of step 1. Then in the subsequent iterations, we only want to do the simple back-substitution according to the above equation. This method is called modified Newton method, shown in Figure 15.6.

The modified Newton method saves more computing time per iteration, although the rate of convergence in iterative processes decreases, the total time of computation is saved compared with Newton method in most cases. In order to improve the rate of convergence in the modified Newton method, $[K_t]$ may be recalculated after each k iteration, which means that $[K_t^0]$ in Eq. (15.21) will be changed to

$$[K_t^j] = [K_t(\delta_j)], \quad j = k, 2k, \ldots \tag{15.22}$$

In addition, some excessive correction technologies can also be employed, which is replacing Eq. (15.21) with the following equation for iterative calculation:

$$\{\delta_{n+1}\} = \{\delta_n\} - \omega^n [K^0]^{-1}\{\psi_n\} \tag{15.23}$$

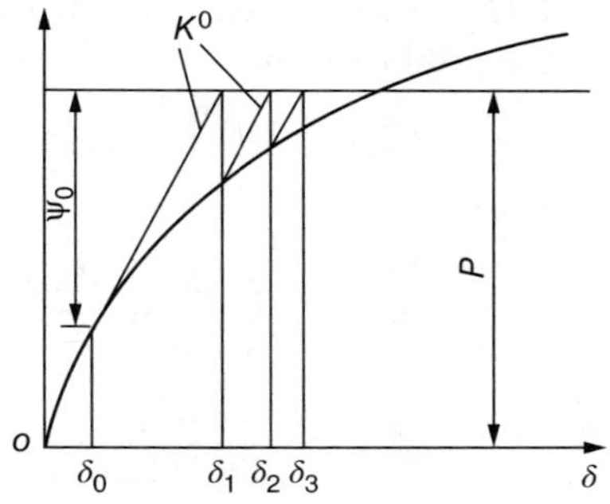

Figure 15.6 Modified Newton method.

where $\omega^n > 1$ is called excessive correction factor and its value is generally determined by trial calculations.

15.2.4 Quasi-Newton Method

The iterative equation of Newton method is

$$\{\delta_{n+1}\} = \{\delta_n\} - [K_t^n]^{-1}\{\psi_n\}$$

In each iteration the stiffness matrix and its inverse must be computed. In the quasi-Newton method, a recursive equation of $[K_t^n]^{-1}$ is established, and $[K_t^{n+1}]^{-1}$ is directly calculated from $[K_t^n]^{-1}$. Let

$$[C_n] = [K_t^n]^{-1} \tag{15.24}$$

Substituting it into Eq. (15.19), we get

$$\{\delta_{n+1}\} = \{\delta_n\} - [C_n]\{\psi_n\} \tag{15.25}$$

As for how to calculate $[C_{n+1}]$ from $[C_n]$, there are the following two algorithms:

1) DFP (Davidon–Fletcher–Power) equation

$$[C_{n+1}] = [C_n] + \frac{\{\Delta\delta_n\}\{\Delta\delta_n\}^{\mathrm{T}}}{\{\Delta\delta_n\}^{\mathrm{T}}\{\Delta\psi_n\}} - \frac{[C_n]\{\Delta\psi_n\}\{\Delta\psi_n\}^{\mathrm{T}}[C_n]}{\{\Delta\psi_n\}^{\mathrm{T}}[C_n]\{\Delta\psi_n\}} \tag{15.26}$$

Here

$$\{\Delta\delta_n\} = \{\delta_{n+1}\} - \{\delta_n\}$$

$$\{\Delta\psi_n\} = \{\psi_{n+1}\} - \{\psi_n\}$$

This is a famous and relatively successful algorithm, but its numerical stability is relatively poor.

2) BFGS (Broyden–Fletcher–Goldfarb–Shanno) equation

$$[C_{n+1}] = \left([I] - \frac{\{\Delta\delta_n\}\{\Delta\psi_n\}^{\mathrm{T}}}{\{\Delta\delta_n\}^{\mathrm{T}}\{\Delta\psi_n\}}\right)[C_n]\left([I] - \frac{\{\Delta\psi_n\}\{\Delta\delta_n\}^{\mathrm{T}}}{\{\Delta\delta_n\}^{\mathrm{T}}\{\Delta\psi_n\}} + \frac{\{\Delta\delta_n\}\{\Delta\delta_n\}^{\mathrm{T}}}{\{\Delta\delta_n\}^{\mathrm{T}}\{\Delta\psi_n\}}\right) \tag{15.27}$$

If $[K_t^n]$ is a symmetric positive definite matrix, BFGS equation can also be written as the product form:

$$[C_{n+1}] = ([I] + \{w_n\}\{v_n\}^\mathrm{T})[C_n]([I] + \{v_n\}\{w_n\}^\mathrm{T}) \tag{15.28}$$

Here

$$\{v_n\} = \pm\left(\frac{\{\Delta\delta_n\}^\mathrm{T}\{\Delta\psi_n\}}{\{\Delta\delta_n\}^\mathrm{T}\{-\psi_n\}}\right)^{1/2}(-\psi_n) - \{\Delta\psi_n\}$$

$$\{w_n\} = \frac{\{\Delta\delta_n\}}{\{\Delta\delta_n\}^\mathrm{T}\{\Delta\psi_n\}} \tag{15.29}$$

BFGS algorithm is one of the most successful algorithms currently, with better numerical stability, and there is ALGOL language program about BFGS algorithm in Ref. [5]. The derivations of DFP and BFGS equations are shown in Ref. [1].

In the quasi-Newton method, in each iteration, instead of calculating the stiffness matrix $[K_t^{n+1}]$ and then computing its inverse, we calculate $[K_t^{n+1}]^{-1}$ directly by recursive equations to improve the computational efficiency.

In the analysis of finite elements, the matrix $[K]$ is sparse, while the inverse matrix $[K]^{-1}$ may be a full matrix. In practical computation, it is best not to directly use Eqs. (15.26), (15.27), and (15.28), but to use the following calculating steps: firstly, calculate $[K_t^0]^{-1}$ and $\{\delta_1\} = \{\delta_0\} - [K_t^0]^{-1}\{\Psi_0\}$ from $\{\delta_0\}$ and at the same time put the triangular matrix $[L]$ and the diagonal matrix $[H]$ of $[K_t^0] = [L][H][L]^\mathrm{T}$ in the storage, and also store $\{v_n\}$ and $\{w_n\}$ in each calculation step. For example, the calculations in the third step are as follows:

$$\begin{aligned}\{\Delta\delta_2\} &= -[C_2]\{\psi_2\}\\ &= -([I] + \{w_1\}\{v_1\}^\mathrm{T})([I] + \{w_0\}\{v_0\}^\mathrm{T})[C_0]\\ &\quad \times ([I] + \{v_0\}\{w_0\}^\mathrm{T})([I] + \{v_1\}\{w_1\}^\mathrm{T})\{\psi_2\}\end{aligned}$$

Here, firstly calculate

$$\{b_1\} = ([I] + \{v_1\}\{w_1\}^\mathrm{T})\{\psi_2\}$$

Then calculate $\{b_0\} = ([I] + \{v_0\}\{w_0\}^\mathrm{T})\{b_1\}$, and call back-substitution process, to calculate $\{s\} = [C_0]\{b_0\}$

Then

$$\{r_0\} = ([I] + \{w_0\}\{v_0\}^\mathrm{T})\{s\}$$

Finally

$$\{r_1\} = ([I] + \{w_1\}\{v_1\}^\mathrm{T})\{r_0\}$$

and

$$\{\Delta\delta_2\} = -\{r_1\}, \{\delta_3\} = \{\delta_2\} + \{r_1\}$$

The processes of calculation aforementioned are just a series of inner product computations. With the increase of iteration steps, the number of $\{v_n\}$ and $\{w_n\}$ to be stored also increases. To reduce the amount of storage, we can recalculate a $[K_0]$ after each k iterative step.

Three kinds of Newton method are described above, namely, Newton method, modified Newton method, and quasi-Newton method. The amount of calculation in each iteration is largest in Newton method, followed by quasi-Newton method, and smallest in modified Newton method. But the overall computational efficiency is not only related to the amount of computation in each iteration but also related to the speed of the convergence. Now it seems that different algorithms may be applied to different problems, numerical experiments are needed to decide which algorithm is the most appropriate to the kind of problems for solving.

15.2.5 The Calculation of $\{\Psi_n\}$ and Initial Stress Method and Initial Strain Method

Equations (15.21) and (15.25) show that $\{\Psi_n\}$ must be used for solving in modified Newton method and quasi-Newton method. Equation (15.18) shows that calculating $\{\Psi_n\}$ needs $[K_n]$, but in modified Newton method and quasi-Newton method, $[K_n]$ is formed only in the first step, not in the subsequent steps. This is the difference between the nonlinear finite elements and general nonlinear equations. For the latter, the equations are given in advance, so that $\{\Psi_n\}$ is calculated without any difficulties, while for the former, the equations are established by $[K_n]$, not given in advance. The following shows how to calculate $\{\Psi_n\}$ in the analysis of nonlinear finite elements:

1) The general equations of $\{\Psi_n\}$
   Assuming that $\{\delta_n\}$ is the nth approximate solution obtained already, the corresponding strain is $\{\varepsilon_n\}$, and the corresponding stress is $[D_n]\{\varepsilon_n\}$, so the corresponding force of element node is

$$\{F(\delta_n)\}^{\mathrm{e}} = \int [B]^{\mathrm{T}}[D_n]\{\varepsilon_n\}\mathrm{d}V$$

   Substituted into Eq. (15.18), we get the unbalanced force $\{\Psi_n\}$ as follows:

$$\{\psi_n\} = \{F(\delta_n)\} - \{P\} = \sum_{\mathrm{e}} \int [B]^{\mathrm{T}}[D_n]\{\varepsilon_n\}\mathrm{d}V - \{P\} \tag{15.30}$$

   where $\sum_{\mathrm{e}}$ is the set of element nodal forces around the node and $[D_n]$ is secant elasticity matrix.
2) Initial stress method
   Assuming that the stress–strain relationship is

$$\{\sigma\} = f(\{\varepsilon\}) \tag{e}$$

   the initial tangent elastic matrix is $[D_0]$, and the stress calculated by $[D_0]$ and linear stress–strain relationship is $[D_0]\{\varepsilon\}$. As shown in Figure 15.7, the initial stress $\{\sigma_0\}$ is so introduced that the sum of the stress $[D_0]\{\varepsilon\}$ calculated according to a linear relationship and the initial stress $\{\sigma_0\}$ is equivalent to

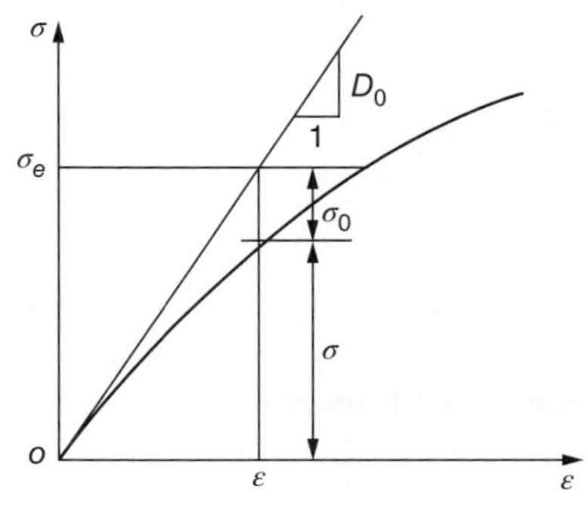

Figure 15.7 Initial stress.

the stress calculated according to the real stress–strain relationship as follows:

$$\{\sigma\} = f(\{\varepsilon\}) = [D_0]\{\varepsilon\} + \{\sigma_0\}$$

It can be seen that the initial stress $\{\sigma_0\}$ should be calculated by the following equation:

$$\{\sigma_0\} = f(\{\varepsilon\}) - [D_0]\{\varepsilon\} = [D_n]\{\varepsilon\} - [D_0]\{\varepsilon_n\} \tag{15.31}$$

Equation (3.10) shows that the element nodal force caused by the initial stress is

$$\{F\}^{\mathrm{e}} = \int [B]^{\mathrm{T}}\{\sigma_0\}\mathrm{d}V$$

Collecting the element nodal forces near the node, we get the nodal unbalance force $\{\psi_n\}$ caused by the initial stress as follows:

$$\{\Psi_n\} = \sum_{\mathrm{e}} \int [B]^{\mathrm{T}}\{\sigma_0\}\mathrm{d}V \tag{15.32}$$

It is easy to prove that Eq. (15.32) is equivalent to Eq. (15.30).

3) Initial strain method
In some problems, it is difficult to express the stress clearly with the strain as Eq. (e), such as creep problems. In contrast, it is easy to express the strain with the stress as follows:

$$\{\varepsilon\} = f(\{\sigma\}) \tag{f}$$

The linear elastic stress–strain relationship is

$$\{\varepsilon_{\mathrm{e}}\} = [D]^{-1}\{\sigma\} \tag{g}$$

As shown in Figure 15.8, the initial strain $\{\varepsilon_0\}$ in the stress–strain relationship is introduced so that the sum of the strain $[D]^{-1}\{\sigma\}$ calculated by linear elastic relationship and the initial strain $\{\varepsilon_0\}$ is equal to the strain calculated by a nonlinear relationship (f) as follows:

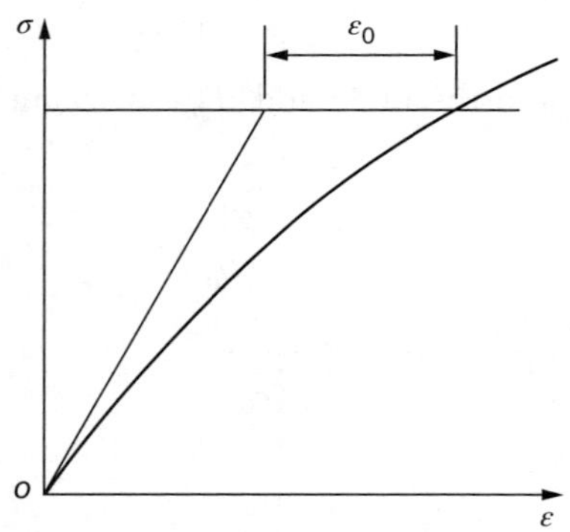

Figure 15.8 Initial strain.

$$\{\varepsilon\} = [D]^{-1}\{\sigma\} + \{\varepsilon_0\}$$

It can be seen that the initial strain $\{\varepsilon_0\}$ should be calculated by the following equation:

$$\{\varepsilon_0\} = \{\varepsilon\} - [D]^{-1}\{\sigma\}$$
$$= f(\{\sigma\}) - [D]^{-1}\{\sigma\} \tag{15.33}$$

Equation (3.15) shows that the nodal unbalance force $\{\Psi\}$ caused by the initial strain should be calculated according to Eq. (15.34):

$$\{\Psi_n\} = \sum_{\mathrm{e}} \int [B]^{\mathrm{T}}[D_n]\{\varepsilon_0\}\mathrm{d}V \tag{15.34}$$

15.3 Mixed Method

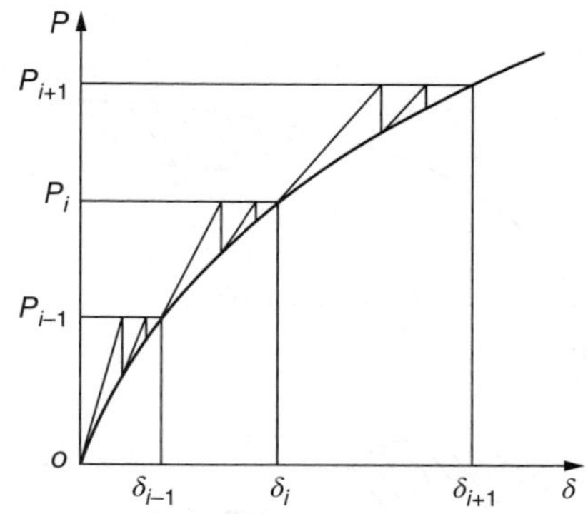

Figure 15.9 Mixed method.

As shown in Figure 15.9, the mixed method takes advantage of both the incremental method and the iterative method. The load is divided into a small number of increments, and iterative calculations are applied to each load increment.

The following is to make a comparison of these calculation methods.

One of the advantages of incremental method is wide range of application, high universality; another advantage is that it can provide load–displacement curve. Incremental method has two disadvantages also: the first is that it usually consumes more computing time than iterative method, and the second is that we do not know how much the approximate solution differs from the true solution.

The calculation amount of iterative method is smaller than incremental method, and the calculation accuracy can be controlled. But iterative method cannot give the load–displacement curve, so its range of application is small, for example, when the deformation properties of the material are related to the process of loading (loading and unloading different in nature), the dynamic problems, and so on, iterative method cannot be applied.

Mixed method has the advantages of both incremental method and iterative method to some extent and avoids their shortcomings, reduces the calculations of each incremental load, and may estimate the degree of approximation because of iterations, but the calculated amount is still large.

15.4 Application of Substructure Method in Nonlinear Analysis

In the structures of practical engineering, generally only the local regions get into the nonlinear range of the stress–strain relationship rather than the entire structure.

As shown in Figure 15.10, the shaded regions are nonlinear, and the others are linear elastic. Usually, we may estimate the boundaries of the nonlinear regions based on historical experiences (for safety reasons, we may estimate largely); then, using the substructure method, we can eliminate all the degrees of freedom within the linear regions, and in nonlinear calculations, only the node freedoms both in nonlinear regions and on the public boundaries of the two regions remain. Thus the nonlinear regions are relatively small, and a lot of computing time can be saved.

Divide the degrees of freedom into three categories:

1) $\{\delta^e\}$ are the degrees of freedom in the linear elastic regions and on the boundaries of nonlinear regions.

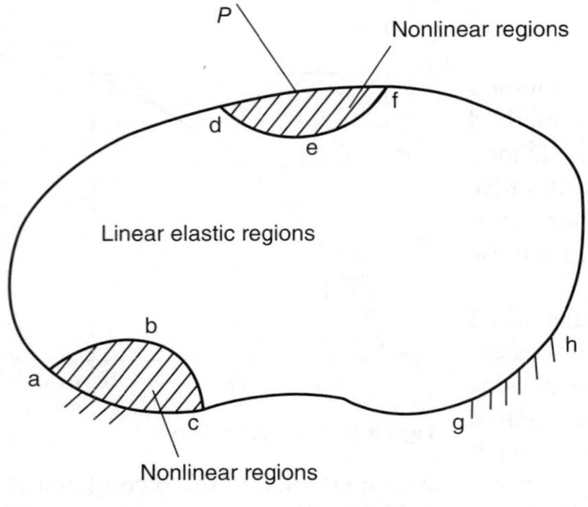

Figure 15.10 The application of substructure method in nonlinear analysis.

2) $\{\delta^p\}$ are the degrees of freedom in the nonlinear regions and on their boundaries.
3) $\{\delta^b\}$ are the degrees of freedom on the public boundaries of the two kinds of regions and on the external constraint boundaries of linear elastic regions, as shown as abc, def, and gh in Figure 15.10.

In linear elastic regions, the relationship between force and displacement is

$$\begin{Bmatrix} F^e \\ F^b \end{Bmatrix} = \begin{bmatrix} k^e_{ee} & k^e_{eb} \\ k^e_{be} & k^e_{bb} \end{bmatrix} \begin{Bmatrix} \delta^e \\ \delta^b \end{Bmatrix} \tag{15.35}$$

Expand Eq. (15.35) and get

$$\{F^e\} = [k^e_{ee}]\{\delta^e\} + [k^e_{eb}]\{\delta^b\} \tag{15.36}$$

$$\{F^b\} = [k^e_{be}]\{\delta^e\} + [k^e_{bb}]\{\delta^b\} \tag{15.37}$$

In Eq. (15.36), replacing $\{F^e\}$ with the nodal load $\{P^e\}$ in elastic regions and on the nonlinear boundaries, the equation that represents the displacement $\{\delta^e\}$ in elastic regions with the boundary displacement $\{\delta^b\}$ is obtained as follows:

$$\{\delta^e\} = [k^e_{ee}]^{-1}\{P^e\} - [k^e_{ee}]^{-1}[k^e_{eb}]\{\delta^b\} \tag{15.38}$$

Substituting Eq. (15.38) into (15.37), we obtain the following boundary nodal force:

$$\{F^b\} = [\bar{k}_b]\{\delta^b\} + \{R^b\} \tag{15.39}$$

where

$$[\bar{k}_b] = [k^e_{bb}] - [k^e_{be}][k^e_{ee}]^{-1}[k^e_{eb}] \tag{15.40}$$

$$\{R^b\} = [k^e_{be}][k^e_{ee}]^{-1}\{P^e\} \tag{15.41}$$

In nonlinear regions, the relationship between force and displacement is

$$\begin{bmatrix} k^p_{pp} & k^p_{pb} \\ k^p_{bp} & k^p_{bb} \end{bmatrix} \begin{Bmatrix} \delta^p \\ \delta^b \end{Bmatrix} = \begin{Bmatrix} P^p \\ P^b \end{Bmatrix} \tag{15.42}$$

where $\{P^p\}$ are the nodal loads in nonlinear regions and $\{P^b\}$ are the nodal loads on the public boundaries of the two kinds of regions.

In the nonlinear regions and on the public boundaries and the constraint boundaries, the node equilibrium condition is

$$\begin{bmatrix} k_{pp}^p & k_{pb}^p \\ k_{bp}^p & k_{bb}^p \end{bmatrix} \begin{Bmatrix} \delta^p \\ \delta^b \end{Bmatrix} + \begin{bmatrix} 0 & 0 \\ 0 & \overline{k}_b \end{bmatrix} \begin{Bmatrix} \delta^p \\ \delta^b \end{Bmatrix} + \begin{Bmatrix} 0 \\ R^b \end{Bmatrix} = \begin{Bmatrix} P^p \\ P^b \end{Bmatrix}$$

That is,

$$\begin{bmatrix} k_{pp}^p & k_{pb}^p \\ k_{bp}^p & k_{bb}^p + \overline{k}_b \end{bmatrix} \begin{Bmatrix} \delta^p \\ \delta^b \end{Bmatrix} = \begin{Bmatrix} P^p \\ P^b - R^b \end{Bmatrix} \tag{15.43}$$

Equation (15.43) is the basic equation of nonlinear analysis. The degrees of freedom in linear regions have been eliminated; only the degrees of freedom in nonlinear regions and on the public boundaries of the two kinds of regions and on the external constraint boundaries remain. If the degrees of freedom on the external constraint boundaries are zero, if necessary, they can be eliminated. When the nonlinear regions are relatively small, the order of equations is significantly reduced, so that the calculation amount is saved. $[\overline{k}_b]$ is given by Eq. (15.40), just once, and recycled in nonlinear iterative calculations. $\{R^b\}$ is given by Eq. (15.41), only recalculated once when the nodal load $\{P^t\}$ is changed, and the calculation of $\{R^b\}$ is more convenient because of recycling of $[k_{be}^e][k_{ee}^e]^{-1}$.

The nodal displacement $\{\delta^p\}$ in nonlinear regions and the nodal displacement $\{\delta^b\}$ on the boundaries are solved by Eq. (15.43) through nonlinear analysis. Substituting the boundary nodal displacements into Eq. (15.38), the nodal displacements $\{\delta^e\}$ can be calculated in the linear regions. Furthermore the stresses in the linear regions can be calculated by $\{\sigma\} = [D][B]\{\delta\}$.

Bibliography

1 Wang, D. (1979) *Solution of Nonlinear Equation Set and Method of Optimization (in Chinese)*, People's Education Press, Beijing.

2 Youqiang, Y. (1987) *Introduction to Nonlinear Finite Elements in Solid Mechanics (in Chinese)*, Beijing University Press, Beijing.

3 Bathe, K.J. and Cimento, A.P. (1980) Some practical procedures for the solution of nonlinear finite element equations. *Journal of Computer Methods in Applied Mechanics and Engineering*, **22**, 59–85.

4 Matthies, H. and Strang, G. (1979) The solution of nonlinear finite element equations. *International Journal for Numerical Methods in Engineering*, **14**, 1613–1626.

5 Fielding, K. (1970) Algorithm 387, function minimization and linear search. *Communications of the ACM*, **13**, 509–510.

16

Problems in Theory of Plasticity

In the problems of solid mechanics, when the strain is small, the stress–strain relation is linear and elastic, and when the strain is large, the relationship is no longer linear and elastic. This kind of problem belongs to the category of theory of plasticity. Application of finite element method in this respect is very successful. This chapter will explain how to use finite element method to solve the problems of theory of plasticity.

16.1 One-Dimensional Stress–Strain Relation

Figure 16.1 shows the simple tensile stress–strain curve of metal under static load and normal temperature, in which point A is the proportional limit. Before this point, the stress–strain relation is linear and it can be expressed by Hooke's law. After point A, the stress–strain relation is nonlinear. But if unloading before point B, the deformation will be fully recovered and point B is the elastic limit.

After point B, there will be a yield (flow) stage where the stress is constant, and the strain increases significantly. Reaching point C, the ability of material to resist deformation is recovered again. More deformation can be produced only after the increase of load. This phenomenon is called the strengthening of the material (or hardening). Point D is the stress under peak load, known as the strength limit. After point D, the stress begins to decline, but due to the necking of the specimen, sectional area reduces; in fact the real stress in the neck is still a slight increase until finally the material is destroyed.

For convenience of study, on the basis of experimental data, it is often abstracted as some simplified models, as shown in Figure 16.2. The simplified models have a common characteristic, namely, the stress–strain relations of the elastic stage and plastic stage must be expressed by different formulas; it is not very convenient for use. For convenience of calculation, the power function is sometimes used to describe the stress–strain relation approximately as follows:

$$\sigma = a|\varepsilon|^m sign\, \varepsilon \tag{16.1}$$

where $a > 0,\ \ 1 \geq m \geq 0$. a and m are the characteristic constants of material. When $m = 0$, it is the ideal rigid plastic model and when $m = 1$, it is the ideal elastic model. The values of sign ε are as follows:

$$\text{sign}\, \varepsilon = \begin{cases} 1 & (\varepsilon > 0) \\ -1 & (\varepsilon < 0) \end{cases}$$

The Finite Element Method: Fundamentals and Applications in Civil, Hydraulic, Mechanical and Aeronautical Engineering, First Edition. Bofang Zhu.

Companion website: www.wiley.com/go/zhufem17

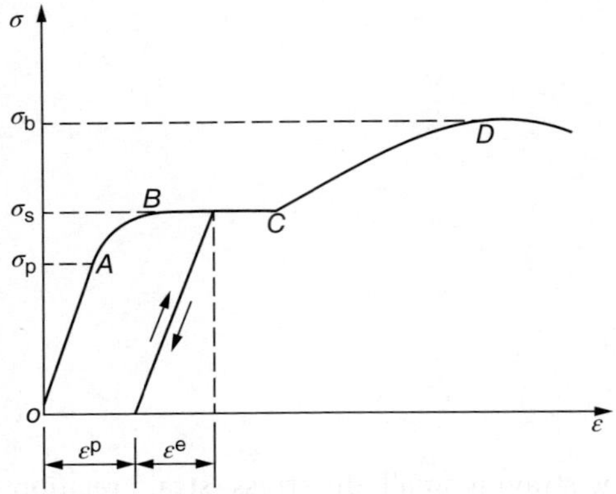

Figure 16.1 Curve of one-dimensional tensile test of metal.

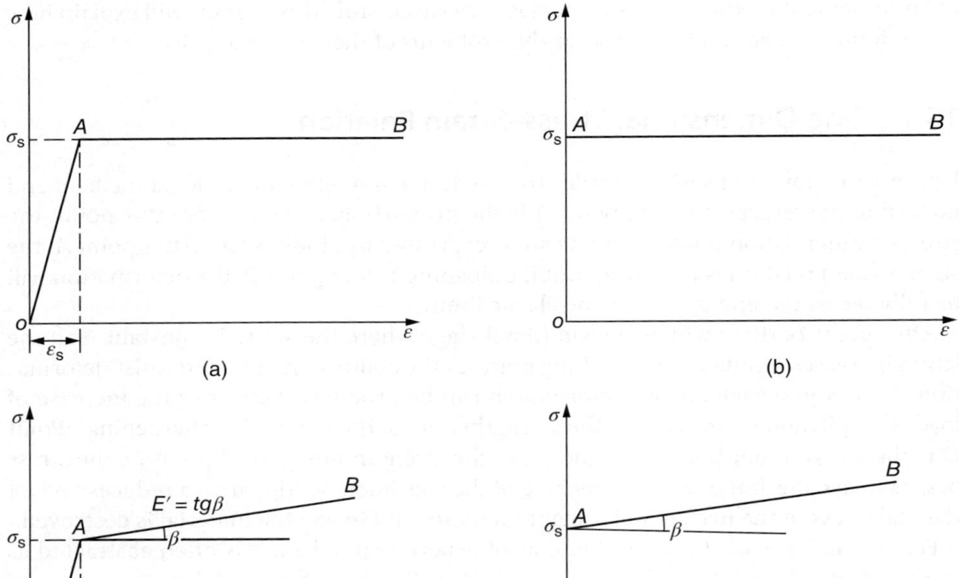

Figure 16.2 Some simplified models of stress–strain relation (a) ideal elastic–plastic model (b) ideal rigid plastic model. (c) linear strengthening elastic–plastic model and (d) linear strengthening rigid-plastic model.

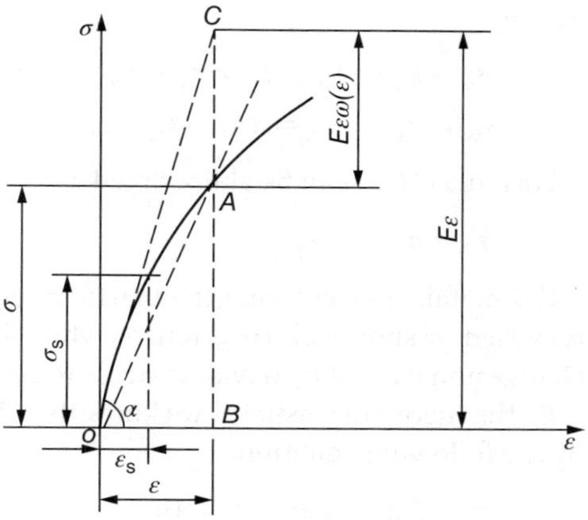

Figure 16.3 Haigh–Westergaard stress space $\sigma = E[1-\omega(\varepsilon)]\varepsilon$.

Sometimes the stress–strain relationship below is also taken (Figure 16.3):

$$\sigma = E[1-\omega(\varepsilon)] \quad \varepsilon = E'\varepsilon$$
$$E' = E[1-\omega(\varepsilon)] \tag{16.2}$$

From Figure 16.3, it is clear that E' is the secant modulus of point A and $\omega(\varepsilon)$ is a nonlinear function of ε.

16.2 Decompose of Stress Tensor and Stress Invariant

Under the action of external force, the strain corresponding with the stress σ_{ij} of one point in the body generally can be divided into two parts, namely, the volume deformation and the shape deformation. Experimental data shows that the change laws of two kinds of the deformation are different. For example, for metal, under the action of uniform pressure, volume deformation is elastic, and there is no plastic deformation. So, in order to study plastic deformation, uniform pressure (or hydrostatic pressure) must be separated from the total stress, making the following decomposition of stress tensor:

$$\begin{bmatrix} \sigma_x & \tau_{xy} & \tau_{xz} \\ \tau_{xy} & \sigma_y & \tau_{yz} \\ \tau_{xz} & \tau_{yz} & \sigma_z \end{bmatrix} = \begin{bmatrix} \sigma_m & 0 & 0 \\ 0 & \sigma_m & 0 \\ 0 & 0 & \sigma_m \end{bmatrix} + \begin{bmatrix} \sigma_x-\sigma_m & \tau_{xy} & \tau_{xz} \\ \tau_{xy} & \sigma_y-\sigma_m & \tau_{yz} \\ \tau_{xz} & \tau_{yz} & \sigma_z-\sigma_m \end{bmatrix} \tag{16.3}$$

$$\sigma_m = \frac{1}{3}(\sigma_x+\sigma_y+\sigma_z) \tag{16.4}$$

In formula (16.3), the first tensor on the right side is called the spherical stress tensor, and the second tensor is called the deviator stress tensor, expressed by $[s_{ij}]$ as follows:

$$[s_{ij}] = \begin{bmatrix} s_x & s_{xy} & s_{xz} \\ s_{xy} & s_y & s_{yz} \\ s_{xz} & s_{yz} & s_z \end{bmatrix} \tag{16.5}$$

where

$$s_x = \sigma_x - \sigma_m, \quad s_y = \sigma_y - \sigma_m, \quad s_z = \sigma_z - \sigma_m$$
$$s_{xy} = \tau_{xy}, \quad s_{yz} = \tau_{yz}, \quad s_{xz} = \tau_{xz}$$

Formula (16.3) can be abbreviated as

$$\sigma_{ij} = \sigma_m \delta_{ij} + s_{ij} \tag{16.6}$$

For metal, after entering the plastic stage, the volume deformation is still elastic and is caused by spherical stress tensor, while the plastic deformation associated with shape change and caused by deviator stress tensor is not elastic.

By the theory of elasticity, we know that the principal stress of any point is determined by the following equation:

$$\sigma_i^3 - I_1\sigma_i^2 - I_2\sigma_i - I_3 = 0 \tag{16.7}$$

When the direction of the axis changes, the components of the stress tensor will change, but the values of the principal stresses remain the same. As a result, the values of the coefficients I_1, I_2, I_3 in formula (16.7) have nothing to do with the orientation of the coordinate axis; they are called three invariants of the stress tensor, where

$$\left.\begin{aligned} I_1 &= \sigma_x + \sigma_y + \sigma_z \\ I_2 &= -(\sigma_x\sigma_y + \sigma_y\sigma_z + \sigma_z\sigma_x) + \tau_{xy}^2 + \tau_{yz}^2 + \tau_{zx}^2 \\ I_3 &= \sigma_x\sigma_y\sigma_z - \sigma_x\tau_{yz}^2 - \sigma_y\tau_{zx}^2 - \sigma_z\tau_{xy}^2 \end{aligned}\right\} \tag{16.8}$$

If we take the principal stress axes for coordinate axes, the stress invariants expressed by the principal stresses can be expressed as follows:

$$\left.\begin{aligned} I_1 &= \sigma_1 + \sigma_2 + \sigma_3 \\ I_2 &= -(\sigma_1\sigma_2 + \sigma_2\sigma_3 + \sigma_3\sigma_1) \\ I_3 &= \sigma_1\sigma_2\sigma_3 \end{aligned}\right\} \tag{16.9}$$

The deviator stress tensor also has three invariants, namely,

$$\left.\begin{aligned} J_1 &= s_x + s_y + s_z = (\sigma_x + \sigma_y + \sigma_z) - 3\sigma_{\mathrm{m}} = 0 \\ J_2 &= -(s_xs_y + s_ys_z + s_zs_x) + s_{xy}^2 + s_{yz}^2 + s_{zx}^2 \\ &= \frac{1}{6}[(\sigma_x - \sigma_y)^2 + (\sigma_y - \sigma_z)^2 + (\sigma_z - \sigma_x)^2 + 6(\tau_{xy}^2 + \tau_{yz}^2 + \tau_{zx}^2)] \\ J_3 &= s_xs_ys_z + 2s_{xy}s_{yz}s_{zx} - s_xs_{yz}{}^2 - s_ys_{zx}{}^2 - s_zs_{xy}{}^2 \end{aligned}\right\} \tag{16.10}$$

If expressed by the principal deviator stresses, they can be written as follows:

$$\left.\begin{aligned} J_1 &= s_1 + s_2 + s_3 = (\sigma_1 - \sigma_m) + (\sigma_2 - \sigma_m) + (\sigma_3 - \sigma_m) = 0 \\ J_2 &= -(s_1s_2 + s_2s_3 + s_3s_1) \\ &= \frac{1}{2}(s_1^2 + s_2^2 + s_3^2) \\ &= \frac{1}{6}[(\sigma_1 - \sigma_2)^2 + (\sigma_2 - \sigma_3)^2 + (\sigma_3 - \sigma_1)^2] \\ J_3 &= s_1s_2s_3 \end{aligned}\right\} \tag{16.11}$$

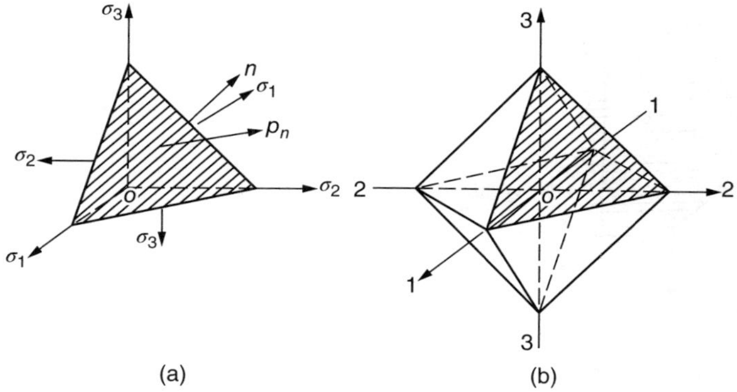

Figure 16.4 Octahedron and the stress on its surfaces.

Then we research the stress on the isoclinal plane. The isoclinal plane is the plane whose normal line has the same angles with the directions of three principal stresses. Totally there are eight planes as such, so-called the surfaces of octahedron (Figure 16.4).

The normal stress on the surfaces of octahedron is

$$\sigma_8 = \sigma_m \tag{16.12}$$

The shear stress on the surfaces of octahedron is

$$\tau_8 = \sqrt{\frac{2}{3}J_2} \tag{16.13}$$

16.3 Haigh–Westergaard Stress Space

The stress state of one point can be expressed by six stress components σ_x, σ_y, σ_z, τ_{xy}, τ_{yz}, τ_{zx}. Taking six stress components as the axes, we get a six-dimensional stress space, where any point represents a state of stress. But the six-dimensional stress space cannot be shown visually and its application is not convenient. Because the stress state of one point can be expressed by three principal stresses, taking three principal stresses as the axes, the principal stress space is obtained. One point $(\sigma_1, \sigma_2, \sigma_3)$ in the principal stress space represents a state of stress. The principal stress space is proposed by Haigh–Westergaard.

16.3.1 Geometric Characteristics of Stress Space

Stress space has the following geometric features:

16.3.1.1 The Hydrostatic Stress Axis

As shown in Figure 16.5, in the stress space, the line L passes through the original point and has the same angles with the coordinate axes. Its equation is $\sigma_1 = \sigma_2 = \sigma_3$; apparently, the point on the line represents that the body is under hydrostatic pressure. The line L is called the hydrostatic stress axis.

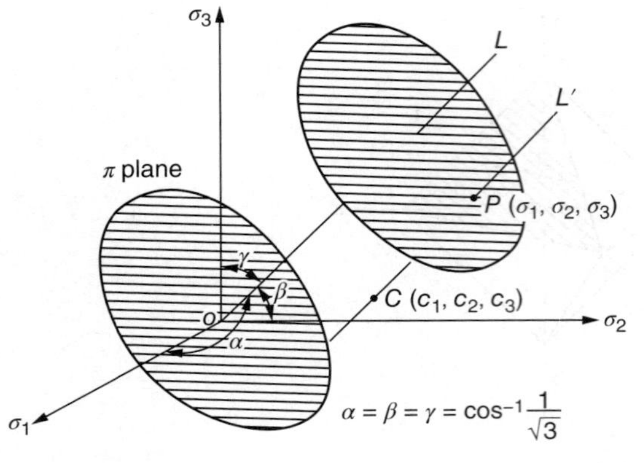

Figure 16.5 Haigh–Westergaard stress space.

16.3.1.2 π Plane

As shown in Figure 16.5, in the stress space, the plane through the original point and normal to the line L is called π plane. The equation of the π plane is $\sigma_1 + \sigma_2 + \sigma_3 = 0$. Because the average stress of each point on the plane is zero, only the deviator stress tensor exists, so the points on π plane represent the stress state not causing volume deformation.

16.3.1.3 Line L' Parallel to the Line L

Assume the line L' is through the point $C(c_1, c_2, c_3)$ and its equation is $\sigma_1 - c_1 = \sigma_2 - c_2 = \sigma_3 - c_3$, where c_1, c_2, c_3 are constants. The average stress of any point $P(\sigma_1, \sigma_2, \sigma_3)$ on the line L' is $\sigma_m = (\sigma_1 + \sigma_2 + \sigma_3)/3$, and the components of the deviator stress tensor of the point are

$$s_1 = \sigma_1 - \sigma_m = \sigma_1 - (\sigma_1 + \sigma_2 + \sigma_3)/3 = (2c_1 - c_2 - c_3)/3 = c_a$$

Likewise

$$s_2 = \sigma_2 - \sigma_m = (2c_2 - c_1 - c_3)/3 = c_b$$
$$s_3 = \sigma_3 - \sigma_m = (2c_3 - c_1 - c_2)/3 = c_c$$

where c_a, c_b, c_c are constants.

So the deviator stress tensor of each point on the line L' is the same, namely, each point on the line L' has the same J_2.

16.3.1.4 The Plane Parallel to π Plane

In Figure 16.5, the equation of the plane parallel to π plane is $\sigma_1 + \sigma_2 + \sigma_3 = c$, where c is a constant. The plane is called deviator plane, in which each point on the plane has the same average stress $\sigma_m = c/3$. As a result, each point on the plane has the same elastic volume deformation.

16.3.2 The Geometric Expression of Any Point

As shown in Figure 16.6, any point $P(\sigma_1, \sigma_2, \sigma_3)$ in the stress space can be expressed by the vector $\overrightarrow{OP}$, and it can be decomposed into two components: one component $\overrightarrow{ON}$

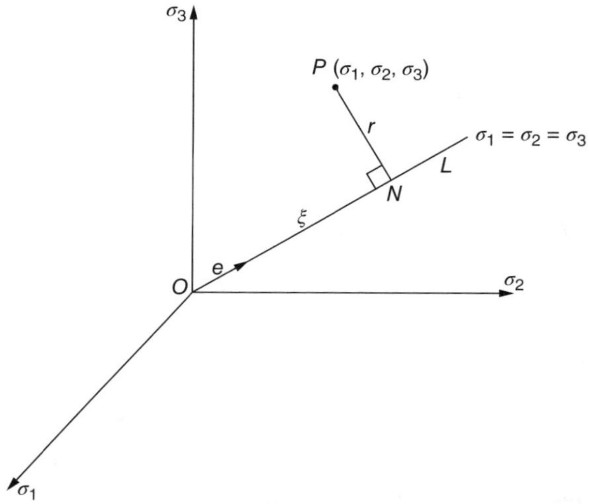

Figure 16.6 The geometric expression of one point in the stress space.

along the hydrostatic stress axis that corresponds with the spherical stress tensor and another component $\overrightarrow{NP}$ in a plane normal to the line L that corresponds with the deviator stress tensor of the point. Then we calculate the length of $\overrightarrow{ON}$ and $\overrightarrow{NP}$, respectively.

Assume $\vec{e}$ is unit length of the hydrostatic stress axis L expressed by the vector as follows:

$$\vec{e} = \frac{1}{\sqrt{3}}\begin{bmatrix}1 & 1 & 1\end{bmatrix}$$

According to the rules of vector operation, the length of $\overrightarrow{ON}$ is

$$\begin{aligned}|ON| &= \xi = \overrightarrow{OP}\cdot\vec{e}\\ &= \begin{bmatrix}\sigma_1 & \sigma_2 & \sigma_3\end{bmatrix}\frac{1}{\sqrt{3}}\begin{bmatrix}1 & 1 & 1\end{bmatrix}\\ &= \frac{1}{\sqrt{3}}I_1\end{aligned}$$

Namely,

$$\xi = \frac{1}{\sqrt{3}}I_1 = \sqrt{3}\sigma_m = \sqrt{3}\sigma_8 \tag{16.14}$$

Components of vector $\overrightarrow{ON}$ on the three axes are

$$\overrightarrow{ON} = \begin{bmatrix}\sigma_m & \sigma_m & \sigma_m\end{bmatrix} = \frac{I_1}{3}\begin{bmatrix}1 & 1 & 1\end{bmatrix}$$

Vector $\overrightarrow{NP}$ is

$$\overrightarrow{NP} = \overrightarrow{OP} - \overrightarrow{ON} = \begin{bmatrix}\sigma_1 & \sigma_2 & \sigma_3\end{bmatrix} - \frac{I_1}{3}\begin{bmatrix}1 & 1 & 1\end{bmatrix} = \begin{bmatrix}s_1 & s_2 & s_3\end{bmatrix}$$

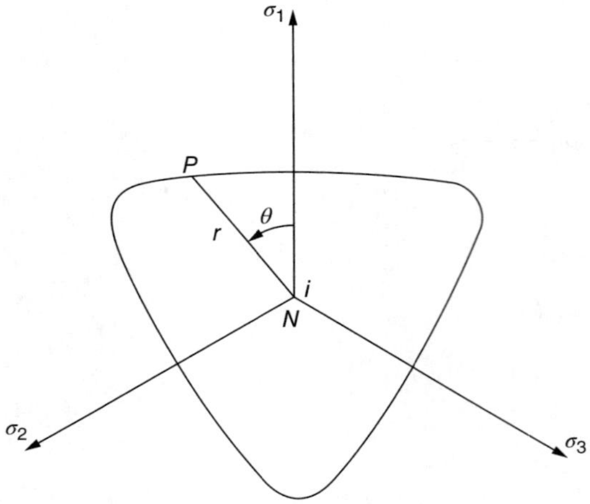

Figure 16.7 Projection of the coordinate axes $\sigma_1, \sigma_2, \sigma_3$ onto the deviator plane.

The length of vector $\overrightarrow{NP}$ is

$$
\begin{aligned}
r &= \sqrt{s_1^2 + s_2^2 + s_3^2} \\
&= \sqrt{2J_2} = \sqrt{3}\tau_8
\end{aligned}
\tag{16.15}
$$

Then we consider the direction of the vector $\overrightarrow{NP}$ on the deviator plane. In Figure 16.7, the paper plane is the deviator plane, and the coordinate axes σ_1, σ_2, σ_3 are projected onto the plane. Because the deviator plane has the same angle with three axes, namely, $\cos^{-1}(1/\sqrt{3}) = 54.7°$, the angles between the projections of three axes onto the deviator plane are equal to $(2\pi/3)$. The components of unit vector $\vec{i}$ that is the projection of coordinate axis σ_1 onto the deviator plane in the coordinate system σ_1, σ_2, σ_3 are

$$
\vec{i} = \frac{1}{\sqrt{6}}\begin{bmatrix} 2 & -1 & -1 \end{bmatrix}
$$

From Figure 16.7, we know

$$
\begin{aligned}
\overrightarrow{NP}\cdot\vec{i} &= r\cos\theta \\
&= \begin{bmatrix} s_1 & s_2 & s_3 \end{bmatrix}\frac{1}{\sqrt{6}}\begin{bmatrix} 2 & -1 & -1 \end{bmatrix} \\
&= \frac{1}{\sqrt{6}}(2s_1 - s_2 - s_3) = 3s_1/\sqrt{6}
\end{aligned}
$$

Then we get

$$
\cos\theta = \frac{\sqrt{3}}{2}\frac{s_1}{\sqrt{J_2}} = \frac{2\sigma_1 - \sigma_2 - \sigma_3}{2\sqrt{3}\sqrt{J_2}}
\tag{16.16}
$$

By trigonometric formula $\cos 3\theta = 4\cos^3\theta - 3\cos\theta$, the formula above can be rewritten as

$$\cos 3\theta = \frac{3\sqrt{3}}{2}\frac{J_3}{J_2^{3/2}} = \frac{\sqrt{2}J_3}{\tau_8^3} \tag{16.17}$$

If $\sigma_1 \geq \sigma_2 \geq \sigma_3$, then we get

$$0 \leq \theta \leq 60°$$

Consider the following several special cases:

1) Pure tension: $\sigma_1 = \sigma_0, \sigma_2 = \sigma_3 = 0, \theta = 0$
2) Pure compression: $\sigma_3 = -\sigma_0, \sigma_1 = \sigma_2 = 0, \theta = 60°$
3) Pure shear: $\sigma_2 = 0, \sigma_1 = \tau, \sigma_3 = -\tau, \theta = 30°$.

In the stress space, the plane containing the hydrostatic stress axis is called the meridian plane. The meridian plane $\theta = 0$ is called tensile meridian plane, and the meridian plane $\theta = 60°$ is called compressive meridian plane.

16.3.3 Principal Stresses

It is inconvenient to solve three principal stresses by formula (16.7) directly. Using trigonometric function relation, it is more convenient to express the principal stresses. Because of $J_1 = 0$, three principal deviator stresses are the three roots of the following equation:

$$s^3 - J_2 s - J_3 = 0 \tag{a}$$

Formula (a) is similar with the following trigonometric function equation:

$$\cos^3\theta - \frac{3}{4}\cos\theta - \frac{1}{4}\cos 3\theta = 0 \tag{b}$$

By substituting $s = \rho\cos\theta$ into formula (a), we get

$$\cos^3\theta - \frac{J_2}{\rho^2}\cos\theta - \frac{J_3}{\rho^3} = 0 \tag{c}$$

Comparing with formula (a), it is clear that

$$\rho = 2\sqrt{J_2}/\sqrt{3} \tag{d}$$

$$\cos 3\theta = \frac{4J_3}{\rho^3} = \frac{3\sqrt{3}}{2}\frac{J_3}{J_2^{3/2}} \tag{e}$$

Noticing the cyclical of $\cos(3\theta \pm 2n\pi)$, if taking $0 \leq \theta \leq \pi/3$, three principal deviator stresses should be

$$\rho\cos\theta, \quad \rho\cos(\theta - 2\pi/3), \quad \rho\cos(\theta + 2\pi/3)$$

namely,

$$\begin{Bmatrix} s_1 \\ s_2 \\ s_3 \end{Bmatrix} = \begin{Bmatrix} \sigma_1 \\ \sigma_2 \\ \sigma_3 \end{Bmatrix} - \begin{Bmatrix} \sigma_m \\ \sigma_m \\ \sigma_m \end{Bmatrix} = \frac{2\sqrt{J_2}}{\sqrt{3}}\begin{Bmatrix} \cos\theta \\ \cos(\theta - 2\pi/3) \\ \cos(\theta + 2\pi/3) \end{Bmatrix} \tag{16.18}$$

where $\sigma_1 \geq \sigma_2 \geq \sigma_3$, $0 \leq \theta \leq \pi/3$.

16.4 Decompose of Strain Tensor

In the condition of small deformation, the relationship between displacement and strain is as follows:

$$\varepsilon_x = \frac{\partial u}{\partial x}, \quad \gamma_{xy} = \frac{\partial u}{\partial y} + \frac{\partial v}{\partial x}$$
$$\varepsilon_y = \frac{\partial v}{\partial y}, \quad \gamma_{yz} = \frac{\partial v}{\partial z} + \frac{\partial w}{\partial y}$$
$$\varepsilon_z = \frac{\partial w}{\partial z}, \quad \gamma_{zx} = \frac{\partial w}{\partial x} + \frac{\partial u}{\partial z}$$

where $\varepsilon_x, \varepsilon_y, \varepsilon_z, \gamma_{xy}, \gamma_{yz}, \gamma_{zx}$ are the strain components.

Let

$$\varepsilon_{xy} = \frac{1}{2}\gamma_{xy}, \quad \varepsilon_{yz} = \frac{1}{2}\gamma_{yz}, \quad \varepsilon_{zx} = \frac{1}{2}\gamma_{zx}$$

and

$$\varepsilon_{xy} = \varepsilon_{yx}, \quad \varepsilon_{yz} = \varepsilon_{zy}, \quad \varepsilon_{zx} = \varepsilon_{xz}$$

Then strain components of a point compose a symmetric strain tensor, namely,

$$[\varepsilon_{ij}] = \begin{bmatrix} \varepsilon_x & \varepsilon_{xy} & \varepsilon_{xz} \\ \varepsilon_{xy} & \varepsilon_y & \varepsilon_{yz} \\ \varepsilon_{xz} & \varepsilon_{yz} & \varepsilon_z \end{bmatrix} = \begin{bmatrix} \varepsilon_x & \frac{1}{2}\gamma_{xy} & \frac{1}{2}\gamma_{xz} \\ \frac{1}{2}\gamma_{xy} & \varepsilon_y & \frac{1}{2}\gamma_{yz} \\ \frac{1}{2}\gamma_{xz} & \frac{1}{2}\gamma_{yz} & \varepsilon_z \end{bmatrix} \tag{16.19}$$

Here, there is a coefficient 1/2 in the front of the shear strain component, because only in this way the formula can meet the rotation rule of the tensor.

Similar to the stress tensor, the strain tensor can also be decomposed into the spherical strain tensor and the deviator strain tensor, namely,

$$\begin{bmatrix} \varepsilon_x & \varepsilon_{xy} & \varepsilon_{xz} \\ \varepsilon_{xy} & \varepsilon_y & \varepsilon_{yz} \\ \varepsilon_{xz} & \varepsilon_{yz} & \varepsilon_z \end{bmatrix} = \begin{bmatrix} \varepsilon_m & 0 & 0 \\ 0 & \varepsilon_m & 0 \\ 0 & 0 & \varepsilon_m \end{bmatrix} + \begin{bmatrix} \varepsilon_x - \varepsilon_m & \frac{1}{2}\gamma_{xy} & \frac{1}{2}\gamma_{xz} \\ \frac{1}{2}\gamma_{xy} & \varepsilon_y - \varepsilon_m & \frac{1}{2}\gamma_{yz} \\ \frac{1}{2}\gamma_{xz} & \frac{1}{2}\gamma_{yz} & \varepsilon_z - \varepsilon_m \end{bmatrix} \tag{16.20}$$

in which

$$\varepsilon_m = (\varepsilon_x + \varepsilon_y + \varepsilon_z)/3$$

where ε_m is the average strain.

The first item on the right side of formula (16.20) is the spherical strain tensor, and the second item is the deviator strain tensor, expressed by $[e_{ij}]$, namely,

$$[e_{ij}] = \begin{bmatrix} e_x & e_{xy} & e_{xz} \\ e_{xy} & e_y & e_{yz} \\ e_{xz} & e_{yz} & e_z \end{bmatrix} = \begin{bmatrix} \varepsilon_x & \frac{1}{2}\gamma_{xy} & \frac{1}{2}\gamma_{xz} \\ \frac{1}{2}\gamma_{xy} & \varepsilon_y & \frac{1}{2}\gamma_{yz} \\ \frac{1}{2}\gamma_{xz} & \frac{1}{2}\gamma_{yz} & \varepsilon_z \end{bmatrix} \tag{16.21}$$

As a result, the strain tensor can be abbreviated as follows:

$$\varepsilon_{ij} = \varepsilon_m \delta_{ij} + e_{ij} \tag{16.22}$$

The spherical strain tensor has the same normal strain in all directions, and it represents the volume change. The sum of three normal strains of the deviator strain tensor is zero, that is to say, it has no volume deformation, but only reflects the change of shape.

Three invariants of the deviator strain tensor are expressed by J_1', J_2', J_3', respectively, as follows:

$$\left.\begin{aligned} J_1' &= e_x + e_y + e_z = 0 \\ J_2' &= \frac{1}{6}[(e_x - e_y)^2 + (e_y - e_z)^2 + (e_z - e_x)^2 + 6(e_{xy}^2 + e_{yz}^2 + e_{zx}^2)] \\ J_3' &= e_x e_y e_z + 2e_{xy}e_{yz}e_{zx} - e_x e_{yz}^2 - e_y e_{zx}^2 - e_z e_{xy}^2 \end{aligned}\right\} \tag{16.23}$$

16.5 Criterion of Yield

Under one-dimensional load, taking the ideal elastic–plastic model as an example, when the stress is less than the yield limit σ_s, the material is in the elastic stage, and when the stress reaches σ_s, the material enters into the plastic stage. Therefore, $\sigma = \sigma_s$ is the yield condition under one-dimensional load.

In the complex stress state, when a point in the body begins to produce plastic deformation, the stress also should meet a certain condition: it is the yield condition in the complicated stress state. Generally speaking, it should be the function of six stress components and can be expressed as follows:

$$F(\sigma_x, \sigma_y, \sigma_z, \tau_{xy}, \tau_{yz}, \tau_{zx}) = C \tag{16.24}$$

where C is a constant related to the material and F is the yield function.

We put the six stress components of one point into formula (16.24): if $F < C$, the point is in the elastic state, and if $F = C$, the point is in the plastic state.

The material we considered is isotropic; the change of the directions of coordinates has no effect on yield condition, so it can be expressed by the principal stresses as follows:

$$F(\sigma_1, \sigma_2, \sigma_3) = C \tag{16.25}$$

It can also be expressed by invariants of the stress tensor I_1, I_2, I_3 or invariants of the deviator stress tensor J_2, J_3.

In some kinds of materials, such as rock, soil, concrete, and so on, their yield conditions are affected by the hydrostatic stress, generally expressed as follows:

$$F(J_1, J_2, J_3) = C \tag{16.26}$$

In other kinds of materials, such as metal, their yield conditions are not affected by the hydrostatic stress and can be expressed as

$$F(J_2, J_3) = C \tag{16.27}$$

Yield condition is often referred to as the yield criterion.

16.5.1 Tresca Yield Criterion

In 1864, Tresca proposed that when the maximum shear stress $\tau_{\max}$ reached a certain value k, the material yielded. This condition can be expressed as follows:

$$\tau_{\max} = k \tag{16.28a}$$

When $\sigma_1 \geq \sigma_2 \geq \sigma_3$, it can be written as

$$\sigma_1 - \sigma_3 = 2k \tag{16.28b}$$

In general, if the order of the principal stresses is unknown, then Tresca yield condition should be expressed as

$$\left.\begin{aligned} \sigma_1 - \sigma_2 &= \pm 2k \\ \sigma_2 - \sigma_3 &= \pm 2k \\ \sigma_3 - \sigma_1 &= \pm 2k \end{aligned}\right\} \tag{16.28c}$$

The general expression is

$$[(\sigma_1 - \sigma_2)^2 - 4k^2][(\sigma_2 - \sigma_3)^2 - 4k^2][(\sigma_3 - \sigma_1)^2 - 4k^2] = 0 \tag{16.28d}$$

The formula above can also be expressed by the invariants of the deviator stress tensor as follows:

$$F = 4J_2^3 - 27J_3^2 - 36k^2J_2^2 + 96k^4J_2 - 64k^2 = 0 \tag{16.28e}$$

In the three-dimensional stress space, $\sigma_1 - \sigma_2 = \pm 2k$ is a pair of planes parallel to the normal line of the deviator plane π and the axis σ_3. Therefore, the yield surfaces built according to formula (16.8) are hexagonal cylinder composed of three pairs of planes parallel to each other and perpendicular to the plane π, as shown in Figure 16.8. The cutting line (yield line) with the deviator plane π is a hexagon, as shown in Figure 16.9(a).The radius of its circumcircle is $2k\sqrt{2/3}$. This is obtained by the projection of yield stress $(2k, 0,0)$ onto the deviator plane π under one-dimensional tensile load.

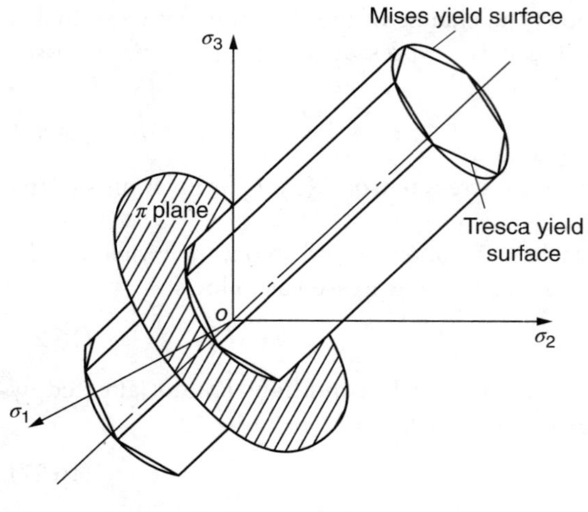

Figure 16.8 The yield surfaces of Tresca and Mises.

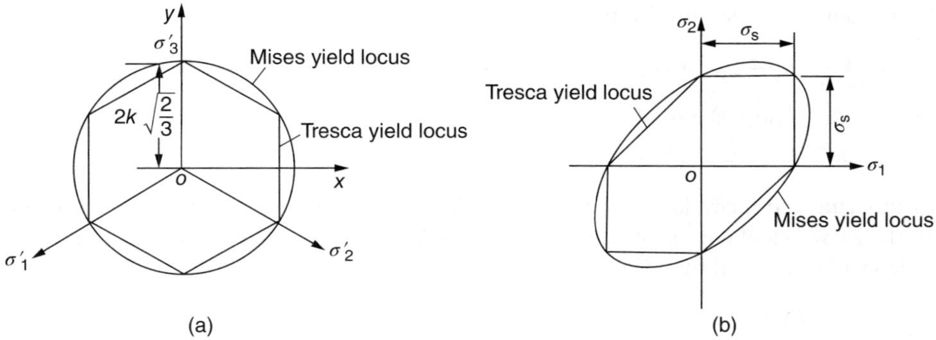

Figure 16.9 The yield lines on deviator plane π and σ_1, σ_2 planes.

If the material is in plane stress state, making $\sigma_3 = 0$, then formula (16.28c) will be

$$\left.\begin{aligned} \sigma_1 - \sigma_2 &= \pm 2k \\ \sigma_2 &= \pm 2k \\ \sigma_1 &= \pm 2k \end{aligned}\right\} \tag{16.29}$$

The yield line in the plane of σ_1 and σ_2 is shown in Figure 16.9(b).

The constant k is determined by one-dimensional tensile test, so $k = \sigma_s/2$. If the constant is determined by pure shear test, then $k = \tau_s$, in which τ_s is pure shear yield limit. According to Tresca yield condition, the relationship between the shear yield limit and the tensile yield limit of the material is as follows:

$$\tau = \sigma_s/2 \tag{16.30}$$

Tresca yield condition is a linear function of the principal stresses and its application is more convenient, which is also consistent with the experiment data of metal materials. But it ignores the effect of the intermediate principal stress, and there are angular points on the yield line, bringing some difficulties to mathematical processing, which is its disadvantage.

16.5.2 Mises Yield Criterion

The yield condition proposed by Mises in 1913 is that the yield line on the deviator plane π is the circumscribed circle of Tresca hexagon, as shown in Figure 16.9(a). The yield surface in stress space is a cylinder perpendicular to the deviator plane π, as shown in Figure 16.8. The cutting line with the plane of principal stress such as plane of σ_1, σ_2 is an ellipse, as shown Figure 16.9(b). Its expression is

$$J_2 = \frac{1}{3}\sigma_s^2 = k^2 \tag{16.31}$$

It shows that as long as the second invariant of the deviator stress tensor reaches a certain value, the material will yield. σ_s is the yield limit under one-dimensional tensile load. Because the octahedral shear stress is $\tau_8 = \sqrt{2J_2/3}$, Mises yield condition can also be expressed as

$$\tau_8 = \sqrt{\frac{2}{3}}k \tag{16.32}$$

In the general stress condition,

$$J_2 = \frac{1}{6}[(\sigma_x - \sigma_y)^2 + (\sigma_y - \sigma_z)^2 + (\sigma_z - \sigma_x)^2 + 6(\tau_{xy}^2 + \tau_{yz}^2 + \tau_{zx}^2)]$$

In the case of pure shear,

$$J_2 = \tau_{xy}^2 = \tau_s^2 = k^2$$

Comparing the formula above with formula (16.31), it is known that in accordance with the Mises yield condition, the relationship between the shear yield limit τ_s and the tensile yield limit σ_s of material is as follows:

$$\tau_8 = \sigma_s/\sqrt{3} \tag{16.33}$$

Taylor and Quinney had tested thin tube made of copper, aluminum, and mild steel under the action of tension T and torque M. The average radius of the tube is R, the thickness is h, and the stress components are

$$\sigma_z = T/(2\pi Rh), \quad \tau_{\theta z} = M/(2\pi R^2 h)$$

The principal stresses are

$$\sigma_1 = \sigma_z/2 + \sqrt{\sigma_z^2 + 4\tau_{\theta z}^2}/2, \quad \sigma_2 = 0$$

$$\sigma_3 = \sigma_z/2 - \sqrt{\sigma_z^2 + 4\tau_{\theta z}^2}/2$$

Noting $\sigma = \sigma_z$, $\tau = \tau_{\theta z}$, for Tresca yield criterion, there is

$$\left(\frac{\sigma}{\sigma_s}\right)^2 + 4\left(\frac{\tau}{\sigma_s}\right)^2 = 1$$

For Mises yield criterion, there is

$$\left(\frac{\sigma}{\sigma_s}\right)^2 + 3\left(\frac{\tau}{\sigma_s}\right)^2 = 1$$

Figure 16.10 shows the theoretical curve and the experimental results. It is evident that Mises criterion accords with the test results better.

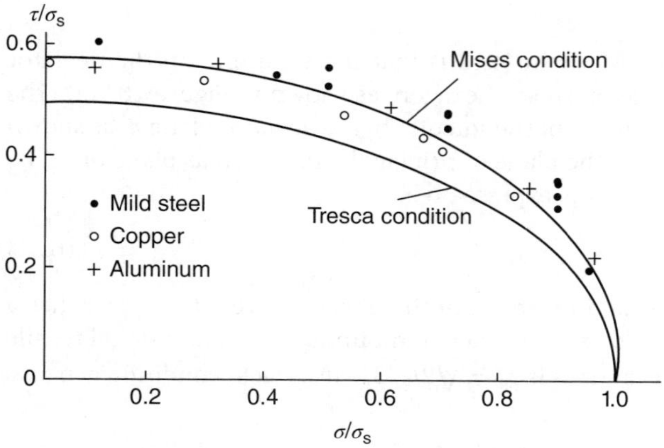

Figure 16.10 Metal test results.

16.5.3 Mohr–Coulomb Yield Criterion

According to Mohr–Coulomb yield criterion, when the stress state reaches the following limit, the material will yield, namely,

$$|\tau| = c - \sigma tg\phi \tag{16.34}$$

where τ is the maximum shear stress, σ is the normal stress in the same plane, c is the cohesion of the material, and ϕ is the internal friction angle of material.

As special cases, such as when the internal friction angle $\phi = 0$, the formula above degrades to maximum shear stress criterion of Tresca, and when $\tau = c$, the cohesive force is equal to the yield limit of material on pure shear condition, $c = \tau_s$.

As shown in Figure 16.11, formula (16.34) represents a tangent of the biggest stress circle. If $\sigma_1 \geq \sigma_2 \geq \sigma_3$, formula (16.34) can be rewritten as follows:

$$-\frac{1}{2}(\sigma_1 - \sigma_3)\cos\phi = c - \left(\frac{\sigma_1 + \sigma_3}{2} - \frac{\sigma_1 - \sigma_3}{2}\sin\phi\right) tg\phi \tag{16.35a}$$

namely,

$$\sigma_1 - \sigma_3 = 2c\cos\phi - (\sigma_1 + \sigma_3)\sin\phi \tag{16.35b}$$

Considering every possible combination of stress producing yield, then the yield surface can be obtained; it is a pyramid, as shown in Figure 16.12. The apex of the pyramid is on the hydrostatic stress axis, $\sigma_1 = \sigma_2 = \sigma_3 = cctg\phi$. Figure 16.13 shows the yield line on plane π.

Mohr–Coulomb yield criterion can be expressed by ξ, r, θ (formulas (16.14), (16.15), (16.16)) as below:

$$F(\xi, r, \theta) = \sqrt{2}\xi\sin\phi + \sqrt{3}r\sin(\theta + \pi/3) + r\cos(\theta + \pi/3)\sin\phi - \sqrt{6}c\cos\phi = 0 \tag{16.36}$$

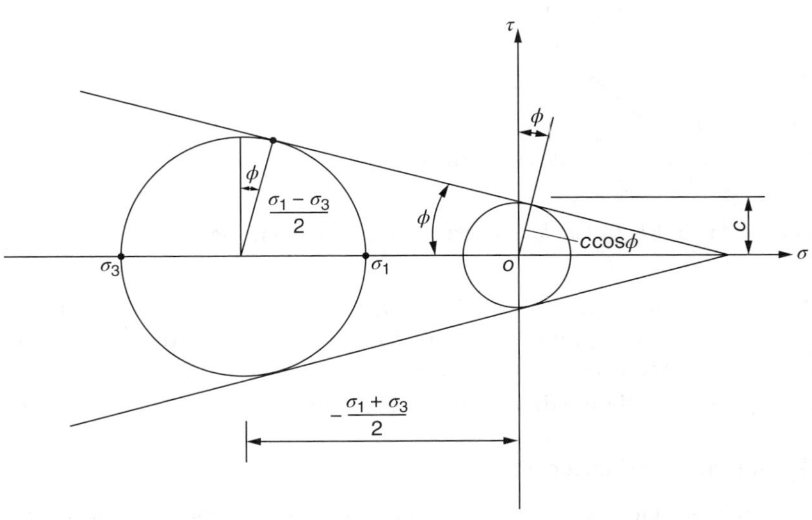

Figure 16.11 Mohr–Coulomb criterion expressed by principle stresses.

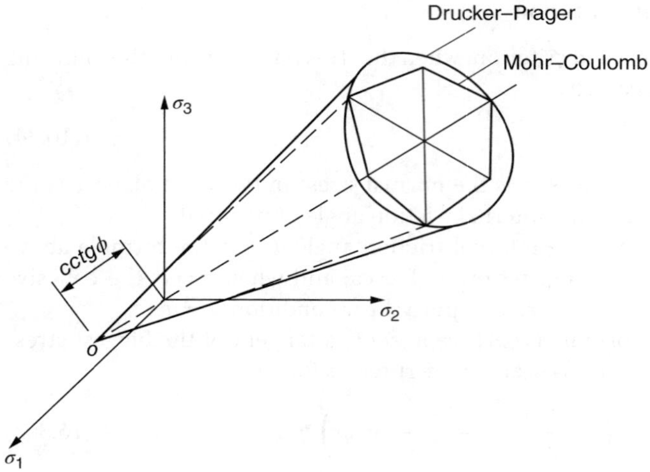

Figure 16.12 Mohr–Coulomb yield surface and Drucker–Prager yield surface.

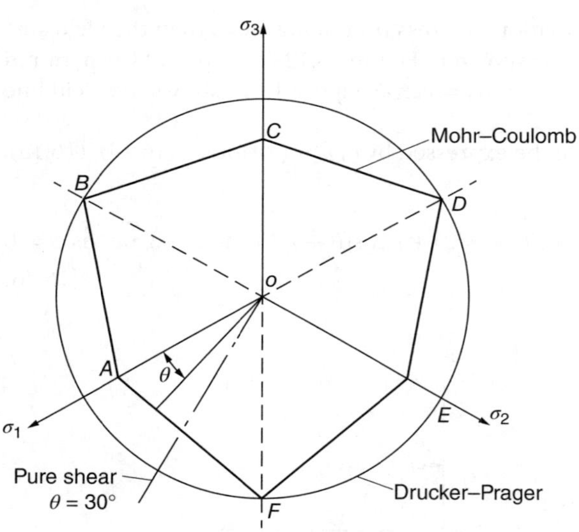

Figure 16.13 Mohr–Coulomb yield surface and Drucker–Prager yield surface on plane π.

in which $0 \leq \theta \leq \pi/3$. In the formula above, if $\theta = 0, \theta = 60°$, we get yield lines on two meridian surfaces as shown in Figure 16.14(a), and then if $\xi = 0$, we get the yield line on the plane π as shown in Figure 16.14(b). In formula (16.36), if $\theta = 0, \xi = 0$, we get r_{t0}, and if $r = 0$, $\xi_0 = \sqrt{3}cctg\phi$ is obtained; by the ratio of r_{t0} and ξ_0, we get $tg\phi_t$.

16.5.4 Drucker–Prager Yield Criterion

Mohr–Coulomb yield surface is an angular cone surface, but the angular points cause inconvenience in numerical calculation. In order to get a smooth yield surface

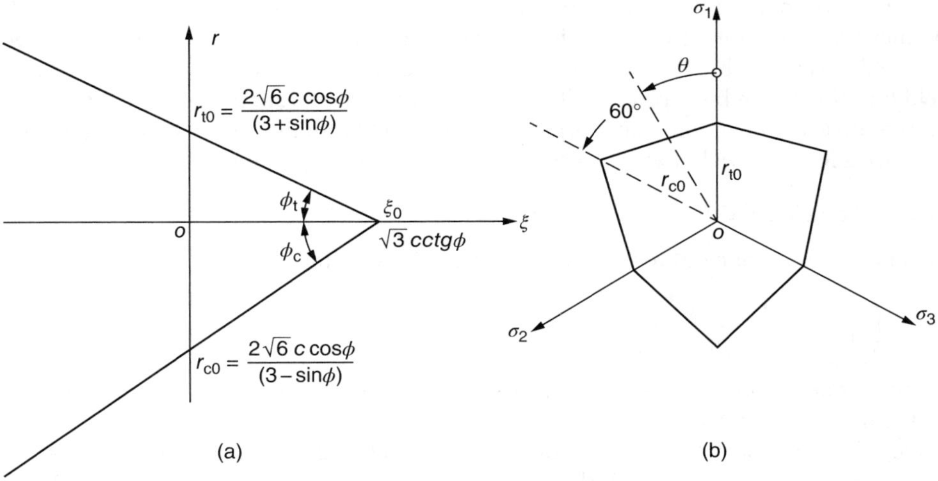

Figure 16.14 Mohr–Coulomb yield criterion (a) meridian plane $\theta = 0$ and $\theta = 60°$ and (b) π plane.

approximate to Mohr–Coulomb surface, in 1952, Drucker–Prager modified Mises yield criterion and put forward the following yield criterion:

$$F = \alpha I_1 + \sqrt{J_2} - k = 0 \tag{16.37}$$

$\xi = I_1/\sqrt{3},\ r = \sqrt{2J_2}$ can be written as

$$F = \sqrt{6}\alpha\xi + r - \sqrt{2}k = 0 \tag{16.38}$$

where α, k are the material constants.

If $\alpha = 0$, the two formulas above degrade to Mises yield criterion. From the formula above, we know that Drucker–Prager yield surface is a conical surface, as shown in Figure 16.12, and its cutting line on the plane π is a circle, as shown in Figure 16.13.

By selecting two constants α and k appropriately, Drucker–Prager yield surface may be close to Mohr–Coulomb yield surface. For example, if

$$\alpha = \frac{2\sin\phi}{\sqrt{3}(3-\sin\phi)}, \quad k = \frac{6c\cos\phi}{\sqrt{3}(3-\sin\phi)} \tag{16.39}$$

in each section, Drucker–Prager yield circle will coincide with external vertexes of Mohr–Coulomb hexagon. If

$$\alpha = \frac{2\sin\phi}{\sqrt{3}(3+\sin\phi)}, \quad k = \frac{6c\cos\phi}{\sqrt{3}(3+\sin\phi)} \tag{16.40}$$

then Drucker–Prager yield circle will coincide with internal vertexes of Mohr–Coulomb hexagon.

Drucker–Prager yield function is the linear function of ξ and r and does not contain θ, so on the meridional plane, the yield line is a straight line, and on the deviator plane, the yield curve is a circle. Test data for the brittle material such as concrete shows that on the meridional plane, the yield line is not straight, but it is a bulge curve outward and on the deviator plane, the yield curve isn't a circle. In three-dimensional tensile zone, it is a triangle with rounded corners. In three-dimensional compressive zone, with the increase of compressive stress, it is gradually close to a circle. In order to describe the yield surface of brittle materials better, in recent years, some new more complex yield functions have been proposed. Tresca and Mises yield functions only contain one parameter. Mohr–Coulomb and Drucker–Prager yield functions have two parameters. Here are a few new yield functions containing three to five parameters.

16.5.5 Lade Yield Criterion

For soil, in 1977, Lade put forward the following yield criterion:

$$\left(\frac{I_1^3}{I_3} - 27\right)\left(\frac{I_1}{p_a}\right)^m = \eta \tag{16.41}$$

where p_a is atmospheric pressure, I_1, I_3 are the first and third stress invariants, and m, η are the material constants.

According to test data, drawing the points of $I_1^3/I_3 - 27$ and p_a/I_1 on the paper of log–log, if a line passes through the majority of points, the intercept on $p_a/I_1 = 1$ is η and the slope is m. On the hydrostatic stress axis, $I_1^3/I_3 = 27$.

For concrete, its tensile strength is

$$R_t = ap_a \tag{16.42}$$

Before substituting it into formula (16.41), make the transformation as follows:

$$\overline{\sigma_x} = \sigma_x + ap_a, \quad \overline{\sigma_y} = \sigma_y + ap_a, \quad \overline{\sigma_z} = \sigma_z + ap_a \tag{16.43}$$

This criterion contains three parameters a, m, η, which should be determined by the experimental data.

16.5.6 Bresler–Pister Yield Criterion

In 1958, Bresler and Pister put forward the following yield criterion:

$$\frac{\tau_8}{R_c} = a - b\frac{\sigma_8}{R_c} + c\left(\frac{\sigma_8}{R_c}\right)^2 \tag{16.44}$$

where τ_8 is the octahedral shear stress, σ_8 is the octahedral normal stress, R_c is the one-dimensional compressive strength of materials, and a, b, c are the three constants determined by the experimental data.

For example, if the one-dimensional tensile strength is R_t, one-dimensional compressive strength is R_c, and equivalent bidirectional compressive strength is R_{bc}, the ratios are as follows:

$$s_1 = R_t/R_c, \quad s_2 = R_{bc}/R_c$$

The stress components of octahedron corresponding to the three experimental conditions are as follows:

Experimental condition	σ_8/R_c	τ_8/R_c
$\sigma_1 = R_t$	$s_1/3$	$\sqrt{2}s_1/3$
$\sigma_3 = -R_c$	$-1/3$	$\sqrt{2}/3$
$\sigma_2 = \sigma_3 = -R_{bc}$	$-2s_2/3$	$\sqrt{2}s_2/3$

By substitution of values of σ_8/R_c, τ_8/R_c above into formula (16.44), we can get the coefficients a, b, c.

Bresler–Pister yield function doesn't contain θ, so the yield curve on the deviator plane is a circle.

16.5.7 Ottosen Yield Criterion

In 1977, Ottosen put forward a yield criterion as follows:

$$\left.\begin{aligned} &F(I_1, J_2, \theta) = a\frac{J_2}{R_c^2} + \lambda(\theta)\frac{\sqrt{J_2}}{R_c} + b\frac{I_1}{R_c} - 1 = 0 \\ &\lambda(\theta) = k_1 \cos\left[\frac{1}{3}\cos^{-1}(k_2 \cos 3\theta)\right] \quad (\cos 3\theta \geq 0) \\ &\lambda(\theta) = k_1 \cos\left[\frac{\pi}{3} - \frac{1}{3}\cos^{-1}(-k_2 \cos 3\theta)\right] \quad (\cos 3\theta \leq 0) \end{aligned}\right\} \tag{16.45}$$

where a, b, k_1, k_2 are four parameters determined by test data.

As $\xi = I_1/3, r = \sqrt{2J_2}$, from formula (16.45), we know that on the meridian line (θ is a constant), the yield curve is a quadratic curve. As for $\lambda(\theta)$, it is derived from the analogy of triangular membrane deformation. Therefore, the yield curve on the plane π, in the low stress zone, is close to a triangle and, in the high stress area, is close to a circle, corresponding to the failure properties of brittle materials such as concrete. Four parameters a, b, k_1, k_2 can be determined according to one-dimensional compressive strength $R_c(\theta = 60°)$, one-dimensional tensile strength $R_t(\theta = 0)$, biaxial compressive strength R_{bc}, and triaxial compressive strength.

For concrete, according to the biaxial test data of Kupfer *et al.*, taking $R_{bc} = 1.16R_c$, based on the triaxial test data of Balmer and Richart, on the pure compressive meridional plane ($\theta = 60°$), taking $(\xi/R_c, r/R_c) = (-5{,}4)$, then the four parameters are listed as shown in Table 16.1.

Figure 16.15 shows the comparison of Ottosen yield criterion and results of concrete triaxial test.

16.5.8 Hsieh–Ting–Chen Four-Parameter Criterion

In 1979, Hsieh, Ting, and Chen put forward the following yield criterion:

$$F = a\frac{J_2}{R_c^2} + b\frac{\sqrt{J_2}}{R_c} + c\frac{\sigma_1}{R_c} + d\frac{I_1}{R_c} - 1 = 0 \tag{16.46}$$

where a, b, c, d are four parameters determined by test data. The yield curve on the meridian plane is a curve and that on the deviator plane is not circle, but close to a triangle. When $a = c = 0$, it degrades to Drucker–Prager yield criterion, and when

Table 16.1 Parameters of Ottosen criterion applied to concrete.

R_t/R_c	a	b	k_1	k_2
0.08	1.8076	4.0962	14.4863	0.9914
0.10	1.2759	3.1962	11.7365	0.9801
0.12	0.9218	2.5969	9.9110	0.9467

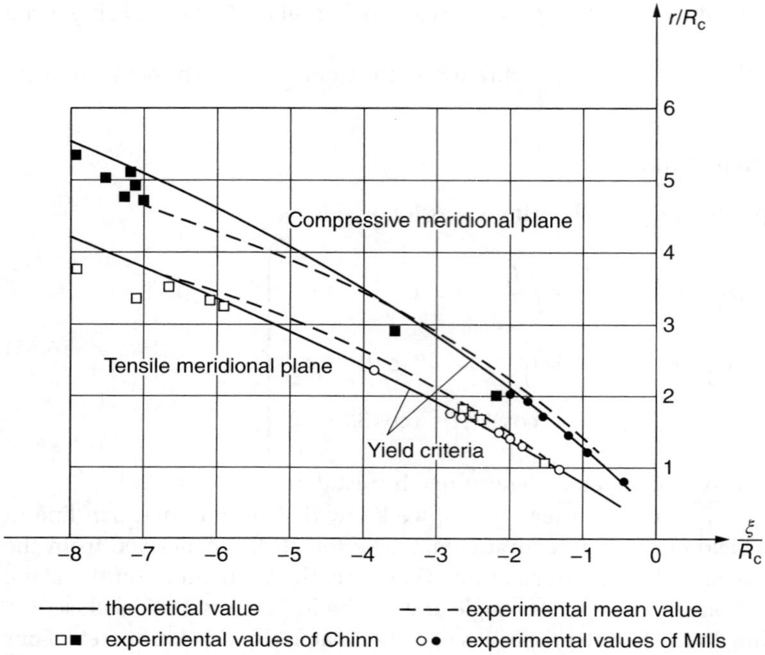

Figure 16.15 Comparison of Ottosen yield criterion and results of concrete triaxial test —— theoretical value, ----- experimental mean value, □■ experimental values of Chinn, and ○● experimental values of Mills.

$a = c = d = 0$, it degrades to Mises yield criterion. The calculation of this criterion is simple, and it is easy to match with the test data, but on the compressive meridional plane, the yield surface has angular points.

16.5.9 Mohr–Coulomb Criterion with the Maximum Tensile Stress

For brittle materials such as concrete and rock, the following mixing rule can be used: when the material produces shear failure, Mohr–Coulomb criterion is applied, as formula (16.36), and when the material produces tensile failure, the maximum tensile stress criterion is applied, namely,

$$F = \sigma_1 - R_t = 0 \tag{c}$$

or the maximum tensile strain criterion, namely,

$$F = \varepsilon_1 - \varepsilon_t = 0 \tag{d}$$

where R_t is the tensile strength and ε_t is the ultimate tensile strain.

Using incremental method, at each incremental step, formula (16.36), formulas (c) and (d) should be tested, respectively. If it is controlled by tensile stress, the yield function expressed by formula (c) should be used in the rest calculation of this incremental step, but the residual tensile strength R'_t (usually taking $R'_t = 0$) should be adopted. If it is controlled by shear stress, the yield function expressed by formula (16.36) should be

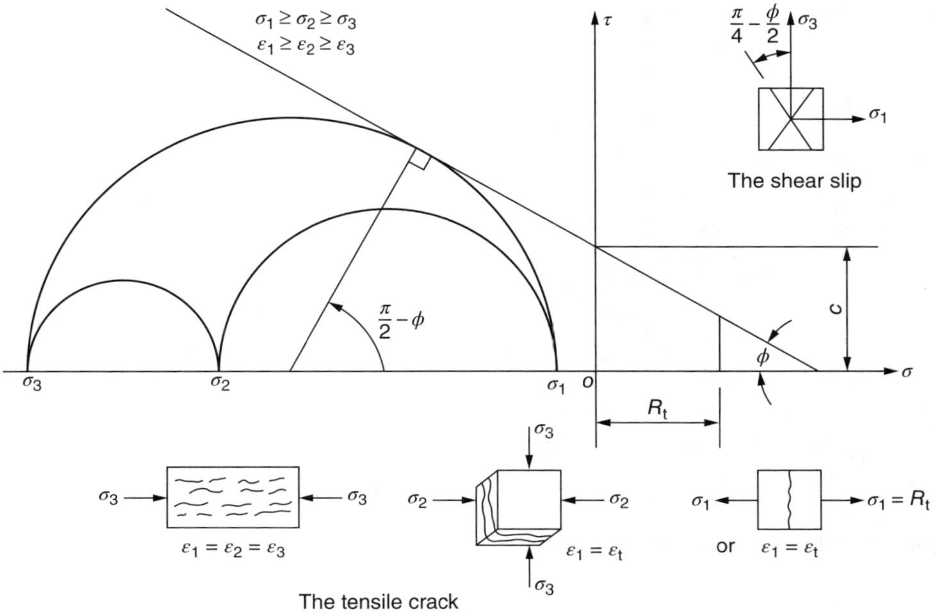

Figure 16.16 Mohr–Coulomb criterion with the maximum tensile stress (strain).

used in the rest calculation of this incremental step, but the residual cohesion c' and residual friction angle ϕ' should be adopted.

Figure 16.16 shows Mohr–Coulomb criterion with the maximum tensile stress, and several kinds of tensile cracks are shown at the bottom of the figure.

16.5.10 Willam–Warnke Criterion with Three and Five Parameters

In 1975, Willam and Warnke put forward a yield criterion with three parameters. The yield curve on the deviator plane is close to a triangle, and that on the meridional plane is a straight line. Shortly afterward a criterion of five parameters was put forward, making the yield curve on the meridian plane a curve, in order to match with experimental data better.

The failure curve on the deviator plane of brittle materials such as concrete is shown in Figure 16.17, similar to an equilateral triangle. In order to get a smooth curve without sharp corners, due to triple symmetry, Willam and Warnke used part of the ellipse to fit within the scope of $0 \le \theta \le 60^\circ$. As shown in Figure 16.18, the center of the ellipse is on the point B, its distance to the center o of the deviator plane is $r_t - b$, the long and short axes of ellipse are a and b, and at the points of P_1 and P_2, the ellipse is orthogonal with two meridian planes $\theta = 0$ and $\theta = 60^\circ$; according to these two conditions, finally, the ellipse expressed in polar coordinate is as follows:

$$r(\theta) = \frac{2r_c(r_c^2 - r_t^2)\cos\theta + r_c(2r_t - r_c)[4(r_c^2 - r_t^2)\cos^2\theta + 5r_t^2 - 4r_t r_c]^{1/2}}{4(r_c^2 - r_t^2)\cos^2\theta + (r_c - 2r_t)^2} \tag{16.47}$$

where r_t and r_c are the values of r on the tensile meridian plane and compressive meridian plane, respectively.

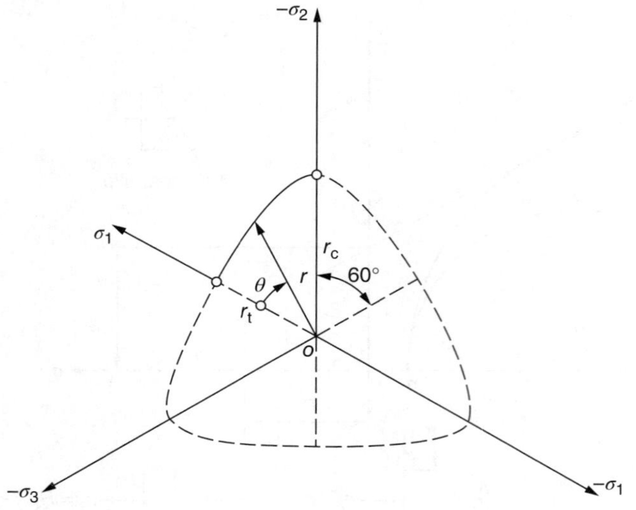

Figure 16.17 The failure curve of brittle materials on the deviator plane.

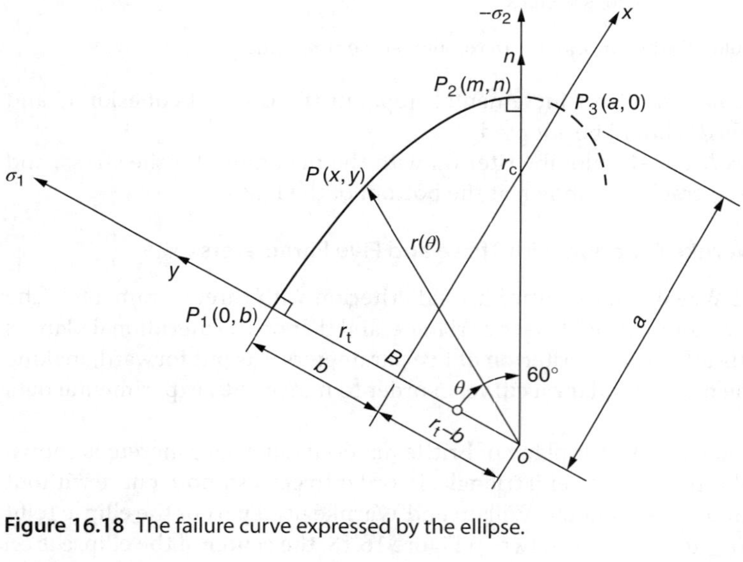

Figure 16.18 The failure curve expressed by the ellipse.

16.5.10.1 Willam–Warnke Criterion with Three Parameters

Using the calculated r_θ above, Willam and Warnke put forward the yield criterion with three parameters as follows:

$$F(\sigma_m, \tau_m, \theta) = \frac{1}{\rho}\frac{\sigma_m}{R_c} + \frac{1}{r(\theta)}\frac{\tau_m}{R_c} - 1 = 0 \tag{16.48}$$

where

$$\left.\begin{aligned} \sigma_m &= I_1/3 = \xi/\sqrt{3} \\ \tau_m &= \sqrt{2J_2/5} = r/\sqrt{5} \end{aligned}\right\} \tag{16.49}$$

Formula (16.48) can also be rewritten as follows:

$$\frac{\tau_m}{R_c} = r(\theta)\left(1 - \frac{1}{\rho}\frac{\sigma_m}{R_c}\right) \tag{16.50}$$

r_θ contains two parameters, r_c, r_t, and adding parameter ρ, there are a total of three parameters. From formula (16.50), we know when θ is a constant; the relationship between τ_m and σ_m is linear; so on the meridional plane of stress space, the yield line is a straight line; if $r_t = r_c$, it degrades to Drucker–Prager yield criterion, and if $\rho \to \infty$, it degrades to Mises yield criterion.

Three parameters, r_c, r_t, ρ, are determined by the test data. For example, knowing one-dimensional tensile strength R_t, one-dimensional compressive strength R_c, and equivalent biaxial compressive strength R_{bc}, if $s_1 = R_t/R_c$, $s_2 = R_{bc}/R_c$, the test data can be expressed as follows:

Test	σ_m/R_c	τ_m/R_c	θ	$r(\theta)$
$\sigma_1 = R_t$	$s_1/3$	$s_1\sqrt{2/15}$	0°	r_t
$\sigma_z = -R_c$	$-1/3$	$\sqrt{2/15}$	60°	r_c
$\sigma_2 = \sigma_3 = -R_{bc}$	$-2s_2/3$	$s_2\sqrt{2/15}$	0°	r_t

By substitution of the test data above into formula (16.50), three parameters are obtained as follows:

$$\rho = \frac{s_1 s_2}{s_2 - s_1}, \quad r_t = \sqrt{\frac{6}{5}\frac{s_1 s_2}{2s_2 + s_1}}, \quad r_c = \sqrt{\frac{6}{5}\frac{s_1 s_2}{3s_1 s_2 + s_2 - s_1}} \tag{16.51}$$

The vertex of yield surface is on the hydrostatic stress axis and its location is

$$\sigma_m/R_c = \rho \tag{16.52}$$

From Figure 16.19, it is clear that in low stress zone, the theoretical and experimental results correspond with each other well, but in high stress zone, the error is bigger.

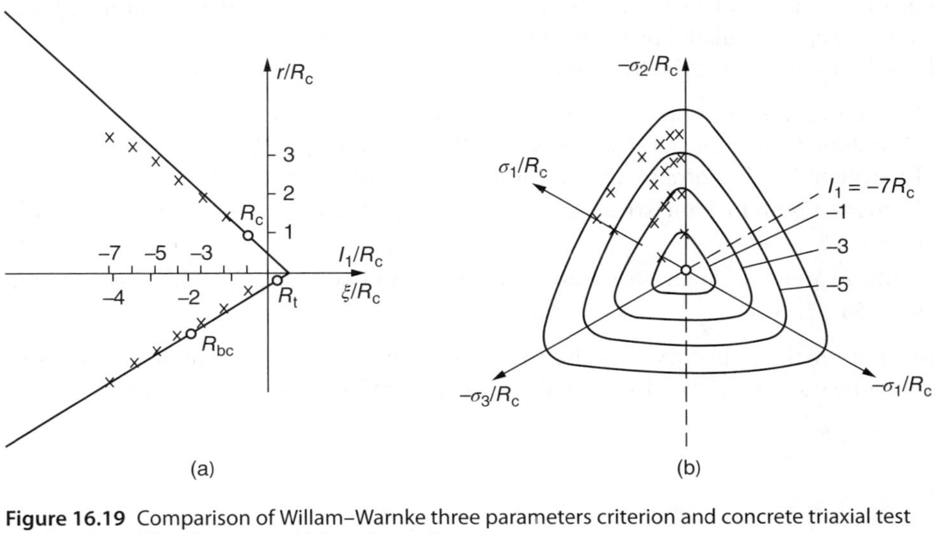

Figure 16.19 Comparison of Willam–Warnke three parameters criterion and concrete triaxial test results (a) meridional plane and (b) deviator plane.

This is due to the assumption in the yield function that the relationship between ξ and r on the meridional plane is linear. According to the experimental data, the relationship between ξ and r is a curve actually.

16.5.10.2 Willam–Warnke Criterion with Five Parameters

In order to accord with experimental data better, assuming that on the tensile and compressive meridional plane, there is a quadratic relation between τ_m and σ_m (viz., r and ξ) as follows:

$$\frac{r_t}{\sqrt{5}R_c} = \frac{\tau_{mt}}{R_c} = a_0 + a_1\frac{\sigma_m}{R_c} + a_2\left(\frac{\sigma_m}{R_c}\right)^2 \quad (\theta = 0^\circ) \tag{16.53a}$$

$$\frac{r_c}{\sqrt{5}R_c} = \frac{\tau_{mc}}{R_c} = b_0 + b_1\frac{\sigma_m}{R_c} + b_2\left(\frac{\sigma_m}{R_c}\right)^2 \quad \theta = 60^\circ \tag{16.53b}$$

There are six parameters in formulas (16.53a) and (16.53b), but the two curves should intersect at the same point on the hydrostatic stress axis, σ_{m0}/R_c; after considering this condition, parameters are reduced to five. An ellipse is used to connect the two meridian planes, and the ellipse is expressed by polar coordinates as follows:

$$r(\sigma_m, \theta) = \frac{2r_c(r_c^2 - r_t^2)\cos\theta + r_c(2r_t - r_c)[4(r_c^2 - r_t^2)\cos^2\theta + 5r_t^2 - 4r_t r_c]^{1/2}}{4(r_c^2 - r_t^2)\cos^2\theta + (r_c - 2r_t)^2} \tag{16.54}$$

In five parameters criterion, calculating r_c and r_t by formula (16.53) first and substituting them into the formula above, then the yield surface is obtained.

Both formulas (16.54) and (16.47) represent ellipses and the forms are exactly same, but the specific numerical values are different. In five-parameter criterion, r_c and r_t are calculated by formula (16.53), their values are related to σ_m, so $r(\sigma_m, \theta)$ calculated by formula (16.47) is not only correlated with θ but also related to σ_m. In three-parameter criterion, r_c and r_t calculated by formula (16.53) are constants and have nothing to do with σ_m, so $r(\theta)$ calculated by formula (16.47) has nothing to do with σ_m.

In order to determine the five parameters, the following test data can be used:

1) One-dimensional compressive strength $R_c(\theta = 60^\circ, R_c > 0)$
2) One-dimensional tensile strength $R_t(\theta = 0^\circ)$, the ratio $s_1 = R_t/R_c$
3) Equivalent biaxial compressive strength R_{bc}, the ratio $s_2 = R_{bc}/R_c$
4) Damage point in high stress area of the triaxial test $(\sigma_m/R_c, \tau_m/R_c) = (-\overline{\xi_1}, \overline{r_1})$, $(\theta = 0^\circ, \overline{\xi_1} > 0)$
5) Damage point in high stress area of the triaxial test $(\sigma_m/R_c, \tau_m/R_c) = (-\overline{\xi_2}, \overline{r_1})$, $(\theta = 60^\circ, \overline{\xi_2} > 0)$

In addition, the yield curves on the tensile and compressive meridian planes intersect at the same point on the hydrostatic stress axis; when $r_t(\rho) = r_c(\rho) = 0$, there is

$$\sigma_{m0}/R_c = \rho \tag{16.55}$$

Table 16.2 Data determining $a_0, a_1, a_2, b_0, b_1, b_2$.

Test data	σ_m/R_c	τ_m/R_c	$\theta(°)$	$r(\sigma_m, \theta)$
1. $\sigma_1 = R_t$	$s_1/3$	$s_1\sqrt{2/15}$	0	$r_t = R_t\sqrt{2/3}$
2. $\sigma_2 = \sigma_3 = -R_{bc}$	$-2s_2/3$	$s_2\sqrt{2/15}$	0	$r_t = R_{bc}\sqrt{2/3}$
3. $(-\overline{\xi_1}, \overline{r_1})$	$-\overline{\xi_1}$	$\overline{r_1}$	0	$r_t = \sqrt{5}\overline{r_1}R_c$
4. $\sigma_3 = -R_c$	$-1/3$	$\sqrt{2/15}$	60	$r_c = R_c\sqrt{2/3}$
5. $(-\overline{\xi_2}, \overline{r_2})$	$-\overline{\xi_2}$	$\overline{r_2}$	60	$r_c = \sqrt{5}\overline{r_2}R_c$
6. Formula (16.55)	ρ	0	0 and 60	$r_t = r_c = 0$

The information above are sorted in Table 16.2. By substitution of three sets of data in Table 16.2 into formula (16.53a), obtaining three equations and solving simultaneously, we get

$$\left.\begin{aligned} a_0 &= \frac{2}{3}s_2a_1 - \frac{4}{9}a_2s_2^2 + \sqrt{\frac{2}{15}}s_2 \\ a_1 &= \frac{1}{3}(2s_2 - s_1)a_2 + \sqrt{\frac{6}{5}}\frac{s_1 - s_2}{2s_2 + s_1} \\ a_2 &= \frac{\sqrt{6/5}\overline{\xi_1}(s_1 - s_2) - \sqrt{6/5}s_1s_2 + \overline{r_1}(2s_2 + s_1)}{(2s_2 + s_1)(\overline{\xi_1^2} - 2s_2\overline{\xi_1}/3 + s_1\overline{\xi_1}/3 - 2s_1s_2/9)} \end{aligned}\right\} \tag{16.56}$$

By formula (16.55), there is

$$a_0 + a_1\rho + a_2\rho^2 = 0$$

Therefore

$$\rho = \left[-a_1 - \sqrt{a_1^2 - 4a_0a_2}\right]/(2a_2) \tag{16.57}$$

By substitution of the three conditions behind ($\theta = 60°$) in Table 16.2 into formula (16.53b), we get

$$\left.\begin{aligned} b_0 &= -\rho b_1 - \rho^2 b_2 \\ b_1 &= \left(\overline{\xi_2} + \frac{1}{3}\right)b_2 + \frac{\sqrt{6/5} - 3\overline{r_2}}{3\overline{\xi_2} - 1} \\ b_2 &= \frac{\overline{r_2}(\rho + 1/3) - \sqrt{2/15}(\rho + \overline{\xi_2})}{(\rho + \overline{\xi_2})(\overline{\xi_2} - 1/3)(\rho + 1/3)} \end{aligned}\right\} \tag{16.58}$$

Figure 16.20 shows that the five-parameter criterion corresponds with the test results well.

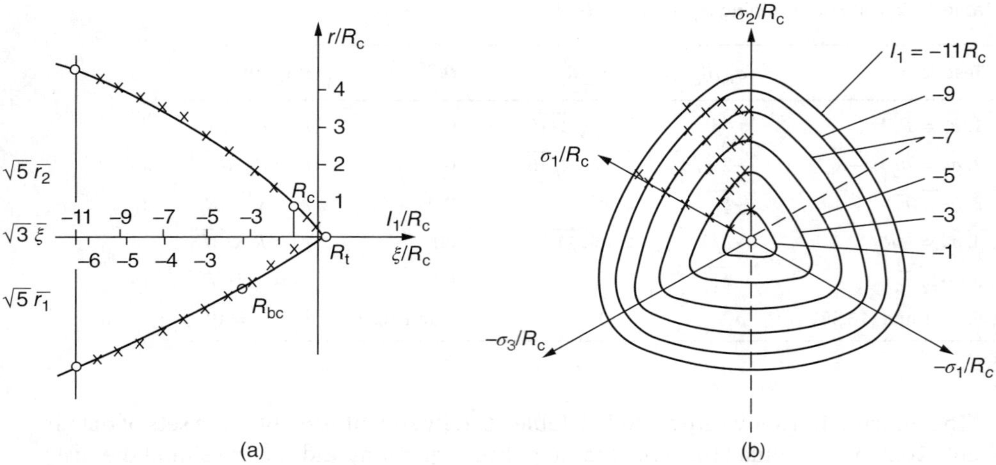

Figure 16.20 Willam–Warnke five parameters criterion and concrete triaxial test results (a) meridian plane and (b) deviator plane.

16.5.11 Zhang–Lu Yield Criterion

Zhang Yuangao and Lu Mingwan assumed that the equations of the cutting lines of the yield surfaces on the two meridian planes $\theta = 0$ and $\theta = 60^\circ$ are, respectively,

$$\left.\begin{array}{l} ar_t^{1.5} + br_t + c\xi = 1 \ (\theta = 0) \\ a_1 r_c^{1.5} + b_1 r_c + c_1 \xi = 1 \ (\theta = 60^\circ) \end{array}\right\} \tag{e}$$

In Ottosen four-parameter criterion and Willam–Warnke five-parameter criterion, the index of r in the first item of formula (a) is taken as 2.0; here it is taken as 1.5, reflecting the weaker nonlinear relationship between r and ξ.

According to the experimental data of concrete, the following conditions are introduced to reduce the number of parameters:

1) When under three-dimensional equivalent tensile load ($r_t = r_c = 0$), the failure surface and hydrostatic pressure axis intersect at the same point; thus there is $c_1 = c$.
2) When the stress is close to three-dimensional equivalent tensile stress ($r_t, r_c \to 0$), the cutting line of the failure surface on the deviator plane is regular triangle, ($r_c = 2r_t$) (because of $\cos 60^\circ = 0.5$), namely, when ($r_t, r_c \to 0, c\xi \to 1$) and $\lim r_c/r_t = 2$. Because when $r_t, r_c \to 0$, the first item of formula (e) relative to the second item is very small, it can be ignored, so by the conditions above, $b_1 = b/2$ is obtained.
3) Under high hydrostatic pressure, the material is changed from brittle failure into ductile failure obeying Mises criterion gradually. The cutting line of the failure surface on the deviator plane is circle $r_c \approx r_t$, namely, when ($r_t, r_c \to \infty, \xi \to \infty$), $\lim r_c/r_t = 1$. The second item of formula (e) relative to the first item is a very small number, and it can be ignored, so there is $a_1 = a$.

Then the equation of the failure surface on the meridional plane is

$$\left.\begin{array}{l} ar_t^{1.5} + br_t + c\xi = 1 \ (\theta = 0) \\ ar_t^{1.5} + 0.5br_t + c\xi = 1 \ (\theta = 60^\circ) \end{array}\right\} \tag{f}$$

Based on the derivation above, Zhang and Lu proposed two kinds of three-parameter yield criterion; the equation of the first proposed yield criterion for the failure curve is

$$ar^{1.5} + br\cos\theta + c\xi = 1 \quad (0 \leq \theta \leq 60^\circ) \tag{g}$$

The advantage of this yield criterion is simple and fundamentally reflecting the characteristics of brittle materials such as concrete. The defect of this yield criterion is that when $\theta = 60^\circ$, the surface is not smooth. In order to overcome the disadvantage of unsmoothness, the second kind of yield criterion suggested by Zhang and Lu is that r_t and r_c are determined by formula (f) and on the deviator plane the Willam–Warnke method is still used to determine the value of r by formula (16.54).

According to the test of Kupfer *et al.*, taking $R_t = 0.1R_c$, $R_{bc} = 1.16R_c$, we get

$$a = 1.671/R_c^{1.5}, \quad b = 7.656/R_c, \quad c = 5.817/R_c$$

16.6 Strain Hardening

As shown in Figure 16.21, under one-dimensional load, when the stress exceeds the initial yield point A into the plastic state, unloading and then loading again, the stress–strain relationship will still change with the elastic law until reaching the highest stress point B before unloading, and then the material enters into plastic state again. The point B is a new yield point after suffering the plastic deformation known as strain hardened point. It is the diacritical point of stress–strain relationship changing with the elastic or plastic law when loading again.

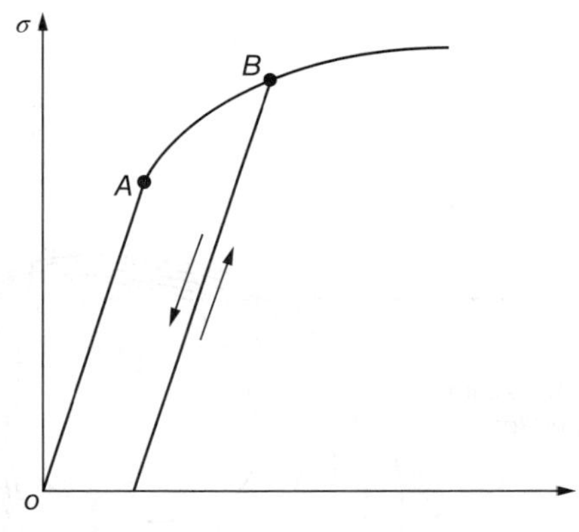

Figure 16.21 Strain hardening under one-dimensional load.

Similarly, unloading when the material has entered into the plastic stage under complex stress state and then loading again, the yield function will change with the previous plastic deformation history. When the stress components satisfy a certain relationship, the materials will be back into the plastic state and produce new plastic deformation; this phenomenon is called strain hardening. When the material enters into plastic state again after the initial yield, the function relation that the stress components must satisfy is called strain hardened condition or load condition – sometimes referred to as subsequent yield condition to distinguish from the initial yield condition. The figure of strain hardened condition in the stress space is called strain hardened surface or loading surface.

Figure 16.22 shows the initial yield surface and failure surface of brittle materials such as concrete in three-dimensional stress space. Figure 16.23 shows the initial yield surface and subsequent yield surface of brittle materials such as concrete in two-dimensional stress plane.

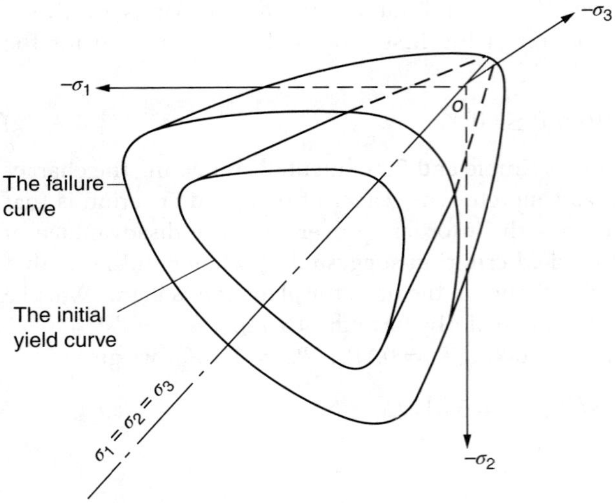

Figure 16.22 The failure surface of brittle materials in three-dimensional stress space.

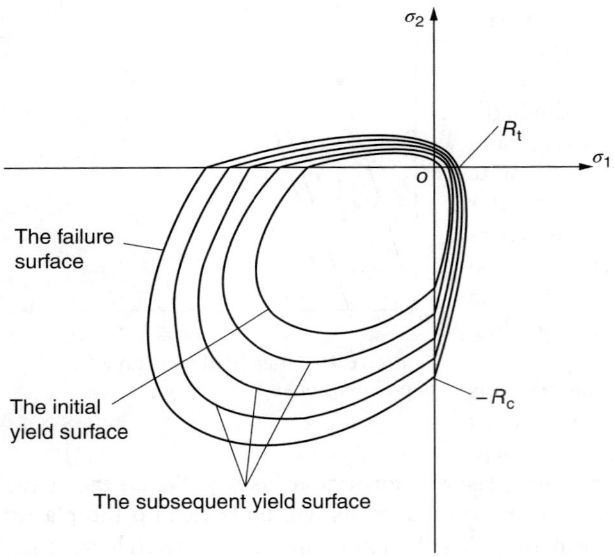

Figure 16.23 The failure curve of brittle materials in two-dimensional stress plane.

16.6.1 Isotropic Strain Hardening Model

Assuming that the shape and center position of the loading surface in the stress space remain invariant, with the increase of degree of strain hardening, the yield surface will expand from the initial yield surface in shape. Loading surface is only decided by the maximum stress points it has ever reached and has nothing to do with the

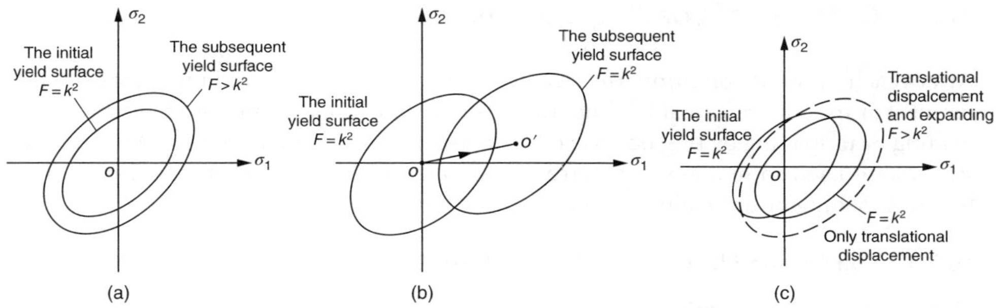

Figure 16.24 Strain hardening model (a) isotropic strain hardening (b) flowing strain hardening, and (c) mixed strain hardening.

loading history, as shown in Figure 16.24(a). The strain hardening condition can be expressed as

$$F(\sigma_{ij}) - k(\varepsilon^{p}) = 0 \tag{16.59}$$

where $k(\varepsilon^{p})$ is the function of the effective plastic strain ε^{p}.

16.6.2 Flowing Strain Hardening Model

It is assumed that in the process of plastic deformation, the shape and size of the yield surface don't change; it just makes rigid translational displacements in the stress space, as shown in Figure 16.24(b). Assuming that in the stress space, coordinate of the internal center of the yield surface is α_{ij}, it is equal to zero at the time of initial yield, so the loading surface of flowing strengthened model can be expressed as

$$F(\sigma_{ij} - \alpha_{ij}) - k = 0 \tag{16.60}$$

Obviously, $F(\sigma_{ij}) - k = 0$ is the initial yield surface; after producing plastic deformation, the loading surface moves with α_{ij}, which is called moving tensor.

16.6.3 Mixed Strain Hardening Model

Combining isotropic strain hardening model and flowing strain hardening model, then mixed strain hardening model is obtained. It is assumed that in the process of plastic deformation, the loading surface not only makes rigid translational displacement but also equably expands in all directions at the same time. The loading surface can be expressed as

$$F(\sigma_{ij} - \alpha_{ij}) - k(\varepsilon^{p}) = 0 \tag{16.61}$$

where α_{ij} is the movement of the center of the yield surface. k is the strain hardening parameter and is a function of effective plastic strain ε^{p}.

In the several kinds of strain hardening models above, the isotropic strain hardening model is most widely used. On one hand, it is convenient for mathematical treatment, and on the other hand, if the directions of stresses (or the ratio of stress components) in the process of loading change little, the calculated results using this model correspond well with the actual situation. The flowing strain hardening model can consider the effect of Bauschinger, so it can be used in the problem of cyclic loading or reverse yield.

16.7 Criterion of Loading and Unloading

After reaching yield condition, the stress–strain laws of loading and unloading are different. For one-dimensional loading, there is only one stress component; we can judge loading or unloading by the increase or decrease of this stress component. For complex stress state, each component in six stress components can increase or decrease; in order to judge loading or unloading, a criterion is needed.

16.7.1 Loading and Unloading of Ideal Plastic Material

There is no strain hardening in ideal plastic material, the loading condition is the same with the yield condition, and stress points can't be located outside the yield surface. When stress points remain on the yield surface, it is called loading, because the plastic deformation can grow. Assuming that the yield condition is $F(\sigma_{ij}) = 0$, when the stress components reach the yield condition, $F(\sigma_{ij}) = 0$, for the stress increment $\mathrm{d}\sigma_{ij}$, if $\mathrm{d}F = F(\sigma_{ij} + \mathrm{d}\sigma_{ij}) - F(\sigma_{ij}) = 0$, it is known that the new stress point remains on the yield surface, called loading. Conversely, if $\mathrm{d}F = F(\sigma_{ij} + \mathrm{d}\sigma_{ij}) - F(\sigma_{ij}) < 0$, it is known that the new stress point is back to the inner of the yield surface, called unloading. Therefore, the loading and unloading criterion of ideal plastic materials can be expressed as

$$\left.\begin{array}{lll} F(\sigma_{ij}) < 0 & & \text{(elastic)} \\ F(\sigma_{ij}) = 0, & \mathrm{d}F = F(\sigma_{ij} + \mathrm{d}\sigma_{ij}) - F(\sigma_{ij}) = \dfrac{\partial F}{\partial \sigma_{ij}}\mathrm{d}\sigma_{ij} = 0 & \text{(loading)} \\ F(\sigma_{ij}) = 0, & \mathrm{d}F = F(\sigma_{ij} + \mathrm{d}\sigma_{ij}) - F(\sigma_{ij}) = \dfrac{\partial F}{\partial \sigma_{ij}}\mathrm{d}\sigma_{ij} < 0 & \text{(unloading)} \end{array}\right\} \tag{16.62}$$

In the stress space, the component of vector n along the direction of the outward normal line of the yield surface is proportional to $\partial F/\partial\sigma_{ij}$, and $\frac{\partial F}{\partial \sigma_{ij}}\mathrm{d}\sigma_{ij} < 0$ shows that stress increment vector points to the inner of the yield surface, called unloading. $\frac{\partial F}{\partial \sigma_{ij}}\mathrm{d}\sigma_{ij} = 0$ indicates $n \cdot \mathrm{d}\sigma = 0$, namely, the stress point can only change along the yield surface, belonging to loading. Because the yield surface can't expand, $\mathrm{d}\sigma$ can't point outside of the yield surface.

16.7.2 Loading and Unloading of Strain Hardened Materials

The loading surface of strain hardened materials can expand, so only when $\mathrm{d}\sigma$ points outside, it belongs to loading, as shown in Figure 16.25(b). When $\mathrm{d}\sigma$ changes along the loading surface, the loading surface does not change; only the stress state changes from one point to another point on the loading surface, but there is no new plastic deformation, and the change process is called neutral variable loading. When $\mathrm{d}\sigma$ points to the inner of loading surface, it is called unloading. The loading and unloading criterion of strain hardened materials can be expressed as

$$\left.\begin{array}{lll} F = 0, & \dfrac{\partial F}{\partial \sigma_{ij}}\mathrm{d}\sigma_{ij} > 0 & \text{(loading)} \\ F = 0, & \dfrac{\partial F}{\partial \sigma_{ij}}\mathrm{d}\sigma_{ij} = 0 & \text{(neutral variable loading)} \\ F = 0, & \dfrac{\partial F}{\partial \sigma_{ij}}\mathrm{d}\sigma_{ij} < 0 & \text{(unloading)} \end{array}\right\} \tag{16.63}$$

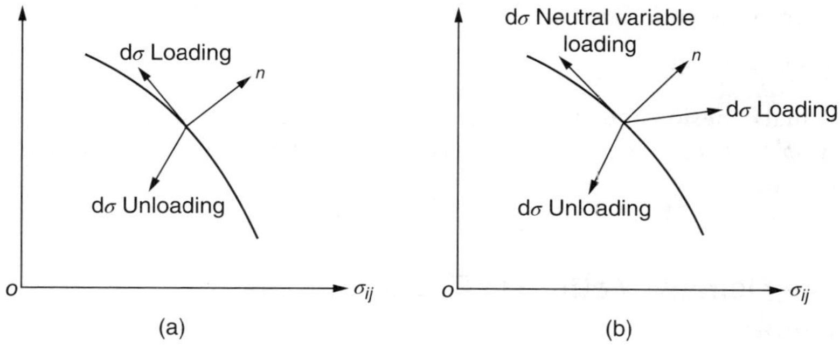

Figure 16.25 Loading and unloading (a) ideal elastic–plastic material and (b) strain hardened material.

16.7.3 Strain Softening, Brittle Failure, and Residual Strength

As shown in Figure 16.26, when the material reaches failure strength, there are two kinds of typical performance: one is brittle material, dropping from failure strength to residual strength suddenly, and another is strain softening material, dropping from failure strength to residual strength gradually. The process that the strength gradually reduces is called strain softening process.

Taking rock, for example, the tensile failure condition is

$$F = \sigma_1 - R_t = 0$$

where R_t is tensile strength.

The condition after damage is

$$F = \sigma_1 - R'_t = 0$$

where R'_t is the residual tensile strength, usually $R'_t = 0$.

The shear failure condition of the rock is

$$F = |\tau| + \sigma tg\phi - c = 0$$

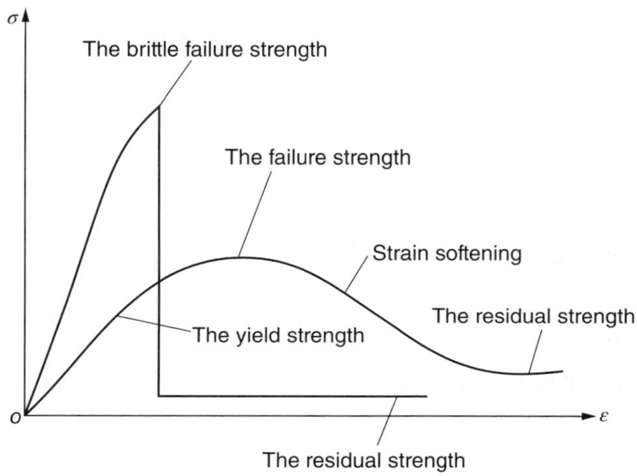

Figure 16.26 The brittle failure, strain softening, and residual strength.

The condition after the shear failure is

$$F = |\tau| + \sigma tg\phi' - c' = 0$$

where c' is the residual cohesion and ϕ' is the residual internal friction angle.

Generally, $c' = 0, \phi' < \phi$. For ideal plastic material, the residual strength is the same with the failure strength.

16.8 The Finite Element Method in Elastic–Plastic Incremental Theory

Within the scope of the initial elastic state, there is a one-to-one corresponding relationship between stress and strain, namely, the generalized Hooke's law. After entering the plastic state, generally speaking, there is no longer one-to-one corresponding relationship between stress and strain, but it is only possible to establish the relationship between the stress increment and the strain increment. The material constitutive relationship expressed in incremental form is called increment theory or flow theory.

16.8.1 The Elastoplastic Matrix of Incremental Theory

The strain increment $\{d\varepsilon\}$ of any point is composed of two parts, namely, the elastic strain increment $\{d\varepsilon_e\}$ and the plastic strain increment $\{d\varepsilon_p\}$:

$$\{d\varepsilon\} = \{d\varepsilon_e\} + \{d\varepsilon_p\} \tag{16.64}$$

where subscript e represents elastic and subscript p represents plastic.

According to the plastic potential theory put forward by Mises, the direction of plastic flow (the direction of plastic strain increment vector) and the gradient direction of the plastic potential function Q are consistent, namely,

$$\{d\varepsilon_p\} = \lambda \left\{ \frac{\partial Q}{\partial \sigma} \right\} \tag{16.65a}$$

where λ is an undetermined parameter. For any stress component, there is

$$d\varepsilon_{ij}^{p} = \lambda \frac{\partial Q}{\partial \sigma_{ij}} \tag{16.65b}$$

By the generalized Hooke's law, the elastic strain increment can be expressed as

$$\{d\varepsilon_e\} = [D]^{-1}\{d\sigma\}$$

So formula (16.64) can be rewritten as

$$\{d\varepsilon\} = [D]^{-1}\{d\sigma\} + \lambda \left\{ \frac{\partial Q}{\partial \sigma} \right\} \tag{16.66}$$

When the stress vector is $\{\sigma\}$, the loading condition is

$$F(\{\sigma\}, \kappa) = 0 \tag{16.67}$$

The value of κ is associated with the plastic strain. When the stress $\{\sigma\}$ is changed to $\{\sigma\} + \{\mathrm{d}\sigma\}$, the plastic strain $\{\varepsilon_{\mathrm{p}}\}$ is changed to $\{\varepsilon_{\mathrm{p}}\} + \{\mathrm{d}\varepsilon_{\mathrm{p}}\}$, and at the same time, when κ is changed to $\kappa + \mathrm{d}\kappa$, the new yield surface is

$$F(\{\sigma\} + \{\mathrm{d}\sigma\}, \kappa + \mathrm{d}\kappa) = 0$$

Subtracting formula above from formula (16.67), we get

$$\mathrm{d}F = F(\{\sigma\} + \{\mathrm{d}\sigma\}, \kappa + \mathrm{d}\kappa) - F(\{\sigma\}, \kappa) = 0$$

According to total differential law, we get

$$\left\{\frac{\partial F}{\partial \sigma}\right\}^{\mathrm{T}} \{\mathrm{d}\sigma\} + \frac{\partial F}{\partial \kappa}\mathrm{d}\kappa = 0$$

The formula above is rewritten as follows:

$$\left\{\frac{\partial F}{\partial \sigma}\right\}^{\mathrm{T}} \{\mathrm{d}\sigma\} + A\lambda = 0 \tag{16.68}$$

$$A = \frac{\partial F}{\partial \kappa}\mathrm{d}\kappa \frac{1}{\lambda} \tag{16.69}$$

Multiplying formula (16.66) by $\left\{\frac{\partial F}{\partial \sigma}\right\}^{\mathrm{T}} [D]$ and eliminating $\{\mathrm{d}\sigma\}$ with formula (16.68), we get

$$\left\{\frac{\partial F}{\partial \sigma}\right\}^{\mathrm{T}} [D]\{\mathrm{d}\varepsilon\} = -A\lambda + \left\{\frac{\partial F}{\partial \sigma}\right\}^{\mathrm{T}} [D] \left\{\frac{\partial Q}{\partial \sigma}\right\} \lambda$$

Getting λ by the formula above, we get

$$\lambda = \frac{\left\{\frac{\partial F}{\partial \sigma}\right\}^{\mathrm{T}} [D]\{\mathrm{d}\varepsilon\}}{\left\{\frac{\partial F}{\partial \sigma}\right\}^{\mathrm{T}} [D] \left\{\frac{\partial Q}{\partial \sigma}\right\} - A} \tag{16.70}$$

Multiplying formula (16.66) by $[D]$ and substituting λ into it, the relationship of stress increment and strain increment is obtained as follows:

$$\{\mathrm{d}\sigma\} = ([D] - [D]_{\mathrm{p}})\{\mathrm{d}\varepsilon\} = [D]_{\mathrm{ep}}\{\mathrm{d}\varepsilon\} \tag{16.71}$$

where

$$[D]_{\mathrm{ep}} = [D] - [D]_{\mathrm{p}} \tag{16.72a}$$

$$[D]_{\mathrm{p}} = \frac{[D]\left\{\frac{\partial Q}{\partial \sigma}\right\}\left\{\frac{\partial F}{\partial \sigma}\right\}^{\mathrm{T}} [D]}{\left\{\frac{\partial F}{\partial \sigma}\right\}^{\mathrm{T}} [D] \left\{\frac{\partial Q}{\partial \sigma}\right\} - A} \tag{16.72b}$$

in which $[D]_{\mathrm{ep}}$ is the elastic–plastic matrix.

For strain hardened material, κ is equal to the plastic work produced in the process of plastic deformation, namely,

$$\mathrm{d}\kappa = \sigma_x \mathrm{d}\varepsilon_x^{\mathrm{p}} + \sigma_y \mathrm{d}\varepsilon_y^{\mathrm{p}} + \cdots = \{\sigma\}^{\mathrm{T}}\{\mathrm{d}\varepsilon_{\mathrm{p}}\} \tag{a}$$

By substitution of formula (16.66) into formula (a), we get

$$d\kappa = \lambda\{\sigma\}^{T}\left\{\frac{\partial Q}{\partial \sigma}\right\}$$

Then substituting it into formula (16.69), we get

$$A = \frac{\partial F}{\partial \kappa}\{\sigma\}^{T}\left\{\frac{\partial Q}{\partial \sigma}\right\} \tag{16.73}$$

For an ideal plastic body, $A = 0$.

From the theory of plasticity, we know that under the condition of Drucker hypothesis, $Q = F$, the plastic constitutive relation is referred to the associated flow rule. Only in this case $[D]_{ep}$ is a symmetric matrix and in general, $Q \neq F$, $[D]_{ep}$ is not a symmetric matrix.

The elastic–plastic matrix of mixed strain hardened model is given in the literature [1].

16.8.2 Symmetric Expression of Nonassociated Elastic–Plastic Stiffness Matrix

Using the nonassociated flowing rule, the elastic–plastic stiffness matrix is asymmetric, the amount of computing increases a lot, and the computer memory capacity that asymmetric stiffness matrix needs is nearly double of that symmetric stiffness matrix needs. Xiong Wenlin derived the symmetric formula of elastic–plastic tangent stiffness matrix under the nonassociated flowing rule and overcame the shortcomings above [2].

Let

$$K_1 = \frac{\left\{\frac{\partial Q}{\partial \sigma}\right\}\{d\sigma\}}{\left\{\frac{\partial F}{\partial \sigma}\right\}^{T}\{d\sigma\}}$$

Multiplying formula (16.68) by K_1, we get

$$\left\{\frac{\partial Q}{\partial \sigma}\right\}^{T}\{d\sigma\} + K_1 A \lambda = 0$$

$$\lambda = \frac{-1}{K_1 A}\left\{\frac{\partial Q}{\partial \sigma}\right\}^{T}\{d\sigma\}$$

By substitution of the formula above into flowing rule formula (16.65a), we get

$$\{d\varepsilon_p\} = \frac{-1}{K_1 A}\left\{\frac{\partial Q}{\partial \sigma}\right\}\left\{\frac{\partial Q}{\partial \sigma}\right\}^{T}\{d\sigma\} = [C_p^s]\{d\sigma\}$$

where

$$[C_p^s] = \frac{-1}{K_1 A}\begin{bmatrix} \frac{\partial Q}{\partial \sigma_x}\frac{\partial Q}{\partial \sigma_x} & \frac{\partial Q}{\partial \sigma_x}\frac{\partial Q}{\partial \sigma_y} & \cdots & \frac{\partial Q}{\partial \sigma_x}\frac{\partial Q}{\partial \tau_{zx}} \\ \frac{\partial Q}{\partial \sigma_y}\frac{\partial Q}{\partial \sigma_x} & \frac{\partial Q}{\partial \sigma_y}\frac{\partial Q}{\partial \sigma_y} & \cdots & \frac{\partial Q}{\partial \sigma_y}\frac{\partial Q}{\partial \tau_{zx}} \\ \vdots & \vdots & \vdots & \vdots \\ \frac{\partial Q}{\partial \tau_{zx}}\frac{\partial Q}{\partial \sigma_x} & \frac{\partial Q}{\partial \tau_{zx}}\frac{\partial Q}{\partial \sigma_y} & \cdots & \frac{\partial Q}{\partial \tau_{zx}}\frac{\partial Q}{\partial \tau_{zx}} \end{bmatrix}$$

in which $[C_{\mathrm{p}}^{\mathrm{s}}]$ is a symmetrical plastic flexible matrix.

Thus we have

$$[C_{\mathrm{ep}}^{\mathrm{s}}] = [C_{\mathrm{e}}^{\mathrm{s}}] + [C_{\mathrm{p}}^{\mathrm{s}}]$$
$$\{\mathrm{d}\sigma\} = [C_{\mathrm{ep}}^{\mathrm{s}}]^{-1}\{\mathrm{d}\varepsilon\} = [D_{\mathrm{ep}}^{\mathrm{s}}]\{\mathrm{d}\varepsilon\}$$

Inversing $[C_{\mathrm{p}}^{\mathrm{s}}]$, the symmetrical elastic–plastic stiffness matrix is obtained, namely,

$$[D_{\mathrm{ep}}^{\mathrm{s}}] = [D]_{\mathrm{e}} - \frac{1}{\beta_{\mathrm{s}}}[D]_{\mathrm{e}}\left\{\frac{\partial Q}{\partial \sigma}\right\}\left\{\frac{\partial Q}{\partial \sigma}\right\}^{\mathrm{T}}[D]_{\mathrm{e}} \tag{16.74}$$

where

$$\beta_{\mathrm{s}} = -K_1 A + \left\{\frac{\partial Q}{\partial \sigma}\right\}^{\mathrm{T}}[D]_{\mathrm{e}}\left\{\frac{\partial Q}{\partial \sigma}\right\}$$

16.8.3 The Calculation of $\left\{\frac{\partial F}{\partial \sigma}\right\}$

For three-dimensional problems,

$$\{\sigma\}^{\mathrm{T}} = \left[\sigma_x\ \sigma_y\ \sigma_z\ \tau_{xy}\ \tau_{yz}\ \tau_{zx}\right]$$

$$\left\{\frac{\partial F}{\partial \sigma}\right\}^{\mathrm{T}} = \left[\frac{\partial F}{\partial \sigma_x}\ \frac{\partial F}{\partial \sigma_y}\ \frac{\partial F}{\partial \sigma_z}\ \frac{\partial F}{\partial \tau_{xy}}\ \frac{\partial F}{\partial \tau_{yz}}\ \frac{\partial F}{\partial \tau_{zx}}\right] \tag{16.75}$$

Usually, the loading function F is expressed with I_1, J_2, J_3 as $F(I_1, J_2, J_3)$; therefore

$$\left\{\frac{\partial F}{\partial \sigma}\right\} = \frac{\partial F}{\partial I_1}\left\{\frac{\partial I_1}{\partial \sigma}\right\} + \frac{\partial F}{\partial J_2}\left\{\frac{\partial J_2}{\partial \sigma}\right\} + \frac{\partial F}{\partial J_3}\left\{\frac{\partial J_3}{\partial \sigma}\right\} \tag{16.76}$$

$$\left\{\frac{\partial I_1}{\partial \sigma}\right\}^{\mathrm{T}} = \left[1\ 1\ 1\ 0\ 0\ 0\right] \tag{16.77a}$$

$$\left\{\frac{\partial J_2}{\partial \sigma}\right\}^{\mathrm{T}} = \left[s_x\ s_y\ s_z\ 2\tau_{xy}\ 2\tau_{yz}\ 2\tau_{zx}\right] \tag{16.77b}$$

$$\left\{\frac{\partial J_3}{\partial \sigma}\right\}^{\mathrm{T}} = \begin{bmatrix} \frac{1}{3}J_2 + s_y s_z - \tau_{yz}^2 & \frac{1}{3}J_2 + s_x s_z - \tau_{zx}^2 & \frac{1}{3}J_2 + s_x s_y - \tau_{xy}^2 \\ 2(\tau_{yz}\tau_{zx} - s_z\tau_{xy}) & 2(\tau_{xy}\tau_{zx} - s_x\tau_{yz}) & 2(\tau_{xy}\tau_{yz} - s_y\tau_{zx}) \end{bmatrix} \tag{16.77c}$$

After calculating $\partial F/\partial I_1, \partial F/\partial J_2, \partial F/\partial J_3$ and substituting them into formula (16.76), we can get $\partial F/\partial \sigma$.

Taking Drucker–Prager criterion, for example, by formula (16.37),

$$\partial F/\partial I = \alpha, \quad \partial F/\partial J_2 = 1/(2\sqrt{J_2}), \quad \partial F/\partial J_3 = 0$$

As for Tresca criterion, it is better to rewrite it by formula (16.18). Originally Tresca criterion is $F = (\sigma_1 - \sigma_3)/2 - k = 0$; from formula (16.18), there is

$$\begin{aligned}\frac{1}{2}(\sigma_1 - \sigma_3) &= \frac{\sqrt{J_2}}{\sqrt{3}}\left[\cos\theta - \cos\left(\theta + \frac{2\pi}{3}\right)\right] \\ &= \frac{\sqrt{J_2}}{\sqrt{3}}[\sin\theta - \sqrt{3}\cos\theta]\end{aligned}$$

As a result, Tresca criterion can be rewritten as follows:

$$F(J_2, \theta) = \frac{\sqrt{J_2}}{2}(\sin\theta + \sqrt{3}\cos\theta) - k = 0 \tag{16.78}$$

Notice that

$$\frac{\partial\theta}{\partial J_2} = \frac{3\sqrt{3}}{4\sin 3\theta}\frac{J_3}{J_2^{5/2}}, \quad \frac{\partial\theta}{\partial J_3} = -\frac{\sqrt{3}}{2\sin 3\theta}\frac{1}{J_2^{3/2}}$$

Working out the partial derivatives by formula (16.78), we get

$$\frac{\partial F}{\partial I_1} = 0 \tag{16.79a}$$

$$\begin{aligned}\frac{\partial F}{\partial J_2} &= \frac{\sin\theta + \sqrt{3}\cos\theta}{4\sqrt{J_2}} + \frac{3\sqrt{3}J_3(\cos\theta - \sqrt{3}\sin\theta)}{8J_2^2\sin 3\theta}\\ &= \frac{\sin\theta + \sqrt{3}\cos\theta}{4\sqrt{J_2}} + \frac{ctg3\theta(\cos\theta - \sqrt{3}\sin\theta)}{4\sqrt{J_2}}\end{aligned} \tag{16.79b}$$

$$\frac{\partial F}{\partial J_3} = \frac{\sqrt{3}(\sqrt{3}\sin\theta - \cos\theta)}{4J_2\sin 3\theta} \tag{16.79c}$$

where the second item of $\partial F/\partial J_2$ is from $\partial\theta/\partial J_2$ and transformed by formula (16.16).

Using formulas (16.18) and (16.35b), Mohr–Coulomb criterion can be rewritten as follows:

$$F = I_1\sin\phi + \frac{1}{2}[3(1 - \sin\phi)\sin\theta + \sqrt{3}(3 + \sin\phi)\cos\theta]\sqrt{J_2} - 3c\cos\phi = 0 \tag{16.80}$$

The partial derivatives of $\partial F/\partial I_1$ can be calculated by formula (16.80). The partial derivatives of several commonly used yield functions are listed in Table 16.3. By formulas (16.76) and (16.77), it is not difficult to find out partial derivatives to stress components of other various yield functions.

Plastic potential function Q often can be expressed as $Q(I_1, J_2, J_3)$ and also can be calculated by formula (16.76), replacing function F with function Q.

Table 16.3 Functions $\partial F/\partial I_1$, $\partial F/\partial J_2$, and $\partial F/\partial J_3$.

Yield criteria	$\partial F/\partial I_1$	$\partial F/\partial J_2$	$\partial F/\partial J_3$
1. Mises	0	$1/(2\sqrt{J_2})$	0
2. Drucker–Prager	α	$1/(2\sqrt{J_2})$	0
3. Tresca	0	$\frac{\sin\theta + \sqrt{3}\cos\theta - \sin\theta ctg3\theta + \cos\theta ctg3\theta}{4\sqrt{J_2}}$	$\frac{\sqrt{3}(\sqrt{3}\sin\theta - \cos\theta)}{4J_2\sin 3\theta}$
4. Mohr–Coulomb	$\sin\phi$	$\frac{\sqrt{3}}{4\sqrt{J_2}}[\sqrt{3}(1 - \sin\phi)(\sin\theta + \cos\theta ctg3\theta) + (3 + \sin\phi)(\cos\theta - \sin\theta ctg3\theta)]$	$\frac{3[(3 + \sin\phi)\sin\theta - \sqrt{3}(1 - \sin\phi)\cos\theta]}{4J_2\sin 3\theta}$

16.8.4 Effective Stress, Effective Plastic Strain, and Calculation of $\partial F/\partial \kappa$

To facilitate the application, the loading condition formula (16.67) should be connected with one-dimensional material test. So it is necessary to define the effective stress and effective plastic strain in such a way that the relationship between effective stress and effective plastic strain just become one-dimensional stress–strain relationship in one-dimensional material test; thus it is easy to find out the related parameters.

16.8.4.1 The Effective Stress σ_i

For the effective stress to become uniaxial stress, the loading function should have the following form:

$$f(\{\sigma\}) = C\sigma_i^n \tag{16.81}$$

where C is a constant.

For example, for Mises materials, $f = J_2$, there is

$$J_2 = C\sigma_i^n$$

or

$$\sigma_i = \left(\frac{J_2}{C}\right)^{1/n} = \left\{\frac{1}{6C}[(\sigma_1-\sigma_2)^2+(\sigma_2-\sigma_3)^2+(\sigma_3-\sigma_1)^2]\right\}^{1/n}$$

For uniaxial test, $\sigma_i = \sigma_1, \sigma_2 = \sigma_3 = 0$, so

$$C = \frac{1}{3},\quad n = 2,\quad \sigma_i = \sqrt{3J_2} \tag{16.82}$$

Similarly, for Drucker–Prager material, $f = \alpha I_1 + \sqrt{J_2}$, thus

$$C = \alpha + \frac{1}{\sqrt{3}},\quad n = 1,\quad \sigma_i = \frac{\sqrt{3}\alpha I_1 + \sqrt{3J_2}}{1+\sqrt{3}\alpha} \tag{16.83}$$

16.8.4.2 The Effective Plastic Strain ε_i

The effective plastic strain increment $\mathrm{d}\varepsilon_i$ is defined by the formula below:

$$\mathrm{d}\varepsilon_i = C[\{\mathrm{d}\varepsilon_\mathrm{p}\}^\mathrm{T}\{\mathrm{d}\varepsilon_\mathrm{p}\}]^{1/2}$$

For example, for incompressible materials, $\mu = 1/2$, in uniaxial tests, there is

$$\mathrm{d}\varepsilon_1^\mathrm{p} = \mathrm{d}\varepsilon_i = C\left[(\mathrm{d}\varepsilon_1^\mathrm{p})^2 + \left(\frac{1}{2}\mathrm{d}\varepsilon_1^\mathrm{p}\right)^2 + \left(\frac{1}{2}\mathrm{d}\varepsilon_1^\mathrm{p}\right)^2\right]^{1/2} = C\sqrt{\frac{3}{2}}\mathrm{d}\varepsilon_1^\mathrm{p}$$

Thus, $C = \sqrt{2/3}$ is obtained. Therefore, the effective plastic strain increment is defined as

$$\mathrm{d}\varepsilon_i = \sqrt{\frac{2}{3}}[\{\mathrm{d}\varepsilon_\mathrm{p}\}^\mathrm{T}\{\mathrm{d}\varepsilon_\mathrm{p}\}]^{1/2} \tag{16.84}$$

In uniaxial tests, the relationship between effective stress and effective plastic strain is

$$\sigma_i = H(\varepsilon_i) \tag{16.85}$$

After differentiating, we get

$$\mathrm{d}\sigma_i = H'\,\mathrm{d}\varepsilon_i \tag{16.86}$$

where H' is the slope at the point σ_i of uniaxial stress–plastic strain curve.

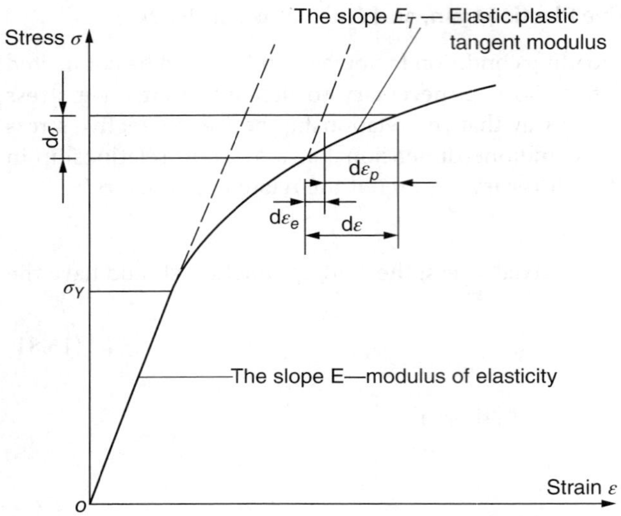

Figure 16.27 Elastic–plastic strain hardened curve under one-dimensional load.

Usually in uniaxial material test curve $\sigma - \varepsilon$, the horizontal ordinate ε includes elastic strain and plastic strain, as shown in Figure 16.27, and the formula below should be used to calculate H':

$$H'(\varepsilon_i) = \frac{d\sigma}{d\varepsilon_p} = \frac{d\sigma}{d\varepsilon - d\varepsilon_e} = \frac{1}{d\varepsilon/d\sigma - d\varepsilon_e/d\sigma}$$

Thus

$$H' = \frac{E_T}{1 - E_T/E} \tag{16.87}$$

where

$$E_T = d\sigma/d\varepsilon, \quad E = d\sigma/d\varepsilon_e$$

in which E_T is the elastic–plastic tangent modulus and E is the modulus of elasticity.

By formula (16.86), taking the development process of strain into account, the magnitude of the effective strain should be calculated by the formula below:

$$\varepsilon_i = \int d\varepsilon_i = \int \frac{d\sigma_i}{H'} \tag{16.88}$$

16.8.4.3 The Calculation of $\partial F/\partial \kappa$

Taking isotropic strain hardened Mises material, for example, by formula (16.31), the loading condition can be expressed as

$$F = \sigma_i - \sigma_s(\kappa) = 0 \tag{b}$$

where σ_i is the effective stress and $\sigma_s(\kappa)$ is the yield stress.

In uniaxial experiments, formula (a) becomes

$$d\kappa = \sigma_s \, d\varepsilon_{1p} \tag{c}$$

By formulas (b) and (c), we get

$$\frac{\partial F}{\partial \kappa} = -\frac{\partial \sigma_s}{\partial \kappa} = -\frac{\partial \sigma_s}{\partial \varepsilon_{1p}} \frac{d\varepsilon_{1p}}{d\kappa} = -\frac{H'}{\sigma_s} \tag{16.89}$$

where H' is shown in formula (16.87).

Assuming that the material obeys the associated flowing rule, $Q = F$, by substitution of formula (16.89) into formula (16.73), we get

$$A = -\frac{H'}{\sigma_s}\{\sigma\}^{\mathrm{T}}\left\{\frac{\partial F}{\partial \sigma}\right\} = -\frac{H'}{\sigma_s}\left\{\frac{\partial F}{\partial \sigma}\right\}^{\mathrm{T}}\{\sigma\} \tag{d}$$

By Euler's theorem, if $F(\{x\})$ is an uniform function of order n; then $\left\{\frac{\partial F}{\partial x}\right\}^{\mathrm{T}}\{x\} = nF$. Now as F is a linear uniform function of order $n = 1$, there is

$$\left\{\frac{\partial F}{\partial \sigma}\right\}^{\mathrm{T}}\{\sigma\} = F = \sigma_s$$

Substituting it into formula (d), we get

$$A = -H' \tag{16.90}$$

16.8.5 Singular Points on the Yield Surface

For Tresca criterion and Mohr–Coulomb criterion, when $\theta = 0$ and $\theta = 60°$, the yield surface has angular points; at this time, the direction of plastic strain is uncertain. As can be seen from Table 16.3, when $\theta = 0$ and $\theta = 60°$, for the two criteria above, $\partial F/\partial J_2$ and $\partial F/\partial J_3$ are infinite. Koiter has given the scope of the plastic strain vector of these angular points. In order to overcome the numerical difficulty, referring to literature [3], the following algorithm is used.

For Tresca criterion, back to formula (16.78), making $\theta = 0$ and $\theta = 60°$, we get

$$\frac{\sqrt{3}}{2}\sqrt{J_2} - \kappa = 0 \tag{e}$$

Thus, when $\theta = 0$ and $\theta = 60°$,

$$\frac{\partial F}{\partial I_1} = 0, \quad \frac{\partial F}{\partial J_2} = \frac{\sqrt{3}}{4\sqrt{J_2}}, \quad \frac{\partial F}{\partial J_3} = 0 \tag{16.91}$$

When $\theta = 0$ and $\theta = 60°$, formula (16.91) is used, while when $1° \leq \theta \leq 59°$, formula (16.79) is used. In fact, formula (e) is Mises criterion; the algorithm above means that on the angular points of Tresca criterion, the direction of plastic strain is given by Mises circle through these points.

In the same way, for Mohr–Coulomb criterion, back to formula (16.80), we get

$$\left.\begin{aligned} F &= I_1 \sin\phi + \frac{\sqrt{3}}{2}(3 + \sin\phi)\sqrt{J_2} - 3C\cos\phi = 0 \quad (\theta = 0) \\ F &= I_1 \sin\phi + \frac{\sqrt{3}}{2}(3 - \sin\phi)\sqrt{J_2} - 3C\cos\phi = 0 \quad (\theta = 60°) \end{aligned}\right\} \tag{f}$$

From formula (b), we get

$$\frac{\partial F}{\partial I_1} = \sin\phi, \quad \frac{\partial F}{\partial J_2} = \frac{\sqrt{3}}{4\sqrt{J_2}}(3+\sin\phi), \quad \frac{\partial F}{\partial J_3} = 0 \quad (\theta = 0) \tag{16.92a}$$

$$\frac{\partial F}{\partial I_1} = \sin\phi, \quad \frac{\partial F}{\partial J_2} = \frac{\sqrt{3}}{4\sqrt{J_2}}(3-\sin\phi), \quad \frac{\partial F}{\partial J_3} = 0 \quad (\theta = 60^\circ) \tag{16.92b}$$

When $\theta = 0$ and $\theta = 60^\circ$, formula (16.92) is used; when $1^\circ \le \theta \le 59^\circ$, the formula in Table 16.3 is used.

The algorithm above can determine the direction of the plastic strain of angular points of Tresca criterion and Mohr–Coulomb criterion, and it conforms to the requirements of Koiter. In fact, this algorithm is equivalent to the smoothing of angular points on the yield surface.

16.8.6 Numerical Calculation Method

For finite element method of elastic–plastic incremental theory, generally the incremental method is used; within each incremental step, several iterative calculations sometimes are needed.

Now we explain how to calculate the stress increment and the displacement increment within each incremental step. The yield surface is

$$F(\{\sigma\}, \varepsilon_i, \kappa) = 0 \tag{16.93}$$

At the beginning of the incremental step, the stress $\{\sigma\}_n$, effective plastic strain ε_i^n, strain hardened parameter κ^n, and deflection $\{\delta\}_n$ are known. The elastic–plastic constitutive equation is

$$\{\Delta\sigma\} = ([D]_e - [D]_p)\{\Delta\varepsilon\} = [D]_{ep}\{\Delta\varepsilon\} \tag{16.94}$$

Our task is to work out the stress increment $\{\Delta\sigma\}_{n+1}$ that satisfies the yield formula (16.93) using the constitutive equation above. All the stresses and strains are checked on Gauss integral points to determine whether this point is in the plastic state. Therefore, part of an element may enter the plastic state and the rest is in the elastic state. For each loading increment, which part is elastic and which part generates plastic strain must be decided, and the stress and strain values should be adjusted until both yield condition and constitutive equation are satisfied.

16.8.6.1 The Displacement Increment

Calculating the elastic–plastic matrix $[D]_{ep}$ and the tangent stiffness matrix of each element by $\{\sigma\}_n$ at the end of last incremental step, forming a global stiffness matrix $[K_t]_n$, the displacement increment of the $n+1$ step is calculated as follows:

$$\{\Delta\delta\}_{n+1} = [K_t]_n^{-1}(\{\Delta P\}_{n+1} - \{\psi_n\}) \tag{16.95}$$

and then the strain increment $\{\Delta\varepsilon\}_{n+1}$ is calculated, and for each Gauss integral point, according to the following steps, its stress $\{\sigma\}_{n+1}$ is calculated.

16.8.6.2 Tentative Stress

After imposing loading increment, for any point, we don't know whether its stress is in the elastic or plastic state, so we have to ignore the plasticity of the material temporarily.

The tentative stress is calculated by the formula below:

$$\{\sigma^{e}\} = \{\sigma\}_{n} + [D]_{e}\{\Delta\varepsilon\}_{n+1} \tag{16.96}$$

By substitution of the tentative stress above into the yield condition formula (16.93), if it is satisfied, it indicates that at this time the behavior of the material is elastic, strain hardening parameters ε_i^n and κ^n remain invariant, and the stress at the end of the $n+1$ increment step is equal to the tentative stress:

$$\{\sigma\}_{n+1} = \{\sigma^{e}\} \tag{16.97}$$

In the calculation above, we assume that the elastic modulus in the $n+1$ step is a constant and is equal to the value at the end of the n step. If the strain increment $\{\Delta\varepsilon\}_{n+1}$ is small, the above assumption has little impact on the calculation accuracy. If the strain increment is bigger or the nonlinear elasticity of the material is stronger, the strain increment can be divided into m parts, and then we do m times calculation in proper order each time the corresponding elastic modulus is used for calculation.

16.8.6.3 The Scale Factor

If the calculation result is that the yield condition was destroyed by tentative stress, it indicates that the integral point has entered the plastic state. At this time, there are two cases: one case is that the stress at the beginning of an incremental step is within the yield surface as shown in Figure 16.28(a), and another case is that the stress at the beginning is on the yield surface as shown in Figure 16.28(b).

For the first case, as shown in Figure 16.28(a), it is assumed that the stress at the beginning of this step is at point A and conforms to the condition

$$F(\{\sigma_A\}, \varepsilon_i^n, \kappa^n) = F_0 < 0 \tag{16.98}$$

The formula above shows that it is in the elastic state. After imposing the load increment, ignoring the plasticity, and calculating according to the elasticity, the stress path goes through the yield surface and reaches point B, and its stress state is $\{\sigma_B\}$. At this time

$$F(\{\sigma_B\}, \varepsilon_i^n, \kappa^n) = F_1 > 0 \tag{16.99}$$

showing that the yield condition has been destroyed. Now, the load increment is divided into two parts: one part is elastic, corresponding with the stress path AC, and the other

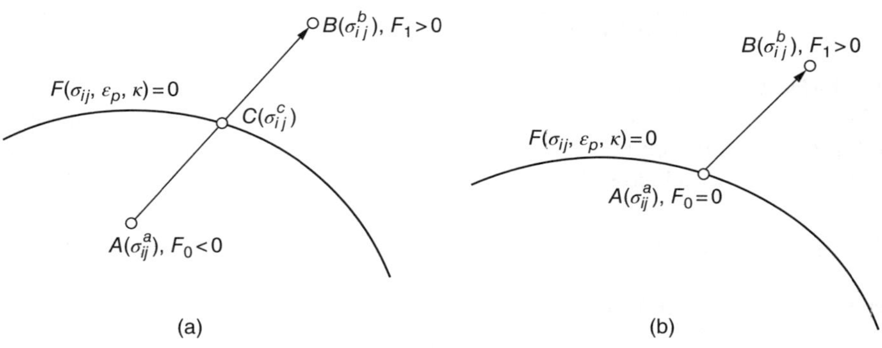

Figure 16.28 The stress adjustment on the yield surface (a) from an elastic state to a plastic state and (b) from a plastic state to another plastic state.

part is plastic, controlling the behavior of point C after it reaches the yield surface. In order to determine the location of point C, let

$$\{\sigma_C\} = \{\sigma_A\} + r\{\Delta\sigma^{\mathrm{e}}\} \tag{16.100}$$

where r is a scale factor; because point C is on the yield surface, it should satisfy the yield condition

$$F(\{\sigma_C\}, \varepsilon_i^n, \kappa^n) = F(\{\sigma_A\} + r\{\Delta\sigma^{\mathrm{e}}\}, \varepsilon_i^n, \kappa^n) = 0 \tag{16.101}$$

Theoretically, r can be calculated by formula (16.101). But in fact, except some very simple yield function, it is difficult to work out r directly, so the approximate algorithm is used as follows. Due to the linear interpolation of F, the first approximate value of r is obtained:

$$r_1 = -\frac{F_0}{F_1 - F_0} \tag{16.102}$$

Actually function F is nonlinear, so

$$F(\{\sigma_A\} + r_1\{\Delta\sigma^{\mathrm{e}}\}, \varepsilon_i^n, \kappa^n) = F_2 \neq 0 \tag{16.103}$$

In order to find out better approximation of r, might as well assuming ε_i^n and κ^n to be constant temporarily, by the first-order Taylor expansion of the yield function, we get

$$\mathrm{d}F = \left\{\frac{\partial F}{\partial \sigma}\right\}^{\mathrm{T}} \{\mathrm{d}\sigma\} \tag{g}$$

Let

$$\{\mathrm{d}\sigma\} = \Delta r_1\{\Delta\sigma^{\mathrm{e}}\} \tag{h}$$

As the change is small, we may take

$$\mathrm{d}F = -F_2 \tag{i}$$

By substitution of formulas (h) and (i) into formula (g), we get

$$F_2 = -\left\{\frac{\partial F}{\partial \sigma}\right\}^{\mathrm{T}} \{\Delta\sigma^{\mathrm{e}}\}\Delta r_1 \tag{j}$$

The improved value of r is obtained as follows:

$$r = r_1 + \Delta r_1 = r_1 - \frac{F_2}{\{\partial F/\partial \sigma\}^{\mathrm{T}}\{\Delta\sigma^{\mathrm{e}}\}} \tag{16.104}$$

16.8.6.4 The Plastic Stress Increment

The strain increment has been divided into two parts – the elastic strain increment $r\{\Delta\varepsilon\}$ and the plastic strain increment $(1 - r)\{\Delta\varepsilon\}$ – so the plastic stress increment can be calculated by the formula below:

$$\{\Delta\sigma^{\mathrm{p}}\} = \int_{r\Delta\varepsilon}^{\Delta\varepsilon} [D]_{\mathrm{p}}\{\mathrm{d}\varepsilon\} \tag{16.105}$$

The elastic stress increment is

$$\{\Delta\sigma^{\mathrm{e}}\} = \int_{0}^{\Delta\varepsilon} [D]\{\mathrm{d}\varepsilon\} = [D]\{\Delta\varepsilon\} \tag{16.106}$$

The stress increment should meet the constitutive equation (16.94):

$$\begin{aligned}\{\Delta\sigma\} &= \int_0^{\Delta\varepsilon}([D]_e - [D]_p)\{d\varepsilon\} = \int_0^{r\Delta\varepsilon}[D]_e\{d\varepsilon\} + \int_{r\Delta\varepsilon}^{\Delta\varepsilon}([D]_e - [D]_p)\{d\varepsilon\} \\ &= \int_0^{\Delta\varepsilon}[D]_e\{d\varepsilon\} - \int_{r\Delta\varepsilon}^{\Delta\varepsilon}[D]_p\{d\varepsilon\} \\ &= \{\Delta\sigma^e\} - \{\Delta\sigma^p\}\end{aligned} \tag{16.107}$$

Because the plastic stiffness matrix $[D]_p$ changes with the stress state, when we use formula (16.105) to calculate $\{\Delta\sigma^p\}$, numerical integration is required. Generally the formula below is used for approximate calculation:

$$\{\Delta\sigma^p\} = (1-r)[D]_p\{\Delta\varepsilon\} \tag{16.108}$$

If the load and strain increment is small, the algorithm above can be directly used. If the load and strain increment is big, more precise algorithm is needed. Usually dividing $(1-r)\{\Delta\varepsilon\}$ into m equal parts, each time the plastic stress increment is calculated by formula (16.108) adopting corresponding new value of $[D]_p$.

As shown in Figure 16.28(b), the stress state is on the yield surface at the beginning of the incremental step, only needing to take $r = 0$ in the calculation above.

16.8.6.5 Stress Back to the Yield Surface

So far, the stress at the end of the $n+1$ incremental step has been calculated as follows:

$$\{\sigma\}_{n+1} = \{\sigma\}_n + \{\Delta\sigma^e\} - \{\Delta\sigma^p\} \tag{16.109}$$

But there are two approximations in the calculation above: one is using approximate formula (16.108) instead of formula (16.105), and the other is using approximate formula (16.104) to solve r instead of formula (16.101).

Therefore $\{\sigma\}_{n+1}$ calculated by formula (16.109) is approximate too. Generally speaking, it can't meet the yield condition strictly; thus

$$F(\{\sigma\}_{n+1}, \varepsilon_i^n, \kappa^n) = F_3 \neq 0 \tag{16.110}$$

In the process of calculation, the error will be accumulated, so further correction is needed to make stress return to the yield surface. Assuming that stress correction is along the direction of the normal line of the yield surface, then we get

$$\{\delta\sigma\} = a\left\{\frac{\partial F}{\partial\sigma}\right\} \tag{16.111}$$

where a is a scalar and $\{\delta\sigma\}$ is a stress correction vector. By substitution of formula (16.110) and formula (16.111) into formula (g), we get

$$dF = -F_3 = \left\{\frac{\partial F}{\partial\sigma}\right\}^T\{\delta\sigma\} = a\left\{\frac{\partial F}{\partial\sigma}\right\}^T\left\{\frac{\partial F}{\partial\sigma}\right\}$$

Thus a is calculated, and substituting it into formula (16.111), the stress correction is obtained:

$$\{\delta\sigma\} = -\frac{\{\partial F/\partial\sigma\}F_3}{\{\partial F/\partial\sigma\}^T\{\partial F/\partial\sigma\}} \tag{16.112}$$

Finally, the stress increment is obtained as follows:

$$\{\sigma\}_{n+1} = \{\sigma\}_n + \{\Delta\sigma^e\} - \{\Delta\sigma^p\} + \{\delta\sigma\} \tag{16.113}$$

16.8.6.6 Calculation Steps

Summing up the algorithm above, the steps to calculate $\{\sigma\}_{n+1}$ are as follows:

1) Calculate $[D]_{ep}$ and the tangent modulus of each element by $\{\sigma\}_n$ at the beginning of the incremental step, form the global tangent stiffness matrix, and then calculate the displacement increment $\{\Delta\delta\}_{n+1} = [K_t]_n^{-1}\{\Delta P\}_{n+1}$ by the loading increment $\{\Delta P\}_{n+1}$.
2) Calculate $\{\Delta\varepsilon\}_{n+1}$ of each Gauss integral point by $\{\Delta\delta\}_{n+1}$.
3) Calculate the elastic stress increment $\{\Delta\sigma^e\} = [D]_e\{\Delta\varepsilon\}_{n+1}$ and the tentative stress $\{\sigma^e\} = \{\sigma\}_n + \{\Delta\sigma^e\}$.
4) Calculate the invariants I_1, J_2, J_3 and θ corresponding to $\{\sigma^e\}$.
5) Calculate $F(\{\sigma^e\}) = F_1$ by the instantaneous yield function.
6) If F_1 is negative, showing that it is in the elastic state, saving the following steps, let $\{\sigma\}_{n+1} = [\sigma^e]$.
7) If $F_1 > 0$ and $F_0 < 0$, go to the ninth step.
8) If $F_0 = 0$, making $r = 0$ and $1 - r = 1$, go to the tenth step.
9) Calculate r by formulas (16.102) and (16.104).
10) Calculate $\{\sigma\}_i = \{\sigma_n\} + r\{\Delta\sigma^e\}$, decide m and calculate the increment $\{d\varepsilon\}_i = [(1-r)/m]\{\Delta\varepsilon\}_{n+1}$ and $\{d\sigma^e\}_i = [(1-r)/m]\{\Delta\sigma^e\}$, and repeat m times from (11) to (16).
11) Calculate the stress invariants I_1, J_2, J_3 of $\{\sigma\}_i$.
12) Using ε_i^n, calculate the strain hardening parameter H' from uniaxial stress–strain curve, such as formula (16.87).
13) Calculate $\{\partial F/\partial\sigma\}_i$ and $[D]_e\{\partial F/\partial\sigma\}_i$.
14) Calculate λ and $[D]_p$ by formulas (16.70) and (16.72); if $\lambda < 0$, let $\lambda = 0$.
15) Calculate $\{d\sigma_i\} = \{d\sigma^e\}_i - [D]_p\{d\varepsilon\}_i$, and update the stress $\{\sigma\}_{i+1} = \{\sigma\}_i + \{d\sigma\}_i$.
16) Calculate the plastic work and the plastic strain increment or effective plastic strain increment of unit volume.
17) Calculate $[\delta\sigma]$ by formula (16.112), and go to the 11th step after correcting the stress, and after repeating m times from 11th step to 16th step, the calculated stress $\{\sigma\}_{i+1}$ is the stress $\{\sigma\}_{n+1}$ at the end of the incremental step.
18) Calculate the unbalanced force $\{\psi_n\}$ at the end of the step and merge it into the loading increment $\{\Delta P_{n+1}\}$ of the next step.

For materials such as concrete and rock, when the stress of Gauss points reach the failure strength, attention should be paid to the change from the failure strength to the residual strength.

16.8.7 Example

Figure 16.29 shows the plastic calculation results of a plane specimen with a breach, because of the bidirectional symmetry, only taking out 1/4 for calculation, with 8 cubic isoparametric elements and 59 nodes; the yield stress $\sigma_Y = 243\,\text{MPa}$, $E = 70\,\text{GPa}$, $\mu = 0.2$ is calculated according to ideal plastic material. The figure shows the expansion of the plastic zone along 4×4 Gauss integral points under different loading conditions.

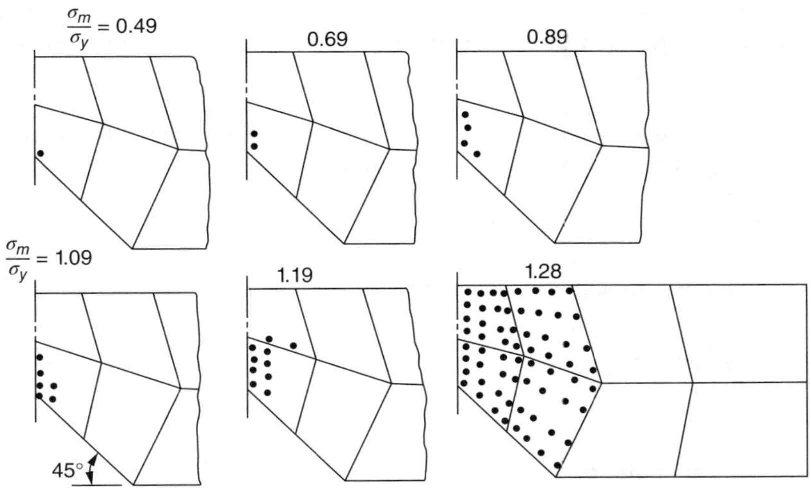

Figure 16.29 The extension of the plastic zone of a plane specimen with a notch through the integral points under different loads. σ_m – the average stress at the notch, σ_y – the yield stress.

16.9 Finite Element Method in the Full Variable Theory of Plasticity

In the state of plastic deformation, there is no one-to-one corresponding relationship between stress and strain, but generally speaking only the relationship between the stress increment and the strain increment can be expressed, and it is called the incremental theory, which has been stated in detail in the last section of this chapter. But under a special loading history condition, the stress and strain relation can be established in full variable form, namely, full variable theory. Its computation is simpler than the incremental theory.

16.9.1 Basic Assumption of Full Variable Theory

1) Assuming that the volume deformation is elastic and is proportional to the hydrostatic stress, but during the plastic deformation, the volume is incompressible, namely,

$$\left.\begin{aligned} \varepsilon_m^{\mathrm{e}} &= \frac{\sigma_m}{3K} = \frac{1-2\mu}{E}\sigma_m \\ \varepsilon_m^{\mathrm{p}} &= 0 \end{aligned}\right\} \tag{16.114}$$

2) Assuming that the deviator strain tensor is similar to the deviator stress tensor and they are coaxial, namely,

$$e_{ij} = \eta s_{ij} \tag{16.115}$$

3) Single curve hypothesis: whatever the stress state is, for the same kind of material, the relationship between equivalent stress σ_i and equivalent strain ε_i is certain, namely,

$$\sigma_i = E(\varepsilon_i)\varepsilon_i$$

in which $E(\varepsilon_i)$ is a function of material properties and it can be obtained from uniaxial test. For strengthened materials, under one-dimensional load, formula (16.2) is applicable, namely,

$$\sigma = E(1-\omega)\varepsilon$$

According to single curve hypothesis, under the complex stress state, the relationship above is

$$\sigma_i = E(1-\omega)\varepsilon_i \tag{16.116}$$

16.9.2 The Stress–Strain Relationship of Yiliuxin

Yiliuxin put forward the stress–strain relationship under elastic–plastic deformation:

$$\left.\begin{aligned} &\text{The elastic stage}: \quad e_{ij} = \frac{s_{ij}}{2G} \\ &\text{The plastic stage}: \quad e_{ij} = \frac{s_{ij}}{2G'} \end{aligned}\right\} \tag{a}$$

where G is the shear elastic modulus and G' is the function of equivalent strain ε_i.

Multiplying formula (a) by itself, summing up, and then taking the square root, we get

$$2G' = \sqrt{\frac{s_{ij}s_{ij}}{e_{kl}e_{kl}}} = \sqrt{\frac{J_2}{J_2'}} = \sqrt{\frac{\sigma_i^2/3}{3\varepsilon_i^2/4}} = \frac{2\sigma_i}{3\varepsilon_i}$$

Back to formula (a), the stress–strain relationship of full variable theory is as follows:

$$e_{ij} = \frac{3\varepsilon_i}{2\sigma_i}s_{ij} \tag{16.117}$$

Comparing with formula (16.115), it is known that

$$\eta = \frac{3\varepsilon_i}{2\sigma_i} \tag{16.118}$$

By substitution of $\mu = 1/2$ into formula (16.116), we get

$$\frac{\sigma_i}{\varepsilon_i} = 2G\left(1+\frac{1}{2}\right)(1-\omega) = 3G(1-\omega)$$

So for the strain hardened material, another expression is obtained, namely,

$$s_{ij} = 2G(1-\omega)e_{ij} \tag{16.119}$$

Full variable theory is applicable to simple loading condition; namely, in the process of loading, the ratio between each component of the deviator stress tensor s_{ij} remains invariant. The research results of Kachanofu *et al.* show that the conditions above can be relaxed a little; as long as the loading path is approaching to a simple loading path, full variable theory can be applied directly.

16.9.3 The Elastic–Plastic Matrix of Full Variable Theory

In order to apply full variable theory into the finite element method, it is necessary to establish the matrix expression between stress and strain. By formulas (16.114) and (16.119), the stress–strain relationship can be written as follows:

$$\left.\begin{aligned}\{s\} &= \frac{2}{3}E(1-\omega)\{e\}\\ \sigma_m &= 3K\varepsilon_m = \frac{E}{1-2\mu}\varepsilon_m\end{aligned}\right\} \tag{16.120}$$

in which $\{s\}$, $\{e\}$ are the deviator stress tensor and deviator strain tensor, respectively, and σ_m, ε_m are the average normal stress and the average normal strain, respectively.

By the formula above, full variable relationship of stress and strain is obtained as follows:

$$\{\sigma\} = [D]_{\text{ep}}\{\varepsilon\} \tag{16.121}$$

in which

$$[D]_{\text{ep}} = \frac{E}{3(1-2\mu)}\begin{bmatrix} 1+2\beta & 1-\beta & 1-\beta & 0 & 0 & 0\\ & 1+2\beta & 1-\beta & 0 & 0 & 0\\ & & 1+2\beta & 0 & 0 & 0\\ & \text{symmetry} & & \frac{3}{2}\beta & 0 & 0\\ & & & & \frac{3}{2}\beta & 0\\ & & & & & \frac{3}{2}\beta \end{bmatrix} \tag{16.122}$$

$$\beta = 2(1-2\mu)(1-\omega)/3$$

where $[D]_{\text{ep}}$ is the elastic–plastic matrix.

For plane strain problems, the third, fourth, and fifth rows and columns in the ranks of $[D]_{\text{ep}}$ can be deleted. For plane stress problems, the elastic–plastic matrix is

$$[D]_{\text{ep}} = \frac{\beta E}{(1-2\mu)(1+2\beta)}\begin{bmatrix} 2+\beta & 1-\beta & 0\\ & 2+\beta & 0\\ \text{symmetry} & & \beta+\frac{1}{2} \end{bmatrix} \tag{16.123}$$

16.10 Practical Simplified Models for Nonlinear Problem of Material

In the front part of this chapter, the finite element method for elastic–plastic theory has been stated in detail. These methods can be used to analyze many engineering problems. For some engineering materials, such as metal, its yield function is simple and its calculation is convenient. For other engineering materials, such as soil, rock, concrete, and so on, their yield functions are more complex, and their strain hardening and softening phenomena are also very obvious, and it is troublesome to apply elastic–plastic finite element method to do fine analysis. In order to determine the parameters of the constitutive relation, complicated tests are needed; for the majority of the engineering, it is not easy to do. In addition, the precision of calculation is related to the importance of engineering. For important projects, accurate model is needed, and for common engineering, we may use simplified model. In fact, in a large number of engineering analyses, simplified models are more used. This section will introduce several practical simplified models for nonlinear problems of materials.

In finite element analysis of nonlinear problems of materials, at present, incremental method is mainly used. Several simplified models below are suitable for incremental method.

16.10.1 Isotropic Model Containing One-Variable Modulus *E*(*t*)

The stress–strain relation of isotropic linear elastic material is

$$\sigma_{ij} = \frac{E}{1+\mu}\varepsilon_{ij} + \frac{\mu E}{(1+\mu)(1-2\mu)}\varepsilon_{kk}\delta_{ij} \tag{16.124}$$

where the elastic modulus E and Poisson's ratio μ are constants. Replacing elastic modulus E with tangent elastic modulus $E(t)$, the relation between the stress increment $\mathrm{d}\sigma_{ij}$ and the strain increment $\mathrm{d}\varepsilon_{ij}$ is as follows:

$$\mathrm{d}\sigma_{ij} = \frac{E(t)}{1+\mu}\mathrm{d}\varepsilon_{ij} + \frac{\mu E(t)}{(1+\mu)(1-2\mu)}\mathrm{d}\varepsilon_{kk}\delta_{ij} \tag{16.125}$$

where Poisson's ratio μ is a constant. For concrete, when the compressive stress is less than 75% of the compressive strength, μ is a constant, and when the stress increases further, the value of μ will increase gradually. In the above equation, all the nonlinear character of the material is reflected by the change of $E(t)$. According to the results of uniaxial test of material, by the use of the relation between equivalent stress and equivalent strain, the tangent elastic modulus can be calculated. The stress–strain relation can be expressed by hyperbolic curve, parabolic curve, or exponential curve. For example, the stress–strain relationship of soil is commonly expressed by a hyperbolic curve, referring to Section 24.1.

For the plane stress problem, formula (16.125) becomes

$$\begin{Bmatrix} \mathrm{d}\sigma_x \\ \mathrm{d}\sigma_y \\ \mathrm{d}\tau_{xy} \end{Bmatrix} = \frac{E(t)}{1-\mu^2}\begin{bmatrix} 1 & \mu & 0 \\ \mu & 1 & 0 \\ 0 & 0 & (1-\mu)/2 \end{bmatrix}\begin{Bmatrix} \mathrm{d}\varepsilon_x \\ \mathrm{d}\varepsilon_y \\ \mathrm{d}\gamma_{xy} \end{Bmatrix} \tag{16.126}$$

where $\gamma_{xy} = 2\varepsilon_{xy}$ is the engineering shear strain.

16.10.2 Isotropic Model Containing Two-Variable Modulus *K*(*t*) and *G*(*t*)

For isotropic linear elastic material, if separating the volume and shape deformation, the stress–strain relationship is

$$\sigma_m = K\varepsilon_{kk}, \quad s_{ij} = 2Ge_{ij} \tag{16.127}$$

where σ_m is the average stress, ε_{kk} is the volume deformation, and s_{ij} is the deviator stress tensor. e_{ij} is the deviator strain tensor, K is the bulk modulus, and G is the shear modulus.

For nonlinear materials, replacing K and G with the variable tangent bulk modulus $K(t)$ and the tangent shear modulus $G(t)$, respectively, the relationship between the stress increment and the strain increment is as follows:

$$d\sigma_m = K(t)\,d\varepsilon_{kk}, \quad ds_{ij} = 2G(t)\,de_{ij} \tag{16.128}$$

In this way, the volume deformation and shape deformation are independent of each other. $K(t)$ and $G(t)$ can be expressed as follows:

$$\left.\begin{aligned} K(t) &= \frac{d\sigma_8}{3\,d\varepsilon_8} = \frac{E(t)}{3[1-2\mu(t)]} \\ G(t) &= \frac{d\tau_8}{d\gamma_8} = \frac{E(t)}{2[1+\mu(t)]} \end{aligned}\right\} \tag{16.129}$$

where the subscript 8 represents an octahedron, for example, σ_8 and τ_8 represent octahedral normal stress and octahedral shear stress, respectively.

According to test data of materials, we can obtain the expressions of $K(t)$ and $G(t)$ and then convert them to $E(t)$ and $\mu(t)$. We can also write the expressions of tangent elastic modulus $E(t)$ and tangent Poisson's ratio $\mu(t)$ directly.

For example, for concrete, according to the test data of Kupfer *et al.*, there is

$$\begin{aligned} \frac{G(t)}{G(t_0)} &= \frac{[1-a(\tau_8/R_c)^{\mathrm{m}}]^2}{1+(m-1)a(\tau_8/R_c)^{\mathrm{m}}} \\ \frac{K(t)}{K(t_0)} &= \frac{G(t)/G(t_0)}{\exp[-(c\gamma_8)^{\mathrm{p}}][1-p(c\gamma_8)^{\mathrm{p}}]} \end{aligned} \tag{16.130}$$

The initial modulus $G(t_0)$, $K(t_0)$ and a, m, c, p mainly depends on the compressive strength R_c. When $R_c = 32.4\,\mathrm{MPa}$, their values are as follows:

$$\left.\begin{aligned} G(t_0)/R_c &= 425, \quad a = 3.5, \quad m = 2.4 \\ K(t_0)/R_c &= 556, \quad c = 210, \quad p = 2.2 \end{aligned}\right\} \tag{16.131}$$

16.10.3 Orthotropic Model and the Equivalent Uniaxial Strain

A three-dimensional (axisymmetric) orthotropic nonlinear model is introduced below. The parameters of this model can be determined by the uniaxial test data of material. Because of adopting the concept of equivalent uniaxial strain, this model not only can be used for simple loading but also can be used in the case of cyclic loading.

16.10.3.1 Orthotropic Constitutive Relations

Considering a three-dimensional axisymmetric structure corresponding to the orthotropic principal axis, the relation between the stress increment and the strain increment is as follows:

$$\begin{Bmatrix} d\varepsilon_1 \\ d\varepsilon_2 \\ d\varepsilon_3 \\ d\gamma_{12} \end{Bmatrix} = \begin{bmatrix} E_1^{-1} & -\mu_{12}E_2^{-1} & -\mu_{13}E_3^{-1} & 0 \\ -\mu_{21}E_1^{-1} & E_2^{-1} & -\mu_{23}E_3^{-1} & 0 \\ -\mu_{31}E_1^{-1} & -\mu_{32}E_2^{-1} & E_3^{-1} & 0 \\ 0 & 0 & 0 & G_{12}^{-1} \end{bmatrix} \begin{Bmatrix} d\sigma_1 \\ d\sigma_2 \\ d\sigma_3 \\ d\tau_{12} \end{Bmatrix} \tag{16.132}$$

where the subscripts 1, 2, 3 represent the orthotropic principal axis.

Due to symmetry, the following conditions are required to be established:

$$\mu_{12}E_1 = \mu_{21}E_2, \quad \mu_{13}E_1 = \mu_{31}E_3, \quad \mu_{23}E_2 = \mu_{32}E_3$$

By substitution of the formula above into formula (16.132), the following symmetric expression is obtained:

$$\begin{Bmatrix} d\varepsilon_1 \\ d\varepsilon_2 \\ d\varepsilon_3 \\ d\gamma_{12} \end{Bmatrix} = \begin{bmatrix} \frac{1}{E_1} & \frac{-\nu_{12}}{\sqrt{E_1E_2}} & \frac{-\nu_{13}}{\sqrt{E_1E_3}} & 0 \\ & \frac{1}{E_2} & \frac{-\nu_{23}}{\sqrt{E_2E_3}} & 0 \\ & & \frac{1}{E_3} & 0 \\ & \text{symmetric} & & \frac{1}{G_{12}} \end{bmatrix} \begin{Bmatrix} d\sigma_1 \\ d\sigma_2 \\ d\sigma_3 \\ 0 \end{Bmatrix} \tag{16.133}$$

Inverting formula (16.133), we get

$$\{d\sigma\} = [C]\{d\varepsilon\} \tag{16.134}$$

The constitutive matrix $[C]$ in the formula above is given by the formula below:

$$[C] = \frac{1}{\phi} \begin{bmatrix} E_1(1-\nu_{32}^2) & \sqrt{E_1E_2}(\nu_{13}\nu_{23}+\nu_{12}) & \sqrt{E_1E_3}(\nu_{12}\nu_{23}+\nu_{13}) & 0 \\ & E_2(1-\nu_{13}^2) & \sqrt{E_2E_3}(\nu_{12}\nu_{13}+\nu_{23}) & 0 \\ & & E_3(1-\nu_{12}^2) & 0 \\ \text{symmetric} & & & \phi G_{12} \end{bmatrix} \tag{16.135}$$

where

$$\left.\begin{aligned} &\nu_{12}^2 = \mu_{12}\mu_{21}, \quad \nu_{23}^2 = \mu_{23}\mu_{32}, \quad \nu_{13}^2 = \mu_{13}\mu_{31} \\ &\phi = 1 - \nu_{12}^2 - \nu_{23}^2 - \nu_{13}^2 - 2\nu_{12}\nu_{23}\nu_{13} \end{aligned}\right\} \tag{16.136}$$

If transforming the matrix $[C]$ into non-orthotropic principal axes $(1', 2', 3'')$ and demanding that after transformation the shear modulus is constant, there is

$$G_{12} = \frac{1}{4\phi}[E_1 + E_2 - 2\nu_{12}\sqrt{E_1E_2} - (\sqrt{E_1}\nu_{23} + \sqrt{E_2}\nu_{13})^2] \tag{16.137}$$

For the plane stress problem, it only needs to let $d\sigma_3 = 0$ in formulas (16.132) and (16.133) and remove the third row and the third column, then we get

$$\begin{Bmatrix} d\sigma_1 \\ d\sigma_2 \\ d\tau_{12} \end{Bmatrix} = \frac{1}{1-\nu_{12}^2} \begin{bmatrix} E_1 & \nu_{12}\sqrt{E_1E_2} & 0 \\ & E_2 & 0 \\ & & (E_1 + E_2 - 2\nu_{12}\sqrt{E_1E_2})/4 \end{bmatrix} \begin{Bmatrix} d\varepsilon_1 \\ d\varepsilon_2 \\ d\gamma_{12} \end{Bmatrix} \tag{16.138}$$

At this time, formula (16.137) becomes

$$(1-\nu_{12}^2)G = \frac{1}{4}(E_1 + E_2 - 2\nu_{12}\sqrt{E_1E_2}) \tag{16.139}$$

16.10.3.2 Equivalent Uniaxial Strain

There are seven material parameters in formula (16.134). In order to decide the relationship between E_1, E_2, E_3 and the change of stress, Ref. [4] puts forward the concept of equivalent uniaxial strain and transform formula (16.134) into the following form:

$$\begin{Bmatrix} d\sigma_1 \\ d\sigma_2 \\ d\sigma_3 \\ d\tau_{12} \end{Bmatrix} = \begin{bmatrix} E_1 b_{11} & E_1 b_{12} & E_1 b_{13} & 0 \\ E_2 b_{21} & E_2 b_{22} & E_2 b_{23} & 0 \\ E_3 b_{31} & E_3 b_{32} & E_3 b_{33} & 0 \\ 0 & 0 & 0 & G_{12} \end{bmatrix} \begin{Bmatrix} d\varepsilon_1 \\ d\varepsilon_2 \\ d\varepsilon_3 \\ d\gamma_{12} \end{Bmatrix} \tag{16.140}$$

Let each item of the matrix on the right side of formula (16.140) equal to the corresponding item of formula (16.135). Then the coefficients b_{ij} of formula (16.140) can be obtained, for example,

$$b_{11} = (1 - \nu_{32}^2)/\phi$$

$$b_{12} = \sqrt{E_2/E_1}(\nu_{13}\nu_{32} + \nu_{12})/\phi$$

$$b_{13} = \sqrt{E_3/E_1}(\nu_{12}\nu_{32} + \nu_{13})/\phi$$

By formula (16.140), we get

$$\left.\begin{aligned} d\sigma_1 &= E_1(b_{11}\,d\varepsilon_1 + b_{12}\,d\varepsilon_2 + b_{13}\,d\varepsilon_3) \\ d\sigma_2 &= E_2(b_{21}\,d\varepsilon_1 + b_{22}\,d\varepsilon_2 + b_{23}\,d\varepsilon_3) \\ d\sigma_3 &= E_3(b_{31}\,d\varepsilon_1 + b_{32}\,d\varepsilon_2 + b_{33}\,d\varepsilon_3) \\ d\tau_{12} &= G_{12}\,d\gamma_{12} \end{aligned}\right\} \tag{a}$$

Let

$$d\varepsilon_{iu} = b_{i1}\,d\varepsilon_1 + b_{i2}\,d\varepsilon_2 + b_{i3}\,d\varepsilon_3 \quad (i = 1, 2, 3) \tag{16.141}$$

Then formula (a) can be written in the following matrix form:

$$\begin{Bmatrix} d\sigma_1 \\ d\sigma_2 \\ d\sigma_3 \\ d\tau_{12} \end{Bmatrix} = \begin{bmatrix} E_1 & 0 & 0 & 0 \\ 0 & E_2 & 0 & 0 \\ 0 & 0 & E_3 & 0 \\ 0 & 0 & 0 & G_{12} \end{bmatrix} \begin{Bmatrix} d\varepsilon_1 \\ d\varepsilon_2 \\ d\varepsilon_3 \\ d\gamma_{12} \end{Bmatrix} \tag{16.142}$$

The vector on the right of formula (16.142) can be defined as the equivalent uniaxial strain increment, and its components can be calculated by formula (16.141). By formula (16.142), the equivalent uniaxial strain increment $d\varepsilon_{iu}$ can be calculated by the following formula:

$$d\varepsilon_{iu} = \frac{d\sigma_i}{E_i} \quad (i = 1, 2, 3) \tag{16.143}$$

Formula (16.143) is the same with the uniaxial stress state in form, so $d\varepsilon_{iu}$ is called equivalent uniaxial strain increment; its value may be positive or negative, depending on the change of the stress state. Integrating formula (16.143), the equivalent uniaxial strain is obtained as follows:

$$\varepsilon_{iu} = \int \frac{d\sigma_i}{E_i} \tag{16.144}$$

By formula (16.132), we know that the strain increment $d\varepsilon_{iu}$ calculated by formula (16.143) is equivalent to the strain increment of orthotropic materials when it is under uniaxial stress increment $d\sigma_i$ and the stress increments in other directions are zero. But $d\varepsilon_{iu}$ depend on the ratio of the stress, and when doing the coordinate transformation of ε_{iu} and $d\varepsilon_{iu}$, they don't transform with the stress state. They are fictional quantities (unless under uniaxial loading); their principal function is used to determine the parameters of material.

What is discussed above is three-dimensional axisymmetric problem, but it is not difficult to popularize to the general three-dimensional problem, only G_{23}, G_{31} need doing the similar treatment like G_{12}.

In order to illustrate the concept of equivalent uniaxial strain further, we may consider an anisotropic elastic material, the parameters of the material are constants at this time, so the equivalent uniaxial strain becomes $\varepsilon_{iu} = \sigma_i/E_i$. For example, $\varepsilon_{1u} = \sigma_1/E_1$, it represents the strain only produced by the stress σ_1 without the influence of Poisson ratio. For nonlinear elastic material of which, a similar explanation can also be made, namely, the equivalent uniaxial strain increment $d\varepsilon_{iu}$ is equal to the strain increment when $d\sigma_i$ acts alone. But when we conduct coordinate transformation, they aren't transformed as the actual strain, so they aren't the actual strain. In addition, by formula (16.144), we know that the equivalent uniaxial strain is obtained according to the integral of the principal stress increment, but under general loading, the directions of the principal stresses are changing. For example, ε_{1u} does not represent the deformation history in a fixed direction 1, but the result of the integral of the principal strain increment $d\sigma_i/E_i$ whose direction is changing. The key point is that the stress–strain relationship is written in the form of uniaxial stress to take advantage of the fact that the form of stress–strain curve is similar to the uniaxial test.

Section 25.1 gives the specific formula of equivalent uniaxial strain of the concrete.

16.10.4 The Approximate Calculation of Strain Softening

For strain softening materials, if we use the full variable method to calculate, whatever the variable stiffness method or the constant stiffness method is adopted, it is sufficient to consider the influence of strain softening when calculating the secant modulus or initial stress; there are no special difficulties of calculation. But if using incremental method to calculate, in the stage of strain softening, the tangent elastic modulus is negative, leading the stiffness matrix to be nonpositive definite, and the calculation will be difficult. In order to overcome this difficulty, the following practical and approximate method can be adopted:

1) When calculating the stiffness matrix, taking $E = 0$ in the strain softening period (in fact taking a small positive number)
2) When calculating the stress increment $\Delta\sigma$ in the strain softening period, taking the actual value of E (negative).

Bibliography

1 Chen, W.F. (1982) *Plasticity in Reinforced Concrete*, McGraw-Hill, New York.
2 Xiong, W.L. (1986) Symmetric expression of non-associated elastic–plastic stiffness matrix. *J. Appl. Math. Mech.*, 7 (11), 1043–1052.

3 Owen, D.R.J. and Hinton, E. (1980) *Finite Elements in Plasticity*, Pineridge Press, Swansea.
4 Darwin, D. and Pecknold, D.A. (1977) Nonlinear biaxial law for concrete. *J. Eng. Mech. Div. ASCE*, **103**, 229–241.
5 Wang, R. (1982) *Fundamentals of Theory of Plasticity*, Science Press, Beijing.
6 Han, D.J. and Chen, W.F. (1987) Constitutive modeling in analysis of concrete structures. *J. Eng. Mech. ASCE*, **113** (4), 577–593.
7 Lade, P.V. (1977) Elasto-plastic stress–strain theory for cohesionless soil with curved yield surface. *Int. J. Solids Struct.*, **13** (11), 1019–1035.
8 Lade, P.V. (1982) Three-parameter failure criterion for concrete. *J. Eng. Mech. Div. ASCE*, **108** (8), 850–863.
9 Ottosen, N.S. (1977) A failure criterion for concrete. *J. Eng. Mech. Div. ASCE*, **103**, 527–534.
10 Elwi, A.A. and Murray, D.W. (1979) A 3D hypoelastic concrete constitutive relationship. *J. Eng. Mech. Div. ASCE*, **105**, 623–641.
11 Bathe, K.J. and Ramaswamy, S. (1979) On three dimensional nonlinear analysis of concrete structures. *Nucl. Eng. Des.*, **52**, 385–409.
12 Bresler, B. and Pister, K.S. (1958) Strength of concrete under combined stresses. *J. ACI*, **55**, 321–345.
13 Hsieh, S.S., Ting, E.C., and Chen, W.F. (1979) *An Elastic Fracture Model for Concrete*. Proc. Third Eng. Mech. Div. Spec. Conf. ASCE. Austin, Texas, 437–440.
14 Willam, K.J. and Warnke, E.P. (1975) Constitutive models for the triaxial behavior of concrete. *Int. Assoc. Bridge Struct. Eng. Proc.*, **19**, 1–30.
15 Koiter, W.T. (1953) Stress–strain relations, uniqueness and variational theorems for elastic–plastic materials with singular surface. *Quart. Appl. Math.*, **11**, 350–354.
16 Kupfer, H.B. and Gestle, K.H. (1973) Behavior of concrete under biaxial stresses. *J. Eng. Mech. Div. ASCE*, **99**, 852–866.
17 Nayak, G.C. and Zienkiewicz, O.C. (1972) Elasto-plastic stress analysis: a generation for various constitutive relations including strain softening. *Int. J. Numer. Methods Eng.*, **5**, 113–135.
18 Wiangao, Z. and Lu, M. (1990) Criterion of rupture of concrete with three parameters. *J. Shanghai Jiaotong Univ.*, **5**, 6.

17

Creep of Concrete and its Influence on Stresses and Deformations of Structures

Creep is the deformation not only related to the stress level but also related to the duration of loading time. Concrete, plastics, wood, clay, and metal under high temperature have significant creep. Particularly, the creep of concrete is relevant to the age of application of load. There are two methods to determine the creep of material, namely, creep test and relaxation test. Creep test is to measure the increasing deformation over time under a fixed stress level. Relaxation test is to maintain a constant strain to measure the decrease of stress. The results of the two test are equivalent in mechanics. Because the creep test is relatively simple, usually creep tests for concrete are done.

In engineering, we are concerned about the effect of creep on the stresses and deformations of the structures. For linear elastocreeping body, if the creep law of the volume deformation is the same with that of the shear deformation, then using several simple theorems, the variation of the stress and deformation can be inferred. If the creep law of the shear deformation and volume deformation are different, for viscoelastic body with coefficients of deformation independent on age, in theory, the variation law of the stress and deformation can be solved by Laplace transformation; except for some relatively simple problems, due to the difficulties of inversion, results from most engineering problems are difficult to obtain. For nonlinear creep problems, except for a few very simple questions, it is difficult to get analytical solution [1–3].

Therefore, for creep problems in the practical engineering, finite element method is an effective solution method.

In general, the creep of the body is not only relevant to stress but also associated with the whole stress history. Therefore, when using the finite element method to analyze creep problems, time is usually divided into a number of time intervals, and the incremental method is used to calculate the stresses and deformations in each time interval.

17.1 Stress–Strain Relation of Concrete

The deformation of concrete is closely related to loading age, and the deformation of concrete in early age is far larger than that in later age. In the calculation of the deformation of concrete, this important feature must be considered.

From the test data, under the long-term loading, Poisson's ratio of concrete creep is a constant and roughly equal to Poisson's ratio of instantaneous deformation [4].

The Finite Element Method: Fundamentals and Applications in Civil, Hydraulic, Mechanical and Aeronautical Engineering, First Edition. Bofang Zhu.

Companion website: www.wiley.com/go/zhufem17

Therefore, in the calculation of the deformation of concrete, it is often assumed that Poisson's ratio of creep is equal to that of instantaneous deformation.

In addition, according to the experimental data, when the compressive stress does not exceed one half of the compressive strength and the tensile stress does not exceed 0.8 times of the tensile strength, there is a linear relationship between the creep and stress of concrete. It can be calculated according to the linear creep theory.

17.1.1 Stress–Strain Relation of Concrete under Action of Stress in One Direction

Assuming that at the age τ, the uniaxial stress $\sigma_x(\tau)$ is applied to concrete, the instantaneous elastic strain is (as shown in Figure 17.1)

$$\varepsilon_x^{\mathrm{e}} = \frac{\sigma_x(\tau)}{E(\tau)} \tag{a}$$

where $E(\tau)$ is instantaneous modulus of elasticity, which is a function of the age τ of concrete.

If the stress remains invariant, the strain will increase with time. This part of strain is known as creep. Its value is

$$\varepsilon_x^{\mathrm{c}} = \sigma(\tau)C(t,\tau) \tag{b}$$

$C(t,\tau)$ is the creep generated under the action of unit stress, known as unit creep. Therefore, if at the age τ the constant stress $\sigma_x(\tau)$ is imposed, to the time t, the total strain is the sum of the elastic strain and creep, that is,

$$\varepsilon_x(t,\tau) = \varepsilon_x^{\mathrm{e}} + \varepsilon_x^{\mathrm{c}} = \frac{\sigma_x(\tau)}{E(\tau)} + \sigma_x(\tau)C(t,\tau) = \sigma_x(\tau)J(t,\tau) \tag{17.1}$$

in which

$$J(t,\tau) = \frac{1}{E(\tau)} + C(t,\tau) \tag{17.2}$$

where $J(t,\tau)$ is the creep compliance.

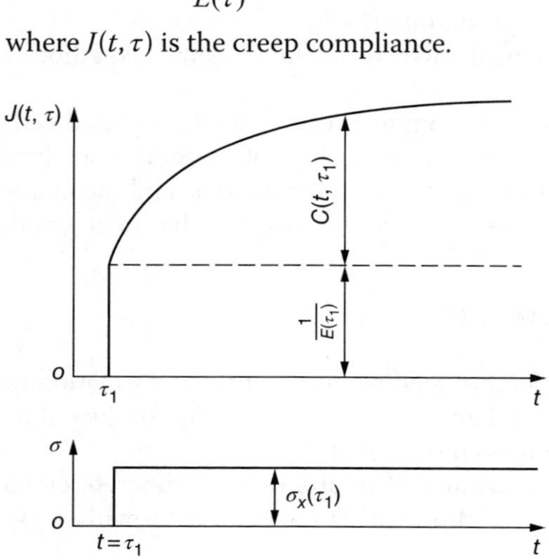

Figure 17.1 The strain of concrete under constant stress.

At the moment of loading, $t = \tau$, creep is zero, so $C(\tau, \tau) = 0$.

It is assumed that the stress does not change with time in the above analysis. If the stress changes with time, based on the principle of superposition, the strain of concrete can be calculated as follows:

$$\varepsilon_x(t) = \Delta\sigma_{x0} J(t, t_0) + \int_{t_0}^{t} J(t, \tau)\frac{\mathrm{d}\sigma_x}{\mathrm{d}\tau}\mathrm{d}\tau \tag{17.3}$$

where t_0 is the age of first loading and $\Delta\sigma_{x0}$ is the stress increment applied at $t = t_0$.

The modulus of elasticity $E(\tau)$ and unit creep $C(t, \tau)$ of concrete are related to the loading age τ. If the loads are applied at different ages, for example, at $t = t_0, t_1, \dots, t_n$, a group of creep compliance curves can be obtained, as shown in Figure 17.2.

The lateral deformation is considered below. At the moment of loading, the lateral instantaneous elastic strain is as follows:

$$\varepsilon_y^{\mathrm{e}} = \varepsilon_z^{\mathrm{e}} = -\mu_1 \frac{\sigma_x(\tau)}{E(\tau)}$$

where $\mu_1(\tau)$ is Poisson's ratio of the instantaneous elastic strain.

If stress σ_x remains constant, the lateral creep is

$$\varepsilon_y^{\mathrm{c}}(t) = \varepsilon_z^{\mathrm{c}}(t) = -\mu_2(t, \tau)\sigma_x(\tau)C(t, \tau) \tag{c}$$

where $\mu_2(t, \tau)$ is Poisson's ratio of creep.

The total lateral strain is calculated by the following formula:

$$\varepsilon_y(t) = \varepsilon_z(t) = -\sigma_x(\tau)\left[\frac{\mu_1(\tau)}{E(\tau)} + \mu_2(t, \tau)C(t, \tau)\right] \tag{17.4}$$

In Ref. [4] the creep test was done using a large concrete specimen with a height of 1.50 m and a diameter of 0.75 m. Results showed that Poisson's ratio of creep is approximately constant and is close to that of elastic strain. Therefore, the following values can be used in engineering:

$$\mu_2(t, \tau) = \mu_1(t, \tau) = \mu = \frac{1}{6} \tag{17.5}$$

By substitution of formula (17.5) into formula (17.4), the total lateral strain can be calculated as follows:

$$\varepsilon_y(t) = \varepsilon_z(t) = -\mu\sigma_x(\tau)J(t, \tau) \tag{17.6}$$

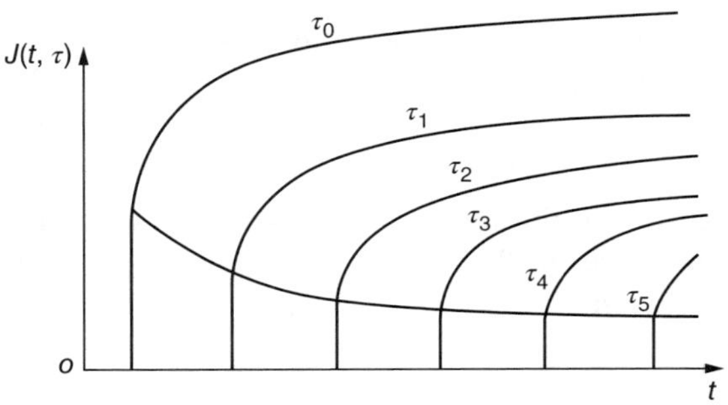

Figure 17.2 Creep compliance of concrete for different loading ages.

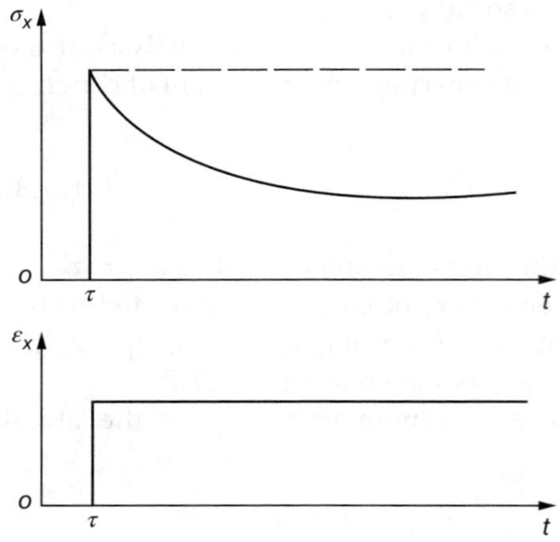

Figure 17.3 Stress relaxation of concrete.

The stress relaxation of concrete is considered below [5–8].

If the concrete is subjected to constrained strain $\varepsilon_x(\tau)$ at age τ, there will be elastic stress:

$$\sigma_x(\tau) = E(\tau)\varepsilon_x(\tau)$$

Assuming that when $t > \tau$, the strain $\varepsilon_x(t)$ remains constant. If it is the ideal elastic body, the stress $\sigma_x(t)$ will also remain unchanged, as the dotted line shown in Figure 17.3. In fact, because concrete is the elastocreeping body, after being subject to the compressive strain, the stress will continue to decay with time, as the solid line shown in Figure 17.3. This phenomenon is known as the stress relaxation of concrete. Assuming that the stress is $\sigma_x(t,\tau)$ at any moment t, the ratio of it and the initial elastic stress is

$$R(t,\tau) = \frac{\sigma_x(t,\tau)}{\sigma_x(\tau)} \tag{17.7}$$

$R(t,\tau)$ is called relaxation coefficient. It can be obtained directly from the relaxation test of concrete. But the relaxation test is troublesome. Generally based on the data of creep test, $R(t,\tau)$ is obtained by the calculation.

Now the relationship between creep compliance $J(t,\tau)$ and relaxation coefficient $R(t,\tau)$ is described. Applying the stress $\sigma_x(\tau_1)$ at age τ_1, the strain can be calculated as follows:

$$\varepsilon_x(t) = \sigma_x(\tau_1)\left[\frac{1}{E(\tau_1)} + C(t,\tau_1)\right] + \int_{\tau_1}^{t} J(t,\tau)\frac{\mathrm{d}\sigma_x}{\mathrm{d}\tau}\mathrm{d}\tau$$

Let

$$\varepsilon_x(t) = \varepsilon_x(\tau_1) = \frac{\sigma_x(\tau_1)}{E(\tau_1)} = \text{constant}$$

By substitution of it into the front formula, after the simplification, we get

$$\sigma_x(\tau_1)C(t,\tau_1) + \int_{\tau_1}^{t} J(t,\tau)\frac{\mathrm{d}\sigma_x}{\mathrm{d}\tau}\mathrm{d}\tau = 0 \tag{d}$$

This is an integral equation. Solving $\sigma_x(t,\tau)$ by the above equation and then substituting it into formula (17.7), the relaxation coefficient $R(t,\tau)$ is obtained. Equation (d) can be solved by the numerical method as follows:

Taking the initial stress $\sigma_x(\tau_1) = 1$, we get

$$C(t,\tau_1) + \int_{\tau_1}^{t} J(t,\tau)\mathrm{d}\sigma = 0 \tag{e}$$

The stress $\sigma(t)$ given by the above equation is the relaxation coefficient. Dividing $t - \tau_1$ into a series of time increments $\Delta\tau_i$, equation (e) is transformed into the following equation:

$$C(\tau_n, \tau_1) + J(\tau_n, \overline{\tau_1})\Delta\sigma_1 + J(\tau_n, \overline{\tau_2})\Delta\sigma_2 + \cdots + J(\tau_n, \overline{\tau_{n-1}})\Delta\sigma_{n-1} + J(\tau_n, \overline{\tau_n})\Delta\sigma_n = 0 \tag{f}$$

Hence, the nth stress increment is

$$\Delta\sigma_n = -\frac{C(\tau_n, \tau_1) + \sum_{i=1}^{n-1} J(\tau_n, \overline{\tau_i})\Delta\sigma_i}{J(\tau_n, \overline{\tau_n})} \tag{g}$$

where $\overline{\tau_i} = (\tau_{i-1} + \tau_i)/2$.

The relaxation coefficient is

$$R(t, \tau) = 1 + \sum \Delta\sigma_i \tag{17.8}$$

Assuming that the strain increment $\Delta\varepsilon_x(\tau)$ is generated at $t = \tau$, the elastic stress increment generated at that time is $\Delta\sigma_x(\tau) = E(\tau)\Delta\varepsilon_x(\tau)$, to the moment t; due to stress relaxation, the stress increment is

$$\Delta\sigma_x(t) = R(t, \tau)\Delta\sigma_x(\tau) = R(t, \tau)E(\tau)\Delta\varepsilon_x(t)$$

Therefore, by formula (17.9), according to the known strain, the stress of the elastocreeping body can be obtained:

$$\sigma_x(t) = E(\tau_0)\Delta\varepsilon_x(\tau_0)R(t, \tau_0) + \int_{\tau_0}^{t} R(t, \tau)E(\tau)\frac{d\varepsilon_x}{d\tau}d\tau \tag{17.9}$$

where $\Delta\varepsilon_x(\tau_0)$ is the strain increment at $t = \tau_0$.

Let

$$K(t, \tau) = E(\tau)R(t, \tau) \tag{17.10}$$

where $K(t, \tau)$ is known as the relaxation modulus.

By substitution of formula (17.10) into formula (17.9), we get

$$\sigma_x(t) = K(t, \tau_0)\Delta\varepsilon_x(\tau_0) + \int_{\tau_0}^{t} K(t, \tau)\frac{d\varepsilon_x}{d\tau}d\tau \tag{17.11}$$

17.1.2 Stress–Strain Relation Under Complex Stress Conditions

As mentioned above, based on test data, it is assumed that Poisson's ratio of the creep of concrete is a constant and is equal to that of the instantaneous elastic strain [8–10]. Therefore, under the complicated stress state, stress–strain relationship of concrete can be expressed as follows:

$$\{\varepsilon(t)\} = [A]\{\Delta\sigma_0\}J(t, \tau_0) + \int_{\tau_0}^{t} J(t, \tau)[A]\left\{\frac{d\sigma(\tau)}{d\tau}\right\}d\tau \tag{17.12}$$

where matrix $[A]$ is used to consider the effect of Poisson's ratio. Referring to formulas (17.13)–(17.15), we get $[A]$ as follows:

1) Plane Stress Problem

$$\{\varepsilon\} = \left[\varepsilon_x \ \varepsilon_y \ \gamma_{xy}\right]^{\mathrm{T}}$$

$$\{\sigma\} = \left[\sigma_x \ \sigma_y \ \tau_{xy}\right]^{\mathrm{T}}$$

$$[A] = \begin{bmatrix} 1 & -\mu & 0 \\ -\mu & 1 & 0 \\ 0 & 0 & 2(1+\mu) \end{bmatrix} \tag{17.13a}$$

$$[A]^{-1} = \frac{1}{1-\mu^2}\begin{bmatrix} 1 & \mu & 0 \\ \mu & 1 & 0 \\ 0 & 0 & 1-\mu/2 \end{bmatrix} \tag{17.13b}$$

2) Plane Strain Problem

$$[A] = (1+\mu)\begin{bmatrix} 1-\mu & -\mu & 0 \\ -\mu & 1-\mu & 0 \\ 0 & 0 & 2 \end{bmatrix} \tag{17.14a}$$

$$[A]^{-1} = \frac{1-\mu}{(1+\mu)(1-2\mu)}\begin{bmatrix} 1 & \dfrac{\mu}{1-\mu} & 0 \\ \dfrac{\mu}{1-\mu} & 1 & 0 \\ 0 & 0 & \dfrac{1-2\mu}{2(1-\mu)} \end{bmatrix} \tag{17.14b}$$

3) Spatial Problem

$$\{\varepsilon\} = \left[\varepsilon_x \ \varepsilon_y \ \varepsilon_z \ \gamma_{xy} \ \gamma_{yz} \ \gamma_{zx}\right]^{\mathrm{T}}$$

$$\{\sigma\} = \left[\sigma_x \ \sigma_y \ \sigma_z \ \tau_{xy} \ \tau_{yz} \ \tau_{zx}\right]^{\mathrm{T}}$$

$$[A] = \begin{bmatrix} 1 & -\mu & -\mu & 0 & 0 & 0 \\ & 1 & -\mu & 0 & 0 & 0 \\ & & 1 & 0 & 0 & 0 \\ & \text{symmetrical} & & 2(1+\mu) & 0 & 0 \\ & & & & 2(1+\mu) & 0 \\ & & & & & 2(1+\mu) \end{bmatrix} \tag{17.15a}$$

$$[A]^{-1} = \frac{1-\mu}{(1+\mu)(1-2\mu)}\begin{bmatrix} 1 & \dfrac{\mu}{1-\mu} & \dfrac{\mu}{1-\mu} & 0 & 0 & 0 \\ & 1 & \dfrac{\mu}{1-\mu} & 0 & 0 & 0 \\ & & 1 & 0 & 0 & 0 \\ & \text{symmetrical} & & \dfrac{1-2\mu}{2(1-\mu)} & 0 & 0 \\ & & & & \dfrac{1-2\mu}{2(1-\mu)} & 0 \\ & & & & & \dfrac{1-2\mu}{2(1-\mu)} \end{bmatrix} \tag{17.15b}$$

The elasticity matrix can be calculated by formula (17.16):

$$[D] = E[A]^{-1} \tag{17.16}$$

17.1.3 Modulus of Elasticity of Concrete $E(\tau)$

The modulus of elasticity of concrete is a function of age. The following expressions are put forward before:

$$E(\tau) = E_0(1 - e^{-a\tau}) \tag{c}$$

$$E(\tau) = E_0(1 - be^{-a\tau}) \tag{d}$$

$$E(\tau) = \sum_{i=1}^{n} E_i(1 - e^{-a_i\tau}) \tag{e}$$

where E_0, E_i, a, b, a_i are the constants determined by the test data. Formula (c) does not accord well with the test results. When $\tau = 0$, $E(0) = E_0(1 - b) \neq 0$ in formula (d). It does not meet the condition $E(0) = 0$, formula (e) is complex, and the coefficients in the formula are difficult to determine.

In order to find a better formula, the author tried the following formula:

$$E(\tau) = \frac{E_0\tau}{c + \tau} \tag{17.17}$$

This is a hyperbolic formula, wherein E_0 and c are constants. When $\tau = \infty$, $E(\tau) = E_0$ and when $\tau = c$, $E(\tau) = E_0/2$. Formula (17.17) also accords badly with test data.

After much exploration, a good formula for the modulus of elasticity of concrete is obtained as follows:

$$E(\tau) = E_0(1 - e^{-a\tau^b}) \tag{17.18}$$

in which E_0, a, b are constants.

Formula (17.18) agrees well with the experimental data and the constants of the formula are relatively easy to determine. Let

$$f(\tau) = a\tau^b = -\ln\left[1 - \frac{E(\tau)}{E_0}\right]$$

Taking logarithm on both sides of the formula above, we get

$$\ln a + b \ln \tau = \ln f(\tau) = \ln\left[-\ln\left(1 - \frac{E(\tau)}{E_0}\right)\right]$$

Taking $\ln(\tau)$ for the horizontal coordinate and $\ln f(\tau)$ for the vertical ordinate and collating the experimental data, draw a straight line through the points of $\ln f(\tau)$. Its intercept is $\ln(a)$, and the slope is b. Figure 17.4 is the experimental value of the concrete of Gongzui gravity dam. It is clear that the testing points are on one line, which illustrates that formula (17.18) is reasonable. The resulting formula is

$$E(\tau) = 38500[1 - \exp(-0.402\tau^{0.385})]\,\text{MPa}$$

The author had proposed the modified logarithmic formula for the modulus of elasticity of concrete as follows:

$$E(\tau) = c\ln(\tau^b + 1) \tag{17.18a}$$

in which c and b are two constants to be determined by test data. The formula agrees well with experimental data.

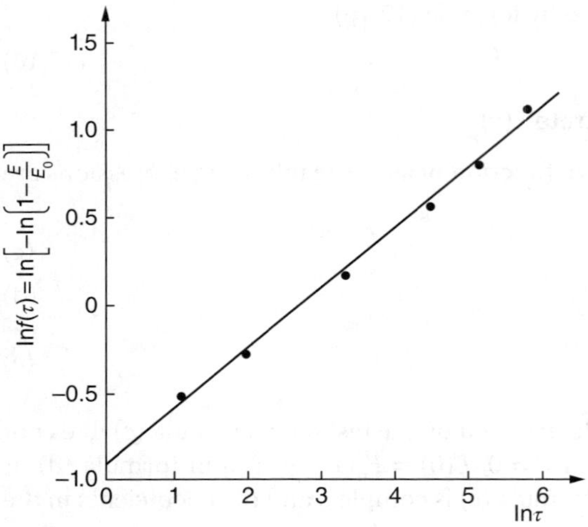

Figure 17.4 The modulus of elasticity of the concrete of Gongzui gravity dam.

17.1.4 Unit Creep of Concrete

According to the test data, the creep deformation of concrete can be divided into two parts – one part is reversible, and the other part is not reversible – so

$$C(t,\tau) = C_1(t,\tau) + C_2(t,\tau) \tag{17.19}$$

where $C_1(t,\tau)$ is reversible creep, that is, at age τ_1 the creep deformation caused by the stress loading increment $\Delta\sigma$ and the stress unloading increment $-\Delta\sigma$ has the same absolute values, and the signs are opposite. $C_2(t,\tau)$ is irreversible, that is, after unloading the creep deformation cannot be restored. In order to reflect the creep deformation of the two parts and save memory in finite element calculation, the author proposed the following expression:

$$C(t,\tau) = \sum_{j=1}^{m} \phi_j(\tau)[1 - e^{-r_j(t-\tau)}] \tag{17.20}$$

For reversible creep, take

$$\phi_j(\tau) = f_j + g_j\tau^{-p_j} \tag{17.21}$$

For irreversible creep, take

$$\phi_j(\tau) = f_j\tau^{-r_j\tau} \tag{17.22}$$

It is obvious that the creep deformation calculated by formula (17.21) is reversible. Next we explain why the creep deformation calculated by formula (17.22) is irreversible. By substitution of formula (17.22) into formula (17.20), we get

$$C_2(t,\tau) = f_j\tau^{-r_j\tau}[1 - e^{-r_j(t-\tau)}] = f_j(e^{-r_j\tau} - e^{-r_jt})$$

If ε_2 represents an irreversible creep deformation, loading at age τ_1, the stress is σ, and the value of ε_2 is

$$\varepsilon_2 = \sigma f_j(e^{-r_j\tau_1} - e^{-r_jt})$$

If removing the load at age τ_2, namely, applying $\Delta\sigma = -\sigma$ at $t = \tau_2$, the deformation when $t > \tau_2$ is

$$\begin{aligned}\varepsilon_2 &= \sigma f_j(e^{-r_j\tau_1} - e^{-r_jt}) + (-\sigma)f_j(e^{-r_j\tau_2} - e^{-r_jt})\\ &= \sigma f_j(e^{-r_j\tau_1} - e^{-r_j\tau_2}) = \text{constant}\end{aligned}$$

As shown in Figure 17.5, ε_2 was unable to recover after unloading and ε_1 is recoverable calculated by formula (17.21). According to the experiment data, in the creep of concrete, reversible creep is about 70–80%. Usually in formula (17.20), reversible creep has two items, and irreversible creep has one item, so there are a total of three items.

How the parameters of formula (17.20) are determined according to test data is explained below. Assume that the creep deformation observed in the test is

$$C'(t,\tau) = C_1'(t,\tau) + C_2'(t,\tau)$$

The parameters of $C_1(t,\tau)$ and $C_2(t,\tau)$ can be determined, respectively, and the methods are similar. Next how to get the parameters of $C_1(t,\tau)$ is explained. Assume that the error of the calculated value and the measured value is

$$Q = C_1(t,\tau) - C_1'(t,\tau)$$

Each point has an error, and F is the sum of squares of error of all observations, that is,

$$F = \sum Q^2$$

For example, taking two large items in $C_1(t,\tau)$ and letting $x_1 = f_1, x_2 = g_1, x_3 = p_1, x_4 = r_1, x_5 = f_2, x_6 = g_2, x_7 = p_2, x_8 = r_2$, then

$$C_1(t,\tau) = (x_1 + x_2\tau^{-x_3})[1 - e^{-x_4(t-\tau)}] + (x_5 + x_6\tau^{-x_7})[1 - e^{-x_8(t-\tau)}]$$

So $F(x)$ is the function of $\{x\} = \begin{bmatrix}x_1 & x_2 & \dots & x_8\end{bmatrix}^T$:

$$F(x) = \sum Q^2 = \sum [C_1(t,\tau) - C'(t,\tau)]^2 \tag{17.23}$$

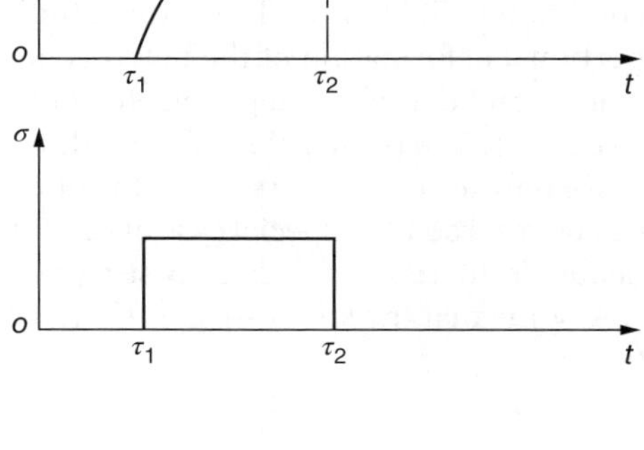

Figure 17.5 The reversible and irreversible creep.

We select $\{x\}$ in this way, and let

$$\left.\begin{aligned} F(x) &\to \text{minimum} \\ \text{subjected to}\quad x_i &\geq 0 \quad (i = 1 \sim 8) \end{aligned}\right\} \tag{17.24}$$

Formula (17.24) is a constrained optimization problem that can be solved by nonlinear programming methods. For example, as the dependent variable is not many, the simple complex method can be used to solve it.

17.1.5 Formula for Preliminary Design

The work of creep test for concrete is big and the test lasts for a long time, so a set of formulas are needed for application in the case of lack of experimental data. After analyzing a lot of test data for hydraulic mass concrete, the author proposed the calculation formulas of modulus of elasticity and unit creep for the preliminary design as follows:

$$E(\tau) = E_0(1 - e^{-0.40\tau^{0.34}}) \tag{17.25}$$

$$\begin{aligned} C(t,\tau) &= C_1(1 + 9.20\tau^{-0.45})[1 - e^{-0.30(t-\tau)}] \\ &\quad + C_2(1 + 1.70\tau^{-0.45})[1 - e^{-0.005(t-\tau)}] \end{aligned} \tag{17.26}$$

where

$$C_1 = \frac{0.23}{E_0}, \quad C_2 = \frac{0.52}{E_0}$$

$$E_0 = 1.05E(360) \quad \text{or} \quad E_0 = 1.20E(90)$$

in which $E(90)$ and $E(360)$ are the instantaneous moduli of elasticity at age of 90 and 360 days, respectively.

17.2 Influence of Creep on Stresses and Deformations of Linear Elastocreeping Body

In engineering, we are concerned about the influence of creep on the stresses and deformations of the structure. This problem is closely related to boundary conditions of structures. The boundary conditions appearing in engineering are summarized in Table 17.1. Alfrey and Qian Xuesen researched the first and second cases in the Table 17.1 and proposed two theorems. In practical engineering, nonhomogeneous structures and mixed boundary conditions often appear, such as the third to the sixth condition in Table 17.1. The author studied this issue, given the conditions for the proportional deformation of the heterogeneous structures, and proved the two theorems, which can be used to judge the effect of creep on the stresses and deformations of the structures. The principal results are described below.

We only have to discuss the sixth condition in Table 17.1, and all other circumstances can be regarded as a special case of it. Without loss of generality, consider the inhomogeneous elastocreeping body consisting of two different materials, as shown in Figure 17.6.

Assume that the stress–strain relation of the material meets the two following conditions:

Table 17.1 The boundary conditions of structures.

Boundary condition	The first kind	The second kind	The mixed condition
Homogeneous body	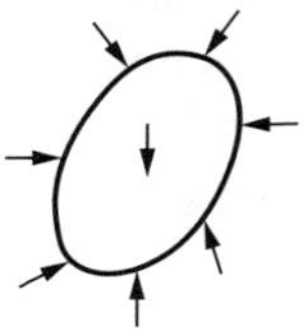	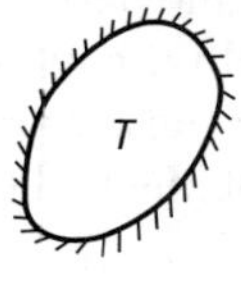	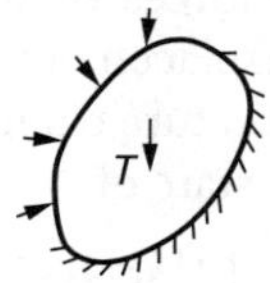
Heterogeneous body	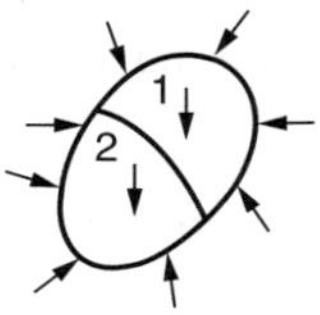	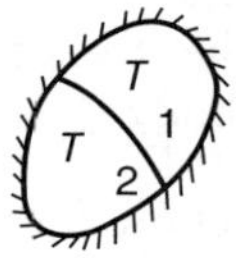	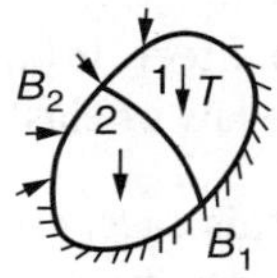

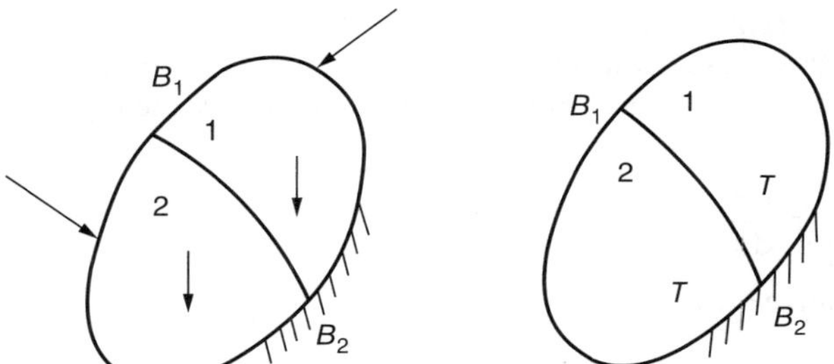

(a) (b)

Figure 17.6 Inhomogeneous elastocreeping body under mixed boundary conditions. (a) Under the action of the load and the body force. (b) Under the action of temperature and forced displacement on the boundary.

1) Poisson's ratio of the creep is constant and is equal to that of the instantaneous elastic strain, which means that at any moment the ratio of the shearing strain to the volumetric strain of the material keeps the same value.
2) The deformation characteristics of the two regions are in accordance with the following relation:

$$E_1(\tau)C_1(t,\tau) = E_2(\tau)C_2(t,\tau) \tag{17.27}$$

or

$$C_1(t,\tau) : C_2(t,\tau) = \frac{1}{E_1(\tau)} : \frac{1}{E_2(\tau)}$$

where subscripts 1 and 2 represent regions 1 and 2.

Formula (17.27) indicates that creep deformations of the two regions are proportional to the corresponding elastic deformations.

The two conditions above are called "proportional deformation condition."

The author has proved the two following theorems [6].

Theorem 17.1 For the homogeneous body whose Poisson's ratio of creep is constant and is equal to the elastic Poisson's ratio or the nonhomogeneous body meeting the proportional deformation condition, the boundary conditions are shown in Figure 17.6(a). Some forces are applied on one part of the border and on the other part of the border, the displacements are zero. In the interior of the body, there are volume forces but no temperature changes. Under the action of the boundary force and the volume force, the stress state of the body is not affected by creep, namely,

$$\{\sigma(x,y,z,t)\} = \{\sigma^{e}(x,y,z,t)\} \tag{a}$$

where $\{\sigma^{e}(x,y,z,t)\}$ is the stress calculated according to elastic body without considering creep effect. $\{\sigma(x,y,z,t)\}$ is the stress calculated taking account of creep and according to the law of elastocreeping body.

In this case, creep has effect on the strain. If the load is applied at $t = t_0$ and remains the same afterward, the strain of any point should be

$$\{\varepsilon(x,y,z,t)\} = \{\Delta\sigma_0(x,y,z)\}J(t,t_0) \tag{b}$$

where $\{\Delta\sigma_0(x,y,z)\}$ is the stress increment caused by external loads at $t = t_0$.

If the load changes with time, the stress components are not affected by creep, so the relationship between any strain component and time can be calculated by the formula under uniaxial stress as follows:

$$\varepsilon(t) = \Delta\sigma_0 J(t,t_0) + \int_{t_0}^{t} J(t,\tau)\frac{d\sigma}{d\tau}d\tau$$

where $J(t,\tau)$ is the creep compliance given by (17.2).

Theorem 17.2 For the homogeneous body whose Poisson's ratio of creep is constant and is equal to the elastic Poisson's ratio or the nonhomogeneous body meeting the proportional deformation condition, the boundary conditions are shown in Figure 17.6(b). One part of the border is free, and on another part, the displacement is given. In the interior of the body, there is no volume force, but there are temperature changes. Under the action of temperature changes and forced displacement on the border, the strain state of objects is not affected by creep, namely,

$$\{\varepsilon(x,y,z,t)\} = \{\varepsilon^{e}(x,y,z,t)\} \tag{c}$$

where $\{\varepsilon^{e}(x,y,z,t)\}$ is the strain calculated according to elastic body without considering creep effect. $\{\varepsilon(x,y,z,t)\}$ is the strain calculated taking account of creep and according to the law of elastocreeping body.

In this case, creep has an impact on the stress. If the temperature changes and the forced displacements on the boundary are applied at $t = t_0$ and they do not change with time, then the stress of the body is

$$\{\sigma(x,y,z,t)\} = \{\Delta\sigma_0(x,y,z)\}R(t,t_0) \tag{d}$$

where $\{\Delta\sigma_0(x,y,z)\}$ is the stress increment caused by the temperature variation and the forced displacement on the border when $t = t_0$ and $R(t,t_0)$ is the relaxation coefficient.

If the temperature and the forced displacements on the border change with time, the strain components are not affected by creep, so the stress components can be calculated according to the elastic strain components and stress relaxation coefficient as follows:

$$\sigma(t) = E(t_0)\Delta\varepsilon(t_0)R(t, t_0) + \int_{t_0}^{t} R(t, \tau)E(\tau)\frac{\mathrm{d}\varepsilon(\tau)}{\mathrm{d}\tau}\mathrm{d}\tau$$

or

$$\sigma(t) = \Delta\sigma_0^{\mathrm{e}}R(t, t_0) + \int_{t_0}^{t} R(t, \tau)\frac{\mathrm{d}\sigma^{\mathrm{e}}}{\mathrm{d}\tau}\mathrm{d}\tau \tag{e}$$

where σ^{e} is the stress calculated according to elastic body without considering the creep.

The two theorem above can be summarized as follows: for the homogeneous body whose Poisson's ratio of creep is constant and is equal to elastic Poisson's ratio or the nonhomogeneous body meeting the proportional deformation conditions, the stress caused by forces is not affected by creep, but the stress caused by temperature changes and forced displacement of boundary is affected by creep and can be calculated by relaxation coefficient as formula (17.9).

For the homogeneous body whose Poisson's ratio of creep is not equal to Poisson's ratio of elastic strain or the nonhomogeneous body that does not meet proportional deformation conditions, the stresses and strains caused by external forces, temperature changes, and forced displacements of boundary will be influenced by creep, and the effect is difficult to calculate by simple methods. Generally the finite element method is used.

17.3 Analysis of Elastocreeping Stresses of Concrete Structure

When calculating the thermal stress and construction stress of concrete structures, the effect of creep and age of concrete must be taken into account. Time t is usually divided into a series of intervals, a set of equilibrium equations is established for each time interval. By formula (17.3), the deformation of concrete structures is related to the whole stress history. In order to calculate concrete deformation at moment t, stress values of all elements at different moments must be stored. For large and medium structures, great store capacity is needed, and the internal memory is not enough, so outside memory is used, and much calculation time will be consumed. So how to reduce the storage volume of stress history is the key of creep stress analysis of concrete structures. Zienkiewicz and Watson took use of the characteristics of exponential function and proposed an explicit solution of equal intervals of time and established the recurrence relation. It is not necessary to store the stress history. The author improved the calculation method, the explicit solution of variable steps are given in the literature [9], and the implicit solution of variable time intervals is given in the literature [5]. This calculation method has been widely used. It is explained in details below.

17.3.1 The Calculation of Strain Increment under Uniaxial Stress

By formulas (17.2) and (17.3), it is known that the strain of concrete can be divided into two parts: elastic strain and creep, namely,

$$\varepsilon(t) = \varepsilon^{\mathrm{e}}(t) + \varepsilon^{\mathrm{c}}(t)$$

$$\varepsilon^{\mathrm{e}}(t) = \frac{\Delta\sigma_0}{E(t_0)} + \int_{t_0}^{t} \frac{1}{E(\tau)}\frac{\mathrm{d}\sigma}{\mathrm{d}\tau}\mathrm{d}\tau \tag{17.28}$$

$$\varepsilon^{\mathrm{c}}(t) = \Delta\sigma_0 C(t, t_0) + \int_{t_0}^{t} C(t, \tau)\frac{\mathrm{d}\sigma}{\mathrm{d}\tau}\mathrm{d}\tau \tag{17.29}$$

where $\varepsilon^{\mathrm{e}}(t)$ is instantaneous elastic strain and $\varepsilon^{\mathrm{c}}(t)$ is creep strain. The time t is divided into a series of time intervals: $\Delta t_1, \Delta t_2, \ldots, \Delta t_n, \ldots$, in which $\Delta t_n = t_n - t_{n-1}$ and the strain increment in each time interval is calculated.

17.3.1.1 The Elastic Strain Increment

By formula (17.28), the elastic strain increment is

$$\Delta\varepsilon_n^{\mathrm{e}} = \varepsilon^{\mathrm{e}}(t_n) - \varepsilon^{\mathrm{e}}(t_{n-1}) = \int_{t_{n-1}}^{t_n} \frac{1}{E(\tau)}\frac{\mathrm{d}\sigma}{\mathrm{d}\tau}\mathrm{d}\tau$$

Taking the mid age

$$t_{n-0.5} = \frac{t_n + t_{n-1}}{2} = t_n - 0.5\Delta t_n$$

when Δt_n is small enough, using the mean value theorem, $E(\tau)$ in the integration sign above can be replaced by the elastic modulus $E(t_{n-0.5})$ at the mid age. Then we get

$$\Delta\varepsilon_n^{\mathrm{e}} = \frac{1}{E(t_{n-0.5})}\int_{t_{n-1}}^{t_n} \frac{\mathrm{d}\sigma}{\mathrm{d}\tau}\mathrm{d}\tau = \frac{1}{E(t_{n-0.5})}\Delta\sigma_n \tag{17.30}$$

17.3.1.2 The Increment of Creep Strain When $C(t, \tau) = \phi(\tau)[1 - \mathrm{e}^{-r(t-\tau)}]$

By formula (17.28), the increment of creep strain is

$$\Delta\varepsilon_n^{\mathrm{c}} = \varepsilon^{\mathrm{c}}(t_n) - \varepsilon^{\mathrm{c}}(t_{n-1}) = \int_{t_{n-1}}^{t_n} C(t, \tau)\frac{\mathrm{d}\sigma}{\mathrm{d}\tau}\mathrm{d}\tau$$

In the integral sign of the right side of the formula above, the age τ is the integration variable. When Δt_n is small enough, replacing $C(t, \tau)$ in the integral sign with unit creep $C(t, t_{n-0.5})$ at the mid age, then

$$\Delta\varepsilon_n^{\mathrm{c}} = C(t, t_{n-0.5})\int_{t_{n-1}}^{t_n} \frac{\mathrm{d}\sigma}{\mathrm{d}\tau}\mathrm{d}\tau = C(t, t_{n-0.5})\Delta\sigma_n \tag{17.31}$$

As shown in Figure 17.7, taking three adjacent moments t_{n-1}, t_n, t_{n+1}, the time intervals are $\Delta t_n = t_n - t_{n-1}, \Delta t_{n+1} = t_{n+1} - t_n$. By equation (17.31), the creep deformation at the

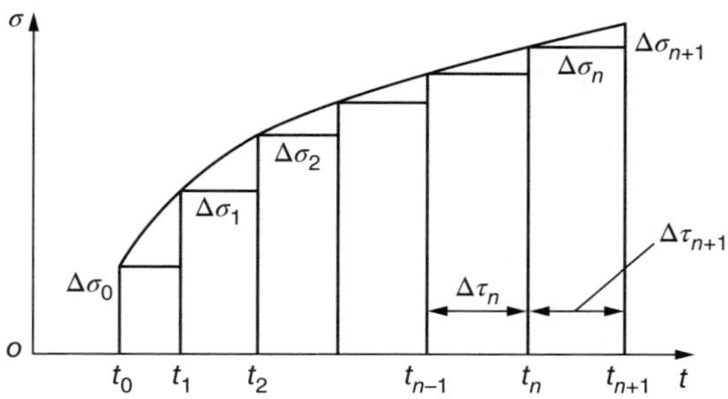

Figure 17.7 The increment of the stress.

three adjacent moments are, respectively,

$$\varepsilon^{c}(t_{n-1}) = \Delta\sigma_0 C(t_{n-1}, t_0) + \Delta\sigma_1 C(t_{n-1}, t_{1-0.5}) + \cdots + \Delta\sigma_{n-1} C(t_{n-1}, t_{n-0.5}) \tag{a}$$

$$\begin{aligned}\varepsilon^{c}(t_n) &= \Delta\sigma_0 C(t_n, t_0) + \Delta\sigma_1 C(t_n, t_{1-0.5}) + \cdots \\ &\quad + \Delta\sigma_{n-1} C(t_n, t_{n-1-0.5}) + \Delta\sigma_n C(t_n, t_{n-0.5})\end{aligned} \tag{b}$$

$$\begin{aligned}\varepsilon^{c}(t_{n+1}) &= \Delta\sigma_0 C(t_{n+1}, t_0) + \Delta\sigma_1 C(t_{n+1}, t_{1-0.5}) + \cdots \\ &\quad + \Delta\sigma_{n-1} C(t_{n+1}, t_{n-1-0.5}) + \Delta\sigma_n C(t_{n+1}, t_{n-0.5}) \\ &\quad + \Delta\sigma_{n+1} C(t_{n+1}, t_{n+1-0.5})\end{aligned} \tag{c}$$

Now considering a simpler case, only take one term in formula (17.20), that is,

$$C(t, \tau) = \phi(\tau)[1 - \mathrm{e}^{-r(t-\tau)}] \tag{d}$$

Subtracting formula (b) from formula (c), we get

$$\Delta\varepsilon^{c}_{n+1} = (1 - \mathrm{e}^{-r\Delta t_{n+1}})\omega_{n+1} + \Delta\sigma_{n+1} C(t_{n+1}, t_{n+1-0.5}) \tag{e}$$

By formula (d)

$$\begin{aligned}&C(t_{n+1}, t_{i-0.5}) - C(t_n, t_{i-0.5}) \\ &\quad = \phi(t_{i-0.5})[1 - \mathrm{e}^{-r(t_{n+1}-t_{i-0.5})}] - \phi(t_{i-0.5})[1 - \mathrm{e}^{-r(t_n-t_{i-0.5})}] \\ &\quad = \phi(t_{i-0.5})[\mathrm{e}^{-r(t_n-t_{i-0.5})} - \mathrm{e}^{-r(t_n+\Delta t_{n+1}-t_{i-0.5})}] \\ &\quad = \phi(t_{i-0.5})\mathrm{e}^{-r(t_n-t_{i-0.5})}[1 - \mathrm{e}^{-r\Delta t_{n+1}}]\end{aligned} \tag{f}$$

By substitution of formula (f) into formula (e), we get

$$\Delta\varepsilon^{c}_{n+1} = (1 - \mathrm{e}^{-r\Delta t_{n+1}})\omega_{n+1} + \Delta\sigma_{n+1} C(t_{n+1}, t_{n+1-0.5}) \tag{17.32}$$

where

$$\begin{aligned}\omega_{n+1} &= \Delta\sigma_0\phi(t_0)\mathrm{e}^{-r(t_n-t_0)} + \Delta\sigma_1\phi(t_{1-0.5})\mathrm{e}^{-r(t_n-t_{1-0.5})} \\ &\quad + \cdots + \Delta\sigma_{n-1}\phi(t_{n-1-0.5})\mathrm{e}^{-r(t_n-t_{n-1-0.5})} \\ &\quad + \Delta\sigma_n\phi(t_{n-0.5})\mathrm{e}^{-r(t_n-t_{n-0.5})}\end{aligned} \tag{g}$$

Similarly, subtracting formulas (a) from (b), we get

$$\Delta\varepsilon^{c}_{n} = \varepsilon^{c}(t_n) - \varepsilon^{c}(t_{n-1}) = (1 - \mathrm{e}^{-r\Delta t_n})\omega_n + \Delta\sigma_n C(t_n, t_{n-0.5}) \tag{17.32a}$$

in which

$$\omega_n = \Delta\sigma_0\phi(t_0)e^{-r(t_{n-1}-t_0)} + \Delta\sigma_1\phi(t_{1-0.5})e^{-r(t_{n-1}-t_{1-0.5})} + \cdots + \Delta\sigma_{n-1}\phi(t_{n-1-0.5})e^{-r(t_{n-1}-t_{n-1-0.5})} \tag{h}$$

Comparing formulas (g) and (h), it is clear that

$$\omega_{n+1} = \omega_n e^{-r\Delta t_n} + \Delta\sigma_n\phi(t_{n-0.5})e^{-0.5r\Delta t_n} \tag{17.33}$$

and

$$\omega_1 = \Delta\sigma_0\phi(t_0)$$

Formulas (17.32) and (h) form a group of recurrence formulas, by means of which it is not necessary to record the history of stress, only storing ω_n, and the increment of creep deformation can be calculated.

17.3.1.3 The Increment of Creep Strain When $C(t, \tau) = \sum \phi_j(\tau)[1 - e^{-r_j(t-\tau)}]$

By substitution of formula (17.20), which is the general expression of creep of concrete into formula (f),

$$C(t_{n+1}, t_{i-0.5}) - C(t_n, t_{i-0.5}) = \sum_{j=1}^{m} \phi_j(t_{i-0.5})e^{-r_j(t_n-t_{i-0.5})}(1 - e^{-r_j\Delta t_{n+1}}) \tag{i}$$

By substitut ion of formula (i) into formula (e),

$$\Delta\varepsilon^{c}_{n+1} = \sum_{j=1}^{m}(1 - e^{-r_j\Delta t_{n+1}})\omega_{j,n+1} + \Delta\sigma_{n+1}C(t_{n+1}, t_{n+1-0.5}) \tag{j}$$

in which

$$\begin{aligned} \omega_{j,n+1} &= \omega_{jn}e^{-r_j\Delta t_n} + \Delta\sigma_n\phi_j(t_{n-0.5})e^{-0.5r_j\Delta t_n} \\ \omega_{j1} &= \Delta\sigma_0\phi_j(t_0) \end{aligned} \tag{k}$$

Therefore, the increment of concrete creep deformation can be calculated as follows:

$$\Delta\varepsilon^{c}_n = \varepsilon^{c}(t_n) - \varepsilon^{c}(t_{n-1}) = \eta_n + q_n\Delta\sigma_n \tag{17.34}$$

$$\left.\begin{aligned} \eta_n &= \sum_{j=1}^{m}(1 - e^{-r_j\Delta t_n})\omega_{jn} \\ q_n &= C(t_n - t_{n-0.5}) \end{aligned}\right\} \tag{17.35}$$

and ω_{jn} can be calculated by the following recursive formula:

$$\begin{aligned} \omega_{jn} &= \omega_{j,n-1}e^{-r_j\Delta t_{n-1}} + \Delta\sigma_{n-1}\phi_j(t_{n-1-0.5})e^{-0.5r_j\Delta t_{n-1}} \\ \omega_{j1} &= \Delta\sigma_0\phi_j(t_0) \end{aligned} \tag{17.36}$$

17.3.2 The Calculation of Strain Increments under Complex Stress Conditions

Taking spatial problems, for example, the stress increment is

$$\{\Delta\sigma\} = \left[\Delta\sigma_x\ \Delta\sigma_y\ \Delta\sigma_z\ \Delta\tau_{xy}\ \Delta\tau_{yz}\ \Delta\tau_{zx}\right]^{\mathrm{T}} \tag{l}$$

The strain increment is

$$\{\Delta\varepsilon\} = \left[\Delta\varepsilon_x\ \Delta\varepsilon_y\ \Delta\varepsilon_z\ \Delta\gamma_{xy}\ \Delta\gamma_{yz}\ \Delta\gamma_{zx}\right]^{\mathrm{T}} \tag{m}$$

By formulas (17.12) and (17.30), the increment of the elastic strain is

$$\{\Delta\varepsilon_n^{\mathrm{e}}\} = \frac{1}{E(t_{n-0.5})}[A]\{\Delta\sigma_n\} \tag{17.37}$$

By formulas (17.12) and (17.34), the increment of creep strain is

$$\{\Delta\varepsilon_n^{\mathrm{c}}\} = \{\eta_n\} + q_n[A]\{\Delta\sigma_n\} \tag{17.38}$$

where

$$q_n = C(t_n, t_{n-0.5})$$

$$\{\eta_n\} = \sum_{j=1}^{m}(1 - \mathrm{e}^{-r_j\Delta t_n})\{\omega_{jn}\} \tag{17.39}$$

$$\{\omega_{jn}\} = \{\omega_{j,n-1}\}\mathrm{e}^{-r_j\Delta t_{n-1}} + [A]\{\Delta\sigma_{n-1}\}\phi_j(t_{n-1-0.5})\mathrm{e}^{-0.5r_j\Delta t_{n-1}} \tag{17.40}$$

$$\{\omega_{j1}\} = [A]\{\Delta\sigma_0\}\phi_j(t_0)$$

Under complicated stress state, $\{\eta\}$ and $\{\omega\}$ are vectors; for example, for spatial problems, there are

$$\{\eta\} = \left[\eta_x\ \eta_y\ \eta_z\ \eta_{xy}\ \eta_{yz}\ \eta_{zx}\right]^{\mathrm{T}} \tag{n}$$

$$\{\omega\} = \left[\omega_x\ \omega_y\ \omega_z\ \omega_{xy}\ \omega_{yz}\ \omega_{zx}\right]^{\mathrm{T}} \tag{o}$$

17.3.3 Equilibrium Equations

The strain increments include the increments of elastic strain, creep strain, and temperature strain:

$$\{\Delta\varepsilon_n\} = \{\Delta\varepsilon_n^{\mathrm{e}}\} + \{\Delta\varepsilon_n^{\mathrm{c}}\} + \{\Delta\varepsilon_n^{\mathrm{I}}\} \tag{17.41}$$

where $\{\Delta\varepsilon_n\}$ is the array of the strain increment, $\{\Delta\varepsilon_n^{\mathrm{e}}\}$ is the array of the elastic strain increment, $\{\Delta\varepsilon_n^{\mathrm{c}}\}$ is the array of the creep strain increment, and $\{\Delta\varepsilon_n^{\mathrm{I}}\}$ is the array of the temperature strain increment.

By formulas (m) and (o), the array of the stress increment can be calculated as follows:

$$\{\Delta\sigma_n\} = [D_n]\{\Delta\varepsilon_n^{\mathrm{e}}\} = [D_n](\{\Delta\varepsilon_n\} - \{\Delta\varepsilon_n^{\mathrm{c}}\} - \{\Delta\varepsilon_n^{\mathrm{I}}\}) \tag{17.42}$$

where

$$[D_n] = E(t_{n-0.5})[A]^{-1} \tag{17.43}$$

in which $[D_n]$ is the elasticity matrix at the mid age.

By substitution of $\{\Delta\varepsilon_n\} = [B]\{\Delta\delta_n\}$ and formula (17.37) into formula (17.41), we get

$$\{\Delta\sigma_n\} = [\overline{D}_n]([B]\{\Delta\delta_n\} - \{\eta_n\} - \{\Delta\varepsilon_n^{\mathrm{I}}\}) \tag{17.44}$$

where

$$[\overline{D}_n]=([I] + q_n[D_n][A])^{-1}[D_n]=\frac{1}{1 + q_nE(t_{n-0.5})}[D_n]=\frac{E(t_{n-0.5})}{1 + q_nE(t_{n-0.5})}[A]^{-1} \tag{17.45}$$

The equilibrium equation of the finite element method is

$$\int [B]^{\mathrm{T}}\{\Delta\sigma_n\}\mathrm{d}V = \{\Delta P_n\} \tag{17.46}$$

where $\{\Delta P_n\}$ is the increment of the external load.

By substitution of formula (17.43) into the formula above, the basic equation of creep analysis of concrete structures is obtained as follows:

$$[K_n]\{\Delta\delta_n\} = \{\Delta P_n\} + \{\Delta P_n^{\mathrm{c}}\} + \{\Delta P_n^{\mathrm{I}}\} \tag{17.47}$$

in which

$$\left.\begin{aligned} [K_n] &= \int [B]^{\mathrm{T}}[\overline{D}_n][B]\mathrm{d}V \\ \{\Delta P_n^{\mathrm{c}}\} &= \int [B]^{\mathrm{T}}[\overline{D}_n][\eta_n]\mathrm{d}V \\ \{\Delta P_n^{\mathrm{I}}\} &= \int [B]^{\mathrm{T}}[\overline{D}_n][\Delta\varepsilon_n^{\mathrm{I}}]\mathrm{d}V \end{aligned}\right\} \tag{17.48}$$

where $[K_n]$ is the stiffness matrix of the structures, $\{\Delta P_n^{\mathrm{c}}\}$ is the equivalent load increment due to creep deformation, and $\{\Delta P_n^{\mathrm{I}}\}$ is the load increment caused by thermal deformation.

By formula (17.47), the displacement increment $\{\Delta\delta_n\}$ can be solved. Substituting it into formula (17.44), then the stress increment can be obtained.

17.4 Compound Layer Element for the Simulation Analysis of Concrete Dams

The massive concrete structures such as concrete dams are often poured in layers. The pouring process goes through several summers and winters, and the construction process has an important influence on dam temperature field and stress field. Therefore, in order to control the temperature field and stress field of concrete dam properly, the process of dam construction must be simulated in the simulation analysis. In 1972, the author and Song Jingting developed the first finite element program for the temperature stress and creep stress of concrete in China. In 1973, this program was used to analyze the temperature stress of Sanmenxia gravity dam, and the actual construction process of the dam was simulated in the calculation. Because concrete dam is higher and higher, the stress is higher and higher, so the simulation analysis of concrete dams is receiving more and more attention. In the simulation calculation of concrete dams, the following problems must be taken into account:

1) The pouring layers are too many. Take a concrete dam of 300 m high, for example, if the thickness of the pouring layer is 1.5 m, there are 200 layers. The stress gradient and temperature gradient in the direction of thickness of the concrete layer in early age are large. Generally a pouring layer should be divided into five layers of finite elements; there are 1000 layers of elements in each dam section. If a layer is divided into 100 nodes, there will be 100,000 nodes only in a dam section. Adding the foundation, the number of nodes will be more. As roller-compacted concrete dam, because the thickness of the pouring layer is only 0.30 m, strictly simulating the construction process, more elements and nodes will be required.
2) The time increment is short. For early-age concrete, because the modulus of elasticity, unit creep, and adiabatic temperature rise change rapidly with age, shorter time increment is required to guarantee the accuracy. For example, taking the time increment $\tau = 0.5$ days, there are 730 steps in 1 year. If the construction time is 3 years, there are a total of 2119 steps. If considering the influence of sunshine and the daily temperature variations, time intervals should be reduced to 1–2 h. If the construction time is 3 years, there are 13,000–26,000 steps. Two very large linear systems of equations (one for temperature field and one for stress field) must be solved in each time interval.
3) The calculation of cooling water pipes is difficult. The radius of the cooling water pipe is 1–2 cm; intensive computing grid must be taken around the water pipes. The size of the element should be 1–3 cm, although the size outward can be larger gradually, yet the total number of elements is large. For example, if the distance between pipes is 1.5 m × 1.5 m, using the linear element, 90 nodes are needed. If the dimension of the pouring layer is 18 m × 1.5 m × 30 m, taking the spacing as 3 m in the direction of the length, there are 12,000 nodes in a pouring layer, and if the dam is 300 m high, there are 2.4 million nodes in a dam section. As far as the current level of computer hardware is concerned, in fact, it is difficult to compute.

The author proposed a series of new methods, including compound layer method, joint elements for compound layer method, the method of different time increments in different regions for the stress field, the method of different time increments in different regions for the temperature field, the equivalent heat conduction equation for pipe cooling, and so on. They overcame the difficulties mentioned above, and the efficiency and precision of calculation are significantly improved, referring to literature [21–27]. Next we focus on explaining the compound layer method for analyzing the stress.

The basic principle of the finite element mesh is that the element size must adapt to the changes of the temperature gradient and the stress gradient. For example, using the linear element, the variation of the temperature and stress in the element are linear. When dividing the grid, this feature must be fully taken into account.

As shown in Figure 17.8, when the dam gradually rises, the dam is divided into four zones from top to bottom. In the new pouring area R_1, the changes of temperature and stress in the direction of the thickness are relatively fast. Each pouring

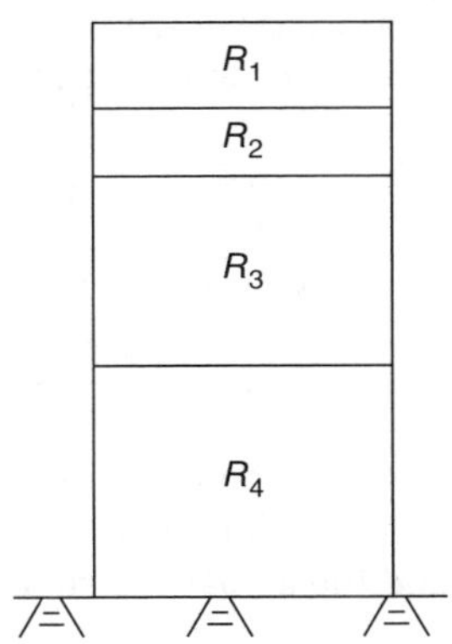

Figure 17.8 Dam section.

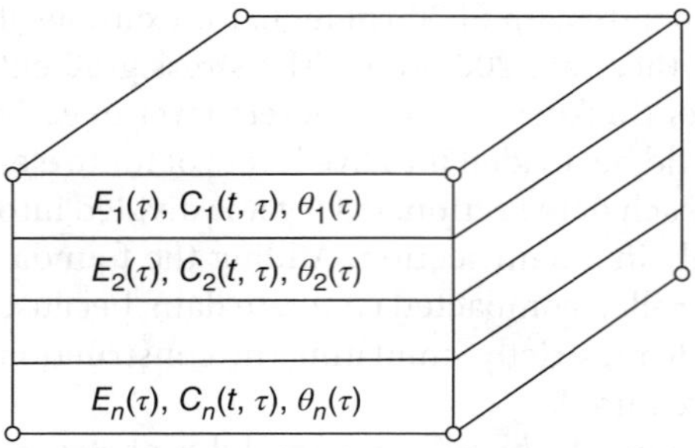

Figure 17.9 Compound layer element.

layer shall be divided into n layers of finite elements, such as $n = 3–5$. In the region R_2 under R_1, in each pouring layer, the increment of temperature and stress in the direction of thickness is linear, and n layers of finite elements can be merged into one layer of finite element in each original pouring layer, namely, a pouring layer is a layer of finite element. In the region R_3, the changes of the temperature increment and strain increment along vertical direction are further flat, and several pouring layers can be merged into one compound layer element of a dam section. Each pouring layer retains the mechanical characteristics and thermal characteristics, respectively, within the compound element as shown in Figure 17.9. In the area R_4, at a later stage, several pouring layers can be merged into one homogeneous compound layer element. At this moment, the mechanical and thermal characteristics of each pouring layer are fully close. We may take an average value of them, as shown in Figure 17.10.

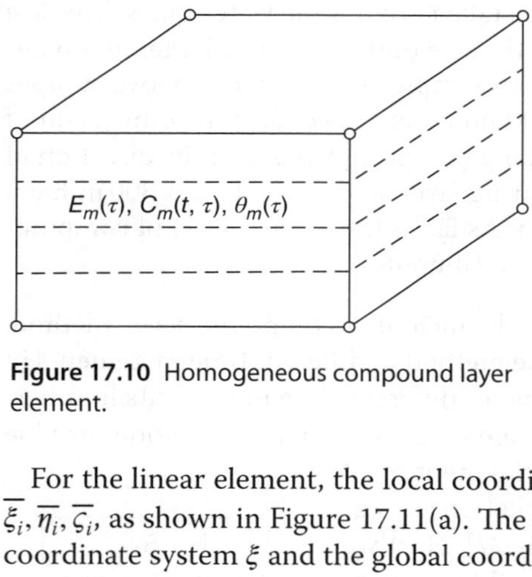

Figure 17.10 Homogeneous compound layer element.

After merging layers, within the compound layer element, each pouring layer has the original modulus of elasticity $E_i(\tau)$, creep $C_i(t, \tau)$, and adiabatic temperature rise $\theta_i(\tau)$, taking concrete as elastocreeping body. Using the finite element incremental method for analysis, the basic equations in the nth time increment Δt_n is formula (17.43) to formula (17.47).

As shown in Figure 17.11, there are several layers of different media within the compound element, and before calculation, the coordinate transformation must be solved within the element.

For the linear element, the local coordinate system of the ith layer before merging is $\overline{\xi}_i, \overline{\eta}_i, \overline{\varsigma}_i$, as shown in Figure 17.11(a). The corresponding relationship between the local coordinate system ξ and the global coordinate system z is as follows:

$$\overline{\xi} = +1 \rightarrow z_{i+1}, \quad \overline{\xi}_i = -1 \rightarrow z_i, \quad \overline{\xi}_i = 0 \rightarrow \frac{z_{i+1} + z_i}{2}$$

The corresponding z to arbitrary $\overline{\varsigma}_i$ is

$$z(\overline{\varsigma}_i) = \frac{z_i + z_{i+1}}{2} + \frac{z_{i+1} - z_i}{2}\overline{\varsigma}_i \tag{17.49}$$

In the compound element, the local coordinate system is ξ, η, ς, as shown in Figure 17.11(b). The corresponding relationship between the local coordinate system and the global coordinate system is as follows:

$$\varsigma = +1 \rightarrow z_{i+1}, \quad \varsigma = -1 \rightarrow z_i, \quad \varsigma = 0 \rightarrow \frac{z_{i+1} + z_i}{2}$$

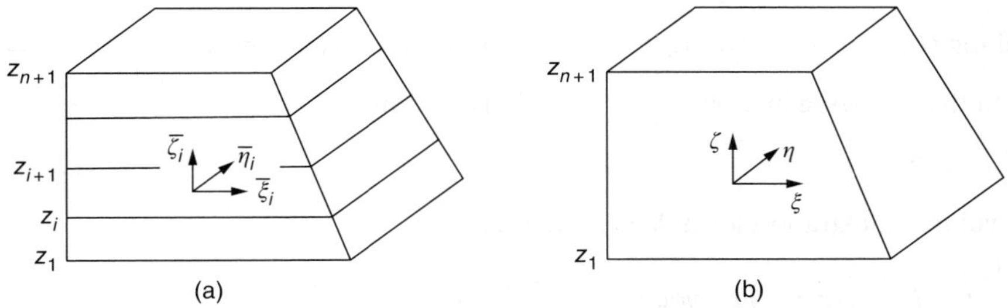

Figure 17.11 The coordinate transformation of the compound layer element. (a) The local coordinate system of each layer of the compound element. (b) The local coordinate system of the global element.

The corresponding z to arbitrary ς is

$$z(\varsigma) = \frac{z_1 + z_{n+1}}{2} + \frac{z_{n+1} - z_1}{2}\varsigma \tag{17.50}$$

That is,

$$\varsigma = \frac{2}{z_{n+1} - z_1}\left[z(\varsigma) - \frac{z_1 + z_{n+1}}{2}\right] \tag{17.51}$$

Let $z(\varsigma) = z(\overline{\varsigma}_i)$ in the formula above, and the relationship between ς and z is obtained as follows:

$$\varsigma = \frac{2}{z_{n+1} - z_1}\left[\frac{z_i + z_{i+1}}{2} + \frac{z_{i+1} - z_i}{2}\overline{\varsigma}_i - \frac{z_1 + z_{n+1}}{2}\right] \tag{17.52}$$

The value of ς corresponding to $\overline{\varsigma}_i$ of the original ith layer can be calculated by the formula above, while the values of ξ, η are the same with $\overline{\xi}_i, \overline{\eta}_i$.

Geometry matrix $[B]$ is divided into blocks as follows:

$$[B] = \begin{bmatrix} B_1 & B_2 & \ldots & B_i & \ldots & B_m \end{bmatrix}$$

Substituting it into formula (17.47), the submatrix is obtained:

$$[k_{rs}] = \int_{-1}^{1}\int_{-1}^{1}\int_{-1}^{1} G_{rs}(\xi, \eta, \varsigma)\mathrm{d}\xi\mathrm{d}\eta\mathrm{d}\varsigma \tag{17.53}$$

where

$$G_{rs}(\xi, \eta, \varsigma) = [B_r]^{\mathrm{T}}[\overline{D}][B_s]|J|$$

in which $[\overline{D}]$ is the equivalent elasticity matrix, $|J|$ is the determinant of the Jacobian matrix J, and the matrix $[B_r]$ contains ξ, η, ς.

Applying Gauss numerical integration to formula (17.53), the stiffness matrix can be calculated.

The basic formula for Gaussian integration is

$$\int_{-1}^{1} f(\xi)\mathrm{d}\varsigma = \sum H_i f(\xi_i) \tag{17.54}$$

Now taking $f(\xi) = 1$, substituting it into the formula above, and getting $\int_{-1}^{1} \mathrm{d}\xi = 2 = \sum H_i$, then the following important relationship is obtained:

$$\sum H_i = 2 \tag{17.55}$$

For the volume integration formula (17.53), there is

$$\int_{-1}^{1}\int_{-1}^{1}\int_{-1}^{1} G_{rs}(\xi,\eta,\varsigma)\mathrm{d}\xi\mathrm{d}\eta\mathrm{d}\varsigma = \sum_i\sum_j\sum_m G_{rs}(\xi_i,\eta_j,\varsigma_m)H_iH_jH_m$$

By formula (17.54), there is

$$\sum_i H_i = 2,\quad \sum_j H_j = 2,\quad \sum_m H_m = 2 \tag{17.56}$$

For three-dimensional linear element, take two integration points, respectively, in the direction of ξ, η, so $H_i = H_j = 1$. In the direction of ς, n layers are merged into one layer. There are a total of $2n$ points, so

$$\sum_{m=1}^{2n} H_m = H_1 + H_2 + \cdots + H_{2n} = 2$$

If the thickness of each layer is same, take

$$H_m = \frac{1}{n}$$

If the thickness of each layer is different, take

$$H_m = \frac{\Delta\varsigma_m}{2} = \frac{\varsigma_{m+1} - \varsigma_m}{2} \tag{17.57}$$

where $\varsigma_{m+1}, \varsigma_m$ are the values ς of the upper and lower surface of the mth layer, respectively.

By numerical integration, the stiffness matrix and load matrix are obtained by formula (17.48), and the displacement increment is obtained by formula (17.47). Then by formula (17.42), the stress increment $\Delta\sigma_n$ can be calculated.

The merging method has the following advantages:

1) High efficiency of computation. After merging, the number of elements is greatly reduced. For example, for a concrete dam, the thickness of the pouring layer is 1.5 m. Originally each pouring layer is divided into 5 layers of finite elements. After merging, four pouring layers of the old concrete are merged into one layer of finite element. A compound layer element contains the original 20 layers of finite elements. As another example, for roller-compacted concrete dam, the thickness of the pouring layer is 0.30 m. Dividing into 4 layers of finite elements, after merging, the height of the element of the old concrete is 6.0 m. A layer contains the original 80 layers of finite elements. Therefore, the calculation efficiency has been greatly improved.
2) High accuracy. Using the compound layer method, not only the calculation efficiency is greatly improved, but also the calculation precision is improved. In the past simulation calculation, from the beginning to the end, using the same network grid, the thickness of the element cannot be too thin to control the calculation scale.

In the calculation of roller-compacted concrete dam, the pouring layer of 0.30 m thick sometimes is divided into one layer of element; even at the beginning several pouring layers are merged into a layer of homogeneous element. Actually it is far from the reality. And using the compound layer method, very thin element can be used in the new concrete, for example, roller-compacted concrete layer of 0.30 m thick can be divided into 4–5 layers of elements. The accuracy is greatly improved, at mid and later ages, merging layers gradually to improve the efficiency of computation.

Combining the compound layer method and the method of different time increments in different regions for the stress field and temperature field in literature [19, 20], for new concrete using a thin element and a short time increment, and for old concrete using a thick element and a long time increment, not only the calculation efficiency but also the calculation precision is greatly improved.

In general, the compound layer method and the method of different time increments in different regions are a set of methods taking account of the evolution of mechanical properties and thermal properties of concrete during the construction process. It not only greatly improves the simulation but also greatly improves the efficiency of computation and changes the simulation computation of high concrete dam completely.

Bibliography

1 Zhu, B.F. (1964) The stresses and deformations in the nonhomogeneous visco-elastic media under mixed boundary conditions. *Journal of Mechanics*, **2**.

2 Zhu, B.F. (1965) *Mountain Rock Pressure Carried by Underground Buildings in Visco-elastic Media*. Proceedings of IWHR(5), China Building Industry Press, Peking.

3 Zhu, B.F. (1985) The modulus of elasticity, creep and coefficient of stress relaxation of concrete. *Journal of Hydraulic Engineering*, **9**, 54–61.

4 Zhu, B.F. (1982) Some problems in the theory of creep of concrete. *Journal of Hydraulic Engineering*, **3**, 35–40.

5 Zhu, B.F. (1983) An implicit method for the stress analysis of concrete structures considering the effect of creep. *Journal of Hydraulic Engineering*, **5**, 40–46.

6 Zhu, B.F. (1984) Substructure method for stress analysis of nonhomogeneous elasto-creeping solids. *Journal of Hydraulic Engineering*, **2**, 20–24.

7 Zhu, B.F. (1986) Effective modulus method and equivalent temperature method for simple harmonic temperature creep stress analysis of late-age concrete and visco-elastic body. *Journal of Hydraulic Engineering*, **8**, 61–67.

8 Zhu, B.F. (1987) Creep analysis of reinforced concrete and pre-stressed concrete member. *Journal of Hydraulic Engineering*, **9**, 53–62.

9 Zhu, B.F. and Wang, T.S. (1976) *Thermal Stress and Temperature Control of Hydraulic Concrete Structures*, China Water Power Press, Peking.

10 Aruchunion (1962) *Some Problems in the Theory of Creep*, Science Press, Peking (Translated from Russian).

11 Alfrey, T. (1944) Non-homogeneous stresses in visco-elastic bodies. *Quarterly of Applied Mathematics*, **2**, 113.

12 Tsien, H.S. (1950) A generation of Alfrey's theorem for visco-elastic media. *Quarterly of Applied Mathematics*, **8**, 104.

13 Zienkiewicz, O.C. and Watson, M. (1996) Some creep effects in stress analysis with particular reference to concrete pressure vessels. *Nuclear Engineering and Design*, **4**.

14 Zienkiewicz, O.C., Watson, M. and King, I.P. (1968) A numerical method of visco-elastic stress analysis. *International Journal of Mechanical Sciences*, **10** (10).

15 Kimishima H. Study on Dam Concrete Creep. Japanese Electric Power Research Institute Report, 10(5), 10(6).

16 Zhu, B.F. (1992) The constrained extreme value method for parameter fitting of concrete creep equations. *Journal of Hydraulic Engineering*, **7**, 75–76.

17 Zhu, B.F., Li, Z.M. and Zhang, B.C. (1984) *Optimal Design of Structures, Principles and Applications*, China Water Power Press, Peking.

18 Dong, F.P. and Zhu, B.F. (1987) Study on creep stress of RCC dams. *Water Resources and Hydropower Engineering*, **10**, 24–32.

19 Zhu, B.F. (1995) Compound layer method for stress analysis simulating construction process. *Dam Engineering*, **VI**(2), 157–178.

20 Zhu, B.F. (1995) A numerical method using different time increments in different regions for analysing stresses in elasto-creeping solids. *Journal of Hydraulic Engineering*, **7**, 23–27.

21 Zhu, B.F. (1994) Mixed-layer method for analysis of stresses by simulating the construction process of multilayered high concrete structures. *Journal of Hydroelectric Engineering*, **3**, 21–30.

22 Zhu, B.F. (1995) Joint elements in mixed-layer method for stress analysis simulating construction process of high concrete dams. *Journal of Hydroelectric Engineering*, **3**, 14–21.

23 Zhu, B.F. (1995) A method using different time increments in different regions for solving unsteady temperature field by numerical method. *Journal of Hydraulic Engineering*, **8**, 46–52.

24 Zhu, B.F. (1991) Equivalent equation of heat conduction in mass concrete considering the effect of pipe cooling. *Journal of Hydraulic Engineering*, **3**, 28–34.

25 Zhu BF. Thermal Stress and Temperature Control of Mass Concrete[M]. Peking: China Electric Power Press, 1999 (in Chinese).

26 Zhu, B.F. and Song, J.T. (1977) *Analysis of Concrete Temperature Field and Thermal Creep Stress by Finite Element*. Selected Papers of Water Resources and Hydropower Engineering Application of Electronic Computer[C]. Water Resources and Electric Power Press, Peking.

27 Zhu, B.F. (2014) *Thermal Stresses and Temperature Control of Mass Concrete[M]*. Elsevier, New York.

18

Stress Analysis for Viscoelastic and Visco-Plastic Bodies

For many engineering materials, such as plastic, rock, soil, and so on, the strain is not only related to the stress but also associated with time. Actually, it is the stress history that affects its strain. These kinds of materials are named as viscoelastic and visco-plastic materials.

18.1 The Stress–Strain Relation of Viscoelastic Body under the Action of Unidirectional Stress

18.1.1 The Stress–Strain Relation of Ideal Elastic Body (Hooke Body)

As shown in Figure 18.1, the stress–strain relation of ideal elastic body (Hooke body) is

$$\sigma = E\varepsilon \tag{18.1}$$

in which E is the modulus of elasticity.

Under the action of a constant stress, the strain is

$$\varepsilon = \frac{\sigma}{E} \tag{18.2}$$

18.1.2 The Stress–Strain Relation of Ideal Plastic Body: The Dashpot

Figure 18.2 shows the model of ideal plastic body, the dashpot. The stress–strain relation of the dashpot is

$$\sigma = f\frac{\mathrm{d}\varepsilon}{\mathrm{d}\tau} \tag{18.3}$$

in which f is coefficient of viscosity.

Under the action of a constant stress, the strain is

$$\varepsilon = \frac{\sigma}{f}t \tag{18.4}$$

18.1.3 Maxwell Body

Figure 18.3 shows the model of Maxwell body. The stress–strain relation of Maxwell body is

$$\sigma = E\varepsilon_1 = f\frac{\mathrm{d}\varepsilon_2}{\mathrm{d}t} \tag{a}$$

$$\varepsilon = \varepsilon_1 + \varepsilon_2 \tag{b}$$

The Finite Element Method: Fundamentals and Applications in Civil, Hydraulic, Mechanical and Aeronautical Engineering, First Edition. Bofang Zhu.

Companion website: www.wiley.com/go/zhufem17

Take the derivative of Eq. (b) and we can get

$$\frac{d\varepsilon}{d\tau} = \frac{d\varepsilon_1}{d\tau} + \frac{d\varepsilon_2}{d\tau} = \frac{1}{E}\frac{d\sigma}{dt} + \frac{\sigma}{f} \tag{c}$$

or

$$\sigma + \frac{f}{E}\frac{d\sigma}{d\tau} = f\frac{d\varepsilon}{d\tau} \tag{18.5}$$

Figure 18.1 Hooke body.

Figure 18.2 Ideal plastic body, the dashpot.

or written as

$$\sigma + p_1\dot{\sigma} = q_1\dot{\varepsilon} \tag{18.5a}$$

in which $p_1 = \frac{f}{E}$, $q_1 = f$.

Under the action of a constant stress σ_0, the strain is

$$\varepsilon = \frac{\sigma_0}{q_1}(p_1 + t) = \sigma_0\left(\frac{1}{E} + \frac{1}{f}t\right) \tag{18.6}$$

If instantaneous constant strain ε is applied at the initial time, the stress is

$$\sigma = E\varepsilon e^{-Et/f} \tag{18.7}$$

18.1.4 Kelvin Body

Figure 18.4 shows the model of Kelvin body. In some literatures it is called Voigt body. The stress–strain relation of Kelvin body is

$$\sigma = E\varepsilon + f\frac{d\varepsilon}{dt} \tag{18.8}$$

or written as

$$\sigma = q_0\varepsilon + q_1\frac{d\varepsilon}{dt} \tag{18.8a}$$

in which $q_0 = E$, $q_1 = f$.

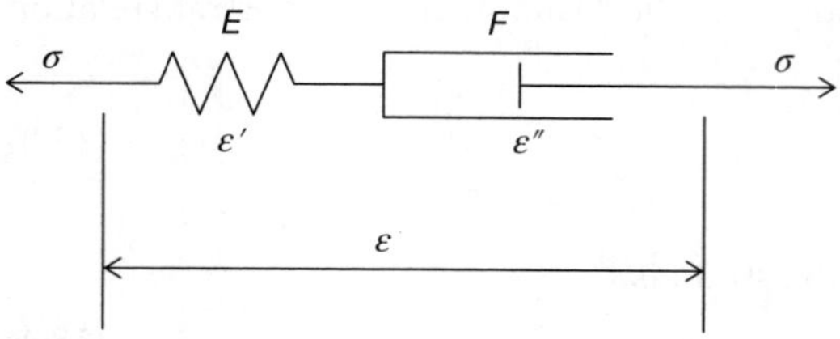

Figure 18.3 Maxwell body.

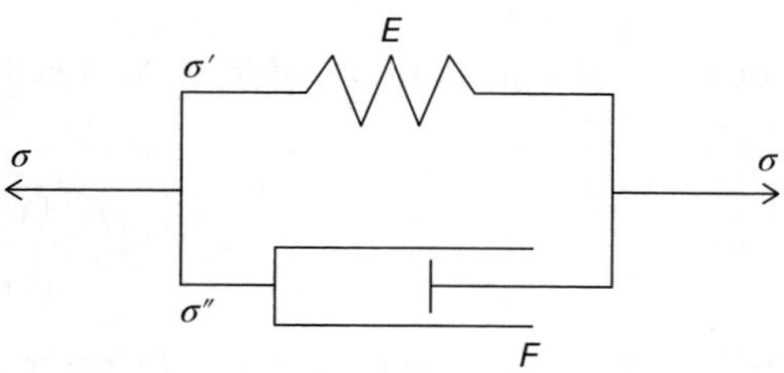

Figure 18.4 Kelvin body.

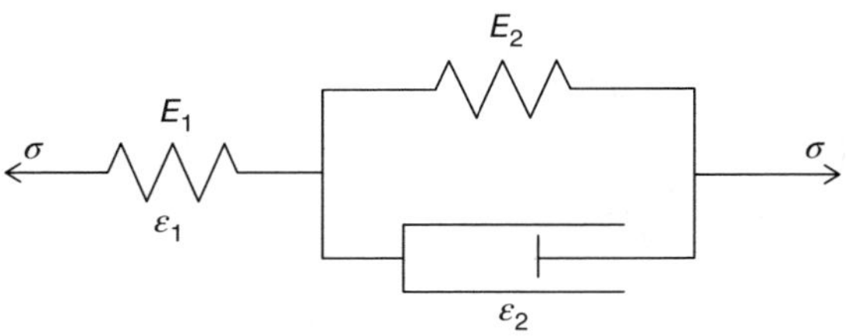

Figure 18.5 Standard three-component viscoelastic body.

When $\tau = 0$, the body is subjected to the action of a constant stress, and the solution of Eq. (18.8) is

$$\varepsilon(t) = \frac{\sigma}{E}(1 - e^{-Et/f}) \tag{18.9}$$

18.1.5 Standard three-Component Viscoelastic Body

Figure 18.5 shows the model of standard three-component viscoelastic body. The stress–strain relation of standard three-component viscoelastic body is

$$\left.\begin{aligned} &\sigma = E_1 \cdot \varepsilon_1 \\ &\sigma = E_2 \cdot \varepsilon_2 + f_2 \frac{d\varepsilon_2}{dt} \\ &\varepsilon = \varepsilon_1 + \varepsilon_2 \end{aligned}\right\} \tag{18.10}$$

Since $\frac{d\varepsilon}{dt} = \frac{d\varepsilon_1}{dt} + \frac{d\varepsilon_2}{dt}$, Eq. (18.10) can be written as

$$\sigma + \frac{\eta_2}{E_1 + E_2}\frac{d\sigma}{d\tau} = \frac{E_1 E_2}{E_1 + E_2} \cdot \varepsilon + \frac{\eta_2 E_1}{E_1 + E_2}\frac{d\varepsilon}{d\tau}$$

or

$$p_0\sigma + p_1\frac{d\sigma}{d\tau} = q_0\varepsilon + q_1\frac{d\varepsilon}{d\tau} \tag{18.10a}$$

in which $p_0 = 1$, $p_1 = \eta_2/(E_1 + E_2)$, $q_0 = E_1E_2/(E_1 + E_2)$, and $q_1 = \eta_2 E_1/(E_1 + E_2)$.

Assume a constant stress is applied when $t = 0$, then the strain is

$$\varepsilon = \frac{\sigma}{q_0}\left[1 - \left(1 - \frac{p_1 q_0}{q_1}\right)e^{-q_0 t/q_1}\right] \tag{18.10b}$$

18.1.6 Kelvin Chain

Figure 18.5 shows the model of Kelvin chain. The stress–strain relation of Kelvin chain under the action of a constant stress is

$$\varepsilon = \sigma\left[\frac{1}{E_0} + \sum \frac{1}{E_i}(1 - e^{-E_i t/f_i})\right] = \sigma J(t) \tag{18.11}$$

in which

$$J(t) = \frac{1}{E_0} + \sum \frac{1}{E_i}(1 - e^{-E_i t/f_i}) \tag{18.12}$$

$J(t)$ is called creep compliance.

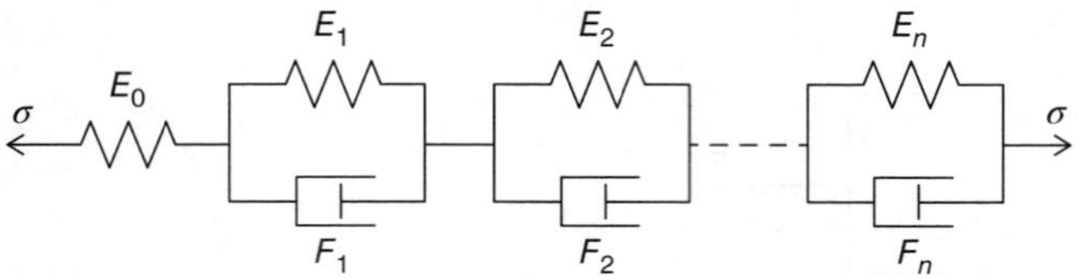

Figure 18.6 Kelvin chain.

18.1.7 The Stress–Strain Relation When Stress Changes with Time

For the Kelvin chain as shown in Figure 18.6, when the stress changes with time, the strain should be calculated with the equation below:

$$\varepsilon(t) = \sigma_0 J(t) + \int_0^t J(t-\tau)\frac{\mathrm{d}\sigma}{\mathrm{d}\tau}\mathrm{d}\tau \tag{18.13}$$

18.2 The Stress–Strain Relation under the Action of Complex Stresses

There are six components of stress and six components of strain for every point in a body under the action of complex stresses that can be written as

$$\{\sigma\} = \{\sigma_x\, \sigma_y\, \sigma_z\, \tau_{xy}\, \tau_{yz}\, \tau_{zx}\}^{\mathrm{T}} \tag{18.14}$$

$$\{\varepsilon\} = \{\varepsilon_x\, \varepsilon_y\, \varepsilon_z\, \gamma_{xy}\, \gamma_{yz}\, \gamma_{zx}\}^{\mathrm{T}} \tag{18.15}$$

18.2.1 The Stress–Strain Relation When Poisson's Ratio Is Constant

The experimental results indicate that Poisson's ratio of some materials can be seen as a constant, such as concrete. The stress and strain usually changes with time, and the component of strain can be calculated as below:

$$\left.\begin{aligned}
\varepsilon_x(t) &= \left[\sigma_x(0) - \mu\sigma_y(0) - \mu\sigma_z(0)\right] J(t) + \int_0^t J(t-\tau)\left[\frac{\mathrm{d}\sigma_x}{\mathrm{d}\tau} - \mu\frac{\mathrm{d}\sigma_y}{\mathrm{d}\tau} - \mu\frac{\mathrm{d}\sigma_z}{\mathrm{d}\tau}\right]\mathrm{d}\tau \\
\varepsilon_y(t) &= \left[\sigma_y(0) - \mu\sigma_x(0) - \mu\sigma_z(0)\right] J(t) + \int_0^t J(t-\tau)\left[\frac{\mathrm{d}\sigma_y}{\mathrm{d}\tau} - \mu\frac{\mathrm{d}\sigma_x}{\mathrm{d}\tau} - \mu\frac{\mathrm{d}\sigma_z}{\mathrm{d}\tau}\right]\mathrm{d}\tau \\
&\ \ \vdots \\
\gamma_{xy}(t) &= \frac{\tau_{xy}(0)}{2(1+\mu)}J(t) + \int_0^t \frac{J(t-\tau)}{2(1+\mu)}\frac{\mathrm{d}\tau_{xy}}{\mathrm{d}\tau}\mathrm{d}\tau
\end{aligned}\right\} \tag{18.16}$$

in which $J(t)$ is creep compliance. The equation set above can be written in matrix form as below:

$$\{\varepsilon(t)\} = [A]\{\sigma_0\}J(t) + \int_0^t J(t-\tau)[A]\left\{\frac{\mathrm{d}\sigma(\tau)}{\mathrm{d}\tau}\right\}\mathrm{d}\tau \tag{18.17}$$

in which $[A]$ is shown in Eqs. (17.13)–(17.15).

18.2.2 Different Law for Volume Deformation and Shear Deformation

Decompose the components of stress and strain as below:

$$\begin{bmatrix} \sigma_x & \tau_{xy} & \tau_{xz} \\ \tau_{yx} & \sigma_y & \tau_{yz} \\ \tau_{zx} & \tau_{zy} & \sigma_z \end{bmatrix} = \begin{bmatrix} \sigma & 0 & 0 \\ 0 & \sigma & 0 \\ 0 & 0 & \sigma \end{bmatrix} + \begin{bmatrix} \sigma_x - \sigma & \tau_{xy} & \tau_{xz} \\ \tau_{yx} & \sigma_y - \sigma & \tau_{yz} \\ \tau_{zx} & \tau_{zy} & \sigma_z - \sigma \end{bmatrix} \tag{18.18}$$

$$\begin{bmatrix} \varepsilon_x & \gamma_{xy} & \gamma_{xz} \\ \gamma_{yx} & \varepsilon_y & \gamma_{yz} \\ \gamma_{zx} & \gamma_{zy} & \varepsilon_z \end{bmatrix} = \begin{bmatrix} e & 0 & 0 \\ 0 & e & 0 \\ 0 & 0 & e \end{bmatrix} + \begin{bmatrix} \varepsilon_x - e & \gamma_{xy} & \gamma_{xz} \\ \gamma_{yx} & \varepsilon_y - e & \gamma_{yz} \\ \gamma_{zx} & \gamma_{zy} & \varepsilon_z - e \end{bmatrix} \tag{18.19}$$

in which

$$\sigma = \frac{1}{3}(\sigma_x + \sigma_y + \sigma_z) \tag{18.20}$$

$$e = \frac{1}{3}(\varepsilon_x + \varepsilon_y + \varepsilon_z) \tag{18.21}$$

σ is called the spherical tensor of stress, and e is called the spherical tensor of strain.

For isotropic elastic medium, the spherical tensor of stress is in direct proportion to the spherical tensor of strain, and the deviator of stress is in direct proportion to the deviator of strain:

$$\begin{bmatrix} \sigma & 0 & 0 \\ 0 & \sigma & 0 \\ 0 & 0 & \sigma \end{bmatrix} = 3K \begin{bmatrix} e & 0 & 0 \\ 0 & e & 0 \\ 0 & 0 & e \end{bmatrix} \tag{18.22}$$

$$\begin{bmatrix} \sigma_x - \sigma & \tau_{xy} & \tau_{xz} \\ \tau_{yx} & \sigma_y - \sigma & \tau_{yz} \\ \tau_{zx} & \tau_{zy} & \sigma_z - \sigma \end{bmatrix} = 2G \begin{bmatrix} \varepsilon_x - e & \gamma_{xy} & \gamma_{xz} \\ \gamma_{yx} & \varepsilon_y - e & \gamma_{yz} \\ \gamma_{zx} & \gamma_{zy} & \varepsilon_z - e \end{bmatrix} \tag{18.23}$$

Using s_{ij} to represent the deviator of stress,

$$s_{ij} = \sigma_{ij} - \sigma\delta_{ij} \tag{18.24}$$

that is,

$$s_{xx} = \sigma_x - \sigma$$
$$s_{xy} = \tau_{xy}$$
$$s_{xz} = \tau_{xz}$$

Using e_{ij} to represent the deviator of strain,

$$e_{ij} = \varepsilon_{ij} - e\delta_{ij} \tag{18.25}$$

that is,

$$e_{xx} = \varepsilon_x - e$$
$$e_{xy} = \gamma_{xy}$$
$$e_{xz} = \gamma_{xz}$$

The stress–strain relation of elastic body is

$$\left.\begin{aligned}\sigma &= 3Ke \\ s_{ij} &= 2Ge_{ij}\end{aligned}\right\} \tag{18.26}$$

in which K is the bulk modulus and G is the shear modulus.

The stress–strain relation of viscoelastic body is

$$P' s_{ij} = Q' e_{ij} \tag{18.27}$$

$$P'' \sigma = Q'' e \tag{18.28}$$

in which

$$\left.\begin{aligned}P' &= \sum p'_i \frac{d^i}{dt^i} \\ Q' &= \sum q'_i \frac{d^i}{dt^i} \\ P'' &= \sum p''_i \frac{d^i}{dt^i} \\ Q'' &= \sum q''_i \frac{d^i}{dt^i}\end{aligned}\right\} \tag{18.29}$$

p'_i, q'_i, p''_i, q''_i are coefficients.

18.3 Stress Analysis of Viscoelastic Body

The fundamental formula of stress analysis of viscoelastic body in the nth time interval is

$$[K_n]\{\delta_n\} = \{\Delta P_n\} + \{\Delta P_n{}^{\mathrm{c}}\} + \{\Delta P_n{}^{\mathrm{I}}\} \tag{18.30}$$

in which

$$\left.\begin{aligned}[K_n] &= \int [B]^{\mathrm{T}}[\overline{\mathrm{D}}_n]\{B\}\mathrm{d}V \\ \{\Delta P_n{}^{\mathrm{c}}\} &= \int [B]^{\mathrm{T}}[\overline{\mathrm{D}}_n]\{\eta_n\}\mathrm{d}V \\ \{\Delta P_n{}^{\mathrm{I}}\} &= \int [B]^{\mathrm{T}}[\overline{\mathrm{D}}_n]\{\Delta\varepsilon_n{}^{i}\}\mathrm{d}V\end{aligned}\right\} \tag{18.31}$$

where

$[K_n]$ = the stiffness matrix of the structure.

$\{\Delta P_n{}^{\mathrm{c}}\}$ = the load increment caused by creep.

$\{\Delta P_n{}^{\mathrm{I}}\}$ = the load increment caused by temperature deformation.

$\{\Delta P_n\}$ = the load increment due to external load.

Inversion of Eq. (18.30) will give the displacement increment:

$$\{\Delta\delta_n\} = [K_n]^{-1}(\{\Delta P_n\} + \{\Delta P_n{}^{\mathrm{c}}\} + \{\Delta P_n{}^{\mathrm{I}}\}) \tag{18.30a}$$

Then we can get the stress increment by Eq. (17.44) with the displacement increment. We can also find that Eqs (18.30) and (18.31) are similar to Eqs (17.47) and (17.48) for the concrete structure. There are some differences between their $[\overline{D}_n]$ that are due to different creep deformation.

18.3.1 Stress Analysis of Viscoelastic Body with Constant Poisson's Ratio

Assume the creep Poisson's ratio of the body was constant and do not change with time; its numerical value is equal to that of elastic strain. Under the action of complex stress, the array of incremental elastic strain in period $\Delta t_n = t_n - t_{n-1}$ is

$$\{\Delta \varepsilon_n{}^{\mathrm{e}}\} = \frac{1}{E_0}[A]\{\Delta \sigma_n\} \tag{18.32}$$

The array of incremental creep strain is

$$\{\Delta \varepsilon_n{}^{\mathrm{c}}\} = \{\Delta \eta_n\} + q_n[A]\{\Delta \sigma_n\} \tag{18.33}$$

in which

$$\left.\begin{aligned} &\{\Delta \eta_n\} = \sum_{j=1}^{m}(1 - e^{-r_j \Delta t_n})\{\Delta \omega_{jn}\} \\ &q_n = \sum_{j=1}^{m} \phi_j(1 - e^{-0.5 r_j \Delta t_n}) \\ &\{\Delta \omega_{jn}\} = \{\Delta \omega_{j,n-1}\}e^{-r_j \Delta t_{n-1}} + [A]\{\Delta \sigma_n\}\phi_j e^{-0.5 r_j \Delta t_{n-1}} \\ &\{\Delta \omega_{j1}\} = [A]\{\Delta \sigma_0\}\phi_j \end{aligned}\right\} \tag{18.34}$$

$[A]$ is given in Eqs (17.13)–(17.15). We can get the displacement increment $\{\Delta \delta_n\}$ by Eq. (18.30) and then get the stress increment $\{\Delta \sigma_n\}$ by Eq. (17.44).

18.3.2 Stress Analysis of Viscoelastic Body with Different Laws for Volume Deformation and Shear Deformation

In this case, the fundamental formula of stress analysis is still Eq. (18.30), but the elasticity matrix $[\overline{D}_n]$ and the equivalent load increment $\{\Delta P_n{}^{\mathrm{c}}\}$ are changed. They should be calculated as follows.

Let the creep compliance of shear deformation be

$$J_1(t) = \frac{1}{G_0} + \sum_{j=1}^{m} \phi_j(1 - e^{-r_j t}) \tag{18.35}$$

The creep compliance of volume deformation is

$$J_2(t) = \frac{1}{K_0} + \sum_{j=1}^{m} \psi_j(1 - e^{-s_j t}) \tag{18.36}$$

The stress–strain relation of elastic body is

$$e_{xy} = \gamma_{xy}/2 = s_{xy}/2G, \; \varepsilon_m = \sigma_m/3K$$

The deviatoric tensor of strain of viscoelastic body is

$$2e_{ij}(t) = \Delta s_{ij0} J_1(t) + \int_0^t J_1(t-\tau)\frac{\mathrm{d}s_{ij}}{\mathrm{d}\tau}\mathrm{d}\tau \tag{18.37}$$

The volume deformation is

$$3\varepsilon_m(t) = \Delta \sigma_{m0} J_2(t) + \int_0^t J_2(t-\tau)\frac{\mathrm{d}\sigma_m}{\mathrm{d}\tau}\mathrm{d}\tau \tag{18.38}$$

The definitions of σ_m, ε_m, s_{ij}, and e_{ij} can be found in Chapter 17.

In period Δt_n, the increment of elastic strain deviator is

$$\Delta e^{\mathrm{e}}_{ijn} = e^{\mathrm{e}}_{ij}(t_n) - e^{\mathrm{e}}_{ij}(t_{n-1}) = \frac{1}{2G_0}\Delta s_{ijn} \tag{18.39}$$

The increment of creep strain deviator is

$$\Delta e^{\mathrm{c}}_{ijn} = e^{\mathrm{c}}_{ij}(t_n) - e^{\mathrm{c}}_{ij}(t_{n-1}) = \frac{1}{2}\eta_{ijn} + \frac{1}{2}q_n\Delta s_{ijn} \tag{18.40}$$

in which

$$\left.\begin{aligned} \eta_{ijn} &= \sum_{j=1}^{m}(1 - e^{-r_j\Delta t_n})\Delta\omega_{jn} \\ q_n &= \sum_{j=1}^{m}\phi_j(1 - e^{-0.5r_j\Delta t_n}) \\ \omega_{ijn} &= \Delta\omega_{ij,n-1}e^{-r_j\Delta t_{n-1}} + \Delta s_{ij,n-1}\phi_j e^{-0.5r_j\Delta t_{n-1}} \\ \omega_{ij1} &= \Delta s_{ij0}\phi_j \end{aligned}\right\} \tag{18.41}$$

In period Δt_n, the increment of elastic volume deformation is

$$\Delta\varepsilon^{\mathrm{c}}_{m,n} = \varepsilon^{\mathrm{c}}_m(t_n) - \varepsilon^{\mathrm{c}}_m(t_{n-1}) = \frac{1}{3K_0}\Delta\sigma_{m,n} \tag{18.42}$$

The increment of creep volume deformation is

$$\Delta\varepsilon^{\mathrm{c}}_{m,n} = \frac{1}{3}\zeta_n + \frac{1}{3}p_n\Delta\sigma_{m,n} \tag{18.43}$$

in which

$$\left.\begin{aligned} \zeta_n &= \sum_{j=1}^{m}(1 - e^{-s_j\Delta t_n})\rho_{jn} \\ p_n &= \sum_{j=1}^{m}\psi_j(1 - e^{-0.5s_j\Delta t_n}) \\ \rho_{jn} &= \rho_{j,n-1}e^{-s_j\Delta t_{n-1}} + \Delta\sigma_{m,n-1}\phi_j e^{-0.5s_j\Delta t_{n-1}} \\ \rho_{j1} &= \Delta\sigma_{m0}\psi_j \end{aligned}\right\} \tag{18.44}$$

The fundamental formulae of stress analysis of viscoelastic body are Eqs (18.30) and (18.31). The elasticity matrix $[\overline{D}_n]$ and the equivalent load increment $\{\Delta P_n{}^{\mathrm{c}}\}$ caused by creep are calculated as below.

First, get $\overline{G}_n$ and $\overline{K}_n$:

$$\overline{G}_n = \frac{G_0}{1 + q_n G_0},\ \overline{K}_n = \frac{K_0}{1 + p_{nK_0}} \tag{18.45}$$

in which q_n is shown in Eq. (18.41) and p_n is shown in Eq. (18.44).

Calculate $\overline{E}_n$ and $\overline{\mu}_n$ by the equation below:

$$\left.\begin{aligned} \overline{E}_n &= \frac{9\overline{K}_n\overline{G}_n}{3\overline{K}_n + \overline{G}_n} \\ \overline{\mu}_n &= \frac{3\overline{K}_n - 2\overline{G}_n}{2(3\overline{K}_n + \overline{G}_n)} \end{aligned}\right\} \tag{18.46}$$

Then replace E with $\overline{E}_n$ and μ with $\overline{\mu}_n$, and we can get $[\overline{D}_n]$ by Eqs (2.21), (2.24), and (7.14). Finally we get the stiffness matrix with Eq. (12.31).

Load increment caused by creep strain is calculated by the equation below:

$$\{\Delta P_n{}^{\mathrm{c}}\} = \int [B]^{\mathrm{T}}[\overline{D}_n]\{\Delta\varepsilon_n{}^{\mathrm{c}}\}\mathrm{d}V \tag{18.47}$$

The increment of creep strain is computed as follows:

$$\{\Delta\varepsilon_n{}^{\mathrm{c}}\} = \left[\frac{1}{2}\eta_{xx,n} + \frac{1}{3}\zeta_n,\ \frac{1}{2}\eta_{yy,n} + \frac{1}{3}\zeta_n,\ \frac{1}{2}\eta_{zz,n} + \frac{1}{3}\zeta_n,\ \eta_{xy,n},\ \eta_{yz,n},\ \eta_{zx,n}\right] \tag{18.48}$$

in which $\eta_{ij,n}$ is given in Eq. (18.41) and ζ_n is given in Eq. (18.44).

The above method for analyzing stress of viscoelastic body is implicit method. It has a better efficiency than the method mentioned in Ref. [5] that used explicit method. A bigger time interval Δt_n can be used in this way.

18.4 Effective Modulus Method and Equivalent Temperature Method for Simple Harmonic Temperature Creep Stress Analysis of Concrete at Late Ages and Viscoelastic Body

Periodical change of external temperature, such as annual variation, will lead to cyclic temperature stress in the structure. This kind of temperature stress may be remarkable in statically indeterminate structures such as arch bridges, arch dams, and so on. The influence of creep on temperature stress is significant, and it must be considered in structural calculation. The author proposed the effective modulus method that made the calculation simple and convenient.

Transform Eq. (18.13) and get

$$\left.\begin{aligned}\varepsilon(t) &= \frac{\sigma(t)}{E(t)} - \int_{t_0}^{t} \sigma(\tau)\frac{\partial J(t,\tau)}{\partial\tau}\,\mathrm{d}\tau \\ J(t,\tau) &= \frac{1}{E(\tau)} + C(t,\tau)\end{aligned}\right. \tag{18.49}$$

in which $J(t, \tau)$ is creep compliance.

For late-age concrete and general viscoelastic bodies, their modulus of elasticity and unit creep can be given as

$$\left.\begin{aligned}E(\tau) &= E = \text{const} \\ C(t,\tau) &= \sum_{i=1}^{m} C_i[1 - e^{-r_i(t-\tau)}]\end{aligned}\right\} \tag{18.50}$$

in which C_i is constant.

Generally, it is only necessary to compute the temperature stresses in quasi-steady state without the initial effect in practical engineering. So let $t_0 \to -\infty$ and then we get the equation below from Eq. (18.49):

$$\varepsilon(t) = \frac{\sigma(t)}{E} - \int_{-\infty}^{t} \sigma(\tau)\frac{\partial J(t,\tau)}{\partial\tau}\mathrm{d}\tau \tag{18.51}$$

Substitute Eq. (18.50) into Eq. (18.51) and get

$$\varepsilon(t) = \frac{\sigma(t)}{E} - \int_{-\infty}^{t} \sigma(\tau) \left[\sum_{i=1}^{m} C_i r_i e^{-r_i(t-\tau)} \right] \mathrm{d}\tau \tag{18.52}$$

In general, cosine functions of time can be used to simulate the change of ambient temperature. When the initial effect has already disappeared and the structure reaches its quasi-steady state, the temperature, stress, and strain of every point of the structure are simple harmonic function of time, which can be described by cosine functions. Suppose the stress state of one point of the structure is

$$\begin{aligned} \sigma(t) &= \sigma_0 \cos \omega(t + \eta) \\ \omega &= \frac{2\pi}{P} \end{aligned} \tag{18.53}$$

in which ω is the circular frequency, P is the period of temperature and stress variation, σ_0 is the stress amplitude, and η is the phase difference. Obviously, both σ_0 and η are functions of coordinates.

Substitute Eq. (18.53) into Eq. (18.52) and get

$$\varepsilon(t) = \frac{\sigma_0}{\rho E} \cos \omega(t + \eta - \xi) \tag{18.54}$$

in which

$$\left. \begin{aligned} &\rho = \frac{1}{\sqrt{a^2 + b^2}}, \quad \xi = \frac{1}{\omega} \tan^{-1}\left(\frac{a}{b}\right) \\ &a = 1 + \sum_{i=1}^{m} \frac{E C_i r_i^2}{r_i^2 + \omega^2}, \quad b = 1 + \sum_{i=1}^{m} \frac{E C_i r_i \omega}{r_i^2 + \omega^2} \end{aligned} \right\} \tag{18.55}$$

where ξ is the phase difference of the peak stress behind the peak strain.

Let

$$E^* = \rho E, \; t^* = t - \xi \tag{18.56}$$

and substitute them into Eq. (18.54), we get

$$\varepsilon(t) = \frac{\sigma_0}{E^*} \cos \omega(t^* + \eta) \tag{18.57}$$

This equation shares the same form with the stress–strain relation of general elastic bodies. If the creep Poisson's ratio of a body is a constant and equal to Poisson's ratio of instantaneous elastic deformation and ignores the effect of inertia force, we may use an equivalent elastic body to simulate a viscoelastic body and the concrete of late age. The equivalent elastic modulus is

$$E^* = \rho E \tag{18.58}$$

We can use the equivalent elastic modulus to calculate the stress and the strain of concrete as it is an elastic body. The stress and the strain are approximatively equal to the peak stress and the peak strain of the original viscoelastic body. We called this computing method effective modulus method.

The temperatures need not be transformed when we using the effective modulus method to solve harmonic stress problems of a viscoelastic body. We just need to substitute the modulus of elasticity by the equivalent modulus of elasticity and then calculate its stress state using the method for elastic bodies. Since the temperature stress lies on $\rho E\alpha T$, we can also transform the temperature into the equivalent temperature instead of changing the modulus of elasticity. The equivalent temperature is

$$T^* = \rho T \tag{18.58a}$$

Then we can compute the stress state as an elastic body. This calculating method is called equivalent temperature method.

For example, let the modulus of elasticity of late-age concrete be E, and its unit creep is

$$C(t,\tau) = c_1[1 - e^{-0.030(t-\tau)}]$$

in which $c_1 = 1/E$, $r_1 = 0.030(1/d)$, the period $P = 365\text{d}$, and $\omega = 2\pi/P = 0.017214$.

From Eq. (18.55) we can get

$$a = 1.752, b = 0.4317, \rho = 0.554, \omega\xi = 13°50'$$

Using the equivalent elastic modulus $E^* = \rho E$ to get the stress state means the effect of creep is taken into account.

18.5 Stress Analysis for Visco-Plastic Bodies

The visco-plasticity theory studies the effects of time on variation of plastic deformation. The stress and strain are functions of time after the initial yielding of the body. The degree of time effect relates to material characteristics, actual conditions such as temperature and loads. As an example, time effect of metal materials under normal temperature condition is not remarkable, and methods of plastic mechanics can be used to solve the nonlinear deformation in this case. But when working under the environment of high temperature, metal materials are heavily influenced by time effect, and the visco-plastic theory should be applied in this case. Time effect of some materials, such as clay, is remarkable even when they are working at normal temperatures.

18.5.1 Viscoelastic–Plastic Problems under Action of One-Dimensional Stress

The stress–strain relation of viscoelastic–plastic bodies under one-dimensional stress can be shown in Figure 18.7, which is a one-dimensional rheological model, consisting of three components: a spring, a dashpot, and a friction slipper. The spring undertakes the total stress σ, with the elastic strain ε^{e}. The stress–strain relation is

$$\sigma^{\text{e}} = \sigma = E\varepsilon^{\text{e}} \tag{18.59}$$

where E is the modulus of elasticity. The stress carried by the friction slipper is σ^{p}. When $\sigma^{\text{p}} < \sigma^{\text{s}}$, the friction slipper will not move and σ^{s} is the yield stress. When $\sigma^{\text{p}} = \sigma^{\text{s}}$, the friction slipper starts to move, while its rate of deformation is controlled by the viscous component. Stress carried by the viscous component is $\sigma^{\text{d}} = \sigma - \sigma^{\text{p}}$, and its stress–strain relation is

$$\sigma^{\text{d}} = b\frac{\text{d}\varepsilon^{\text{vp}}}{\text{d}t} \tag{18.60}$$

in which ε^{vp} is the visco-plastic strain and b is the coefficient of viscosity.

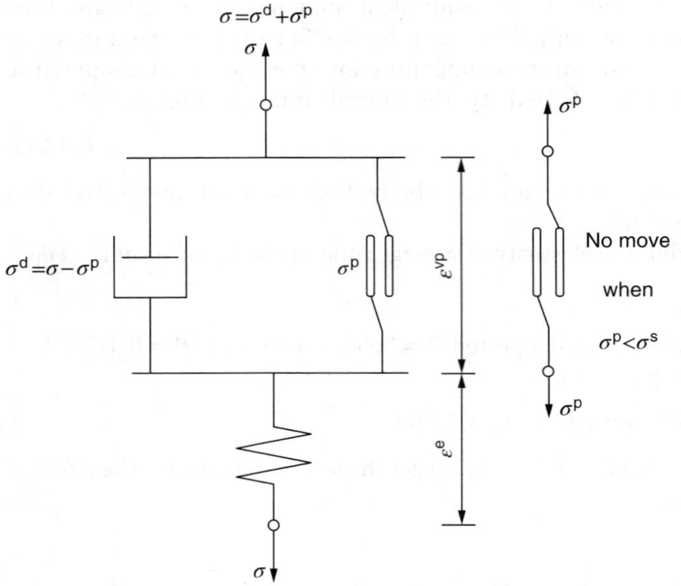

Figure 18.7 One-dimensional viscoelastic–plastic body.

The strain contains two parts: the elastic strain ε^{e} and the visco-plastic strain $\varepsilon^{\mathrm{vp}}$. That is,

$$\varepsilon = \varepsilon^{\mathrm{e}} + \varepsilon^{\mathrm{vp}} \tag{18.61}$$

Suppose a material subjected to linear hardening rule, as shown in Figure 18.8.

$$\sigma^{\mathrm{s}} = \sigma_Y + H'\varepsilon^{\mathrm{vp}} \tag{18.62}$$

in which H' is a constant and

$$H' = \frac{\mathrm{d}\sigma}{\mathrm{d}\varepsilon^{\mathrm{vp}}} = \frac{\mathrm{d}\sigma}{\mathrm{d}\varepsilon - \mathrm{d}\varepsilon^{e}} = \frac{E_T}{1 - E_T/E} \tag{18.63}$$

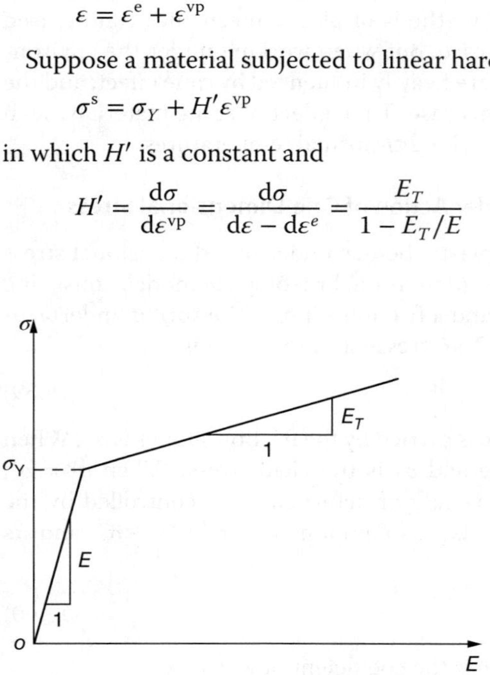

Figure 18.8 Linear elastic hardening.

The stress carried by the friction slipper is

$$\left.\begin{aligned} \sigma^{\mathrm{p}} &= \sigma\,(\sigma^{\mathrm{p}} < \sigma^{\mathrm{s}}) \\ \sigma^{\mathrm{p}} &= \sigma^{\mathrm{s}}\,(\sigma^{\mathrm{p}} \geq \sigma^{\mathrm{s}}) \end{aligned}\right\} \tag{18.64}$$

From Figure 18.7 we have

$$\sigma = \sigma^{\mathrm{d}} + \sigma^{\mathrm{p}} \tag{18.65}$$

Before visco-plastic yield comes into being, $\varepsilon^{\mathrm{vp}} = 0$, $\sigma^{\mathrm{d}} = 0$, and $\sigma^{\mathrm{p}} = \sigma$. The strain is $\varepsilon = \varepsilon^{\mathrm{e}} = \sigma/E$. After producing the visco-plastic yield, the strain consists of elastic strain ε^{e} and visco-plastic strain $\varepsilon^{\mathrm{vp}}$ as Eq. (18.61). By substituting Eqs (18.60), (18.62), (18.64) into Eq. (18.65), we get

$$\sigma = \sigma_{\mathrm{Y}} + H'\varepsilon^{\mathrm{vp}} + b\frac{\mathrm{d}\varepsilon^{\mathrm{vp}}}{\mathrm{d}t} \tag{18.66}$$

Substitution of Eqs (18.59) and (18.61) into Eq. (18.66) gives

$$H'E\varepsilon + bE\frac{\mathrm{d}\varepsilon}{\mathrm{d}t} = H'\sigma + E(\sigma - \sigma_{\mathrm{Y}}) + b\frac{\mathrm{d}\sigma}{\mathrm{d}t} \tag{18.67}$$

This is the first-order ordinary differential equation that determines the stress–strain relation under viscoelastic–plastic condition. Let

$$\gamma = 1/b \tag{18.68}$$

where γ is the flow coefficient. Substitution of Eq. (18.68) into Eq. (18.67) gives

$$\dot{\varepsilon} = \frac{\dot{\sigma}}{E} + \gamma[\sigma - (\sigma_{\mathrm{Y}} + H'\varepsilon^{\mathrm{vp}})] \tag{18.69}$$

The equation above can also be written as

$$\dot{\varepsilon} = \dot{\varepsilon}^{\mathrm{e}} + \dot{\varepsilon}^{\mathrm{vp}} \tag{18.70}$$

in which

$$\dot{\varepsilon}^{\mathrm{e}} = \frac{\dot{\sigma}}{E} \tag{18.71}$$

$$\dot{\varepsilon}^{\mathrm{vp}} = \gamma[\sigma - (\sigma_{\mathrm{Y}} + H'\varepsilon^{\mathrm{vp}})] \tag{18.72}$$

When under the action of constant stress $\sigma = \sigma_{\mathrm{A}}$, due to $\frac{\mathrm{d}\sigma}{\mathrm{d}t} = 0$, the equation changes into

$$\gamma H'\varepsilon + \frac{\mathrm{d}\varepsilon}{\mathrm{d}t} = \frac{\gamma H'}{E}\sigma_{\mathrm{A}} + \gamma(\sigma_{\mathrm{A}} - \sigma_{\mathrm{Y}}) \tag{18.73}$$

The right-hand side of the equation above is a constant. When $H' \neq 0$, the solution of Eq. (18.73) is

$$\varepsilon = \frac{\sigma_{\mathrm{A}}}{E} + \frac{(\sigma_{\mathrm{A}} - \sigma_{\mathrm{Y}})}{H'}(1 - e^{-H'\gamma t}) \tag{18.74}$$

The stress–strain is shown in Figure 18.9(a). The strain rate decreases gradually and finally reaches a stable value.

For ideal visco-plastic bodies, $H' = 0$. Let $H' \to 0$ in the Eq. (18.74) and apply L'Hospital's rule to get

$$\varepsilon = \frac{\sigma_{\mathrm{A}}}{E} + (\sigma_{\mathrm{A}} - \sigma_{\mathrm{Y}})\gamma t \tag{18.75}$$

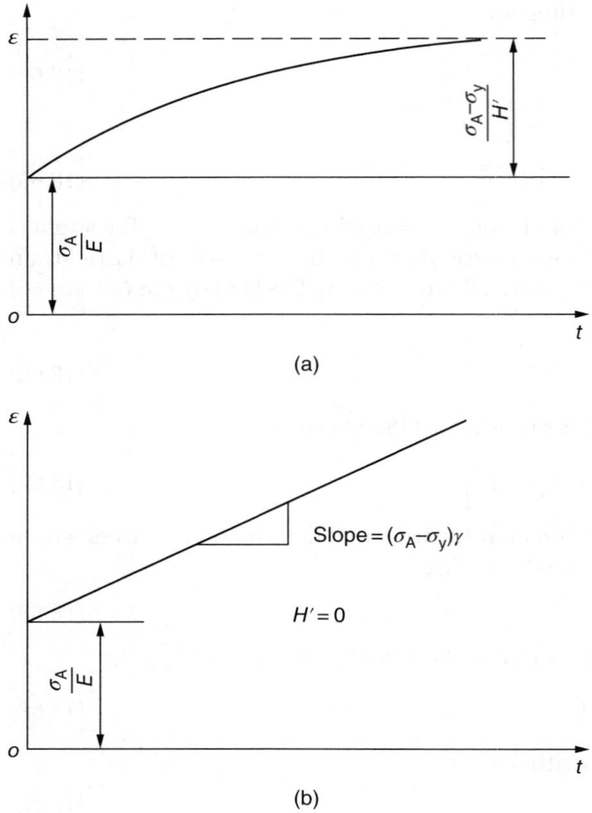

Figure 18.9 Strain response of viscoelastic–plastic bodies for constant stress (a) linear hardening material and (b) perfect-plastic material.

The relation between strain ε and time t is shown in Figure 18.9(b). The strain is increasing with time without limit. The differences between Figure 18.9(a) and (b) are that yield stress σ^s of strain-hardening materials is increasing with visco-plastic strain ε^{vp}, while when σ^s reaches σ_A, visco-plastic strain rate changes to 0 and the strain no more increases. For ideal viscoelastic–plastic bodies, $H' = 0$, stress carried by the viscous component is $\sigma^d = \sigma_A - \sigma_Y$ so that strain will be increasing with time with no limit.

18.5.2 Viscoelastic–Plastic Problems with Complex Stress States

When working under complex stress states, the total strain rate consists of elastic strain rate and visco-plastic strain rate:

$$\{\dot{\varepsilon}\} = \{\dot{\varepsilon}^e\} + \{\dot{\varepsilon}^{vp}\} \tag{18.76}$$

in which $\{\dot{\varepsilon}\}$ is the total strain rate, $\{\dot{\varepsilon}^e\}$ is the elastic strain rate, and $\{\dot{\varepsilon}^{vp}\}$ is the visco-plastic strain rate.

The stress rate is obtained by

$$\{\dot{\sigma}\} = [D]\{\dot{\varepsilon}\} \tag{18.77}$$

in which $[D]$ is the elasticity matrix.

When working under the action of stress in one direction, the condition that increase visco-plastic strain is $\sigma^{p} - \sigma^{s} = 0$. Similarly, when working under complex stress states, the condition for visco-plastic strain begins to arise, which can be considered as

$$F(\sigma, \varepsilon^{vp}) - F_0 = 0 \tag{18.78}$$

When the equation above is satisfied, visco-plastic strain appears. As to the numerical value of visco-plastic strain rate, it certainly refers to the current stress state. For hardened materials, it may be in relation to visco-plastic strain state. In order to simplify the calculation, generally it is assumed that visco-plastic strain rate only refers to the current stress state, that is,

$$\{\dot{\varepsilon}^{vp}\} = f\{\sigma\} \tag{18.79}$$

The visco-plastic flow rule written below gets widespread usage:

$$\{\dot{\varepsilon}^{vp}\} = \gamma\langle \emptyset(f)\rangle \left\{\frac{\partial Q}{\partial \sigma}\right\} \tag{18.80}$$

in which $Q = Q\{\{\sigma\}, \{\varepsilon^{vp}\}, k\}$. Q is the plastic potential, and γ is the flow coefficient.

The meaning of the symbol $\langle\,\rangle$ is

$$\left.\begin{aligned} &\langle \emptyset(x)\rangle = \emptyset(x) \quad (x > 0) \\ &\langle \emptyset(x)\rangle = 0 \qquad\;\; (x \le 0) \end{aligned}\right\} \tag{18.81}$$

Compare the two Eqs (18.80) and (18.65), and we will find that the nonaffiliated plastic potential theory is similar to the function of visco-plastic strain rate so far. If the associated flow rule was adopted and suppose

$$Q \equiv F$$

then Eq. (18.80) would change into

$$\{\dot{\varepsilon}^{vp}\} = \gamma\langle \emptyset(f)\rangle \left\{\frac{\partial Q}{\partial \sigma}\right\} = \gamma\langle \emptyset(f)\rangle\{a\} \tag{18.82}$$

in which

$$\{a\} = \left\{\frac{\partial F}{\partial \sigma}\right\} = \left[\frac{\partial F}{\partial \sigma_x}, \frac{\partial F}{\partial \sigma_y}, \dots, \frac{\partial F}{\partial \tau_{zx}}\right]^{T}$$

Different functions for $\emptyset$ had been adopted and the two most commonly used now are

$$\emptyset(F) = e^{\alpha\left(\frac{F-F_0}{F_0}\right)} - 1 \tag{18.83}$$

$$\emptyset(F) = \left(\frac{F - F_0}{F_0}\right)^{\beta} \tag{18.84a}$$

in which α and β are constants.

Equation (18.84a) is sometimes simply considered as

$$\emptyset(F) = F \tag{18.84b}$$

18.5.3 Visco-Plastic Strain Increment

Divide time t into a series of intervals of time, and the visco-plastic strain increment appearing in interval $\Delta t_n = t_{n+1} - t_n$ is obtained by Eq. (18.85):

$$\{\Delta\varepsilon_n^{\mathrm{vp}}\} = \{\varepsilon^{\mathrm{vp}}(t_{n+1})\} - \{\varepsilon^{\mathrm{vp}}(t_n)\} = \Delta t_n[(1-s)\{\dot{\varepsilon}_n^{\mathrm{vp}}\} + s\{\dot{\varepsilon}_{n+1}^{\mathrm{vp}}\}] \tag{18.85}$$

In Eq. (18.85), in case of $s = 0$, it becomes forward difference formula such that all the strain increments depend on the state of $t = t_n$, the beginning of period, and this is explicit solution. In case of $s = 1$, it becomes backward difference formula such that the strain increments involve the state of $t = t_{n+1}$, the ending of period, and this is implicit method. In case of $s = 0.5$, it turns into midpoint differentiation formula.

On the basis of Eq. (18.79), expand $\{\dot{\varepsilon}^{\mathrm{vp}}\}$ and ignore high-order terms, and we can get

$$\{\dot{\varepsilon}_{n+1}^{\mathrm{vp}}\} = \{\dot{\varepsilon}^{\mathrm{vp}}\} + [H_n]\{\Delta\sigma_n\} \tag{18.86}$$

in which

$$[H_n] = \left.\frac{\partial\{\dot{\varepsilon}^{\mathrm{vp}}\}}{\partial\{\sigma\}}\right|_{t=t_n} \tag{18.87}$$

Substitute Eq. (18.86) into Eq. (18.85) and get

$$\{\Delta\varepsilon_n^{\mathrm{vp}}\} = \{\dot{\varepsilon}^{\mathrm{vp}}\}\Delta t_n + [C_n]\{\Delta\sigma_n\} \tag{18.88}$$

in which

$$[C_n] = s\Delta t_n[H_n] \tag{18.89}$$

18.5.4 Stress Analysis of Viscoelastic–Plastic Bodies

The stress increments producing in period $\Delta t_n = t_{n+1} - t_n$ is

$$\{\Delta\sigma_n\} = [D]\{\Delta\varepsilon_n^e\} = [D](\{\Delta\varepsilon_n\} - \{\Delta\varepsilon_n^{\mathrm{vp}}\}) \tag{18.90}$$

Substitute $\{\Delta\varepsilon_n\} = [B]\{\Delta\delta_n\}$ into the equation above and get

$$\{\Delta\sigma_n\} = [D]([B]\{\Delta\delta_n\} - \{\Delta\varepsilon_n^{\mathrm{vp}}\}) \tag{18.91}$$

Then substitute Eq. (18.88) into Eq. (18.91) and get

$$\{\Delta\sigma_n\} = [D]([B]\{\Delta\delta_n\} - \{\dot{\varepsilon}^{\mathrm{vp}}\}\Delta t_n - [C_n]\{\Delta\sigma_n\})$$

Transform Eq. (18.90) and get

$$\{\Delta\sigma_n\} = [\overline{D}_n]([B]\{\Delta\delta_n\} - \{\dot{\varepsilon}^{\mathrm{vp}}\}\Delta t_n) \tag{18.92}$$

in which

$$[\overline{D}_n] = ([I] + [D][C_n])^{-1}[D] = ([D]^{-1} + [C_n])^{-1} \tag{18.93}$$

The equilibrium condition during period Δt_n is

$$\int [B]^{\mathrm{T}}\{\Delta\sigma_n\}\mathrm{d}V = \{\Delta P_n\} \tag{18.94}$$

Substitute Eq. (18.92) into Eq. (18.94) and get the equilibrium equation set:

$$[K_n]\{\Delta\delta_n\} = \{\Delta P_n\} + \{\Delta P_n^{\text{vp}}\} \tag{18.95}$$

in which

$$[K_n] = \int [B]^{\text{T}}[\overline{D}_n][B]\text{d}V \tag{18.96}$$

$$\{\Delta P_n^{\text{vp}}\} = \int [B]^{\text{T}}[\overline{D}_n]\{\dot{\varepsilon}_n^{\text{vp}}\}\Delta t_n \text{d}V \tag{18.97}$$

where $[K_n]$ is the stiffness matrix, $\{\Delta P_n^{\text{vp}}\}$ is the equivalent load increment caused by visco-plastic strain increment, and $\{\Delta P_n\}$ is the external load increment.

Substitute the strain increment obtained from Eq. (18.95) into Eq. (18.92) to get the stress increment $\{\Delta\sigma_n\}$. Further we can get

$$\{\delta_{n+1}\} = \{\delta_n\} + \{\Delta\delta_n\} \tag{18.98}$$

$$\{\sigma_{n+1}\} = \{\sigma_n\} + \{\Delta\sigma_n\} \tag{18.99}$$

We have already known from Eq. (18.90) that

$$\{\Delta\varepsilon_n^{\text{vp}}\} = \{\Delta\varepsilon_n\} - [D]^{-1}\{\Delta\sigma_n\} = [B]\{\Delta\delta_n\} - [D]^{-1}\{\Delta\sigma_n\} \tag{18.100}$$

As a result,

$$\{\varepsilon_{n+1}^{\text{vp}}\} = \{\varepsilon_n^{\text{vp}}\} + \{\Delta\varepsilon_n^{\text{vp}}\} \tag{18.101}$$

The visco-plastic strain rate $\{\dot{\varepsilon}^{\text{vp}}\}$ is calculated by Eq. (18.92). When the visco-plastic strain rate of every point is 0, it means the deformation is stable so that the calculation can be stopped.

The stress state should meet the equilibrium equation below at any moment t:

$$\int [B]^{\text{T}}\{\sigma\}\text{d}V - \{P\} = 0 \tag{18.102}$$

Because of the error caused by linearization, the stress $\{\sigma_{n+1}\}$ we have already calculated may not fit the equilibrium equation above, and this kind of error will accumulate. To avoid the accumulation of error, we can get the out-of-balance forces by the equation below:

$$\psi_{n+1} = \int [B]^{\text{T}}\{\sigma_{n+1}\}\text{d}V - \{P_{n+1}\} \tag{18.103}$$

Then we superimpose the out-of-balance forces above to the load increment in the next calculation step.

Calculation steps for each time interval are as follows.

Step 1: $\{\delta_n\}$, $\{\Delta\sigma_n\}$, $\{\varepsilon_n\}$, $\{\varepsilon_n^{\text{vp}}\}$, and F_n are known, then calculate the factors following:

$$[C_n] = s\Delta t_n[H_n]$$

$$[\overline{D}_n] = ([D]^{-1} + [C_n])^{-1}$$

$$[K_n] = \int [B]^{\text{T}}[\overline{D}_n][B]\text{d}V$$

$$\{\dot{\varepsilon}_n^{\text{vp}}\} = \gamma\langle\emptyset\rangle\{a\}$$

$$\{\Delta P_n^{\mathrm{vp}}\} = \int [B]^{\mathrm{T}}[\overline{D}_n]\{\dot{\varepsilon}_n^{\mathrm{vp}}\}\Delta t_n \mathrm{d}V$$

$$\{\psi_n\} = \int [B]^{\mathrm{T}}\{\sigma_n\}\mathrm{d}V$$

Step 2: Calculate the displacement increment by the equation below:

$$\{\Delta\delta_n\} = [K_n]^{-1}(\{\Delta P_n\} + \{\Delta P_n^{\mathrm{vp}}\} + \{\Psi_n\})$$

Then calculate $\{\Delta\sigma_n\}$, $\{\delta_{n+1}\}$, and $\{\sigma_{n+1}\}$ as follows:

$$\{\Delta\sigma_n\} = [\overline{D}_n]([B]\{\Delta\delta_n\} - \{\dot{\varepsilon}_n^{\mathrm{vp}}\}\Delta t_n)$$
$$\{\sigma_{n+1}\} = \{\sigma_n\} + \{\Delta\sigma_n\}$$
$$\{\delta_{n+1}\} = \{\delta_n\} + \{\Delta\delta_n\}$$

Step 3: Calculate $\dot{\varepsilon}_{n+1}^{\mathrm{vp}}$ as follows:

$$\{\dot{\varepsilon}_{n+1}^{\mathrm{vp}}\} = \gamma\langle\emptyset\rangle\{a_{n+1}\}$$

Step 4: Examine the visco-plastic strain rate $\{\dot{\varepsilon}_{n+1}^{\mathrm{vp}}\}$ of every point. If every numerical value is close to 0, stop calculating. Otherwise go back to step 1 and continue to calculate.

18.5.5 The Choice of Time Interval Δt_n

In Eq. (18.85), in case of $s < 0.5$, it is conditional stability so that time interval Δt_n must be less than a critical value, or the calculation will be unstable. In case of $s \geq 0.5$, the calculation will be stable without qualification, no matter how long the time interval Δt_n value is. However, in order to guarantee a good calculation accuracy, actually time interval Δt_n is going to have a limitation even $s \geq 0.5$.

Usually the conditions below are formulated for controlling the length of the time interval:

$$(\dot{\varepsilon}_{\mathrm{un}}^{\mathrm{vp}})^{\frac{1}{2}}\Delta t_n \leq \rho(\varepsilon_{\mathrm{iin}})^{\frac{1}{2}} \tag{18.104}$$

in which $\varepsilon_{\mathrm{iin}}$ is the first invariant of total strain, $\dot{\varepsilon}_{\mathrm{un}}^{\mathrm{vp}}$ is the first invariant of visco-plastic strain rate, and ρ is a coefficient.

Every point should satisfy Eq. (18.104) so that we can get

$$\Delta t_n \leq \rho[\varepsilon_{\mathrm{iin}}/\dot{\varepsilon}_{\mathrm{un}}^{\mathrm{vp}}]^{\frac{1}{2}}{}_{\min} \tag{18.105}$$

It means that the upper limit of Δt_n is the minimum of numerical values that are the results of Δt_n fitting calculations for every point. According to the experiences, we can use $\rho = 0.01$–0.15 for explicit solution. As to implicit method, we can give ρ a large number. $\rho = 10$ was used for calculation and it was still stable, but computational accuracy was bad.

When we apply variable time intervals for calculating, there is a limitation below for two adjacent time intervals:

$$\Delta t_{n+1} \leq k\Delta t_n \tag{18.106}$$

in which k is a constant given advance.

There is no criterion for the value of the constant k. Experience shows that $k = 1.5$ is appropriate.

The limitations for time intervals above are experimental. Reference [10] gave a theoretical value for upper limitation of time interval Δt_n to explicit solution. For associated visco-plastic flow problems $Q \equiv F$ and linear function $\emptyset(F) = F$, limitation for time intervals is

$$\left.\begin{array}{ll} \text{Tresca materials:} & \Delta t_n \le \dfrac{(1+\mu)F_0}{\gamma E} \\ \text{Mises materials:} & \Delta t_n \le \dfrac{4(1+\mu)F_0}{3\gamma E} \\ \text{Mohr-Coulomb materials:} & \Delta t_n \le \dfrac{4(1+\mu)(1-2\mu)F_0}{\gamma(1-2\mu+\sin^2\emptyset)E} \end{array}\right\} \tag{18.107}$$

in which γ is flow coefficient, μ is Poisson's ratio, and $\emptyset$ is internal friction angle. For Tresca and Mises materials, F_0 is axial yield stress; for Mohr–Coulomb materials, $F_0 = c\cos\emptyset$, and c is cohesive force.

18.6 Combined Viscoelastic–Plastic Models

As shown in Figure 18.10, combine m sets of viscoelastic–plastic models in parallel into one model and its total stress should be calculated by Eq. (18.108):

$$\sigma = \sum_{i=1}^{m} s_j \sigma_j \tag{18.108}$$

in which σ_j is the stress carried by viscoelastic–plastic model № j. s_j is its participation coefficient, and that is weighting coefficient actually.

s_j should meet condition

$$\sum_{i=1}^{m} s_j = 1 \tag{18.109}$$

Actually we can multiply E_j, σ_i^s, and b_i by s_j.

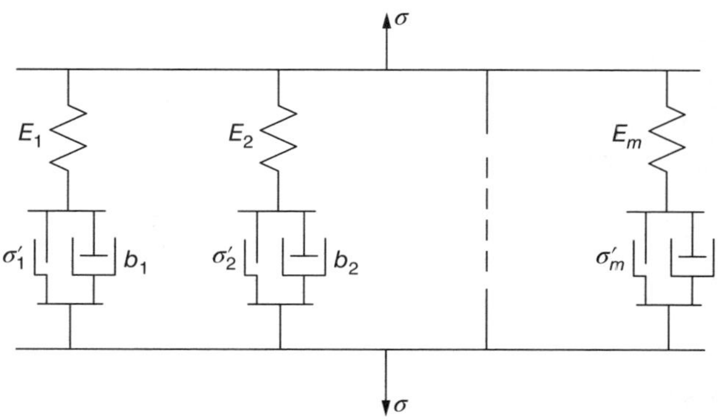

Figure 18.10 Combined model.

The stress increment during time interval $\Delta t_n = t_{n+1} - t_n$ is

$$\Delta\sigma_n = \sum_{i=1}^{m} s_j \Delta\sigma_{jn} \tag{18.110}$$

Substitute Eq. (18.110) into Eq. (18.94) and get the equilibrium condition

$$\int [B]^{\mathrm{T}} \left(\sum_{i=1}^{m} s_j \{\Delta\sigma_{jn}\} \right) \mathrm{d}V = \{\Delta P_n\} \tag{18.111}$$

Thus

$$[K_n]\{\Delta\delta_n\} = \{\Delta P_n\} + \{\Delta P_n^{\mathrm{vp}}\} \tag{18.112}$$

in which

$$[K_n] = \sum_{i=1}^{m} s_j [K_{jn}] = \sum_{i=1}^{m} s_j \int [B]^{\mathrm{T}} \left[[\overline{D}_{jn}] \right] [B] \mathrm{d}V \tag{18.113}$$

$$\{\Delta P_n^{\mathrm{vp}}\} = \sum_{i=1}^{m} s_j \{\Delta P_{jn}^{\mathrm{vp}}\} = \sum_{i=1}^{m} s_j \int [B]^{\mathrm{T}} [\overline{D}_{jn}] \{\Delta \dot{\varepsilon}_{jn}^{\mathrm{vp}}\} \Delta t_n \mathrm{d}V \tag{18.114}$$

It is clear that we can get all the computational formulae for combined model by simple superposition of involved items in the upper segment, and the calculation is convenient.

This kind of combined model shares a strong simulation ability. For example, take one model and let $\sigma_1^s = 0$, and we will get Maxwell model as shown in Figure 18.3. Take two submodels and let $\sigma_1^{\mathrm{s}} = E_2 = \infty$, $\sigma_2^{\mathrm{s}} = 0$, and we will get Voigt model as shown in Figure 18.4. Take two submodels and let $\sigma_1^{\mathrm{s}} = \infty$, $\sigma_2^{\mathrm{s}} = 0$, and we will get a standard three-element model as shown in Figure 18.5. Take two models and let $\sigma_1^{\mathrm{s}} = \sigma_2^{\mathrm{s}} = 0$, and we will get a four-element model.

Certainly, this kind of combined model is used to simulate the stress–strain–time relation of practical engineering materials more frequently. Figure 18.11 shows an example that uses a model combined with two submodels to simulate the stress–strain–time relation of rock salt [9]. The calculating parameters are as follows:

For submodel 1:

$$\begin{aligned}
&s_1 = 0.5 \\
&E_1 = 470\,\mathrm{MPa} \\
&\mu_1 = 0.24 \\
&\text{cohesive force } c_1 = 0 \\
&\gamma_1 = 2.0 \times 10^{-3}/(\mathrm{MPa} \cdot \mathrm{d}) \\
&\text{frictional angle } \phi_1 = 0
\end{aligned}$$

For submodel 2:

$$\begin{aligned}
&s_2 = 0.5 \\
&E_2 = 1540\,\mathrm{MPa} \\
&\mu_2 = 0.24 \\
&\text{cohesive force } c_2 = 5.5\,\mathrm{MPa} \\
&\gamma_2 = 1.0 \times 10^{-4}/(\mathrm{MPa} \cdot \mathrm{d}) \\
&\text{frictional angle } \phi_1 = 0
\end{aligned}$$

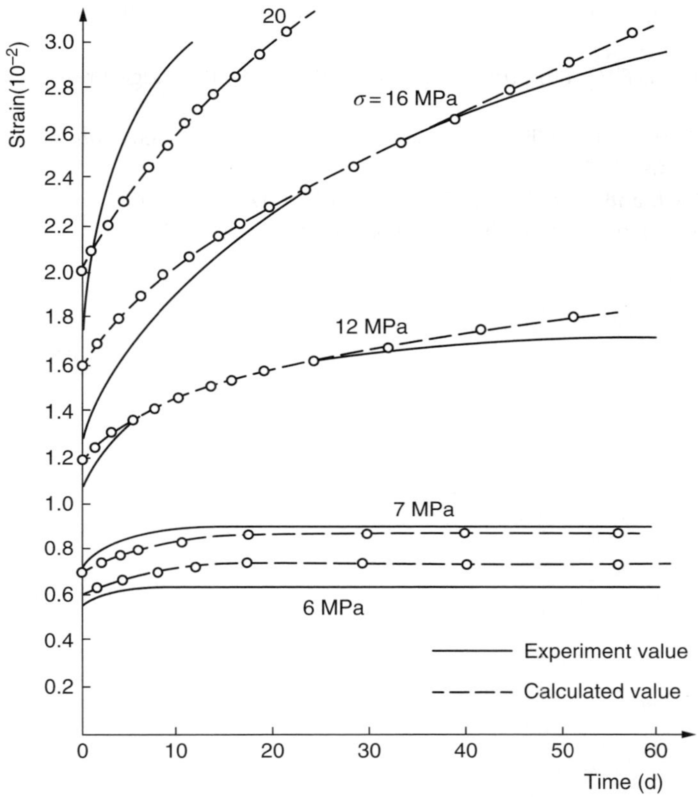

Figure 18.11 Practical creep curve simulated by combined models.

It can be seen from Figure 18.11 that the calculated value agrees well with experimental value. If more kinds of submodels were combined, the result would be better.

Bibliography

1 Zhu, B.F. (1964) The stresses and deformations in the nonhomogeneous visco-elastic media under mixed boundary conditions. *J. Hydraul. Eng.*, **2**, 74–79.

2 Zhu, B.F. (1984) Substructure method for stress analysis of nonhomogeneous elasto-creeping solids. *J. Hydraul. Eng.*, **2**, 22–26.

3 Alfrey, T. (1944) Non-homogeneous stresses in visco-elastic bodies. *Q. Appl. Math.*, **2**, 113.

4 Tsien, H.S. (1950) A generalization of Alfrey's theorem for visco-elastic media. *Q. Appl. Math.*, **8**, 114.

5 Zienkiewicz, O.C., Watson, M. and King, I.P. (1968) A numerical method of visco-elastic stress analysis. *Intern. J. Mech. Sci.*, **10** (10), 807–827.

6 Bland, D.R. (1960) *The Theory of Linear Visco-Elasticity*, Pergamon Press, Oxford.

7 Christensen, R.M. (1982) *Theory of Visco-Elasticity*, Academic Press, New York.

8 Cristescu, N. and Suliciu, I. (1982) *Visco-Elasticity*, Martinus Nijhoff Publishers, Hague.
9 Owen, D.R.J. and Hinton, E. (1980) *Finite Elements in Plasticity*, Pineridge Press, Swansca, UK.
10 Cormeau, I. (1975) Numerical stability in quasistatic elasto-visco-plasticity. *Int. J. Numer. Meth. Eng.*, **9**, 109–127.
11 Pande, G.N., Wen, D.R.J. and Zienkiewicz, O.C. (1977) Overlay models in time-dependent nonlinear material analysis. *Comput. Struct.*, **7**, 435–443.

19

Elastic Stability Problem

In the design of thin-walled structures, the elastic structural stability must be taken into consideration. The finite element method is widely applied in this area. The analysis of elastic stability is often divided into two steps. The first step is to calculate the structural internal force distribution with the linear method. The second step is to calculate the critical load of instability. Method of the first step has been explained in the chapters before. In this chapter, we will focus on how to calculate the critical load by finite element method when the initial distribution of the internal forces is given. The principle of minimum potential energy is applied in the derivation of formulas.

This chapter focuses on the linear elastic stability problem. Here, "linear" implies that (1) the bar axial force and the plate membrane force are determined by the linear elastic analysis, and (2) on the occurrence of the infinitesimal displacement caused by buckling, the axial or membrane forces stay constant. For plates, the membrane forces are calculated with linear elastic plane stress analysis and remain unchanged in the buckling state. The nonlinear buckling problem and the post-buckling problem, that is, the geometrical nonlinear problems will be discussed in the next chapter.

19.1 Geometrical Stiffness Matrix of the Beam Element

As shown in Figure 19.1, an axial force F_x is applied on the beam element *ij*, which has constant cross section, and it is assumed that the axial force will not be affected by the deflection w.

The deflection w leads to the additional strain at the neutral axis of the beam. Take an infinitesimal segment AB from the beam element as shown Figure 19.2; its length is $\mathrm{d}x$. With the deflection w in the beam, the length becomes

$$\overline{\mathrm{d}x} = \sqrt{(\mathrm{d}x)^2 + \left(\frac{\mathrm{d}w}{\mathrm{d}x}\mathrm{d}x\right)^2} = \mathrm{d}x\sqrt{1 + \left(\frac{\mathrm{d}w}{\mathrm{d}x}\right)^2}$$

Expand the right part of the equation above by binomial theorem

$$\overline{\mathrm{d}x} = \mathrm{d}x\left[1 + \frac{1}{2}\left(\frac{\mathrm{d}w}{\mathrm{d}x}\right)^2 + \cdots\right]$$

The Finite Element Method: Fundamentals and Applications in Civil, Hydraulic, Mechanical and Aeronautical Engineering, First Edition. Bofang Zhu.

Companion website: www.wiley.com/go/zhufem17

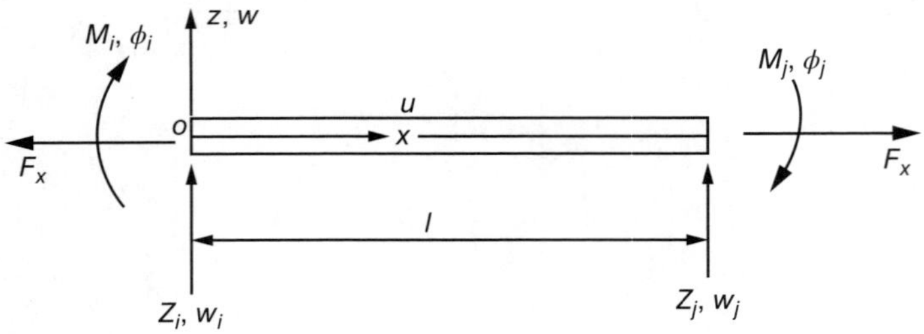

Figure 19.1 Beam element.

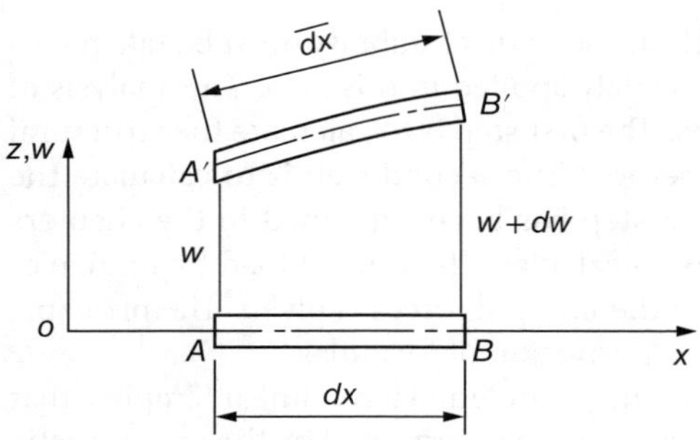

Figure 19.2 A differential part.

In the right part of the above equation, ignore the high-order term after the third term, so that the additional strain in the neutral axis caused by the deflection w is

$$\varepsilon_x^{\mathrm{I}} = \frac{1}{2}\left(\frac{\mathrm{d}w}{\mathrm{d}x}\right)^2$$

With the lateral deflection w, the curvature of the neutral axis is $\mathrm{d}^2w/\mathrm{d}x^2$, and the flexure strain of the beam is

$$\varepsilon_x^{\mathrm{II}} = -z\left(\frac{\mathrm{d}^2w}{\mathrm{d}x^2}\right)$$

The strain in the beam due to the axial displacement u is

$$\varepsilon_x^{\mathrm{III}} = \frac{\mathrm{d}u}{\mathrm{d}x}$$

Superposing the three terms above, we obtain the total strain in the beam element as

$$\varepsilon_x = \frac{\mathrm{d}u}{\mathrm{d}x} - z\left(\frac{\mathrm{d}^2w}{\mathrm{d}x^2}\right) + \frac{1}{2}\left(\frac{\mathrm{d}w}{\mathrm{d}x}\right)^2 \tag{19.1}$$

The strain energy of the beam element can be calculated by the Eq. (19.2):

$$U^{\mathrm{e}} = \frac{1}{2}\int E\varepsilon_x^2\mathrm{d}\,(\mathrm{vol}) \tag{19.2}$$

where vol represents the volume. Substituting Eq. (19.1) into Eq. (19.2), and letting d (vol) = dAdx, we get

$$U^{\mathrm{e}} = \frac{1}{2}\int_L\int_A E\left\{\begin{matrix}\left(\frac{\mathrm{d}u}{\mathrm{d}x}\right)^2 + z^2\left(\frac{\mathrm{d}^2w}{\mathrm{d}x^2}\right)^2 + \frac{1}{4}\left(\frac{\mathrm{d}w}{\mathrm{d}x}\right)^4 - 2z\left(\frac{\mathrm{d}u}{\mathrm{d}x}\right)\left(\frac{\mathrm{d}^2w}{\mathrm{d}x^2}\right) \\ -z\left(\frac{\mathrm{d}^2w}{\mathrm{d}x^2}\right)\left(\frac{\mathrm{d}w}{\mathrm{d}x}\right)^2 + \left(\frac{\mathrm{d}u}{\mathrm{d}x}\right)\left(\frac{\mathrm{d}w}{\mathrm{d}x}\right)^2\end{matrix}\right\}\mathrm{d}A\mathrm{d}x \tag{19.3}$$

Integrating through the thickness direction, and putting the origin of coordinates at the centroid of the beam cross section, we have

$$\int_A \mathrm{d}A = A, \quad \int_A z\mathrm{d}A = 0, \quad \int_A z^2\mathrm{d}A = I$$

Hence, from the Eq. (19.3), we have

$$U^{\mathrm{e}} = \frac{1}{2}\int_L E\left\{A\left(\frac{\mathrm{d}u}{\mathrm{d}x}\right)^2 + I\left(\frac{\mathrm{d}^2w}{\mathrm{d}x^2}\right)^2 + A\left(\frac{\mathrm{d}u}{\mathrm{d}x}\right)\left(\frac{\mathrm{d}w}{\mathrm{d}x}\right)^2 + \frac{A}{4}\left(\frac{\mathrm{d}w}{\mathrm{d}x}\right)^4\right\}\mathrm{d}x \tag{19.4}$$

The relation between axial force F_x and the axial displacement u is

$$F_x = EA\frac{\mathrm{d}u}{\mathrm{d}x} \tag{19.5}$$

Here for F_x, tension is positive and compression is negative. Ignoring the high-order term, and substituting Eq. (19.5) into Eq. (19.4), we get

$$U^{\mathrm{e}} = \frac{1}{2}\int_L\left\{EA\left(\frac{\mathrm{d}u}{\mathrm{d}x}\right)^2 + EI\left(\frac{\mathrm{d}^2w}{\mathrm{d}x^2}\right)^2 + F_x\left(\frac{\mathrm{d}w}{\mathrm{d}x}\right)^2\right\}\mathrm{d}x \tag{19.6}$$

From Eq. (19.6), the beam element strain energy can be divided into the axial strain energy and the flexure strain energy as

$$U^{\mathrm{e}} = U_{\mathrm{a}}^{\mathrm{e}} + U_{\mathrm{b}}^{\mathrm{e}} \tag{19.7}$$

$$U_{\mathrm{a}}^{\mathrm{e}} = \frac{1}{2}\int_L EA\left(\frac{\mathrm{d}u}{\mathrm{d}x}\right)^2\mathrm{d}x \tag{19.8}$$

$$U_{\mathrm{b}}^{\mathrm{e}} = \frac{1}{2}\int_L\left\{EI\left(\frac{\mathrm{d}^2w}{\mathrm{d}x^2}\right)^2 + F_x\left(\frac{\mathrm{d}w}{\mathrm{d}x}\right)^2\right\}\mathrm{d}x \tag{19.9}$$

As discussed above, the analysis of elastic stability is divided into two steps. First is to calculate the structural internal force distribution with the linear method. The axial strain energy can be considered in this step. Second is to calculate the critical load and the bending strain energy should be taken into consideration.

The nodal displacement of the beam element can also be divided into two parts, that is, the axial displacement and the bending displacement:

$$\{\delta_{\mathrm{a}}^{\mathrm{e}}\} = \begin{Bmatrix} u_i \\ u_j \end{Bmatrix}, \quad \{\delta_{\mathrm{b}}^{\mathrm{e}}\} = \begin{Bmatrix} w_i \\ \varphi_i \\ w_j \\ \varphi_j \end{Bmatrix}$$

The deflection of the beam element w can be expressed by the nodal bending displacement:

$$w = \begin{bmatrix} N_1 & N_2 & N_3 & N_4 \end{bmatrix} \begin{Bmatrix} w_i \\ \varphi_i \\ w_j \\ \varphi_j \end{Bmatrix} = [N] \left\{ \delta_b^e \right\} \tag{19.10}$$

By the Eq. (19.10), we can get the first- and second-order derivatives of w with respect to x, namely,

$$\frac{dw}{dx} = [g] \left\{ \delta_b^e \right\} \tag{19.11}$$

$$\frac{d^2w}{dx^2} = [B] \left\{ \delta_b^e \right\} \tag{19.12}$$

where

$$[g] = \left[\frac{dN}{dx} \right] = \left[\frac{dN_1}{dx} \quad \cdots \quad \frac{dN_4}{dx} \right] \tag{19.13}$$

$$[B] = \left[\frac{d^2N}{dx^2} \right] = \left[\frac{d^2N_1}{dx^2} \quad \cdots \quad \frac{d^2N_4}{dx^2} \right] \tag{19.14}$$

By matrix algebra, we get

$$\left(\frac{dw}{dx} \right)^2 = \left\{ \delta_b^e \right\}^T [g]^T [g] \left\{ \delta_b^e \right\}$$

$$\left(\frac{d^2w}{dx^2} \right)^2 = \left\{ \delta_b^e \right\}^T [B]^T [B] \left\{ \delta_b^e \right\}$$

By substituting Eqs (19.11) and (19.12) into Eq. (19.9), we get the bending strain energy of the beam element as

$$U_b^e = \frac{1}{2} \left\{ \delta_b^e \right\}^T [k_b] \left\{ \delta_b^e \right\} + \frac{1}{2} \left\{ \delta_b^e \right\}^T [k_\sigma] \left\{ \delta_b^e \right\} \tag{19.15}$$

where

$$[k_b] = \int_L [B]^T EI [B] \, dx \tag{19.16}$$

$$[k_\sigma] = \int_L [g]^T F_x [g] \, dx \tag{19.17}$$

Here, matrix $[k_b]$ is a usual bending stiffness matrix of the beam element. Matrix $[k_\sigma]$ is the additional stiffness matrix owing to the axial force F_x.

In other words, when we take the axial force F_x into consideration, the stiffness matrix of the beam element is

$$[k] = [k_b] + [k_\sigma] \tag{19.18}$$

When the axial force F_x is tension, $[k_\sigma]$ is positive, and the stiffness matrix of the beam element increases with F_x. Conversely, when the axial force F_x is compression, $[k_\sigma]$ is negative, and the stiffness matrix of the beam element decreases with F_x. The Eq. (19.17) indicates that matrix $[k_\sigma]$ is independent on the material physical constants, such as E, and is merely influenced by the elemental geometry so it is called geometrical stiffness matrix. Matrix $[k_\sigma]$ is only related to the initial force F_x, so it is called initial stress stiffness matrix.

19.2 Geometrical Stiffness Matrix of Plate Elements

Similar to the analysis in the last section, in a thin plate, the additional strain in the plate caused by the deflection w is (Figure 19.3)

$$\varepsilon_x^{\mathrm{I}} = \frac{1}{2}\left(\frac{\partial w}{\partial x}\right)^2, \quad \varepsilon_y^{\mathrm{I}} = \frac{1}{2}\left(\frac{\partial w}{\partial y}\right)^2, \quad \gamma_{xy}^{\mathrm{I}} = \frac{\partial w}{\partial x}\frac{\partial w}{\partial y}$$

The strain caused by the plate flexure is

$$\varepsilon_x^{\mathrm{II}} = -z\frac{\partial^2 w}{\partial x^2}, \quad \varepsilon_y^{\mathrm{II}} = -z\frac{\partial^2 w}{\partial y^2}, \quad \gamma_{xy}^{\mathrm{II}} = -2z\frac{\partial^2 w}{\partial x \partial y}$$

The strain caused by the displacement u, v in the plate is

$$\varepsilon_x^{\mathrm{II}} = \frac{\partial u}{\partial x}, \quad \varepsilon_y^{\mathrm{II}} = \frac{\partial v}{\partial y}, \quad \gamma_{xy}^{\mathrm{II}} = \frac{\partial u}{\partial y} + \frac{\partial v}{\partial x}$$

Superimposing the terms above, we get the strain–displacement relation of the thin plate:

$$\left.\begin{aligned}
\varepsilon_x &= \frac{\partial u}{\partial x} - z\frac{\partial^2 w}{\partial x^2} + \frac{1}{2}\left(\frac{\partial w}{\partial x}\right)^2 \\
\varepsilon_y &= \frac{\partial v}{\partial y} - z\frac{\partial^2 w}{\partial y^2} + \frac{1}{2}\left(\frac{\partial w}{\partial y}\right)^2 \\
\gamma_{xy}^{\mathrm{II}} &= \frac{\partial u}{\partial y} + \frac{\partial v}{\partial x} - 2z\frac{\partial^2 w}{\partial x \partial y} + \frac{\partial w}{\partial x}\frac{\partial w}{\partial y}
\end{aligned}\right\} \tag{19.19}$$

Similar to the equation derivation in the last section, the flexure strain energy of an isotropic thin plate is

$$\begin{aligned}
U_{\mathrm{b}}^{\mathrm{e}} = {} & \frac{D}{2}\int_A \left\{\left(\frac{\partial^2 w}{\partial x^2}\right)^2 + \left(\frac{\partial^2 w}{\partial y^2}\right)^2 + 2\mu\frac{\partial^2 w}{\partial x^2}\frac{\partial^2 w}{\partial y^2} + 2(1-\mu)\left(\frac{\partial^2 w^2}{\partial x \partial y}\right)^2\right\}\mathrm{d}A \\
& + \frac{1}{2}\int_A \left\{\sigma_x t\left(\frac{\partial w}{\partial x}\right)^2 + \sigma_y t\left(\frac{\partial w}{\partial y}\right)^2 + 2\tau_{xy} t\frac{\partial w}{\partial x}\frac{\partial w}{\partial y}\right\}\mathrm{d}A
\end{aligned} \tag{19.20}$$

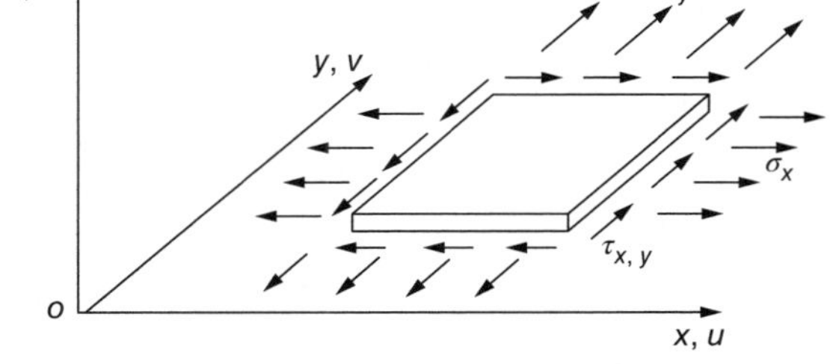

Figure 19.3 Plate element.

where $D = \frac{Et^3}{12(1-\mu^2)}$. In the above equation, t represents thickness of the plate, and $\sigma_x t, \sigma_y t, \tau_{xy} t$ are the internal forces in the plane of the plate.

The second term in right part of Eq. (19.20) represents strain energy induced by the initial plane stress of the plate, which can be written by matrix as

$$\frac{1}{2}\int_A \begin{Bmatrix} \frac{\partial w}{\partial x} \\ \frac{\partial w}{\partial y} \end{Bmatrix}^{\mathrm{T}} \begin{bmatrix} \sigma_x & \tau_{xy} \\ \tau_{xy} & \sigma_y \end{bmatrix} \begin{Bmatrix} \frac{\partial w}{\partial x} \\ \frac{\partial w}{\partial y} \end{Bmatrix} t\mathrm{d}A$$

By the shape function $[N]$ and the nodal flexure displacement $\{\delta_{\mathrm{b}}^{\mathrm{e}}\}$ of the plate element, we can express the deflection of the plate as

$$w = [N]\{\delta_{\mathrm{b}}^{\mathrm{e}}\} \tag{19.21}$$

Let

$$\{\psi\} = \begin{Bmatrix} -\frac{\partial^2 w}{\partial x^2} \\ -\frac{\partial^2 w}{\partial y^2} \\ -2\frac{\partial^2 w}{\partial x \partial y} \end{Bmatrix} = [B]\{\delta_{\mathrm{b}}^{\mathrm{e}}\} \tag{19.22}$$

$$\begin{Bmatrix} \frac{\partial w}{\partial x} \\ \frac{\partial w}{\partial y} \end{Bmatrix} = [g]\{\delta_{\mathrm{b}}^{\mathrm{e}}\} \tag{19.23}$$

where

$$[B] = -\begin{bmatrix} \frac{\partial^2 N}{\partial x^2} \\ \frac{\partial^2 N}{\partial y^2} \\ \frac{\partial^2 N}{\partial x \partial y} \end{bmatrix}, \quad [g] = \begin{bmatrix} \frac{\partial N}{\partial x} \\ \frac{\partial N}{\partial y} \end{bmatrix} \tag{19.24}$$

Substituting Eqs (19.22) and (19.23) into Eq. (19.20), we get the flexure strain energy of the plate element as

$$U_{\mathrm{b}}^{\mathrm{e}} = \frac{1}{2}\{\delta_{\mathrm{b}}^{\mathrm{e}}\}^{\mathrm{T}}[k_{\mathrm{b}}]\{\delta_{\mathrm{b}}^{\mathrm{e}}\} + \frac{1}{2}\{\delta_{\mathrm{b}}^{\mathrm{e}}\}^{\mathrm{T}}[k_{\sigma}]\{\delta_{\mathrm{b}}^{\mathrm{e}}\} \tag{19.25}$$

Here,

$$[k_{\mathrm{b}}] = \int_L [B]^{\mathrm{T}}[D][B]\,\mathrm{d}x \tag{19.26}$$

$$[k_{\sigma}] = \int_A [g]^{\mathrm{T}} \begin{bmatrix} \sigma_x & \tau_{xy} \\ \tau_{xy} & \sigma_y \end{bmatrix} [g]\, t\mathrm{d}A \tag{19.27}$$

$$[D] = D\begin{bmatrix} 1 & \mu & 0 \\ \mu & 1 & 0 \\ 0 & 0 & \frac{1-\mu}{2} \end{bmatrix}$$

In the above equations, $[k_\mathrm{b}]$ is the usual flexure stiffness matrix of the plate and $[k_\sigma]$ is the geometrical stiffness matrix of the thin plate.

According to Eq. (19.20), the geometrical stiffness matrix can also be written as

$$[k_\sigma] = \sigma_x [k_{\sigma x}] + \sigma_y [k_{\sigma y}] + \tau_{xy} [k_{\sigma xy}] \tag{19.28}$$

Here,

$$\left.\begin{aligned}
[k_{\sigma x}] &= t \int_A \left[\frac{\partial N}{\partial x}\right]^\mathrm{T} \left[\frac{\partial N}{\partial x}\right] \mathrm{d}A \\
[k_{\sigma y}] &= t \int_A \left[\frac{\partial N}{\partial y}\right]^\mathrm{T} \left[\frac{\partial N}{\partial y}\right] \mathrm{d}A \\
[k_{\sigma xy}] &= 2t \int_A \left[\frac{\partial N}{\partial x}\right]^\mathrm{T} \left[\frac{\partial N}{\partial y}\right] \mathrm{d}A
\end{aligned}\right\} \tag{19.29}$$

According to Eq. (19.27), the geometrical stiffness matrix of the plate element is independent of the material physical constant. It can be applied to the isotropic plate and the anisotropic plate. Besides, if the internal plane stresses of the plate are not known in advance, we can calculate these stresses by the finite element method with the same mesh.

19.3 Global Analysis

Assuming there are two groups of loads, shown in the Figure 19.4, one group is the longitudinal load acting in the plane of the plate, which will induce internal stress. The other group is the transverse load $\{P\}$, which will lead to the flexure of the plate.

It is assumed that we have already calculated the stress $\{\sigma\}$ induced by the longitudinal load acting on the plate. Thus, only the flexure behavior induced by the transverse load will be discussed.

The strain energy is a scalar; the strain energy of the whole structure equals the sum of that of each element. So under the transverse load, the flexure strain energy of the plate is

$$U = \sum U^\mathrm{e} = \frac{1}{2}\{\delta_\mathrm{b}\}^\mathrm{T} [K_\mathrm{b}] \{\delta_\mathrm{b}\} + \frac{1}{2}\{\delta_\mathrm{b}\}^\mathrm{T} [K_\sigma] \{\delta_\mathrm{b}\} \tag{19.30}$$

where

$$[K_\mathrm{b}] = \left[\sum k_\mathrm{b}\right] \tag{19.31}$$

$$[K_\sigma] = \left[\sum k_\sigma\right] \tag{19.32}$$

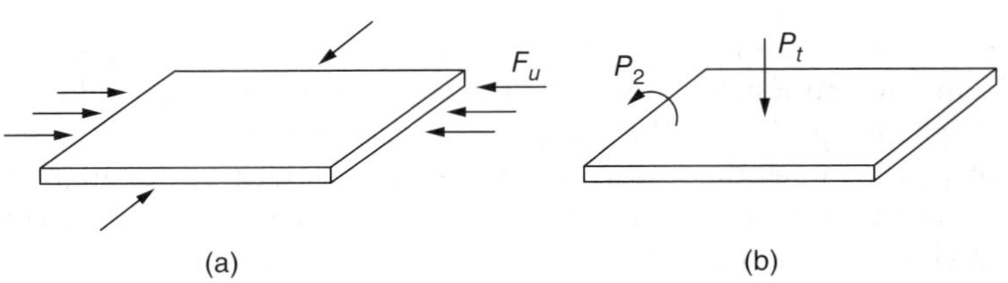

Figure 19.4 Load acting on the plate. (a) Longitudinal load and (b) transverse load.

With the transverse load {P}, we assume that the external potential energy of the plate is

$$V = \{\delta_b\}^T \{P\}$$

The potential energy of the whole structure is

$$\Pi_p = U - V = \frac{1}{2}\{\delta_b\}^T [K_b] \{\delta_b\} + \frac{1}{2}\{\delta_b\}^T [K_\sigma] \{\delta_b\} - \{\delta_b\}^T \{P\}$$

According to the principle of minimum potential energy, the first-order variation of the potential energy is equal to zero. So we get Eq. (19.33) as follows:

$$([K_b] + [K_\sigma]) \{\delta_b\} = \{P\} \tag{19.33}$$

When the longitudinal load is given, and the transverse load is nonzero, we can calculate the displacement of the structure with the equation above. Here, the geometric stiffness matrix reflects the influence of the longitudinal load on the transverse deflection.

According to the analysis of the last two sections, the geometric stiffness matrix is proportional to the longitudinal load. If the longitudinal load increases with the factor λ as

$$\{F\} = \lambda \{F_o\}$$

Then the geometric stiffness matrix increases correspondingly, so Eq. (19.33) turns to

$$([K_b] + \lambda [K_\sigma]) \{\delta_b\} = \{P\}$$

When the longitudinal load is small, $\{\delta_b\}$ is proportional to $\{P\}$. Especially when $\{P\} = 0$, $\{\delta_b\} = 0$ and when the longitudinal load gets to the critical load, that is,

$$\{F_{cr}\} = \lambda_{cr} \{F_o\} \tag{19.34}$$

The structure loses elastic stability. At this moment, even though the transverse load $\{P\} = 0$, the structure can still keep a certain displacement, that is,

$$([K_b] + \lambda [K_\sigma]) \{\delta_b\} = 0 \tag{19.35}$$

This equation is a system of homogeneous equations. To get a solution other than zero, the determinant of the coefficient matrix must be zero, that is,

$$\left| [K_b] + \lambda [K_\sigma] \right| = 0 \tag{19.36}$$

By this equation we can get the minimum eigenvalue λ. Then substituting it into Eq. (19.34), we can get the critical load of buckling.

As for the large-scale problem, it is not easy to calculate eigenvalue directly by Eq. (19.36). Iterative method is often used to solve the problem. Transfer Eq. (19.35) to

$$\frac{1}{\lambda} \{\delta_b\} = -[K_b]^{-1} [K_\sigma] \{\delta_b\} \tag{19.37}$$

By iteration, we can get the minimum eigenvalue λ as well as the corresponding eigenvector by Eq. (19.37). In engineering, we are interested is the minimum eigenvalue λ. Compared with other dynamic problems, the calculation cost is always smaller.

Nodal displacement $\{\delta_b\}$ includes the linear displacement w as well as the angular displacement φ. Merely by linear displacement, we can usually calculate the critical load with enough accuracy. Hence, we can decrease the freedom of nodes similarly as in the calculation of the natural frequency of vibration, which will be explained in Section 22.4.

19.4 Cases of Beam System

For beam element shown in Figure 19.1, we take the displacement function as

$$w = \left[L_1^2(3-2L_1) \quad -L_1^2L_2l \quad L_2^2(3-2L_2) \quad L_1^2L_2l\right]\begin{Bmatrix} w_i \\ \phi_i \\ w_j \\ \phi_j \end{Bmatrix} \tag{19.38}$$

where $L_1 = 1 - \frac{x}{l}, \quad L_2 = \frac{x}{l}$

The flexure stiffness matrix of the beam element is shown in Eq. (1.24). By Eq. (19.38), we get

$$[g] = \left[\frac{\mathrm{d}N}{\mathrm{d}x}\right] = \left[-\frac{6L_1L_2}{l} \quad -L_1(L_1-2L_2) \quad \frac{6L_1L_2}{l} \quad -L_1(L_2-2L_1)\right]$$

According to Eq. (19.17), we get the geometric stiffness matrix of the beam element as

$$[k_\sigma] = \frac{F_x}{30l}\begin{array}{c} \begin{array}{cccc} w_i & \phi_i & w_j & \phi_j \end{array} \\ \begin{bmatrix} 36 & -3l & -36 & -3l \\ -3l & 4l^2 & 3l & -l^2 \\ -36 & 3l & 36 & 3l \\ -3l & -l^2 & 3l & 4l^2 \end{bmatrix} \end{array} \tag{19.39}$$

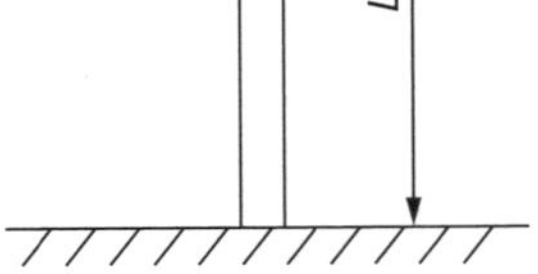

Figure 19.5 Column.

By the geometric stiffness matrix above, we can calculate the critical load of the beam system. We take the column shown in Figure 19.5 as an example, whose length is L and the load is P, that is,

$$F_x = -P$$

Here, only one element is used as a simple example. According to boundary conditions, we have

$$w_j = \phi_j = 0$$

There are only two degrees of freedom in the column, namely, w_i and ϕ_i. According to Eqs (1.24) and (19.39), the stiffness matrix is

$$[k] = [k_\mathrm{b}] + [k_\sigma] = \frac{EI}{l^3}\begin{bmatrix} 12 & -6l \\ -6l & 4l^2 \end{bmatrix} - \frac{P}{30l}\begin{bmatrix} 36 & -3l \\ -3l & 4l^2 \end{bmatrix} \tag{a}$$

Let

$$\omega = \frac{Pl^2}{30EI} \tag{b}$$

then

$$[k] = \frac{EI}{l^3}\begin{bmatrix} 12-36\omega & -6l-3\omega l \\ -6l-3\omega l & 4l-4\omega l^2 \end{bmatrix}$$

By the condition that the determinant of the stiffness matrix equals to zero, we have

$$\begin{vmatrix} 12(1-3\omega) & -3l(2+\omega) \\ -3l(2+\omega) & 4l(1-\omega l) \end{vmatrix} = 0$$

Two solutions of this equation are $\omega_1 = 0.0829, \omega_2 = 1.072$. Substituting the smaller solution ω_1 into Eq. (19.40b), the critical load of the column is

$$P_{cr} = 0.251\frac{\pi^2 EI}{l^2}$$

Compared with the theoretical solution $P_{cr} = 0.250\frac{\pi^2 EI}{l^2}$, the error is 0.4%. If the number of elements is increased, even higher computational accuracy can be obtained.

19.5 Computing Examples of Elastic Stability of Thin Plate System

When the deflection function of the thin plate element $w = [N]\{\delta_b\}^e$ is given, we can calculate the geometric stiffness matrix by Eq. (19.27). (The geometric stiffness of rectangular and triangular elements is given in Refs [2] and [5], respectively.) By these element geometric stiffness matrices, we can easily get the critical load of longitudinal buckling of each kind of thin plates. Some examples are shown in the following.

19.5.1 Rectangular Thin-Plate Element

For the rectangular thin plate shown in Figure 19.6, the plane size is $2a \times 2b$ and the thickness is t. In Ref. [2], the critical load of buckling is computed by rectangular element. Results of different element subdivision are shown in Table 19.1. Critical load is written as

$$\sigma_{cr} = \frac{k\pi^2 D}{4b^2 t}$$

When it is compressed in one direction, σ_{cr} in the above equation represents the critical value of σ_x. When it is pure shear on the four boundaries, σ_{cr} represents the critical value of shear stress τ. The last case in the Table 19.1 is a rectangular plate with reinforcing rib. The area of the rib is A_r, and the moment of inertia is I. Besides,

$$\frac{A_r}{2bt} = 0.20, \quad \frac{EI}{2bD} = 5$$

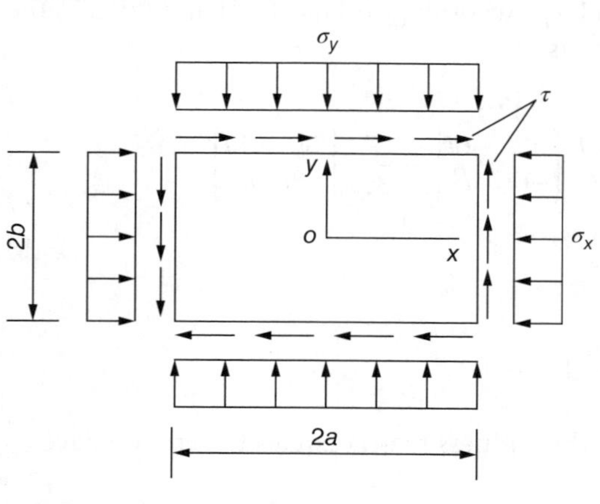

Figure 19.6 Rectangular thin plate.

Table 19.1 Longitudinal buckling load parameter k in different element division of rectangular thin plate (size 2a × 2b).

a/b	Load	Boundary conditions	Accurate k	4 × 4		6 × 6		8 × 8		10 × 10	
				k	Error	k	Error	k	Error	k	Error
1.0	Simple compress σ_x	Simply supported four sides	4	3.77	−5.75%	3.89	−2.80%	3.93	−1.68%	3.96	−1.00%
1.0	Simple compress σ_x	Fixed on four sides	10.07	9.28	−7.80%	9.61	−4.56%	9.78	−2.86%	9.89	−1.79%
1.0	Biaxial compress $\sigma_x = \sigma_y$	Fixed on four sides	5.315	4.98	−6.38%	5.08	−4.46%	5.16	−2.92%	5.22	−1.84%
1.25	Sheared in four sides	Simply supported four sides	7.71	6.95	−9.92%	7.25	−6.00%	7.45	−3.37%	—	—
1.0	Simple compress σ_x	Simply supported on four sides with longitudinal reinforced rib with A_r, EI $\frac{A_r}{2bt} = 0.20$ $\frac{EI}{2BD} = 5$	9.72	9.46	−2.66%	9.58	−1.48%	9.63	0.98%	9.65	0.71%

According to the Table 19.1, when the computational grid gets denser, the calculation error decreases rapidly.Ref. [2] also solved a longitudinal buckling problem of orthotropic rectangular plate (a/b = 2). It is compressed in the x direction, and the material constants and the size of the plate are

$$E_x = 211\,\text{GPa},\ \ E_y = 35.1\,\text{GPa},\ \ G = 70.2\,\text{GPa},$$
$$\mu_{xy} = 0.3,\ \ \mu_{yx} = 0.05,\ \ 2a = 61\,\text{cm},\ \ t = 0.254\,\text{cm}$$

Now in equation $\{M\} = [D]\{\psi\}$, the flexure elasticity matrix of the thin plate is

$$[D] = \begin{bmatrix} D_{11} & D_{12} & 0 \\ D_{12} & D_{22} & 0 \\ 0 & 0 & D_{33} \end{bmatrix}$$

$$D_{11} = \frac{E_x t^3}{12\left(1-\mu_{xy}\mu_{yx}\right)},\qquad D_{22} = \frac{E_y t^3}{12\left(1-\mu_{xy}\mu_{yx}\right)},$$

$$D_{33} = \frac{Gt^3}{12}, D_{12} = \mu_{yx} D_{11}$$

With a 10×10 grid, by finite elemental method the critical load is 1830 N, and the accurate result is 1841 N, with the error 0.6%.

19.5.2 Triangular Thin-Plate Elements

The triangular plate element can adapt to irregular geometric shapes, so its practical value is big. In Ref. [10], with the grid shown in Figure 19.7, two kinds of triangular elements are used to calculate the buckling load of a rectangular thin plate in unidirectional compression. The calculated and exact values are shown in Table 19.2, where T_9 represents triangular element with 9 degrees of freedom; T_{18} represents triangular element with 18 degrees of freedom (there are 6 degrees of freedom in each node); R_{12} represents rectangular element with 12 degrees of freedom. Due to symmetry, only 1/4 needs to be taken into calculation, in which the calculated values of rectangular element are taken from Ref. [2]. According to this table, T_{18} has a higher accuracy. Certainly, stiffness of this element is more complex.

Figure 19.8 shows longitudinal buckling problems of isosceles triangular plate under uniform compressive stress σ, each side of which is simply supported. By T_{18} triangular element with the grid in the figure, the critical load factor $k = 4.648$ [10] is obtained, and the exact value $k = 4.50$, which differs by 3.3%. The critical load is

$$\sigma_{\text{cr}} = k\frac{\pi^2 D}{a^2 t}$$

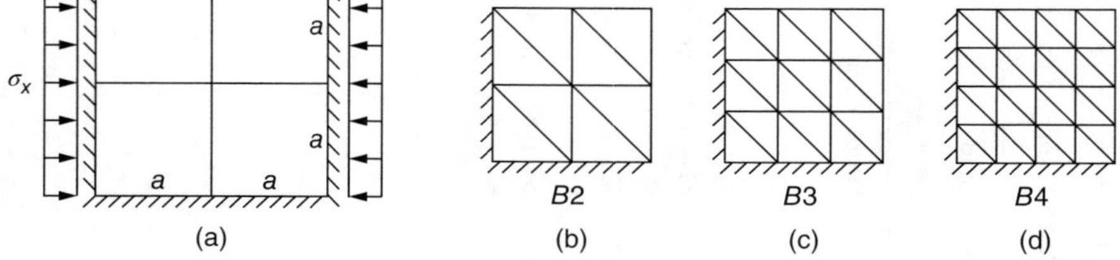

Figure 19.7 Grid for calculating buckling load of rectangular thin plate.

Table 19.2 Ratio of calculating value to exact value of buckling load of thin plate under compression in one direction.

Boundary		Simply supported on four sides			Fixed on four sides		
Element		R_{12}	T_9	T_{18}	R_{12}	T_9	T_{18}
Grid	B_2	0.943	0.934	1.001	0.923	0.914	1.001
	B_3	0.972	—	1.001	0.956	—	—
	B_4	0.983	0.968	—	0.972	—	—

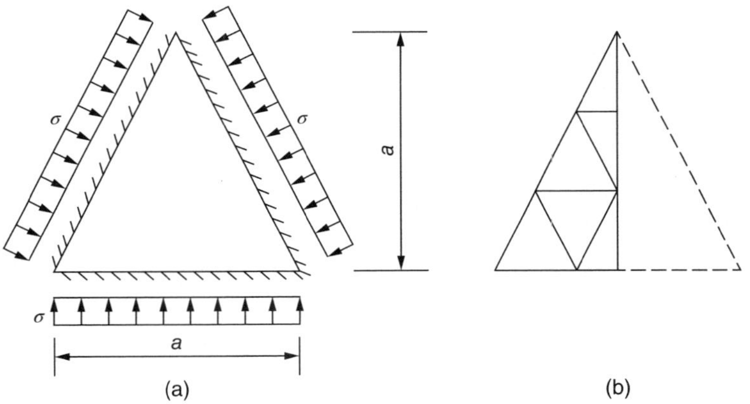

Figure 19.8 Calculation of elastic stability of triangular thin plate. (a) Load and (b) grid.

For all the plate elements above, the displacement conforming element T_{18} gives the best result of the critical loads; the precision of computation of the nonconforming element T_9 and R_{12} are somewhat 1ower.

Bibliography

1 Timoshenko, S.P. and Gere, J.M. (1961) *Theory of Elastic Stability*, 2nd edn, Mc Graw-Hill, New York.
2 Kapur, K.K. and Hartz, B.J. (1966) Stability of thin plates using the finite element method. *Proceedings of American Society of Civil Engineers*, **EM2**, 177–195.
3 Gallagher, R.H. and Padlog, J. (1963) Discrete element approach to structural instability analysis. *Journal of AIAA*, **1**, 1537–1539.
4 Anderson, R.G. and Newton, R.E. (1969) Plate buckling analysis using a fully compatible finite element. *Journal of AIAA*, **8**, 527–529.
5 Gallagher, R.H., Gellatly, R.A., Mallett, R.H. and Padlog, J. (1967) A discrete element procedure for thin shell instability analysis. *Journal of AIAA*, **5**, 138–145.

6 Gallagher, R.H. and Yang, H.T.Y. (1968) Elastic instability predictions for doubly curved shells, in *Air Force Institute of Technology, Proceedings 2nd Conference on Matrix Methods in Structural Mechanics*, Wright-Patterson A. F. Base, Ohio.

7 Matsui, T. and Matsuoka, O. (1976) A new finite element scheme for instability analysis of thin shells. *International Journal of Numerical Methods in Engineering*, **10**, 145–170.

8 Zienkiewicz, O.C. and Taylor, R.L. (2009) *Finite Element Method for Solid and Structural Mechanics*, Elsevier, New York,London.

9 Gallagher, R.H. (1975) *Finite Element Analysis Fundamentals*, Prentice-Hall, Englewood Cliffs.

10 Holland, I. and Bell, K. (1972) *Finite Element Methods in Stress Analysis*, Tapir, Trondheim(Norway).

20

Problems in Analysis of Structures with Large Displacement

Nonlinear problems of structures can be divided into two categories: material nonlinearity and geometric nonlinearity. Material nonlinear problems have been discussed from Chapters 16 to 18 of this book. Geometric nonlinear problems, namely, the problems with large displacement, will be discussed in this chapter.

In most structures with large displacement, the strains of structures are still small, which indicates that; the stress–strain relation of materials still remains linear, but the strain–displacement relation is nonlinear, namely, geometric nonlinearity. How to solve geometric nonlinear problems by the finite element method will be explained in the following sections of this chapter.

If both the displacement–strain relation and the stress–strain relation are nonlinear, then that is a double nonlinear (material nonlinearity and geometric nonlinearity) problem. How to solve double nonlinear problems by the finite element method will be illustrated in the last section of this chapter.

20.1 The Basic Method for Geometrical Nonlinear Problems

20.1.1 Basic Formulas

First, the equilibrium equations are established by the principle of virtual displacements. The array $\{\psi\}$ represents the sum of the vectors of generalized external force and internal force of each node; according to the principle of virtual displacements, the work done by external forces due to virtual displacements is equal to the strain energy resulting from the virtual strain, so there is

$$\{d\delta\}^{\mathrm{T}}\{\psi\} = \int \{d\varepsilon\}^{\mathrm{T}}\{\sigma\}dV - \{d\delta\}^{\mathrm{T}}\{P\} = 0 \tag{20.1}$$

where $\{d\delta\}$ is the virtual displacement, $\{d\varepsilon\}$ is the virtual strain, and $\{P\}$ is the load array.

The relationship between the displacement and the strain is expressed by the strain increment

$$\{d\varepsilon\} = [\overline{B}]\{d\delta\} \tag{20.2}$$

The Finite Element Method: Fundamentals and Applications in Civil, Hydraulic, Mechanical and Aeronautical Engineering, First Edition. Bofang Zhu.

Companion website: www.wiley.com/go/zhufem17

Eliminating $\{d\delta\}^T$ by formula (20.2), the equilibrium equations of the nonlinear problems are as follows:

$$\{\psi(\delta)\} = \int [\overline{B}]^T\{\sigma\}dV - \{P\} = 0 \tag{20.3}$$

whether there are big displacements or small displacements, formula (20.3) is applicable.

In the situation of big displacement, the strain–displacement relationship is nonlinear; matrix $[\overline{B}]$ is the function of $\{\delta\}$. For operational convenience, it can be written as

$$[\overline{B}] = [B_0] + [B_L] \tag{20.4}$$

where $[B_0]$ is the matrix for the analysis of linear strain, unrelated to $\{\delta\}$; $[B_L]$ is caused by the nonlinear deformation, related to $\{\delta\}$, usually $[B_L]$ is the linear function of $\{\delta\}$.

In most cases, although the displacement is great, the strain is not big; the stress–strain relationship is linear and elastic, so there is

$$\{\sigma\} = [D](\{\varepsilon\} - \{\varepsilon_0\}) + \{\sigma_0\} \tag{20.5}$$

where $[D]$ is the elasticity matrix of materials.

If the strain is relatively large, belonging to the nonlinear stress–strain relation, the elastoplastic matrix $[D]_{ep}$ should be used, which will be discussed in Section 20.4.

Usually Newton–Raphson method is used to solve formula (20.3), so the relationship between $\{d\delta\}$ and $\{d\psi\}$ should be established, taking the differential of formula (20.3), we get

$$\{d\psi\} = \int [d\overline{B}]^T\{\sigma\}dV + \int [\overline{B}]^T\{d\sigma\}dV \tag{20.6}$$

If the initial strain and stress is ignored in formula (20.5), we get

$$\{d\sigma\} = [D]\{d\varepsilon\} = [D][\overline{B}]\{d\delta\}$$

In formula (20.4), $[B_0]$ is not related to $\{\delta\}$, so

$$[d\overline{B}] = [dB_L]$$

Substituting it into formula (20.6), there is

$$\{d\psi\} = \int [dB_L]^T\{\sigma\}dV + [\overline{K}]\{d\delta\} \tag{20.7}$$

where

$$[\overline{K}] = \int [\overline{B}]^T[D][\overline{B}]dV = \int [B_0 + B_L]^T[D][B_0 + B_L]dV = [K_0] + [K_L] \tag{20.8}$$

$$[K_0] = \int [B_0]^T[D][B_0]dV \tag{20.9}$$

$$[K_L] = \int \left([B_0]^T[D][B_L] + [B_L]^T[D][B_L] + [B_L]^T[D][B_0]\right)dV \tag{20.10}$$

where $[K_0]$ is the linear stiffness matrix for small displacement and $[K_L]$ is the matrix for the initial displacement or big displacement.

The first term in the right side of formula (20.7) can be written as follows:

$$\int [dB_L]^T\{\sigma\}dV = [K_\sigma]\{d\delta\} \tag{20.11}$$

in which $[K_\sigma]$ is the symmetric matrix of the stress $\{\sigma\}$, called the initial stress matrix or geometric stiffness matrix.

Thus, formula (20.7) can be written as

$$\{d\psi\} = ([K_0] + [K_\sigma] + [K_L])\{d\delta\} = [K_T]\{d\delta\} \tag{20.12}$$

where

$$[K_T] = [K_0] + [K_\sigma] + [K_L] \tag{20.13}$$

in which $[K_T]$ is the tangent stiffness matrix.

20.1.2 The Solution

The problems with large displacement usually are solved by Newton–Raphson method; the iterative formulas are

$$\{\Delta\delta\}_n = -[K_T]^{-1}\{\psi\}_n \tag{20.14}$$

$$\{\delta\}_{n+1} = \{\delta\}_n + \{\Delta\delta\}_n \tag{20.15}$$

Calculation procedures are as follows:

1) Solving the linear elastic solution and using it as the first approximate value $\{\delta\}_1$
2) Calculating $[\overline{B}]$ by formula (20.4), stress $\{\sigma\}$ by formula (20.5), and the imbalanced force $\{\psi\}$ by formula (20.3).
3) Calculating the tangent stiffness matrix $[K_T]$.
4) Calculating by formulas (20.14) and (20.15), then

   $$\{\Delta\delta\}_1 = -[K_T]^{-1}\{\psi\}_1$$

   and

   $$\{\delta\}_2 = \{\delta\}_1 + \{\Delta\delta\}_1$$

5) Going to the second step, repeat the computation iteratively until $\{\psi\}_n$ is small enough.

As described in Chapter 16, there are three methods to solve nonlinear problems: the integral variable method, the incremental method, and the mixed method. For the problem with large displacement, the mixed method is often used, namely, giving several levels of incremental loads, for each level of the incremental load, using an iterative method to solve. The variable stiffness iteration is available and the constant stiffness iteration can also be used. The more effective method is that in the first and second iterations, the variable stiffness iteration is adopted, and after the second iteration, the constant stiffness iteration is used.

When the load increment is small enough, for each level of load increment, the iteration is only once, which is an ordinary incremental method.

It must be noted that when deriving formula (20.7) the influence of the displacement on the load is ignored. In some cases, such as the flutter of the wing, the load is related to the displacement, $(\partial\{P\}/\partial\{\delta\})\{d\delta\}$ should be added in the formula (20.7).

20.1.3 The Elastic Stability Problem

By formulas (19.17) and (19.27), it is clear that the stiffness matrix $[K_\sigma]$ of the initial stress does not depend on the nodal displacements and only contains the stress $\{\sigma\}$; therefore, as the first approximation, assuming that $[K_L] = 0$, calculating the stress $\{\sigma\}$ by the linear elastic method, by formula (20.12), we get

$$\{d\psi\} = ([K_0] + [K_\sigma])\{d\delta\}$$

If the load increases by the proportional factor λ, then

$$\{d\psi\} = ([K_0] + \lambda[K_\sigma])\{d\delta\}$$

Thus, the critical condition of the elastic instability is

$$|[K_0] + \lambda[K_\sigma]| = 0 \tag{20.16}$$

By formula (20.16) the eigenvalue λ_{cr} is obtained, namely, the critical load of the elastic stability.

It shows that the preconditions of the elastic stability analysis is $[K_L] = 0$; $[K_L]$ is given by equation (20.10), meaning that the displacement is not big. If the displacement is large, the critical load of the elastic stability analysis is often too big. In order to get the correct answer, it should be solved by formulas (20.14) and (20.15) step by step according to the problem with the larger displacement. Figure 20.1 shows the elastic stability solution and the finite element solution of the large displacement of the arch under the action of concentrated load.

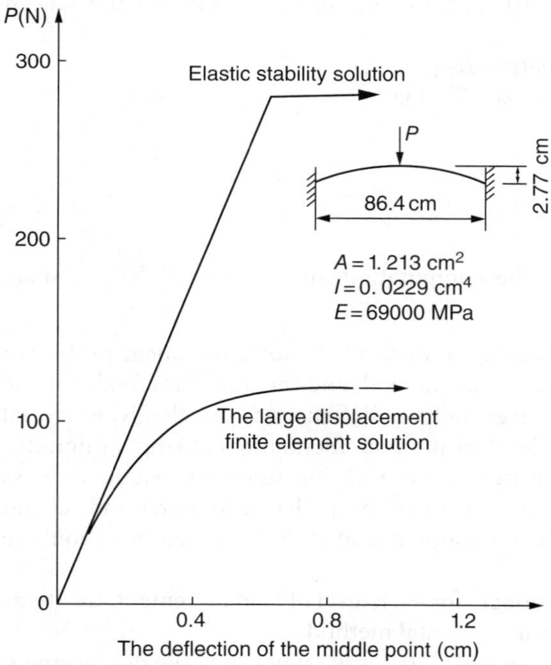

Figure 20.1 The elastic stability solution and the finite element solution of the large displacement of the arch under the action of concentrated load.

This example shows that the critical load obtained by the large displacement solution sometimes can be lower than the critical load of the elastic stability.

20.2 The Plate Element of Large Deflection

For the thin plate under the transverse load, if the deflection is small, the membrane stress in the plate can be ignored, but if the deflection is larger, the membrane stress in the plate cannot be ignored and the membrane strain caused by the lateral displacement should be taken into account in calculation; the deformation of the middle face and the bending deformation of the thin plate are no longer independent of each other but coupled with each other.

Let plane oxy coincide with the middle surface of plates. The strain of the plate can be described by the displacement of the middle surface. $\{\varepsilon\}$ represents the array of the strain and the curvature of the middle face, while $\{\sigma\}$ represents the array of the membrane stress and bending forces. Then

$$\{\varepsilon\} = \begin{Bmatrix} \varepsilon^p \\ \varepsilon^b \end{Bmatrix} = \begin{bmatrix} \varepsilon_x & \varepsilon_y & \gamma_{xy} & \dfrac{\partial^2 w}{\partial x^2} & \dfrac{\partial^2 w}{\partial y^2} & 2\dfrac{\partial^2 w}{\partial x \partial y} \end{bmatrix}^{\mathrm{T}}$$

$$\{\sigma\} = \begin{Bmatrix} \sigma^{\mathrm{p}} \\ \sigma^{\mathrm{b}} \end{Bmatrix} = \begin{bmatrix} N_x & N_y & N_{xy} & M_x & M_y & M_{xy} \end{bmatrix}^{\mathrm{T}} \tag{20.17}$$

where the superscript b represents bending, the superscript p represents the plane.

By the theory of the large deflection of plates, it is known that (Figure 20.2)

$$\varepsilon_x = \frac{\partial u}{\partial x} + \frac{1}{2}\left(\frac{\partial w}{\partial x}\right)^2, \quad \varepsilon_y = \frac{\partial v}{\partial y} + \frac{1}{2}\left(\frac{\partial w}{\partial y}\right)^2, \quad \gamma_{xy} = \frac{\partial u}{\partial y} + \frac{\partial v}{\partial x} + \frac{\partial w}{\partial x}\frac{\partial w}{\partial y}$$

Therefore formula (20.17) can be written as

$$\{\varepsilon\} = \begin{Bmatrix} \dfrac{\partial u}{\partial x} \\ \dfrac{\partial v}{\partial y} \\ \dfrac{\partial u}{\partial y} + \dfrac{\partial v}{\partial x} \\ \dfrac{\partial^2 w}{\partial x^2} \\ \dfrac{\partial^2 w}{\partial y^2} \\ 2\dfrac{\partial^2 w}{\partial x \partial y} \end{Bmatrix} + \begin{Bmatrix} \dfrac{1}{2}\left(\dfrac{\partial w}{\partial x}\right)^2 \\ \dfrac{1}{2}\left(\dfrac{\partial w}{\partial y}\right)^2 \\ \dfrac{\partial w}{\partial x}\dfrac{\partial w}{\partial y} \\ 0 \\ 0 \\ 0 \end{Bmatrix} = \begin{Bmatrix} \varepsilon_0^{\mathrm{p}} \\ \varepsilon_0^{\mathrm{b}} \end{Bmatrix} + \begin{Bmatrix} \varepsilon_{\mathrm{L}}^{\mathrm{p}} \\ 0 \end{Bmatrix} \tag{20.18}$$

in which the first term of the right side is the linear strain and the second term is the nonlinear strain.

Assuming that the material is elastic (the elastoplastic material is discussed in Section 20.4), the elasticity matrix of plates is

$$[D] = \begin{bmatrix} D^{\mathrm{p}} & 0 \\ 0 & D^{\mathrm{b}} \end{bmatrix} \tag{20.19}$$

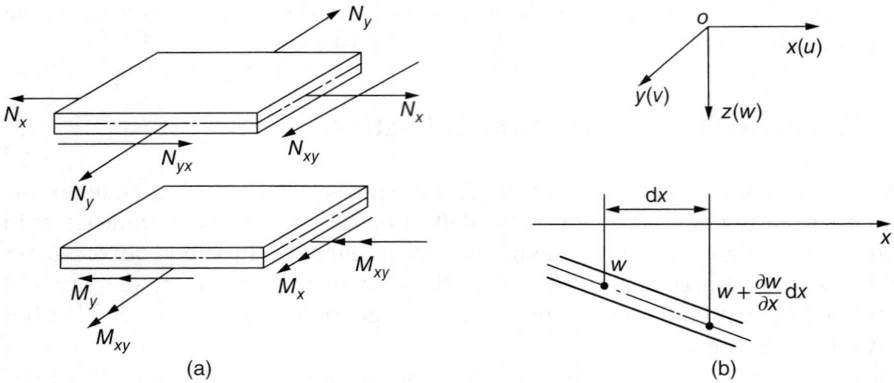

Figure 20.2 The internal force and the deflection of the plate.

where

$$[D^{\mathrm{p}}] = \frac{Et}{1-\mu^2}\begin{bmatrix} 1 & \mu & 0 \\ \mu & 1 & 0 \\ 0 & 0 & (1-\mu)/2 \end{bmatrix}$$

and

$$[D^{\mathrm{b}}] = \frac{Et^3}{12(1-\mu)^2}\begin{bmatrix} 1 & \mu & 0 \\ \mu & 1 & 0 \\ 0 & 0 & (1-\mu)/2 \end{bmatrix}$$

The displacement of any point of the plate element can be expressed by the shape function and the nodal displacement as follows:

$$\begin{Bmatrix} u \\ v \\ w \end{Bmatrix} = [N]\{\delta\}^{\mathrm{e}} \tag{20.20}$$

For operational convenience, the nodal displacement can be divided into two categories: the plane displacement and the bending displacement, namely,

$$\{\delta_i\} = \begin{Bmatrix} \delta_i^{\mathrm{p}} \\ \delta_i^{\mathrm{b}} \end{Bmatrix}$$

$$\{\delta_i^{\mathrm{p}}\} = \begin{bmatrix} u_i & v_i \end{bmatrix}^{\mathrm{T}}, \quad \{\delta_i^{\mathrm{b}}\} = \begin{bmatrix} w_i & \left(\dfrac{\partial w}{\partial x}\right)_i & \left(\dfrac{\partial w}{\partial y}\right)_i \end{bmatrix}^{\mathrm{T}}$$

The shape function can also be divided into

$$[N_i] = \begin{bmatrix} N_i^{\mathrm{b}} & 0 \\ 0 & N_i^{\mathrm{b}} \end{bmatrix}$$

In the definition above, except for the nonlinear strain term $\{\varepsilon_{\mathrm{L}}^{\mathrm{p}}\}$, the remaining terms are identical to that described in Chapter 11.

Matrix $[\overline{B}]$ is calculated in the following, by formula (20.4), there is

$$[\overline{B}] = [B_0] + [B_L]$$

$$[B_0] = \begin{bmatrix} B_0^p & 0 \\ 0 & B_0^b \end{bmatrix}, \quad [B_L] = \begin{bmatrix} 0 & B_L^b \\ 0 & 0 \end{bmatrix} \tag{20.21}$$

where $[B_0]$ is the matrix for the linear strain analysis and $[B_L]$ is the matrix for the nonlinear strain analysis.

By formula (20.18), the nonlinear strain $\{\varepsilon_L^b\}$ can be expressed as follows:

$$\{\varepsilon_L^b\} = \frac{1}{2}\begin{bmatrix} \frac{\partial w}{\partial x} & 0 \\ 0 & \frac{\partial w}{\partial y} \\ \frac{\partial w}{\partial y} & \frac{\partial w}{\partial x} \end{bmatrix}\begin{Bmatrix} \frac{\partial w}{\partial x} \\ \frac{\partial w}{\partial y} \end{Bmatrix} = \frac{1}{2}[A]\{\theta\} \tag{20.22}$$

where

$$\{\theta\} = \begin{Bmatrix} \frac{\partial w}{\partial x} \\ \frac{\partial w}{\partial y} \end{Bmatrix} = [g]\{\delta^b\}^e \tag{20.23}$$

$$[g] = \begin{bmatrix} \frac{\partial N^b}{\partial x} & \frac{\partial N^b}{\partial x} & \cdots \\ \frac{\partial N^b}{\partial y} & \frac{\partial N^b}{\partial y} & \cdots \end{bmatrix} \tag{20.24}$$

It is clear that matrix $[g]$ depends on the derivative of the shape function with respect to the coordinate.

The properties of matrix $[A]$ and array $\{\theta\}$ are proved in the following.

Property 1: The differential of matrix $[A]$ multiplied by $\{\theta\}$ is equal to matrix $[A]$ multiplied by the differential of array $\{\theta\}$. The proof is in the following:

$$[dA]\{\theta\} = \begin{bmatrix} d\left(\frac{\partial w}{\partial x}\right) & 0 \\ 0 & d\left(\frac{\partial w}{\partial y}\right) \\ d\left(\frac{\partial w}{\partial y}\right) & d\left(\frac{\partial w}{\partial x}\right) \end{bmatrix}\begin{Bmatrix} \frac{\partial w}{\partial x} \\ \frac{\partial w}{\partial y} \end{Bmatrix} = \begin{bmatrix} \frac{\partial w}{\partial x} & 0 \\ 0 & \frac{\partial w}{\partial y} \\ \frac{\partial w}{\partial y} & \frac{\partial w}{\partial x} \end{bmatrix}\begin{Bmatrix} d\left(\frac{\partial w}{\partial x}\right) \\ d\left(\frac{\partial w}{\partial y}\right) \end{Bmatrix} = [A]\{d\theta\} \tag{20.25}$$

Property 2: If $\{y\} = \lfloor y_1 \ y_2 \ y_3 \rfloor^T$, there is

$$[dA]^T\{y\} = \begin{bmatrix} d\left(\frac{\partial w}{\partial x}\right) & 0 & d\left(\frac{\partial w}{\partial y}\right) \\ 0 & d\left(\frac{\partial w}{\partial y}\right) & d\left(\frac{\partial w}{\partial x}\right) \end{bmatrix}\begin{Bmatrix} y_1 \\ y_2 \\ y_3 \end{Bmatrix} = \begin{bmatrix} y_1 & y_3 \\ y_3 & y_2 \end{bmatrix}\{d\theta\} \tag{20.26}$$

The expression $[B_L^b]$ is derived below, taking the differential of formula (20.22), and by formula (20.25), we get

$$\{d\varepsilon_L^b\} = \frac{1}{2}[dA]\{\theta\} + \frac{1}{2}[A]\{d\theta\} = [A]\{d\theta\} = [A][g]\{d\delta^b\}^e$$

By formulas (20.2) and (20.4), it is known that

$$[B_L^b] = [A][g] \tag{20.27}$$

The tangent stiffness matrix $[k_T] = [k_0] + [k_\sigma] + [k_L]$ of the plate element is calculated below; $[k_0]$ is the stiffness matrix of the linear small deflection

$$[k_0] = \begin{bmatrix} k_0^p & 0 \\ 0 & k_0^b \end{bmatrix} \tag{20.28}$$

where $[k_0^p]$ is seen in Chapter 2 and $[k_0^b]$ is seen in Chapter 10. Substituting formula (20.21) into formula (20.10), we get

$$[k_L] = \int \begin{bmatrix} 0 & B_0^{pT} D^p B_L^b \\ B_L^{bT} D^p B_0^p & B_L^{bT} D^p B_L^b \end{bmatrix} dxdy \tag{20.29}$$

Then solving $[k_\sigma]$ below, taking the differential of formula (20.21), we can obtain

$$[dB_L]^T = \begin{bmatrix} 0 & 0 \\ dB_L^{bT} & 0 \end{bmatrix}$$

Substituting the formula above into formula (20.11) and by formulas (20.17) and (20.27), we get

$$[k_\sigma]\{d\delta\}^e = \int \begin{bmatrix} 0 & 0 \\ g^T dA^T & 0 \end{bmatrix} \begin{Bmatrix} N_x \\ N_x \\ N_{xy} \\ M_x \\ M_y \\ M_{xy} \end{Bmatrix} dxdy$$

By formula (20.26), there is

$$[dA^T]\begin{Bmatrix} N_x \\ N_y \\ N_{xy} \end{Bmatrix} = \begin{bmatrix} N_x & N_{xy} \\ N_{xy} & N_y \end{bmatrix} \{d\theta\} = \begin{bmatrix} N_x & N_{xy} \\ N_{xy} & N_y \end{bmatrix} [g]\{d\delta^b\}^e$$

So the initial stress stiffness matrix is as follows:

$$[k_\sigma] = \begin{bmatrix} 0 & 0 \\ 0 & k_\sigma^b \end{bmatrix} \tag{20.30}$$

$$[k_\sigma^b] = \int [g]^T \begin{bmatrix} N_x & N_{xy} \\ N_{xy} & N_y \end{bmatrix} [g]dxdy \tag{20.31}$$

An example is listed below. The square plate simply supported is under uniformly distributed load; in the middle surface, the normal displacement is constrained along the border, resulting in the secondary membrane stress. The plane dimension is 40.6 cm × 40.6 cm, the thickness is 0.254 cm, the elastic modulus is 207 GPa, the Poisson's ratio is 0.316, and the lateral load is $\Delta q = 103.5$ kPa, divided into six levels of loading, the first two levels of load increments are $q = 10.35$ kPa and the latter four levels of load increments are $\Delta q = 20.70$ kPa. For the symmetry, taking one-fourth of the plate, using the rectangular plate element of 4 × 4, solved by Newton–Raphson method, the calculation result is shown in Figure 20.3; the result of the finite element analysis is very close to Levy theoretical solution, and the displacement error is less than 1%.

As for the solution of the large deflection problem of the shell, one way is to assemble the plate element into the shell, as in Chapter 11.2; of course, the nonlinear method in this section is used for the calculation of the stiffness matrix of plates. Another method is to use a curved shell element, using the basic equation of the shell and taking the nonlinear factors into account, to calculate the tangent stiffness matrix $[K_T]$ of the shell element. For the flat shells, because the slopes $\partial z/\partial x$ and $\partial z/\partial y$ of the coordinate lines are small, in comparison with 1, the quadratic term of the slopes can be neglected, and the calculation can be simplified. Due to the space limitation, the large deflection problem of the shell is not introduced in detail in this book, and the readers can refer to Refs [4, 7, 8, 11].

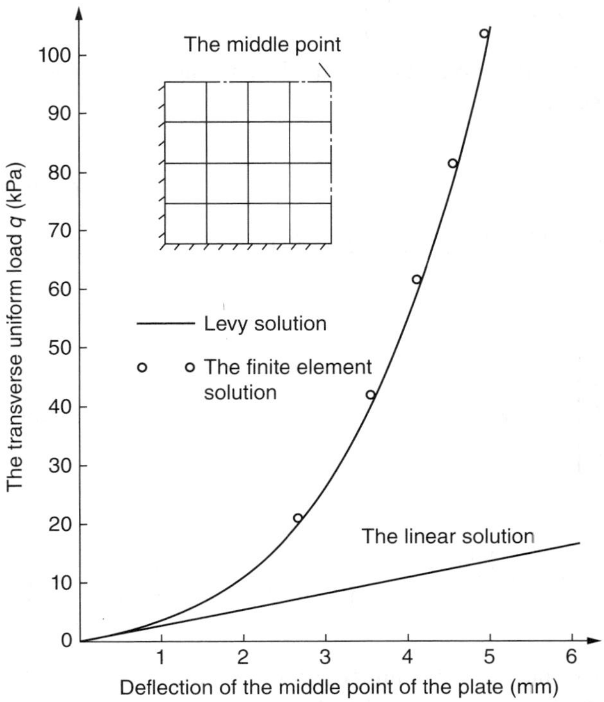

Figure 20.3 The analysis of large deflection of the simply supported square plate.

20.3 Three-Dimensional Solid Element of Large Displacement

Defining the strain components by Lagrange's view, namely, defining the strain components by the coordinate (x, y, z) before deformation as independent variables, we get

$$\left.\begin{aligned}
\varepsilon_x &= \frac{\partial u}{\partial x} + \frac{1}{2}\left[\left(\frac{\partial u}{\partial x}\right)^2 + \left(\frac{\partial v}{\partial x}\right)^2 + \left(\frac{\partial w}{\partial x}\right)^2\right] \\
\varepsilon_y &= \frac{\partial v}{\partial y} + \frac{1}{2}\left[\left(\frac{\partial u}{\partial y}\right)^2 + \left(\frac{\partial v}{\partial y}\right)^2 + \left(\frac{\partial w}{\partial y}\right)^2\right] \\
\varepsilon_z &= \frac{\partial w}{\partial z} + \frac{1}{2}\left[\left(\frac{\partial u}{\partial z}\right)^2 + \left(\frac{\partial v}{\partial z}\right)^2 + \left(\frac{\partial w}{\partial z}\right)^2\right] \\
\gamma_{yz} &= \frac{\partial v}{\partial z} + \frac{\partial w}{\partial y} + \left(\frac{\partial u}{\partial y}\frac{\partial u}{\partial z} + \frac{\partial v}{\partial y}\frac{\partial v}{\partial z} + \frac{\partial w}{\partial y}\frac{\partial w}{\partial z}\right) \\
\gamma_{zx} &= \frac{\partial w}{\partial x} + \frac{\partial u}{\partial z} + \left(\frac{\partial u}{\partial x}\frac{\partial u}{\partial z} + \frac{\partial v}{\partial x}\frac{\partial v}{\partial z} + \frac{\partial w}{\partial x}\frac{\partial w}{\partial z}\right) \\
\gamma_{xy} &= \frac{\partial u}{\partial y} + \frac{\partial v}{\partial x} + \left(\frac{\partial u}{\partial x}\frac{\partial u}{\partial y} + \frac{\partial v}{\partial x}\frac{\partial v}{\partial y} + \frac{\partial w}{\partial x}\frac{\partial w}{\partial y}\right)
\end{aligned}\right\} \tag{20.32}$$

The relationship between the strain components and the displacement components mentioned above can be expressed as

$$\{\varepsilon\} = \{\varepsilon_0\} + \{\varepsilon_{\mathrm{L}}\} \tag{20.33}$$

$$\{\varepsilon\} = \left\{\varepsilon_x \quad \varepsilon_y \quad \varepsilon_z \quad \gamma_{yz} \quad \gamma_{zx} \quad \gamma_{xy}\right\} \tag{20.34}$$

$$\{\varepsilon_0\} = \left[\frac{\partial u}{\partial x} \quad \frac{\partial v}{\partial y} \quad \frac{\partial w}{\partial z} \quad \frac{\partial v}{\partial z} + \frac{\partial w}{\partial y} \quad \frac{\partial w}{\partial x} + \frac{\partial u}{\partial z} \quad \frac{\partial u}{\partial y} + \frac{\partial v}{\partial x}\right]^{\mathrm{T}} \tag{20.35}$$

$$\{\varepsilon_{\mathrm{L}}\} = \frac{1}{2}\begin{bmatrix} \theta_x^{\mathrm{T}} & 0 & 0 \\ 0 & \theta_y^{\mathrm{T}} & 0 \\ 0 & 0 & \theta_z^{\mathrm{T}} \\ 0 & \theta_z^{\mathrm{T}} & \theta_y^{\mathrm{T}} \\ \theta_z^{\mathrm{T}} & 0 & \theta_x^{\mathrm{T}} \\ \theta_y^{\mathrm{T}} & \theta_x^{\mathrm{T}} & 0 \end{bmatrix} \begin{Bmatrix} \theta_x \\ \theta_y \\ \theta_z \end{Bmatrix} = \frac{1}{2}[A]\{\theta\} \tag{20.36}$$

$$\{\theta_x\} = \left[\frac{\partial u}{\partial x} \quad \frac{\partial v}{\partial x} \quad \frac{\partial w}{\partial x}\right]^{\mathrm{T}}, \quad \{\theta_y\} = \left[\frac{\partial u}{\partial y} \quad \frac{\partial v}{\partial y} \quad \frac{\partial w}{\partial y}\right]^{\mathrm{T}}, \quad \{\theta_z\} = \left[\frac{\partial u}{\partial z} \quad \frac{\partial v}{\partial z} \quad \frac{\partial w}{\partial z}\right]^{\mathrm{T}} \tag{20.37}$$

where $[A]$ is a 6×9 matrix. It is clear that $\{\varepsilon_0\}$ is the linear part of the strain components, while $\{\varepsilon_{\mathrm{L}}\}$ is the nonlinear part; matrix $[A]$ and array $\{\theta\}$ also satisfy the two properties described in the previous section, so

$$\{\mathrm{d}\varepsilon_{\mathrm{L}}\} = \frac{1}{2}[\mathrm{d}A]\{\theta\} + \frac{1}{2}[A]\{\mathrm{d}\theta\} = [A]\{\mathrm{d}\theta\} \tag{20.38}$$

The displacement of any point of the element is expressed as

$$u = \sum N_i u_i, \quad v = \sum N_i v_i, \quad w = \sum N_i w_i$$

so

$$\{\theta\} = \begin{Bmatrix} \frac{\partial u}{\partial x} \\ \frac{\partial v}{\partial x} \\ \frac{\partial w}{\partial x} \\ \vdots \\ \frac{\partial w}{\partial z} \end{Bmatrix} = \begin{bmatrix} \frac{\partial N_1}{\partial x} & 0 & 0 & \frac{\partial N_2}{\partial x} & \cdots \\ 0 & \frac{\partial N_1}{\partial x} & 0 & 0 & \cdots \\ 0 & 0 & \frac{\partial N_1}{\partial x} & 0 & \cdots \\ \vdots & \vdots & \vdots & \vdots & \vdots \\ 0 & 0 & \frac{\partial N_1}{\partial z} & 0 & \cdots \end{bmatrix} \begin{Bmatrix} u_1 \\ v_1 \\ w_1 \\ u_2 \\ \vdots \end{Bmatrix} = [g]\{\delta\}^e \tag{20.39}$$

By formula (20.38), there is

$$\{\mathrm{d}\varepsilon_{\mathrm{L}}\} = [A][g]\{\mathrm{d}\delta\}^{\mathrm{e}}$$

so we get

$$\{B_{\mathrm{L}}\} = [A][g] \tag{20.40}$$

The matrix $[B_0]$ of linear strain analysis can refer to formula (8.20); by substitution of $[B_0]$ and $[B_{\mathrm{L}}]$ into formula (20.8), $[K_0] + [K_{\mathrm{L}}]$ can be calculated. By formula (20.11), we get

$$[K_\sigma]\{\mathrm{d}\delta\}^{\mathrm{e}} = \int [\mathrm{d}B_L]^{\mathrm{T}}\{\sigma\}\mathrm{d}V = \int [g]^{\mathrm{T}}[\mathrm{d}A]^{\mathrm{T}}\{\sigma\}\mathrm{d}V \tag{20.41}$$

According to the derivation in the previous section, using property two of matrix $[A]$ and array $\{\theta\}$ and $\{\mathrm{d}\theta\} = [g]\{\mathrm{d}\delta\}^{\mathrm{e}}$, we get

$$[\mathrm{d}A]^{\mathrm{T}}\{\sigma\} = \begin{bmatrix} \sigma_x I & \text{symmetrical} & \\ \tau_{xy} I & \sigma_y I & \\ \tau_{zx} I & \tau_{zy} I & \sigma_z I \end{bmatrix} \{\mathrm{d}\theta\} = [M]\{\mathrm{d}\theta\} = [M][g]\{\mathrm{d}\delta\}^{\mathrm{e}} \tag{20.42}$$

where I is a third-order unit matrix and $[M]$ is a 9×9 matrix with six stress components.

By substitution of formula (20.42) into formula (20.41), the stiffness matrix of the initial stress is as follows:

$$[K_\sigma] = \int [g]^{\mathrm{T}}[M][g]\mathrm{d}V \tag{20.43}$$

Thus, the tangent stiffness matrix of three-dimensional solid element with large displacement is in the following:

$$[K_{\mathrm{T}}] = [K_\sigma] + [K_0] + [K_{\mathrm{L}}] \tag{20.44}$$

Experience has shown that in the problem with large displacement, it is better to use the isoparametric element. Moreover, in the formulas above, as far as omitting some items, the corresponding formulas of two-dimensional and one-dimensional problems can be obtained.

20.4 Double Nonlinearity: Elastoplastic Large Displacement Problem

The elastic large deflection problem is discussed in the previous sections of this chapter, namely, it is a geometrical nonlinear and the material properties are linear elastic. The double nonlinear problems are discussed in this section, that is, elastoplastic large displacement problem.

For the double nonlinear problems, the nonlinear geometric relation derived in the previous section of this chapter is still available; the key step is that the stress–strain relationship is given by the following equation instead:

$$\{\Delta\sigma\} = [D]_{\mathrm{ep}}\{\Delta\varepsilon\} = ([D] - [D]_{\mathrm{p}})\{\Delta\varepsilon\} \tag{20.45}$$

where $[D]_{\mathrm{ep}}$ is the elastoplastic matrix.

In formulas (20.9) and (20.10) for calculating $[K_0]$ and $[K_{\mathrm{L}}]$, change $[D]$ into $[D]_{\mathrm{ep}}$, we get

$$[K_0]_{\mathrm{p}} = \int [B_0]^{\mathrm{T}}([D] - [D]_{\mathrm{p}})[B_0]\mathrm{d}V = [K_0] - \int [B_0]^{\mathrm{T}}[D]_p[B_0]\mathrm{d}V$$

$$[K_{\mathrm{L}}]_{\mathrm{p}} = [K_{\mathrm{L}}] - \int ([B_0]^{\mathrm{T}}[D]_{\mathrm{p}}[B_{\mathrm{L}}] + [B_{\mathrm{L}}]^{\mathrm{T}}[D]_{\mathrm{p}}[B_{\mathrm{L}}] + [B_{\mathrm{L}}]^{\mathrm{T}}[D]_{\mathrm{p}}[B_0])\mathrm{d}V$$

Therefore, for the double nonlinear problems, the tangent stiffness matrix of the structures is

$$[K_{\mathrm{T}}] = [K_0] + [K_{\mathrm{L}}] + [K_\sigma] - [K_{\mathrm{R}}] \tag{20.46}$$

where

$$[K_{\mathrm{R}}] = \int ([B_0]^{\mathrm{T}}[D]_{\mathrm{p}}[B_0] + [B_0]^{\mathrm{T}}[D]_{\mathrm{p}}[B_{\mathrm{L}}] + [B_{\mathrm{L}}]^{\mathrm{T}}[D]_{\mathrm{p}}[B_{\mathrm{L}}] + [B_{\mathrm{L}}]^{\mathrm{T}}[D]_{\mathrm{p}}[B_0])\mathrm{d}V \tag{20.47}$$

in which $[K_{\mathrm{R}}]$ is called the load correction matrix.

The double nonlinear problem generally is solved by Newton–Raphson method; the calculating steps are basically the same with that of the geometric nonlinear problem, but in each load increment, in order to calculate $[K_{\mathrm{T}}]$ by formula (20.46), $[D]_{\mathrm{p}}$ needs to be calculated by the method described in Section 17.8.

Bibliography

1 Bathe, K.J. and Ozdemir, H. (1976) Elastic–plastic large deformation static and dynamic analysis. *Computers and Structures*, **6** (2), 81–92.

2 Haisler, W.E., Stricklin, J.A. and Stebbins, F.J. (1972) Development and evaluation of solution procedures for geometrically non-linear analysis. *AIAA Journal*, **10**, 264–272.

3 Marcal, P.V. (1967) Effect of Initial Displacement on Problem of Large Deflection and Stability. Technical Report ARPA E54, Brown University.

4 Brebbia, C.A. and Conner, J. (1969) Geometrically non-linear finite element analysis. *Journal of Engineering Mechanics Division, Proceedings ASCE*, **95** (EM2), 463–483.

5 Kawai, T. and Yoshimura, N. (1969) Analysis of large deflection of plates by finite element method. *International Journal for Numerical Methods in Engineering*, **1**, 123–133.
6 Murray, D.W. and Wilson, E.L. (1968) Finite element large deflection analysis of plates. *Journal of the Engineering Mechanics Division Proceedings ASCE*, **94** (EM1), 143–165.
7 Bergan, P.G. and Clough, R.W. (1973) Large deflection analysis of plates and shallow shells using the finite element method. *International Journal for Numerical Methods in Engineering*, **5**, 543–555.
8 Batoz, J.L., Chattopadhyay, A. and Dhatt, G. (1976) Finite element large deflection analysis of shallow shells. *International Journal for Numerical Methods in Engineering*, **10**, 35–38.
9 Walker, A.C. (1969) A non-linear finite element analysis of shallow circular arches. *International Journal of Solids and Structures*, **5**, 97–107.
10 Oden, J.T. (1967) Numerical formulation of non-linear elasticity problems. *Proceeding Journal of Engineering Structural Division, ASCE*, **93** (ST3), 235–255.
11 Mescall, J.F. (1965) Large deflections of spherical shells under concentrated loads. *Journal of Applied Mechanics*, **32**, 936–938.
12 Da Deppo, D.A. and Schmidt, R. (1975) Instability of clamped-hinged circular arches subjected to a point load. *Transactions of the American Society of Mechanical Engineers, Journal of Applied Mechanics*, **12**, 894–896.
13 Mc Meeking, R.M. and Rice, J.R. (1975) Finite element formulation for problems of large elasto-plastic deformation. *International Journal of Solids and Structures*, **11**, 601–616.
14 Bathe, K.J. (1982) *Finite Element Procedures in Engineering Analysis*, Prentice-Hall, Englewood Cliffs.
15 Zienkiewicz, O.C. and Nayak, G.C. (1973) A General Approach to Problems of Plasticity and Large Deformation Using Isoparametric Elements, in Wright-Patterson Air Force Base. Proceedings of the 3rd Conference Matrix Methods in Structural Mechanics, Ohio.
16 Zienkiewicz, O.C., Taylor, R.L. and Zhu, J.Z. (2005) *The Finite Element Method, its Basis and Fundamentals*, 6th edn, Elsevier, New York.

21

Problems in Fracture Mechanics

21.1 Introduction

Fracture mechanics is a new subject developed in recent decades. With promotion of productivity, in a short period of time, it had made a great development both experimentally and theoretically.

Early in 1921, Griffith explained that the actual strength of a material is only one thousandth of theoretical strength due to flaws inside the material. Meanwhile, he believed that when cracked body is under load, if the surface energy needed for crack developing is less than the release value of elastic energy, the crack will develop and lead to a final break. This theory is proved in glass. While it only works in the perfect elastic body, that is, perfect elastic material, the theory has no development.

During World War II, nearly 2000 American fully welded ships encountered more than 1000 brittle failures, in which 238 ships were totally destroyed. In 1950, the shell of American North Star missile with solid fuels exploded in experiment. The shell is made of high strength steel whose yield strength is 1400 MPa. It had passed the traditional toughness test, which estimated that the breaking stress is less than half of the yield strength. Afterward, lots of water pressure test for shell were carried out, which showed that brittle failures are always caused by macro cracks (0.1–10 mm). These cracks may be metallurgy faults or may come from processing or application. For many construction members, the existence of these macro cracks is inevitable. Fracture mechanics becomes a new subject focusing on studying the stress, strain, and crack extending pattern near the crack acme.

Crack extending problems also exist in concrete structures. For instance, concrete dams often have some surface cracks as a result of dramatic temperature changing during construction process. The depth of the crack is merely several centimeters. Among these surface cracks, some develop into deep ones and even penetrate the whole body, causing serious influence.

In general stress field, there are three modes of crack extending: Mode I is opening mode. Mode II is sliding mode. Mode III is tearing mode. They are shown in Figure 21.1. Based on the analysis of theory of elasticity, stress and strain near the crack acme can

The Finite Element Method: Fundamentals and Applications in Civil, Hydraulic, Mechanical and Aeronautical Engineering, First Edition. Bofang Zhu.

Companion website: www.wiley.com/go/zhufem17

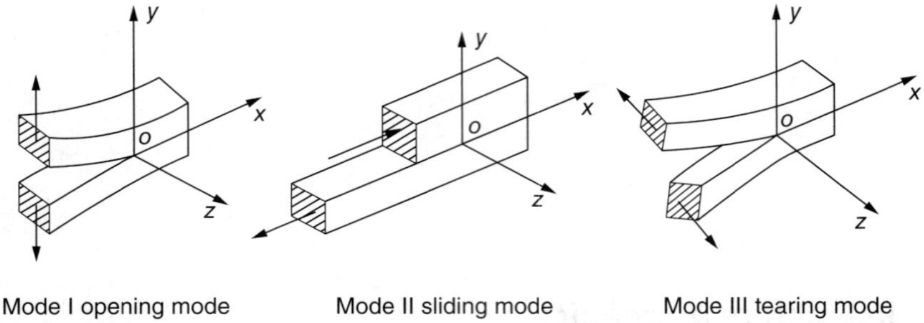

Figure 21.1 Types of crack extending.

be represented as

$$\left.\begin{aligned}\sigma_{ij} &= \frac{K}{\sqrt{r}} f_{ij}(\theta) \\ u_i &= \frac{K}{\sqrt{r}} f_i(\theta)\end{aligned}\right\} \tag{21.1}$$

in which r,θ are polar coordinates using crack acme as the original point and $f_{ij}(\theta)$ and $f_i(\theta)$ are definite functions for each cracks, as shown in Eq. (21.2).

Take the most common as well as the most dangerous opening modes as an example, the stress component and displacement component near the crack acme can be represented as

$$\left.\begin{aligned}\sigma_x &= \frac{K_\mathrm{I}}{\sqrt{2\pi r}} \cos\frac{\theta}{2}\left[1 - \sin\frac{\theta}{2}\sin\frac{3\theta}{2}\right] \\ \sigma_y &= \frac{K_\mathrm{I}}{\sqrt{2\pi r}} \cos\frac{\theta}{2}\left[1 + \sin\frac{\theta}{2}\sin\frac{3\theta}{2}\right] \\ \tau_{xy} &= \frac{K_\mathrm{I}}{\sqrt{2\pi r}} \sin\frac{\theta}{2}\cos\frac{\theta}{2}\cos\frac{3\theta}{2}\end{aligned}\right\} \tag{21.2}$$

$$\begin{aligned}u &= \frac{(1+\mu)K_\mathrm{I}}{2E}\sqrt{\frac{r}{2\pi}}\left[(2s-1)\cos\frac{\theta}{2} - \cos\frac{3\theta}{2}\right] \\ v &= \frac{(1+\mu)K_\mathrm{I}}{2E}\sqrt{\frac{r}{2\pi}}\left[(2s+1)\sin\frac{\theta}{2} - \sin\frac{3\theta}{2}\right]\end{aligned} \tag{21.3}$$

As for the sliding mode, that is, mode II, stress component and displacement component near the crack acme are

$$\begin{aligned}\sigma_x &= \frac{-K_\mathrm{II}}{\sqrt{2\pi r}} \sin\frac{\theta}{2}\left[2 + \cos\frac{\theta}{2}\cos\frac{3\theta}{2}\right] \\ \sigma_y &= \frac{K_\mathrm{II}}{\sqrt{2\pi r}} \cos\frac{\theta}{2}\sin\frac{\theta}{2}\cos\frac{3\theta}{2}\end{aligned}$$

$$\tau_{xy} = \frac{K_{II}}{\sqrt{2\pi r}} \cos\frac{\theta}{2}\left[1 - \sin\frac{\theta}{2}\sin\frac{3\theta}{2}\right]$$

$$u = \frac{(1+\mu)K_{II}}{2E}\sqrt{\frac{r}{2\pi}}\left[(2s+3)\sin\frac{\theta}{2} + \sin\frac{3\theta}{2}\right]$$

$$v = \frac{-(1+\mu)K_{II}}{2E}\sqrt{\frac{r}{2\pi}}\left[(2s-2)\cos\frac{\theta}{2} + \cos\frac{3\theta}{2}\right]$$

Plane stress:

$$s = \frac{3-\mu}{1+\mu}, w = -\frac{\mu}{E}\int(\sigma_x + \sigma_y)\mathrm{d}z, \sigma_z = 0$$

Plane strain:

$$s = 3 - 4\mu, \sigma_x = \mu(\sigma_x + \sigma_y), w = 0$$

As for tearing mode, that is, mode III, the displacement is vertical to the xy plane, $u = v = 0, w \neq 0$, stress component and displacement component around the crack tip are

$$\tau_{xx} = -\frac{K_{III}}{\sqrt{2\pi r}} \sin\frac{\theta}{2}$$

$$\tau_{yx} = \frac{K_{III}}{\sqrt{2\pi r}} \cos\frac{\theta}{2}$$

$$w = \frac{2(1+\mu)K_{III}}{E}\sqrt{\frac{2r}{\pi}} \sin\frac{\theta}{2}$$

in which E and μ are modulus of elasticity and Poisson's ratio.

Therefore, for mode I cracks, stress field (as well as displacement field) in the crack front is determined by factor K_I. It depends on the amount and distribution of external forces, geometric conditions of objects, and shape and location of the cracks. Meanwhile, for mode II and III cracks, stress field in crack front is determined by their own factors K_{II} and K_{III}. K_I, K_{II}, and K_{III} are called stress intensity factors.

Experiments have indicated that when K_I, K_{II}, or K_{III} reaches critical values K_{Ic}, K_{IIc}, and K_{IIIc}, cracks lose their stabilities and extend. K_I, K_{II}, and K_{III} all depend on the property of materials. Thus different materials and conditions have various values. They represent the resistance of materials toward crack developing, which are new indexes to measure the tenacity derived from fracture mechanics.

Among three crack modes, opening mode is the most regular, and also the most dangerous ones, which has been best studied. K_{Ic} is called plane strain fracture toughness of opening development mode. The critical condition of crack developing is

$$K_I = K_{Ic} \tag{21.4}$$

In other words, when stress intensity factor K_I reaches fracture toughness K_{Ic}, cracks will develop, leading to the break of the structure. K_{Ic} is measured by material tests, while stress intensity factor K_I is calculated by structural analysis. One of the most important content of fracture mechanics is to calculate the stress intensity factor for different structural conditions and different types of cracks. Regular methods include the complex variable method, the integral transform method, the boundary collocation method, and the finite element method. However, the former methods can work out results only when the given conditions are simple. For the complex conditions in practical engineering, it can only be worked out by the finite element method. Thus, the finite element method has a wide use in analysis of fracture mechanics.

21.2 Direct Method

In linear elastic fracture mechanics, the simplest method to calculate stress intensity factor of crack tip by the finite element method is the direct method. That is, firstly use the finite element method to find out the numerical solution of components of stress or displacement around the crack tip. Then use the expression (Formula (21.2) or (21.3)) of the stress field of the crack tip to calculate the stress intensity factor directly.

To calculate the stress intensity factor, we could use the displacement value of the crack tip, as well as the stress value, which are introduced as follows.

21.2.1 Displacement Method

Put the coordinate original point to the crack tip, as shown in Figure 21.2. Letting $\theta = \pi$, for plane stress problems, we can get the vertical displacement around the crack tip as

$$v = \frac{4(1-\mu^2)}{\sqrt{2\pi}E} K_I \sqrt{r}$$

In this formula, we ignore higher-order terms of r. If we retain the higher terms, the formula turns to

$$v = \frac{4(1-\mu^2)}{\sqrt{2\pi}E} K_I \sqrt{r} \left(1 + a_1 \frac{r}{l} + \cdots\right)$$

Rearrange this formula as

$$\frac{\sqrt{2\pi}Ev}{4(1-\mu^2)\sqrt{r}} = K_I \left(1 + a_1 \frac{r}{l} + \cdots\right) \tag{21.5}$$

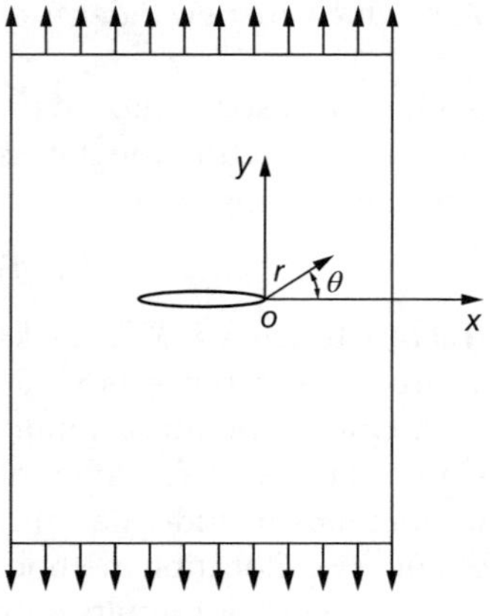

Figure 21.2 Plane crack.

Taking the limit $r \to 0$ of this formula, we can get the stress intensity factor as

$$K_I = \lim_{r \to 0} \left(\frac{\sqrt{2\pi}Ev}{4(1-\mu^2)\sqrt{r}} \right) \tag{21.6}$$

The finite element method can only get approximate displacement v^* of each node. Meanwhile, close to the crack tip $r=0$, the error of the finite element method is large, where it cannot be used directly. In order to overcome this problem, method is used as follows. Assume v^* is the displacement value around the crack tip by the finite element method. Let

$$K_I^* = \frac{\sqrt{2\pi}Ev^*}{4(1-\mu^2)\sqrt{r}} \tag{21.7}$$

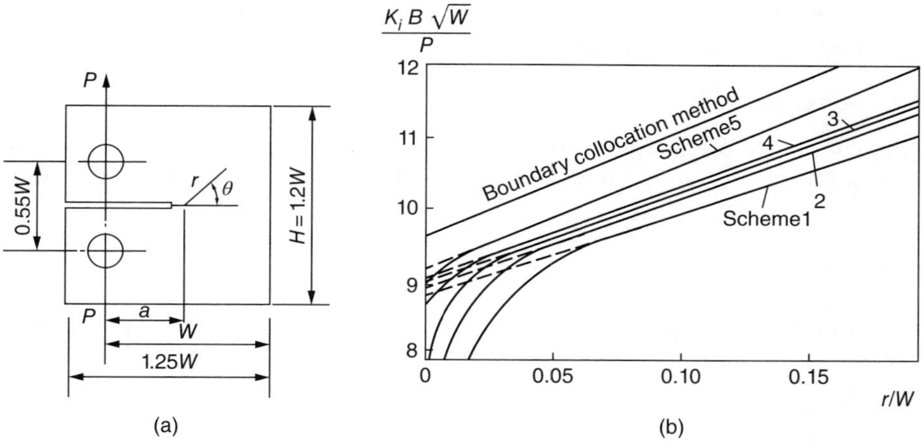

Figure 21.3 K_i of compact tensile specimen calculated by finite element method. (a) Compact tensile specimen; (b) the result of finite element method.

As shown in Formula (21.3), when $\frac{r}{l}$ is small, K_I^* can be treated as linear function of r approximately. Let K_I^* be the vertical coordinate and r the horizontal coordinate, and plot the related points. Then draw the fitted line by least squares method. The intersection of this line and the vertical axis is the estimated value of stress intensity factor K_I.

Because we can only get the approximate solution by the finite element method, the estimated value of K_I calculated by the aforementioned method is apparently related to the intensity of the meshes. The thinner the meshes are, the displacement value as well as the estimated value of K_I is more accurate.

Figure 21.3(a) shows a compact tensile specimen. Using the finite element method of 5 different meshes to this specimen, we get a group of curves as Figure 21.3(b) of the calculated displacements by the aforementioned method, where the vertical and horizontal coordinates are dimensionless term $K_i^* B\sqrt{W}/P$ and r/W, B represents the thickness of the specimen. The straight line on the top is a more accurate solution calculated by boundary collocation method. The other 5 straight lines are approximate solutions calculated by the finite element method. The density of mesh of these 5 schemes is in Table 21.1. As shown in Table 21.1 and the straight lines of scheme 4 and 5 in

Table 21.1 Density of meshes.

Scheme	Around the crack tip $(A/a^2)*10^6$	Around the outside $(A/a^2)*10^6$
1	312	2
2	78	2
3	21	2
4	1.2	2
5	1.2	1

A, the area of the finite element (mm^2); a, the length of crack (mm).

Figure 21.3(b), the accuracy of the calculated K_I is also influenced by the area of the finite element mesh on the boundary of the specimen.

21.2.2 Stress Method

Put the coordinate original point to the crack tip, as in Figure 21.2. Let $\theta = 0$, by Eq. (21.2), for plane strain problems, we can get the vertical stress around the crack tip as

$$\sigma_y = \frac{K_I}{\sqrt{2\pi r}}$$

In this formula, we also ignore higher-order terms of r. If we retain the higher terms, the formula turns to

$$\sigma_y = \frac{K_I}{\sqrt{2\pi r}}\left(1 + b_1\frac{r}{l} + \cdots\right)$$

Rearrange this formula as

$$\sqrt{2\pi r}\sigma_y = K_I\left(1 + b_1\frac{r}{l} + \cdots\right) \tag{21.8}$$

Taking the limit $r \to 0$ of this formula, we can get the stress intensity factor as follows:

$$K_I = \lim_{r\to 0}(\sqrt{2\pi r}\sigma_y) \tag{21.9}$$

Just as mentioned before, the finite element method can only get approximate stress σ_y^* of each element. Meanwhile, close to the crack tip, the error is big; thus the limit process cannot be used directly. Let

$$K_I^* = \sqrt{2\pi r}\sigma_y^* \tag{21.10}$$

By Formula (21.8), when $\frac{r}{l}$ is small, K_I^* can be treated as linear function of r approximately. Similar to the displacement method, let K_I^* be the vertical coordinate and r the horizontal coordinate, and plot the related points. Then draw the fitted line through most of the points and extrapolate it to the place where $r = 0$. Then by the intersection of this line and the vertical axis, the estimated value of stress intensity factor K_I can be computed.

Stress method and displacement method have no significant differences. Now we usually adopt the displacement method. As stress is the first-order derivative of displacement, the accuracy of displacement is higher. So displacement method is better.

In a word, although to calculate the stress intensity factor of the crack tip by direct method has a simpler procedure, and it can take advantage of the general programs of the finite element method, its accuracy is influenced by the size of the finite element mesh. In order to ensure that the calculation result has enough accuracy, it is necessary to apply very thin mesh around the crack tip. So it is not an ideal method.

21.3 *J-Integral* Method

In continuum mechanics, some integrals with characteristics of conservation are always used in order to analyze stress field around the incision or crack tip. For plane crack problems, the so-called *J-integral* method put forward by J. R. Rice is a linear integral that has nothing to do with the path of integration. It is one kind of the integral with characteristics of conservation, which is defined as

$$J = \int_{\Gamma} \left[W\,\mathrm{d}y - \vec{T} \cdot \frac{\partial \vec{u}}{\partial x}\,\mathrm{d}s \right] \tag{21.11}$$

where W is the strain energy of unit volume, as for linear elastic body, $W = \frac{1}{2}\{\sigma\}^{\mathrm{T}}\{\varepsilon\} = \frac{1}{2}(\sigma_x\varepsilon_x + \sigma_y\varepsilon_y + \tau_{xy}\gamma_{xy})$; $\vec{T}$ is the tension vector defined by direction of the outer normal of integral curve Γ; $\vec{u}$ is the corresponding displacement vector; the symbol '·' between $\vec{T}$ and $\frac{\partial \vec{u}}{\partial x}$ represents "dot multiply" of two vectors; ds is one part of differential arc length along the integral curve Γ; and Γ is a contour started with lower boundary and ended with upper boundary of the crack (Figure 21.4).

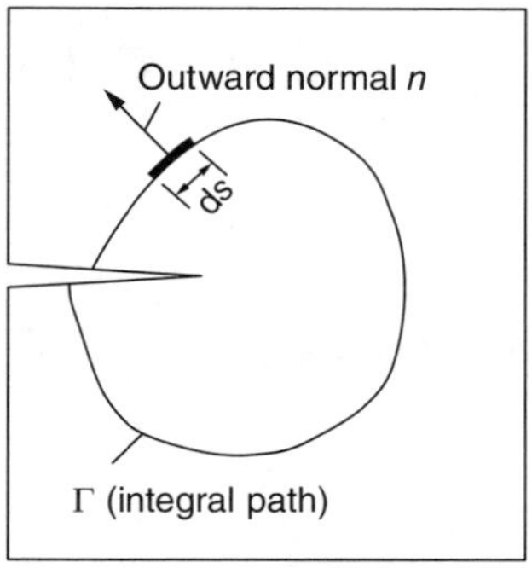

Figure 21.4 Contour integral.

Based on Green's theory and equilibrium equation, Rice proved that in small-scale yielding, in the area around the crack tip, *J-integral* has nothing to do with the path of integration and the value is constant along any integral path. He also proved that in linear elasticity, the value of *J-integral* equals the release rate of strain energy G around the crack tip. As a result, J and stress intensity factor are connected as

$$\left.\begin{aligned} &\text{for plane strain problem :} \quad K_{\mathrm{I}} = \sqrt{\frac{JE}{1-\mu^2}} \\ &\text{for plane stress problem :} \quad K_{\mathrm{I}} = \sqrt{JE} \end{aligned}\right\} \tag{21.12}$$

In order to calculate the value of *J-integral,* the finite element method is used to get the components of stress, strain, and displacement around the crack tip. Optionally pick out an integral path Γ that is convenient to calculate the linear integral and get its value by the definition of *J-integral* (21.11). Then K_{I} can be calculated by one of the two formulas mentioned earlier. As the irrelevance of the value of the *J-integral* to integral path, the minor area around crack tip that changes drastically can be avoided; thus it has a better computational accuracy. Also different integral paths are optional to check each other. *J-integral* reflects the overall energy relation of the cracks, which does not request a particular high local accuracy around the crack tip, so the elements are unnecessary to be much thin.

A plate is shown in Figure 21.5, which is made of two materials of different moduli of elasticity. There is a penetrating crack whose length is $2a$ on the left and uniform vertical displacement v_0 on the up and down sides of the plate. The question is how to calculate the stress intensity factor K_{I}. In order to illustrate the application of *J-integral,* the right of the crack tip is chosen in the following. The stress intensity factor of the left of the crack tip can be calculated by the same method.

Due to the vertical symmetry, it is feasible to analyze merely the upper half. By the symmetry, the vertical displacement on the cut plane $v = 0$. Triangular meshes of different densities are used based on the different stress gradients, as shown in Figure 21.6(a). Work out each component of stress, strain, and displacement by the finite element method. Then the value of the *J-integral* can be calculated by any integral path around the right side of the crack tip. For the calculation mesh used now, the rectangle integral path coincident with the mesh edge is the simplest.

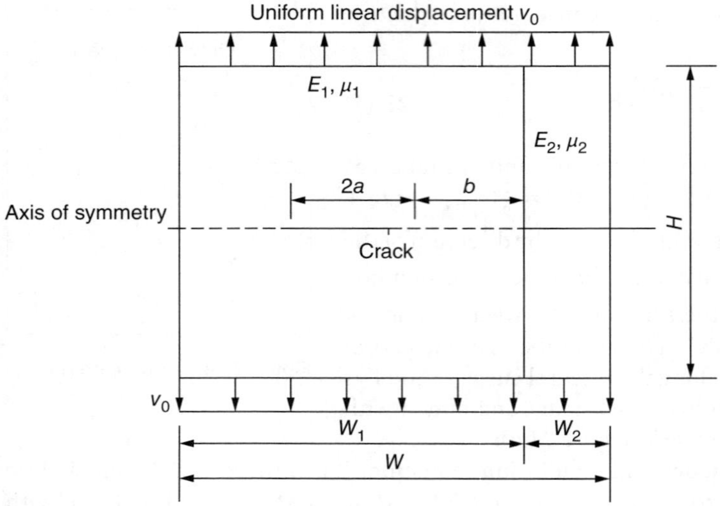

Figure 21.5 Example.

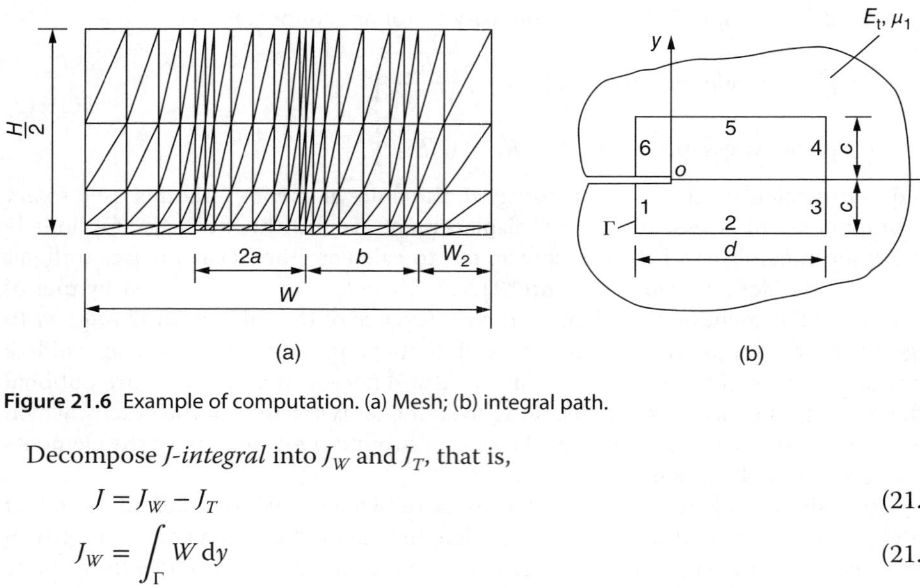

Figure 21.6 Example of computation. (a) Mesh; (b) integral path.

Decompose *J-integral* into J_W and J_T, that is,

$$J = J_W - J_T \tag{21.13}$$

$$J_W = \int_\Gamma W\,\mathrm{d}y \tag{21.14}$$

$$J_T = \int_\Gamma \vec{T} \cdot \frac{\partial \vec{u}}{\partial x}\,\mathrm{d}s \tag{21.15}$$

As shown in Figure 21.6(b), integral path Γ can be divided into straight lines of Line 1, 2, 3, 4, 5, and 6. For the linear integrals of Eqs (21.14) and (21.15), started with Line 1, and integrate them to Line 6 in counterclockwise direction. Thus,

$$J_W = \int_0^{-c} W_1\,\mathrm{d}y + \int_{-c}^{-c} W_2\,\mathrm{d}y + \int_{-c}^{0} W_3\,\mathrm{d}y + \int_0^{c} W_4\,\mathrm{d}y + \int_c^{c} W_5\,\mathrm{d}y + \int_c^{0} W_6\,\mathrm{d}y \tag{a}$$

where W_i represents the value of W in Line i. Due to symmetry, $W_1 = W_6$, $W_3 = W_4$. Besides,

$$\int_{-c}^{-c} W_2\,\mathrm{d}y = 0, \int_{c}^{c} W_5\,\mathrm{d}y = 0$$

(because $\mathrm{d}y = 0$ along these two lines). Formula (a) can be simplified to

$$J_W = 2\left[\int_{-c}^{-c} W_4\,\mathrm{d}y + \int_{c}^{0} W_6\,\mathrm{d}y\right] \tag{b}$$

The next integral J_T is calculated by the same integral path:

$$\begin{aligned} J_T = &\int_0^c \vec{T}\cdot\frac{\partial\vec{u}}{\partial x}\,\mathrm{d}s + \int_c^{c+d} \vec{T}\cdot\frac{\partial\vec{u}}{\partial x}\,\mathrm{d}s + \int_{c+d}^{2c+d} \vec{T}\cdot\frac{\partial\vec{u}}{\partial x}\,\mathrm{d}s + \int_{2c+d}^{3c+d} \vec{T}\cdot\frac{\partial\vec{u}}{\partial x}\,\mathrm{d}s \\ &+ \int_{3c+d}^{3c+2d} \vec{T}\cdot\frac{\partial\vec{u}}{\partial x}\,\mathrm{d}s + \int_{3c+2d}^{4c+2d} \vec{T}\cdot\frac{\partial\vec{u}}{\partial x}\,\mathrm{d}s \end{aligned} \tag{c}$$

Due to symmetry, this formula can be simplified to

$$J_T = 2\left[\int_0^c \left(\vec{T}\cdot\frac{\partial\vec{u}}{\partial x}\right)^{(4)}\mathrm{d}s + \int_0^d \left(\vec{T}\cdot\frac{\partial\vec{u}}{\partial x}\right)^{(5)}\mathrm{d}s + \int_0^c \left(\vec{T}\cdot\frac{\partial\vec{u}}{\partial x}\right)^{(6)}\mathrm{d}s\right] \tag{d}$$

In Line 4, $\vec{T}$ can be divided to σ_x and τ_{xy}, and the corresponding displacement vector $\vec{u}$ is u and v. Therefore,

$$\left(\vec{T}\cdot\frac{\partial\vec{u}}{\partial x}\right)^{(4)} = \sigma_x^{(4)}\frac{\partial u^{(4)}}{\partial x} + \tau_{xy}^{(4)}\frac{\partial v^{(4)}}{\partial x} = \sigma_x^{(4)}\varepsilon_x^{(4)} + \tau_{xy}^{(4)}\frac{\partial v^{(4)}}{\partial x}$$

In Line 5, $\vec{T}$ can be divided to σ_y and τ_{xy}, and the corresponding displacement vector $\vec{u}$ is u and v. Therefore,

$$\left(\vec{T}\cdot\frac{\partial\vec{u}}{\partial x}\right)^{(5)} = \sigma_y^{(5)}\frac{\partial v^{(5)}}{\partial x} + \tau_{xy}^{(5)}\frac{\partial u^{(5)}}{\partial x} = \sigma_y^{(5)}\frac{\partial v^{(5)}}{\partial x} + \tau_{xy}^{(5)}\varepsilon_x^{(5)}$$

In Line 6, $\vec{T}$ can be divided to σ_y and τ_{xy}, and the corresponding displacement vector $\vec{u}$ is $-u$ and $-v$. Therefore,

$$\left(\vec{T}\cdot\frac{\partial\vec{u}}{\partial x}\right)^{(6)} = -\sigma_x^{(6)}\frac{\partial u^{(6)}}{\partial x} - \tau_{xy}^{(6)}\frac{\partial v^{(6)}}{\partial x} = -\sigma_x^{(6)}\varepsilon_x^{(6)} - \tau_{xy}^{(6)}\frac{\partial v^{(6)}}{\partial x}$$

Substitute the previous three formulas into Formula (g) to obtain

$$\begin{aligned} J_T = 2\Bigg[&\int_0^c \left(\sigma_x^{(4)}\varepsilon_x^{(4)} + \tau_{xy}^{(4)}\frac{\partial v^{(4)}}{\partial x}\right)\mathrm{d}s + \int_0^d \left(\sigma_y^{(5)}\frac{\partial v^{(5)}}{\partial x} + \tau_{xy}^{(5)}\varepsilon_x^{(5)}\right)\mathrm{d}s \\ &+ \int_0^c \left(-\sigma_x^{(6)}\varepsilon_x^{(6)} - \tau_{xy}^{(6)}\frac{\partial v^{(6)}}{\partial x}\right)\mathrm{d}s\Bigg] \end{aligned} \tag{e}$$

By substituting the values of stress, strain, and displacement along the integral path calculated by the finite element method into Formula (b), (e), J_W and J_T can be determined. Then work out *J-integral* by Formula (21.13). The stress intensity factor K_I is determined by Formula (21.12). As the crack lies in the first material, factors E_1 and μ_1 are used.

J-integral is widely used because it has higher accuracy and the mesh around the crack tip is unnecessary to be too thin.

It will be pointed out in the latter that not only it is feasible to calculate the stress intensity factor $K_{\rm I}$ in small-scale yield conditions by *J-integral*, but also it is used as factors to determine brittle rupture in elastic–plastic fracture mechanics in large-scale yield conditions.

21.4 Energy Method, Flexibility Method, and Bueckner Formula

Several energy methods to calculate the stress intensity factor are introduced in this chapter.

21.4.1 Energy Release Rate *G* and the Related Formulas

Energy release rate G is defined as the energy released by the system when crack develops unit area. As shown in Figure 21.7, the load is P. The relative displacement of points of force application is v. The area of the crack is A. Assume that the crack area virtually increases by $\mathrm{d}A$ and the strain energy increases by $\mathrm{d}U$. Thus $\mathrm{d}U + G\mathrm{d}A = P\mathrm{d}v$. That is,

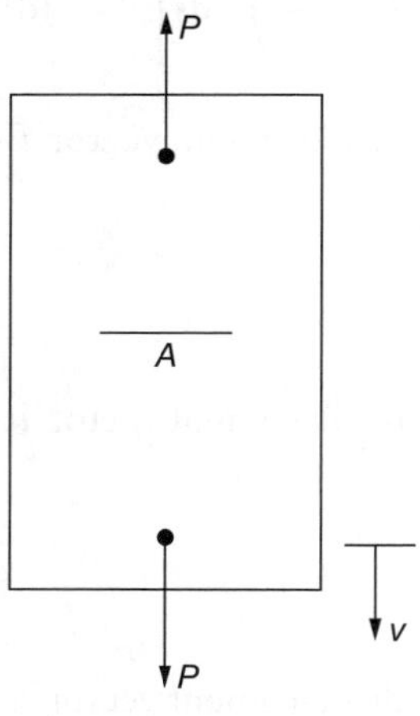

Figure 21.7 Flexibility.

$$G\mathrm{d}A = P\mathrm{d}v - \mathrm{d}U \tag{a}$$

As $\mathrm{d}(Pv) = P\mathrm{d}v + v\mathrm{d}P$,

$$G\mathrm{d}A = \mathrm{d}(Pv - U) - v\mathrm{d}P = -\mathrm{d}\Pi - v\mathrm{d}P \tag{b}$$

where $\Pi = U - Pv$, in which Π is the potential energy of the subject.

From Formula (b),

$$G = P\frac{\partial v}{\partial A} - \frac{\partial U}{\partial A} = -\frac{\partial \Pi}{\partial A} - v\frac{\partial P}{\partial A} \tag{21.16}$$

Let

$$\lambda = \frac{v}{P} \tag{21.17}$$

where λ is the flexibility of the subject.

For linear elastic body, stress energy $U = Pv/2$. Substitute this formula into Formula (21.16):

$$G = P\frac{\partial v}{\partial A} - \frac{1}{2}\frac{\partial (Pv)}{\partial A} = \frac{P}{2}\frac{\partial v}{\partial A} - \frac{v}{2}\frac{\partial P}{\partial A} \tag{c}$$

As $\lambda = v/P$,

$$\frac{\partial \lambda}{\partial A} = \left(P\frac{\partial v}{\partial A} - v\frac{\partial P}{\partial A}\right)/P^2$$

Therefore,

$$P\frac{\partial v}{\partial A} = P\frac{\partial \lambda}{\partial A} + v\frac{\partial P}{\partial A} \tag{d}$$

Substituting Formula (d) into Formula (c), then we have the well-known Irwin–Kies formula:

$$G = \frac{1}{2}P^2\frac{\partial \lambda}{\partial A} \tag{21.18}$$

Consider two special cases in the following:

1) When the displacement is fixed (the points of force application are fixed), $dv = 0$. From Formula (a),

$$G = -\left(\frac{\partial U}{\partial A}\right)_v \tag{21.19}$$

where the subscript v represents that v is fixed.

For linear elastic body,

$$U = \frac{1}{2}Pv = \frac{1}{2}\lambda P^2 = \frac{1}{2\lambda}v^2$$

From Formula (21.19),

$$G = -\frac{\partial}{\partial A}\left(\frac{v^2}{2\lambda}\right) = \frac{P^2}{2}\frac{\partial \lambda}{\partial A} \tag{e}$$

2) When the load is fixed, $dP = 0$. From Formula (b),

$$G = -\left(\frac{\partial \Pi}{\partial A}\right)_P \tag{21.20}$$

Because $U = Pv/2$, $\Pi = U - Pv = -U$. Thus

$$G = -\left(\frac{\partial \Pi}{\partial A}\right)_P = \left(\frac{\partial U}{\partial A}\right)_P = \frac{\partial}{\partial A}\left(\frac{1}{2}\lambda P^2\right) = \frac{1}{2}P^2\frac{\partial \lambda}{\partial A} \tag{21.21}$$

Formulas (e), (21.21), and (21.18) are the same. So whether the displacement is fixed, or the load is fixed, or neither of the displacement and load is fixed, Formula (21.18) is workable.

There is a one-to-one correspondence between the energy release rate G and the stress intensity factor. For mode I cracks,

$$G_I = \frac{K_I^2}{E'} \tag{21.22}$$

$$E' = \begin{cases} E(\text{plane stress}) \\ E/(1-\mu^2)\ (\text{plane strain}) \end{cases}$$

For mode II and III cracks,

$$\left.\begin{aligned} G_{II} &= \frac{K_{II}^2}{E'} \\ G_{III} &= \frac{(1+\mu)K_{III}^2}{E} \end{aligned}\right\} \tag{21.23}$$

21.4.2 Flexibility Method

In Formula (21.18), replacing the differential quotient $\partial\lambda/\partial A$ by difference quotient $\Delta\lambda/\Delta A$, there is

$$G = \frac{1}{2}P^2\frac{\Delta \lambda}{\Delta A} \tag{21.24}$$

Carry out two finite element calculations. In the first calculation, the area of the crack is A. Then determine λ by Formula (21.17). In the second calculation, the area of the crack is $A + \Delta A$, and the flexibility is $\lambda + \Delta\lambda$. Putting $\Delta\lambda/\Delta A$ into Formula (21.24), we can get the energy release rate G.

21.4.3 Energy Method

When the structure bears several concentrated loads and distributed load as shown in Figure 21.8, the strain energy is determined by Formula (21.25):

$$U = \frac{1}{2}\Sigma P_i \delta_i \tag{21.25}$$

When the load is fixed, by Formula (21.21), there is

$$G = \frac{\partial U}{\partial A} \tag{21.26}$$

Thus, by two finite element calculations, ΔU is determined. Substituting it into the previous formula, we can calculate G.

21.4.4 Bueckner Formula

Bueckner put forward a formula to calculate energy release rate. Considering a three-dimensional subject as Figure 21.9(a), the volume is V, and the volume force is X_i. There is an external force T_i on the boundary S_σ. Give a fixed displacement on the boundary S_u. There was a crack inside the subject, of which the up boundary is C_1 and the down boundary is C_2. Figure 21.9(b) shows the situation when the virtual crack develops. The load and boundary conditions remain the same. New crack surfaces C_1' and C_2' appear. Adding these two conditions together, we can get the sum condition as Figure 21.9(c), in which the volume force is $2X_i$ and the external force on the boundary S_σ is $2T_i$. There is no surface force on C_1 and C_2. There is surface force T_i^* on C_1' and C_2'. Give a fixed displacement on the boundary S_u. Surface force T_i^* is the required force that makes the virtual crack surface C_1' and C_2' closely matched. Then subtracting condition (a) from (b), we can get the difference condition as Figure 21.9(d). Difference condition is that the same body develops its virtual crack and there is no volume force. Also there is no surface force on S_σ, C_1 and C_2. The fixed displacement on S_u is zero. There is a surface force on C_1' and C_2'. Bueckner proved that the elastic potential energy caused by crack developing only depends on strain energy in difference condition, which is calculated by Formula (21.27):

$$U = -\frac{1}{2}\int_{C_1'+C_2'} T_i^* v_i \, ds \tag{21.27}$$

where v_i is the displacement of crack surface in difference condition. This integral is calculated along the surface of virtual crack.

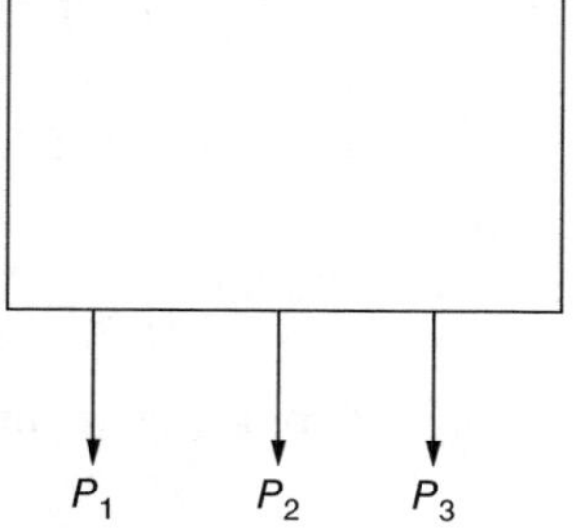

Figure 21.8 Structure subjected to multi-loads.

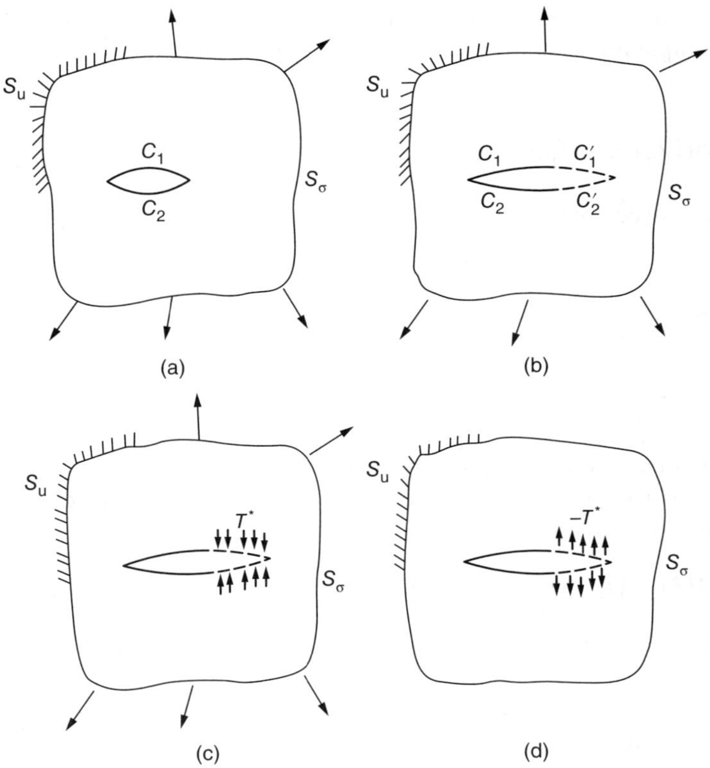

Figure 21.9 Several stress conditions of any crack. (a) before the crack virtually extends, (b) after the crack virtually extends, (c) sum condition, and (d) difference condition.

During the deduction of Formula (21.27), the virtual crack does not have to be infinitely small. Therefore, it can be viewed that before the virtual crack develops, the subject has no cracks at all. In this condition, T_i^* equals to the stress of corresponding location in crack C_1 and C_2 of certain load when there is no crack in the subject. It can be written as T_i^0. Thus, the difference condition equals the condition of no volume force, no surface force on S_σ, fixed zero displacement on S_u, and surface force T_i^0 on the crack surface. The energy release rate of crack is determined by Formula (21.28):

$$G = \frac{\partial U}{\partial A} = -\frac{1}{2}\frac{\partial}{\partial A}\int_{C_1'+C_2'} T_i^0 v_i \, ds \tag{21.28}$$

where A is the area of the crack and the length of the crack for plane problems, T_i^0 is the stress of the body with no crack under given certain load and boundary displacement, and v_i is the displacement of the crack surface when the external force of cracked subject is zero and there is a surface force T_i^0 on the crack surface.

Given different sizes of cracks, G can be calculated by Formula (21.28). Then the stress intensity factor is determined by Formula (21.22). In the process of using Formula (21.28), considering the normal stress and shearing stress individually, K_{I} and K_{II} can be calculated separately.

Reference [10] gives some examples of calculation. For plane problems, by mesh of about 500 nodes, the computational accuracy of the stress intensity factor is 3–5%.

21.5 Stiffness Derivative Method

When the external load is fixed, by Formula (21.21),

$$G = -\frac{\partial \Pi}{\partial A} \tag{21.29}$$

where Π is the potential energy.

By Formula (3.42),

$$\Pi = \frac{1}{2}\{\delta\}^{\mathrm{T}}[K]\{\delta\} - \{\delta\}^{\mathrm{T}}\{P\} \tag{21.30}$$

where $[K]$ is the stiffness matrix and $\{P\}$ is the nodal load.

Taking the derivative of Formula (21.30) with respect to crack length a and considering the symmetry of $[K]$, we get

$$\frac{\partial \Pi}{\partial a} = \frac{\partial \{\delta\}^{\mathrm{T}}}{\partial a}([K]\{\delta\} - \{P\}) + \frac{1}{2}\{\delta\}^{\mathrm{T}}\frac{\partial [K]}{\partial a}\{\delta\} - \{\delta\}^{\mathrm{T}}\frac{\partial \{P\}}{\partial a}$$

But in the finite element method, according to the equilibrium condition, $[K]\{\delta\} - \{P\} = 0$. Therefore,

$$G = -\frac{\partial \Pi}{\partial a} = -\frac{1}{2}\{\delta\}^{\mathrm{T}}\frac{\partial [K]}{\partial a}\{\delta\} + \{\delta\}^{\mathrm{T}}\frac{\partial \{P\}}{\partial a} \tag{21.31}$$

If there is no volume force and no external force on the crack surface, $\partial\{P\}/\partial a = 0$. Therefore,

$$G = -\frac{1}{2}\{\delta\}^{\mathrm{T}}\frac{\partial [K]}{\partial a}\{\delta\} \tag{21.32}$$

Given a virtual extension Δa, then the strain energy release rate G is determined by Formulas (21.31) and (21.32). In the next part we will discuss it in three conditions.

21.5.1 Plane Problem

As shown in Figure 21.10, draw two curves Γ_0 and Γ_1 around the crack tip. When the crack extends virtually, nodes above and nodes outside Γ_1 keep constant. In the meanwhile nodes above and nodes outside Γ_0 move right at a distance of Δa. After the virtual extension, the element stiffness matrix outside Γ_1 and inside Γ_0 stays the same. Only the element stiffness matrix between Γ_0 and Γ_1 has changed. Thus,

$$\frac{\partial [K]}{\partial a} = \sum_e \frac{\partial [K^{\mathrm{e}}]}{\partial a} \tag{21.33}$$

where $\sum_e$ is the sum of the elements between Γ_0 and Γ_1.

$\partial[K^{\mathrm{e}}]/\partial a$ can be calculated by difference method:

$$\frac{\partial [K^{\mathrm{e}}]}{\partial a} = \frac{1}{\Delta a}([K^{\mathrm{e}}]_{a+\Delta a} - [K^{\mathrm{e}}]_a) \tag{21.34}$$

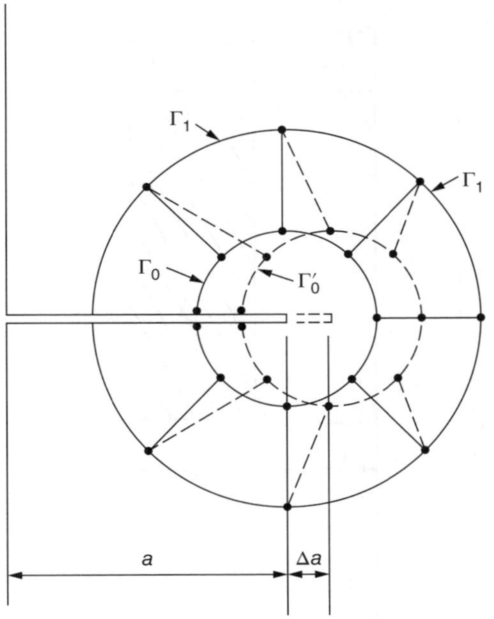

Figure 21.10 Virtual extension of plane crack.

Similarly,

$$\frac{\partial\{P\}}{\partial a} = \frac{1}{\Delta a}(\{P\}_{a+\Delta a} - \{P\}_a) \tag{21.35}$$

The shape and area of contours Γ_0 and Γ_1 are arbitrary. There are two contours including the crack tip and Γ_0 is inside Γ_1. In practice, Γ_0 can shrink into a point of the crack tip. At that time, the sum in Formula (21.33) only involved several elements near the crack tip.

Reference [6] gives an example as shown in Figure 21.11, a double-notched plane strain tensile specimen. The ratio of the depth of the notch to specimen width is $2a/W = 0.5$. By symmetry, it is only necessary to calculate 1/4 of the specimen. Nodes around the crack tip are in the arcs of radius 0, 0.2a, 0.5a, and a. Each arc has 12 elements of included angle 15°. The innermost elements are always constant strain triangle elements. Others are quadrilateral elements. In total there are 94 elements and 106 nodes. Four inner rings are chosen as contour Γ_0 in turns. Calculate G and the stress intensity factor by Formula (21.32). The results are listed in Table 21.2. To compare two methods, the results determined by conformal mapping by Bowie are listed in the same table. The errors of Bowie's calculation are less than 1%. As shown in the table, the errors of stiffness derivative method are 4.3%, 0.6%, 2.8%, and 2.6% correspondingly.

It should be noted that if singular isoparametric elements in the next section are adopted around the crack tip, the accuracy can still increase.

21.5.2 Axial Symmetrical Problem

Stiffness derivative method is useful for axial symmetrical problem. In axial symmetrical problem, the circumferential strain $\varepsilon_\theta = u/r$. So by Formula (6.12), stiffness matrix is related to radius r. After the elements inside Γ_0 have rigid-body displacement, the

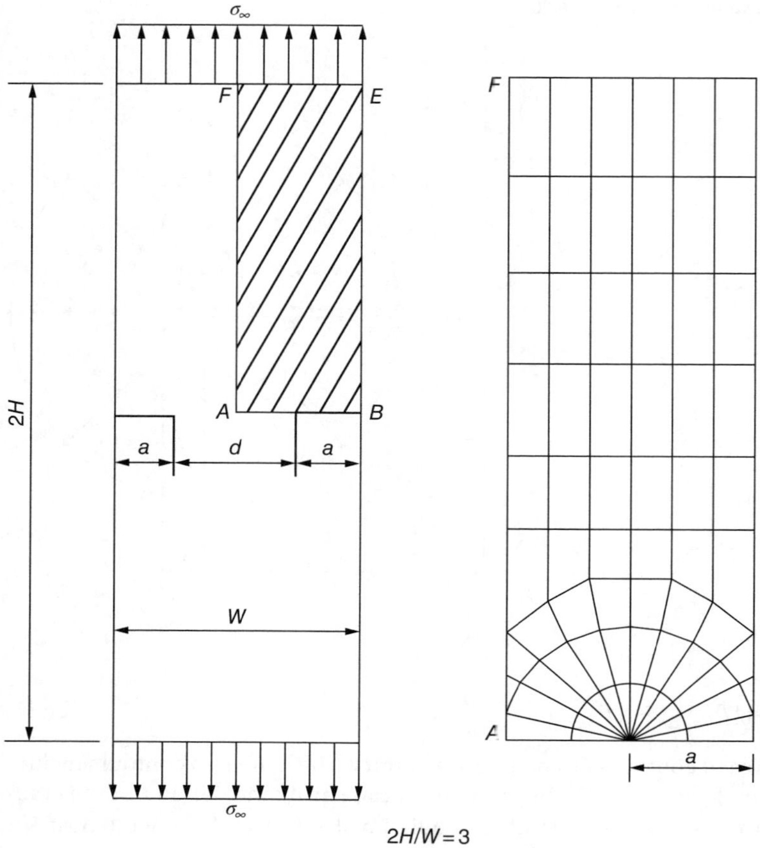

Figure 21.11 Example.

Table 21.2 Examples of stiffness derivative method (plane problems).

Method	Bowie	Stiffness derivative method			
		$\Gamma_0 : r/a = 0$	0.1	0.2	0.5
$K_i/\sigma_\infty\sqrt{a}$	1.028	0.985	1.022	0.999	1.001
Error (%)		4.3	0.6	2.8	2.6

stiffness matrix changes. Therefore, the sum in Formula (21.33) involved all the elements inside Γ_1 rather than merely the elements between Γ_0 and Γ_1, which is different from plane problem.

Reference [6] offers an example: a circular tensile specimen with ring notch. The external diameter is D. The diameter in the notch tip is d. $D/d = 2$. By symmetry, it is necessary only to calculate 1/4 of the specimen. The finite element mesh is the same as that in Figure 21.11. When the crack has a virtual extend, nodes on and inside Γ_0 move radial

Table 21.3 Cases of stiffness derivative method (axisymmetric problems).

Method	Bueckner	Stiffness derivative method			
		$r_{\Gamma_0}/(d/2) = 0$	0.1	0.2	0.5
$K_i/\sigma_\infty\sqrt{\pi D}$	0.240	0.228	0.238	0.232	0.234
Error (%)		5.0	0.8	3.3	2.5

distance of Δa. The results are listed in Table 21.3. The results determined by approximate conformal mapping by Bueckner are listed in the same table, of which errors are regarded as less than 1%.

21.5.3 Space Problem

Consider a crack in a three-dimensional subject as Figure 21.12(a). Assuming that node i in crack front has a virtual extension in the radial direction at a distance of Δa, stiffness matrix related to node i change from $[K^e]_a$ to $[K^e]_{a+\Delta a}$. The increment is

$$[\Delta K^e] = [K^e]_{a+\Delta a} - [K^e]_a$$

Sum up all the $[\Delta K^e]$ of related elements:

$$[\Delta K] = \sum([K^e]_{a=\Delta a} - [K^e]_a) \tag{21.36}$$

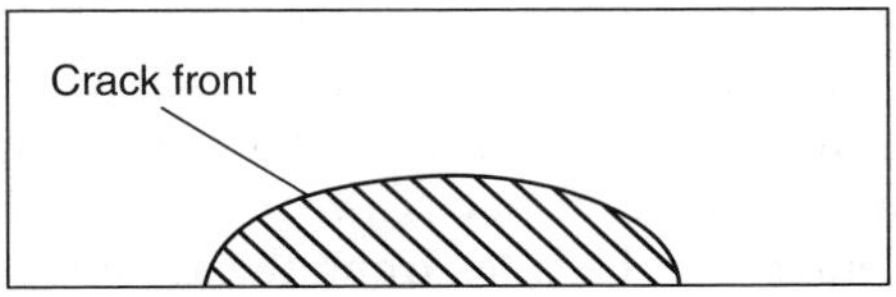

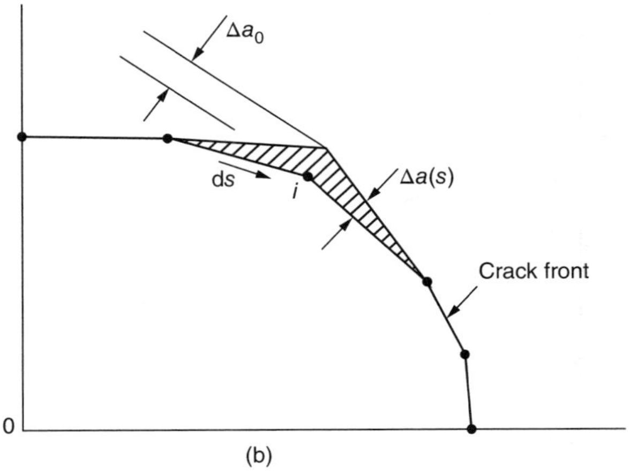

Figure 21.12 Crack virtually extends of space problems. (a) Crack; (b) extension.

Thus,

$$\int G(s)\Delta a(s)\mathrm{d}s = -\Delta\Pi = -\frac{1}{2}\{\delta\}^{\mathrm{T}}[\Delta K][\delta] + \{\delta\}^{\mathrm{T}}\{\Delta P\} \tag{21.37}$$

Reference [6] recommended to linearly interpolate $G(s)$ among adjacent nodes in the crack front. Thus,

$$G(s) = \alpha s + \beta$$

If there are $n+1$ nodes in the crack front, there are n factors α and n factors β, $2n$ unknown factors in total. In each intermediate node, $G(s)$ need to be continuous; there are $n-1$ continuous conditions in total. Each time when a node is moved, an energy equation shown as Eq. (21.37) is obtained, and there are $n+1$ energy equations in total. Together with $n-1$ continuous conditions, there are $2n$ equations. Solve these equations, and n factors α and n factors β are determined.

This method is troublesome, and the author recommended simplifying it appropriately. By mean value theorem,

$$\int G(s)\Delta a(s)\mathrm{d}s \cong G_i\int \Delta a(s)\mathrm{d}s = G_i\Delta A_s$$

Substituting Formula (21.37),

$$G_i = -\frac{1}{2\Delta A_i}\{\delta\}^{\mathrm{T}}[\Delta K]\{\delta\} + \frac{1}{\Delta A_i}\{\delta\}^{\mathrm{T}}\{\Delta P\} \tag{21.38}$$

in which $\Delta A_i = \int \Delta a_s \,\mathrm{d}s$

where ΔA_i is the area of crack extension and G_i is the energy release rate of node i.

From the aforementioned examples for plane and axisymmetrical problems, it is clear that the accuracy of computation is poor if only some points near the tip of crack are removed. Now in three-dimensional problems, when the crack has a virtual extension, if the range of the disturbance is limited to elements in the crack front, the accuracy is

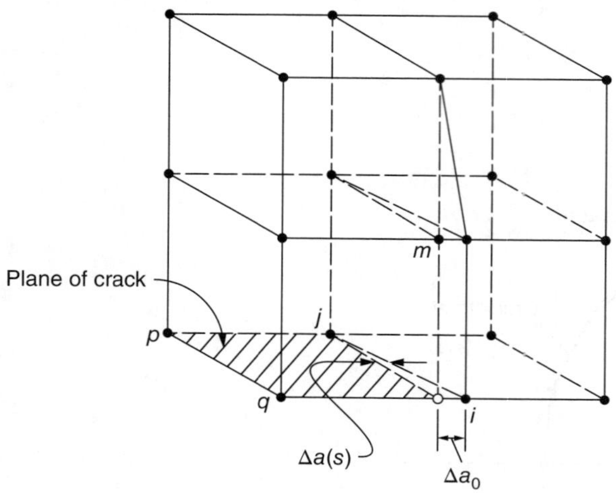

Figure 21.13 Crack virtually extends of space problems.

relatively low. In order to get a higher accuracy, the range of the disturbance needs to be larger.

A case is shown in Figure 21.13: there are four 8-noded isoparametric elements on the crack plane *ijpq*, where side *ij* is the crack front. If only node i moves $\Delta a(s)$ alongside ij, while other nodes stay still, the calculation in Formula (21.38) only involves two elements below. On the contrary, if the full line shown in Figure 21.13 has virtual crack extension, not only node i moves Δa_0, but also node m moves Δa_0. The calculation in Formula (21.38) involves four elements. In this way the accuracy is higher.

21.6 Singular Element of the Crack Tip

The stress field around the crack tip has singularity. As shown in Eq. (21.1), each stress component close to crack tip is direct proportional to $r^{-1/2}$. When $r \to 0$, stress increases rapidly (Figure 21.14). In regular finite element method, the stress and displacement inside the element are represented by polynomial, so it cannot reflect the change of the stress near the singular point. In order to overcome this drawback, before mid-1970s, two ways are adopted. One is to divide the mesh around the crack tip into extremely thin. The result of this method leads to increase of nodes and calculated amount. The other way is to set special singular elements close to the crack tip. Outside these special elements, others are still regular. The following are cases of singular element of the crack tip:

21.6.1 Triangular Singular Element

As shown in 21.14, circle the crack tip by a group of triangular elements. One vertex of the triangular element coincides with the crack tip. The displacement function of triangular singular element is

$$\begin{Bmatrix} u \\ v \end{Bmatrix} = \begin{Bmatrix} u_0 \\ v_0 \end{Bmatrix} + \frac{\theta_j - \theta}{\theta_j - \theta_i} \begin{Bmatrix} u_i \\ v_i \end{Bmatrix} + \frac{\theta - \theta_i}{\theta_j - \theta_i} \begin{Bmatrix} u_i \\ v_i \end{Bmatrix} \sqrt{\frac{r}{r_0}} \tag{21.39}$$

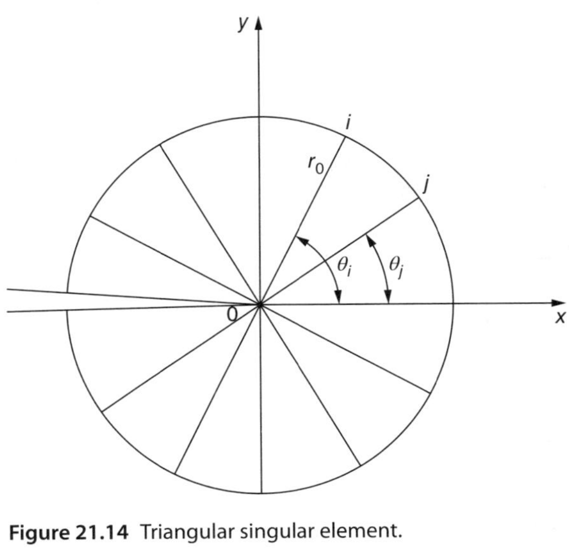

Figure 21.14 Triangular singular element.

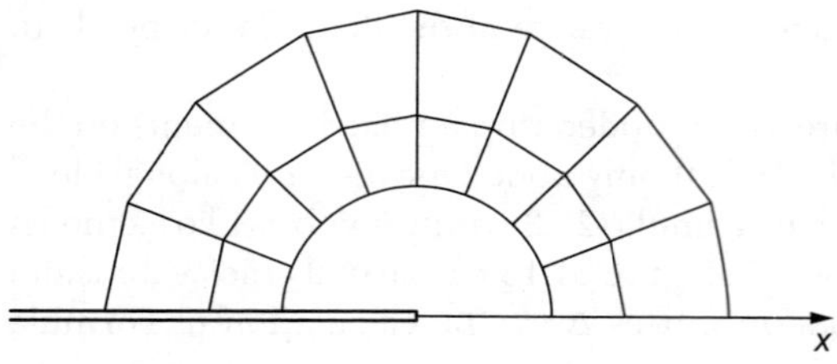

Figure 21.15 Circle singular element.

where u_0, v_0 are displacements of the crack tip. The displacement shown earlier is related to $r^{1/2}$, so the strain has a factor $r^{-1/2}$.

21.6.2 Circle Singular Element

Set a circle singular element in the crack tip; the others outside are still regular ones. For plane strain tensile crack, due to symmetry, only half is required as Figure 21.15 shows. The displacements in the semicircle element are

$$u = u_0 + \frac{(1+\mu)K_I}{2E}\sqrt{\frac{r}{2\pi}}\left[(5-8\mu)\cos\frac{\theta}{2} - \cos\frac{3\theta}{2}\right]$$
$$v = v_0 + \frac{(1+\mu)K_I}{2E}\sqrt{\frac{r}{2\pi}}\left[(7-8\mu)\sin\frac{\theta}{2} - \sin\frac{3\theta}{2}\right] \quad (21.40)$$

In fact, this formula is the first term deduced from Williams's stress function.

21.6.3 Hybrid Singular Element

As shown in Figure 21.16, hybrid singular element is used in the crack tip; elements in other part usually are continuous displacement elements. For plane hybrid singular

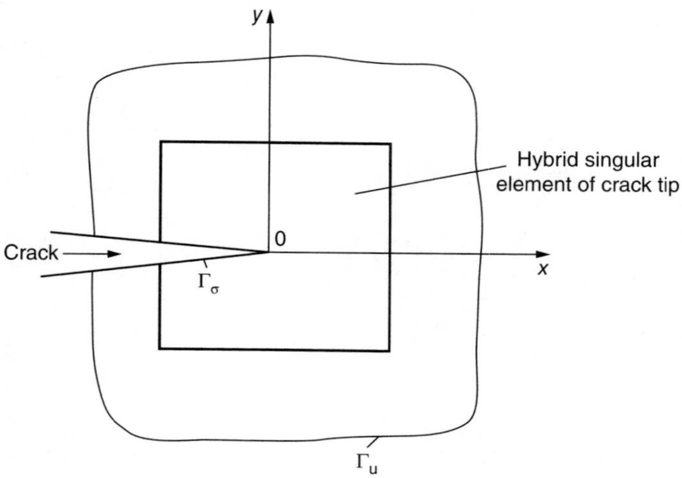

Figure 21.16 Hybrid singular element of the crack tip.

element of the crack tip, by Formula (3.68), define the functional as

$$\Pi = \iint \left(\{\sigma\}^{\mathrm{T}}\{\varepsilon\} - \frac{1}{2}\{\sigma\}^{\mathrm{T}}[D]^{-1}\{\sigma\} \right) \mathrm{d}x\,\mathrm{d}y - \int_{\Gamma_\sigma} \{r\}^{\mathrm{T}}\{\overline{p}\}\mathrm{d}s - \int_{\Gamma_u} \{p\}^{\mathrm{T}}\{r - \overline{r}\}\mathrm{d}s \tag{21.41}$$

where $\{\overline{p}\}$ is the given surface force on the boundary Γ_σ. $\{\overline{r}\}$ is the given displacement on the other boundary Γ_u. $\{\sigma\}$ and $\{\varepsilon\}$ are stress and strain inside the element. $\{r\}$ is the boundary displacement of the element. $\{p\}$ is the surface force on the boundary.

Now use complex variables functions to represent the stress inside the element.

Let

$$z = x + iy$$

where x, y are rectangular coordinates, $i = \sqrt{-1}$.

Stress and strain of any points in the plane can be expressed as

$$\left.\begin{aligned} &\sigma_y + \sigma_x = 2[\Phi'(z) + \overline{\Phi'(z)}] \\ &\sigma_y - \sigma_x + 2i\tau_{xy} = 2\left[\overline{z}\Phi''(z) + \Psi'(z)\right] \\ &2G(u + iv) = \eta\Phi(z) - z\overline{\Phi'(z)} - \overline{\Psi(z)} \end{aligned}\right\} \tag{21.42}$$

Plane strain:

$$\eta = 3 - 4\mu$$

Plane stress:

$$\eta = (3 - \mu)/(1 + \mu)$$

where G is shear modulus, μ is Poisson's ratio, $(\ )'$ represents derivative, and $\overline{(\ )}$ represents conjugate. Φ and Ψ both are analytic functions.

By mapping

$$z = \omega(\zeta) = \zeta^2$$

the area inside the element is transferred to the right half plane, where the crack is on the imaginary axis. Stress functions Φ and Ψ are both analytic functions of ζ. Use simple polynomial of ζ to represent stress function $\psi(\zeta)$. By the condition that the stresses on the surface of the crack are zero (or equal to given surface force), express $\psi(\zeta)$ by $\Phi(\zeta)$. Then functions $\Phi(\zeta)$ and $\psi(\zeta)$ are obtained. Inside the element, equilibrium and continuity conditions are strictly satisfied. On the surface of the crack, the condition that stress is zero is strictly satisfied. On other boundaries of the element, the surface forces $\{p\}$ and surface displacement $\{r\}$ are given. Express the given displacement $\{\overline{r}\}$ on the boundary Γ_u by nodal displacement $\{\delta\}$. In the last, take the stationary value of functional Π given by Formula (21.41) to determine the polynomial coefficients of stress function $\psi(\zeta)$. The stiffness matrix $[K^e]$ is calculated.

By direct method, when the total degrees of freedom are about 1500, the error of K_{I} is 3–5%. By flexibility method, when the total degrees of freedom are about 500, the error of K_{II} is 3–5% as well. By hybrid singular element method in this section, when the total degrees of freedom are 15, the result is nothing else but accurate. Thus, its accuracy is pretty high, but its shortcoming is the complex computational procedure.

Since the mid-1970s, thanks to the singular isoparametric elements in the next section, the singularity of the crack tip appears if the midpoints of the boundary are moved to 1/4 length, which is convenient to put into use. The several singular elements of the crack tip mentioned in this section have been less used. By brief introduction, this section is aimed at giving a complete picture of the use of the finite element method in fracture mechanics to the reader.

21.7 Singular Isoparametric Element (1/4 Length Midpoint Method)

21.7.1 Rectangular Singular Isoparametric Element

In the mid-1970s, Barsoum and Henshell individually put forward a convenient method reflecting the stress singularity in the crack tip: if the midpoint of 8-nodal isoparametric element is moved to the 1/4 point, there is stress singularity of $r^{-1/2}$ around the angular point. These elements are called singular isoparametric elements. In the crack tip, by putting 4 singular isoparametric elements, the stress field around the crack tip can be well reflected.

In the following, the change law of element stress field is analyzed when the midpoint is moved to the 1/4 point from the regular location. As Figure 21.17, the transform mapping the side 123 of the child element to side $\eta = 1$ of the parent element is

$$\begin{aligned} x &= N_1(\xi - 1)x_1 + N_2(\xi - 1)x_2 + N_3(\xi - 1)x_3 \\ &= -\frac{1}{2}(1 - \xi)\xi x_1 + (1 - \xi^2)x_2 + \frac{1}{2}\xi(1 + \xi)x_3 \end{aligned}$$

In this formula, let $x_1 = 0, x_2 = h/4, x_3 = h$, and we get

$$x = \frac{1}{4}(1 - \xi^2)h + \frac{1}{2}\xi(1 + \xi)h = \frac{1}{4}(1 + \xi^2)h$$

Determine ξ by this formula:

$$\xi = \sqrt{4x/h - 1}$$

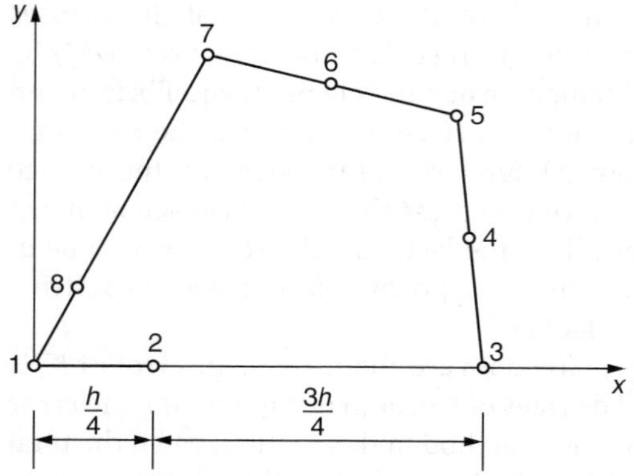

Figure 21.17 Rectangle singular isoparametric element.

In side 123, because η = constant, displacement u has the following form:

$$u = \alpha_0 + \alpha_1 \xi + \alpha_2 \xi^2$$

Therefore,

$$\varepsilon_x = \frac{\partial u}{\partial x} = \frac{\partial u}{\partial \xi}\frac{d\xi}{dx} = (\alpha_1 + 2\alpha_2 \xi)\frac{1}{\sqrt{hx}} \tag{21.43}$$

By Formula (21.43), in side 123, strain has singularity of $r^{-1/2}$. Further research shows that there are two shortcomings of rectangle singular isoparametric element: ① Only sides 123 and 187 have singularity, but other radials started by the crack tip do not have. ② By strict integration, strain energy of the element becomes infinite (Figure 21.18). Therefore the calculated stiffness matrix is unbounded. As the stiffness matrix is computed by integration on Gaussian integration points, the computed results practically are finite. From the cases listed in the following, it shows that the accuracy increases quite a lot by adopting singular isoparametric element. Besides, considering that the calculated local stress and displacement around the crack tip are not ideal, when stress intensity factor is determined by displacement method, the nodal displacement close to the crack tip should be ignored and replaced by the value extrapolated by other nodal displacements.

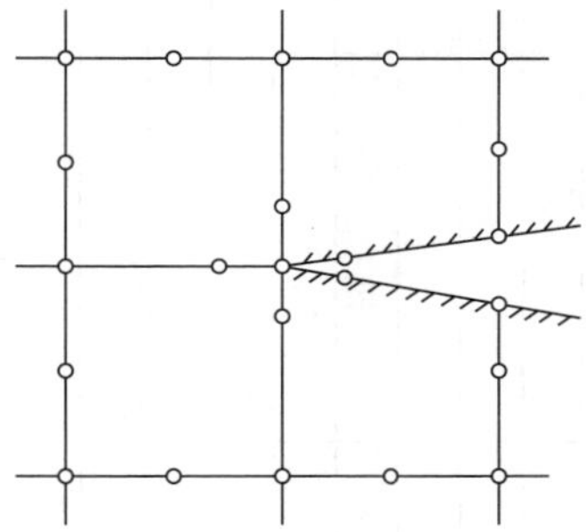

Figure 21.18 Arrangement of singular isoparametric elements.

There are two examples.

Example 1 Figure 21.19 shows the computed result of K_I of a strip with central crack. The mesh is shown in (a). The result is shown in (b). The figure shows, with $\theta = 0$, how K_I^* changes over r. As shown in the figure, the result by singular isoparametric element is better than that by normal element and is close to the accurate value. 4 square points in the figure represent results of 4 cases, respectively.

As shown here, although singular isoparametric element merely uses 147 nodes and 274 degrees of freedom, its accuracy is better than that of any other methods mentioned earlier. Meanwhile the fourth method uses 893 nodes and almost 1700 degrees of freedom.

Example 2 Calculation of side crack K_{II} is shown in Figure 21.20. Mode II cracks will develop under antisymmetric shearing force. In order to keep the moment balance, moment $M_0 = V_0 L_1$ is required besides shearing force V_0. The computational mesh is in Figure 21.21(a) and the result is in Figure 21.21(b). The result of K_{II} by singular element is close to Wilson's result.

21.7.2 Triangular Degenerated Singular Isoparametric Element

Barsoum put forward triangular degenerated singular isoparametric element. Combine nodes 1, 7, and 8 in 8-nodal isoparametric element to one node coincided with the crack tip. Then the midpoints of sides 123 and 165 are moved to the 1/4 length as in Figure 21.21. In this case, there is singularity not only in elements in two sides but also in elements in other radials started by the crack tip. The strain energy of the element becomes bounded.

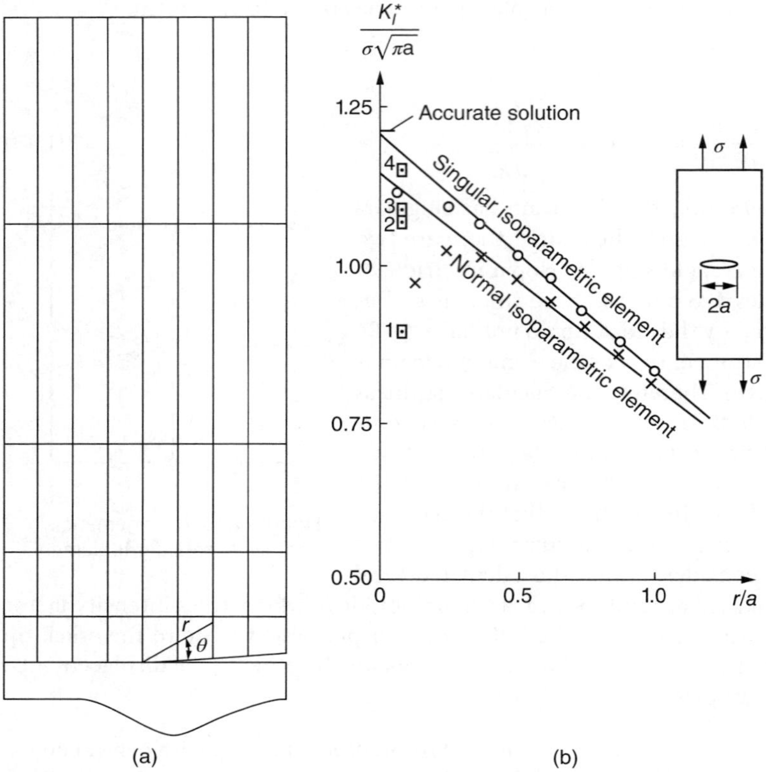

Figure 21.19 The calculation of K_I for strip with central crack. (a) Mesh (147 nodes); (b) result of K_I. 1□- 414 triangle isoparametric element, 240 nodes. 2□- 414 triangle isoparametric element, 240 nodes; special elements are used in the crack tip. 3□- 414 triangle linear strain element, 893 nodes. 4□- 414 triangle linear strain element, 893 nodes; special elements are used in the crack tip.

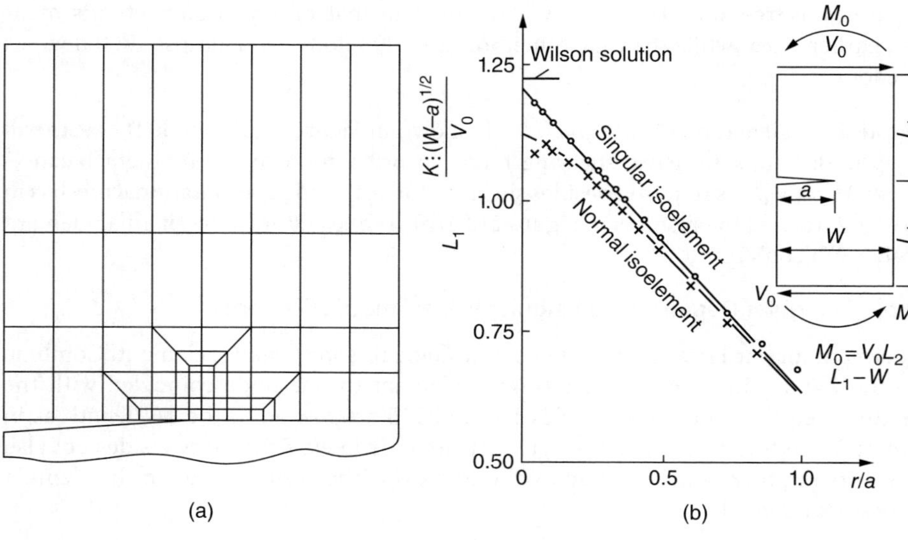

Figure 21.20 The calculation of side crack K_{II}. (a) Mesh; (b) result.

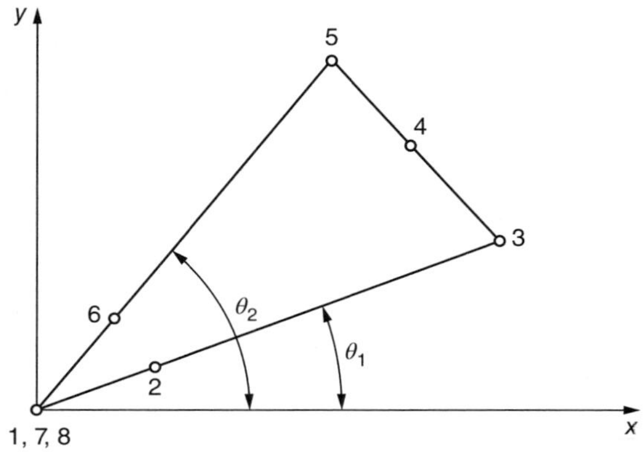

Figure 21.21 Triangular degenerated singular isoparametric element.

As shown in Figure 21.21, nodes 2 and 6 are on 1/4 length of two sides, respectively. Node 4 is in the midpoint of side 35. Therefore, $x_1 = x_7 = x_8 = 0, x_2 = x_3/4, x_4 = (x_3 + x_5)/2, x_6 = x_5/4$. By isoparametric element coordinate transformation $x = \sum N_i x_i$,

$$\begin{aligned} x &= \left(\frac{1}{4}N_2 + N_3 + \frac{1}{2}N_4\right)x_3 + \left(\frac{1}{2}N_4 + N_5 + \frac{1}{4}N_6\right)x_5 \\ &= \frac{1}{8}\left[(1-\eta)x_3 + (1+\eta)x_5\right](1+\xi)^2 \end{aligned}$$

from which

$$\xi = \frac{2\sqrt{2}}{\sqrt{(1-\eta)x_3 + (1+\eta)x_5}}\sqrt{x} - 1 \tag{21.44}$$

As shown in Figure 21.21,

$$\eta = \frac{\alpha}{\theta_2 - \theta_1}\left[\theta - \frac{\theta_1 + \theta_2}{2}\right] \quad (\theta_1 \le \theta \le \theta_2) \tag{21.45}$$

where θ_1 and θ_2 represent the polar angles of edges 13 and 15, respectively.

By Formula (21.45), when $\theta =$ constant, $\eta =$ constant. This indicated that the radial started by node 1 of degenerated element is line of iso-η.

Substitute shape function $N_i(\xi, \mu)$ into displacement mode of isoparametric element $u = \sum N_i u_i$ and $v = \sum N_i v_i$,

$$\begin{aligned} u &= \alpha_0 + \alpha_1\xi + \alpha_2\eta + \alpha_3\xi\eta + \alpha_4\xi^2 + \alpha_5\eta^2 + \alpha_6\xi^2\eta + \alpha_7\xi\eta^2 \\ v &= \beta_0 + \beta_1\xi + \beta_2\eta + \beta_3\xi\eta + \beta_4\xi^2 + \beta_5\eta^2 + \beta_6\xi^2\eta + \beta_7\xi\eta^2 \end{aligned}$$

In degenerated element any strain component in radials $\theta =$ constant (correspondingly $\eta =$ constant) started with node 1 can be determined by the previous formula and Formula (21.44). For example,

$$\begin{aligned} \varepsilon_x &= \frac{\partial u}{\partial x} = \frac{\partial u}{\partial \xi}\frac{\partial \xi}{\partial x} \\ &= (\alpha_1 + \alpha_3\eta + 2\alpha_4\xi + 2\alpha_6\xi\eta + \alpha_7\eta^2) \\ &\quad \times \frac{\sqrt{2}}{\sqrt{(1-\eta)x_3 + (1+\eta)x_5}} \times \frac{1}{\sqrt{x}} \end{aligned} \tag{21.46}$$

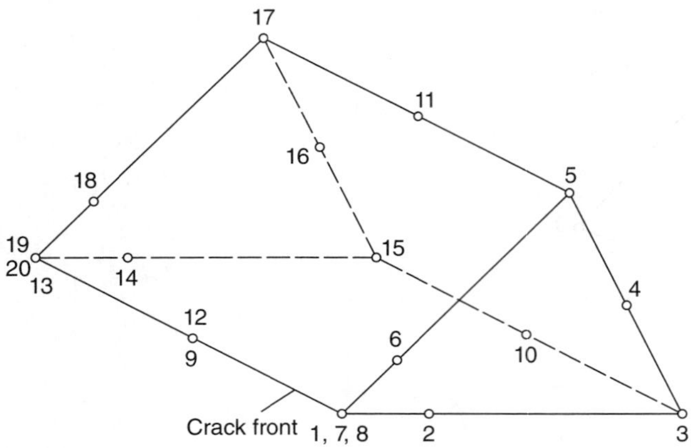

Figure 21.22 Three-dimensional degeneration singular isoparametric element.

Substitute $x = r \cos \theta$ into Eq. (21.46). There is term $r^{-1/2}$ in ε_x. It is the same for other components.

For space problem as in Figure 21.22, degenerate plane 1–7–8–12–19–21–13–9 into a line 1–9–13; then move four side midpoints 2, 6, 14, and 18 to 1/4 position. The triangular degenerated singular isoparametric element is obtained. Parent element coordinate system is ξ, η, ζ. By taking sections along $\zeta =$ constant, it is not difficult to prove that each section has stress singularity in the crack front.

21.8 Blunt Crack Zone Model

In the calculation earlier, when crack extends, the crack tip changes from one node to two nodes, which is not convenient for the finite element calculation. Now distributed crack model is of more use, which simulates the action of crack in mechanics. When crack develops, in the new cracked element, it is assumed that the modulus of elasticity normal to the direction of the cracks is zero, but there is nonzero modulus of elasticity parallel to the direction of the cracks. As it is not necessary to modify the computation mesh after the crack extending, only the mechanical parameters need to be changed, which is more convenient in program designing. But the aforementioned methods for calculating stress intensity factors are not applicable in this moment. In order to calculate stress intensity factors by distributed crack model, Bazant put forward blunt crack zone model. However, his original version did not consider aggregate interlocking. The author improves his method, and in the following deduction, this aspect is also considered.

A plane problem is shown in Figure 21.23. The crack is parallel to the x-axis (if it is not parallel to the x-axis, transformation of coordinates is required). The length is a. The volume is V. The surface acted by the external force (including free surface) is S. The external force on the surface is $\{p^0\} = \{p_x \ p_y\}^{\mathrm{T}}$. Before crack extending, the stress is $\{\sigma^0\} = [\sigma_x^0 \ \sigma_y^0 \ \tau_{xy}^0]^{\mathrm{T}}$ and the strain $\{\varepsilon^0\} = [\varepsilon_x^0 \ \varepsilon_y^0 \ \gamma_{xy}^0]^{\mathrm{T}}$. The extended length of the crack is Δa. The volume of extending element is ΔV. The extended area of the surface is ΔS.

Figure 21.23 Blunt crack zone model.

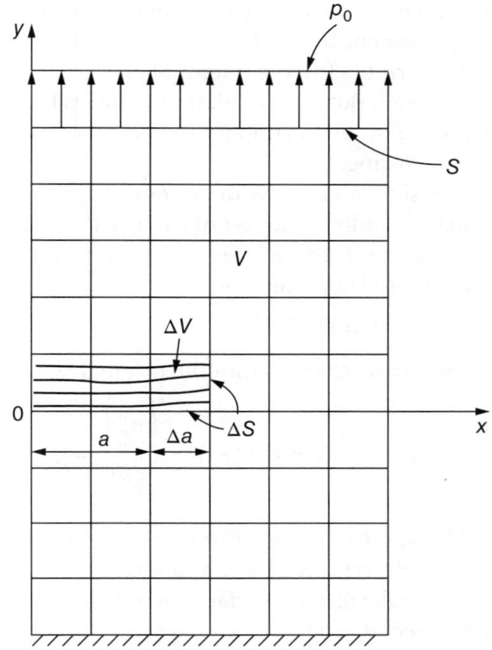

After crack extending, the component of stress in element ΔV is

$$\{\sigma\} = [D_c]\{\varepsilon\} \tag{a}$$

where

$$[D_c] = \begin{bmatrix} E & 0 & 0 \\ 0 & 0 & 0 \\ 0 & 0 & \beta G \end{bmatrix} \tag{b}$$

in which $[D_c]$ is the element elasticity matrix after cracking and G is the shearing modulus.

During the crack extending, the strain energy change in element ΔV is

$$\frac{1}{2}\int_{\Delta V} \left(\{\varepsilon\}^{\mathrm{T}}[D_c]\{\varepsilon\} - \{\sigma^0\}^{\mathrm{T}}\{\varepsilon^0\}\right) \mathrm{d}V \tag{c}$$

The potential energy change of the whole subject during the cracking is

$$\begin{aligned} \Delta\Pi =& \frac{1}{2}\int_{\Delta V} \left(\{\varepsilon\}^{\mathrm{T}}[D_c]\{\varepsilon\} - \{\sigma^0\}^{\mathrm{T}}\{\varepsilon^0\}\right) \mathrm{d}V \\ &+ \frac{1}{2}\int_{V-\Delta V} \left(\{\sigma\}^{\mathrm{T}}\{\varepsilon\} - \{\sigma^0\}^{\mathrm{T}}\{\varepsilon^0\}\right) \mathrm{d}V - \int_S \{p^0\}^{\mathrm{T}}(\{r\} - \{r^0\})\mathrm{d}S \end{aligned} \tag{d}$$

where

$$\{r\} = \begin{bmatrix} u & v \end{bmatrix}^{\mathrm{T}}$$
$$\{r^0\} = \begin{bmatrix} u^0 & v^0 \end{bmatrix}^{\mathrm{T}}$$

in which $\{r\}$ is the surface displacement after the crack extending and $\{r^0\}$ is the surface displacement before the crack extending.

By Formula (d), the potential energy change of the subject during the cracking can be calculated. But the calculation involves the whole subject and the calculating amount is large. In order to simplify the calculation, $\Delta\Pi$ is expressed by elements $\Delta V$and ΔS in the following.

Consider an intermediate state. Apply surface force $\{\Delta P\}$ on the surface ΔS. After the crack extending, the rest of subject $V - \Delta V$ still keeps the original stress $\{\sigma^0\}$ and strain $\{\varepsilon^0\}$. While in the element ΔV strain equals the original strain $\{\varepsilon^0\}$, the elastic matrix changes to $[D_c]$. The stress is

$$\{\sigma'\} = [D_c]\{\varepsilon^0\} \tag{e}$$

The stress change from the original state to intermediate state of element ΔV is

$$\{\Delta\sigma\} = \{\sigma^0\} - \{\sigma'\} = \begin{Bmatrix} \sigma_x^0 \\ \sigma_y^0 \\ \tau_{xy}^0 \end{Bmatrix} - [D_c]\{\varepsilon^0\} \tag{f}$$

The applied surface force $\{\Delta p\}$ on the surface ΔS in order to keep the intermediate state is determined by $\{\Delta\sigma\}$ above.

Now take out the surface force $\{\Delta p\}$ on the surface ΔS. The subject changes from the intermediate state to the final state after crack extending. In element $V - \Delta V$, the stress is $\{\sigma\}$, and the strain is $\{\varepsilon\}$. In element ΔV, the stress is $\{\sigma\}$ as shown in Formula (a), and the strain is $\{\varepsilon\}$. The energy change from the intermediate state to final state of the subject is

$$\begin{aligned} &-\frac{1}{2}\int_{\Delta V} \left(\{\varepsilon\}^{\mathrm{T}}[D_c]\{\varepsilon\} - \{\sigma^0\}^{\mathrm{T}}\{\varepsilon^0\}\right) \mathrm{d}V + \frac{1}{2}\int_{V-\Delta V} \left(\{\sigma\}^{\mathrm{T}}\{\varepsilon\} - \{\sigma^0\}^{\mathrm{T}}\{\varepsilon^0\}\right) \mathrm{d}V \\ &= \int_S \{p^0\}^{\mathrm{T}} \left(\{r\} - \{r^0\}\right) \mathrm{d}S + \frac{1}{2}\int_{\Delta S} \{\Delta p\}^{\mathrm{T}} \left(\{r\} - \{r^0\}\right) \mathrm{d}S \end{aligned} \tag{g}$$

Substitute Formula (g) into Formula (d). The potential energy change of the whole subject during the crack extending is

$$\Delta\Pi = \frac{1}{2}\int_{\Delta V} \left(\{\varepsilon^0\}^{\mathrm{T}}[D_c]\{\varepsilon^0\} - \{\sigma^0\}^{\mathrm{T}}\{\varepsilon^0\}\right) \mathrm{d}V + \frac{1}{2}\int_{\Delta S} \{\Delta p\}^{\mathrm{T}} \left(\{r\} - \{r^0\}\right) \mathrm{d}S \tag{21.47}$$

$\{r\} = [N]\{\delta\}$, where $[N]$ is shape function matrix. Formula (21.47) can be rewritten as

$$\Delta\Pi = \frac{1}{2}\int_{\Delta V} \left(\{\varepsilon^0\}^{\mathrm{T}}[D_c]\{\varepsilon^0\} - \{\sigma^0\}^{\mathrm{T}}\{\varepsilon^0\}\right) \mathrm{d}V + \frac{1}{2}\left(\{\delta\}^{\mathrm{T}} - \{\delta^0\}^{\mathrm{T}}\right)\{\Delta P\} \tag{21.48}$$

$$\{\Delta P\} = \int_{\Delta S} [N]^{\mathrm{T}}\{\Delta p\}\mathrm{d}S \tag{21.49}$$

So the potential energy change Δ during the crack extending is determined by Formula (21.49). Then calculate energy release rate G by $G = \Delta\Pi/\Delta a$. The stress intensity factor is determined by Eq. (21.22).

In Eq. (21.48), the first term represents the energy loss of the elastic matrix change of ΔV. The second term represents effect caused by ΔP of the whole subject V. In fact, the force $\{\Delta p\}$ on the surface ΔS can be worked out by the stiffness matrix and displacement in ΔV, which is shown in Formula (f).

21.9 Elastic–Plastic Fracture

As previously mentioned, when we consider the cracked body as linear elastic body, there is singularity in crack tip as Formula (21.1). When $r \to 0$, the stress becomes infinite. Therefore, there is always a stress yielding zone in the crack tip. For small-scale yielding, because the large region around the plastic zone of the crack is still elastic zone, the result of linear fracture mechanics is still representative. The fracture is judged by $K = K_{\text{Ic}}$, which is still workable. Most high strength steel and ultrahigh strength steel belongs to this situation. When the size of plastic zone is close to or more than the size of the crack, it is called large-scale yielding. The fracture of large-scale yielding is called plastic–elastic fracture. Most intermediate steel and low strength steel show plastic–elastic fracture. In this situation, the fracture criterion $K = K_{\text{Ic}}$ is invalid. But $J = J_{\text{Ic}}$ is still workable.

The stress–strain relationship of material is

$$\tau \propto \gamma^N$$

where τ is shearing stress and γ is shearing strain.

According to the well-known HRR singular solution (Hutchinson, Rice, and Rosengren), in small-scale conditions, the stress and strain around the crack tip have singularity as

$$\left.\begin{aligned} \varepsilon_{ij} &\to r^{-1/(1+N)}F_{ij}(\theta, N) \\ \sigma_{ij} &\to r^{-N/(1+N)}G_{ij}(\theta, N) \end{aligned}\right\} \tag{21.50}$$

Apparently, when $N = 1$, the formula returns to linear elastic fracture; when $N = 0$, the stress singularity disappears, while the strain has singularity of r^{-1}. The solution is called Prandtl plastic solution, as shown in Figure 21.24, where τ_0 is shearing yield strength.

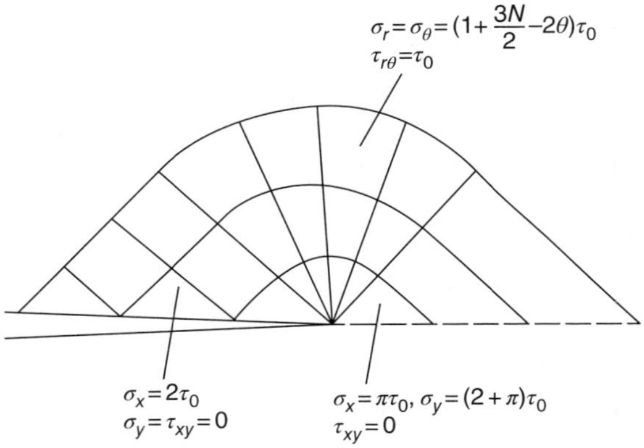

Figure 21.24 Plastic solution of Prandtl.

In elastic–plastic fracture, *J-integral* is still defined by Formula (21.11). But W is determined by

$$W = W(\varepsilon_{mn}) = \int_0^{\varepsilon_{mn}} \sigma_{ij}\, \mathrm{d}\varepsilon_{ij} \tag{21.51}$$

As mentioned before, in linear elastomer, *J-integral* is independent on integral path and $J = G$. In elastic–plastic fracture, strictly speaking, J-integral is not independent on integral path. However, the result shows that in small-scale conditions, *J-integral* is approximately independent on integral path. For example, when hardening exponent $N = 0.1–0.3$, comparing the *J-integral* along the plastic zone of the crack tip with *J-integral* of far field, the value of the former is 96–98%. Thus, *J-integral* is a criterion for fracture in large-scale yielding. When $J = J_c$, the crack extends. J_c reflecting material character is measured by experiment. The value of *J-integral* that depends on structure shape, the size, the location of the crack and the load is calculated by the finite element method.

For plane problems, substituting W defined by Formula (21.51) into Formula (21.11), we can calculate *J-integral*. But this method is not applicable for axisymmetric problems and three-dimensional problems. The virtual crack extending method by Parks is introduced in the following, which can be used for calculating *J-integral* of plane problems and axis symmetric problems as well as three-dimensional problems.

Firstly, consider plane problems of unit thickness. According to deformation theory of plasticity, after discreted by the finite element method, the potential of the structure is

$$\begin{aligned}\Pi &= \sum_e \int W\,\mathrm{d}V_i - \{P\}^{\mathrm{T}}\{\delta\} \\ &= \sum_e \overline{W}_i(\overline{X}, \delta) - \{P\}^{\mathrm{T}}\{\delta\}\end{aligned} \tag{21.52}$$

where

$$W = W(\varepsilon_{ij})$$

in which V_i is the volume of element i, W is the strain energy density (which is a function of total strain), $\{P\}$ is the nodal load, $\{\delta\}$ is the nodal displacement, $\{\overline{X}\}$ is the nodal coordinate array, and $\overline{W}_i$ is the volume integral of the strain energy density of element i; apparently, $\overline{W}_i$ is a function of $\{\overline{X}\}$ and $\{\delta\}$.

Assume that the crack is coincided with x-axis and the length is a. As shown in Figure 21.10, Γ_0 is a contour surround the crack tip. Assume that all the nodes in and inside Γ_0 move right Δa in the direction of crack length, meanwhile the nodes outside Γ_1 stay still. Then the length of the crack extending is Δa, and the increment of the potential energy is

$$\begin{aligned}\Delta\Pi &= \sum_e \left[\left(\frac{\partial \overline{W}_i}{\partial x}\right)^{\mathrm{T}}\{\Delta\overline{x}\} + \left(\frac{\partial \overline{W}_i}{\partial \delta}\right)^{\mathrm{T}}\{\Delta\delta\}\right] - \{P\}^{\mathrm{T}}\{\Delta\delta\} - \{\delta\}^{\mathrm{T}}\left\{\frac{\partial P}{\partial \overline{x}}\right\}\{\Delta\overline{x}\} \\ &= \left[\sum_e \left(\frac{\partial \overline{W}_i}{\partial \delta}\right)^{\mathrm{T}} - \{P\}^{\mathrm{T}}\right]\{\Delta\delta\} + \left[\sum_e \left(\frac{\partial \overline{W}_i}{\partial \overline{x}}\right)^{\mathrm{T}} - \{\delta\}^{\mathrm{T}}\left\{\frac{\partial P}{\partial \overline{x}}\right\}\right]\{\Delta\overline{x}\}\end{aligned} \tag{21.53}$$

As $\sum_e \left(\frac{\partial \overline{W}_i}{\partial \delta}\right)^{\mathrm{T}} - \{P\}^{\mathrm{T}} = 0$, the first term in the right of Formula (21.53) is zero. Rice proved that for plane problems in plastic deformation theory, we have

$$J = -\frac{\partial \Pi}{\partial a} \tag{21.54}$$

By Formulas (21.53) and (21.54), we get

$$-\Delta\Pi = J\Delta a = -\sum_e \left(\frac{\partial \overline{W}_i}{\partial \overline{x}}\right)^{\mathrm{T}} \{\Delta\overline{x}\} + \{\delta\}^{\mathrm{T}} \left\{\frac{\partial P}{\partial \overline{x}}\right\} \{\Delta\overline{x}\} \tag{21.55}$$

By Formula (21.55), the crack tip's *J-integral* of elastic–plastic plane problems is calculated. For linear elastic problems, $\overline{W}_i = \frac{1}{2}\{\delta\}^{\mathrm{T}} \times [\Delta K_i]\{\delta\}$, $[\Delta K_i] = (\partial[K_i]/\partial\overline{x})\{\Delta\overline{x}\}$. Therefore,

$$\left(\frac{\partial \overline{W}_i}{\partial \overline{x}}\right)^{\mathrm{T}} \{\Delta\overline{x}\} = \frac{1}{2}\{\delta\}^{\mathrm{T}}[\Delta K_i]\{\delta\}$$

Here Formula (21.55) degenerates to Formula (21.31) of stiffness derivative method.

In the process of calculating the crack tip's *J-integral* of elastic–plastic plane problems by Formula (21.55), we only need to calculate elements between Γ_0 and Γ_1. As for axisymmetric problems, all the elements in Γ_1 need to be calculated.

For space problems, assume that in crack plane, along the crack tip, there is virtual crack extension $\Delta a(s)$. The potential energy increases by

$$-\Delta\Pi = \int_{\text{Front}} J(s)\Delta a(s)\mathrm{d}s = -\sum_e \left(\frac{\partial \overline{W}_i}{\partial \overline{x}}\right)^{\mathrm{T}} \{\Delta\overline{x}\} + \{\delta\}^{\mathrm{T}} \left\{\frac{\partial P}{\partial \overline{x}}\right\} \{\Delta\overline{x}\} \tag{21.56}$$

If there are no external force on the crack surface and no volume force inside the subject, then $\partial P/\partial\overline{x} = 0$. The second term in the right of Formulas (21.55) and (21.56) is zero. Formula (21.56) turns to

$$-\Delta\Pi = \int_{\text{Front}} J(s)\Delta a(s)\mathrm{d}s = -\sum_e \left(\frac{\partial \overline{W}_i}{\partial \overline{x}}\right)^{\mathrm{T}} \{\Delta\overline{x}\} \tag{21.57}$$

In three-dimensional crack plane, as shown in Figure 21.12, along the crack front, by moving node i in turns, we can get *J-integral* of node i as

$$J_i = -\frac{1}{\Delta A_i}\sum_e \left(\frac{\partial \overline{W}_i}{\partial \overline{x}}\right)^{\mathrm{T}} \{\Delta\overline{x}\} \tag{21.58}$$

where

$$\Delta A_i = \int \Delta a(s)\mathrm{d}s$$

in which ΔA_i is the crack area extension by moving node i.

Similar as calculating the energy release rate G of elastic fracture by stiffness derivative method, when virtual crack extends, the area of disturbance should be larger as shown in full lines of Figure 21.12 in order to improve accuracy.

The virtual crack extending method mentioned earlier not only can be used in plastic deformation (full value) theory but also can be used in plastic increment theory. But in the latter conditions, *J-integral* is not strictly independent on integral path. The values of *J-integral* calculated by different contours differ somewhat. Experience shows that the results of both theories have good accuracy.

21.10 Extended Finite Element Method for Fracture Analysis

In the extended finite element method, first the usual finite element mesh is produced, and then a few degrees of freedom are added to represent the discontinuities of a crack without remeshing the computing mesh.

Moes and Mohammadi suggested that the displacements be expressed as follows [19]

$$U(x) = \sum_{j=1}^{n} N_j(x)U_j + \sum_{h=1}^{m} N_h(x)\left[H(\xi(x)) - H(\xi(x_{kh}))\right] a_h$$
$$+ \sum_{k=1}^{m1} N_k(x)\left[\sum_{c=1}^{mf}(F_c^1(x) - F_c^1(x_k))b_{k1}\right]$$
$$+ \sum_{k=1}^{m2} N_k(x)\left[\sum_{c=1}^{mf}(F_c^2(x) - F_c^2(x_k))b_{k2}\right] \tag{21.59}$$

In the right part of the previous equation, the first term represents the conventional finite element mesh, and the second to the fourth term represent the discontinuities of the crack surface and the two crack tips.

U_j are the nodal displacements of the usual finite element mesh, and a_h, b_{k1}, and, b_{k2} are vectors of additional nodal degrees of freedom for modeling crack faces and the two crack tips, respectively. The displacements near the tip of crack are expressed by the following functions with the crack tip coordinates (r,θ) and with the crack tip as origin

$$\{F(r,\theta)\} = \left\{\sqrt{r}\sin\frac{\theta}{2}, \sqrt{r}\cos\frac{\theta}{2}, \sqrt{r}\sin\theta\sin\frac{\theta}{2}, \sqrt{r}\sin\theta\cos\frac{\theta}{2}\right\}$$

The Heaviside function $H(\xi)$ can be defined as a step function:

$$H(\xi) = \begin{cases} 1 \text{ when } \xi \geq 0 \\ 0 \text{ when } \xi < 0 \end{cases} \tag{21.60}$$

or a signed function:

$$H(\xi) = \text{sign}\,\xi = \begin{cases} 1 \text{ when } \xi \geq 0 \\ -1 \text{ when } \xi < 0 \end{cases} \tag{21.61}$$

In order to avoid numerical instabilities, the following smoothed Heaviside function can also be used for a small value of B less than the element size:

$$H(\xi) = \begin{cases} 0 & \text{when } \quad \xi < -\chi \\ \dfrac{1}{2} + \dfrac{\xi}{2\beta} + \dfrac{1}{2\pi}\sin\dfrac{n\xi}{\beta}, & \beta < \xi < \beta \\ 1 & \xi > \beta \end{cases} \tag{21.62}$$

The derivative of the Heaviside function is the Dirac delta function $\delta(\xi)$:

$$H'(\xi) = \delta(\xi)$$

which may be approximated by the following smoothed function:

$$\delta(\xi) = \begin{cases} \dfrac{1}{2\beta} + \dfrac{1}{2\beta}\sin\dfrac{\pi\xi}{\beta}, & -\beta < \xi < \beta \\ 0, & \text{otherwise} \end{cases} \tag{21.63}$$

A numerical example is given as follows.

A tensile plate with a central crack as shown in Figure 21.25 is calculated [23]. The elements that are fully cut by a crack are enriched by Heaviside enrichment, whereas the elements containing a crack tip are enriched by tip enrichment functions. The element matrices are integrated over the Gauss points as shown in Figure 21.26.

Table 21.4 compares the normalized stress intensity factors for various meshes as well as different ratios of crack length a to plate width b. There is a close agreement between the numerical results and the exact solution of Irwin.

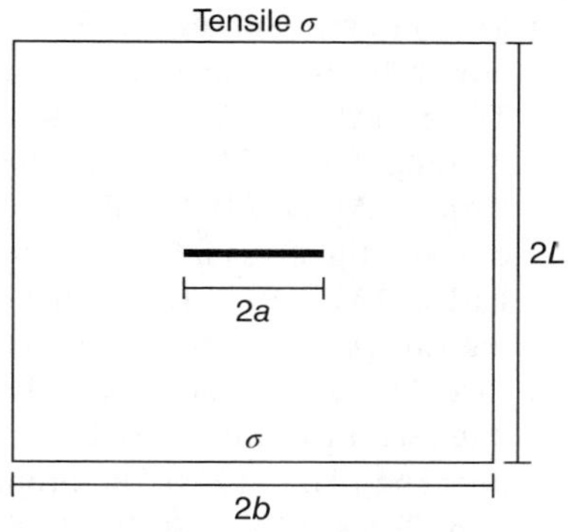

Figure 21.25 Tensile plate with a central crack.

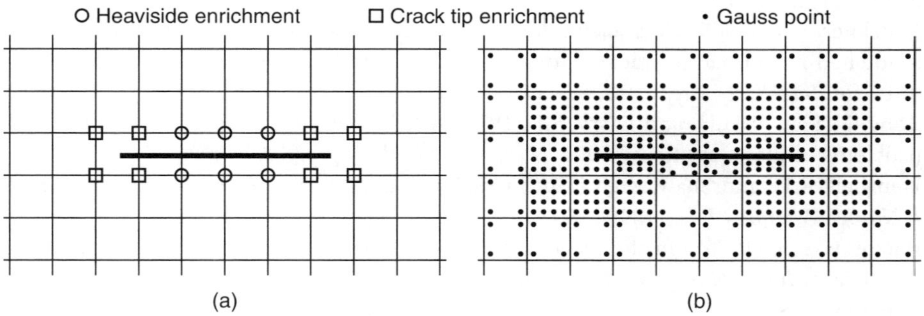

Figure 21.26 (a) Selection of enrichment and (b) position of Gauss points around the central crack.

Table 21.4 Normalized values of stress intensity factors.

a/b	Irwin $\overline{K}_I$	Mesh 24 × 50		Mesh 48 × 90	
		$\overline{K}_I$	Error (%)	$\overline{K}_I$	Error (%)
1/8	1.040	1.033	0.72	1.035	0.56
1/6	1.075	1.066	0.78	1.069	0.59
1/4	1.189	1.180	0.81	1.183	0.56

Bibliography

1 Chan, S.K., Tuba, I.S. and Wilson, W.K. (1970) On the finite element method in linear fracture mechanics. *Eng. Fract. Mech.*, **2**, 1–17.

2 Rice, J.R. (1968) A path independent integral and the approximate analysis of strain concentration by notches and cracks. *J. Appl. Mech.*, **35**, 379–386.

3 Hutchinson, J.W. (1968) Singular behavior at the end of a tensile crack. *J. Mech. Phys. Solids*, **16**, 13–31.

4 Rice, J.R. and Rosengren, G.F. (1968) Plane strain deformation near a crack tip in a power law hardening material. *J. Mech. Phys. Solids*, **16**, 1–12.

5 Tracey, D.M. (1976) Finite element solutions for crack tip behavior in small-scale yielding. *J. Eng. Mater. Tech.*, **98**, 146–151.

6 Parks, D.M. (1974) A stiffness derivative finite element technique for determination of crack tip stress intensity factors. *Int. J. Fract.*, **10** (4), 487–502.

7 Parks, D.M. (1977) The virtual crack extension method for nonlinear material behavior. *Comput. Methods Appl. Mach Eng.*, **12**, 353–364.

8 Rice, J.R. (1968) Mathematical analysis in the mechanics of fracture, in *Fracture*, vol. 2 (ed. H. Liebowitz), Academic Press, pp. 191–311.

9 Bueckner, H.F. (1958) The propagation of cracks and the energy of elastic deformation. *Trans. ASME*, **80** (6), 1225–1230.

10 Hayes, D.J. (1972) A practical application of bueckner's formulation for determining stress intensity factors for cracked bodies. *Int. J. Fract. Mech.*, **8** (2), 157–165.

11 Bowie, O.L. (1964) Rectangular tensile sheet with symmetric edge cracks. *J. Appl. Mech.*, **31**, 208–212.

12 Bueckner, H.F. (1965) Coefficients for computation of stress intensity factor K_I for a notched round bar, in fracture toughness testing and its applications. ASTM STP 381. *Am. Soc. Test. Matls.*, **23**, 82–84.

13 Tong, P., Pian, T.H.H. and Lasry, S.J. (1973) A hybrid element approach to crack problems in plane elasticity. *Int. J. Numer. Methods Eng.*, **7**, 297–308.

14 Henshell, R.D. and Shaw, K.G. (1975) Crack tip finite elements are unnecessary. *Int. J. Numer. Methods Eng.*, **9**, 495–507.

15 Barsoum, R.S. (1976) On the use of isoparametric finite elements in linear fracture mechanics. *Int. J. Numer. Methods Eng.*, **10**, 25–37.

16 Barsoum, R.S. (1977) Triangular quarter-point elements as elastic and perfectly-plastic crack tip elements. *Int. J. Numer. Methods Eng.*, **11**, 85–98.

17 Hibbitt, H.D. (1977) Some properties of singular isoparametric elements. *Int. J. Numer. Methods Eng.*, **11**, 180–184.

18 Bazant, Z.P. and Cedolin, L. (1979) Blunt crack band propagation in finite element analysis. *J. Eng. Mech. Div. ASCE*, **105** (EM2), 297–315.

19 Mohammadi, S. (2008) *Extended finite element method for fracture analysis of structures [M]*. Blackwell Publishing.

22

Problems in Structural Dynamics

Dynamic analysis, especially the one that considers the effects of earthquake and blast, is an important problem in engineering design. Dynamic analysis of complex structures is very difficult in the past. Now with the finite element method and high-speed computer, it is possible to perform dynamic analysis of various complex structures. This chapter will explain how to apply the finite element method to dynamic analysis of structures.

22.1 Equations of Motion

After the discretization of structures, the dynamic equilibrium equations of each node in the state of motion are as follows:

$$\{F_\mathrm{i}\} + \{F_\mathrm{d}\} + \{P(t)\} = \{F_\mathrm{e}\} \tag{a}$$

where $\{F_\mathrm{i}\}$, $\{F_\mathrm{d}\}$, $\{P(t)\}$, and $\{F_\mathrm{e}\}$ are the inertia force, the damping force, the dynamic loads, and the elastic force, which all are vectors.

The vector of elastic force can be expressed by the nodal displacement $\{\delta\}$ and the stiffness matrix $[K]$ as follows:

$$\{F_\mathrm{e}\} = [K]\{\delta\} \tag{b}$$

Here the element K_{ij} in the stiffness matrix $[K]$ is the elastic force at the node i caused by the unit displacement of the node j.

According to d'Alembert principle, the inertia force can be expressed by the mass matrix $[M]$ and the node acceleration $\frac{\partial^2\{\delta\}}{\partial t^2}$ as follows:

$$\{F_\mathrm{i}\} = -[M]\frac{\partial^2\{\delta\}}{\partial t^2} \tag{c}$$

Here the element M_{ij} of the mass matrix is the inertia force at the node i caused by the unit acceleration of the node j.

The structure is assumed to have viscous damping; the damping force can be expressed by the damping matrix $[C]$ and the node speed $\frac{\partial\{\delta\}}{\partial t}$ as follows:

$$\{F_\mathrm{d}\} = -[C]\frac{\partial\{\delta\}}{\partial t} \tag{d}$$

Here the element C_{ij} of the damping matrix is the damping force at the node i caused by the unit speed of the node j.

The Finite Element Method: Fundamentals and Applications in Civil, Hydraulic, Mechanical and Aeronautical Engineering, First Edition. Bofang Zhu.

Companion website: www.wiley.com/go/zhufem17

Substituting Eqs (b), (c), and (d) into (a), we obtain the equation of motion as follows:

$$[M]\frac{\partial^2\{\delta\}}{\partial^2 t} + [C]\frac{\partial\{\delta\}}{\partial t} + [K]\{\delta\} = \{P(t)\} \tag{22.1}$$

Let

$$\{\dot{\delta}\} = \frac{\partial\{\delta\}}{\partial t}, \quad \{\ddot{\delta}\} = \frac{\partial^2\{\delta\}}{\partial^2 t}$$

then the equation of motion can be expressed as:

$$[M]\{\ddot{\delta}\} + [C]\{\dot{\delta}\} + [K]\{\delta\} = \{P(t)\} \tag{22.2}$$

When the earthquake happens, let the ground acceleration be a, the acceleration of the structure relative to the ground be $\ddot{\delta}$, then the actual acceleration of each node in the structure is equal to $a + \ddot{\delta}$, which will replace $\ddot{\delta}$ in Eq. (22.2) to calculate the inertial force. As for the elastic force and the damping force, they depend on the strain and the strain rate of the structure, so they depend on the displacement δ and $\dot{\delta}$, not on the ground acceleration. If there are no other dynamic loads, using $a + \ddot{\delta}$ instead of $\ddot{\delta}$ in the first item on the left of Eq. (22.2), the motion equation of the structure during an earthquake is

$$[M]\{\ddot{\delta}\} + [C]\{\dot{\delta}\} + [K]\{\delta\} = -[M]\{a\} \tag{22.3}$$

The advantage of dynamic analysis using finite element method is that the constants of material properties of each element can be arbitrarily changed, so that after calculating the stiffness matrix, mass matrix, and damping matrix of each element, the global stiffness matrix, mass matrix, and damping matrix of the structure can be obtained by simple data processing. Another important advantage is that the stiffness matrix, mass matrix, and damping matrix obtained are highly sparse and easy to be arranged in banded matrices, so they are easy to calculate.

It can be seen from Eqs (22.2) and (22.3) that the following two problems must be solved to perform the structural dynamic calculation using finite element method: (1) the establishment of the mass matrix $[M]$, the damping matrix $[C]$, and the stiffness matrix $[K]$ of the structure and (2) equations of motion (22.2) and (22.3) are second-order differential equations, including second derivatives of displacement with respect to time, so we must find effective methods to solve such a large group of second-order differential equations.

In the following sections of this chapter, we will gradually solve the previous two problems. The stiffness matrix $[K]$ in dynamic problem is the same as in static problem, as previously described in detail and not repeated here.

22.2 Mass Matrix

Let $[m]$ represent the element mass matrix and $[M]$ represent the global mass matrix. After obtaining element mass matrix, the global mass matrix will be obtained with appropriate combination. The method of combination is similar to the method of computing the global stiffness matrix from element stiffness matrix.

Two types of mass matrices can be used in dynamic calculation, the consistent mass matrix and the lumped mass matrix, which will be introduced as follows.

22.2.1 Consistent Mass Matrix

According to d'Alembert principle, the inertial force that acts on the elementary volume of a tiny part of moving structure is

$$\{p_{\mathrm{i}}\} = -\rho \frac{\partial^2 \{r\}}{\partial t^2} \tag{a}$$

Here ρ is density of the material.

After the discretization of structures, consider one element of the structure, and take the following form of displacement function:

$$\{r\} = [N]\{\delta\}^{\mathrm{e}} \tag{b}$$

Substituting the previous equation into Eq. (a), we get

$$\{p_{\mathrm{i}}\} = -\rho[N] \frac{\partial^2 \{\delta\}^{\mathrm{e}}}{\partial t^2} \tag{c}$$

Then using the formula (3.14) of load transposition, we can calculate the inertial force acted on the node of element e as follows:

$$\{F_{\mathrm{i}}\}^{\mathrm{e}} = \int [N]^{\mathrm{T}}\{p_{\mathrm{i}}\}\,\mathrm{d}V = -\int [N]^{\mathrm{T}}\rho[N]\,\mathrm{d}V \frac{\partial^2 \{\delta\}^{\mathrm{e}}}{\partial t^2}$$

or

$$\{F_{\mathrm{i}}\}^{\mathrm{e}} = -[m]\{\ddot{\delta}\}^{\mathrm{e}} \tag{d}$$

It is clear that the element mass matrix is

$$[m] = \int [N]^{\mathrm{T}}\rho[N]\,\mathrm{d}V \tag{22.4}$$

When we calculate the element mass matrix by the previous method, the kinetic energy and the potential energy of the element are consistent, so it is called consistent mass matrix.

22.2.2 Lumped Mass Matrix

It is assumed that the mass of the element is lumped on its nodes. The mass matrix obtained is a diagonal matrix.

The lumped mass matrix of the element is defined as follows:

$$[m] = \int \rho[\psi]^{\mathrm{T}}[\psi]\,\mathrm{d}V \tag{22.5}$$

Here $[\psi]$ is matrix of function ψ_i, ψ_i takes 1 in the region that is assigned to the node i, and 0 out of the region.

Because the regions assigned to each node cannot be intersected, the mass matrix calculated previously is diagonal.

22.2.3 Several Typical Element Mass Matrices

Several typical element mass matrixes are given as follows:

22.2.3.1 Beam Element

1) *Lumped mass matrix*. Assign $l/2$ to each node, neglect the rotational term, and get

$$[m] = \frac{W}{2g}\begin{bmatrix} 1 & 0 & 0 & 0 \\ 0 & 0 & 0 & 0 \\ 0 & 0 & 1 & 0 \\ 0 & 0 & 0 & 0 \end{bmatrix}$$

Here W is the weight of the beam element.

2) *Consistent mass matrix*. Using cubic interpolation function (Figure 22.1),

$$w = \left[L_1^2(3-2L_1)\ L_1^2L_2 l\ L_2^2(3-2L_2)\ -L_1L_2^2 l\right]\{\delta\}^{\mathrm{e}}$$
$$\{\delta\}^{\mathrm{e}} = \left[w_1\ \theta_1\ w_2\ \theta_2\right]^{\mathrm{T}}$$
$$L_1 = 1 - x/l, L_2 = x/l$$

Element mass matrix given by Eq. (22.4) is as follows:

$$[m] = \frac{W}{420g}\begin{bmatrix} 156 & -22l & 54 & 13l \\ & 4l^2 & -13l & -3l^2 \\ \text{symmetric} & & 156 & 22l \\ & & & 4l^2 \end{bmatrix}$$

22.2.3.2 Plane Constant Strain Triangular Elements

1) *Lumped mass matrix*. It is assumed that the weight of element is W, which is divided into three equal parts and assigned to each node, then we get element mass matrix as follows:

$$[m] = \frac{W}{3g}\left[\begin{array}{cc:cc:cc} 1 & 0 & 0 & 0 & 0 & 0 \\ 0 & 1 & 0 & 0 & 0 & 0 \\ \hdashline 0 & 0 & 1 & 0 & 0 & 0 \\ 0 & 0 & 0 & 1 & 0 & 0 \\ \hdashline 0 & 0 & 0 & 0 & 1 & 0 \\ 0 & 0 & 0 & 0 & 0 & 1 \end{array}\right]$$

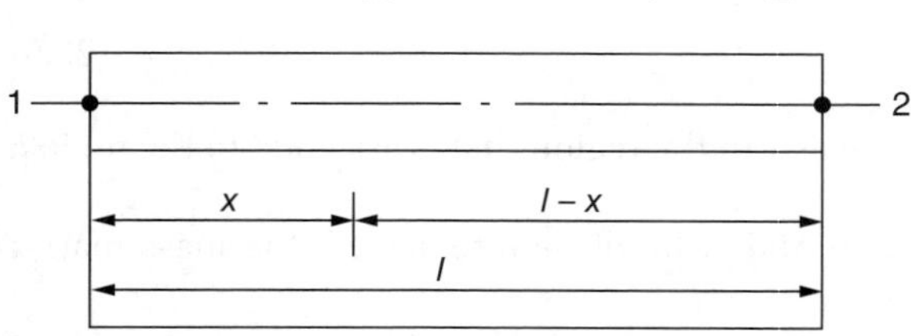

Figure 22.1 Beam element.

2) *Consistent mass matrix.* Using displacement function as Eq. (b), the shape function $[N]$ can be expressed by area coordinates as

$$[N] = \lfloor IL_i \ IL_j \ IL_m \rfloor$$

where

$$I = \begin{bmatrix} 1 & 0 \\ 0 & 1 \end{bmatrix}, \quad L_i = (a_i + b_i x + c_i y)/2A, \ldots$$

Let the thickness of the element be t, according to Eq. (22.4), we get

$$\begin{aligned}
[m] &= \rho t \iint [N]^{\mathrm{T}}[N]\,\mathrm{d}x\,\mathrm{d}y \\
&= \rho t \iint \begin{bmatrix} IL_i \\ IL_j \\ IL_m \end{bmatrix} \lfloor IL_i \ IL_j \ IL_m \rfloor \,\mathrm{d}x\,\mathrm{d}y \\
&= \rho t \iint \begin{bmatrix} L_iL_i & L_iL_j & L_iL_m \\ L_jL_i & L_jL_j & L_jL_m \\ L_mL_i & L_mL_j & L_mL_m \end{bmatrix} \mathrm{d}x\,\mathrm{d}y
\end{aligned}$$

as

$$\iint L_r L_s \,\mathrm{d}x\,\mathrm{d}y = \begin{cases} \dfrac{A}{12} & (r \neq s) \\ \dfrac{A}{6} & (r = s) \end{cases}$$

Let the weight of the element be W, as $\rho t A = W/g$, we get the mass matrix of the element as follows:

$$[m] = \frac{W}{3g}\left[\begin{array}{cc:cc:cc}
\frac{1}{2} & 0 & \frac{1}{4} & 0 & \frac{1}{4} & 0 \\
0 & \frac{1}{2} & 0 & \frac{1}{4} & 0 & \frac{1}{4} \\ \hdashline
\frac{1}{4} & 0 & \frac{1}{2} & 0 & \frac{1}{4} & 0 \\
0 & \frac{1}{4} & 0 & \frac{1}{2} & 0 & \frac{1}{4} \\ \hdashline
\frac{1}{4} & 0 & \frac{1}{4} & 0 & \frac{1}{2} & 0 \\
0 & \frac{1}{4} & 0 & \frac{1}{4} & 0 & \frac{1}{2}
\end{array}\right]$$

22.2.3.3 Rectangular Plate Element

1) *Lumped mass matrix.* Let the weight of the element be W, which is divided into four equal parts and assigned to each node, and neglect the rotational term, we get lumped mass matrix as follows:

$$[m] = \frac{W}{4g}\begin{bmatrix} 1 & & & & & & & & & & & \\ & 0 & & & & & & & & & & \\ & & 0 & & & & & & & & & \\ & & & 1 & & & & & 0 & & & \\ & & & & 0 & & & & & & & \\ & & & & & 0 & & & & & & \\ & & & & & & 1 & & & & & \\ & & & & & & & 0 & & & & \\ & 0 & & & & & & & 0 & & & \\ & & & & & & & & & 1 & & \\ & & & & & & & & & & 0 & \\ & & & & & & & & & & & 0 \end{bmatrix}$$

2) *Consistent mass matrix.* Substituting the shape function (10.13) of rectangular plate element described in Chapter 10 into Eq. (22.4), we get the consistent mass matrix of rectangular plate element as follows:

$$[m] = \frac{W}{25{,}200g}$$

$$\times\begin{bmatrix} 3454 & & & & & & & & & & & \\ 922b & 320b^2 & & & & & & & & & & \\ -922a & -252ab & 320a^2 & & & & & & & & & \\ 1226 & 398b & -548a & 3454 & & & & & \text{symmetrical} & & & \\ 398b & 160b^2 & -168ab & 922b & 320b^2 & & & & & & & \\ 548a & 168ab & -240a^2 & 922a & 252ab & 320a^2 & & & & & & \\ 394 & 232b & -232a & 122b & 548b & 398a & 3454 & & & & & \\ -232b & -120b^2 & 112ab & -548b & -240b^2 & -168ab & -922b & 320b^2 & & & & \\ 232a & 112ab & -120a^2 & 398a & 168ab & 160a^2 & 922a & -252ab & 320a^2 & & & \\ 1226 & 548b & -398a & 394 & 232b & 232a & 1226 & -398b & 548a & 3454 & & \\ -548b & -240b^2 & 168ab & -232b & -120b^2 & -112ab & -398b & 160b^2 & -168ab & -922b & 320b^2 & \\ -398a & -168ab & 160a^2 & -232a & -112ab & -120a^2 & -548a & 168ab & -240a^2 & -922a & 252ab & 320a^2 \end{bmatrix}$$

The node number and the significance of a and b are shown in Figure 10.4.

22.2.4 Comparison of Two Mass Matrices

The experience of calculation has shown that under the conditions of the same number of elements, the calculation accuracy given by two mass matrices is almost the same. The lumped mass matrix not only is easy to calculate itself but also can simplify a lot of dynamic computing because it is a diagonal matrix. For some problems, such as beams, plates, shells, and so on, because the rotation items are neglected, the number of degrees of freedom in equations of motion can be significantly reduced.

Figure 22.2 is the computing grid of a simply supported square plate; due to the symmetry in two directions, we can only take 1/4 to calculate. When calculating the stiffness matrix, four triangular elements are combined into a quadrilateral element. We compare the calculations of two mass matrices. The results are shown in Table 22.1. It is

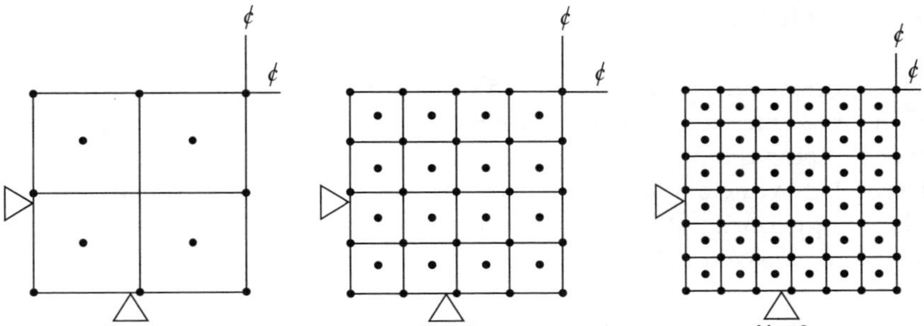

Figure 22.2 Simply supported square plate (1/4).

Table 22.1 Calculated natural frequency of simply supported square plate (two kinds of mass matrices).

λ_{mn}	Vibration type	Theoretical value m^2+n^2	$N=2$ Lumped mass	$N=2$ Consistent	$N=4$ Lumped mass	$N=4$ Consistent mass	$N=6$ Lumped mass	$N=6$ Consistent mass
λ_{11}	Symmetric–symmetric	2.000	2.034	2.031	2.007	2.007	2.001	2.001
λ_{12}	Antisymmetric–symmetric	5.000	5.142	5.135	5.032	5.031	5.013	5.012
λ_{13}	Symmetric–symmetric	10.000	10.185	10.390	10.076	10.078	10.032	10.032
λ_{23}	Antisymmetric–symmetric	13.000	13.499	14.028	13.317	13.289	13.131	13.125
λ_{14}	Antisymmetric–symmetric	17.000	—	19.014	17.136	17.176	17.059	17.065
λ_{33}	Symmetric–symmetric	18.000	17.117	19.747	18.733	18.647	18.305	18.283
λ_{34}	Antisymmetric–symmetric	25.000	23.672	29.880	26.283	26.153	25.557	25.509
λ_{15}	Symmetric–symmetric	26.000	30.267	28.954	26.186	26.390	26.095	26.127

Note: The theoretical natural frequency of square plate is $f=\frac{\pi}{2l^2}\sqrt{\frac{D}{\rho}\lambda_{nm}}$, $\lambda_{nm}=m^2+n^2$, $D=Eh^3/12(1-\mu^2)$, where l is side length of plate and D is flexural rigidity of plate. Those listed in the table are the values of λ_{mn}.

clear that under the conditions of the same mesh size, the calculation results of the two mass matrices are very close, but the number of degrees of freedom solving with lumped mass matrix is far less than that with consistent mass matrix.

It must be noted that when using high-order element, it is difficult to derive a lumped mass matrix. Furthermore, as long as the continuity between the discrete elements is maintained, the frequency calculated by consistent mass matrix represents the upper limit of the natural frequency of the structure.

22.3 Damping Matrix

As mentioned before, the mass matrix $[M]$ and stiffness matrix $[K]$ of the structure are consisted of the mass matrices $[m]$ and stiffness matrices $[k]$ of elements. Relatively, the problem of damping is more complicated, because the damping matrix of structure $[C]$ is not assembled by the damping matrix of element, and the approximate value is decided by the energy consumption of the whole structure during the vibration process according to existing measured data.

22.3.1 Damping of Single Freedom System

The free vibration equation of single freedom system is

$$m\ddot{\delta} + c\dot{\delta} + k\delta = 0 \tag{a}$$

Here m is the mass, c is the damping coefficient, k is the stiffness coefficient, and δ is the displacement.

After Eq. (a) divided by m, we get

$$\ddot{\delta} + 2\zeta\omega\dot{\delta} + \omega^2\delta = 0 \tag{b}$$

where

$$\omega = \sqrt{k/m} \tag{c}$$

$$\zeta = c/(2m\omega) \tag{d}$$

Here ζ is named as the damping ratio; ω is the natural frequency of system (this point will be described later).

Let the initial conditions be the following: when $t = 0$, $\delta = \delta_0$, $\dot{\delta} = v_0$. The solution satisfying these initial conditions and Eq. (b) is

$$\delta = \exp(-\zeta\omega t)\left(\delta_0 \cos\omega_\text{d} t + \frac{v_0 + \zeta\omega\delta_0}{\omega_\text{d}}\sin\omega_\text{d} t\right) \tag{e}$$

$$\omega_\text{d} = \omega\sqrt{1-\zeta^2} \tag{f}$$

Equation (e) shows the natural frequency of the system is ω_d and the amplitude of the system decays with time.

Let $\zeta = 0$ in Eq. (e), we get undamped free vibration as follows:

$$\delta = \delta_0 \cos\omega t + \frac{v_0}{\omega}\sin\omega t \tag{g}$$

Equation (g) shows the natural frequency without damp is ω.

We can analyze the influence of damping on the natural frequency from $\omega_\text{d} = \omega\sqrt{1-\zeta^2}$. The damping ratios of most structures are small number, $\zeta = 0.01 - 0.10$, generally less than 0.20. For example, according to the analysis of microseismic observations to dozens of dams in China, the damping ratios ζ of different types of dams are 0.03–0.05 for arch dam, 0.05–0.10 for gravity dam and buttress dam, and 0.10–0.20 for earth dam and rock-fill dam. During earthquake, the value of ζ not only will increase but also is a small value. According to Eq. (f), when $\zeta = 0.05$, $\omega_\text{d} = 0.9987\omega$; when $\zeta = 0.10$, $\omega_\text{d} = 0.995\omega$; and when $\zeta = 0.20$, $\omega_\text{d} = 0.9798\omega$. It can be clearly seen that the influence of the damp on the natural frequency is small; usually we can take $\omega_\text{d} = \omega$.

The damping ratio ζ can be determined by the measured data of the amplitude attenuation of free vibration. From Eq. (e), the amplitude at the time t_i is $\delta_i = A\exp(-\zeta\omega t_i)$; after a period $T_d = 2\pi/\omega_d$, the amplitude becomes $\delta_{i+1} = A\exp[-\zeta\omega(t_i + T_d)]$, and A is the value of the right brackets in formula (e). Let the natural logarithm of the ratio of the two adjacent amplitudes be ρ_1, and called the logarithm decline rate of amplitude, then

$$\rho_1 = \ln\frac{\delta_i}{\delta_{i+1}} = \zeta\omega T_d = \zeta\omega\frac{2\pi}{\omega_d} = \frac{2\pi\zeta}{\sqrt{1-\zeta^2}} \tag{h}$$

So we can get the damping ratio

$$\zeta = \rho_1/\sqrt{(2\pi)^2 + \rho_1^2} \tag{22.6}$$

Therefore, as long as the two adjacent amplitudes are measured, the damping ratio ζ is calculated by substituting the natural logarithm ρ_1 of the ratio into Eq. (22.6).

For example, taking $\sqrt{1-\zeta^2} \approx 1$, then we get from Eq. (h)

$$\zeta = \frac{\rho_1}{2\pi} = \frac{1}{2\pi}\ln\frac{\delta_i}{\delta_{i+1}} \tag{22.7}$$

For system of multidegree of freedom, there are different damping ratios corresponding to the different vibration modes, which can also be measured by experiments.

22.3.2 Damping of System of Multidegree of Freedom

If it is assumed that the damping force is proportional to the particle velocity, the damping force acting on unit volume of the tiny part of the moving structure is

$$\{p_d\} = -\alpha\rho\frac{\partial}{\partial t}\{r\} = -\alpha\rho[N]\{\dot{\delta}\}^e$$

Here α is the constant of ratio, ρ is the material density, and N is the shape function.

We can calculate the damping force acting on the node of element e by the general formula (3.14) of load displacement as follows:

$$\{F_d\}^e = \int [N]^T\{p_d\}\,dV = -\alpha\int [N]^T\rho[N]\,dV\{\dot{\delta}\}^e$$

as

$$\{F_d\}^e = -[c]\{\dot{\delta}\}^e$$

and

$$[c] = \alpha\int [N]^T\rho[N]\,dV = \alpha[m]$$

It can be seen that the element damping matrix is proportional to the element mass matrix. If it is assumed that the damping force is proportional to the strain velocity, the damping stress can be expressed as

$$\{\sigma_d\} = -\beta[D]\frac{\partial\{\varepsilon\}}{\partial t} = -\beta[D][B]\{\dot{\delta}\}^e$$

According to Eq. (3.16), the damping force acting on the node of element e is

$$\{F_d\}^e = \int [B]^T\{\sigma_d\}\,dV = -\beta\int [B]^T[D][B]\,dV\{\dot{\delta}\}^e = -[c]\{\dot{\delta}\}^e$$

where

$$[c] = \beta \int [B]^{\mathrm{T}}[D][B]\,\mathrm{d}V = \beta[k]$$

It can be seen that the element damping matrix is proportional to the element stiffness matrix $[k]$.

As already said, the approximate value of damp is usually decided by the overall energy consumption of structure based on measured data, so the overall damping matrix is calculated directly and the element damping matrix is not computed. The following linear formula is usually adopted to compute the damping matrix and called the Rayleigh damping, as

$$[C] = \alpha[M] + \beta[K] \tag{22.8}$$

The coefficients α and β here are decided by measured data.

Now explain how to calculate α and β. It is assumed that $\{\phi\}_i$ and $\{\phi\}_j$ are two vibration modes. After multiplying both sides of Eq. (22.8) by $\{\phi\}_j^{\mathrm{T}}$ and $\{\phi\}_j$, we get

$$\{\phi\}_j^{\mathrm{T}}[C]\{\phi\}_i = \alpha\{\phi\}_j^{\mathrm{T}}[M]\{\phi\}_i + \beta\{\phi\}_j^{\mathrm{T}}[K]\{\phi\}_i \tag{i}$$

According to the modal orthogonality mentioned later as Eqs (22.31) and (22.34), we get from Eq. (i)

$$\left.\begin{aligned} &\{\phi\}_j^{\mathrm{T}}[C]\{\phi\}_i = 0 \quad (i \neq j) \\ &\{\phi\}_j^{\mathrm{T}}[C]\{\phi\}_j = (\alpha + \beta\omega_j^2)m_{pj} \quad (i = j) \end{aligned}\right\} \tag{j}$$

where

$$m_{pj} = \{\phi\}_j^{\mathrm{T}}[M]\{\phi\}_j$$

let

$$\alpha + \beta\omega_j^2 = 2\zeta_j\omega_j \tag{22.9}$$

so

$$\{\phi\}_j^{\mathrm{T}}[C]\{\phi\}_j = 2\zeta_j\omega_j m_{pj} \tag{22.10}$$

From Eq. (22.9), we get

$$\zeta_i = \frac{\alpha}{2\omega_i} + \frac{\beta\omega_i}{2} \tag{22.11}$$

According to the two measured damping ratios, α and β can be calculated from Eq. (22.9); for example, substituting the two damping ratios ζ_i and ζ_j into Eq. (22.9), we get

$$\left.\begin{aligned} \alpha + \beta\omega_i^2 &= 2\zeta_i\omega_i \\ \alpha + \beta\omega_j^2 &= 2\zeta_j\omega_j \end{aligned}\right\} \tag{k}$$

After simultaneous solution, we get

$$\left.\begin{aligned} \alpha &= \frac{2(\zeta_i\omega_j - \zeta_j\omega_i)}{\omega_j^2 - \omega_i^2}\omega_i\omega_j \\ \beta &= \frac{2(\zeta_j\omega_j - \zeta_i\omega_i)}{\omega_j^2 - \omega_i^2} \end{aligned}\right\} \tag{22.12}$$

If $\zeta_i = \zeta_j = \zeta$, then

$$\alpha = \frac{2\omega_i\omega_j}{\omega_i + \omega_j}\zeta, \quad \beta = \frac{2\zeta}{\omega_i + \omega_j} \tag{22.13}$$

If there are more than two damping ratios in measured data, the two damping ratios $\overline{\zeta}_1$ and $\overline{\zeta}_2$ can be determined by graphical method. As shown in Figure 22.3, taking the frequency ω as x-axis and the damping ratio ζ as the y-axis, drawing a straight line through the measured points on the graph, taking two points (ω_1, ζ_1) and (ω_2, ζ_2) on the line, substituting them into Eq. (22.12), and then α and β are calculated.

For better consideration of the effect of damping on vibration of structure, the damping matrix can be taken as the following form:

$$[C] = [M]\sum_{c=0}^{p-1} \alpha_c([M]^{-1}[K])^c \tag{22.14}$$

Here α_c is an undetermined coefficient.

If just taking $c = 0$ and $c = 1$ in the sum symbol, $[C] = \alpha_0[M] + \alpha_1[K]$ is Rayleigh damping matrix. If there are p measured damping ratios, p items can be taken in Eq. (22.14).

Equation (22.37) shows that, when $i \neq j$, we get

$$\{\phi\}_j^{\mathrm{T}}[C]\{\phi\}_i = \{\phi\}_j^{\mathrm{T}}[M]\sum_{c=0}^{p-1} \alpha_c([M]^{-1}[K])^c\{\phi\}_i = 0 \tag{22.15}$$

When $i = j$, from Eqs (22.39) and (22.9), we get

$$\begin{aligned}\{\phi\}_i^{\mathrm{T}}[C]\{\phi\}_i &= \sum_{c=0}^{p-1} \alpha_c\{\phi\}_i^{\mathrm{T}}[M]([M]^{-1}[K])^c\{\phi\}_i \\ &= \sum_{c=0}^{p-1} \alpha_c\omega_i^{2c} m_{pi} = 2\zeta_i\omega_i m_{pi}\end{aligned} \tag{22.16}$$

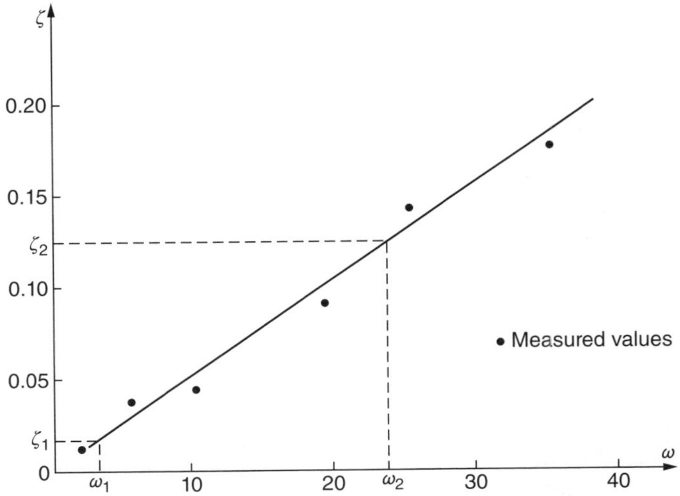

Figure 22.3 Relationship of the damping ratio ζ and frequency ω.

Then we obtain

$$\zeta_i = \frac{1}{2\omega_i}\sum_{c=0}^{p-1}\alpha_c\omega_i^{2c}$$

$$= \frac{1}{2}\left(\frac{\alpha_0}{\omega_i} + \alpha_1\omega_i + \alpha_2\omega_i^3 + \cdots + \alpha_{p-1}\omega_i^{2p-1}\right) \quad (i = 1,2,\ldots,p) \tag{22.17}$$

Substituting p measured damping ratios $\zeta_1, \zeta_2, \ldots, \zeta_p$ and the corresponding natural frequencies $\omega_1, \omega_2, \ldots, \omega_p$ into the previous equation, we get p linear equations, and p coefficients $\alpha_0, \alpha_1, \ldots, \alpha_{p-1}$ are obtained after simultaneously solving them. Thus, the assuming damping of each order can be matched up with measured results. But generally, when $p > 2$, the matrix $[C]$ calculated by Eq. (22.14) is often full, so the calculation increases a lot when solving the forced vibration of structure by direct integration method. Therefore, when using direct integration method, the Rayleigh damping matrix is usually used.

The damping of structure is mainly related to properties of materials, and the damping of foundation is larger than the damping of structure in the interaction problems of structure and foundation, so the different values of α and β are given to structure and foundation. In this case, the different vibration modes are no longer orthogonal to damping matrix $[C]$, so it cannot be decoupled between the vibration modes. Then, the direct integration method should be used, in which the orthogonal condition of vibration modes is not used.

22.4 Natural Frequency and Vibration Mode of Structure

22.4.1 Natural Frequency and Vibration Mode

Letting $P(t) = 0$ in Eq. (22.2), the free vibration equation is obtained. In practical engineering, the damping has little effect on the natural frequency and vibration mode, so the nondamping free vibration equations are obtained by neglecting the damping force:

$$[K]\{\delta\} + [M]\{\ddot{\delta}\} = 0 \tag{22.18}$$

Let the structure have simple harmonic motion as follows:

$$\{\delta\} = \{\phi\}\cos\omega t$$

Substituting the previous equation into Eq. (22.18), the homogeneous equation is obtained as follows:

$$[K]\{\phi\} = \omega^2[M]\{\phi\} \tag{22.19a}$$

or

$$([K] - \omega^2[M])\{\phi\} = 0 \tag{22.19b}$$

In free vibration, the amplitudes $\{\phi\}$ of each node in the structure are not all zero, so the determinant of the matrix in the brackets of Eq. (22.19b) must be equal to zero, and the natural frequency equation of structure is obtained as

$$|[K] - \omega^2[M]| = 0 \tag{22.20}$$

The stiffness matrix $[K]$ and mass matrix $[M]$ of structure are all n order matrices, where n is the number of degrees of freedom of nodes, so the previous equation is nth algebraic equation of ω^2, and n natural frequencies of structure are as follows:

$$\omega_1 \leq \omega_2 \leq \omega_3 \leq \cdots \leq \omega_n$$

For each natural frequency, a group of amplitudes of each node $\{\phi\}_i = [\phi_{i1}, \phi_{i2}, \ldots, \phi_{in}]^{\mathrm{T}}$ can be determined by Eq. (22.19), which should maintain a fixed ratio, but absolute values can be changed arbitrarily; they form a characteristic vector, often called the natural vibration modes of the structure in engineering.

Because in each mode, the amplitude of each node is relative, the absolute value can be any number. In practice, one of the following two methods is often used to determine the specific number of vibration mode:

1) *Regular mode*. Take one item of $\{\phi\}_i$, such as nth item as 1, that is, $\phi_{in} = 1$, so

$$\{\phi\}_i = [\phi_{i1}, \phi_{i2}, \ldots, 1]^{\mathrm{T}} \tag{22.21}$$

2) N*ormalized mode*. Select the value of ϕ_{ij} to make

$$\{\phi\}_i^{\mathrm{T}}[M]\{\phi\}_i = 1 \tag{22.22}$$

Assuming that a vibration mode $\{\overline{\phi}\}_i = [\overline{\phi}_{i1}, \overline{\phi}_{i2}, \ldots, \overline{\phi}_{in}]$ is calculated, if letting

$$\phi_{ij} = \overline{\phi}_{ij}/\overline{\phi}_{in} \tag{22.23}$$

then the mode $\{\phi\}_i = [\phi_{i1}, \phi_{i2}, \ldots, \phi_{in}]^{\mathrm{T}}$ obtained is regular mode. If letting

$$\begin{aligned} \phi_{ij} &= \overline{\phi}_{ij}/c \\ c &= [\{\overline{\phi}\}_i^{\mathrm{T}}[M]\{\overline{\phi}\}_i]^{1/2} \end{aligned} \tag{22.24}$$

then the mode $\{\phi\}_i = [\phi_{i1}, \phi_{i2}, \ldots, \phi_{in}]^{\mathrm{T}}$ obtained is normalized mode.

Let

$$m_{pi} = \{\phi\}_i^{\mathrm{T}}[M]\{\phi\}_i \tag{22.25}$$

When $[M]$ is the lumped mass matrix, then

$$m_{pi} = \begin{bmatrix} \phi_{i1} & \phi_{i2} & \cdots & \phi_{in} \end{bmatrix} \begin{bmatrix} m_1 & 0 & \cdots & 0 \\ 0 & m_2 & \cdots & 0 \\ \vdots & \vdots & & \vdots \\ 0 & 0 & \cdots & m_n \end{bmatrix} \begin{Bmatrix} \phi_{i1} \\ \phi_{i2} \\ \vdots \\ \phi_{in} \end{Bmatrix} = \sum_{s=1}^{n} m_s \phi_{is}^2 \tag{22.26a}$$

When $\{\phi\}_i$ is the normalized mode, we have

$$m_{pi} = 1 \tag{22.26b}$$

Let

$$k_{pi} = \{\phi\}_i^{\mathrm{T}}[K]\{\phi\}_i = \{\phi\}_i^{\mathrm{T}}\omega_i^2[M]\{\phi\}_i = \omega_i^2 m_{pi} \tag{22.27}$$

Here m_{pi} and k_{pi} are called generalized mass and generalized stiffness corresponding to ith mode.

From Eq. (22.27), we obtain

$$\omega_i = \sqrt{k_{pi}/m_{pi}} \tag{22.28}$$

Equation (22.28) is the generalization of the computation formula of frequency in single degree-of-freedom system.

Example 1 Solving $[K]\{\phi\} = \omega^2[M]\{\phi\}$, where

$$[K] = \begin{bmatrix} 2 & -1 & 0 \\ -1 & 4 & -1 \\ 0 & -1 & 2 \end{bmatrix}, \quad [M] = \begin{bmatrix} 0.5 & 0 & 0 \\ 0 & 1 & 0 \\ 0 & 0 & 0.5 \end{bmatrix}$$

From Eq. (22.20), the frequency equation is

$$|[K] - \omega^2[M]| = \begin{vmatrix} 2-0.5\omega^2 & -1 & 0 \\ -1 & 4-\omega^2 & -1 \\ 0 & -1 & 2-0.5\omega^2 \end{vmatrix} = 0$$

After expansion, the frequency equation obtained is

$$(\omega^2 - 2)(\omega^2 - 4)(\omega^2 - 6) = 0$$

Three natural frequencies obtained are as follows:

$$\omega^2 = 2, \quad \omega^2 = 4, \quad \omega^2 = 6$$

Substituting $\omega_1^2 = 2$ into Eq. (22.19a), equations that the first vibration mode must be met are obtained as follows:

$$\begin{aligned} \phi_{11} - \phi_{12} + 0 &= 0 \\ -\phi_{11} + 2\phi_{12} - \phi_{13} &= 0 \\ \phi_{11} - \phi_{12} + \phi_{13} &= 0 \end{aligned}$$

After simultaneous solution of the first and second equations, we obtain ϕ_{11}, ϕ_{12} represented by ϕ_{13} as follows:

$$\phi_{11} = \phi_{13}, \quad \phi_{12} = \phi_{13}$$

If taking $\phi_{13} = 1$, the first vibration mode of normalized mode is as follows:

$$\{\phi\}_1 = \begin{bmatrix} 1 & 1 & 1 \end{bmatrix}^{\mathrm{T}}$$

Also, the second and third vibration modes of normalized mode are as follows:

$$\{\phi\}_2 = \begin{bmatrix} -1 & 0 & 1 \end{bmatrix}^{\mathrm{T}}$$
$$\{\phi\}_3 = \begin{bmatrix} 1 & -1 & 1 \end{bmatrix}^{\mathrm{T}}$$

From Eq. (22.24), the normalized mode is obtained as follows:

$$\{\phi\}_1 = \begin{bmatrix} 1/\sqrt{2}, & 1/\sqrt{2}, & 1/\sqrt{2} \end{bmatrix}^{\mathrm{T}}$$
$$\{\phi\}_2 = \begin{bmatrix} -1 & 0 & 1 \end{bmatrix}^{\mathrm{T}}$$
$$\{\phi\}_3 = \begin{bmatrix} 1/\sqrt{2}, & -1/\sqrt{2}, & 1/\sqrt{2} \end{bmatrix}^{\mathrm{T}}$$

22.4.2 Orthogonality of Modes

If a set of loads are applied to each node of structure,

$$\{P_i\} = \omega_i^2[M]\{\phi\}_i \tag{22.29}$$

Here $\{P_i\}$ is called the characteristic load of the ith vibration mode.

The structural displacement $\{\delta\}$ caused by $\{P_i\}$ is determined by the following equation:

$$[K]\{\delta\} = \{P_i\} = \omega_i^2[M]\{\phi\}_i$$

Comparing the previous equation with Eq. (22.19), we know $\{\delta\} = \{\phi\}_i$, so in characteristic loads, the structural displacement is equal to the characteristic vector of the vibration mode $\{\phi\}_i$.

Now assume that the characteristic load of ith vibration mode is applied to the structure, and let the virtual work done by the virtual displacement corresponding to jth vibration mode $\{\phi\}_j$ caused by structure be

$$W_1 = \{\phi\}_j^{\mathrm{T}}\omega_i^2[M]\{\phi\}_i$$

If the characteristic load of jth vibration mode is applied to the structure, then let the virtual work done by the virtual displacement corresponding to ith vibration mode $\{\phi\}_i$ caused by structure be

$$W_2 = \{\phi\}_i^{\mathrm{T}}\omega_j^2[M]\{\phi\}_j$$

According to the reciprocal theorem, $W_1 = W_2$, we obtain

$$(\omega_i^2 - \omega_j^2)\{\phi\}_i^{\mathrm{T}}[M]\{\phi\}_j = 0$$

When $i \neq j$, $\omega_i^2 - \omega_j^2 \neq 0$, it can be deduced from the equation

$$\{\phi\}_i^{\mathrm{T}}[M]\{\phi\}_j = 0 \quad (i \neq j) \tag{22.30}$$

The previous property is called orthogonality of natural vibration mode.

It is known from Eqs (22.22) and (22.30) that the properties of normalized mode are as follows:

$$\{\phi\}_i^{\mathrm{T}}[M]\{\phi\}_j = \begin{cases} 0 & (i \neq j) \\ 1 & (i = j) \end{cases} \tag{22.31}$$

A matrix formed by the characteristic vectors $\{\phi\}_i$ of each vibration mode is as follows:

$$[\Phi] = \left[\{\phi\}_1 \ \{\phi\}_2 \ \cdots \ \{\phi\}_n\right] \tag{22.32}$$

It is known from Eq. (22.23) that the properties of $[\Phi]$ are as follows:

$$[\Phi]^{\mathrm{T}}[M][\Phi] = [I] \quad \text{(unit matrix)} \tag{22.33}$$

Premultiplying both sides of Eq. (22.19a) by $\{\phi\}_i^{\mathrm{T}}$, we obtain

$$\{\phi\}_i^{\mathrm{T}}[K]\{\phi\}_j = \omega_j^2\{\phi\}_i^{\mathrm{T}}[M]\{\phi\}_j = \begin{cases} 0 & (i \neq j) \\ \omega_i^2 & (i = j) \end{cases} \tag{22.34}$$

It can be seen that

$$[\Phi]^{\mathrm{T}}[K][\Phi] = [\Omega^2] \tag{22.35}$$

$$[\Omega^2] = \begin{bmatrix} \omega_1^2 & 0 & \cdots & 0 \\ 0 & \omega_2^2 & \cdots & 0 \\ \vdots & \vdots & & \vdots \\ 0 & 0 & \cdots & \omega_n^2 \end{bmatrix} \tag{22.36}$$

Now introduce the more generalized orthogonal relations derived by Caughey. From Eqs (22.19a), (22.30), (22.34), we have

$$[K]\{\phi\}_i = \omega_i^2[M]\{\phi\}_i \tag{a}$$

$$\{\phi\}_j^{\mathrm{T}}[M]\{\phi\}_i = 0, \quad (i \neq j) \tag{b}$$

$$\{\phi\}_j^{\mathrm{T}}[K]\{\phi\}_i = 0, \quad (i \neq j) \tag{c}$$

Premultiplying both sides of Eq. (a) by $\{\phi\}_j^{\mathrm{T}}[K][M]^{-1}$, we obtain

$$\{\phi\}_j^{\mathrm{T}}[K][M]^{-1}[K]\{\phi\}_i = \omega_i^2\{\phi\}_j^{\mathrm{T}}[K]\{\phi\}_i = 0$$

namely,

$$\{\phi\}_j^{\mathrm{T}}[K][M]^{-1}[K]\{\phi\}_i = 0 \tag{d}$$

Premultiplying both sides of Eq. (a) by $\{\phi\}_j^{\mathrm{T}}[K][M]^{-1}[K][M]^{-1}$ and taking into account Eq. (d), we obtain

$$\{\phi\}_j^{\mathrm{T}}[K][M]^{-1}[K][M]^{-1}[K]\{\phi\}_i = \omega_i^2\{\phi\}_j^{\mathrm{T}}[K][M]^{-1}[K]\{\phi\}_i = 0$$

namely,

$$\{\phi\}_j^{\mathrm{T}}[K]([M]^{-1}[K])^2\{\phi\}_i = 0 \tag{e}$$

Continuing the previous calculations, we get a more general equation of orthogonal relation as

$$\{\phi\}_j^{\mathrm{T}}[K]([M]^{-1}[K])^a\{\phi\}_i = 0 \quad (a = 0, 1,2, \ldots, n) \tag{f}$$

Premultiplying both sides of Eq. (a) by $\frac{1}{\omega_i^2}\{\phi\}_j^{\mathrm{T}}[M][K]^{-1}$, we obtain

$$\{\phi\}_j^{\mathrm{T}}[M][K]^{-1}[M]\{\phi\}_i = 0 \tag{g}$$

Premultiplying both sides of Eq. (a) by $\frac{1}{\omega_i^2}\{\phi\}_j^{\mathrm{T}}[M][K]^{-1}[M][K]^{-1}$ and taking into account Eq. (g), we obtain

$$\{\phi\}_j^{\mathrm{T}}[M][K]^{-1}[M][K]^{-1}[M]\{\phi\}_i = \frac{1}{\omega_i^2}\{\phi\}_j^{\mathrm{T}}[M][K]^{-1}[M]\{\phi\}_i = 0$$

namely,

$$\{\phi\}_j^{\mathrm{T}}[M]([K]^{-1}[M])^2\{\phi\}_i = 0$$

Continuing the previous calculations, we get another more general equation of orthogonal relation as

$$\{\phi\}_j^{\mathrm{T}}[M]([K]^{-1}[M])^b\{\phi\}_i = 0 \quad (b = 0, 1,2, \ldots, n) \tag{h}$$

If c is an integer, it is the inverse matrix when c is the power of negative, so integrated Eq. (f) and (h), we get a more general equation of orthogonal relation as

$$\{\phi\}_j^{\mathrm{T}}[M]([M]^{-1}[K])^c\{\phi\}_i = 0 \quad (-\infty < c < \infty) \tag{22.37}$$

From Eq. (22.37) it is not difficult to verify that the previous equations of orthogonal relation are all its special circumstances. For example, when $c=0$, Eq. (22.30) is obtained and when $c=1$ and -1, Eqs (22.34) and (g) are obtained, respectively.

Now introduce the calculations when $i=j$. Premultiplying both sides of Eq. (a) by $\{\phi\}_i^{\mathrm{T}}[K][M]^{-1}$ and multiplying after $\{\phi\}_i^{\mathrm{T}}$ on the left part of the equation by $[I] = [M][M]^{-1}$, we obtain

$$\{\phi\}_i[M][M]^{-1}[K][M]^{-1}[K]\{\phi\}_i = \omega_i^2\{\phi\}_i^{\mathrm{T}}[K]\{\phi\}_i = \omega_i^2(\omega_i^2\{\phi\}_i^{\mathrm{T}}[M]\{\phi\}_i)$$

namely,

$$\{\phi\}_i^{\mathrm{T}}[M]([M]^{-1}[K])^2\{\phi\}_i = \omega_i^{2\times 2}m_{pi}, \quad m_{pi} = \{\phi\}_i^{\mathrm{T}}[M]\{\phi\}_i$$

Premultiplying both sides of Eq. (a) by $\{\phi\}_i^{\mathrm{T}}([K][M]^{-1})^2$ and multiplying after $\{\phi\}_i^{\mathrm{T}}$ on the left part of the equation by $[I] = [M][M]^{-1}$, we obtain

$$\{\phi\}_i^{\mathrm{T}}[M]([M]^{-1}[K])^3\{\phi\}_i = \omega_i^{2\times 3}m_{pi}$$

Repeating such calculations, we get the relationship as follows:

$$\{\phi\}_i^{\mathrm{T}}[M]([M]^{-1}[K])^c\{\phi\}_i = \omega_i^{3c}m_{pi} \tag{22.38}$$

Premultiplying both sides of Eq. (22.38) by α_c and summing from $c=0$ to $c=p-1$, we obtain

$$\sum_{c=0}^{n-1}\alpha_c\{\phi\}_i^{\mathrm{T}}[M]([M]^{-1}[K])^c\{\phi\}_i = \sum_{c=0}^{p-1}\alpha_c\omega_i^{2c}m_{pi} \tag{22.39}$$

The coefficient α_c in general expression (22.14) of damping matrix is calculated by Eq. (22.39).

22.4.3 Free Vibration Equation of Structure Represented by Flexibility Matrix

Premultiplying both sides of Eq. (22.19a) by $[K]^{-1}$, the free vibration equation of structure represented by flexibility matrix is obtained as follows:

$$[f][M]\{\phi\} = \lambda\{\phi\} \tag{22.40}$$

where

$$[f] = [K]^{-1} \tag{22.41}$$

$$\lambda = 1/\omega^2 \tag{22.42}$$

Here $[f]$ is flexibility matrix; λ is eigenvalue.

There are n eigenvalues for structure with n degrees of freedom as follows:

$$\lambda_1 \geq \lambda_2 \geq \lambda_3 \geq \cdots \geq \lambda_n$$

Each eigenvalue λ_i and its corresponding characteristic vector $\{\phi\}_i$ are known as eigenpair.

22.4.4 Effects of Zero Mass

If there are some zero elements in the diagonal of lumped mass matrix $[M]$, we analyze their effects on the natural vibration frequency and vibration mode.

Rewrite Eq. (22.19a) in the following form:

$$[M]\{\phi\} = \lambda[K]\{\phi\} \qquad \lambda = 1/\omega^2 \tag{22.43}$$

If the diagonal element $m_{kk} = 0$ of lumped mass matrix $[M]$, its corresponding eigenpair is

$$\lambda_i = 0, \{\phi\}_i = [0 \;\; 0 \;\; \cdots \;\; 0 \;\; \underset{\uparrow\, k\text{th element}}{1} \;\; 0 \;\; \cdots \;\; 0]^{\mathrm{T}} \tag{22.44}$$

It is not difficult to verify that the previous eigenpair satisfies Eq. (22.43). Because $\lambda_i = 1/\omega_i^2$, the eigenpair to meet $[K]\{\phi\} = \omega^2[M]\{\phi\}$ is

$$\omega_i = \infty, \quad \{\phi\}_i = \begin{bmatrix}0 \; 0 \; \cdots \; 0 \; 1 \; 0 \; \cdots \; 0\end{bmatrix}^{\mathrm{T}} \tag{22.45}$$

Example 2 In the eigenvalue problem

$$\begin{bmatrix} 2 & -1 & 0 & 0 \\ -1 & 2 & -1 & 0 \\ 0 & -1 & -1 & 0 \\ 0 & 0 & -1 & 1 \end{bmatrix}\{\phi\} = \omega^2\begin{bmatrix} 0 & 0 & 0 & 0 \\ 0 & 2 & 0 & 0 \\ 0 & 0 & 0 & 0 \\ 0 & 0 & 0 & 1 \end{bmatrix}\{\phi\} \tag{i}$$

There are two diagonal elements equal to zero in the mass matrix, so $\omega_3 = \infty, \omega_4 = \infty$, and the two corresponding eigenvectors are

$$\{\phi\}_3 = \begin{bmatrix}1 \; 0 \; 0 \; 0\end{bmatrix}^{\mathrm{T}}, \quad \{\phi\}_4 = \begin{bmatrix}0 \; 0 \; 1 \; 0\end{bmatrix}^{\mathrm{T}}$$

Attention must be payed to that, when $i = 3, 4$, $\{\phi\}_i^{\mathrm{T}}[M]\{\phi\}_i = 0$, so the characteristic values of $\{\phi\}_3$ and $\{\phi\}_4$ cannot be decided by orthogonal conditions of $[M]$.

22.4.5 Static Condensation

In structural analysis with finite element method, the grid density is generally determined by the need of accuracy of stress analysis, so the order of stiffness matrix is also decided by the need of stress analysis. In most cases, the determination of deformation of structure requires much less degrees of freedom. The distribution of mass is only related to the deformation, so in the analysis of vibration mode, the degrees of freedom in mass matrix are much smaller than in stiffness matrix. In the dynamic analysis of complex structures, meshing is to do first on the needs of stress analysis, then it can be reduced to a dynamic problem with less degrees of freedom.

Relatively simple reduction method assumes that the mass is concentrated on a small number of selected nodes. Let the degree of freedom of nonzero mass of whole structure be ϕ_1 and the degree of freedom of zero mass be ϕ_2; Eq. (22.19) can be divided into the following blocks:

$$\begin{bmatrix} K_{11} & K_{12} \\ K_{21} & K_{22} \end{bmatrix}\begin{Bmatrix} \phi_1 \\ \phi_2 \end{Bmatrix} = \omega^2\begin{bmatrix} M_{11} & 0 \\ 0 & 0 \end{bmatrix}\begin{Bmatrix} \phi_1 \\ \phi_2 \end{Bmatrix} \tag{22.46}$$

From the lower part of Eq. (22.46), $\{\phi_2\}$ can be expressed by $\{\phi_1\}$ as follows:

$$\{\phi_2\} = -[K_{22}]^{-1}[K_{21}]\{\phi_1\} \tag{22.47}$$

And then substituting back into the upper part of Eq. (22.46), the condensed equation obtained is as follows:

$$[\overline{K}_{11}]\{\phi_1\} = \omega^2[M_{11}]\{\phi_1\} \tag{22.48}$$

where

$$[\overline{K}_{11}] = [K_{11} - K_{12}K_{22}^{-1}K_{21}] \tag{22.49}$$

$\{\phi_1\}$ can be calculated from Eq. (22.48) and then substituted into Eq. (22.47) to compute $\{\phi_2\}$.

In fact, Gauss elimination method can also be used to reduce the order; the right part of Eq. (22.46) can be understood as the load vector as follows:

$$\{P\} = \begin{Bmatrix} \omega^2[M_{11}]\{\phi_1\} \\ 0 \end{Bmatrix} \tag{22.50}$$

Using the Gauss elimination method, the degrees of freedom corresponding to the zero mass are all eliminated.

It must be pointed out that after the free vibration equation of structure (22.19) is reduced to Eq. (22.48), while the degrees of freedom are reduced, the bandwidth of stiffness matrix has increased, so only when the degrees of freedom are reduced much, it can bring obvious benefits in the calculation.

Figure 22.4 shows the vibration of a cantilever square plate. Before the reduced order, the plate has 90 degrees of freedom and then gradually reduced to 6 degrees of freedom; as the figure shows, the first four natural frequencies are slightly changed [1].

Example 3 For Eq. (i) of the eigenvalue problem, the third and fourth eigenpairs are solved earlier; now we calculate the other two eigenpairs by static condensation method.

Rearranging the rows and columns of matrix in Eq. (i), we obtain

$$\begin{bmatrix} 2 & 0 & -1 & -1 \\ 0 & 1 & 0 & -1 \\ -1 & 0 & 2 & 0 \\ -1 & -1 & 0 & 2 \end{bmatrix} \begin{Bmatrix} \phi_1 \\ \phi_2 \end{Bmatrix} = \omega^2 \begin{bmatrix} 2 & 0 & 0 & 0 \\ 0 & 1 & 0 & 0 \\ 0 & 0 & 0 & 0 \\ 0 & 0 & 0 & 0 \end{bmatrix} \begin{Bmatrix} \phi_1 \\ \phi_2 \end{Bmatrix}$$

From Eq. (22.49), we obtain

$$[\overline{K}_{11}] = \begin{bmatrix} 2 & 0 \\ 0 & 1 \end{bmatrix} - \begin{bmatrix} -1 & -1 \\ 0 & -1 \end{bmatrix} \begin{bmatrix} 0.5 & 0 \\ 0 & 0.5 \end{bmatrix} \begin{bmatrix} -1 & 0 \\ -1 & -1 \end{bmatrix} = \begin{bmatrix} 1 & -0.5 \\ -0.5 & 0.5 \end{bmatrix}$$

Substituting into Eq. (22.48), the reduced equation are obtained as

$$\begin{bmatrix} 1 & -0.5 \\ -0.5 & 0.5 \end{bmatrix} \{\phi_1\} = \omega^2 \begin{bmatrix} 2 & 0 \\ 0 & 1 \end{bmatrix} \{\phi_1\}$$

The natural frequency equation after reduction is

$$|[\overline{K}_{11}] - \omega^2[M_{11}]| = 2\omega^4 - 2\omega^2 + \frac{1}{4} = 0$$

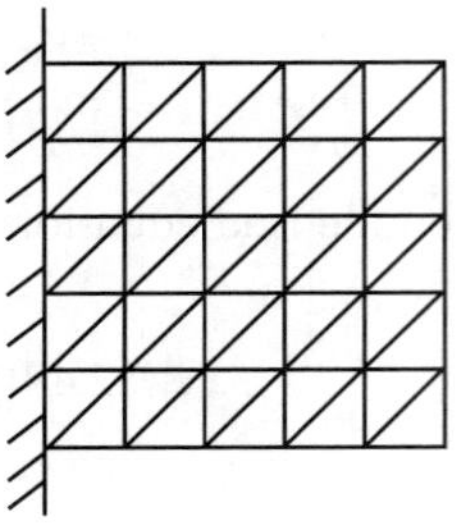

Vibration mode	$\omega\sqrt{D/\rho t a^4}$
1	3.469
2	8.535
3	21.450
4	27.059

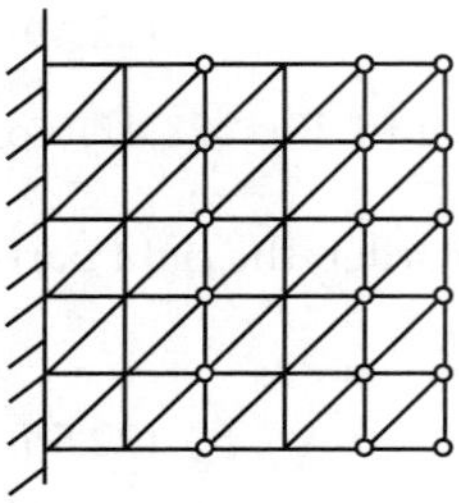

Vibration mode	$\omega\sqrt{D/\rho t a^4}$
1	3.470
2	8.540
3	21.559
4	27.215

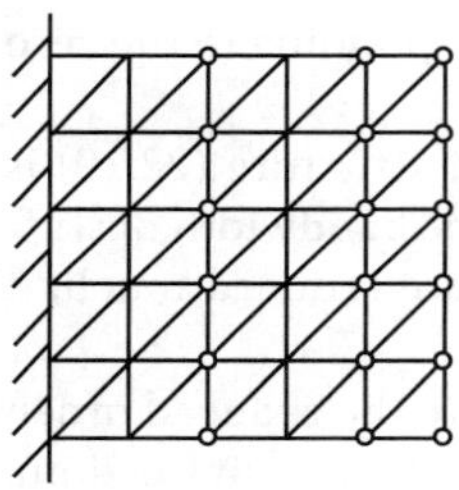

Vibration mode	$\omega\sqrt{D/\rho t a^4}$
1	3.470
2	8.543
3	21.645
4	27.296

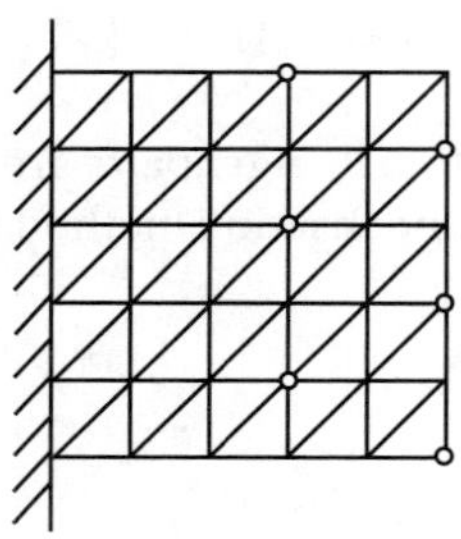

Vibration mode	$\omega\sqrt{D/\rho t a^4}$
1	3.473
2	8.604
3	22.690
4	29.490

Figure 22.4 Reducing degrees of freedom in vibration of a cantilever square plate.

And then

$$\omega_1^2 = \frac{1}{2} - \frac{\sqrt{2}}{4}, \quad \omega_2^2 = \frac{1}{2} + \frac{\sqrt{2}}{4}$$

$$\{\phi_1\}_1 = \begin{bmatrix} \frac{1}{2} & \frac{\sqrt{2}}{2} \end{bmatrix}^{\mathrm{T}}, \quad \{\phi_1\}_2 = \begin{bmatrix} -\frac{1}{2} & \frac{\sqrt{2}}{2} \end{bmatrix}^{\mathrm{T}}$$

Substituting into Eq. (22.33), we obtain

$$\{\phi_2\}_1 = \begin{bmatrix} \frac{1}{4} & \frac{1+\sqrt{2}}{4} \end{bmatrix}^{\mathrm{T}}, \quad \{\phi_2\}_2 = \begin{bmatrix} -\frac{1}{4} & \frac{-1+\sqrt{2}}{4} \end{bmatrix}^{\mathrm{T}}$$

After adding two eigenpairs corresponding to zero quality, we know that the four eigenpairs of eigenvalue problem equation (i) are

$$\omega_1^2 = \frac{1}{2} - \frac{\sqrt{2}}{4}, \quad \{\phi\}_1 = \begin{bmatrix} \frac{1}{4} & \frac{1}{2} & \frac{1+\sqrt{2}}{4} & \frac{\sqrt{2}}{2} \end{bmatrix}^{\mathrm{T}}$$

$$\omega_2^2 = \frac{1}{2} + \frac{\sqrt{2}}{4}, \quad \{\phi\}_2 = \begin{bmatrix} -\frac{1}{4} & -\frac{1}{2} & \frac{-1+\sqrt{2}}{4} & \frac{\sqrt{2}}{2} \end{bmatrix}^{\mathrm{T}}$$

$$\omega_3^2 = \infty, \quad \{\phi\}_3 = \begin{bmatrix} 1 & 0 & 0 & 0 \end{bmatrix}^{\mathrm{T}}$$

$$\omega_4^2 = \infty, \quad \{\phi\}_4 = \begin{bmatrix} 0 & 0 & 1 & 0 \end{bmatrix}^{\mathrm{T}}$$

22.5 Mode Superposition Method for Analyzing the Structure of Forced Vibration

Currently, there are two methods used to analyze the forced vibration of the structure, the mode superposition method and the direct integration method. This section describes the mode superposition method.

We represent the structural displacement vector $\{\delta\}$ in the state of motion with a linear superposition of modes $\{\phi\}_i$ as follows:

$$\{\delta\} = \{\phi\}_1\eta_1(t) + \{\phi\}_2\eta_2(t) + \cdots + \{\phi\}_n\eta_n(t) = \sum_{i=1}^{n} \{\phi\}_i\eta_i(t) \tag{22.51}$$

Premultiplying on both sides of Eq. (22.51) by $\{\phi\}_j^{\mathrm{T}}[M]$, because of modal orthogonality, n items on the right side of equality are all zero when $i \neq j$; only $i = j$ is left as

$$\{\phi\}_j^{\mathrm{T}}[M]\{\delta\} = \eta_j(t)\{\phi\}_j^{\mathrm{T}}[M]\{\phi\}_j = m_{pj}\eta_j(t)$$

Then we obtain

$$\eta_i(t) = \frac{\{\phi\}_i^{\mathrm{T}}[M]\{\delta\}}{m_{pi}} \tag{22.52}$$

The initial values of η_i and $\dot{\eta}_i$ may be expressed as follows:

$$\eta_i(0) = \frac{\{\phi\}_i^{\mathrm{T}}[M]\{\delta(0)\}}{m_{pi}} \tag{22.53}$$

$$\dot{\eta}_i(0) = \frac{\{\phi\}_i^{\mathrm{T}}[M]\{\dot{\delta}(0)\}}{m_{pi}} \tag{22.54}$$

Now consider the following equations of motion:

$$[M]\{\ddot{\delta}\} + [C]\{\dot{\delta}\} + [K]\{\delta\} = \{P(t)\}$$

Substituting Eq. (22.51) into the previous equation, we obtain

$$[M]\sum_{i=1}^{n} \{\phi\}_i\ddot{\eta}_i + [C]\sum_{i=1}^{n} \{\phi\}_i\dot{\eta}_i + [K]\sum_{i=1}^{n} \{\phi\}_i\eta_i = \{P(t)\}$$

Premultiplying both sides of the previous equation by $\{\phi\}_j^{\mathrm{T}}$ and letting $[C] = \alpha[M] + \beta[K]$, we obtain

$$\sum_{i=1}^{n}\{\phi\}_j^{\mathrm{T}}[M]\{\phi\}_i\ddot{\eta}_i + \sum_{i=1}^{n}\{\phi\}_j^{\mathrm{T}}(\alpha[M] + \beta[K])\{\phi\}_j\dot{\eta}_i + \sum_{i=1}^{n}\{\phi\}_j^{\mathrm{T}}[K]\{\phi\}_i\eta_i$$
$$= \{\phi\}_j^{\mathrm{T}}\{P(t)\} \tag{a}$$

Because of modal orthogonality, the items with $i \neq j$ on the right side of Eq. (a) are all zero; only the item with $i = j$ left, so we obtain

$$m_{pi}\ddot{\eta}_i + (\alpha + \beta\omega_i^2)m_{pi}\dot{\eta}_i + \omega_i^2 m_{pi}\eta_i = \{\phi\}_i^{\mathrm{T}}\{P(t)\} \tag{b}$$

From Eq. (22.9), $\alpha + \beta\omega_i^2 = 2\zeta_i\omega_i$, substituting it into Eq. (b), we obtain

$$\ddot{\eta}_i + 2\zeta_i\omega_i\dot{\eta}_i + \omega_i^2\eta_i = \frac{1}{m_{pi}}\{\phi\}_i^{\mathrm{T}}\{P(t)\} \quad (i = 1,2, \ldots, n) \tag{22.55}$$

This is a second-order ordinary differential equation; there are n such equations, which are independent of each other. The form of Eq. (22.55) is the same as the motion equation of single degree-of-freedom system. Equation (22.55) can be solved by numerical integral method, also by the Duhamel integral method as follows:

$$\eta_i(t) = \frac{1}{\omega_{di}m_{pi}}\int_0^t P^*(\tau)\,\mathrm{e}^{-\zeta_i\omega_i(t-\tau)}\sin\omega_{di}(t-\tau)\,\mathrm{d}\tau$$
$$+\,\mathrm{e}^{-\zeta_i\omega_i t}\left\{\eta_i(0)\cos\omega_{di}t + \frac{\dot{\eta}_i(0) + \zeta_i\omega_i\eta_i(0)}{\omega_{di}}\sin\omega_{di}t\right\} \tag{22.56}$$

where

$$\omega_{di} = \omega_i\sqrt{1 - \zeta_i^2} \tag{c}$$

$$P^*(t) = \{\phi\}_i^{\mathrm{T}}\{P(t)\} = \phi_{11}P_1(t) + \phi_{12}P_2(t) + \cdots + \phi_{1n}P_n(t) \tag{d}$$

m_{pi} is seen in Eq. (22.55).

After substituting $\eta_i(t)$ into Eq. (22.51), the required answer is obtained. In structural dynamic analysis with finite element method, degrees of freedom n can reach hundreds or even thousands, but because of the influences of higher modes on the structural dynamic responses are generally small, usually the calculation of a portion of the low modes is enough. For example, for the earthquake load, only the first 5–20 vibration modes are generally calculated. For blast and impact loads, we need to take more vibration modes, sometimes as many as $2n/3$ modes for calculation. For the dynamic responses induced by vibration, sometimes only a portion of the intermediate vibration modes are working, so it is necessary to take only vibration modes of $\omega_{\mathrm{l}} \leq \omega \leq \omega_{\mathrm{u}}$ for analysis, where ω_{l} and ω_{u} are the lower and upper vibration frequencies, respectively.

22.6 Dynamic Response of Structure under the Action of Earthquake Solving by Vibration Mode Superposition Method

During an earthquake, it is assumed that the ground accelerations in x, y, z directions are $\ddot{x}_0$, $\ddot{y}_0$, $\ddot{z}_0$; the deformation of node i relative to ground is $\{\delta_i\} = \begin{bmatrix} u_i & v_i & w_i \end{bmatrix}^{\mathrm{T}}$, and the actual acceleration of node i is

$$\begin{Bmatrix} \ddot{u}_i \\ \ddot{v}_i \\ \ddot{w}_i \end{Bmatrix} + \ddot{x}_0 \begin{Bmatrix} 1 \\ 0 \\ 0 \end{Bmatrix} + \ddot{y}_0 \begin{Bmatrix} 0 \\ 1 \\ 0 \end{Bmatrix} + \ddot{z}_0 \begin{Bmatrix} 0 \\ 0 \\ 1 \end{Bmatrix} \tag{a}$$

Therefore, the dynamic load of the structure during an earthquake is

$$\{P\} = -[M]\{a\} = -[M](\{A_x\}\ddot{x}_0 + \{A_y\}\ddot{y}_0 + \{A_z\}\ddot{z}_0) \tag{22.57}$$

$$\{A_x\} = \begin{Bmatrix} 1 \\ 0 \\ 0 \\ \vdots \\ 1 \\ 0 \\ 0 \end{Bmatrix}, \quad \{A_y\} = \begin{Bmatrix} 0 \\ 1 \\ 0 \\ \vdots \\ 0 \\ 1 \\ 0 \end{Bmatrix}, \quad \{A_z\} = \begin{Bmatrix} 0 \\ 0 \\ 1 \\ \vdots \\ 0 \\ 0 \\ 1 \end{Bmatrix} \tag{22.58}$$

Let

$$\{M_x\} = [M]\{A_x\}, \quad \{M_y\} = [M]\{A_y\}, \quad \{M_z\} = [M]\{A_z\} \tag{22.59}$$

then Eq. (22.57) can be written as

$$\{P\} = -\{M_x\}\ddot{x}_0 - \{M_y\}\ddot{y}_0 - \{M_z\}\ddot{z}_0 \tag{b}$$

For the lumped mass matrix, we get

$$\left.\begin{aligned} \{M_x\} &= \begin{bmatrix} m_1 & 0 & 0 & m_4 & 0 & 0 & \cdots & m_{n-2} & 0 & 0 \end{bmatrix}^{\mathrm{T}} \\ \{M_y\} &= \begin{bmatrix} 0 & m_2 & 0 & 0 & m_5 & 0 & \cdots & 0 & m_{n-1} & 0 \end{bmatrix}^{\mathrm{T}} \\ \{M_z\} &= \begin{bmatrix} 0 & 0 & m_3 & 0 & 0 & m_6 & \cdots & 0 & 0 & m_n \end{bmatrix}^{\mathrm{T}} \end{aligned}\right\} \tag{c}$$

Substituting Eq. (22.57) into Eq. (22.55), we know that during an earthquake η_i will meet the following equations:

$$\ddot{\eta}_i + 2\zeta_i\omega_i\dot{\eta}_i + \omega_i^2\eta_i = -\gamma_{ix}\ddot{x}_0 - \gamma_{iy}\ddot{y}_0 - \gamma_{iz}\ddot{z}_0, \quad (i = 1,2,\ldots,n) \tag{22.60}$$

where

$$\left.\begin{aligned} \gamma_{ix} &= \frac{1}{m_{pi}}\{\phi\}_i^{\mathrm{T}}\{M_x\} \\ \gamma_{iy} &= \frac{1}{m_{pi}}\{\phi\}_i^{\mathrm{T}}\{M_y\} \\ \gamma_{iz} &= \frac{1}{m_{pi}}\{\phi\}_i^{\mathrm{T}}\{M_z\} \end{aligned}\right\} \tag{22.61}$$

Here $\gamma_{ix}, \gamma_{iy}, \gamma_{iz}$ are called modal participation factors.

For the lumped mass matrix, from Eqs (c) and (22.26a), we know

$$\left.\begin{aligned}\gamma_{ix} &= \frac{m_1\phi_{i1} + m_4\phi_{i4} + \cdots + m_{n-2}\phi_{i,n-2}}{\sum_{i=1}^{n} m_s\phi_{is}^2}\\ \gamma_{iy} &= \frac{m_2\phi_{i2} + m_5\phi_{i5} + \cdots + m_{n-1}\phi_{i,n-1}}{\sum_{i=1}^{n} m_s\phi_{is}^2}\\ \gamma_{iz} &= \frac{m_3\phi_{i3} + m_6\phi_{i6} + \cdots + m_n\phi_{in}}{\sum_{i=1}^{n} m_s\phi_{is}^2}\end{aligned}\right\} \tag{d}$$

During an earthquake, three components of the ground acceleration $\ddot{x}_0, \ddot{y}_0, \ddot{z}_0$ are often different, but when substituting them into Eq. (22.60) simultaneously, there is no difficulty in computing. The three acceleration components can also be calculated separately and then superimposed.

When calculating the response of structure under the action of $\ddot{x}_0$, let

$$\eta_i(t) = \gamma_{ix}u_{ix}(t) \tag{e}$$

Substituting into Eq. (22.60), we obtain

$$\ddot{u}_{ix} + 2\omega_i\zeta_i\dot{u}_{ix} + \omega_i^2 u_{ix} = -\ddot{x}_0 \tag{f}$$

The previous equation is actually an equation of motion of structure of single degree of freedom with frequency ω_i under the effect of ground acceleration $\ddot{x}_0$; using numerical integration method or Duhamel integral method, its answer $u_{ix}(t)$ can be found. Similarly, for $\ddot{y}_0$ and $\ddot{z}_0$, answers $u_{iy}(t)$ and $u_{iz}(t)$ may be obtained. Substituting into Eq. (22.51), it is known that under the effects of $\ddot{x}_0$, $\ddot{y}_0$, and $\ddot{z}_0$, the displacement of the structure is

$$\{\delta\} = \sum_{i=1}^{n} \{\phi\}_i(\gamma_{ix}u_{ix} + \gamma_{iy}u_{iy} + \gamma_{iz}u_{iz}) \tag{22.62}$$

Because u_{ix}, u_{iy}, u_{iz} are responses of the structure of single degree of freedom with frequency ω_i under the effect of ground acceleration in $\ddot{x}_0$, $\ddot{y}_0$, and $\ddot{z}_0$. So, we can use the displacement response spectrum or acceleration response spectrum, which is already known to simplify the calculation.

22.7 Vector Iteration Method for Computing the Natural Frequency and Vibration Mode

By using the mode superposition method to calculate the forced vibration of structure, the main computational work is the computation of natural frequency and vibration mode of structure. Usually only a few frequencies and modes need to be calculated, not all the frequencies and modes, so it is convenient to use the vector iteration method.

Vector iteration method is divided into two kinds, namely, the inverse iteration method and the positive iterative method. The inverse iteration method converges to the lowest frequency and vibration mode of structure, while the positive iterative method converges to the highest frequency and vibration mode of structure.

22.7.1 Inverse Iteration Method: The Calculation of Lowest Frequency and Vibration Mode

The inverse iteration method is to calculate a free vibration Eq. (22.40) of structure represented by the flexibility matrix. Let

$$[A] = [f][M] = [K]^{-1}[M] \tag{22.63}$$

Substituting it into Eq. (22.40), we obtain

$$[A]\{\phi\} = \lambda\{\phi\} \tag{22.64a}$$

where

$$\lambda = 1/\omega^2$$

Here λ is called eigenvalue.

Equation (22.64a) can be rewritten as

$$\{\phi\} = \frac{1}{\lambda}[A]\{\phi\} \tag{22.64b}$$

By means of Eq. (22.64b), an iterative algorithm may be established; the kth step is calculated as follows:

$$\{y^{(k+1)}\} = [A]\{\phi^{(k)}\} = \begin{bmatrix} y_1^{(k+1)} & \cdots & y_n^{(k+1)} \end{bmatrix}^{\mathrm{T}} \tag{22.65a}$$

$$\lambda^{(k+1)} = y_n^{(k+1)} \tag{22.65b}$$

$$\{\phi^{(k+1)}\} = \frac{1}{\lambda^{(k+1)}}\{y^{(k+1)}\} = \begin{bmatrix} \dfrac{y_1^{(k+1)}}{y_n^{(k+1)}} & \cdots & \dfrac{y_{n-1}^{(k+1)}}{y_n^{(k+1)}} & 1 \end{bmatrix}^{\mathrm{T}} \tag{22.65c}$$

Usually, we may take the initial value $\{\phi^{(0)}\} = \begin{bmatrix} 1 & 1 & \cdots & 1 \end{bmatrix}^{\mathrm{T}}$ and substitute into Eq. (22.65a). We will get $\{y^{(1)}\}$, then $\lambda^{(1)}$ from Eq. (22.65b), and $\{\phi^{(1)}\}$ from Eq. (22.65c). Repeat the calculations, until the results of nth step and $(n+1)$th step are sufficiently close.

Example 1 $[f]$, $[M]$ are as follows:

$$[f] = \begin{bmatrix} 54 & 28 & 8 \\ 28 & 16 & 5 \\ 8 & 5 & 2 \end{bmatrix}, \quad [M] = \begin{bmatrix} 4 & 0 & 0 \\ 0 & 1 & 0 \\ 0 & 0 & 1 \end{bmatrix}, \quad [A] = [f][M] = \begin{bmatrix} 216 & 28 & 8 \\ 112 & 16 & 5 \\ 32 & 5 & 2 \end{bmatrix}$$

Taking the initial value $\{\phi^{(0)}\} = \begin{bmatrix} 1 & 1 & 1 \end{bmatrix}^{\mathrm{T}}$, we get $\{y^{(1)}\} = [A]\{\phi^{(0)}\} = \begin{bmatrix} 252 & 133 & 39 \end{bmatrix}^{\mathrm{T}}$ from Eq. (22.65a), $\lambda^{(1)} = 39$ from Eq. (22.65b), and $\{\phi^{(1)}\} = \begin{bmatrix} 6.4615 & 3.4103 & 1 \end{bmatrix}^{\mathrm{T}}$ from Eq. (22.65c). Repeat the calculations as shown in Table 22.2. After five iterations, answers are as follows:

$$\lambda = 231.83, \quad \{\phi\} = \begin{bmatrix} 6.6404 & 3.4690 & 1.0000 \end{bmatrix}^{\mathrm{T}}$$

The advantages of iterative method are that the procedure of calculation is simple and the error generated in the calculation process can be automatically revised in calculation.

Table 22.2 Examples of inverse iteration.

k	$\{\phi\}$			$\{y\}$			λ
	ϕ_1	ϕ_2	ϕ_3	y_1	y_2	y_3	
0	1	1	1	252	133	39	39
1	6.4615	3.4103	1.0000	1499.17	783.25	225.82	225.82
2	6.6388	3.4685	1.0000	1539.43	804.04	231.78	231.78
3	6.6403	3.4689	1.0000	1539.43	804.22	231.83	231.83
4	6.6403	3.4690	1.0000	1539.44	804.22	231.83	231.83
5	6.6404	3.4690	1.0000				

Now we shall prove that the results obtained in inverse iteration method are the lowest natural frequency ω_1 and corresponding vibration mode $\{\phi\}_1$ of the structure.

After the given initial mode $\{\phi^{(0)}\}$, inverse iteration method is actually computing a series of vectors as follows:

$$[A]\{\phi^{(0)}\}, [A]([A]\{\phi^{(0)}\}) = [A]^2\{\phi^{(0)}\}, \ldots$$
$$[A]([A]^{k-1}\{\phi^{(0)}\}) = [A]^k\{\phi^{(0)}\}$$

The initial vibration mode $\{\phi^{(0)}\}$ can be expressed as

$$\{\phi^{(0)}\} = \eta_1\{\phi\}_1 + \eta_2\{\phi\}_2 + \cdots + \eta_n\{\phi\}_n$$

in which η_i is constant and $\{\phi\}_i$ is i-order vibration mode.

Premultiplying $[A]$ on the previous equation and considering $[A]\{\phi\}_i = \lambda_i\{\phi\}_i$, we get

$$[A]\{\phi^{(0)}\} = \eta_1[A]\{\phi\}_1 + \eta_2[A]\{\phi\}_2 + \cdots + \eta_n[A]\{\phi\}_n$$
$$= \eta_1\lambda_1\{\phi\}_1 + \eta_2\lambda_2\{\phi\}_2 + \cdots + \eta_n\lambda_n\{\phi\}_n$$

Similarly

$$[A]^2\{\phi^{(0)}\} = \eta_1\lambda_1^2\{\phi\}_1 + \eta_2\lambda_2^2\{\phi\}_2 + \cdots + \eta_n\lambda_n^2\{\phi\}_n$$
$$[A]^k\{\phi^{(0)}\} = \eta_1\lambda_1^k\{\phi\}_1 + \eta_2\lambda_2^k\{\phi\}_2 + \cdots + \eta_n\lambda_n^k\{\phi\}_n$$

Because $\lambda_1 > \lambda_2 > \cdots > \lambda_n > 0$, when k is sufficiently large, we get

$$\lambda_1^k \gg \lambda_2^k \gg \cdots \gg \lambda_n^k$$

where the symbol "≫" means "much greater than." Compared with λ_1^k, other items can be ignored, so as long as $\eta_1 \neq 0$, we get

$$[A]^k\{\phi^{(0)}\} \approx \eta_1\lambda_1^k\{\phi\}$$

The previous equation shows that, after k iterations, the difference between vectors $[A]\{\phi^{(0)}\} \approx \eta_1\lambda_1^k\{\phi\}$ and $\{\phi\}_1$ is only a constant or vector $[A]^k\{\phi^{(0)}\}$ converges to $\eta_1\lambda^k\{\phi\}_1$. After k iterations the inverse iteration method converges to the first mode of the structure.

It can be proved that the convergence rate for the inverse iteration method is λ_2/λ_1. So the bigger the difference between λ_2 and λ_1, the smaller the ratio, and the faster the convergence. Conversely, when λ_2 and λ_1 are close, λ_2/λ_1 is close to 1, and the convergence is slow.

Back to the basic equations of the free vibration of structure, we get

$$[K]\{\phi\} = R[M]\{\phi\} \tag{22.66}$$

where $R = \omega^2 = 1/\lambda$; in order to obtain the regularization mode, the iterative calculations can also be done according to the following equations [2]:

$$\{x^{(k+1)}\} = [K]^{-1}\{y^{(k)}\} \tag{22.67a}$$

$$\{\overline{y}^{(k+1)}\} = [M]\{x^{(k+1)}\} \tag{22.67b}$$

$$R^{(k+1)} = \frac{\{x^{(k+1)}\}^{\mathrm{T}}\{y^{(k)}\}}{\{x^{(k+1)}\}^{\mathrm{T}}\{\overline{y}^{(k+1)}\}} \tag{22.67c}$$

$$\{y^{(k+1)}\} = \frac{\{\overline{y}^{(k+1)}\}}{(\{x^{(k+1)}\}^{T}\{\overline{y}^{(k+1)}\})^{1/2}} \tag{22.67d}$$

As long as $\{y_1\}^{\mathrm{T}}\{\phi\}_1 \neq 0$, when $k \to \infty$, we get

$$\{y^{(k+1)}\} \to [M]\{\phi\}_1, \quad R^{(k+1)} \to \omega_1^2 = 1/\lambda_1$$

when

$$\frac{|R^{(k+1)} - R^{(k)}|}{R^{(k+1)}} \leq \varepsilon$$

where ε is allowable error, after the previous calculations are over, and taking

$$\omega_1^2 = R^{(k+1)} \tag{22.67e}$$

$$\{\phi\}_1 = \frac{\{x^{(k+1)}\}}{(\{x^{(k+1)}\}^{\mathrm{T}}\{\overline{y}^{(k+1)}\})^{1/2}} \tag{22.67f}$$

22.7.2 Mode Clearance: Calculation of Other Frequencies and Modes

As mentioned previously, the inverse iteration method converges to the first vibration mode, and its basic iterative equation is Eq. (22.65). After the first mode is obtained, the effects of the first vibration mode must be cleared from $\{\phi^{(k)}\}$; after iterations, it will converge to the second vibration mode. After the second mode is obtained, the effects of the first and the second vibration modes must be cleared from $\{\phi^{(k)}\}$; it will converge to the third mode. Generally, after obtaining the rth vibration mode, the effects of the first to the rth vibration modes must be cleared from $\{\phi^{(k)}\}$, and then it will converge to the $(r+1)$th mode.

Now we explain how to clear the effects of the first r modes after they are obtained. $\{\phi^{(k)}\}$ is expressed as a linear combination of each mode as

$$\{\phi^{(k)}\} = \sum_{j=1}^{n} \eta_j\{\phi\}_j \tag{a}$$

Premultiplying both sides of the previous equation by $\{\phi\}_i^{\mathrm{T}}[M]$ and considering the orthogonal condition of vibration mode (22.30), we obtain

$$\{\phi\}_i^{\mathrm{T}}[M]\{\phi^{(k)}\} = \{\phi\}_i^{\mathrm{T}}[M]\sum_{j=1}^{n} \eta_j\{\phi\}_j = \{\phi\}_i^{\mathrm{T}}[M]\{\phi\}_i\eta_i$$

Thus,

$$\eta_i = \frac{\{\phi\}_i^{\mathrm{T}}[M]\{\phi^{(k)}\}}{\{\phi\}_i^{\mathrm{T}}[M]\{\phi\}_i} \tag{b}$$

In order to clear the influence of the first r vibration modes from $\{\phi^{(k)}\}$, we can take

$$\{\phi^{(k)}\} - \sum_{j=1}^{r} \eta_j\{\phi\}_j = \{\phi^{(k)}\} - \sum_{j=1}^{r} \{\phi\}_j \frac{\{\phi\}_j^{\mathrm{T}}[M]\{\phi^{(k)}\}}{\{\phi\}_j^{T}[M]\{\phi\}_j}$$

$$= \left([I] - \sum_{j=1}^{r} \frac{\{\phi\}_j\{\phi\}_j^{\mathrm{T}}[M]}{\{\phi\}_j^{T}[M]\{\phi\}_j}\right)\{\phi^{(k)}\} = [Q]_r\{\phi^{(k)}\}$$

where

$$[Q]_r = [I] - \sum_{j=1}^{r} \frac{\{\phi\}_j\{\phi\}_j^{\mathrm{T}}[M]}{\{\phi\}_j^{\mathrm{T}}[M]\{\phi\}_j} \tag{c}$$

Here $[Q]_r$ is called mode clearing matrix.

In each iteration step, replacing $\{\phi^{(k)}\}$ with $[Q]_r\{\phi^{(k)}\}$ means that in Eq. (22.65a), taking

$$\begin{aligned} \{y^{(k+1)}\} &= [A][Q]_r\{\phi^{(k)}\} = [A]_r\{\phi^{(k)}\} \\ [A]_r &= [A][Q]_r \end{aligned} \tag{22.68}$$

Here $[A]_r$ is called shrinkage matrix.

Then we get the new equations of k iterative step as follows:

$$\{y^{(k+1)}\} = [A]_r\{\phi^{(k)}\} = \left[y_1^{(k+1)} \ \cdots \ y_n^{(k+1)}\right]^{\mathrm{T}} \tag{22.69a}$$

$$\lambda^{(k+1)} = y_n^{(k+1)} \tag{22.69b}$$

$$\{\phi^{(k+1)}\} = \frac{1}{\lambda^{(k+1)}}\{y^{(k+1)}\} = \left[\frac{y_1^{(k+1)}}{y_n^{(k+1)}} \ \cdots \ \frac{y_{n-1}^{(k+1)}}{y_n^{(k+1)}} \ \ 1\right]^{\mathrm{T}} \tag{22.69c}$$

Equation (22.69c) means the basic equation $\lambda\{\phi\} = [A]\{\phi\}$ has been replaced by Eq. (22.70a):

$$\lambda\{\phi\} = [A]_r\{\phi\} = [A][Q]_r\{\phi\} \tag{22.70a}$$

Substituting Eq. (c) into Eq. (22.70a) and noting $[A]\{\phi\}_j = \lambda_j\{\phi\}_j$, we obtain

$$\lambda\{\phi\} = [A]_r\{\phi\} = \left([A] - \sum_{j=1}^{r} \frac{\lambda_j\{\phi\}_j\{\phi\}_j^{\mathrm{T}}[M]}{\{\phi\}_j^{\mathrm{T}}[M]\{\phi\}_j}\right)\{\phi\} \tag{22.70b}$$

It is not difficult to prove that the first r eigenvectors of the previous equation are same as Eq. (22.64), but the corresponding eigenvalues are zero. From $r+1$, Eqs (22.70b) and (22.64) have the same eigenvalues and eigenvectors. The previous conclusions will be proved in the following in two steps.

Firstly, let λ_i and $\{\phi\}_i (i \le r)$ be the eigenvalues and eigenvectors of $\lambda\{\phi\} = [A]\{\phi\}$. Substituting $\{\phi\}_i$ into Eq. (22.70b), we obtain

$$\lambda\{\phi\}_i = [A]\{\phi\}_i - \sum_{j=1}^{r} \frac{\lambda_j\{\phi\}_j\{\phi\}_j^{\mathrm{T}}[M]\{\phi\}_i}{\{\phi\}_j^{\mathrm{T}}[M]\{\phi\}_j}$$

Because of the orthogonal conditions of the vibration mode, when $i \neq j$, $\{\phi\}_j^{\mathrm{T}}[M]\{\phi\}_i = 0$, so we get

$$\lambda\{\phi\}_i = [A]\{\phi\}_i - \lambda_i\{\phi\}_i = \lambda_i\{\phi\}_i - \lambda_i\{\phi\}_i = 0\cdot\{\phi\}_i$$

Then

$$\lambda = 0$$

It can be seen that $\{\phi\}_i$ is the eigenvector of Eq. (22.70), and its corresponding eigenvalue is zero.

Next, letting λ_s and $\{\phi\}_s$ $(s > r)$ be the eigenvalues and eigenvectors of $\lambda\{\phi\} = [A]\{\phi\}$ and substituting $\{\phi\}_s$ into Eq. (22.70), we obtain

$$\lambda\{\phi\}_s = [A]\{\phi\}_s - \sum_{j=1}^{r} \frac{\lambda_j\{\phi\}_j\{\phi\}_j^{\mathrm{T}}[M]\{\phi\}_s}{\{\phi\}_j^{\mathrm{T}}[M]\{\phi\}_j}$$

Because of $s > r$, according to the orthogonal conditions of vibration mode, item Σ on the right side of the previous equation is equal to zero, so we get

$$\lambda\{\phi\}_s = [A]\{\phi\}_s = \lambda_s\{\phi\}_s$$

It can be seen that

$$\lambda = \lambda_s$$

So, λ_s and $\{\phi\}_s$ are also the eigenvalues and eigenvectors of Eq. (22.70). In a word, through the iterative calculations by Eq. (22.70), we obtain λ_{r+1} and $\{\phi\}_{r+1}$.

Recursive formula of shrinkage matrix $[A]_r$ is given in the following text. From Eqs (c) and (22.68), and noting $[A]\{\phi\}_j = \lambda_j\{\phi\}_j$, we get

$$[A]_r = [A][Q]_r = [A] - \sum_{j=1}^{r} \frac{[A]\{\phi\}_j\{\phi\}_j^{\mathrm{T}}[M]}{\{\phi\}_j^{\mathrm{T}}[M]\{\phi\}_j} = [A] - \sum_{j=1}^{r} \frac{\lambda_j\{\phi\}_j\{\phi\}_j^{\mathrm{T}}[M]}{\{\phi\}_j^{\mathrm{T}}[M]\{\phi\}_j}$$
$$= [A] - \sum_{j=1}^{r-1} \frac{\lambda_j\{\phi\}_j\{\phi\}_j^{\mathrm{T}}[M]}{\{\phi\}_j^{\mathrm{T}}[M]\{\phi\}_j} - \lambda_r\frac{\{\phi\}_r\{\phi\}_r^{\mathrm{T}}[M]}{\{\phi\}_r^{\mathrm{T}}[M]\{\phi\}_r}$$

The recursive formula is obtained as follows:

$$[A]_r = [A]_{r-1} - \lambda_r\frac{\{\phi\}_r\{\phi\}_r^{\mathrm{T}}[M]}{\{\phi\}_r^{\mathrm{T}}[M]\{\phi\}_r} \tag{22.71}$$

Example 2 $\lambda_1 = 231.83$ and $\{\phi\}_1 = \lfloor 6.6404\ 3.4690\ 1.0000 \rfloor^{\mathrm{T}}$ are obtained in Example 1. Now the second eigenpair is calculated using the iterative method. From Eq. (22.71), we get

$$[A]_1 = [A] - \lambda_1\frac{\{\phi\}_1\{\phi\}_1^{\mathrm{T}}[M]}{\{\phi\}_1^{\mathrm{T}}[M]\{\phi\}_1}$$

$$= \begin{bmatrix} 216 & 28 & 8 \\ 112 & 16 & 5 \\ 32 & 5 & 2 \end{bmatrix} - \begin{bmatrix} 215.8772 & 28.1938 & 8.12738 \\ 112.7756 & 14.7288 & 4.2458 \\ 32.5095 & 4.2458 & 1.22393 \end{bmatrix}$$

$$= \begin{bmatrix} 0.12279 & -0.1938 & -0.12738 \\ -0.7756 & 1.2712 & 0.7542 \\ -0.5095 & 0.7542 & 0.77607 \end{bmatrix}$$

The second eigenpair is obtained in the following by iterative calculations using Eq. (22.69):

$$\lambda_2 = 1.9354, \quad \{\phi\}_2 = \begin{bmatrix} -0.2188 & 1.3899 & 1.0000 \end{bmatrix}^{\mathrm{T}}$$

22.7.3 Shifting: To Improve the Convergence Speed

As mentioned before, the convergence speed of the inverse iteration method is λ_2/λ_1. When λ_2/λ_1 is very small, for example, $\lambda_2/\lambda_1 = 0.01$, convergence is fast. On the contrary, when λ_2/λ_1 is close to 1, for example, $\lambda_2/\lambda_1 = 0.99$, convergence is very slow. Then the convergence speed can be improved by shifting.

It is known from Eq. (22.19a), the basic equation of free vibration of the structure is

$$[K]\{\phi\} = R[M]\{\phi\} \tag{22.72}$$

$$R = \omega^2 = 1/\lambda \tag{d}$$

Let

$$R = \eta + s \tag{22.73}$$

Substituting it into Eq. (22.72), we obtain

$$([K] - s[M])\{\phi\} = \eta[M]\{\phi\} \tag{e}$$

Let

$$[A] = ([K] - s[M])^{-1}[M] \tag{f}$$

Substituting it into Eq. (22.72), we obtain

$$\{\phi\} = \eta[A]\{\phi\} \tag{g}$$

Assuming that the eigenvalues for Eq. (22.72) are $R_1 < R_2 < \cdots < R_n$, without mode clearance, the iterative calculation is performed by the inverse iteration method to Eq. (g), which can be proved to converge to η_j and $\{\phi\}_j$; the eigenpair of the original problem formula (22.72) is λ_j, $\{\phi\}_j$, and

$$\lambda_j = \omega_j^2 = \frac{1}{R_j} = \frac{1}{\eta_j + s}$$

where η_j takes the value of R_j closest to μ. The iterative convergence speed is

$$r = \max\left|\frac{R_j - s}{R_p - s}\right| \quad (p \neq j)$$

Because R_j is closest to s, the speed of convergence is actually the larger one of two ratios:

$$\left|\frac{R_j - s}{R_{j-1} - s}\right| \quad \text{or} \quad \left|\frac{R_j - s}{R_{j+1} - s}\right| \tag{h}$$

Obviously, the closer the values of s to R_j, the faster the convergence is. The question is how to choose a value of s. Generally the value of s can be estimated by Rayleigh quotient (22.74) $R = \omega^2$ as equation in Section 22.8.

22.7.4 Positive Iterative Method: Calculation of the Maximum Frequency and Vibration Mode

The basic equation of free vibration of structure represented by the stiffness matrix is

$$\lambda[K]\{\phi\} = [M]\{\phi\} \tag{i}$$

Premultiplying both sides of Eq. (i) by $[M]^{-1}$, we obtain

$$\{\phi\} = \lambda[B]\{\phi\} \tag{j}$$

where

$$[B] = [M]^{-1}[K] \tag{k}$$

In the positive iterative method, a set of iterative formulas as follows is established by Eq. (j):

$$\{y^{(k+1)}\} = [B]\{\phi^{(k)}\} \tag{l}$$

$$\lambda^{(k+1)} = 1/y_n^{(k+1)} \tag{m}$$

$$\{\phi^{(k+1)}\} = \lambda\{y^{(k+1)}\} = \begin{bmatrix} \frac{y_1^{(k+1)}}{y_n^{(k+1)}} & \cdots & \frac{y_{n-1}^{(k+1)}}{y_n^{(k+1)}} & 1 \end{bmatrix} \tag{n}$$

After iterations, the maximum natural vibration frequency and the corresponding vibration mode are obtained finally as

$$\{\phi^{(k+1)}\} \to \{\phi\}_n, \quad \lambda^{(k+1)} \to \lambda_n, \quad (k \to \infty) \tag{i}$$

Usually, some of the lowest frequencies and modes are needed by the engineering, rather than the highest frequencies and vibration modes, so it is rare to use the positive iterative method.

22.8 Energy Method for Computing the Natural Frequencies of Structure

There are two kinds of energy methods for calculating the natural frequency of structure, namely, the Rayleigh method and the Ritz method.

22.8.1 Rayleigh Energy Method

According to the conservation of energy, if ignoring the energy dissipation of structure during vibration, at any instant, the sum of kinetic energy V and strain energy U of structure remains unchanged, namely, $V(t) + U(t) = \text{constant}$.

The displacement of the node in free vibration of structure is expressed as follows:

$$\{\delta\} = \{\phi\}\sin(\omega t + \theta) \tag{a}$$

The velocity of the node is

$$\{\dot{\delta}\} = \{\phi\}\omega\cos(\omega t + \theta) \tag{b}$$

The kinetic energy of structure is

$$V = \frac{1}{2}\{\dot{\delta}\}^{\mathrm{T}}[M]\{\dot{\delta}\} = \frac{1}{2}\{\phi\}^{\mathrm{T}}[M]\{\phi\}\omega^2\cos^2(\omega t + \theta) \tag{c}$$

The strain energy of structure is

$$V = \frac{1}{2}\{\delta\}^{\mathrm{T}}[K]\{\delta\} = \frac{1}{2}\{\phi\}^{\mathrm{T}}[K]\{\phi\}\sin^2(\omega t + \theta) \tag{d}$$

When $\omega t + \theta = \pi/2$, $\sin(\omega t + \theta) = 1$, the displacement reaches the maximum value $\{\phi\}$, and the strain energy of structure reaches the maximum value

$$U_{\max} = \frac{1}{2}\{\phi\}^{\mathrm{T}}[K]\{\phi\} \tag{e}$$

At this time, $\cos(\omega t + \theta) = 0$, so the kinetic energy of structure $V = 0$.

When $\omega t + \theta = 0$, $\cos(\omega t + \theta) = 1$, $\sin(\omega t + \theta) = 0$, at this time the strain energy of structure $U = 0$, and the kinetic energy of structure reaches the maximum value:

$$V_{\max} = \frac{1}{2}\omega^2\{\phi\}^{\mathrm{T}}[M]\{\phi\} \tag{f}$$

According to the conservation of energy, we get

$$U_{\max} = V_{\max}$$

So there is

$$R = \omega^2 = \frac{\{\phi\}^{\mathrm{T}}[K]\{\phi\}}{\{\phi\}^{\mathrm{T}}[M]\{\phi\}} \tag{22.74}$$

R is called Rayleigh quotient.

Substituting the accurate ith vibration mode $\{\phi\}_j$ into Eq. (22.74), the accurate jth vibration frequency ω_j is obtained. Usually, the approximate value $\{\phi\}$ is substituted into Eq. (22.74) to calculate the approximate self-vibration frequency. Because it is difficult to estimate the high-order mode, the first natural frequency of structure is generally calculated by this equation. For this reason, the approximate value of $\{\phi\}$ needs to be given. One approach is to give the approximate value of $\{\phi\}$ based on the boundary conditions of structure; another approach is to use the displacement under the action of a static load (such as gravity) instead of $\{\phi\}$, which is substituted into Eq. (22.74) to calculate ω.

The result of calculating the lowest natural frequency of structure by Eq. (22.74) is good. Experience shows that although a rough value $\{\phi\}$ is given, the accuracy of the lowest natural frequency calculated is satisfactory.

22.8.2 Ritz Energy Method

The vibration mode of structure is expressed in the following form:

$$\begin{aligned}\{\phi\} &= a_1\{\psi\}_1 + a_2\{\psi\}_2 + \cdots + a_s\{\psi\}_s \\ &= \left[\{\psi\}_1\ \{\psi\}_2\ \cdots\ \{\psi\}_s\right]_{n\times s}\{a\}_{s\times 1} = [\psi]_{n\times s}\{a\}_{s\times 1}\end{aligned} \tag{22.75}$$

Here $\{\psi\}_1$, $\{\psi\}_2$, …, $\{\psi\}_s$ are s vectors, which are given in advance according to the boundary conditions of structure; $a_1, a_2, \ldots, a_s$ are s undetermined coefficients.

Substituting Eq. (22.75) into Rayleigh quotient (22.74) and letting the numerator and denominator of the formula, respectively, be $U(a)$ and $V(a)$, we get

$$R = \frac{\{\phi\}^{\mathrm{T}}[K]\{\phi\}}{\{\phi\}^{\mathrm{T}}[M]\{\phi\}} = \frac{\{a\}^{\mathrm{T}}\{\psi\}^{\mathrm{T}}[K][\psi]\{a\}}{\{a\}^{\mathrm{T}}\{\psi\}^{\mathrm{T}}[M][\psi]\{a\}} = \frac{U(a)}{V(a)} \tag{g}$$

The value of a_j is so selected that $\omega^2 = R$ will take the minimum value; let the partial derivative of R to a_j be equal to zero:

$$\frac{\partial R}{\partial a_j} = \frac{1}{V^2}\left(V(a)\frac{\partial U}{\partial a_j} - U(a)\frac{\partial V}{\partial a_j}\right) = 0 \quad (j = 1,2,\ldots,s) \tag{h}$$

From the Rayleigh quotient (22.74) and Eq. (g), we get $U(a) = \omega^2 V(a)$; substituting it into Eq. (h), we obtain

$$\frac{\partial U}{\partial a_j} - \omega^2\frac{\partial V}{\partial a_j} = 0 \quad (j = 1,2,\ldots,s) \tag{i}$$

From Eq. (g), we get

$$\begin{aligned}\frac{\partial U}{\partial a_j} &= \frac{\partial}{\partial a_j}(\{a\}^{\mathrm{T}}[\psi]^{\mathrm{T}}[K][\psi]\{a\}) \\ &= \frac{\partial\{a\}^{\mathrm{T}}}{\partial a_j}[\psi]^{\mathrm{T}}[K]\{\psi\}\{a\} + \{a\}^{\mathrm{T}}\{\psi\}^{\mathrm{T}}[K][\psi]\frac{\partial\{a\}}{\partial a_j}\end{aligned}$$

Two terms on the right side of the previous equation are all scalars, whose transposes are still scalars. All the terms in the second part of the previous equation are transposed and then multiplied in reserve order; then we get

$$\frac{\partial U}{\partial a_j} = 2\frac{\partial\{a\}^{\mathrm{T}}}{\partial a_j}[\psi]^{\mathrm{T}}[K][\psi]\{a\} \tag{j}$$

But

$$\frac{\partial\{a\}^{\mathrm{T}}}{\partial a_j} = \frac{\partial}{\partial a_j}\,[a_1\ a_2\ \ldots\ a_j\ \ldots\ a_s] = [0\ \ 0\ \ \cdots\ \ 0\ \ \underset{\substack{\uparrow \\ j\text{th term}}}{1}\ \ 0\ \ \cdots\ \ 0]$$

So

$$\frac{\partial\{a\}^{\mathrm{T}}}{\partial a_j} = \frac{\partial}{\partial a_j}[\psi]^{\mathrm{T}} = \{\psi\}_j^{\mathrm{T}}$$

Substituting into Eq. (j), we get

$$\frac{\partial U}{\partial a_j} = 2\{\psi\}_j^{\mathrm{T}}[K][\psi]\{a\}$$

Similarly, we have

$$\frac{\partial V}{\partial a_j} = 2\{\psi\}_j^{\mathrm{T}}[M][\psi]\{a\}$$

Therefore, Eq. (i) can be written as

$$\{\psi\}_j^{\mathrm{T}}[K][\psi]\{a\} - \omega^2\{\psi\}_j^{\mathrm{T}}[M][\psi]\{a\} = 0 \quad (j = 1,2, \ldots, s) \tag{22.76}$$

This is a set of homogeneous linear equations about a_1, a_2, …, a_s, expanding the determinant composed of the coefficients of a_1, a_2, …, a_s, and, making it equal to zero, we get the frequency equation and obtain s frequencies by solving it: ω_1^2, ω_2^2, …, ω_s^2.

In fact, linear equations represented by Eq. (22.76) can be expressed as follows:

$$[\psi]^{\mathrm{T}}[K][\psi]\{a\} - \omega^2[\psi]^{\mathrm{T}}[M][\psi]\{a\} = 0 \tag{22.77}$$

Let

$$\left.\begin{aligned} [\overline{K}]_{s\times s} &= [\psi]^{\mathrm{T}}[K][\psi] \\ [\overline{M}]_{s\times s} &= [\psi]^{\mathrm{T}}[M][\psi] \end{aligned}\right\} \tag{22.78}$$

Equation (22.77) can be written as

$$[\overline{K}]_{s\times s}\{a\}_{s\times 1} = \omega^2[\overline{M}]_{s\times s}\{a\} \tag{22.79}$$

Here $[\overline{K}]$ is called generalized stiffness matrix; $[\overline{M}]$ is called generalized mass matrix.

So, the problem is attributed to the eigenvalue problem of matrix; but here is the eigenvalue problem of $s \times s$ matrix, instead of the eigenvalue problem of original $n \times n$ matrix. Thus, the Ritz method plays a role in reducing the degrees of freedom, and the problem of $n \times n$ is reduced to $s \times s$. The s eigenvalues calculated are the approximations of the first s values of frequencies ω^2 of structure, and the s corresponding approximate vibration modes are

$$\{\phi\}_j = [\psi]\{a\}_j \quad (j = 1,2, \ldots, s)$$

$\{a\}_j$ is orthogonal to generalized mass matrix as

$$\{a\}_i^{\mathrm{T}}[\overline{M}]\{a\}_j = 0 \quad (i \neq j) \tag{k}$$

So, the approximate mode of each order is orthogonal to mass matrix as

$$\{\phi\}_i^{\mathrm{T}}[M]\{\phi\}_j = \{a\}_i^{\mathrm{T}}[\psi]^{\mathrm{T}}[M][\psi]\{a\}_j = \{a\}_i^{\mathrm{T}}[\overline{M}]\{a\}_j = 0 \quad (i \neq j) \tag{l}$$

22.9 Subspace Iteration Method for Computing the Natural Frequencies and Vibration Modes of Structure

As mentioned earlier, the inverse iteration method is to assume an initial vibration mode and to obtain the lowest frequency and mode by the basic equations iteratively. After mode clearance, the higher order frequencies and vibration modes are obtained by iterative calculation. When the degrees of freedom are small, the iterative method is effective, but when the degrees of freedom are large, it will take quite a long time to compute, with low efficiency.

The Ritz method is to give s initial vectors $\{\psi\}_j$, $j = 1\sim s$. The vibration mode of structure is expressed as a linear combination of the s vectors: $\{\phi\} = a_1\{\psi\}_1 + a_2\{\psi\}_2 + \cdots + a_s\{\psi\}_s$; after calculating the coefficients $a_1, \ldots, a_s$, s approximate natural frequencies ω_j and vibration modes $\{\phi\}_j$ are obtained, $j = 1\sim s$. Ritz method reduces the scale of the problem, from n order to s order. However, the accuracy of the calculation is related to the accuracy of the given s initial vectors. The closer the initial vector to the vibration mode of structure, the higher the accuracy of the calculation.

The basic idea of subspace iteration method [3] is to combine the Ritz method and the inverse iterative method, not only reducing the degrees of freedom by using the Ritz method but also making the vibration mode gradually close to its exact value by using the inverse iterative method in calculation. Its calculation result is quite good because of absorbing advantages of these two methods. Experience shows that this is one of the most effective methods for solving the natural vibration frequency and vibration mode of large-scale structure.

The modified subspace iteration method improves the computational efficiency of subspace iteration method by shifting.

22.9.1 Subspace Iteration Method

The basic equation for free vibration of structure is

$$[K]\{\phi\}_i = \omega_i^2[M]\{\phi\}_i \quad (i = 1,2, \ldots n) \tag{a}$$

or

$$[f][M]\{\phi\}_i = \lambda_i\{\phi\}_i \quad (i = 1,2, \ldots, n) \tag{b}$$

There are n eigenvectors in the previous problem. Selecting the first s ($s < n$) characteristic vectors of n characteristic vectors in n-dimensional space, the space defined by these s vectors is called subspace of the original n-dimensional space.

We select s ($s < n$) n-dimensional initial vectors $\{\psi^{(0)}\}_j$ ($j = 1,2, \ldots, s$) to form a matrix of $n \times s$ as

$$[\psi^{(0)}] = \left[\{\psi^{(0)}\}_1 \ \{\psi^{(0)}\}_2 \ \cdots \ \{\psi^{(0)}\}_s\right] \tag{c}$$

Taking it as the zero-order approximation of the first s-order vibration modal matrix $[\phi]$ of structure, that is,

$$[\phi^{(0)}] = \left[\{\phi^{(0)}\}_1 \ \{\phi^{(0)}\}_2 \ \cdots \ \{\phi^{(0)}\}_s\right] = [\psi^{(0)}] \tag{d}$$

Premultiplying $[\psi^{(0)}]$ by $[f][M]$, that is, premultiplying each column of $[\psi^{(0)}]$ by $[f][M]$, we obtain

$$[\tilde{\psi}^{(1)}] = [f][M][\psi^{(0)}] \tag{e}$$

Letting the maximum norm of the displacement of each mode in $[\tilde{\psi}^{(1)}]$ be 1, we obtain $[\psi^{(1)}]$.

The first-order approximation of ith order vibration mode of structure is expressed as

$$\{\phi^{(1)}\}_i = a_{i1}\{\psi^{(1)}\}_1 + a_{i2}\{\psi^{(1)}\}_2 + \cdots + a_{is}\{\psi^{(1)}\}_s \tag{f}$$

where $a_{i1}, a_{i2}, \ldots, a_{is}$ are s undetermined coefficients. So the first-order approximation of the first s-order vibration modal matrix of structure is expressed as

$$[\phi^{(1)}]_{n\times s} = [\psi^{(1)}]_{n\times s}[a]_{s\times s} \tag{g}$$

where $[a] = [\{a\}_1, \{a\}_2, \ldots, \{a\}_s]$ is the matrix of undetermined coefficients. According to the Ritz method, the generalized mass matrix and stiffness matrix are as follows:

$$[\overline{M}^{(1)}]_{s\times s} = [\psi^{(1)}]^{\mathrm{T}}[M][\psi^{(1)}] \tag{h}$$

$$[\overline{K}^{(1)}]_{s\times s} = [\psi^{(1)}]^{\mathrm{T}}[K][\psi^{(1)}] \tag{i}$$

So, the problem is attributed to the eigenvalue problem of $s \times s$ order as

$$[\overline{K}^{(1)}]_{s\times s}\{a\}_{s\times 1} = \omega^2[\overline{M}^{(1)}]_{s\times s}\{a\} \tag{j}$$

or

$$[\overline{f}^{(1)}][\overline{M}]\{a\} = \lambda\{a\} \tag{k}$$

where

$$[\overline{f}^{(1)}] = [\overline{K}^{(1)}]^{-1}, \quad \lambda = 1/\omega^2$$

Usually $s \ll n$, so the eigenvalue problem of $s \times s$ order is relatively easy to solve. The eigenvalue $\lambda_i^{(1)}$ $(i = 1,2, \ldots, s)$ and its corresponding characteristic vector $\{a^{(1)}\}_i$ are calculated, and substituted into Eq. (g); then we obtain the first-order approximation of the first s-order vibration modal matrix of structure as

$$[\phi^{(1)}]_{n\times s} = [\psi^{(1)}]_{n\times s}[a^{(1)}]_{s\times s} \tag{l}$$

Now begin the second step of iterations. Premultiplying $[\phi^{(1)}]$ by $[f][M]$, we obtain

$$[\tilde{\psi}^{(2)}] = [f][M][\phi^{(1)}]$$

Let the maximum norm of the displacement of each mode in $[\tilde{\psi}^{(2)}]$ be 1, we obtain $[\psi^{(2)}]$; the second-order approximation of the first s-order vibration modal matrix of structure is expressed as

$$[\phi^{(2)}] = [\psi^{(2)}][a^{(2)}] \tag{m}$$

Repeating the calculations of (h)–(l) in the first iteration as

$$[\overline{M}^{(2)}] = [\psi^{(2)}]^{\mathrm{T}}[M][\psi^{(2)}] \tag{n}$$

$$[\overline{K}^{(2)}] = [\psi^{(2)}]^{\mathrm{T}}[K][\psi^{(2)}] \tag{o}$$

$$[\overline{K}^{(2)}]\{a^{(2)}\} = \omega^2[\overline{M}^{(2)}]\{a^{(2)}\} \tag{p}$$

$$[\overline{f}][\overline{M}]\{a^{(2)}\} = \lambda\{a^{(2)}\} \tag{q}$$

After $[a^{(2)}]$ is calculated, the second-order approximation of the first s-order vibration modal matrix is obtained as

$$[\phi^{(2)}] = [\psi^{(2)}][a^{(2)}] \tag{r}$$

Then the next iteration is performed, until the results of two iterations are sufficiently close.

Since the first s-order vibration modes are needed to calculate, usually taking more initial presumed modes in the calculation, for example, the first p-order ($p > s$) presumed modes, the iterative calculations are performed until the first s vibration modes meet the required accuracy. The increased p–s-order presumed modes are to speed up the convergence of the first s-order vibration modes. Of course, this will also increase the computation in each iteration. We must weigh the gain and loss and select a reasonable value for p. Reference [3] suggests taking the small one of $p = 2s$ and $p = s + 8$. Reference [4] suggests taking $p = \sqrt{b}$ according to the actual computing experiences, where b is the bandwidth of the stiffness matrix.

Example 1 Subspace iteration method for solving the first two-order eigenpairs.

$$[K] = k\begin{bmatrix} 2 & -1 & 0 & 0 \\ -1 & 2 & -1 & 0 \\ 0 & -1 & 2 & -1 \\ 0 & 0 & -1 & 1 \end{bmatrix}, \quad [f] = \frac{1}{k}\begin{bmatrix} 1 & 1 & 1 & 1 \\ 1 & 2 & 2 & 2 \\ 1 & 2 & 3 & 3 \\ 1 & 2 & 3 & 4 \end{bmatrix}$$

$$[M] = m\begin{bmatrix} 1 & 0 & 0 & 0 \\ 0 & 1 & 0 & 0 \\ 0 & 0 & 1 & 0 \\ 0 & 0 & 0 & 1 \end{bmatrix}, \quad [f][M] = \frac{m}{k}\begin{bmatrix} 1 & 1 & 1 & 1 \\ 1 & 2 & 2 & 2 \\ 1 & 2 & 3 & 3 \\ 1 & 2 & 3 & 4 \end{bmatrix}$$

Taking the zero-order approximation of the first two-order modal matrix,

$$[\phi^{(0)}] = [\psi^{(0)}] = \begin{bmatrix} 0.25 & 0.50 & 0.75 & 1 \\ 1 & 1 & 0 & -0.90 \end{bmatrix}^{\mathrm{T}}$$

From Eq. (e), we obtain

$$[\tilde{\psi}^{(1)}] = [f][M][\psi^{(0)}] = \frac{m}{k}\begin{bmatrix} 2.5 & 4.75 & 6.5 & 7.5 \\ 1.1 & 1.2 & 0.3 & -0.6 \end{bmatrix}^{\mathrm{T}}$$

Letting the maximum norm of the displacement of each mode in $[\tilde{\psi}^{(0)}]$ be 1, we obtain

$$[\psi^{(1)}] = \begin{bmatrix} 0.3333333 & 0.9166667 \\ 0.6333333 & 1.0000000 \\ 0.8666667 & 0.2500000 \\ 1.0000000 & -0.5000000 \end{bmatrix}$$

From Eqs (h) and (i), we calculate the generalized mass matrix and stiffness matrix as

$$[\overline{M}^{(1)}] = [\psi^{(1)}]^{\mathrm{T}}[M][\psi^{(1)}] = m\begin{bmatrix} 2.2633333 & 0.6555556 \\ 0.6555556 & 2.1527778 \end{bmatrix}$$

$$[\overline{K}^{(1)}] = [\psi^{(1)}]^{\mathrm{T}}[K][\psi^{(1)}] = k\begin{bmatrix} 0.2733333 & 0.0555556 \\ 0.0555556 & 1.9722222 \end{bmatrix}$$

From Eq. (j), we obtain the Ritz eigenvalue problem $[\overline{K}^{(1)}]\{a\} = \omega^2[\overline{M}^{(1)}]\{a\}$ as follows:

$$\begin{bmatrix} 0.2733333 - 2.2633333\beta & 0.0555556 - 0.6555556\beta \\ 0.0555556 - 0.6555556\beta & 1.9722222 - 2.1527778\beta \end{bmatrix}\begin{Bmatrix} a_1 \\ a_2 \end{Bmatrix} = \begin{Bmatrix} 0 \\ 0 \end{Bmatrix}$$

where

$$\beta = m\omega^2/k$$

The frequency equation is obtained by the determinant $|[\overline{K}^{(1)}] - \omega^2[\overline{M}^{(1)}]| = 0$ as follows

$$4.4427006\beta^2 - 4.97938271\beta - 0.535987653 = 0$$

Solving it results in

$$\beta_1 = 0.120623117, \quad \beta_2 = 1.000177752$$

Substituting $\beta = \beta_1, \omega_1^2 = \beta_1 k/m$ into $([\overline{K}^{(1)}] - \omega_1^2[\overline{M}^{(1)}])\{a\} = \{0\}$, we get

$$a_2 = 0.0137374a_1$$

From $\omega_2^2 = \beta_2 k/m$, we get $a_1 = -0.3015051a_2$ and

$$[a^{(1)}] = \left[\{a_1^{(1)}\}\ \{a_2^{(1)}\}\right] = \begin{bmatrix} 1.0000000 & -0.3015051 \\ 0.0137374 & 1.0000000 \end{bmatrix}$$

From Equation (l), the approximate value of the first and second mode of natural vibration are

$$[\phi^{(1)}] = [\psi^{(1)}][a^{(1)}] = \begin{bmatrix} 0.3459259 & 0.8161650 \\ 0.6470707 & 0.8090468 \\ 0.8701010 & -0.0113044 \\ 0.9931313 & -0.8015051 \end{bmatrix}$$

Taking the maximum norm of the vectors as 1, the normalized modes of vibration are

$$[\phi^{(1)}] = \begin{bmatrix} 0.3483183 & 1.0000001 \\ 0.6515459 & 0.9912784 \\ 0.8761186 & -0.0138505 \\ 1.0000000 & -0.9820379 \end{bmatrix}$$

Now we consider the second iterations

$$[\tilde{\psi}^{(2)}] = [f][M][\phi^{(1)}] = \frac{m}{k}\begin{bmatrix} 2.8759828 & 0.9953900 \\ 5.4036473 & 0.9907800 \\ 7.2797659 & -0.0051080 \\ 8.2797659 & -0.9871459 \end{bmatrix}$$

After normalization, we get

$$[\tilde{\psi}^{(2)}] = \frac{m}{k}\begin{bmatrix} 0.3473507 & 1.0000000 \\ 0.6526328 & 0.9953686 \\ 0.8792236 & -0.0051316 \\ 1.0000000 & -0.9917177 \end{bmatrix}$$

The second approximations of the generalized mass matrix and stiffness matrix are

$$[\overline{M}^{(2)}] = [\psi^{(2)}]^{\mathrm{T}}[M][\psi^{(2)}] = m\begin{bmatrix} 2.3196125 & 0.0007313 \\ 0.0007313 & 2.9742888 \end{bmatrix}$$

$$[\overline{K}^{(2)}] = [\psi^{(2)}]^{\mathrm{T}}[K][\psi^{(2)}] = k\begin{bmatrix} 0.2797799 & 0.0000765 \\ 0.0000765 & 2.9743741 \end{bmatrix}$$

The Ritz eigenvalue problem $([\overline{K}^{(2)}] - \omega^2[\overline{M}^{(2)}])\{a\} = \{0\}$ is obtained as follows

$$\begin{bmatrix} 0.2797799 - 2.3196125\beta & 0.0000765 - 0.0007313\beta \\ 0.0000765 - 0.0007313\beta & 2.9743741 - 2.9742888\beta \end{bmatrix}\begin{Bmatrix} a_1 \\ a_2 \end{Bmatrix} = \begin{Bmatrix} 0 \\ 0 \end{Bmatrix}$$

The frequency equation is

$$6.8991969\beta^2 - 7.7315414\beta + 0.8321700 = 0$$

Solving it, we get

$$\beta_1 = 0.1206149, \quad \beta_2 = 1.0000287$$

and we get

$$[a^{(2)}] = \begin{bmatrix} 1 & -0.0003 \\ 0.0001 & 1 \end{bmatrix}$$

So, the second approximation of the first two-order vibration modal matrix is obtained as

$$[\phi^{(2)}] = [\psi^{(2)}][a^{(2)}] = \begin{bmatrix} 0.3474507 & 0.9998958 \\ 0.6527323 & 0.9951728 \\ 0.8792231 & -0.0053954 \\ 0.9999008 & -0.9920177 \end{bmatrix}$$

After normalization, we get

$$[\phi^{(2)}] = \begin{bmatrix} 0.3474852 & 1.000000 \\ 0.6527970 & 0.9952765 \\ 0.8793102 & -0.0053959 \\ 1.00000 & -0.9921210 \end{bmatrix}$$

The second approximations of the squares of the first two-order natural frequencies are $\omega_1^2 = \beta_1 k/m = 0.1206149k/m$, $\omega_2^2 = \beta_2 k/m = 1.0000287k/m$.

In this example the exact solutions of the first two-order frequencies are $\omega_1^2 = 0.12061k/m$, $\omega_2^2 = 1.00000k/m$.

22.9.2 Modified Subspace Iteration Method

When the volume of the eigenpairs calculated is large, the computation of the subspace iteration method is relatively large. Reference [4] suggests to improve the subspace iteration method by shifting.

According to a lot of experiences of calculation, Ref. [4] suggests the best width for the subspace is

$$p = \max[4, \sqrt{b}] \tag{s}$$

Here p is the width of subspace, that is, the number of initial vectors $\{\psi^{(0)}\}$ taken in the subspace iteration method; b is the bandwidth of the stiffness matrix.

Assuming that s characteristic vectors need to be calculated, if $s \le \sqrt{b}/2$, conventional subspace iteration method can be used for calculating. Conversely, if $s > \sqrt{b}/2$, the characteristic vectors to be calculated can be divided into several groups. Each group contains about p characteristic vectors. From the second group, shifting firstly is performed, and then iterative calculations are performed near the shifted point to determine the characteristic vectors of this group.

The value of shifting D is calculated by the following equation:

$$D = R_n + 0.9(R^*_{n+1} - R_n) \tag{22.80a}$$

If $R^*_{n+1} - R_n < 0.01s$, taking

$$D = 0.99R_n \tag{22.80b}$$

$$R_n = \omega_n^2$$

Here R_n is the last eigenvalue calculated; R^*_{n+1} is the estimate value of the next eigenvalue to be calculated.

The number of iterations for each shifting is calculated by the following formula:

$$I = \max\left[2, \quad \frac{nb^2/2}{3np^2 + 2npb + 10p^3}\right] \tag{22.81}$$

22.10 Ritz Vector Superposition Method for Solving Forced Vibration of Structure

In the mode superposition method for solving forced vibrations of structures, the orthogonality of vibration modes is used to make the problem decoupled, so the vibration mode is completely independent of the distribution of load on structure. For some specific loads, some vibration modes may contribute small, even nothing, but in the analysis of vibration modes, they will still be calculated, which is a waste of time of the computer. Wilson *et al.* [5] proposed the Ritz vector superposition method, which produces a set of orthogonal Ritz vectors related to the spatial distribution of loads, and then use them to solve the forced vibration of structures. Experiences show that it often has a better computational efficiency than the subspace iteration method.

Let the equation of motion be

$$[M]\{\ddot{\delta}\} + [C]\{\dot{\delta}\} + [K]\{\delta\} = \{S\}r(t) \tag{22.82}$$

Here $r(t)$ is function of time t, independent of the position of the node; $\{S\}$ is the load vector, providing the space distribution of dynamic load.

The first base vector of Ritz vector superposition method is the static response due to load $\{S\}$. The next few Ritz vectors are to reflect the impact of dynamic inertia force. Then, the final group of Ritz vectors orthogonal to each other is obtained through one subspace iteration. They can be used to solve equations of motion (22.82).

The calculation steps of Ritz vector superposition method are as follows.

Step 1 Generating p ($p \ll n$) M-Orthogonal Base Vectors

1) Generating the first base vector $\{\psi\}_1$:

$$\{\overline{\psi}\}_1 = [K]^{-1}\{S\} \tag{22.83}$$

Regularizing

$$\{\psi\}_1 = \{\overline{\psi}\}_1/(\{\overline{\psi}\}_1[M]\{\overline{\psi}\}_1)^{1/2}$$

2) Generating $\{\psi\}_2$–$\{\psi\}_p$:

$$\{\overline{\psi}^*\}_{i+1} = [K]^{-1}[M]\{\psi\}_i \tag{a}$$

$$\beta_i = \{\overline{\psi}^*\}_{i+1}[M]\{\psi\}_i \tag{b}$$

M-orthogonalization

$$\{\overline{\psi}\}_{i+1} = \{\overline{\psi}^*\}_{i+1} - \sum_{j=1}^{i} \beta_j\{\psi\}_j \tag{22.84}$$

Regularizing

$$\{\psi\}_{i+1} = \{\overline{\psi}\}_{i+1}/(\{\overline{\psi}\}_{i+1}[M]\{\overline{\psi}\}_{i+1})^{1/2}$$

Let

$$[\psi] = \left[\{\psi\}_1 \ \{\psi\}_2 \ \cdots \ \{\psi\}_p\right]$$

Step 2 Generating the Vector Orthogonal to [*K*]

Let

$$[\overline{K}] = [\psi]^{\mathrm{T}}[K][\psi] \tag{c}$$

$$[\overline{M}] = [\psi]^{\mathrm{T}}[M][\psi] \tag{d}$$

Solving

$$[\overline{K}]\{a\} = \omega^2[\overline{M}]\{a\} \tag{22.85a}$$

or

$$\lambda\{a\} = [K]^{-1}[\overline{M}]\{a\} \tag{22.85b}$$

and we get λ_i and $\{a\}_i$, $i = 1,2, \ldots, p$.

The first p-order characteristic vectors of structure are

$$\{\phi\}_i = [\psi]\{a\}_i \quad (i = 1,2,\ldots,p) \tag{22.86}$$

Step 3 Solving Equations of Motion (22.82)

Let

$$\{\delta\} = \sum_{i=1}^{p} \{\phi\}_i \eta_i(t) \tag{22.87}$$

Substituting into Eq. (22.82), we obtain

$$\ddot{\eta}_i + 2\zeta_i\omega_i\dot{\eta}_i + \omega_i^2\eta_i = \{\phi\}_i^{\mathrm{T}}\{S\}r(t) \tag{22.88}$$

Here since the $\{\phi\}_i$ is already regularized, so $m_{pi} = 1$. η_i is obtained from the previous equation, and substituted into Eq. (22.87), to get the solution of the problem.

The first p-order eigenvectors obtained by the Ritz vector superposition method have automatically excluded the eigenvectors that do not work, so it improves the efficiency of calculation. If the structure is under the dynamic loads from more than one directions at the same time, for example, from the directions of x, y, z, then

$$\{P(t)\} = \{S_x\}r_1(t) + \{S_y\}r_2(t) + \{S_z\}r_3(t)$$

In this case, there are two kinds of calculation methods: the first method is to compute each group of eigenvectors for $\{S_x\}$, $\{S_y\}$, $\{S_z\}$, respectively; the second method [6] is to take in formula (22.83):

$$\{S\} = \{S_x\} + \{S_y\} + \{S_z\} \tag{e}$$

It is necessary to compute only one group of eigenvectors that will consider the dynamic loads in three directions. Obviously, the computational efficiency of second method is higher.

In the calculation of the dynamic responses of structures under earthquake, if there are seismic loads from two directions at the same time, we can take in Eq. (22.83):

$$\{S\} = \{M_x\} + \{M_y\} \tag{f}$$

If there are seismic loads from three directions at the same time, we can take

$$\{S\} = \{M_x\} + \{M_y\} + \{M_z\} \tag{g}$$

Here $\{M_x\}$, $\{M_y\}$, $\{M_z\}$ are seen in Eq. (22.59).

22.11 Modified Ritz Vector Superposition Method

As mentioned before, in the Ritz vector superposition method, some frequencies may leak, such as the vibration modes independent of given loads. In order to prevent the leakage of frequency, Ref. [7] suggests that the first step in the algorithm described in the last segment is amended as follows.

1) Generating the first base vector $\{\psi\}_1$:

$$\{\overline{\psi}\}_1 = [K]^{-1}\{S\} \tag{a}$$

$$\{y\} = [K]^{-1}[M]\{\overline{\psi}\}_1 \tag{b}$$

Do m cycles to Eqs (a), (b), and regularize finally

$$\{\psi\}_1 = \{\overline{\psi}\}_1/(\{\overline{\psi}\}_1^{\mathrm{T}}[M]\{\overline{\psi}\}_1)^{1/2} \tag{c}$$

2) Generating the second to the pth base vectors:
For $i = 1, 2, \ldots, p-1$, there are

$$\{y^*\}_{i+1} = [K]^{-1}[M]\{\psi\}_i \tag{d}$$

$$\beta_i = \{y^*\}_{i+1}[M]\{\psi\}_i \tag{e}$$

$$\{y\}_{i+1} = \{y^*\}_{i+1} - \sum_{j=1}^{i} \beta_j\{\psi\}_j \tag{f}$$

$$\{\psi\}_{i+1} = \{y\}_{i+1}/(\{y\}_{i+1}[M]\{y_{i+1}\})^{1/2} \tag{g}$$

$$\{\psi\}_{i+1} \Rightarrow \{\psi\}_i \tag{h}$$

Equation (h) assigns $\{\psi\}_{i+1}$ to $\{\psi\}_i$.

For each i, do m cyclic iterations of Eqs (d)–(h). But in the last iteration, Eq. (*h*) does not work.

Steps 2 and 3 of original algorithm are unchanged.

It can be proved that [7] after such modifications, the results obtained will be the first p natural vibration frequencies and modes of structure and no frequency and mode of vibration for $i \leq p$ will leak. But the calculating speed is very fast, because there are no iterations.

The author and Rao Bin applied the Ritz vector method to seismic stress analysis of arch dam; the calculating speed is 4–12 times faster than the subspace iteration method [8]. The more the modes taken in calculation, the greater the gap of the computational efficiency, because in the Ritz vector method, there are no iterations.

22.12 Dynamic Substructure Method

Combining the Ritz vector method with substructure method, the dynamic substructure method [7, 9] will be formed, which is favorable to the dynamic analysis of large structures on a microcomputer.

1) Solving $[K]\{\overline{\psi}\}_1$,

$$[K]\{\overline{\psi}\}_1 = \{S\} \tag{22.89}$$

As mentioned before, this step is static calculation – it can be solved by static substructure method. The original structure is divided into m substructures, and the ith substructure is divided into blocks as follows ($i = 1, 2, \ldots, m$):

$$\begin{bmatrix} K^i_{\text{cc}} & K^i_{\text{cb}} \\ K^i_{\text{bc}} & K^i_{\text{bb}} \end{bmatrix} \begin{Bmatrix} \overline{\psi}^i_{1\text{c}} \\ \overline{\psi}^i_{1\text{b}} \end{Bmatrix} = \begin{Bmatrix} S^i_{\text{c}} \\ S^i_{\text{b}} \end{Bmatrix} \tag{a}$$

Here the subscript "c" represents the interior of substructures, the subscript "b" represents the boundaries of substructures, $\{\overline{\psi}^i_{1\text{c}}\}$ is the degrees of freedom of the nodes in the interior of substructures, and $\{\overline{\psi}^i_{1\text{b}}\}$ is the degrees of freedom of the nodes on the boundaries of substructures.

According to the static condensation theory, from Eq. (a), we obtain

$$[\overline{K}^i_{\text{b}}]\{\overline{\psi}^i_{1\text{b}}\} = \{S^i_{\text{b}}\} - \{R^i_{\text{b}}\} \tag{b}$$

where

$$[\overline{K}^i_{\text{b}}] = [K^i_{\text{bb}}] - [K^i_{\text{bc}}][K^i_{\text{cc}}]^{-1}[K^i_{\text{cb}}] \tag{c}$$

$$\{R^i_{\text{b}}\} = [K^i_{\text{bc}}][K^i_{\text{cc}}]^{-1}\{S^i_{\text{c}}\} \tag{d}$$

All the substructures are assembled, and we obtain

$$[\overline{K}_{\text{b}}]\{\overline{\psi}_{1\text{b}}\} = \{S_{\text{b}}\} - \{R_{\text{b}}\} \tag{22.90}$$

where

$$[\overline{K}_{\text{b}}] = \sum_{i=1}^{m}[\overline{K}^i_{\text{b}}], \quad \{S_{\text{b}}\} = \sum_{i=1}^{m}[S^i_{\text{b}}], \quad \{R_{\text{b}}\} = \sum_{i=1}^{m}[R^i_{\text{b}}] \tag{22.91}$$

The degrees of freedom of the nodes on the boundaries of each substructure $\{\overline{\psi}^i_{1\text{b}}\}$ can be solved by Eq. (22.90), and the degrees of freedom of the nodes in each substructure $\{\overline{\psi}^i_{1\text{c}}\}$ can be solved by the following equation:

$$\{\overline{\psi}^i_{1\text{c}}\} = [K^i_{\text{cc}}]^{-1}(\{S^i_{\text{c}}\} - [K^i_{\text{cb}}]\{\overline{\psi}^i_{1\text{b}}\}) \tag{22.92}$$

So, $\{\overline{\psi}\}_1$ on all of the nodes are solved.

2) Solving $[K]\{\overline{\psi}^*\}_{j+1}$,

$$[K]\{\overline{\psi}^*\}_{j+1} = [M]\{\psi\}_j \tag{22.93}$$

Let

$$\{P\} = [M]\{\psi\}_j \tag{22.94}$$

$\{P\}$ is regarded as equivalent load vector, instead of $\{S\}$ on the right side of Eq. (22.89). Replacing $\{\overline{\psi}\}_1$ on the left side of Eq. (22.89) with $\{\overline{\psi}^*\}_{j+1}$, following the approach of the previous section, we can get the analysis formula of substructure.

After solving all the regularization vectors, let

$$[\psi] = \left[\{\psi\}_1 \ \{\psi\}_2 \ \cdots \ \{\psi\}_p\right] \tag{e}$$

3) Subspace iteration

Let

$$[\overline{K}] = [\psi]^{\mathrm{T}}[K][\psi] \tag{f}$$

$$[\overline{M}] = [\psi]^{\mathrm{T}}[M][\psi] \tag{g}$$

Obtain the equation

$$[\overline{K}]\{a\} = \omega^2[\overline{M}]\{a\} \tag{h}$$

ω_i^2 and $\{a\}_i$ ($i = 1, 2, \ldots, p$) are solved by Eq. (h). The first p-order eigenvectors of structure are

$$\{\phi\}_i = [\psi]\{a\}_i \quad (i = 1,2,\ldots,p) \tag{i}$$

Analysis of dynamic substructure can also use the modified Ritz vector superposition method, only when generating the basis vectors $\{\psi\}_i$, as described in Section 22.11, some corresponding iterative steps are increased.

Following the analysis of static multiple substructures, the analysis of multiple dynamic substructures can also be performed.

For the analysis of dynamic substructures, Guyan method was used in the past, which uses static condensation to each substructure; due to ignoring the effect of dynamic inertia force, the computational accuracy is poor. Kuhar method considers the inertial force ignored previously, but, because of iterations, the computational efficiency is poor. The dynamic substructure method described in this section is based on the Ritz vector method, with better computational accuracy and efficiency. The method does not make any approximation in the process of dividing the whole structure into substructure; in this sense, it is an accurate method.

Example 1 As shown in Figure 22.5, a straight bar with uniform cross-section fixed at both ends, whose length is L, total mass is M, cross section is F, and the modulus of elasticity is E, calculate its longitudinal free vibration. It is divided into 14 equal parts, where each short rod can be regarded as a spring, the stiffness coefficient is $k = EF/l = 14EF/L$, and the mass of each segment is $m = M/14$ and divided into four substructures, as shown in Figure 22.6(b).

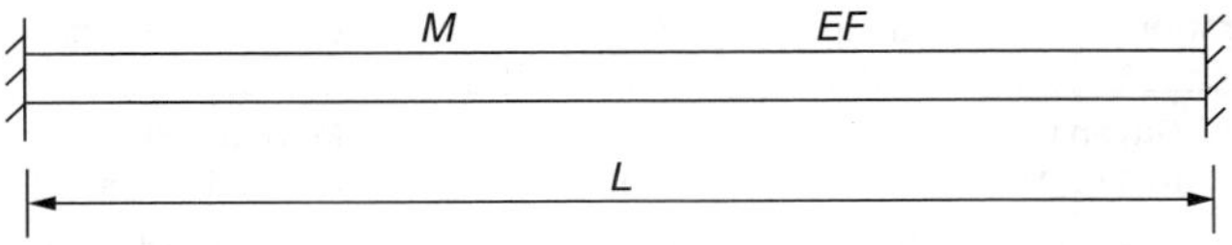

Figure 22.5 An example.

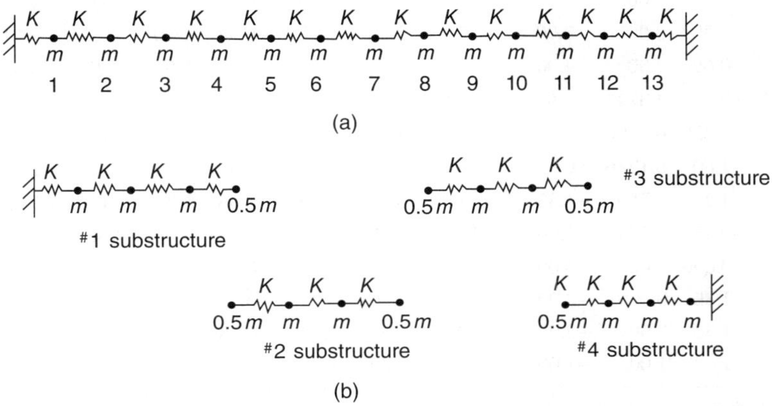

Figure 22.6 An example. (a) Divisions of elements and (b) divisions of substructures.

Table 22.3 Frequencies calculated by various dynamic substructure methods.

Frequency method	$\omega_1\sqrt{\frac{LM}{EF}}$	$\omega_2\sqrt{\frac{LM}{EF}}$	$\omega_3\sqrt{\frac{LM}{EF}}$	**Computational error**		
				ω_1	ω_2	ω_3
Exact solution	3.1416	6.2832	9.4248	0	0	0
Guyan method in Ref. [10]	3.1960	6.8580	11.5030	1.7%	9.1%	22%
Ritz method (adding symmetric load)	3.1350		9.2480	−0.2%		−1.8%
Ritz method (adding antisymmetric load)		6.231			−0.8%	
Improved Ritz method (two iterations)	3.1350	6.3310	9.2510	−0.2%	−0.8%	−1.8%

Reference [7] conducted a substructure calculation with both Ritz vector method and modified Ritz vector method. The results of frequency calculation are shown in Table 22.3; the results of vibration modal calculation are shown in Table 22.4. The following points can be seen from the results:

1) From the results through Guyan method in Ref. [10], only the first-order frequencies and modes can meet the requirements of engineering calculations, and the computational accuracy of second- and third-order frequencies and vibration modes are poor.

Table 22.4 Comparison of the vibration modes calculated by various dynamic substructure methods.

	Exact solution			Guyan method in Ref. [10]			Ritz method[a] in Ref. [7]			Modified Ritz method in Ref. [7]		
Node	First	Second	Third	First	Second	Third	First	Second	Third	First	Second	Third
1	1.00	1.00	1.00	1.00	1.00	1.00	1.00	1.00	1.00	1.00	1.00	1.00
2	1.95	1.802	1.56	2.00	2.00	2.00	1.95	1.80	0.08	1.95	1.80	1.56
3	2.80	2.25	1.35	3.00	3.00	3.30	2.80	2.25	−0.58	2.80	2.25	1.44
4	3.51	2.25	0.696	4.00	4.00	4.00	3.51	2.25	−0.36	3.51	2.25	0.69
5	4.05	1.802	−0.357	4.3744	2.6667	0.4403	4.05	1.80	0.35	4.05	1.80	−0.36
6	4.38	1.00	−1.250	4.7490	1.3333	−3.1195	4.38	1.00	0.58	4.38	1.00	−1.25
7	4.49	0	−1.60	5.1236	0	−6.6795	4.49	0	−0.98	4.49	0	−1.95
8	4.38	−1.00	−1.25	4.7492	−1.3333	−3.1195	4.38	−1.00	0.58	4.38	−1.00	−1.25
9	4.05	−1.82	−0.357	4.3744	−2.6667	0.4403	4.05	−1.80	0.34	4.05	−1.80	−0.36
10	3.51	−2.25	0.696	4.00	−4.00	4.00	3.51	−2.25	−0.36	3.51	−2.25	0.69
11	2.80	−2.25	1.35	3.00	−3.00	3.00	2.80	−2.25	−0.58	2.80	−2.25	1.44
12	1.95	−1.802	1.56	2.00	−2.00	2.00	1.95	−1.80	0.08[a]	1.95	−1.80	1.56
13	1.00	−1.00	1.00	1.00	−1.00	1.00	1.00	−1.00	1.00	1.00	−1.00	1.00

a) The first and third vibration modes are obtained by Ritz method with symmetric load, and the second vibration mode is obtained by Ritz method with antisymmetric load. The accuracy of third vibration mode is poor as seen from the calculation.

2) When calculating substructures with Ritz vector method, if symmetric load is applied, symmetric vibration mode is obtained, as ω_1, $\{\phi_1\}$ and ω_3, $\{\phi_3\}$. If antisymmetric load is applied, antisymmetric vibration mode is obtained, as ω_2, $\{\phi_2\}$. The accuracy of three frequencies and the first two vibration modes are better, but the accuracy of the third vibration mode is poor.
3) When analyzing with modified Ritz vector method, the accuracy of third-order natural frequencies and vibration modes are good, there are only two iterations, and the computational efficiency is high, without leaking frequency.

22.13 Direct Integration Method for Solving the Equation of Motion

Equation of motion (22.2) is a two-order ordinary differential equation that can be solved by the numerical integration method. There are many direct integration methods applied to dynamic problems. Now we introduce several effective methods that can be used in the dynamic analysis of large structure by finite element method.

When the initial displacement $\{\delta_0\}$, velocity $\{\dot{\delta}_0\}$, and acceleration $\{\ddot{\delta}_0\}$ are given, it is required to solve the equation of motion as follows:

$$[M]\{\ddot{\delta}\} + [C]\{\dot{\delta}\} + [K]\{\delta\} = \{P(t)\}$$

Dividing t into a series of Δt, the approximate solutions at time 0, Δt, $2\Delta t$, $3\Delta t$, …, t, $t + \Delta t$, … are calculated by numerical method. The main difference between the different calculation methods is how to assume the variation of displacement, velocity, and acceleration at Δt.

22.13.1 Linear Acceleration Method

As shown in Figure 22.7(a), assuming that acceleration components of each node in the period Δt are varied linearly with time t as

$$\ddot{\delta}(t) = \ddot{\delta}(t_1) + \frac{\ddot{\delta}(t_2) - \ddot{\delta}(t_1)}{t_2 - t_1}(t - t_1) \tag{a}$$

So the speeds and displacements of the nodes at time t can be obtained by two integrals as follows:

$$\dot{\delta}(t) = \dot{\delta}(t_1) + \int_{t_1}^{t} \ddot{\delta}(t)\,\mathrm{d}t \tag{b}$$

$$\delta(t) = \delta(t_1) + \int_{t_1}^{t} \dot{\delta}(t)\,\mathrm{d}t \tag{c}$$

Substituting Eq. (a) into the previous Eqs (b) and (c) and letting $t = t_2$, we obtain

$$\dot{\delta}(t) = \dot{\delta}(t_1) + \frac{\Delta t}{2}\ddot{\delta}(t_1) + \frac{\Delta t}{2}\ddot{\delta}(t) \tag{d}$$

$$\delta(t) = \delta(t_1) + \Delta t\dot{\delta}(t_1) + \frac{\Delta t^2}{3}\ddot{\delta}(t_1) + \frac{\Delta t^2}{6}\ddot{\delta}(t) \tag{e}$$

From Eq. (e), we obtain

$$\ddot{\delta}(t) = \frac{6}{\Delta t^2}\delta(t) - \frac{6}{\Delta t^2}\delta(t_1) - \frac{6}{\Delta t}\dot{\delta}(t_1) - 2\ddot{\delta}(t_1) \tag{f}$$

Substituting Eq. (f) into Eq. (d), we obtain

$$\dot{\delta}(t) = \frac{3}{\Delta t}\delta(t) - \frac{3}{\Delta t}\delta(t_1) - 2\dot{\delta}(t_1) - \frac{\Delta t}{2}\ddot{\delta}(t_1) \tag{g}$$

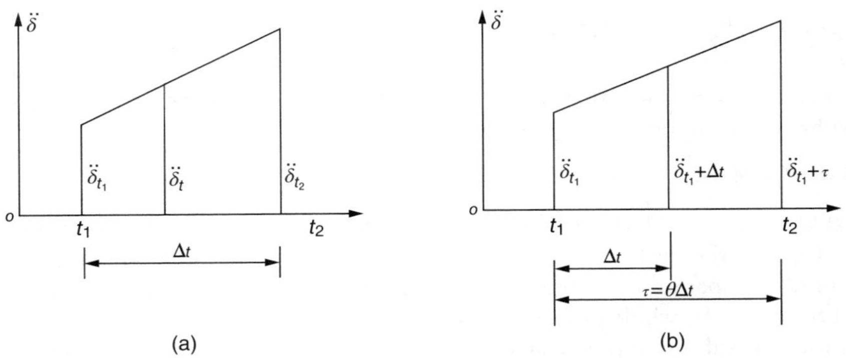

Figure 22.7 Linear acceleration method and θ method. (a) Linear acceleration method and (b) θ method.

Substituting Eqs (f) and (g) into equation of motion (22.2) and denoting $\delta(t)$ and $\delta(t_1)$ by δ_t and δ_{t_1}, we obtain

$$
\begin{aligned}
&[M]\left\{\frac{6}{\Delta t^2}\delta_t-\frac{6}{\Delta t^2}\delta_{t_1}-\frac{6}{\Delta t}\dot{\delta}_{t_1}-2\ddot{\delta}_{t_1}\right\}\\
&+(\alpha[M]+\beta[K])\left\{\frac{3}{\Delta t}\delta_t-\frac{3}{\Delta t}\delta_{t_1}-2\dot{\delta}_{t_1}-\frac{\Delta t}{2}\ddot{\delta}_{t_1}\right\}\\
&+[K]\{\delta_t\}=\{P_t\}
\end{aligned}
$$

After further simplification, we get the basic equation of dynamic analysis as follows:

$$[K_t]\{\overline{\delta}_t\}=\{\overline{P}_t\} \tag{h}$$

where

$$[K_t]=[K]+a_0[M] \tag{i}$$

$$\{\overline{P}_t\}=\{P_t\}+[M]\{a_1\delta_{t_1}+a_2\dot{\delta}_{t_1}+a_3\ddot{\delta}_{t_1}\} \tag{j}$$

$$\overline{\delta}_t=\left(\frac{3\beta}{\Delta t}+1\right)\delta_t-\frac{3\beta}{\Delta t}\delta_{t_1}-2\beta\dot{\delta}_{t_1}-\frac{\Delta t}{2}\beta\ddot{\delta}_{t_1} \tag{k}$$

or

$$\delta_t=\frac{\Delta t}{3\beta+\Delta t}\overline{\delta}_t+\frac{3\beta}{3\beta+\Delta t}\delta_{t_1}+\frac{2\beta}{3\beta+\Delta t}\dot{\delta}_{t_1}+\frac{\beta\Delta t^2}{6\beta+2\Delta t}\ddot{\delta}_{t_1} \tag{l}$$

Substituting Eq. (l) into Eqs (f) and (g), we obtain

$$\ddot{\delta}_t=a_4\overline{\delta}_t+a_5\delta_{t_1}+a_6\dot{\delta}_{t_1}+a_7\ddot{\delta}_{t_1} \tag{m}$$

$$\dot{\delta}_t=\dot{\delta}_{t_1}+\frac{\Delta t}{2}[a_4\overline{\delta}_t+a_5\delta_{t_1}+a_6\dot{\delta}_{t_1}+(1+a_7)\ddot{\delta}_{t_1}] \tag{n}$$

Coefficients $a_0, a_1, \ldots$ are calculated by the following equations:

$$
\left.
\begin{aligned}
&a_0=\frac{6+3\alpha\Delta t}{\Delta t^2+3\beta\Delta t},\quad a_1=\frac{6}{\Delta t^2}+\frac{3}{\Delta t}(\alpha-\beta a_0)\\
&a_2=\frac{6}{\Delta t}+2(\alpha-\beta a_0),\quad a_3=2+\frac{\Delta t}{2}(\alpha-\beta a_0)\\
&a_4=\frac{6}{3\beta\Delta t+\Delta t^2},\quad a_5=\frac{3\beta}{\Delta t}a_4-\frac{6}{\Delta t^2}\\
&a_6=2\beta a_4-\frac{6}{\Delta t},\quad a_7=\frac{\beta\Delta t}{2}a_4-2
\end{aligned}
\right\} \tag{o}
$$

The previous equations show that the term $\overline{P}_t$ on the right side of basic equation (h) is known; solving it we obtain

$$\{\overline{\delta}_t\}=[K_t]^{-1}\{\overline{P}_t\} \tag{p}$$

Substituting into Eqs (l), (m), (n), we calculate $\delta_t, \dot{\delta}_t, \ddot{\delta}_t$ at $t=t_2$. Substituting these values into Eq. (j), $\overline{P}_t$ at the next time can be calculated, and substituting into Eq. (p), the displacement of next period is obtained. Repeating such calculations step by step, it is not difficult to obtain the whole process of the dynamic response.

Equation (o) shows if the step Δt is fixed in the whole calculation process, a_0 is constant, and $[K_t]$ is also constant. Only once reverse of $[K_t]$ is required; the following calculation in each step is just simply back substitutions, so it is convenient to calculate.

The Taylor formula can prove that the truncation error of displacement in the previous method is four-order. This calculation method is stable when the time step is smaller than the minimum vibration period of discrete structure; otherwise, it is unstable. The magnitude of the minimum period is the time for the stress wave goes through 2 nodes, and it is extremely short. For example, for triangle mesh of constant strain plane problem, ignoring the effect of damping, the upper bound of the time step may be estimated from the following equation:

$$(\Delta t)^2 \leq \frac{\rho(1+\mu)}{E}\Delta x \Delta y$$

Here E is the modulus of elasticity, μ is Poisson's ratio, ρ is density, and Δx and Δy are minimum grid spacing.

For example, for the concrete structure, if we take $\Delta x = \Delta y = 1m$, the upper bound of the time step obtained by the previous equation is 0.0003 s.

According to the experience of calculation, when the time step is over the threshold, the calculation is not stable. At this time, although the high frequency components of dynamic load input is very small, the high frequency of the structural dynamic responses will grow infinitely, to make the calculation result meaningless. In order to eliminate the instability, we can use the following θ method.

22.13.2 Wilson Method (θ Method)

Assuming that the acceleration in the time interval $\tau = \theta\Delta t$ is linear, according to the reason before, it is known that

$$\dot{\delta}_{t_1+\tau} = \dot{\delta}_{t_1} + \frac{\tau}{2}[\ddot{\delta}_{t_1} + \ddot{\delta}_{t_1+\tau}]$$

$$\delta_{t_1+\tau} = \delta_{t_1} + \tau\dot{\delta}_{t_1}\frac{\tau^2}{6}[2\ddot{\delta}_{t_1} + \ddot{\delta}_{t_1+\tau}]$$

Simultaneous solution of the previous two equations yields the acceleration and velocity at time $t_1 + \tau$ as follows:

$$\left.\begin{aligned} \ddot{\delta}_{t_1+\tau} &= \frac{6}{\tau^2}[\delta_{t_1+\tau} - \delta_{t_1}] - \frac{6}{\tau}\dot{\delta}_{t_1} - 2\ddot{\delta}_{t_1} \\ \dot{\delta}_{t_1+\tau} &= \frac{3}{\tau}[\delta_{t_1+\tau} - \delta_{t_1}] - 2\dot{\delta}_{t_1} - \frac{\tau}{2}\ddot{\delta}_{t_1} \end{aligned}\right\} \tag{q}$$

Substituting Eq. (q) into equation of motion at time $t_1 + \tau$:

$$[M]\{\ddot{\delta}_{t_1+\tau}\} + [C]\{\dot{\delta}_{t_1+\tau}\} + [K]\{\delta_{t_1+\tau}\} = \{P_{t_1+\tau}\} \tag{r}$$

In the previous formula the terms without $\delta_{t_1+\tau}$ are moved to the right side, and we obtain the equation of motion as

$$[\overline{K}]\{\delta_{t_1+\tau}\} = \{\overline{P}_{t_1+\tau}\} \tag{22.95}$$

where

$$[\overline{K}] = [K] + \frac{3}{\tau}[C] + \frac{6}{\tau^2}[M]$$

$$\begin{aligned} \{\overline{P}_{t_1+\tau}\} = \{P_{t_1+\tau}\} &+ [M]\left(2\{\ddot{\delta}_{t_1}\} + \frac{6}{\tau}\{\dot{\delta}_{t_1}\} + \frac{6}{\tau^2}\{\delta_{t_1}\}\right) \\ &+ [C]\left(\frac{\tau}{2}\{\ddot{\delta}_{t_1}\} + 2\{\dot{\delta}_{t_1}\} + \frac{3}{\tau}\{\delta_{t_1}\}\right) \end{aligned}$$

Table 22.5 Calculation of the linear structure with θ method.

1. Original calculation
   (1) Calculate the following constants
   $$\tau = \theta\Delta t\ (\theta \geq 1.37)$$
   $$b_0 = \frac{6}{\tau^2}, b_1 = \frac{3}{\tau}, b_2 = 2b_1, b_3 = \frac{\tau}{2}, b_4 = \frac{b_0}{\theta}$$
   $$b_5 = -\frac{b_2}{\theta}, b_6 = 1 - \frac{3}{\theta}, b_7 = \frac{\Delta t}{2}, b_8 = \frac{\Delta t^2}{2}$$
   (2) Establish an effective stiffness matrix
   $$[\overline{K}] = [K] + b_1[C] + b_0[M]$$
   (3) Decompose $[\overline{K}]$
   $$[\overline{K}] = [L][D][L]^{\mathrm{T}}$$
2. Calculation of each step
   (1) Calculate the effective load array $\{\overline{P}_{t_1+\tau}\}$
   $$\{\overline{P}_{t_1+\tau}\} = \{P_{t_1+\tau}\} + [M](b_0\{\delta_{t_1}\} + b_2\{\dot{\delta}_{t_1}\} + 2\{\ddot{\delta}_{t_1}\}) + [C](b_1\{\delta_{t_1}\} + 2\{\dot{\delta}_{t_1}\} + b_3\{\ddot{\delta}_{t_1}\})$$
   (2) Solve $\{\delta_{t_1+\tau}\}$ by the equation of motion
   $$\{\delta_{t_1+\tau}\} = [\overline{K}]^{-1}\{\overline{P}_{t_1+\tau}\}$$
   (3) Calculate the acceleration, velocity, and displacement at $t_1 + \Delta t$
   $$\ddot{\delta}_{t_1+\Delta t} = b_4(\delta_{t_1+\Delta t} - \delta_{t_1}) + b_5\dot{\delta}_{t_1} + b_6\ddot{\delta}_{t_1}$$
   $$\dot{\delta}_{t_1+\Delta t} = \dot{\delta}_{t_1} + b_7(\ddot{\delta}_{t_1+\Delta t} + \ddot{\delta}_{t_1})$$
   $$\delta_{t_1+\Delta t} = \delta_{t_1} + \Delta t\dot{\delta}_{t_1} + b_8(\ddot{\delta}_{t_1+\Delta t} + 2\ddot{\delta}_{t_1})$$
   (4) Calculate the stress (if necessary)

From Eq. (22.95), the displacement vector $\delta_{t_1+\tau}$ is solved, and $\ddot{\delta}_{t_1+\tau}$ is computed by Eq. (q). Finally, the acceleration at time $t_1 + \Delta t$ will be obtained from Eq. (22.96):

$$\ddot{\delta}_{t_1+\Delta t} = \left(1 - \frac{1}{\theta}\right)\ddot{\delta}_{t_1} + \frac{1}{\theta}\ddot{\delta}_{t_1+\tau} \tag{22.96}$$

Substituting the previous accelerations into Eqs (d) and (e), we can obtain the velocity and displacement at time $t_1 + \Delta t$. The calculation steps are shown in Table 22.5.

If taking $\theta = 1$, we obtain the general linear acceleration method previously; as noted before, it is conditionally stable. If taking $\theta \geq 1.37$, the calculation is unconditionally stable.

The difference of the previous two kinds of calculation methods is that θ method is according to the equilibrium condition at $t + \theta\Delta t$, as Eq. (22.2), to solve the displacement of structure at $t + \Delta t$, while the ordinary linear acceleration method is according to the balance condition at $t + \Delta t$ to solve the displacement of structure.

22.13.3 Newmark Method

The Newmark method also can be regarded as the extension of linear acceleration method. Using the following assumptions:

$$\dot{\delta}_{t_1+\Delta t} = \dot{\delta}_{t_1} + [(1-\beta)\ddot{\delta}_{t_1} + \beta\ddot{\delta}_{t_1+\Delta t}]\Delta t \tag{s}$$

$$\delta_{t_1+\Delta t} = \delta_{t_1} + \dot{\delta}_{t_1}\Delta t + \left[\left(\frac{1}{2} - \alpha\right)\ddot{\delta}_{t_1} + \alpha\ddot{\delta}_{t_1+\Delta t}\right]\Delta t^2 \tag{t}$$

Table 22.6 Calculation procedure of Newmark method.

1. Original calculation

(1) Calculate the constants

$\alpha \geq 0.25(0.5+\beta)^2$ $(\beta \geq 0.50)$

$a_0 = \dfrac{1}{\alpha \Delta t^2}, a_1 = \dfrac{\beta}{\alpha \Delta t}, a_2 = \dfrac{1}{\alpha \Delta t}, a_3 = \dfrac{1}{2\alpha} - 1$

$a_4 = \dfrac{\beta}{\alpha} - 1, a_5 = \dfrac{\Delta t}{2}\left(\dfrac{\beta}{\alpha} - 2\right), a_6 = (1-\beta)\Delta t$

$a_7 = \beta \Delta t$

(2) Establish an effective stiffness matrix

$[\overline{K}] = [K] + a_0[M] + a_1[C]$

(3) Decompose $[\overline{K}]$

$[\overline{K}] = [L][D][L]^{\mathrm{T}}$

2. Calculation of each step

(1) Calculate the effective load array $\{\overline{P}_{t_1+\tau}\}$

$\{\overline{P}_{t_1+\tau}\} = \{P_{t_1+\tau}\} + [M](a_0\{\delta_{t_1+\Delta t}\} + a_2\{\dot{\delta}_{t_1}\} + a_3\{\ddot{\delta}_{t_1}\}) + [C](a_1\{\delta_{t_1}\} + a_4\{\dot{\delta}_{t_1}\} + a_5\{\ddot{\delta}_{t_1}\})$

(2) Solve $\{\delta_{t_1+\Delta t}\}$

$[L][D][L]^{\mathrm{T}}\{\delta_{t_1+\Delta t}\} = \{\overline{P}_{t_1+\Delta t}\}$

(3) Calculate the acceleration and velocity at $t_1 + \Delta t$

$\{\ddot{\delta}_{t_1+\Delta t}\} = a_0(\{\delta_{t_1+\Delta t}\} - \{\delta_{t_1}\}) - a_2\{\dot{\delta}_{t_1}\} - a_3\{\ddot{\delta}_{t_1}\}$

$\{\dot{\delta}_{t_1+\Delta t}\} = \{\dot{\delta}_{t_1}\} + a_6\{\ddot{\delta}_{t_1}\} + a_7\{\ddot{\delta}_{t_1+\Delta t}\}$

Here α, β are two parameters; if taking $\alpha = 1/6$, $\beta = 1/2$, we get linear acceleration method, and if taking $\alpha = 1/4$, $\beta = 1/2$, we get the average acceleration method (taking the average acceleration on Δt and calculating the velocity according to trapezoid formula). The calculation steps are shown in Table 22.6.

Simultaneous solution of Eqs (r) and (s) yields

$$\ddot{\delta}_{t_1+\Delta t} = \frac{1}{\alpha \Delta t^2}(\delta_{t_1+\Delta t} - \delta_{t_1}) - \frac{1}{\alpha \Delta t}\dot{\delta}_{t_1} - \left(\frac{1}{2\alpha} - 1\right)\ddot{\delta}_{t_1} \tag{u}$$

$$\dot{\delta}_{t_1+\Delta t} = \dot{\delta}_{t_1} + (1-\beta)\Delta t \ddot{\delta}_{t_1} + \beta \Delta t \ddot{\delta}_{t_1+\Delta t} \tag{v}$$

Substituting Eqs (u) and (v) into the equation of motion at the following $t_1 + \Delta t$ moment:

$$[M]\{\ddot{\delta}_{t_1+\Delta t}\} + [C]\{\dot{\delta}_{t_1+\Delta t}\} + [K]\{\delta_{t_1+\Delta t}\} = \{P_{t_1+\Delta t}\}$$

After simplification, we get

$$[\overline{K}]\{\delta_{t_1+\Delta t}\} = \{\overline{P}_{t_1+\Delta t}\} \tag{22.97}$$

where

$$[\overline{K}] = [K] + a_0[M] + a_1[C] \tag{22.98}$$

$$\{\overline{P}_{t_1+\Delta t}\} = \{P_{t_1+\Delta t}\} + [M](a_0\{\delta_{t_1}\} + a_2\{\dot{\delta}_{t_1}\} + a_3\{\ddot{\delta}_{t_1}\}) + [C](a_1\{\delta_{t_1}\} + a_4\{\dot{\delta}_{t_1}\} + a_5\{\ddot{\delta}_{t_1}\}) \tag{22.99}$$

and

$$\begin{aligned} a_0 &= \frac{1}{\alpha \Delta t^2}, \quad a_1 = \frac{\beta}{\alpha \Delta t}, \quad a_2 = \frac{1}{\alpha \Delta t} \\ a_3 &= \frac{1}{2\alpha} - 1, \quad a_4 = \frac{\beta}{\alpha} - 1, \quad a_5 = \frac{\Delta t}{2}\left(\frac{\beta}{\alpha} - 2\right) \\ a_6 &= (1-\beta)\Delta t, \quad a_7 = \beta \Delta t \\ \alpha &\geq 0.25(0.5+\beta)^2 \quad (\beta \geq 0.50) \end{aligned} \tag{22.100}$$

Comparing Table 22.5 with Table 22.6, it can be seen that θ method and Newmark method are similar in program structure; the only difference is in the calculation formulas of a_i and b_i. Therefore, we can use these two kinds of different calculation methods in the same program.

22.13.4 Calculation Stability, Precision, and the Selection of Time Step

Using direct integral method for dynamic analysis, the CPU time is inversely proportional to the time step Δt, so it is important to choose the value of Δt. To reduce computational cost, it is better to take the bigger Δt. But in order to ensure the appropriate precision, Δt cannot be too large; this involves two issues, namely, the computational stability and accuracy.

Our task is to solve the equation of motion as

$$[M]\{\ddot{\delta}\} + [C]\{\dot{\delta}\} + [K]\{\delta\} = \{P(t)\}$$

As described before, the displacement vector $\{\delta\}$ is represented by the linear superposition of mode $\{\phi\}_i$ as

$$\{\delta\} = \sum_{i=1}^{n} \{\phi\}_i \eta_i(t)$$

Substituted into Eq. (22.2), we obtain n independent two-order ordinary differential equations as follows:

$$\ddot{\eta}_i + 2\zeta_i \omega_i \dot{\eta}_i + \omega_i^2 \eta_i = p_i(t) \quad (i = 1,2,\ldots,n)$$

where

$$p_i(t) = \frac{1}{m_{pi}} \{\phi\}_i^{\mathrm{T}} \{P(t)\}$$

Here ζ_i is the damping ratio; ω_i is natural frequency; m_{pi} is seen in Eq. (22.25).

Equation (22.55) can be solved by the Duhamel integral, or by the linear acceleration method, θ method, or Newmark method mentioned before. Because the vibration period of structure $T_i = 2\pi/\omega_i$ is known, $i = 1 - n$, using vibration modal superposition method for each equation, we can choose the appropriate step Δt according to the corresponding natural vibration period T_i to ensure the necessary accuracy. If using unified Δt to solve all n equations in Eq. (22.55), then the mode superposition method is completely equivalent to direct integration method. The important conclusion is that in examining the accuracy of the direct integration method, we only need to analyze how to get the integral of Eq. (22.55) with unified Δt, to replace the analysis of the formula (22.2). In this case, when examining computational stability and accuracy of the direct integration method, the constants we need to consider are Δt, ω_i, and $\zeta_i (i = 1,2, \ldots, n)$, instead of elements of stiffness matrix, damping matrix, and mass matrix. Moreover, the n equations in Eq. (22.55) are independent of each other, each of which is equivalent to an equation of motion of the structure of single degree of freedom, so the problem is greatly simplified.

22.13.4.1 Computational Stability

Let $p_i(t) = 0$ in Eq. (22.55). We obtain free vibration equation of the structure of single degree of freedom. Under given initial conditions, if no damping, the amplitude should remain constant. If there is damping, the amplitude of displacement should gradually decay and finally becomes to zero. However, when using the numerical integration method for solving this equation, if the time step Δt is too large, the displacement does not decay, but may extend progressively. This phenomenon is called the instability of calculation. In addition, when using numerical methods, there is always rounding error in the calculation process. Because of the damping, the impact of a particular rounding error on calculation accuracy should be smaller with the extended time t. However, when Δt is too large, the impact of rounding error may be larger with the extension of the time t; this is a kind of computational instability. Therefore, the calculation stability has two layers of meaning: (1) the impact of initial conditions will not be gradually enlarged and (2) the impact of rounding error will not be gradually enlarged.

If for any Δt, especially when $\Delta t/T$ (natural period) is larger, the calculation is stable. This calculation method is called the unconditional stability. Conversely, if only when $\Delta t/T$ is less than a certain value, the calculation is stable; otherwise the calculation is unstable, this method is called the conditional stability.

The results of research [2] show that for θ method, when $\theta \geq 1.37$, the calculation is unconditionally stable. Usually $\theta = 1.40$ is taken in the calculation. Linear acceleration method is equivalent to $\theta = 1.0$; it is conditionally stable.

For Newmark method, when $\beta \geq 0.5$ and $\alpha \geq 0.25(\beta + 0.5)^2$, it is unconditionally stable. Usually $\beta = 0.5$ and $\alpha = 0.25$ are taken.

22.13.4.2 Calculation Accuracy

For the calculation methods of conditional stability, such as linear acceleration method, when time step Δt is less than the critical time step Δt_{cr}, the computation is stable, and the remaining space for option of Δt is generally small. For the calculation methods of unconditional stability, for any Δt, the computation is stable, the selection of Δt mainly depends on the calculation accuracy under the condition to meet the necessary accuracy, and the time step Δt is as large as possible, in order to save the computing time.

The so-called calculation accuracy is the accuracy of the numerical solution compared with the exact solution. As already mentioned, in order to analyze the impact of Δt on calculation accuracy of direct integration method, according to the principle of vibration mode superposition, it is enough to analyze the impact of the size of Δt on Eq. (22.55). In fact, to get some of the main conclusions, we only need to analyze the following undamped free vibration problem of single degree of freedom:

$$\left.\begin{array}{ll} & \ddot{\eta} + \omega^2 \eta = 0 \\ \text{Initial value} & \eta(0) = 1.0, \quad \dot{\eta}(0) = 0.0, \quad \ddot{\eta}(0) = -\omega^2 \end{array}\right\} \tag{w}$$

The exact solution of Eq. (w) is a cosine function:

$$\eta = \cos \omega t \tag{x}$$

The solution of Eq. (w) is calculated by numerical method, whose variation period and amplitude will be different from the exact solution. The calculation error can be represented by the following two variables:

1) $\Delta T/T$, where $T = 2\pi/\omega$ is the natural vibration period of the exact solution, $T + \Delta T$ is the natural vibration period of the numerical solution, and $\Delta T/T$ is called period extension ratio.
2) $\Delta A/A$, where $A = 1.0$ is the amplitude of the exact solution, $A - \Delta A$ is the amplitude of the numerical solution, and $\Delta A/A$ is called amplitude attenuation ratio.

Figure 22.8 shows that the relation between $\Delta t/T$ and the calculation errors of θ method and Newmark method, where Δt is the time step and T is natural vibration period. The following points can be seen from the figure:

1) When $\Delta t/T \leq 0.01$, the calculation errors of these two methods are small.
2) For θ method, the calculation error when $\theta = 1.4$ is less than that when $\theta = 2.0$. When $\Delta t/T = 0.10$, the calculation error is less than 6.8%; when $\Delta t/T = 0.05$, the calculation error is less than 1.3%.
3) The calculation error of Newmark method ($\beta = 0.5$, $\alpha = 0.25$) is small, and there is only period extension, but no amplitude attenuation.

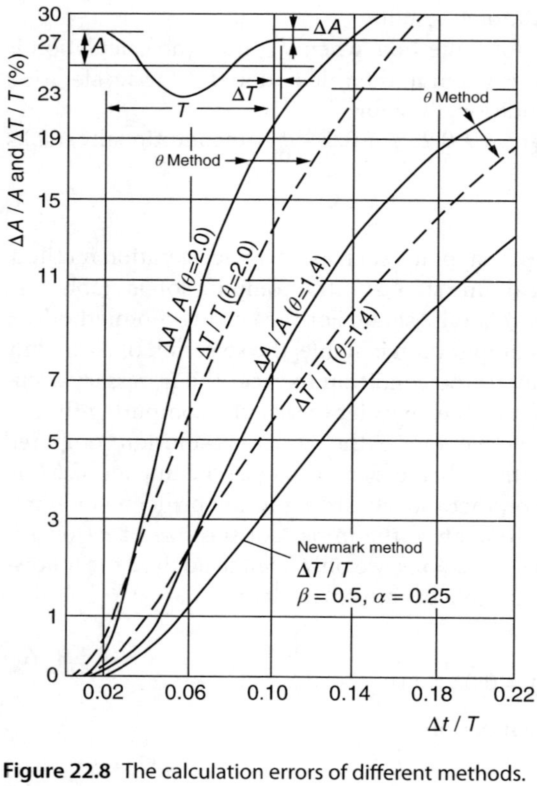

Figure 22.8 The calculation errors of different methods.

22.13.4.3 The Selection of the Time Step Δt

Experience shows that, in the dynamic responses of structure, in fact only the first s vibration modes work, the influence of the next n–s high modes is negligible, so when using the modal superposition method, it only needs to calculate the first s vibration modes and can neglect the next higher modes.

Solving by direct integral method is equivalent to getting the integrals of n equations in formula (22.55) with a uniform time step. Let the sth self-vibration period be $T_s = 2\pi/\omega_s$. If taking the time step $\Delta t = 0.05T_s$, for the first s vibration modes, $\Delta t/T_s \leq 0.05$, $i = 1 - s$; as shown in Figure 22.8, the calculation accuracy is fine. For the next n–s higher modes, T_i is getting smaller; sometimes T_s/T_n can reach the order of 1000, so $\Delta t/T_s$ is getting larger with the increase of i, and the relative calculation error is also getting bigger. If the method is unconditionally stable, because the contribution of higher modes to the dynamic responses of structures is very small, the calculation error can be still negligible. If the calculation method is conditionally stable, for low order vibration modes, the time step selected satisfies the stability conditions $\Delta t \leq \Delta t_{cr}$. But for the higher modes, it may not satisfy the stability condition, the calculation is not stable, and the errors in the computation may grow with time t, although the contribution of higher modes on the dynamic responses of the structure is small, but the calculation error may also be unacceptable. So, generally speaking, when solved by direct integration method, the unconditionally stable method should be used for calculation.

It must be noted that, for the problems of structural vibration and wave propagation, it must take different time steps and finite element meshes.

For the structural vibration problems, usually only a few low vibration modes (or a few middle modes close to load frequency) are excited. If performing Fourier analysis to dynamic load input, the results show that when the maximum frequency contained in load is ω_u, the frequency simulated accurately by the finite element meshes can be taken as

$$\omega_{co} = 4\omega_u$$

It is unnecessary to calculate more accurate high-order vibration modes using finer meshes, because their contribution to the structural dynamic responses is small, so they can be neglected.

Accordingly, the time step of the direct integration method (the algorithm of unconditional stability) can be taken as

$$\Delta t = 0.05T_{co} = 0.05(2\pi/\omega_{co})$$

The main difference of the problems of wave propagation and structural vibration is that, in wave propagation problems, there will be more frequencies excited.

Assuming that L_w is wavelength, the time of the wave passing through the length L_w is

$$t_w = \frac{L_w}{c}$$

Here c is the wave velocity.

If we must use n times of the time step to represent the wave propagation, the time step should be taken as

$$\Delta t = \frac{t_w}{n}$$

and the effective length of the finite element should be

$$L_e = c\Delta t = L_w/n$$

When we use the unconditionally stable implicit method, L_e can be regarded as the shortest distance between 2 nodes in the direction of the wave propagation.

22.14 Coupled Vibration of Solid and Fluid

When the solid is contacting with the fluid, under dynamic load, they will affect each other. This section will explain the calculation method for coupled vibration of solid and fluid.

For small amplitude vibration of compressible fluid, the pressure p must satisfy the following differential equation:

$$\frac{\partial^2 p}{\partial x^2} + \frac{\partial^2 p}{\partial y^2} + \frac{\partial^2 p}{\partial z^2} + \frac{1}{\bar{c}^2}\frac{\partial^2 p}{\partial t^2} = 0 \tag{22.101}$$

Here $\bar{c}$ is the speed of sound; in the derivation of the equation, the damping (viscous) effect has been neglected.

On the free boundary of the fluid, $p = 0$, and on the interface between fluid and solid, the boundary condition is

$$\frac{\partial p}{\partial n} = -\rho \frac{\partial^2}{\partial t^2}(u_n) \tag{22.102}$$

Here u_n is the displacement component of solid boundary in the normal direction.

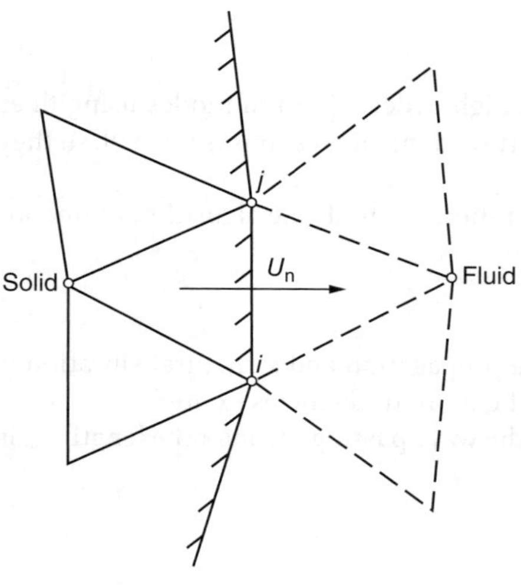

Figure 22.9 The interface of fluid and solid.

After discretization of Eq. (22.101) by finite element method, the following equation is obtained:

$$[H]\{p\} + [G]\{\ddot{p}\} + \{F\} = 0 \tag{22.103}$$

where $[H]$, $[G]$ are the matrices obtained by the general method, the vector $\{F\}$ is obtained by surface integral on the boundary. The motion of boundary depends on the movement of solid. The finite element method is also used to discrete the solid. The displacement components of solid boundary in the normal direction can be expressed by appropriate shape functions $[\overline{N}]$ and the displacement of node $\{\delta\}$ (Figure 22.9) as follows:

$$u_n = [\overline{N}]\{\delta\} \tag{a}$$

From Eq. (14.22), we have

$$\{F\} = [S]\{\ddot{\delta}\} \tag{22.104}$$

where

$$[S] = \int_s [N]^{\mathrm{T}} \rho [\overline{N}]\, \mathrm{d}s \tag{b}$$

Here $[N]$ is the shape function of the specified pressure p; s is the interface of solid and fluid.

For solid, the finite element method is used to discrete, and we get the following equation:

$$[M]\{\ddot{\delta}\} + [C]\{\dot{\delta}\} + [K]\{\delta\} + \{\overline{F}\} + \{P\} = 0 \tag{22.105}$$

Here $\{P\}$ is dynamic loads acting on the solid; $\{\overline{F}\}$ is the nodal load produced by fluid pressure.

Let $p = [N]\{p\}$, from virtual work principle, we get

$$\{\overline{F}\} = \int_s [\overline{N}]^{\mathrm{T}} p\, \mathrm{d}s = \frac{1}{\rho}[S]^{\mathrm{T}}\{p\} \tag{22.106}$$

From Eqs (22.103), (22.104), (22.105), and (22.106), the differential equations of the coupling movement of solid and fluid is obtained as follows:

$$\left.\begin{aligned} &[H]\{p\} + [G]\{\ddot{p}\} - [S]\{\ddot{\delta}\} = 0 \\ &[M]\{\ddot{\delta}\} + [C]\{\dot{\delta}\} + [K]\{\delta\} + \frac{1}{\rho}[S]^{\mathrm{T}}\{p\} + \{P\} = 0 \end{aligned}\right\} \tag{22.107}$$

The pressure of fluid p and the displacement of solid δ, velocity $\dot{\delta}$, and acceleration $\ddot{\delta}$ can be solved by Eq. (22.107).

If the fluid is incompressible, the velocity of sound $\overline{c} = \infty$, so

$$[G] = 0$$

So the first term in Eq. (22.107) can be solved directly, and we get

$$\{p\} = [H]^{-1}[S]\{\ddot{\delta}\} \tag{22.108}$$

Substituting Eq. (22.108) into the second item of Eq. (22.107), we get

$$([M] + [M_p])\{\ddot{\delta}\} + [C]\{\dot{\delta}\} + [K]\{\delta\} + \{P\} = 0 \tag{22.109}$$

where

$$[M_p] = \frac{1}{\rho}[S]^{\mathrm{T}}[H]^{-1}[S] \tag{22.110}$$

Thus, if the fluid is incompressible, the form of the equation of motion of solid (22.109) is consistent with the motion equation of the general elastic structure (22.2), just adding an additional matrix $[M_p]$ to mass matrix $[M]$. Because $[M_p]$ reflects the influence of the mass of fluid on the solid motion, so it can be called the added mass matrix.

22.15 Seismic Stress of Gravity Dam

In the previous sections, we describe the principles and methods of structural dynamic calculation by finite element method. In the following sections we will give some examples of applications; as can be seen, the finite element method is powerful in solving structural dynamic problems in engineering. In this section, we will illustrate some important results of the earthquake stress of gravity dams calculated by finite element method.

According to traditional dam design specifications, in construction of gravity dams in the earthquake region, the horizontal acceleration 0.1*g* or 0.05*g* (*g* is the acceleration of gravity) is generally taken as the seismic load, and by calculating the dam stress by static method, the values of the stresses obtained are often very small and do not play any roles in the design. But practically some gravity dams had suffered much damage after the earthquake, and this calculation method is completely divorced from reality; only the dynamic calculation using the finite element method can realistically reflect the seismic stress of gravity dam.

For example, after the earthquake in Koyina on December 11, 1967, several serious cracks appeared in the dam. The dam was built in 1954–1963, by the cyclopean concrete, with length of 853 m, height of 103 m, and bottom width of 70.2 m. It is designed by the static method; with an earthquake load of 0.05*g*, there is no tensile stress in the dam. The maximum ground acceleration measured at the time of the earthquake along the dam axis is 0.63*g*, in the direction of flow is 0.49*g*, and in the vertical direction is 0.34*g*. After the earthquake, there were cracks on the upstream and downstream surface of dam; the main horizontal cracks happened at the place of dam slope changing (629 m height). Although the dam has not failed, the damage is very serious. Later, it was decided to construct buttress on the downstream surface to reinforce dam body. This case shows that in the past the calculations of dam seismic design are divorced from reality, causing adverse consequences.

The seismic stress of Koyina gravity dam was calculated by two-dimensional finite element method [11], ignoring the dynamic action of the water. Assuming that the dam is homogeneous, $E = 31.7$ GPa, $\gamma = 2.64$ t/m^3, and $\mu = 0.20$, the damping ratio is 0.05. The first four vibration periods calculated by finite element method are 0.326, 0.122, 0.093, 0.063 s; the maximum tensile stress is shown in Figure 22.10, including gravity and hydrostatic pressure; the maximum tensile stress on the upstream face of the dam is up to 4.0 MPa, on the downstream face to 7.0 MPa. In a place where the dam slope changes, there was obvious stress concentration and the maximum compressive stress at this elevation was 8.8 MPa (not shown in Figure 22.10).

The Figure 22.11 shows the stresses of Pine Flat gravity Dam under Taft earthquake in 1952: $E = 35.1$ GPa, $\gamma = 2.48$ t/m^3, $\mu = 0.20$, and the damping ratio is 0.05. The results

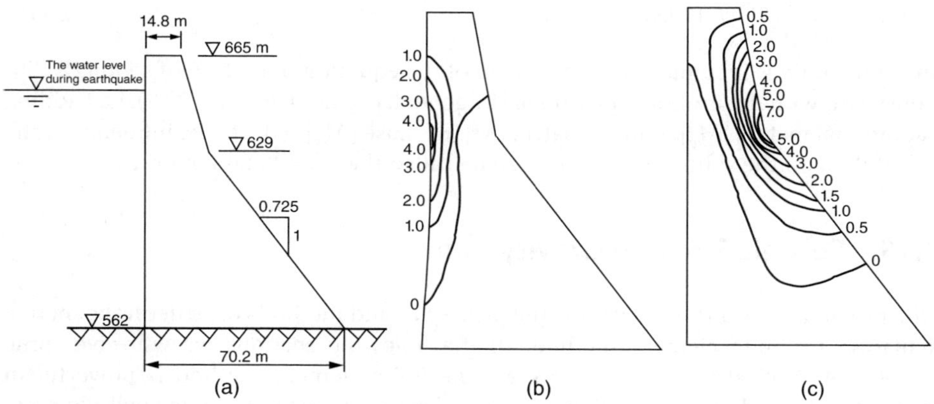

Figure 22.10 Stresses of Koyina gravity dam caused by the horizontal and vertical earthquake (MPa). (a) The dam section; (b) the tensile stress at $t = 4.425$ s; and (c) the tensile stress at $t = 4.250$ s.

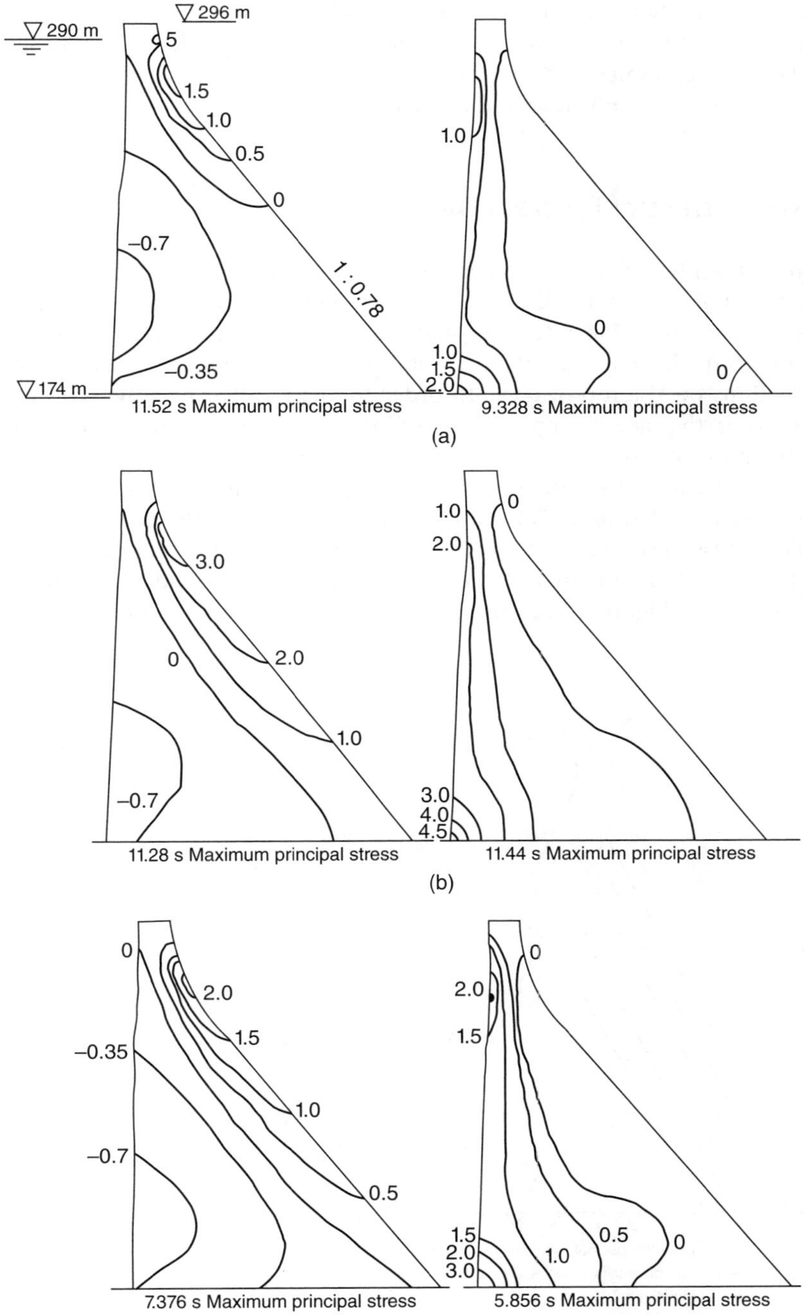

Figure 22.11 Seismic stress of Pine Flat gravity dam (MPa). (a) During the horizontal and vertical earthquake, not including the dynamic water pressure; (b) during the horizontal and vertical earthquake, including the dynamic water pressure; and (c) during the horizontal earthquake, including the dynamic water pressure.

show that, under the empty reservoirs, the vertical earthquake has little effect on the maximum stress, but, under the full reservoirs, the vertical earthquake makes tensile stress within the dam increased by 40–70% because the hydrodynamic pressure caused by vertical earthquake is perpendicular to the face of the dam.

22.16 Seismic Stress of Buttress Dam

The seismic stress of Xinfengjiang buttress dam in China [12] is shown in Figure 22.12. The maximum height of the dam is 105 m; after water retaining, earthquakes frequently happened in reservoir area. The intensity of the most serious earthquake reached VIII$^+$; after the earthquake, continuous horizontal penetrating cracks occurred near the changing sections on the top of dam. In order to analyze the dynamic stress in the dam under the earthquake, China Institute of Water Resources and Hydropower Research and Institute of Computing Technology of Chinese Academy of Sciences calculated dynamic stress by two-dimensional finite element method. There are 260 triangular elements, with 160 nodes in the dam in total; the θ method was used to calculate the dam stress under action of the measured horizontal ground acceleration of the dam site. The maximum dynamic stress of the upstream and downstream dam surface is shown in Figure 22.12. The results agree well with the experimental results.

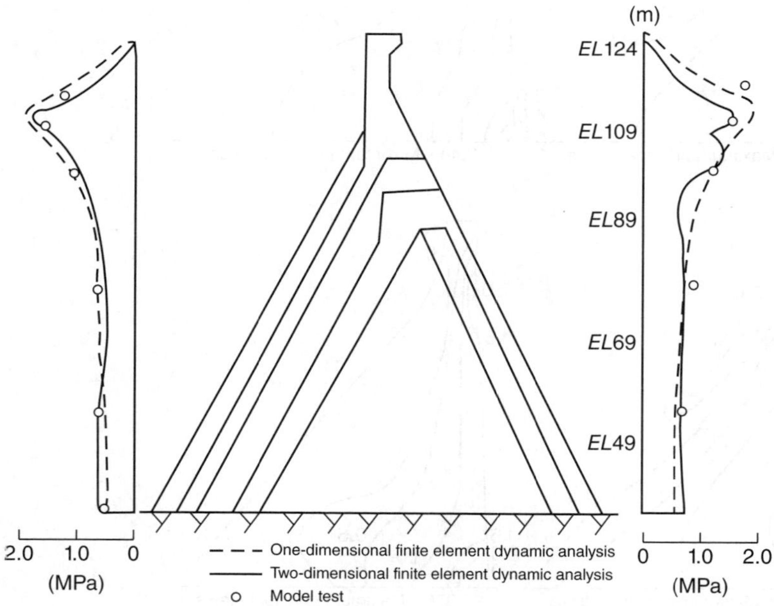

Figure 22.12 Maximum dynamic stress of the upstream (left) and downstream (right) surface under earthquake of Xinfengjiang buttress dam.

22.17 Vibration of Arch Dam

Arch dam is a complex spatial shell structure, and it is very difficult to analyze its dynamic stress by classical methods. But this problem can be easily solved by finite element method. Figure 22.13 shows the calculation results of an arch dam vibration by the finite element method [13]. The valley is symmetrical, so we can only take a half to calculate, assuming that the foundation is rigid; using the shell element for calculation, the mesh of discretization is shown in Figure 22.13(i). The first six vibration modes and corresponding natural frequencies are shown.

22.18 Seismic Stress of Earth Dam

In the past, the seismic stress of the earth dam was calculated as a cantilever, only producing the horizontal shear deformation. Obviously, this is a very rough calculating model. In fact, earth dam is a three-dimensional continuum with anisotropic, heterogeneous, and nonelastic materials, and there is also water in voids in the soil. The seismic stress problem of earth dam is very complicated but can be solved by the finite element method and large high-speed electronic computer.

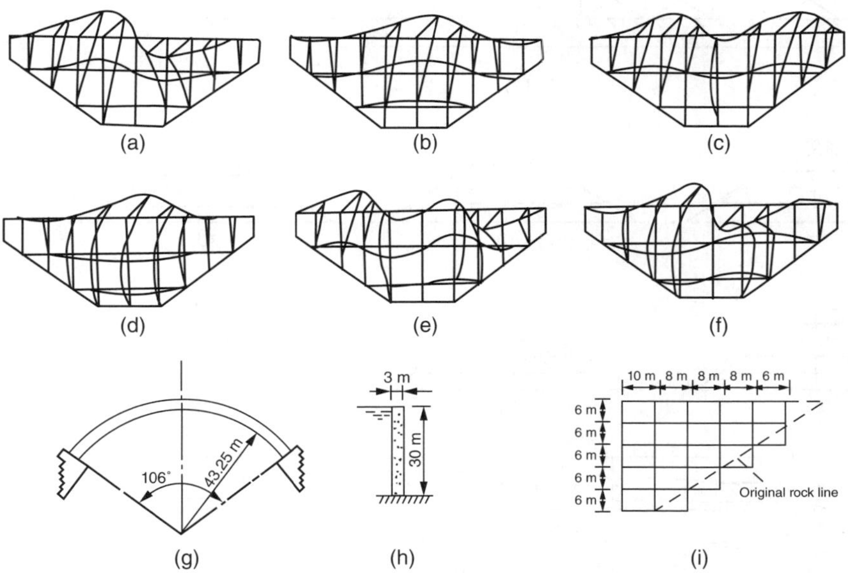

Figure 22.13 Radial free vibration of arch dam (empty reservoirs). (a) The first vibration mode: 8.50 times per second; (b) the second vibration mode: 8.77 times per second; (c) the third vibration mode: 12.36 times per second; (d) the fourth vibration mode: 15.75 times per second; (e) the fifth vibration mode: 16.18 times per second; (f) the sixth vibration mode: 20.33 times per second; (g) the plan; (h) the section of arch crown; (i) computational meshes.

For the earth dam in broad valley, the seismic stress along the direction of flow can be calculated according to plane strain problem. Figure 22.14 shows a simple example of seismic stress analysis of a dam [14]. The section of dam is a triangular, homogeneous, linear elastic body. $E = 572\,\text{GPa}$, $\mu = 0.45$, $\gamma = 2.08\,\text{t/m}^3$, and damping is assumed to be 20% of the critical damping of each vibration mode. After the discretization of the dam section, there are 100 elements, 66 nodes (including 11 nodes on the bottom, which are assumed to be fixed), the remaining 55 nodes, and 110 degrees of freedom in total. The first 15 vibration modes and frequencies are calculated first; Figure 22.13 shows the first six vibration modes where only the first vibration mode is similar to the pure shear deformation obtained by the triangular shear beam theory. There are obvious vertical motions, swings, and so on in all other modes. After obtaining all the modes, the seismic stress is calculated by modal superposition method. Assume that the dam is also under the effects of two components of ground acceleration, one horizontal component and one vertical component. These two components are the actual records of El Centro

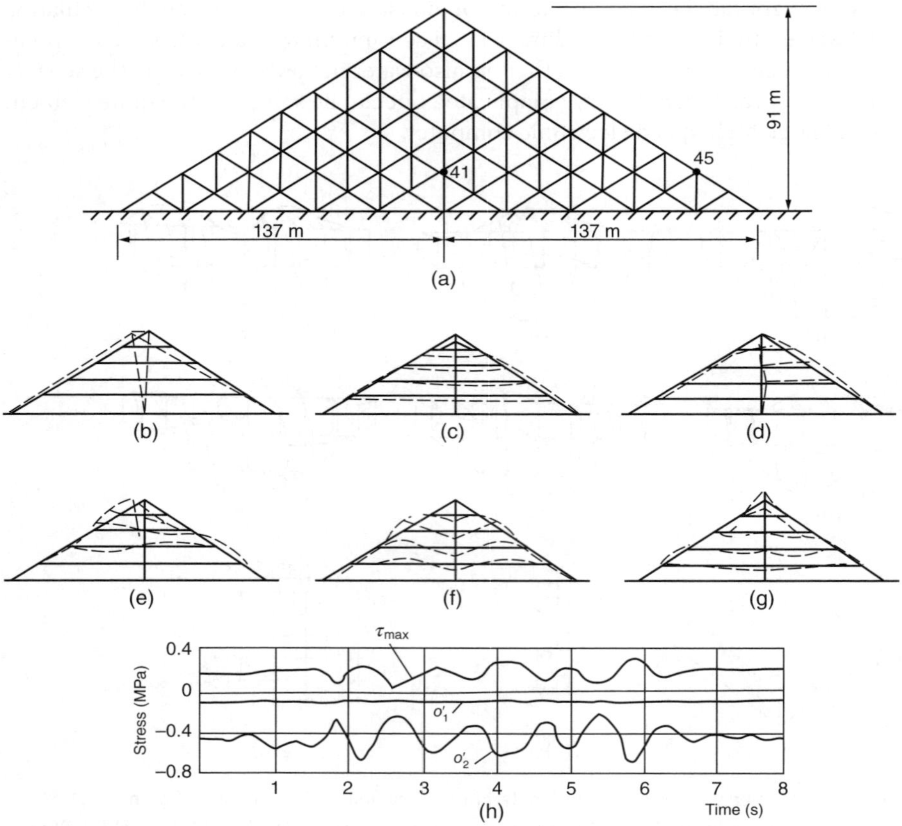

Figure 22.14 Seismic stress analysis of the earth dam. (a) Size and meshes of dam body; (b) the first vibration mode: $\omega = 7.71$ rad/s; (c) the second vibration mode: $\omega = 12.52$ rad/s; (d) the third vibration mode: $\omega = 14.60$ rad/s; (e) the fourth vibration mode: $\omega = 19.31$ rad/s; (f) the fifth vibration mode: $\omega = 20.12$ rad/s; (g) the sixth vibration mode: $\omega = 23.10$ rad/s; and (h) stress–time relationship (node 45).

earthquake in May 18, 1940. Figure 22.14(h) shows the stress–time relationship at 45th node on the dam slope. Initial stress at $t = 0$ is the static stress obtained by the finite element method.

The slope of the earth dam during the earthquake is easy to slide, so only when the stress states near the dam surface are calculated by finite element method, it is possible to estimate accurately the sliding stability of the slope. In addition, when the size of the dam is relatively large, due to the phase difference, the ground motion during the earthquake is not uniform; this effect can also be considered by the finite element method. When the width of dam bottom is larger than the wavelength of seismic waves, this effect cannot be ignored.

22.19 Seismic Stresses of Cylindrical Shell

Cylindrical shell has been widely adopted in practical engineering. Now we introduce the results of the dynamic response of a cylindrical shell roof with finite element method. The shape and size of the thin shell are shown in Figure 22.15(a), the variation of vertical displacement of the midpoint of the free edge of the shell in a practical earthquake is shown in Figure 22.15(b), and the vertical components of the first and second vibration

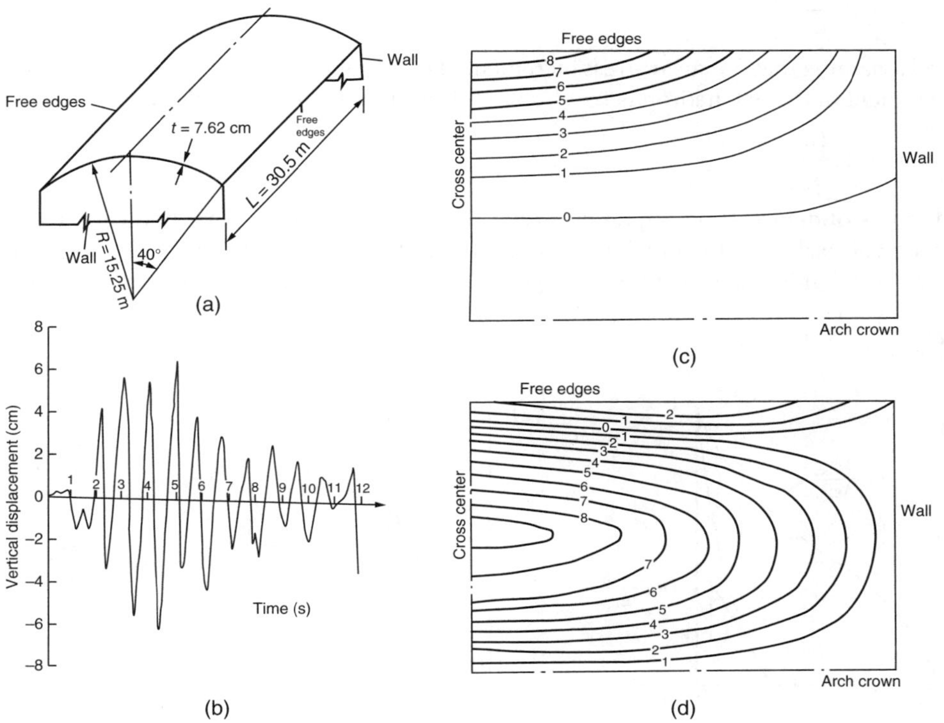

Figure 22.15 Dynamic stress of cylindrical shell. (a) The size of roof of cylindrical shell; (b) vertical displacement of the midpoint of the free edge under Ell Centro earthquake in 1944; (c) first mode of free vibration (vertical component); and (d) second mode of free vibration (vertical component).

modes of shell are shown in Figure 22.15(c) and (d). Because of bidirectional symmetry, only a quarter is necessary to be calculated. The shell is divided into the triangular plate elements, the membrane stress of the shell is the plane stress of triangular element, and the bending stresses of the shell are the bending stresses of triangular element [15].

22.20 Nonlinear Dynamic Responses of Underground Structures

When calculating the dynamic responses of underground structures, we must consider the interaction between the structure and the surrounding medium, which is typically soil or rock. Its stress–strain relationship is often nonlinear and should be obtained by dynamic tests in the field or in the laboratory, and the deformation laws of loading and unloading are inconsistent, as shown in Figure 22.16. For calculating the nonlinear dynamic responses of underground structures, load increment method and numerical integration method of equations of motion should be adopted. The stress–strain relationship of soil and rock is usually expressed in stress and strain of the octahedron. For example, the normal stress increment of the octahedron is $\Delta\sigma_{oct}$, the normal strain increment of the octahedron is $\Delta\varepsilon_{oct}$, so the volume deformation modulus is

$$K \cong \frac{\Delta\sigma_{oct}}{\Delta\varepsilon_{oct}}$$

And assuming that the shear stress increment of the octahedron is $\Delta\tau_{oct}$, the shear strain increment of the octahedron is $\Delta\gamma_{oct}$, so the shear deformation modulus is

$$G \cong \frac{\Delta\tau_{oct}}{\Delta\gamma_{oct}}$$

Modulus of deformation required in each step of load increment method can be calculated according to the previous two equations. Typically, the deformation moduli in loading and unloading are different. Figure 22.17 shows an axisymmetric computational

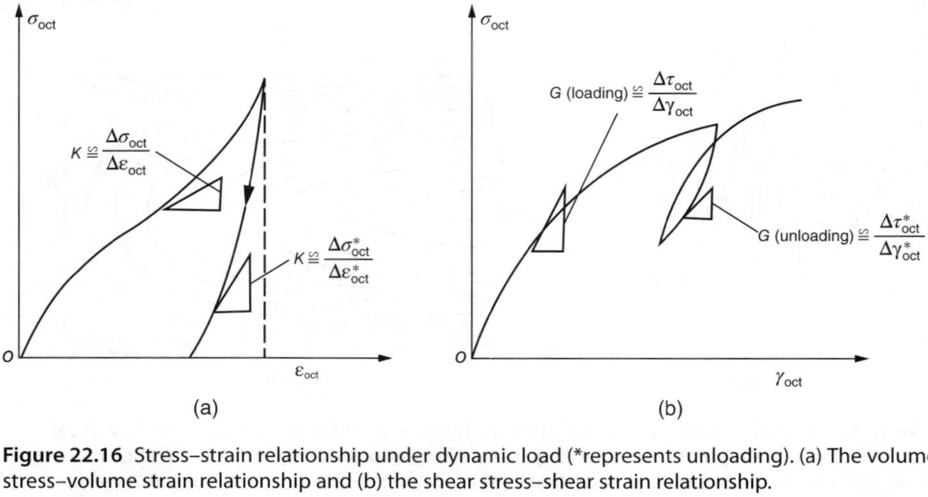

Figure 22.16 Stress–strain relationship under dynamic load (*represents unloading). (a) The volume stress–volume strain relationship and (b) the shear stress–shear strain relationship.

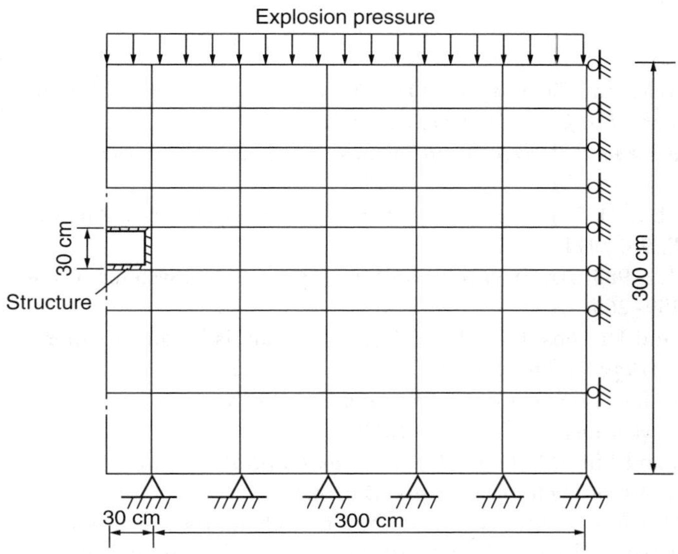

Figure 22.17 Computational meshes of axisymmetric underground structures.

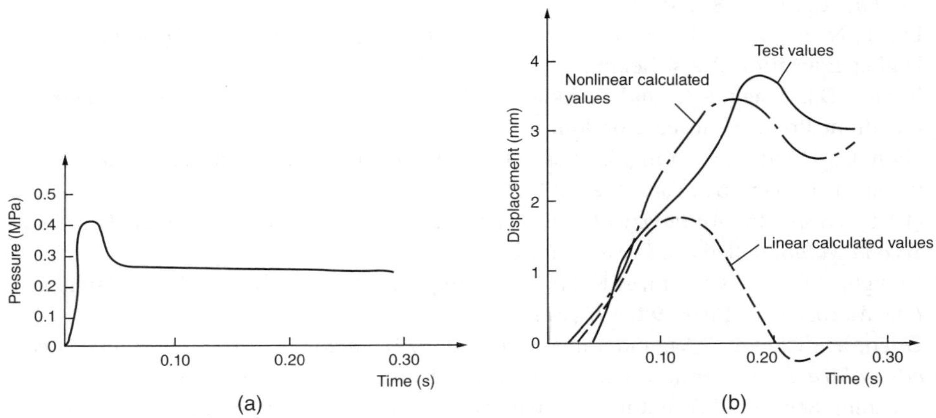

Figure 22.18 Explosion pressure and dynamic response of the solid medium. (a) Explosion pressure changing with time and (b) displacement of a point of the solid medium.

grid of underground structures. Using isoparametric axisymmetric finite elements, the surrounding soil is divided into 47 quadrilateral elements; the structure itself is divided into three thin plate elements, 63 nodes in total. Calculated by the linear acceleration method, nonlinear stress–strain relationship is obtained from the uniaxial dynamic test. The same structure was tested; the explosion pressure is shown in Figure 22.18(a).The test and calculation results are shown in Figure 22.18(b). The figure shows that the nonlinear calculation results and experimental data are close, with residual deformations. The agreement between linear calculation results and the experimental data is poor, and an important drawback is that there is no residual deformation.

Bibliography

1 Anderson, R.G., Irons, B.M. and Zienkiewicz, O.C. (1968) Vibration and stability of plates using finite elements. *Int. J. Solids Struct.*, **4**, 1031–1055.

2 Bathe, K.J. (1982) *Finite Element Procedures in Engineering Analysis*, Prentice-Hall, Englewood Cliffs.

3 Bathe, K.J. and Wilson, E.L. (1972) Large Eigenvalue Problems in Dynamic Analysis. *J. Eng. Mech. Div. ASCE*, **98**, 1471–1485.

4 Wilson, E.L. and Itoh, T. (1983) An Eigensolution Strategy for Large Systems. *Comput. Struct.*, **16** (1–4), 259–265.

5 Wilson, E.L., Yuan, M. and Dickens, J.M. (1982) Dynamic analysis by direct superposition of Ritz vectors. *Earthquake Eng. Struct. Dyn.*, **10**, 813–821.

6 Ruiming, D. and Wilson, E.L. (1984) An effective modified Ritz vector direct superposition method. *Earthquake Eng. Eng. Vib.*, **4** (2), 36–48.

7 Huanchun, S., Naisi, Q. and Lin, J. (1992) *Advanced Computational Structural Dynamics*, Dalian University of Technology Press, Dalian.

8 Bin, R. and Zhu, B. (1994) Wilson-Ritz vector method for solving the eigenvalue of the nonsymmetric matrix and its application to the stress analysis of arch dam under the action of seismic load. *Technol. Comput. Comput. Appl.*, **2**, 12–18.

9 Yuan, M., Xiong, S. and Chen, X. (1991) Multiple level dynamic substructure analysis. *Eng. Comput.*, **8**, 231–244.

10 Lin, J., Naisi, Q. and Huanchun, S. (1989) *Computational Structural Dynamics*, Higher Education Press, Beijing.

11 Naylor, D.J., Stagg, K.G. and Zienkiewicz, O.C. (1975) *Numerical Analysis of Dams*, Quadrant Press, Swansea, London.

12 Shen, C., Houqun, C., Yang, Z. *et al.* (1974) Xinfengjiang reservoir earthquake and its effect on dam. *Sci. China*, **2**, 239–272.

13 (1975) *Translated Anthology of Finite Element Method in Solid Mechanics (First and Second Volume)*, Science Press, Beijing.

14 Clough, R.W. and Chopra, A.K. (1996) Earthquake stress analysis in Earth dams. *J. Eng. Mech. Proc. ASCE*, **92**, 197–212.

15 Oden, J.T., Clough, R.W. and Yamamoto, Y. (1972) *Advances in Computational Methods in Structural Mechanics and Design*, UAH Press, Huntsville, Alabama.

16 Teaching Section of Structural Mechanics of Water Conservancy Institute of East China (1983) *Structural Mechanics. Second Volume*, Water and Power Press, Beijing.

23

Problems in Rock Mechanics

Rock mechanics is an important but a young discipline. Since the character of rock is quite complicated, the theoretical methods are helpless in most cases, while the finite element method (FEM) can be widely used. FEM could be applied to compute the stress, deformation, and stability of rock slope, dam foundation, tunnel, lane, and mine. It can also be used to analyze the cause of fault, joint, fold, and other geological structures.

Though rock mechanics resembles soil mechanics to some extent, there are still some significant differences. For example, it is always assumed that soil is isotropic, while rock is usually anisotropic. Soil mass can be taken as homogeneous, but in rock mass, there exist joints, cracks, and faults, which play an important role in the deformation and stability of rock.

Rock and soil are both porous materials and sometimes are saturated. In saturated rock and soil, the total stress in rock and soil can be decomposed into two parts: one part is pore water pressure and the other part is the stress between particles of rock and soil, referred as effective stress. The principle of effective stress can be expressed as follows:

$$\begin{aligned}
\{\sigma\} &= \{\sigma'\} - [m]p \\
\{\sigma\}^{\mathrm{T}} &= \left\{\sigma_x \ \sigma_y \ \sigma_z \ \tau_{xy} \ \tau_{yz} \ \tau_{zx}\right\} \\
\{\sigma'\}^{\mathrm{T}} &= \left\{\sigma_x' \ \sigma_y' \ \sigma_z' \ \tau_{xy}' \ \tau_{yz}' \ \tau_{zx}'\right\} \\
[m]^{\mathrm{T}} &= \left[1 \ 1 \ 1 \ 0 \ 0 \ 0\right]
\end{aligned}$$

where $\{\sigma\}$ is total stress vector, $\{\sigma'\}$ is effective stress vector, and p is pore water pressure (scalar).

The one that dominates the yield and failure of rock and soil is effective stress $\{\sigma'\}$. In this chapter and the next, all equations are expressed with effective stress. For convenience, the apostrophes are omitted. Besides, the symbols in the theory of elasticity are used. For normal stress, tensile stress is taken as positive.

23.1 Structure of Rock

In the view of engineering, the structure of rock can be divided into several parts, as shown in Figure 23.1.

The Finite Element Method: Fundamentals and Applications in Civil, Hydraulic, Mechanical and Aeronautical Engineering, First Edition. Bofang Zhu.

Companion website: www.wiley.com/go/zhufem17

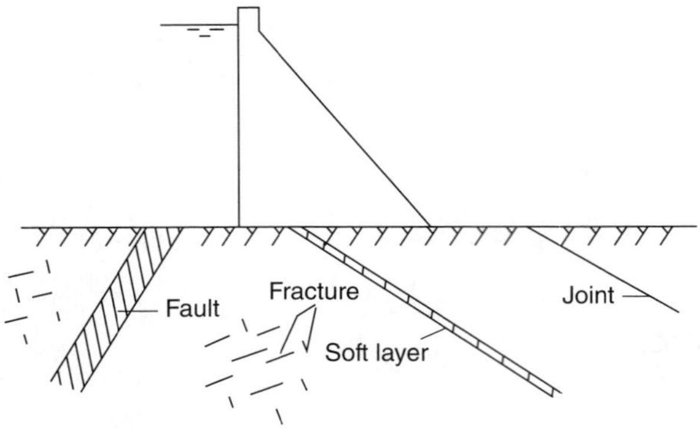

Figure 23.1 Structure of rock.

23.1.1 Rock Block

Integral block can be represented by ordinary physical element in FEM computation. For hard rock, since its strength is high, while the stress level of general project is usually not very high, it could be taken as elastomer. For soft rock, it should be taken as nonlinear material in some cases.

23.1.2 Fault

The thickness of fault varies from 10 cm to several meters or even more than 10 m. In FEM computing, fault is usually represented by solid element. The material in faults is fragmented rocks or fault gouge, the characters of which are similar to those of soil, with strong nonlinearity. Major faults should be computed as nonlinear media. While for minor faults, such as those far away from buildings, their deformation has little effect on the stress and deformation of structure, and they can also be computed as elastomer, using corresponding elastic modulus.

23.1.3 Soft Layer

Soft layer has a thickness ranging from several centimeters to more than 10 cm. The material in the layers is fragmented rocks or soft soil. For major soft layers, such as those adjacent to buildings or those having great effects on the safety of buildings, layer elements are usually used in computing, considering nonlinear constitutive relation. For minor soft layer, equivalent elements could be used.

23.1.4 Joint

Joint is a discontinuous plane in rock, and its thickness is close to zero. Therefore, the relation between stress and deformation is not expressed with elastic modulus, but with stiffness coefficient as follows:

$$\sigma_{\mathrm{n}} = \lambda_{\mathrm{n}} \Delta v, \quad \tau_{\mathrm{s}} = \lambda_{\mathrm{s}} \Delta u \tag{23.1}$$

where λ_{n} is the normal stiffness coefficient, λ_{s} is the tangential stiffness coefficient, Δv is the difference of the normal displacement at both sides of the joint, Δu is the difference of the tangential displacement at both sides of the joint, σ_n is normal stress, and τ_s is shear stress.

For major joints, joint elements are usually used and for minor joints, equivalent elements could be used.

23.1.5 Crack

Crack is a small discontinuous plane with near-zero thickness. Since there are many cracks whose areas are small and thicknesses close to zero, generally it is unnecessary to use specific elements to simulate cracks, while the influence of cracks is considered when determining the elastic modulus of rock block.

23.2 Equivalent Deformation Modulus

There are usually groups of soft layers in bedrock. It will be troublesome to take all of them as special elements. The commonly used method is to take crucial parts near the building as special elements while using an equivalent orthotropic body for simulation in other parts. Assume that the two elastic principal axes are parallel and orthogonal to the soft layer, respectively. This kind of element that considers comprehensively rock mass, soft layers, and joints is referred as equivalent rock element, or geology element. Pan suggested the composite modulus of rock with soft layers [3]. Gao suggested the modulus of rock with joints [4]. Based on their work, we will give the equivalent elastic constant of rock with both soft layers and joints.

As Figure 23.2 shows, take out an element of micro-block $dx\,dy\,dz$ ($d_z = 1$). Suppose the element contains several soft layers and the total thickness is $\beta\,dz = \beta$. It also contains some joints whose thickness h is near zero. Therefore, the thickness of rock block is $1 - \beta$. Assume the elastic modulus and Poisson's ratio of rock block and soft layers are E_r, μ_r, E_f, μ_f, respectively, and the normal and tangential stiffness coefficient of joints are λ_n and λ_s, respectively. The equivalent elastic constants of the medium are as follows: E_z, $E_x = E_y$, $G_{zx} = G_{zy}$, G_{xy}, $\mu_{zx} = \mu_{zy} = \mu_{xy}$.

Under the action of vertical stress, the deflection of the element in direction of z is

$$\Delta w = \frac{(1-\beta)\sigma_z}{E_r} + \frac{\beta\sigma_z}{E_f} + \sum \frac{\sigma_z}{\lambda_n} = \frac{\sigma_z}{E_z}$$

Thus we have the equivalent elastic modulus of the geology element:

$$E_z = 1 \Big/ \left[\frac{1-\beta}{E_r} + \frac{\beta}{E_f} + \sum \frac{1}{\lambda_n} \right] \tag{23.2}$$

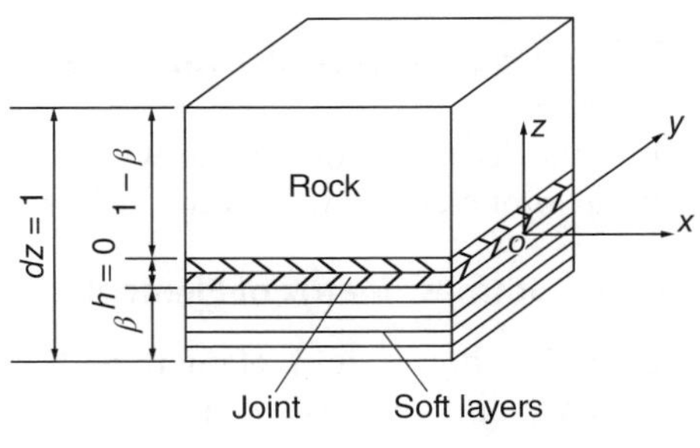

Figure 23.2 Micro-unit of rock with joints and soft intercalated layers.

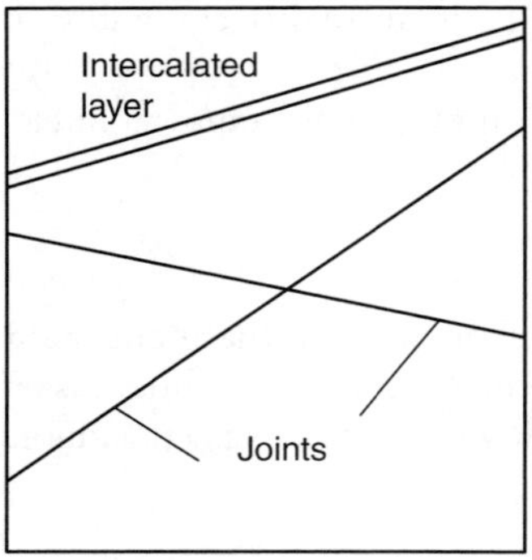

Figure 23.3 Rock that is not parallelly cut.

Suppose the horizontal strain is ε_x. Then the horizontal stress is

$$\sigma_x = (1-\beta)E_r\varepsilon_x + \beta E_f\varepsilon_x = E_x\varepsilon_x$$

Therefore,

$$E_x = E_y = (1-\beta)E_r + \beta E_f \tag{23.3}$$

Under the action of shear stress τ_{zx}, the horizontal deflection is

$$\Delta u = \frac{(1-\beta)\tau_{zx}}{G_r} + \frac{\beta\tau_{zx}}{G_f} + \sum\frac{\tau_{zx}}{\lambda_s} = \frac{\tau_{zx}}{G_{zx}}$$

Then we have

$$G_{zx} = G_{zy} = 1\Bigg/\left[\frac{1-\beta}{G_r} + \frac{\beta}{G_f} + \Sigma\frac{1}{\lambda_s}\right] \tag{23.4}$$

where

$$G_r = E_r/2(1+\mu_r), \quad G_f = E_f/2(1+\mu_f)$$

Suppose the shear strain is γ_{xy}. Then the shear stress τ_{xy} is

$$\tau_{xy} = (1-\beta)G_r\gamma_{xy} + \beta G_f\gamma_{xy} = G_{xy}\gamma_{xy}$$

Therefore,

$$G_{xy} = (1-\beta)G_r + \beta G_f \tag{23.5}$$

With similar methods, Poisson's ratio can be derived as follows:

$$\mu = (1-\beta)\mu_r + \beta\mu_f \tag{23.6}$$

For rock that is not parallelly cut, as Figure 23.3 shows, we can firstly compute the equivalent flexibility matrix:

$$[C] = [C_r] + \Sigma[R]^T[C_j][R] \tag{23.7}$$

where $[C]$ is the equivalent flexibility matrix, $[C_r]$ is the flexibility matrix of rock, $[C_j]$ is the flexibility matrix of joints or soft layers, and $[R]$ is the coordinate transformation matrix. The details can be seen in Appendix A.

Then compute the inversion and we can get the equivalent elastic matrix of the medium, that is,

$$[D] = [C]^{-1} \tag{23.8}$$

23.3 Two-Dimensional Linear Joint Element

Joint is a discontinuous plane whose thickness is almost zero. The two-dimensional 4-node joint element with zero thickness is applied widely in practical engineering.

23.3.1 Stiffness Matrix of Element

Figure 23.4 shows a joint element with length of l, width of e, thickness of t (vertical to x, y plane), and 4 nodes i, j, m, r. Locate the origin of coordinates at the centroid.

Figure 23.4 Joint element in rock.

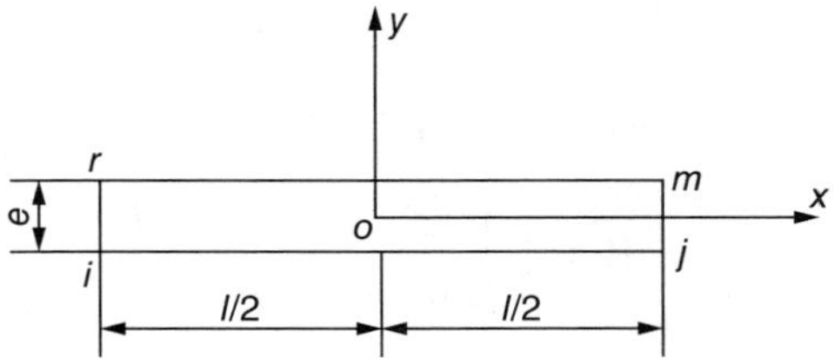

The nodal force of the element is

$$\{F\}^{e} = \left[U_i\ V_i\ U_j\ V_j\ U_m\ V_m\ U_r\ V_r\right]^{T} \tag{23.9}$$

The nodal displacement of the element is

$$\{\delta\}^{e} = \left[u_i\ v_i\ u_j\ v_j\ u_m\ v_m\ u_r\ v_r\right]^{T} \tag{23.10}$$

Assume the displacements of upper border rm and lower border ij are distributed linearly, that is,

$$u_{\text{upper}} = \frac{1}{2}\left(1 - \frac{2x}{l}\right)u_r + \frac{1}{2}\left(1 + \frac{2x}{l}\right)u_m$$

$$u_{\text{lower}} = \frac{1}{2}\left(1 - \frac{2x}{l}\right)u_i + \frac{1}{2}\left(1 + \frac{2x}{l}\right)u_j$$

Therefore we can get the difference between the horizontal displacements of upper and lower borders of the element

$$\Delta u = u_{\text{upper}} - u_{\text{lower}} = \frac{1}{2}\left[(u_r - u_i)\left(1 - \frac{2x}{l}\right) + (u_m - u_j)\left(1 + \frac{2x}{l}\right)\right] \tag{a}$$

Similarly, the difference between the vertical displacements of upper and lower borders of the element is

$$\Delta v = v_{\text{upper}} - v_{\text{lower}} = \frac{1}{2}\left[(v_r - v_i)\left(1 - \frac{2x}{l}\right) + (v_m - v_j)\left(1 + \frac{2x}{l}\right)\right] \tag{b}$$

Assume the shear stress in the element is proportional to the difference of horizontal displacement, that is,

$$\tau_s = \lambda_s \Delta u + \tau_{s0} \tag{23.11}$$

and the normal stress in the element is proportional to the difference of vertical displacement, that is,

$$\sigma_n = \lambda_n \Delta v + \sigma_{n0} \tag{23.12}$$

where λ_s, λ_n are tangential and normal stiffness coefficient of the joint, respectively, and τ_{s0}, σ_{n0} are initial shear stress and normal stress in the element.

Let

$$\{\Delta\delta\} = \begin{Bmatrix} \Delta u \\ \Delta v \end{Bmatrix}, \quad [\lambda] = \begin{bmatrix} \lambda_s & 0 \\ 0 & \lambda_n \end{bmatrix}, \quad \{\sigma_0\} = \begin{Bmatrix} \tau_{s0} \\ \sigma_{n0} \end{Bmatrix} \tag{23.13}$$

Then according to Eqs (23.11) and (23.12), we have

$$\{\sigma\} = \begin{Bmatrix} \tau_s \\ \sigma_n \end{Bmatrix} = [\lambda]\{\Delta\delta\} + \{\sigma_0\} \tag{23.14}$$

From Eqs (a) and (b), we get

$$\{\Delta\delta\} = \begin{Bmatrix} \Delta u \\ \Delta v \end{Bmatrix} = [M]\{\delta\}^{\mathrm{e}} \tag{23.15}$$

$$[M] = \frac{1}{2}\begin{bmatrix} -z_1 & 0 & -z_2 & 0 & z_2 & 0 & z_1 & 0 \\ 0 & -z_1 & 0 & -z_2 & 0 & z_2 & 0 & z_1 \end{bmatrix} \tag{23.16}$$

where

$$z_1 = 1 - \frac{2x}{l}, \quad z_2 = 1 + \frac{2x}{l} \tag{c}$$

If there is a virtual displacement $\{\delta^*\}^{\mathrm{e}}$ on each node, then the difference of virtual displacement in the element is

$$\{\Delta\delta^*\} = [M]\{\delta^*\}^{\mathrm{e}}$$

For unit length of element, the virtual work of the stress is

$$t\{\Delta\delta^*\}^{\mathrm{T}}\{\delta\} = \{\delta^{*\mathrm{e}}\}^{\mathrm{T}} t[M]^{\mathrm{T}}([\lambda]\{\Delta\delta\} + \{\sigma_0\})$$

After integrating along the length of element, we get the virtual work of the stress of the element as follows:

$$\{F\}^{\mathrm{e}} = \left(t\int_{-l/2}^{l/2} [M]^{\mathrm{T}}[\lambda][M]\,\mathrm{d}x\right)\{\delta\}^{\mathrm{e}} + \int_{-l/2}^{l/2} [M]^{\mathrm{T}}\{\sigma_0\}\,\mathrm{d}x \tag{d}$$

The second part on the right end of Eq. (d) is the nodal force resulting from initial stress. By changing the sign we can get the nodal load due to initial stress. The first part on the right end of Eq. (d) is the nodal force caused by nodal displacement. Thus from Eq. (d) we get the nodal force of the element:

$$[F]^{\mathrm{e}} = [k]^{\mathrm{e}}\{\delta\}^{\mathrm{e}} \tag{23.17}$$

$$[k]^{\mathrm{e}} = t\int_{-l/2}^{l/2} [M]^{\mathrm{T}}[\lambda][M]\,\mathrm{d}x \tag{23.18}$$

The load caused by initial stress is

$$\{P\}_{\sigma_0}^{\mathrm{e}} = -t\int_{-l/2}^{l/2} [M]^{\mathrm{T}}\{\sigma_0\}\,\mathrm{d}x \tag{23.19}$$

Substitute the expression of $[M]$ (23.16) into Eq. (23.18) and notice that

$$\int_{-l/2}^{l/2} z_1^2\,\mathrm{d}x = \frac{4l}{3}, \quad \int_{-l/2}^{l/2} z_1 z_2\,\mathrm{d}x = \frac{2l}{3}, \quad \int_{-l/2}^{l/2} z_2^2\,\mathrm{d}x = \frac{4l}{3}$$

We get the stiffness matrix of the element as follows [1]:

$$[k]^{\mathrm{e}} = \frac{lt}{6}\begin{bmatrix} 2\lambda_{\mathrm{s}} & 0 & \lambda_{\mathrm{s}} & 0 & -\lambda_{\mathrm{s}} & 0 & -2\lambda_{\mathrm{s}} & 0 \\ 0 & 2\lambda_{\mathrm{n}} & 0 & \lambda_{\mathrm{n}} & 0 & -\lambda_{\mathrm{n}} & 0 & -2\lambda_{\mathrm{n}} \\ \lambda_{\mathrm{s}} & 0 & 2\lambda_{\mathrm{s}} & 0 & -2\lambda_{\mathrm{s}} & 0 & -\lambda_{\mathrm{s}} & 0 \\ 0 & \lambda_{\mathrm{n}} & 0 & 2\lambda_{\mathrm{n}} & 0 & -2\lambda_{\mathrm{n}} & 0 & -\lambda_{\mathrm{n}} \\ -\lambda_{\mathrm{s}} & 0 & -2\lambda_{\mathrm{s}} & 0 & 2\lambda_{\mathrm{s}} & 0 & \lambda_{\mathrm{s}} & 0 \\ 0 & -\lambda_{\mathrm{n}} & 0 & -2\lambda_{\mathrm{n}} & 0 & 2\lambda_{\mathrm{n}} & 0 & \lambda_{\mathrm{n}} \\ 2\lambda_{\mathrm{s}} & 0 & -\lambda_{\mathrm{s}} & 0 & \lambda_{\mathrm{s}} & 0 & 2\lambda_{\mathrm{s}} & 0 \\ 0 & -2\lambda_{\mathrm{n}} & 0 & -\lambda_{\mathrm{n}} & 0 & \lambda_{\mathrm{n}} & 0 & 2\lambda_{\mathrm{n}} \end{bmatrix} \tag{23.20}$$

23.3.2 Nodal Force Due to Initial Stress

According to Eq. (23.14), the stress of the element in the direction of thickness (the direction of y) is uniform, while that in the direction of length (the direction of x) is linearly distributed. Correspondingly, assume the initial stress of the element is also linearly distributed in the direction of x, that is,

$$\tau_{s0} = \frac{1}{2}\left(1 - \frac{2x}{l}\right)\tau_{si0} + \frac{1}{2}\left(1 + \frac{2x}{l}\right)\tau_{sj0}$$

$$\sigma_{n0} = \frac{1}{2}\left(1 - \frac{2x}{l}\right)\sigma_{ni0} + \frac{1}{2}\left(1 + \frac{2x}{l}\right)\sigma_{nj0}$$

where τ_{si0}, σ_{ni0} and τ_{sj0}, σ_{nj0} are the initial stress of node i and j, respectively.

From Eq. (23.19), we can get the nodal load caused by initial stress as below:

$$\{P\}^{e}_{\sigma_0} = -\frac{tl}{6}\begin{bmatrix} -2 & 0 & -1 & 0 \\ 0 & -2 & 0 & -1 \\ -1 & 0 & -2 & 0 \\ 0 & -1 & 0 & -2 \\ 1 & 0 & 2 & 0 \\ 0 & 1 & 0 & 2 \\ 2 & 0 & 1 & 0 \\ 0 & 2 & 0 & 1 \end{bmatrix}\begin{Bmatrix} \tau_{si0} \\ \sigma_{ni0} \\ \tau_{sj0} \\ \sigma_{nj0} \end{Bmatrix} \tag{23.21}$$

If the initial stress in the element is uniform, that is, τ_{s0} and σ_{n0}, then according to Eq. (23.19), the nodal load caused by initial stress is

$$\{P\}^{e}_{\sigma_0} = \frac{tl}{2}\left[\tau_{s0} \quad \sigma_{n0} \quad \tau_{s0} \quad \sigma_{n0} \quad -\tau_{s0} \quad -\sigma_{n0} \quad -\tau_{s0} \quad -\sigma_{n0}\right]^{T} \tag{23.22}$$

If the local coordinate system of the element is (x', y') and the global coordinate system is (x, y), as Figure 23.5 shows, the coordinate system should be transformed. Please refer to Appendix A for more details.

23.4 Stiffness Coefficients of Joint Element

When computing joint element in rock, the stiffness coefficients of joints λ_n and λ_s are required. Obviously, these coefficients can be obtained by measured data. We

Figure 23.5 Coordinate transformation.

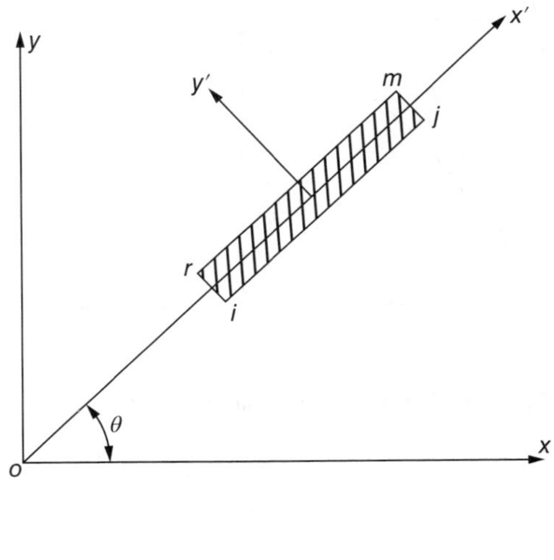

will introduce the computing method proposed by the author in 1976 (see Ref. [22]). According to the load–displacement curve from field test, the stiffness coefficient may be calculated by

$$\lambda_{\rm n} = \frac{\partial \sigma}{\partial v}, \quad \lambda_{\rm s} = \frac{\partial \tau}{\partial u} \tag{23.23}$$

where u and v are the difference of horizontal and vertical displacements, respectively, between the upper and lower edges of joint.

In general, joint in rock cannot sustain normal tension; thus we have

$$\lambda_{\rm n} = \lambda_{\rm s} = 0 \quad (\sigma > 0) \tag{23.24}$$

In actual computing, values close to zero should be used instead of zero to avoid overflow during the computing.

When $\sigma < 0$, $\lambda_{\rm n}$, and $\lambda_{\rm s}$ do not equal to zero, according to measured data, the normal stiffness coefficient $\lambda_{\rm n}$ can be approximately denoted with a constant, while the tangential stiffness coefficient $\lambda_{\rm s}$ shows strong nonlinearity, varying with tangential displacement difference u and normal stress σ. Then we will mainly discuss the tangential stiffness coefficient (if the normal stiffness coefficient is not a constant, it could be calculated with similar method).

According to actual data, we found that the relation between shear stress and tangential displacement of joint in rock could be expressed with hyperbola as follows:

$$\tau = \frac{u}{a + bu} \tag{23.25}$$

where τ is the shear stress on joint, u is the displacement difference between the upper and lower borders of the joint, and a and b are constants determined by measured data.

a and b have apparent physical significance. As shown in Figure 23.6(a), a is the reciprocal of initial stiffness coefficient λ_0, while b is the reciprocal of the maximum shear stress (shear strength) $\tau_{\rm m}$, that is,

$$\frac{1}{a} = \lambda_0, \quad \frac{1}{b} = \tau_{\rm m} \tag{23.26}$$

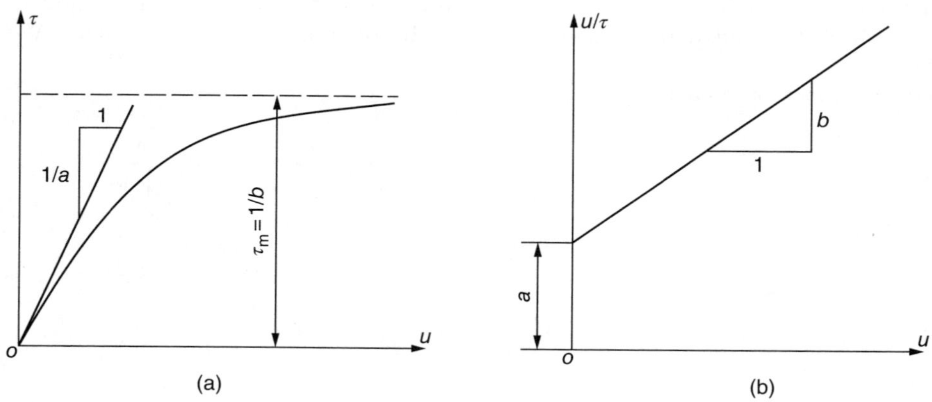

Figure 23.6 $\tau - u$ relationship of rock joint.

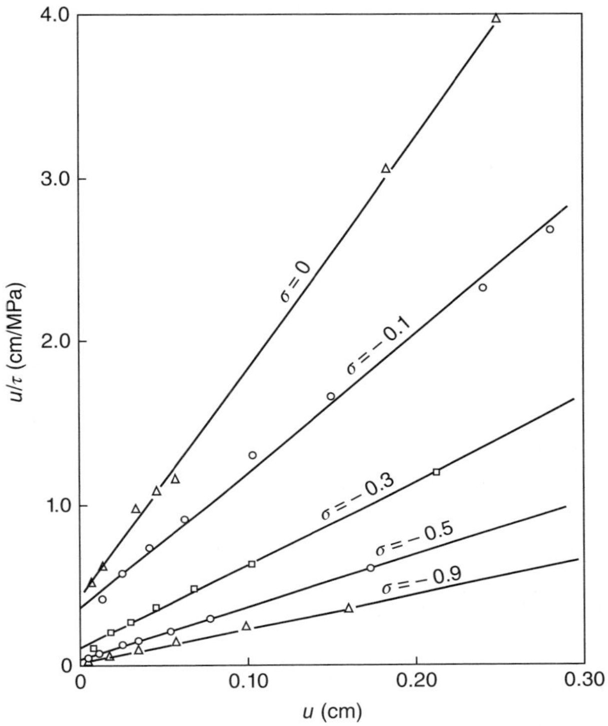

Figure 23.7 Field test data of rock joint of a hydropower station.

If we transform the coordinate system, taking u/τ as the vertical coordinate, as Figure 23.6(b) shows, then Eq. (23.25) could be written as

$$\frac{u}{\tau} = a + bu \tag{23.27}$$

Therefore $u/\tau - u$ is a line and a and b are the intercept and slope of the line, respectively. Loading under different normal stress, we can obtain a group of these lines. Figure 23.7 shows the data of field tests on joint in rock of a hydropower station, arranged by the author according to Eq. (23.27). It is shown that almost all measured points locate on these lines, which highlight the effectiveness of Eq. (23.25) in describing the relation between stress and displacement of joint in rock.

From Eqs (23.23) to 23.25, we get

$$\lambda_s = \frac{\partial \tau}{\partial u} = \frac{a}{(a + bu)^2} \tag{a}$$

According to Eq. (23.25), $u = a\tau/(1 - b\tau)$, substitute it into Eq. (a) and obtain

$$\lambda_s = (1 - b\tau)^2/a \tag{23.28}$$

Then substitute $1/a = \lambda_0$, $1/b = \tau_m$ into Eq. (23.28), and we have

$$\lambda_s = \lambda_0 \left(1 - \frac{\tau}{\tau_m}\right)^2 \tag{23.29}$$

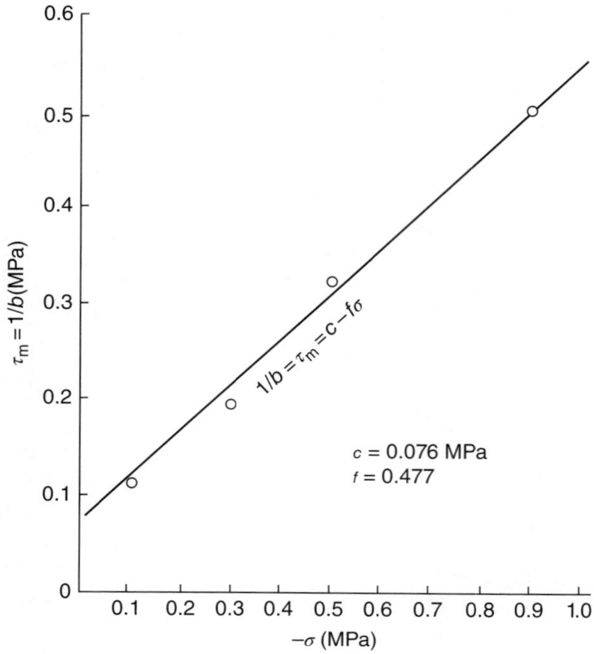

Figure 23.8 $\tau_{\mathrm{m}} - \sigma$ relationship of rock joint.

The relation between shear strength τ_{m} and normal stress could be expressed with the Coulomb equation as follows:

$$\tau_{\mathrm{m}} = \frac{1}{b} = c - f\sigma \tag{23.30}$$

Put the τ_{m} under different normal stress calculated in Figure 23.7 into Figure 23.8, from which we can see that σ and τ_{m} almost satisfy Eq. (23.30). For this group of data, $c = 0.076\,\mathrm{MPa}, f = 0.477$, where c is equivalent to cohesive force and f is equivalent to friction coefficient. It should be noted that due to the sign of theory of elasticity that is applied here, for σ, tensile stress is positive and compressive stress is negative. Thus the σ in Eq. (23.30) should be negative.

Draw the initial tangential stiffness coefficient λ_0 of joint and corresponding normal stress σ on log–log paper, as Figure 23.9 shows. With an appropriate constant s, almost all points locate on a line. Therefore,

$$\lambda_0 = \frac{1}{a} = F(s - \sigma)^n \tag{23.31}$$

where F and n are constants. For data in Figure 23.9, $F = 28.3, n = 1.758, s = 0.20\,\mathrm{MPa}$.

Substitute Eqs (23.30) and (23.31) into Eq. (23.29), consider the Eq. (23.24), and we get the expression of tangential stiffness coefficient of joint as follows:

$$\begin{aligned} &\lambda_{\mathrm{s}} = 0 && (\sigma > 0) \\ &\lambda_{\mathrm{s}} = F(s - \sigma)^n \left(1 - \frac{\tau}{c - f\sigma}\right)^2 && (\sigma \le 0) \end{aligned} \tag{23.32}$$

where F, s, n, c, f are constants determined by measured data.

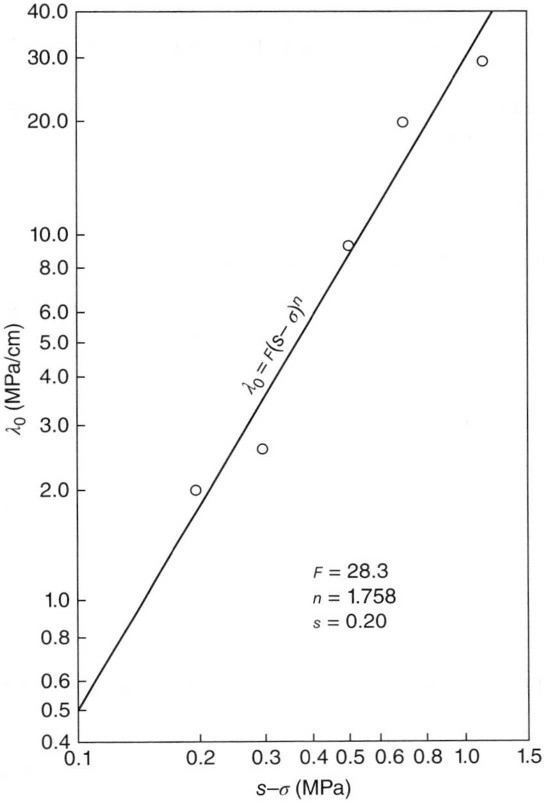

Figure 23.9 $\lambda_0 - \sigma$ relationship of rock joint.

In view of physical significance, s is actually equivalent to tensile stress of joint. In order to take it into consideration and bring in atmospheric pressure p_a and water density γ_w, to make parameter F to be dimensionless, the equation above could be rewritten in the following:

$$\lambda_s = 0 \qquad (\sigma > s)$$

$$\lambda_s = F\gamma_w \left(\frac{s-\sigma}{p_a}\right)^n \left(1 - \frac{\tau}{c - f\sigma}\right)^2 \qquad (\sigma \le s) \tag{23.32a}$$

When the relation of $\tau - u$ is relatively complex, spline function could be used. For soft layer with limit thickness, similar methods can be used to express shear modulus G and elastic modulus E.

23.5 Layer Element

The thickness of joint in rock is near zero. In practical work, it is difficult to determine the thickness of joint according to geological survey data. While things are different for soft intercalated layer, whose thickness varies from several centimeters to more than ten centimeters, the thickness could be determined with geological survey data, and the influence of the thickness of soft intercalated layer is also expected to be considered in engineering analysis. Therefore, it is necessary to reflect the impact of thickness in soft layer element.

Take a two-dimensional element, as shown in Figure 23.4, with length of l, thickness of e, and 4 nodes of i, j, m, r. The strain in the element is

$$\{\varepsilon\} = \begin{Bmatrix} \gamma_s \\ \varepsilon_n \end{Bmatrix} = \begin{Bmatrix} \Delta u/e \\ \Delta v/e \end{Bmatrix} = \frac{1}{e}\{w\} \tag{23.33}$$

According to Eq. (23.15), $\{\Delta\delta\} = [M]\{\delta\}^e$, substitute it into Eq. (23.33) and get

$$\{\varepsilon\} = [B]\{\delta\}^e \tag{23.34}$$

where

$$[B] = \frac{1}{e}[M]$$

The stress in the element is

$$\{\sigma\} = \begin{Bmatrix} \tau_s \\ \sigma_n \end{Bmatrix} = \begin{bmatrix} G & 0 \\ 0 & E \end{bmatrix} \begin{Bmatrix} \gamma_s \\ \varepsilon_n \end{Bmatrix} = [D]\{\varepsilon\} \tag{23.35}$$

where

$$[D] = \begin{bmatrix} G & 0 \\ 0 & E \end{bmatrix} \tag{23.36}$$

According to the universal Eq. (3.11) for stress analysis, the stiffness matrix of joint element is calculated with Eq. (23.37):

$$[k]^e = \int [B]^T[D][B]\,d(\text{vol}) \tag{23.37}$$

Since

$$d(\text{vol}) = tl\,dx, \quad [B] = \frac{1}{e}[M]$$

$$[k]^e = \frac{t}{e}\int_{-l/2}^{l/2} [M]^T[D][M]\,dx \tag{a}$$

Let

$$\lambda_s = \frac{G}{e}, \quad \lambda_n = \frac{E}{e} \tag{23.38}$$

Then

$$[D] = e[\lambda]$$

Substitute it into Eq. (23.37) and we get

$$[k]^e = t\int_{-l/2}^{l/2} [M]^T[\lambda][M]\,dx$$

The equation above is absolutely same with Eq. (23.18).

Thus it can be seen that if we define the stiffness coefficients λ_n and λ_s of joint as Eq. (23.38), joint element with zero thickness and soft layer element with limit thickness share the same stiffness matrix. Therefore, we can use uniform equations to compute joint element and soft layer element. For joint element, λ_n and λ_s obtained from measured data could be used, which include the influence of thickness. For soft layer element, use the λ_n and λ_s defined with Eq. (23.38), which reflect the influence of thickness e. Equation (23.38) was firstly proposed by the author in Ref. [21].

23.6 Two-Dimensional High-Order Joint Element

Joint and soft layers have crucial influence on the stress, strain, and stability of rock. For rock mechanics, joint element plays an important role. In Section 23.3, we introduced the commonly used 4-node joint element. In the next two sections, we will derive several other kinds of joint element. In this section, several kinds of two-dimensional high-order joint element will be introduced.

23.6.1 6-Node Quadratic Joint Element

If we use 6-node triangular element in integral rock, then correspondingly, we could use 6-node joint element, as Figure 23.10 shows. For joint element, take length as l, width as e, and thickness (vertical to the paper) as t. Locate the origin of local coordinate system (x, y) at the centroid of the element. The nodal force $\{F\}^e$ and nodal displacement $\{\delta\}^e$ of the element are, respectively,

$$\{F\}^e = \begin{Bmatrix} U_1 \\ V_1 \\ \vdots \\ U_6 \\ V_6 \end{Bmatrix}, \quad \{\delta\}^e = \begin{Bmatrix} u_1 \\ v_1 \\ \vdots \\ u_6 \\ v_6 \end{Bmatrix} \tag{23.39}$$

Suppose the horizontal displacement along the lower border *123* of the element is

$$u_{\text{lower}} = N_1 u_1 + N_2 u_2 + N_3 u_3$$

The displacement along the upper border *456* is

$$u_{\text{upper}} = N_1 u_4 + N_2 u_5 + N_3 u_6$$

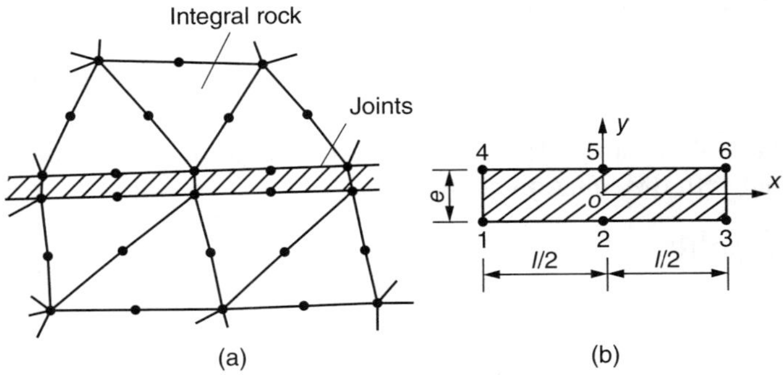

Figure 23.10 6-Node quadratic joint element.

where

$$N_1 = -\frac{x}{l}\left(1 - \frac{2x}{l}\right), \quad N_2 = 1 - \left(\frac{2x}{l}\right)^2, \quad N_3 = \frac{x}{l}\left(1 + \frac{2x}{l}\right) \tag{23.40}$$

in which N_1, N_2, N_3 are shape functions.

Then, the difference between the horizontal displacement of upper and lower borders is

$$\Delta u = u_{\text{upper}} - u_{\text{lower}} = N_1(u_4 - u_1) + N_2(u_5 - u_2) + N_3(u_6 - u_3)$$

Similarly, the difference between the vertical displacements of upper and lower borders is

$$\Delta v = v_{\text{upper}} - v_{\text{lower}} = N_1(v_4 - v_1) + N_2(v_5 - v_2) + N_3(v_6 - v_3)$$

The strain of any point in the element is

$$\{\varepsilon\} = \begin{Bmatrix} \gamma_s \\ \varepsilon_n \end{Bmatrix} = \begin{Bmatrix} \dfrac{\Delta u}{e} \\ \dfrac{\Delta v}{e} \end{Bmatrix} = [B]\{\delta\}^e \tag{23.41}$$

$$[B] = \frac{1}{e}[N] \tag{23.42}$$

$$[N] = \begin{bmatrix} -N_1 & 0 & -N_2 & 0 & -N_3 & 0 & N_1 & 0 & N_2 & 0 & N_3 & 0 \\ 0 & -N_1 & 0 & -N_2 & 0 & -N_3 & 0 & N_1 & 0 & N_2 & 0 & N_3 \end{bmatrix} \tag{23.43}$$

The stress of any point in the element is

$$\{\sigma\} = \begin{Bmatrix} \tau_s \\ \sigma_n \end{Bmatrix} = \begin{bmatrix} G & 0 \\ 0 & E \end{bmatrix} \begin{Bmatrix} \gamma_s \\ \varepsilon_n \end{Bmatrix} + \begin{Bmatrix} \tau_{s0} \\ \sigma_{n0} \end{Bmatrix} = [D]\{\varepsilon\} + \{\sigma_0\} \tag{23.44}$$

$$[D] = \begin{bmatrix} G & 0 \\ 0 & E \end{bmatrix} \tag{23.45}$$

where $\{\sigma_0\}$ is the initial stress.

According to the universal equation of stress analysis, the element stiffness matrix should be calculated with the equation below:

$$[k]^e = \int [B]^T[D][B]\,d(\text{vol})$$

Let $d(\text{vol}) = te\,dx$ and $[B] = \frac{1}{e}[N]$, thus we have

$$[k]^e = \frac{t}{e}\int_{-l/2}^{l/2} [N]^T[D][N]\,dx \tag{23.46}$$

When the width of joint $e \to 0$, substitute $[D] = e[\lambda]$ into Eq. (23.46), and the element stiffness matrix is calculated with the equation below:

$$[k]^e = t\int_{-l/2}^{l/2} [N]^T[\lambda][N]\,dx$$

where

$$[\lambda] = \begin{bmatrix} \lambda_s & 0 \\ 0 & \lambda_n \end{bmatrix}$$

in which λ_s, λ_n are stiffness coefficients.

The load caused by initial stress $\{\sigma_0\}$ can be calculated with Eq. (23.47)

$$\{P\}^{\mathrm{e}}_{\sigma_0} = -t\int_{-l/2}^{l/2} [N]^{\mathrm{T}}\{\sigma_0\}\,\mathrm{d}x \tag{23.47}$$

The calculation above is conducted in local coordinate system. In order to get the stiffness matrix and nodal load of the element in global coordinate system, it is necessary to transform the coordinate system. The process is similar to that in Section 23.1 and is omitted here.

23.6.2 6-Node Curved Joint Element

If we use 8-node curved element for integral rock, then we can use 6-node curved element for joint correspondingly, as Figure 23.11 shows. Suppose the width of joint element is constant e and the upper and lower borders of joint are both curves, the coordinate should be transformed. For example, the coordinates of the lower border of joint are

$$x = N_1x_1 + N_2x_2 + N_3x_3$$
$$y = N_1y_1 + N_2y_2 + N_3y_3$$

where x_i, y_i are coordinates of node i, $i = 1, 2, 3$.

The coordinates of the center line of joint are

$$\left.\begin{aligned} x &= N_1\overline{x_1} + N_2\overline{x_2} + N_3\overline{x_3} \\ y &= N_1\overline{y_1} + N_2\overline{y_2} + N_3\overline{y_3} \end{aligned}\right\} \tag{23.48}$$

where

$$\overline{x_1} = \frac{1}{2}(x_1 + x_4),\quad \overline{x_2} = \frac{1}{2}(x_2 + x_5),\quad \overline{x_3} = \frac{1}{2}(x_3 + x_6)$$

The shape functions N_i are

$$N_1 = -\xi\frac{(1-\xi)}{2},\quad N_2 = 1-\xi^2,\quad N_3 = \xi\frac{(1+\xi)}{2} \tag{23.49}$$

The horizontal displacement along the lower border of the joint element is

$$u_{\text{lower}} = N_1u_1 + N_2u_2 + N_3u_3$$

The horizontal displacement along the upper border of the joint element is

$$u_{\text{upper}} = N_1u_4 + N_2u_5 + N_3u_6$$

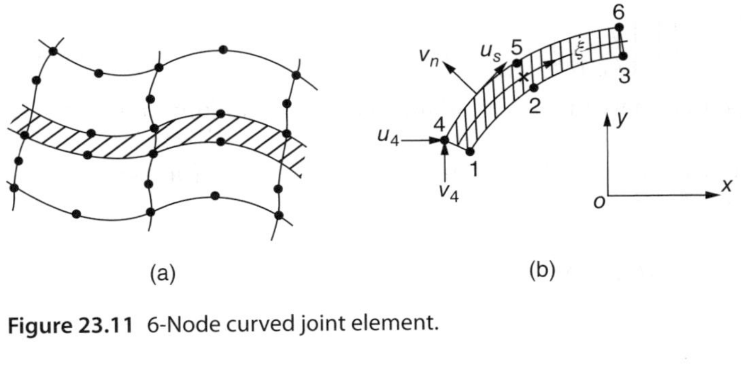

Figure 23.11 6-Node curved joint element.

The difference between the horizontal displacements of the upper and lower borders of the joint element is

$$\Delta u = N_1(u_4 - u_1) + N_2(u_5 - u_2) + N_3(u_6 - u_3)$$

Similarly, the difference between the vertical displacements of the upper and lower borders of the joint element is

$$\Delta v = N_1(v_4 - v_1) + N_2(v_5 - v_2) + N_3(v_6 - v_3)$$

Therefore,

$$\begin{Bmatrix} \Delta u \\ \Delta v \end{Bmatrix} = [N]\{\delta\}^{\mathrm{e}} \tag{23.50}$$

$$[N] = \begin{bmatrix} -N_1 & 0 & -N_2 & 0 & -N_3 & 0 & N_1 & 0 & N_2 & 0 & N_3 & 0 \\ 0 & -N_1 & 0 & -N_2 & 0 & -N_3 & 0 & N_1 & 0 & N_2 & 0 & N_3 \end{bmatrix}$$

Suppose the tangential displacement difference between the upper and lower borders is Δu_{s}, and the normal displacement difference between the upper and lower borders is Δv_{n}. Then Δu_{s} and Δv_{n} are calculated according to Δu and Δv after coordinate transformation as follows:

$$\begin{Bmatrix} \Delta u_{\mathrm{s}} \\ \Delta v_{\mathrm{n}} \end{Bmatrix} = [L]\begin{Bmatrix} \Delta u \\ \Delta v \end{Bmatrix} = [L][N]\{\delta\}^{\mathrm{e}} \tag{23.51}$$

where

$$[L] = \begin{bmatrix} \alpha & \beta \\ -\beta & \alpha \end{bmatrix}$$

$$\alpha = \frac{\dfrac{\partial x}{\partial \xi}}{\sqrt{\left(\dfrac{\partial x}{\partial \xi}\right)^2 + \left(\dfrac{\partial y}{\partial \xi}\right)^2}}, \quad \beta = \frac{\dfrac{\partial y}{\partial \xi}}{\sqrt{\left(\dfrac{\partial x}{\partial \xi}\right)^2 + \left(\dfrac{\partial y}{\partial \xi}\right)^2}}$$

Thus the strain of any point in the element can be expressed by

$$\{\varepsilon\} = \begin{Bmatrix} \Delta u_{\mathrm{s}}/e \\ \Delta v_{\mathrm{n}}/e \end{Bmatrix} = [B]\{\delta\}^{\mathrm{e}} \tag{23.52}$$

where

$$[B] = \frac{1}{e}[L][N] \tag{23.53}$$

The stress of any point in the element can be expressed as

$$\{\sigma\} = \begin{Bmatrix} \tau_{\mathrm{s}} \\ \sigma_{\mathrm{n}} \end{Bmatrix} = \begin{bmatrix} G_{\mathrm{s}} & 0 \\ 0 & E_{\mathrm{n}} \end{bmatrix}\begin{Bmatrix} \gamma_{\mathrm{s}} \\ \varepsilon_{\mathrm{n}} \end{Bmatrix} + \{\sigma_0\} = [D]\{\varepsilon\} + \{\sigma_0\} \tag{23.54}$$

where G_{s}, E_{n} are the tangential shear modulus and normal elastic modulus, respectively, of joint element.

According to the universal equation of stress analysis, the element stiffness matrix is calculated with the Eq. (23.55)

$$[k]^{\mathrm{e}} = \frac{t}{e}\int_{-1}^{1} [N]^{\mathrm{T}}[L]^{\mathrm{T}}[D][L][N]S\,\mathrm{d}\xi \tag{23.55}$$

where

$$S = \sqrt{\left(\frac{\partial x}{\partial \xi}\right)^2 + \left(\frac{\partial y}{\partial \xi}\right)^2}$$

$\frac{\partial x}{\partial \xi}$ and $\frac{\partial y}{\partial \xi}$ should be calculated with Eq. (23.48).

The nodal load caused by initial stress can be calculated with the equation below:

$$\{P\}^{e}_{\sigma_0} = -t \int_{-1}^{1} [N]^{T}[L]^{T}[\sigma_0] S \, d\xi \tag{23.56}$$

Equations (23.55) and (23.56) can be computed with numerical integration. It should be noted that the element matrix and nodal load computed with these two equations are in global coordinate system (x, y), and there is no need for coordinate transformation.

23.7 Three-Dimensional Joint Element

In this section, we will introduce several kinds of three-dimensional joint element for solving spatial problem in rock mechanics.

23.7.1 6-Node Three-Dimensional Joint Element

If we use tetrahedron element for integral rock, then 6-node three-dimensional joint element could be used for joint, as Figure 23.12 shows. There are 6 nodes for each element, with thickness of e. The nodal force $\{F\}^{e}$ and nodal displacement $\{\delta\}^{e}$ of the element are, respectively,

$$\{F\}^{e} = \begin{Bmatrix} U_1 \\ V_1 \\ W_1 \\ \vdots \\ U_6 \\ V_6 \\ W_6 \end{Bmatrix}, \quad \{\delta\}^{e} = \begin{Bmatrix} u_1 \\ v_1 \\ w_1 \\ \vdots \\ u_6 \\ v_6 \\ w_6 \end{Bmatrix}$$

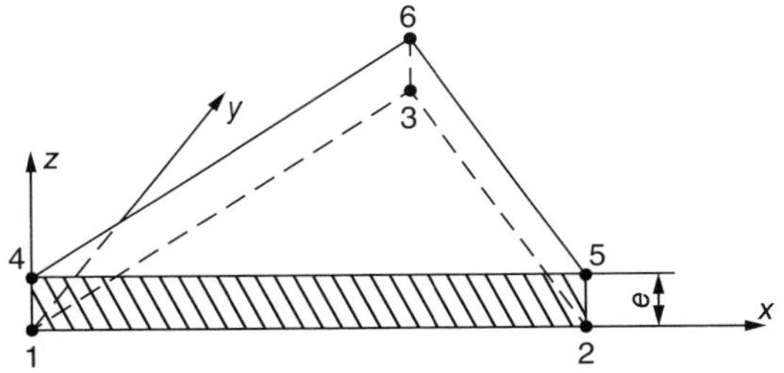

Figure 23.12 6-Node three-dimensional joint element.

The horizontal displacement of any point on the lower surface *123* of the element can be expressed as

$$u_{\text{lower}} = N_1u_1 + N_2u_2 + N_3u_3$$

The horizontal displacement of any point on the upper surface *456* of the element can be expressed as

$$u_{\text{upper}} = N_1u_4 + N_2u_5 + N_3u_6$$

Therefore, for any point (x, y) on the upper and lower surface of the element, the difference of the horizontal displacement in the direction of x is

$$\Delta u = N_1(u_4 - u_1) + N_2(u_5 - u_2) + N_3(u_6 - u_3)$$

where N_1, N_2, N_3 are shape functions below:

$$N_1 = \frac{a_1 + b_1x + c_1y}{2A}, \quad N_2 = \frac{a_2 + b_2x + c_2y}{2A}, \quad N_3 = \frac{a_3 + b_3x + c_3y}{2A} \tag{23.57}$$

where A is the area of triangle *123* and a_i, b_i, c_i are coefficients calculated with Eq. (2.5), $i = 1, 2, 3$.

Similarly, the difference of the horizontal displacements of any point on the upper and lower surface in the direction of y is

$$\Delta v = N_1(v_4 - v_1) + N_2(v_5 - v_2) + N_3(v_6 - v_3)$$

The difference of the vertical displacement of any point on the upper and lower surface in the direction of z is

$$\Delta w = N_1(w_4 - w_1) + N_2(w_5 - w_2) + N_3(w_6 - w_3)$$

Since the thickness e of the element is quite small, the strain of any point in the element can be calculated from the displacement difference of the upper and lower surface as follows:

$$\{\varepsilon\} = \begin{Bmatrix} \gamma_{xz} \\ \gamma_{yz} \\ \varepsilon_{\text{n}} \end{Bmatrix} = \frac{1}{e}\begin{Bmatrix} \Delta u \\ \Delta v \\ \Delta w \end{Bmatrix} = [B]\{\delta\}^{\text{e}} \tag{23.58}$$

where $[B] = \frac{1}{e}[N]$

$$[N] = \begin{bmatrix} -N_1 & 0 & 0 & -N_2 & 0 & 0 & -N_3 & 0 & 0 & N_1 & 0 & 0 & N_2 & 0 & 0 & N_3 & 0 & 0 \\ 0 & -N_1 & 0 & 0 & -N_2 & 0 & 0 & -N_3 & 0 & 0 & N_1 & 0 & 0 & N_2 & 0 & 0 & N_3 & 0 \\ 0 & 0 & -N_1 & 0 & 0 & -N_2 & 0 & 0 & -N_3 & 0 & 0 & N_1 & 0 & 0 & N_2 & 0 & 0 & N_3 \end{bmatrix} \tag{23.59}$$

The stress of any point (x, y) in the element can be expressed as

$$\{\sigma\} = [D]\{\varepsilon\} + \{\sigma_0\} \tag{23.60}$$

where

$$[D] = \begin{bmatrix} G & 0 & 0 \\ 0 & G & 0 \\ 0 & 0 & E \end{bmatrix} \tag{23.61}$$

in which G is shear modulus, E is elastic modulus, and $\{\sigma_0\}$ is initial stress.

According to the principle of virtual work, the stiffness matrix of the element can be calculated as below:

$$[k]^{\mathrm{e}} = \frac{1}{e}\iint [N]^{\mathrm{T}}[D][N]\,\mathrm{d}x\,\mathrm{d}y \tag{23.62}$$

The nodal load caused by the initial stress $\{\sigma_0\}$ can be calculated with Eq. (23.63):

$$\{P\}^{\mathrm{e}}_{\sigma_0} = -\iint [N]^{\mathrm{T}}\{\sigma_0\}\,\mathrm{d}x\,\mathrm{d}y \tag{23.63}$$

According to the feature of shape function N_i, there are following relationships:

$$\iint N_i\,\mathrm{d}x\,\mathrm{d}y = \frac{A}{3}$$

$$\iint N_i N_j \mathrm{d}x\mathrm{d}y = \begin{cases} \dfrac{A}{6} & (i = j) \\ \dfrac{A}{12} & (i \neq j) \end{cases}$$

where A is the area of triangle *123*. Based on the relationships above, we can easily calculate each element in $[k]^{\mathrm{e}}$ and $\{P\}^{\mathrm{e}}_{\sigma_0}$.

If the thickness of joint $e \to 0$, it can be assumed that the component of stress in the element is proportional to the displacement difference, that is,

$$\tau_{xz} = \lambda_{\mathrm{s}}\Delta u + \tau_{xz0}$$

$$\tau_{yz} = \lambda_{\mathrm{s}}\Delta v + \tau_{yz0}$$

$$\sigma_{\mathrm{n}} = \lambda_{\mathrm{n}}\Delta w + \sigma_{\mathrm{n}0}$$

that is,

$$\{\sigma\} = [\lambda]\begin{Bmatrix} \Delta u \\ \Delta v \\ \Delta w \end{Bmatrix} + \{\sigma_0\} \tag{23.64}$$

$$[\lambda] = \begin{bmatrix} \lambda_{\mathrm{s}} & 0 & 0 \\ 0 & \lambda_{\mathrm{s}} & 0 \\ 0 & 0 & \lambda_{\mathrm{n}} \end{bmatrix} \tag{23.65}$$

In Eq. (23.62), substitute $[D]$ with $e[\lambda]$, and we can get the following equation when the thickness of joint $e \to 0$:

$$[k]^{\mathrm{e}} = \iint [N]^{\mathrm{T}}[\lambda][N]\,\mathrm{d}x\,\mathrm{d}y \tag{23.66}$$

The calculation above is conducted within local coordinate system. Coordinate systems should be transformed when establishing the global nodal equilibrium equation set.

23.7.2 Three-Dimensional Curved Joint Element

If we use 20-node isoparametric curved element for integral rock, for joint, we must correspondingly use a 16-node curved joint element, as Figure 23.13 shows. Assume

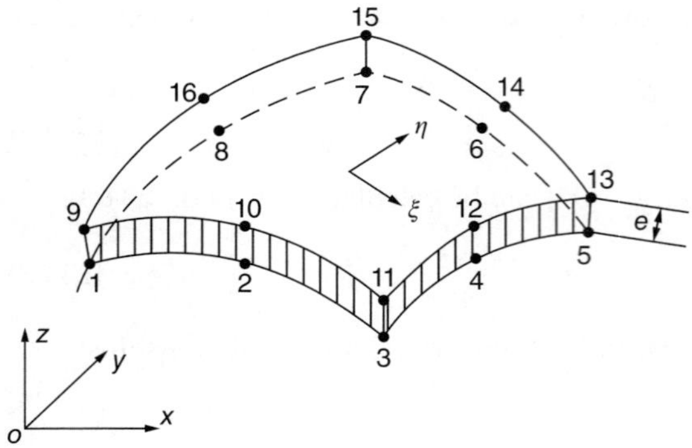

Figure 23.13 Three-dimensional curved joint element.

the thickness e of the joint element is a constant and the upper and lower surfaces of the join are curved surface. Then we need to transform the coordinate system. For example, the coordinates of the lower surface are

$$x = N_1 x_1 + N_2 x_2 + \cdots + N_8 x_8$$

$$y = N_1 y_1 + N_2 y_2 + \cdots + N_8 y_8$$

$$z = N_1 z_1 + N_2 z_2 + \cdots + N_8 z_8$$

$$\text{For corner point}: \quad N_i = \frac{1}{4}(1+\xi_0)(1+\eta_0)(\xi_0+\eta_0-1) \quad (i = 1,3,5,7)$$

$$\text{For middle point}: \quad N_i = \frac{1}{2}(1-\xi^2)(1+\eta_0) \quad (i = 2,6)$$

$$N_i = \frac{1}{2}(1-\eta^2)(1+\xi_0) \quad (i = 4,8)$$

$$\xi_0 = \xi_i \xi, \quad \eta_0 = \eta_i \eta$$

where x_i, y_i, z_i are the coordinates of node i and N_i is shape function.

The middle surface of the joint element is also curved surface, whose coordinates could be calculated with Eq. (23.67):

$$x = \sum_1^8 N_i \overline{x_i}, \quad y = \sum_1^8 N_i \overline{y_i}, \quad z = \sum_1^8 N_i \overline{z_i} \tag{23.67}$$

where

$$\overline{x_i} = \frac{1}{2}(x_i + x_{i+8}), \quad \overline{y_i} = \frac{1}{2}(y_i + y_{i+8}), \quad \overline{z_i} = \frac{1}{2}(z_i + z_{i+8})$$

The displacement difference between upper and lower surfaces in the direction of x, y, z can be expressed as

$$\begin{Bmatrix} \Delta u \\ \Delta v \\ \Delta w \end{Bmatrix} = [N]\{\delta\}^{\mathrm{e}} \tag{23.68}$$

$$[N] = \begin{bmatrix} -N_1 & 0 & 0 & -N_2 & 0 & 0 & \cdots & N_1 & 0 & 0 & \cdots & N_8 & 0 & 0 \\ 0 & -N_1 & 0 & 0 & -N_2 & 0 & \cdots & 0 & N_1 & 0 & \cdots & 0 & N_8 & 0 \\ 0 & 0 & -N_1 & 0 & 0 & -N_2 & \cdots & 0 & 0 & N_1 & \cdots & 0 & 0 & N_8 \end{bmatrix}$$

Now we establish a local coordinate system (x', y', z') at any point (ξ_0, η_0) in the element, and the axis z' is required to be perpendicular to the middle surface of the element. At the point (ξ_0, η_0), there are two curves on the middle surface: one is $\xi = \xi_0$ and the other is $\eta = \eta_0$. Take two tangent vectors $\overrightarrow{d\xi}$ and $\overrightarrow{d\eta}$ on the middle surface of the element. Among them, $\overrightarrow{d\xi}$ is tangent to curve $\eta = \eta_0$ and $\overrightarrow{d\eta}$ is tangent to curve $\xi = \xi_0$. From this we know that

$$\overrightarrow{d\xi} = \left(\vec{i}\frac{\partial x}{\partial \xi} + \vec{j}\frac{\partial y}{\partial \xi} + \vec{k}\frac{\partial z}{\partial \xi}\right) d\xi$$

$$\overrightarrow{d\eta} = \left(\vec{i}\frac{\partial x}{\partial \eta} + \vec{j}\frac{\partial y}{\partial \eta} + \vec{k}\frac{\partial z}{\partial \eta}\right) d\eta$$

where $\vec{i}$, $\vec{j}$, $\vec{k}$ are unit vectors in direction of x, y, z, respectively.

Let z' be vertical to $\overrightarrow{d\xi}$ and $\overrightarrow{d\eta}$, that is, vertical to the middle surface of the element. According to vector algebra,

$$z' = \overrightarrow{d\xi} \times \overrightarrow{d\eta} = \begin{vmatrix} \vec{i} & \vec{j} & \vec{k} \\ \frac{\partial x}{\partial \xi} & \frac{\partial y}{\partial \xi} & \frac{\partial z}{\partial \xi} \\ \frac{\partial x}{\partial \eta} & \frac{\partial y}{\partial \eta} & \frac{\partial z}{\partial \eta} \end{vmatrix}$$

Thus we can calculate the direction cosines l_3, m_3, n_3 of axis z'. The other two axes y' and x' are tangent to the middle surface of the element and may be computed with following method. Let

$$y' = z' \times x' = \begin{vmatrix} \vec{i} & \vec{j} & \vec{k} \\ l_3 & m_3 & n_3 \\ 1 & 0 & 0 \end{vmatrix}$$

From this we can get the direction cosines l_2, m_2, n_2 of axis y'. If axis z' is parallel to axis x, $y' = z' \times y$ could be used. A subprogram ought to be written to verify if z' is parallel to x. Finally, take x' orthogonal to y' and z', that is,

$$x' = y' \times z'$$

Thus we can calculate the direction cosines l_1, m_1, n_1 of axis x'. Then the matrix of direction cosines for local coordinate system (x', y', z') is as below:

$$[L] = \begin{bmatrix} l_1 & l_2 & l_3 \\ m_1 & m_2 & m_3 \\ n_1 & n_2 & n_3 \end{bmatrix} \tag{23.69}$$

For any point (ξ, η) in the joint element, the displacement difference between the upper and lower surfaces in the direction of x', y', z' can be calculated by Eq. (23.68) with coordinate transformation:

$$\begin{Bmatrix} \Delta u' \\ \Delta v' \\ \Delta w' \end{Bmatrix} = [L]\begin{Bmatrix} \Delta u \\ \Delta v \\ \Delta w \end{Bmatrix} = [L][N]\{\delta\}^{\mathrm{e}} \tag{23.70}$$

Considering the thickness e of joint element is quite small, the strain of any point in the local coordinate systems can be expressed as follows:

$$\{\varepsilon'\} = \begin{Bmatrix} \gamma_{x'z'} \\ \gamma_{y'z'} \\ \varepsilon_{z'} \end{Bmatrix} = \frac{1}{e} \begin{Bmatrix} \Delta u' \\ \Delta v' \\ \Delta w' \end{Bmatrix} = [B']\{\delta\}^{\mathrm{e}} \tag{23.71}$$

$$[B'] = \frac{1}{e}[L][N] \tag{23.72}$$

In local coordinate system, the stress of any point is

$$\{\sigma'\} = \begin{Bmatrix} \tau_{x'y'} \\ \tau_{y'z'} \\ \sigma_{z'} \end{Bmatrix} = [D']\{\varepsilon'\} + \{\sigma_0'\} \tag{23.73}$$

where

$$[D'] = \begin{bmatrix} G_{\mathrm{s}} & 0 & 0 \\ 0 & G_{\mathrm{s}} & 0 \\ 0 & 0 & E_{\mathrm{n}} \end{bmatrix} \tag{23.74}$$

in which G_{s}, E_{n} are, respectively, tangential shear modulus and normal elastic modulus of joint element.

The stiffness matrix of the element can be calculated by the equation below:

$$[k]^{\mathrm{e}} = \frac{1}{e}\int_{-1}^{1}\int_{-1}^{1} [N]^{\mathrm{T}}[L]^{\mathrm{T}}[D'][L][N]|J|\,\mathrm{d}\xi\,\mathrm{d}\eta \tag{23.75}$$

where

$$|J| = \left\{ \left(\frac{\partial x}{\partial \xi}\frac{\partial y}{\partial \eta} - \frac{\partial x}{\partial \eta}\frac{\partial y}{\partial \xi}\right)^2 + \left(\frac{\partial y}{\partial \xi}\frac{\partial z}{\partial \eta} - \frac{\partial y}{\partial \eta}\frac{\partial z}{\partial \xi}\right)^2 + \left(\frac{\partial z}{\partial \xi}\frac{\partial x}{\partial \eta} - \frac{\partial z}{\partial \eta}\frac{\partial x}{\partial \xi}\right)^2 \right\}^{1/2}$$

The nodal load caused by initial stress is

$$\{P\}_{\sigma_0}^{\mathrm{e}} = -\int_{-1}^{1}\int_{-1}^{1} [N]^{\mathrm{T}}[L]^{\mathrm{T}}\{\sigma_0'\}|J|\,\mathrm{d}\xi\,\mathrm{d}\eta \tag{23.76}$$

The stiffness matrix and nodal load of the element calculated in this way belong to the global coordinate system.

In the two sections above, the equations of two-dimensional high-order joint element and three-dimensional joint element are first given by the author in [21].

23.8 Infinite Joint Element

Infinite joint element is used to simulate the joint extending to infinite distance. Ge Xiurun suggested a formula for the computing of three-dimensional 10-node infinite joint element [8]. We will introduce the formula for 4-node joint element in plane problem and 8-node and 10-node joint elements in space problem.

23.8.1 Infinite Joint Element in Plane Problem

Figure 23.14 shows a 4-node joint element in plane problem. The thickness e of the element is zero or a very small number. The thickness vertical to the paper is t. The coordinates of any point on the middle surface of the element can be expressed as follows:

$$x = \sum_{i=1}^{2} N_i x_i, \quad y = \sum_{i=1}^{2} N_i y_i \qquad \text{(a)}$$

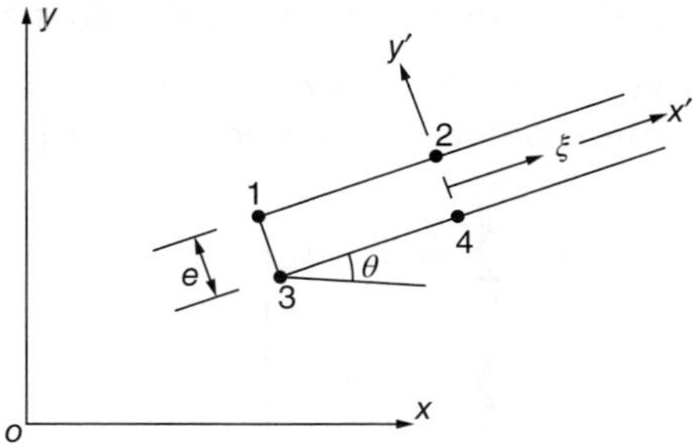

Figure 23.14 Infinite joint element in plane problem.

The shape functions are as follows:

$$\left.\begin{aligned} N_1 &= -\xi, \quad N_2 = 1+\xi && (\xi \le 0) \\ N_1 &= -\xi/(1-\xi), \quad N_2 = 1+\xi/(1-\xi) && (\xi > 0) \end{aligned}\right\} \qquad (23.77)$$

In global coordinate system, the displacement difference between the upper and lower walls of the joint is

$$\{\Delta\delta\} = \begin{Bmatrix} \Delta u \\ \Delta v \end{Bmatrix} = [M]\{\delta\}^{\mathrm{e}} \qquad (23.78)$$

$$\{\delta\}^{\mathrm{e}} = \begin{bmatrix} u_1 & v_1 & u_2 & v_2 & u_3 & v_3 & u_4 & v_4 \end{bmatrix}^{\mathrm{T}}$$

$$[M] = \begin{bmatrix} M_1 & 0 & M_2 & 0 & -M_1 & 0 & -M_2 & 0 \\ 0 & M_1 & 0 & M_2 & 0 & -M_1 & 0 & -M_2 \end{bmatrix} \qquad (23.79)$$

where the shape function M_i is as follows:

$$M_i = M_i^0 f(r_i/r)$$

in which M_i^0 is the shape function N_i for $\xi \le 0$ and $f(r_i/r)$ is attenuation function, as Eq. (8.77) shows.

In local coordinate system (x', y'), the displacement difference between the upper and lower surfaces of the joint is

$$\{\Delta\delta'\} = \begin{Bmatrix} \Delta u' \\ \Delta v' \end{Bmatrix} = [L]\{\Delta\delta\} = [L][M]\{\delta\}^{\mathrm{e}} \qquad (23.80)$$

$$[L] = \begin{bmatrix} \cos\theta & \sin\theta \\ -\sin\theta & \cos\theta \end{bmatrix}$$

In local coordinate system, the stress of any point is

$$\{\sigma'\} = \begin{Bmatrix} \tau_{\mathrm{s}} \\ \tau_{\mathrm{n}} \end{Bmatrix} = \begin{bmatrix} \lambda_{\mathrm{s}} & 0 \\ 0 & \lambda_{\mathrm{n}} \end{bmatrix} \begin{Bmatrix} \Delta u \\ \Delta v \end{Bmatrix} = [\lambda][L][M]\{\delta\}^{\mathrm{e}} \qquad (23.81)$$

The stiffness matrix of the element is

$$[k]^{\mathrm{e}} = \int_{-1}^{1} [M]^{\mathrm{T}}[L]^{\mathrm{T}}[\lambda][L][M] S \,\mathrm{d}\xi \qquad (23.82)$$

$$S = \sqrt{\left(\frac{\partial x}{\partial \xi}\right)^2 + \left(\frac{\partial y}{\partial \xi}\right)^2} = \left[\left(\sum \frac{\mathrm{d}N_i}{\mathrm{d}\xi} x_i\right)^2 + \left(\sum \frac{dN_i}{d\xi} y_i\right)^2\right]^{1/2} \qquad \text{(b)}$$

23.8.2 Infinite Joint Element in Spatial Problem

Figure 23.15 shows two kinds of infinite joint element in spatial problem.

The coordinates of any point on the middle surface of the element can be expressed by

$$\left.\begin{aligned} x &= \sum_{i=1}^{s} N_i(\xi,\eta)x_i \\ y &= \sum_{i=1}^{s} N_i(\xi,\eta)y_i \\ z &= \sum_{i=1}^{s} N_i(\xi,\eta)z_i \end{aligned}\right\} \tag{c}$$

For 8-node element, $s = 4$; for 10-node element, $s = 5$. For 8-node element, the shape functions are

$$\left.\begin{aligned} \eta \le 0: \quad & N_1 = -\frac{1}{2}(1-\xi)\eta, \quad N_2 = -\frac{1}{2}(1+\xi)\eta \\ & N_3 = \frac{1}{2}(1+\xi)(1+\eta), \quad N_4 = \frac{1}{2}(1-\xi)(1+\eta) \\ \eta > 0: \quad & N_1 = -\frac{(1-\xi)\eta}{2(1-\eta)}, \quad N_2 = -\frac{(1+\xi)\eta}{2(1-\eta)} \\ & N_3 = \frac{1+\xi}{2}\left(1+\frac{\eta}{1-\eta}\right), \quad N_4 = \frac{1-\xi}{2}\left(1+\frac{\eta}{1-\eta}\right) \end{aligned}\right\} \tag{23.83}$$

For 10-node element, the shape functions are

$$\left.\begin{aligned} \eta \le 0: \quad & N_i = -(1+\xi_i\xi)\eta(\xi_i\xi-\eta-1)/2 \qquad (i = 1,3) \\ & N_i = (1+\xi_i\xi)(1-\eta)^2 \qquad (i = 4,5) \\ & N_2 = -(1-\xi^2)\eta \\ \eta > 0: \quad & N_i = -0.5(1+\xi_i\xi)\eta/(1-\eta) \qquad (i = 1,3) \\ & N_i = (1+\xi_i\xi)/2(1-\eta) \qquad (i = 4,5) \\ & N_2 = 0 \end{aligned}\right\} \tag{23.84}$$

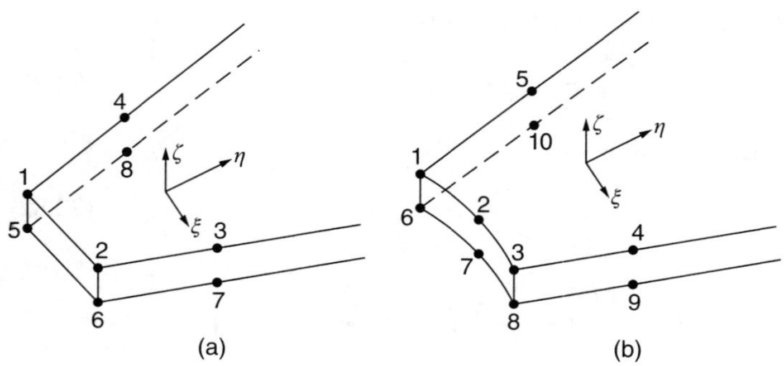

Figure 23.15 Infinite joint element in spatial problem (a) 8-node element and (b) 10-node element.

The displacement difference between the upper and lower surfaces is

$$\{\Delta\delta\} = \left[\Delta u \ \Delta v \ \Delta w\right]^{\mathrm{T}} = [M]\{\delta\}^{\mathrm{e}} \tag{d}$$

$$\{\delta\}^{\mathrm{e}} = \left[u_1 \ v_1 \ w_1 \ u_2 \ v_2 \ w_2 \ \cdots \ u_{2s} \ v_{2s} \ w_{2s}\right]^{\mathrm{T}} \tag{e}$$

$$[M] = \begin{bmatrix} M_1 & 0 & 0 & M_2 & 0 & 0 & \cdots & -M_{2s} & 0 & 0 \\ 0 & M_1 & 0 & 0 & M_2 & 0 & \cdots & 0 & -M_{2s} & 0 \\ 0 & 0 & M_1 & 0 & 0 & M_2 & \cdots & 0 & 0 & -M_{2s} \end{bmatrix} \tag{f}$$

$$M_i = M_i^0 f(r_i/r)$$

where M_i^0 is the shape function N_i for $\eta < 0$.

Suppose z' is the normal direction of the middle surface of the joint. Then in local coordinate system (x', y', z'), the displacement difference between the upper and lower surfaces of the joint is

$$\{\Delta\delta'\} = \left[\Delta u' \ \Delta v' \ \Delta w'\right]^{\mathrm{T}} = [L]\{\Delta\delta\} = [L][M]\{\delta\}^{\mathrm{e}} \tag{g}$$

where

$$[L] = \begin{bmatrix} l_1 & m_1 & n_1 \\ l_2 & m_2 & n_2 \\ l_3 & m_3 & n_3 \end{bmatrix}$$

in which $l_i, m_i, n_i \ (i = 1, 2, 3)$ are direction cosines of x', y', z'.

In local coordinate system (x', y', z'), the stress of any point is

$$\{\sigma'\} = \left[\tau_{z'x'} \ \tau_{z'y'} \ \sigma_{z'}\right]^{\mathrm{T}} = [\lambda]\{\Delta\delta'\} = [\lambda][L][M]\{\delta\}^{\mathrm{e}} \tag{23.85}$$

where

$$[\lambda] = \begin{bmatrix} \lambda_{\mathrm{s}} & 0 & 0 \\ 0 & \lambda_{\mathrm{s}} & 0 \\ 0 & 0 & \lambda_{\mathrm{n}} \end{bmatrix} \tag{h}$$

The stiffness matrix of the element is

$$[k]^{\mathrm{e}} = \int_{-1}^{1}\int_{-1}^{1} [M]^{\mathrm{T}}[L]^{\mathrm{T}}[\lambda][L][M]A \,\mathrm{d}\xi \,\mathrm{d}\eta \tag{23.86}$$

in which A is given in Eq. (8.61).

23.9 Choice of Method for Stress Analysis in Rock

The methods for stress analysis in rock can be classified into two main types: the linear elastic method and the nonlinear method.

There are several kinds of nonlinear method: (1) nonlinear elastic computing method, (2) initial stress method and no tension method, (3) elastic–plastic method, and (4) viscoelastic–plastic method.

We will expound the nonlinear methods in Sections 23.10–23.13, and the amount of their calculation differs a lot.

When choosing method for stress analysis in rock, the following factors ought to be considered.

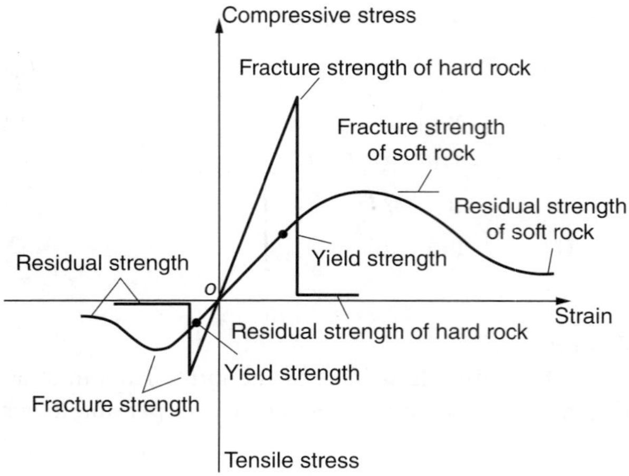

Figure 23.16 Stress–strain curves of hard rock and soft rock.

23.9.1 The Objective and Importance of Analysis

Obviously, refined method should be used for important analysis, while relatively sketchy method might be used for less important analysis.

23.9.2 Character of Rock

Figure 23.16 shows typical stress–strain curves of rock. It is clear that there is great difference between the stress–strain curves of hard rock (such as granite) and soft rock (such as mudstone, coal).

23.9.3 Buried Depth of the Project

For excavation near the surface or shallow-buried tunnels or caverns, due to the joint with low strength in rock, the yield and failure of rock are prevented. Thus the rock can be computed as elastomer while considering the nonlinearity in the computing of joint. In contrast, for deep-buried mines, due to the high ground stress and small amount of joints, the rock itself might yield or fail.

23.9.4 Original Data

For more refined computing method, the more original data (computing parameters) are required. Therefore, the availability of original data also must be taken into consideration when choosing appropriate computing method.

23.10 Elastic Increment Method for Nonlinear Stress Analysis

Compute with increment method using the tangential elastic modulus in each step. The fundamental equation is

$$[K_i]\{\Delta\delta_{i+1}\} = \{\Delta P_{i+1}\} - \{\psi_i\} \tag{23.87}$$

$$\{\delta_{i+1}\} = \{\delta_i\} + \{\Delta\delta_{i+1}\}, \quad \{\sigma_{i+1}\} = \{\sigma_i\} + \{\Delta\sigma_{i+1}\} \tag{23.88}$$

where $\{\Delta\delta_{i+1}\}$ is array of displacement increment, $\{\Delta P_{i+1}\}$ is array of load increment, $\{\psi_i\}$ is array of out-of-balance force, and $\{K_i\}$ is stiffness matrix.

23.10.1 Fracture and Slide of Rock

As Figure 23.17 shows, take the rock as stratified anisotropy material. In local coordinate system, the elastic modulus in the direction of x' is E_i, the elastic modulus in the direction of y' is E_2, and the shear modulus is G_2. The relationships of stress–strain are as follows:

$$\begin{aligned}
\varepsilon_{x'} &= \frac{\sigma_{x'}}{E_1} - \frac{\mu_2\sigma_{y'}}{E_2} - \frac{\mu_1\sigma_{z'}}{E_1}\\
\varepsilon_{y'} &= \frac{\sigma_{y'}}{E_2} - \frac{\mu_2\sigma_{x'}}{E_2} - \frac{\mu_2\sigma_{z'}}{E_2}\\
\varepsilon_{z'} &= \frac{\sigma_{z'}}{E_1} - \frac{\mu_1\sigma_{x'}}{E_1} - \frac{\mu_2\sigma_{y'}}{E_2}\\
\gamma_{x'y'} &= \frac{\tau_{x'y'}}{G_2}
\end{aligned}$$

Corresponding elasticity matrix can be seen in Eqs (2.25) and (2.26).

Using the anisotropy elastic matrix, we can consider the fracture and slide of rock. For example, when $\sigma_{y'}$ exceeds the tensile strength and the rock fractures, we can take $E_2 = G_2 = 0$ (in actual computing, a value near zero should be used to avoid overflow during matrix inversion). If the normal stress is compressive stress, while the shear stress $\tau_{x'y'}$ exceeds the shear strength, the rock will slide, and we can take $G_2 = 0$.

Take the axis of x' in local coordinate system parallel to stratification. The stress in local coordinate system should satisfy the following condition:

$$\sigma_{y'} \le 0, \quad |\tau_{x'y'}| \le c - f\sigma_{y'} \tag{a}$$

Check each element, and the criterions are as follows:

1) If $\sigma_{y'} > 0$, the layers will be pulled apart; thus we should take $E_2 = G_2 = 0$ in the computing in the next step. Meanwhile, $\sigma_{y'}$ and $\tau_{x'y'}$ on the layer should be eliminated, that is,

$$\{\sigma_0'\} = \begin{Bmatrix} 0 \\ -\sigma_{y'} \\ -\tau_{x'y'} \end{Bmatrix}$$

2) If $\sigma_{y'} < 0$, $|\tau_{x'y'}| > c - f\sigma_{y'}$, the layer will slide; thus we should take $G_2 = 0$ and

$$\{\sigma_0'\} = \begin{Bmatrix} 0 \\ 0 \\ \eta f'|\sigma_{y'}| - \tau_{x'y'} \end{Bmatrix}$$

When $\tau_{x'y'}$ is positive, $\eta = +1$; when $\tau_{x'y'}$ is negative, $\eta = -1$. f' is residual friction coefficient.

Calculate $\{\sigma_0'\}$ with the out-of-balance force $\{\psi_i\}$.

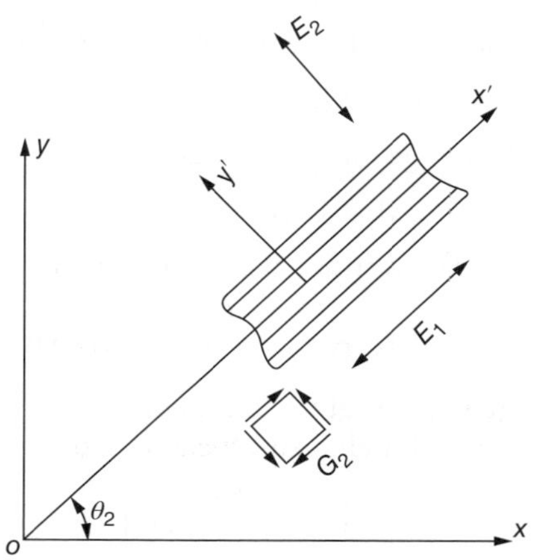

Figure 23.17 Anisotropy rock.

3) If $\sigma_{y'} < 0$, $|\tau_{x'y'}| < c - f\sigma_{y'}$, then the layer will neither be pulled apart nor slide and no adjustment is necessary.

Although the initial stress appears during the computing, the corresponding modulus has been taken as zero and will be eliminated after a single step; thus it is unnecessary for repeated iteration as the initial stress method in the next section.

23.10.2 Fracture and Slide of Joint and Soft Layer

If we use a hyperbola to express the tangential stiffness coefficient of joint, the criterion could be taken as follows:

1) If $\sigma'_{\rm n} > 0$, the joint surface will be pulled apart. Let $\lambda_{\rm n} = \lambda_{\rm s} = 0$ and take

$$\{\sigma'_0\} = -\begin{Bmatrix} \tau'_{\rm s} \\ \sigma'_{\rm n} \end{Bmatrix}$$

2) If $\sigma'_{\rm n} < 0$, the joint surface will not be pulled apart. In this case, $\lambda_{\rm n}$ can be taken as specified constant, while $\lambda_{\rm s}$ can be calculated with Eq. (23.32).

If test data is insufficient and $\lambda_{\rm s}$ cannot be expressed with hyperbola, when $|\tau'_{\rm s}| > -f\sigma'_{\rm n}$, we can take $\lambda_{\rm s} = 0$.

The author and Song applied the methods mentioned in the above sections to the computing of Shuangpai dam and other projects and achieved satisfactory results [16].

23.11 Initial Stress Method and No Tension Method

Using the full variable method, apply all of the load at once.

For given initial elastic modulus E_0 of rock, the initial elasticity matrix is $[D_0]$. Solve this problem linearly and get the linear stress below:

$$\{\sigma_{\rm e}\} = [D_0]\{\varepsilon\} \tag{23.89}$$

Suppose the allowable stress provided by the stress–strain relation of rock is $\{\sigma_{\rm b}\}$. Since the stress–strain relation of rock is nonlinear, the allowable stress $\{\sigma_{\rm b}\}$ will be different with the linear stress $\{\sigma_{\rm e}\}$. We introduce the initial stress $\{\sigma_0\}$ as follows:

$$\{\sigma_{\rm b}\} = [D_0]\{\varepsilon\} + \{\sigma_0\} = \{\sigma_{\rm e}\} + \{\sigma_0\} \tag{23.90}$$

Hence the initial stress is

$$\{\sigma_0\} = \{\sigma_{\rm b}\} - \{\sigma_{\rm e}\} \tag{23.91}$$

For conventional element, the nodal load caused by initial stress is

$$\{P\}_{\sigma_0} = -t\iint [B]^{\rm T}\{\sigma_0\}\,{\rm d}x\,{\rm d}y \tag{23.92}$$

where $[B]$ is given in Eq. (4.11).

For joint element, the nodal load caused by initial stress is

$$\{P\}_{\sigma_0} = -t\int_{-l/2}^{l/2} [M]^{\rm T}\{\sigma_0\}\,{\rm d}x \tag{23.93}$$

where $[M]$ is given in Eq. (23.16).

After computing the load caused by initial stress, we can make an adjustment. After successive adjustments, we will reach the final result.

From the analysis above, it is clear that after the allowable stress $\{\sigma_b\}$ is derived from the stress–strain relation of rock, it will be easy to compute with initial stress method.

23.11.1 No Tension Method

This is the easiest way for nonlinear analysis of rock, proposed by Zienkiewicz *et al.* [2]. Considering there are many fractures distributed randomly in rock, it could be assumed that the rock cannot bear any tensile stress.

Take plane problem as example. Suppose the normal directions of linear stress are $x'y'$, the linear stress is

$$\{\sigma'_e\} = \left[\sigma_{x'} \ \ \sigma_{y'} \ \ 0\right]^T$$

If $\sigma_{x'}$ is tensile stress, then the initial stress is

$$\{\sigma'_0\} = \left[-\sigma_{x'} \ \ 0 \ \ 0\right]^T$$

If $\sigma_{x'}$ and $\sigma_{y'}$ are both tensile stress, then the initial stress is

$$\{\sigma'_0\} = \left[-\sigma_{x'} \ \ -\sigma_{y'} \ \ 0\right]^T$$

Transform the initial stress into the global coordinate system, compute the nodal load caused by the initial stress, and conduct one iteration. In this way, compute iteratively until normal tensile stress is eliminated.

Figure 23.18 is an example computed with no tension method [13]. This is a gravity dam on rock foundation. The linear computation result is shown in Figure 23.18(a), indicating large area with tensile stress at the dam heel. Figure 23.18(b) shows the result of nonlinear computation, which assumes rock is unbearable for tensile stress, that is, the normal stress cannot be greater than zero. A cracked area appears in the rock of the dam heel.

23.11.2 Fracture and Slide of Stratified Rock

When a large amount of joints with similar direction exists, it can be approximately taken as stratified material. Assume that the rock cannot bear tensile stress in the direction vertical to stratification while can sustain shear stress, not greater than friction force, in the direction parallel to stratification.

As Figure 23.19 shows, suppose the axis x' of local coordinate system is parallel to stratification. In local coordinate system, the stress should satisfy the following conditions:

$$\sigma_{y'} \le 0, \quad |\tau_{x'y'}| \le c - f\sigma_{y'} \tag{23.94}$$

where c is cohesive force on layer's surface and f is friction coefficient.

The computing procedure is as follows:

1) Compute elastic displacement:

$$\{\delta\} = [K]^{-1}\{P\}$$

Then compute elastic stress $\{\sigma\}$.

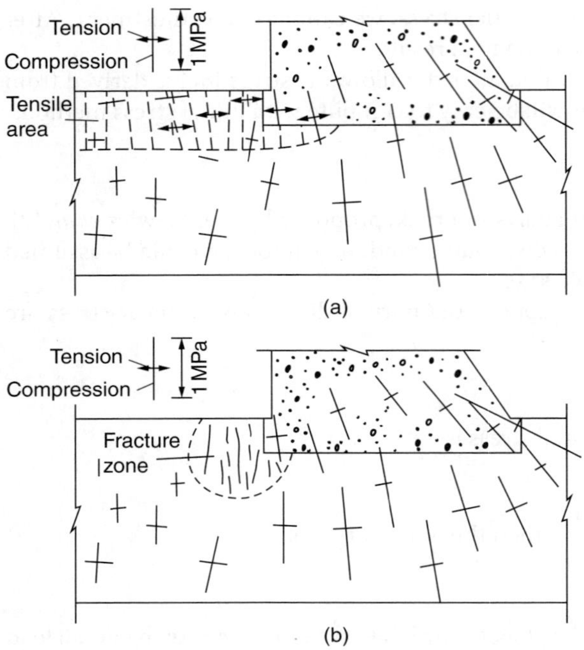

Figure 23.18 Gravity dam on rock foundation (a) elastic stress and (b) stress when rock is a no-tension material.

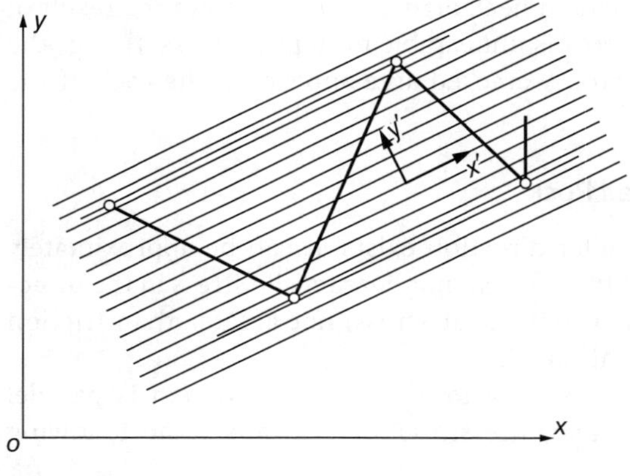

Figure 23.19 Stratified rock.

2) Compute linear stress in local coordinate system:

$$\{\sigma'\} = [\theta]^{-1}\{\sigma\}$$

where $[\theta]$ is stress transform matrix given in Eq. (2.19).

3) Determine if the stress condition of the element satisfies Eq. (23.94) and compute allowable stress $\{\sigma_b\}$.

1) If $\sigma_{y'} > 0$, the stratifications will be pulled apart. The normal stress $\sigma_{y'}$ and shear stress $\tau_{x'y'}$ should be zero. The allowable stress is

$$\{\sigma_b'\} = \begin{Bmatrix} \sigma_{x'} \\ 0 \\ 0 \end{Bmatrix} \tag{23.95}$$

2) If $\sigma_{y'} \leq 0$ and $|\tau_{x'y'}| > c - f\sigma_{y'}$, the stratifications will slide. The allowable stress is

$$\{\sigma_b'\} = \begin{Bmatrix} \sigma_{x'} \\ \sigma_{y'} \\ \eta|f'\sigma_{y'}| \end{Bmatrix} \tag{23.96}$$

When $\tau_{x'y'}$ is positive, $\eta = +1$; when $\tau_{x'y'}$ is negative, $\eta = -1$. Eq. (23.96) indicates that $\sigma_{x'}, \sigma_{y'}$ need not to be adjusted, while the value of shear stress equals to friction force $|f'\sigma_{y'}|$, in the same direction of $\tau_{x'y'}$. f' is the residual friction coefficient.
3) If $\sigma_{y'} \leq 0$ and $|\tau_{x'y'}| \leq f - c\sigma_{y'}$, the stratifications will neither be pulled apart nor slide. The allowable stress equals to the original linear stress.
4) Compute initial stress in local coordinate system:

$$\{\sigma_0'\} = \{\sigma_b'\} - \{\sigma'\}$$

Then transform it into a global coordinate system and compute the nodal load with the initial stress.
5) Compute displacement increment caused by the initial stress:

$$\{\Delta\delta\} = [K]^{-1}\{P\}_{\sigma 0}$$

Compute the stress increment $\{\Delta\sigma\}$. The element stress is

$$\{\sigma\} + \{\sigma_0\} + \{\Delta\sigma\}$$

Repeat the procedure above until $\{\Delta\delta\}$ is small enough.

23.11.3 Fracture and Slide of Joint and Soft Layer

In general, joints and soft layers in rock cannot sustain tensile stress, and their shear stress also satisfies Eq. (23.55). Therefore, the stress in local coordinate system should satisfy the following conditions:

$$\sigma_n' \leq 0 \tag{23.97}$$

$$\tau_s' = \frac{\Delta u'}{a + b\Delta u'} \tag{23.98}$$

During computing, the criterions are as below:

1) If $\sigma_n' > 0$, the joint surfaces or soft layers will be pulled apart. The allowable stress is zero and the initial stress is

$$\{\sigma_0'\} = -\{\sigma'\} = -\begin{Bmatrix} \tau_s' \\ \sigma_n' \end{Bmatrix} \tag{23.99}$$

2) If $\sigma_n' \leq 0$, the joint surfaces or soft will not be pulled apart; thus there is no need to adjust the normal stress. The allowable shear stress is shown as Eq. (23.98); therefore

the initial stress is

$$\{\sigma_0'\} = \begin{Bmatrix} \Delta\tau' \\ 0 \end{Bmatrix} \tag{23.100}$$

$$\Delta\tau' = \frac{\Delta u'}{a + b\Delta u'} - \tau_s' \tag{23.101}$$

If the measured data is insufficient and Eq. (23.98) is hard to derive, approximate computation with friction coefficient could be applied. In this case, the $\Delta\tau'$ in Eq. (23.100) should be calculated with the equation below:

$$\Delta\tau' = \eta|f'\sigma_n'| - \tau_s' \tag{23.102}$$

When τ_s' is positive, $\eta = +1$; when τ_s' is negative, $\eta = -1$.

23.12 Elastic–Plastic Increment Method

The fundamental equations of elastic–plastic increment method are Eqs (17.94) and (17–95). The computing procedure has been elaborated in Section 17.8. Next we will explain some features of the application of elastic–plastic increment method to the computation in rock mechanics.

23.12.1 Elastic–Plastic Computation for Integral Rock

Use the mixed failure criteria: Mohr–Coulomb criteria with maximum tensile stress (for shear failure) and maximum tensile stress criteria (for tensile failure) as Figure 23.16 shows.

The yield function of Mohr–Coulomb criteria is

$$F_s = |\tau| + \sigma_n tg\phi - c = 0 \tag{23.103}$$

It is inconvenient to apply Eq. (23.103) directly to numerical computation. Thus we need to find out the direction of failure surface first. Expressing with normal stress, substitute $\tau = (\sigma_1 - \sigma_3)/2$, $\sigma_n = (\sigma_1 + \sigma_3)/2$ into the equation above, and we get

$$F_s = \sigma_1(1 + \sin\phi) - \sigma_3(1 - \sin\phi) - 2c\cos\phi = 0 \tag{23.104}$$

Also, $\xi(= I_1/\sqrt{3})$, $r(= \sqrt{2J_2})$ and θ (Figure 17.7) can be used to express Mohr–Coulomb criteria, as Eq. (17.36) shows.

The maximum tensile stress criteria is

$$F_n = \sigma_1 - R_t = 0 \tag{23.105}$$

where R_t is the tensile strength and σ_1 is the maximum normal tensile stress.

After tensile failure, the residual tensile strength is R_t'. Generally, we take $R_t' = 0$. After shear failure, the residual cohesive force is c' and the residual friction angle is ϕ'. These values can be determined by tests. We usually take $c' = 0$, $\phi' < \phi$.

Before reaching the yield, it can be computed as elastomer. In order to calculate the deformation after yield, we need to define the flow rule. Since in practical engineering, yield hardly occurs in integral rock, corresponding test data is limited. In actual computation, associated flow rule can be applied, that is, to assume $Q = F$.

23.12.2 Elastic–Plastic Computation for Rock with Weak Surface

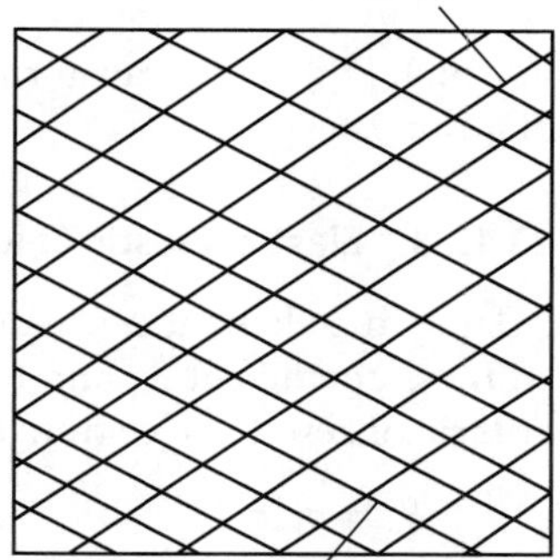

Figure 23.20 Rock with weak surfaces.

Suppose there are several weak surfaces that consist of joints and soft layers (Figure 23.20). In this case, the yield of rock is mainly controlled by these weak surfaces, while the rock itself can hardly yield. To guarantee the integrality of computation, we can verify the yield of rock with Eqs (23.104) and (23.105). The yield functions of weak surface of group j are as follows:

$$F_{\mathrm{s}}^{j} = |\tau_j| + \sigma_{\mathrm{n}}^{j} tg\phi_j - c_j = 0 \tag{23.106}$$

$$F_{\mathrm{n}}^{j} = \sigma_{\mathrm{n}}^{j} - R_{\mathrm{t}}^{j} = 0 \tag{23.107}$$

where $\sigma_{\mathrm{n}}^{j}, \tau_j$ are the normal stress and shear stress, respectively, on the weak surface of group j; $c_j, \phi_j, R_{\mathrm{t}}^{j}$ are cohesive force, friction angle, and tensile strength on the weak surface of group j; and in spatial problems, τ_j is not a component of shear stress on weak surface, but the compound shear stress on the weak surface.

Using associated flow rule, in local coordinate system, the plastic strain increment of joints in group j is

$$\{\mathrm{d}\bar{\varepsilon}_j^{\mathrm{p}}\} = \begin{Bmatrix} \mathrm{d}\varepsilon_{nj}^{\mathrm{p}} \\ \mathrm{d}\gamma_j^{\mathrm{p}} \end{Bmatrix} = \lambda_j \begin{Bmatrix} \dfrac{\partial F}{\partial \sigma_{nj}} \\ \dfrac{\partial F}{\partial \tau_j} \end{Bmatrix} \tag{23.108}$$

In global coordinate system, the plastic strain increment contributed by joints in group j is

$$\mathrm{d}\varepsilon_j^{\mathrm{p}} = \begin{Bmatrix} \mathrm{d}\varepsilon_{xj}^{\mathrm{p}} \\ \mathrm{d}\varepsilon_{yj}^{\mathrm{p}} \\ \mathrm{d}\varepsilon_{zj}^{\mathrm{p}} \\ \mathrm{d}\gamma_{xyj}^{\mathrm{p}} \\ \vdots \end{Bmatrix} = \lambda_j [T_j] \begin{Bmatrix} \dfrac{\partial F}{\partial \sigma_{nj}} \\ \dfrac{\partial F}{\partial \tau_j} \end{Bmatrix} \tag{23.109}$$

where

$$[T_j] = \left[\frac{\partial \hat{\sigma}_j}{\partial \sigma}\right] = \begin{bmatrix} \dfrac{\partial \sigma_{nj}}{\partial \sigma_x} & \dfrac{\partial \tau_j}{\partial \sigma_x} \\ \dfrac{\partial \sigma_{nj}}{\partial \sigma_y} & \dfrac{\partial \tau_j}{\partial \sigma_y} \\ \dfrac{\partial \sigma_{nj}}{\partial \sigma_z} & \dfrac{\partial \tau_j}{\partial \sigma_z} \\ \dfrac{\partial \sigma_{nj}}{\partial \tau_{xy}} & \dfrac{\partial \tau_j}{\partial \tau_{xy}} \\ \vdots & \vdots \end{bmatrix} \tag{23.110}$$

$$\{\hat{\sigma}_j\}^{\mathrm{T}} = \begin{bmatrix}\sigma_{nj} & \tau_j\end{bmatrix}$$

in which $[T_j]$ is the transform matrix and $\{\hat{\sigma}_j\}^{\mathrm{T}}$ is the stress vector on the joint surface of group j.

23.12.3 Elastic–Plastic Calculation for Joint Element

Assume the element is elastic then elastic deformation can be computed using normal stiffness coefficient λ_{n} and tangential stiffness coefficient λ_{s}. Use the Mohr–Coulomb criteria of zero tensile strength ($R_{\mathrm{t}} = 0$) as yield condition, that is,

$$F_{\mathrm{n}} = \sigma_{\mathrm{n}} = 0 \tag{23.111}$$

$$F_{\mathrm{s}} = |\tau| + \sigma_{\mathrm{n}} tg\phi - c = 0 \tag{23.112}$$

If associated flow rule is applied, when tensile failure occurs, the plastic strain increment should be computed with the following equation:

$$\begin{Bmatrix} \mathrm{d}\varepsilon_{\mathrm{n}}^{\mathrm{p}} \\ \mathrm{d}\gamma^{\mathrm{p}} \end{Bmatrix} = \lambda \begin{Bmatrix} \dfrac{\partial F}{\partial \sigma_{\mathrm{n}}} \\ \dfrac{\partial F}{\partial \tau} \end{Bmatrix} = \lambda \begin{Bmatrix} 1 \\ 0 \end{Bmatrix} \tag{23.113}$$

When shear yield occurs, the plastic strain increment should be computed with Eq. (23.114):

$$\begin{Bmatrix} \mathrm{d}\varepsilon_{\mathrm{n}}^{\mathrm{p}} \\ \mathrm{d}\gamma^{\mathrm{p}} \end{Bmatrix} = \lambda \begin{Bmatrix} \dfrac{\partial F}{\partial \sigma_{\mathrm{n}}} \\ \dfrac{\partial F}{\partial \tau} \end{Bmatrix} = \lambda \begin{Bmatrix} tg\phi \\ 0 \end{Bmatrix} \tag{23.114}$$

In this case, the ratio of plastic normal strain increment and plastic shear strain increment is

$$\frac{\mathrm{d}\varepsilon_{\mathrm{n}}^{\mathrm{p}}}{\mathrm{d}\gamma^{\mathrm{p}}} = tg\phi \tag{23.115}$$

The right end of Eq. (23.115) is positive, indicating that after shear yield, together with plastic shear deformation, the element will become dilatant, which is also called as dilatation. The mechanical schematic diagram of joint dilatation is shown in Figure 23.21. Since the surface of joint is fairly coarse, the climbing effect to go across the coarse bumps on the joint surface will lead to the dilatation of joint, together with shear displacement.

The joint dilatation observed in actual measurement is much smaller than that calculated with Eq. (23.104). In order to match the calculated joint dilatation with the observed values, we can use the nonassociated flow rule. Take the flow potential function as

$$Q = |\tau| + \sigma_{\mathrm{n}} tg\psi = 0 \tag{23.116}$$

In this case,

$$\frac{\mathrm{d}\varepsilon_{\mathrm{n}}^{\mathrm{p}}}{\mathrm{d}\gamma^{\mathrm{p}}} = tg\psi \tag{23.117}$$

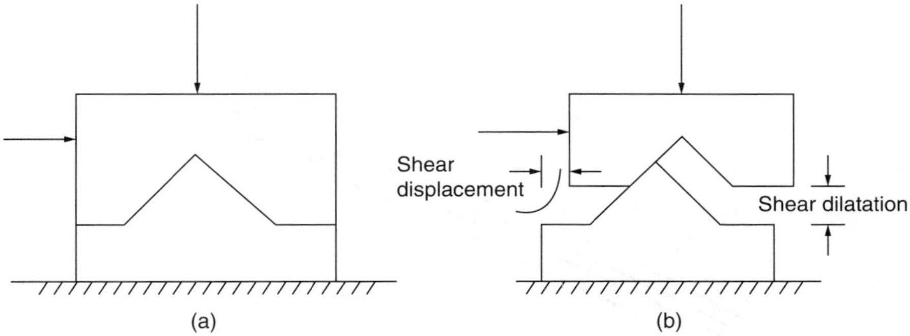

Figure 23.21 Mechanical schematic diagram of joint dilatation.

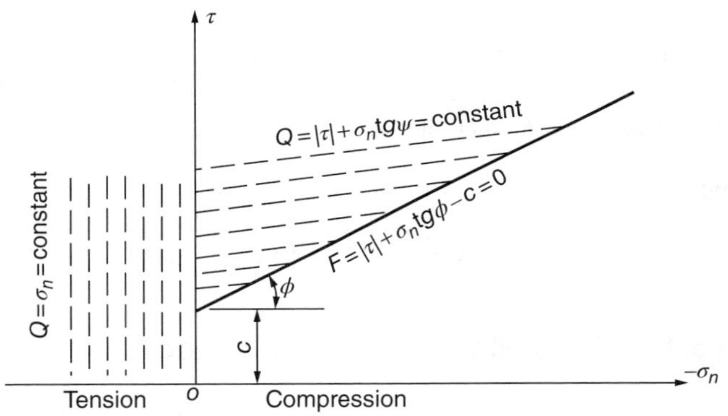

Figure 23.22 Mohr–Coulomb criteria of zero tensile strength.

where ψ is called as dilatation angle, whose value could be determined according to test result.

This kind of Mohr–Coulomb criteria of zero tensile strength can be seen in Figure 23.22.

Strictly speaking, the relation between the shear strength and normal stress of joint is nonlinear. Based on some test data on artificial joints and natural joints, Barton and Choubey [17] proposed the following shear yield criteria:

$$F = |\tau| + \sigma_n tg\left[s\ln(-R_c/\sigma_n) + \phi'\right] \tag{23.118}$$

where s is the coarse coefficient of joint, R_c is the compressive strength of joint, and ϕ' is the residual friction angle.

Considering the $\ln(-R_c/\sigma_n)$ in Eq. (23.118), the joint must not have tensile strength. The parameters s, R_c, and ϕ' could be determined by field tests. The diagram of Eq. (23.118) is shown in Figure 23.23.

If associated flow rule is applied, the dilatation angle at peak strength can be calculated with Eq. (23.115) while not consistent with test data. In order to let the dilatation angle close to test data, Pande and Xiong [18] proposed following plastic potential function:

$$Q = |\tau| - \frac{\sigma_n tg\lambda_1}{K_1} - \frac{s}{263.86}\frac{\sigma_n^2}{R_c} = \text{constant} \tag{23.119}$$

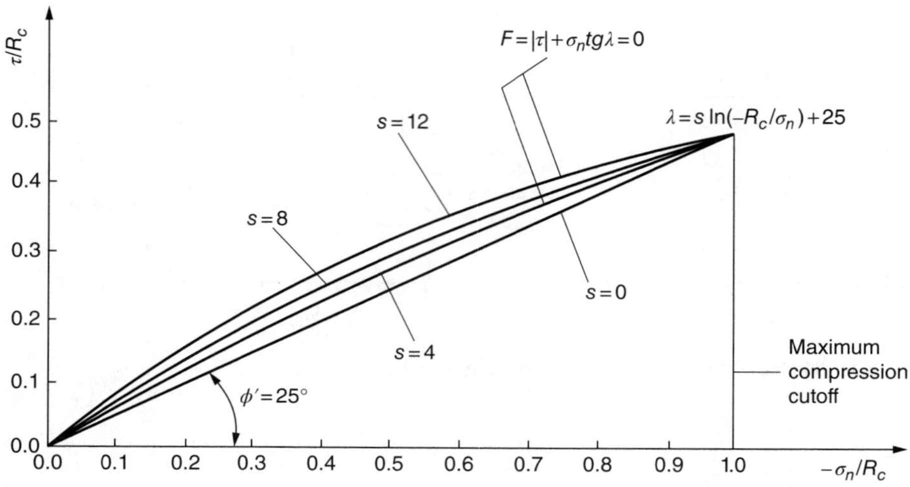

Figure 23.23 Barton and Choubey's fracture criteria for rock joint ($\phi' = 25°$).

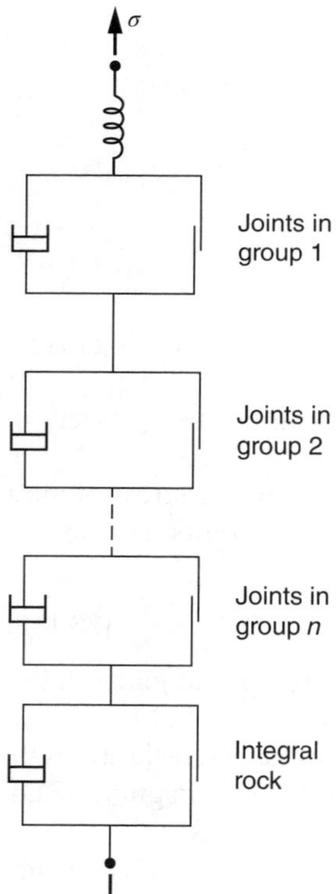

Figure 23.24 Multilaminate viscoelastic–plastic model of rock.

where

$$K_1 = 1 - tg\lambda_1 tg\phi', \quad \lambda_1 = s\ln\left(-\frac{R_c}{\sigma_n}\right)$$

23.13 Viscoelastic–Plastic Method

There are two cases where viscoelastic–plastic method should be used in rock mechanics: one is to consider the time effect of rock deformation physically and the other one is to take it as a computational tool. Experience shows that [18] compared with elastic–plastic method, viscoelastic–plastic method shows higher computational accuracy. The result curve of load–displacement is smoother and the distribution of stress is more reasonable. When it is near to structural failure, the curves computed with elastic–plastic method usually show a jump and is not smooth enough. Thus it will be difficult to calculate the safety coefficient for structural failure. The load–displacement curve computed with viscoelastic–plastic method is smoother, which means more accurate safety coefficient can be obtained.

Suppose there are n groups of joints in the rock and each joint has a failure criterion. The rheological model is shown in Figure 23.24, with $n + 1$ tandem visco-plastic elements. Among them, the first n elements represent n groups of joints successively, and the last visco-plastic element represents integral rock. Combine all the deformation of the n groups of joints and the integral rock and denote it with a spring. This kind of computational model is called as multilaminate viscoelastic–plastic model, which is also called as multilaminate framework model in some literature.

The basic assumptions of multilaminate viscoelastic–plastic model are as follows:

1) Assume there are n groups of joints in rock. Joints in a group are parallel to each other, as Figure 23.20 shows.
2) Assume the volume of joints is quite small compared with the gross volume of rock.
3) Assume that joints and rock are both elastic–plastic, that is, they are elastic before yield. The failure criteria are Mohr–Coulomb criteria of zero tensile strength, as shown in Figure 23.22 (or Mohr–Coulomb criteria with maximum tensile stress (Figure 17.16)).

We will describe the computational model in three-dimensional space. Denote the stress of any point in global coordinate system with $\{\sigma\}^{\mathrm{T}} = \left[\sigma_x\ \sigma_y\ \sigma_z\ \tau_{xy}\ \tau_{yz}\ \tau_{zx}\right]$. $\{\hat{\sigma}_j\}^{\mathrm{T}} = \left[\sigma_{nj}\ \tau_j\right]$ denotes the normal and tangential stress, respectively, of joints in group j, which can be calculated by transforming $\{\sigma\}^{\mathrm{T}}$ into local coordinate system (n, s, t), where n is the normal direction of joint surface and s, t locate on joint surface.

The yield functions of joints in group j are

$$F_{\mathrm{s}}^{j} = |\tau_j| + \sigma_{\mathrm{n}}^{j} tg\phi_j - c_j = 0$$

$$F_{\mathrm{n}}^{j} = \sigma_{\mathrm{n}}^{j} = 0$$

For tensile failure, use the associated flow rule and the potential function is

$$Q_{\mathrm{n}}^{j} = \sigma_{\mathrm{n}}^{j} \tag{23.120}$$

For shear failure, use nonassociated flow rule and the potential function is

$$Q_{\mathrm{s}}^{j} = |\tau_j| + \sigma_{\mathrm{n}}^{j} \tan \psi_j - \text{constant} = 0 \tag{23.121}$$

The flow equation of joints in group j is

$$\{\dot{\varepsilon}_j^{\mathrm{vp}}\} = \gamma_j \langle F_j \rangle \left\{ \frac{\partial Q_j}{\partial \hat{\sigma}_j} \right\} \left\{ \frac{\partial \hat{\sigma}_j}{\partial \sigma} \right\} \tag{23.122}$$

where $\dot{\varepsilon}_j^{\mathrm{vp}}$ is the contribution of joints in group j to the visco-plastic strain rate of rock in global coordinate system. If this group of joints does not yield ($F_j < 0$), the contribution will be zero. γ_j is the flow parameter of joints in group j. When analyzing the time effect of rock deformation, γ_j should be determined according to test data; when using it as a kind of computational tool, γ_j can be taken as any constant, usually 1.

Use Mohr–Coulomb rule with maximum tensile stress for integral rock, as Eqs (23.94) and (23.95) show. Suppose the yield function is F_{r} and plastic potential function is Q_{r}, then the flow equation will be

$$\{\dot{\varepsilon}_{\mathrm{r}}^{\mathrm{vp}}\} = \gamma_{\mathrm{r}} \langle F_{\mathrm{r}} \rangle \left\{ \frac{\partial Q_{\mathrm{r}}}{\partial \sigma_j} \right\} \tag{23.123}$$

where γ_{r} is the flow parameter of integral rock.

The gross visco-plastic strain rate $\dot{\varepsilon}^{\mathrm{vp}}$ is the sum of contribution of integral rock and each group of joints (Figure 23.24), that is,

$$\dot{\varepsilon}^{\mathrm{vp}} = \gamma_{\mathrm{r}} \langle F_{\mathrm{r}} \rangle \left\{ \frac{\partial Q_{\mathrm{r}}}{\partial \sigma} \right\} + \sum_{j=1}^{n} \gamma_j \langle F_j \rangle \left\{ \frac{\partial Q_j}{\partial \hat{\sigma}_j} \right\} \left\{ \frac{\partial \hat{\sigma}_j}{\partial \sigma} \right\} \tag{23.124}$$

The first term on the right end of the equation above is the contribution of integral rock and the second item the contribution of n groups of joints.

The yield functions and potential functions of each group of joints can also be expressed with Eqs (23.118) and (23.119).

Joint surfaces cannot sustain tensile stress. When normal stress is tensile stress, the joints will open. During next stage of loading, if the normal stress turns to be compressive stress, the joints will close. Generally, we can determine the open and close of joints with normal strain. Suppose the maximum gross tensile strain of joints in the loading history is $\varepsilon_{\mathrm{n}}^{\max}$. In this increment step, the normal tensile strain of joints becomes ε_{n}. Let

$$\varepsilon^* = \varepsilon_{\mathrm{n}} + \varepsilon_{\mathrm{n}}^{\max} \tag{23.125}$$

Obviously, $\varepsilon^* > 0$ means joints are open, while $\varepsilon^* \leq 0$ means joints are closed.

In the analysis above, we considered a general condition, that is, rock with n groups of joints. Only taking the first term on the right hand of Eq. (23.124), we can get the computational equation for integral rock. If we take a joint-related term, it can be used to compute joint element, though of course, some computing details need to be modified correspondingly. To improve computational accuracy, Barton and Choubey's shear yield criteria can also be used for each group of joints.

23.14 Computation of Anchor Bolt in Rock Foundation

Anchor bolts are often used to strengthen rocks with developed joints. There are many forms of anchor bolt in rocks. Some are full-length grouted, some are merely anchored at the end, prestressed, or not prestressed.

If there contains only a small amount of anchor bolts in rocks, link elements can be used. The effect of prestressed anchor bolt could also be considered with a similar method, that is, loading prestressed force on appropriate nodes.

If there are a great number of anchor bolts, it will be tedious to use link elements. In this case, we can compute by an equivalent method as in the computation of steel bar in reinforced concrete structure.

Figure 23.25 shows a viscoelastic–plastic computational model for rocks strengthened by non-prestressed anchor bolts. Suppose there are m group of anchor bolts in rocks. When the anchor bolts are grouted for full length, the strain of the bolts and the rock should be consistent. In the figure, the first column represents rock, the second column represents bolts in group 1, and the last column represents bolts in group m. The sum of strain in each column should be equal.

Use subscript i to denote the column of number i and subscript ij to denote the rheological element of number j in column i. Others without subscripts represent equivalent material. Therefore we have

$$\{\Delta\varepsilon\} = \{\Delta\varepsilon_i\} = \sum_j \{\Delta\varepsilon_{ij}\} \tag{23.126}$$

$$\{\Delta\sigma\} = \sum_i p_i \Delta\sigma_i \tag{23.127}$$

$$\{\Delta\sigma_i\} = \{\Delta\sigma_{ij}\} \tag{23.128}$$

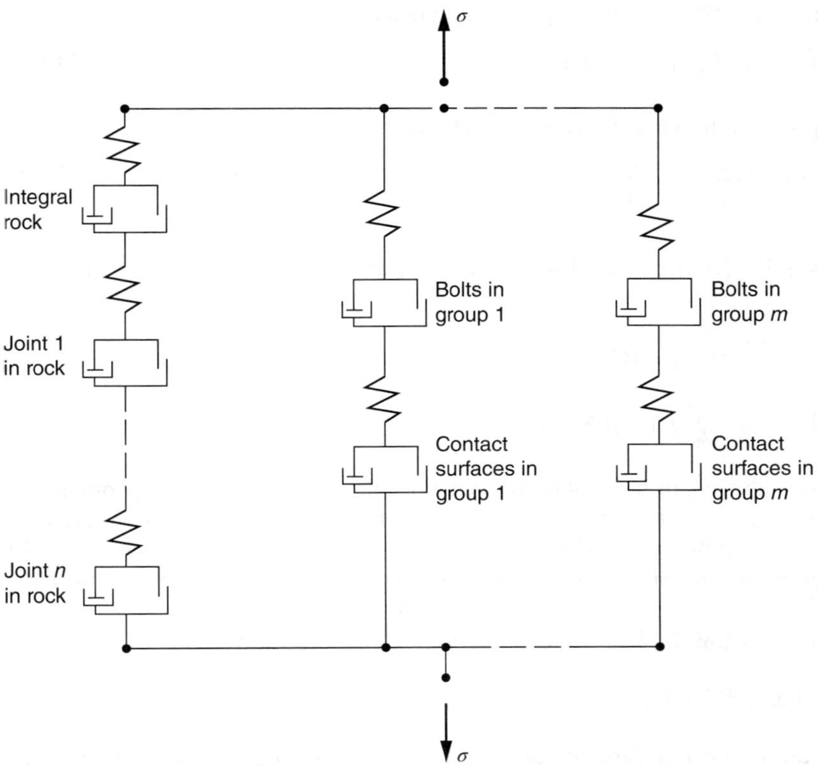

Figure 23.25 Viscoelastic–plastic computational model for rock strengthened by bolts.

where p_i is the volume ratio of column i. For rock, $p_i = 1$, and for bolts, p_i equals to the ratio of section areas.

For a single rheological element, the relation between the vector of stress increment and the vector of strain increment is

$$\{\Delta\varepsilon_{ij}\} = [D_{ij}]^{-1}\{\Delta\sigma_{ij}\} + \{\Delta\varepsilon_{ij}^{\text{vp}}\} = [C_{ij}]\{\Delta\sigma_{ij}\} + \{\Delta\varepsilon_{ij}^{\text{vp}}\} \tag{23.129}$$

or

$$\{\Delta\sigma_{ij}\} = [D_{ij}](\{\Delta\varepsilon_{ij}\} - \{\Delta\varepsilon_{ij}^{\text{vp}}\}) \tag{23.130}$$

Substitute Eq. (23.129) into Eq. (23.126), we get the stress–stain relation of column i, that is,

$$\{\Delta\varepsilon\} = [D_i]^{-1}\{\Delta\sigma_i\} + \{\Delta\varepsilon_i^{\text{vp}}\} = [C_i]\{\Delta\sigma_i\} + \{\Delta\varepsilon_i^{\text{vp}}\} \tag{23.131}$$

$$\{\Delta\sigma_i\} = [D_i](\{\Delta\varepsilon_i\} - \{\Delta\varepsilon_i^{\text{vp}}\}) \tag{23.132}$$

where

$$[C_i] = [D_i]^{-1} = \sum_j [D_{ij}]^{-1} = \sum_j [C_{ij}] \tag{23.133}$$

$$\{\Delta\varepsilon_i^{\text{vp}}\} = \sum_j \{\Delta\varepsilon_{ij}^{\text{vp}}\} \tag{23.134}$$

Substitution of Eq. (23.132) into Eq. (23.127) yields

$$\{\Delta\sigma\} = \sum_i p_i [D_i](\{\Delta\varepsilon\} - \{\Delta\varepsilon_i^{\mathrm{vp}}\}) \tag{23.135}$$

Hence we get the following stress–strain relation:

$$\{\Delta\sigma\} = [D](\{\Delta\varepsilon\} - \{\Delta\varepsilon^{\mathrm{vp}}\}) \tag{23.136}$$

or

$$\{\Delta\varepsilon\} = [D]^{-1}\{\Delta\sigma\} + \{\Delta\varepsilon^{\mathrm{vp}}\} = [C]\{\Delta\sigma\} + \{\Delta\varepsilon^{\mathrm{vp}}\} \tag{23.137}$$

where

$$[D] = \sum_i p_i [D_i] = [C]^{-1} \tag{23.138}$$

$$\{\Delta\varepsilon^{\mathrm{vp}}\} = [D]^{-1} \sum_i p_i [D_i]\{\Delta\varepsilon_i^{\mathrm{vp}}\} \tag{23.139}$$

The equations above can be applied to space problem as well as plane problem. To simplify the computation, it is often assumed that there is no relative movement between bolts and the surrounding rock. Thus we can ignore the contact surface between bolts and the surrounding rock shown in Figure 23.25, assuming the bolts and rock have the same strain.

Substitute Eq. (23.136) into the following equilibrium condition:

$$\int [B]^{\mathrm{T}}\{\Delta\sigma\}\,\mathrm{d}V = \{\Delta P\}$$

We obtain the equilibrium function of surrounding rock strengthened by anchor bolts as follows:

$$[K]\{\Delta\delta\} = \{\Delta P\} + \{\Delta P^{\mathrm{vp}}\} \tag{23.140}$$

where

$$[K] = \int [B]^{\mathrm{T}}[D][B]\,\mathrm{d}V \tag{23.141}$$

$$\{\Delta P^{\mathrm{vp}}\} = \sum_i p_i \int [B]^{\mathrm{T}}[D_i]\{\Delta\varepsilon_i^{\mathrm{vp}}\}\,\mathrm{d}V \tag{23.142}$$

For more details of calculation, please refer to Section 18.6. It should be noted that only the components of stress and strain in the same coordinate system can be added together. When the direction of anchor bolt and axis in global coordinate system are different, coordinate transformation is required for all of the stress, strain, flexibility matrix, elastic matrix, and so on. Please refer to Appendix A for more details.

Anchor bolts in rock are usually made of steel. Thus the Mises yield criteria could be applied.

Now we will summarize the computation method:

1) Compute the $[C_i]$ of each column with Eq. (23.133) and get $[D_i]$ after matrix inversion.
2) Compute $\{\Delta\varepsilon_i^{\mathrm{vp}}\}$ with Eq. (23.134).
3) Compute $[D]$ with Eq. (23.138), $[K]$ with Eq. (23.141), and $\{\Delta P^{\mathrm{vp}}\}$ with Eq. (23.142).

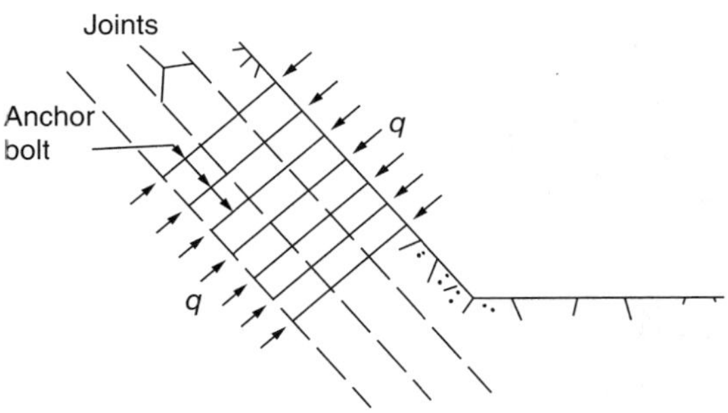

Figure 23.26 Prestressed anchor bolts in rock.

4) Compute $\{\Delta\delta\}$ with Eq. (23.140).
5) Compute the stress increment of each column $\{\Delta\sigma_i\}$ with Eq. (23.132).

Next we will explain the computation method for prestressed anchor bolts in rock (Figure 23.26). Let the prestress in bolt be σ_0. The equivalent pressure on the surface of the rock of both sides of the bolt is

$$q = -p\sigma_0 \tag{23.143}$$

where p is the volume ratio of anchor bolt.

In computation, q is regarded as a part of external load on rock and σ_0 is regarded as the initial stress in anchor bolt.

23.15 Computing Examples in Rock Mechanics

Practically, many computations of rock mechanics are three-dimensional. But three-dimensional problems are usually complex. It is necessary to use too many figures and words to explain the condition and results clearly. Since the computational principles of two-dimensional problem and three-dimensional problem are essentially the same, for saving space, we only give two examples of two-dimensional problem.

23.15.1 Computing of Rock Slope

The rock slope, with height of 70 m, contains two groups of weak surfaces, as Figure 23.27 shows.

Assume the initial stress condition before excavation is as follows [15]:

$$\sigma_y = \rho g h, \quad \sigma_x = \sigma_z = k_0 \sigma_y$$

Parameters are as follows:

$$\text{Rock}: \quad \rho = 2.5\,\text{t/m}^3, \quad E = 28.0\,\text{GPa}, \quad \mu = 0.20, \quad k_0 = 1/3, \quad \phi = 45°, \quad c = 5.6\,\text{MPa}$$

Two groups of joints: $\phi = 45°$, c = variable value, tensile strength = $c/10$.

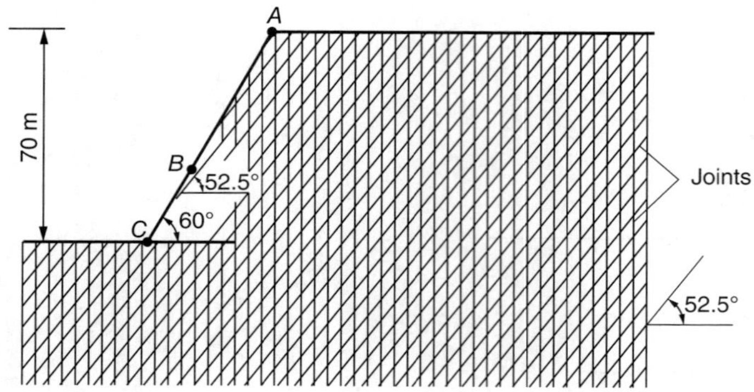

Figure 23.27 Rock slope.

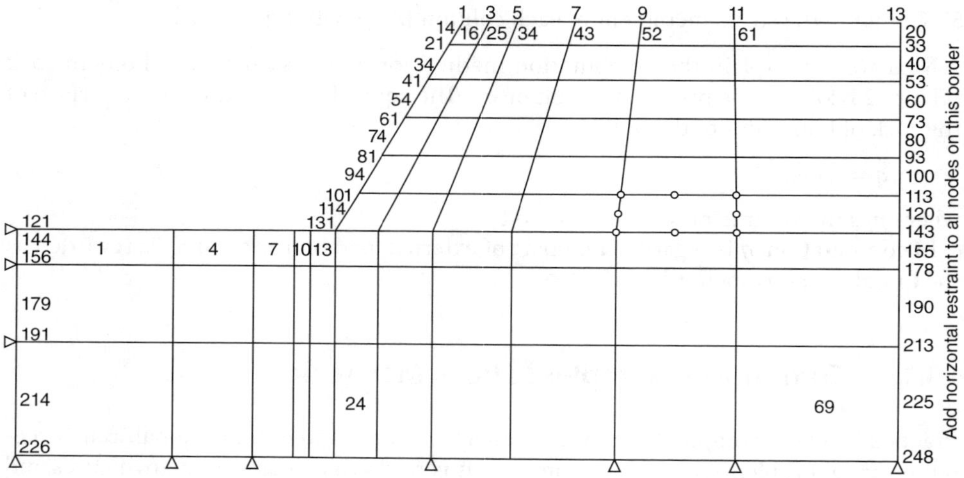

Figure 23.28 Computing grid.

During computing, the value of c for joints decreases step by step, from high to low, until rock failure occurs (more reasonable method is to decrease c and ϕ simultaneously to calculate safety coefficient). Assume excavation is finished in one step (obviously it is greatly simplified).

Use the 8-node plane isoparametric element. At first, a series of linear analysis should be applied to different grids to select a reasonable computing grid. The finally selected computing grid contains 450 degrees of freedom, as Figure 23.28 shows.

Figure 23.29 shows the variation of horizontal displacement with decreasing c of joint surface. The final value of c at failure (no convergence) using associated flow rule is 21 kPa and that using nonassociated flow rule is 23 kPa. Figures 23.30 and 23.31 illustrate the convergent solution and contour of effective shear strain at failure using the two flow rules. The maximum effective shear zone locates about $0.1H$ above the foot of the slope.

Figure 23.32 illustrates the Gauss points, tensile fractures, and slide surfaces that are still creeping after 450 time steps.

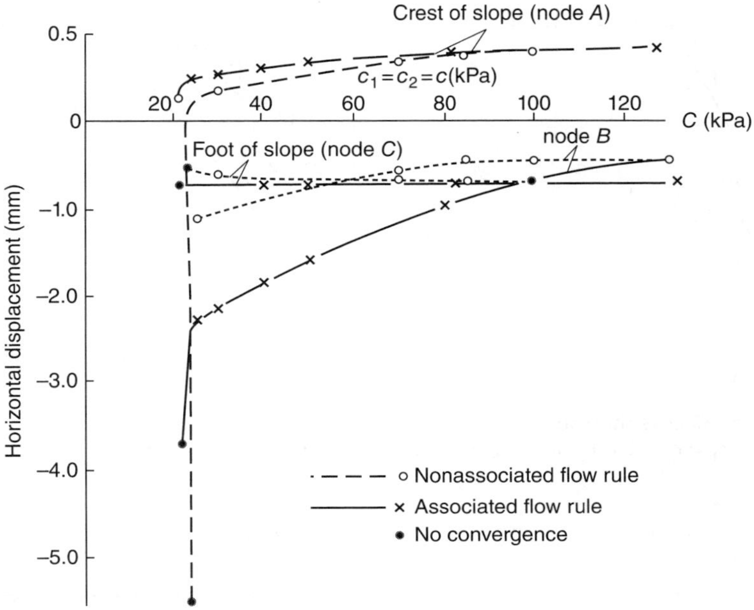

Figure 23.29 Variation of horizontal displacement with decreasing c of joint surface.

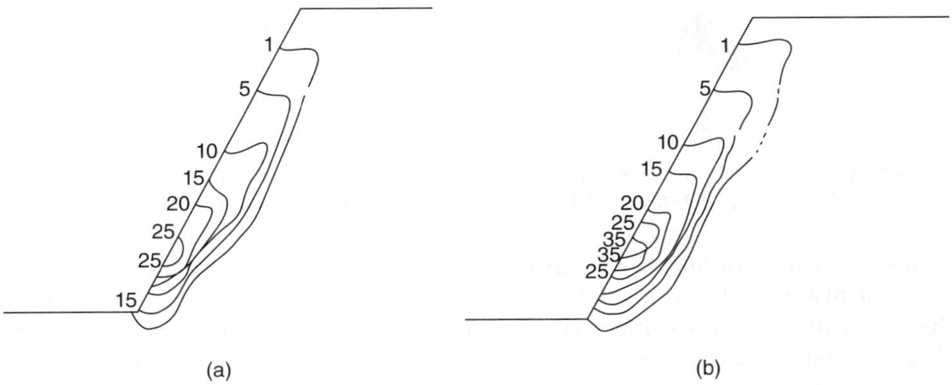

Figure 23.30 Contour of effective shear strain (associated flow rule) (a) convergent solution ($c_1 = c_2 = 23$ kPa, $\phi_1 = \phi_2 = 45°$) and (b) failure ($c_1 = c_2 = 21$ kPa, $\phi_1 = \phi_2 = 45°$).

23.15.2 Antisliding Stability of Gravity Dam on Rock Foundation

When there are soft intercalated layers in rock foundation, the antisliding stability of gravity dam is a crucial issue. For cases with two or more slide surfaces, it is difficult to calculate accurate safety coefficient with limit equilibrium method of rigid body, and FEM computation is needed in most cases. Calculate the overload safety coefficient K_p with overload method. Or calculate the safety coefficient K_f under the meaning of strength reservation by decreasing strength indexes of material. To be more specific, compute the load–displacement curve with FEM, and then determine the critical

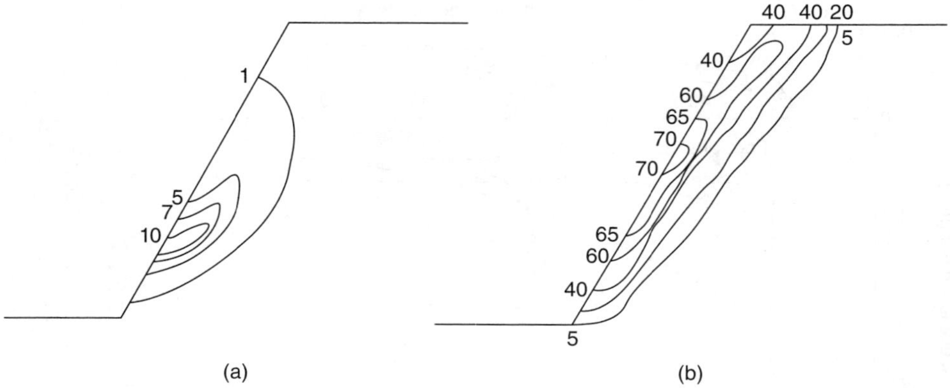

Figure 23.31 Contour of effective shear strain (nonassociated flow rule, $\psi = 0$, no dilatation) (a) convergent solution ($c_1 = c_2 = 25$ kPa, $\phi_1 = \phi_2 = 45°$) and (b) failure ($c_1 = c_2 = 23$ kPa, $\phi_1 = \phi_2 = 45°$).

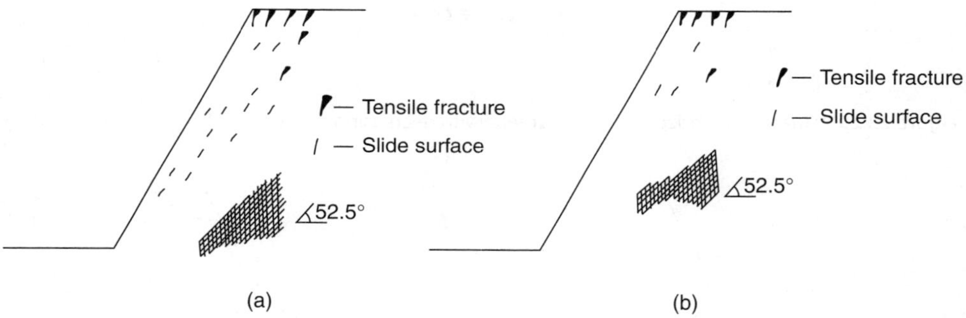

Figure 23.32 Gauss points, tensile fractures, and slide surfaces creeping after 450 time steps (a) associated flow rule ($c = 21$ kPa) and (b) nonassociated flow rule ($c = 23$ kPa).

collapsing load according to the variation of the curve. Experience has shown that when calculating with elastic–plastic increment method, the load–displacement curve changes gently and is insensitive to the increment of load during the stage of collapsing. Therefore the safety coefficient of collapsing can hardly be obtained accurately. However, this shortage could be avoided by computing with viscoelastic–plastic increment theory.

Figure 23.33 shows a gravity dam with two slide surfaces in rock foundation. The height of the dam is 71.5 m. The antisliding parameters of the two slide surfaces are $f_1 = 0.65, c_1 = 1$ kPa and $f_2 = 0.50, c_2 = 1$ kPa, respectively. Meanwhile, compute with both elastic–plastic FEM and viscoelastic–plastic FEM [19]. In the computing of viscoelastic–plastic FEM, use the Bingham model to simulate deformation characters of the two slide surfaces. Results are shown in Figure 23.34. From the figure we can see that when computing with viscoelastic–plastic theory, an apparent deflection of the relative displacement of all representative points appears when the load ratio

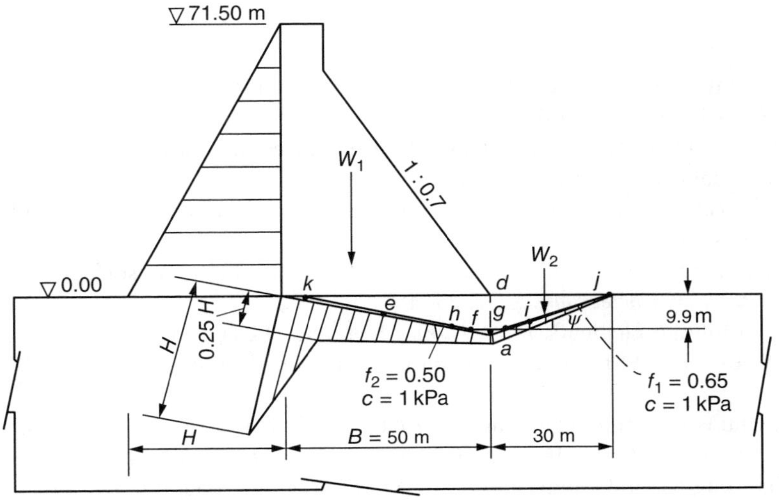

Figure 23.33 Gravity dam with two slide surfaces in rock foundation.

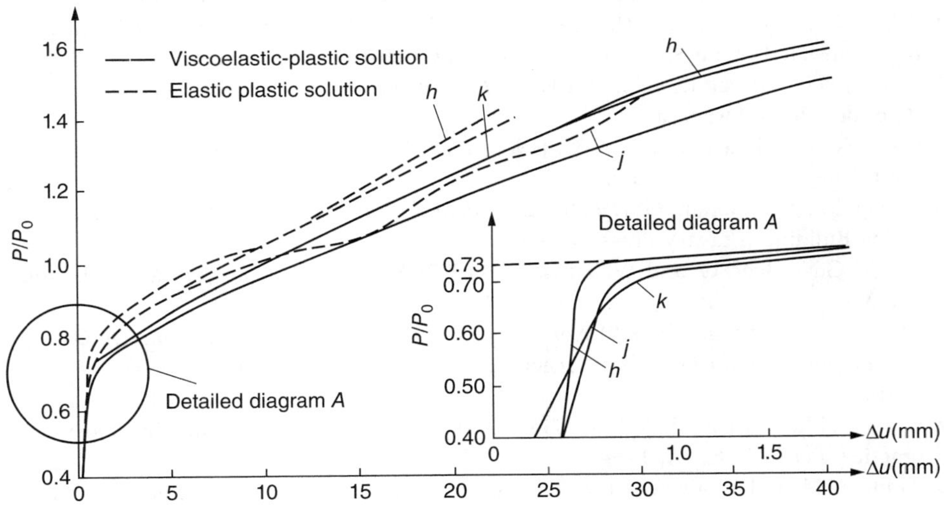

Figure 23.34 Viscoelastic–plastic solution and elastic–plastic solution to calculate K_f in two slide surfaces.

increases from 0.70 to 0.73. All elements on the slide surface yield at this time. We can take $K_p = 0.73$. However, for elastic–plastic solution, this change is not apparent. The viscoelastic–plastic solution also shows more reasonable stress distribution. To save computing time, these two methods could be used together. The elastic–plastic method can be performed during the first several stages of loading increment, while the viscoelastic–plastic method is used when it is near to the collapsing load.

Bibliography

1 Goodman, R.E., Taylor, R.L. and Brekke, T. (1968) A model of the mechanics of jointed rock. *Proc. ASCE*, **94** (SM3), 298.

2 Zienkiewicz, O.C., Vallianppian, S. and King, I.P. (1968) Stress analysis of rock as a "no tension" material. *Geotechnigue*, **18**, 56–66.

3 Pan, J. (1981) *Collected Papers of Hydraulic Structure Analysis*, Water Resources and Electric Power Press, Beijing.

4 Gao, A. and Xianhong, Z. (1983) On the stress–strain relationship of anisotropic jointed rock mass. *J. Hydroelectric Eng.*, **2**, 25–46.

5 Bettess, P. (1977) Infinite Elements. *Int. J. Num. Meth. Eng.*, **11** (1), 53–64.

6 Beer, G. and Meek, J.L. (1981) Infinite domain elements. *Int. J. Num. Meth. Eng.*, **17** (1), 43–52.

7 Chongbin, Z., Chuhan, Z. and Guangdou, Z. (1986) Simulation of semi-infinite plane elastic foundation using infinite elements. *J. Tsinghua Univ.*, **26** (3), 51–64.

8 Xiurun, G., Xianrong, G. and Dingxiang, F. (1986) Three-dimensional infinite domain elements and joint infinite domain elements. *J. Geotech. Eng*, 9–20.

9 Jun, S. and Xueyuan, H. (1987) *Underground Structure*, Science Press, Beijing.

10 Gudehus, G. (1977) *Finite Elements in Geomechanics*, John Wiley, London.

11 Desay, C.S. and Christian, J.T. (1981) *Numerical Methods in Geotechnical Engineering*. Translated by Lu Shishen, China Building Industry Press, Beijing.

12 Naylor, D.J. and Pande, G.N. (1981) *Finite Elements in Geotechnical Engineering*, Pineridge Press, Swansea.

13 Stagg, K.G. and Zienkiewicz, O.C. (1969) *Rock Mechanics in Engineering Practice*, John Wiley and sons, London.

14 Zheng, Y. and Xiaonan, G. (1989) *Fundamental Geotechnical Plastic Mechanics*, China Building Industry Press, Beijing.

15 Pande, G.N., Beer, G. and Williams, J.R. (1990) *Numerical Methods in Rock Mechanics*, Wiley, New York.

16 Zhu, B. and Jingting, S. (1979) *Nonlinear Finite Element Analysis of Concrete Dam on Complex Foundation*. See: Proceedings of Dam Research, China Waterpower Press.

17 Barton, N.R. and Choubey, V. (1977) The shear strength of rock joints in theory and practice. *Rock Mech.*, **10**, 1–54.

18 Pande, G.N. and Xiong, W. (1982) An improved multilaminate model of jointed rock masses, in *Numerical Models in Geomechanics* (eds R. Dungar, G.N. Pande and J.A. Studer), A. A. Bulkema, Rotterdam, pp. 218–226.

19 Zhonghua, C. and Zhu, B. (1984) Analysis of slide stability of concrete dams by means of the elasto-visco-plastic FEM. *J. Hydraulic Eng.*, **7**, 24–34.

20 Gerrard, C.M. and Pande, G.N. (1979) Numerical model of reinforced jointed rock masses. *Comput. Geotech.*, **1**, 293–318.

21 Zhu, B. (1997) *Collected Papers of Zhu Bofang Academician*, China Electric Power Press, Beijing.

22 Zhu, B. (1979, 3rd ed. 2009) *The Finite Element Method, Theory and Applications (in Chinese)*, 1rst edn, Water Resources and Hydropwer Press, Beijing.

24

Problems in Soil Mechanics

There are many factors affecting the stress and strain of soil, such as soil structure, pores, density, stress history, features of load, pore water, time effect, etc. These factors lead to complex behaviors of soil under the action of forces, which are nonlinear in most cases. It is difficult to solve the practical problems by theoretical methods. However, with the finite element method (FEM), many practical problems can be readily solved.

We can solve the following problems in soil mechanics with FEM: (1) analysis of deformation and stability of soil slope, excavation, earth dam, soil dike, and bank; (2) computation of bearing capacity and settlement of foundation; and (3) interactions between soil and structure.

Plane element and space element are mostly used when solving soil mechanics problems. The key is to appropriately describe the stress–strain relation of soil. It must tally with the results of experiments and observation, as well as be suitable for FEM computation.

24.1 Nonlinear Elastic Model

The stress–strain relation of soil is fairly complex, with characteristic of nonlinear, inelastic, compressive hardening, dilatancy, etc. It is also affected by pore water and time effect. In this chapter, we will describe the stress–strain relation of soil with consideration of the effects caused by nonlinear, inelastic, compressive hardening and pore water. Time effect has been discussed in Chapter 18.

For ordinary sand and clay, Kondner suggested to describe the stress–strain relation with hyperbola as follows [1]:

$$\sigma_1 - \sigma_3 = \frac{\varepsilon}{a + b\varepsilon} = \frac{1}{(a/\varepsilon) + b} \tag{24.1}$$

in which σ_1 is major principal stress, σ_3 is minor principal stress, ε is axial strain, and a, b are constants.

Let $\varepsilon \to \infty$ in the equation above and we get

$$\frac{1}{b} = (\sigma_1 - \sigma_3)_u \tag{24.2}$$

in which $(\sigma_1 - \sigma_3)_u$ is asymptotic value of stress difference, which is a little bigger than the compressive strength of soil. It is reasonable because hyperbola is always below the

The Finite Element Method: Fundamentals and Applications in Civil, Hydraulic, Mechanical and Aeronautical Engineering, First Edition. Bofang Zhu.

Companion website: www.wiley.com/go/zhufem17

asymptotic line when the strain is finite. Denote the ratio of compressive strength and asymptotic value of stress difference with R, and then we have

$$(\sigma_1 - \sigma_3)_f = R(\sigma_1 - \sigma_3)_u \tag{24.3}$$

in which $(\sigma_1 - \sigma_3)_f$ is the compressive strength of soil and R is the break ratio, whose value is less than 1, usually between 0.75 and 1.00, and is irrelevant with confining pressure.

From Eqs (24.2) and (24.3), we can get constant b, that is,

$$b = \frac{R}{(\sigma_1 - \sigma_3)_f} \tag{24.4}$$

By taking the derivative of Eq. (24.1), we can obtain the tangent modulus of soil, that is,

$$E_t = \frac{d(\sigma_1 - \sigma_3)}{d\varepsilon} = \frac{a}{(a + b\varepsilon)^2} \tag{24.5}$$

Let $\varepsilon = 0$ in the equation above, and we get

$$E_0 = \frac{1}{a} \tag{24.6}$$

in which E_0 is initial tangent modulus and constant a is equal to the reciprocal of the initial tangent modulus.

Transform Eq. (24.1) and we have

$$\frac{\varepsilon}{\sigma_1 - \sigma_3} = a + b\varepsilon \tag{24.7}$$

As Figure 24.1(b) shows, taking $\frac{\varepsilon}{\sigma_1 - \sigma_3}$ as vertical ordinate and ε as horizontal ordinate, the equation above will be a line, whose intercept and slope are a and b, respectively.

Experiment has shown that the tangent modulus of soil varies with confining pressure. Duncan and Zhang suggested expressing the relation between the tangent modulus and confining pressure with Eq. (24.8) [2]:

$$E_0 = Kp_a \left(\frac{\sigma_3}{p_a} \right)^n \tag{24.8}$$

where E_0 is the initial tangent modulus, σ_3 is the minor principal stress, K, n are the parameters, and p_a is the atmospheric pressure.

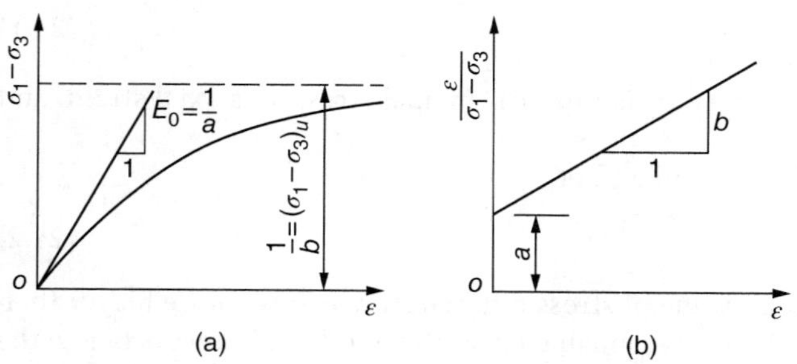

Figure 24.1 Stress–strain relation of soil expressed with hyperbola.

p_a is introduced to make $\frac{\sigma_3}{p_a}$ a dimensionless value. In order to take tensile strength into consideration, Ref. [4] suggested Eq. (24.9) to calculate initial tangent modulus, that is,

$$E_0 = Kp_a\left(\frac{\sigma_1 - \sigma_3}{p_a}\right)^n \tag{24.9}$$

$$\sigma_t = -2c \times tg\left(45^\circ - \frac{\phi}{2}\right) \tag{24.10}$$

where σ_t is the tensile strength of soil, c is the cohesive force of soil, and ϕ is the internal friction angle.

Taking logarithm of Eq. (24.9), we know there will be a straight line on log–log paper. With this relation, we can determine parameter K and n according to experiments.

According to Mohr–Coulomb failure criterion, the compressive strength can be denoted as Eq. (24.11):

$$(\sigma_1 - \sigma_3)_f = \frac{2c\cos\phi + 2\sigma_3\sin\phi}{1 - \sin\phi} \tag{24.11}$$

Substituting the expression of a and b into Eq. (24.5), we can have the tangent modulus E_t as follows:

$$E_t = \frac{\frac{1}{E_0}}{\left[\frac{1}{E_0} + \frac{R\varepsilon}{(\sigma_1 - \sigma_3)_f}\right]^2} \tag{24.12}$$

Although the tangent modulus can be calculated with Eq. (24.12), it contains both stress difference and strain. It will be more convenient to be applied to FEM with strain eliminated.

Transform Eq. (24.1) into

$$\varepsilon = \frac{(\sigma_1 - \sigma_3)a}{1 - b(\sigma_1 - \sigma_3)} = \frac{\sigma_1 - \sigma_3}{E_0\left[1 - \frac{R(\sigma_1 - \sigma_3)}{(\sigma_1 - \sigma_3)_f}\right]} = \frac{\sigma_1 - \sigma_3}{E_0(1 - R_S)} \tag{a}$$

in which

$$s = \frac{\sigma_1 - \sigma_3}{(\sigma_1 - \sigma_3)_f} \tag{24.13}$$

where s is called as stress intensity.

Substituting Eq. (a) into Eq. (24.12), we can eliminate the strain ε in the equation and get

$$E_t = (1 - R_s)^2 E_0 \tag{24.14}$$

If the expression of $E_0, s, (\sigma_1 - \sigma_3)_f$ is substituted into Eq. (24.14), the tangent modulus can be expressed as follows:

$$E_t = \left[1 - \frac{R(1 - \sin\phi)(\sigma_1 - \sigma_3)}{2c\cos\phi + 2\sigma_3\sin\phi}\right]^2 \times Kp_a\left(\frac{\sigma_1 - \sigma_3}{p_a}\right)^n \tag{24.15}$$

This equation can be applied easily into the finite element analysis of soil. There are five parameters contained, that is, c, ϕ, R, K, n, which should be determined by experiments.

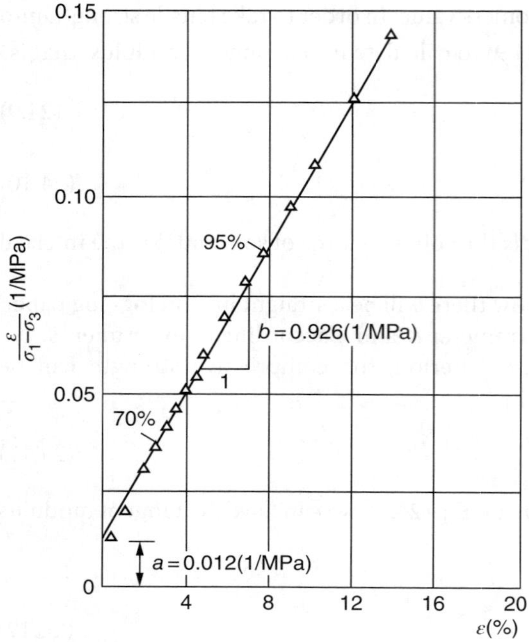

Figure 24.2 Stress–strain relation of quartz sand ($\sigma_3 = 0.50$ MPa).

The stress–strain relation of quartz sand under confining pressure of $\sigma_3 = 0.5$ MPa is shown in Figure 24.2. We can see that the experiment results greatly correspond with the hyperbola equation (24.7). With these materials, we can calculate the parameters a and b under condition of $\sigma_3 = 0.5$ MPa and then E_0 and $(\sigma_1 - \sigma_3)_u$. According to experiments of each confining pressure, a group of parameters can be determined. Figure 24.3 shows the tangent modulus E_0 of quartz sand under different confining pressure, from which

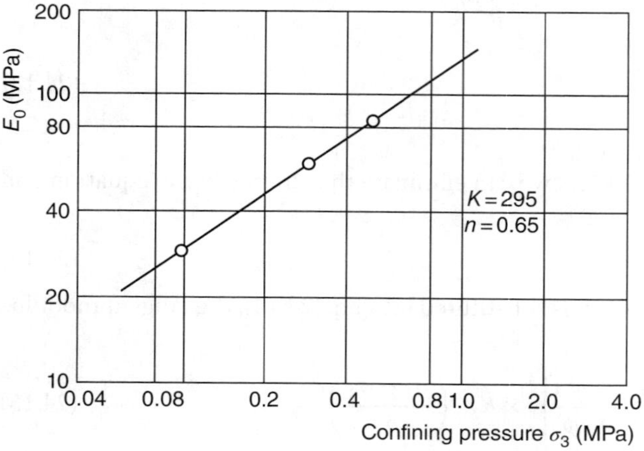

Figure 24.3 Relation between initial tangent modulus E_0 and confining pressure σ_3 of quartz.

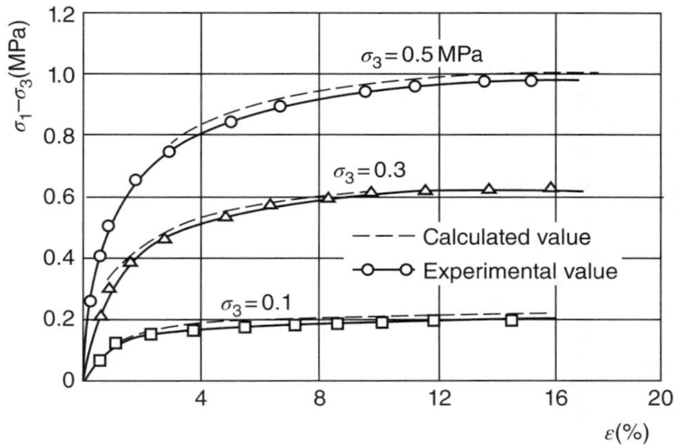

Figure 24.4 Calculated and experimental stress–strain relation of quartz sand.

parameters K and n can be determined. Figure 24.4 compared the stress–strain relation obtained from experiments and that calculated by substituting relevant parameters into Eq. (24.1). From the figure we can see that the calculated curve corresponds quite well with the experimental curve. To obtain good correspondence on a whole, Ref. [2] suggested to let the line go through the points of $s = 70\%$ and $s = 95\%$ when taking the value of a and b, as shown in Figure 24.2.

The stress–strain relation under load was given previously. In practical projects, there may be conditions of unloading or loading after unloading. Experiments have shown that it is impossible for soil after unloading to return to the strain condition before loading, while it can only partly recover the strain caused by the initial loading. There will be a bit of hysteresis in the cycle of unloading–loading. But it is nearly linear and elastic. Moreover, as Figure 24.5 shows, the tangent modulus in the two cycles of unloading–loading are approximate, despite the different stress levels in both cycles. It can be assumed that the stress–strain relation of soil in unloading–loading cycles can be expressed with a uniform tangent modulus E_u with enough accuracy. According to experiments, the tangent modulus in unloading–loading cycles has little relation with the stress level, though it is only relevant to the confining pressure, which could be expressed as

$$E_u = K_u p_a \left(\frac{\sigma_1 - \sigma_3}{p_a} \right)^n \tag{24.16}$$

where E_u is the tangent modulus of loading after unloading and K_u and n are parameters.

Actually, n here can take the value of n in the equation (24.8) calculating initial tangent modulus. K_u here, however, is usually greater than the parameter K for the initial tangent modulus. For instance, for the quartz sand mentioned before, $K = 295$, $K_u = 1090$, $n = 0.65$.

When calculating the stress and strain of soil, besides the tangent modulus E, we also need Poisson's ratio. According to experiment results, the relation between lateral and longitudinal deformation may be also expressed with hyperbola as follows [3]:

$$\varepsilon_a = \frac{\varepsilon_r}{\mu_0 + m\varepsilon_r} \tag{24.17}$$

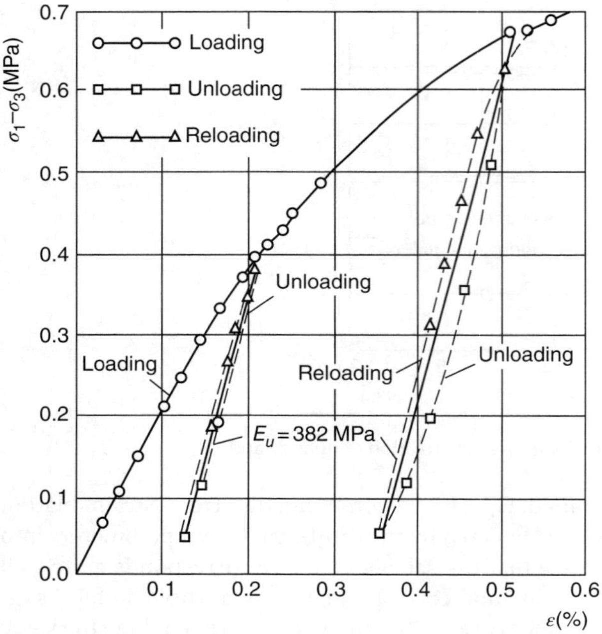

Figure 24.5 Stress–strain relation of sand under unloading and reloading ($\sigma_3 = 0.3$ MPa).

or

$$\frac{\varepsilon_r}{\varepsilon_a} = \mu_0 + m\varepsilon_r \tag{24.18}$$

where ε_a is the axial strain, ε_r is the radial strain (triaxial test), μ_0 is the initial Poisson's ratio with zero strain, and m is the parameter.

From Eq. (24.18), we have

$$\varepsilon_r = \frac{\mu_0 \varepsilon_a}{1 - m\varepsilon_a} \tag{24.19}$$

According to the definition, Poisson's ratio could be derived from Eq. (24.20), that is,

$$\mu_t = \frac{d\varepsilon_r}{d\varepsilon_a} \tag{24.20}$$

Substituting Eq. (24.19) into Eq. (24.20), we get

$$\mu_t = \frac{\mu_0}{(1 - m\varepsilon_a)^2} \tag{24.21}$$

Experiments have shown that the initial Poisson's ratio μ_0 decreases with the increase of confining pressure σ_3, denoted as Eq. (24.22):

$$\mu_0 = g - h \lg \left(\frac{\sigma_3 - \sigma_t}{p_a} \right) \tag{24.22}$$

in which g, h are the parameters depending on experiments, σ_3 is the confining pressure, and σ_t is the tensile strength.

Substituting Eq. (24.22) into Eq. (24.21), we can obtain Poisson's ratio as follows:

$$\mu_t = \frac{g - h\lg\left(\dfrac{\sigma_3 - \sigma_t}{p_a}\right)}{(1 - m\varepsilon_a)^2} \tag{24.23}$$

where ε_a, σ_t could be calculated with Eq. (a) and Eq. (24.10). In order to calculate Poisson's ratio, we need three parameters g, h, m, which ought to be determined by experiment.

According to all above, we can know that the stress–strain relation expressed by hyperbola was firstly proposed by Kondner, while Eqs (24.15) and (24.23) calculating E_t and μ_t based on experimental materials are proposed by Duncan. These equations have been applied universally, usually called as Duncan's model.

Poisson's ratio μ_t calculated with Eq. (24.23) is sometimes larger than its actual value. In 1980, Duncan and Wang *et al.* turned to take the volumetric deformation modulus K_t as parameter as follows [5]:

$$K_t = \frac{dp}{d\varepsilon_v} = K_b p_a \left(\frac{\sigma_3}{p_a}\right)^m \tag{24.24}$$

in which

$$p = (\sigma_1 + \sigma_2 + \sigma_3)/3$$

where ε_v is the volumetric deformation, K_b, m are the experimental parameters, and the value of m is usually 0–1.0.

After having K_t and E_t, Poisson's ratio could be calculated with Eq. (24.25):

$$\mu_t = \frac{3K_t - E_t}{6K_t} \tag{24.25}$$

24.2 Elastic–Plastic Model with Two Yield Surfaces

Duncan's model mentioned in the last section is intuitive and easy to be accepted by engineers because its parameters are obtained by matching experimental curves. For the problem that the direction of principal stress does not change obviously, the result is usually acceptable, though the calculated displacement is larger at times. Therefore, Duncan's model is applied universally in the computation of practical projects. Nevertheless, when applied to problems with obvious variation of direction of principal stress, such as impoundment of earth dam, Duncan's model's computation error is always large. Elastic–plastic models, represented by the Cambridge model [7], derive the stress–strain relation from assumed yield

Figure 24.6 Two-yield-surface function.

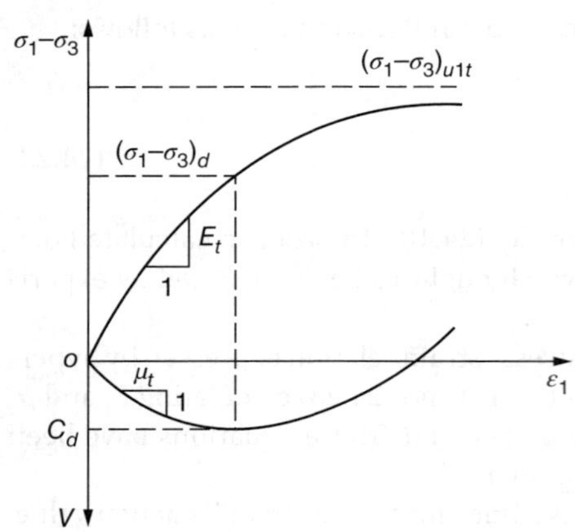

Figure 24.7 Relation between principal stress difference, principal strain, and volumetric strain.

surface. In spite of the occasionally larger calculated displacement, the result is qualitatively reasonable, especially for cases where the direction of principal stress deflects apparently. Based on the advantages of both models, Shen Zhujiang proposed the two-yield-surface elastic–plastic model [6], whose stress–strain relation shares the form of Cambridge model, while the parameters are obtained from matching the experimental stress–strain relation, just as that in Duncan's model. This model uses two yield surfaces and overcomes the shortage that the two models mentioned above are inadaptable to cases when confining pressure decreases (Figures 24.6 and 24.7).

24.2.1 Yield Function and Elastic–Plastic Matrix

Divide the strain increment into two parts (the elastic strain increment $\{\Delta\varepsilon^e\}$ and plastic strain increment $\{\Delta\varepsilon^p\}$), that is,

$$\{\Delta\varepsilon\} = \{\Delta\varepsilon^e\} + \{\Delta\varepsilon^p\} \tag{a}$$

Then divide the plastic strain increment into two parts, that is,

$$\{\Delta\varepsilon^p\} = \{\Delta\varepsilon_1^p\} + \{\Delta\varepsilon_2^p\} \tag{b}$$

Assuming there is a yield surface for each part of plastic strain, we can calculate the strain increment with orthogonal flow rule:

$$\Delta\varepsilon_{ij} = \Delta\varepsilon_{ij}^e + A_1\Delta f_1\frac{\partial f_1}{\partial\sigma_{ij}} + A_2\Delta f_2\frac{\partial f_2}{\partial\sigma_{ij}} \tag{24.26}$$

where A_1, A_2 are plastic coefficients for yield surface f_1, f_2, respectively, which are non-negative and equal to zero only when the yield surface is unloading or neutral loading.

Suppose

$$p = \frac{1}{3}(\sigma_1+\sigma_2+\sigma_3) \quad \tau = \frac{1}{3}\left[(\sigma_1-\sigma_2)^2+(\sigma_2-\sigma_3)^2+(\sigma_3-\sigma_1)^2\right]^{1/2}$$

in which p, τ are normal stress and shear stress of octahedron, respectively.

Shen Zhujiang suggests to express the first and second yield function with ellipse and power function respectively, that is,

$$f_1 = p^2 + r^2\tau^2, \quad f_2 = \frac{\tau^s}{p} \tag{24.27}$$

in which r is the ratio of major and minor axis of the ellipse and s is the power.

Suppose $v = \varepsilon_1+\varepsilon_2+\varepsilon_3$ is volumetric strain, $\gamma = (2/3)[(\varepsilon_1-\varepsilon_2)^2+(\varepsilon_2-\varepsilon_3)^2+(\varepsilon_3-\varepsilon_1)^2]^{1/2}$ is shear strain of octahedron. From Eqs (24.26) and (24.27), we can derive their increment as follows:

$$\left.\begin{aligned}\Delta v &= \frac{\Delta p}{K} + A_1\frac{\partial f_1}{\partial p}\Delta f_1 + A_2\frac{\partial f_2}{\partial p}\Delta f_2\\ \Delta \gamma &= \frac{\Delta\tau}{G} + \frac{2}{3}\left(A_1\frac{\partial f_1}{\partial \tau}\Delta f_1 + A_2\frac{\partial f_2}{\partial \tau}\Delta f_2\right)\end{aligned}\right\} \tag{24.28}$$

where K, G are elastic volumetric modulus and shear modulus, respectively.

From Eq. (24.27), we have

$$\Delta f_1 = 2p\Delta p + 2r^2\tau\Delta\tau, \quad \Delta f_2 = -\frac{\tau^s}{p^2}\Delta p + \frac{s\tau^{s-1}}{p}\Delta\tau$$

Substitute the equation above into Eq. (24.28) and we get

$$\Delta v = \frac{\Delta p}{K} + A\Delta p + C\Delta\tau \tag{c}$$

$$\Delta\gamma = \frac{\Delta\tau}{G} + \frac{2}{3}(B\Delta\tau + C\Delta p) \tag{d}$$

where

$$A = 4p^2A_1 + \frac{\tau^{2s}}{p^4}A_2, \quad B = 4r^2\tau^2A_1 + \frac{s^2\tau^{2s}}{p^2\tau^2}A_2, \quad C = 4r^2p\tau A_1 - \frac{s\tau^{2s}}{p^3\tau}A_2$$

Applying the Prandtl–Reuss flow rule on π plane, Eq. (d) could be extended as

$$\Delta e_{ij} = \frac{\Delta s_{ij}}{2G} + \frac{1}{3}(B\Delta\tau + C\Delta p)\frac{s_{ij}}{\tau} \tag{24.29}$$

$$\Delta e_{ij} = \Delta\varepsilon_{ij} - \frac{1}{3}\Delta v\delta_{ij}, \quad s_{ij} = \sigma_{ij} - p\delta_{ij}$$

where δ_{ij} is the Kronecker unit function.

Considering $\Delta\tau = \frac{1}{3}\frac{s_{ij}}{\tau}\Delta s_{ij}$, $\Delta\tau$ can be solved by multiplying s_{ij} to the two sides of Eq. (24.29). Then substituting it back into Eq. (c) and Eq. (24.29), we get

$$\left.\begin{aligned}\Delta p &= K_{\mathrm{p}}\Delta v - P\frac{1}{\tau}\{s\}^{\mathrm{T}}\{\Delta e\}\\ \Delta s_{ij} &= 2G\Delta e_{ij} - P\frac{s_{ij}}{\tau}\Delta v - Q\frac{s_{ij}}{\tau^2}\{s\}^{\mathrm{T}}\{\Delta e\}\end{aligned}\right\} \tag{24.30}$$

in which

$$K_{\mathrm{p}} = \frac{K}{1+KA}\left(1 + \frac{2}{3}\frac{KGC^2}{1+KA+GD}\right)$$

$$P = \frac{2}{3}\frac{KGC}{1+KA+GD}$$

$$Q = \frac{2}{3}\frac{G^2D}{1+KA+GD}$$

$$D = \frac{2}{3}(B + KAB - KC^2)$$

For plane problems, Eq. (24.30) can be transformed to

$$\{\Delta\sigma\} = [D]_{\mathrm{ep}}\{\Delta\varepsilon\}$$

in which

$$\{\Delta\sigma\}^{\mathrm{T}} = \left[\Delta\sigma_x, \Delta\sigma_y, \Delta\sigma_z, \Delta\tau_{xy}\right], \quad \{\Delta\varepsilon\}^{\mathrm{T}} = \left[\Delta\varepsilon_x, \Delta\varepsilon_y, \Delta\gamma_{xy}\right] \tag{24.31}$$

$$[D]_{\mathrm{ep}} = \begin{bmatrix} M_1 - P\dfrac{s_x + s_x}{\tau} - Q\dfrac{s_x{}^2}{\tau^2}, & M_2 - P\dfrac{s_x + s_y}{\tau} - Q\dfrac{s_x s_y}{\tau^2}, & -P\dfrac{s_{xy}}{\tau} - Q\dfrac{s_x s_{xy}}{\tau^2} \\ M_2 - P\dfrac{s_y + s_x}{\tau} - Q\dfrac{s_y s_x}{\tau^2}, & M_1 - P\dfrac{s_y + s_z}{\tau} - Q\dfrac{s_y{}^2}{\tau^2}, & -P\dfrac{s_{xy}}{\tau} - Q\dfrac{s_y s_{xy}}{\tau^2} \\ M_2 - P\dfrac{s_z + s_x}{\tau} - Q\dfrac{s_z s_x}{\tau^2}, & M_2 - P\dfrac{s_z + s_y}{\tau} - Q\dfrac{s_z s_y}{\tau^2}, & -P\dfrac{s_{xy}}{\tau} - Q\dfrac{s_z s_{xy}}{\tau^2} \\ -P\dfrac{s_{xy}}{\tau} - Q\dfrac{s_{xy} s_x}{\tau^2}, & -P\dfrac{s_{xy}}{\tau} - Q\dfrac{s_{xy} s_y}{\tau^2}, & G - Q\dfrac{s_{xy}{}^2}{\tau^2} \end{bmatrix} \tag{24.32}$$

$$M_1 = K_{\mathrm{p}} + 4G/3, \quad M_2 = K_{\mathrm{p}} - 2G/3$$

where $[D]_{\mathrm{ep}}$ is symmetrical elastic–plastic matrix.

24.2.2 Plastic Coefficient

Assume plastic coefficient A_1 and A_2 are merely functions of stress state while irrelevant with stress path. Thus the results measured indoor with simple stress path can be directly applied to complex stress conditions on the spot. Like Duncan's model, we can use the results of ordinary triaxial test. In this case, $\Delta\gamma = \sqrt{2}(3\Delta\varepsilon_1 - \Delta\upsilon)/3$, $\Delta p = \Delta\sigma_1/3$, $\Delta\tau = \sqrt{2}\Delta\sigma_1/3$. Substitute it into Eq. (c) and define $E_{\mathrm{t}} = \Delta\sigma_1/\Delta\varepsilon_1$, $\mu_{\mathrm{t}} = \Delta\upsilon/\Delta\varepsilon_1$ (Figure 24.7), and then we will have

$$\left.\begin{aligned} \frac{9}{E_{\mathrm{t}}} &= \frac{1}{K} + \frac{3}{G} + 4(p + \sqrt{2}r^2\tau)^2 A_1 + \frac{\tau^{2s}}{p^2}\left(\frac{1}{p} - \frac{\sqrt{2}s}{\tau}\right)^2 A_2 \\ \frac{3\mu_{\mathrm{t}}}{E} &= \frac{1}{K} + 4p(p + \sqrt{2}r^2\tau)A_1 + \frac{\tau^{2s}}{p^2}\left(\frac{1}{p} - \frac{\sqrt{2}s}{\tau}\right)A_2 \end{aligned}\right\} \tag{e}$$

A_1 and A_2 can be solved from Eq. (e):

$$\left.\begin{aligned} A_1 &= \frac{\tau\left(\dfrac{9}{E_{\mathrm{t}}} - \dfrac{3\mu_{\mathrm{t}}}{E_{\mathrm{t}}} - \dfrac{3}{G}\right) + \sqrt{2}sp\left(\dfrac{3\mu_{\mathrm{t}}}{E_{\mathrm{t}}} - \dfrac{1}{K}\right)}{4\sqrt{2}(p + \sqrt{2}r^2\tau)(sp^2 + r^2\tau^2)} \\ A_2 &= \frac{p^4\tau^2}{\tau^{2s}} \frac{p\left(\dfrac{9}{E_{\mathrm{t}}} - \dfrac{3\mu_{\mathrm{t}}}{E_{\mathrm{t}}} - \dfrac{3}{G}\right) - \sqrt{2}r^2\tau\left(\dfrac{3\mu_{\mathrm{t}}}{E_{\mathrm{t}}} - \dfrac{1}{K}\right)}{\sqrt{2}(\sqrt{2}sp - \tau)(sp^2 + r^2\tau^2)} \end{aligned}\right\} \tag{24.33}$$

in which E_{t} could be solved with Eq. (24.15) given by Duncan and the five parameters c, ϕ, R, K, n can be calculated from experimental materials. Shen obtained the $\upsilon - \varepsilon_1$ relation curve via matching experiments with parabola. According to the definition, $\mu_{\mathrm{t}} = \Delta\upsilon/\Delta\varepsilon_1$ (Figure 24.7), we have

$$\mu_{\mathrm{t}} = 2C_d\left(\frac{\sigma_3}{p_{\mathrm{a}}}\right)^d \frac{E_0 R_{\mathrm{s}}}{\sigma_1 - \sigma_3} \frac{1 - R_{\mathrm{d}}}{R_{\mathrm{d}}}\left(1 - \frac{R_{\mathrm{s}}}{1 - R_{\mathrm{s}}} \frac{1 - R_{\mathrm{d}}}{R_{\mathrm{d}}}\right) \tag{24.34}$$

$$R_d = \frac{(\sigma_1 - \sigma_3)_d}{(\sigma_1 - \sigma_3)_{\mathrm{ult}}}$$

where C_d is the maximum volumetric strain when $\sigma_3 = 1$at; d is the power when volumetric strain varies with σ_3; and R_d is the stress ratio (dilatancy ratio) when the maximum volumetric strain occurs. Obviously, these three parameters are easy to determine.

The tangent modulus of unloading–loading could be calculated with Eq. (24.16). Assuming that Poisson's ratio μ is constant, the elastic volumetric modulus K and shear modulus G in the equations mentioned above can be calculated as follows:

$$K = \frac{E_{\mathrm{u}}}{3(1-2\mu)}, \quad G = \frac{E_{\mathrm{u}}}{2(1+\mu)} \tag{24.35}$$

See Eq. (24.16) for E_{u}. For most kinds of soil, we can take $\mu = 0.30$.

As for the yield surface parameters r and s, we can take $r = 2$, $s = 3$ for soil according to the results in Ref. [6] and $r = 2$, $s = 2$ for rockfill according to Ref. [8].

24.3 Interaction between Soil and Structure: Contact Element

Most structures are built on soil foundation. Therefore, the interactions between soil and structure will directly affect the stress and displacement of structure. In the past, limited by calculating method, people could only make some coarse assumptions about the interactions between soil and structure when designing structures, leading to great difference between computation result and reality. After taking FEM, we can consider nonlinear stress–strain relation as well as stage construction process. Thus the reality could be well reflected in calculating, leading to great improvement in precision, and the computing result gets closer to actual observation as shown in Figure 24.10.

When analyzing interactions between soil and structure with FEM, besides using nonlinear stress–strain relation according to structure and features of soil, the contact surface of soil and structure deserves special concern. In the past, one of the following two extreme simplified assumptions is applied when analyzing interactions between soil and structure: (1) the contact surface is coarse enough to prevent any slide between soil and structure, and (2) the contact surface is so smooth that there won't be shear stress to restrain the relative slide between soil and structure. The reality, of course, is not so simple. In order to reflect the stress distribution on contact surface well, we will use a specified contact element.

Referring to the rock joint element mentioned in last chapter, we take the contact element as Figure 24.8 shows, with length of l, width of $e = 0$, and thickness (perpendicular to xy plane) of t. The element has 4 nodes i, j, m, r. The origin of coordinates is on the

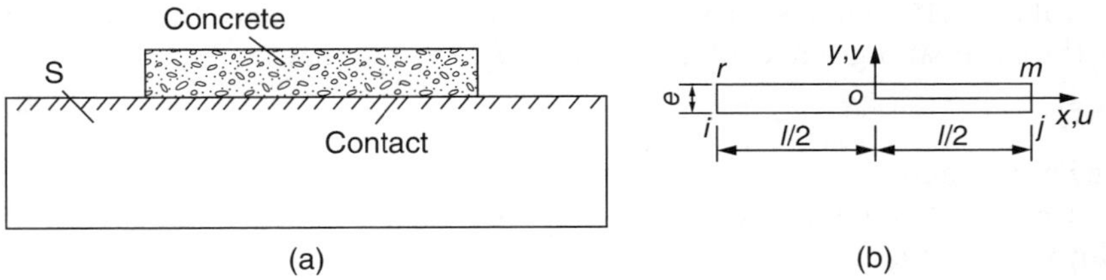

Figure 24.8 Foundation beam and contact element. (a) Foundation and beam; (b) contact element.

element's centroid. One side of the element is soil and the other is structure's material, such as concrete.

Assume that the shear stress τ_s in the element is proportional to the displacement difference of the two sides and the normal stress σ_n is proportional to the vertical displacement difference of the two sides, that is,

$$\tau_s = \lambda_s \Delta u, \quad \sigma_n = \lambda_n \Delta v \tag{a}$$

where λ_s, λ_n are, respectively, the tangent and normal stiffness coefficient of contact element and Δu, Δv are the tangent and normal displacement difference of two sides of the element.

To measure the stiffness coefficient, experiments ought to be conducted. Test the composite sample with half soil and half structure material on direct shear apparatus. For instance, if the structure material is concrete, the lower half of the shear box should be filled with concrete and the upper half should be filled with soil and then tamp it simulating that in construction procedure. It will be better to keep the interval between the upper half and concrete smaller, so that we can assume that the relative displacement between the upper and lower part in the shear box is completely the result of movement between contact surfaces.

The shear stress and displacement difference curve of contact element could be expressed with hyperbola:

$$\tau_i = \frac{\Delta u}{a + b\Delta u} \tag{b}$$

in which τ_s is the shear stress on contact surface, Δu is the tangent displacement difference on contact surface, and a, b are the parameters depending on experimental materials. After transformation, Eq. (b) will become

$$\frac{\Delta u}{\tau_s} = a + b\Delta u \tag{c}$$

The reciprocal of coefficient a is the initial slope of shear stress and displacement difference curve, called as initial tangent stiffness coefficient λ_{s0}. It is relevant with normal stress σ_n on the contact surface. According to experiments, it can be expressed as

$$\lambda_{s0} = \frac{1}{a} = k_s \gamma_w \left(\frac{\sigma_n}{p_a}\right)^n \tag{d}$$

where λ_{s0} is the initial tangent stiffness coefficient, σ_n is the normal stress, γ_w is the density of water, p_a is the atmospheric pressure, and k_s, n are the parameters depending on experimental materials.

γ_w and p_a are introduced to make parameters k and n dimensionless (the unit of λ_{s0} is kN/cm^3, same as that of γ_w).

The reciprocal of coefficient b is the asymptotic value of $\tau_s - \Delta u$ curve, denoted with τ_{su}. Suppose the shear strength of contact surface is τ_f. Then we have

$$\tau_f = R\tau_{su} \tag{e}$$

where R is the break ratio.

The shear strength of contact surface τ_f is proportional to normal stress, expressed with the following equation:

$$\tau_f = \sigma_n tg\delta \tag{f}$$

where δ is friction angle of the contact surface between soil and structure, which is less than the inner friction angle of soil ϕ in most cases, generally $\delta = 0.8\phi$.

According to Eqs (e) and (f), we can get

$$b = \frac{1}{\tau_{su}} = \frac{R}{\tau_f} = \frac{R}{\sigma_n tg\delta} \tag{g}$$

By taking the derivative of Eq. (b) and eliminating Δu, we can obtain the following tangent stiffness coefficient:

$$\lambda_s = k_s \gamma_w \left(\frac{\sigma_n}{p_a}\right)^n \left(1 - \frac{R\tau_s}{\sigma_n tg\delta}\right)^2 \tag{24.36}$$

When the normal stress σ_n is tensile stress, the contact surface will be pulled apart, and thus there should be $\lambda_n = \lambda_s = 0$. When σ_n is compressive stress, λ_s should be calculated with the equation above, while λ_n should be taken as a number large enough. Since the thickness of contact element e is equal to zero, from last chapter we know that $\lambda_n = E/e$. Therefore as long as E is finite, when the width $e \to 0$, there will be $\lambda_n \to \infty$. In practical computation it can be taken as around $\lambda_n = 1 \times 10^3\,\text{kN/cm}^3$.

The experimental materials are shown in Figure 24.9, from which we can see that the experimental points suit quite well with the hyperbola equation [10].

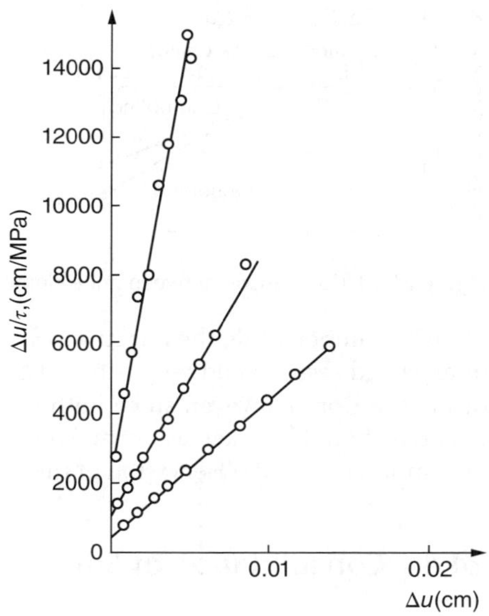

Figure 24.9 Experimental curve of contact surface between soil and concrete.

Figure 24.10 shows the result of finite element analysis of a ship lock [9]. This lock is a reinforced concrete structure, built on soil foundation and analyzed with linear strain quadrangle element. At the beginning, it was assumed the soil was linear elastic and the load was added instantly without consideration of construction process. The computing result was inconsistent with actual materials. Thus it was turned to analyze with increment method. The process was divided into ten stages with step-by-step calculation to simulate construction process. In the computation, foundation excavation, drainage, concreting by stages, backfill, variation of underground water level, seasonal air temperature change, impoundment of lock, and other factors are considered. For concrete, long-term elastic modulus was used to account for creep. The stress–strain relation of soil was expressed with hyperbola. On the contact surface between soil and concrete, contact element was used to consider relative displacement. A specialized test was conducted to measure the stiffness coefficient of the element. The calculated foundation settlement, soil pressure, and structure deflection all suited well with the observations. Some observations appeared as abnormal were shown reasonable with the result of finite element analysis. For example, both computation and observation have shown that with the impoundment of lock, lock wall moves toward the center line

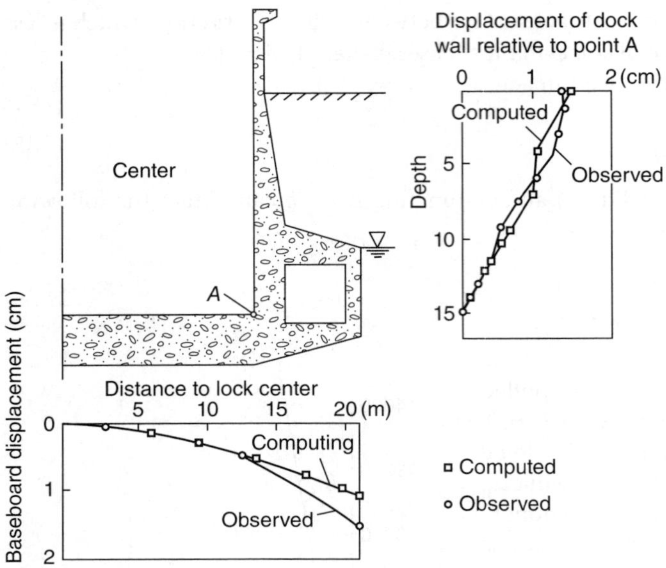

Figure 24.10 Comparison between finite element computing result and observations of lock.

of lock chamber while the effective soil pressure on the top of lock wall increases. This unexpected result could be attributed to some error if it is shown only in computation or observation. However, since both observation and computation arrive at the same result, it should be taken as normal phenomenon. The comparison between deflections of computation and observation of the lock when completed are shown in Figure 24.10.

24.4 Consolidation of Soil

Under the effect of load, water in soil will gradually seep out and the volume will decrease. This phenomenon is called consolidation of soil. It results in compressive deformation and the strength of soil increases at the same time. Therefore, consolidation is an important issue in soil mechanics because it not only leads to foundation settlement but also controls the stability of foundation.

Next we will first introduce the approximate consolidation theory of Terzaghi and then describe the more accurate consolidation theory of Biot. Both of the theories are applied in engineering at present.

24.4.1 Terzaghi's Consolidation Theory

Figure 24.11 shows an elementary volume $dxdydz$ taken from soil. The seepage velocities in the direction of x, y, z are v_x, v_y, v_z, respectively. The inward and outward flows of the elementary volume in unit time are, respectively,

$$q_{\text{in}} = v_x \mathrm{d}y\mathrm{d}z + v_y \mathrm{d}z\mathrm{d}x + v_z \mathrm{d}x\mathrm{d}y$$

$$q_{\text{out}} = \left(v_x + \frac{\partial v_x}{\partial x}\mathrm{d}x\right)\mathrm{d}y\mathrm{d}z + \left(v_y + \frac{\partial v_y}{\partial y}\mathrm{d}y\right)\mathrm{d}z\mathrm{d}x + \left(v_z + \frac{\partial v_z}{\partial z}\mathrm{d}z\right)\mathrm{d}x\mathrm{d}y$$

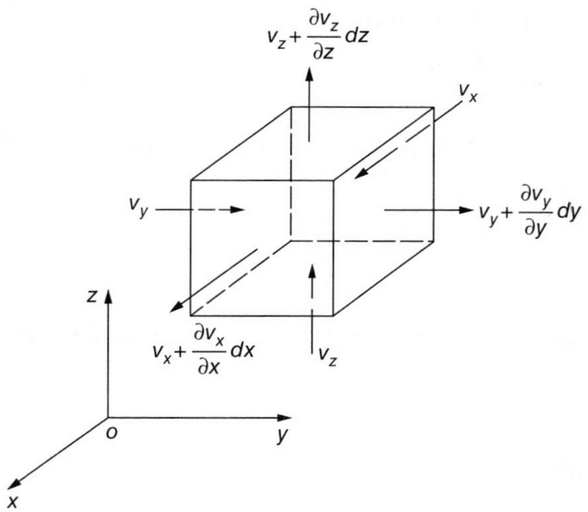

Figure 24.11 Micro unit of soil.

Therefore the net inflow in unit time is

$$q_{in} - q_{out} = -\left(\frac{\partial v_x}{\partial x} + \frac{\partial v_y}{\partial y} + \frac{\partial v_z}{\partial z}\right) dxdydz \tag{a}$$

Suppose the soil is completely saturated and compression of soil particles and water is negligible. Then the volume change of the elementary volume $\partial V/\partial t$ is equal to the net inflow, that is,

$$\frac{\partial V}{\partial t} = q_{in} - q_{out} \tag{b}$$

Substitute Eq. (a) into Eq. (b) and get

$$\frac{\partial V}{\partial t} = -\left(\frac{\partial v_x}{\partial x} + \frac{\partial v_y}{\partial y} + \frac{\partial v_z}{\partial z}\right) dxdydz \tag{c}$$

The volume of the elementary volume V can be expressed with soil particle volume V_s and void ratio e as follows:

$$V = dxdydz = (1 + e)V_s \tag{d}$$

Since the volume change of soil particle is negligible, V_s can be taken as constant. Therefore the volume change of the elementary volume mainly comes from change in void ratio. From Eq. (d) we may have

$$\frac{\partial V}{\partial t} = V_s \frac{\partial e}{\partial t} = \frac{dxdydz}{1+e}\frac{\partial e}{\partial t} \tag{e}$$

From Eqs (c) and (e) we can get

$$\frac{\partial e}{\partial t} = -(1+e)\left(\frac{\partial v_x}{\partial x} + \frac{\partial v_y}{\partial y} + \frac{\partial v_z}{\partial z}\right) \tag{f}$$

Suppose the pore pressure in soil is p. According to Darcy's law, we have

$$v_x = -\frac{k_x}{\gamma}\frac{\partial p}{\partial x}, \quad v_y = -\frac{k_y}{\gamma}\frac{\partial p}{\partial y}, \quad v_z = -\frac{k_z}{\gamma}\frac{\partial p}{\partial z} \tag{24.37}$$

where γ is the density of water and k_x, k_y, k_z are the percolate coefficients in direction of x, y, z, respectively.

Substituting Eq. (24.37) into Eq. (f), we will obtain

$$\frac{\partial e}{\partial t} = \frac{1+e}{\gamma}\left(k_x\frac{\partial^2 p}{\partial x^2} + k_y\frac{\partial^2 p}{\partial y^2} + k_z\frac{\partial^2 p}{\partial z^2}\right) \tag{24.38}$$

The effective stress σ' in any point of soil is equal to the total stress σ minus pore pressure p, that is,

$$\sigma = \sigma' - p \tag{24.39}$$

If the total stress σ' is constant, then

$$\frac{\partial \sigma}{\partial t} = -\frac{\partial p}{\partial t} \tag{g}$$

$$\frac{\partial \sigma}{\partial e} = -\frac{\partial p}{\partial e} \tag{h}$$

Define compression coefficient as

$$a_{\mathrm{v}} = \frac{\partial e}{\partial \sigma} \tag{i}$$

Therefore

$$\frac{\partial e}{\partial t} = \frac{\partial e}{\partial p}\frac{\partial p}{\partial t} = a_{\mathrm{v}}\frac{\partial p}{\partial t} \tag{24.40}$$

Substitute Eq. (24.40) into Eq. (24.38) and get

$$\frac{\partial e}{\partial t} = C_{\mathrm{v}x}\frac{\partial^2 p}{\partial x^2} + C_{\mathrm{v}y}\frac{\partial^2 p}{\partial y^2} + C_{\mathrm{v}z}\frac{\partial^2 p}{\partial z^2} \tag{24.41}$$

where

$$C_{\mathrm{v}x} = \frac{(1+e)k_x}{\gamma a_{\mathrm{v}}}, \quad C_{\mathrm{v}y} = \frac{(1+e)k_y}{\gamma a_{\mathrm{v}}}, \quad C_{\mathrm{v}z} = \frac{(1+e)k_z}{\gamma a_{\mathrm{v}}} \tag{24.42}$$

in which $C_{\mathrm{v}x}, C_{\mathrm{v}y}, C_{\mathrm{v}z}$ are called consolidate coefficients.

Equation (24.41) is Terzaghi's consolidation equation.

Suppose the three directions share the same permeate coefficient, $k_x = k_y = k_z$. Then we have

$$C_{\mathrm{v}x} = C_{\mathrm{v}y} = C_{\mathrm{v}z} = C_{\mathrm{v}} = (1+e)k/\gamma a_{\mathrm{v}}$$

Equation (24.41) can be rewritten as follows:

$$\frac{\partial p}{\partial t} = C_{\mathrm{v}}\left(\frac{\partial^2 p}{\partial x^2} + \frac{\partial^2 p}{\partial y^2} + \frac{\partial^2 p}{\partial z^2}\right) \tag{24.41a}$$

Consolidation equations (24.41) and (24.41a) share the same form with the heat conduction equation (14.2). The finite element discretization is also the same, which won't be repeated here.

24.4.2 Biot's Consolidation Theory

Based on stricter foundation, Biot derived the three-dimensional (3D) consolidation equation that can correctly reflect dissipation of pore pressure and deformation of soil skeleton, which is usually called as "true 3D consolidation theory," and Terzaghi's 3D consolidation equation is called as "pseudo 3D consolidation theory."

The basic assumptions of Biot's consolidation theory are the following: (1) deformation of soil skeleton is linear elastic, (2) deformation is small, (3) seepage follows Darcy's law, and (4) water is incompressible.

The volume force on soil skeleton due to pore pressure is

$$\begin{bmatrix} X \\ Y \\ Z \end{bmatrix} = -\begin{bmatrix} \dfrac{\partial p}{\partial x} \\ \dfrac{\partial p}{\partial y} \\ \dfrac{\partial p}{\partial z} \end{bmatrix} \tag{j}$$

Take the direction of z axis straight up. In rectangular coordinate system, the balance equation expressed with displacement is [12]

$$\left.\begin{aligned} G\nabla^2 u + \frac{G}{1-2\mu}\frac{\partial}{\partial x}\left(\frac{\partial u}{\partial x}+\frac{\partial v}{\partial y}+\frac{\partial w}{\partial z}\right) - \frac{\partial p}{\partial x} = 0 \\ G\nabla^2 v + \frac{G}{1-2\mu}\frac{\partial}{\partial y}\left(\frac{\partial u}{\partial x}+\frac{\partial v}{\partial y}+\frac{\partial w}{\partial z}\right) - \frac{\partial p}{\partial y} = 0 \\ G\nabla^2 w + \frac{G}{1-2\mu}\frac{\partial}{\partial z}\left(\frac{\partial u}{\partial x}+\frac{\partial v}{\partial y}+\frac{\partial w}{\partial z}\right) - \frac{\partial p}{\partial z} - \gamma = 0 \end{aligned}\right\} \tag{24.43}$$

where

$$\nabla^2 = \frac{\partial^2}{\partial x^2} + \frac{\partial^2}{\partial y^2} + \frac{\partial^2}{\partial z^2}$$

in which γ is the density of soil.

The volume deformation of soil is

$$\varepsilon_{\mathrm{v}} = \frac{\partial u}{\partial x} + \frac{\partial v}{\partial y} + \frac{\partial w}{\partial z} \tag{k}$$

According to deformation compatibility condition, the rate of volume deformation should equal to the net inflow of unit volume in unit time:

$$\frac{\partial \varepsilon_{\mathrm{v}}}{\partial t} = \frac{\partial}{\partial t}\left(\frac{\partial u}{\partial x}+\frac{\partial v}{\partial y}+\frac{\partial w}{\partial z}\right) = -\frac{\partial v_{\mathrm{x}}}{\partial x} - \frac{\partial v_{\mathrm{y}}}{\partial y} - \frac{\partial v_{\mathrm{z}}}{\partial z} \tag{l}$$

Suppose x, y, z are the principal directions of seepage. According to Darcy's law, we have

$$v_x = -\frac{k_{\mathrm{x}}}{\gamma}\frac{\partial p}{\partial x}, \quad v_y = -\frac{k_y}{\gamma}\frac{\partial p}{\partial y}, \quad v_z = -\frac{k_z}{\gamma}\frac{\partial p}{\partial z} \tag{m}$$

Substituting Eq. (m) into Eq. (l), we can get the deformation compatibility condition as follows:

$$\frac{\partial}{\partial t}\left(\frac{\partial u}{\partial x}+\frac{\partial v}{\partial y}+\frac{\partial w}{\partial z}\right) - \frac{k_x}{\gamma}\frac{\partial^2 p}{\partial x^2} - \frac{k_y}{\gamma}\frac{\partial^2 p}{\partial y^2} - \frac{k_z}{\gamma}\frac{\partial^2 p}{\partial z^2} = 0 \tag{24.44}$$

If $k_x = k_y = k_z = k$, then Eq. (24.44) will become

$$\frac{\partial}{\partial t}\left(\frac{\partial u}{\partial x}+\frac{\partial v}{\partial y}+\frac{\partial w}{\partial z}\right)-\frac{k}{\gamma}\nabla^2 p = 0 \tag{24.44a}$$

Equations (24.43) and (24.44) are Biot's 3D consolidation equation. There are four partial differential equations that are just enough to solve the four variables: u, v, w, p.

Although Biot's consolidation equation was proposed early in 1941, it was not applied widely in engineering because it is difficult to solve. Until 1970s, attributing to FEM, it began to be universally applied.

Next we will describe how to solve Biot's consolidation equation with FEM.

Express the pore pressure p and soil displacement u, v, w of each element with shape functions as follows:

$$p = [N]\{p\}^{\mathrm{e}}, \quad \begin{Bmatrix} u \\ v \\ w \end{Bmatrix} = \left[\overline{N}\right]\{\delta\}^{\mathrm{e}} \tag{n}$$

$$[N] = \left[N_i \ N_j \ N_m \ \cdots\right], \quad \left[\overline{N}\right] = \begin{bmatrix} N_i & 0 & 0 & N_j & 0 & 0 & N_m & 0 & 0 & \cdots \\ 0 & N_i & 0 & 0 & N_j & 0 & 0 & N_m & 0 & \cdots \\ 0 & 0 & N_i & 0 & 0 & N_j & 0 & 0 & N_m & \cdots \end{bmatrix}$$

$$\{p\}^{\mathrm{e}} = \left[p_i \ p_j \ p_m \ \cdots\right]^{\mathrm{T}}, \quad \{\delta\}^{\mathrm{e}} = \left[u_i \ v_i \ w_i \ u_j \ v_j \ w_j \ u_m \ v_m \ w_m \ \cdots\right]^{\mathrm{T}}$$

From virtual work principle, we know that the nodal load on soil skeleton caused by pore pressure is

$$\{P\}_{\mathrm{p}} = \int \left[\overline{N}\right]^{\mathrm{T}} \begin{Bmatrix} \dfrac{\partial p}{\partial x} \\ \dfrac{\partial p}{\partial y} \\ \dfrac{\partial p}{\partial z} \end{Bmatrix} \mathrm{d}V = [L]\{p\}^{\mathrm{e}} \tag{o}$$

in which

$$[L] = \int \left[\overline{N}\right]^{\mathrm{T}} \begin{Bmatrix} \dfrac{\partial}{\partial x} \\ \dfrac{\partial}{\partial y} \\ \dfrac{\partial}{\partial z} \end{Bmatrix} [N]\mathrm{d}V \tag{p}$$

Thus the node equilibrium equation of soil skeleton is shown as follows:

$$[K]\{\delta\} + [L]\{p\} - \{P\} = 0 \tag{24.45}$$

in which $[K]$ is the stiffness matrix and $\{P\}$ is the nodal force caused by other loads except for pore pressure.

The pore pressure must satisfy continuous equation, that is, Eq. (14.9) in Chapter 14, where the inner source Q is the volume of fluid that flows in unit volume in unit time.

Comparing Eqs (14.9) and (l) in this section, we know that in this case

$$Q = -\frac{\partial \varepsilon_v}{\partial t} = -\frac{\partial}{\partial t}\left(\frac{\partial u}{\partial x} + \frac{\partial v}{\partial y} + \frac{\partial w}{\partial z}\right) = -\frac{\partial}{\partial t}[C]\{\delta\}^{\mathrm{e}} \tag{q}$$

$$[C] = \left[\frac{\partial N_i}{\partial x}\ \frac{\partial N_i}{\partial y}\ \frac{\partial N_i}{\partial z}\ \frac{\partial N_j}{\partial x}\ \frac{\partial N_j}{\partial y}\ \frac{\partial N_j}{\partial z}\ \frac{\partial N_m}{\partial x}\ \frac{\partial N_m}{\partial y}\ \frac{\partial N_m}{\partial z}\ \ldots\right] \tag{r}$$

Substitute Eq. (q) into Eqs (14.22) and (14.25) and we obtain

$$[H]\{p\} + [S]\frac{\partial\{\delta\}}{\partial t} = 0 \tag{24.46}$$

where

$$[S] = \int [N]^{\mathrm{T}}[C]\mathrm{d}V \tag{24.47}$$

We can solve soil displacement u, v, w and pore pressure p with simultaneous equations (24.45) and (24.46).

Equation (24.46) contains partial differentiation of displacement to time thus it is necessary to discrete in time domain. Now we discretize with two-point form in time domain. Take the shape function of time domain as

$$\begin{aligned} N_n(t) &= 1 - t/\Delta t = 1 - \xi, \quad N_{n+1}(t) = t/\Delta t = \xi \\ \dot{N}_n &= \mathrm{d}N_n/\mathrm{d}t = -1/\Delta t, \quad \dot{N}_{n+1} = \mathrm{d}N_{n+1}/\mathrm{d}t = 1/\Delta t \end{aligned} \tag{s}$$

Therefore we have

$$\{\delta\} = N_n(t)\{\delta_n\} + N_{n+1}(t)\{\delta_{n+1}\} = (1-\xi)\{\delta_n\} + \xi\{\delta_{n+1}\} \tag{24.48}$$

$$\{p\} = N_n(t)\{p_n\} + N_{n+1}(t)\{p_{n+1}\} = (1-\xi)\{p_n\} + \xi\{p_{n+1}\} \tag{24.49}$$

The weighted residual of Eq. (24.46) is

$$\begin{aligned} \int_0^1 W_j\{R\}\mathrm{d}\xi = \int_0^1 W_j \Big([H]((1-\xi)\{p_n\} + \xi\{p_{n+1}\}) \\ + [S]\left(-\frac{1}{\Delta t}\{\delta_n\} + \frac{1}{\Delta t}\{\delta_{n+1}\}\right)\Big)\mathrm{d}\xi = 0 \end{aligned}$$

Divide the equation above by $\int_0^1 W_j\mathrm{d}\xi$ so that we get

$$[H]((1-\theta)\{p_n\} + \theta\{p_{n+1}\}) + \frac{1}{\Delta t}[S](\{\delta_{n+1}\} - \{\delta_n\}) = 0 \tag{24.50}$$

in which

$$\theta = \int_0^1 W_j\xi\mathrm{d}\xi \Big/ \int_0^1 W_j\mathrm{d}\xi$$

θ is s in Figure 15.4. For forward difference $\theta = 0$, for center point difference $\theta = 1/2$, and for backward difference $\theta = 1$. Consolidation problems are usually solved with increment method. For this reason, Eq. (24.50) can be better rewritten in increment form as

$$\theta[H]\{\Delta p_n\} + \frac{1}{\Delta t}[S]\{\Delta\delta_n\} = -[H]\{p_n\} \tag{24.51}$$

in which

$$\{\Delta\delta_n\} = \{\delta_{n+1}\} - \{\delta_n\}, \quad \{\Delta p_n\} = \{p_{n+1}\} - \{p_n\} \tag{t}$$

Equation (24.45) could also be rewritten in increment form as well:

$$[K]\{\Delta\delta_n\} + [L]\{\Delta p_n\} = \{\Delta P_n\} \tag{24.52}$$

Equations (24.51) and (24.52) are basic equations to solve Biot's consolidation problems with FEM. They can also be combined as follows:

$$\begin{bmatrix} K & L \\ \frac{1}{\Delta t}S & \theta H \end{bmatrix} \begin{Bmatrix} \Delta\delta_n \\ \Delta p_n \end{Bmatrix} = \begin{Bmatrix} F_1 \\ F_2 \end{Bmatrix} \tag{24.53}$$

where

$$\{F_1\} = \{\Delta P_n\}, \quad \{F_2\} = -[H]\{p_n\} \tag{24.54}$$

In order to decrease the band width of coefficient matrix in Eq. (24.53) to save storage, it would be better to have hybrid coding for the degree of freedom for each node, that is, to take the degree of freedom of element node as follows:

$$\{a\}^{\mathrm{e}} = \begin{bmatrix} u_i & v_i & w_i & p_i & u_j & v_j & w_j & p_j & \cdots \end{bmatrix}^{\mathrm{T}} \tag{u}$$

Taking plane strain problem as example, we use triangular element with nodes i, j, m. Let

$$p = [N]\{p\}^{\mathrm{e}}, \quad \begin{Bmatrix} u \\ v \end{Bmatrix} = \begin{bmatrix} \overline{N} \end{bmatrix} \{\delta\}^{\mathrm{e}} \tag{v}$$

$$[N] = \begin{bmatrix} N_i & N_j & N_m \end{bmatrix}, \quad \begin{bmatrix} \overline{N} \end{bmatrix} = \begin{bmatrix} N_i & 0 & N_j & 0 & N_m & 0 \\ 0 & N_i & 0 & N_j & 0 & N_m \end{bmatrix}$$

in which

$$N_i = \frac{1}{2\Delta}(a_i + b_i x + c_i y)$$

where Δ is the area of triangle.

The volume force caused by seepage is

$$\begin{Bmatrix} \dfrac{\partial p}{\partial x} \\ \dfrac{\partial p}{\partial y} \end{Bmatrix} = \frac{1}{2\Delta} \begin{bmatrix} b_i & b_j & b_m \\ c_i & c_j & c_m \end{bmatrix} \begin{Bmatrix} p_i \\ p_j \\ p_m \end{Bmatrix} \tag{w}$$

Since $\int N_i \mathrm{d}A = \Delta/3$, substitute Eq. (w) into Eq. (p) and we get

$$[L] = \frac{1}{6} \begin{bmatrix} b_i & b_j & b_m \\ c_i & c_j & c_m \\ b_i & b_j & b_m \\ c_i & c_j & c_m \\ b_i & b_j & b_m \\ c_i & c_j & c_m \end{bmatrix} \tag{x}$$

From Eq. (r) we have

$$[C] = \frac{1}{2\Delta} \begin{bmatrix} b_i & c_i & b_j & c_j & b_m & c_m \end{bmatrix}$$

Substitute it into Eq. (24.47). Thus

$$[S] = \frac{1}{6}\begin{bmatrix} b_i & c_i & b_j & c_j & b_m & c_m \\ b_i & c_i & b_j & c_j & b_m & c_m \\ b_i & c_i & b_j & c_j & b_m & c_m \end{bmatrix} \tag{y}$$

The stiffness matrix $[K]$ is shown in Eq. (2.36). From Eq. (14.23) we know the conductive matrix is

$$[H]^e = \frac{k_x}{4\Delta}\begin{bmatrix} b_i^2 & b_ib_j & b_ib_m \\ b_jb_i & b_j^2 & b_jb_m \\ b_mb_i & b_mb_j & b_m^2 \end{bmatrix} + \frac{k_x}{4\Delta}\begin{bmatrix} c_i^2 & c_ic_j & c_ic_m \\ c_jc_i & c_j^2 & c_jc_m \\ c_mc_i & c_mc_j & c_m^2 \end{bmatrix} \tag{z}$$

Substitute $[K], [H], [L], [S]$ into Eq. (24.53), and we get the basic equation to solve Biot's consolidation problem. u, v, p at different times can be calculated by solving this equation.

Figure 24.12 shows the comparison between FEM computing result and theoretical solution of elastic soil consolidation. The theoretical solution is the settlement of half-space surface with uniform load with width of L. In FEM computation, a block with a width of $2.5L$ and a depth of $1.25L$ is used instead of half-space, bringing in some error. For pore pressure we use linear function and 3-node triangular element. For elastic displacement, we use quadratic function and 6-node triangular element. In the figure, $\xi = \sqrt{et}/L$, where e is consolidation coefficient, t is time, and L is loading width. The d in the figure is the surface settlement when $\xi = 1/8, L = (-\infty, \infty)$. It is clear in the figure that the FEM computing result corresponds very well with theoretical solution.

24.4.3 Nonlinear Consolidation Problem

In Biot's consolidation theory, the soil skeleton is assumed to be linear elastic. In reality, the stress–strain relation of soil is usually nonlinear. Next we will illustrate how to consider the nonlinear constitutive relation of soil when solving consolidation problems with FEM.

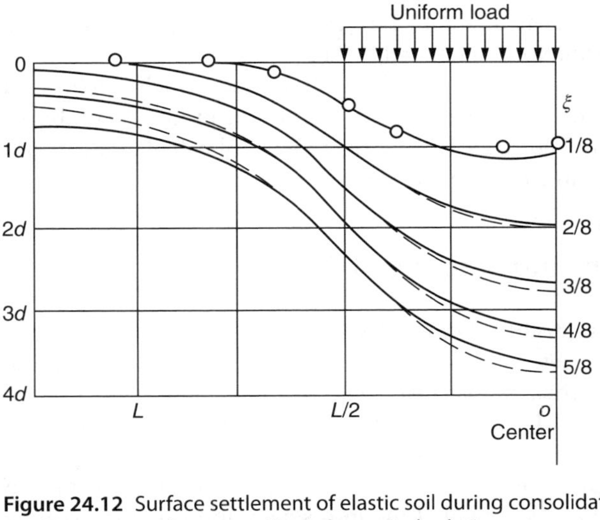

Figure 24.12 Surface settlement of elastic soil during consolidation. — FEM computation; – – – Biot's theoretical solution.

Equations (24.51) and (24.52) are basic equations for solving consolidation problems. Continuous equation (24.51) is irrelevant with the stress–strain relation of soil. Thus for nonlinear problems, we merely need to properly modify the balance equation (24.52).

24.4.3.1 Nonlinear Elastic Consolidation Problem

The equilibrium equation (24.52) could be rewritten as

$$[K_n]\{\Delta\delta_n\} + [L]\{\Delta p_n\} = \{\Delta P_n\} \tag{24.55}$$

in which $[K_n]$ is the tangent stiffness matrix in step n.

Usually we use Duncan's model to calculate tangent elastic modulus E_t and tangent Poisson's ratio μ_t with Eqs (24.15) and (24.23) and then get the tangent stiffness matrix $[K_n]$.

24.4.3.2 Elastic–Plastic Consolidation Problem

Assume the soil skeleton is elastic–plastic and choose a yield function, such as Mohr–Coulomb yield function or Shen Zhujiang's two-yield-surface function. Then we can solve the elastic–plastic matrix $[D]_{ep}$ and establish the equilibrium equation (24.52), solving together with Eq. (24.51).

24.4.3.3 Viscoelastic Consolidation Problem

Assume the soil skeleton is viscoelastic. The balance equation could be rewritten as

$$[K_n]\{\Delta\delta_n\} + [L]\{\Delta P_n\} = \{\Delta P_n\} + \{\Delta P_n^c\} \tag{24.56}$$

in which $\{\Delta P_n^c\}$ is the load increment caused by viscous deformation, which can be calculated by Eq. (18.79), and $[K_n]$ is the stiffness matrix, calculated by Eq. (18.48).

For soil, the three-element viscoelastic model is often used, shown as Figure 26.11.

24.4.3.4 Viscoelastic–Plastic Consolidation Problem

Assume the soil skeleton is viscoelastic–plastic. The equilibrium equation is

$$[K_n]\{\Delta\delta_n\} + [L]\{\Delta P_n\} = \{\Delta P_n\} + \left\{\Delta P_n^{vp}\right\} \tag{24.57}$$

in which $[K_n]$ is the stiffness matrix, calculated by Eq. (18.96), and $\{\Delta P_n{}^{vp}\}$ is the nodal load caused by visco-plastic deformation, calculated by Eq. (18.97).

For soil, the five-element viscoelastic–plastic model is often used, which can be seen in Figure 26.12. Take Poisson's ratio μ as constant. For important projects, combined model may also be used, shown in Figure 24.15. The reference [15] calculated consolidation settlement of a tunnel with five-element viscoelastic–plastic model.

24.5 Stress, Deformation, and Stability of Earth Dam

As mentioned before, the stress–strain relation of soil is highly nonlinear, so it is impossible to calculate stress and deformation of earth dam with traditional method. These years, as earth dams grow higher, people are concerned more about the stress and deformation of earth dam. With FEM, nonlinear stress–strain relations, construction by stages, and other complicated factors can be taken into account, which can give us more accurate stress and deformation of soil dam. As shown in Figure 24.13, according to practical construction procedures, we divide time into stages and calculate step by

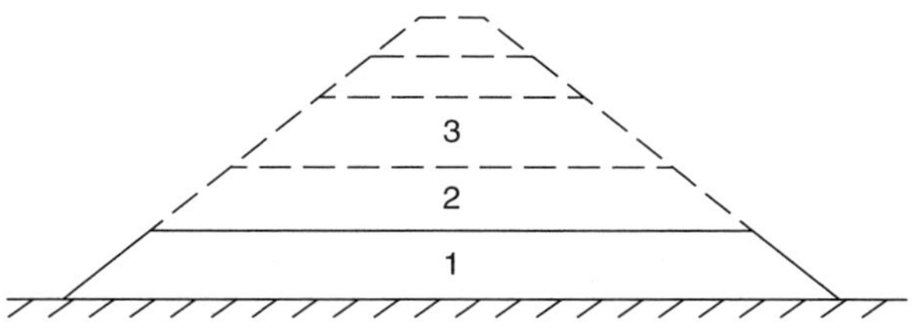

Figure 24.13 Simulating construction process.

step. The stress increment in each step is $\{\Delta\sigma\}$, and then the total stress is

$$\{\sigma\} = \Sigma\{\Delta\sigma\}$$

Figure 24.14 shows the computing result of a 260 m high earth dam [14]. Clay was used in the core of this dam, and the dam shell materials came from floodplains, consisting of silt, sand, cobble, and boulder. Computing grids are shown in Figure 24.14(a). Use 4-node quadrangle linear element for the upstream and downstream dam shell and 8-node quadrangle quadratic element for the core due to its large gradient. Divide it into 10 layers and calculate by increment method, with the stress–strain relation expressed with hyperbola. Take the lateral pressure coefficient in dam shell and inverted filter as $k = 0.5$–0.7 and that in the core as $k = 0.8$–0.9. The calculated deformation and stress are shown in Figure 24.14(b–d).

According to the stress calculated with FEM, assuming the slide surface, we can calculate the slide stability coefficient K. The results are shown in Table 24.1.

From the table above, we can see that the result computed with Bishop's method is close to that of FEM.

24.6 Computation of Rockfill Dam with Concrete Face Slab

In rockfill dam with concrete face slab, the concrete slab is thin, thus having little effect on dam deformation. Both the displacement of rockfill and the stress of slab depend mainly on the deformation features of rockfill. In the finite element computation of rockfill dam with concrete face slab, Duncan's nonlinear elastic model was mainly used in the past. Compared to observations, however, the results show the following two main problems: (1) According to calculated results, there is great tensile stress in the slab, while the observed stress is mainly compressive stress. (2) The calculated lateral displacement is apparently larger, causing the downstream dam slope to extend outward, which is not observed in actual.

Duncan's model reflects the compressive hardening and nonlinear deformation rule of rockfill but not the shear shrinkage and anisotropy caused by stress. The two-yield-surface elastic–plastic model proposed by Shen Zhujiang comprehensively reflects the compressive hardening, nonlinearity, shear shrinkage, and anisotropy caused by stress, which will be more reasonable when applied in calculating rockfill dam with concrete face slab [8].

The concrete slab itself can be calculated with elastic solid element. As for whether contact element should be added between slab and rockfill, Ref. [8] showed an experiment. According to the result, if concrete is poured after building the rockfill, then

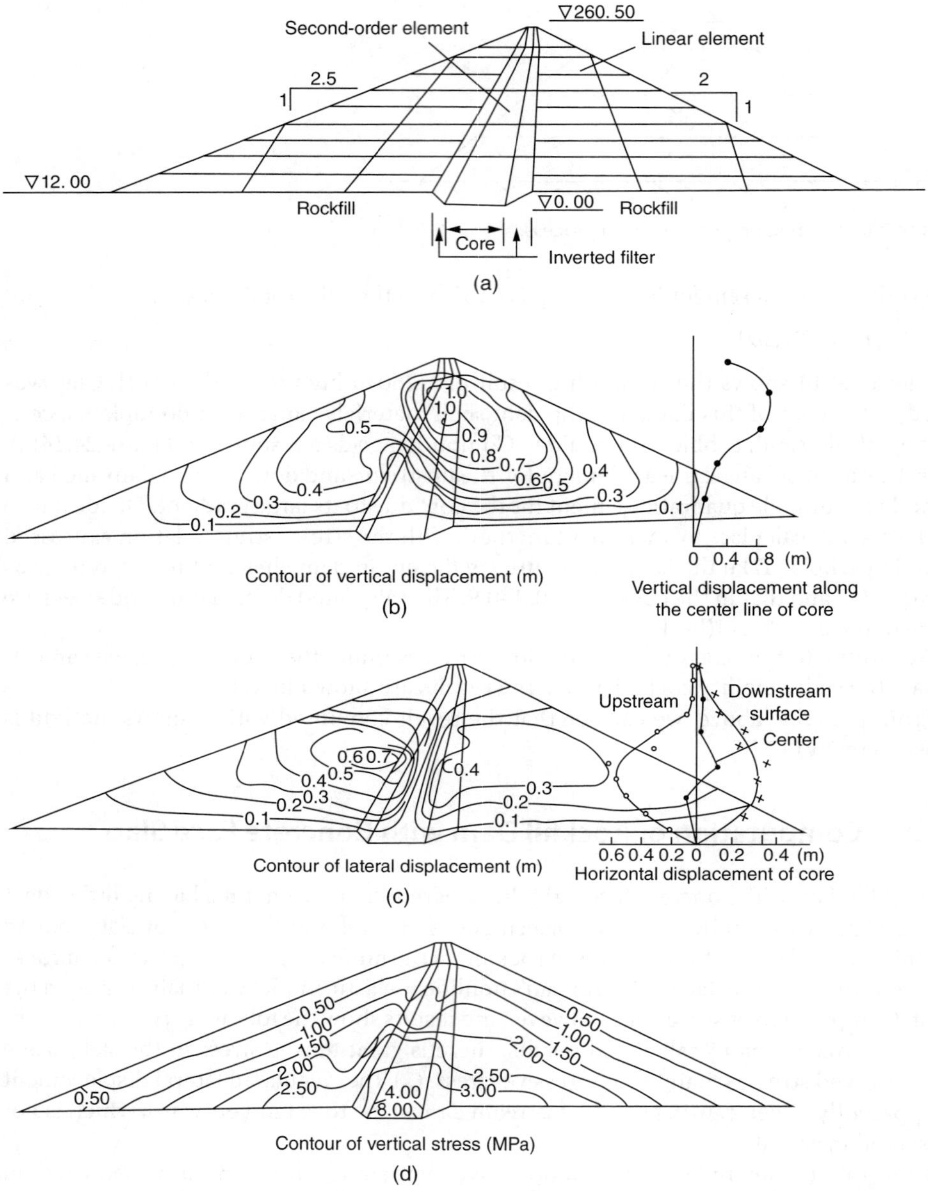

Figure 24.14 Stress and deformation of earth dam. (a) Computing grids (dam height of 260 m); (b) contour of vertical displacement (m); (c) contour of lateral displacement (m); (d) contour of vertical stress σ_y (MPa).

the rock on the surface will combine with concrete, forming a coarse surface. The sliding surface at failure will occur inside the rockfill, so there is no need to add contact element.

There are many joints in the slab. The peripheral joint is often pulled apart and thus can be taken as free boundary. Theoretically, the downward slope joints in slab could be simulated with thin-layer element, but it is troublesome. One way is to share the deformation of joint to the whole length of slab and calculate the average elastic modulus. In this way, the slab turns to be orthogonal anisotropic, with two elastic moduli: the concrete elastic modulus E_1 in direction of downward slope and the equivalent average elastic modulus E_2 in direction of dam axis. Copper sheets, plastic water stop strips, wood boards, and other materials are often embedded in joints. In Ref. [8], it is assumed

Table 24.1 Sliding stability coefficient.

Computing method	Sliding stability coefficient	
	Vertical core	Slant core
Fellenius' method	2.50	2.38
Bishop's method	2.89	2.85
Finite element method	2.84	2.83

Table 24.2 Parameters in computation of Tianshengqiao rockfill dam with concrete face slab.

Material	ϕl	$\Delta\phi$	k_{ur}	k	n	R_f	k_b	m	c_d	n_d	R_d
Subcrust	50.6	7	2210	1050	0.354	0.706	480	0.236	0.0028	0.79	0.67
Transition layer	52.5	8	2000	970	0.361	0.760	440	0.193	0.0045	0.70	0.70
Limestone rockfill	54.0	13	1980	940	0.350	0.849	340	0.18	0.0013	1.49	0.75
Limestone rockfill	54.0	13.5	1550	720	0.303	0.798	800	−0.18	0.0028	1.13	0.68
Mudstone	48.0	10	1050	500	0.250	0.727	250	0	0.014	0.57	0.68

Table 24.3 Computing result of Tianshengqiao rockfill dam with concrete face slab.

Computing methods	Displacement in construction (cm)		Displacement in impoundment (cm)		Tensile stress at bottom of slab (MPa)
	Max settlement	Max displacement	Face slab	Peripheral joint	
2 dimensional D	227	77.6	48.5	6.4	10.52
2 dimensional D_a	229	74.7	42.8	9.8	8.77
2 dimensional N	163	42.8	33.9	3.7	2.23
3 dimensional D	165	64.5	67.2	11.1	18.3
3 dimensional D_b	165	64.5	66.1	10.4	19.0
3 dimensional N	136	13.4	29.5	2.1	1.0

D, Duncan model; N, two-yield-surface elastic-plastic model.

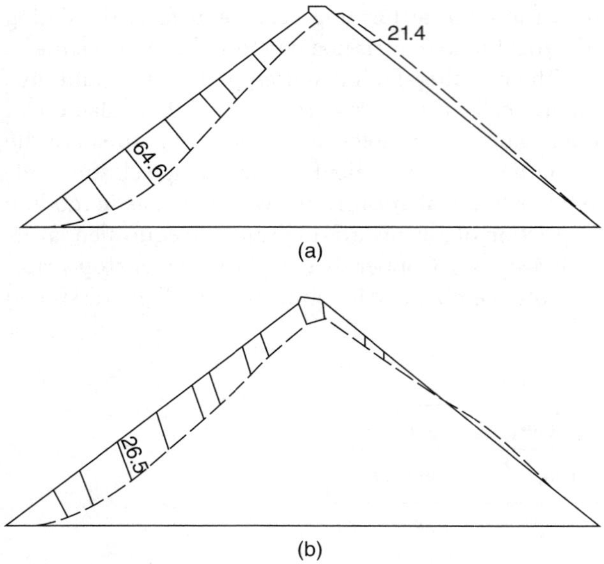

Figure 24.15 Displacement after impoundment at typical cross section of Tianshengqiao rockfill dam with concrete face slab (cm).

that wood boards bear the pressure in joints and plastic water stop strips bear the tensile force. The computing result shows that the equivalent elastic modulus in tension is $E_2 \cong 100\,\text{MPa}$, about 1/200 of elastic modulus of concrete. In compression, E_2 is close to elastic modulus of concrete.

Tianshengqiao rockfill dam with concrete face slab is constructed in a V-shaped valley, with a height of 180 m and a top length of 1178 m. Reference [8] conducted 2D and 3D computation with Duncan's model and two-yield-surface elastic–plastic model. The computation was divided into 20 stages to simulate construction and impoundment process. The material parameters used are shown in Table 24.2. Computing results are shown in Table 24.3 and Figures 24.15 and 24.16. It is clear that both of the dam displacement and slab tensile stress calculated with two-yield-surface elastic–plastic model are less than those of Duncan's model, which seems to be reasonable.

24.7 Limit Analysis in Rock and Soil Mechanics

In geotechnical engineering, we often need to analyze slope stability and limit bearing capacity of foundation. In the past, slices method and limit equilibrium method were widely used. However, since many assumptions were introduced in computation, there were many shortages. Zienkiewicz proposed to use FEM for limit analysis in geotechnical engineering [19], which has been greatly developed these years and can overcome many shortages in traditional methods.

24.7.1 Computation Methods

There are two methods for limit analysis in rock and soil mechanics using FEM.

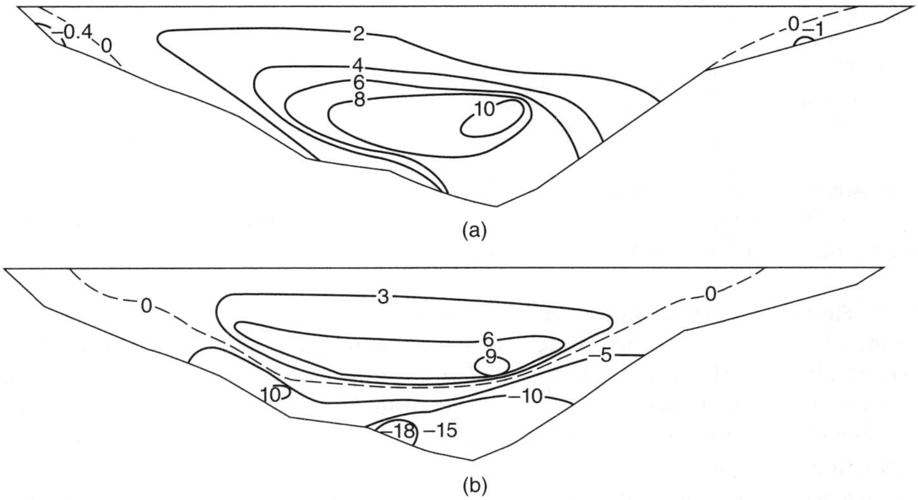

Figure 24.16 Stresses in the slab of Tianshengqiao rockfill dam in the direction parallel to upstream slope after impoundment (MPa). (a) Two-yield-surface model; (b) Duncan's model.

24.7.1.1 Finite Element Strength Discount Method

The finite element strength discount method may be called as strength discount method for short. In geotechnical engineering, failure is mainly caused by shear and tension fracture. Generally, it is appropriate to use Mohr–Coulomb strength criterion with maximum tensile stress, that is,

$$\left.\begin{array}{l} \tau \le c + f\sigma \\ \sigma \le R_{\mathrm{t}} \end{array}\right\} \tag{24.58}$$

where

$$f = tg\phi$$

in which c is the cohesive force, f is the friction coefficient, and R_{t} is the tensile strength.

In computation, discount material strength by k times, that is, to take the strength criterion as

$$\left.\begin{array}{l} \tau \le \dfrac{c}{k} + \dfrac{f}{k}\sigma \\ \sigma \le \dfrac{R_t}{k} \end{array}\right\} \tag{24.59}$$

Gradually increase the value of k until the ultimate failure occurs. The value of k at failure is safe coefficient or strength storage coefficient.

In geotechnical engineering, the friction coefficient f is relatively stable, while the cohesive force c is relatively dispersive. Considering this, the author suggests to use the modified strength criteria, that is,

$$\left.\begin{array}{l} \tau \le \dfrac{c}{k_{\mathrm{c}}} + \dfrac{f}{k_{\mathrm{f}}}\sigma \\ \sigma \le \dfrac{R_{\mathrm{t}}}{k_{\mathrm{t}}} \end{array}\right\} \tag{24.60}$$

where

$$k_c = 1 + \omega$$
$$k_f = 1 + s_1\omega$$
$$k_t = 1 + s_2\omega$$

Gradually increase the overloading coefficient ω from 0; s_1 and s_2 are two coefficients unequal to 1. For example, we may take $s_1 = 0$, meaning the friction coefficient f remains constant and merely decreases the value of c.

24.7.1.2 Finite Element Increment Loading Method

In computation with finite element increment loading method (overloading method for short), load is gradually increased. The rock and soil will begin from the initial elastic state, then gradually change to plastic state, and finally, come to the ultimate failure state, when the corresponding load is limit load. The ratio of limit load and design load is overloading coefficient.

Cohesive force c is dimensional, while friction coefficient is dimensionless. Therefore, the values of strength storage coefficient and overloading coefficient are different. Next we will analyze it.

Calculating with strength discount method, suppose the limit state occurs when strength is discounted by k times. The limit condition is

$$\tau = \frac{c}{k} + \frac{f}{k}\sigma \tag{24.61}$$

Calculating with overloading method, the shear stress τ and normal stress σ are both increasing with load. The limit state occurs when the load comes to ω times of the original value. The stress condition is

$$\left.\begin{aligned} &\omega\tau = c + f\omega\sigma \\ \text{or} \quad &\tau = \frac{c}{\omega} + f\sigma \end{aligned}\right\} \tag{24.62}$$

Comparing Eqs (24.61) and (24.62), it is clear that in overloading method, the friction resistance $f\sigma$ never changes while the adhesive force decreases. But in strength discount method, the friction resistance and the adhesive force decrease at the same time. Therefore, for the same material and structure, the strength storage coefficient k is less than the overloading coefficient ω.

24.7.2 Failure Criteria

The three failure criteria used at present are as follows:

1) A failure surface is formed thoroughly through the whole structure, and each point on the failure surface arrives at limit state.
2) The displacement of node on failure surface shows abrupt change.
3) The computation is not convergent.

In most cases, the results given by the three failure criteria above are close. Since nonlinear calculating method is relatively complex, non-convergence of computation may be related to calculating method. Thus some scholars disagree to use non-convergence of computation as a failure criterion.

24.7.3 Advantage of Finite Element Limit Analysis Method

The advantage of traditional limit equilibrium method lies in simple calculation as well as the long-term use. But it does have some significant shortages:

1) We have to know the position and shape of failure surface beforehand. For plane problems in uniform soil, we can use optimal methods to search for dangerous slide surface. But for plane problems in heterogeneous soil or space problems, it is too difficult to find out the most dangerous failure surface. As for bedrock, due to large quantity of joints, cracks, interlayers, fault, and other discontinuous structural surfaces, it will be even more difficult to find out the most dangerous failure surface and thus difficult to calculate minimum safe coefficient.
2) For soil where there is a dam or supporting structure on bedrock, the limit equilibrium method cannot take account of the interaction between foundation and dam or supporting structure.
3) It cannot consider seepage field in foundation and crustal stress field in bedrock.

The finite element limit analysis method can overcome the shortages above, automatically find the most dangerous failure surface, considering interactions between foundation and dam, and considering the effects of seepage field in foundation and crustal stress field in bedrock.

24.7.4 Calculation Examples

Example 1 Stability analysis of soil slope

Reference [20] applies the finite element strength discount method to computation in uniform and heterogeneous soil slopes and earth dams, all achieving good results. For soil slope shown in Figure 24.17, when c_2/c_1 is taken three different values, the failure mode is shown in Figure 24.18. When $c_2/c_1 = 0.60$, the foundation is soft so that the failure surface crosses the foundation, as Figure 24.18(a) shows; when $c_2/c_1 = 1.5$, there are two failure surfaces: one crosses the slope toe and the other crosses the foundation, as Figure 24.18(b) shows; when $c_2/c_1 = 2.0$, the failure surface crosses slope toe, as Figure 24.18(c) shows.

This example indicates that the finite element strength discount method could automatically find out the failure surface and safe coefficient under different conditions. Reference [21] uses finite element strength discount method to calculate a group of stability problems of rock slope, which also obtain fine result.

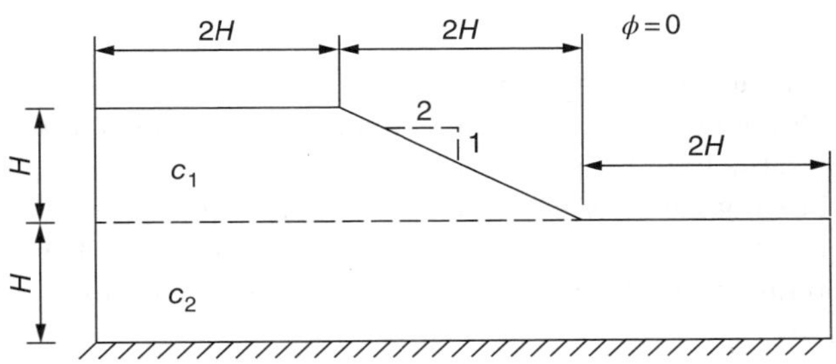

Figure 24.17 Undrained clay slope ($c_1/\gamma H = 0.25, \phi = 0$).

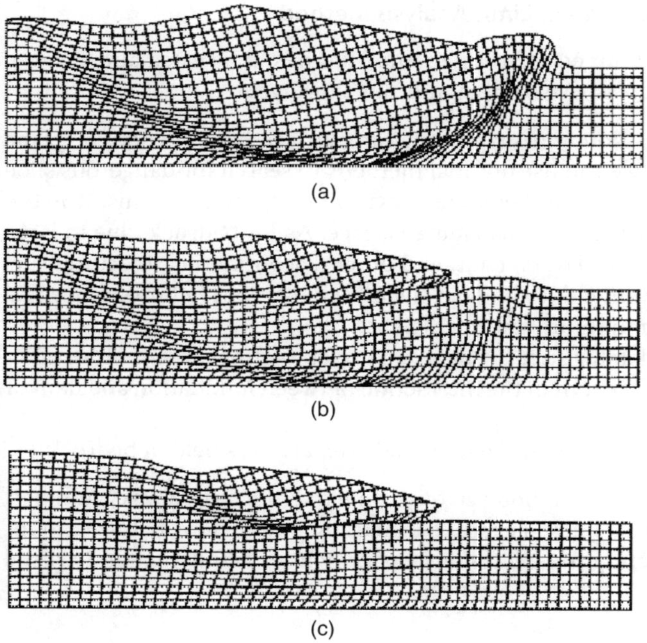

Figure 24.18 Failure form of undrained clay slope. (a) $c_2/c_1 = 0.60$; (b) $c_2/c_1 = 1.5$; (c) $c_2/c_1 = 2.0$.

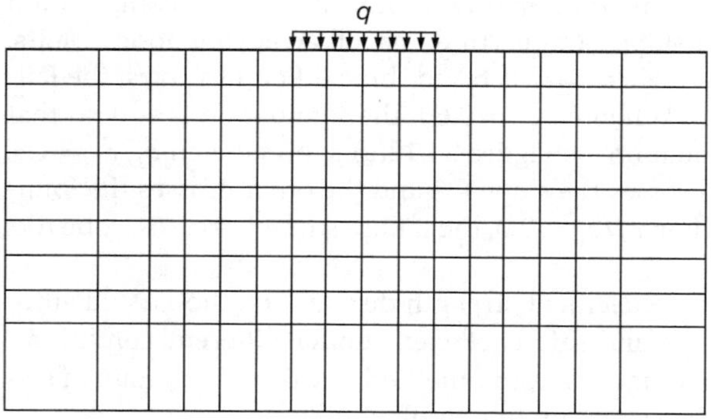

Figure 24.19 Foundation bearing capacity computation.

Example 2 Computation of foundation bearing capacity

For a semi-infinite weightless foundation bearing uniform perpendicular load q in a range of B_0, as Figure 24.19 shows, the limit bearing capacity have a Prandtl theoretical solution. Comparison of limit bearing capacity between results calculated with finite element overloading method and the theoretical solution is shown in Table 24.4. We can see that the two results are quite close. We take $c = 10\,\text{kPa}$ [21] in computation.

Table 24.4 Compare of limit bearing capacity (kPa).

Friction angle, ϕ(∘)	Theoretical solution	FEM	Relative error (%)	Friction angle, ϕ(°)	Theoretical solution	FEM	Relative error (%)
0	51.42	52.19	1.50	20	148.35	151.75	2.29
10	83.45	84.98	1.83	30	301.40	309.96	2.85

Bibliography

1 Kondner, R.L. (1963) Hyperbolic stress–strain response of cohesive soils. *Proceedings of the American Society of Civil Engineers*, **89** (SM1), 115–143.
2 Duncan, J.M. and Chang, C.Y. (1970) Nonlinear analysis of stress and strain in soils. *Proceedings of the American Society of Civil Engineers*, **96** (SM5), 1629–1653.
3 Kulhavy, F.H. and Duncan, J.M. (1972) Stresses and movements in Oroville Dam. *Proceedings of the American Society of Civil Engineers*, **98** (SM7), 653–655.
4 Wang, F. (1979) Improvement of Duncan's model. *Chinese Journal of Geotechnical Engineering*, (1), 80–86.
5 Duncan J M et al. *Strength, Stress–Strain and Bulk Modulus Parameters for Finite Element Analysis of Stress and Movements in Soil Masses.* Berkeley: University of California, 1980, UCB/GT/80-01.
6 Shen, Z. (1990) *A New Model for Stress–Strain Analysis in Soil Masses*, China Architecture and Building Press, Beijing.
7 Roscoe, K.H. *et al.* (1963) Yielding of clays in states wetter than critical. *Geotechnique*, **13** (3), 211–240.
8 Zhu, B. and Shen, Z. (1990) *Computational Soil Mechanics*, Shanghai Science and Technology Press, Shanghai.
9 Duncan, J.M. and Clough, G.W. (1971) Finite element analysis of port Allen lock. *Proceedings of the American Society of Civil Engineers*, **97** (SM8), 1053–1067.
10 Clough, G.W. and Duncan, J.M. (1971) Finite element analysis of retaining wall behavior. *Proceedings of the American Society of Civil Engineers*, **97** (SM10), 1657–1673.
11 Biot, A. (1941) General theory of three dimensional consolidation. *Journal of Applied Physics*, **12**, 155–164.
12 Xu, Z. (1982) *Theory of Elasticity*, 2nd edn, Higher Education Press, Beijing.
13 Zhang, M., Qian, J. and Chen, X. (1993) Visco-elastic–plastic FEM and its application in tunnel analysis. *China Civil Engineering Journal*, **6** (3), 26.
14 Naylor, D.J., Stagg, K.G. and Zienkiewicz, O.C. (1975) Numerical Analysis of Dams.
15 Desai, C.S. and Christian, J.T. (eds) (1981) *Numerical Method in Geotechnical Engineering* Lu Shishen et al. translated, China Architecture and Building, Beijing.
16 Huang, W. (1979) Theory of elastoplastic stress–strain models for soil. *Journal of Tsinghua University*, **19** (1), 1–26.
17 Jiang, P. (1982) *Constitutive Low of Soil*, Science Press, Beijing.

18 Zhang, W. and Bo, S. (1991) Rockfill feature analysis and calculation in concrete faced rockfill dam. *Water Resources and Hydropower Engineering,* (1), 17–23.

19 Zienkiewicz, O.C., Humpheson, C. and Lewis, R.W. (1975) Associated and non-associated visco-plasticity and plasticity in soil mechanics. *Geotechnique,* **25** (4), 671–689.

20 Griffiths, D.V. and Lane, P.A. (1999) Slope stability analysis by finite elements. *Geotechnique,* **49** (3), 387–403.

21 Zheng, Y., Zhao, S. and Deng, C. (2007) Development of finite element limit analysis method and its application in geotechnical engineering. *Engineering Science,* (3), 10–26.

25

Plain and Reinforced Concrete Structures

Plain and reinforced concrete structures have been widely used in engineering practice such as dams and pavements that are built of plain concrete and buildings, bridges, offshore oil platforms, and nuclear reactor containers that are built of reinforced concrete. Finite element methods (FEM) become more and more popular in the design and research of these structures.

FEM have been widely applied in the analysis of concrete structures like dams. Taking arch dams as examples, the structural mechanics like the arch-cantilever methods were usually used. However, many complex factors can hardly be taken into account, such as the holes in the dams, the interaction between dam and foundation, the effect of seepage water, the construction process, the effect of joints on the dam stress, and the local stress redistribution after cracking, whereas they can be well simulated by FEM. Hence, FEM has been recommended for the arch dam design in some countries instead of structural mechanics, while in other countries, FEM has been used for important arch dam designs though structural mechanics is still used for normal designs. In general, methods based on strength of materials will still be adopted in normal designs for simple concrete structures, such as concrete retaining wall, concrete gravity dams, and small- and medium-sized concrete arch dams, while FEM will be more and more widely used for more important and more complex concrete structures.

As for reinforced concrete structures, the structural responses against mechanical loads are rather complex as they are comprised of two types of materials: steel bar and concrete. Even normal design loads can lead to nonlinear behaviors like concrete cracking and localized damage of steel bar in a simply supported beam. These behaviors can hardly be computed by traditional structural mechanics. As a matter of fact, the present design methods of reinforced concrete structures are empirical methods based on vast experimental data.

D. Ngo and A.C. Scordelis were the first to apply FEM for reinforced concrete structural analysis; since then plenty of research work has been done and lots of literatures have been published. FEM has been proved a powerful tool for reinforced concrete structural analysis, which can simulate those complex factors such as concrete cracking and nonlinear multiaxial stress–strain relationship and interactions between steel bar and concrete. Two advantages can be expected:

1) Systematic analysis can be applied for normal reinforced concrete members, which can be used to study the effects of important parameters on structural response and

The Finite Element Method: Fundamentals and Applications in Civil, Hydraulic, Mechanical and Aeronautical Engineering, First Edition. Bofang Zhu.

Companion website: www.wiley.com/go/zhufem17

lead to a deeper understanding of these members. Based on that, the design codes for normal reinforced concrete members can be improved.

2) It can be used for the design and analysis of box girders, offshore oil platform, and nuclear reactor container, for which the normal computing methods can hardly be used.

It is believed that the normal design codes based on large amount of experimental data will still be used for simple reinforced concrete members like beams, columns, and frames, whereas the analysis of offshore structures, nuclear reactor container, and volutes of hydropower stations needs FEM.

25.1 Constitutive Models of Concrete

25.1.1 Uniaxial Stress–Strain Relationship of Concrete

The typical uniaxial stress–strain relationship of concrete is shown in Figure 25.1. The relation is linear before point A in compression; after that, it becomes nonlinear increasing (hardening) and turns nonlinear decreasing (softening) beyond C; finally at D, it fails and loses the capacity. In tension, the linear percentage is relatively larger in the curve. The proportion limit σ_t is about 80% of the tensile strength. The crack appears when the strain reaches ε_t. A hysteresis loop will be produced when the concrete is unloaded at B and reloaded. Approximately, a line parallel to oA is assumed instead of the hysteresis loop.

1) Monotonic stress–strain relationship in uniaxial compression.
The Saenz equation is usually used:

$$\sigma = \frac{E\varepsilon}{a + b\left(\dfrac{\varepsilon}{\varepsilon_c}\right) + c\left(\dfrac{\varepsilon}{\varepsilon_c}\right)^2 + d\left(\dfrac{\varepsilon}{\varepsilon_c}\right)^3}$$

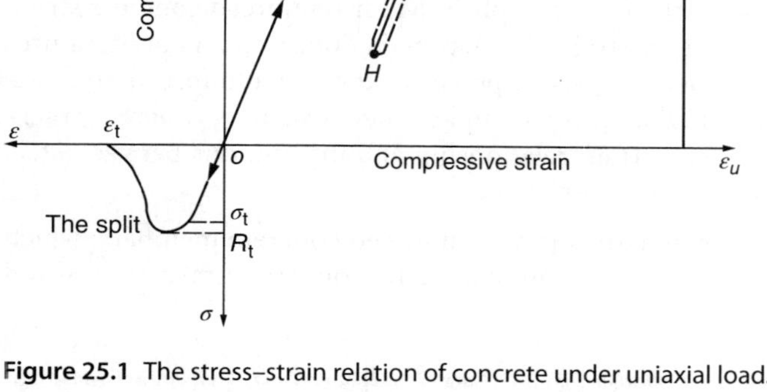

Figure 25.1 The stress–strain relation of concrete under uniaxial load.

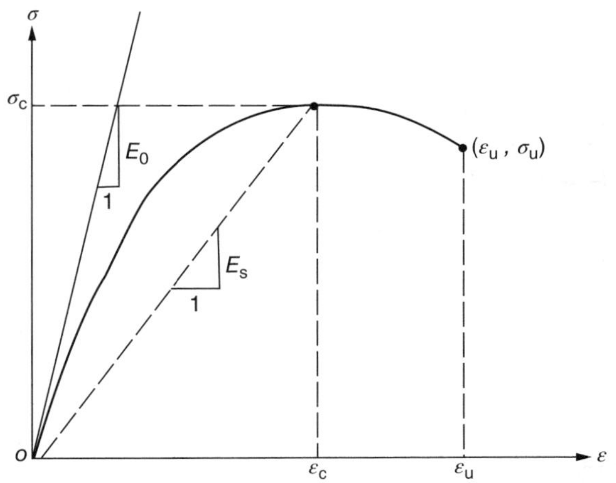

Figure 25.2 The stress–strain relation of concrete under the axial compression.

in which E is modulus of elasticity and a, b, c, d are the constants, which are determined as follows (Figure 25.2):

1) $\varepsilon = 0$, $d\sigma/d\varepsilon = E_0$.
2) $\varepsilon = \varepsilon_c$, $\sigma = R_c$ (the maximum stress in the figure and related strain).
3) When $\varepsilon = \varepsilon_c$, $d\sigma/d\varepsilon = 0$ ($E=0$ at peak point).
4) When $\varepsilon = \varepsilon_u$ (ultimate strain), $\sigma = \sigma_u$.

With a, b, c, d determined by the above conditions, it can be written as

$$\sigma = \frac{E_0\varepsilon}{1 + \left(R + \frac{E_0}{E_s} - 2\right)\frac{\varepsilon}{\varepsilon_c} + (2R-1)\left(\frac{\varepsilon}{\varepsilon_c}\right)^2 + R\left(\frac{\varepsilon}{\varepsilon_c}\right)^3} \tag{25.1}$$

where

$$R = \frac{(E_0/E_s)(R_c/\sigma_u - 1)}{(\varepsilon_u/\varepsilon_c - 1)^2} - \frac{\varepsilon_c}{\varepsilon_u}$$

If (4) is neglected, that is, the softening stage of the curve is neglected, we have

$$\sigma = \frac{E_0\varepsilon}{1 + \left(\frac{E_0}{E_s} - 2\right)\frac{\varepsilon}{\varepsilon_c} + \left(\frac{\varepsilon}{\varepsilon_c}\right)^2} \tag{25.2}$$

2) Poisson's ratio.
   Figure 25.3 shows the experimental data of Poisson's ratio of concrete. It stays at the constant 0.2 when the stress is less than $0.8R_c$, and it increases rapidly when the stress is beyond $0.8R_c$.
3) Stress–strain relationship in uniaxial tension.
   Figure 25.4 shows the uniaxial tensile stress–strain relations of concrete with five different mix portions. As the tensile strength is relative small, and the proportional limit is relatively large compared with tensile strength, a linear stress–strain relation is assumed for concrete in tension until rupture. The concrete will be cracked when the stress reaches the strength. Also we assume the tensile modulus of elasticity equals the compressive modulus of elasticity.

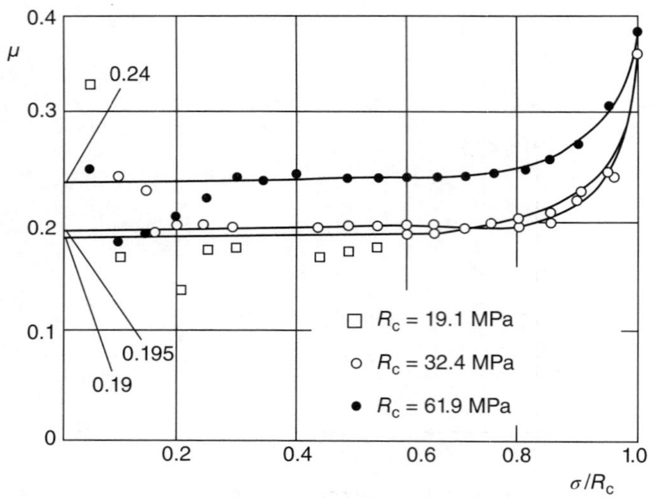

Figure 25.3 Poisson's ratio μ of concrete.

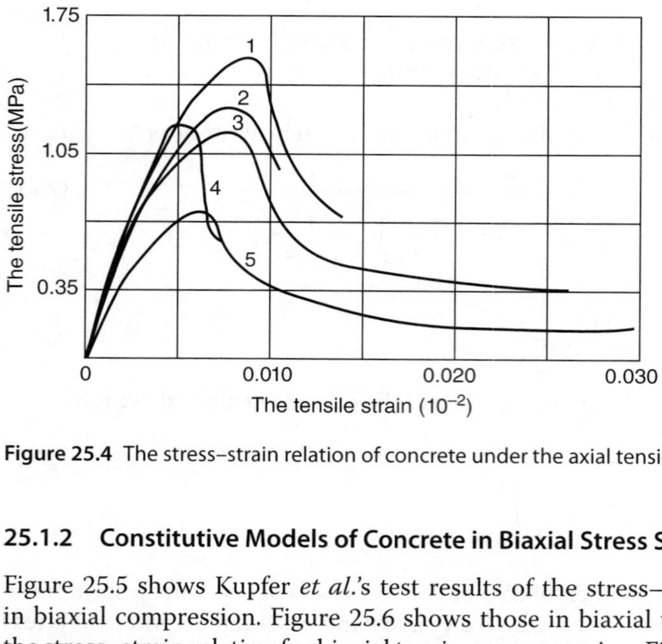

Figure 25.4 The stress–strain relation of concrete under the axial tension.

25.1.2 Constitutive Models of Concrete in Biaxial Stress State

Figure 25.5 shows Kupfer *et al.*'s test results of the stress–strain relation of concrete in biaxial compression. Figure 25.6 shows those in biaxial tension. Figure 25.7 shows the stress–strain relation for biaxial tension–compression. Figure 25.8 shows the biaxial strength of concrete.

It can be found that the difference of the stress–strain curves in Figures 25.5–25.7 is mainly due to the effect of Poisson's ratio. For example, the stress–strain relationship of elastic material in biaxial stress state is

$$\sigma = \frac{E_0\varepsilon}{1-\mu\alpha} = E_0'\varepsilon$$

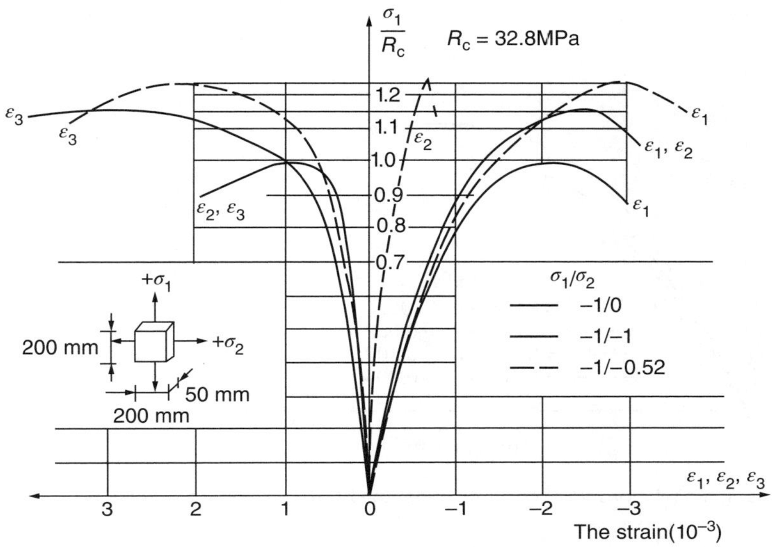

Figure 25.5 The stress–strain relation of concrete under the biaxial compression.

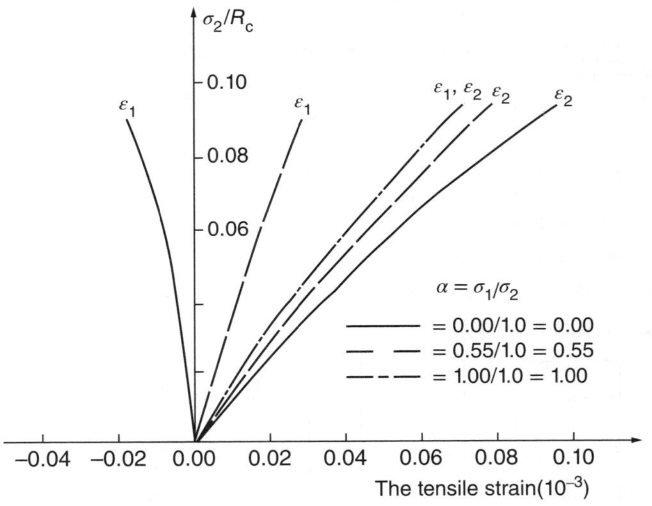

Figure 25.6 The stress–strain relation of concrete under the biaxial tension.

where

$$E_0' = \frac{E_0}{1 - \mu\alpha}$$

Here, α means the ratio of lateral stress to the target stress, for example, if σ_1 is the target stress, the lateral stress is σ_2. Then we have $\alpha = \sigma_2/\sigma_1$. μ means Poisson's ratio for biaxial stress state; E_0 is the initial tangent modulus in uniaxial stress state; E_0' is the initial tangent modulus in biaxial stress state.

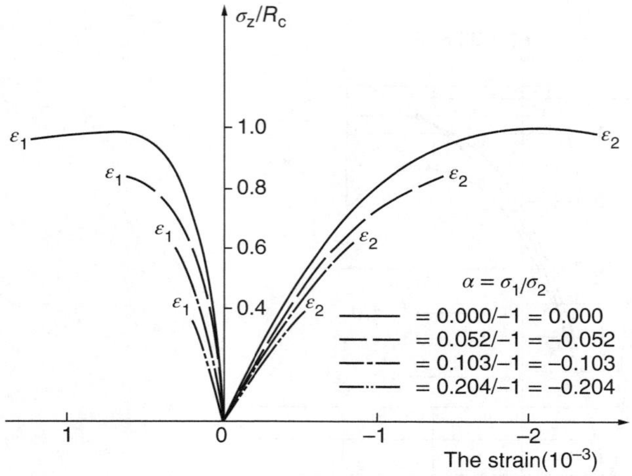

Figure 25.7 The stress–strain relationship of concrete under the biaxial tension–compression.

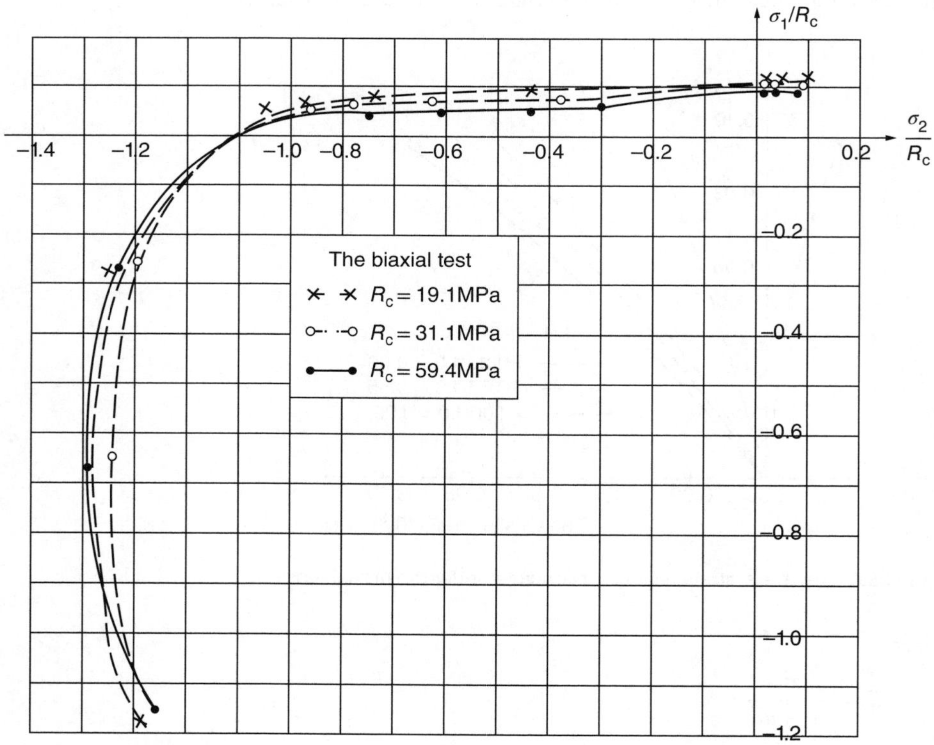

Figure 25.8 The biaxial strength of concrete.

Similar to Eq. (25.2) for uniaxial stress state, the stress–strain relationship for biaxial stress state can be written as

$$\sigma = \frac{E_0\varepsilon}{(1-\mu\alpha)\left[1+\left(\frac{E_0'}{E_s'}-2\right)\frac{\varepsilon}{\varepsilon_p}+\left(\frac{\varepsilon}{\varepsilon_p}\right)^2\right]} \tag{25.3}$$

where $E_s' = \sigma_p/\varepsilon_p$.

Here, E_s' is the initial secant modulus at maximum stress σ_p, and ε_p is related strain at peak stress σ_p.

According to Kupfer's experiment data, the maximum stress σ_p and related strain ε_p have the following data (tension is positive and the principal stress is ranked as the algebraic value $\sigma_1 \geq \sigma_2$) for four different regions in Figure 25.8:

1) Compression–compression zone ($\sigma_1 < 0, \sigma_2 < 0, 0 \leq \alpha \leq 1$)

$$\left.\begin{aligned}
\sigma_{2p} &= \frac{1+3.65\alpha}{(1+\alpha)^2}R_c \\
\varepsilon_{2p} &= \varepsilon_c\left(\frac{3\sigma_{2p}}{R_c}-2\right) \\
\sigma_{1p} &= \alpha\sigma_{2p}, \varepsilon_{1p} = \varepsilon_c\left[0.35\frac{\sigma_{1p}}{R_c}+2.25\left(\frac{\sigma_{1p}}{R_c}\right)^2-1.6\left(\frac{\sigma_{1p}}{R_c}\right)^3\right]
\end{aligned}\right\} \tag{25.4}$$

2) Compression–tension zone ($\sigma_1 < 0, \sigma_2 > 0, -0.17 \leq \alpha \leq 0$)

$$\left.\begin{aligned}
\sigma_{2p} &= \frac{(1+3.28\alpha)R_c}{(1+\alpha)^2} \\
\varepsilon_{2p} &= \varepsilon_c\left[4.42-8.38\left(\frac{\sigma_{2p}}{R_c}\right)+7.54\left(\frac{\sigma_{2p}}{R_c}\right)^2-2.58\left(\frac{\sigma_{2p}}{R_c}\right)^3\right] \\
\sigma_{1p} &= \alpha\sigma_{2p}, \varepsilon_{1p} = \frac{\sigma_{1p}}{E_0}.
\end{aligned}\right\} \tag{25.5}$$

3) Tension–compression zone ($\sigma_1 > 0, \sigma_2 < 0, -\infty < \alpha < -0.17$)

$$\left.\begin{aligned}
\sigma_{2p} &\leq 0.65R_c \\
\varepsilon_{2p} &= \varepsilon_c\left[4.42-8.38\left(\frac{\sigma_{2p}}{R_c}\right)+7.54\left(\frac{\sigma_{2p}}{R_c}\right)^2-2.58\left(\frac{\sigma_{2p}}{R_c}\right)^3\right] \\
\sigma_{1p} &= R_t, \varepsilon_{1p} = \frac{\sigma_{1p}}{E_0}
\end{aligned}\right\} \tag{25.6}$$

4) Tension–tension zone ($\sigma_1 > 0, \sigma_2 > 0, 1 < \alpha < +\infty$)

$$\left.\begin{aligned}
\sigma_{1p} &= R_t \geq \sigma_{2p} \\
\varepsilon_{1p} &= \frac{R_t}{E_0} \geq \varepsilon_{2p} = \frac{\sigma_{2p}}{E_0}
\end{aligned}\right\} \tag{25.7}$$

In Eqs (25.4)–(25.7), R_c and R_t are uniaxial compressive and tensile strength, respectively; ε_c is the related strain to the maximum stress R_c; and $\mu = 0.2$ in the above analysis (Figure 25.9).

All these equations can be used for plane problems such as beams, slabs, shells, and shear walls, but cannot be used for three-dimensional problems.

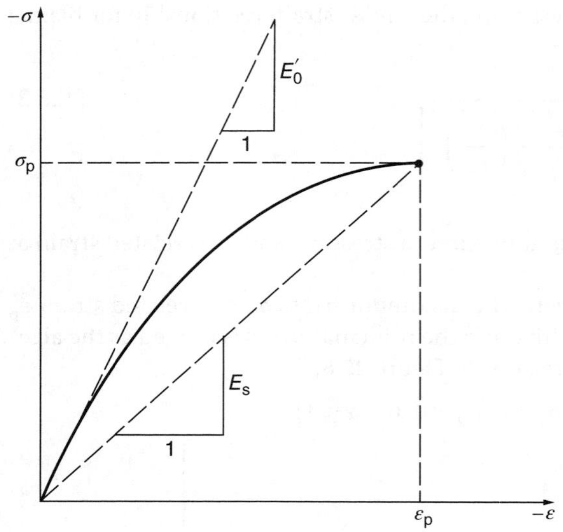

Figure 25.9 The stress–strain relationship of concrete under the biaxial load.

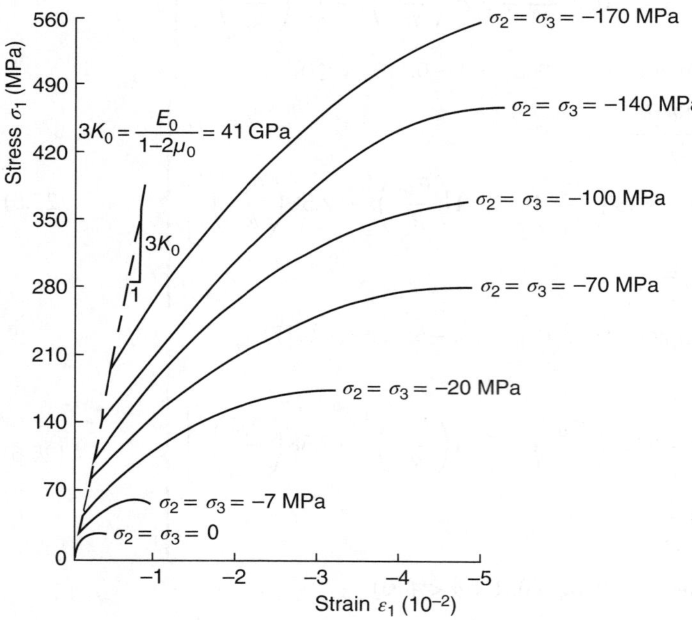

Figure 25.10 The stress–strain relation of concrete under the triaxial load.

25.1.3 Constitutive Models of Concrete in Triaxial Stress State

Figure 25.10 shows the typical stress–strain relations for concrete in triaxial compression. It is clear that the compressive strength and deformability are remarkably enhanced by the increase of the lateral stress.

1) Failure surface of concrete
   The following models have been proposed for failure surface of concrete:
   1) Lade's criteria, referred to Eqs (17.41) and (17.43)
   2) Bresler–Pister criteria, referred to Equation (17.44)
   3) Ottosen's four-parameter criteria, referred to Equation (17.45)
   4) Hsieh–Ting–Chen's four-parameter criteria, referred to (17.46)
   5) Mohr–Coulomb's criteria with maximum tensile stress, referred to Equation 17.16
   6) Willam–Warnke's three- and five-parameter criteria, referred to Eqs (17.47), (17.48), and (17.53)

   Willam–Warnke's five-parameter criteria fits the experimental data well, as shown in Equation 17.20. It can be used for important analysis. Mohr–Coulomb's criteria with maximum tensile stress can be used for relative simple analysis.
2) Hardening of concrete
   Isotropic hardening models are used in general plastic theory, that is, it is assumed that the initial yielding surface and successor yielding surface have similar shape but different sizes, as shown in Figure 25.11(a). These hardening models have been applied for concrete; the disadvantages are the underestimation of plastic deformation in triaxial compression.
   In fact, nonuniform hardening behavior is observed in concrete, as shown in Figures 25.11(b) and 25.12. Reference [3] proposed the following models as Figure 25.13; four regions are divided in the meridian plane:
   1) The yielding surface coincides with the failure surface in tension–tension zone, that is, no plasticity before failure.
   2) The yielding surface is separated from failure surface in tension–compression zone and compression–tension zone.
   3) The yielding surface is similar to failure surface but with smaller size in the compression zone with low lateral stress.
   4) The failure surface closes at the hydro pressure axis in the compression zone with high lateral stress.

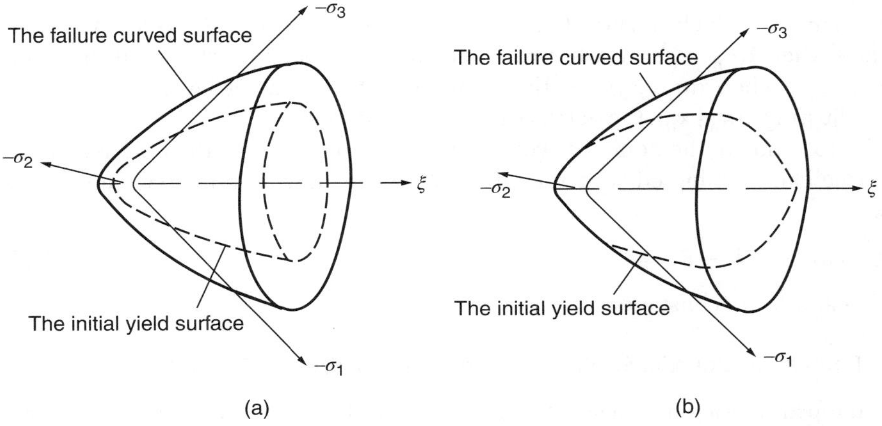

Figure 25.11 The yield surface and the failure surface. (a) The isotropic hardening. (b) Inhomogeneous hardening.

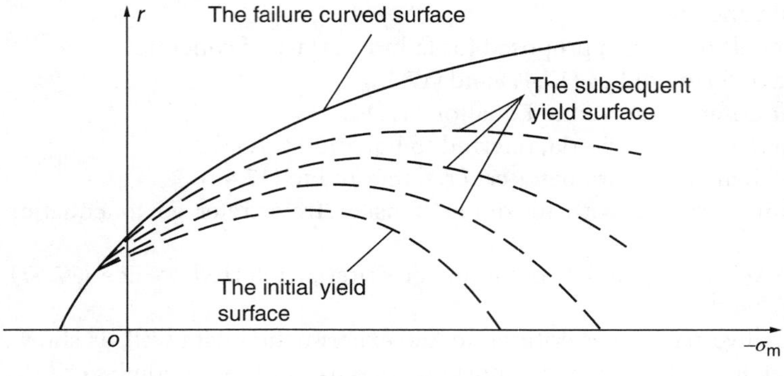

Figure 25.12 Inhomogeneous hardening model.

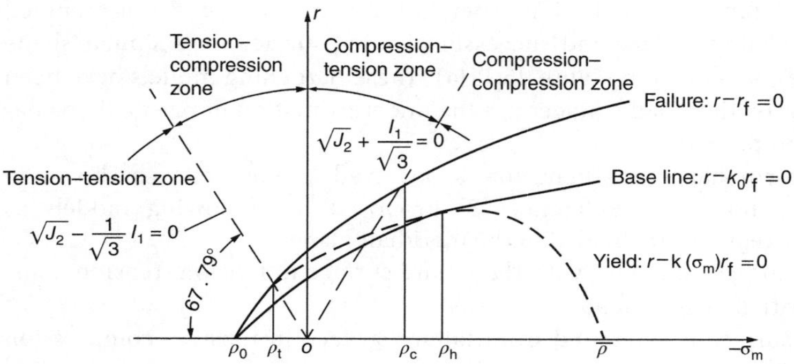

Figure 25.13 The hardening model for concrete.

The yielding surface and successor yielding surface are expressed as follows:

$$F = r - k(\sigma_m, \kappa_0)r_f = 0 \tag{25.8}$$

where $r_f = r(\sigma_m, \theta)$, which can be referred to Eq. (17.47), and $r = \sqrt{2J_2}$, which is the failure surface when $k(\sigma_m, \kappa_0) = 1$, whereas it becomes the yielding surface and successor yielding surface when $k(\sigma_m, \kappa_0) < 1$. The nonuniform hardening behavior can be simulated by adjusting $k(\sigma_m, \kappa_0)$; the details can be referred to Ref. [3].

In order to simulate the nonlinear volumetric strain of concrete, the non-associated flowing rule can be adopted with the Drucker–Prager flowing potential function as follows:

$$Q = \alpha I_1 + \sqrt{J_2} - K^* = 0 \tag{25.9}$$

in which α and K^* are constants.

25.1.4 Equivalent Uniaxial Strain and Orthotropic Model for Concrete

The nonuniform hardening constitutive relations can be well simulated by the incremental theory of plasticity; however, the computation becomes very complex. Several simplified practical models have been introduced in Eq. (17.10), among which the equivalent uniaxial strain and orthotropic model is a better one. Therefore, the detail is explained as follows:

1) Equivalent uniaxial stress–strain relation
   1) Biaxial loading: the biaxial stress–strain relations of concrete are shown in Figures 17.5–17.7. If every stress–strain curve with different stress ratios is standardized with the peak stress σ_c and related strain ε_c, these curves will coincide with each other. Hence, we can use a unique expression to simulate the stress–strain relations for concrete in compression. As the widely used Saenz equation for concrete in uniaxial compression, a similar formulation is adopted for the equivalent uniaxial stress–strain relation subjected to biaxial loading as follows:

$$\sigma_i = \frac{E_0\varepsilon_{\text{iu}}}{1+[(E_0/E_s)-2](\varepsilon_{\text{iu}}/\varepsilon_{\text{ic}})+(\varepsilon_{\text{iu}}/\varepsilon_{\text{ic}})^2} \tag{25.10}$$

$$E_s = \frac{\sigma_{\text{ic}}}{\varepsilon_{\text{ic}}}$$

   where ε_{iu} is the equivalent uniaxial strain, referred to Eq. (17.41); E_0 is the initial tangent modulus when stress is zero; σ_{ic} is the peak compressive stress; ε_{ic} is the equivalent uniaxial strain at the maximum compressive stress; and E_s is the secant modulus.
   According to Eq. (17.43),

$$E_i = \frac{d\sigma_i}{d\varepsilon_{\text{iu}}}$$

   The tangent modulus at the biaxial state is computed with the derivative of the Eq. (25.10):

$$E_i = \frac{E_0[1-(\varepsilon_{\text{iu}}/\varepsilon_{\text{ic}})^2]}{\{1+[(E_0/E_s)-2](\varepsilon_{\text{iu}}/\varepsilon_{\text{ic}})+(\varepsilon_{\text{iu}}/\varepsilon_{\text{ic}})^2\}^2} \tag{25.11}$$

   2) Triaxial loading: Elwi and Murray [14] extended the above stress–strain relation from biaxial loading condition to triaxial compression and tension. The equivalent uniaxial strain and stress relation under triaxial loading is as follows:

$$E_i = \frac{E_0\varepsilon_{\text{iu}}}{1+\left(R+\dfrac{E_0}{E_s}-2\right)(\varepsilon_{\text{iu}}/\varepsilon_{\text{ic}})-(2R-1)(\varepsilon_{\text{iu}}/\varepsilon_{\text{ic}})^2+R(\varepsilon_{\text{iu}}/\varepsilon_{\text{ic}})^3} \tag{25.12}$$

   where

$$R = \frac{E_0(\sigma_{\text{ic}}/\sigma_{\text{if}}-1)}{E_s(\varepsilon_{\text{if}}/\varepsilon_{\text{ic}}-1)^2}-\frac{\varepsilon_{\text{ic}}}{\varepsilon_{\text{if}}} \tag{25.13}$$

   Figure 25.14 shows the curve representing Eq. (25.12) and the meaning of some parameters in R formulations. Here, E_0 is the initial modulus of elasticity, σ_{ic} is the maximum stress, ε_{ic} is the related strain to σ_{ic}, and σ_{if} and ε_{if} are the stress and strain at the softening stage of the equivalent stress–strain curve, respectively.

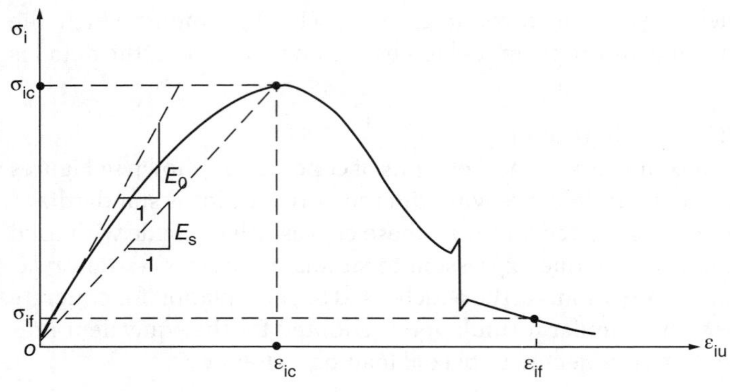

Figure 25.14 The typical curve of the compressive stress – the equivalent uniaxial strain for concrete.

Getting the derivative of ε_{iu} from Eq. (17.12), we could obtain $E_i = d\sigma_i/d\varepsilon_i$ as follows:

$$E_i = \frac{\left[1 + (2R-1)(\varepsilon_{iu}/\varepsilon_{ic})^2 - 2R(\varepsilon_{iu}/\varepsilon_{ic})^3\right] E_0}{\left[1 + \left(R + \frac{E_0}{E_s} - 2\right)(\varepsilon_{iu}/\varepsilon_{ic}) - (2R-1)(\varepsilon_{iu}/\varepsilon_{ic})^2 + R(\varepsilon_{iu}/\varepsilon_{ic})^3\right]^2} \tag{25.14}$$

3) Poisson's ratio: According to the experiment data of Kupfer [4], Poisson's ratio of concrete can be expressed as follows:

$$\mu = \mu_0 \left[1 + 1.3763\frac{\varepsilon}{\varepsilon_u} - 5.360\left(\frac{\varepsilon}{\varepsilon_u}\right)^2 + 8.586\left(\frac{\varepsilon}{\varepsilon_u}\right)^3\right] \tag{25.15}$$

where μ_0 is the initial Poisson's ratio, ε is the strain at uniaxial test, and ε_u is the ε_{ic} in the uniaxial test.

Due to different stress paths, three different Poisson's ratios are obtained for three directions. The upper limit of μ_i should be set as 0.50. When μ_i reaches 0.5, the volumetric increment should be zero. In fact, it means the start of the propagation of the unstable micro cracks in concrete. Although the dilatancy effect has been observed in the tests without lateral restricts when concrete approaching failure, it might give more reliable estimation for stress in real structures when the upper limit is set as 0.50.

2) Ultimate strength surface

Here, let us explain how to determine the parameters σ_{ic}, ε_{ic}, σ_{if}, and ε_{if} in the formulation of R. σ_{ic} is the ultimate strength. For softening material, it has not failed when the stress reaches σ_{ic}. Therefore, ultimate strength surface is used in the context instead of failure surface. With all the yielding surface functions in Figure 17.5, the ultimate strength surface can be described. For example, with the five-parameter criteria, the ultimate strength surface can be determined by five material parameters.

According to different stress ratios, three different σ_{ic} corresponding to three principal stresses can be calculated based on the ultimate strength surface.

To compute the equivalent uniaxial strain ε_{ic} corresponding to the σ_{ic}, an equivalent surface is assumed in the equivalent uniaxial strain space. It has the same shape as the ultimate strength surface in the stress space. ε_{1u}, ε_{2u}, ε_{3u} are used to replace σ_1, σ_2, σ_3 in the formulation of the ultimate strength surface, and then ε_{1c}, ε_{2c}, ε_{3c} are computed.

The softening part of the stress–strain curve is significantly affected by the experimental setup; it cannot be obtained with usual simple facilities. In general, it can be deduced from the flexure stress distribution. It is assumed by Elwi and Murray as follows:

$$\varepsilon_{if} = 4\varepsilon_{ic}, \quad \sigma_{if} = \frac{\sigma_{ic}}{4} \tag{25.16}$$

Up to now, all the material parameters used in the analysis have been determined. Here, anisotropy is induced by the nonlinear stress–strain relationship.

For practical analysis, a simplified approach is suggested by Bathe and Ramaswamy [9]:

1) The concrete is treated as isotropic material when it is in tension or loaded by low compressive stress. Let three principal stresses be σ_1, σ_2, σ_3, and make tension as positive. The order of the three principal stresses is $\sigma_3 \le \sigma_2 \le \sigma_1$. According to stress–strain relation as Eq. (25.14), three corresponding moduli of elasticity E_1, E_2, E_3 can be computed. When $\sigma_3 \ge kR_c$ (the typical value of k is 0.4), the material is loaded by low compressive stress and can be treated as isotropic material. The modulus of elasticity is computed by the following weighted average value:

$$E = \frac{|\sigma_1|E_1 + |\sigma_2|E_2 + |\sigma_3|E_3}{|\sigma_1| + |\sigma_2| + |\sigma_3|} \tag{25.17}$$

2) When the concrete is loaded by high compressive stress, that is, $\sigma_3 < kR_c$, it is treated as anisotropic material. The elastic matrix is computed as

$$D = \frac{1}{(1+\mu)(1-2\mu)} \begin{bmatrix} (1-\mu)E_1 & \mu E_{12} & \mu E_{13} & 0 & 0 & 0 \\ & (1-\mu)E_2 & \mu E_{23} & 0 & 0 & 0 \\ & & (1-\mu)E_3 & 0 & 0 & 0 \\ & & & \dfrac{(1-2\mu)E_{12}}{2} & 0 & 0 \\ \text{symmetric} & & & & \dfrac{(1-2\mu)E_{13}}{2} & 0 \\ & & & & & \dfrac{(1-2\mu)E_{23}}{2} \end{bmatrix} \tag{25.18}$$

where μ is Poisson's ratio.

The shear modulus G_{ij} in the coordinate plane ij is calculated according to the weighted average modulus E_{ij} in the coordinate plane ij as follows:

$$G_{ij} = \frac{E_{ij}}{2(1+\mu)} = \frac{1}{2(1+\mu)} \times \frac{|\sigma_i|E_i + |\sigma_j|E_j}{|\sigma_i| + |\sigma_j|} \tag{25.19}$$

3) In order to avoid the nonpositive stiffness matrix in the strain softening region, E_i is set as 0 for stiffness matrix computation (a small positive number is used for practical analysis). To compute incremental stress, the real negative modulus E_i is used.

25.2 Finite Element Models for Cracks in Concrete

The structural response of concrete and reinforced concrete is affected by many factors, among which cracks are the most important. The cross section of the member will be weakened after cracking in concrete, and the stress of concrete and reinforcing bars will be redistributed. Therefore, the effect of cracks in concrete has to be reasonably simulated to obtain accurate structural response of the concrete and reinforced concrete.

25.2.1 Crack Inducement in Concrete

Two criteria are usually applied to judge the crack inducement in concrete:

1) Maximum principal stress criterion: Cracks will appear when the maximum principal stress is larger than a limit.
2) Maximum principal strain criterion: Cracks will appear when the maximum principal strain is larger than a limit.

Figure 25.15(a) shows the two criteria for biaxial stress state and the experimental data by Kupfer *et al.* [4]. It can be found that the maximum principal stress criterion fits the experimental data better. Therefore, it is used for general analysis for engineering practice, that is, the concrete will be cracked when

$$\sigma_1 = R_t \tag{25.20}$$

where R_t is the concrete uniaxial tensile strength. The tensile plastic deformation is rather small, and the proportional limit is about $0.8R_t$. Therefore, in practical analysis, a brittle stress–strain relation can be used as shown in Figure 25.15(b). It states that the concrete is treated as elastic material when $\sigma_1 < R_t$, while the stress drops to zero at the occurrence of crack when $\sigma_1 = R_t$.

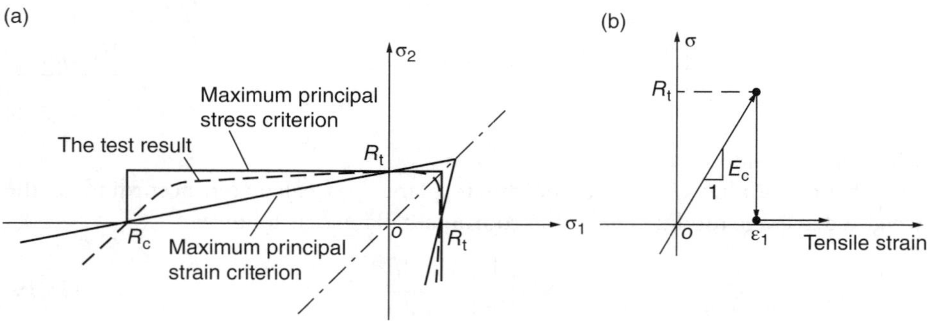

Figure 25.15 The crack of concrete. (a) The fracture criterion of the tension zone under the biaxial tension. (b) The stress–strain relationship before and after the brittle crack.

It can be implied from detailed experiment data that the concrete will not rupture immediately when $\sigma_1 = R_t$ and a softening stage is included in the stress–strain relation. Due to the low tensile strength of concrete, the softening stage is not a significant factor for the whole structural response. Therefore, the usual analysis will not take it into account.

There are five finite element models for concrete cracks: ① discrete crack model, ② smeared crack model, ③ thin-layer crack model, ④ no-tension model, and ⑤ fracture crack model. All the five models will be explained in the following sections.

25.2.2 Discrete Crack Model

The crack position and direction are set in the finite element mesh. As shown in Figure 25.16, two nodes are set at the same coordinate for single directional crack, each of which represents one side of the crack. For the cross point of two cracks, four nodes will be set at the same coordinate. In the early finite element literature, the nodes at different sides of the crack are independent to each other. Crack will numerically appear when the tensile principal stress reaches the concrete tensile strength in the load incremental analysis. Then the node is modified as two different nodes, and all the nodes are renumbered for next step computation.

The crack surface is not smooth for normal concrete. Shear stress can still be transferred through crack surface when slip is occurred along the surface due to the interlock of the aggregates. Only for high strength concrete the interlock effect of the aggregates turns unimportant as the mortar strength is close to the aggregate strength. It will produce a smoother crack surface, and the crack will propagate directly across the aggregates rather than along the interface between aggregate and mortar (Figure 25.17).

Figure 25.18 shows the modified discrete crack model. Two springs are used between the two nodes at the two sides of the crack. One is parallel to the crack surface, and the other is orthogonal to the crack surface. These springs work elastically when the stress in the concrete element is less than tensile strength; this is the normal condition. The springs' stiffness will be decreased when the stress in the concrete element reaches the tensile strength; then the crack will open. Small residual stiffness has to be remained to represent the interlock effect of aggregates. As the crack width h has to be zero before cracking, large stiffness is set for the springs (theoretically speaking, it is

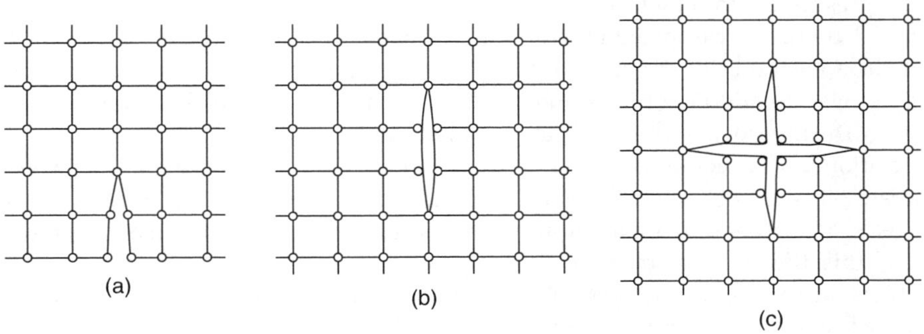

Figure 25.16 The discrete crack model for concrete. (a) The external crack. (b) The internal cracks. (c) The biaxial crack.

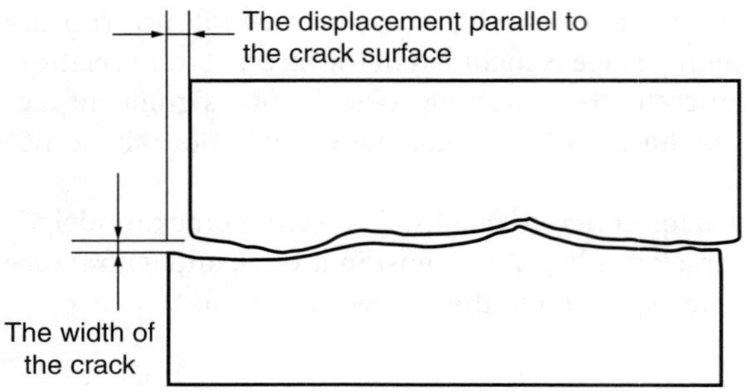

Figure 25.17 The interlocking of the aggregates.

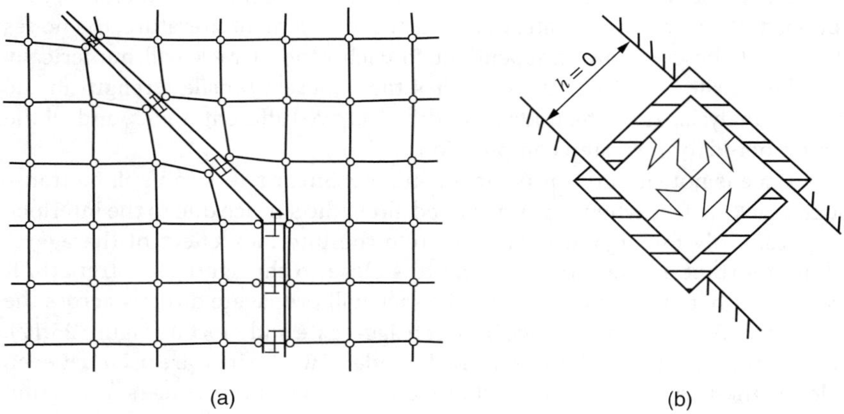

Figure 25.18 The discrete crack model with springs. (a) The computing grid. (b) The spring.

infinity). The stress state of the concrete is obtained from the solid element instead of the springs. Thus, the springs' stiffness has to be sufficiently large. Comparatively, the detailed number is not significant to the computational results. But for post-cracking, a proper value has to be set for the spring stiffness so that it can simulate both the interlock effect of aggregate and the crack opening. Compared with the complete separate discrete crack model, there are two advantages for this type of discrete crack model with springs: ① After the appearance of the crack, only the spring stiffness has to be modified, and there is no need to add new nodes and modify the mesh, which lead to much easier analysis. ② The interlock of the aggregates can be simulated.

The position and direction of the cracks have to be assumed before analysis when the discrete crack model is used; this is a significant disadvantage. Except for special simple cases, it is rather difficult to know the precise position and direction of the crack. Because of this, the discrete crack model is not widely used.

Still, there are some advantages: ① The discrete crack model can also simulate the interlock effect of the reinforcement and the bond between reinforcement and surrounding concrete. ② As the crack is a discontinuous surface inside the structure, discrete crack model can explicitly express the real discontinuous effect (Figure 25.19).

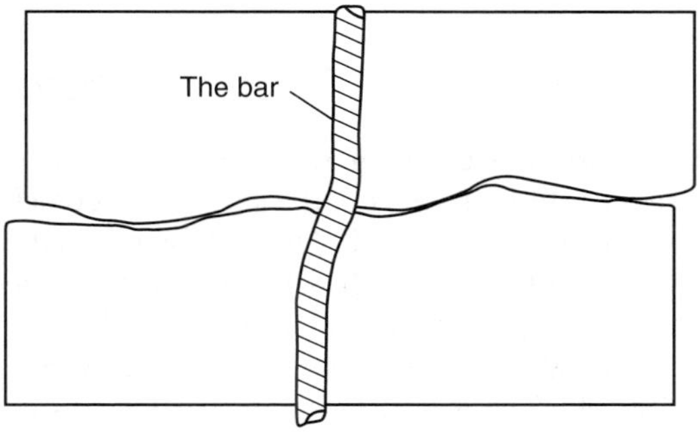

Figure 25.19 The interlocking effect of the bar.

25.2.3 Smeared Crack Model

Smeared crack model is not to explicitly simulate the crack but to simulate the mechanical effect. The influence of the crack will be considered in the stiffness matrix. Obviously, the main influence of the crack is that the crack surface cannot bear the tensile stress anymore, but the tensile stress will be sustained parallel to the crack surface. As shown in Figure 25.20, infinite numbers of cracks are assumed inside the element, which makes orthotropic material. Orthogonal to the crack surface, the stiffness become zero, while original stiffness is used in the parallel direction. To simulate the interlock effect, a small shear modulus has to be used in the direction parallel to the crack. Taken the plane stress problem as an example, and the relation between concrete stress increment and strain increment is as follows:

$$\begin{Bmatrix} \Delta\sigma'_x \\ \Delta\sigma'_y \\ \Delta\tau'_{xy} \end{Bmatrix} = [D'_c] \begin{Bmatrix} \Delta\varepsilon'_x \\ \Delta\varepsilon'_y \\ \Delta\gamma'_{xy} \end{Bmatrix} \tag{25.21}$$

where

$$[D'_c] = \begin{bmatrix} E & 0 & 0 \\ 0 & 0 & 0 \\ 0 & 0 & \beta G \end{bmatrix} \tag{25.22}$$

in which $[D'_c]$ is the elemental tangent elastic stiffness matrix.

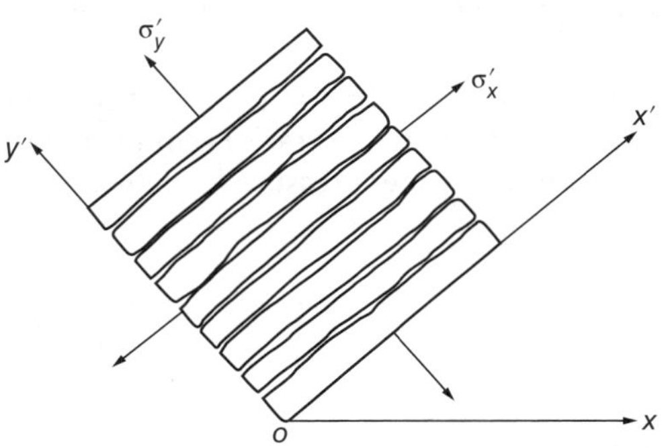

Figure 25.20 The smeared crack model.

The incremental dilatancy is

$$\Delta\varepsilon_z = -\frac{\mu}{E}\Delta\sigma'_x \tag{25.23}$$

In Eq. (25.22), the concrete modulus of elasticity is set to zero in the direction orthogonal to the crack surface. In order to simulate the interlock effect of the aggregate at crack surface, a small shear stiffness βG is set along the crack surface, $0 \le \beta \le 1$. Obviously, the value of β is affected by the roughness and width of the crack. It will decrease with the increase of the crack width. When the crack opening is sufficiently large, there will be no interlock effect and $\beta = 0$. In addition, the crack will be closed when the strain normal to the crack surface turns in compression. The closed crack surface is a weak face inside the concrete, but it can still sustain compressive stress, and a similar interlock effect of aggregates contributes to the shear transfer, that is, β can be set as a suitable value. For practical analysis, the modulus for y' direction is set as E, and β is 1 for the closed crack. It is assumed the crack is completely healed. When crack opens, $\beta < 1$, and its value can be a decreasing function of the crack opening strain. When the crack opening strain is sufficiently large, β should be zero but is recommended to set as a small positive value instead of zero to avoid numerical difficulties.

After cracking, there is no interaction between the orthogonal direction and parallel direction of the crack surface; the effect of Poisson's ratio is ignored in Eq. (25.23). Certainly, due to the roughness of the crack surface, not only shear stress but also dilatancy stress will be induced by the shear slip along the crack. Strictly speaking, the non-diagonal element of $[D'_c]$ cannot be set as zero. Therefore, Poisson's ratio μ is still used in $[D'_c]$ in some literatures.

For constant strain elements, when crack appears, it represents the whole element is cracked. For high-order elements, the crack direction will be different according to different stress states in each node. Although the detailed propagating path of crack is rather complicated, only the direction and the cracking region need to be known. Therefore, for practical analysis, the computation is based on each Gauss point.

The advantage of using smeared crack model is no need to know the exact position and direction of the crack, which are automatically obtained in the analysis according to the real stress state with the incremental loading method. However, it can hardly consider the dowel action and bond between concrete and reinforcement. Fortunately, these two effects only affect the local behavior around the crack, such as the bond stress distribution, crack width, and so on, and they hardly affect the global structural behavior. Compared with the discrete crack model, the smeared crack model is much easier for analysis, and thus it is widely used.

According to local coordinates, the stress before cracking is $\{\sigma'\} = \left[\sigma'_x,\ \sigma'_y,\ \tau'_{xy}\right]^{\mathrm{T}}$, and the stress on surface of crack turns to zero after cracking. If the crack is connected to pressure water, the stress on surface of crack will become water pressure ($-p$); therefore, the released stress is

$$\{\sigma'_{\mathrm{R}}\} = \begin{Bmatrix} \sigma'_x \\ \sigma'_y \\ \tau'_{xy} \end{Bmatrix} - \begin{Bmatrix} \sigma'_x \\ -p \\ 0 \end{Bmatrix} \tag{25.24}$$

where $-p$ denotes the water pressure on surface of crack.

For analysis, the released stress can be transformed to the nodal load, which will be balanced with global equilibrium equations. Here, we assume the normal stress and shear stress are zero on surface of crack at the transient state of the cracking. After stress redistribution, the shear stress at cracking surface may be nonzero according to the aggregate–aggregate interaction.

25.2.4 Thin-Layer Element for Crack

The thin-layer element is a kind of solid element with very small thickness. It was once applied to simulate the weak layer inside the rock mass, the joint inside the concrete dams, and contacting surface between concrete dams and rock foundation. Up to now, it has been seldom used for reinforced concrete. As a matter of fact, it can also be applied to the discrete crack inside the reinforced concrete and might have better results than general discrete crack model, according to the author. It behaves the same as the normal concrete element before cracking, while it can also simulate the dowel action and bond performance like the discrete element. As shown in Figure 25.21, (a) illustrates the usage of three-dimensional thin-layer elements for the joint of an arch dam and interface between the dam and rock foundation, (b) two-dimensional thin-layer elements for the contact surface between a gravity dam and the rock foundation, and (c) two-dimensional thin-layer elements to simulate the main cracks of a reinforced concrete beam, of which both reinforcing bar and concrete are simulated with two-dimensional solid element, while contact element is applied for the interface between concrete and the steel bar.

There is no theoretical rule for the thickness of the thin-layer element. Generally, it is set as the 1/10–1/30 of the element length.

Two schemes are proposed for the computation of the nodal force of the thin-layer elements:

1) Goodman Type
   Only one normal stress component is considered here, as shown in Figure 25.22. For plane problem, the stress–strain relation becomes

$$\{\sigma'\} = \begin{Bmatrix} \sigma'_y \\ \tau'_{xy} \end{Bmatrix} = [D'_t] \begin{Bmatrix} \varepsilon'_y \\ \gamma'_{xy} \end{Bmatrix} = [D'_t]\{\varepsilon'\} \tag{25.25}$$

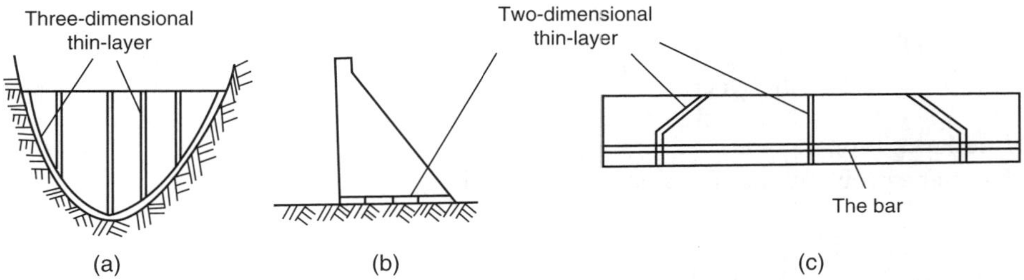

Figure 25.21 The thin-layer element for crack model. (a) The arch dam. (b) The gravity dam. (c) The reinforced concrete beam.

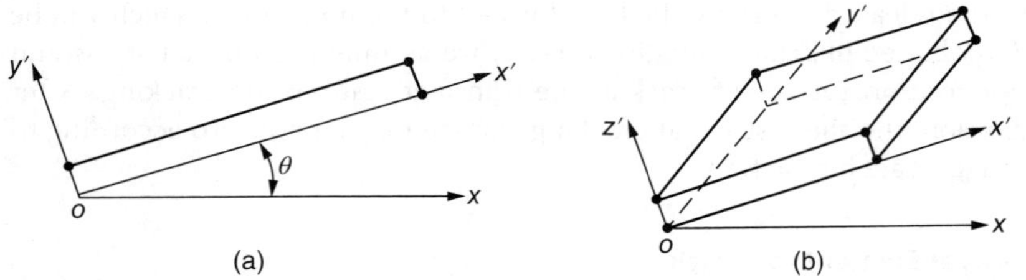

Figure 25.22 The thin-layer element for crack model. (a) The plane problem. (b) The spatial problem.

The element strain ε'_y and γ'_{xy} are computed according to the local displacement difference – the normal component $\Delta v'$ and the shear component $\Delta u'$, as Eq. (23.33) – and the tangential elasticity matrix $[D'_t]$ is set as follows:
Before cracking

$$[D'_t] = \begin{bmatrix} E & 0 \\ 0 & G \end{bmatrix} \tag{25.26}$$

After cracking

$$[D'_t] = \begin{bmatrix} 0 & 0 \\ 0 & \beta G \end{bmatrix} \tag{25.27}$$

For three-dimensional problem, the stress–strain relation becomes (Figure 25.22(b))

$$\{\sigma'\} = \begin{Bmatrix} \sigma'_z \\ \tau'_{zx} \\ \tau'_{xy} \end{Bmatrix} = [D'_t] \begin{Bmatrix} \varepsilon'_z \\ \gamma'_{yz} \\ \gamma'_{xy} \end{Bmatrix} = [D'_t]\{\varepsilon'\} \tag{25.28}$$

Before cracking

$$[D'_t] = \begin{bmatrix} E & 0 & 0 \\ 0 & G & 0 \\ 0 & 0 & G \end{bmatrix} \tag{25.29}$$

After cracking

$$[D'_t] = \begin{bmatrix} E & 0 & 0 \\ 0 & \beta G & 0 \\ 0 & 0 & \beta G \end{bmatrix} \tag{25.30}$$

2) Interlayer Type
For plane problem, the stress–strain relation is

$$\{\sigma'\} = \begin{Bmatrix} \sigma'_z \\ \sigma'_y \\ \tau'_{xy} \end{Bmatrix} = [D'_t] \begin{Bmatrix} \varepsilon'_z \\ \varepsilon'_y \\ \gamma'_{xy} \end{Bmatrix} = [D'_t]\{\varepsilon'\} \tag{25.31}$$

For plane stress problem, Eq. (2.21) is used for pre-cracking conditions.

After cracking, $\sigma'_y = 0$, $[D'_t]$ is computed as follows:

$$[D'_t] = \begin{bmatrix} E & 0 & 0 \\ 0 & 0 & 0 \\ 0 & 0 & \beta G \end{bmatrix} \tag{25.32}$$

For plane strain problem, Eq. (2.25) is used for pre-cracking conditions.
After cracking, $\sigma'_y = 0$ and $\varepsilon'_z = 0$. $[D'_t]$ is computed as follows:

$$[D'_t] = \begin{bmatrix} E/(1-\mu^2) & 0 & 0 \\ 0 & 0 & 0 \\ 0 & 0 & \beta G \end{bmatrix} \tag{25.33}$$

For three-dimensional problem, before cracking, $[D'_t]$ is computed as Eq. (7.14). After cracking, as $\sigma'_z = 0$, stress can be calculated according to Eq. (25.34):

$$\begin{Bmatrix} \sigma'_x \\ \sigma'_y \\ \sigma'_z \\ \tau'_{xy} \\ \tau'_{yz} \\ \tau'_{zx} \end{Bmatrix} = \frac{E}{1-\mu^2} \begin{bmatrix} 1 & \mu & 0 & 0 & 0 & 0 \\ \mu & 1 & 0 & 0 & 0 & 0 \\ 0 & 0 & 0 & 0 & 0 & 0 \\ 0 & 0 & 0 & \frac{1-\mu}{2} & 0 & 0 \\ 0 & 0 & 0 & 0 & \frac{\beta(1-\mu)}{2} & 0 \\ 0 & 0 & 0 & 0 & 0 & \frac{\beta(1-\mu)}{2} \end{bmatrix} \begin{Bmatrix} \varepsilon'_x \\ \varepsilon'_y \\ \varepsilon'_z \\ \gamma'_{xy} \\ \gamma'_{yz} \\ \gamma'_{zx} \end{Bmatrix} \tag{25.34}$$

If we compare the above two methods, it is clear that the Goodman-type method is easier, especially for three-dimensional problem, in which only three stress components are considered, and Mohr–Coulomb criterion is often used for post-cracking nonlinear analysis. As the stress computation accuracy might be a little lower (only one normal stress component is computed), the stress in the element aside can also be used for cracking judgment (by extending the stress from the Gauss points to cracking surface).

As for interlayer-type method, it is in fact the same as the smeared cracking element (only less Gauss points will be used in the thickness direction) with almost same computation cost. The difference is the crack can only occur in a specific region; it can more reliably simulate the cracking effect on the structural response, such as the bonding properties between steel bar and concrete, dowel action, the interaction between reinforcement and concrete between two cracks, and so on.

The boundary of the element do not need to be a straight line or a plane; general curve and surface can also be applied, as discussed in Chapter 23.

25.2.5 No-Tension Crack Model

No-tension model is mainly used in geotechnical engineering; so far, it seems that it has not been used in reinforced concrete structures. The author thinks that, in fact, it can also be used for analysis of reinforced concrete structures.

Taking the initial stress $\{\sigma_0\}$ as follows:

$$\{\sigma_0\} = -[T]\{\sigma'_R\} \tag{25.35}$$

where $\{\sigma'_R\}$ is the release stress in the local coordinate system, referring to formula (25.24), and $[T]$ is the coordinate transformation matrix, referring to Appendix A; it transforms the release stress in the local coordinate system to that in the global coordinate system.

Imbalanced force caused by the initial stress can be calculated according to formula (17.32), and then it is solved by the iterative method. The constant stiffness method is available and it is relatively simple. For the element (or Gaussian integral points), if the principal tensile stress is less than R_t, we think it does not crack, and the initial stress is zero.

In a variety of models for concrete cracks, no-tension model is the most simple of calculation. In this model, the bond stress and the interlocking of the reinforcement cannot be considered, but the most important factor, namely, the cracking of concrete, is considered.

25.2.6 Fracture Mechanics Model

Practical experience has shown that concrete is a cutting sensitive material; when the concrete structure has an incision, due to the stress concentration at the tip, it is easy to generate to a crack. When taking the principal tensile stress reaching the tensile strength as the criterion of crack, because of ignoring the stress concentration at the crack tip, it may be rather unsafe.

Therefore, from 1960s, many studies on fracture mechanics of concrete have been done.

The FEM is an important tool for concrete fracture study. At present, there are two kinds of models:

25.2.6.1 The Sharp Crack Model

The sharp crack model is the traditional fracture mechanics model, as shown in Figure 25.23(a). Assuming that the crack is discrete and develops along the side of elements, arranging several quadratic isoparametric elements around the crack tip, the midpoints of several sides near the crack tip are close to the crack tip. The distance to the crack tip is equal to 1/4 of the length of the side to reflect the stress concentration at crack tip. Then calculate the stress intensity factor at the crack tip and compare it

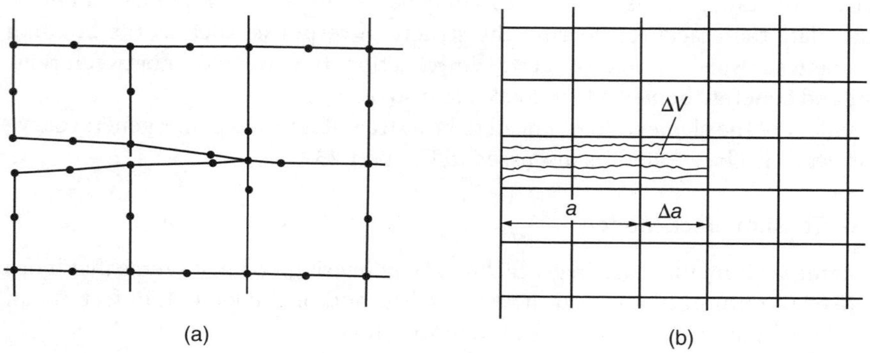

Figure 25.23 Fracture mechanics model for concrete. (a) The sharp crack model. (b) The blunt crack band model.

with the fracture toughness of concrete to determine whether the crack will extend (for details refer to Chapter 21).

For this computational model, once the crack extends, it is necessary to modify the finite element mesh, and the calculation is cumbersome.

25.2.6.2 The Blunt Crack Band Model

As shown in Figure 25.23(b), calculating by the smeared crack model, assuming that the crack length is a, research the conditions that the crack extends to the next element (the length is Δa, the volume is ΔV). Considering that before the crack extends, there are a large of micro cracks in the concrete in the front of the crack tip, Bazant proposed a criterion without taking the stress intensity factor as crack extension, but according to the ratio $\Delta U/\Delta a$ of the change of the energy ΔU of the new cracked element to the change of the length Δa of the crack to judge whether the crack will extend (for details refer to Section 21.8).

By the sharp crack model, the nonlinear constitutive relation of concrete can be considered, and the change of the energy can also be used to determine whether the crack will extend. But once the crack extends, it is necessary to change the mesh, and it is not convenient. By the blunt crack band model, the crack extension is realized by changing the material parameters of the crack element; it does not need to change the mesh after the crack extension, and it is convenient.

25.2.6.3 Comparison of Concrete Crack Models

Five kinds of concrete crack models are introduced above. Thin-layer element model and no-tension model are not used in reinforced concrete previously, but the author thinks that they can be used in reinforced concrete.

For traditional discrete crack model without springs, after the crack extension, it needs to modify the calculation mesh, and it is not convenient. They should not be used in future. For the discrete crack model and thin-layer crack element model with springs, when the crack extends, it only needs to modify spring constants or material constants of thin-layer element, it does not need to modify the calculation mesh (if using the fracture mechanical criterion, it needs to modify the location of the midpoints of sides of isoparametric elements in the front of the crack), and it is also convenient. But it needs to assume the location and orientation of the crack first; for more complex structures, it is difficult. Therefore, this kind of model is now seldom used. But the two kinds of crack models also have some advantages: they can reflect the interlocking of the reinforcement bar and the changes of the bond stress better. In addition, the cracks are the discontinuities in the interior of the structure; such discrete cracks could better reflect the effect of the discontinuity on the structural response. So the two kinds of crack models can be used in the following conditions:

1) There are cracks in the structure, so it needs to analyze its effect.
2) The weak surface in the structure, such as concrete joints and the interface between concrete dam and the rock.
3) It needs to study the local effects, such as the interlocking, the changes of the bond stress of the bar, and the width of the crack.

Between the two kinds of models, the accuracy of the thin-layer crack model is better. Before cracking, it is the same with the common concrete, and after cracking, it is the same with the discrete crack.

The smeared crack model reflects the effect of the crack by modifying the parameters of the material, without assuming the location and direction of the crack, and it is more convenient.

But it cannot consider the local effects such as the change of the bond stress and the interlocking of the bar. If you only need to calculate the global properties such as the bearing capacity and the displacement of the structure, there is no doubt that the smeared crack model is the best choice. So the smeared model has been widely used at present. No-tension crack model is similar to the smeared model, and the calculation is simpler; of course, it is rough, but it can be used to calculate the global response of the concrete and reinforced concrete structures.

The main function of the fracture mechanics model is that it makes the condition of crack extension more precise. It can be used for the analysis of crack propagation conditions of plain concrete structures such as the concrete blocks, the gravity dams, and the buttress dams. As for reinforced concrete structures, the tension on the section is mainly born by the steel bar, and the accuracy of the crack range has little influence on the bearing capacity of the structure, so the fracture mechanics model is rarely used.

25.3 The Calculation of the Smeared Crack Model

In the last section, the basic concepts of the smeared crack model has been illustrated; since it has been widely used in practice, in this section we will give a detailed description of the calculation method.

25.3.1 Modes of the Concrete Failure and Constitutive Relations Before and After Failure

As shown in Figure 25.24, concrete has two modes of damage: crushing and split. Crushing is the disintegration and the complete loss of the strength of materials under compressive stress. Split is that cracks generate in a direction perpendicular to the principal stress. After the split, the material cannot withstand the tensile stress in the direction perpendicular to the crack, but in the direction parallel to the crack, the strength has no loss. During the loading process in the future, if the joint is under pressure again in the normal direction, the cracks may close. If the concrete were split by two groups (plane problems) or three groups (spatial problems) of cracks in different directions, the strength of the material is completely lost.

The stress–strain relation of concrete before and after the failure is shown in Figure 24.25. The gradients of line 1–2 and 3–4 represent the elastic matrices before and after the destruction, respectively, and line 2–3 represents the release stress $\{\sigma'_{\mathrm{R}}\}$ at the failure moment. The release stress will be redistributed to the material nearby. The stress–strain incremental relationship (in the local coordinate system) after the destruction is

$$\{\mathrm{d}\sigma'\} = [D'_{\mathrm{c}}]\{\mathrm{d}\varepsilon'\} \tag{25.36}$$

where $[D'_{\mathrm{c}}]$ is the elasticity matrix of concrete after destruction.

In Figure 25.25, point 4 represents the state after stress redistribution. From the beginning of the failure to the end of the stress redistribution, in the whole process, the stress increment $\{\Delta\sigma\}$ is calculated by formula (25.37):

$$\begin{aligned}\{\Delta\sigma'\} &= \{d\sigma'\} - \{\sigma_R'\}\\ &= [D_c']\{d\varepsilon'\} - \{\sigma_R'\}\end{aligned} \tag{25.37}$$

The formula above is the relationship in the local coordinate system where x' axis is parallel to the crack surface, and the relationship in the global coordinate system is

$$\{\Delta\sigma\} = [D_c]\{d\varepsilon\} - \{\sigma_R\} \tag{25.38}$$

$$[D_c] = [T][D_c'][T]^T \tag{25.39}$$

$$\{\sigma_R\} = [T]\{\sigma_R'\} \tag{25.40}$$

where $[T]$ is the coordinate transformation matrix, $[D_c]$ is the elasticity matrix of concrete after failure, and σ_R is the release stress at the failure moment.

Formula (25.38) is the stress–strain model during the process from the beginning of failure to the end of the stress redistribution. As for the stress–strain relation of concrete before and after damage can be linear or nonlinear, the elasticity matrices $[D]$ and $[D_c]$ can be calculated according to different circumstances; the nodal loads caused by the release stress of the crack is

$$\{\Delta P_R\} = \int [B]^T\{\sigma_R\}dV \tag{25.41}$$

It should be adjusted by the global equilibrium equations and transferred to other elements.

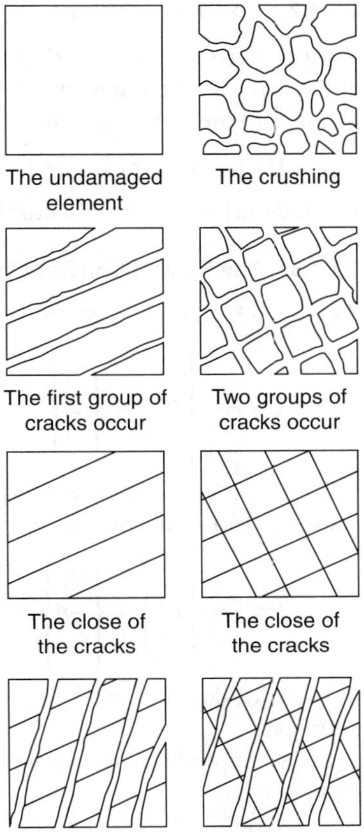

Figure 25.24 The failure modes of the concrete.

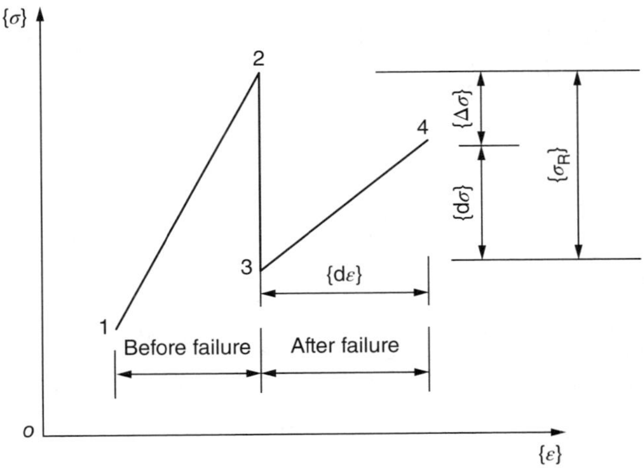

Figure 25.25 The stress–strain relation of concrete before and after failure.

25.3.2 Concrete Crushing

Concrete crushing means that the material is completely disintegrated, so the original stress is released completely; after crushing, concrete is no longer able to withstand any stress, and the elasticity matrix is zero, namely, after crushing

$$[D_c] = 0, \quad \{\sigma_R\} = \{\sigma\} \tag{25.42}$$

in which $\{\sigma\}$ is the stress vector before crushing.

25.3.3 The Split of Concrete Under the Plane Stress

As shown in Figure 25.26, the release stress in the local coordinate system (x', y') is

$$\{\sigma'_R\} = \begin{Bmatrix} \sigma'_x \\ \sigma'_y \\ \tau'_{xy} \end{Bmatrix} - \begin{Bmatrix} \sigma'_x \\ 0 \\ 0 \end{Bmatrix} \tag{25.43}$$

If the cracks connect with artesian water, after the split, the joint surface bears the water pressure $-p$, and the releasing stress is

$$\{\sigma'_R\} = \begin{Bmatrix} \sigma'_x \\ \sigma'_y \\ \tau'_{xy} \end{Bmatrix} - \begin{Bmatrix} \sigma'_x \\ -p \\ 0 \end{Bmatrix} \tag{25.44}$$

In the local coordinate system, the elasticity matrix is represented by formula (25.22) after split.

The stress and strain coordinate transformation relationships of plane problems are as follows:

$$\left.\begin{aligned} \{\sigma\} = [t]\{\sigma'\}, \quad \{\sigma'\} = [t']\{\sigma\} \\ \{\varepsilon\} = [t']^{\mathrm{T}}\{\varepsilon'\}, \quad \{\varepsilon'\} = [t]^{\mathrm{T}}\{\varepsilon\} \end{aligned}\right\} \tag{25.45}$$

in which

$$[t] = \begin{bmatrix} c^2 & s^2 & -2sc \\ s^2 & c^2 & 2sc \\ sc & -sc & -s^2 \end{bmatrix}, \quad [t'] = \begin{bmatrix} c^2 & s^2 & 2sc \\ s^2 & c^2 & -2sc \\ -sc & sc & c^2 - s^2 \end{bmatrix}, \quad c = \cos\theta,\ s = \sin\theta \tag{25.46}$$

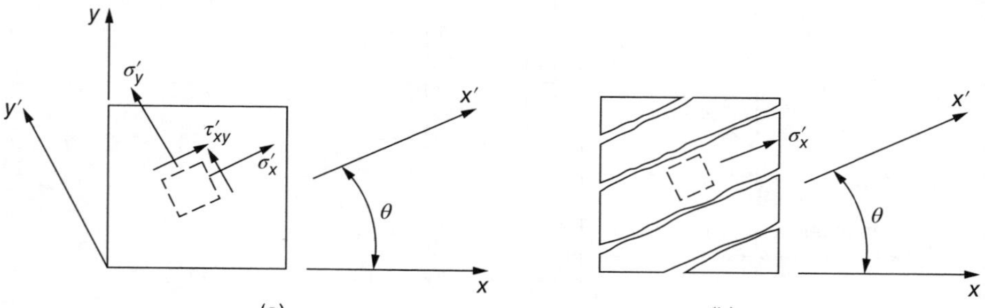

Figure 25.26 The stress status of concrete before and after cracking. (a) Before cracking. (b) After cracking.

By formulas (25.39) and (25.46), the elasticity matrix after split in the global coordinate system is as follows:

$$[D_c] = E\begin{bmatrix} c^4 + 4\overline{\beta}s^2c^2 & c^2s^2 - 4\overline{\beta}s^2c^2 & sc^3 - 2\overline{\beta}sc(c^2 - s^2) \\ c^2s^2 - 4\overline{\beta}sc & s^4 + 4\overline{\beta}s^2c^2 & s^3c + 2\overline{\beta}sc(c^2 - s^2) \\ sc^3 - 2\beta sc(c^2 - s^2) & s^3c + 2\overline{\beta}sc(c^2 - s^2) & s^2c^2 + \overline{\beta}(c^2 - s^2)^2 \end{bmatrix} \tag{25.47}$$

where $\overline{\beta} = \beta/[2(1+\mu)]$.

25.3.4 Concrete Split Under the Plane Strain State

The release stress is the same with the plane stress problems, expressed by formula (25.44).

After there are cracks perpendicular to axis y', the following stress–strain relationship holds:

$$\mathrm{d}\varepsilon'_x = \frac{\mathrm{d}\sigma'_x}{E} - \frac{\mu\mathrm{d}\sigma_z}{E}, \quad \mathrm{d}\varepsilon_z = -\frac{\mu\mathrm{d}\sigma'_x}{E} + \frac{\mathrm{d}\sigma_z}{E} = 0$$

After eliminating $\mathrm{d}\sigma_z$, we get

$$\mathrm{d}\sigma'_x = \frac{E}{1-\mu^2}\mathrm{d}\varepsilon'_x$$

Therefore, the elasticity matrix of concrete after the split in the local coordinate system is

$$[D'_c] = \frac{E}{1-\mu^2}\begin{bmatrix} 1 & 0 & 0 \\ 0 & 0 & 0 \\ 0 & 0 & \beta' \end{bmatrix} \tag{25.48}$$

in which $\beta' = \beta(1-\mu)/2$.

In formula (25.47), changing E into $E/(1-\mu^2)$ and changing $\overline{\beta}$ into β', the elasticity matrix $[D_c]$ in the global coordinate system after the split for the plane strain problems is obtained.

25.3.5 The Split of Concrete of the Spatial Problems

As shown in Figure 25.27, cracks occur along the direction perpendicular to axis z'. After the split, $\sigma'_z = 0$ or $\sigma'_z = -p$, the elasticity matrix in the local coordinate system and the stress–strain relationship are as follows:

$$\begin{Bmatrix} \sigma'_x \\ \sigma'_y \\ \sigma'_z \\ \tau'_{xy} \\ \tau'_{yz} \\ \tau'_{zx} \end{Bmatrix} = \frac{E}{1-\mu^2}\begin{bmatrix} 1 & \mu & 0 & 0 & 0 & 0 \\ \mu & 1 & 0 & 0 & 0 & 0 \\ 0 & 0 & 0 & 0 & 0 & 0 \\ 0 & 0 & 0 & (1-\mu)/2 & 0 & 0 \\ 0 & 0 & 0 & 0 & \beta(1-\mu)/2 & 0 \\ 0 & 0 & 0 & 0 & 0 & \beta(1-\mu)/2 \end{bmatrix}\begin{Bmatrix} \varepsilon'_x \\ \varepsilon'_y \\ \varepsilon'_z \\ \gamma'_{xy} \\ \gamma'_{yz} \\ \gamma'_{zx} \end{Bmatrix} \tag{25.49}$$

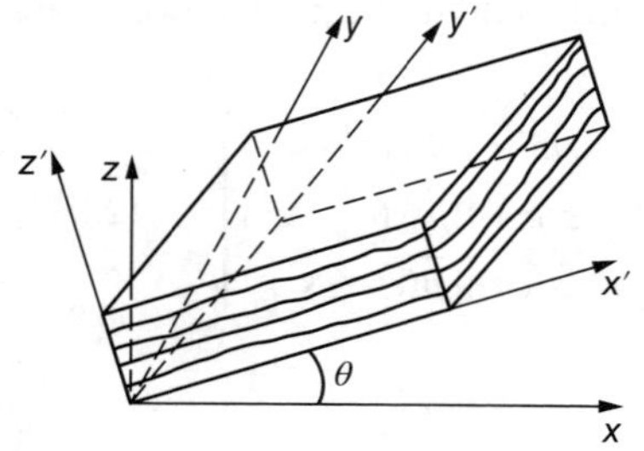

Figure 25.27 The cracking of three-dimensional concrete.

The release stress caused by the crack is

$$\{\sigma_R'\} = \begin{Bmatrix} \sigma_x' \\ \sigma_y' \\ \sigma_z' \\ \tau_{xy}' \\ \tau_{yz}' \\ \tau_{zx}' \end{Bmatrix} - \begin{Bmatrix} \sigma_x' \\ \sigma_y' \\ -p \\ \tau_{xy}' \\ 0 \\ 0 \end{Bmatrix} \tag{25.50}$$

Using the tensor notation, the transformation relationships of the stress tensor and the strain tensor between the local coordinate system (x_1', x_2', x_3') and the global coordinate system (x_1, x_2, x_3) are as follows:

$$\sigma_{ij}' = l_{im} l_{jn} \sigma_{mn}, \quad \sigma_{ij} = l_{mi} l_{nj} \sigma_{mn}' \tag{25.51}$$

$$\varepsilon_{ij}' = l_{im} l_{jn} \varepsilon_{mn}, \quad \varepsilon_{ij} = l_{mi} l_{nj} \varepsilon_{mn}' \tag{25.52}$$

$$l_{ij} = \cos(x_i', x_j)$$

where (x_i', x_j) is the angle between axis x_i' and axis x_j.

It is noticed that the shear strain ε_{ij} in the strain tensor is half of the shear strain γ_{ij} in the engineering, that is, $\varepsilon_{ij} = \gamma_{ij}/2$.

25.3.6 The Behavior of Concrete After Split

The behavior of concrete after split depends on the state of the stress and the strain, which may cause new cracks or crushing, and the old cracks may continue to open or close back, using the strain ε_n along the normal direction of the joint to determine. If $\varepsilon_n \geq 0$, the crack is open. If $\varepsilon_n \leq 0$, the crack is close. ε_n can be calculated based on the coordinate transformation relation.

For example, for plane problems, by formula (25.45), it is known that the strain in the normal direction of the joint surface is

$$\varepsilon_y' = \begin{bmatrix} \sin^2\theta & \cos^2\theta & -\sin\theta\cos\theta \end{bmatrix} \begin{Bmatrix} \varepsilon_x \\ \varepsilon_y \\ \gamma_{xy} \end{Bmatrix} \tag{25.53}$$

25.3.7 The Stress Adjustment and the Calculation Procedure when Concrete Splits

Now we consider the calculation of a typical incremental step. As shown in Figure 25.28, point 1 is the start of an incremental step. According to the known states of the stress and strain, calculating the initial stiffness matrix $[K]_n$ of the step by the load increment $\{\Delta P\}_n$ and calculating the displacement increment, point 2 is obtained. Then calculating the stress and strain of the Gaussian points, if finding some points split, calculating the stiffness matrix after split and the nodal loads caused by the release stress of the cracks, solving again and checking each Gaussian point, if new crack appears, repeat the calculation above, until there are no new cracks, the stress adjustment of this incremental step

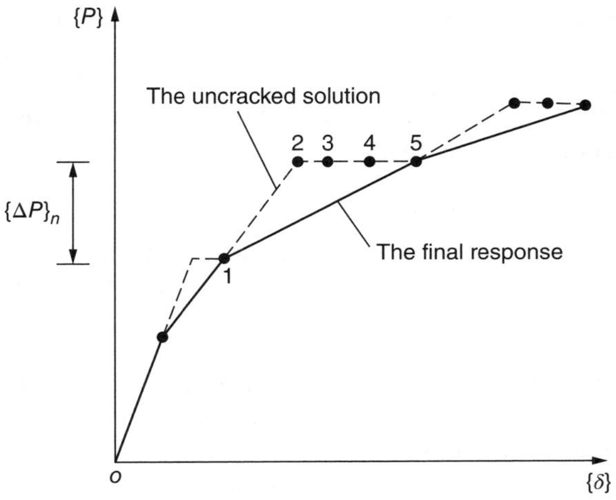

Figure 25.28 The load–displacement increment relationship considering the split of concrete.

is ended, and point 5 is obtained. Then, going to the calculation of the next incremental step. When the change of the stress is intense or the load increment $\{\Delta P\}_n$ is large, several iterations are needed in each incremental step. Therefore, at the beginning of loading, the load increment can be larger but when the nonlinearity of the material is intense, or when the structure will be destroyed, the load increment should be smaller.

The calculation of each incremental step is as follows:

1) According to the known stress and strain at the beginning of this step, calculate $[K]_n$, giving $\{\Delta P\}_n$. By formula (25.54), the displacement increment is obtained:

$$\{\Delta\delta\}_n = [K]_n^{-1}\{\Delta P\}_n \tag{25.54}$$

2) Calculating the stress $\{\sigma\}_n$ and the strain $\{\varepsilon\}_n$ of each Gaussian point, checking every point whether the points that do not split originally will split, whether the points that split originally produce a new crack, and whether the original crack will be close. According to the check results, calculating the new global stiffness matrix $[K_t]_n$ and the nodal load $\{\Delta P_R\}_n$ caused by the new crack, the external load does not change, by formula (25.55), calculating the displacement increment caused by the stress adjustment:

$$\{\Delta\delta_R\}_n = [K_t]_n^{-1}\{\Delta P_R\}_n \tag{25.55}$$

and calculating the corresponding stress increment $\{d\sigma_R\}_n$ and strain increment $\{d\varepsilon_R\}_n$.
3) For each point without failure, calculating the scaling factor r, making the total stress $\{\sigma\}_n + r\{d\sigma_R\}_n$ meeting the failure condition.
4) Finding the minimum r_m of all scale factors.
5) If $r_m \geq 1$, representing that the nodal loads $\{\Delta P_R\}_n$ caused by the release stress have been allocated, $\{\sigma\}_n + \{d\sigma_R\}_n$, $\{\varepsilon\}_n + \{d\varepsilon_R\}_n$, $\{\delta\}_n + \{d\delta_R\}_n$ are the final responses of this incremental step, such as point 5 in Figure 25.27.

6) If $r_m < 1$, suggesting that there are other points to split before $\{\Delta P_R\}_n$ is redistributed well, dividing $\{\Delta P_R\}_n$ into two parts, the first part is $r_m\{\Delta P_R\}_n$, adjusted by formula (25.55),and the corresponding responses are $r_m\{\Delta\delta_R\}_n$, $r_m\{d\sigma_R\}_n$, $r_m\{d\varepsilon_R\}_n$, and so on.
7) The rest nodal load is $(1 - r_m)\{\Delta P_R\}_n$, plus the nodal load $\{\Delta P_R\}_{n+1}$ caused by the new cracks; by formula (25.56), compute again, that is,

$$[K_t]_{n+1}\{\Delta\delta_R\}_{n+1} = (1 - r_m)\{\Delta P_R\}_n + \{\Delta P_R\}_{n+1} \tag{25.56}$$

where $[K_t]_{n+1}$ is the new tangent stiffness matrix considering the new cracks.
8) Repeating step (3) to step (7), until $r_m \geq 1$, the calculation of this incremental step is finished.
An example is given as follows.
There is a plain concrete container, the top of which is hemispherical, the thickness is 5.8 cm, the length is 163 cm, and it is bearing external uniform hydrostatic pressure. The uniaxial compressive strength is $R_c = 42.3$ MPa, and the stress–strain relationship is shown in Figure 25.29. The ultimate tensile strain is $\varepsilon_t = 0.08\%$ and the ultimate compressive strain is 0.35%.

The relationship between the radial displacement of point *A* of the container and the external pressure *p* is shown in Figure 25.30. At the load step H, one element on the internal margin begins to yield. When load continues to increase, more elements yield. When one element is destroyed, the external pressure *p* keeps invariable, and the stress of the destroyed element is released and distributed to the elements nearby, so the yield range on the internal margin expands further, the number of the destroyed elements gradually increases, to load step 9; the stiffness matrix of the structure is singular, the structure is damaged, and the external pressure is the same with the load step 1 when the first destroyed element appeared. The failure process is shown in Figure 25.31.

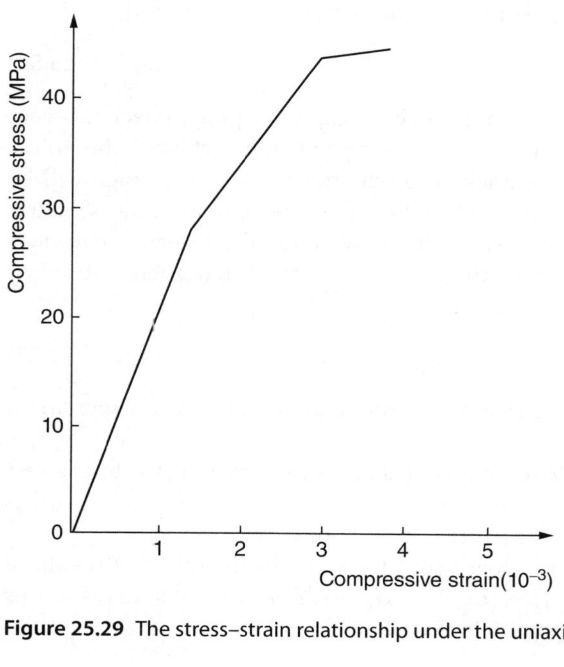

Figure 25.29 The stress–strain relationship under the uniaxial compression.

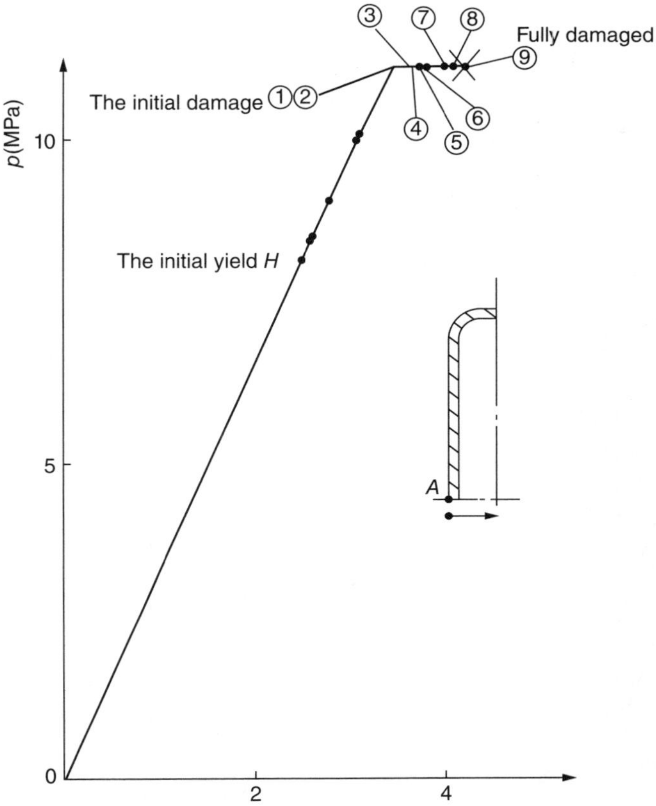

Figure 25.30 The relationship between the displacement of point *A* and the external pressure.

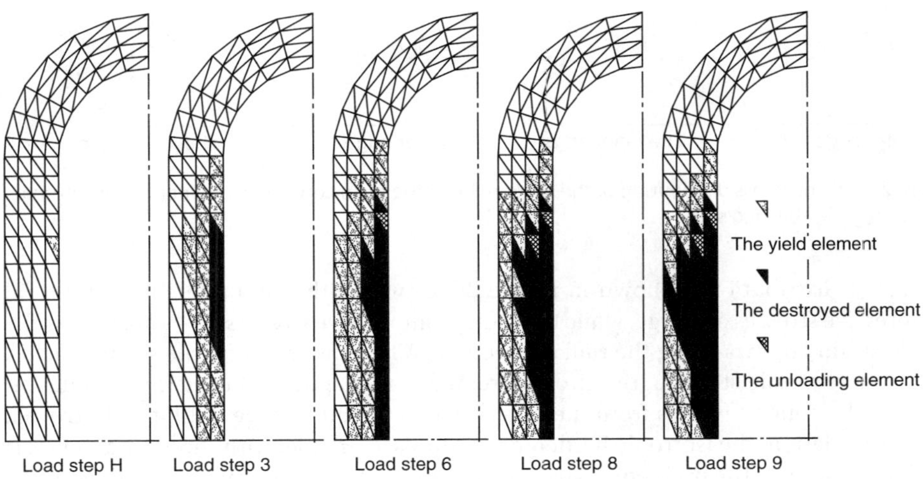

Figure 25.31 The yield and failure process of the plain concrete container under external pressure.

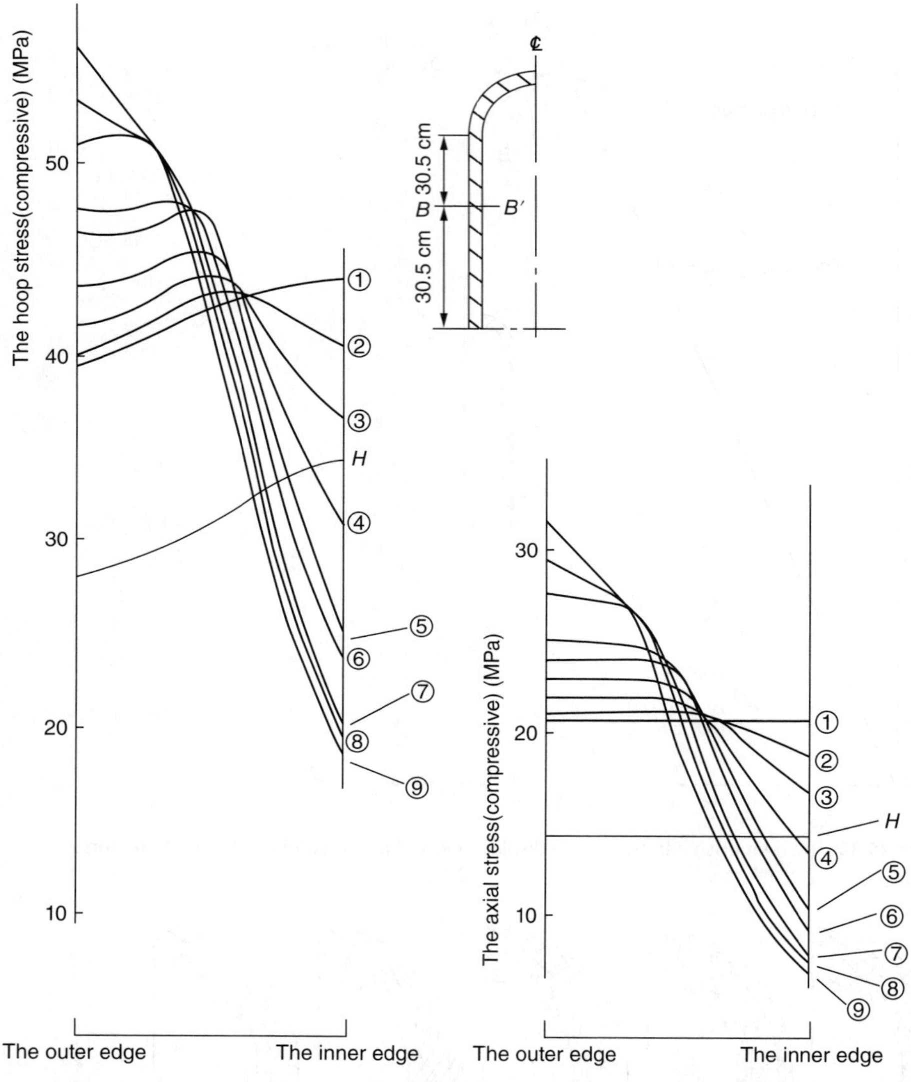

Figure 25.32 The stress distribution of the section BB′ of the plain concrete container under external pressure ①–⑨ the load step.

The stress distribution is shown in Figure 25.32, where the circumferential and axial compressive stresses are large, while the radial compressive stress is very small. Thus the tensile strain appears along the radial direction. When the radial tensile strain reaches the ultimate tensile strain ε_t, the circumferential cracks appear. During the destruction process, the concrete spalls from the inner edge to the outer edge until it is destroyed. In the calculation, the isotropic hardening elastoplastic model with three parameters is used for concrete, The problem is analyzed according to an axisymmetric problem.

25.4 The Constitutive Relation and the Stress Calculation of the Steel

25.4.1 The Constitutive Relation of the Steel Bar

In the reinforced concrete structure, the steel is in uniaxial stress state, and the stress–strain relationship is relatively simple. There are three constitutive models that are often used, as shown in Figure 25.33:

25.4.1.1 The Ideal Elastic–Plastic Model

As shown in Figure 25.33(a), when $\sigma < \sigma_y$, the material is elastic, where σ_y is the yield stress and E_s is the modulus of elasticity. When $\sigma = \sigma_y$, the material is ideal plastic, in which the modulus of elasticity is zero.

25.4.1.2 The Trilinear Model

As shown in Figure 25.33(b), the stress–strain relationship of the steel bar is represented by three lines. When $\sigma < \sigma_y$, it is linear and elastic, the modulus of elasticity is E_s; when $\varepsilon_y \le \varepsilon \le \varepsilon_s$, there is a plastic platform, $\sigma = \sigma_y$, and the modulus of elasticity is zero; when $\varepsilon > \varepsilon_s$, it is linear hardening, and the modulus of elasticity is E_s'.

25.4.1.3 The Complete Model

As shown in Figure 25.33(c), when $\sigma < \sigma_y$, it is linear and elastic, and the modulus of elasticity is E_s. When $\varepsilon_y \le \varepsilon \le \varepsilon_s$, there is a plastic platform, $\sigma = \sigma_y$, and the modulus of elasticity is zero; when $\varepsilon > \varepsilon_s$, the stress–strain relationship is expressed by the exponential function as follows:

$$\sigma = \sigma_y + (\sigma_u - \sigma_y)[1 - e^{-b(\varepsilon - \varepsilon_s)}] \tag{25.57}$$

in which σ_y is the yield stress, σ_u is the ultimate strength, and b is a constant.

The modulus of elasticity is

$$E_s' = \frac{d\sigma}{d\varepsilon} = b(\sigma_u - \sigma_y)e^{-b(\varepsilon - \varepsilon_s)} \tag{25.58}$$

Some steel products, such as high strength steel and the prestressing cables after cold drawing, may be no platform. Let $\varepsilon_s = \varepsilon_y$ in Figure 25.33(b) and (c).

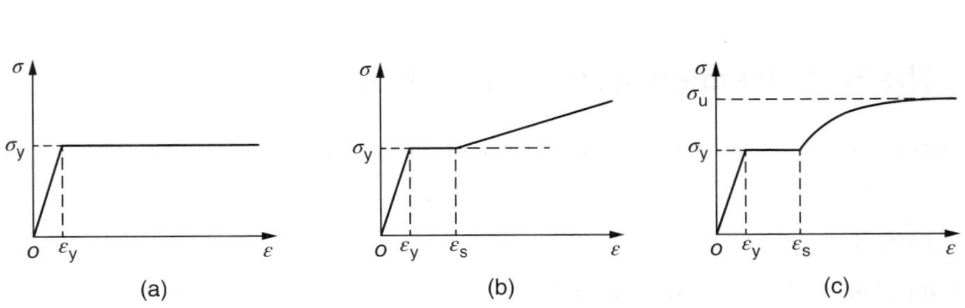

Figure 25.33 The constitutive model of the steel bar. (a) The ideal elastic–plastic model. (b) The trilinear model. (c) The complete model.

25.4.2 The Calculation of the Stress of the Steel Bar

The factors considered in the formula for the calculation of the stress of steel are related with the calculation purposes. For example, in the calculation of seismic response, because the loading time is very short, only the effect of reverse sign of the nonlinear strain needs considering, regardless of the temperature and time (creep) effect. In contrast, in the analysis of the shell of the nuclear reactor, the effects of temperature and time must be considered, regardless of the change of the stress symbol. Generally speaking, according to the nature of the problem, a formula as simple as possible should be used to simulate the properties of the steel.

When the bar is in the elastic state, the stress should be calculated by formula (25.59):

$$\sigma_s = E_s(\varepsilon_s - \varepsilon^T) \tag{25.59}$$

where ε_s is the total strain of the bar, ε^T is the temperature strain of the bar, and E_s is the modulus of elasticity of the bar.

After the bar enters into the plastic state, generally calculated according to the ideal elastoplastic model, the iterative method is used. Usually calculating a test elastic stress, that is,

$$\sigma_s^e = E_s(\varepsilon_s - \varepsilon_{i-1}^p - \varepsilon^T) \tag{25.60}$$

in which ε_{i-1}^p is the plastic deformation at the end of the previous step, and then check the yield condition, namely,

$$-\sigma_y \le \sigma_s^e \le \sigma_y \tag{25.61}$$

If the condition above is satisfied, in this increment step, the stress of the bar is elastic and the plastic strain increment is zero, that is,

$$\sigma_s = \sigma_s^e, \quad \varepsilon_i^p = \varepsilon_{i-1}^p \tag{25.62}$$

If formula (25.61) cannot be establish, then

$$\sigma_s = \sigma_y \frac{\sigma_s^e}{|\sigma_s^e|} \tag{25.63}$$

The formula above shows that when σ_s^e is positive, $\sigma_s = \sigma_y$, whereas $\sigma_s = -\sigma_y$. The plastic deformation at the end of this step is

$$\varepsilon_i^p = \varepsilon_s - \frac{\sigma_s}{E_s} \tag{25.64}$$

25.5 The Finite Element Model of the Steel Bar

In the reinforced concrete structures, the calculation models of the steel bar are as follows:

25.5.1 Line Element

As shown in Figure 25.34(a), the bar withstanding the axial force is as follows:

$$N_s = A_s\sigma_s \tag{25.65}$$

where N_s is the axial force of the bar, A_s is the area of the reinforcing bar, and σ_s is the stress of the bar determined by the stress–strain relationship.

The line element can be a line (linear element) or a curve (higher-order element).

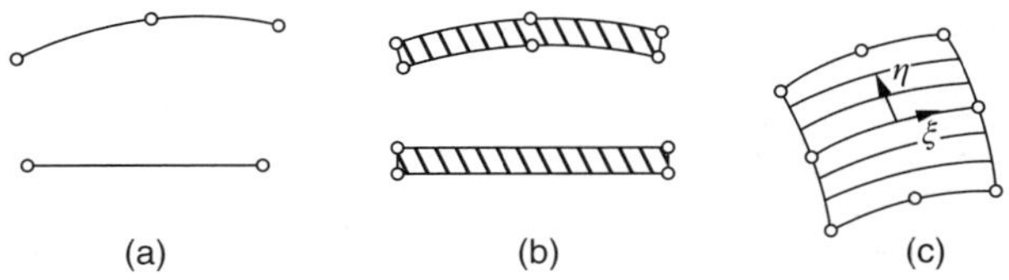

Figure 25.34 The calculation model of the steel bar. (a) The line element. (b) The solid element. (c) The thin membrane element.

25.5.2 Solid Element

As shown in Figure 25.34(b), converting the circular area into a square section with equal area and then dividing it into rectangular solid elements along the direction of the length, it can be a linear element or a higher-order element. So, the bar can withstand not only the axial forces but also the bending moments and shear forces.

25.5.3 Thin Membrane Element

As shown in Figure 25.34(c), for spatial problems, the steel bar can be regarded as thin membrane, the stress component along the direction s of the length is $\sigma_s = E_s\varepsilon_s$, and the stresses in the other directions are all zeros. In the local coordinate system, the elasticity matrix is

$$[D_s] = \begin{bmatrix} E_s & 0 & 0 \\ 0 & 0 & 0 \\ 0 & 0 & 0 \end{bmatrix} \tag{25.66}$$

where E_s is the modulus of elasticity of the steel, taking $E_s = 0$ after the bar yields.

The thickness of the thin membrane is

$$t = \frac{A_s}{a} \tag{25.67}$$

in which t is the thickness of the thin membrane, a is the spacing of the bars, and A_s is the area of single bar.

For the steel mesh, if the bars are orthogonal, it can be converted into orthotropic steel membrane. But a better approach is to calculate according to two layers of steel membrane, which is favorable for nonlinear analysis.

25.6 The Connection of the Steel Bar and Concrete

The steel bars are laid in concrete. In the calculation models of reinforced concrete structures, there are three models for the connection between the bars and concrete.

25.6.1 Fixed Connection

Assuming that the steel bar and concrete are fully connected, there is no relative displacement. The strain of the bar is equal to that of the concrete around it.

25.6.2 The Linking Spring Element

As shown in Figure 25.35(a), the connection elements are set between the steel bar and the concrete. The contact forces between the bar and concrete are represented by a set of

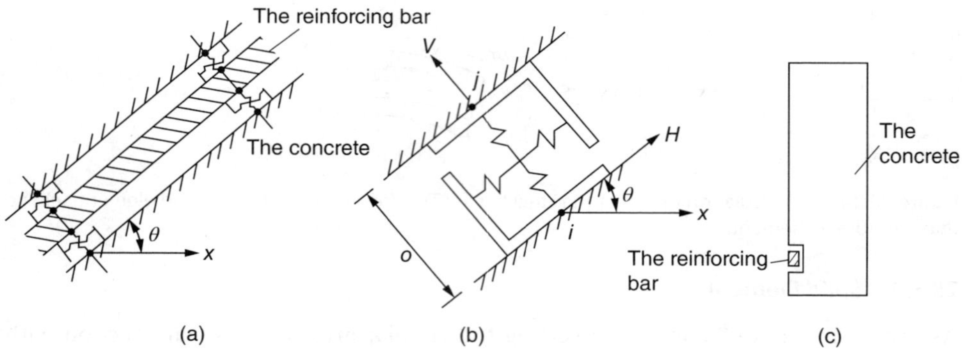

Figure 25.35 The linking element.

springs as shown in Figure 25.35(b). The distance between the surface of the bar and that of the concrete is zero. Assuming that the coefficient of the tangential spring is k_h and the coefficient of the normal spring is k_v, in local coordinate system (h, v), the relation between the nodal forces and the nodal displacement is

$$\{\overline{F}\} = \begin{Bmatrix} \overline{X_i} \\ \overline{Y_i} \\ \overline{X_j} \\ \overline{Y_j} \end{Bmatrix} = \begin{bmatrix} k_\mathrm{h} & 0 & -k_\mathrm{h} & 0 \\ 0 & k_\mathrm{v} & 0 & -k_\mathrm{v} \\ -k_\mathrm{h} & 0 & k_\mathrm{h} & 0 \\ 0 & -k_\mathrm{v} & 0 & k_\mathrm{v} \end{bmatrix} \begin{Bmatrix} \overline{u_i} \\ \overline{v_i} \\ \overline{u_j} \\ \overline{v_j} \end{Bmatrix} = [\overline{k^e}]\{\overline{\delta^e}\} \tag{25.68}$$

where $[\overline{k^e}]$ is the stiffness matrix in the local coordinate system.

After the coordinate transformation, the stiffness matrix of the spring in the global coordinate system is as follows:

$$[k^e] = [T]^T[\overline{k^e}][T] = \begin{bmatrix} k_\mathrm{h}c^2 + k_\mathrm{v}s^2 & k_\mathrm{h}sc - k_\mathrm{v}sc & -k_\mathrm{h}c^2 - k_\mathrm{v}s^2 & -k_\mathrm{h}cs + k_\mathrm{v}sc \\ & k_\mathrm{h}s^2 + k_\mathrm{v}c^2 & -k_\mathrm{h}sc + k_\mathrm{v}sc & -k_\mathrm{h}s^2 - k_\mathrm{v}c^2 \\ & & k_\mathrm{h}c^2 + k_\mathrm{v}s^2 & k_\mathrm{h}sc - k_\mathrm{v}sc \\ \text{symmetrical} & & & k_\mathrm{h}s^2 + k_\mathrm{v}c^2 \end{bmatrix} \tag{25.69}$$

in which $c = \cos\theta, s = \sin\theta$.

The coordinate transformation matrix is formula (1.15).

As shown in Figure 25.35(c), the concrete section is continuous, but in the position corresponding to the steel, deducting the section of the steel bar, leaving a rectangular notch, embedding the steel bars, a set of springs are set to connect the steel and concrete. The solid element is adopted for the bar.

If the line element is adopted for the bar, usually assuming that the bar is fixed with the concrete, while the line element can also be connected with the concrete by the spring, then only a spring parallel to the steel is needed to connect the nodes of the bar and those of the adjacent concrete. In formula (25.69) for the stiffness matrix of the spring element, we let

$$k_\mathrm{v} = 0$$

In the actual calculation, k_v should be retained, taking a small constant.

25.6.3 The Contact Element

The contact element can be used between the steel bar and the concrete as shown in Figure 25.36(a). If the solid element is used for the bar, two contact elements can be used on both sides. Compared with the axial stress, the lateral stress of the bar is small. Actually it is possible to use contact element on one side of the bar. If the line element is used, as shown in Figure 25.36(b), the contact element is used only on one side of the bar, and the calculation method of the contact element is given in Section 24.3.

There are several different suppositions about the relationship between the steel bar and the concrete, as shown in Figure 25.37. One supposition is that the bar is embedded in concrete as shown in Figure 25.37(a); another supposition is that one or two rigid arms are fixed on the concrete surface (as shown in Figure 25.37(b) and (c)), which connect the bar through the contact element or the spring element.

Within the contact element, the contact force between the bar and concrete is continuously distributed. Within the link spring element, the contact force is concentrated. Therefore, the contact element is better than the spring element. In order to simulate

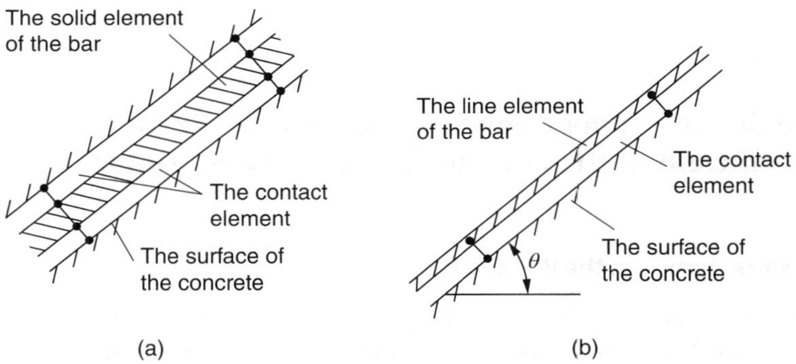

Figure 25.36 The contact element between the bar and concrete. (a) The contact elements on both sides. (b) The contact elements on one side.

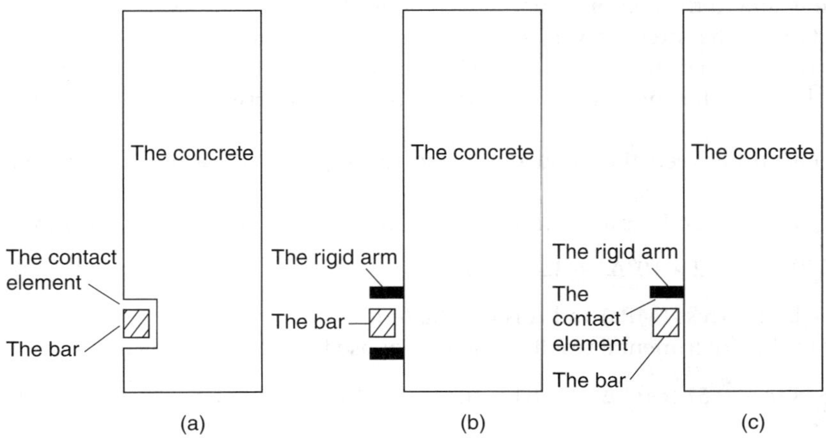

Figure 25.37 Connection of the steel bar and concrete.

the nonlinear characteristics and the damage of the bond stress of the reinforcement, more Gaussian integration points can be arranged within the contact element, doing the nonlinear analysis point by point.

When dividing the computational grid, we should notice that ① the concrete must be continuous and cannot be cut off by the contact elements and ② different numbers should be given to the nodes of the concrete and those of the steel bar. The nodes of the bar are connected with those of the concrete through the contact element or the link spring.

For the spatial problems, if necessary, the contact element can also be used to connect the concrete and the steel bar.

Among the three ways of the connection above, of course, the fixed connection is the simplest.

Generally speaking, if the objective is to understand the bearing capacity and the overall response of the structure, the fixed connection between the steel and concrete can be assumed. If it is necessary to know the local responses such as the width of the crack, the distribution of the bond stress, and the interlocking of the bar, the contact element or the link spring element must be used, and the contact element is better.

25.7 The Bond Stress between the Steel Bar and Concrete: The Stiffness Coefficient of the Linking Spring and the Contact Element

25.7.1 The Bond Stress between the Bar and the Concrete

Bond stress between the plain bar and the concrete is due to chemical bonding and the friction. When the bar bears the tension, due to the effect of Poisson's ratio, there is tensile stress in the direction perpendicular to the surface, making the bond stress unreliable.

In the engineering, corrugated bar is mostly used. The bonding stress between bars and concrete is due to the compression between the ribs and the keyway of concrete. The actual state of the stress is very complex. In the calculation, usually assuming that there is the nominal shear stress, namely, the bond stress on the contact surface between the bar and the concrete, it concerns with the slip between the bar and the concrete.

The relationship between the bond stress and the slip should be obtained by the experiment.

According to Bresler and Bertero's test data, Nilson proposed the following relation:

$$\tau = 9781\Delta - 5.72\times 10^6\Delta^2 + 8.35\times 10^8\Delta^3 \tag{25.70}$$

where τ is the bond stress (MPa) and Δ is the slip (cm).

According to the experimental results, Houde proposed the following formula:

$$\tau = (5289\Delta - 2.51\times 10^6\Delta^2 + 5.84\times 10^8\Delta^3 - 5.46\times 10^{10}\Delta^4)\sqrt{R_c/42.3} \tag{25.71}$$

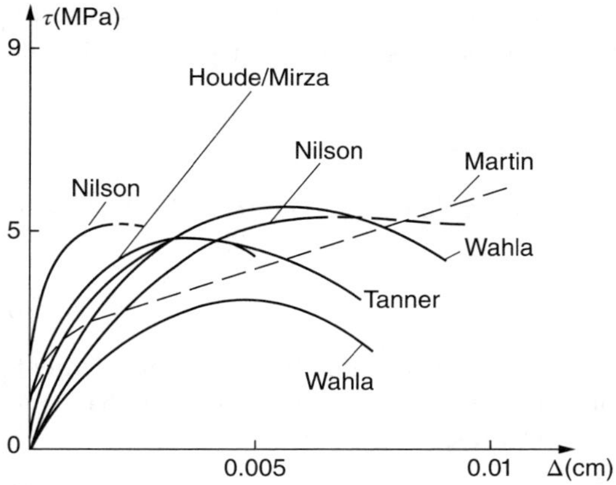

Figure 25.38 The relationship between the bond stress and the slip (the test results of different scholars).

in which R_c is the compressive strength of concrete (MPa) and other symbols are same with those mentioned before.

The test results of different scholars are shown in Figure 25.38 for reference.

25.7.2 The Stiffness Coefficient of the Linking Spring

As shown in Figure 25.35, the tangential stiffness coefficient of the link spring elements can be calculated by formula (25.72):

$$k_h = A\frac{d\tau}{d\Delta} \tag{25.72}$$

in which A is the surface area allocated to the spring.

For example, when a spring represents the contact area between m bars and concrete, the spacing of the springs is l, the diameter of the bar is d, and A is calculated by formula (25.73):

$$A = \frac{m\pi dl}{2} \tag{25.73}$$

where 2 appears in the denominator, because usually the spring elements are set above and below the bar element.

By formula (25.70),

$$\frac{d\tau}{d\Delta} = 9781 - 11.44\times10^6\Delta + 25.05\times10^8\Delta^2 \tag{25.74}$$

By formula (25.71),

$$\frac{d\tau}{d\Delta} = (5289 - 5.06\times10^6\Delta + 17.53\times10^8\Delta^2 - 21.85\times10^{10}\Delta^3)\sqrt{R_c/42.3} \tag{25.75}$$

Regarding the normal stiffness coefficient k_v, the experimental data is lacking, and usually in the calculation, a large value is taken, which has the same magnitude with the modulus of elasticity of concrete.

25.7.3 The Stiffness Coefficient of the Contact Element

The tangential stiffness coefficient λ_s of the contact element in Figure 25.36 can be calculated by formula (25.76):

$$\lambda_s = a\frac{d\tau}{d\Delta} \tag{25.76}$$

where a is the surface area of the bar per unit length represented by the contact element.

For example, when the contact element represents the interface between m bars and concrete and there are contact elements on both sides of the bar,

$$a = \frac{m\pi d}{2} \tag{25.77}$$

$d\tau/d\Delta$ can be calculated by formulas (25.74) and (25.75).

The normal stiffness coefficient λ_n of the contact element usually takes a large number that has the same magnitude with the modulus of elasticity of concrete.

25.8 The Stiffness Matrix of the Reinforced Concrete Structure

In the reinforced concrete structures, there are three kinds of elements: the concrete element, the reinforcement element, and the contact element. Combining the stiffness matrices of these three kinds of elements, the global stiffness matrix is obtained as follows:

$$[K] = \sum_e [K_c^e] + \sum_e [K_s^e] + \sum_e [K_f^e] \tag{25.78}$$

where $[K]$ is the global stiffness matrix, $[K_c^e]$ is the stiffness matrix of concrete element, $[K_s^e]$ is the stiffness matrix of the reinforcement element, and $[K_f^e]$ is the stiffness matrix of the contact element (or the linking spring).

The incremental method is usually used in the calculation of the reinforced concrete structures, and the global equilibrium equation is

$$[K]\{\delta\} = \{\Delta P\} \tag{25.79}$$

After the displacement increment is calculated by the formula above, in accordance with their constitutive relations, the stresses of the concrete, bar, and contact elements are calculated, respectively.

25.9 The Calculation of Steel Bar in the Isoparametric Element

In the calculation of reinforced concrete structures, the isoparametric element is widely used, as shown in Figure 25.39. In this section, the calculation method of the steel bar in the isoparametric element is described.

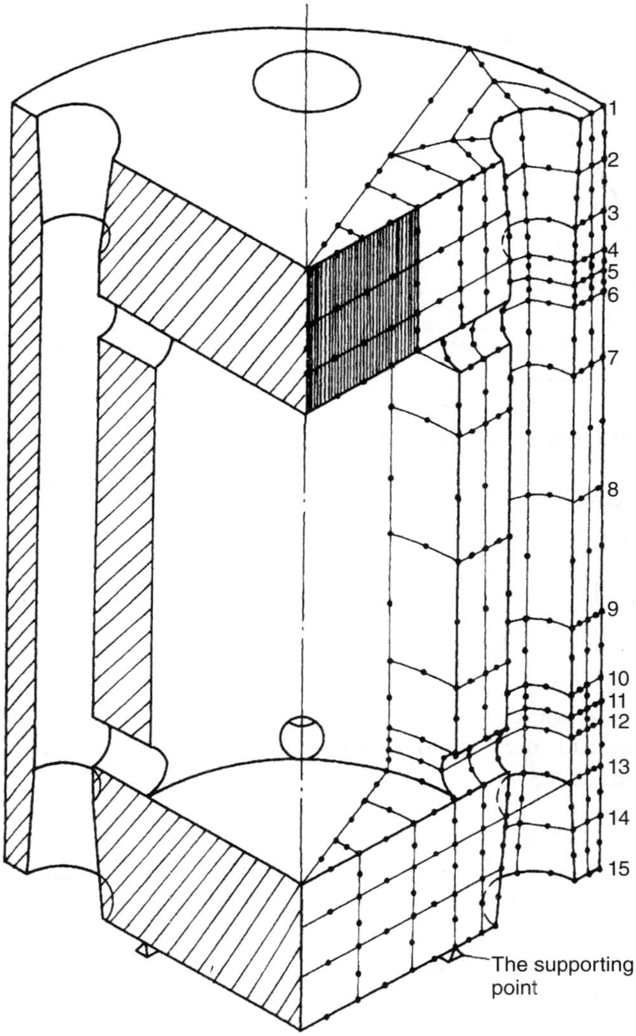

Figure 25.39 The computing grid of the pressure container of the nuclear power station.

25.9.1 Plane Problem

The plane isoparametric element is shown in Figure 25.40, assuming that the bar is located on the curve $\eta = \eta_c$, in which η_c is a constant. Assuming that the deformation of the bar is equivalent to that of the concrete, namely, there is no slip. The displacement of any point within the element is expressed as

$$u = \sum N_i u_i, \quad v = \sum N_i v_i$$
$$N_i = N_i(\xi, \eta)$$

where N_i is the shape function and u_i, v_i are the nodal displacements.

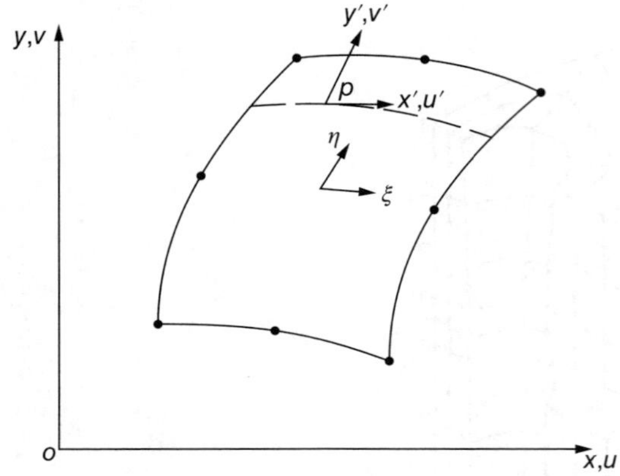

Figure 25.40 The calculation of the bar in the plane isoparametric element.

In the global coordinate system (x, y), the deformation matrix is

$$[j] = \begin{Bmatrix} \dfrac{\partial u}{\partial x} & \dfrac{\partial v}{\partial x} \\ \dfrac{\partial u}{\partial y} & \dfrac{\partial v}{\partial y} \end{Bmatrix} = \begin{bmatrix} \dfrac{\partial [N]}{\partial x} \\ \dfrac{\partial [N]}{\partial y} \end{bmatrix} \begin{bmatrix} u_1 & v_1 \\ u_2 & v_2 \\ \vdots & \vdots \end{bmatrix} \tag{25.80}$$

in which $[N] = \lfloor N_1 \ \ N_2 \ \ \ldots \rfloor$.

Because $[j]$ is a second-order tensor, the deformation matrix $[j']$ in the local coordinate system (x', y') should be transformed from the deformation matrix $[j]$ in the global coordinate system (x, y) by formula (25.81):

$$[j'] = \begin{Bmatrix} \dfrac{\partial u'}{\partial x'} & \dfrac{\partial v'}{\partial x'} \\ \dfrac{\partial u'}{\partial y'} & \dfrac{\partial v'}{\partial y'} \end{Bmatrix} = [R][j][R]^{\mathrm{T}} \tag{25.81}$$

where $[R]$ is the direction cosine matrix of point P.

Because axis x' coincides with ξ,

$$[R] = \begin{bmatrix} \dfrac{\partial x}{\partial x'} & \dfrac{\partial y}{\partial x'} \\ \dfrac{\partial x}{\partial y'} & \dfrac{\partial y}{\partial y'} \end{bmatrix} = \frac{1}{\sqrt{\left(\dfrac{\partial x}{\partial \xi}\right)^2 + \left(\dfrac{\partial y}{\partial \xi}\right)^2}} \begin{bmatrix} \dfrac{\partial x}{\partial \xi} & \dfrac{\partial y}{\partial \xi} \\ -\dfrac{\partial y}{\partial \xi} & \dfrac{\partial x}{\partial \xi} \end{bmatrix} \tag{25.82}$$

By formulas (25.80), (25.81), and (25.82), the axial strain of the bar at point P is as follows:

$$\begin{aligned} \varepsilon_p &= \frac{\partial u'}{\partial x'} \\ &= \frac{1}{h^2} \left\{ \left(\frac{\partial x}{\partial \xi}\right)^2 \sum \frac{\partial N_i}{\partial x} u_i + \frac{\partial x}{\partial \xi}\frac{\partial y}{\partial \xi} \left(\sum \frac{\partial N_i}{\partial x} v_i + \sum \frac{\partial N_i}{\partial y} u_i \right) + \left(\frac{\partial y}{\partial \xi}\right)^2 \sum \frac{\partial N_i}{\partial y} v_i \right\} \end{aligned} \tag{25.83a}$$

or

$$\varepsilon_p = \frac{\partial u'}{\partial x'} = \frac{1}{h^2}\left[C_1\frac{\partial N_1}{\partial x} + C_2\frac{\partial N_1}{\partial y} \quad C_2\frac{\partial N_1}{\partial x} + C_3\frac{\partial N_1}{\partial y} \quad \cdots\right]\begin{Bmatrix} u_1 \\ v_1 \\ u_2 \\ v_2 \\ \vdots \end{Bmatrix}$$

$$= [B][\delta]^e \tag{25.83b}$$

in which

$$[B] = \frac{1}{h^2}\left[C_1\frac{\partial N_1}{\partial x} + C_2\frac{\partial N_1}{\partial y} \quad C_2\frac{\partial N_1}{\partial x} + C_3\frac{\partial N_1}{\partial y} \quad \cdots\right] \tag{25.84}$$

$$\left.\begin{aligned} &C_1 = \left(\frac{\partial x}{\partial \xi}\right)^2, \quad C_2 = \frac{\partial x}{\partial \xi}\frac{\partial y}{\partial \xi}, \quad C_3 = \left(\frac{\partial y}{\partial \xi}\right)^2 \\ &h = \sqrt{\left(\frac{\partial x}{\partial \xi}\right)^2 + \left(\frac{\partial y}{\partial \xi}\right)^2} \end{aligned}\right\} \tag{25.85}$$

$\partial N_i/\partial x$ and $\partial N_i/\partial y$ of formula (25.84) are the values of point P, referring to formula (8.24).

Because only the axial strain of the bar needs considering, the elasticity matrix $[D]$ is given by formula (25.86):

$$[D] = \begin{bmatrix} E_s & 0 & 0 \\ 0 & 0 & 0 \\ 0 & 0 & 0 \end{bmatrix} \tag{25.86}$$

in which E_s is the tangent modulus of elasticity of the bar.

The stiffness matrix of the bar is

$$[k] = \int [B]^T A_s E_s [B] \mathrm{d}s$$

in which A_s is the section area of the bar and d_s is the differential arc length of the bar. Because the bar is located on the curve $\eta = \eta_c$,

$$\mathrm{d}s = \sqrt{\left(\frac{\partial x}{\partial \xi}\right)^2 + \left(\frac{\partial y}{\partial \xi}\right)^2}\mathrm{d}\xi = h\mathrm{d}\xi$$

Therefore, the stiffness matrix of the bar is

$$[k] = A_s \int [B]^T E_s [B] h \mathrm{d}\xi \tag{25.87}$$

25.9.2 The Axisymmetric Problems

Using the cylindrical coordinate (r, θ, z), due to the axial symmetry, the circumferential displacement is zero. At any point there are only two displacement components, namely, the radial displacement u and the axial displacement w.

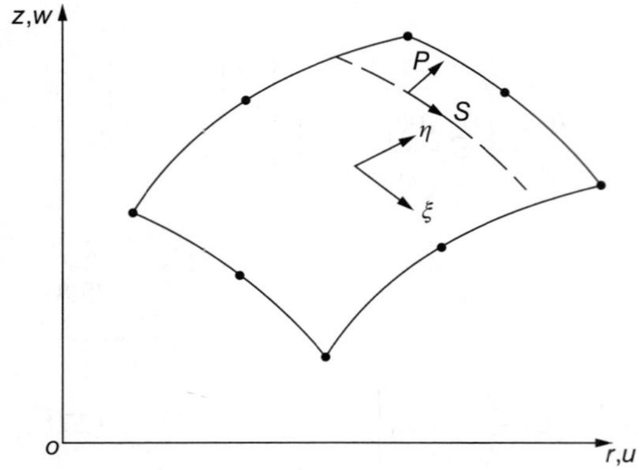

Figure 25.41 The calculation of the bar in the isoparametric element of the axisymmetric problems.

25.9.2.1 The Calculation of the Bar in the Radial Plane

As shown in Figure 25.41, assuming that the bar is located on the curve $\eta = \eta_c$ and it is similar to the plane problem, the normal strain of point P along the tangent direction S can be calculated by formula (25.88):

$$\varepsilon_s = \frac{1}{h^2}\left[C_1\frac{\partial N_1}{\partial r} + C_2\frac{\partial N_2}{\partial z} \quad C_2\frac{\partial N_1}{\partial r} + C_3\frac{\partial N_1}{\partial z} \quad \dots\right]\begin{Bmatrix} u_1 \\ w_1 \\ u_2 \\ w_2 \\ \vdots \end{Bmatrix} = [B][\delta]^e \tag{25.88}$$

in which C_1, C_2, C_3, h can be calculated by formula (25.85), simply replacing x, y with r, z.

The stiffness matrix of the steel element is calculated by formula (25.89):

$$[k] = 2\pi \int [B]^{\mathrm{T}} A_s E_s r [B] h \mathrm{d}\xi \tag{25.89}$$

where A_s is the section area of the bar per unit length in the circumferential direction.

25.9.2.2 The Calculation of the Circumferential Bar

The strain of the circumferential bar is

$$\varepsilon_\theta = \frac{u}{r} = \left[\frac{N_1}{r} \quad 0 \quad \frac{N_2}{r} \quad 0 \quad \dots\right]\begin{Bmatrix} u_1 \\ w_1 \\ u_2 \\ w_2 \\ \vdots \end{Bmatrix} = [B]\{\delta^e\} \tag{25.90}$$

The stiffness matrix is still calculated by formula (25.87), only calculating $[B]$ by formula (25.90).

25.9.2.3 The Calculation of the Steel Plate Lining

The strain of the axisymmetric steel lining is (the shear strain is zero)

$$\{\varepsilon\} = \begin{Bmatrix} \varepsilon_s \\ \varepsilon_\theta \end{Bmatrix} = \sum_{i=1}^{n} [B_i] \begin{Bmatrix} u_i \\ w_i \end{Bmatrix} \tag{25.91}$$

in which

$$[B_i] = \begin{bmatrix} a_i & b_i \\ c_i & 0 \end{bmatrix}, \quad a_i = \frac{1}{h^2}\left(C_1\frac{\partial N_i}{\partial r} + C_2\frac{\partial N_i}{\partial z}\right), \quad b_i = \frac{1}{h^2}\left(C_2\frac{\partial N_i}{\partial r} + C_3\frac{\partial N_i}{\partial z}\right)$$

$$c_i = \frac{N_i}{r} \tag{25.92}$$

The stiffness matrix is calculated by formula (25.93)

$$[k] = 2\pi \int_{-1}^{1} [B]^T [D_s] tr [B] h \mathrm{d}\xi \tag{25.93}$$

$$[D_s] = \frac{E_s}{1-\mu^2}\begin{bmatrix} 1 & \mu \\ \mu & 1 \end{bmatrix}$$

where t is the thickness of the steel plate.

25.9.2.4 The Circumferential Point Element

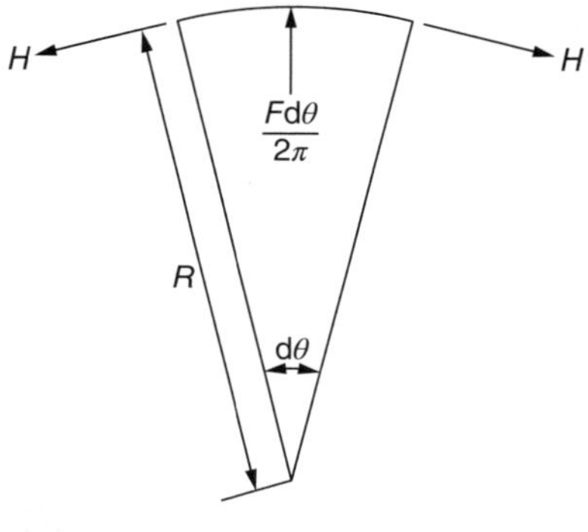

Figure 25.42 The circumferential point element.

For the circular prestressed steel cable with concentrated sections, the *circumferential point element is available.* Assuming that, under the action of the radial force F, the radial displacement is u, the tension in the cable is

$$H = \frac{E_s A_s u}{R}$$

As shown in Figure 25.42, the radial equilibrium conditions of arc $R\mathrm{d}\theta$ is

$$FR\mathrm{d}\theta/2\pi R = 2H \sin\left(\frac{1}{2}\mathrm{d}\theta\right) = 2H\frac{1}{2}\mathrm{d}\theta$$

Then $F = 2\pi H$ is obtained, so

$$F = ku \tag{25.94}$$

$$k = \frac{2\pi E_s A_s}{R} \tag{25.95}$$

Therefore, the circumferential cable can be simulated by point element (at a node), where the radial stiffness coefficient is k, as the formula above.

25.9.2.5 The Spatial Problems

As shown in Figure 25.43, assuming that the equivalent film element of the bar is located on the curved surface $\varsigma = \varsigma_c$, the directions of the bar are the same with that of the local coordinates ξ, η, and the equivalent thickness in the two directions are t_ξ and t_η,

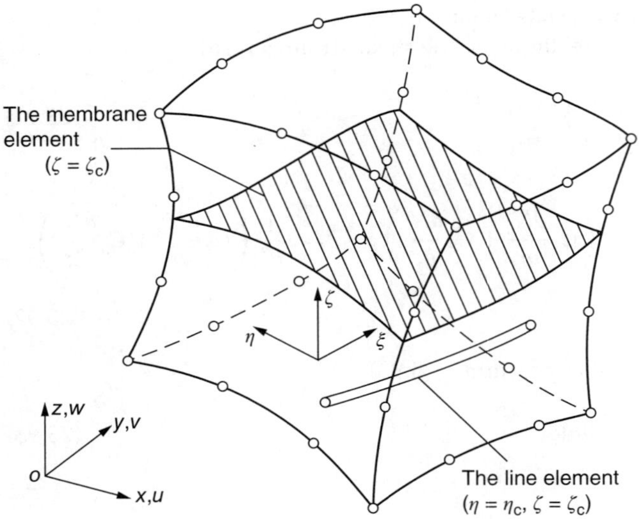

Figure 25.43 The line element and the equivalent membrane element in the spatial solid element.

respectively, only considering the axial stresses in the directions of ξ, η, the elasticity matrix of the equivalent membrane element can be taken as

$$[D] = \begin{bmatrix} E_s t_\xi & 0 & 0 \\ 0 & E_s t_\eta & 0 \\ 0 & 0 & 0 \end{bmatrix} \tag{25.96}$$

In the local coordinate system (x', y', z'), the strain should be

$$\{\varepsilon'\} = \begin{Bmatrix} \dfrac{\partial u'}{\partial x'} \\ \dfrac{\partial v'}{\partial y'} \\ \dfrac{\partial u'}{\partial y'} + \dfrac{\partial v'}{\partial x'} \end{Bmatrix} \tag{25.97}$$

where u' and v' are the displacements in the local coordinate system and axis z' is orthogonal to the membrane.

The deformation matrix in the global coordinate system (x, y, z) is

$$[j] = \begin{bmatrix} \dfrac{\partial u}{\partial x} & \dfrac{\partial v}{\partial x} & \dfrac{\partial w}{\partial x} \\ \dfrac{\partial u}{\partial y} & \dfrac{\partial v}{\partial y} & \dfrac{\partial w}{\partial y} \\ \dfrac{\partial u}{\partial z} & \dfrac{\partial v}{\partial z} & \dfrac{\partial w}{\partial z} \end{bmatrix} = \begin{bmatrix} \dfrac{\partial N_1}{\partial x} & \cdots & \dfrac{\partial N_n}{\partial x} \\ \dfrac{\partial N_1}{\partial y} & \cdots & \dfrac{\partial N_n}{\partial y} \\ \dfrac{\partial N_1}{\partial z} & \cdots & \dfrac{\partial N_n}{\partial z} \end{bmatrix} \begin{bmatrix} u_1 & v_1 & w_1 \\ \vdots & \vdots & \vdots \\ u_n & v_n & w_n \end{bmatrix} \tag{25.98}$$

By formula (8.24), it is known that

$$[j] = [J]^{-1} \begin{bmatrix} \frac{\partial N_1}{\partial \xi} & \cdots & \frac{\partial N_n}{\partial \xi} \\ \frac{\partial N_1}{\partial \eta} & \cdots & \frac{\partial N_n}{\partial \eta} \\ \frac{\partial N_1}{\partial \varsigma} & \cdots & \frac{\partial N_n}{\partial \varsigma} \end{bmatrix} \begin{bmatrix} u_1 & v_1 & w_1 \\ \vdots & \vdots & \vdots \\ u_n & v_n & w_n \end{bmatrix} \tag{25.99}$$

Through the coordinate transformation, the deformation matrix in the local coordinate system can be obtained as follows:

$$[j'] = \begin{bmatrix} \frac{\partial u'}{\partial x'} & \frac{\partial v'}{\partial x'} & \frac{\partial w'}{\partial x'} \\ \frac{\partial u'}{\partial y'} & \frac{\partial v'}{\partial y'} & \frac{\partial w'}{\partial y'} \\ \frac{\partial u'}{\partial z'} & \frac{\partial v'}{\partial z'} & \frac{\partial w'}{\partial z'} \end{bmatrix} = [R][j][R]^{\mathrm{T}} \tag{25.100}$$

in which

$$[R] = \begin{bmatrix} \frac{\partial x}{\partial x'} & \frac{\partial x}{\partial y'} & \frac{\partial x}{\partial z'} \\ \frac{\partial y}{\partial x'} & \frac{\partial y}{\partial y'} & \frac{\partial y}{\partial z'} \\ \frac{\partial z}{\partial x'} & \frac{\partial z}{\partial y'} & \frac{\partial z}{\partial z'} \end{bmatrix} \tag{25.101}$$

where $[R]$ is the direction cosine matrix for the coordinate rotation.

By formulas (25.97), (25.99), and (25.100), the strain of the membrane in the local coordinate system can be written as formula (25.102):

$$\{\varepsilon'\} = \sum_{i=1}^{n} [B_i] \begin{Bmatrix} u_i \\ v_i \\ w_i \end{Bmatrix} \tag{25.102}$$

Therefore, the stiffness matrix of the membrane element can be calculated as follows:

$$[K_{ij}] = \int [B_i]^{\mathrm{T}} [D][B_j] \mathrm{d}A \tag{25.103}$$

The area dA of the microelement can be calculated by formula (25.104):

$$\mathrm{d}A = \sqrt{a^2 + b^2 + c^2} \mathrm{d}\xi \mathrm{d}\eta \tag{25.104}$$

in which

$$a = \det \begin{vmatrix} \frac{\partial y}{\partial \xi} & \frac{\partial y}{\partial \eta} \\ \frac{\partial z}{\partial \xi} & \frac{\partial z}{\partial \eta} \end{vmatrix}, \quad b = \det \begin{vmatrix} \frac{\partial z}{\partial \xi} & \frac{\partial z}{\partial \eta} \\ \frac{\partial x}{\partial \xi} & \frac{\partial x}{\partial \eta} \end{vmatrix}, \quad c = \det \begin{vmatrix} \frac{\partial x}{\partial \xi} & \frac{\partial x}{\partial \eta} \\ \frac{\partial y}{\partial \xi} & \frac{\partial y}{\partial \eta} \end{vmatrix} \tag{25.105}$$

where det represents the determinant and a, b, c are calculated at $\varsigma = \varsigma_{\mathrm{c}}$.

For the line element of the single bar as shown in Figure 25.43, $\eta = \eta_c, \varsigma = \varsigma_c$, only the strain ε'_x along the direction of ξ needs calculating, repeating the steps above, and remembering that η, ς are constants. Therefore $[B_i]$, $[B_j]$ can be calculated, and the stiffness matrix can be calculated as follows:

$$[K_{ij}] = \int [B_i]^{\mathrm{T}} E_s A_s [B_j] h \mathrm{d}\xi \tag{25.106}$$

$$h = \sqrt{\left(\frac{\partial x}{\partial \xi}\right)^2 + \left(\frac{\partial y}{\partial \xi}\right)^2 + \left(\frac{\partial z}{\partial \xi}\right)^2}$$

Because the ratio of the erection bar is small, usually it is not calculated. If calculated, it can be calculated as follows. Setting the ratio of the bar p, the direction is orthogonal to the equivalent membrane, namely, it is parallel to the local coordinate ς, so that only the strain along the direction ς in the local coordinate system needs calculating, that is,

$$\varepsilon'_z = \frac{\partial w'}{\partial z'}$$

Thus $[B_i]$, $[B_j]$ can be obtained. Assuming that the erection bar is distributed in the volume evenly, the stiffness matrix can be calculated by the formula as follows:

$$[k_{ij}] = \int_{-1}^{1}\int_{-1}^{1}\int_{-1}^{1} [B_i]^{\mathrm{T}} p E_s [B_j] |J| \mathrm{d}\xi \mathrm{d}\eta \mathrm{d}\varsigma$$

where $|J|$ is the determinant of the Jacobian matrix.

25.10 The Layered Element of the Reinforced Concrete Plates and Shells

As shown in Figure 25.44, dividing the reinforced concrete plate or shell into several thin layers, each layer can have a different property. While assuming that the material property of each layer in the direction of the thickness is constant and each layer is in plane stress state, the stress–strain relationship is two-dimensional nonlinear.

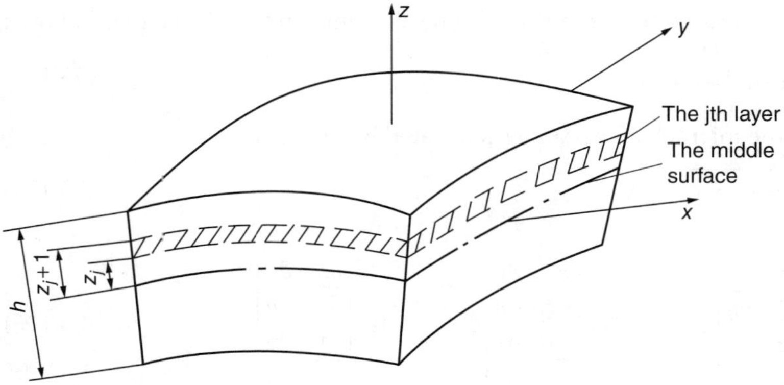

Figure 25.44 The layered element of the reinforced concrete plates and shells.

Due to the nonlinear stress–strain relation, especially the tensile property of concrete is different from the compressive property, and the characteristics of the material are asymmetric for the middle surface. Not only for the reinforced concrete shell the internal forces of the middle surface are coupled with the bending forces, but also for the reinforced concrete plates under lateral loads the free degrees of the middle surface also must be taken into account in the stress analysis. Therefore, in the nonlinear analysis of the reinforced concrete plates and shells, the same nodal degrees of freedom should be used.

In the jth layer, the relationship between the stress increments and the strain increments is

$$\{\Delta\sigma\} = [D]_j\{\Delta\varepsilon\} \tag{25.107}$$

$$\{\Delta\sigma\} = \left[\Delta\sigma_x,\ \Delta\sigma_y,\ \Delta\tau_{xy}\right]^{\mathrm{T}},\quad \{\Delta\varepsilon\} = \left[\Delta\varepsilon_x,\ \Delta\varepsilon_y,\ \Delta\gamma_{xy}\right]^{\mathrm{T}}$$

According to Kirchhoff assumption, the strain increment is calculated as follows:

$$\{\Delta\varepsilon\} = \{\Delta\varepsilon_0\} + z\{\Delta\psi\} \tag{25.108}$$

in which $\{\Delta\varepsilon_0\}$ is the strain increment of the middle surface, $\{\Delta\psi\}$ is the curvature increment, $\{\psi\}$ refers to formula (10.1), and z is the distance to the middle surface.

The increment of the internal forces is calculated by formula (25.109):

$$\{\Delta N\} = \int\{\Delta\sigma\}\mathrm{d}z,\quad \{\Delta M\} = \int z\{\Delta\sigma\}\mathrm{d}z \tag{25.109}$$

in which

$$\{\Delta N\} = \left[\Delta N_x,\ \Delta N_y,\ \Delta N_{xy}\right]^{\mathrm{T}},\quad \{\Delta M\} = \left[\Delta M_x,\ \Delta M_y,\ \Delta M_{xy}\right]^{\mathrm{T}}$$

By substitution of formulas (25.107) and (25.108) into formula (25.109), we get

$$\begin{Bmatrix}\Delta N\\ \Delta M\end{Bmatrix} = \sum_j \begin{bmatrix}(z_{j+1}-z_j)[D]_j & \frac{1}{2}(z_{j+1}^2-z_j^2)[D]_j\\ \frac{1}{2}(z_{j+1}^2-z_j^2)[D]_j & \frac{1}{3}(z_{j+1}^3-z_j^3)[D]_j\end{bmatrix}\begin{Bmatrix}\Delta\varepsilon_0\\ \Delta\psi\end{Bmatrix} + \sum_j \begin{bmatrix}t_{sj}[D]_{sj} & z_j t_{sj}[D]_{sj}\\ z_j t_{sj}[D]_{sj} & z_j^2 t_{sj}[D]_{sj}\end{bmatrix}\begin{Bmatrix}\Delta\varepsilon_0\\ \Delta\psi\end{Bmatrix} \tag{25.110}$$

The first item in the right side of formula (25.110) represents the contributions of concrete, where z_{j+1} and z_j are the distance of the top and the bottom of the jth layer of concrete to the middle surface, respectively. The second item represents the contribution of the reinforced bar, where t_{sj} is the equivalent thickness of the jth layer bar (per unit length) and z_j is the distance of the reinforcing bars to the middle surface.

The original stiffness method is used for calculation. According to the state of the stress and the crack at the nth step, by formula (25.110) and the normal line assumption, the stiffness matrix $[K_t]_n$ and the unbalanced force $\{\psi_n\}$ can be calculated, and by the formula below, the displacement increment of the $n+1$th step can be calculated:

$$\{\Delta\delta\}_{n+1} = [K_t]_n^{-1}(\{\Delta P\}_{n+1} - \{\psi_n\}) \tag{25.111}$$

Examples For a concrete shell, the plane size is 10.7 m × 12.4 m, the thickness is 5.1 cm, and the height of rise is 0.76 m. Under uniform load, during the test, when loading to 6.5 kPa, the shell is destroyed. There are side beams on the four sides, wherein the dimension of the beam is 34.cm × 35.6 cm.

During the calculations, assume that the four sides are fixed. Due to symmetry, taking out 1/4 for calculation, the calculation grid is shown in Figure 25.45, and dividing into nine layers in the direction of the thickness, the thickness of each layer is 0.56 cm. The results of the deflection of the midpoint are shown in Figure 25.46.

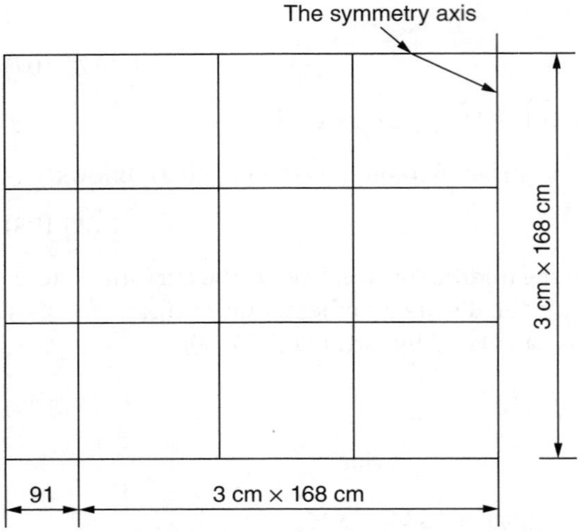

Figure 25.45 The mesh of the shell.

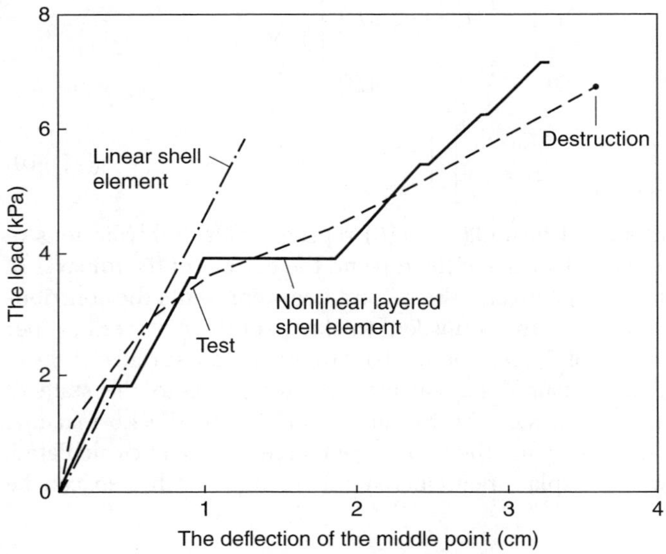

Figure 25.46 The results of the deflection of the midpoint of the concrete shell.

Bibliography

1 Ngo, D. and Scordelis, A.C. (1967) Finite element analysis of reinforced concrete beams. *ACI Journal*, **3**, 152–163.
2 ASCE Task Committee on Finite Element Analysis of Reinforced Concrete Structures (1982) *State of the Art Report on Finite Element Analysis of Reinforced Concrete*, ASCE, New York.
3 Han, D.J. and Chen, W.F. (1987) Constitutive modeling in analysis of concrete structures. *Journal of the Engineering Mechanics, ASCE*, **113** (EM4), 577–593.
4 Kupfer, H., Hilsdore, H.K. and Rush, H. (1966) Behavior of concrete under biaxial stresses. *Journal of ACI*, **66** (8), 656–666.
5 Bazant, Z.P. and Cedolin, L. (1979) Blunt crack band propagation in finite element analysis. *Journal of the Engineering Mechanics Division, ASCE*, **105** (EM2), 297–315.
6 Bresler, B. and Bertero, V.V. (1968) Behavior of reinforced concrete under repeated loads. *Journal of the Structural Divison, ASCE*, **94** (ST6), 1567–1590.
7 Nilson, A.H. (1968) Nonlinear analysis of reinforced concrete by the finite element method. *ACI Journal*, **65** (9), 757–766.
8 Houde J. (1973) Study of Force-Displacement Relationships for the Finite Element Analysis of Reinforced Concrete. In: Department of Civil Engineering and Applied Mechanics. Report No. 73-2 Montreal: Mc Gill University, 12.
9 Bathe, K.J. and Ramaswamy, S. (1979) On three-dimensional nonlinear analysis of concrete structures. *Journal of Nuclear Engineering and Design*, **52**, 385–409.
10 Argyris, J.H. *et al.* (1974) Recent developments in finite element analysis of prestressed concrete reactor vessels. *Nuclear Engineering and Design*, **28**, 42–75.
11 Zienkiewicz, O.C. *et al.* (1972) Finite element methods in the analysis of reactor vessels. *Nuclear Engineering and Design*, **20**, 507–541.
12 Hand, F.R., Pecknold, D.A. and Schnobrich, W.C. (1973) Nonlinear analysis of RC plates and shells. *Journal of the Structural Division, ASCE*, **99** (ST7), 1491–1505.
13 Saenz, L.P. (1964) Discussion of equation for the stress–strain curve of concrete by Desayi and Krishman. *Journal of the American Concrete Institute*, **61**, 1229–1235.
14 Elwi, A.A. and Murray, D.W. (1979) A 3D hypoelastic concrete constitutive relationship. *Journal of the Engineering Mechanics Division, ASCE*, **105** (EM4), 623–641.
15 Chen, W.F. (1982) *Plasticity in Reinforced Concrete*, McGraw-Hill, New York.
16 Jianjing, J. (1983) *Finite Element Techniques for Static Analysis of Structures in Reinforced Concrete*, Chalmers University of Technology, Goteborg.
17 Zienkiewicz, O.C., Valliappian, S. and King, I.P. (1968) Stress analysis of rock as a "no-tension" material. *Geotechnique*, **18**, 56–66.

26

Back Analysis of Engineering

So far, what we discussed are all positive analysis problems. For example, giving the material properties, initial conditions, and boundary and load conditions of the body, through calculation, the reaction quantities of the body are obtained, such as displacement, stress, vibration frequency, temperature, seepage, and so on.

There are three kinds of back analysis problems in engineering: which are proposed in engineering design stage, in the construction process, and after the completion of the structure. For the first kind, because of the complex situation, it is difficult to determine the natural deformation modulus, permeability coefficient, and initial ground stress of rock mass through indoor test. Generally according to some reaction quantities obtained on site, they are reckoned by back analysis.

For the second kind, in the process of excavation of tunnel or underground workshop, getting data of the displacement or stress release, through back analysis, the deformation modulus and the initial ground stress of the rock mass are obtained.

For the third kind, after completion of the concrete dam, according to the measured displacements, we may compute the elastic modulus of the dam body, rock foundation, and so on.

Generally speaking, back analysis problems are more difficult than positive analysis problems. Experience shows that the finite element method is a powerful tool for back analysis.

26.1 General Principles of Back Analysis

Engineering analysis has two characteristics:

1) The known measured reaction quantities are the comprehensive reflection of various factors.
2) The calculation amount of the back analysis is often very large.

Therefore when doing back analysis, the following principles should be paid attention to:

1) Seize the main factors and ignore the secondary factors as much as possible.
   For example, in displacement back analysis, the influence of Poisson's ratio of the object on the stress and displacement field in general is not big; the variation range of its value is very small, so usually it does not have to do inverse analysis on Poisson's ratio.

The Finite Element Method: Fundamentals and Applications in Civil, Hydraulic, Mechanical and Aeronautical Engineering, First Edition. Bofang Zhu.

Companion website: www.wiley.com/go/zhufem17

For another instance, the weight of the object can be determined accurately and its variation range is usually not big; also generally we do not have to do inverse analysis on the weight of the object.

In back analysis, the main contradiction should be seized. For example, in the underground engineering, the deformation modulus and initial stress of the rock mass are the main factors; they should be included in the back analysis, while the secondary factors, such as gravity and Poisson's ratio, can be calculated according to the known value.

2) Segment many factors for the back analysis.
   For example, the displacements of concrete dam are affected by the water level, temperature, and time comprehensively. So we should use the displacement difference of the two times, when the water level changes sharply and the temperature field is very close, for back analysis as far as possible, in order to reduce the number of unknown variables.
3) Analyze the conditions carefully and reduce the unknowns.
   For example, when the ground layer is deep, the vertical component of the ground stress, which generally can be calculated according to the law of proportionality to the depth, does not have to be included in the back analysis. For another instance, when a very deep gulley cuts through the nearby underground caverns, it can be judged that the horizontal component of the ground stress from the side of the gulley is not too big and does not have to be included in the analysis.
   For a long tunnel, using the measured values near the working face in the process of excavation to do the back analysis, if the measured values are mainly from the local reactions before and after the excavation, and if the rock of this section is single, then we can do the back analysis according to the homogeneous mass.
   In the back analysis of viscoelastic mass, sometimes we can separate the instantaneous elastic effect and time effect in the measured values; in this case, we should do the back analysis for the instantaneous elastic displacement first, in order to solve the instantaneous elastic modulus, the initial ground stress, and so on. Then taking these quantities as the known values, do the back analysis of the effect of time. In this way, the amount of calculation of the back analysis can be reduced a lot.
4) Choose the constitutive model carefully.
   In back analysis, whether the constitutive model is appropriate is very important. Although the raw data taken from the actual observation data, if the constitutive model used is not appropriate, the results cannot reflect the real situation and lose the practical value. The constitutive model on one hand should conform to the actual situation basically and on the other hand should be easy to calculate.

26.2 Back Analysis of the Seepage Field

In the exploration stage of the hydropower projects, there are usually a certain number of drill holes. Using the drilling pressure water test, the permeability coefficient of rock mass of a small region near the drill holes can be calculated.

If using the observed values of underground water level in many drilling holes, by the back analysis of seepage field, the permeability coefficient of a large region of rock mass can be calculated.

26.2.1 The Optimization Method

Take the waterhead function as

$$\varphi = z + \frac{p}{\gamma} \tag{26.1}$$

where p is the groundwater pressure, γ is the density of water, and z is the height above a base level, and the z axis points upward.

After the finite element discretization, the basic equation of the stable seepage is

$$[H]\{\phi\} = \{F\} \tag{26.2}$$

where $[H]$ is the seepage matrix and $[F]$ is the right item; for details, refer to Chapter 14.

According to the actual hydrogeologic conditions, rock mass can be divided into several regions. In region j, if the seepage is isotropic, a permeability coefficient is k_j, and if it is anisotropic, there will be more than one permeability coefficient. In order to reduce the number of unknown variables, when the conditions permit, we should try to assume the fixed ratio of the coefficients of permeability of different directions in the region and keep only one or two permeability coefficients as variables. Assuming that there are a total of m permeability coefficients to be solved,

$$x_j = k_j \quad (j = 1,2, \dots, m) \tag{26.3}$$

Our task is to solve $\{x\}^T = \lfloor x_1 \; x_2 \; \dots \; x_m \rfloor$ and make the sum of squares of the weighted errors to take minimum value, namely,

$$S = \sum_{i=1}^{n} w_i(\phi_i - \phi_i^*)^2 \rightarrow \text{minimum} \tag{26.4}$$

in which w_i is the weight coefficient, ϕ_i and ϕ_i^* are the calculated value and measured value of waterhead function of point i, respectively, and n is the number of the measured values.

According to the actual hydrogeologic conditions, the scope of the permeability coefficient can be given, so as to get a set of constraints

$$\underline{x}_j \le x_j \le \overline{x}_j (j = 1,2, \dots, m) \tag{26.5}$$

Formulas (26.4) and (26.5) form a constrained minimization problem and can be solved by the mathematical programming method.

Taking $x_j(j = 1 \sim m)$ as coordinate axes, an m-dimensional space is formed. $\{x\}$ is a point of the space. According to the experience, if solving formulas (26.4) and (26.5) by the nonlinear programming method, dozens or even hundreds of iterations are needed. If building matrices $[H]$ and $\{F\}$ for each $\{x\}$, solving $\{\phi\}$ by formula (26.2), the amount of calculation is very large, and this is the difficulty of seepage back analysis.

26.2.2 The Approximate Reanalysis

In order to simplify the calculation, the author suggests that giving an initial value $\{x^0\}$, and then expanding ϕ_i into Taylor series in the neighborhood of $\{x^0\}$, ignoring the higher-order term, we get

$$\phi_i = \phi_i^0 + \sum_{j=1}^{m} \left(\frac{\partial \phi_i}{\partial x_j}\right)_0 (x_j - x_j^0) = c_{i0} + \sum_{j=1}^{m} c_{ij} x_j \tag{26.6}$$

in which

$$c_{i0} = \phi_i^0 - \sum_{j=1}^{m} c_{ij}x_j^0, \quad c_{ij} = \left(\frac{\partial \phi_i}{\partial x_j}\right)_0 = \left.\frac{\partial \phi_i}{\partial x_j}\right|_{x_j=x_j^0} \tag{26.7}$$

By substitution of formula (26.6) into formula (26.4), we get

$$S = \sum_{i=1}^{n} w_i \left(c_{i0} + \sum_{j=1}^{m} c_{ij}x_j - \phi_i^* \right)^2 = \text{minimum} \tag{26.8}$$

By the condition of S taking minimum, there is

$$\frac{\partial S}{\partial x_j} = 2\sum_{i=1}^{n} w_i c_{ij} \left(c_{i0} + \sum_{j=1}^{m} c_{ij}x_j - \phi_i^* \right) = 0$$

thus linear equations of m order are obtained as follows:

$$[a]\{x\} = \{b\} \tag{26.9}$$

where the elements of the matrix $[a]$ and the array $[b]$ can be calculated as follows:

$$a_j = \sum_{i=1}^{n} w_i c_{ir} c_{ij} \quad b_j = \sum_{i=1}^{n} w_i c_{ij}(c_{i0} - \phi_i^*) \tag{26.10}$$

By inversing formula (26.9), we get

$$\{x\} = [a]^{-1}\{b\} \tag{26.11}$$

Considering the constraint condition of formula (26.5), namely,

$$\left.\begin{aligned} x_j &= \overline{x}_j \quad (x_j > \overline{x}_j) \\ x_j &= \underline{x}_j \quad (x_j < \underline{x}_j) \end{aligned}\right\} \tag{26.12}$$

So we get the first approximation $\{x^1\}$, which ignore the higher-order term in the Taylor expansion, the solution is approximate. Substitute $\{x^1\}$ for $\{x^0\}$ next and then do Taylor expansion to ϕ_i in the neighborhood of $\{x^1\}$. Repeat the above calculation until the first $\{x^n\}$ and the second $\{x^{n+1}\}$ are fully close to each other. Iteration usually takes about ten times.

The derivation $\partial\phi_i/\partial x_j$ is calculated below. The simplest method is to calculate with the difference method as follows:

$$\frac{\partial \phi_i}{\partial x_j} = \frac{\phi_i(x_j + \Delta x_j) - \phi_i(x_j)}{\Delta x_j} \tag{26.13}$$

More efficient algorithm is given below. By computing partial derivatives with respect to x_j on both sides of formula (26.2), we get

$$\frac{\partial \phi_i}{\partial x_j} = [H]^{-1}\left(\frac{\partial\{F\}}{\partial x_j} - \frac{\partial[H]}{\partial x_j}\{\phi\}\right) \tag{26.14}$$

$\partial[F]/\partial x_j$ and $\partial[H]/\partial x_j$ is calculated by difference method as follows:

$$\left.\begin{aligned} \frac{\partial\{F\}}{\partial x_j} &= \frac{1}{\Delta x_j}(\{F(x_j + \Delta x_j)\} - \{F(x_j)\}) \\ \frac{\partial[H]}{\partial x_j} &= \frac{1}{\Delta x_j}(\{H(x_j + \Delta x_j)\} - \{H(x_j)\}) \end{aligned}\right\} \tag{26.15}$$

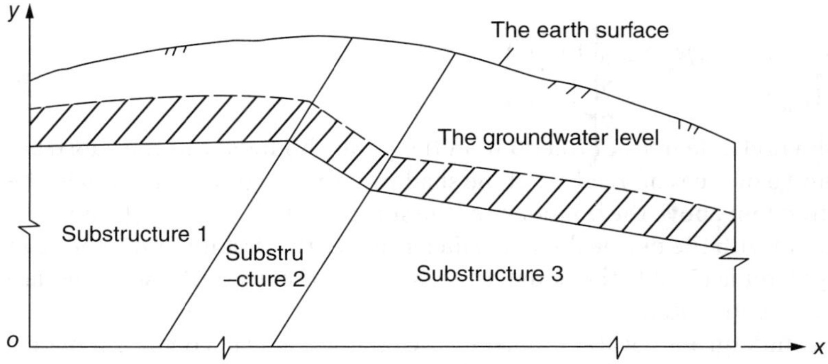

Figure 26.1 Vertical profiles of mountain massif.

By substitution of above formula into formula (26.14), $\partial \phi_i / \partial x_j$ can be calculated, because of the decomposed $[H]$; it requires only simple back substitution. So for each iteration, it only needs to solve formula (26.2) once.

26.2.3 Application of the Substructure Method

Using the substructure method, calculation can be simplified further.

When the permeability coefficient changes, the underground water level will fluctuate, we may estimate a rough range of fluctuation. According to the actual underground water level of the known n points, the lower boundary of underground water level fluctuation can be determined, considering it as the upper limit of classification of substructure of each rock mass. Under the upper limit, degrees of freedom of internal nodes of each substructure can be condensed off entirely, as shown in Figure 26.1.

Considering the area of the same permeability coefficient as a substructure, in the j substructure, the basic equation is

$$x_j \begin{bmatrix} \lambda_{jj} & \lambda_{jb} \\ \lambda_{bj} & \lambda_{bb} \end{bmatrix} \begin{Bmatrix} \phi_j \\ \phi_b \end{Bmatrix} = \begin{Bmatrix} F_j \\ F_b \end{Bmatrix} \tag{26.16}$$

where ϕ_j and ϕ_b are the nodal degrees of freedom for the inside and boundary of a substructure, respectively.

By the first formula of formula (26.16), we get

$$\{\phi_j\} = \frac{1}{x_j}[\lambda_{jj}]^{-1}\{F_j\} - [\lambda_{jj}]^{-1}[\lambda_{jb}]\{\phi_b\}$$

By substitution of the formula above into the second formula of formula (26.16), we get

$$x_j[\lambda_b^*]\{\phi_b\} = \{F_b^*\} \tag{26.17}$$

in which

$$\left.\begin{aligned} [\lambda_b^*] &= [\lambda_{bb}] - [\lambda_{bj}][\lambda_{jj}]^{-1}[\lambda_{jb}] \\ \{F_b^*\} &= \{F_b\} - [\lambda_{bj}][\lambda_{jj}]^{-1}\{F_j\} \end{aligned}\right\} \tag{26.18}$$

Assembling the contribution of each substructure, we get

$$\begin{bmatrix} H_{bb} & H_{ba} \\ H_{ab}^T & H_{aa} \end{bmatrix} \begin{Bmatrix} \phi_b \\ \phi_a \end{Bmatrix} = \begin{Bmatrix} F_b \\ F_a \end{Bmatrix} \tag{26.19}$$

where

$$\left.\begin{aligned} [H_{bb}] &= \sum x_j[\lambda_b^*], \quad [H_{ba}] = x_a[\lambda_{ba}] \\ [H_{aa}] &= x_a[\lambda_{aa}], \qquad \{F_b\} = \sum\{F_b^*\} \end{aligned}\right\} \tag{26.20}$$

in which $\{\phi_b\}$ is the nodal degrees of freedom on the public boundary of the substructures. $\{\phi_a\}$ is the nodal degrees of freedom of the shadow part in Figure 26.1, namely, the nodal degrees of freedom above the boundary of the substructures, $[\lambda_b^*]$, $[\lambda_{ba}]$, and $[\lambda_{aa}]$ all have nothing to do with the permeability coefficient; after the first iteration, they can be reused later. By formula (26.20), the nodal degrees of freedom inside the substructure have been condensed completely.

According to the underground water level, only the relative value of coefficient of permeability of various regions can be determined; in order to determine the absolute value of coefficient of permeability, a condition would be added.

One way is that, in the area where the information of water pressure test in borehole is complete, using the data of water pressure test, the absolute value of the permeability coefficient of a region can be determined. Another method is that, using the measured flow of the groundwater dew point, a discharge condition can be added.

26.3 Elastic Displacement Back Analysis of Homogeneous Body and Proportional Deformation Heterogeneous Body

Considering the foundation and building as elastic body, by back analysis according to a set of measured displacement either during or after the construction, the elastic modulus of the foundation and building or the initial ground stress can be calculated.

In this section, we consider two kinds of circumstances: One is homogenate elastic body, and its elastic modulus is E, as shown in Figure 26.2(a). The other is proportional deformation heterogeneous elastic body, and it is divided into several regions, with different elastic modulus, and the elastic modulus are keep a known fixed proportion, as shown in Figure 26.2(b), namely,

$$E_1 = k_1E, \quad E_2 = k_2E, \quad E_3 = k_3E \tag{a}$$

The coefficients k_1, k_2, k_3 are known, and only one unknown elastic modulus E needs to be calculated. As for the method of inversing several different elastic modulus at the same time, we will introduce it in the next section.

Due to the back analysis of the proportional deformation, inhomogeneous elastic body is easier than that of the inhomogeneous elastic body that does not keep proportional

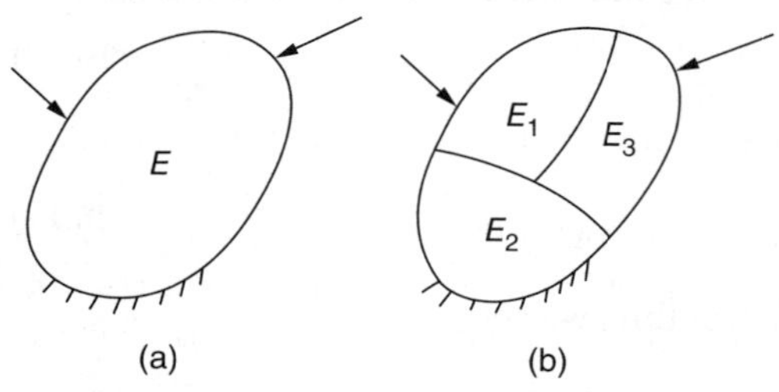

Figure 26.2 Elastic body. (a) Homogeneous. (b) Heterogeneous.

deformation; therefore, as long as the scale factor k_i can be determined approximately, the method of this section can be used. First calculating according to the approximate value of k_i, getting the value of E, and then modifying k_i appropriately, the error is further reduced.

26.3.1 Inversion of Elastic Modulus

First assuming an initial elastic modulus E_0, calculate the displacement of each measuring point under known load by the finite element method:

$$\delta_{i0} = c_i \quad (i = 1,2, \ldots, n)$$

Assuming that the actual elastic modulus of the object is

$$E = E_0/\eta \tag{26.21}$$

then the displacement of each measuring point is

$$\delta_i = c_i\eta \quad (i = 1,2, \ldots, n)$$

Assuming that the measured displacement of each measuring point is δ_i^*, the sum of error square is

$$S = \sum_{i=1}^{n} (\delta_i - \delta_i^*)^2 = \sum_{i=1}^{n} (c_i\eta - \delta_i^*)^2 \tag{b}$$

Choosing η to get minimum S, by

$$\frac{\partial S}{\partial \eta} = 2\sum_{i=1}^{n}(c_i\eta - \delta_i^*)c_i = 0$$

we get

$$\eta = \frac{\sum c_i\delta_i^*}{\sum c_i^2} \tag{26.22}$$

26.3.2 The Inversion of Initial Ground Stress in a Small Area

In the design of underground structure, we are interested in the initial ground stress in a small area, as shown in Figure 26.3. One is the vertical component q of ground stress, and the other is the horizontal component p perpendicular to the cave walls. Setting some measuring points near the working plane, by the displacement difference before and after the excavation of ΔL, p, q, and elastic modulus E of rock mass can be inversed.

By finite element method, the displacement difference of the measuring point is as follows:

$$\delta_i = \frac{a_i p}{E} + \frac{b_i q}{E} \tag{26.23}$$

where coefficients a_i, b_i are the known values calculated by the finite element method, in the range of ΔL, making $\sigma_x = -1$ and $\sigma_y = -1$, respectively, on cave walls. Dividing into three different conditions, we may do analysis as follows.

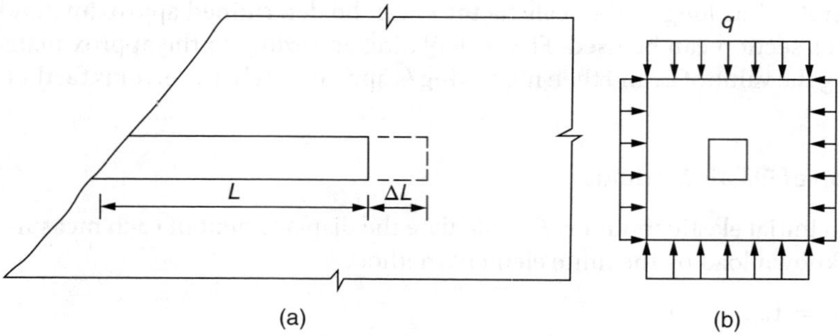

Figure 26.3 Back analysis of the initial ground stress in a small area. (a) Excavation. (b) The component of ground stress.

26.3.2.1 *E* Is Known: Inversion of *q* and *p*

Let

$$x_1 = p, \quad x_2 = q$$

By formula (26.23), we know

$$\delta_i = \alpha_i x_1 + \beta_i x_2 \qquad \text{(c)}$$
$$\alpha_i = a_i/E, \beta_i = b_i/E$$

The sum of error square is

$$S = \sum_{i=1}^{n} (\delta_i - \delta_i^*)^2 = \sum_{i=1}^{n} (\alpha_i x_1 + \beta_i x_2 - \delta_i^*)^2 \qquad \text{(d)}$$

Choosing x_1 and x_2 to minimize S, by

$$\left.\begin{aligned} \frac{\partial S}{\partial x_1} &= 2\sum(\alpha_i x_1 + \beta_i x_2 - \delta_i^*)\alpha_i = 0 \\ \frac{\partial S}{\partial x_2} &= 2\sum(\alpha_i x_1 + \beta_i x_2 - \delta_i^*)\beta_i = 0 \end{aligned}\right\} \qquad \text{(e)}$$

we get

$$\left.\begin{aligned} Q_{11}x_1 + Q_{12}x_2 &= P_1 \\ Q_{21}x_1 + Q_{22}x_2 &= P_2 \end{aligned}\right\} \qquad (26.24)$$

where

$$Q_{11} = \sum \alpha_i^2, \quad Q_{12} = Q_{21} = \sum \alpha_i \beta_i, \quad Q_{22} = \sum \beta_i^2$$
$$P_1 = \sum \alpha_i \delta_i^*, \quad P_2 = \sum \beta_i \delta_i^*$$

solving x_1 and x_2 by formula (26.24), and p and q are obtained.

26.3.2.2 *q* Is Known: Inversion of *p* and *E*

The vertical component *q* of ground stress often can be calculated according to the depth and density of rock mass, if inversing *p* and *E*, making

$$x_1 = p, \quad x_2 = 1/E$$

By formula (26.23), we know

$$\delta_i = a_i x_1 x_2 + b_i q x_2 \tag{f}$$

According to the method often used in regression analysis, let

$$y_1 = x_1 x_2 \quad y_2 = x_2 \tag{g}$$

We get

$$\delta_i = \alpha_i y_1 + \beta_i y_2 \tag{h}$$

where

$$\alpha_i = a_i \quad \beta_i = b_i q$$

Formulas (h) and (c) are in the same form; calculating y_1 and y_2 by the least-square method and substituting them into formula (g), $x_1 = p$ and $x_2 = 1/E$ can be calculated.

26.3.2.3 Inversion of *E*, *p*, and *q* at the Same Time

Let

$$x_1 = p, \quad x_2 = q, \quad x_3 = 1/E \tag{i}$$

By formula (26.23), we have

$$\delta_i = a_i x_1 x_3 + b_i x_2 x_3 \tag{26.25}$$

Now the question is to solve x_1, x_2, and x_3. Let

$$s = \sum_{i=1}^{n} (a_i x_1 x_3 + b_i x_2 x_3 - \delta_i^*)^2 = \text{minimum} \tag{26.26}$$

This is an unconstrained extreme problem, and the nonlinear programming method is available; for details see literature [7]. Because the variables are not many, Powell method or simplex method can be used. Usually the amount of calculation is little.

The following algorithm also can be used. Taking the initial value $\{x^0\}^T = \left[x_1^0 \ x_2^0 \ x_3^0\right]$, expanding δ_i into Taylor series in the neighborhood of $\{x^0\}$, and neglecting higher-order terms, we get

$$\begin{aligned} \delta_i &= \delta_i^0 + \sum_{j=1}^{3} \left(\frac{\partial \delta_i}{\partial x_j}\right)_0 (x_j - x_j^0) \\ &= c_{i0} + c_{i1} x_1 + c_{i2} x_2 + c_{i3} x_3 \end{aligned} \tag{26.27}$$

where

$$c_{i0} = \delta_i^0 - \sum_{j=1}^{3} c_{ij} x_j^0 \quad c_{i1} = \left(\frac{\partial \delta_i}{\partial x_1}\right)_0 = a_i x_3^0$$

$$c_{i2} = \left(\frac{\partial \delta_i}{\partial x_2}\right)_0 = b_i x_3^0 \quad c_{i3} = \left(\frac{\partial \delta_i}{\partial x_3}\right)_0 = a_i x_1^0 + b_i x_2^0$$

By substitution of formula (26.27) into formula (26.26), we get

$$s = \sum_{i=1}^{n} (c_{i0} + c_{i1} x_1 + c_{i2} x_2 + c_{i3} x_3 - \delta_i^*)^2 \to \text{minimum} \tag{26.28}$$

By the extreme conditions, there are

$$\frac{\partial S}{\partial x_1} = 0 \quad \frac{\partial S}{\partial x_2} = 0 \quad \frac{\partial S}{\partial x_3} = 0 \tag{26.29}$$

Solving the above linear equations, the first approximation $\{x^1\} = \lfloor x_1^1 \ x_2^1 \ x_3^1 \rfloor$ is obtained. Making Taylor series expanding in the neighborhood of $\{x^1\}$ and repeating the above calculation, the second approximation $\{x^2\}$ is obtained, doing so iteratively, until two calculation results are fully close.

26.3.3 Back Analysis of Initial Ground Stress in a Wide Range

By the theory of stress release, the initial ground stress can be measured on-site, but the cost is expensive and it is hard to conduct a large number of measuring points. In the construction of large underground structure, the wide range of initial ground stress field often is necessary. The method of combining the measurement and calculation can be used, namely, measuring the ground stresses at some points and then doing back analysis, calculating a wide range of ground stress field. Two methods have been proposed in China that can be used for plane problems as well as spatial problems.

26.3.3.1 The First Method

This is the method put forward by Professor Guo Huaizhi. Figure 26.4 shows a plane problem, taking the straight border of calculation section in the valley or the symmetric position of mountain watershed; in order to simplify the boundary condition, let the shear stress of the boundary be equal to zero. Presupposing the elastic modulus of the calculation domain by finite element method, the gravity stress field and tectonic stress field are calculated, respectively. When calculating the tectonic stress field, assuming that one side is fixed and the other side has a uniform horizontal displacement, namely,

$$u = u_0 \tag{j}$$

Then assuming that the actual stress field is a linear combination of the two kinds of stress field, namely,

$$\sigma_i = b_1\sigma_{i1} + b_2\sigma_{i2} \tag{26.30}$$

where σ_i is the ground stress of i point, σ_{i1} is the gravity stress of i point, σ_{i2} is the tectonic stress of i point, and b_1 and b_2 are the undetermined coefficients.

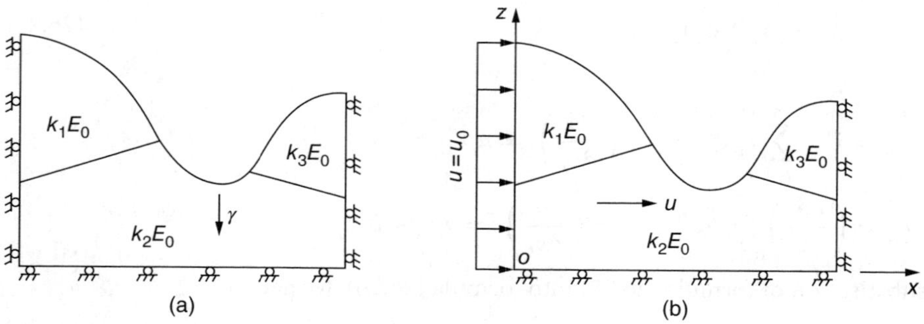

Figure 26.4 Back analysis of ground stress field in a big region. (a) The gravity stress field. (b) The tectonic stress field.

Assuming that the measured ground stress component of the i point is σ_i^*, the sum of error squares is

$$S = \sum_{i=1}^{n} (b_1\sigma_{i1} + b_2\sigma_{i2} - \sigma_i^*)^2 \tag{26.31}$$

Coefficients b_1 and b_2 can be obtained by the condition of $\partial S/\partial b_1 = 0$ and $\partial S/\partial b_2 = 0$ to get minimum S; substituting them into formula (26.30), the ground stress of any point can be obtained.

26.3.3.2 The Second Method

This is the author's method, which is improved on the basis of Guo's method.

The calculation precision of gravity stress and tectonic stress is completely different. The bulk density of rock mass can be determined more accurately, and the scope of its change is not large. If the rock is homogeneous, gravity stress is not affected by elastic modulus. For proportional deformation heterogeneous elastic body, as long as the ratio k_i keeps invariant, E_0 also has no effect on gravity stress. Of course, strictly speaking, the change of elastic modulus ratio k_i has a certain influence on gravity stress distribution, but its influence is much smaller than that on tectonic stress, as long as the change range is not big, the effect on gravity stress is not big. Experience shows that vertical component of ground stress calculated according to formula $q = \gamma H$ is very close to the measured value. Based on the actual topography geological conditions, the precision of gravity stress calculated by the finite element method is more accurate than that calculated by the formula above. Therefore, the precision of gravity stress calculated by the finite element method can meet the requirements of engineering design; we do not have to inverse it during the design phase.

The situation of the tectonic stress is different. First of all, the boundary displacement values u_0 of formula (j) are completely unknown; tectonic stresses are more likely affected by elastic modulus. According to the principle of stress release when measuring ground stress, we use elastic modulus of rock block obtained by the indoor test; while solving the tectonic stress field, we use the elastic modulus of rock mass, which reflects the influence of joints and fissures in rock mass. The two kinds of elastic modulus are different. For homogeneous elastic body and the proportional deformation inhomogeneous elastic body, the influence of u_0 and E_0 on tectonic stress is reflected by the product E_0u_0.

According to the analysis above, because the accuracy of gravity stress field calculated by the finite element method is far higher than that of tectonic stress field, it is not fair to put them in the same level for regression analysis. The main contradiction of back analysis of ground stress is the inversion of tectonic stress, and the precision of the gravity stress field calculated by the finite element method can basically meet the requirements of engineering design, so we may consider it as known.

For plane problems, as shown in Figure 26.4, assuming that rock mass is homogeneous elastic body or proportional deformation inhomogeneous elastic body, according to the actual condition of rock mass, given coefficients k_i and elastic modulus E_0, gravity stress σ_{i1} and tectonic stress σ_{i2} are calculated, respectively, by finite element method.

Keeping k_i invariant, when the elastic modulus changes from E_0 to E, gravity stress is constant, and when the boundary displacement changes from u_0 to u', ground stress can be calculated as follows:

$$\sigma_i = \sigma_{i1} + \frac{Eu'}{E_0 u_0}\sigma_{i2}$$

where σ_i is ground stress, σ_{i1} is gravity stress, and σ_{i2} is tectonic stress corresponding to the elastic modulus E_0 and boundary displacement u_0 (both refer to a particular stress component).

Let

$$x = \frac{Eu'}{E_0 u_0} \tag{26.32}$$

We get

$$\sigma_i = \sigma_{i1} + x\sigma_{i2} \tag{26.33}$$

The first item on the right side of the formula above is gravity stress, and the second item is tectonic stress. Assuming that the measured ground stress is σ_i^*, the sum of error square is

$$S = \sum_{i=1}^{n} (\sigma_{i1} + x\sigma_{i2} - \sigma_i^*)^2 \tag{26.34}$$

By formula $\partial S/\partial x = 0$, we get

$$x = \frac{\sum \sigma_{i2}(\sigma_i^* - \sigma_i^1)}{\sum \sigma_{i2}^2} \tag{26.35}$$

By substitution of x into formula (26.33), the ground stress field is obtained.

For the space problems, we can cut a cylinder whose horizontal sectional plane is rectangular as an analysis object. When calculating tectonic stress, we assume that $u = u_0$ on the plane $x = 0$, $v = v_0$ on the plane $y = 0$, and $u = 0$ and $v = 0$ on the other two relative planes. So ground stress can be calculated as follows:

$$\sigma_i = \sigma_{i1} + x\sigma_{i2} + y\sigma_{i3} \tag{26.36}$$

$$x = \frac{Eu'}{E_0 u_0}, \quad y = \frac{Ev'}{E_0 v_0}$$

where σ_{i1} is gravity stress, σ_{i2} is tectonic stress caused by $u = u_0$, and σ_{i3} is tectonic stress caused by $v = v_0$. x and y can be calculated by least-squares method.

26.4 Back Analysis of Material Parameters of Heterogeneous Elastic Body

This section will clarify how to do back analysis of material parameters of heterogeneous elastic body.

26.4.1 The Difference State and Its Inverse Analysis

For a building, from start, construction, and completion to operation, the change process of its stress and displacement is very complicated. In order to simplify the problem, it is better to analyze the difference status. Taking the back analysis of a

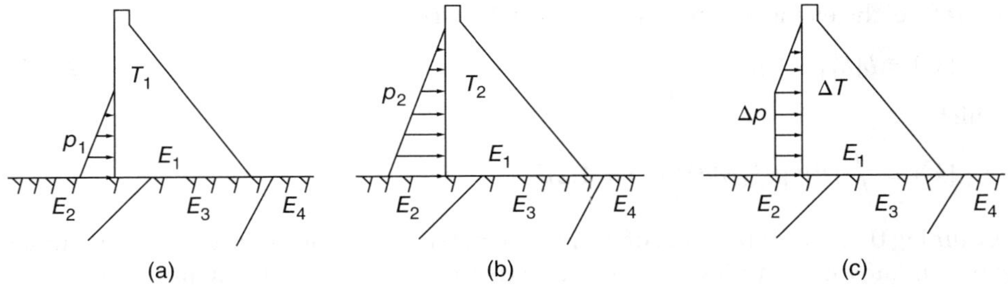

Figure 26.5 The difference state: (c) = (b) − (a).

dam, for example, as shown in Figure 26.5, it would be best to take out the difference (c) = (b) − (a) of two states where water level rapidly change for analysis. At this time, external load is the water pressure difference $\Delta p = p_2 - p_1$, the temperature field is the difference between the two temperature field $\Delta T = T_2(x, y, z) - T_1(x, y, z)$, and the displacement difference of the ith point is $\delta_i = \delta_{i2} - \delta_{i1}$.

The benefits of taking difference status for back analysis are as follows: first the complex nonlinear initial displacement during the construction and operation can be removed, second the influence of gravity can be removed, and third we can choose the difference of two states where water level rapidly change but temperature field change little, namely, Δp is larger and ΔT is smaller, in order to highlight the effect of Δp.

Generally speaking, in order to get a larger Δp, time difference won't be too small; during the calculation it is best not to make $\Delta T = 0$, because the temperature field calculation is easy and has high precision. ΔT can be calculated according to the actual condition.

The basic equation of doing back analysis for difference status is

$$[K]\{\delta\} = \{P\} \tag{26.37}$$

where the load array is

$$\{P\} = \{P_1\} + \{P_2\} = \int_S [N]^{\mathrm{T}}\{\Delta P\}\mathrm{d}S + \int_V [B]^{\mathrm{T}}[D]\{\varepsilon^1\}\mathrm{d}V \tag{26.38}$$

where $\{P_1\}$ is the nodal load caused by external load, $\{P_2\}$ is the nodal load caused by the temperature variation, and $\{\varepsilon^I\}$ is the free temperature deformation.

26.4.2 The Stiffness Matrix Decomposition Method

Assuming that b is the bulk deformation modulus and G is the shear deformation modulus, their relationship with E and μ are as follows:

$$E = \frac{9bG}{3b + G} \quad \mu = \frac{3b - 2G}{6b + 2G} \tag{26.39}$$

Elastic matrix can be written as follows:

$$[D] = b[d_1] + G[d_2] \tag{26.40}$$

For example, for the plane strain problem, there is

$$[d_1] = \begin{bmatrix} 1 & 1 & 0 \\ 1 & 1 & 0 \\ 0 & 0 & 0 \end{bmatrix} \quad [d_2] = \begin{bmatrix} 4/3 & -2/3 & 0 \\ -2/3 & 4/3 & 0 \\ 0 & 0 & 1 \end{bmatrix} \tag{a}$$

Therefore, the element stiffness matrix can be written as

$$[k^e] = b[k_1^e] + G[k_2^e] \tag{26.41}$$

in which

$$[k_i^e] = \int [B]^{\mathrm{T}}[d_i][B]\mathrm{d}V \quad (i = 1,2)$$

Assuming that there are a total of m kinds of materials, as isotropic body, each material has two unknown quantities, E_j and μ_j, and there are $2m$ unknown quantities. Let

$$\{x\}^{\mathrm{T}} = \begin{bmatrix} x_1 & x_2 & \dots & x_{2m} \end{bmatrix} = \begin{bmatrix} E_1, & \mu_1, & E_2, & \mu_2, & \dots \end{bmatrix} \tag{b}$$

The global stiffness matrix is

$$[K] = \sum_{j=1}^{2m} x_j[K_j] \tag{26.42}$$

Assuming that there are a total of n observed values of displacement components, the measuring points coincide with the nodes of the finite element mesh and the global equilibrium equation can be partitioned as follows:

$$\begin{bmatrix} K_{11} & K_{12} \\ K_{21} & K_{22} \end{bmatrix} \begin{Bmatrix} \delta_1^* \\ \delta_2 \end{Bmatrix} = \begin{Bmatrix} P_1 \\ P_2 \end{Bmatrix} \tag{26.43}$$

where $\{\delta_i^*\}$ is the known displacement measurements.

Working out $\{\delta_2\}$ by the second formula of formula (26.43) and then substituting it into the first formula, we get

$$\left([K_{11}] - [Q][K_{21}]\right)\{\delta_i^*\} = \{P_1\} - [Q]\{P_2\} \tag{26.44}$$

in which

$$[Q] = [K_{12}][K_{22}^{-1}] \tag{c}$$

Considering formula (26.42), formula (26.44) can be rewritten as

$$\sum_{j=1}^{2m} x_j[C_j] = \{P_1\} - [Q]\{P_2\} \tag{d}$$

where

$$[C_j] = [K_{11,j}] - [Q][K_{21,j}] \tag{e}$$

Formula (d) can be rewritten as

$$[C]\{x\} = \{P_1\} - [Q]\{P_2\} \tag{26.45}$$

in which

$$[C] = \begin{bmatrix} C_1 & C_2 & \dots & C_{2m} \end{bmatrix}$$

Assuming that the number of observations is over that of unknown quantities, solving by the least-square method, the sum of error square is

$$S = ([C]\{x\} - \{P_1\} + [Q]\{P_2\})^{\mathrm{T}}([C]\{x\} - \{P_1\} + [Q]\{P_2\}) \tag{f}$$

By the extreme condition $\partial S/\partial\{x\} = 0$, we get

$$[C]^{\mathrm{T}}[C]\{x\} = [C]^{\mathrm{T}}(\{P_1\} - [Q]\{P_2\}) \tag{26.46}$$

By formulas (c) and (e), we know both $[Q]$ and $[C]$ contain $[x]$, so formula (26.46) is essentially an iterative formula of a nonlinear equation. It should be written in the following form:

$$[C^k]^{\mathrm{T}}[C^k]\{x^{k+1}\} = [C^k]^{\mathrm{T}}(\{P_1\} - [Q^k]\{P_2\}) \tag{26.47}$$

where k is the number of iterations.

$[Q^k]$ and $[C^k]$ are computed by Equations (c), (26.45) and $\{x^k\}$, and then $\{x^{k+1}\}$ is calculated by formula (26.47).

26.4.3 Optimization Method

Giving material parameter $\{x\}$, such as formula (b), the global stiffness matrix $[K]$ is established; inversing formula (26.37) we get $\{\delta\}$. The sum of error squares S with the measured values can be calculated. According to the actual situation, the upper limit $\overline{x}_j$ and lower limit $\underline{x}_j$ of x_j can be given. The question is how to solve $\{x\}$, making the sum of error squares

$$S = \sum_{i=1}^{n} (\delta_i - \delta_i^*)^2 = \text{minimum} \tag{26.48}$$

subjected the constraint conditions

$$\underline{x}_j \leq x_j \leq \overline{x}_j \tag{26.49}$$

This is a constrained extreme problem, and we can use nonlinear programming method to solve it. When variables are not many, we can use complex method and the procedure is simpler. If not considering constraint condition formula (26.49), simplex method can be used to solve it.

In literature [10], an example was calculated using the two methods above. The finite element mesh is shown in Figure 26.6(a), and the calculation results are shown in Figure 26.6(b). Since each simplex may also be iterative several times, the number of the total iterations may be more than that of the simplices.

26.4.4 Improvement of the Optimization Method

Expanding δ_i into Taylor form in the neighborhood of $\{x^0\}$ and ignoring the higher-order term, we get

$$\delta_i = \delta_i^0 + \sum_{j=1}^{2m} \left(\frac{\partial \delta_i}{\partial x_j}\right)_0 (x_j - x_j^0) = c_i^0 + \sum_{j=1}^{2m} c_{tj} x_j \tag{26.50}$$

where

$$c_i^0 = \delta_i^0 - \sum_{j=1}^{2m} c_{ij} x_j^0, \quad c_{ij} = \left(\frac{\partial \delta_i}{\partial x_j}\right)_0 = \left.\frac{\partial \delta_i}{\partial x_j}\right|_{x_j = x_j^0}$$

By substitution of formula (26.50) into formula (26.48), formula (26.9) is obtained; by inversing we get $\{x\}$ as formula (26.11), and then considering constraint condition as

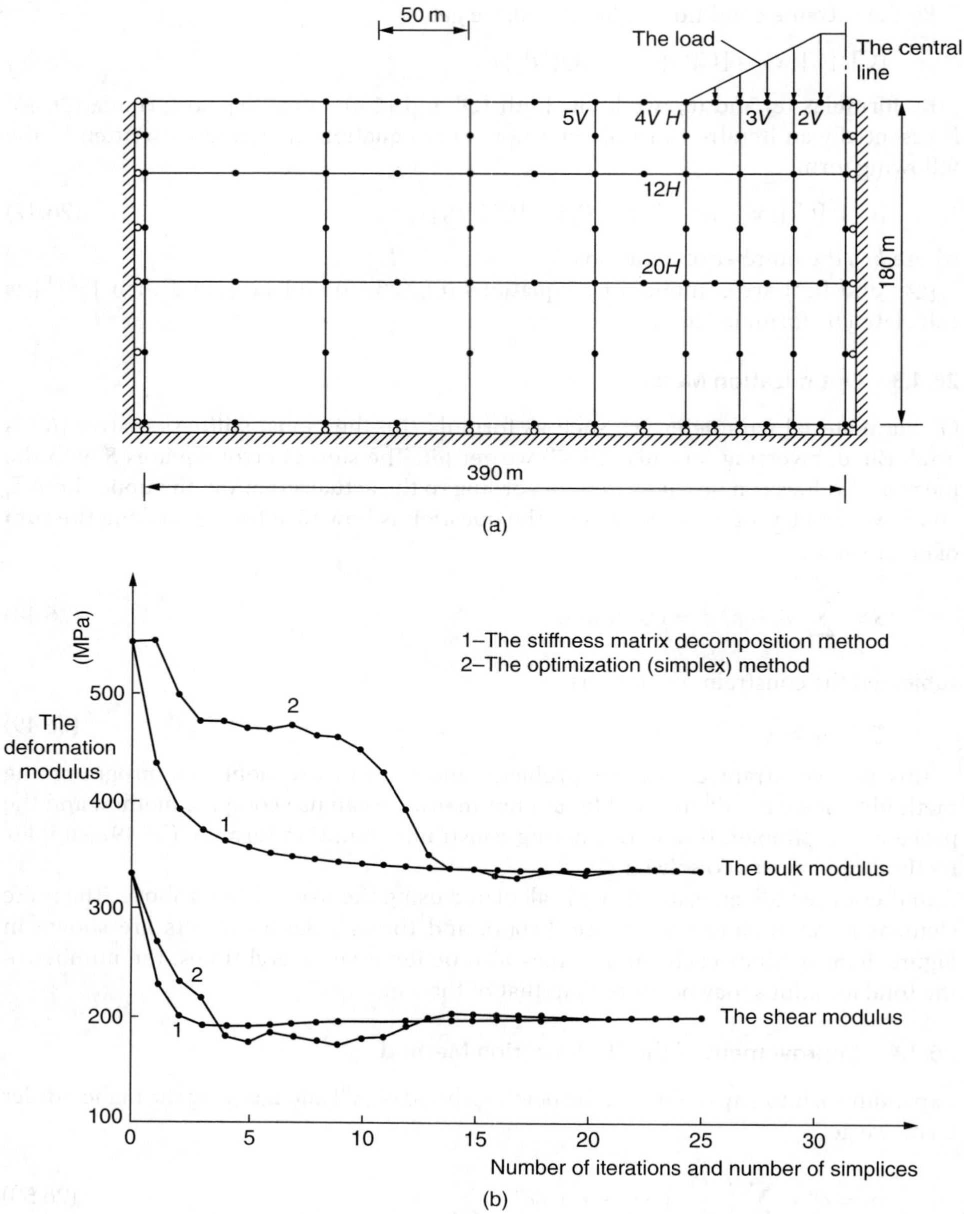

Figure 26.6 Back analysis of the bulk modulus and the shear modulus of bedrock. (a) The calculation mesh. (b) The calculation results.

formula (26.12), the first approximate solution $\{x^1\}$ is obtained. So calculate step by step, until two calculation results before and after are fully close.

Because $\{\delta\} = [K]^{-1}\{P\}$, in order to improve the precision of the Taylor expansion formula (26.50) to speed up the convergence speed, we should take the reciprocal of elastic modulus as inversion variable, namely, replacing formula (b), taking

$$\{x\}^{\mathrm{T}} = \begin{bmatrix} x_1 & x_2 & \dots & x_{2m} \end{bmatrix} = \begin{bmatrix} 1/E_1 & \mu_1 & 1/E_2 & \mu_2 & \dots \end{bmatrix} \tag{g}$$

We illustrate how to calculate the displacement sensitivity $\partial\delta_i/\partial x_j$ as follows. The simplest way is to use the difference method to calculate, similar to formula (26.13). A more effective way is to compute partial derivative of both sides of formula (26.37) on x_j, we get

$$\frac{\partial\{\delta\}}{\partial x_j} = [K]^{-1}\left(\frac{\partial\{P\}}{\partial x_j} - \frac{\partial\{K\}}{\partial x_j}\{\delta\}\right) \tag{26.51}$$

By formula (18.43), $[D] = E[A]^{-1}$, the element stiffness matrix is

$$[k^{\mathrm{e}}] = \int [B]^{\mathrm{T}}[D][B]\mathrm{d}V = E\int [B]^{\mathrm{T}}[A]^{-1}[B]\mathrm{d}V = E[\lambda^{\mathrm{e}}] \tag{h}$$

where $[A]^{-1}$ is a matrix related to μ, referring to formulas (18.13)–(18.15).

$$\frac{\partial\{k^{\mathrm{e}}\}}{\partial x_j} = E\int [B]^{T}\frac{[\Delta A]^{-1}}{\Delta x_j}\mathrm{d}V \quad (x_j = \mu_j) \tag{i}$$

$$\frac{\partial\{k^{\mathrm{e}}\}}{\partial x_j} = -\frac{1}{x_j^2}[\lambda]^{\mathrm{e}} = -\frac{1}{x_j}[k]^{\mathrm{e}} \quad \left(x_j = \frac{1}{E_j}\right) \tag{j}$$

Then we calculate $\partial\{P\}/\partial x_j$. By formula (26.38), we know $\{P_1\}$ has nothing to do with E and μ, so we only need to calculate $\partial\{P_2\}/\partial x_j$. By substitution of $[D] = E[A]^{-1}$ into it, we get

$$\{P_2^{\mathrm{e}}\} = E\int [B]^{\mathrm{T}}[A]^{-1}\{\varepsilon^I\}\mathrm{d}V \tag{k}$$

in which $[A]$ is a matrix associated with μ; thus we know

$$\frac{\partial\{P_2^{\mathrm{e}}\}}{\partial x_j} = E\int [B]^{\mathrm{T}}\frac{[A(x_j+\Delta x_j)]^{-1} - [A(x_j)]^{-1}}{\Delta x_j}\{\varepsilon^I\}\mathrm{d}V \quad (x_j = \mu_j) \tag{l}$$

$$\frac{\partial\{P_2^{\mathrm{e}}\}}{\partial x_j} = -\frac{1}{x_j^2}\int [B]^{\mathrm{T}}[A]^{-1}[B]\mathrm{d}V = -\frac{1}{x_j}\{P_2^{\mathrm{e}}\} \quad \left(x_j = \frac{1}{E_j}\right) \tag{m}$$

By substitution of formulas (h) to (k) into formula (26.51), we get $\partial\delta_i/\partial x_j$.

In order to improve the calculation efficiency further, the substructure method can be adopted. Through static condensation, all the nodal degrees of freedom inside the substructure can be eliminated, only those on the public side are left.

Literature [8] made back analysis of the elastic modulus for Baishan arch dam by improved optimization method.

Inversion objects are four different elastic moduli of the dam body, the right bank, the riverbed, and the left bank. Stress analysis uses the three-dimensional isoparametric element, a total of 732 nodes. After twelve iterations, satisfactory results are obtained.

26.5 Back Analysis of Interaction of Elastic Structure with the Surrounding Medium

For structures as the tunnel lining and steel sheet pile shown in Figure 26.7, there is a group of observed data of the deformation of structure; it is required to solve the interaction forces between structure and the surrounding media.

26.5.1 The Displacement Method

Dividing the structure into several elements, at node i, on the contact surface of the structure and the surrounding media, the forces in the normal and tangential direction per unit area are q_{ni} and $q_{ti,}$ respectively. Let

$$\{q_i\} = \begin{Bmatrix} q_{ni} \\ q_{ti} \end{Bmatrix} \tag{a}$$

On the contact surface, the interaction force of any point in the normal and tangential direction per unit area can be represented as

$$\{\overline{p}\} = \begin{Bmatrix} \overline{p}_n \\ \overline{p}_t \end{Bmatrix} = [\overline{N}]\{q^e\} \tag{b}$$

$$\{q^e\} = \begin{bmatrix} q_i & q_j & q_m & \cdots \end{bmatrix}^T$$

where $[\overline{N}]$ is the shape function and i, j, and m are the nodal numbers of the element.

In the global coordinate system, the interaction force is

$$\{p\} = \begin{Bmatrix} p_x \\ p_y \end{Bmatrix} = [T]\{\overline{p}\} = [T][\overline{N}]\{q^e\} \tag{c}$$

in which $[T]$ is the coordinate transformation matrix.

For the lining or the sheet pile, the interacting force $\{p\}$ is external load, and by formula (3.15), the nodal load of the element caused by $\{p\}$ is

$$\{p^e\} = \int [N]^T[p]\mathrm{d}s = [\alpha^e][q^e] \tag{26.52}$$

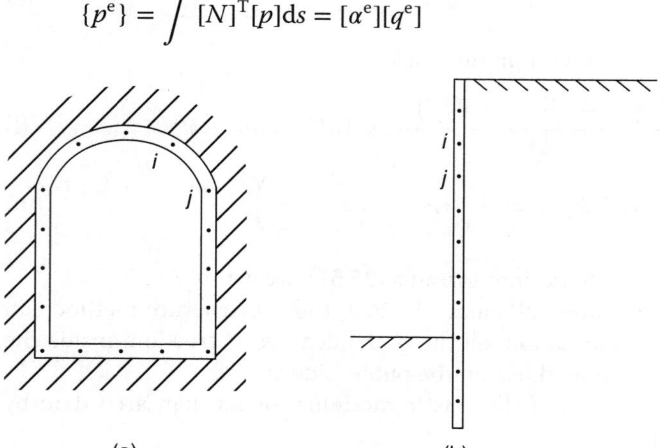

Figure 26.7 The structure interacting with the surrounding media. (a) The tunnel lining. (b) The steel sheet pile.

where

$$[\alpha^{e}] = \int [N]^{T}[T][\overline{N}]dS \tag{d}$$

Assembling the stiffness matrix and the nodal load of each element, we get the global equilibrium equation of the structure as follows:

$$\begin{bmatrix} K_{11} & K_{12} & K_{13} \\ K_{21} & K_{22} & K_{23} \\ K_{31} & K_{32} & K_{33} \end{bmatrix} \begin{Bmatrix} \delta_1^0 \\ \delta_2^* \\ \delta_3 \end{Bmatrix} = \begin{bmatrix} \alpha_1 \\ \alpha_2 \\ \alpha_3 \end{bmatrix} \{q\} \tag{26.53}$$

in which $\{\delta_1^0\}$ is the known constraint displacements for preventing the rigid movements of the structure. To keep generality, we make $\{\delta_1^0\} = 0$, $\{\delta_2^*\}$ is the known measured displacement, and $\{\delta_3\}$ is the displacement of other nodes.

Expanding formula (26.53), we get

$$[K_{11}]\{\delta_1^0\} + [K_{12}]\{\delta_2^*\} + [K_{13}]\{\delta_3\} = [\alpha_1]\{q\} \tag{e}$$

$$[K_{21}]\{\delta_1^0\} + [K_{22}]\{\delta_2^*\} + [K_{23}]\{\delta_3\} = [\alpha_2]\{q\} \tag{f}$$

$$[K_{31}]\{\delta_1^0\} + [K_{32}]\{\delta_2^*\} + [K_{33}]\{\delta_3\} = [\alpha_3]\{q\} \tag{g}$$

Let $\{\delta_1^0\} = 0$, by formula (g), we get

$$\{\delta_3\} = [K_{33}]^{-1}([\alpha_3]\{q\} - [K_{32}]\{\delta_2^*\}) \tag{h}$$

By substitution of formula (h) into formula (f), we get

$$[C]\{q\} = \{\delta_2^*\} \tag{26.54}$$

where

$$[C] = ([K_{22}] - [K_{23}][K_{33}]^{-1}[K_{32}])^{-1}([\alpha_2] - [K_{23}][K_{33}]^{-1}[\alpha_3]) \tag{i}$$

By substitution of formula (h) into formula (e), we get

$$[R]\{q\} = \{0\} \tag{26.55}$$

where

$$[R] = [K_{12}][C] - [K_{13}][K_{33}]^{-1}([\alpha_3] - [K_{32}][C]) - [\alpha_1]$$

If besides a set of measured displacement $\{\delta_2^*\}$, the contact stress $\{\overline{p}^*\}$ of a group of contact surfaces is observed by pressure box, and by formula (b), we get

$$[L]\{q\} = \{\overline{p}^*\} \tag{26.56}$$

If the number of measured values and the number of constraint condition of structure's rigid body displacement are just equal to the number of unknown variables, by formulas (26.54)–(26.56), $\{q\}$ can be solved. Usually, the number of measured values is over than that of the unknowns, so we need to use least-square method to solve. The sum of weighted square error is

$$S = ([C]\{q\} - \{\delta_2^*\})^{T}([C]\{q\} - \{\delta_2^*\}) + \omega_1([L]\{q\} - \{\overline{P}^*\})^{T}([L]\{q\} - \{\overline{p}^*\}) \\ + \omega_2([R]\{q\})^{T}([R]\{q\})$$

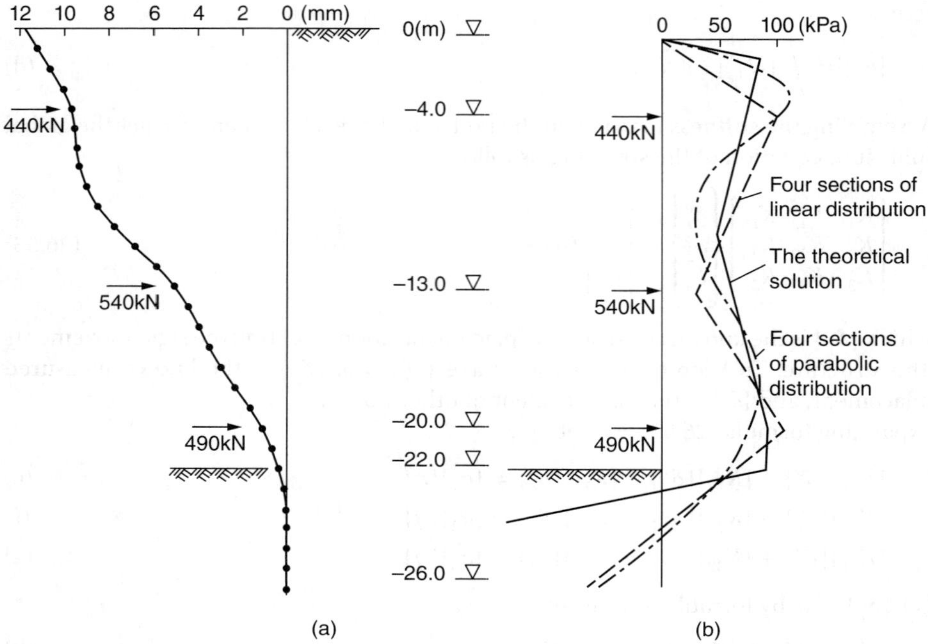

Figure 26.8 Back analysis of earth pressure. (a) The measured displacement of the steel sheet pile (dot) and the tensile stress of the bolt. (b) The earth pressure distribution of back analysis.

Because the dimensions of the deflection and stress are different, it is necessary to introduce the weight coefficient ω_1 and ω_2, by $\partial S/\partial\{q\} = 0$, we get

$$[C]^{\mathrm{T}}([C]\{q\} - \{\delta_2^*\}) + \omega_1[L^{\mathrm{T}}]([L]\{q\} - \{\overline{p}^*\}) + \omega_2[R]^{\mathrm{T}}[R]\{q\} = 0$$

Thus the obtained unknown variable is as follows:

$$\{q\} = ([C]^{\mathrm{T}}[C] + \omega_1[L]^{\mathrm{T}}[L] + \omega_2[R]^{\mathrm{T}}[R])^{-1}([C]^{\mathrm{T}}\{\delta_2^*\} + \omega_1[L]^{\mathrm{T}}\{\overline{p}^*\}) \tag{26.57}$$

The literature [10] gives an example of inverting the earth pressure by the measured deformation of steel sheet pile, which is 28 m high. There are three bolts to support. The measured displacement of the steel sheet pile and the tension in the bolt are shown in Figure 26.8(a). Assuming that the earth pressure is divided into four intervals, the elevations of the division surfaces are the same with the bolts; only considering horizontal soil pressure, two back analyses are conducted. One assumes that the earth pressure is a linear distribution for partition, while the other assumes that it is parabolic distribution in each interval. The results of the back analysis are shown in Figure 26.8(b).

In field measuring, sometimes we get the relative displacement, while, in the back analysis, we use the absolute displacement; therefore we need a transformation, as shown in Figure 26.9, namely,

$$\{\Delta\delta'\} = \begin{Bmatrix} u_j' - u_i' \\ v_j' - v_i' \end{Bmatrix} = \begin{bmatrix} -\cos\theta & -\sin\theta & \cos\theta & \sin\theta \\ \sin\theta & -\cos\theta & -\sin\theta & \cos\theta \end{bmatrix} \begin{Bmatrix} u_i \\ v_i \\ u_j \\ v_j \end{Bmatrix}$$

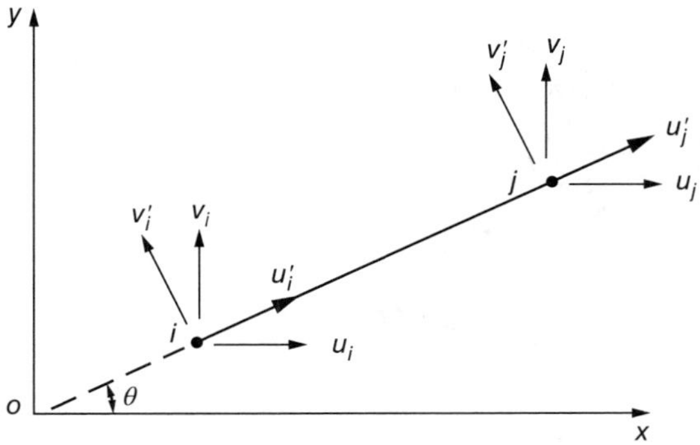

Figure 26.9 The transformation of the relative displacement and the absolute displacement.

Thus obtaining

$$\{\Delta\delta'\} = [T][\delta] \tag{j}$$

where $[T]$ is the transformation matrix.

Considering the relations above, formula (26.54) should be rewritten as follows:

$$\begin{aligned}[c']\{q\} &= [T][c]\{q\} \\ &= [T]\{\delta_2^*\} = \{\Delta\delta'^*_2\}\end{aligned}$$

26.5.2 The Hybrid Method

By formula (26.53) and formula (i), in the back analysis of the structure interacting with the surrounding medium by the displacement method above, a series of complex matrix operations such as the matrix decomposition, inversion, and so on are needed; it is rather complex. The hybrid method proposed by the authors is introduced as follows; the calculation program is much more simple.

Figure 26.10 represents a lining of underground powerhouse, taking m nodes along the axis direction, assuming that the normal stress and shear stress between the structure and the surrounding medium are a linear distribution between the nodes. At node k, the normal stress is $\sigma_n = x_{2k}$, and the shear stress is $\sigma_t = x_{2k-1}$, and they are unknown quantities to be solved with a total of $2m$ number.

Let

$$x_{j-1} = x_{2k-1}, \quad x_j = x_{2k} \ (j = 1 \sim 2m)$$

In addition, there are six unknown quantities, namely, the horizontal displacement u_0, the vertical displacement v_0, the rotation ϕ_0 of the left side of the lining, and three reaction forces of the right side of the lining: the horizontal reaction force x_{2m+1}, the vertical reaction force x_{2m+2}, and the bending moment x_{2m+3}.

The horizontal displacement u_i and the vertical displacement v_i at point i of the lining can be calculated as follows:

$$\left.\begin{aligned} u_i &= \sum_{j=1}^{2m+3} c_{ij}x_j + u_0 + \phi_0 y_i + u_{ip} \\ v_i &= \sum_{j=1}^{2m+3} d_{ij}x_j + v_0 - \phi_0 x_i + v_{ip} \end{aligned}\right\} \tag{26.58}$$

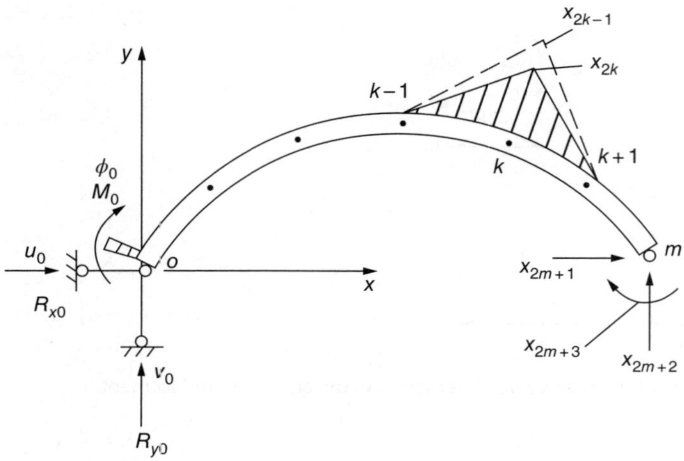

Figure 26.10 The hybrid method of back analysis for structure interacting with the surrounding media.

where c_{ij} and d_{ij} are the deformation coefficients for the lining calculated according to the cantilever structure when $x_j = 1$, x_i, and y_i are the coordinates of point i. u_{ip} and v_{ip} are the horizontal and vertical displacements caused by other known static loads (including temperature change).

Assuming that there are n_1 horizontal displacement observations u_i^* and n_2 vertical displacement observations v_i^*, the sum of weighted squares error is

$$S = \sum_{i=1}^{n_1} \omega_i (u_i - u_i^*)^2 + \sum_{i=1}^{n_2} \omega_i (v_i - v_i^*)^2 = \text{minimum} \tag{26.59}$$

By the extreme condition, we get

$$\frac{\partial S}{\partial x_j} = 0 \quad (j = 1,2, \dots , 2m) \tag{26.60}$$

Equilibrium conditions are

$$\sum X = 0, \quad \sum Y = 0, \quad \sum M = 0 \tag{26.61}$$

There are three deformation continuity conditions on the right side of the lining. Assuming that the lining is the cantilever structure, which is elastically supported on the left side, by formula (26.58), the horizontal displacement u_m and the vertical displacement v_m of the right side can be calculated and its rotation is

$$\phi_m = \sum_{j=1}^{2m+3} e_{ij} x_j + \phi_0 + \phi_{ip}$$

u_m, v_m, and ϕ_m must be equal to the displacements of the right side support, namely,

$$u_m = x_{2m+1}/k'_x, \quad v_m = x_{2m+2}/k'_y, \quad \phi_m = x_{2m+3}/k'_\phi \tag{26.62}$$

where k'_x, k'_y, and k'_ϕ are the deformation coefficients of the right support.

There are $2m + 6$ conditions of formulas (26.60), (26.61), and (26.62). $2m + 6$ unknown quantities can be solved. The relationship between the deformation and the reaction force of the left bearing is

$$R_{x0} = k_x u_0, R_{y0} = k_y v_0, M_0 = R_\phi \phi_0$$

These bearing forces should be included in the equilibrium condition formula (26.61). To the left, k_x, k_y, and k_ϕ are the deformation coefficients of the left side. The deformation coefficients k_x, k_x^1… of both sides can be calculated by the finite element method.

26.6 Nonlinear Solid Back Analysis

The stress–strain relationship of many engineering structures is nonlinear; compared with the structure of linear elastic material, the difficulty of back analysis is big. The main difficulty lies in that the amount of calculation of the stress analysis is too big.

26.6.1 The Solving Method

It is usually to use the optimization method for the back analysis of nonlinear solids. Assuming that there are n_1 measured displacement values and n_2 measured stress values (or the soil pore water pressure), back analysis problems can be attributed as follows:

Solving $\{x\}^{\mathrm{T}} = \lfloor x_1 \ x_2 \ \dots \ x_m \rfloor$ to make sum of the weighted squares error,

$$S = \sum_{i=1}^{n_1} \omega_i \left(\frac{\delta_i}{\delta_i^*} - 1 \right)^2 + \sum_{i=1}^{n_2} \omega_i \left(\frac{\sigma_i}{\sigma_i^*} - 1 \right)^2 = \text{minimum} \quad (26.63)$$

with the constraints

$$\underline{x}_j \leq x_j \leq \overline{x}_j \quad (j = 1,2, \dots, m) \quad (26.64)$$

where δ_i^* is the measured displacement component value, σ_i^* is the measured stress component value, $\underline{x}_j$ is the lower limit of x_j, $\overline{x}_j$ is the upper limit of x_j, and ω_i is the weighted coefficient.

Inversion variable x_j can be material parameters of stress–strain relation; also it can be other quantities, such as initial ground stress of rock mass.

Usually stress analysis uses the incremental method, and the basic equation is

$$\left. \begin{aligned} &[K]\{\Delta\delta\} = \{\Delta P\} \\ &\delta = \sum \Delta\delta, \sigma = \sum \Delta\sigma \end{aligned} \right\} \quad (26.65)$$

The following are methods of solution.

26.6.1.1 The Direct Search Method

It is not necessary to calculate the sensitivity of δ and σ; giving a group of $\{x\}$, by formula (26.65), δ and σ are calculated, and substituting them into formula (26.63), S can be obtained. When variables are not many, relatively simple complex method can be used. When variables are many (more than 10), the complex method has low efficiency, and we can use the penalty function method or sequential quadratic programming method.

26.6.1.2 The First-Order Taylor Expansion

In the neighborhood of $\{x^0\}$ doing the first-order Taylor expansion, the expansion of δ_i is expressed in formula (26.50); accordingly the stress is calculated by formula (26.66):

$$\sigma_i = \sigma_i^0 + \sum_{j=1}^{m} \left(\frac{\partial \sigma_i}{\partial x_j}\right)_0 (\sigma_j - \sigma_j^0) \tag{26.66}$$

The derivative of the displacement and stress normally only can be calculated by the finite difference method, namely,

$$\frac{\partial \delta_i}{\partial x_j} = \frac{\delta_i(x_j + \Delta x_j) - \delta_i(x_j)}{\Delta x_j}, \quad \frac{\partial \sigma_i}{\partial x_j} = \frac{\sigma_i(x_j + \Delta x_j) - \sigma_i(x_j)}{\Delta x_j} \tag{26.67}$$

Doing expansion once, $m+1$ times stress analysis is required.

By substitution of the first-order Taylor expansion of the deflection and stress into formula (26.63), by the extreme condition $\partial S/\partial x_j = 0$, linear equations (26.9) can be obtained. Inversing and considering formula (26.64), the first approximate solution $\{x^1\}$ can be obtained. Repeat the calculation above, until the iteration results are fully close to each other.

26.6.1.3 The Second-Order Taylor Expansion

Using second-order Taylor expansion with no cross terms to represent δ_i and σ_i, the calculation formula of δ_i is [the calculation formula of σ_i is similar to formula (26.68)]

$$\delta_i(x) = \delta_i(x_j^0) + \sum_{j=1}^{m} b_i(x_j - x_j^0) + \sum_{j=1}^{m} c_j(x_j - x_j^0)^2 \tag{26.68}$$

$$\left.\begin{aligned} b_j &= \frac{\delta_i(x_j^0 + \Delta x_j) - \delta_i(x_j^0 - \Delta x_j)}{2\Delta x_j} \\ c_j &= \frac{\delta_i(x_j^0 + \Delta x_j) + \delta_i(x_j^0 - \Delta x_j) - 2\delta_i(x_j^0)}{\Delta x_j^2} \end{aligned}\right\} \tag{26.69}$$

Doing second-order expansion once, $2m+1$ times stress analysis is required. The penalty function method, the gradient projection method, sequential quadratic programming method, and reduced gradient method may be used to solve the problem.

26.6.2 The Constitutive Model

In back analysis, it is very important to select the constitutive model. If the constitutive model is improper, calculation results cannot reflect the real situation, and actually they are worthless.

For example, literature [6] is a paper widely cited in the inverse analysis, but the author thinks that in this paper the constitutive model is wrong. Its analysis object is test results of a deep circular tunnel under axisymmetric internal water pressure. The author of this paper assumes that it is an isotropic axisymmetric problem; the initial ground stress is uniform hydrostatic pressure. He assumes that material constitutive model is ideal elastic–plastic Mohr–Coulomb criteria. According to the measured displacement of rock mass, the cohesion, internal friction angle, and the initial ground stress of rock mass are inversed. The problem is that since it is the axisymmetric problem, the shear stress of various points is zero, the destruction of the rock mass is caused by hoop tensile stress, the maximum tensile stress criterion should be used, and the inversion objects should be the tensile strength and the initial ground stress of rock mass. As it is known to all, the tensile strength of rock mass reckoned by Mohr–Coulomb theory is not real, so the inversed cohesion and internal friction angle of rock mass in literature [6] is not true.

This example tells us that though the original data of back analysis is the actual observation, if the constitutive model is not appropriate, the results of back analysis is of no significance. So in the back analysis, material constitutive model should be selected very carefully.

When choosing a constitutive model, we should consider the following factors: (1) material properties, (2) the stress state of object, (3) the approximate scope of the stress and strain, (4) the structure characteristics of the building and the foundation, (5) the indoor test result, and (6) the experience of similar projects.

Besides the linear elastic model, there are several commonly used constitutive models.

26.6.2.1 The Nonlinear Elastic Model

The more famous model is Duncan model, wherein the tangent modulus E_t is calculated by formula (24.15), which has five parameters. The tangent Poisson's ratio is calculated by formula (24.23), which has three parameters. Taking the eight parameters as inversion objects, there is a total of eight inversion variables. If Poisson's ratio uses the laboratory test results, three variables can be reduced.

For the joint element of rock mass, we can use the hyperbolic equation given by the author, which contains five parameters. If the tensile strength is zero, there are only four parameters.

26.6.2.2 Elastic–Plastic Model

For metal, concrete, rock, and soil the mises yield criterion (see formula (17.31)), five-parameter yield criterion (see formula (17.53)), maximum tensile stress criterion (Figure 17.16), and Shen Zhujiang double yield surface elastic–plastic model (see Section 24.2), respectively, are available.

26.6.2.3 Viscoelastic and Viscoelastic–Plastic Model

For viscoelastic material, usually the three-component standard viscoelastic model shown in Figure 26.11 can be used. There are three parameters. If the volume deformation and the shear deformation use different models, there are a total of six parameters. If the volume deformation is elastic, expressed with the bulk modulus K, the shear deformation is expressed with three components viscoelastic model, and there is a total of four parameters. If Poisson's ratio μ is constant (indicating that the volume deformation is proportional to the shear formation), the axial deformation uses three-component visco-elastic model, with four parameters.

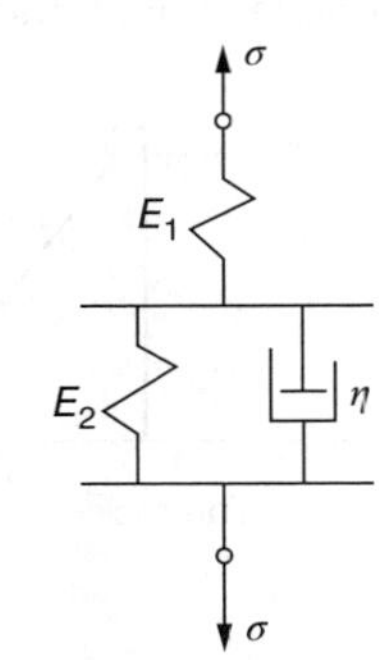

Figure 26.11 Viscoelastic body.

For viscoelastic–plastic material, the five-component model shown in Figure 26.12 can be used. Because the yield function of the plastic component may contain more than one parameter, therefore, the material parameters of the model may be more than 5. In addition, the volume deformation and shear deformation may need to use different models.

In the most general case, a combined viscoelastic–plastic model shown in Figure 18.10 can be used.

26.6.2.4 Partition Composite Model

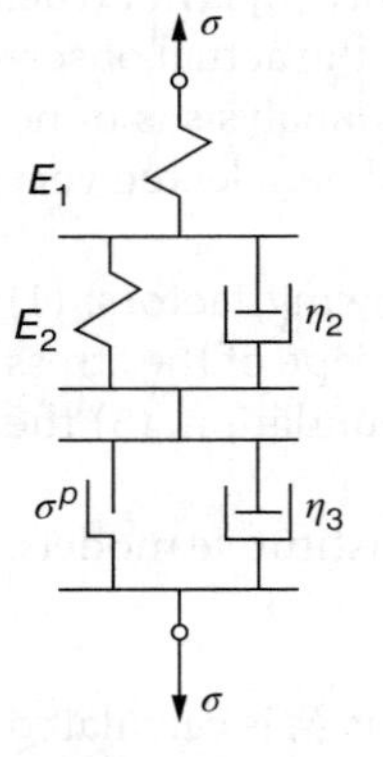

Figure 26.12 Viscoelastic–plastic body.

Practical engineering structures are made up of different materials. For different materials, different constitutive models should be adopted, respectively, for example, for concrete gravity dam as shown in Figure 26.13(a), the dam body and rock (because the stress is not big) can use the linear elastic models, and the joint uses the hyperbolic model. For concrete-faced rock-fill dam shown in Figure 26.13(b), the concrete face uses linear elastic model, and usually the elastic modulus and Poisson's ratio can be solved by indoor test and do not have to be the inversion objects. The part of rock fill uses Shen Zhujiang's double yield surfaces elastic–plastic model.

A numerical example is listed below. Literature [9] did back analysis for the dam shown in the Figure 26.14. Dam body includes rubble slope protection, and clay core and sand dam shell, and the constitutive model of three kinds of materials is the Duncan–Zhang model. The tangent elastic modulus and Poisson's ratio are expressed with formula (24.15) and formula (24.23), respectively. There are a total of eight parameters, namely, $c, \phi, R, K, n, g, h, m$. Considering that c, ϕ are more important, in order to reduce the amount of calculation, we only do back analysis for c, ϕ. There are six variables of three materials, and other parameters use test values.

Using Biot consolidation theory and the finite element method (see Section 24.6), the dam displacement, stress, and pore water pressure can be calculated. The sum of squares error is

$$S = \sum_{i=1}^{L}\left[\sum_{j=1}^{n_1}\left(\frac{u_j}{u_j^*} - 1\right)^2 + \sum_{k=1}^{n_2}\left(\frac{p_k}{p_k^*} - 1\right)^2\right] = \text{minimum} \tag{26.70}$$

where L is the number of construction stages, n_1 is the number of displacement observation points, n_2 is the number of pore pressure observation points, u_j and p_k are the calculated displacement and pore pressure, respectively, and u_j^* and p_k^* are the measured displacement and pore pressure.

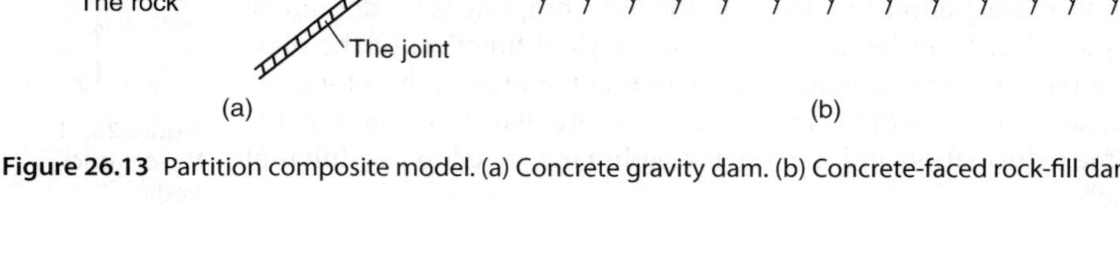

Figure 26.13 Partition composite model. (a) Concrete gravity dam. (b) Concrete-faced rock-fill dam.

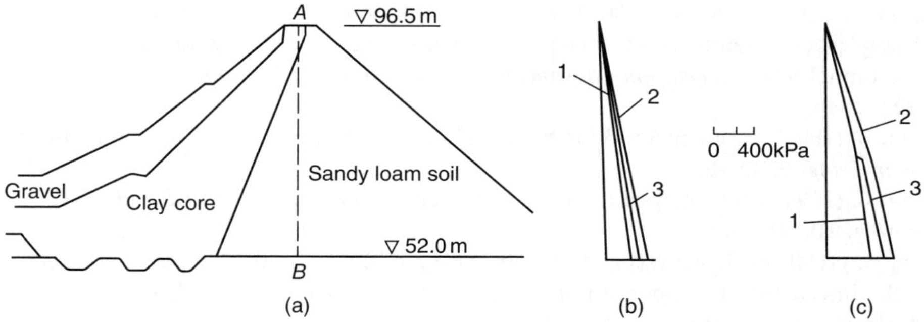

Figure 26.14 Back analysis of earth dam. (a) The profile of dam. (b) The comparison of σ_x. (c) The comparison of σ_y. 1, The measured stress; 2, the stress calculated by the indoor test parameters; 3, the stress calculated by back analysis parameters.

Table 26.1 The comparison of indoor test values and back analysis values.

Material	Sandy soil		Ball clay		Gravel	
Parameter	Test value	Back analysis	Test value	Back analysis	Test value	Back analysis
c (kPa)	15.3	16.8	42.9	45.4	10.2	11.1
ϕ (∘)	35.5	36.1	29.5	28.9	35.3	34.8

In the process of back analysis, c and ϕ of various materials are given the following restrictions on experience:

$$0 < c < a, 0 < \phi < b \tag{a}$$

On the premise of meeting the constraint condition (a), making S the minimum, this is a constrained extreme problem. Literature [9] solves by the simplex method. The comparison of back analysis results and indoor test is listed in Table 26.1.

Figure 26.14(b) and (c) represents the comparison of the measured stress and the computed stress. Curve 1 is the measured stress, curve 2 is the stress calculated according to the test parameters, and curve 3 is the stress calculated according to the parameter obtained by back analysis,

The deviation of curve 2 and curve 1 is about 10%, while the deviation of curve 3 and 1 is about 3–5%. By Table 26.1 and Figure 26.14, overall, for this example, indoor test parameters are close to the result of back analysis.

Bibliography

1 Guo, H. (1983) Method for analysing the initial stresses in rocks. *Journal of Rock and Soil*, **3**.

2 Kavanagh, K.T. and Clough, R.W. (1971) Finite element application in the characterization of elastic solids. *International Journal of Solids and Structures*, **7**, 11–23.

3 Gioda, G. and Maier, G. (1980) Direct search solution of an inverse problem in elastoplasticity identification of cohesion, friction angle and in situ stress by pressure tunnel test. *International Journal for Numerical Methods in Engineering*, **15**, 1823–1848.
4 Zhu, B. (1994) A new method for back analysis of seepage field. *Journal of Hydraulic Engineering*, **9**, 42–46.
5 Zhu, B. (1994) Back analysis of initial stresses in rocks. *Journal of Hydraulic Engineering*, **10**, 30–35.
6 Liu, H. (1993) Back analysis and its application in Tiesan Earth Dam. Proceedings of the International Symposium on Application of Computer Methods in Rock Mechanics and Engineering, Xian, China.
7 Gioda, G. and Jurina, L. (1981) Numerical identification of soil-structure interaction pressures. *International Journal for Numerical and Analytical Methods in Geomechanics*, **5**, 33–56.
8 Gioda, G. and Sakurai, S. (1987) Back analysis procedures for the interpretation of field measurements in geomechanics. *International Journal for Numerical and Analytical Methods in Geomechanics*, **11**, 555–583.
9 Zhu, B. (1984) Engineering back analysis. *Journal of Computing Technique and Computer Applications*, **2**, 14–24.
10 Zhu, B. (1995) Feedback design of hydraulic structures in construction stage. *Journal of Hydropower*, **2**, 74–82.

27

Automatic Mesh Generation, Error Estimation, and Auto-adaptation Technique

Practical engineering structures are complex, so it may be troublesome to divide the computing grid and to input the node coordinates and loads manually. Therefore, it is necessary to establish an automatic mesh generation method, which means that if you divide the practical engineering structures into several subfields and input the corresponding data, then a detailed computing grid will be created by the computer and numerical values such as nodal coordinates and loads will be calculated automatically.

In practical engineering structures, stress gradients of some parts may be so large that intensive computing grids for these parts are needed. In contrast, sparse computing grids can be used for the parts where stress gradients are small. However, it is difficult to estimate the rule of stress variation beforehand for complex practical engineering structures. This problem may be solved by two steps:

Step 1: Divide a grid based on experiences first and analyze.
Step 2: Estimate errors for the calculated result and modify the grid where errors are large or the predetermined calculation precisions are not met, and then reanalyze the stress.

The iterative method can be applied in the second step when it is necessary until a satisfactory result is obtained. This automatic modification method of computing grid is called auto-adaptation technique.

There are three ways to modify the computing grid:

1) *h mode*: Refine the grids where stress precisions are not enough by decreasing the maximum size h of the element to improve the calculation accuracy.
2) *p mode*: Increase the orders of shape functions where stress precisions are not enough to improve the calculation accuracy, while the computing grid does not change.
3) *mixed mode*: Use both h mode and p mode.

In the three methods above, h mode is convenient and applied more frequently.

The Finite Element Method: Fundamentals and Applications in Civil, Hydraulic, Mechanical and Aeronautical Engineering, First Edition. Bofang Zhu.

Companion website: www.wiley.com/go/zhufem17

27.1 Automatic Generation of Computing Grid

The methods for automatic generation of computing grid are shown below.

27.1.1 Isoparametric Transformation Method

For plane problems as shown in Figure 27.1, panel (a) is the practical region. Use 8-node isoparametric elements to transform the coordinates, that is,

$$x = \sum N_i(\xi,\eta)x_i, \quad y = \sum N_i(\xi,\eta)y_i \tag{a}$$

Divide equidistant grids (or non-equidistance grids) in the parent element shown as Figure 27.1(b), and a computing grid will be obtained in the child element as shown in Figure 27.1(c).

For complex geometrical shapes, a coarse mesh should be made to divide the area into several subfields, and then input the corresponding data. Afterward, isoparametric transformations are made for every subfield so that a refined computational grid is created automatically.

The isoparametric transformation method can also be applied for space problems. The transformation formulas of the coordinates are

$$x = \sum N_i(\xi,\eta,\zeta)x_i, \quad y = \sum N_i(\xi,\eta,\zeta)y_i, \quad z = \sum N_i(\xi,\eta,\zeta)z_i \tag{b}$$

27.1.2 Composite Function Method

Any complex geometric figure can be expressed in one element by composite function method.

Now construct the composite functions. Consider a square area $-1 \le \xi \le 1, -1 \le \eta \le 1$ to give function ϕ for its four edges, that is,

$$\phi(-1,\eta), \quad \phi(1,\eta), \quad \phi(\xi,-1), \quad \phi(\xi,1) \tag{c}$$

We need to contrive a function $\phi(\xi,\eta)$ that can form a smooth surface in the square area and take given values like Eq. (c) on the edges. Use linear interpolation functions,

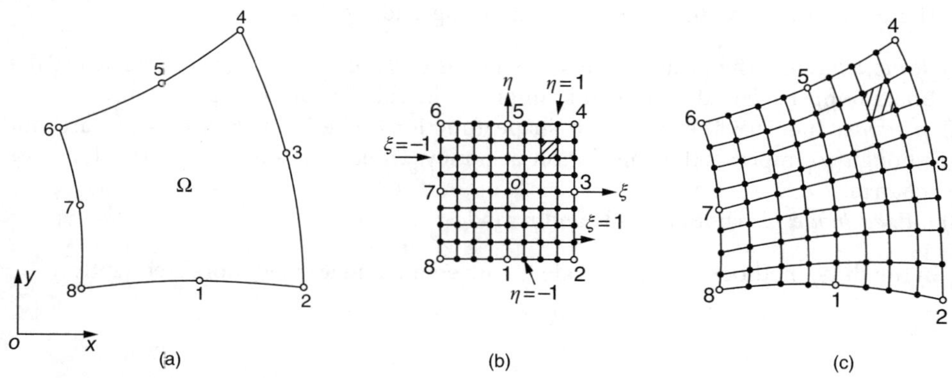

Figure 27.1 Mesh created automatically by isoparametric transformation. (a) The actual area Ω; (b) the parent element; (c) mesh created in Ω.

written as

$$N_1(\xi) = (1 - \xi)/2, \quad N_2(\xi) = (1 + \xi)/2$$
$$N_1(\eta) = (1 - \eta)/2, \quad N_2(\eta) = (1 + \eta)/2 \tag{27.1}$$

Let

$$\left.\begin{aligned} P_1 &= N_2(\eta)\phi(\xi, 1) + N_1(\eta)\phi(\xi, -1) \\ P_2 &= N_2(\xi)\phi(1, \eta) + N_1(\xi)\phi(-1, \eta) \\ P_3 &= N_2(\xi)N_2(\eta)\phi(1, 1) + N_2(\xi)N_1(\eta)\phi(1, -1) \\ &\quad + N_1(\xi)N_2(\eta)\phi(-1,1) + N_1(\xi)N_1(\eta)\phi(-1, -1) \end{aligned}\right\} \tag{27.2}$$

Then take

$$\phi = P_1 + P_2 - P_3 \tag{27.3}$$

Obviously, function $\phi(\xi, \eta)$ can form a smooth surface in the rectangular area and take given values like Eq. (c) on the edges; P_1 and P_2 are linear interpolation functions in the η and ξ direction, respectively, (Figure 27.2).

Any geometric figure can be mapped to the square area R: $-1 \le \xi \le 1$, $-1 \le \eta \le 1$. Divide grids in the square area R, and the computing mesh for actual geometric figures can be obtained by the mapping relations.

Figure 27.3 shows 1/4 of a square plate with circular hole. Figure 27.3(a) shows that the two divided subfields and computing grid can be created automatically by the isoparametric transformation method. Figure 27.3(b) shows the case that no subfield is divided, but the computing grid is created directly by composite functions.

For the methods creating finite element mesh automatically, the readers may refer to Ref. [1–11].

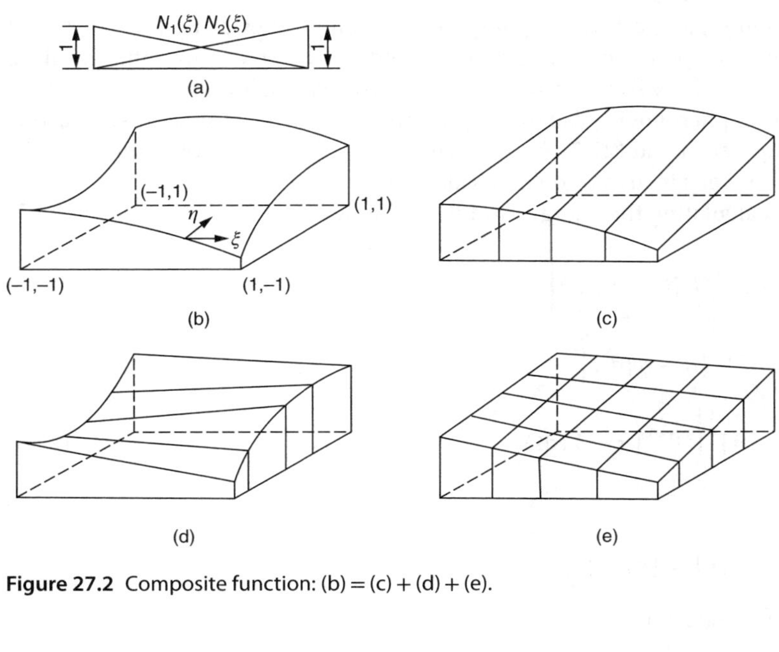

Figure 27.2 Composite function: (b) = (c) + (d) + (e).

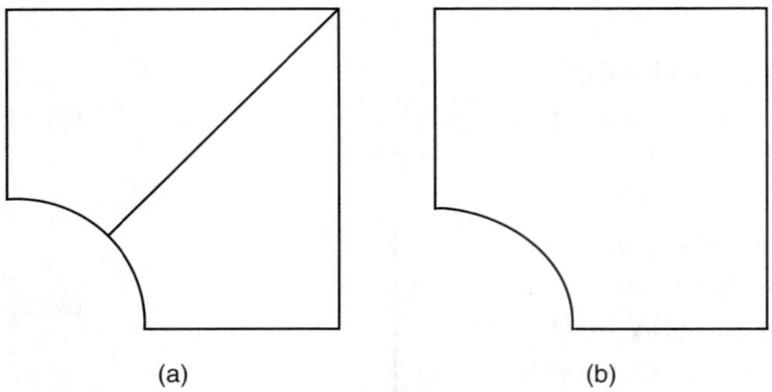

Figure 27.3 1/4 of a square plate with circular hole. (a) Two divided subfields; (b) no subfields.

27.2 Error Estimation

Solutions given by the finite element method (FEM) are approximate, which have differences with exact solutions, and these differences are errors. For example, the displacement error is

$$e_u = u - \hat{u} \tag{27.4}$$

in which u is the exact displacement and $\hat{u}$ is the approximate displacement.

The stress error is

$$e_\sigma = \sigma - \hat{\sigma} \tag{27.5}$$

in which σ is the exact stress and $\hat{\sigma}$ is the approximate stress.

There are stress concentration phenomena in practical engineering structures. In the stress concentration region, errors are going to be infinite values using Eqs (27.4) and (27.5). For example, for points carrying concentrated forces, errors are infinite values using Eqs (27.4) and (27.5). But as a whole, its approximate solutions are acceptable. Stress concentration phenomena also exist in the concave places of structures. For this reason, usually Eqs (27.4) and (27.5) are not used to calculate errors of points directly, but error moduli for certain areas would be calculated.

One method is calculating the energy module as follows:

$$\begin{aligned} e &= \left[\int_\Omega \{\varepsilon - \hat{\varepsilon}\}^{\mathrm{T}}[D]\{\varepsilon - \hat{\varepsilon}\}\mathrm{d}\Omega\right]^{1/2} \\ &= \left[\int_\Omega \{\varepsilon - \hat{\varepsilon}\}^{\mathrm{T}}\{\sigma - \hat{\sigma}\}\mathrm{d}\Omega\right]^{1/2} \\ &= \left[\int_\Omega \{\sigma - \hat{\sigma}\}^{T}[D]^{-1}\{\sigma - \hat{\sigma}\}\mathrm{d}\Omega\right]^{1/2} \end{aligned} \tag{27.6}$$

in which

$$\{\varepsilon\} = [B]\{u\},\ \{\hat{\varepsilon}\} = [B]\{\hat{u}\}$$

Ω is the region of integration.

Another method is calculating the module of displacement error $||e_u||_L$ and the module of stress error $||e_\sigma||_L$ as follows:

$$||e_u||_L = \left[\int_\Omega \{u - \hat{u}\}^\mathrm{T}\{u - \hat{u}\}\mathrm{d}\Omega\right]^{1/2} \tag{27.7}$$

$$||e_\sigma||_L = \left[\int_\Omega \{\sigma - \hat{\sigma}\}^\mathrm{T}\{\sigma - \hat{\sigma}\}\mathrm{d}\Omega\right]^{1/2} \tag{27.8}$$

We can also calculate the mean squared error. For example, the mean squared error of stress is

$$|\Delta\sigma| = \left(\frac{||e_\sigma||_L^2}{\Omega}\right)^{1/2} \tag{27.9}$$

Approximate solutions $\hat{u}$ and $\hat{\sigma}$ can be obtained by FEM calculation. Here the question is how to get the exact solutions u and σ. Some simple questions have their exact solutions already, but the vast majority of practical engineering problems do not have (it would be not necessary to do FEM calculations if they have exact solutions). Thus a practical method for estimating the exact solutions is needed.

According to the revision method in Section 8.13, the nodal stress after revision is $\tilde{\sigma}_i$, $i = 1\text{–}p$, and the stress of any point in the element is

$$\tilde{\sigma} = \sum_{i=1}^{p} N_i \tilde{\sigma}_i \tag{27.10}$$

The stress $\tilde{\sigma}$ after revision is continuous on the edges of adjacent elements. Experience shows that the stress $\tilde{\sigma}$ after revision shares a good accuracy, so that it can be used to substitute the exact solutions, and then the stress error can be estimated approximately as

$$e_\sigma = \tilde{\sigma} - \hat{\sigma} \tag{27.11}$$

Substitute the equation above into Eqs (27.6), (27.8), and (27.9), and we can get the modules of all kinds of errors.

Assume the module of error calculated by the method mentioned above is $||e||_{\text{approximate}}$ and the accurate result is $||e||_{\text{accurate}}$, and we can get the coefficient of efficiency by the following equation:

$$\theta = \frac{||e||_{\text{approximate}}}{||e||_{\text{accurate}}} \tag{27.12}$$

According to calculation experiences, the module of error calculated by the approximate calculation multiplied by a correction factor k is more close to the module of error calculated by the exact calculation. For elastic problems, the correction factors k are as follows:

Bilinear quadrilateral element	$k = 1.1$
Linear triangular element	$k = 1.3$
Biquadratic 9-node element	$k = 1.6$
Quadratic triangular element	$k = 1.4$

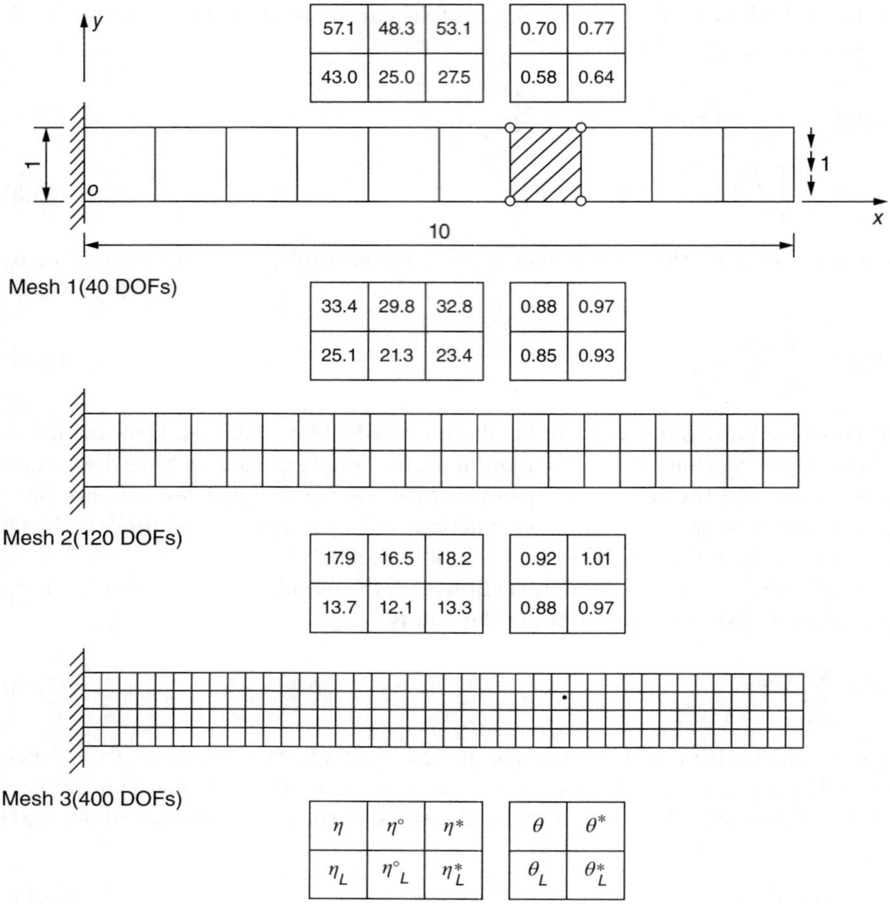

Figure 27.4 Error estimation for a cantilever (bilinear element, plane stress, $E = 10^5$ MPa, $\mu = 0.3$). η, accurate value of energy error module, %; $\eta°$, estimated value of energy error module, %; η^*, estimated value of energy error module adapted with correction factors, %; θ, effective coefficient; θ^*, effective coefficient modified with correction factors; subscript L, error calculated by Eq. (27.8).

The coefficient of efficiency for error estimation is calculated by the following equation:

$$\theta^* = \frac{k\|e\|_{\text{approximate}}}{\|e\|_{\text{accurate}}} \tag{27.13}$$

Figure 27.4 shows the errors calculated by the aforementioned approximate method and the results by the accurate method. It is clear from the figure that these two kinds of errors are very close. The coefficient of efficiency is very close to 1.00 even without adopting the correction factor.

27.3 Auto-adaptation Technique: *h* Method

After stress analysis for initial mesh, values such as the energy module, the stress error module, and the mean squared error of stress can be used to check the accuracy of stress analysis. For example, we can use the mean squared error of stress to do the check. Calculate the error ratio for every element as follows:

$$\xi_i = \frac{|\Delta\sigma|_i}{|\Delta\sigma|_\mathrm{a}} \tag{27.14}$$

in which $|\Delta\sigma|_i$ is the mean squared error of stress by Eq. (27.9) and $|\Delta\sigma|_\mathrm{a}$ is the allowable mean squared error of stress.

If the following equation was satisfied for every element, the calculation satisfied the requirements and the calculation work could stop:

$$\xi_i \leq 1 \tag{a}$$

If the following equation was tenable to some elements, the mesh should be modified to decrease the error:

$$\xi_i > 1 \tag{b}$$

There are two ways to modify the mesh as follows:

Way 1: Local grid refinement
Decrease the sizes of the elements where $\xi_i > 1$, which means local grid refinement, and recalculate the stress. Then calculate the error ratio. If there were still elements where $\xi_i > 1$, do the refinement again. Do this cycle till every element meets the requirement $\xi_i \leq 1$.

Way 2: Design a new mesh
The requirement will finally be satisfied by local grid refinement, but it may call for stress analysis repeatedly so that it is not economic in time. Designing a new mesh is a more effective way to make every element meet the requirement $\xi_i \leq 1$. Obviously, the smaller the element size, the less the calculation error. Generally we may assume that

$$|\Delta\sigma|_i \propto h_i^p \tag{27.15}$$

in which p is the order of shape function.

The size of the new mesh should not be bigger than h to meet the requirement $\xi_i \leq 1$, that is,

$$h = \xi_i^{-1/p} h_i \tag{27.16}$$

The relation between the stress error and the mesh size in the stress concentration area is

$$|\Delta\sigma|_i \propto h_i^\lambda \tag{27.17}$$

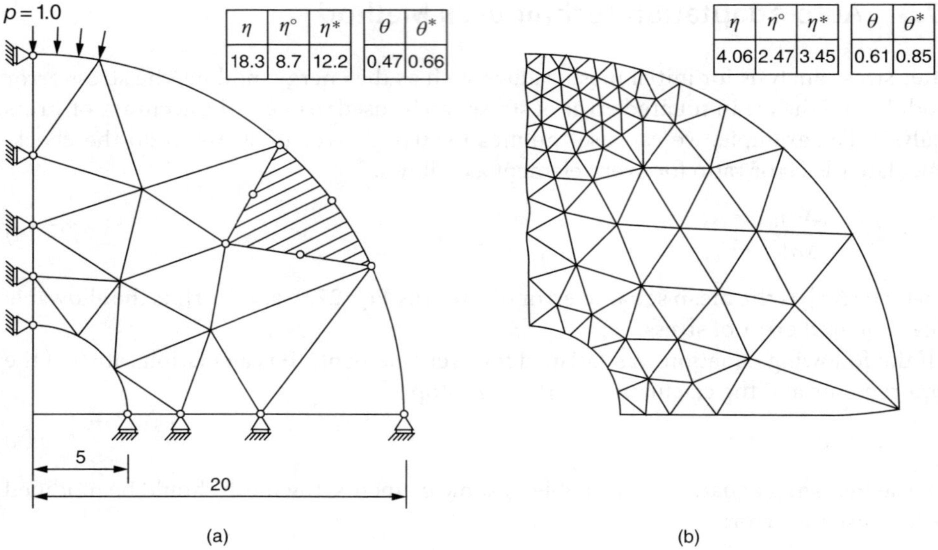

Figure 27.5 Mesh created automatically within 5% calculation error. (a) Original mesh (94 DOFs); (b) new mesh created automatically (482 DOFs).

For elastic problems, at the crack tip, $\lambda = 0.50$ and at the reentrant angle of 90°, $\lambda = 0.71$. The size of new mesh should not be bigger than h, that is,

$$h = \xi_i^{-1/\lambda} h_i \tag{27.18}$$

It is possible to design a new mesh according to Eqs (27.16) and (27.18). Experience shows that the requirement can be met when the new mesh is used to reanalyze the stress.

Figure 27.5 shows a calculation example. Quadratic triangular elements are used to analyze the thick cylinder in this case. Figure 27.5(a) is the initial calculation mesh, and Figure 27.5(b) is the auto-adapted mesh according to the requirement that calculation error is less than 5%. We can find from the figure that actual errors are less than 5%.

27.4 Auto-adaptation Technique: *p* Method

To improve the accuracy of stress analysis, we can increase the order of the shape function instead of changing the mesh where the stress accuracy is not enough after stress analysis. This method is called p method.

Let the formula for interpolation of the original elements be

$$\{u\} = [N]\{\delta\} \tag{a}$$

Assume the formula for interpolation of the new elements after increasing the order of the shape function is

$$\{u'\} = [N_1]\{\delta_1{}'\} + [N_2]\{\delta_2{}'\} \tag{b}$$

in which $\{\delta_1{}'\}$ is the nodal displacement of the original element and $\{\delta_2{}'\}$ is the newly added nodal displacement. For new elements, the equilibrium equations are

$$\begin{bmatrix} K_{11} & K_{12} \\ K_{12} & K_{22} \end{bmatrix} \begin{Bmatrix} \delta_1{}' \\ \delta_2{}' \end{Bmatrix} = \begin{Bmatrix} P_1 \\ P_2 \end{Bmatrix}$$

From the second equation above, we can get

$$\{\delta_2{}'\} = [K_{22}]^{-1}(\{P_2\} - [K_{12}]\{\delta_1{}'\})$$

As an approximation, we may assume

$$\{\delta_1{}'\} = \{\delta\} \tag{c}$$

Then we get

$$\{\delta_2{}'\} = [K_{22}]^{-1}(\{P_2\} - [K_{12}]\{\delta\}) \tag{d}$$

The nodal degree of freedom newly added can be computed one by one approximately as follows. For ith degree of freedom, we can get

$$[K_{22}]^{-1} = 1/k_{ii}$$

so that

$$\delta_i' = (\{P_2\} - [K_{12}]\{\delta_1\})/k_{ii} \tag{e}$$

Displacement differences at ith node between newly added elements and the original elements is

$$\Delta\delta_i = \delta_i' - \delta_i \tag{f}$$

The strain caused by it is

$$\varepsilon_i = [B]\Delta\delta_i \tag{g}$$

Substitute the previous equation into Eqs (27.6)–(27.9) and we can calculate the modification values for errors.

If the allowable maximum of the modification values for errors is $c_{\max}$,

$$c_i \geq \gamma c_{\max} \tag{h}$$

where c_i is the value of the ith element. Increase the order of the shape function for elements, which meet Eq. (h) and reanalyze the stress. γ is an empirical coefficient in the previous equation; its value usually is $\gamma = 0.1–0.5$ (the order of the shape function for every element should be increased in case of $\gamma = 0$). Iterative computations are necessary to reach the expected calculation accuracy.

Figure 27.6 shows an example. There are 72 degrees of freedom (DOF) for the original mesh, and the energy error module is $\eta = 25.8\%$. Applying p auto-adaptation technique,

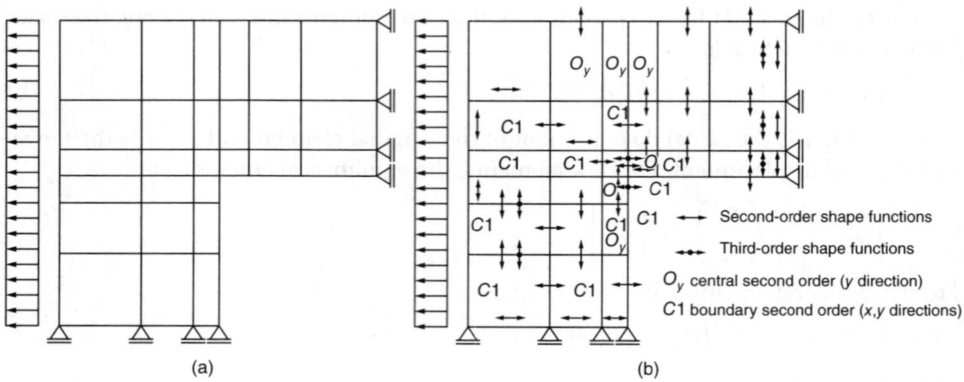

Figure 27.6 An example of p mode. (a) Mesh for the first calculation (72 DOFs, $\eta = 25.8\%$); (b) mesh for the fifth calculation (144 DOFs, $\eta = 7.2\%$).

when the iteration calculation comes to step 5, the DOF are 155 and the energy error module is $\eta = 7.2\%$.

Bibliography

1 Zienkiewicz, O.C. and Phillips, D.V. (1971) An automatic mesh generation scheme for plane and curved surface by isoparametric co-ordinates. *Int. J. Numer. Meth. Eng.*, **3**, 519–528.

2 Hermann, L.R. (1976) Laplacian isoparametric grid generation scheme. *J. Eng. Mech. Div. ASCE*, **102**, 749–756.

3 Cook, W.A. (1974) Body oriented (natural) co-ordinates for generating three dimensional meshes. *Int. J. Numer. Meth. Eng.*, **8**, 27–43.

4 Gordon, W.J. and Hall, C.A. (1973) Construction of curvilinear co-ordinates systems and applications to mesh generation. *Int. J. Numer. Meth. Eng.*, **7**, 461–477.

5 Barnhill, E.E., Birkhoff, T. and Fordon, W.J. (1973) Smooth interpolation in triangles. *J. Approx. Theory*, **8**, 114–128.

6 Cavendish, J.C. (1974) Automatic triangulation of arbitrary planar domains for the finite element method. *Int. J. Numer. Meth. Eng.*, **8**, 679–697.

7 Bukat, A. (1976) Automatic generation of triangular grid. *Int. J. Numer. Meth. Eng.*, **10**, 1329–1342.

8 Sadek, E.A. (1980) A scheme for automatic generation of triangular finite elements. *Int. J. Numer. Meth. Eng.*, **15**, 1813–1822.

9 Cavendish, J.C., Field, D.A. and Frey, W.H. (1985) An approach to automatic 3-D finite element mesh generation. *Int. J. Numer. Meth. Eng.*, **21**, 329–348.

10 Nguyen, N.P. (1982) Automatic mesh generation with tetrahedron elements. *Int. J. Numer. Meth. Eng.*, **18**, 273–280.

11 Yerry, M.A. and Shepherd, M.S. (1984) Automatic 3D mesh generation by the modified-octree technique. *Int. J. Numer. Meth. Eng.*, **20**, 1965–1990.

12 Babuska, I., Zienkiewicz, O.C., Gago, J. and Olivera, E.R. de A. (1986) *Accuracy estimates and adaptive refinements in finite element computations*, Wiley.

13 Zienkiewicz, O.C. and Zhu, J.Z. (1987) A simple error estimator and adaptive procedure for practical engineering analysis. *Int. J. Numer. Meth. Eng.*, **24**, 337–357.

14 Kelly, D.W., Gago, J.P., Zienkiewicz, O.C. and Babushka, I. (1983) A posteriori error analysis and adaptive processes in finite element method. *Int. J. Numer. Meth. Eng.*, **19**, 1593–1619, 1621–1656.

15 Babuska, I. and Miller, A. (1984) The post processing approach in finite element method. *Int. J. Numer. Meth. Eng.*, **20**, 1085–1109, 1111–1129, 2311–2324.

16 Zienkiewicz, O.C., Taylor, R.C. and Zhu, J.Z. (2008) *The Finite Element Method its Basis and Fundamentals*, Elsevier, New York.

28

Matrix

Matrix has been widely used in finite element method in order that many complex formulas can be expressed in a quite compact way. This chapter introduces matrix to engineers who engage in practical engineering and gives the definition of matrix as well as algebraic manipulation rules. Generally, precise deduction and proofs are not given.

28.1 Definition of Matrix

Put $m \times n$ a_{ij} $(i = 1,2,\ldots,m; j = 1,2,\ldots,n)$ in a table with m rows and n columns:

$$[A] = \begin{bmatrix} a_{11} & a_{12} & \cdots & a_{1j} & \cdots & a_{1n} \\ a_{21} & a_{22} & \cdots & a_{2j} & \cdots & a_{2n} \\ \cdots & \cdots & \cdots & \cdots & \cdots & \cdots \\ a_{i1} & a_{i2} & \cdots & a_{ij} & \cdots & a_{in} \\ \vdots & \vdots & \vdots & \vdots & \vdots & \vdots \\ a_{m1} & a_{m2} & \cdots & a_{mj} & \cdots & a_{mn} \end{bmatrix} \tag{28.1}$$

This table is called $m \times n$ matrix. a_{ij} is the element of matrix $[A]$, which can be a real number, a complex numbers or a function of variables. Generally, i and j denote the sequence number of row and column where a_{ij} locates in the matrix $[A]$, respectively. For example, the coefficients of linear equations

$$3x_1 - 2x_2 + 2x_3 = 7$$
$$x_1 + 2x_2 = -1$$
$$x_2 - 2x_3 = -3$$

can be noted as

$$[A] = \begin{bmatrix} 3 & -2 & 2 \\ 1 & 2 & 0 \\ 0 & 1 & -2 \end{bmatrix}$$

This is a 3×3 matrix. It should be noted that $[A]$ has completely different meaning from $|A|$, although they have similar notation. Determinant $|A|$ represents a number whose value can be calculated, whereas matrix $[A]$ is a table that is arrayed by $m \times n$ numbers and cannot be calculated as a determinant.

The Finite Element Method: Fundamentals and Applications in Civil, Hydraulic, Mechanical and Aeronautical Engineering, First Edition. Bofang Zhu.

Companion website: www.wiley.com/go/zhufem17

28.2 Principal Types of Matrix

28.2.1 Square Matrix

If a matrix has the same number of rows as the number of columns (i.e., $m=n$), it is called an nth-order square matrix. For example,

$$[A] = \begin{bmatrix} 1 & 2 \\ 4 & 3 \end{bmatrix}$$

is a second-order square matrix.

28.2.2 Row Matrix

The matrix which contains only one row (i.e., $m=1$) is called row matrix. For example,

$$[A] = \begin{bmatrix} a_{11} & a_{11} & \cdots & a_{1n} \end{bmatrix}$$

The row matrix represents an n dimensional vector that is composed of n elements.

28.2.3 Column Matrix

The matrix that contains only one column (i.e., $n=1$) is called column matrix. For example,

$$= \begin{Bmatrix} a_{11} \\ a_{21} \\ \vdots \\ a_{m1} \end{Bmatrix}$$

The column matrix represents an m dimensional vector that is composed of m elements. The three stress components in the plane stress field can be expressed as

$$\{\sigma\} = \begin{Bmatrix} \sigma_x \\ \sigma_y \\ \tau_{xy} \end{Bmatrix}$$

where σ_x is the normal stress in the x direction, σ_y is the normal stress in the y direction, and τ_{xy} is the shear stress.

28.2.4 Scalar

A 1×1 matrix only contains one single element. In matrix algebra, a single number is called scalar so as to be different from matrix.

28.2.5 Triangular Matrix

The line from the top left, through $a_{11}, a_{22}, \ldots, a_{nn}$, to the right bottom of $n \times n$ matrix $[A]$, is called the main diagonal of matrix $[A]$. A square matrix, of which the elements above or below the main diagonal are all zero, is called triangular matrix. For example,

$$[A_1] = \begin{bmatrix} a_{11} & 0 & 0 & \cdots & 0 \\ a_{21} & a_{22} & 0 & \cdots & 0 \\ a_{31} & a_{32} & a_{33} & \cdots & 0 \\ \vdots & \vdots & \vdots & \vdots & \vdots \\ a_{n,1} & a_{n,2} & a_{n,3} & \cdots & a_{n,n} \end{bmatrix}$$

$$[A_2] = \begin{bmatrix} a_{11} & a_{12} & a_{13} & \cdots & a_{1,n} \\ 0 & a_{22} & a_{23} & \cdots & a_{2,n} \\ 0 & 0 & a_{33} & \cdots & a_{3,n} \\ \vdots & \vdots & \vdots & \vdots & \vdots \\ 0 & 0 & 0 & \cdots & a_{n,n} \end{bmatrix}$$

where $[A_1]$ is called lower triangular matrix and $[A_2]$ is called upper triangular matrix.

28.2.6 Diagonal Matrix

A square matrix, of which all the elements equal zero except for the ones at the main diagonal, is called diagonal matrix. For example,

$$[A] = \begin{bmatrix} a_{11} & & & & \\ & a_{12} & & \mathbf{0} & \\ & & \ddots & & \\ & \mathbf{0} & & \ddots & \\ & & & & a_{nn} \end{bmatrix}$$

In the diagonal matrix and triangular matrix, a bold zero can represent all the zero elements above or below the main diagonal.

28.2.7 Unit Matrix

A square matrix, of which all the elements equal zero except that the ones at the main diagonal equal 1, is called unit matrix and is expressed as $[I]$. For example,

$$[I] = \begin{bmatrix} 1 & 0 & 0 \\ 0 & 1 & 0 \\ 0 & 0 & 1 \end{bmatrix}$$

is a third-order unit matrix.

28.2.8 Zero Matrix

An $m \times n$ matrix, of which all the elements equal zero, is called $m \times n$ zero matrix.

The zero matrix can be expressed as $[\mathbf{0}]$. For example,

$$[0] = \begin{bmatrix} 0 & 0 & 0 \\ 0 & 0 & 0 \end{bmatrix}$$

is a 2×3 zero matrix.

28.2.9 Transpose Matrix

The rows and columns of $m \times n$ matrix $[A]$ are reversed in sequence to get a new $n \times m$ matrix, which is called transpose matrix of $[A]$, noted as $[A]^T$. For example, suppose

$$[A] = \begin{bmatrix} a_{11} & a_{12} & \cdots & a_{1n} \\ a_{21} & a_{22} & \cdots & a_{2n} \\ \vdots & \vdots & \vdots & \vdots \\ a_{m1} & a_{m2} & \cdots & a_{mn} \end{bmatrix}, \quad \text{then } [A]^T = \begin{bmatrix} a_{11} & a_{21} & \cdots & a_{m1} \\ a_{12} & a_{22} & \cdots & a_{m2} \\ \vdots & \vdots & \vdots & \vdots \\ a_{1n} & a_{2n} & \cdots & a_{mn} \end{bmatrix} \tag{28.2}$$

For another example, $[A] = \begin{bmatrix} 2 & 1 & 3 \end{bmatrix}$, then $[A]^T = \begin{Bmatrix} 2 \\ 1 \\ 3 \end{Bmatrix}$

For a third example, $\{x\} = \begin{Bmatrix} x_1 \\ x_2 \\ x_3 \end{Bmatrix}$, then $\{x\}^T = \begin{bmatrix} x_1 & x_2 & x_3 \end{bmatrix}$

If a matrix is transposed twice, then it resumes to its original form, that is,

$$([A]^T)^T = [A] \tag{28.3}$$

For example,

$$[A] = \begin{bmatrix} 2 & 0 & 1 \\ 3 & 1 & 4 \end{bmatrix}, \quad [A]^T = \begin{bmatrix} 2 & 3 \\ 0 & 1 \\ 1 & 4 \end{bmatrix}, \quad ([A]^T)^T = \begin{bmatrix} 2 & 0 & 1 \\ 3 & 1 & 4 \end{bmatrix} = [A]$$

28.2.10 Symmetric Matrix, Antisymmetric Matrix and Skew Symmetric Matrix

If the elements of matrix $[A]$ are symmetric about the main diagonal (i.e., $a_{ij} = a_{ji}$), then $[A]$ is a symmetric matrix. For example,

$$[A] = \begin{bmatrix} 2 & 0 & 3 \\ 0 & 1 & 4 \\ 3 & 4 & 5 \end{bmatrix}$$

is a third-order symmetric matrix. Clearly, the symmetric matrix is the same as its transpose matrix, that is,

$$[A] = [A]^T$$

If all the elements of matrix $[A]$ at the diagonal equal zero and other elements are antisymmetric about the main diagonal (i.e., $a_{ii} = 0,\ \ a_{ij} = -a_{ji}$), then $[A]$ is an antisymmetric matrix. For example,

$$[A] = \begin{bmatrix} 0 & 2 & 1 \\ -2 & 0 & -3 \\ -1 & 3 & 0 \end{bmatrix}$$

is a third-order antisymmetric matrix. The antisymmetric matrix is the same as its minus transpose matrix, that is,

$$[A] = -[A]^T$$

If not all the elements of matrix $[A]$ at the diagonal equal zero and other elements are antisymmetric about the main diagonal (i.e., $a_{ij} = -a_{ji}$), then $[A]$ is a skew symmetric matrix. For example,

$$[A] = \begin{bmatrix} 2 & 3 & -4 \\ -3 & 1 & 5 \\ 4 & -5 & 0 \end{bmatrix}$$

28.2.11 Band Matrix

If the nonzero elements of matrix $[A]$ are all concentrated around the main diagonal, forming a nonzero element band along the main diagonal, then $[A]$ is called band matrix. For example,

$$[A] = \begin{bmatrix} a_{11} & a_{12} & a_{13} & 0 & \cdots & 0 & 0 \\ a_{21} & a_{22} & a_{23} & 0 & \cdots & 0 & 0 \\ 0 & 0 & a_{33} & a_{34} & \cdots & 0 & 0 \\ 0 & 0 & a_{43} & a_{44} & \cdots & 0 & 0 \\ \vdots & \vdots & \vdots & \vdots & \vdots & \vdots & \vdots \\ 0 & 0 & 0 & 0 & \cdots & a_{n-1,n-1} & a_{n-1,n} \\ 0 & 0 & 0 & 0 & \cdots & a_{n,n-1} & a_{n,n} \end{bmatrix}$$

28.3 Equality, Addition, and Subtraction of Matrices

The basic rule of matrix algebra is letting the linear transformation perform in a rather simple way.

28.3.1 The Equality of Matrices

For two matrices $[A]$ and $[B]$ with the same order

$$[A] = \begin{bmatrix} a_{11} & a_{12} & \cdots & a_{1n} \\ a_{21} & a_{22} & \cdots & a_{2n} \\ \vdots & \vdots & \vdots & \vdots \\ a_{m1} & a_{m2} & \cdots & a_{mn} \end{bmatrix}, \quad [B] = \begin{bmatrix} b_{11} & b_{12} & \cdots & b_{1n} \\ b_{21} & b_{22} & \cdots & b_{2n} \\ \vdots & \vdots & \vdots & \vdots \\ b_{m1} & b_{m2} & \cdots & b_{mn} \end{bmatrix}$$

If the corresponding element equals each other, that is,

$$a_{ij} = b_{ij}, \quad 1 \le i \le m, \quad 1 \le j \le n$$

then the two matrices are equal to each other, that is,

$$[A] = [B]$$

For example,

$$[A] = \begin{bmatrix} 2 & 1 & 3 \\ -1 & 0 & 4 \end{bmatrix}, \quad [B] = \begin{bmatrix} 2 & 1 & 3 \\ -1 & 0 & 4 \end{bmatrix}, \quad [C] = \begin{bmatrix} 2 & 1 & 3 \\ 1 & 0 & 4 \end{bmatrix}$$

$[A] = [B]$, but $[A] \neq [C]$, because $a_{21} = -1 \neq c_{21}$.

28.3.2 The Addition and Subtraction of Matrices

If the matrices $[A]$ and $[B]$ have the same order, it is possible to add or subtract the corresponding elements of two matrices and form a new matrix $[C]$ with the same order. The matrix $[C]$ is called the addition or subtraction of the matrices $[A]$ and $[B]$, that is,

$$[C] = [A] \pm [B]$$

where

$$c_{ij} = a_{ij} \pm b_{ij}, \quad 1 \le i \le m, \quad 1 \le j \le n$$

Example 1

$$\begin{bmatrix} a_{11} & a_{12} & a_{13} \\ a_{21} & a_{22} & a_{23} \end{bmatrix} + \begin{bmatrix} b_{11} & b_{12} & b_{13} \\ b_{21} & b_{22} & b_{23} \end{bmatrix} = \begin{bmatrix} a_{11} + b_{11} & a_{12} + b_{12} & a_{13} + b_{13} \\ a_{21} + b_{21} & a_{22} + b_{22} & a_{23} + b_{23} \end{bmatrix}$$

Example 2

$$\begin{bmatrix} a_{11} & a_{12} & a_{13} \\ a_{21} & a_{22} & a_{23} \end{bmatrix} - \begin{bmatrix} b_{11} & b_{12} & b_{13} \\ b_{21} & b_{22} & b_{23} \end{bmatrix} = \begin{bmatrix} a_{11} - b_{11} & a_{12} - b_{12} & a_{13} - b_{13} \\ a_{21} - b_{21} & a_{22} - b_{22} & a_{23} - b_{23} \end{bmatrix}$$

Example 3

$$\begin{bmatrix} 1 & -2 & 0 \\ 2 & 3 & 4 \end{bmatrix} + \begin{bmatrix} 2 & 1 & 3 \\ 0 & 1 & -2 \end{bmatrix} = \begin{bmatrix} 3 & -1 & 3 \\ 2 & 4 & 2 \end{bmatrix}$$

Example 4

$$\begin{bmatrix} 1 & -2 & 0 \\ 2 & 3 & 4 \end{bmatrix} - \begin{bmatrix} 2 & 1 & 3 \\ 0 & 1 & -2 \end{bmatrix} = \begin{bmatrix} -1 & -3 & -3 \\ 2 & 2 & 6 \end{bmatrix}$$

Clearly, the addition and subtraction of matrices obey the commutative law and the associative law, that is,

$$[A] \pm [B] = [B] \pm [A] \quad \text{(commutative law)}$$

and

$$([A] \pm [B]) \pm [C] = [A] + (\pm[B] \pm [C]) \quad \text{(associative law)}$$

28.4 Matrix Multiplied by a Number

A number s multiplying a matrix $[A]$ equals s multiplying each element of $[A]$, that is,

$$s[A] = \begin{bmatrix} sa_{11} & sa_{12} & \cdots & sa_{1n} \\ sa_{21} & sa_{22} & \cdots & sa_{2n} \\ \vdots & \vdots & \vdots & \vdots \\ sa_{m1} & sa_{m2} & \cdots & sa_{mn} \end{bmatrix}$$

Clearly, for the matrices $[A]$ and $[B]$ with the same order, the following formula can be established:

$$s([A] \pm [B]) = s[A] \pm s[B]$$

Example 1

$$2\begin{bmatrix}2 & 1\\ 3 & 4\end{bmatrix}=\begin{bmatrix}4 & 2\\ 6 & 8\end{bmatrix}$$

Example 2

$$3\begin{bmatrix}1 & 2\\ 0 & -1\\ 3 & 2\end{bmatrix}+2\begin{bmatrix}0 & 1\\ 3 & 2\\ 1 & 0\end{bmatrix}-4\begin{bmatrix}1 & 2\\ 0 & 1\\ 2 & 0\end{bmatrix}=\begin{bmatrix}3 & 6\\ 0 & -3\\ 9 & 6\end{bmatrix}+\begin{bmatrix}0 & 2\\ 6 & 4\\ 2 & 0\end{bmatrix}-\begin{bmatrix}4 & 8\\ 0 & 4\\ 8 & 0\end{bmatrix}=\begin{bmatrix}-1 & 0\\ 6 & -3\\ 3 & 6\end{bmatrix}$$

28.5 Multiplication of Matrices

28.5.1 Compatible Matrix

If the number of rows of $[A]$ equals the number of columns of $[B]$, $[A]$ and $[B]$ are called compatible matrix. For example,

$$\underset{m\times p}{[A]}\quad\underset{q\times n}{[B]} = \underset{m\times n}{[C]}$$

Only when $p=q$ the above equation can establish, and the product $[C]$ is an $m\times n$ matrix.

28.5.2 Rules of Matrix Multiplication

The product of compatible matrix $[A]$ and $[B]$ is matrix $[C]$, of which the element c_{ij} equals the sum of each element of ith row in $[A]$ multiplying the corresponding element of jth column in $[B]$, that is,

$$\underset{m\times n}{[C]} = \underset{m\times p}{[A]}\quad\underset{q\times n}{[B]} \tag{28.4}$$

where

$$c_{ij}=a_{i1}b_{1j}+a_{i2}b_{2j}+\cdots+a_{ip}b_{pi}=\sum_{r=1}^{p}a_{ir}b_{rj}$$

For better memorizing, the matrix multiplication can be written as

$$\underset{p\times n}{[B]}=\begin{bmatrix}b_{11} & b_{12} & \cdots & b_{1j} & \cdots & b_{1n}\\ b_{21} & b_{22} & \cdots & b_{2j} & \cdots & b_{2n}\\ \vdots & \vdots & \vdots & \vdots & \vdots & \vdots\\ b_{p1} & b_{p2} & \cdots & b_{pj} & \cdots & b_{pn}\end{bmatrix}$$

$$\underset{m\times p}{[A]}=\begin{bmatrix}a_{11} & a_{12} & \cdots & a_{1p}\\ \vdots & \vdots & \vdots & \vdots\\ a_{i1} & a_{i2} & \cdots & a_{ip}\\ \vdots & \vdots & \vdots & \vdots\\ a_{m1} & a_{m2} & \cdots & a_{mp}\end{bmatrix}\begin{bmatrix}\vdots\\ \vdots\\ \vdots\\ c_{ij}\end{bmatrix}=\underset{m\times n}{[C]}$$

It can be seen that the number of rows in $[C]$ equals the number of rows in $[A]$, and the number of columns in $[C]$ equals the number of columns in $[B]$.

Example 1

$$\begin{bmatrix} a_{11} & a_{12} \\ a_{21} & a_{22} \\ a_{31} & a_{32} \end{bmatrix} \begin{bmatrix} b_{11} & b_{12} \\ b_{21} & b_{22} \end{bmatrix} = \begin{bmatrix} a_{11}b_{11} + a_{12}b_{21} & a_{11}b_{12} + a_{12}b_{22} \\ a_{21}b_{11} + a_{22}b_{21} & a_{21}b_{12} + a_{22}b_{22} \\ a_{31}b_{11} + a_{32}b_{21} & a_{31}b_{12} + a_{32}b_{22} \end{bmatrix}$$

Example 2 If

$$\{x\} = \begin{Bmatrix} x_1 \\ x_2 \\ x_3 \end{Bmatrix}, \quad \{y\} = \begin{Bmatrix} y_1 \\ y_2 \\ y_3 \end{Bmatrix}$$

then

$$\{x\}^{\mathrm{T}}\{y\} = \begin{bmatrix} x_1 & x_2 & x_3 \end{bmatrix} \begin{Bmatrix} y_1 \\ y_2 \\ y_3 \end{Bmatrix} = x_1y_1 + x_2y_2 + x_3y_3$$

and

$$\{x\}\{y\}^{\mathrm{T}} = \begin{Bmatrix} x_1 \\ x_2 \\ x_3 \end{Bmatrix} \begin{bmatrix} y_1 & y_2 & y_3 \end{bmatrix} = \begin{bmatrix} x_1y_1 & x_1y_2 & x_1y_3 \\ x_2y_1 & x_2y_2 & x_2y_3 \\ x_3y_1 & x_3y_2 & x_3y_3 \end{bmatrix}$$

Example 3

$$\begin{bmatrix} 1 & 3 & 2 \\ -2 & 0 & 4 \end{bmatrix} \begin{bmatrix} 2 & 1 \\ 1 & -2 \\ 0 & -1 \end{bmatrix} = \begin{bmatrix} 1\times 2 + 3\times 1 + 2(0) & 1\times 1 + 3\times(-2) + 2\times(-1) \\ -2\times 2 + 0\times 1 + 4(0) & -2\times 1 + 0\times(-2) + 4\times(-1) \end{bmatrix}$$

$$= \begin{bmatrix} 5 & -7 \\ -4 & -6 \end{bmatrix}$$

28.5.3 Properties of Matrix Multiplication

It is easy to verify that the matrix multiplication has the following properties:

1) $s([A][B]) = (s([A])\ [B] = [A]\ (s[B])$, where s is a scalar.
2) $([A][B])[C] = [A]([B][C])$, which means that the matrix multiplication obeys the associative law.

3) $([A]+[B])\,[C]=[A]\,[C]+[B][C]$, $[A]\,([B]+[C])=[A]\,[B]+[A]\,[C]$, which means that the matrix multiplication obeys the distributive law.
4) $$([A]\,[B])^{\mathrm{T}} = [B]^{\mathrm{T}}[A]^{\mathrm{T}} \tag{28.5}$$

   The above equation indicates that the transpose of the product of two matrices equals the product of their respective transpose matrix with the sequence of $[A]\,[B]$ reversed.
5) Generally, the matrix multiplication does not obey the commutative law. However, some cases are exceptional.

Example 1

$$[A] = \begin{bmatrix} 1 & 0 & -1 \\ 1 & 2 & 3 \end{bmatrix}, \quad [B] = \begin{bmatrix} 2 & 0 \\ -1 & 1 \\ 1 & 1 \end{bmatrix}$$

$$[A][B] = \begin{bmatrix} 1 & -1 \\ 3 & 5 \end{bmatrix}, \quad [B][A] = \begin{bmatrix} 2 & 0 & -2 \\ 0 & 2 & 4 \\ 2 & 2 & 2 \end{bmatrix}$$

Example 2

$$\begin{bmatrix} 2 & 1 \\ 1 & 3 \end{bmatrix}\begin{bmatrix} 1 & 0 \\ 0 & 1 \end{bmatrix} = \begin{bmatrix} 1 & 0 \\ 0 & 1 \end{bmatrix}\begin{bmatrix} 2 & 1 \\ 1 & 3 \end{bmatrix} = \begin{bmatrix} 2 & 1 \\ 1 & 3 \end{bmatrix}$$

6) Generally, $[A]=[\mathbf{0}]$ or $[B]=[\mathbf{0}]$ cannot be alleged from equation $[A]\,[B]=[\mathbf{0}]$. For example,

   $$\begin{bmatrix} 3 & 0 \\ 0 & 0 \end{bmatrix}\begin{bmatrix} 0 & 0 \\ 1 & 2 \end{bmatrix} = \begin{bmatrix} 0 & 0 \\ 0 & 0 \end{bmatrix}$$

   In this case, the product of two matrices is zero matrix, but neither of the two matrices is zero matrix.
7) Generally, $[B]=[C]$ cannot be alleged from equation $[A]\,[B]=[A]\,[C]$. For example,

   $$[A] = \begin{bmatrix} 0 & 1 & 0 \\ 0 & 0 & 1 \end{bmatrix}, \quad [B] = \begin{bmatrix} 0 & 0 \\ 1 & 0 \\ 0 & 1 \end{bmatrix}, \quad [C] = \begin{bmatrix} 2 & 1 \\ 1 & 0 \\ 0 & 1 \end{bmatrix}$$

   $$[A][B] = \begin{bmatrix} 1 & 0 \\ 0 & 1 \end{bmatrix}, \quad [A][C] = \begin{bmatrix} 1 & 0 \\ 0 & 1 \end{bmatrix}$$

   Although $[A]\,[B]=[A]\,[C]$, $[B]\neq[C]$.
   It should be noted that the properties in the above (5), (6), and (7) are different from that of the scalar manipulation.
8) If the element a_{ij} of matrix $[A]$ can be any value to make the equation

   $$[A][B] = [A][C] \tag{a}$$

established, $[B]=[C]$ can be alleged. This conclusion can be easily verified. Suppose $[A]$ is an $m \times p$ matrix and $[B]$ and $[C]$ are $p \times n$ matrices. Let

$$\underset{m\times p}{[A]} = \begin{bmatrix} a_{11} & a_{12} & \cdots & a_{1p} \\ a_{21} & a_{22} & \cdots & a_{ip} \\ \vdots & \vdots & \vdots & \vdots \\ a_{m1} & a_{m2} & \cdots & a_{mp} \end{bmatrix}, \quad \underset{p\times n}{[B]} = \begin{bmatrix} b_{11} & b_{12} & \cdots & b_{1n} \\ b_{21} & b_{22} & \cdots & b_{2n} \\ \vdots & \vdots & \vdots & \vdots \\ b_{p1} & b_{p2} & \cdots & b_{pn} \end{bmatrix}, \quad \underset{p\times n}{[C]} = \begin{bmatrix} c_{11} & c_{12} & \cdots & c_{1n} \\ c_{21} & c_{22} & \cdots & c_{2n} \\ \vdots & \vdots & \vdots & \vdots \\ c_{p1} & c_{p2} & \cdots & c_{pn} \end{bmatrix}$$

Firstly, since matrix $[A]$ can be any value, we can take $a_{11}=1$, and let all the other elements be zero. From Eq. (a), we have

$$\begin{bmatrix} b_{11} & b_{12} & \cdots & b_{1n} \\ 0 & 0 & \cdots & 0 \\ \vdots & \vdots & \vdots & \vdots \\ 0 & 0 & \cdots & 0 \end{bmatrix} = \begin{bmatrix} c_{11} & c_{12} & \cdots & c_{1n} \\ 0 & 0 & \cdots & 0 \\ \vdots & \vdots & \vdots & \vdots \\ 0 & 0 & \cdots & 0 \end{bmatrix} \tag{b}$$

Thus $b_{11}=c_{11}, b_{12}=c_{12}, \ldots, b_{1n}=c_{1n}$. Firstly, it can be seen that the element of the 1st row in $[B]$ equals the corresponding element of the 1st row in $[C]$. Secondly, the element of the 2nd row in $[B]$ equals the corresponding element of the 2nd row in $[C]$. Analogically, the elements of each row in $[B]$ equals the corresponding element of each row in $[C]$. Therefore, matrix $[B]$ equals matrix $[C]$.

In the finite element method, this property is often used when deriving the stiffness matrix by principle of virtual work. Clearly, the equality of $[B]$ and $[C]$ is based on randomicity of $[A]$. Therefore, if the elements of $[A]$ cannot be any value, then $[B]=[C]$ cannot be alleged as previously mentioned.

28.5.4 The Positive Power of Square Matrix

The continued product of square matrix $[A]$ by itself is defined as the positive power of square matrix $[A]^n$, that is,

$$[A]^n = \underbrace{[A][A]\cdots[A]}_{n \text{ times}}$$

28.6 Determinant

28.6.1 Definition of Determinant

The nth-order determinant is a scalar that is obtained by $a_{ij}\,(i,j=1,2,\ldots,n)$ of n^2 (n rows by n columns) numbers according to certain operation rules and denoted as

$$|A| = \begin{bmatrix} a_{11} & a_{12} & \cdots & a_{1n} \\ a_{21} & a_{22} & \cdots & a_{2n} \\ \vdots & \vdots & \vdots & \vdots \\ a_{n1} & a_{n2} & \cdots & a_{nn} \end{bmatrix} \tag{28.6}$$

where a_{ij} is the element of $|A|$. A first-order determinant $|A|$ is composed of one element, that is,

$$|A| = |a_{11}| = a_{11}$$

A second-order determinant is defined as

$$|A| = \begin{vmatrix} a_{11} & a_{12} \\ a_{21} & a_{22} \end{vmatrix} = a_{11}a_{22} - a_{12}a_{21} \tag{28.7}$$

that is, the product of the left top element and the right bottom element minus the product of the rest element. The third order as well as higher-order determinant will be defined later.

28.6.2 Minors and Cofactors

The minor of element a_{ij} of $|A|$ is the determinant of $(n-1)$th-order matrix obtained by deleting the ith row and jth column and is denoted as $M_{i,j}$. Multiplying $M_{i,j}$ by $(-1)^{i+j}$ is defined as the cofactor of element a_{ij}, denoted as $A_{i,j}$. For example,

For a third-order determinant

$$|A| = \begin{vmatrix} a_{11} & a_{12} & a_{13} \\ a_{21} & a_{22} & a_{23} \\ a_{31} & a_{32} & a_{33} \end{vmatrix}$$

the minor and cofactor of the element a_{11} are

$$M_{11} = \begin{vmatrix} a_{22} & a_{23} \\ a_{32} & a_{33} \end{vmatrix}, \quad A_{11} = (-1)^{1+1}M_{11} = \begin{vmatrix} a_{22} & a_{23} \\ a_{32} & a_{33} \end{vmatrix}$$

respectively; the minor and cofactor of the element a_{12} are

$$M_{12} = \begin{vmatrix} a_{21} & a_{23} \\ a_{31} & a_{33} \end{vmatrix}, \quad A_{12} = (-1)^{1+2}M_{12} = -\begin{vmatrix} a_{21} & a_{23} \\ a_{31} & a_{33} \end{vmatrix}$$

respectively.

28.6.3 Principal Minors

The minor, which is composed of the elements from the first column to kth column and the first row to kth row, that is, the top left kth-order minor, is called the kth-order principal minor and denoted as D_k.

Example 1 A fourth-order determinant

$$|A| = \begin{vmatrix} a_{11} & a_{12} & a_{13} & a_{14} \\ a_{21} & a_{22} & a_{23} & a_{24} \\ a_{31} & a_{32} & a_{33} & a_{34} \\ a_{41} & a_{42} & a_{43} & a_{44} \end{vmatrix}$$

The leading principal minors of different orders are

$$D_1 = a_{11}, \quad D_2 = \begin{vmatrix} a_{11} & a_{12} \\ a_{21} & a_{22} \end{vmatrix}, \quad D_3 = \begin{vmatrix} a_{11} & a_{12} & a_{13} \\ a_{21} & a_{22} & a_{23} \\ a_{31} & a_{32} & a_{33} \end{vmatrix}, \quad D_4 = \begin{vmatrix} a_{11} & a_{12} & a_{13} & a_{14} \\ a_{21} & a_{22} & a_{23} & a_{24} \\ a_{31} & a_{32} & a_{33} & a_{34} \\ a_{41} & a_{42} & a_{43} & a_{44} \end{vmatrix}$$

respectively.

28.6.4 Expansion of the Determinant by One Row (Column)

Suppose there is a third-order determinant $|A|$, with its three elements a_{11}, a_{12}, a_{13} in the first row and the corresponding cofactors A_{11}, A_{12}, and A_{13}. Then the determinant can be defined as

$$|A| = a_{11}A_{11} + a_{12}A_{12} + a_{13}A_{13} = \sum_{j=1}^{3} a_{1j}A_{1j}$$

In this way, the third-order determinant is defined with the second-order determinant. Analogically, an nth-order determinant can be defined as

$$|A| = a_{11}A_{11} + a_{12}A_{12} + \cdots + a_{1n}A_{1n} = \sum_{j=1}^{n} a_{1j}A_{1j}$$

The above equation gives the definition of determinant and also provides the method of calculating the determinant, that is, denoting the nth-order determinant by $(n-1)$th-order determinant, which is denoted by $(n-2)$th-order determinant. Analogically and finally, the nth-order determinant can be denoted by the second-order determinant.

In fact, the sum of each element multiplying its cofactor in any row (column) of the determinant equals the determinant, that is,

$$|A| = \sum_{j=1}^{n} a_{ij}A_{ij}, \quad (i = 1,2,\ldots,n) \tag{28.8a}$$

$$|A| = \sum_{i=1}^{n} a_{ij}A_{ij}, \quad (j = 1,2,\ldots,n) \tag{28.8b}$$

where the first equation implies that the determinant is expanded along the row and the second equation implies that the determinant is expanded along the column. Thus there are $2n$ ways to expand an nth-order determinant. Anyway, the value of the determinants is certainly all the same. Additionally, the determinant has such properties that the sum of each element in any row (column) multiplying the cofactor of the corresponding element in another row (column) equals zero, that is,

$$|A| = \sum_{j=1}^{n} a_{ij}A_{kj} = 0, \quad (i = 1,2,\ldots,n;\ \ k = 1,2,\ldots,n;\ \ i \neq k) \tag{28.9a}$$

$$|A| = \sum_{i=1}^{n} a_{ij}A_{kj} = 0, \quad (j = 1,2,\ldots,n;\ \ k = 1,2,\ldots,n;\ \ j \neq k) \tag{28.9b}$$

Example 2 Expand the third determinant along the first row:

$$|A| = \begin{vmatrix} a_{11} & a_{12} & a_{13} \\ a_{21} & a_{22} & a_{23} \\ a_{31} & a_{32} & a_{33} \end{vmatrix} = a_{11}\begin{vmatrix} a_{22} & a_{23} \\ a_{32} & a_{33} \end{vmatrix} - a_{12}\begin{vmatrix} a_{21} & a_{23} \\ a_{31} & a_{33} \end{vmatrix} + a_{13}\begin{vmatrix} a_{21} & a_{22} \\ a_{31} & a_{32} \end{vmatrix}$$
$$= a_{11}(a_{22}a_{33}-a_{23}a_{32}) - a_{12}(a_{21}a_{33}-a_{23}a_{31}) + a_{13}(a_{21}a_{32}-a_{22}a_{31})$$

Example 3 Calculate the following third-order determinant (expand along the first column):

$$|A| = \begin{vmatrix} 2 & 0 & 3 \\ 0 & 1 & 2 \\ 1 & 2 & 1 \end{vmatrix} = 2\begin{vmatrix} 1 & 2 \\ 2 & 1 \end{vmatrix} - 0 \times \begin{vmatrix} 0 & 3 \\ 2 & 1 \end{vmatrix} + 1 \times \begin{vmatrix} 0 & 3 \\ 1 & 2 \end{vmatrix}$$
$$= 2(1 \times 1 - 2 \times 2) - 0 + (0 \times 2 - 1 \times 3) = -9$$

28.6.5 Properties of Determinant

From the definition of the determinant, it has the following basic properties:

1) The determinant equals zero if all the elements of one row (or column) of the determinant are zero. In fact, the determinant must equal zero if expanding along this row (column).
2) Exchanging the row and column of the determinant, we obtain the transpose determinant, which equals the original determinant.
3) If any two rows (or columns) are exchanged, the determinant changes its sign.
4) If any two rows (or columns) are equivalent, the determinant equals zero.
5) Multiplying all the elements of any row (or column) of the determinant by the same constant equals to the determinant multiplied by this constant.
6) Multiplying each element of one row (or column) of the determinant by the same number and adding them to the corresponding element of another row (or column), respectively, the value of the determinant remains unchanged.

28.7 Inverse Matrix

In the algebra, if

$$ab = 1 \tag{a}$$

b is called the inverse of a and is denoted as

$$b = \frac{1}{a} = a^{-1} \tag{b}$$

In the matrix algebra, the unit matrix $[I]$ is similar to 1 in the algebra. Suppose $[A]$ is an nth-order square matrix, whereas $[I]$ is an nth-order unit matrix. Then

$$[A]\,[I] = [I]\,[A] = [A] \tag{c}$$

The above equation is equivalent to $a \cdot 1 = 1 \cdot a = a$.

28.7.1 The Definition of Inverse Matrix

Suppose $[A]$ is an nth-order square matrix. If there is another square matrix $[B]$ satisfying

$$[A][B] = [B][A] = [I] \tag{28.10}$$

where $[I]$ is an nth-order unit matrix, then $[B]$ is the inverse matrix of $[A]$ and can be denoted as $[B] = [A]^{-1}$. It can be seen that the above two equations are similar to Eqs. (a) and (b). From Eq. (28.10), we have

$$[A][A]^{-1} = [A]^{-1}[A] = [I] \tag{28.11}$$

28.7.2 The Adjoint Matrix

Suppose $[A]$ is an nth-order square matrix and the cofactor of the corresponding element a_{ij} is A_{ij}. A new square matrix $\widehat{A}$ can be made with the element of ith row and jth column A_{ij}. that is,

$$[\widehat{A}] = \begin{bmatrix} A_{11} & A_{12} & \cdots & A_{1n} \\ A_{21} & A_{22} & \cdots & A_{2n} \\ \vdots & \vdots & \vdots & \vdots \\ A_{n1} & A_{n2} & \cdots & A_{nn} \end{bmatrix}$$

where $[\widehat{A}]$ is called the cofactor matrix of $[A]$.

The transpose matrix of $[\widehat{A}]$ is called the adjoint matrix of $[A]$ and is denoted as

$$[\overset{*}{A}] = [\widehat{A}]^{\mathrm{T}} = \begin{bmatrix} A_{11} & A_{21} & \cdots & A_{n1} \\ A_{12} & A_{22} & \cdots & A_{n2} \\ \vdots & \vdots & \vdots & \vdots \\ A_{1n} & A_{2n} & \cdots & A_{nn} \end{bmatrix} \tag{28.12}$$

From the matrix multiplication and the properties of determinant, we have

$$[A][\overset{*}{A}] = [\overset{*}{A}][A] = \begin{bmatrix} |A| & 0 & \cdots & 0 \\ 0 & |A| & \cdots & 0 \\ \vdots & \vdots & \vdots & \vdots \\ 0 & 0 & \cdots & |A| \end{bmatrix} = |A|[I] \tag{28.13}$$

The above equation is not difficult to verify. Let $[C] = [A][\overset{*}{A}]$. Then

$$[C] = [A][\overset{*}{A}] = \begin{bmatrix} a_{11} & a_{12} & \cdots & a_{1n} \\ a_{21} & a_{22} & \cdots & a_{2n} \\ \vdots & \vdots & \vdots & \vdots \\ a_{n1} & a_{n2} & \cdots & a_{nn} \end{bmatrix} \begin{bmatrix} A_{11} & A_{12} & \cdots & A_{1n} \\ A_{21} & A_{22} & \cdots & A_{2n} \\ \vdots & \vdots & \vdots & \vdots \\ A_{n1} & A_{n2} & \cdots & A_{nn} \end{bmatrix}$$

From the matrix multiplication and Eq. (28.8), the main diagonal elements are

$$c_{ii} = a_{i1}A_{i1} + a_{i2}A_{i2} + \cdots + a_{in}A_{in} = |A|$$

From the matrix multiplication and Eq. (28.9), the non-main diagonal elements are

$$c_{ij} = a_{i1}A_{j1} + a_{i2}A_{j2} + \cdots + a_{in}A_{jn} = 0, \quad i \neq j$$

It can be seen that the main diagonal elements are all $|A|$, and the non-main diagonal elements are all zero, that is, Eq. (28.13) is established.

28.7.3 Inverse Matrix

Both sides of Eq. (28.13) are divided by the determinant $|A|$. Then

$$[A]\frac{[\overset{*}{A}]}{|A|} = \frac{[\overset{*}{A}]}{|A|}[A] = [I]$$

is obtained. From the definition of the inverse matrix, we have

$$[A]^{-1} = \frac{[\overset{*}{A}]}{|A|} \tag{28.14}$$

The inverse matrix can be calculated by the above equation. If $|A| = 0$, $[A]^{-1}$ does not exist, and $[A]$ is called singular matrix. If $|A| \neq 0$, the unique inverse matrix can be calculated, and $[A]$ is called nonsingular matrix.

Example 1 Given the matrix

$$[A] = \begin{bmatrix} 1 & 2 & 0 \\ 0 & 1 & 1 \\ 1 & 0 & 2 \end{bmatrix}$$

with its cofactors

$$A_{11} = 2,\quad A_{12} = 1,\quad A_{13} = -1,\quad A_{21} = -4,\quad A_{22} = 2,\quad A_{23} = 2,\quad A_{31} = 2,$$
$$A_{32} = -1,\quad A_{33} = 1$$

The cofactor matrix is

$$[\hat{A}] = \begin{bmatrix} 2 & 1 & -1 \\ -4 & 2 & 2 \\ 2 & -1 & 1 \end{bmatrix}$$

From Eq. (28.12), the adjoint matrix $[\overset{*}{A}]$ is

$$[\overset{*}{A}] = [\hat{A}]^{\mathrm{T}} = \begin{bmatrix} 2 & -4 & 2 \\ 1 & 2 & -1 \\ -1 & 2 & 1 \end{bmatrix}$$

If $[\overset{*}{A}]$ is divided by $|A| = 4$, we can obtain

$$[A]^{-1} = \begin{bmatrix} 0.5 & -1 & 0.5 \\ 0.25 & 0.5 & -0.25 \\ -0.25 & 0.5 & 0.25 \end{bmatrix}$$

The readers can check: $[A][A]^{-1} = [I]$.

28.7.4 The Inverse Matrix of the Diagonal Matrix

The inverse matrix of the diagonal matrix is still a diagonal matrix; the nonzero elements at the main diagonal of it equal the corresponding reciprocals of the original matrix.

$$\begin{bmatrix} a_{11} & 0 & \cdots & 0 \\ 0 & a_{22} & \cdots & 0 \\ \vdots & \vdots & \vdots & \vdots \\ 0 & 0 & \cdots & a_{nn} \end{bmatrix}^{-1} = \begin{bmatrix} \dfrac{1}{a_{11}} & 0 & \cdots & 0 \\ 0 & \dfrac{1}{a_{22}} & \cdots & 0 \\ \vdots & \vdots & \vdots & \vdots \\ 0 & 0 & \cdots & \dfrac{1}{a_{nn}} \end{bmatrix}$$

28.7.5 The Properties of the Inverse Matrix

The inverse matrix has the following properties:

1) $([A]^{-1})^{\mathrm{T}} = ([A]^{\mathrm{T}})^{-1}$.
2) $([A]^{-1})^{-1} = [A]$.
3) $([A][B])^{-1} = [B]^{-1}[A]^{-1}$.
4) If $[A]$ is a symmetric matrix, then $[A]^{-1}$ is also a symmetric matrix.
5) The inverse matrix of the triangular matrix is still the same type of triangular matrix.
6) The negative power of the square matrix.

Suppose $[A]$ is a nonsingular matrix, then there is an inverse matrix $[A]^{-1}$. The negative power of $[A]$ can be defined as

$$[A]^{-n} = ([A]^{-1})^n = \underbrace{[A]^{-1}[A]^{-1}\cdots[A]^{-1}}_{n\ \text{times}}$$

28.8 Partitioned Matrix

28.8.1 Definition of the Partitioned Matrix

For the convenience of denoting the matrix or meeting the requirement of computation, any matrix $[A]$ can be divided into small blocks with some vertical lines and horizontal lines. The divided matrix is called partitioned matrix, and those small partitioned matrices are called submatrices. For example,

$$\begin{bmatrix} 1 & 2 & 1 & -1 \\ 3 & 0 & 1 & 2 \\ -1 & 2 & 0 & 4 \\ 0 & 2 & 1 & 3 \end{bmatrix} = \left[\begin{array}{cc:cc} 1 & 2 & 1 & -1 \\ 3 & 0 & 1 & 2 \\ \hdashline -1 & 2 & 0 & 4 \\ 0 & 2 & 1 & 3 \end{array}\right] = \left[\begin{array}{cc:cc} 1 & 2 & 1 & -1 \\ \hdashline 3 & 0 & 1 & 2 \\ -1 & 2 & 0 & 4 \\ \hdashline 0 & 2 & 1 & 3 \end{array}\right] = \left[\begin{array}{c:cc:c} 1 & 2 & 1 & -1 \\ 3 & 0 & 1 & 2 \\ \hdashline -1 & 2 & 0 & 4 \\ 0 & 2 & 1 & 3 \end{array}\right] = \cdots$$

The partitioned matrix can be denoted as

$$[A] = \begin{bmatrix} A_{11} & A_{12} & \cdots & A_{1n} \\ A_{21} & A_{22} & \cdots & A_{2n} \\ \vdots & \vdots & \vdots & \vdots \\ A_{m1} & A_{m2} & \cdots & A_{mn} \end{bmatrix} \tag{28.15}$$

Clearly, the submatrix in the same row, such as $[A_{11}]$, $[A_{12}]$, …, $[A_{1n}]$, has the same rows, while the submatrix in the same column, such as $[A_{11}]$, $[A_{21}]$, …, $[A_{m1}]$, has the same columns. When calculating the partitioned matrix, first, $[A]$ is calculated by taking the submatrix as "element" of the matrix $[A]$, and then the submatrix is calculated.

28.8.2 Addition and Subtraction of the Partitioned Matrix

Suppose the partitioned matrices are as the following:

$$[A] = \begin{bmatrix} A_{11} & A_{12} & \cdots & A_{1n} \\ A_{21} & A_{22} & \cdots & A_{2n} \\ \vdots & \vdots & \vdots & \vdots \\ A_{m1} & A_{m2} & \cdots & A_{mn} \end{bmatrix}, \quad [B] = \begin{bmatrix} B_{11} & B_{12} & \cdots & B_{1n} \\ B_{21} & B_{22} & \cdots & B_{2n} \\ \vdots & \vdots & \vdots & \vdots \\ B_{m1} & B_{m2} & \cdots & B_{mn} \end{bmatrix} \tag{28.16}$$

If the corresponding submatrix $[A_{ij}]$ and $[B_{ij}]$ have the same rows and columns, then $[A] \pm [B]$ equals the matrix obtained from the addition or subtraction of the corresponding sub matrix, that is,

$$[A] \pm [B] = \begin{bmatrix} A_{11} \pm B_{11} & A_{12} \pm B_{12} & \cdots & A_{1n} \pm B_{1n} \\ A_{21} \pm B_{21} & A_{22} \pm B_{22} & \cdots & A_{2n} \pm B_{2n} \\ \vdots & \vdots & \vdots & \vdots \\ A_{m1} \pm B_{m1} & A_{m2} \pm B_{m2} & \cdots & A_{mn} \pm B_{mn} \end{bmatrix} \tag{28.17}$$

Clearly, if the corresponding submatrices do not have the same rows and columns, the addition and subtraction cannot be done.

28.8.3 The Multiplication of the Partitioned Matrix

$$[A] = \begin{bmatrix} A_{11} & A_{12} & \cdots & A_{1p} \\ A_{21} & A_{22} & \cdots & A_{2p} \\ \vdots & \vdots & \vdots & \vdots \\ A_{m1} & A_{m2} & \cdots & A_{mp} \end{bmatrix}, \quad [B] = \begin{bmatrix} B_{11} & B_{12} & \cdots & B_{1n} \\ B_{21} & B_{22} & \cdots & B_{2n} \\ \vdots & \vdots & \vdots & \vdots \\ B_{q1} & B_{q2} & \cdots & B_{qn} \end{bmatrix} \tag{28.18}$$

If $p = q$ and the columns of the sub matrix $[A_{ij}]$ equals the rows of the submatrix $[B_{ij}]$, then

$$[A][B] = \begin{bmatrix} C_{11} & C_{12} & \cdots & C_{1n} \\ C_{21} & C_{22} & \cdots & C_{2n} \\ \vdots & \vdots & \vdots & \vdots \\ C_{m1} & C_{m2} & \cdots & C_{mn} \end{bmatrix} \tag{28.19}$$

where

$$[C_{ik}] = \sum_{i=1}^{p} [A_{ij}][B_{jk}]$$

Example 1

$$[A] = \left[\begin{array}{cc:cc} a_{11} & a_{12} & a_{13} & a_{14} \\ a_{21} & a_{22} & a_{23} & a_{24} \\ \hdashline a_{31} & a_{32} & a_{33} & a_{34} \\ a_{41} & a_{42} & a_{43} & a_{44} \end{array}\right] = \begin{bmatrix} A_{11} & A_{12} \\ A_{21} & A_{22} \end{bmatrix}$$

$$[B] = \left[\begin{array}{cc:cc} b_{11} & b_{12} & b_{13} & b_{14} \\ b_{21} & b_{22} & b_{23} & b_{24} \\ \hdashline b_{31} & b_{32} & b_{33} & b_{34} \\ b_{41} & b_{42} & b_{43} & b_{44} \end{array}\right] = \begin{bmatrix} B_{11} & B_{12} \\ B_{21} & B_{22} \end{bmatrix}$$

$$[A][B] = \begin{bmatrix} A_{11} & A_{12} \\ A_{21} & A_{22} \end{bmatrix} \begin{bmatrix} B_{11} & B_{12} \\ B_{21} & B_{22} \end{bmatrix} = \begin{bmatrix} A_{11}B_{11} + A_{12}B_{21} & A_{11}B_{12} + A_{12}B_{22} \\ A_{21}B_{11} + A_{22}B_{21} & A_{21}B_{12} + A_{22}B_{22} \end{bmatrix}$$

Example 2 When dividing the matrices, the partitioned matrices must be consistent ($p = q$), and the order of the submatrices must obey the rule of multiplication. Otherwise, the matrices cannot multiply by each other. For example, for the following matrices multiplication, (a) is correct, whereas (b) is incorrect, as the submatrices of the matrix for multiplication in (b) are not consistent with each other.

$$\left[\begin{array}{cc:cc} 1 & 0 & 1 & 2 \\ 2 & 1 & 0 & 3 \\ \hdashline 1 & 1 & 2 & 4 \\ 2 & 1 & 1 & 3 \end{array}\right]\left[\begin{array}{cc} 1 & 2 \\ 0 & -1 \\ \hdashline 1 & 0 \\ 3 & 1 \end{array}\right] \tag{a}$$

$$\left[\begin{array}{cc:cc} 1 & 0 & 1 & 2 \\ 2 & 1 & 0 & 3 \\ \hdashline 1 & 1 & 2 & 4 \end{array}\right]\left[\begin{array}{cc} 1 & 2 \\ 0 & -1 \\ \hdashline 1 & 0 \\ 3 & 1 \end{array}\right] \tag{b}$$

28.8.4 The Inverse of the Partitioned Matrix

The inverse of a higher-order matrix can be transferred to the inverse of the lower-order partitioned matrix by dividing the matrix.

For a matrix

$$[A] = \begin{bmatrix} A_{11} & A_{12} \\ A_{21} & A_{22} \end{bmatrix}$$

let

$$[A]^{-1} = [B] = \begin{bmatrix} B_{11} & B_{12} \\ B_{21} & B_{22} \end{bmatrix}$$

According to the definition of the inverse matrix, $[A][B] = [I]$, that is,

$$\begin{bmatrix} A_{11} & A_{12} \\ A_{21} & A_{22} \end{bmatrix}\begin{bmatrix} B_{11} & B_{12} \\ B_{21} & B_{22} \end{bmatrix} = \begin{bmatrix} I & 0 \\ 0 & I \end{bmatrix}$$

the following equations can be obtained.

$$[A_{11}][B_{11}] + [A_{12}][B_{21}] = [I] \tag{a}$$

$$[A_{21}][B_{11}] + [A_{22}][B_{21}] = [0] \tag{b}$$

$$[A_{11}][B_{12}] + [A_{12}][B_{22}] = [0] \tag{c}$$

$$[A_{21}][B_{12}] + [A_{22}][B_{22}] = [I] \tag{d}$$

From Eq. (c),

$$[B_{12}] = -[A_{11}]^{-1}[A_{12}][B_{22}] \tag{e}$$

Substitute Eq. (e) into Eq. (d),

$$([A_{22}] - [A_{21}][A_{11}]^{-1}[A_{12}])[B_{22}] = [I]$$

can be obtained.

Thus,

$$[B_{22}] = ([A_{22}] - [A_{21}][A_{11}]^{-1}[A_{12}])^{-1} \tag{f}$$

From $[B][A] = [I]$,

$$[B_{21}][A_{11}] + [B_{22}][A_{21}] = 0$$

can be derived, thus

$$[B_{21}] = -[B_{22}][A_{21}][A_{11}]^{-1} \tag{g}$$

Finally, Eq. (h) can be obtained from Eq. (a).

$$[B_{11}] = [A_{11}]^{-1} - [A_{11}]^{-1}[A_{12}][B_{21}] \tag{h}$$

All of the submatrices of inverse matrix $[B]$ can be computed by Eqs. (e)–(h), from which the inverse computation of the partitioned matrix only includes the inverse of the matrix with the same order as the submatrix of $[A]$. Although the lengths of the equations are long, the inverse calculation of the partitioned matrix can save more time than that of the original matrix, because the multiplication, addition, and subtraction calculation of the matrices is much faster than the inverse calculation of the matrix.

To begin with the first term of the left top part of the given matrix, the higher-order matrix is obtained by adding one row and column after the inverse calculation. Repeating the above calculation, the inverse matrix of any order matrix can be obtained. For example, there is a fourth-order square matrix

$$[A] = \begin{bmatrix} a_{11} & a_{12} & a_{13} & a_{14} \\ a_{21} & a_{22} & a_{23} & a_{24} \\ a_{31} & a_{32} & a_{33} & a_{34} \\ a_{41} & a_{42} & a_{43} & a_{44} \end{bmatrix}$$

First of all, take a_{11} as the first-order matrix:

$$[A_1] = [a_{11}]$$

of which the inverse matrix is

$$[A_1]^{-1} = \left[\frac{1}{a_{11}}\right]$$

Adding one row and one column and taking

$$[A_2] = \left[\begin{array}{c:c} a_{11} & a_{12} \\ \hdashline a_{21} & a_{22} \end{array}\right]$$

the inverse matrix of $[A_2]^{-1}$ can be calculated through Eqs. (e)–(h). Adding another row and column and taking

$$[A_3] = \left[\begin{array}{cc:c} a_{11} & a_{12} & a_{13} \\ a_{21} & a_{22} & a_{23} \\ \hdashline a_{31} & a_{32} & a_{33} \end{array}\right]$$

$[A_3]^{-1}$ can be calculated by $[A_2]^{-1}$ and Eqs. (e)–(h). Finally, taking the partitioned matrix

$$[A_4] = \left[\begin{array}{ccc:c} a_{11} & a_{12} & a_{13} & a_{14} \\ a_{21} & a_{22} & a_{23} & a_{24} \\ a_{31} & a_{32} & a_{33} & a_{34} \\ \hdashline a_{41} & a_{41} & a_{43} & a_{44} \end{array}\right]$$

$[A_4]^{-1}$ can be computed by $[A_3]^{-1}$ and Eqs. (e)–(h).

28.9 Orthogonal Matrix

For a square matrix $[A]$, if its transpose matrix $[A]^T$ multiplied by itself is a unit matrix, that is,

$$[A][A]^T = [A]^T[A] = [I] \tag{28.20}$$

then this matrix is called the orthogonal matrix. According to the definition of the inverse matrix Eqs. (28.10) and (28.20), the transpose of the orthogonal matrix equals its inverse matrix, that is,

$$[A]^T = [A]^{-1} \tag{28.21}$$

From Eq. (28.20) we know that

$$\begin{bmatrix} a_{11} & a_{12} & \cdots & a_{1n} \\ a_{21} & a_{22} & \cdots & a_{2n} \\ \vdots & \vdots & & \vdots \\ a_{n1} & a_{n2} & \cdots & a_{nn} \end{bmatrix} \begin{bmatrix} a_{11} & a_{21} & \cdots & a_{n1} \\ a_{12} & a_{22} & \cdots & a_{n2} \\ \vdots & \vdots & & \vdots \\ a_{1n} & a_{2n} & \cdots & a_{nn} \end{bmatrix} = \begin{bmatrix} 1 & 0 & \cdots & 0 \\ 0 & 1 & \cdots & 0 \\ \vdots & \vdots & & \vdots \\ 0 & 0 & \cdots & 1 \end{bmatrix}$$

Based on the matrix multiplication, we have

$$a_{i1}^2 + a_{i2}^2 + \cdots + a_{in}^2 = 1, \quad i = 1,2,\ldots,n \tag{28.22a}$$

$$a_{i1}a_{j1} + a_{i2}a_{j2} + \cdots + a_{in}a_{jn} = 0, \quad \text{if } i \neq j, \; i,j = 1,2,\ldots,n \tag{28.22b}$$

Therefore, if $[A]$ is an orthogonal matrix, then the quadratic sum of elements in each row (or column) equals 1, while the sum of the product of element and its corresponding element from different row (or column) equals 0. If the elements in matrix $[A]$ do not satisfy Eq. (28.22), then $[A]$ is not an orthogonal matrix (Figure 28.1).

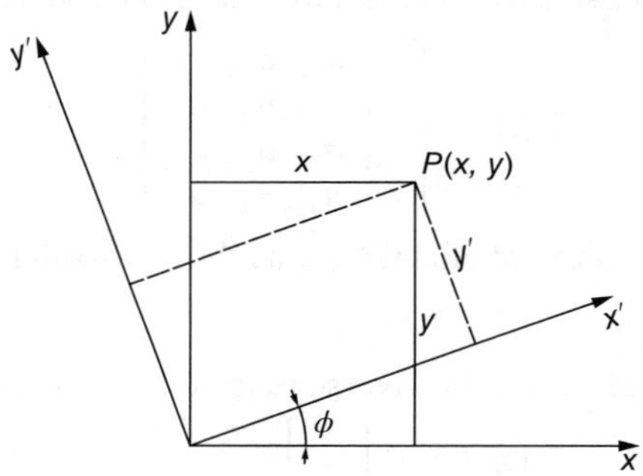

Figure 28.1 Coordinate transform.

Example 1 For any point $P(x, y)$ in the plane rectangular coordinate system, when the axis rotates around the original point with an angle ϕ, the new coordinates are

$$x' = x\cos\phi + y\sin\phi$$
$$y' = -x\sin\phi + y\cos\phi$$

The new coordinates can also be presented in the matrix pattern as

$$\begin{Bmatrix} x' \\ y' \end{Bmatrix} = [L] \begin{Bmatrix} x \\ y \end{Bmatrix}$$

where

$$[L] = \begin{bmatrix} \cos\phi & \sin\phi \\ -\sin\phi & \cos\phi \end{bmatrix}$$

Since

$$\cos^2\phi + \sin^2\phi = 1$$
$$(-\sin\phi)^2 + \cos^2\phi = 1$$
$$\cos\phi(-\sin\phi) + \sin\phi\cos\phi = 0$$

and the elements of $[L]$ satisfy Eq. (28.22), $[L]$ is an orthogonal matrix, that is,

$$[L]^{-1} = [L]^{\mathrm{T}} = \begin{bmatrix} \cos\phi & -\sin\phi \\ \sin\phi & \cos\phi \end{bmatrix}$$

Thus,

$$\begin{Bmatrix} x \\ y \end{Bmatrix} = [L]^{-1}\begin{Bmatrix} x' \\ y' \end{Bmatrix} = [L]^{\mathrm{T}}\begin{Bmatrix} x' \\ y' \end{Bmatrix}$$

28.10 Positive Definite Matrix

For the real matrix

$$[A] = \begin{bmatrix} a_{11} & a_{12} & \cdots & a_{1n} \\ a_{21} & a_{22} & \cdots & a_{2n} \\ \vdots & \vdots & \vdots & \vdots \\ a_{n1} & a_{n2} & \cdots & a_{nn} \end{bmatrix}$$

if the elements of $[A]$ are taken as the coefficients of quadratic homogenous polynomial

$$\sum_{i=1}^{n}\sum_{j=1}^{n} a_{ij}x_ix_j = a_{11}x_1^2 + a_{12}x_1x_2 + \cdots + a_{1n}x_1x_n$$
$$+a_{21}x_2x_1 + a_{22}x_2^2 + \cdots + a_{2n}x_2x_n + \cdots +$$
$$a_{n1}x_nx_1 + a_{n2}x_nx_2 + \cdots + a_{nn}x_n^2$$

and for any group with not all zero numbers $x_1, x_2, \ldots, x_n$ always satisfying

$$\sum_{i=1}^{n}\sum_{j=1}^{n} a_{ij}x_ix_j > 0 \tag{28.23}$$

then matrix $[A]$ is positive definite. The sufficient and necessary condition for a matrix $[A]$ to be positive definite is all the principal minor determinants of $[A]$ are greater than zero, that is,

$$a_{11} > 0, \quad \begin{vmatrix} a_{11} & a_{12} \\ a_{21} & a_{22} \end{vmatrix} > 0, \cdots \begin{vmatrix} a_{11} & a_{12} & \cdots & a_{1n} \\ a_{21} & a_{22} & \cdots & a_{2n} \\ \vdots & \vdots & \vdots & \vdots \\ a_{n1} & a_{n2} & \cdots & a_{nn} \end{vmatrix} > 0 \tag{28.24}$$

It can be seen that the positive definite matrix must be nonsingular.

For example, matrix

$$[A] = \begin{bmatrix} 4 & 2 & 1 \\ 2 & 3 & 0 \\ 1 & 0 & 2 \end{bmatrix}$$

of which the leading principal minors are

$$4>0,\quad \begin{vmatrix}4&2\\2&3\end{vmatrix}=8>0,\quad \begin{vmatrix}4&2&1\\2&3&2\\1&0&2\end{vmatrix}=13>0$$

thus $[A]$ is positive definite.

28.11 Derivative of Matrix

If the element $a_{i,j}$ of matrix $[A]$ is the function of parameter t, the derivative of matrix with respect to t is

$$\frac{d}{dt}[A]=[\dot{A}]=\begin{bmatrix}\frac{da_{11}}{dt} & \frac{da_{12}}{dt} & \cdots & \frac{da_{1n}}{dt}\\ \frac{da_{21}}{dt} & \frac{da_{22}}{dt} & \cdots & \frac{da_{2n}}{dt}\\ \vdots & \vdots & \vdots & \vdots\\ \frac{da_{n1}}{dt} & \frac{da_{n2}}{dt} & \cdots & \frac{da_{mn}}{dt}\end{bmatrix} \tag{28.25}$$

The derivative of matrix $[A]$ can be obtained by calculating the derivative of each element. The higher-order derivative of the matrix can be defined in the same way. For example, the second-order derivative of the matrix is obtained by differentiating the element twice.

Example 1

$$[A]=\begin{bmatrix}t^2+1 & 2t\\ 4 & t^3+t\end{bmatrix}$$

The first derivative of $[A]$ is

$$\frac{d}{dt}[A]=[\dot{A}]=\begin{bmatrix}2t & 2\\ 0 & 3t^2+1\end{bmatrix}$$

The second derivative of $[A]$ is

$$\frac{d^2}{dt^2}[A]=[\ddot{A}]=\begin{bmatrix}2 & 0\\ 0 & 6t\end{bmatrix}$$

Example 2 As shown in Figure 28.2, suppose there is an elastic beam with two concentrated mass m_1 and m_2. Two external forces $f_1(t)$ and $f_2(t)$ are applied at m_1 and m_2, respectively. Figure 28.2(b) shows the equilibrium of the mass m_1. The inertia of the beam can be ignored compared to the concentrated mass. According to the equilibrium condition, the motion equation of the elastic system can be written as

$$\begin{aligned} m_1\frac{d^2x_1}{dt^2}+k_{11}x_1+k_{12}x_2 &= f_1(t)\\ m_2\frac{d^2x_2}{dt^2}+k_{21}x_1+k_{22}x_2 &= f_2(t) \end{aligned} \tag{a}$$

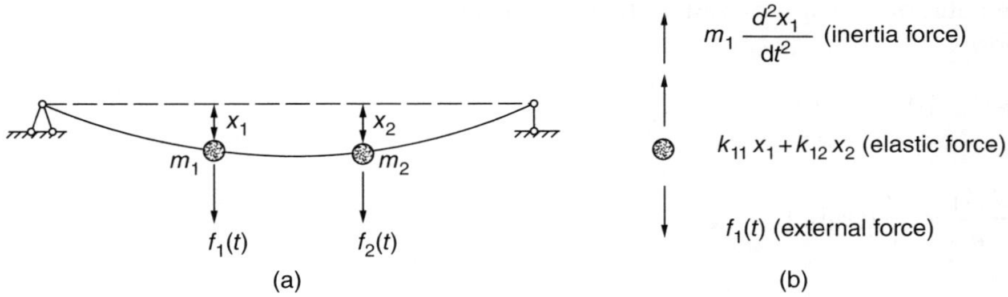

Figure 28.2 Vibration of beam.

where k_{11}, k_{12}, etc. are stiffness coefficient. The matrix form of the above equations is

$$\begin{bmatrix} m_1 & 0 \\ 0 & m_2 \end{bmatrix} \begin{Bmatrix} \dfrac{d^2x_1}{dt^2} \\ \dfrac{d^2x_2}{dt^2} \end{Bmatrix} + \begin{bmatrix} k_{11} & k_{12} \\ k_{21} & k_{22} \end{bmatrix} \begin{Bmatrix} x_1 \\ x_2 \end{Bmatrix} = \begin{Bmatrix} f_1(t) \\ f_2(t) \end{Bmatrix} \tag{b}$$

Let $\{x\} = \begin{Bmatrix} x_1 \\ x_2 \end{Bmatrix}$. Then

$$\begin{Bmatrix} \dfrac{d^2x_1}{dt^2} \\ \dfrac{d^2x_2}{dt^2} \end{Bmatrix} = \frac{d^2}{dt^2}\{x\} = \{\ddot{x}\}$$

Thus Eq. (b) can be noted as

$$[M]\{\ddot{x}\} + [K]\{x\} = \{f\} \tag{c}$$

where

$$[M] = \begin{bmatrix} m_1 & 0 \\ 0 & m_2 \end{bmatrix}, \quad [K] = \begin{bmatrix} k_{11} & k_{12} \\ k_{21} & k_{22} \end{bmatrix}, \quad \{f\} = \begin{Bmatrix} f_1(t) \\ f_2(t) \end{Bmatrix}$$

The derivative of a scalar with respect to each element of the column matrix $\{x\}$ can also constitute a column matrix and can be noted as

$$\begin{Bmatrix} \dfrac{\partial I}{\partial x_1} \\ \dfrac{\partial I}{\partial x_2} \\ \vdots \\ \dfrac{\partial I}{\partial x_n} \end{Bmatrix} = \frac{\partial I}{\partial \{x\}}$$

where

$$\{x\} = \begin{Bmatrix} x_1 \\ x_2 \\ \vdots \\ x_n \end{Bmatrix}$$

and I is a scalar.

These notations are often used in the finite element method.
Suppose

$$[A] = [B][C]$$

Then

$$\frac{d[A]}{dt} = \frac{d}{dt}([B][C])$$

but

$$\frac{d[A]}{dt} \neq \left(\frac{d}{dt}[B]\right)\left(\frac{d}{dt}[C]\right)$$

28.12 Integration of Matrix

If the element a_{ij} of matrix $[A]$ is the function of parameter t, the definition of the integral of matrix $[A]$ is

$$\int [A]dt = \begin{bmatrix} \int a_{11}dt & \int a_{12}dt & \cdots & \int a_{1n}dt \\ \int a_{21}dt & \int a_{22}dt & \cdots & \int a_{2n}dt \\ \vdots & \vdots & \vdots & \vdots \\ \int a_{m1}dt & \int a_{m2}dt & \cdots & \int a_{mn}dt \end{bmatrix} \tag{28.26}$$

The multiple integral of matrix can be defined in the similar way.

Example 1 Suppose there is a matrix

$$[A] = \begin{bmatrix} x & 2x^2 \\ 3x^3 & 1 \end{bmatrix}$$

Then

$$\int [A]dx = \begin{bmatrix} \frac{x^2}{2} & \frac{2x^3}{3} \\ \frac{3x^4}{4} & x \end{bmatrix}$$

Suppose

$$[A] = [B][C]$$

Then

$$\int [A]dt = \int ([B][C])dt$$

and

$$\int [A]dt \neq \int [B]dt \int [C]dt$$

In the finite element method, it is often required to calculate the integral of the product of several matrices as follows:

$$[K] = \iiint [B]^{\mathrm{T}}[D][B]\mathrm{d}x\mathrm{d}y\mathrm{d}z$$

where the element of $[B]$ is the function of x, y, and z. In order to calculate the above integral, the matrix multiplication $[B]^{\mathrm{T}}[D][B] = [C]$ has to be accomplished first, and then the triple integral is done with respect to each element of matrix $[C]$.

Bibliography

1 Pipes, L.A. (1963) *Matrix Methods for Engineering*, Prentice-Hall, Englewood Cliffs, NJ.

2 Puzemieniecki, J.S. (1965) Matrix Methods in Structural Mechanics. Proceedings of the Conference Held at Wright-Patterson Air Force Base, Ohio.

3 Martin, H.C. (1963) *Introduction to Matrix Methods of Structural Analysis*, McGraw-Hill.

4 Jenkins, W.M. (1969) *Matrix and Digital Computer Methods in Structural Analysis*, McGraw-Hill.

29

Linear Algebraic Equation Set

When the finite element method is applied to solve engineering problems, it is necessary to solve a linear algebraic equation set eventually.

In this chapter, the methods used on electronic computers for the solution of linear algebraic equation set will be introduced. The methods only applicable for manual calculation are ignored.

29.1 Linear Algebraic Equation Set

The general form of linear algebraic equation set is

$$\left.\begin{aligned} a_{11}x_1 + a_{12}x_2 + \cdots + a_{1n}x_n &= b_1 \\ a_{21}x_1 + a_{22}x_2 + \cdots + a_{2n}x_n &= b_2 \\ &\vdots \\ a_{n1}x_1 + a_{n2}x_2 + \cdots + a_{nn}x_n &= b_n \end{aligned}\right\} \tag{29.1}$$

Its matrix form is

$$\begin{bmatrix} a_{11} & a_{12} & \cdots & a_{1n} \\ a_{21} & a_{22} & \cdots & a_{2n} \\ \vdots & \vdots & \vdots & \vdots \\ a_{n1} & a_{n2} & \cdots & a_{nn} \end{bmatrix} \begin{Bmatrix} x_1 \\ x_2 \\ \vdots \\ x_n \end{Bmatrix} = \begin{Bmatrix} b_1 \\ b_2 \\ \vdots \\ b_n \end{Bmatrix} \tag{29.1a}$$

The equation above can be written as

$$[A]\{x\} = \{b\} \tag{29.2}$$

in which

$$[A] = \begin{bmatrix} a_{11} & a_{12} & \cdots & a_{1n} \\ a_{21} & a_{22} & \cdots & a_{2n} \\ \vdots & \vdots & \vdots & \vdots \\ a_{n1} & a_{n2} & \cdots & a_{nn} \end{bmatrix} \tag{29.3}$$

The Finite Element Method: Fundamentals and Applications in Civil, Hydraulic, Mechanical and Aeronautical Engineering, First Edition. Bofang Zhu.

Companion website: www.wiley.com/go/zhufem17

$[A]$ is the coefficient matrix of the equation set:

$$\{x\} = \begin{Bmatrix} x_1 \\ x_2 \\ \vdots \\ x_n \end{Bmatrix}, \quad \{b\} = \begin{Bmatrix} b_1 \\ b_2 \\ \vdots \\ b_n \end{Bmatrix}$$

where $\{x\}$ is the unknown array and $\{b\}$ is the free term array.

The necessary and sufficient condition that the unique solution exists for the equation set is that the coefficient matrix $[A]$ is nonsingular, which means the determinant $|A|$ is not equal to zero.

Obviously, if $|A| \neq 0$, there is an inverse matrix $[A]^{-1}$. Multiplying both sides of Eq. (29.2) by $[A]^{-1}$, we get

$$\{x\} = [A]^{-1}\{b\} \tag{29.4}$$

The solution $\{x\}$ calculated by the equation above is unique because $[A]^{-1}$ is uniquely determined by $[A]$.

29.2 Simple Iterative Method

We take the following linear equation set, for example, to introduce the simple iterative method:

$$\left.\begin{aligned} a_{11}x_1 + a_{12}x_2 + a_{13}x_3 &= b_1 \\ a_{21}x_1 + a_{22}x_2 + a_{23}x_3 &= b_2 \\ a_{31}x_1 + a_{32}x_2 + a_{33}x_3 &= b_3 \end{aligned}\right\} \tag{29.5}$$

Suppose the absolute values of the elements of leading diagonal a_{11}, a_{22}, and a_{33} are much bigger than the absolute values of other elements. Get x_1, x_2, and x_3 in sequence as follows:

$$\left.\begin{aligned} x_1 &= \frac{1}{a_{11}}(b_1 - a_{12}x_2 - a_{13}x_3) \\ x_2 &= \frac{1}{a_{22}}(b_2 - a_{21}x_1 - a_{23}x_3) \\ x_3 &= \frac{1}{a_{33}}(b_3 - a_{31}x_1 - a_{32}x_2) \end{aligned}\right\} \tag{29.6}$$

The initial value may be given as

$$x_1^{(0)} = x_2^{(0)} = x_3^{(0)} = 0 \tag{29.7}$$

Substitute the equation above into Eq. (29.6), and we get the first approximate value as

$$x_1^{(1)} = \frac{b_1}{a_{11}}, \quad x_2^{(1)} = \frac{b_2}{a_{22}}, \quad x_3^{(1)} = \frac{b_3}{a_{33}} \tag{29.8}$$

Substitution of the approximate value set into the right-hand side of Eq. (29.6) will give the second approximate values as

$$\left.\begin{aligned} x_1^{(2)} &= \frac{1}{a_{11}}(b_1 - a_{12}x_2^{(1)} - a_{13}x_3^{(1)}) \\ x_2^{(2)} &= \frac{1}{a_{22}}(b_2 - a_{21}x_1^{(1)} - a_{23}x_3^{(1)}) \\ x_3^{(2)} &= \frac{1}{a_{33}}(b_3 - a_{31}x_1^{(1)} - a_{32}x_2^{(1)}) \end{aligned}\right\} \tag{29.9}$$

If the iteration is convergent when calculating repeatedly, the results obtained in sequence will be more accurate. The iterative computation will finish when the stipulated accuracy is reached. Let

$$\Delta x_i^{(r)} = x_i^{(r)} - x_i^{(r-1)},\ i = 1,2,3 \tag{29.10}$$

Let $\max|\Delta x_i^{(r)}|$ be the maximum value of Δx_i. Obviously, the smaller the $\max|\Delta x_i^{(r)}|$ is, the more accurate the result is so that we can use it as the scale of the calculation error. Let the allowable error be ε, and then if

$$\max|\Delta x_i^{(r)}| < \varepsilon \tag{29.11}$$

the calculation stops and the array $x_i^{(r)}$ is the approximate solution of the equation set.

For an n-order equation set (29.1), the iterative formula is

$$x_i^{(r)} = \frac{1}{a_{ii}}\left(b_i - \sum_{j=1}^{i-1} a_{ij}x_j^{(r-1)} - \sum_{j=i+1}^{n} a_{ij}x_j^{(r-1)}\right),\ i = 1,2,3,\dots,n \tag{29.12}$$

Substitution of $x_j^{(r-1)}$ into the above equation will give the rth approximate result $x_i^{(r)}$, $i = 1,2,\dots,n$.

According to Eq. (29.10), the difference between the two approximate values is

$$\Delta x_i^{(r)} = x_i^{(r)} - x_i^{(r-1)} = \frac{1}{a_{ii}}\left(\sum_{j=1}^{i-1} a_{ij}\Delta x_j^{(r-1)} + \sum_{j=i+1}^{n} a_{ij}\Delta x_j^{(r-1)}\right) \tag{29.13}$$

Then by Eq. (29.11) it can be judged whether the results meet the accuracy requirement.

Calculation steps for the simple iterative method are as follows:

1) Select an arbitrary initial value $x_i^{(0)}$, $i = 1,2,\dots,n$; usually the zero vector is used.
2) Substitute $x_i^{(0)}$ into the right-hand side of Eq. (29.12) and calculate the first approximate value $x_i^{(1)}$, $i = 1,2,\dots,n$.
3) Judge whether the approximate value meets the accuracy requirement. If the accuracy is not met, substitute $x_i^{(1)}$ into the right-hand side of Eq. (29.12) and calculate the second approximate value $x_i^{(2)}$, $i = 1,2,\dots,n$.

Repeat the steps above and stop until Eq. (29.11) is satisfied.

When the calculation steps above are used in calculating, the sufficient and necessary condition of convergence is that each absolute value of the element of main diagonal of the coefficient matrix $[A]$ is, respectively, greater than the sum of the absolute values of the other elements in the same row, namely,

$$|a_{ii}| > \sum_{j=1}^{i-1} |a_{ij}| + \sum_{j=i+1}^{n} |a_{ij}|, \quad i = 1,2,\ldots,n \tag{29.14}$$

When this condition is met, the calculation must be convergent. But when the condition is not met, the calculation may not necessarily be divergent.

29.3 Seidel Iterative Method

We still take the third-order linear equation set, for example, to introduce the Seidel iterative method. Substitute the initial value $x_1^{(0)} = x_2^{(0)} = x_3^{(0)} = 0$ into the right-hand side of the first equation in equation set (29.6) and obtain the first approximate value $x_1^{(1)}$. When the first approximate value $x_2^{(1)}$ is calculated, use $x_1^{(1)}$, which is already obtained instead of $x_1^{(0)}$. Similarly, use $x_1^{(1)}$ and $x_2^{(1)}$ when calculating the first approximate value $x_3^{(1)}$. It can be written as follows:

$$x_1^{(1)} = \frac{b_1}{a_{11}}$$

$$x_2^{(1)} = \frac{1}{a_{22}}\left(b_2 - a_{21}x_1^{(1)}\right)$$

$$x_3^{(1)} = \frac{1}{a_{33}}\left(b_3 - a_{31}x_1^{(1)} - a_{32}x_2^{(1)}\right)$$

In the next step, $x_2^{(1)}$, and $x_3^{(1)}$ are used to calculate $x_1^{(2)}$; $x_1^{(2)}$, and $x_3^{(1)}$ are used to calculate $x_2^{(2)}$; and $x_1^{(2)}$, $x_2^{(2)}$, are used to calculate $x_3^{(2)}$. The second approximate values are obtained as follows:

$$\left.\begin{aligned} x_1^{(2)} &= \frac{1}{a_{11}}\left(b_1 - a_{12}x_2^{(1)} - a_{13}x_3^{(1)}\right) \\ x_2^{(2)} &= \frac{1}{a_{22}}\left(b_2 - a_{21}x_1^{(2)} - a_{23}x_3^{(1)}\right) \\ x_3^{(2)} &= \frac{1}{a_{33}}\left(b_3 - a_{31}x_1^{(2)} - a_{32}x_2^{(2)}\right) \end{aligned}\right\} \tag{29.15}$$

Repeat the calculating steps until the calculation precision meets the requirements.

We can find the differences between the simple iterative method and the Seidel iterative method by comparing Eqs (29.15) and (29.9). When the rth approximate values are calculated, the $r - 1$th approximate values are used in the right-hand side for the simple iterative method. But for the Seidel iterative method, the rth approximate values already obtained are used as possible in the right-hand side of the equation, and the $(r - 1)$th approximate values are used in case the rth approximate values are unknown provisionally. This is the reason why the Seidel iterative method shares a faster rate of convergence.

For an n-order equation set (24.1), the computational formula of the Seidel iterative method is

$$x_i^{(r)} = \frac{1}{a_{ii}}\left(b_i - \sum_{j=1}^{i-1} a_{ij}x_j^{(r)} - \sum_{j=i+1}^{n} a_{ij}x_j^{(r-1)}\right), \qquad i = 1,2,3,\ldots,n \tag{29.16}$$

The difference between the two adjacent approximate calculations is

$$\Delta x_i^{(r)} = x_i^{(r)} - x_i^{(r-1)} = \frac{1}{a_{ii}}\left(\sum_{j=1}^{i-1} a_{ij}\Delta x_j^{(r)} - \sum_{j=i}^{n} a_{ij}\Delta x_j^{(r-1)}\right) \tag{29.17}$$

Suppose that the permissible error is ε, when

$$\max|\Delta x_i^{(r)}| < \varepsilon$$

is satisfied, the calculation stops and the rth approximate values are used as the approximate solution for the equation set.

The sufficient condition for convergence of the Seidel iterative method is that if the coefficient matrix $[A]$ is symmetric and positive definite, the Seidel iterative calculation will be convergent with an arbitrary initial value $x_i^{(0)}$.

29.4 Over-Relaxation Iterative Method

The rth approximate values are given by Eq. (29.17) when the Seidel iterative method is applied to solve equation set (29.1):

$$x_i^{(r)} = x_i^{(r-1)} + \Delta x_i^{(r)} = x_i^{(r-1)} + \frac{1}{a_{ii}}\left(\sum_{j=1}^{i-1} a_{ij}\Delta x_j^{(r)} + \sum_{j=i}^{n} a_{ij}\Delta x_j^{(r-1)}\right)$$

To speed up the rate of convergence, we multiply $\Delta x_i^{(r)}$ by a factor ω and then plus $x_i^{(r-1)}$ to calculate the rth approximate value, that is,

$$x_i^{(r)} = x_i^{(r-1)} + \omega\Delta x_i^{(r)} = x_i^{(r-1)} + \frac{\omega}{a_{ii}}\left(\sum_{j=1}^{i-1} a_{ij}\Delta x_j^{(r)} + \sum_{j=i}^{n} a_{ij}\Delta x_j^{(r-1)}\right) \tag{29.18}$$

in which ω is a constant. It can be proved that if the coefficient matrix $[A]$ of the equation set is symmetric and positive definite, when and only when

$$0 < \omega < 2,$$

the iterative formula (29.18) is convergent with an arbitrary initial value $x_i^{(0)}$.

If $\omega = 1$, it turns to be the Seidel iterative method. In the case of $\omega < 1$, it becomes the under-relaxation iteration method, and in case of $1 < \omega < 2$, it is the over-relaxation iteration method, which is used more often. ω is called the over-relaxation factor.

The value of the over-relaxation factor has a great influence on the convergence rate of the iterative calculation. The optimal over-relaxation factor changes with the structural style and the loading conditions. From the existing experiences, a better result will be obtained when $\omega = 1.8 - 1.9$ for elastic plane problems, and a smaller value ω is better for the elastic thin plate and shell.

29.5 Block Over-Relaxation Iterative Method

The block over-relaxation iterative method can be applied to speed up the calculation further. Take an n-order equation set (29.1), for example, the coefficient matrix $[A]$ can be divided into $m \times m$ submatrices, and the unknown array $\{x\}$ and the free term array $\{b\}$ are also divided into m subarrays, that is,

$$\begin{bmatrix} a_{11} & a_{12} & \cdots & | & \cdots & | & \cdots & \cdots & a_{1n} \\ a_{21} & a_{22} & \cdots & | & \cdots & | & \cdots & \cdots & a_{2n} \\ \cdots & \cdots & \cdots & | & \cdots & | & \cdots & \cdots & \cdots \\ -- & -- & -- & | & - & | & -- & -- & -- \\ \cdots & \cdots & \cdots & | & \cdots & | & \cdots & \cdots & \cdots \\ \cdots & \cdots & \cdots & | & \cdots & | & \cdots & \cdots & \cdots \\ -- & -- & -- & | & - & | & -- & -- & -- \\ \cdots & \cdots & \cdots & | & \cdots & | & \cdots & \cdots & \cdots \\ a_{n1} & a_{n2} & \cdots & | & \cdots & | & \cdots & \cdots & a_{nn} \end{bmatrix} \begin{Bmatrix} x_1 \\ x_2 \\ \cdots \\ -- \\ \cdots \\ \cdots \\ -- \\ \cdots \\ x_n \end{Bmatrix} = \begin{Bmatrix} b_1 \\ b_2 \\ \cdots \\ -- \\ \cdots \\ \cdots \\ -- \\ \cdots \\ b_n \end{Bmatrix}$$

Take $[A_{ij}]$,$\{X_i\}$, and $\{B_i\}$ to represent the coefficient submatrix, the unknown subarray, and the free term subarray in succession, and the equation above can be written as

$$\begin{bmatrix} A_{11} & A_{12} & \cdots & A_{1m} \\ A_{21} & A_{22} & \cdots & A_{2m} \\ \vdots & \vdots & \vdots & \vdots \\ A_{m1} & A_{m2} & \cdots & A_{mm} \end{bmatrix} \begin{Bmatrix} X_1 \\ X_2 \\ \vdots \\ X_m \end{Bmatrix} = \begin{Bmatrix} B_1 \\ B_2 \\ \vdots \\ B_m \end{Bmatrix} \tag{29.19}$$

m matrix equations can be obtained from the equation above:

$$[A_{i1}]\{X_1\} + [A_{i2}]\{X_2\} + \cdots + [A_{im}]\{X_m\} = \{B_i\}, \ i = 1,2, \ldots, m.$$

Keeping the ith item in the left side and moving the other items to the right side of the equation, then we get

$$[A_{ii}]\{X_i\} = \{B_i\} - \sum_{j=1}^{i-1}[A_{ij}]\{X_j\} - \sum_{j=i+1}^{m}[A_{ij}]\{X_j\}$$

Multiply both sides of the equation by $[A_{ii}]^{-1}$ to get

$$\{X_i\} = [A_{ii}]^{-1}\left(\{B_i\} - \sum_{j=1}^{i-1}[A_{ij}]\{X_j\} - \sum_{j=i+1}^{m}[A_{ij}]\{X_j\}\right)$$

Substitute the known initial value into the right-hand side of the equation above and $\{X_i\}$ can be calculated. Now introduce the relaxation factor ω and use similar steps for deriving Eq. (29.18) in the previous chapter, we can obtain the computational formula of the block over-relaxation iterative method as follows:

$$\begin{aligned} \{X_i^{(r)}\} &= \{X_i^{(r-1)}\} + \omega\{\Delta X_i^{(r)}\} \\ &= \{X_i^{(r-1)}\} + \omega[A_{ii}]^{-1}\left(\sum_{j=1}^{i-1}[A_{ij}]\{\Delta X_j^{(r)}\} + \sum_{j=i}^{m}[A_{ij}]\{\Delta X_j^{(r-1)}\}\right), \quad 1 \le i \le m \end{aligned} \tag{29.20}$$

Substitute the known approximate value into the right-hand side of the equation and the rth approximate value $\{X_i^{(r)}\}$, $i = 1,2, \ldots, m$ can be calculated.

If the explicit formulation of the inverse matrix $[A_{ii}]^{-1}$ cannot be displayed, transform Eq. (29.20) as follows:

$$[A_{ii}]\{X^{(r)}\} = [A_{ii}]\{X_i^{(r-1)}\} + \omega\left(\{B_i\} - \sum_{j=1}^{i-1}[A_{ij}]\{X_j^{(r)}\} - \sum_{j=i}^{m}[A_{ij}]\{X_j^{(r-1)}\}\right),\ 1 \le i \le m \tag{29.21}$$

The direct solution method can be used to solve the equation above.

Now we explain the condition for convergence of the block over-relaxation iterative method. If the coefficient matrix $[A]$ is symmetric and positive definite and the matrix

$$\begin{bmatrix} A_{11} & & 0 \\ & A_{12} & \\ & & \ddots \\ 0 & & & A_{mm} \end{bmatrix}$$

is positive definite and when and only when $0 < \omega < 2$, the block over-relaxation iterative calculation will be convergent with an arbitrary initial value $\{x^{(0)}\}$.

29.6 Direct Solution Method

The direct solution methods for linear equations generally used are elimination method, pivoting elimination method, decomposing coefficient matrix method, and so on. Here we only discuss the decomposing coefficient matrix method, which is used more often.

For the equation set

$$[A]\{x\} = \{b\} \tag{29.2}$$

we introduce, respectively, the decomposing coefficient matrix methods for the equation set when the coefficient matrix $[A]$ is symmetrical or asymmetrical.

Case 1 The coefficient matrix $[A]$ is asymmetrical

Do LU decomposition to the coefficient matrix $[A]$:

$$[A] = [L][M] \tag{29.22}$$

where $[L]$ is a lower triangular matrix and $[M]$ is an upper triangular matrix of which values of the elements of main diagonal are all 1:

$$[L] = \begin{bmatrix} l_{11} & 0 & 0 & \cdots & 0 \\ l_{21} & l_{22} & 0 & \cdots & 0 \\ l_{31} & l_{32} & l_{33} & \cdots & 0 \\ \vdots & \vdots & \vdots & \vdots & \vdots \\ l_{n1} & l_{n2} & l_{n3} & \cdots & l_{nn} \end{bmatrix},\ [M] = \begin{bmatrix} 1 & m_{12} & m_{13} & \cdots & m_{1n} \\ 0 & 1 & m_{23} & \cdots & m_{2n} \\ 0 & 0 & 1 & \cdots & m_{3n} \\ \vdots & \vdots & \vdots & \vdots & \vdots \\ 0 & 0 & 0 & \cdots & 1 \end{bmatrix} \tag{29.23}$$

Now we take a four-order matrix, for example, to introduce the way to calculate the elements of the matrix [L] and [M]:

$$\begin{bmatrix} a_{11} & a_{12} & a_{13} & a_{14} \\ a_{21} & a_{22} & a_{23} & a_{24} \\ a_{31} & a_{32} & a_{33} & a_{34} \\ a_{41} & a_{42} & a_{43} & a_{44} \end{bmatrix} = \begin{bmatrix} l_{11} & 0 & 0 & 0 \\ l_{21} & l_{22} & 0 & 0 \\ l_{31} & l_{32} & l_{33} & 0 \\ l_{41} & l_{42} & l_{43} & l_{44} \end{bmatrix} \begin{bmatrix} 1 & m_{12} & m_{13} & m_{14} \\ 0 & 1 & m_{23} & m_{24} \\ 0 & 0 & 1 & m_{34} \\ 0 & 0 & 0 & 1 \end{bmatrix} \tag{a}$$

From Eq. (24.22), according to the matrix multiplication rule, the elements of the matrix [L] and [M] can be calculated line by line as follows:

Line 1: According to the matrix multiplication rule,

$$a_{11} = l_{11},\ a_{12} = l_{11}m_{12},\ a_{13} = l_{11}m_{13},\ a_{14} = l_{11}m_{14}$$

Then we get

$$l_{11} = a_{11},\ m_{12} = \frac{a_{12}}{l_{11}},\ m_{13} = \frac{a_{13}}{l_{11}},\ m_{14} = \frac{a_{14}}{l_{11}}$$

Line 2: According to the matrix multiplication rule,

$$a_{21} = l_{21},\ a_{22} = l_{21}m_{12} + l_{22},\ a_{23} = l_{21}m_{13} + l_{22}m_{23},\ a_{24} = l_{21}m_{14} + l_{22}m_{24}$$

Then we get

$$l_{21} = a_{21},\ l_{22} = a_{22} - l_{21}m_{12},\ m_{23} = (a_{23} - l_{21}m_{13})/l_{22},\ m_{24} = (a_{24} - l_{21}m_{14})/l_{22}$$

Line 3: According to the matrix multiplication rule,

$$a_{31} = l_{31},\ a_{32} = l_{31}m_{12} + l_{32},$$
$$a_{33} = l_{31}m_{13} + l_{32}m_{23} + l_{33},\ a_{34} = l_{31}m_{14} + l_{32}m_{24} + l_{33}m_{34}$$

Then we get

$$l_{31} = a_{31},\ l_{32} = a_{32} - l_{31}m_{12},$$
$$l_{33} = a_{33} - l_{31}m_{13} - l_{32}m_{23},\ m_{34} = (a_{34} - l_{31}m_{14} - l_{32}m_{24})/l_{33}$$

Line 4: According to the matrix multiplication rule,

$$a_{41} = l_{41},\ a_{42} = l_{41}m_{12} + l_{42},$$
$$a_{43} = l_{41}m_{13} + l_{42}m_{23} + l_{43},\ a_{44} = l_{41}m_{14} + l_{42}m_{24} + l_{43}m_{34} + l_{44}$$

Then we get

$$l_{41} = a_{41},\ l_{42} = a_{42} - l_{41}m_{12},$$
$$l_{43} = a_{43} - l_{41}m_{13} - l_{42}m_{23},\ l_{44} = a_{44} - l_{41}m_{14} - l_{42}m_{24} - l_{43}m_{34}$$

Concluding from the above, we obtain the general formulas of l_{ij} and m_{ij} as follows:

$$\begin{aligned} l_{ij} &= a_{ij} - l_{i1}m_{1j} - l_{i2}m_{2j} - \cdots - l_{i,j-1}m_{j-1,i} \\ &= a_{ij} - \sum_{k=1}^{j-1} l_{ik}m_{kj} \quad i \geq j \end{aligned} \tag{29.24}$$

$$m_{ij} = (a_{ij} - l_{i1}m_{1j} - l_{i2}m_{2j} - \cdots - l_{i,i-1}m_{i-1,j}/l_{ii})$$
$$= \left(a_{ij} - \sum_{k=1}^{j-1} l_{ik}m_{kj}\right)/l_{ii}, \quad i < j \tag{29.25}$$

It becomes easy to solve the equation set after decomposing the coefficient matrix into two triangular matrices. In fact, the equation set $[A]\{x\} = \{b\}$ has already transformed into another form as follows:

$$[L][M]\{x\} = \{b\} \tag{29.26}$$

Let

$$[M]\{x\} = \{y\} \tag{29.27}$$

and substitute it into Eq. (29.26) to get

$$[L]\{y\} = \{b\} \tag{29.28}$$

Solve the array $\{y\}$ first from the above equation and then substitute them into Eq. (29.27); the array $\{x\}$ is obtained, and we obtain the solution of the equation set $[A]\{x\} = \{b\}$.

Rewrite Eq. (29.28) and Eq. (29.27) in explicit expression, that is,

$$\left.\begin{aligned} &l_{11}y_1 = b_1 \\ &l_{21}y_1 + l_{22}y_2 = b_1 \\ &\quad\vdots \\ &l_{i1}y_1 + l_{i2}y_2 + \cdots + l_{ii}y_i = b_i \\ &\quad\vdots \\ &l_{n1}y_1 + l_{n2}y_2 + \cdots + l_{nn}y_n = b_n \end{aligned}\right\} \tag{29.28a}$$

$$\left.\begin{aligned} &x_1 + m_{12}x_2 + m_{13}x_3 + \cdots + m_{1n}x_n = y_1 \\ &x_2 + m_{23}x_3 + \cdots + m_{2n}x_n = y_2 \\ &\qquad\qquad\vdots \\ &x_i + \cdots + m_{in}x_n = y_i \\ &\qquad\qquad\vdots \\ &x_n = y_n \end{aligned}\right\} \tag{29.27a}$$

Solve $y_1, y_2, \ldots, y_n$ from above to below by Eq. (29.28a) and substitute them into Eq. (29.27a) as the free item array, and then $x_n, x_{n-1}, \ldots, x_2, x_1$ can be solved from below to above.

The sufficient condition that *LU* decomposition $[A] = [L][M]$ can be done is that all the principal minors of $|A|$ do not equal to 0.

In engineering calculations, the coefficient matrix $[A]$ of the equation set is banded frequently, which is in favor of compressing the computer storage capacity, because it is not necessary to store zero elements outside the band. We may have a question whether the matrices $[L]$ and $[M]$ are still banded after the *LU* decomposition and where is the edge of the band. We take the matrix below to illustrate this problem.

(The upper edge of the band)

$$[A]=\begin{bmatrix} a_{11} & a_{12} & a_{13} & 0 & 0 & 0 & 0 & 0 \\ a_{21} & a_{22} & a_{23} & a_{24} & a_{25} & 0 & 0 & 0 \\ 0 & a_{32} & a_{33} & a_{34} & 0 & 0 & 0 & 0 \\ 0 & 0 & a_{43} & a_{44} & a_{45} & a_{46} & 0 & 0 \\ 0 & 0 & 0 & a_{54} & a_{55} & 0 & 0 & a_{58} \\ 0 & 0 & a_{63} & a_{64} & 0 & a_{66} & a_{67} & 0 \\ 0 & 0 & 0 & 0 & a_{75} & a_{76} & a_{77} & 0 \\ 0 & 0 & 0 & 0 & 0 & a_{86} & a_{87} & a_{88} \end{bmatrix} \tag{b}$$

(The lower edge of the band)

Take the fifth line out of the matrix, and the left three elements are all zero elements, that is,

$$a_{51} = a_{52} = a_{53} = 0, a_{54} \neq 0$$

We can get from Eq. (29.24) that

$$l_{51} = l_{52} = l_{53} = 0, l_{54} = a_{54} \neq 0$$

It can be found that the location of the first nonzero element from the left does not change.

If we call the ligature of the locations of the first nonzero element from the left in each line the lower edge of the band, then the lower triangular matrix $[L]$ and the matrix $[A]$ share the same lower edge of the band.

Then take the fifth row out of the matrix. The upper three elements are all zero elements, that is,

$$a_{16} = a_{26} = a_{36} = 0, a_{46} \neq 0$$

We can get from Eq. (29.25) that

$$m_{16} = m_{26} = m_{36} = 0, m_{46} = \frac{a_{46}}{l_{46}} \neq 0$$

If we call the ligature of the locations of the first nonzero element from above to below in each row the upper edge of the band, then the upper triangular matrix $[M]$ and the matrix $[A]$ share the same upper edge of the band. It must be pointed that if we call the ligature of the locations of the last nonzero element from the left in each line as shown by dotted line in Eq. (b) the "upper edge of the band," this "upper edge of the band" will change after the decomposition.

Case 2 The coefficient matrix [*A*] is symmetrical

In case that the coefficient matrix [*A*] is a symmetric matrix,

$$a_{ij} = a_{ji},\ i \neq j$$

From Eqs. (29.24) and (29.25), we know that

$$l_{i1} = a_{i1},\ 1 \leq i \leq n$$

$$m_{1i} = \frac{a_{1i}}{l_{11}} = \frac{l_{i1}}{l_{11}}$$

$$l_{i2} = a_{i2} - l_{i1}m_{12} = a_{i2} - \frac{l_{i1}l_{21}}{l_{11}}$$

$$m_{2i} = \left(\frac{a_{i2} - l_{21}m_{1i}}{l_{22}}\right) = \frac{a_{i2} - \frac{l_{i1}l_{21}}{l_{11}}}{l_{22}} = \frac{l_{i2}}{l_{22}},\quad 1 \leq i \leq n$$

By parity of reasoning, it can be obtained:

$$\left.\begin{aligned} l_{ij} &= a_{ij} - \sum_{k=1}^{j-1} \frac{l_{ik}l_{jk}}{l_{kk}} \\ m_{ij} &= \frac{l_{ij}}{l_{jj}},\quad i \geq j, 1 \leq i \leq n, 1 \leq j \leq n \end{aligned}\right\} \tag{29.29}$$

Thus

$$[M] = [L_0][L]^{\mathrm{T}} \tag{29.30}$$

in which

$$[L_0] = \begin{bmatrix} \frac{1}{l_{11}} & 0 & \cdots & 0 \\ 0 & \frac{1}{l_{22}} & \cdots & 0 \\ \vdots & \vdots & \vdots & \vdots \\ 0 & 0 & \cdots & \frac{1}{l_{nn}} \end{bmatrix} \tag{29.31}$$

So the equation set $[A]\{x\} = \{b\}$ is transformed to be

$$[L][L_0][L]^{\mathrm{T}}\{x\} = \{b\} \tag{29.32}$$

Let

$$[L_0][L]^{\mathrm{T}}\{x\} = \{y\} \tag{29.33}$$

Then

$$[L]\{y\} = \{b\} \tag{29.34}$$

Get the array $\{y\}$ by Eq. (29.34):

$$y_i = \left(b_i - \sum_{k=1}^{i-1} l_{ik} y_k\right) \Big/ l_{ii}, \quad i = 1,2,\dots,n \tag{29.35}$$

Substitute $\{y\}$ into Eq. (29.33), and the array $\{x\}$ can be solved:

$$x_i = y_i - \left(\sum_{k=i+1}^{n} l_{ki} x_k\right) \Big/ l_{ii}, \quad i = n, n-1, \dots, 1 \tag{29.36}$$

29.7 Conjugate Gradient Method

Consider the linear equation set

$$[A]\{x\} = \{b\}$$

where $[A]$ is a symmetric positive definite matrix. The question to solve the equation set is equivalent to find an array $\{x\}$, which makes the quadratic functional below to obtain its minimum value:

$$H(\{x\}) = ([A]\{x\}, \{x\}) - 2(\{x\}, \{b\}) \tag{29.37}$$

in which $(\{x\}, \{b\})$ represent the inner product of vector $\{x\}$ and vector $\{b\}$. Suppose $\{x\} = [x_1, x_2, \dots, x_n]^{\mathrm{T}}$ and $\{b\} = [b_1, b_2, \dots, b_n]^{\mathrm{T}}$, and then the inner product can be written as

$$(\{x\}, \{b\}) = b_1 x_1 + b_2 x_2 + \cdots + b_n x_n$$

The relation of equivalence mentioned above can be proved. See Reference [1].

The conjugate gradient method is a successive approximation to find the array $\{x\}$, which makes the quadratic functional (29.37) to obtain its minimum value. If there is no roundoff error in the calculation, the exact solution will be obtained within no more than n steps by the conjugate gradient method for an n-order equation set.

Take array $\{x^{(0)}\}$ to be the initial approximation. Generally, this approximation will not meet the equation set (29.2). Then according to the following formula, calculate the first approximate value $\{x^{(1)}\}$, the second approximate value $\{x^{(2)}\}$, and so on:

$$\{q^{(0)}\} = \{r^{(0)}\} = \{b\} - [A]\{x^{(0)}\} \tag{29.38}$$

$$\{p^{(i)}\} = [A]\{q^{(i)}\} \tag{29.39}$$

$$\alpha^{(i)} = \frac{(\{q^{(i)}\}, \{r^{(i)}\})}{(\{q^{(i)}\}, \{p^{(i)}\})} \tag{29.40}$$

$$\{x^{(i+1)}\} = \{x^{(i)}\} + \alpha^{(i)}\{q^{(i)}\} \tag{29.41}$$

$$\{r^{(i+1)}\} = \{r^{(i)}\} - \alpha^{(i)}\{q^{(i)}\} \tag{29.42}$$

$$\beta^{(i)} = -\frac{(\{r^{(i+1)}\}, \{p^{(i)}\})}{(\{q^{(i)}\}, \{p^{(i)}\})} \tag{29.43}$$

$$\{q^{(i+1)}\} = \{r^{(i+1)}\} + \beta^{(i)}\{q^{(i)}\} \tag{29.44}$$

If the matrix $[A]$ is not symmetric positive definite, when the matrix $[A]$ is nonsingular, matrix $[A]^T[A]$ is symmetric and positive definite. Multiply both sides of equation $[A]\{x\} = \{b\}$ by $[A]^T$ and get

$$[A]^T[A]\{x\} = [A]^T\{b\} \tag{29.45}$$

Then the conjugate gradient method can be applied to this equation.

Comparing the conjugate gradient method and the direct solution method, we can find a common property that limited computing steps are necessary to obtain the exact solution of the equation set if there is no roundoff error in the calculation. The difference is that the conjugate gradient method uses iterative computations while the direct solution method does not, so that the program design for the conjugate gradient method is much easier when calculating with computers, and the zero elements in the matrix $[A]$ are not necessary to store, which can save a lot of computer storage.

Example 1 Solve the equation set below by the conjugate gradient method:

$$\begin{bmatrix} 2 & -1 \\ -1 & 1 \end{bmatrix} \begin{pmatrix} x_1 \\ x_2 \end{pmatrix} = \begin{pmatrix} 0 \\ 1 \end{pmatrix}$$

The coefficient matrix is symmetric and positive definite so that the conjugate gradient method can be applied to the equation set:

$$\{x^{(0)}\} = [0, 0]^T$$
$$\{q^{(0)}\} = \{r^{(0)}\} = \{b\} - [A]\{x^{(0)}\} = [0, 1]^T$$
$$\{p^{(0)}\} = [A]\{q^{(0)}\} = [-1,1]^T$$
$$\alpha^{(0)} = \frac{(\{q^{(0)}\}, \{r^{(0)}\})}{(\{q^{(0)}\}, \{p^{(0)}\})} = 1$$
$$\{x^{(1)}\} = \{x^{(0)}\} + \alpha^{(0)}\{q^{(0)}\} = [0, 1]^T$$

Check whether the first approximate value $\{x^{(1)}\}$ meets the allowable error and find out that the array $\{x^{(1)}\}$ does not meet the requirement, so go on calculating the second approximate value $\{x^{(2)}\}$:

$$\{r^{(1)}\} = \{r^{(0)}\} - \alpha^{(0)}\{q^{(0)}\} = [1, 0]^T$$
$$\beta^{(0)} = -\frac{(\{r^{(1)}\}, \{p^{(0)}\})}{(\{q^{(0)}\}, \{p^{(0)}\})} = 1$$
$$\{q^{(1)}\} = \{r^{(1)}\} + \beta^{(0)}\{q^{(0)}\} = \begin{Bmatrix} 1 \\ 1 \end{Bmatrix}$$
$$\{p^{(1)}\} = [A]\{q^{(1)}\} = [1, 0]^T$$
$$\alpha^{(1)} = \frac{(\{q^{(1)}\}, \{r^{(1)}\})}{(\{q^{(1)}\}, \{p^{(1)}\})} = 1$$
$$\{x^{(2)}\} = \{x^{(1)}\} + \alpha^{(1)}\{q^{(1)}\} = [1, 2]^T$$

The array $\{x^{(2)}\}$ meets the allowable error so that the calculation ends up. There is no roundoff error in the calculation and the obtained result is the exact solution.

29.8 Comparison of Several Kinds of Commonly Used Method

In the calculations of engineering problems, the algebraic equations to be solved are often very large. Certainly, the order n of the equations relates to the nature of the problems, the number of elements, and the element form. Generally speaking, for plane problems of fluid mechanics and heat transfer problems, $n = 300–600$. For plane stress problems and plate and shell problems, $n = 600–2000$. For space stress problems, $n = 5000–15,000$. In fact, most of the computing time is to solve the equation set when the finite element method is applied to analysis engineering problems. Thus it is important to choose a suitable equation solution method.

When choosing the equation solution method, we must consider two principal factors: the storage capacity of the computer and the computing time.

There are n^2 elements in the coefficient matrix $[A]$, in which the zero elements out of the band are not necessary to store. But when decomposition of the coefficient matrix is needed, the elements inside the band are all needed to store. Such iterative method calls for storing all the nonzero elements, while there is no need to store the zero elements inside the band. Suppose that the band width is b, and the number of the average nonzero elements inside the band is c, then the memory spaces for different methods are shown as follows:

Method	Direct solution	Iteration method	Conjugate gradient method
Memory space	nb	nc	$nc + 3n$

Taking advantage of the symmetry of the matrix will halve the memory space. Because $b > c$, the iteration method uses the least memory space. Taking plane stress problems, for example, the memory space for the iteration method is only a quarter of the memory space needed for the direct solution method.

As for computing speed, if an appropriate relaxation factor ω is applied, the computing times for the iteration method and the direct method may differ little. But in most cases, the iteration method needs more computing time, and the convergence rate of the iteration method may be very slow in some cases. When the memory space of the computer is enough, the direct solution method is more appropriate, because the computing time is less and definite, while the computing time for iterative calculation will depend on the load conditions and structure style and may be very long. In case the memory space is not enough, the over-relaxation iterative method and the conjugate gradient method are appropriate.

When the finite element method is applied to solve the space stress problems, the memory space of the computer may be not enough because the equation set is very huge. However, the ordinary iteration method will spend a lot of computing time. So the blocked over-relaxation iterative method is used more often. According to the memory space of the computer, 100–200 nodes can be included in a block and then conduct the block iterative computations.

In recent years, as the memory space of the computer gets larger and larger, the memory space of computer is not a problem for common engineering problems.

29.9 Homogeneous Linear Equations

If the free item array $\{b\}$ is a zero vector in Eq. (29.2), it becomes a homogeneous linear equation:

$$[A]\{x\} = \{0\} \tag{29.46}$$

If the matrix $[A]$ is nonsingular, which means $|A| \neq 0$, the inverse matrix $[A]^{-1}$ exists and the unique solution of the equation above is

$$\{x\} = [A]^{-1}\{0\} = \{0\} \tag{29.47}$$

In other words, there only exists the meaningless solution $x_1 = x_2 = \cdots = x_n = 0$.

When and only when

$$|A| = 0 \tag{29.48}$$

the homogeneous linear equation has an untrivial solution.

Consider the equation set

$$[A]\{x\} = \lambda\{x\}$$

which can be written as

$$\begin{bmatrix} a_{11}-\lambda & a_{12} & \cdots & a_{1n} \\ a_{21} & a_{22}-\lambda & \cdots & a_{2n} \\ \vdots & \vdots & \vdots & \vdots \\ a_{n1} & a_{n2} & \cdots & a_{nn}-\lambda \end{bmatrix} \begin{Bmatrix} x_1 \\ x_2 \\ \vdots \\ x_n \end{Bmatrix} = 0 \tag{29.49}$$

The condition of existence of the untrivial solution is that the determinant below equals zero, that is,

$$\begin{vmatrix} a_{11}-\lambda & a_{12} & \cdots & a_{1n} \\ a_{21} & a_{22}-\lambda & \cdots & a_{2n} \\ \vdots & \vdots & \vdots & \vdots \\ a_{n1} & a_{n2} & \cdots & a_{nn}-\lambda \end{vmatrix} = 0 \tag{29.50}$$

It comes to be an n-order equation of λ after the expansion of the formula above, and it is called the characteristic equation. It has n solutions $\lambda_1, \lambda_2, \ldots, \lambda_n$, which are called eigenvalues. There is an untrivial solution corresponding to every eigenvalue, and these vectors are called eigenvectors. The methods for solving the eigenvalues and the eigenvectors are given in Chapter 19.

Example 2 Consider the homogeneous linear equation below:

$$\begin{bmatrix} 3-\lambda & -2 \\ -1 & 4-\lambda \end{bmatrix} \begin{Bmatrix} x_1 \\ x_2 \end{Bmatrix} = 0$$

The characteristic equation is

$$\begin{vmatrix} 3-\lambda & -2 \\ -1 & 4-\lambda \end{vmatrix} = \lambda^2 - 7\lambda + 10 = 0$$

There are two roots of the above equation, $\lambda_1 = 2,\ \lambda_2 = 5$.

Substituting $\lambda_1 = 2$ into the equation set and letting $x_1 = 1$, we can get $x_2 = \frac{1}{2}$. The first eigenvector is $x_1 = 1, x_2 = \frac{1}{2}$. Then substituting $\lambda_2 = 5$ into the equation set and letting $x_1 = 1$, we can get $x_2 = -1$. The second eigenvector is $x_1 = 1, x_2 = -1$.

Bibliography

1 Yuangda, Z. and Qianyang, X. (1964) *Linear Algebra*, People's Education Press, Beijing, China.

30

Variational Method

Variational method is the method for studying the maxima and minima of functionals. It has been widely used in the finite element method. In this chapter, we will describe the fundamentals of variational method and some skills for its application to the finite element method.

30.1 The Extrema of Functions

Variational method is used to study the extrema of functionals. Because the extrema of functionals have a lot of connections and similarities with the extrema of the functions, here we recapitulate the extrema of functions.

30.1.1 The Extrema of One-Variable Functions

The maxima and minima of function are known as extrema of a function. The points where the functions obtain the extrema are known as extreme points.

Let $f = f(x)$ be a continuous and differentiable function in region $a \leq x \leq b$, and if $f(x)$ has an extremum at $x = x_0$, the derivative at this point must be equal to zero:

$$\frac{df(x)}{dx} = 0$$

As at the extreme point $x = x_0$ the first derivative of the function $f(x)$ is equal to zero, the differential of the function must also be equal to zero, that is,

$$df = \frac{df(x)}{dx}dx = 0$$

We called the points where the first derivative is zero the stationary points of function $f(x)$. The two formulas above show that the extreme points of functions must be the stationary points, but the stationary points are not necessarily extreme points. In order to determine whether the function has extreme value at the stationary point, the second derivative criterion should be used as follows:

If $\dfrac{d^2f(x)}{dx^2} > 0$, the function has minima.

The Finite Element Method: Fundamentals and Applications in Civil, Hydraulic, Mechanical and Aeronautical Engineering, First Edition. Bofang Zhu.

Companion website: www.wiley.com/go/zhufem17

If $\dfrac{d^2f(x)}{dx^2} < 0$, the function has maxima.

If $\dfrac{d^2f(x)}{dx^2} = 0$, it is uncertain.

For the third case above, $f''(x) = 0$, there may be the extreme or no extreme, and the following theorem can be used to judge.

Assume that the first-order, the second-order, and the $(k-1)$th-order derivative of the continuous differentiable function $f(x)$ at point $x = x_0$ are equal to zero, but the kth-order derivative is not equal to zero, that is,

$$f'(x) = f''(x) = \cdots = f^{(k-1)}(x) = 0$$

but $f^{(k)}(x) \neq 0$.

If k is an odd number, then $f(x)$ has no extremum at this point, and if k is an even number, then $f(x)$ has an extremum at this point. When $f^{(k)}(x) > 0$, the function has a minimum, and when $f^{(k)}(x) < 0$, the function has a maximum.

30.1.2 The Extrema of a Function with Several Variables

Now we consider the necessary condition when a function with several variables has extrema.

Assume that $f = f(x_1, x_2, \ldots, x_n)$ is a continuous and differentiable function, if the function $f(x_1, x_2, \ldots, x_n)$ has the extreme value at the point $(x_1^0, x_2^0, \ldots, x_n^0)$, the full differential df must be equal to zero at this point, that is,

$$\begin{aligned} df &= \frac{\partial f}{\partial x_1}dx_1 + \frac{\partial f}{\partial x_2}dx_2 + \cdots + \frac{\partial f}{\partial x_n}dx_n \\ &= \sum_{i=1}^{n} \frac{\partial f}{\partial x_i}dx_i = 0 \end{aligned}$$

In the formula above the increment dx_i is arbitrary. Among increments of n independent variables, if we only retain $dx_k \neq 0$, leaving the rest $n-1$ dx_i equal to zero, from the formula above, we know

$$df = \frac{\partial f}{\partial x_k}dx_k = 0$$

As $dx_k \neq 0$, from the formula above, we have

$$\frac{\partial f}{\partial x_k} = 0$$

Because dx_k is arbitrary, we can make $k = 1,2, \ldots, n$ in turn, so for every x_k the formula above must be tenable. In other words, if the function $f(x_1, x_2, \ldots, x_n)$ has the extreme value at the point $(x_1^0, x_2^0, \ldots, x_n^0)$, then all the first-order partial derivatives at the point must be equal to zero, that is,

$$\frac{\partial f}{\partial x_i} = 0, \quad i = 1,2, \ldots, n$$

As with one-variable functions, extreme points of function with several variables must also be stationary points.

30.2 The Extrema of Functionals

So-called functional refers to a variable whose value is determined by a function $y = y(x)$ or by several functions $y_1(x), y_2(x)\ldots$.

For example, in Figure 30.1, given two points 1 and 2, the length L of the curve connecting the two points is associated with curve shape. If the curve shape is different, the length L is different, so the length L is defined by equation $y = y(x)$ of the curve. As long as the equation of the curve is given, the length L can be calculated by the following formula:

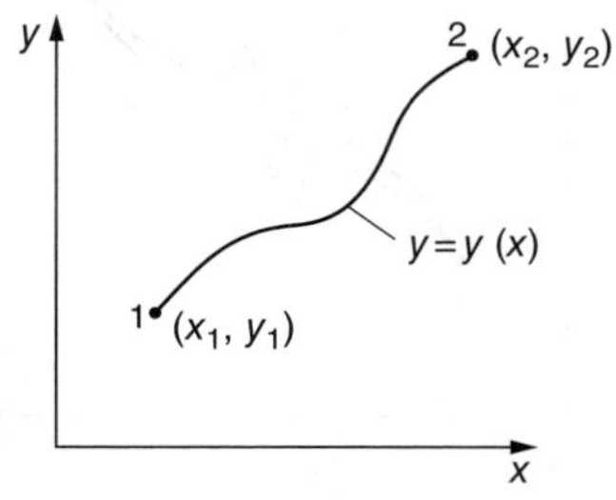

Figure 30.1 Curve in plane.

$$L[y(x)] = \int_{x_1}^{x_2} \sqrt{1 + \left(\frac{dy}{dx}\right)^2}\, dx$$

Therefore, the curve length L is a functional, recorded as $L[y(x)]$.

As shown in Figure 30.2, given a surface $z = z(x, y)$ in space, the surface area S shall be determined by the following formula:

$$S[z(x, y)] = \iint_D \sqrt{1 + \left(\frac{\partial z}{\partial x}\right)^2 + \left(\frac{\partial z}{\partial y}\right)^2}\, dxdy$$

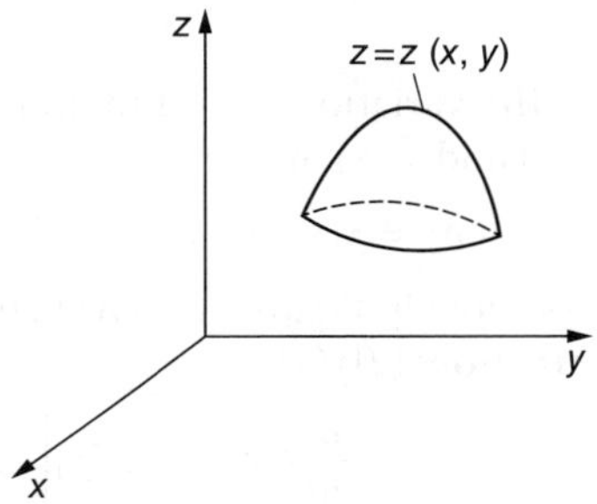

Figure 30.2 Surface in space.

Apparently, the surface area S depends on the shape of the surface. If the shape changes, the surface area also changes, so the surface area S is also a functional, expressed by $S[z(x, y)]$.

The comparison of function and functional are given in the following:

Function	$f(x)$	Functional	$I[y(x)]$
Variable	f	Variable	I
Independent variable	x	Function	$y(x)$
The increment of x	Δx	The variation of $y(x)$	δy
The differential of function	df	The variation of functional	δI

The variational method is to study the problem of maxima and minima of functional. The solution to this question is very similar to the solution to the extreme value of the function. Now omitting the mathematical proof, briefly describe the result and compare with the method for determining extreme value of the function.

Function $f = f(x)$

1) If, for each x in a particular domain of the variable x, f has a value corresponding with it, the variable f is called the function of variable x, recorded as $f(x)$.
2) If, for minor change of x, there is minor change of function $f(x)$ corresponding with it, then the function $f(x)$ is continuous.
3) If the differentiable function $f(x)$ reaches the maximum or minimum value at $x = x_0$, then at this point there is

$$df = 0$$

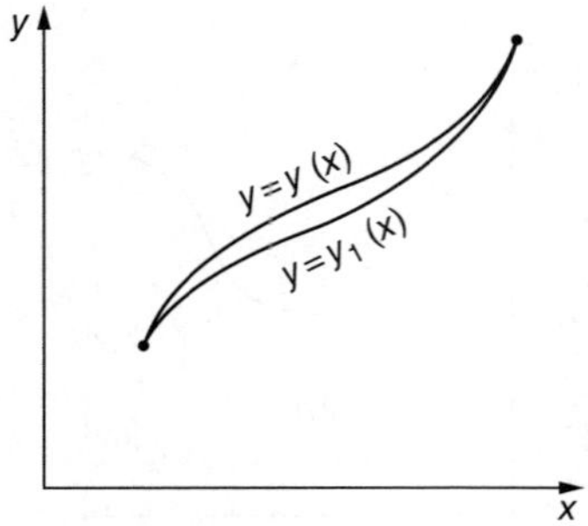

Figure 30.3 Two curves in plane.

Functional $I[y(x)]$

1) If, for each function $y(x)$ in one category of function $y(x)$, I has a value corresponding to it, the variable I is called the functional depending on the function $y(x)$, expressed as $I[y(x)]$.
2) If, for minor change of $y(x)$, there is minor change of functional $I[y(x)]$ corresponding to it, then the functional is continuous.
3) If the variational functional $I[y(x)]$ reaches the maximum or minimum value at $y = y_0(x)$, then at $y = y_0(x)$ there is

$$\delta I = 0$$

The variation δy of function $y(x)$ refers to the difference between the two functions $y(x)$ and $y_1(x)$, namely,

$$\delta y = y(x) - y_1(x)$$

As show in Figure 30.3, $y(x)$ and $y_1(x)$ are very close to each other. The variation of the functional $I[y(x)]$ is

$$\delta I = \frac{\partial}{\partial \alpha} I[y(x) + \alpha \delta y]\bigg|_{\alpha=0} \tag{30.1}$$

This is correspondent to the differential of a function:

$$df = \frac{\partial}{\partial \alpha} f(x + \alpha \Delta x)\bigg|_{\alpha=0} \tag{30.2}$$

For the function, if $f(x)$ has the maximum or the minimum value at $x = x_0$, we can search for x_0 according to the criteria $df = 0$. Equally, for the functional, if $I[f(x)]$ has the maximum or minimum value at $y = y_0(x)$, we can search for $y_0(x)$ according to the criteria $\delta I = 0$.

30.3 Preliminary Theorems

In order to derive Euler's equation in later sections, here we describe a theorem first.

Assume that $\Phi(x)$ is a continuous function in the interval $[a,b]$, if, for any continuous function $\eta(x)$ satisfying $\eta(a) = \eta(b) = 0$, there is

$$\int_a^b \eta(x) \ \Phi(x) = 0$$

Then in the interval $[a, b]$, there is

$$\Phi(x) \equiv 0$$

This theorem is not difficult to prove. Assuming that at the point $x = x_1$ in the interval $[a, b]$, there is $\Phi(x) \neq 0$, and we can get contradictory results. In fact, because the function $\Phi(x)$ is continuous, it does not change sign in the neighborhood of x_1 $(x_0 \leq x_1 \leq x_2)$.

Since the function $\eta(x)$ is arbitrary, we can select the function $\eta(x)$ in such a way that it does not change sign in the neighborhood of x_1 ($x_0 \leq x_1 \leq x_2$) and it is greater than zero, while outside the neighborhood, it is constantly equal to zero, as shown in Figure 30.4. Then we get

$$\int_a^b \eta(x)\ \Phi(x)\mathrm{d}x = \int_{x_0}^{x_2} \eta(x)\ \Phi(x)\mathrm{d}x \neq 0$$

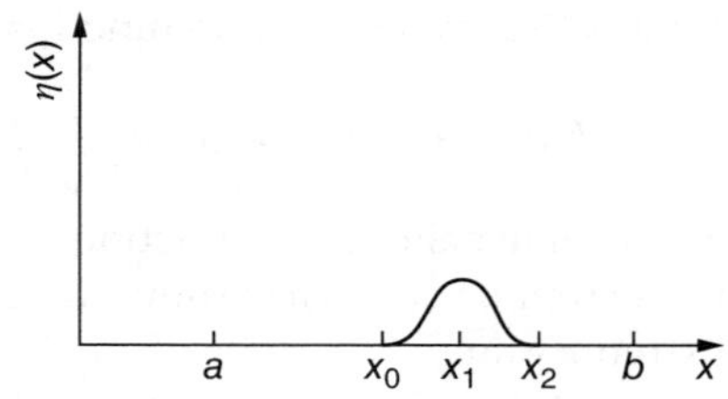

Figure 30.4 Function $\eta(x)$.

This is due to the product $\eta(x)\ \Phi(x)$ that does not change sign in the interval $[x_0, x_2]$, and outside the interval it is equal to zero. This leads to contradictions, and then we know $\Phi(x) \equiv 0$.

For spatial problem, a similar conclusion can be obtained.

Assuming that $\Phi(x, y, z)$ is a continuous function in the domain D of the space (x, y, z), while $\eta(x, y, z)$ is an arbitrary continuous function that is zero on the boundary of D, if

$$\iiint_D \eta(x, y, z)\ \Phi(x, y, z) = 0$$

then $\Phi(x, y, z) \equiv 0$ in the domain D.

30.4 Euler's Equation of One-Dimensional Problems

Now we study the necessary condition for a simplest functional to achieve its extreme value that there is only one independent variable x.

Assume that the functional is

$$I[y(x)] = \int_a^b F[x, y(x), y'(x)]\mathrm{d}x \tag{30.3}$$

where F is the function of $x, y(x)$, and $y'(x) = \frac{\mathrm{d}y}{\mathrm{d}x}$.

Assuming that functional I has extreme value and achieves its extreme value at

$$y = y(x)$$

boundary conditions are as follows:

$$\left.\begin{array}{ll}\text{when} & x = a, y = c \\ \text{when} & x = b, y = d\end{array}\right\} \tag{a}$$

For arbitrary continuous function $\eta(x)$, when the parameter ε is small enough,

$$\overline{y}(x) = y(x) + \varepsilon\eta(x) \tag{b}$$

will be a continuous function in the neighborhood of $y(x)$. In order to satisfy the boundary condition (a), the function $\eta(x)$ must satisfy the following boundary conditions:

$$\left.\begin{array}{ll}\text{when} & x = a,\ \eta = 0 \\ \text{when} & x = b,\ \eta = 0\end{array}\right\} \tag{c}$$

By substitution of $\bar{y}(x)$ into functional (30.3), we get

$$I[y(x) + \varepsilon\eta(x)] = I(\varepsilon) = \int_a^b F[x, y + \varepsilon\eta, y' + \varepsilon\eta']\mathrm{d}x \tag{d}$$

The formula above is a function of parameter ε, and when $\varepsilon = 0$, functional I reaches the extreme value. According to the necessary condition for the function to reach its extreme value,

$$\left.\frac{\mathrm{d}I}{\mathrm{d}\varepsilon}\right|_{\varepsilon=0} = 0 \tag{e}$$

According to the method of differentiation in integral sign, we can know

$$\frac{\mathrm{d}I}{\mathrm{d}\varepsilon} = \int_a^b \left[\frac{\partial F}{\partial \bar{y}}\frac{\mathrm{d}\bar{y}}{\mathrm{d}\varepsilon} + \frac{\partial F}{\partial \bar{y}'}\frac{\mathrm{d}\bar{y}'}{\mathrm{d}\varepsilon}\right]\mathrm{d}x \tag{f}$$

where

$$\frac{\partial F}{\partial y'} = \frac{\partial F}{\partial(\mathrm{d}y/\mathrm{d}x)}$$

By formula (b)

$$\frac{\mathrm{d}\bar{y}}{\partial\varepsilon} = \eta(x), \quad \frac{\mathrm{d}\bar{y}'}{\partial\varepsilon} = \eta'(x)$$

By substituting them into formula (f) and making $\varepsilon = 0$, we get

$$\left.\frac{\mathrm{d}I}{\mathrm{d}\varepsilon}\right|_{\varepsilon=0} = \int_a^b \left[\frac{\partial F}{\partial y}\eta + \frac{\partial F}{\partial y'}\eta'\right]\mathrm{d}x \tag{g}$$

Doing partial integration of the second item on the right side, we get

$$\int_a^b \frac{\partial F}{\partial y'}\eta'\mathrm{d}x = \left[\frac{\partial F}{\partial y'}\eta\right]_a^b - \int_a^b \eta\left[\frac{\mathrm{d}}{\mathrm{d}x}\left(\frac{\partial F}{\partial y'}\right)\right]\mathrm{d}x$$

According to the boundary condition formula (c), $\eta(a) = \eta(b) = 0$, by the formula above, we get

$$\int_a^b \frac{\partial F}{\partial y'}\eta'\mathrm{d}x = -\int_a^b \eta\left[\frac{\mathrm{d}}{\mathrm{d}x}\left(\frac{\partial F}{\partial y'}\right)\right]\mathrm{d}x$$

By substitution of it into formula (g) and by formula (e), when the functional achieves the extreme value,

$$\left.\frac{\mathrm{d}I}{\mathrm{d}\varepsilon}\right|_{\varepsilon=0} = 0$$

Thus

$$\left.\frac{\mathrm{d}I}{\mathrm{d}\varepsilon}\right|_{\varepsilon=0} = \int_a^b \eta\left[\frac{\partial F}{\partial y} - \frac{\mathrm{d}}{\mathrm{d}x}\left(\frac{\partial F}{\partial y'}\right)\right]\mathrm{d}x = 0$$

Because the item in the brackets on the right of the formula above is continuous, while $\eta(x)$ is arbitrary, according to the preliminary theorems, from the formula above, the necessary condition for functional $I[y(x)]$ to reach its extreme value is

$$\frac{\partial F}{\partial y} - \frac{\mathrm{d}}{\mathrm{d}x}\left(\frac{\partial F}{\partial y'}\right) = 0 \tag{30.4}$$

The formula above is called Euler's equation.

Note

$$F_{y'} = \frac{\partial F}{\partial y'}$$

The total derivative is (remembering that $F = F(x, y, y')$ the function of x, y, y')

$$\frac{\mathrm{d}}{\mathrm{d}x}F_{y'} = \frac{\partial F_{y'}}{\partial x} + \frac{\partial F_{y'}}{\partial y}\frac{\mathrm{d}y}{\mathrm{d}x} + \frac{\partial F_{y'}}{\partial y'}\frac{\mathrm{d}y'}{\mathrm{d}x}$$

$$= F_{xy'} + y'F_{yy'} + y''F_{y'y'}$$

By substitution of it into formula (30.4), we get the expanded form of Euler's equation:

$$F_y - F_{xy'} - y'F_{yy'} - y''F_{y'y'} = 0 \tag{30.4a}$$

The necessary condition for functional $I[y(x)]$ to achieve the extreme value on $y(x)$ is that $y(x)$ satisfies Euler's equation (30.4). The integral curve of Euler's equation is called extreme curves. It is clear that only on the extremum curves the functional reaches the extreme value.

Example 1 ***Functional***

$$I(y(x)) = \int_0^{\pi/2} [(y')^2 - y^2]\mathrm{d}x$$

boundary condition

$$y(0) = 0, \quad y\left(\frac{\pi}{2}\right) = 1$$

On what kind of curve the functional above reaches extreme value? Here

$$F = (y')^2 - y^2$$

$$\frac{\partial F}{\partial y} = -2y, \quad \frac{\partial F}{\partial y'} = 2y', \quad \frac{\mathrm{d}}{\mathrm{d}x}\left(\frac{\partial F}{\partial y'}\right) = 2y''$$

By substitution of them into formula (30.4), Euler's equation is

$$y'' + y' = 0$$

The general solution is $y = c_1 \cos x + c_2 \sin x$, and by boundary conditions, $c_1 = 0$ and $c_2 = 1$ are obtained, so $y = \sin x$ is the extremum curve.

In other words, only on the curve $y = \sin x$ functional reaches extreme value.

Example 2 ***Functional***

$$I(y(x)) = \int_0^1 [(y')^2 + 2xy]\,\mathrm{d}x$$

boundary condition $y(0) = 0,\ y(1) = 1$

On what kind of curve the functional will reach extreme value?

Here

$$F = (y')^2 + 2xy, \quad \frac{\partial F}{\partial y} = 2x, \quad \frac{\partial F}{\partial y'} = 2y', \quad \frac{\mathrm{d}}{\mathrm{d}x}\left(\frac{\partial F}{\partial y'}\right) = 2y''$$

Euler's equation is

$$y'' - x = 0$$

Its general solution is

$$y = \frac{x^3}{6} + c_1 x + c_2$$

By boundary conditions, $c_1 = \frac{5}{6}$ and $c_2 = 0$ are obtained, so only on the curve $y = \frac{5}{6}x + \frac{x^3}{6}$ functional reaches extreme value.

Example 3 ***Functional***

$$L(y(x)) = \int_{x_1}^{x_2} \sqrt{1 + (y')^2}\mathrm{d}x$$

represents the length of curve $y = y(x)$ on a plane:

$$F = \sqrt{1 + (y')^2}$$

Euler's equation is

$$y'' = 0$$

Its general solution is $y = c_1 x + c_2$, which is the equation of a line. It is known that on the plane functional $L(y(x))$ gets the extreme value on the line. In other words, when $y = y(x)$ is a line, the length of the curve reaches the extreme value.

30.5 Euler's Equation for Plane Problems

Assume that function

$$\phi = \phi(x, y)$$

is continuous and two order differentiable in region R. The boundary of region R is divided into two parts, B and C. On boundary B, the value of function ϕ has been given, namely,

$$\phi = \phi_B \tag{30.5}$$

while on boundary C, the value of function ϕ is not given in advance.

Taking functional into account,

$$I(\phi) = \iint\limits_R F(\phi, \phi_x, \phi_y)\mathrm{d}x\mathrm{d}x + \int_G G(\phi)\mathrm{d}s \tag{30.6}$$

where $\phi_x = \frac{\partial \varphi}{\partial x}$, $\phi_y = \frac{\partial \varphi}{\partial y}$. Here value of function $G(\phi)$ is taken along boundary C, without giving the value of ϕ on the boundary C previously.

Assume functional I has extreme values and gets its extremes on $\phi = \phi(x, y)$. On the neighborhood of extreme surface $\phi(x, y)$, function can be written as

$$\overline{\phi}(x, y) = \phi(x, y) + \varepsilon\eta(x, y) \tag{a}$$

where ε is in an arbitrarily small scalar quantity and $\eta(x, y)$ is an arbitrary function in region R. On boundary B, because function $\overline{\phi} = \phi_B$, η should meet the following boundary condition:

On boundary B

$$\eta = 0 \tag{b}$$

By substitution of formula (a) into formula (30.6), we get

$$I(\phi + \varepsilon\eta) = I(\varepsilon) = \iint_R F(\phi + \varepsilon\eta, \phi_x + \varepsilon\eta_x, \phi_y + \varepsilon\eta_y)\mathrm{d}x\mathrm{d}x + \int_G G(\phi + \varepsilon\eta)\mathrm{d}s \tag{c}$$

The formula above is a function of parameter ε. When $\varepsilon = 0$, functional I obtains the extreme value, according to the necessary condition for functional I to obtain the extreme value:

$$\left.\frac{\mathrm{d}I}{\mathrm{d}\varepsilon}\right|_{\varepsilon=0} = 0 \tag{d}$$

According to the method of differentiation under the integral sign,

$$\frac{\mathrm{d}I(\varepsilon)}{\mathrm{d}\varepsilon} = \iint_R \left[\frac{\partial F}{\partial \phi}\frac{\mathrm{d}\phi}{\mathrm{d}\varepsilon} + \frac{\partial F}{\partial \phi_x}\frac{\mathrm{d}\phi_x}{\mathrm{d}\varepsilon} + \frac{\partial F}{\partial \phi_y}\frac{\mathrm{d}\phi_y}{\mathrm{d}\varepsilon}\right]\mathrm{d}x\mathrm{d}y + \int_G \frac{\partial G}{\partial \phi}\frac{\mathrm{d}\phi}{\mathrm{d}s}\mathrm{d}s = 0 \tag{e}$$

By formula (a)

$$\frac{\partial \phi}{\partial \varepsilon} = \eta, \quad \frac{\partial \phi_x}{\partial \varepsilon} = \eta_x, \quad \frac{\partial \phi_y}{\partial \varepsilon} = \eta_y$$

By substitution of the formula above into formula (e) and when $\varepsilon = 0$, $\phi = \phi$, $\phi_x = \phi_x, \phi_y = \phi_y$; therefore when the functional obtains the extreme value,

$$\delta I = \left.\frac{\mathrm{d}I(\varepsilon)}{\mathrm{d}\varepsilon}\right|_{\varepsilon=0} = \iint_R \left[\frac{\partial F}{\partial \phi}\eta + \frac{\partial F}{\partial \phi_x}\eta_x + \frac{\partial F}{\partial \phi_y}\eta_y\right]\mathrm{d}x\mathrm{d}y + \int_G \frac{\partial G}{\partial \phi}\eta\mathrm{d}s = 0 \tag{f}$$

Note

$$\frac{\partial F}{\partial \phi} = F_\phi, \quad \frac{\partial F}{\partial \phi_x} = F_{\phi_x}, \quad \frac{\partial F}{\partial \phi_y} = F_{\phi_y} \tag{g}$$

By partial integration, we know

$$\left.\begin{aligned} F_{\phi_x}\eta_x &= \frac{\partial}{\partial x}(F_{\phi_x}\eta) - \frac{\partial F_{\phi_x}}{\partial x}\eta \\ F_{\phi_y}\eta_y &= \frac{\partial}{\partial y}(F_{\phi_y}\eta) - \frac{\partial F_{\phi_y}}{\partial y}\eta \end{aligned}\right\} \tag{h}$$

By substitution of formulas (g) and (h) into formula (f), we get

$$\begin{aligned} \delta I &= \iint_R \eta\left[\frac{\partial F}{\partial \phi} - \frac{\partial F_{\phi_x}}{\partial x} - \frac{\partial F_{\phi_y}}{\partial y}\right]\mathrm{d}x\mathrm{d}y \\ &\quad + \iint_R \left[\frac{\partial}{\partial x}(F_{\phi_x}\eta) + \frac{\partial}{\partial y}(F_{\phi_y}\eta)\right]\mathrm{d}x\mathrm{d}y \\ &\quad + \int_G \frac{\partial G}{\partial \phi}\eta\mathrm{d}s \\ &= 0 \end{aligned} \tag{i}$$

According to Green's theorem,

$$\iint \left(\frac{\partial Q}{\partial x} + \frac{\partial P}{\partial y}\right) \mathrm{d}x\mathrm{d}y = \int (Q\mathrm{d}y - P\mathrm{d}x)$$

Making $Q = F_{\phi_x}\eta, \quad P = F_{\phi_y}\eta$, the second item on the right side of formula (i) can be transformed as follows:

$$\iint_R \left[\frac{\partial}{\partial x}(F_{\phi_x}\eta) + \frac{\partial}{\partial y}(F_{\phi_y}\eta)\right] \mathrm{d}x\mathrm{d}y = \int_{B+G} \eta(F_{\phi_x}\mathrm{d}y - F_{\phi_y}\mathrm{d}x)$$

By formula (b), on boundary B, $\eta = 0$, there is only line integral along boundary C on the right side of the formula above and $\mathrm{d}y = l_x\mathrm{d}s$, $\mathrm{d}x = -l_y\mathrm{d}s$, where l_x and l_y are the direction cosines along the outward normal of the boundary. Thus

$$\iint_R \left[\frac{\partial}{\partial x}(F_{\phi_x}\eta) + \frac{\partial}{\partial y}(F_{\phi_y}\eta)\right] \mathrm{d}x\mathrm{d}y = \int_G \eta(F_{\phi_x}l_x + F_{\phi_y}l_y)\mathrm{d}s$$

By substitution of the formula above into formula (i), we get

$$\begin{aligned}\delta I = &\iint_R \eta \left[\frac{\partial F}{\partial \phi} - \frac{\partial F_{\phi_x}}{\partial x} - \frac{\partial F_{\phi_y}}{\partial y}\right] \mathrm{d}x\mathrm{d}y \\ &+ \int_C \eta \left[\frac{\partial G}{\partial \phi} + F_{\phi_x}l_x + F_{\phi_y}l_y\right] \mathrm{d}s \\ = &\ 0\end{aligned} \tag{j}$$

In the region R and on boundary C, η of the formula above is arbitrary, and according to preliminary theorems, on the right side of the formula above, in either the double integral or the line integral, the factor in the bracket behind η must be equal to zero. Otherwise we can always choose η to make $\delta I \neq 0$. Therefore by $\delta I = 0$, we can infer

$$\left.\begin{aligned}&\text{in the region } R: \quad \frac{\partial F}{\partial \phi} - \frac{\partial}{\partial x}\left(\frac{\partial F}{\partial \phi_x}\right) - \frac{\partial}{\partial y}\left(\frac{\partial F}{\partial \phi_y}\right) = 0 \\ &\text{on boundary } C: \quad \frac{\partial G}{\partial \phi} + l_x\frac{\partial F}{\partial \phi_x} + l_y\frac{\partial F}{\partial \phi_y} = 0\end{aligned}\right\} \tag{30.7}$$

These are Euler's equations of plane problems.

In short, in plane problems, for the functional $I(\phi)$ expressed by formula (30.6), if its extreme value is achieved on $\phi = \phi(x, y)$, where ϕ meets the boundary condition (30.5), then the function $\phi(x, y)$ must satisfy Euler's equations (30.7):

Example 1 ***Functional***

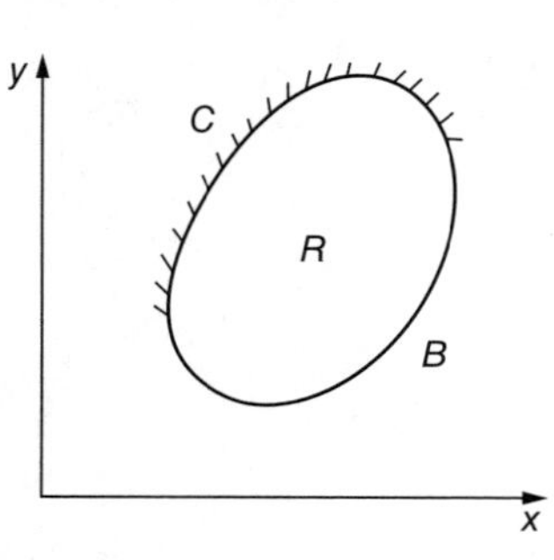

Figure 30.5 A plane problem.

$$\begin{aligned}I[\phi(x,y)] = &\iint_R \frac{1}{2}\left[\left(\frac{\partial \phi}{\partial x}\right)^2 + \left(\frac{\partial \phi}{\partial y}\right)^2\right] \mathrm{d}x\mathrm{d}y \\ &+ \int_C \beta\phi\left(\frac{1}{2}\phi - t\right) \mathrm{d}s\end{aligned} \tag{k}$$

Referring to Figure 30.5, where the boundary conditions of ϕ are on boundary B

$$\phi = \phi_B \tag{l}$$

in what circumstances the functional above reaches the extreme values?

Here

$$F = \frac{1}{2}\left[\left(\frac{\partial \phi}{\partial x}\right)^2 + \left(\frac{\partial \phi}{\partial y}\right)^2\right]$$

$$G = \beta\phi\left(\frac{1}{2}\phi - t\right)$$

$$\frac{\partial F}{\partial \phi} = 0, \quad \frac{\partial F}{\partial \phi_x} = \frac{\partial \phi}{\partial x}, \quad \frac{\partial}{\partial x}\left(\frac{\partial F}{\partial \phi_x}\right) = \frac{\partial^2 \phi}{\partial x^2}$$

$$\frac{\partial F}{\partial \phi_y} = \frac{\partial \phi}{\partial y}, \quad \frac{\partial}{\partial y}\left(\frac{\partial F}{\partial \phi_y}\right) = \frac{\partial^2 \phi}{\partial y^2}, \quad \frac{\partial G}{\partial \phi} = \beta(\phi - t)$$

By Euler's equation formula (30.7), we know that when the functional achieves its extreme, function ϕ must satisfy the following equation:

$$\text{in region } R: \quad \frac{\partial^2 \phi}{\partial x^2} + \frac{\partial^2 \phi}{\partial y^2} = 0 \tag{m}$$

$$\text{on boundary } C: \quad l_x\frac{\partial \phi}{\partial x} + l_y\frac{\partial \phi}{\partial y} + \beta(\phi - t) = 0 \tag{n}$$

When functional $I(\phi)$ achieves its extreme values, function $\phi(x, y)$ satisfies Laplace's equation in region R, and on boundary C it meets formula (n).

Example 2 ***Functional***

$$I[\phi(x, y)] = \iint\limits_R \left[\left(\frac{\partial \phi}{\partial x}\right)^2 + \left(\frac{\partial \phi}{\partial y}\right)^2 + 2\phi f(x, y)\right] \mathrm{d}x\mathrm{d}y$$

On the boundary of region R, ϕ is known, that is, $\phi = \phi_B$. In what circumstances functional I achieves the extreme values?

Here

$$F = \left(\frac{\partial \phi}{\partial x}\right)^2 + \left(\frac{\partial \phi}{\partial y}\right)^2 + 2\phi f, \quad G = 0$$

$$\frac{\partial F}{\partial \phi} = 2f, \quad \frac{\partial}{\partial x}\left(\frac{\partial F}{\partial \phi_x}\right) = 2\frac{\partial^2 \phi}{\partial x^2}, \quad \frac{\partial}{\partial y}\left(\frac{\partial F}{\partial \phi_y}\right) = 2\frac{\partial^2 \phi}{\partial y^2}$$

In region R, Euler's equation is

$$\frac{\partial^2 \phi}{\partial x^2} + \frac{\partial^2 \phi}{\partial y^2} = f(x, y)$$

This is a Poisson equation. Therefore, when functional obtains its extrema, function $\phi(x, y)$ should satisfy the Poisson equation.

30.6 Euler's Equations of Spatial Problems

As shown in Figure 30.6, R is a region of space (x, y, z). Assume that function $\phi(x, y, z)$ is continuous and two order differentiable in region R. On boundary B, the value of ϕ is given:

$$\phi = \phi_B$$

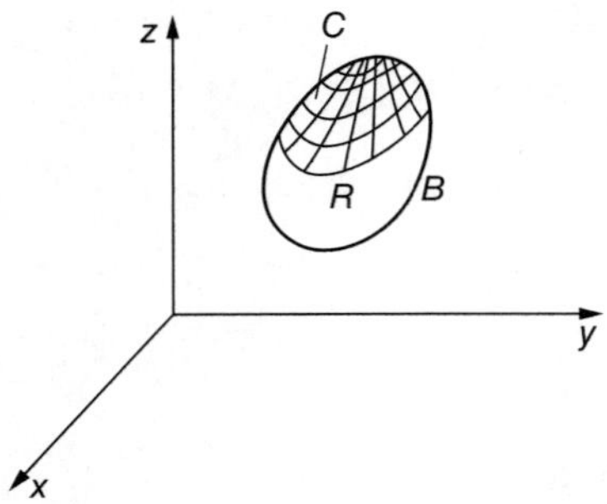

Figure 30.6 A spatial problem.

On boundary C, the value of ϕ is not given.

Taking the functional into account

$$I(\phi) = \iiint_R F(\phi, \phi_x, \phi_y, \phi_z)\mathrm{d}x\mathrm{d}y\mathrm{d}z + \int_C G(\phi)\mathrm{d}s \tag{30.8}$$

the value of function $G(\phi)$ is taken along boundary C where the value of ϕ is not given previously.

Assuming that functional I achieves extreme on $\phi = \phi(x, y, z)$, in the neighborhood of $\phi = \phi(x, y, z)$, the function can be written as

$$\overline{\phi}(x, y, z) = \phi(x, y, z) + \varepsilon\eta(x, y, z) \tag{a}$$

where ε is an arbitrarily small parameter and $\eta(x, y, z)$ is an allowed arbitrary function. On boundary B, $\phi = \phi_B$,

$$\eta = 0 \tag{b}$$

By substitution of formula (a) into formula (30.8), we get

$$I(\phi + \varepsilon\eta) = \iiint_R F(\phi + \varepsilon\eta, \phi_x + \varepsilon\eta_x, \phi_y + \varepsilon\eta_y, \phi_z + \varepsilon\eta_z)\mathrm{d}x\mathrm{d}y\mathrm{d}z + \int_C G(\phi + \varepsilon\eta)\mathrm{d}s \tag{c}$$

The formula above is the function of parameter ε. When $\varepsilon = 0$, functional I obtains the extreme value, according to the necessary condition for functional I to obtain the extreme value:

$$\begin{aligned}\delta I &= \left.\frac{\mathrm{d}I}{\mathrm{d}\varepsilon}\right|_{\varepsilon=0} \\ &= \iiint_R \left[\frac{\partial F}{\partial \phi}\eta + \frac{\partial F}{\partial \phi_x}\eta_x + \frac{\partial F}{\partial \phi_y}\eta_y + \frac{\partial F}{\partial \phi_z}\eta_z\right] \mathrm{d}x\mathrm{d}y\mathrm{d}z + \iint_C \frac{\partial G}{\partial \phi}\mathrm{d}s = 0\end{aligned} \tag{d}$$

By integration by parts, the formula above can be transformed into

$$\begin{aligned}\delta I &= \iiint_R \eta \left[\frac{\partial F}{\partial \phi} - \frac{\partial F_{\phi_x}}{\partial x} - \frac{\partial F_{\phi_y}}{\partial y} - \frac{\partial F_{\phi_z}}{\partial z}\right] \mathrm{d}x\mathrm{d}y\mathrm{d}z \\ &+ \iiint_R \left[\frac{\partial}{\partial x}(F_{\phi_x}\eta) + \frac{\partial}{\partial x}(F_{\phi_y}\eta) + \frac{\partial}{\partial x}(F_{\phi_z}\eta)\right] \mathrm{d}x\mathrm{d}y\mathrm{d}z + \iint_C \frac{\partial G}{\partial \phi}\mathrm{d}s \\ &= 0\end{aligned} \tag{e}$$

Using Green's theorem of spatial problems, the second item on the right side of the formula above can be transformed as follows:

$$\begin{aligned}&\iiint_R \left[\frac{\partial}{\partial x}(F_{\phi_x}\eta) + \frac{\partial}{\partial y}(F_{\phi_y}\eta) + \frac{\partial}{\partial z}(F_{\phi_z}\eta)\right] \mathrm{d}x\mathrm{d}y\mathrm{d}z \\ &= \iint_{B+C} \eta(F_{\phi_x}l_x + F_{\phi_y}l_y + F_{\phi_z}l_z)\mathrm{d}s\end{aligned} \tag{f}$$

By substitution of the formula above into formula (e), on boundary B, $\eta = 0$, we get

$$\delta I = \iiint\limits_R \eta \left[\frac{\partial F}{\partial \phi} - \frac{\partial F_{\phi_x}}{\partial x} - \frac{\partial F_{\phi_y}}{\partial y} - \frac{\partial F_{\phi_z}}{\partial z} \right] \mathrm{d}x\mathrm{d}y\mathrm{d}z$$
$$+ \iint\limits_C \eta \left(\frac{\partial G}{\partial \phi} + F_{\phi_x} l_x + F_{\phi_y} l_y + F_{\phi_z} l_z \right) \mathrm{d}s$$
$$= 0 \tag{g}$$

Because of the arbitrariness of η, according to the preliminary theorems, by the formula above, when functional $I(\phi)$ obtains the extreme value, there must be

$$\left. \begin{aligned} &\text{In region } R \quad : \frac{\partial F}{\partial \phi} - \frac{\partial}{\partial x}\left(\frac{\partial F}{\partial \phi_x}\right) - \frac{\partial}{\partial y}\left(\frac{\partial F}{\partial \phi_y}\right) - \frac{\partial}{\partial z}\left(\frac{\partial F}{\partial \phi_z}\right) = 0 \\ &\text{On boundary } C : \frac{\partial G}{\partial \phi} + l_x \frac{\partial F}{\partial \phi_x} + l_y \frac{\partial F}{\partial \phi_y} + l_z \frac{\partial F}{\partial \phi_z} = 0 \end{aligned} \right\} \tag{30.9}$$

This is Euler's equation of spatial problems, where l_x, l_y, l_z are the directional cosines of the outward normal of boundary surface.

In spatial problems, for the functional $I(\phi)$ expressed by formula (30.8), if it obtains the extreme values on $\phi(x, y, z)$, then function $\phi(x, y, z)$ must satisfy Euler's equation formula (30.9).

In formula (30.8), functional I is only related with the first-order partial derivatives ϕ_x, ϕ_y, ϕ_z of $\phi(x, y, z)$, if, except for these first-order partial derivatives, functional I is also associated with higher-order partial derivatives. Using similar methods, it is easy to infer the appropriate Euler equation. In addition, in formula (30.8), there are only one independent function ϕ and partial derivatives. If functional includes several independent functions $\phi, \Psi, \ldots$ and their partial derivatives, using similar steps, Euler's equation derived is a partial differential equation.

Example 1 ***Referring to Figure 30.6,***

$$I(\phi) = \iiint\limits_R \frac{1}{2} \left[k_x \left(\frac{\partial \phi}{\partial x}\right)^2 + k_y \left(\frac{\partial \phi}{\partial y}\right)^2 + k_z \left(\frac{\partial \phi}{\partial z}\right)^2 - Q\phi \right] \mathrm{d}x\mathrm{d}y\mathrm{d}z$$
$$+ \iint\limits_C \left(q\phi + \frac{1}{2}\alpha\phi^2 \right) \mathrm{d}s$$

Boundary conditions are on boundary B, $\phi = \phi_B$ and on the boundary C, ϕ is not provided. In what conditions the functional reaches extreme values?

Let

$$F = \frac{1}{2} \left[k_x \left(\frac{\partial \phi}{\partial x}\right)^2 + k_y \left(\frac{\partial \phi}{\partial y}\right)^2 + k_z \left(\frac{\partial \phi}{\partial z}\right)^2 \right] - Q\phi$$

$$\frac{\partial F}{\partial \phi} = -Q, \quad \frac{\partial F}{\partial \phi_x} = k_x \frac{\partial \phi}{\partial x}, \quad \frac{\partial}{\partial x}\left(\frac{\partial F}{\partial \phi_x}\right) = \frac{\partial}{\partial x}\left(k_x \frac{\partial \phi}{\partial x}\right), \ldots$$

$$G = q\phi + \frac{1}{2}\alpha\phi^2, \quad \frac{\partial G}{\partial \phi} = q + \alpha\phi$$

By formula (30.9), we know that when functional $I(\phi)$ obtains extreme values, ϕ should satisfy the following conditions:

In region R

$$\frac{\partial}{\partial x}\left(k_x\frac{\partial \phi}{\partial x}\right)+\frac{\partial}{\partial y}\left(k_y\frac{\partial \phi}{\partial y}\right)+\frac{\partial}{\partial z}\left(k_z\frac{\partial \phi}{\partial z}\right)+Q=0$$

On boundary C

$$k_x\frac{\partial \phi}{\partial x}l_x+k_y\frac{\partial \phi}{\partial y}l_y+k_z\frac{\partial \phi}{\partial z}l_z+q+\alpha\phi=0$$

30.7 Ritz Method for Solving Variational Problems

In actual projects, it is often necessary to solve the differential equation with given boundary conditions. Because function $\phi(x, y, z)$ making functional achieves its extreme value must meet corresponding Euler equation, so we can construct a functional, making its Euler equation be the differential equation we need to solve. In this way we transform the problem to solve differential equation into the problem to solve extreme value of the functional. When we obtain function $\phi(x, y, z)$ that makes the functional achieve extreme value, it also meets the differential equation; thus it is the solution we need.

For example, when we need to solve the plane Laplace equation

$$\frac{\partial^2\phi}{\partial x^2}+\frac{\partial^2\phi}{\partial y^2}=0 \tag{a}$$

on all boundary

$$\phi=\phi_B \tag{b}$$

We consider formula (a) as Euler's equations, and the corresponding functional is

$$I(\phi)=\iint\limits_R \frac{1}{2}\left[\left(\frac{\partial \phi}{\partial x}\right)^2+\left(\frac{\partial \phi}{\partial y}\right)^2\right]\mathrm{d}x\mathrm{d}y \tag{c}$$

and we find the function $\phi(x, y)$ that makes the functional $I(\phi)$ obtain its extreme value. It is the solution of the Laplace equation.

The problem now is how to find the function $\phi(x, y)$ that makes the functional achieve extreme value. In this section, we introduce the classical Ritz method first, and in the next section, we will explain the finite element method.

The concept of the Ritz method is that function $\phi(x, y)$ depending on functional $I(\phi)$ has the form as follows (here taking plane problems, as example, for one- and three-dimensional problems, the method is similar):

$$\begin{aligned}\phi(x,y)&=\alpha_1 w_1(x,y)+\alpha_2 w_2(x,y)+\cdots+\alpha_n w_n(x,y)\\&=\sum_{i=1}^{n}\alpha_i w_i(x,y)\end{aligned} \tag{30.10}$$

where $w_1(x, y), w_2(x, y), w_3(x, y)\ldots$ are a set of function sequence that satisfy the boundary conditions, called the coordinate functions, and $\alpha_1, \alpha_2, \ldots, \alpha_n$ are the undetermined coefficients.

By substitution of the formula above into functional $I(\phi)$, the functional $I(\phi)$ is transformed into the function of coefficients $\alpha_1, \alpha_2, \ldots, \alpha_n$ and can be written as

$$I[\phi(x, y)] = I(\alpha_1, \alpha_2, \ldots, \alpha_n)$$

If functional $I[\phi(x, y)]$ obtains the extreme value on $\phi(x, y)$, coefficients $\alpha_1, \alpha_2, \ldots, \alpha_n$ should satisfy the equations

$$\frac{\partial I}{\partial \alpha_1} = 0, \frac{\partial I}{\partial \alpha_2} = 0, \ldots, \frac{\partial I}{\partial \alpha_n} = 0 \tag{30.11}$$

Solving the equations above, we will obtain coefficients $\alpha_1, \alpha_2, \ldots, \alpha_n$, and substitution of them into formula (30.10) will give the solution $\phi(x, y)$.

Therefore, using the Ritz method to solve an engineering problem, the computational steps are as follows:

1) Take differential equations from the project as Euler's equations and find their corresponding functional $I(\phi)$.
2) Assume a set of coordinate function sequence $w_1(x, y), w_2(x, y), \ldots, w_n(x, y)$ that satisfy the boundary conditions, let $\phi = \sum_{i=1}^{n} a_i w_i$, and substitute this formula into functional, so the functional will be function $I(\alpha_1, \alpha_2, \ldots, \alpha_n)$ that take coefficients $\alpha_1, \alpha_2, \ldots, \alpha_n$ as variables.
3) Solve the coefficients $\alpha_1, \alpha_2, \ldots, \alpha_n$ by Eq. (30.11).
4) Substitute the coefficients obtained into formula (30.10) and get a solution that makes the functional achieve extreme values, that is, the approximate solution of differential equations.

Example 1

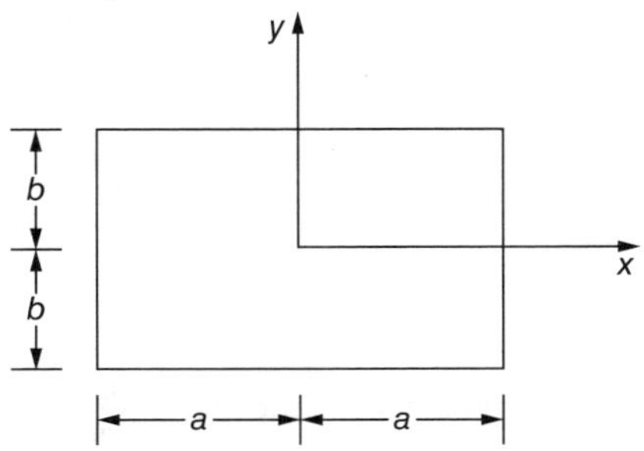

Figure 30.7 Torsion of a rod.

Consider the torsion problem of an elastic cylinder with uniform section. The cross section is a $2a \times 2b$ rectangle, as shown in Figure 30.7. When analyzing the stress induced by torque, the following differential equation must be solved:

$$\frac{\partial^2 \phi}{\partial x^2} + \frac{\partial^2 \phi}{\partial y^2} = -2 \tag{d}$$

Here $\phi = \phi(x, y)$ is the stress function on the boundary

$$\phi(x, y) = 0 \tag{e}$$

Equation (d) is a special case of Poisson's equation with $f(x, y) = -2$, taking it as Euler's equations, and the corresponding functional is

$$I(\phi) = \int_{-a}^{a} \int_{-b}^{b} \left[\left(\frac{\partial^2 \phi}{\partial x^2} \right)^2 + \left(\frac{\partial^2 \phi}{\partial y^2} \right)^2 - 4\phi \right] \mathrm{d}x\mathrm{d}y \tag{f}$$

Take the approximate expression of solution as

$$\phi = (x^2 - a^2)(y^2 - b^2) \sum_{i=0}^{n} \sum_{j=0}^{m} \alpha_{ij} x^i y^j \tag{g}$$

Obviously the formula above meets the boundary condition formula (e). Because the cross section is symmetrical to x and y axes, in the formula above, only the even items are necessary.

Take one item from formula (g), that is,

$$\phi_1 = \alpha_1(x^2 - a^2)(y^2 - b^2)$$

$$\left(\frac{\partial \phi_1}{\partial x}\right)^2 + \left(\frac{\partial \phi_1}{\partial y}\right)^2 - 4\phi_1 = 4\alpha_1^2[x^2(y^2 - b^2)^2 + y^2(x^2 - a^2)^2] - 4\alpha_1(x^2 - a^2)(y^2 - b^2)$$

$$I_1 = \int_{-a}^{a}\int_{-b}^{b}\left[\left(\frac{\partial \phi_1}{\partial x}\right)^2 + \left(\frac{\partial \phi_1}{\partial y}\right)^2 - 4\phi_1\right] \mathrm{d}x\mathrm{d}y$$

$$= 4\int_{-a}^{a}\int_{-b}^{b}\{\alpha_1^2[x^2(y^2 - b^2)^2 + y^2(x^2 - a^2)^2] - \alpha_1(x^2 - a^2)(y^2 - b^2)\}\mathrm{d}x\mathrm{d}y$$

$$\frac{\partial I_1}{\partial \alpha_1} = 4\int_{-a}^{a}\int_{-b}^{b}\{2\alpha_1[x^2(y^2 - b^2)^2 + y^2(x^2 - a^2)^2] - (x^2 - a^2)(y^2 - b^2)\}\mathrm{d}x\mathrm{d}y$$

After computing the integral of the formula above, by $\frac{\partial I_1}{\partial \alpha_1} = 0$, we get

$$\frac{128}{45}a^3b^3(a^2 + b^2)\alpha_1 - \frac{32}{9}a^3b^3 = 0$$

Then we get

$$\alpha_1 = \frac{5}{4}\frac{1}{a^2 + b^2}$$

The approximate solution of the stress function is

$$\phi_1 = \frac{5}{4}\frac{1}{a^2 + b^2}(x^2 - a^2)(y^2 - b^2)$$

Take three items from formula (g), that is,

$$\phi_3 = (x^2 - a^2)(y^2 - b^2)(\alpha_1 + \alpha_2x^2 + \alpha_3y^2)$$

Substituting it into functional expression (f), by the necessary conditions $\frac{\partial I_3}{\partial \alpha_1} = 0$, $\frac{\partial I_3}{\partial \alpha_2} = 0$, $\frac{\partial I_3}{\partial \alpha_3} = 0$ for making functional obtain the extreme value, after calculating, the following equations can be obtained:

$$\frac{128}{45}a^3b^3(a^2 + b^2)\alpha_1 + \frac{128}{45}a^5b^3\left(\frac{a^2}{7} + \frac{b^2}{5}\right)\alpha_2 + \frac{128}{45}a^3b^5\left(\frac{a^2}{5} + \frac{b^2}{7}\right)\alpha_3 = \frac{16a^3b^3}{9}$$

$$\frac{128}{45}a^5b^3\left(\frac{a^2}{7} + \frac{b^2}{5}\right)\alpha_1 + \frac{128}{45 \times 7}a^5b^5\left(\frac{11b^2}{5} + \frac{a^2}{3}\right)\alpha_2 + \frac{128}{45 \times 35}a^5b^5(a^2 + b^2)\alpha_3 = \frac{16a^5b^3}{45}$$

$$\frac{128}{45}a^3b^3\left(\frac{a^2}{5}+\frac{b^2}{7}\right)\alpha_1+\frac{128}{45\times 35}a^5b^5(a^2+b^2)\alpha_2$$

$$+\frac{128}{45\times 7}\left(\frac{11}{5}a^2+\frac{1}{3}b^2\right)\alpha_3=\frac{16a^3b^5}{45}$$

Solving them, we get

$$\alpha_1=\frac{35(9a^4+130a^2b^2+9b^4)}{16(45a^6+509a^4b^2+509a^2b^4+45b^6)}$$

$$\alpha_2=\frac{105(9a^2+b^2)}{16(45a^6+509a^4b^2+509a^2b^4+45b^6)}$$

$$\alpha_3=\frac{105(a^2+9b^2)}{16(45a^6+509a^4b^2+509a^2b^4+45b^6)}$$

30.8 Finite Element Method for Solving the Variational Problems

Using the Ritz method to solve variational problems, the coordinate function w_i should satisfy the boundary conditions in the original solving region. It is clear that only when the solving region is regular (such as circles, rectangles, ellipses) and boundary conditions are relatively simple that such a coordinating function can be found. In real projects, the solving area tends to be irregular, and the boundary conditions also tend to be more complex. Thus it is difficult to find the coordinate function w_i that satisfy the boundary conditions, so it cannot be solved by the Ritz method.

Using the finite element method to solve engineering problems, the solving region R is divided into a number of triangular elements. According to variational principle, function $\phi(x, y)$ that makes the functional achieve extreme value can be found, because the solving region was divided into limited triangular elements, which can adapt to irregular shapes and complex boundary conditions. Thus it is a great progress in the analysis of engineering problems, and it can be used to solve various complex problems in actual engineering.

Taking two-dimensional Laplace equations as an example, the problem is as follows:

In region R

$$\frac{\partial^2\phi}{\partial x^2}+\frac{\partial^2\phi}{\partial y^2}=0 \tag{a}$$

On boundary B

$$\phi=\phi_{\mathrm{B}} \tag{b}$$

On boundary C

$$l_x\frac{\partial\phi}{\partial x}+l_y\frac{\partial\phi}{\partial y}+\beta(\phi-t)=0 \tag{c}$$

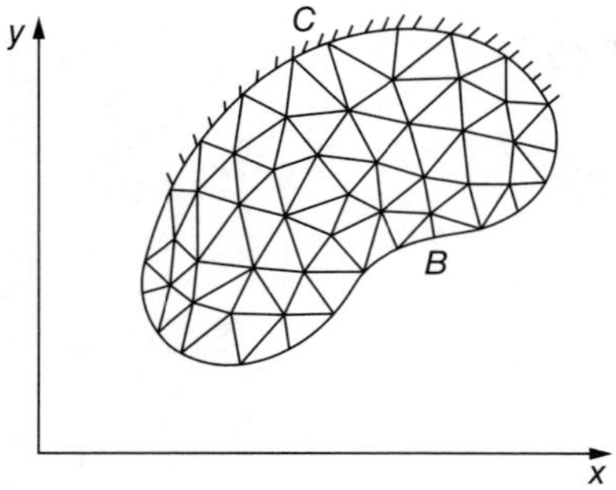

Figure 30.8 Plane problem.

By Example 1 of Section 30.5, the corresponding functional is

$$I[\phi(x,y)] = \iint_R \frac{1}{2}\left[\left(\frac{\partial \phi}{\partial x}\right)^2 + \left(\frac{\partial \phi}{\partial y}\right)^2\right] dxdy + \int_C \beta\left(\frac{1}{2}\phi^2 - t\phi\right) ds \quad \text{(d)}$$

As shown in Figure 30.8, dividing the solving region into m elements, each element constitutes a subdomain ΔR, and these subdomains compose the original region R:

$$R = \sum \Delta R \quad \text{(e)}$$

Accordingly, functional $I(\phi)$ becomes the sum of integrals in these subdomains:

$$I(\phi) = \sum I^e \quad (30.12)$$

where I^e is the integral in the subdomain ΔR of element e:

$$I^e = \iint_{\Delta R} \frac{1}{2}\left[\left(\frac{\partial \phi}{\partial x}\right)^2 + \left(\frac{\partial \phi}{\partial y}\right)^2\right] dxdy + \int_{\Delta C} \beta\left(\frac{1}{2}\phi^2 - t\phi\right) ds \quad (30.13)$$

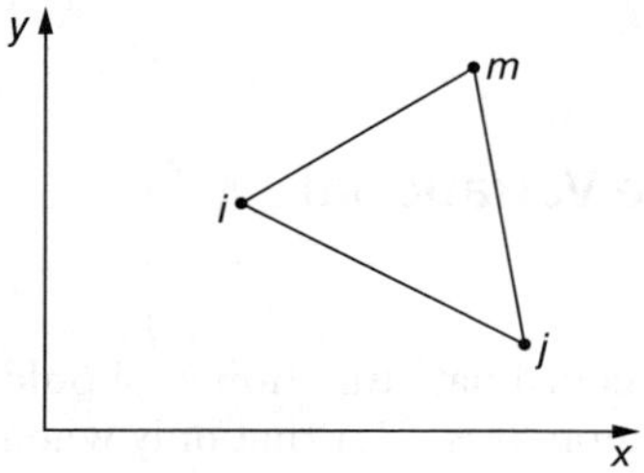

Figure 30.9 Triangular element.

The second item on the right of the formula above is integrated along boundary C; only those elements close to boundary C will have this item.

Taking out one element e, as shown in Figure 30.9, assuming that the three nodes are i, j, m, in element e, function $\phi(x, y)$ can be expressed approximately as follows:

$$\phi^e(x,y) = N_i\phi_i + N_j\phi_j + N_m\phi_m = [N]\{\phi\}^e \quad (30.14)$$

where ϕ_i, ϕ_j, ϕ_m are the values of ϕ on the nodes i, j, m and N_i, N_j, N_m are the shape functions, $[N] = [N_i, N_j, N_m]$.

By substitution of formula (30.14) into formula (30.13), I^e can be expressed by three nodal functions ϕ_i, ϕ_j, ϕ_m of element e. By formula (30.12), accumulating I^e of each element, functional $I(\phi)$ throughout the region is obtained. It is clear that functional $I(\phi)$ has become the function of the nodal function values $\phi_1, \phi_2, \ldots, \phi_n$ (a total of n nodes):

$$I[\phi(x,y)] = I(\phi_1, \phi_2, \ldots, \phi_n)$$

According to the necessary conditions for functional to reach extreme values, we get the equation set as follows:

$$\frac{\partial I}{\partial \phi_1} = 0, \frac{\partial I}{\partial \phi_2} = 0, \ldots, \frac{\partial I}{\partial \phi_n} = 0$$

Because $I = \sum I^e$, the formula above can be simplified as follows:

$$\sum \frac{\partial I^e}{\partial \phi_1} = 0, \sum \frac{\partial I^e}{\partial \phi_2} = 0, \ldots, \sum \frac{\partial I^e}{\partial \phi_n} = 0 \quad (30.15)$$

The equation set include n equations, the solution of which will give n nodal function value $\phi_1, \phi_2, \ldots, \phi_n$. By formula (30.14), the value ϕ of each element can be obtained, and specific formulas can be seen in Chapter 13, not listed here.

Now let us consider the convergence condition for the finite element method to solve variational problems. For the problems discussed currently, because formula (d) only contains the first derivative of ϕ, only the continuity of ϕ must be guaranteed when choosing the shape functions. In addition, when the size of the element shrinks infinitely, the first derivative $\frac{\partial \phi}{\partial x}, \frac{\partial \phi}{\partial y}$ must approach to a constant. If these conditions are satisfied, the solution of the finite element method will converge to real solution.

We will explore the boundary conditions of the problem. There are two kinds of boundary conditions, namely, imposed boundary conditions and natural boundary conditions. On boundary B, the given value ϕ must satisfy formula (b) first, namely, $\phi = \phi_B$, called imposed boundary conditions. On boundary C, it is not required that a trial function should satisfy formula (c) first, but it is satisfied progressively during functional minimization process (see Euler equations formula (30.7)), called natural boundary conditions.

Bibliography

1 Washizu, K. (1975) *Variational Methods in Elasticity and Plasticity*, 2nd edn, Pergamon Press, Oxford.

2 Courant, R. (1943) Variational methods for solution of problems of equilibrium and vibration. *Bulletin of the Amercian Mathematical Society*, **49**, 1–43.

31

Weighted Residual Method

When the variational method is used to find an approximate solution of a problem, at first the corresponding functional of the problem must be given; for some problems it is difficult to find the corresponding functional, or the corresponding functional does not exist. In this case, it is impossible to solve the problem by variational method, but the problem may be solved by the weighted residual method. In this chapter, we shall explain the basic idea of the weighted residual method at first and then show how to use it in the finite element method.

31.1 Introduction to Weighted Residual Method

The weighted residual method is an efficient method to find the approximate solution of a differential equation. Let u satisfy the following differential equation in domain D:

$$L(u) = p \tag{31.1}$$

and satisfy the following boundary condition on boundary C:

$$B(u) = 0 \tag{31.2}$$

where $L(\,)$ and $B(\,)$ are differential operators.

An exact solution u must satisfy the above differential equation at any point in domain D and satisfy the above boundary condition at any point on boundary C. It is difficult to give such exact solution for some complicated engineering problems, so we shall find an approximate solution.

Let the approximate solution of the above problem be

$$u = \beta_1 u_1 + \beta_2 u_2 + \cdots + \beta_n u_n = \sum_{i=1}^{n} \beta_i u_i \tag{31.3}$$

where β_i are coefficients to be determined and u_i are known functions that must be linearly independent and must be taken from a complete set of functions, which means that any function may be expressed by this set.

As u is an approximate solution, substitution of Eq. (31.3) into Eqs (31.1) and (31.2) will give the following residuals:

$$\text{internal residual} \quad R_a = L(u) - p \tag{31.4}$$

$$\text{boundary residual} \quad R_b = B(u) \tag{31.5}$$

The Finite Element Method: Fundamentals and Applications in Civil, Hydraulic, Mechanical and Aeronautical Engineering, First Edition. Bofang Zhu.

Companion website: www.wiley.com/go/zhufem17

For an exact solution, at any point in the solution region D $R_a = 0$ and at any point on the boundary C $R_b = 0$, in the weighted residual methods, these requirements are satisfied approximately by the following methods:

i) Internal residual method
The trial function u satisfies the boundary condition, that is, on the boundary $B(u) = 0$, but in the solution region, $R_a \neq 0$, we choose the coefficients β_i in (31.3) so that the weighted integral

$$\int_D W_a R_a \mathrm{d}V = 0 \tag{31.6}$$

where W_a is the internal weight function.

ii) Boundary residual method
The trial function u satisfies Eq. (31.1), that is, $R_a = L(u) - p = 0$ in the solution region but does not satisfy the boundary condition (31.2), and we choose the coefficients in (31.3) so that the weighted integral of boundary residual equals zero, that is,

$$\int_C W_b R_b \mathrm{d}S = 0 \tag{31.7}$$

iii) Mixed residual method
The trial function does not satisfy both Eqs (31.1) and (31.2). Choose the coefficients β_i so that the following two equations are satisfied:

$$\int_D W_a R_a \mathrm{d}V = 0 \tag{31.8}$$

$$\int_C W_b R_b \mathrm{d}S = 0 \tag{31.9}$$

After the coefficients β_i in Eq. (31.3) are determined by Eqs (31.6)–(31.9), the approximate solution is obtained.

31.2 Weight Function for Internal Residual Method

In the following we shall explain how to choose the weight function in the internal residual method.

31.2.1 Collocation Method

Take an approximate solution $u = \sum \beta_i u_i$, where u_i satisfies the boundary condition but does not satisfy the differential equation; the residual is $R_a = L(u) - p \neq 0$. It is required that Eq. (31.6) is satisfied.

Take the weight function W_i as follows:

$$\left.\begin{array}{ll} \text{on the } n \text{ discrete points} & W_i = 1 \\ \text{in the else part of region } D & W_i = 0 \end{array}\right\} \tag{31.10}$$

Practically, it is required to satisfy the differential equation (31.1) on the n discrete points; in other words, the residuals are equal to zero on these n points:

$$R_i = 0, \quad i = 1,2, \ldots, n \tag{31.11}$$

The n undetermined coefficients β_i can be computed by the above n equations, and the approximate solution is given by Eq. (31.3).

Example 1 Solve the following differential equation:

$$\frac{d^2u}{dx^2} + u + x = 0; \quad 0 \le x \le 1 \tag{a}$$

and boundary condition:

$$\left.\begin{array}{l} u = 0 \text{ when } x = 0 \\ u = 0 \text{ when } x = 1 \end{array}\right\} \tag{b}$$

Take the approximate solution as follows:

$$u = x(1-x)(\beta_1 + \beta_2 x + \beta_3 x^2 + \cdots) \tag{c}$$

It is clear that the above equation satisfies the boundary condition (b) but does not satisfy the differential equation (a).

If only one term is taken in Eq. (c), we obtain the first approximate solution:

$$u = \beta_1 x(1-x)$$

Substituting it into Eq. (a), the residual is

$$R(x) = x + \beta_1(-2 + x - x^2)$$

Take $x = 1/2$ as the collocation point

$$R\left(\frac{1}{2}\right) = \frac{1}{2} - \frac{7}{4}\beta_1 = 0$$

Hence, we get $\beta_1 = 2/7$, so the first approximate solution is

$$u = \frac{2}{7}x(1-x)$$

If two terms are taken in Eq. (c), we obtain the second approximate solution:

$$u = x(1-x)(\beta_1 + \beta_2 x)$$

The residual is

$$R(x) = x + \beta_1(-2 + x - x^2) + \beta_2(2 - 6x + x^2 - x^3)$$

Taking $x = 1/3$ and $x = 2/3$ as collocation points, we have

$$R\left(\frac{1}{3}\right) = \frac{1}{3} - \frac{16}{9}\beta_1 + \frac{2}{27}\beta_2 = 0,$$

$$R\left(\frac{2}{3}\right) = \frac{2}{3} - \frac{16}{9}\beta_1 - \frac{50}{27}\beta_2 = 0$$

Thus we obtain $\beta_1 = 0.1948$, $\beta_2 = 0.1731$, so the second approximate solution is

$$u = x(1-x)(0.1948 + 0.1731x)$$

Table 31.1 Computing results of collocation method.

x	First approximate solution	Second approximate solution	Exact solution
0.25	0.0536	0.0446	0.0440
0.50	0.0713	0.0704	0.0697
0.75	0.0536	0.0619	0.0601

The exact solution of this problem is

$$u = \frac{\sin x}{\sin 1} - x$$

The comparison of the approximate solution and the exact solution is given in Table 31.1. It is clear that the second approximate solution is close to the exact solution and the error is 3%. If more terms are taken, the precision will be higher.

31.2.2 Least Squares Method

For the differential equation $L(u) = p$ and the boundary condition $B(u) = 0$, take the approximate solution $u = \sum \beta_i u_i$, where u_i satisfies the boundary condition but does not satisfy the differential equation; the residual is $R = L(u) - p$. After integrating R^2 in region D, we get

$$I = \int R^2 \,\mathrm{d}V \tag{31.12}$$

Choose the coefficients β_i so as to make the integral I to take minimum value; thus

$$\frac{\partial I}{\partial \beta_i} = 2\int R\frac{\partial R}{\partial \beta_i}\,\mathrm{d}V = 0, \quad i = 1,2,\ldots,n \tag{31.13}$$

Hence we have n equations that may be used to compute the n coefficients β_i. Comparing Eqs (31.9) and (31.13), it is apparent that now the weight function is

$$W_i = \frac{\partial R}{\partial \beta_i} \tag{31.14}$$

Example 2 Solve the problem in Example 1 by least squares method. The first approximate solution is $u = \beta_1 x(1 - x)$:

$$R = x + \beta_1(-2 + x - x^2)$$

$$\frac{\partial R}{\partial \beta_1} = -2 + x - x^2$$

By Eq. (31.13),

$$\int_0^1 R\frac{\partial R}{\partial \beta_1}\mathrm{d}x_1 = \int_0^1 [x + \beta_1(-2 + x - x^2)](-2 + x - x^2)\mathrm{d}x = 0$$

Table 31.2 Computing results of the least squares method.

x	First approximate solution	Second approximate solution	Exact solution
0.25	0.0506	0.0434	0.0440
0.50	0.0681	0.0683	0.0697
0.75	0.0506	0.0592	0.0601

After integrating, we have

$$\frac{101}{5}\beta_1 - \frac{11}{2} = 0$$

Thus $\beta_1 = 55/202 = 0.272$, so $u = 0.272\ x(1-x)$.

Take the second approximate solution as

$$u = x(1-x)(\beta_1 + \beta_2 x)$$
$$R = x + \beta_1(-2 + x - x^2) + \beta_2(2 - 6x + x^2 - x^3)$$

By Eq. (31.13), we get

$$\int_0^1 R\frac{\partial R}{\partial \beta_1}\,\mathrm{d}x = \int_0^1 [x + \beta_1(-2 + x - x^2) + \beta_2(2 - 6x + x^2 - x^3)](-2 + x - x^2)\,\mathrm{d}x = 0$$

$$\int_0^1 R\frac{\partial R}{\partial \beta_2}\,\mathrm{d}x = \int_0^1 [x + \beta_1(-2 + x - x^2) + \beta_2(2 - 6x + x^2 - x^3)](2 - 6x + x^2 - x^3)\,\mathrm{d}x = 0$$

After integration, we have

$$202\beta_1 + 101\beta_2 = 55$$
$$101\beta_1 + 1532\beta_2 = 393$$

Solution of the above equations gives $\beta_1 = 0.192,\ \ \beta_2 = 0.165$. The second approximate solution is $u = x(1-x)(0.192 + 0.165x)$. The comparison of approximate solutions and the exact solution is shown in Table 31.2.

31.2.3 Moment Method

For the differential equation $L(u) - p = 0$ and boundary condition $B(u) = 0$, take the approximate solution $u = \sum \beta_i u_i$, where u_i satisfies the boundary condition but does not satisfy the differential equation. Substituting u into the differential equation, the residual is $R = L(u) - p$, and the coefficients β_i are determined by $\int W_i R\,\mathrm{d}V = 0$. For the one-dimensional problem, take the weight functions as follows:

$$W_1 = 1,\ W_2 = x,\ W_3 = x^2,\ \ldots,\ W_n = x^{n-1}$$

Substituting them in Eq. (31.9), we have

$$\left.\begin{aligned}&\int 1\cdot R\,\mathrm{d}V = 0\\&\int x\cdot R\,\mathrm{d}V = 0\\&\int x^2\cdot R\,\mathrm{d}V = 0\\&\quad\vdots\\&\int x^{n-1}\cdot R\,\mathrm{d}V = 0\end{aligned}\right\}\tag{31.15}$$

Solution of the above equations will give the coefficients β_i. The left parts of Eq. (31.15) represent the moment of residual R of zero order, first order, second order, ..., so this method is called the moment method.

Example 3 Solve the problem of Example 1 by moment method. The first approximate solution is $u = \beta_1 x(1-x)$. Take the weight function $W_1 = 1$ from Eq. (31.15), and we obtain

$$\int_0^1 1\cdot R\mathrm{d}x = \int_0^1 [x + \beta_1(-2 + x - x^2)]\,\mathrm{d}x = 0$$

or

$$1 - \frac{11}{3}\beta_1 = 0$$

Hence

$$\beta_1 = \frac{3}{11} = 0.273,\quad u = 0.273x(1-x)$$

Take the second approximate solution as $u = x(1-x)(\beta_1 + \beta_2 x)$, and the residual is $R = x + \beta_1(-2 + x - x^2) + \beta_2(2 - 6x + x^2 - x^3)$. Given the weight functions $W_1 = 1$, $W_2 = x$, from Eq. (31.15), we have

$$\int_0^1 R\,\mathrm{d}x = 0,\quad \int_0^1 xR\,\mathrm{d}x = 0$$

After integration, we get

$$\frac{11}{6}\beta_1 + \frac{11}{12}\beta_2 = \frac{1}{2}$$

$$\frac{11}{12}\beta_1 + \frac{19}{12}\beta_2 = \frac{1}{3}$$

Hence $\beta_1 = 0.1878$, $\beta_2 = 0.1693$, and $u = x(1-x)(0.1878 + 0.1693x)$.

The comparison of the approximate solution and the exact solution is shown in Table 31.3.

31.2.4 Galerkin Method

For differential equation $L(u) - p = 0$ and boundary condition $B(u) = 0$, let the approximate solution be

$$u = \sum_{i=1}^{n} \beta_i u_i \tag{31.3}$$

Table 31.3 Computing results of moment method.

x	First approximate solution	Second approximate solution	Exact solution
0.25	0.0512	0.0432	0.0440
0.50	0.0682	0.0682	0.0697
0.75	0.0512	0.0591	0.0601

where u_i satisfies the boundary condition but does not satisfy the differential equation; the coefficients β_i are determined by the following equation:

$$\int W_i R \, dV = 0 \tag{31.6a}$$

where R is the residual and W_i is the weight function. In Galerkin method, take

$$W_i = u_i \tag{31.16}$$

After substituting in Eq. (31.9), we get

$$\left.\begin{aligned} &\int u_1 R \, dV = 0 \\ &\int u_2 R \, dV = 0 \\ &\quad\vdots \\ &\int u_n R \, dV = 0 \end{aligned}\right\} \tag{31.17}$$

where $R = L(u) - p$ is the residual. Thus we obtain n equations by means of which we can solve the n coefficients β_i.

The precision of Galerkin method is very good.

Example 4 Solve the problem in Example 1 by Galerkin method. Let the approximate solution be

$$u = \beta_1 u_1 + \beta_2 u_2 + \cdots = \beta_1 x(1-x) + \beta_2 x^2(1-x) + \cdots$$

If only one term is taken, we get the first approximate solution $u = \beta_1 u_1 = \beta_1 x(1-x)$. Let the weight function be $W_1 = u_1 = x(1-x)$. By Eq. (31.17),

$$\int_0^1 u_1 R \, dx = \int_0^1 x(1-x)[x + \beta_1(-2 + x - x^2)] dx = 0$$

Namely,

$$\frac{1}{12} - \frac{3}{10}\beta_1 = 0$$

Hence

$$\beta_1 = 5/18 = 0.278, \quad u = 0.278x(1-x)$$

Table 31.4 Computing results of Galerkin method.

x	First approximate solution	Second approximate solution	Exact solution
0.25	0.0521	0.0440	0.0440
0.50	0.0695	0.0698	0.0697
0.75	0.0512	0.0600	0.0601

Let the second approximate solution be $u = \beta_1 x(1 - x) + \beta_2 x^2(1 - x)$ Taking the weight functions $W_1 = x(1 - x)$, $W_2 = x^2(1 - x)$ and substituting them into Eq. (31.13), we get

$$\int_0^1 x(1-x)[x + \beta_1(-2 + x - x^2) + \beta_2(2 - 6x + x^2 - x^3)]\,dx = 0$$

$$\int_0^1 x^2(1-x)[x + \beta_1(-2 + x - x^2) + \beta_2(2 - 6x + x^2 - x^3)]\,dx = 0$$

Namely,

$$\frac{3}{10}\beta_1 + \frac{3}{20}\beta_2 = \frac{1}{12}$$
$$\frac{3}{20}\beta_1 + \frac{13}{105}\beta_2 = \frac{1}{12}$$

Solving the above simultaneous equations, we get $\beta_1 = 0.1924, \beta_2 = 0.1707$; thus $u = x(1 - x)(0.1924 + 0.1707x)$. The comparison of the approximate and the exact solution is shown in Table 31.4.

In the preceding paragraphs, the trial function (31.3) satisfies the boundary condition, so they belong to the internal residual methods.

The methods of solution for the boundary residual method and the mixed residual methods are similar; they will not be repeated here, and the readers may refer to Refs [1, 2].

31.3 Establish Fundamental Equations of Finite Element Method by Weighted Residual Method

The problem to be solved is that in region D, u satisfies the differential equation

$$L(u) - p = 0 \tag{31.18}$$

and on boundary C,

$$B(u) = 0 \tag{31.19}$$

where $L(\,)$ and $B(\,)$ are differential operators.

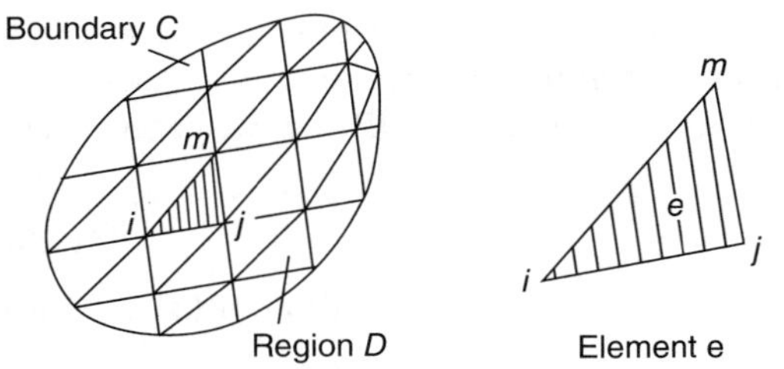

Figure 31.1 Solution region D divided into elements.

Divide the region D into M elements, as shown in Figure 31.1. From which one element e is taken, whose nodes are $i\ j\ m$, the unknowns are the displacements of the nodes, namely,

$$\{u\}^e = \begin{Bmatrix} u_i \\ u_j \\ u_m \end{Bmatrix} \tag{31.20}$$

The displacement u of any point in the element may be expressed as follows:

$$u = \begin{bmatrix} N_i & N_j & N_m \end{bmatrix} \begin{Bmatrix} u_i \\ u_j \\ u_m \end{Bmatrix} \tag{31.21}$$

where N_i, N_j, N_m are shape functions. In the whole region D, u may be expressed by

$$u = [N]\{u\} \tag{31.22}$$

in which $[N]$ is shape function matrix and $\{u\}$ is vector of nodal displacements that satisfy the boundary condition.

As u expressed by (31.22) is an approximate solution that does not satisfy the differential equation, substitution of it into Eq. (31.18) will give the residual

$$R = L(u) - p \tag{31.23}$$

Now it is required that the integral of the weighted residual R in region D is equal to zero:

$$\int_D W_i R\,\mathrm{d}V = \int_D W_i[L(N\{u\} - p)]\,\mathrm{d}V = \sum_e \int_{\Delta D} W_i[L(N\{u\} - p)]\,\mathrm{d}V = 0 \tag{31.24}$$

where $\sum_e$ indicates the summation of elements and ΔD is the subregion corresponding to element e. The above equation indicates that the summation of integrals of weighted residual of each element must be equal to zero. If there are n nodal unknowns to be determined, give n weighted functions W_i, $i = 1,2,\ldots,n$, and equations can be derived from Eq. (31.24). Simultaneous solution of these n equations will give the n nodal displacements $\{u_i\}$.

If different weight function W_i is given in Eq. (31.24), we will obtain a different computing method. For example, take $W_i = 1$ on n discrete point, while $W_i = 0$ elsewhere, we get the collocation method. If we take weight function $W_i = \partial R/\partial u_i$, we get the least squares method. If we take weight function

$$W_i = N_i \tag{31.25}$$

where N_i is the shape function, we get Galerkin method. Substitution of (31.25) into (31.24) yields

$$\int_D W_i R\,\mathrm{d}V = \int_D N_i(L[N]\{u\} - p)\,\mathrm{d}V = 0 \tag{31.26}$$

The computing precision of Galerkin method is good. If there is corresponding functional, the computing results of Galerkin method may be the same as those of variational method. So, Galerkin method has been widely applied.

There is a defect in the abovementioned method. The differential operator $L(\,)$ appears in the integral of Eq. (31.26). Compared with the functional of variational method, the order of differentiation is higher. In order to avoid contribution on the common boundaries of adjacent elements, there must be higher order of continuation of the shape functions, and difficulty may appear in the construction of shape functions. In order to avoid this difficulty, we may take integration by parts for Eq. (31.26) to reduce the order of differentiation of the term in the integral.

In the following, we take a bar under tension as a simple example to show how to establish fundamental equations of finite element method by weighted residual method.

As shown in Figure 31.2, in a bar with length L, cross-sectional area A, horizontal displacement u, horizontal strain $\varepsilon_x = \frac{\mathrm{d}u}{\mathrm{d}x}$, stress $\sigma_x = E\frac{\mathrm{d}u}{\mathrm{d}x}$, and condition of equilibrium $A\frac{\mathrm{d}\sigma_x}{\mathrm{d}x} + q = 0$, after substitution of $\sigma_x = E\frac{\mathrm{d}u}{\mathrm{d}x}$ into the above equation, we get the equilibrium equation expressed by displacement as follows:

$$EA\frac{\mathrm{d}^2u}{\mathrm{d}x^2} + q = 0 \tag{31.27}$$

where q is the horizontal distributed load. Let the left end be fixed and a horizontal force F is acting on the right end, so the boundary conditions are

$$u = 0 \quad \text{when } x = 0 \tag{31.28}$$

$$AE\frac{\mathrm{d}u}{\mathrm{d}x} = F \quad \text{when } x = L \tag{31.29}$$

Now the weighted residual method is used to solve the problem. The displacement u must satisfy the following equation:

$$\int_0^L W_i\left(EA\frac{\mathrm{d}^2u}{\mathrm{d}x^2} + q\right)\mathrm{d}x = 0 \tag{31.30}$$

where W_i is the weight function. For the term $EA\frac{\mathrm{d}^2u}{\mathrm{d}x^2}$, after integrating by parts, we have

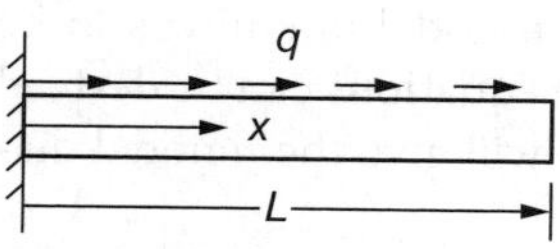

Figure 31.2 A bar.

$$\int_0^L EAW_i\frac{\mathrm{d}^2u}{\mathrm{d}x^2}\mathrm{d}x = \left[EAW_i\frac{\mathrm{d}u}{dx}\right]_0^L - \int_0^L EA\frac{\mathrm{d}W_i}{\mathrm{d}x}\frac{\mathrm{d}u}{\mathrm{d}x}\mathrm{d}x$$

Substitute into Eq. (31.30) to get

$$\int_0^L EA\frac{\mathrm{d}W_i}{\mathrm{d}x}\frac{\mathrm{d}u}{\mathrm{d}x}\mathrm{d}x = \int_0^L W_i q\,\mathrm{d}x + \left(EAW_i\frac{\mathrm{d}u}{\mathrm{d}x}\right)_{x=L} - \left(EAW_i\frac{\mathrm{d}u}{\mathrm{d}x}\right)_{x=0} \quad \text{(a)}$$

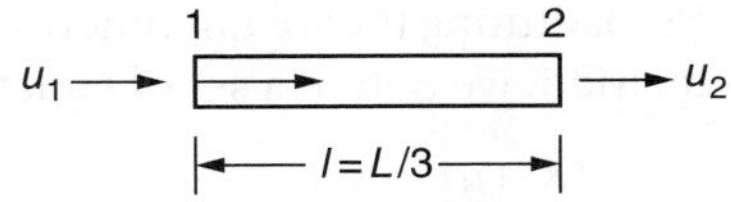

Figure 31.3 Elements 1–2.

Now, divide the bar into three elements with length $l = L/3$. The element 1–2 is shown in Figure 31.3. Let the origin of the local coordinate be node 1, and the displacement of any point of the element may be expressed by

$$u = \begin{bmatrix} N_1 & N_2 \end{bmatrix}\begin{Bmatrix} u_1 \\ u_2 \end{Bmatrix} \quad \text{(b)}$$

in which the shape function $N_1 = 1 - \frac{x}{l}, N_2 = \frac{x}{l}$.

Taking a derivative of Eq. (b), we get the strain as follows:

$$\frac{\mathrm{d}u}{\mathrm{d}x} = \begin{bmatrix} \frac{\mathrm{d}N_1}{\mathrm{d}x} & \frac{\mathrm{d}N_2}{\mathrm{d}x} \end{bmatrix}\begin{Bmatrix} u_1 \\ u_2 \end{Bmatrix} = \begin{bmatrix} -\frac{1}{l} & \frac{1}{l} \end{bmatrix}\begin{Bmatrix} u_1 \\ u_2 \end{Bmatrix} \quad \text{(c)}$$

By Galerkin method, take the weight function as follows:

$$W_i = N_i$$

For the element 1–2, we have

$$W_1 = N_1 = 1 - \frac{x}{l}, \quad W_2 = N_2 = \frac{x}{l}$$

For the whole structure, the weight functions are shown in Figure 31.4.

It is known that $u_0 = 0$ for node 0. As u_0 is not unknown, the weight function W_0 is not necessary. We need only W_1, W_2, and W_3. Substitution of them in Eq. (a) will yield three equations that may be used to determine u_1, u_2, and u_3.

From Figure 31.4, at node 0, all the values of weight functions W_1, W_2, and W_3 are equal to zero, namely, $W_{x=0} = 0$; thus the last term of the right part of Eq. (a) is equal to zero, so Eq. (a) is simplified as follows:

$$\int_0^L EA\frac{\mathrm{d}W_i}{\mathrm{d}x}\frac{\mathrm{d}u}{\mathrm{d}x}\mathrm{d}x = \int_0^L W_i q\,\mathrm{d}x + \left(EAW_i\frac{\mathrm{d}u}{\mathrm{d}x}\right)_{x=L} \quad \text{(d)}$$

Substituting Eq. (b) into the above equation, we have

$$\sum_e \int_0^L EA\frac{\mathrm{d}W_i}{\mathrm{d}x}\left(\frac{\mathrm{d}N_i}{\mathrm{d}x}u_i + \frac{\mathrm{d}N_i}{\mathrm{d}x}u_j\right)\mathrm{d}x = \sum_e \int_0^L W_i q\,\mathrm{d}x + \left(EAW_i\frac{\mathrm{d}u}{\mathrm{d}x}\right)_{x=L} \quad (31.31)$$

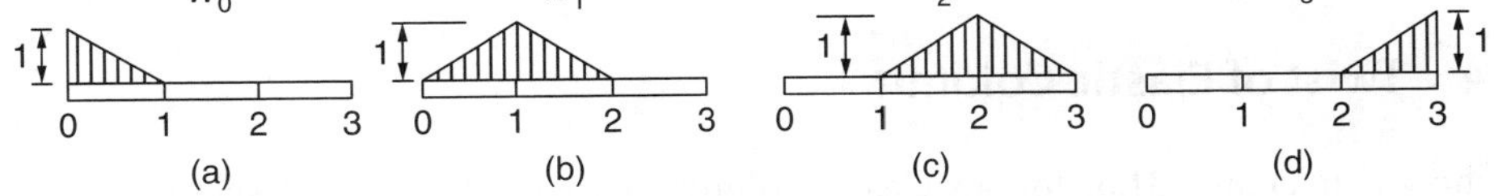

Figure 31.4 Division of elements and weight functions W_i.

Substituting the weight functions W_1, W_2, and W_3 shown in Figure 31.4 into the above equation, we obtain a set of equations:

$$[K]\{u\} = \{P\} \tag{31.32}$$

in which $\{u\} = \left[u_1 \ u_2 \ u_3\right]^{\mathrm{T}}$ and the elements of $[K]$ and $\{P\}$ are

$$K_{ij} = \sum_e k_{ij}^{\mathrm{e}}, \ P_i = \sum_e p_i \tag{e}$$

For element $i\,j$, $W_i = N_i = 1 - \frac{x}{l}$, $W_j = N_j = \frac{x}{l}$, from (31.31), we have

$$k_{ij}^{\mathrm{e}} = \int_0^l EA \frac{\mathrm{d}N_i}{\mathrm{d}x}\frac{\mathrm{d}N_j}{\mathrm{d}x}\mathrm{d}x = \begin{cases} \dfrac{EA}{l}, & \text{when } i = j \\ -\dfrac{EA}{l}, & \text{when } i = j \end{cases} \tag{f}$$

$$p_i^{\mathrm{e}} = \int_0^l N_i q \,\mathrm{d}x = \frac{ql}{2} \tag{g}$$

From Eq. (31.31) it is clear that on node 3, we must add the contribution of $\left(EAW_i\frac{du}{dx}\right)_{x=L}$. As $W_3 = 1$ at $x = L$, from boundary condition (31.29), we get

$$\left(EAW_3\frac{\mathrm{d}u}{\mathrm{d}x}\right)_{x=L} = EA\frac{\mathrm{d}u}{\mathrm{d}x}\bigg|_{X=L} = F$$

Hence

$$p_3^{\mathrm{e}} = \frac{ql}{2} + F \tag{h}$$

Substituting Eqs (e)–(h) and $l = L/3$ into Eq. (31.32), we get

$$\frac{3EA}{L}\begin{bmatrix} 2 & -1 & 0 \\ -1 & 2 & -1 \\ 0 & -1 & 1 \end{bmatrix}\begin{Bmatrix} u_1 \\ u_2 \\ u_3 \end{Bmatrix} = \begin{Bmatrix} \dfrac{qL}{3} \\ \dfrac{qL}{3} \\ \dfrac{qL}{6} + F \end{Bmatrix} \tag{i}$$

Solution of the above equation will give the nodal displacements u_1, u_2, and u_3. The above equation is the same as that derived by energy method.

The above is a simple example. In the following sections, we shall further explain how to establish the fundamental equations of the finite element method for the twist of elastic column, the transient temperature field, and the dynamic response of structure.

31.4 Twist of Elastic Column

As shown in Figure 31.5, for an elastic column with irregular cross section, twist moments M_t are applied at the two ends.

In this case, the normal stress $\sigma_x, \sigma_y, \sigma_z$ and shearing stress τ_{xy} are all equal to zero; only the shearing stresses τ_{zx} and τ_{yz} are not equal to zero. Define the stress function ϕ as follows:

$$\tau_{zx} = \frac{\partial \phi}{\partial y}, \tau_{yz} = \frac{\partial \phi}{\partial x} \tag{31.33}$$

The stress function ϕ must satisfy the following differential equation:

$$\frac{\partial}{\partial x}\left(\frac{1}{G}\frac{\partial \phi}{\partial x}\right) + \frac{\partial}{\partial y}\left(\frac{1}{G}\frac{\partial \phi}{\partial y}\right) + 2\theta = 0 \tag{31.34}$$

where G is shear modulus and θ is the angle of twist in unit length along the axial direction. The boundary of the column is free, so on the boundary we have

$$\frac{\mathrm{d}\phi}{\mathrm{d}s} = 0, \quad \text{namely, } \phi = \text{constant} \tag{31.35}$$

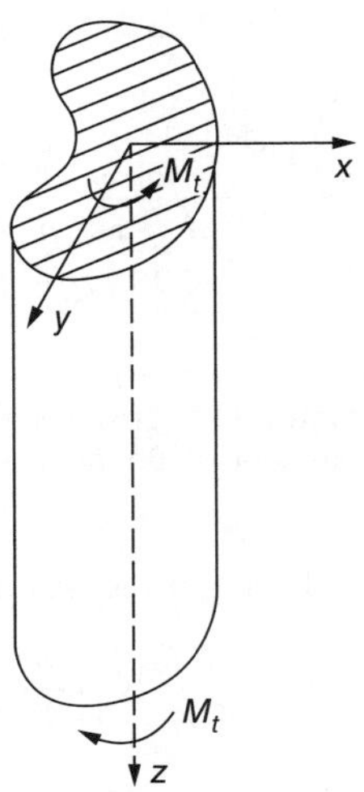

Figure 31.5 A column.

From the definition of stress function, the shearing stress is proportional to the gradient of ϕ and independent of the absolute value of ϕ, so we may let the constant in Eq. (31.35) be zero, namely,

$$\text{on boundary } C \quad \phi = 0 \tag{31.36}$$

By the condition of equilibrium, we have

$$M_t = 2\iint \phi \,\mathrm{d}x\,\mathrm{d}y \tag{31.37}$$

The derivation of the above equations is given in [3].

When the geometry of the cross section is irregular, it is difficult to solve the problem of twist by analytic methods, but it is easy to solve it by the finite element method. We shall explain how to establish the fundamental equations of finite element method by the weighted residual method in the following.

When the problem is solved by weighted residual method, the stress function ϕ must satisfy the following equation:

$$\iint_D W_i \left[\frac{\partial}{\partial x}\left(\frac{1}{G}\frac{\partial \phi}{\partial x}\right) + \frac{\partial}{\partial y}\left(\frac{1}{G}\frac{\partial \phi}{\partial x}\right) + 2\theta\right] \mathrm{d}x\,\mathrm{d}y = 0 \tag{31.38}$$

in which W_i is the weight function. Taking integration by parts for the first term, we have

$$\begin{aligned}
&\iint_D W_i \frac{\partial}{\partial x}\left(\frac{1}{G}\frac{\partial \phi}{\partial x}\right) \mathrm{d}x\,\mathrm{d}y \\
&= \int_C \left| W_i \frac{1}{G}\frac{\partial \phi}{\partial x}\right|_1^2 \mathrm{d}y - \iint_D \frac{1}{G}\frac{\partial W_i}{\partial x}\frac{\partial \phi}{\partial x}\mathrm{d}x\,\mathrm{d}y \\
&= \int_C \frac{1}{G} W_i \frac{\partial \phi}{\partial x} l_x \,\mathrm{d}s - \iint_D \frac{1}{G}\frac{\partial W_i}{\partial x}\frac{\partial \phi}{\partial x}\mathrm{d}x\,\mathrm{d}y
\end{aligned}$$

where l_x is the direction cosine of the normal to the boundary (Figure 31.6). The line integration is taken along the whole boundary C.

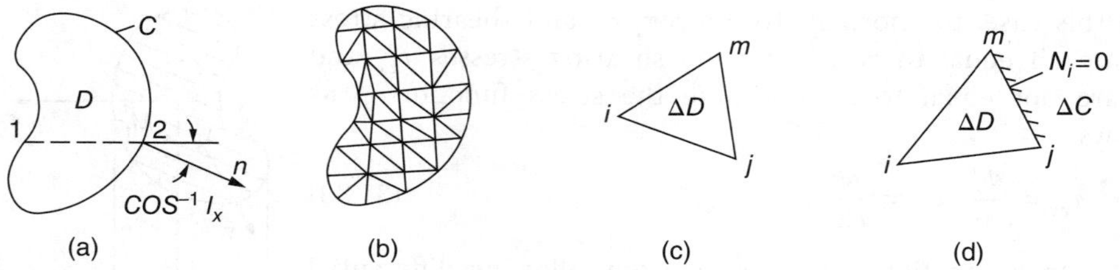

Figure 31.6 Cross section of a column. (a) Cross section of column. (b) Division of elements. (c) Interior element. (d) Boundary element.

Taking integration by parts for the second term of Eq. (31.38), we get

$$\iint_D W_i\left[\frac{\partial}{\partial x}\left(\frac{1}{G}\frac{\partial \phi}{\partial x}\right)+\frac{\partial}{\partial y}\left(\frac{1}{G}\frac{\partial \phi}{\partial y}\right)\right]\mathrm{d}x\,\mathrm{d}y$$

$$=\int_C \frac{1}{G}W_i\left(\frac{\partial \phi}{\partial x}l_x+\frac{\partial \phi}{\partial y}l_y\right)\mathrm{d}s$$

$$-\iint_D \frac{1}{G}\left(\frac{\partial W_i}{\partial x}\frac{\partial \phi}{\partial x}+\frac{\partial W_i}{\partial y}\frac{\partial \phi}{\partial y}\right)\mathrm{d}x\,\mathrm{d}y$$

$$=\int_C \frac{1}{G}W_i\frac{\partial \phi}{\partial n}\mathrm{d}s-\iint_D \frac{1}{G}\left(\frac{\partial W_i}{\partial x}\frac{\partial \phi}{\partial x}+\frac{\partial W_i}{\partial y}\frac{\partial \phi}{\partial y}\right)\mathrm{d}x\,\mathrm{d}y$$

Substituting the above two equations into Eq. (31.38), we obtain

$$\iint_D \frac{1}{G}\left(\frac{\partial W_i}{\partial x}\frac{\partial \phi}{\partial x}+\frac{\partial W_i}{\partial y}\frac{\partial \phi}{\partial y}\right)\mathrm{d}x\,\mathrm{d}y$$

$$=\iint_D 2W_i\theta\,\mathrm{d}x\,\mathrm{d}y+\int_C \frac{1}{G}W_i\frac{\partial \phi}{\partial n}\mathrm{d}s \tag{31.39}$$

Now, there will be no contribution on the boundary of adjacent elements if ϕ is continuous. Because these are the derivatives of weight function $\frac{\partial W_i}{\partial x}$ and $\frac{\partial W_i}{\partial y}$, W_i must also be continuous in region D. The weight function for collocation method is not continuous, so it cannot be applied. The weight functions for Galerkin method and least squares method are continuous, so they can be used.

As shown in Figure 31.6(c), the problem is discretized by finite element method. The stress function ϕ in the interior of any element e is expressed by the shape functions as follows:

$$\phi=\begin{bmatrix}N_i & N_j & N_m\end{bmatrix}\begin{Bmatrix}\phi_i\\ \phi_i\\ \phi_m\end{Bmatrix} \tag{31.40}$$

where N_i is the shape function.

Now use Galerkin method, and take the weight function

$$W_i=N_i$$

Substitute the above two equations into Eq. (31.39) to get

$$\sum_e \iint_{\Delta D} \frac{1}{G} \left[\frac{\partial N_i}{\partial x} \left(\frac{\partial N_i}{\partial x} \phi_i + \frac{\partial N_j}{\partial x} \phi_i + \frac{\partial N_m}{\partial x} \phi_m \right) + \frac{\partial N_i}{\partial y} \left(\frac{\partial N_i}{\partial y} \phi_i + \frac{\partial N_j}{\partial y} \phi_j + \frac{\partial N_m}{\partial y} \phi_m \right) \right] \mathrm{d}x\,\mathrm{d}y$$
$$= \sum_e \iint_{\Delta D} 2N_i\theta\,\mathrm{d}x\,\mathrm{d}y + \sum_e \int_{\Delta C} \frac{1}{G} N_i \frac{\partial \phi}{\partial n} \mathrm{d}s \tag{31.41}$$

where ΔD is the subregion of element e and ΔC is the boundary of element e that coincides with boundary C. $\sum_e$ indicates the summation of relevant elements. On the boundary C, $\phi = 0$. If there are n interior nodes, n equations can be established by Eq. (31.41) by means of which we can solve the n unknowns. As we know $\phi = 0$ on the boundary, so it is not necessary to establish equation for the boundary nodes. When computing the ϕ_i of interior node near the boundary, the shape function on boundary C is $N_i = 0$; thus the last term in the above equation is equal to zero. For example, let the jm boundary of element ijm be located on boundary C. As shown in Figure 31.6(a), ϕ_j and ϕ_m are equal to zero, so they are not unknown. For interior node i, ϕ_i is unknown, but on boundary jm, shape function $N_i = 0$, so

$$\int_{\Delta C} \frac{1}{G} N_i \frac{\partial \phi}{\partial n} \mathrm{d}s = 0$$

Thus, the matrix equation derived from Eq. (31.41) is

$$[H]\{\phi\} = \{P\} \tag{31.42}$$

The elements of matrix $[H]$ and $\{P\}$ are given as follows:

$$H_{ij} = \sum_e h_{ij}^{\mathrm{e}}, \quad P_i = \sum_e p_i^{\mathrm{e}} \tag{31.43}$$

From (31.41), we have

$$\left.\begin{aligned} h_{ij}^{\mathrm{e}} &= \iint_{\Delta D} \frac{1}{G} \left(\frac{\partial N_i}{\partial x}\frac{\partial N_j}{\partial x} + \frac{\partial N_i}{\partial y}\frac{\partial N_j}{\partial y} \right) \mathrm{d}x\,\mathrm{d}y, \\ p_i^{\mathrm{e}} &= \iint_{\Delta D} 2N_i\theta\,\mathrm{d}x\,\mathrm{d}y \end{aligned}\right\} \tag{31.44}$$

The above equations are the same as those derived by variational method.

It is clear that there is θ in the above equation, but θ is determined by Eq. (31.41). θ is unknown before the solution of ϕ. In order to overcome this contradiction, we may let $\theta = 1$, compute the ϕ for $\theta = 1$, and then compute the actual θ by Eq. (31.41).

Let

$$\phi = \theta\Phi$$

Substituting in (31.34), we get

$$\frac{\partial}{\partial x}\left(\frac{1}{G}\frac{\partial \Phi}{\partial x}\right) + \frac{\partial}{\partial y}\left(\frac{1}{G}\frac{\partial \Phi}{\partial y}\right) + 2 = 0$$

From Eq. (31.37), we have

$$\theta = \frac{M_t}{2\iint \Phi \, dx \, dy}$$

Thus, we may let $\theta = 1$, compute Φ by Eqs (31.42)–(31.44), and compute θ by the above equation.

31.5 Unsteady Temperature Field

For a two-dimensional unsteady temperature field, the temperature T must satisfy the differential equation of conduction of heat and the initial and boundary conditions:

$$\text{in region } D \quad \frac{\partial^2 T}{\partial x^2} + \frac{\partial^2 T}{\partial y^2} + \frac{1}{a}\left(\frac{\partial \bar{\theta}}{\partial \tau} - \frac{\partial T}{\partial \tau}\right) = 0, \tag{31.45}$$

$$\text{initial condition, when } \tau = 0, T = T_0 \tag{31.46}$$

$$\left.\begin{array}{ll} \text{boundary condition, on } C_1, & T = T_b \\ \text{on } C_2, & \lambda \dfrac{\partial T}{\partial n} + \beta(T - T_a) = 0 \end{array}\right\} \tag{31.47}$$

in which a is the diffusivity coefficient, $\bar{\theta}$ is the temperature rise due to internal source of heat, τ is time, λ is the conductivity coefficient, and β is the surface heat transfer coefficient.

In the following, we shall go in two steps: first, discrete in space domain, and second, discrete in time domain. Now consider how to discrete in space domain. The approximate solution of temperature field must satisfy the following equation:

$$\iint_D W_i \left[\frac{\partial^2 T}{\partial x^2} + \frac{\partial^2 T}{\partial y^2} + \frac{1}{a}\left(\frac{\partial \bar{\theta}}{\partial \tau} - \frac{\partial T}{\partial \tau}\right)\right] dx \, dy = 0 \tag{31.48}$$

in which W_i is the weight function. For the two terms with second derivatives in the integral, taking integration by parts, we get

$$\iint_D W_i \left(\frac{\partial^2 T}{\partial x^2} + \frac{\partial^2 T}{\partial y^2}\right) dx \, dy = -\iint_D \left(\frac{\partial W_i}{\partial x}\frac{\partial T}{\partial x} + \frac{\partial W_i}{\partial y}\frac{\partial T}{\partial y}\right) dx \, dy + \int_{C_1+C_2} W_i \frac{\partial T}{\partial n} ds$$

Substituting the above equation into Eq. (31.48), we get

$$\iint_D \left[\frac{\partial W_i}{\partial x}\frac{\partial T}{\partial x} + \frac{\partial W_i}{\partial y}\frac{\partial T}{\partial y} - \frac{W_i}{a}\left(\frac{\partial \theta}{\partial \tau} - \frac{\partial T}{\partial \tau}\right)\right] dx \; dy = \int_{C_1+C_2} W_i \frac{\partial T}{\partial n} ds \tag{31.49}$$

The region D is divided into a group of elements, as shown in Figure 31.7. The temperature in each element is expressed in the following:

$$T(x, y, \tau) = N_i(x, y)T_i(\tau) + N_i(x, y)T_j(\tau) + N_m(x, y)T_m(\tau) \tag{31.50}$$

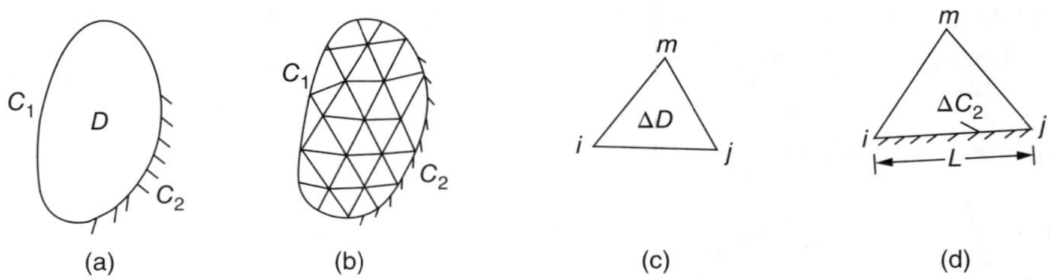

Figure 31.7 Division into elements. (a) Solution region. (b) Division of elements. (c) Interior element. (d) Element near boundary C_2.

in which $N_i(x, y)$ is the shape function in space domain and $T_i(\tau)$ is the temperature of node i, which is a function of time. By Galerkin method, take the weight function as

$$W_i = N_i(x, y)$$

Substituting into Eq. (31.49), we get

$$\sum_e \iint_{\Delta D} \left[\frac{\partial N_i}{\partial x} \left(\frac{\partial N_i}{\partial x} T_i + \frac{\partial N_j}{\partial x} T_j + \frac{\partial N_m}{\partial x} T_m \right) + \frac{\partial N_i}{\partial y} \left(\frac{\partial N_i}{\partial y} T_i + \frac{\partial N_j}{\partial y} T_j + \frac{\partial N_m}{\partial y} T_m \right) + \frac{N_i}{a} \left(N_i \frac{\partial T_i}{\partial \tau} + N_j \frac{\partial T_j}{\partial \tau} + N_m \frac{\partial T_m}{\partial \tau} - \frac{\partial \theta}{\partial \tau} \right) \right] \mathrm{d}x\,\mathrm{d}y = \sum_e \int_{\Delta C_1} N_i \frac{\partial T}{\partial n} \mathrm{d}s + \sum_e \int_{\Delta C_2} N_i \frac{\partial T}{\partial n} \mathrm{d}s \tag{31.51}$$

where ΔD is a subregion of element e. The right part of the above equation is line integral along the boundary $C_1 + C_2$. On the boundary C_1, $T = T_b$, the nodal temperature is known. When computing the nodal temperature near C_1, the shape function on C_1 is equal to zero. For example, if the boundary jm of element e is boundary C_1, when the temperature of node i is computed, the shape function of boundary jm $N_i = 0$, so the first term of the right part of the above equation is equal to zero. On the boundary C_2, from boundary condition (31.47), $\frac{\partial T}{\partial n} = -\frac{\beta}{\lambda}(T - T_a)$, substituting into the above equation, we get

$$\sum_e \iint_{\Delta D} \left[\frac{\partial N_i}{\partial x} \left(\frac{\partial N_i}{\partial x} T_i + \frac{\partial N_j}{\partial x} T_j + \frac{\partial N_m}{\partial x} T_m \right) + \frac{\partial N_i}{\partial y} \left(\frac{\partial N_i}{\partial y} T_i + \frac{\partial N_j}{\partial y} T_j + \frac{\partial N_m}{\partial y} T_m \right) + \frac{N_i}{a} \left(N_i \frac{\partial T_i}{\partial \tau} + N_j \frac{\partial T_j}{\partial \tau} + N_m \frac{\partial T_m}{\partial \tau} - \frac{\partial \theta}{\partial \tau} \right) \right] \mathrm{d}x\,\mathrm{d}y = -\sum_e \int_{\Delta C_2} \frac{\beta}{\lambda} N_i (N_i T_i + N_j T_j - T_a) \mathrm{d}s \tag{31.52}$$

Hence, we get system of equations as follows:

$$[H]\{T\} + [C]\frac{\partial}{\partial \tau}\{T\} = \{P\} \tag{31.53}$$

The elements of matrices $[H]$, $[C]$, and, $\{P\}$ are computed in the following:

$$H_{ij} = \sum_e h^e_{ij},\ \ C_{ij} = \sum_e C^e_{ij},\ \ P_i = \sum_e P^e_i \tag{31.54}$$

For element e, it is clear from Eq. (31.52) that

$$\left.\begin{aligned} h^e_{ij} &= \iint_{\Delta D} \left(\frac{\partial N_i}{\partial x}\frac{\partial N_j}{\partial x} + \frac{\partial N_i}{\partial y}\frac{\partial N_j}{\partial y}\right) \mathrm{d}x\,\mathrm{d}y + \int_{\Delta C_2} \frac{\beta}{\lambda} N_i N_j\,\mathrm{d}s \\ c^e_{ij} &= \iint_{\Delta D} \frac{1}{a} N_i N_j\,\mathrm{d}x\,\mathrm{d}y \\ p^e_j &= \iint_{\Delta D} \frac{1}{a} N_i \frac{\partial \theta}{\partial \tau}\,\mathrm{d}x\,\mathrm{d}y + \int_{\Delta D_2} \frac{\beta}{\lambda} N_i T_a\,\mathrm{d}s \end{aligned}\right\} \tag{31.55}$$

For isoparametric elements, they can be computed by numerical integration. For simple elements, they can be integrated. For example, for a triangular element, shape functions are

$$N_i = \frac{1}{2A}(a_i + b_i x + c_i y) \quad (i, j, m)$$

As

$$\frac{\partial N_i}{\partial x} = \frac{b_i}{2A},\ \frac{\partial N_i}{\partial y} = \frac{c_i}{2A}$$

$$\iint_{\Delta D} N_i N_j\,\mathrm{d}x\,\mathrm{d}y = \begin{cases} \dfrac{A}{6}, & \text{when } i = j \\ \dfrac{A}{12}, & \text{when } i \neq j \end{cases}$$

$$\iint_{\Delta D} N_i\,\mathrm{d}x\,\mathrm{d}y = \frac{A}{3}$$

$$\iint_{\Delta \sigma_2} N_i N_j\,\mathrm{d}s = \begin{cases} \dfrac{L}{3}, \text{when } i = j \\ \dfrac{L}{6}, \text{when } i \neq j \end{cases}$$

$$\int_{\Delta C_2} N_i\,\mathrm{d}s = \frac{L}{2}$$

where A is the area of the triangle and L is the length of boundary ij (on C_2). From Eq. (31.55), for each element e, we get the matrices as follows:

$$[h]^e = \frac{1}{4A}\begin{bmatrix} b_i^2 + c_i^2 & b_i b_j + c_i c_j & b_i b_m + c_i c_m \\ b_j b_i + c_j c_i & b_j^2 + c_j^2 & b_j b_m + c_j c_m \\ b_m b_i + c_m c_i & b_m b_j + c_m c_j & b_m^2 + c_m^2 \end{bmatrix} + \frac{\beta L}{6\lambda}\begin{bmatrix} 2 & 1 & 0 \\ 1 & 2 & 0 \\ 0 & 0 & 0 \end{bmatrix} \tag{31.56}$$

$$[c]^e = \frac{A}{12a}\begin{bmatrix} 2 & 1 & 1 \\ 1 & 2 & 1 \\ 1 & 1 & 2 \end{bmatrix} \tag{31.57}$$

$$\{p\}^e = \frac{1}{3a}\frac{\partial\theta}{\partial\tau}\begin{Bmatrix}1\\1\\1\end{Bmatrix} + \frac{\beta L}{2\lambda}\begin{Bmatrix}1\\1\\0\end{Bmatrix} \tag{31.58}$$

The second terms of the right part of Eqs (31.56) and (31.58) exist only when the boundary *ij* locates on C_2.

Now we return to Eq. (31.53) and consider how to discrete in time domain. Let $\{T\}_0, \{T\}_1, \ldots, \{T\}_n$ be the nodal temperatures at time $\tau = 0, \tau_1, \tau_2, \ldots, \tau_n$

By interpolating in time domain, the nodal temperature at any time τ may be expressed as

$$\{T\} = N_0(\tau)\{T\}_0 + N_1(\tau)\{T\}_1 + \cdots + N_n(\tau)\{T\}_n$$

in which $N_i(\tau)$ is the interpolating function in time domain. $\{T\}_0$ is the known initial nodal temperature. $\{T\}_1, \ldots, \{T\}_n$ are the nodal temperatures to be determined. By Galerkin method, let the weight function in time domain be

$$W_1(\tau) = N_1(\tau), W_2(\tau) = N_2(\tau), \ldots, W_n(\tau) = N_n(\tau)$$

By Eq. (31.53), we have

$$\left.\begin{aligned}&\int N_1(\tau)\left([H]\{T\} + [C]\frac{\partial}{\partial\tau}\{T\} - \{P\}\right) d\tau = 0\\&\int N_2(\tau)\left([H]\{T\} + [C]\frac{\partial}{\partial\tau}\{T\} - \{P\}\right) d\tau = 0\\&\qquad\vdots\\&\int N_n(\tau)\left([H]\{T\} + [C]\frac{\partial}{\partial\tau}\{T\} - \{P\}\right) d\tau = 0\end{aligned}\right\} \tag{31.59}$$

Substitute $N_i(\tau)$ into the above equation and take integration with respect to time. *a* system of equations will be obtained, the solution of which will give $\{T\}_1$, $\{T\}_2, \ldots, \{T\}_n$.

For example, use linear interpolation function. In the time interval $0 \le \tau \le \Delta\tau$, the nodal temperature is expressed by

$$\{T\} = \left[N_0(\tau)\ N_1(\tau)\right]\begin{Bmatrix}\{T\}_0\\\{T\}_1\end{Bmatrix} \tag{31.60}$$

where

$$\left.\begin{aligned}N_0(\tau) &= \frac{1-\tau}{\Delta\tau},\\N_1(\tau) &= \frac{\tau}{\Delta\tau}\end{aligned}\right\} \tag{31.61}$$

As $\frac{\partial N_0}{\partial\tau} = -\frac{1}{\Delta\tau}, \frac{\partial N_1}{\partial\tau} = \frac{1}{\Delta\tau}$, so the time derivative of nodal temperature is

$$\frac{\partial}{\partial\tau}\{T\} = \left[\frac{\partial N_0}{\partial\tau}\ \frac{\partial N_1}{\partial\tau}\right]\begin{Bmatrix}\{T\}_0\\\{T\}_1\end{Bmatrix} = \left[-\frac{1}{\Delta\tau}\quad \frac{1}{\Delta\tau}\right]\begin{Bmatrix}\{T\}_0\\\{T\}_1\end{Bmatrix} \tag{31.62}$$

The initial nodal temperature $\{T\}_0$ is known. Unknown is the nodal temperature when $\tau = \Delta\tau$. Take the weight function in time domain as

$$W_i(\tau) = N_i(\tau)$$

Substituting into Eq. (31.53), we get

$$\int_0^{\Delta\tau} N_1(\tau)\left([H]\{T\} + [C]\frac{\partial}{\partial\tau}\{T\} - \{P\}\right)\mathrm{d}\tau = 0 \tag{31.63}$$

By substituting Eqs (31.60) and (31.62) into the above equation, we have

$$\int_0^{\Delta\tau} \frac{\tau}{\Delta\tau}\left([H][N_0 N_1]\begin{Bmatrix}\{T\}_0\\\{T\}_1\end{Bmatrix} + [C]\begin{bmatrix}\frac{\partial N_0}{\partial\tau} & \frac{\partial N_1}{\partial\tau}\end{bmatrix}\begin{Bmatrix}\{T\}_0\\\{T\}_1\end{Bmatrix} - \{P\}\right)\mathrm{d}\tau = 0$$

After integration with respect to time τ and simplification, we get

$$\left(\frac{2}{3}[H] + \frac{1}{\Delta\tau}[C]\right)\{T\}_1 + \left(\frac{1}{3}[H] - \frac{1}{\Delta\tau}[C]\right)\{T\}_0 = \frac{2}{\Delta\tau}\int_0^{\Delta\tau}\frac{\tau}{\Delta\tau}\{P\}\,\mathrm{d}\tau \tag{31.64}$$

In order to obtain the integral of the right part of the above equation, $\{P\}$ is expressed as follows:

$$\{P\} = \begin{bmatrix}N_0 & N_1\end{bmatrix}\begin{Bmatrix}\{P\}_0\\\{P\}_1\end{Bmatrix} \tag{31.65}$$

where $\{P\}_0$ and $\{P\}_1$ are the values of $\{P\}$ at $\tau = 0$ and $\tau = \Delta\tau$, respectively; thus we get

$$\frac{2}{\Delta\tau}\int_0^{\Delta\tau}\frac{\tau}{\Delta\tau}\{P\}\mathrm{d}\tau = \frac{1}{3}\{P\}_0 + \frac{2}{3}\{P\}_1$$

Substitution of the above equation into Eq. (31.65) yields

$$\left(\frac{2}{3}[H] + \frac{1}{\Delta\tau}[C]\right)\{T\}_1 + \left(\frac{1}{3}[H] - \frac{1}{\Delta\tau}[C]\right)\{T\}_0 = \frac{1}{3}\{P\}_0 + \frac{2}{3}\{P\}_1 \tag{31.66}$$

The inverse of the above equation gives the nodal temperature $\{T\}_1$ at $\tau = \Delta\tau$ as follows:

$$\{T\}_1 = \left(\frac{2}{3}[H] + \frac{1}{\Delta\tau}[C]\right)^{-1}\left(\frac{1}{3}\{P\}_0 + \frac{2}{3}\{P\}_1 - \left(\frac{1}{3}[H] - \frac{1}{\Delta\tau}[C]\right)\{T\}_0\right) \tag{31.67}$$

Repeating the above computation step by step, we will obtain the temperatures at any time. If the interpolation function of higher order is used, the precision of computation may be better, but experience shows that the precision of computation is satisfactory with the linear interpolation function as Eq. (31.60).

31.6 Dynamic Response of Structure

After discretion in space domain, the problem of dynamic response of structure may be expressed by the following equation:

$$[K]\{\delta\} + [C]\frac{\partial}{\partial\tau}\{\delta\} + [M]\frac{\partial^2}{\partial\tau^2}\{\delta\} = \{P\} \tag{31.68}$$

where

$\{\delta\}$ = vector of nodal displacements

$[K]$ = stiffness matrix

$[C]$ = damping matrix

$[M]$ = mass matrix

$[P]$ = vector of dynamic loads

Two initial conditions must be given for $\tau = 0$, namely, the initial nodal displacements $\{\delta\}_0$ and the initial nodal velocity $\{\dot{\delta}\}_0 = \frac{\partial}{\partial\tau}\{\delta\}$.

Now we explain how to discrete in time domain by Galerkin method.

Let the nodal displacement and the nodal velocity at $\tau = \tau_1 = \Delta\tau$ be $\{\delta\}_1$ and $\{\dot{\delta}\}_1 = \left\{\frac{\partial}{\partial\tau}\delta_1\right\}$, and the nodal displacement between $\tau = 0$ and $\tau = \Delta\tau$ is expressed by

$$\{\delta\} = \left[N_{00}(\tau)\ N_{01}(\tau)\ N_{10}(\tau)\ N_{11}(\tau)\right]\begin{Bmatrix}\delta_0\\ \dot{\delta}_0\\ \delta_1\\ \dot{\delta}_1\end{Bmatrix} \tag{31.69}$$

where

$$\left.\begin{aligned} N_{00}(\tau) &= 1 - 3s^2 + 2s^3\\ N_{01}(\tau) &= (s - 2s^2 + s^3)\Delta\tau\\ N_{10}(\tau) &= 3s^2 - 2s^3\\ N_{11}(\tau) &= (-s^2 + s^3)\Delta\tau\\ s &= \frac{\tau}{\Delta\tau} \end{aligned}\right\} \tag{31.70}$$

The polynomial N_i possesses the following properties:

i) when $s = 0, N_{00} = \frac{\partial}{\partial\tau}N_{01} = 1$, all the others are equal to zero.

ii) when $s = 1, N_{10} = \frac{\partial}{\partial\tau}N_{11} = 1$, all the others are equal to zero.

By Eq. (31.69), when $\tau = 0$, the nodal displacement and nodal velocity are $\{\delta\}_0$ and $\{\dot{\delta}\}_0$; when $\tau = \Delta\tau$, the nodal displacement and nodal velocity are $\{\delta\}_1$ and $\{\dot{\delta}\}_1$. As $\{\delta\}_1$ and $\{\dot{\delta}\}_1$ are unknown, by Galerkin method, the weight functions are N_{10} and N_{11}. The equation of weighted residual is

$$\left.\begin{aligned} \int_0^{\Delta\tau} N_{10}\left([K]\{\delta\} + [C]\frac{\partial}{\partial\tau}\{\delta\} + [M]\frac{\partial^2}{\partial\tau^2}\{\delta\} - \{P\}\right)\mathrm{d}\tau = 0\\ \int_0^{\Delta\tau} N_{11}\left([K]\{\delta\} + [C]\frac{\partial}{\partial\tau}\{\delta\} + [M]\frac{\partial^2}{\partial\tau^2}\{\delta\} - \{P\}\right)\mathrm{d}\tau = 0 \end{aligned}\right\} \tag{31.71}$$

Solution of the above equation will give the dynamic displacement $\delta(t)$ of the problem.

After substituting Eqn. (31.69) and (31.70) into the above equation and integrating with respect to time τ, we get the following equation set for solving $\{\delta\}_1$ and $\{\dot{\delta}\}_1$:

$$\begin{bmatrix} \alpha_{11} & \alpha_{12} \\ \alpha_{21} & \alpha_{22} \end{bmatrix} \begin{Bmatrix} \delta_1 \\ \dot{\delta}_1 \end{Bmatrix} + \begin{bmatrix} \beta_{11} & \beta_{12} \\ \beta_{21} & \beta_{22} \end{bmatrix} \begin{Bmatrix} \delta_0 \\ \dot{\delta}_0 \end{Bmatrix} = \begin{Bmatrix} \gamma_1 \\ \gamma_2 \end{Bmatrix} \tag{31.72}$$

where $[\alpha_{ij}]$, $[\beta_{ij}]$, $\{\gamma_i\}$ are got by integration of Eq. (31.71) as follows:

$$[\alpha_{11}] = \int_o^{\Delta\tau} N_{10}\left([K]N_{10} + [C]\frac{\partial N_{10}}{\partial \tau} + [M]\frac{\partial^2 N_{10}}{\partial \tau^2}\right) d\tau,$$

$$[\alpha_{12}] = \int_o^{\Delta\tau} N_{10}\left([K]N_{11} + [C]\frac{\partial N_{11}}{\partial \tau} + [M]\frac{\partial^2 N_{11}}{\partial \tau^2}\right) d\tau,$$

$$[\beta_{11}] = \int_o^{\Delta\tau} N_{10}\left([K]N_{00} + [C]\frac{\partial N_{00}}{\partial \tau} + [M]\frac{\partial^2 N_{00}}{\partial \tau^2}\right) d\tau,$$

$$[\beta_{12}] = \int_o^{\Delta\tau} N_{10}\left([K]N_{01} + [C]\frac{\partial N_{01}}{\partial \tau} + [M]\frac{\partial^2 N_{01}}{\partial \tau^2}\right) d\tau$$

$$[\gamma_1] = \int_o^{\Delta\tau} N_{10}\{p\}\ d\tau$$

$$[\alpha_{21}] = \int_o^{\Delta\tau} N_{11}\left([K]N_{10} + [C]\frac{\partial N_{10}}{\partial C} + [M]\frac{\partial^2 N_{12}}{\partial \tau^2}\right) d\tau$$

$$\{\gamma_2\} = \int_o^{\Delta\tau} N_{11}\{p\}d\tau$$

The equation (31.72) is a formula of circulation, by means of which the displacement and velocity at the end of the time increment $\Delta\tau$ may be computed by the displacement and velocity at the beginning of the time increment.

Bibliography

1 Finlayson, B.A. (1972) *The Method of Weighted Residuals and Variational Principles*, Academic Press.

2 Bofang, Z. and The Weighted Residual Methods (1986) *Handbook of Modern Engineering Mathematics*, Vol. II, Middle China Engineering College Press.

3 Timoshenko, S. and Goodier, J.N. (1970) *Theory of Elasticity*, McGraw-Hill, New York.

Appendix A

Transformation of Stresses, Strains, Flexibility, and Stiffness between Two Coordinate Systems

1) **Transformation of stresses**
Two coordinate systems (x, y, z) and (x', y', z') are linked by Eq. (A.1)

$$\begin{Bmatrix} x' \\ y' \\ z' \end{Bmatrix} = \begin{bmatrix} l_1 & m_1 & n_1 \\ l_2 & m_1 & n_1 \\ l_2 & m_1 & n_1 \end{bmatrix} \begin{Bmatrix} x \\ y \\ z \end{Bmatrix} = [L] \begin{Bmatrix} x \\ y \\ z \end{Bmatrix} \tag{A.1}$$

where l_i, m_i, n_i $(i = 1, 2, 3)$ are the direction cosines of x', y', z'.
Let $\{\sigma\} = \begin{bmatrix} \sigma_x & \sigma_y & \sigma_z & \tau_{xy} & \tau_{yz} & \tau_{zx} \end{bmatrix}^{\mathrm{T}}$ be the stress in the (x', y, z) coordinate system. The stress in $(x', y', ,z')$ coordinate system is given by Eqs. (A.2) and (A.3):

$$\{\sigma'\} = [T]\{\sigma\} \tag{A.2}$$

$$[T_\sigma] = \begin{bmatrix} l_1^2 & m_1^2 & n_1^2 & 2l_1m_1 & 2m_1n_1 & 2l_1n_1 \\ l_2^2 & m_2^2 & n_2^2 & 2l_2m_2 & 2m_2n_2 & 2l_2n_2 \\ l_3^2 & m_3^2 & n_3^2 & 2l_3m_3 & 2m_3n_3 & 2l_3n_3 \\ l_1l_2 & m_1m_2 & n_1n_2 & l_1m_2 + l_2m_1 & m_1n_2 + m_2n_1 & l_1n_2 + l_2n_1 \\ l_2l_3 & m_2m_3 & n_2n_3 & l_2m_3 + l_3m_2 & m_2n_3 + m_3n_2 & l_2n_3 + l_3n_2 \\ l_1l_3 & m_1m_3 & n_1n_3 & l_1m_3 + l_3m_1 & m_1n_3 + m_3n_1 & l_1n_3 + l_3n_1 \end{bmatrix} \tag{A.3}$$

where $[T_\sigma]$ is the matrix of transformation.
If only one part of stress components is considered, the matrix of transformation $[T_\sigma]$ may be reduced; for example, the normal stress and shearing stress on the plane of joint in rock mass may be given by:

$$\begin{Bmatrix} \sigma'_x \\ \tau'_{yz} \\ \tau'_{zx} \end{Bmatrix} = \begin{bmatrix} l_3^2 & m_3^2 & n_3^2 & 2l_3m_3 & 2m_3n_3 & 2l_3n_3 \\ l_2l_3 & m_2m_3 & n_2n_3 & l_2m_3 + l_3m_2 & m_2n_3 + m_3n_2 & l_2n_3 + l_3n_2 \\ l_1l_3 & m_1m_3 & n_1n_3 & l_1m_3 + l_3m_1 & m_1n_3 + m_3n_1 & l_1n_3 + l_3n_1 \end{bmatrix} \{\sigma\} \tag{A.4}$$

where the z' axis is normal to the plane of joint, the axis x' and y' are in the plane of joint, and the principal sheaing stress on the plane of joint is given by Eq. (A.5):

$$\tau = [(\tau'_{yz})^2 + (\tau'_{zx})^2]^{1/2} \tag{A.5}$$

The Finite Element Method: Fundamentals and Applications in Civil, Hydraulic, Mechanical and Aeronautical Engineering, First Edition. Bofang Zhu.

Companion website: www.wiley.com/go/zhufem17

2) **Transformation of strains**
Let $\{\varepsilon\} = [\varepsilon_x, \varepsilon_y, \varepsilon_z, \gamma_{xy}, \gamma_{yz}, \gamma_{zx}]^T$ be the strain in (x, y, z) coordinate system. The strain in (x', y', z') coordinate system is given by Eq. (A.6) and (A.7):

$$\{\varepsilon'\} = [T_\varepsilon]\{\varepsilon\} \tag{A.6}$$

$$[T_\varepsilon] = \begin{bmatrix} l_1^2 & m_1^2 & n_1^2 & l_1m_1 & m_1n_1 & l_1n_1 \\ l_2^2 & m_2^2 & n_2^2 & l_2m_2 & m_2n_2 & l_2n_2 \\ l_3^2 & m_3^2 & n_3^2 & l_3m_3 & m_3n_3 & l_3n_3 \\ 2l_1l_2 & 2m_1m_2 & 2n_1n_2 & l_1m_2+l_2m_1 & m_1n_2+m_2n_1 & l_1n_2+l_2n_1 \\ 2l_2l_3 & 2m_2m_3 & 2n_2n_3 & l_2m_3+l_3m_2 & m_2n_3+m_3n_2 & l_2n_3+l_3n_2 \\ 2l_1l_3 & 2m_1m_3 & 2n_1n_3 & l_1m_3+l_3m_1 & m_1n_3+m_3n_1 & l_1n_3+l_3n_1 \end{bmatrix} \tag{A.7}$$

3) **Transformation of flexibility**
In coordinate system (x, y, z), the stress–strain relation is

$$\{\varepsilon\} = [C]\{\sigma\} \tag{A.8}$$

In coordinate system (x', y', z'),

$$\{\varepsilon'\} = [C']\{\sigma'\} \tag{A.9}$$

The work done in two coordinate systems must be equal, so

$$\{\sigma'\}^T\{\varepsilon'\} = [\sigma]^T\{\varepsilon\} \tag{A.10}$$

Similarily,

$$\{\sigma'\}\{\varepsilon'\}^t = [\varepsilon]^T\{\sigma\} \tag{A.11}$$

By Eqs. (A.2), (A.6), (A.8), and (A.9), we have

$$[C']\{\sigma'\} = \{\varepsilon'\} = [T_\varepsilon]\{\varepsilon\} = [T_\varepsilon][C]\{\sigma\} = [T_\varepsilon][C][T_\sigma]^{-1}\{\sigma'\}$$

As $\{\sigma'\}$ is arbitrary, so

$$[C'] = [T_\varepsilon][C][T_\sigma]^{-1} \tag{A.12}$$

4) **Transformation of displacement, load, nodal force, and stiffness matrix of element**
In coordinate system (x', y', z'), the displacement of node i $\{\delta_i'\} = \left[u_i' \ v_i' \ w_i'\right]^T$ is given by Eq. (A.13):

$$\{\delta_i'\} = [L]\{\delta_i\}$$
$$\{\delta_i\} = \left[u_i \ v_i \ w_i\right]^T \tag{A.13}$$

where $\{\delta_i\}$ is the displacement of node i in the coordinate system (x, y, z) and $[L]$ is the matrix of direction cosines as (A.1).
In the coordinate system (x', y', z'), the nodal displacement of the element $\{\delta'\}^e = \left[\delta_1' \ \delta_2' \ \cdots\right]^T$ is given by Eqs. (A.14) and (A.15):

$$\{\delta'\}^e = [\lambda]\{\delta\}^e \tag{A.14}$$

$$[\lambda] = \begin{bmatrix} L & 0 & 0 & 0 \\ 0 & L & 0 & 0 \\ 0 & 0 & L & 0 \\ 0 & 0 & 0 & L \end{bmatrix} \tag{A.15}$$

Similarily, we have the following equations:

$$\{F'\}^{\mathrm{e}} = [\lambda]\{F\}^{\mathrm{e}}$$

$$\{P'\}^{\mathrm{e}} = [\lambda]\{P\}^{\mathrm{e}} \tag{A.16}$$

$$\{k'\}^{\mathrm{e}} = [\lambda][k][\lambda]^{\mathrm{T}} \tag{A.17}$$

As $[L]$ is orthogonal matrix, we have

$$[L]^{-1} = [L]^{\mathrm{T}}, \quad [\lambda]^{-1} = [\lambda]^{\mathrm{T}} \tag{A.18}$$

So, we have the following relations:

$$\{\delta\}^{\mathrm{e}} = [\lambda]^{\mathrm{T}}\{\delta'\}^{\mathrm{e}} \tag{A.19}$$

$$\{F\}^{\mathrm{e}} = [\lambda]^{\mathrm{T}}\{F'\}^{\mathrm{e}} \tag{A.20}$$

$$\{P\}^{\mathrm{e}} = [\lambda]^{\mathrm{T}}\{P'\}^{\mathrm{e}} \tag{A.21}$$

$$[k]^{\mathrm{e}} = [\lambda]^{\mathrm{T}}[k']^{\mathrm{e}}[\lambda] \tag{A.22}$$

where $\{\delta\}^{\mathrm{e}}, \{F\}^{\mathrm{e}}, \{P\}^{\mathrm{e}}, [k]^{\mathrm{e}}$ are the nodal displacement, nodal force, nodal load, and stiffness matrix of the element in the coordinate system (x, y, z), respectively, and $\{\delta'\}^{\mathrm{e}}, \{F'\}^{\mathrm{e}}, \{P'\}^{\mathrm{e}}, [k]^{\mathrm{e}}$ are those in the coordinate system (x', y', z').

Appendix B

Important Formulas

The displacement of any point in an element

$$\{r\} = \begin{Bmatrix} u \\ v \\ w \end{Bmatrix} = [N]\{\delta\}^{\mathrm{e}} = \begin{bmatrix} N_i & N_j & N_m & \cdots \end{bmatrix} \begin{Bmatrix} \delta_i \\ \delta_j \\ \delta_m \\ \vdots \end{Bmatrix} \tag{3.4}$$

The strain of any point in an element

$$\{\varepsilon\} = [B]\{\delta\}^{\mathrm{e}} \tag{3.5}$$

The stress of any point in an element

$$\{\sigma\} = [D](\{\varepsilon\} - \{\varepsilon_0\}) + \{\sigma_0\} \tag{3.6}$$

The stiffness matrix of an element

$$[k]^{\mathrm{e}} = \iiint [B]^{\mathrm{T}}[D][B]\,\mathrm{d}x\,\mathrm{d}y\,\mathrm{d}z \tag{3.11}$$

The nodal force of an element

$$\{F\} = [k]^{\mathrm{e}}\{\delta\}^{\mathrm{e}} \tag{3.12}$$

The nodal load due to body force $\{q\}$

$$[P]_q^{\mathrm{e}} = \iiint [N]^{\mathrm{T}}\{q\}\,\mathrm{d}x\,\mathrm{d}y\,\mathrm{d}z \tag{3.14}$$

The nodal load due to surface force $\{p\}$

$$\{P\}_p^{\mathrm{e}} = \int_S [N]^{\mathrm{T}}\{p\}\,\mathrm{d}S \tag{3.15}$$

The nodal load due to initial strain $\{\varepsilon_0\}$

$$[P]_{\varepsilon_0}^{\mathrm{e}} = \iiint [B]^{\mathrm{T}}[D][\varepsilon_0]\,\mathrm{d}x\,\mathrm{d}y\,\mathrm{d}z \tag{3.16}$$

The nodal load due to initial stress $\{\sigma_0\}$

$$[P]_{\sigma_0}^{\mathrm{e}} = -\iiint [B]^{\mathrm{T}}[\sigma_0]\,\mathrm{d}x\,\mathrm{d}y\,\mathrm{d}z \tag{3.17}$$

The Finite Element Method: Fundamentals and Applications in Civil, Hydraulic, Mechanical and Aeronautical Engineering, First Edition. Bofang Zhu.

Companion website: www.wiley.com/go/zhufem17

The equilibrium equation

$$\text{or} \quad [K]\{\delta\} = \{P\} \tag{4.1}$$

$$\int_V [B]^{\mathrm{T}}\{\sigma\}\,\mathrm{d}V = \{P\} \tag{4.1a}$$

Index

The Finite Element Method: Fundamentals and Applications in Civil, Hydraulic, Mechanical and Aeronautical Engineering, First Edition. Bofang Zhu.

Companion website: www.wiley.com/go/zhufem17

y